R. Gupta's®

Popular Master Guide

BSNL

Bharat Sanchar Nigam Ltd.

J.T.O.

Junior Telecom Officer

(TELECOM)

Recruitment Examination

2017
EDITION

RAMESH PUBLISHING HOUSE, New Delhi

Published by
O.P. Gupta *for* Ramesh Publishing House

Admin. Office
12-H, New Daryaganj Road, Opp. Officers' Mess,
New Delhi-110002 ✆ 23261567, 23275224, 23275124

E-mail: info@rameshpublishinghouse.com
Website: www.rameshpublishinghouse.com

Showroom:
- Balaji Market, Nai Sarak, Delhi-6 ✆ 23253720, 23282525
- 4457, Nai Sarak, Delhi-6, ✆ 23918938

© Reserved with the Publisher

No Part of this book may be reproduced or transmitted in any form or by any means, electronic or mechanical including photocopying, recording or by any transformation storage and retrieval system without written permission from the Publisher.

Indemnification Clause: ***This book is being sold/distributed subject to the exclusive condition that neither the author nor the publishers, individually or collectively, shall be responsible to indemnify the buyer/user/possessor of this book beyond the selling price of this book for any reason under any circumstances. If you do not agree to it, please do not buy/accept/use/possess this book.***

Book Code: R-139

ISBN: 978-93-5012-839-8

Scheme of Examination & Syllabus

The pattern of the question paper will be multiple choice objective types with negative marking. All candidates will undergo a common multiple-choice type examination of three hours consisting of the following three sections:

Components	Number of Questions	Marks
Section–I : Engineering Stream	50	100
Section–II : Engineering Stream	50	100
Section–III : General Ability Test	20	40
Total	**120**	**240**

SYLLABUS FOR EACH OF THE THREE SECTIONS

Section–I

• Materials and components • Physical Electronics, Electron Devices and ICs • Network Theory • Electromagnetic Theory • Electronic Measurements and Instrumentation • Power Electronics.

Section–II

• Analog Electronic Circuits • Digital Electronic Circuits • Control Systems • Communication Systems • Microwave Engineering • Computer Engineering • Microprocessors

Section–III

General Ability Test

The candidate's comprehension and understanding of General English shall be tested through simple exercises. Questions on knowledge of current events and of such matter of everyday observation and experience in his/her scientific aspects as may be expected of an educated person. Questions will also be included on events and developments in Telecommunications, History of India and Geography.

CONTENTS

ENGINEERING STREAM (SECTION–I & II)

Important Multiple Choice Questions

GENERAL ABILITY TEST (SECTION–III)

Previous Paper (Solved)

BHARAT SANCHAR NIGAM LTD.

BSNL—Junior Telecom Officers (JTO) Recruitment Exam

Section-I

1. Which of the following is true about Address Resolution Protocol (ARP)?
A. It is send across networks
B. It is used to find out the Layer 2 address, given a layer 3 address
C. It is used to find out a Layer 3 address, given a Layer 2 address
D. It is used to find own layer 3 address

2. Which of the following is NOT a valid DNS record?
A. Mail Exchange Record
B. Date stamp Record
C. Address Record
D. Pointer record

3. In which region(s) of operation, can a transistor be used as a switching element?
A. Active region
B. Active, Cut off and Saturation regions
C. Active and Saturation region
D. Cut off and saturation region

4. Race around condition happens in the J-K flip flop when :
A. both inputs are 1
B. both inputs are 0
C. the inputs are 0 and 1
D. output is 0

5. The flip-flop used as a delay flip-flop is :
A. JK flip-flop B. RS flip-flop
C. D flip-flop D. T flip-flop

6. To realize a full adder using half adders, the minimum number of half adders required is :
A. 1 B. 2
C. 3 D. 4

7. In a A/D converter, the analog input is given to :
A. ground of OPAMP
B. virtual ground of OPAMP
C. inverting terminal of OPAMP
D. non inverting terminal of OPAMP

Directions (Qs. Nos. 8-10) : *The maximum deviation in a FM system is 75 KHz with the modulating signal being 10 KHz sinusoid.*

8. The bandwidth of the above system is :
A. 75 KHz B. 150 KHz
C. 170 KHz D. 190 KHz

9. If the modulating frequency is doubled to 20 KHz sinusoid, the new bandwidth of the above system is :
A. 95 KHz B. 190 KHz
C. 170 KHz D. 210 KHz

10. If the modulating signal amplitude is doubled, the new bandwidth of the above system will be :
A. 320 KHz B. 190 KHz
C. 210 KHz D. 100 KHz

11. The relation between Maximum Usable Frequency (MUF), the critical frequency (f_c) and angle of incidence θ is :
A. MUF = $f_c \cos\theta$ B. MUF = $f_c \sec\theta$
C. MUF = $f_c \sin\theta$ D. MUF * $f_c = \cos\theta$

12. Which of the following bands experiences maximum rain attenuation?
A. S Band B. C Band
C. L Band D. Ku band

13. Thevenin equivalent voltage is :
A. equal to source voltage
B. equal to load voltage
C. the open voltage
D. voltage at any point in the circuit

14. To get maximum power transferred to a load, the load impedance should be equal to :
A. source impedance
B. complex conjugate of source impedance
C. source resistance only
D. source reactance only

15. Norton's theorem gives :
A. an equivalent current source in parallel with an equivalent impedance
B. an equivalent current source in series with an equivalent impedance
C. an equivalent voltage source in parallel with an equivalent impedance
D. an equivalent voltage source in series with an equivalent impedance

16. A sine wave is applied across an RL circuit. If the frequency of the voltage is increased, the current :
A. increases
B. decreases
C. remains the same
D. momentarily goes to zero

17. In a series RLC circuit, for a critically damped operation, damping factor ζ =
A. 0 B. infinity
C. 1 D. 0.5

18. Which of the following is a Low Pass Filter?

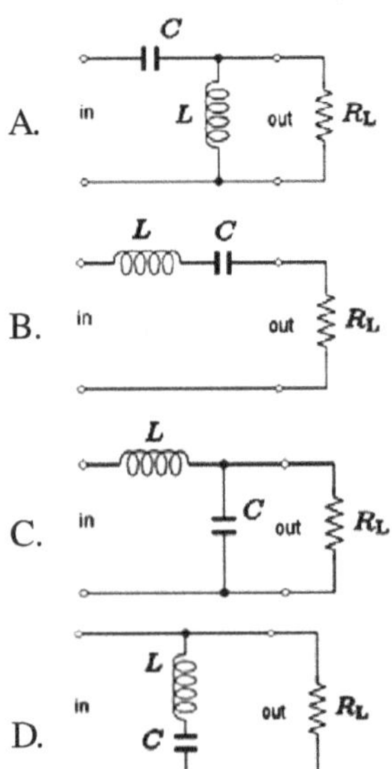

19. In a series RLC circuit at resonance, the current is determined by :
A. Resistance only B. Inductance only
C. Capacitance only D. All R, L, C

20. If for a network, $Z_{11} = Z_{22}$, then the network is :
A. Reciprocal B. symmetric
C. Loss less D. Active

21. The Inductance L is related to the number of turns (N) as :
A. $L \propto N$ B. $L \propto N^{1/2}$
C. $L \propto N^3$ D. $L \propto N^2$

22. For a linear transformer of two coupled inductors of L_1 and L_2, the mutual inductance M :
A. $<= \sqrt{(L_1L_2)}$ B. $= \sqrt{\left(\frac{L_1}{L_2}\right)}$
C. $= \frac{L_1}{L_2}$ D. $= L_1L_2$

23. An ideal voltage source has :
A. terminal voltage in proportion to load
B. terminal voltage in proportion to current
C. zero internal resistance
D. open circuit voltage equal to full load voltage

24. Hystersis loss varies with frequency (f) as :
A. f B. $\sqrt{f}$
C. f^2 D. $f^{2.6}$

25. Tap changing facility is normally provided in :
A. Current transformers
B. Distribution transformers
C. Step Up transformers
D. High voltage transformers

26. In an induction motor, with the increase in input voltage, which of the following increases?
A. Power factor B. Slip
C. Torque D. All of the above

27. For a 6 pole three phase Induction motor operating at 50Hz, the synchronous speed is :
A. 1200 rpm B. 600 rpm
C. 500 rpm D. 1000 rpm

28. The purpose of a commutator in a DC generator is to :
A. increase o/p voltage
B. reduce sparking
C. smoothen o/p
D. convert induced AC into DC

29. For a stepper motor, the number of rotor and stator teeth are 40 and 10 respectively. The step angle is :
A. 54° B. 27°
C. 12° D. 10°

30. For a delta connection, which of the following is true?
A. $E_{ph} = E_{line}$ B. $E_{ph} = \sqrt{3}x\, E_{line}$
C. $E_{ph} = 0.5x\, E_{line}$ D. $E_{ph} = \frac{E_{line}}{\sqrt{3}}$

31. Which of the following is NOT used as a refrigerant?
A. Ammonia B. R22
C. Neon D. Freon

32. A power system whose input and output are DC is called :
A. Rectifier B. Chopper
C. Inverter D. Transformer

33. Which of the following is used for high voltage (>3 KV) control applications?
A. Power diode B. BJT Transistor
C. MOSFET D. SCR

34. Which of the following is an integrating instrument?
A. Voltmeter B. Ammeter
C. Wattmeter D. Watthour meter

35. A moving coil permanent magnet instrument can be used as a __________ by using a low resistance shunt.
A. Ammeter B. Voltmeter
C. Galvanometer D. Flux meter

36. Which of the following is used for measuring the insulation resistance?
A. Tangent Galvanometer
B. Meggar
C. Current transformer
D. Voltmeter

37. In an Anderson bridge, the unknown inductance is measured in terms of :
A. Known Inductance and resistance
B. Known capacitance and resistance
C. Known Resistance
D. Known inductance

38. The resolution of the Digital to Analog Converter (DAC) of '*n*' digital inputs is :
A. $2n$ B. $2/n$
C. 2^n D. $\sqrt{2^n}$

39. A single channel digital storage oscilloscope uses a 12 bit, 10^8 samples/s ADC. For a 10 kHz sine wave input, what is the number of samples taken per cycle of input?
A. 10^8 B. 10^{12}
C. 10^2 D. 10^4

40. A n-type semi conductor :
A. is intrinsic
B. has trivalent impurity atoms added
C. has pentavalent impurity atoms added
D. requires no doping

41. Forward voltage across a conducting silicon diode is :
A. 0.3 V B. –0.3 V
C. 0.5 V D. 0.7 V

42. Normal operating region for a Zener diode is :
A. forward bias region
B. reverse bias region
C. reverse break down region
D. zero crossing region

43. Each diode in a center tapped Full Wave rectifier is in ________ region and conducts for ________ of the input cycle :
A. forward, 180° B. reverse, 90°
C. forward, 90° D. reverse, 180°

44. In the active region of a BJT, the base-emitter junction is ________ biased and the collector base junction is ______________ .
A. forward, forward
B. forward, reverse
C. reverse, forward
D. reverse, reverse

45. β_{dc} =

A. $\frac{I_B}{I_C}$ B. $\frac{I_E}{I_C}$

C. $\frac{I_C}{I_B}$ D. $\frac{I_C}{I_E}$

46. Which of the following configuration has lowest output impedance?

A. Darlington B. Emitter follower
C. Voltage follower D. Fixed bias

47. Which of the following parameters has the units of ohms?

A. h_{ie} B. h_{oe}
C. h_{re} D. h_{fe}

48. A relation is said to be in BCNF when :

A. it has overlapping composite keys
B. it has no composite keys
C. it has no multi valued dependencies
D. it has no overlapping composite keys which have related attributes

49. The language used to by application programs to request data from the data bases is :

A. DML B. DDL
C. Query language D. Any of the above

50. An audit trail is used to :

A. make back up copies
B. record history of operations on a file
C. can be used to restore lost data
D. list the contents of the file

Section-II

51. Representation of 12 (decimal) in two's complement notation is :

A. 0100 B. 0101
C. 0011 D. 1100

52. In which file of DOS is the path defined :

A. SYSTEM.INI B. IO.SYS
C. CONFIG.SYS D. AUTOEXEC.BAT

53. What is the file system used in DOS?

A. UFS B. NFS
C. FAT D. NTFS

54. In DOS, the notation "cd .." denotes change to :

A. Present directory B. Parent directory
C. Home directory D. Child directory

55. What is the error code in HTTP for "Requested Resource not found"?

A. 402 B. 220
C. 404 D. 403

56. Java source code is compiled into :

A. byte code B. .exe file
C. .obj file D. source code

57. Javascript statements are executed by :

A. Compiler B. JVM
C. Server D. Browser

58. The "===" operator will :

A. check the value
B. check the value and type of value
C. equate both sides
D. result in error as it does not exist

59. Which of the following is incorrect with reference to php?

A. Variables should start with $
B. Variables names are case sensitive
C. No command is there to define a variable's data type
D. Variables can start with a number

60. Which of the following method is not related to threading?

A. Wait() B. Notify
C. Clear D. NotifyAll()

61. What is % operator used for in Java?

A. Text formatting
B. Remainder/Modulo operator
C. To give integer value of a division
D. As a division operator

62. Which of the following regarding Java naming convention is correct?

A. The file name should exactly match the class name

B. Java is case insensitive
C. Class name should start with upper/lower case letter
D. Method names should start with upper/ lower case letter

63. The pattern sent on HDLC to denote start and end of data frame is :
A. 0x7E B. 0x7C
C. 0x00 D. 0x7F

64. The FTP server listens for incoming connections on port _______ :
A. 80 B. 23
C. 21 D. 8080

65. Ethernet is associated with IEEE _________ protocol.
A. 802.3 B. 802.4
C. 802.5 D. 802.11

66. The maximum number of 10 Base T Ethernet cable segments that are allowed between two points in the network are :
A. 2 B. 5
C. 10 D. No limit

67. Which of the following IP address is not a part of a private network address pool?
A. 10.x.x.x B. 192.x.x.x
C. 172.x.x.x D. 200.x.x.x

68. The maximum value of TTL field in a IP datagram is :
A. 3 B. 63
C. 255 D. 1023

69. Which of the following is NOT a property of TCP?
A. It ensures packets arrive in order
B. It is connectionless
C. It provides flow control
D. It uses 3 way handshake before start of data transmission

70. In Unix, which file is used to map service names to server names?
A. /etc/hosts B. /etc/protocols
C. /etc/inetd.conf D. /etc/services

71. Which of the following is a Universal gate?
A. NAND B. AND
C. OR D. NOT

72. Which of the following has the lowest power dissipation?
A. CMOS B. TTL
C. ECL D. RTL

73. Which of the following Boolean algebra rules is correct?
A. A.A = 1 B. A + A' B = A + B
C. A + A B = A + B D. A(A + B) = B

74. The correct sequence of the following PCM coding steps is :
1. Encoding 2. Quantising
3. Sampling 4. Low pass filtering
A. 3, 1, 2, 4 B. 4, 3, 2, 1
C. 1, 2, 3, 4 D. 4, 2, 3, 1

75. The Low Pass filter cut-off in a telephone exchange analog subscriber line card is set at :
A. 5 KHz B. 8 KHz
C. 4 KHz D. 16 KHz

76. In PCM, quantization noise depends on :
A. sampling rate
B. number of quantization levels
C. signal power
D. bandwidth of the signal

77. In a Optical fiber, the propagation of signal happens due to :
A. total internal reflection
B. refraction
C. dispersion
D. attenuation

78. Which of the following windows is not used in optical fiber communication?
A. 850 nm B. 1310 nm
C. 1400 nm D. 1550 nm

79. Longest hop radio communication is possible due to ____________ of the Ionosphere.
A. F1 region B. D region
C. F2 region D. E region

80. Which of the following operations can result in an empty stack?
A. PUSH B. POP
C. TOP D. EMPTY

81. Which of the following is incorrect regarding a Queue?
A. Queue is FIFO

B. Insertions go at the end of the list
C. Deletions go at the front of the list
D. Queue is LIFO

82. The Heap sort algorithm for n elements has a complexity of the order :
A. O(n) B. O(n^2)
C. O(n log n) D. O(log n)

83. Binary search in an array of n elements has a complexity of :
A. O(n) B. O(n^2)
C. O(n log n) D. O(log n)

84. Which of the following is required before doing binary search?
A. The array should be a sorted one
B. The array should consist of only +ve numbers
C. The array should not consist of zero
D. The array should consist of only ones and zeroes

85. Of the following, the most efficient algorithm to sort elements is _________ .
A. Simple sort B. Insertion sort
C. Bubble sort D. Heap sort

86. Which of the following statements is FALSE regarding malloc() call :
A. It is used for dynamic memory allocation
B. It allocates the requisite block after searching the free list for a usable memory block
C. If the available memory size is not available, it returns that size which is possible
D. It returns a pointer to the memory block

87. Which of the following functions allocates memory and initializes it also?
A. free() B. malloc()
C. realloc() D. calloc()

88. Principle of Locality is used to justify :
A. DMA B. Interrupt
C. Cache memory D. Polling

89. A class is :
A. a group of objects
B. template for objects of a particular type
C. a class of objects
D. a classification of object

90. Inheritance in object-oriented system is used to :
A. create new classes from existing classes
B. add new operations to existing operations
C. add new attributes to existing attributes
D. add new states to existing states

91. By encapsulation in object-oriented modelling we mean :
A. encapsulating data and programs
B. hiding attributes of an object from users
C. hiding operations on object from users
D. hiding implementation details of methods from users of objects

92. In 8086 microprocessor the following has the highest priority among all type interrupts.
A. NMI B. DIV 0
C. TYPE 255 D. OVER FLOW

93. In 8086 the overflow flag is set when :
A. The sum is more than 16 bits
B. Signed numbers go out of their range after an arithmetic operation
C. Carry and sign flags are set
D. During subtraction

94. 8088 microprocessor differs with 8086 microprocessor in :
A. Data width on the output
B. Address capability
C. Support of coprocessor
D. Support of MAX / MIN mode

95. In ADC 0808 if _______ pin high enables output.
A. EOC B. I/P0-I/P7
C. SOC D. OE

96. The 8279 normally provides a maximum of _____ seven segment display interface with CPU.
A. 8 B. 16
C. 32 D. 18

97. In 8086 microprocessor one of the following instructions is executed before an arithmetic operation :
A. AAM B. AAD
C. DAS D. DAA

98. In FM broadcasting, what is the frequency deviation used?

A. 75 KHz B. 20 KHz
C. 4 KHz D. 48 KHz

99. The gain of a parabolic antenna is proportional to its diameter D by :

A. D B. D^2
C. $\frac{1}{D^2}$ D. $\frac{1}{D}$

100. Antenna gain for a dipole is :

A. 1.5 dBi B. 1 dBi
C. 1.76 dBi D. 2 dBi

Section-III

Directions (Qs. 101-102) : *Each of the following questions consist of one capitalized word followed by 4 words. Choose the option which is most similar in meaning to the capitalized word.*

101. PHILANTHROPY

A. Generosity B. Perversity
C. Flaunting D. Hatred

102. INSCRUTABLE

A. Sublime B. Mysterious
C. Strange D. Happy

Directions (Qs. 103-104) : *Each of the following questions consist of one capitalized word followed by 4 words. Choose the option which is most opposite in meaning to the capitalized word.*

103. EXPLOIT

A. Utilise B. Support
C. Neglect D. Alert

104. CONDEMN

A. Censure B. Approve
C. Recommend D. Prais

Directions (Qs. 105-108) : *Each sentence below has three underlined separate parts (A), (B), (C). Select the underlined part of each sentence where there is an error. If there is no error in the given sentence, select option (D).*

105. Have you gone through (A)/ either of the (B)/ three chapters in this book? (C)/ No error (D)

106. Many a star (A)/ are twinkling (B)/ in the sky. (C)/ No error (D)

107. The speaker was (A)/ not only slow (B)/ but also inaudible as well.(C)/ No error (D)

108. Let us congratulate him (A)/ for his success (B)/ in the examination.(C)/ No error (D)

Directions (Qs. 109-111) : *There are passages given below. After each passage, there are a few questions based on the passage before it. Read the passage first and then answer the questions based on it. You are required to answer the questions based on the contents of the passage and the opinion of the author only.*

It was a very cold evening and so very few people were on the street. I did not go out myself although it was my habit not to keep indoors after sunset. So I closed all the doors and windows and took the book which was lying open on the table and tried to read it. The cold was getting so severe that I started shivering and so I wrapped myself with another blanket. But I could not continue reading the book because I was nearly rendered incapable of turning the pages.

109. The author's habit was to :

A. read in the evening
B. play in the evening
C. go out in the evening
D. sleep in the evening.

110. If it was not cold, the author would have liked to :

A. work
B. sit and look into the street
C. read
D. take a stroll

111. There were not many people on the street because, it was :

A. very cold B. snowing
C. raining D. very dark

112. Clotting of blood is helped by :

A. Vitamin A B. Vitamin K
C. Vitamin D D. Vitamin C

113. The element available in thermometer is :
A. Platinum B. Gold
C. Silver D. Mercury

114. What is the approximate time taken for light from Sun to reach Earth?
A. 8 minutes B. 8 seconds
C. 8 hours D. 2 seconds

115. Where was India's first oil refinery set up?
A. Digboi B. Bombay
C. Barauni D. Vishakhapatnam

116. Which of the following is not a coal field?
A. Jharia B. Singrauli
C. Raniganj D. Khetri

117. From which country is the India's aircraft carrier INS Vikramaditya purchased?
A. Russia B. USA
C. UK D. France

118. Which of the following does not contribute to greenhouse effect?
A. Methane B. Ozone
C. Carbon Dioxide D. Oxygen

119. The boiling point of water is :
A. 100°(degree) F B. 100°(degree) K
C. 100°(degree) C D. 0°(degree) C

120. Which state has the largest coast line in India?
A. Tamil Nadu B. Andhra Pradesh
C. Gujarat D. Kerala

ANSWERS

1	2	3	4	5	6	7	8	9	10
B	B	C	A	C	B	D	C	B	A
11	**12**	**13**	**14**	**15**	**16**	**17**	**18**	**19**	**20**
B	D	C	B	A	B	C	C	A	B
21	**22**	**23**	**24**	**25**	**26**	**27**	**28**	**29**	**30**
D	A	C	A	B	C	D	D	B	A
31	**32**	**33**	**34**	**35**	**36**	**37**	**38**	**39**	**40**
C	B	D	D	A	B	B	C	D	C
41	**42**	**43**	**44**	**45**	**46**	**47**	**48**	**49**	**50**
D	C	A	B	C	B	A	D	A	B
51	**52**	**53**	**54**	**55**	**56**	**57**	**58**	**59**	**60**
A	D	C	B	C	A	D	B	D	C
61	**62**	**63**	**64**	**65**	**66**	**67**	**68**	**69**	**70**
B	A	A	C	A	B	D	C	B	C
71	**72**	**73**	**74**	**75**	**76**	**77**	**78**	**79**	**80**
A	A	B	B	C	B	A	C	C	B
81	**82**	**83**	**84**	**85**	**86**	**87**	**88**	**89**	**90**
D	C	D	A	D	C	D	C	B	A
91	**92**	**93**	**94**	**95**	**96**	**97**	**98**	**99**	**100**
D	A	B	A	D	B	B	A	B	C
101	**102**	**103**	**104**	**105**	**106**	**107**	**108**	**109**	**110**
A	B	B	D	B	B	C	D	C	D
111	**112**	**113**	**114**	**115**	**116**	**117**	**118**	**119**	**120**
A	B	D	A	A	D	A	D	C	C

ENGINEERING STREAM

1

NETWORK THEORY

NETWORK FUNCTIONS

A function relating currents or voltages at different parts of the network is called transfer function. Mathematically transfer function is found to be similar to transfer impedance or admittance and is included in the broader category of functions called network functions.

Driving Point Impedance : For a two port network in zero state with no internal energy sources the driving point impedance is defined as the ratio of the transform voltage at any point to the transform current at the same point. It is normally denoted by Z(s).

Similarly for a two port network in zero state with no internal energy source, driving point admittance is defined as the ratio of the transform current at any port to the transform voltage at the same port. It is normally denoted by Y(s).

Fig. 1 : *A two port network*

Therefore, driving point admittance is simply the reciprocal of the driving point impedance at the same port.

A two port network with reference directions of port voltages and port current is shown in Fig. 1.

The driving point impedance $Z(s)$ at port 1 as per above definition is given by

$$Z(s) = \frac{V_1(s)}{I_1(s)}$$

while the driving point admittance at port 1 as per above definition is given by

$$Z(s) = \frac{I_1(s)}{V_1(s)}$$

Transfer Functions

By convention, transfer function is defined as the ratio of an output quantity to an input quantity. It is normally denoted by H(s). These quantities can be either voltage or current or impedance or admittance. Thus transfer function can take the following four forms.

(i) **Voltage transfer ratio :** This is defined as the ratio of voltage transform at one port of voltage transform to that at another port. It is usually denoted by G(s). Subscript 1, 2 is affixed to G to denote the port number. Using the notations of Fig. 1, we have

$$G_{12}(s) = \frac{V_1(s)}{V_2(s)}$$

and

$$G_{21}(s) = \frac{V_2(s)}{V_1(s)}$$

(ii) **Current transfer ratio :** This is defined as the ratio of current transform at one port to the current transform so that at another port. It is usually denoted by $\alpha(s)$; Subscript 1, 2 is affixed to α to denote the port number. Using the notations of Fig. 1, we get

$$\alpha_{12}(s) = \frac{I_2(s)}{I_1(s)}$$

and

$$\alpha_{21}(s) = \frac{I_1(s)}{I_2(s)}$$

(iii) **Transfer impedance :** This is defined as the ratio of voltage transform at one port to the current transform at another port. It is usually denoted by Z(s). Subscript 1, 2 is affixed to Z to denote the port number. Using again the notations of Fig. 1, we have

$$Z_{12}(s) = \frac{V_1(s)}{I_2(s)}$$

and

$$Z_{21}(s) = \frac{V_2(s)}{I_1(s)}$$

(iv) **Transfer admittance :** It is defined as the ratio of current transform at one port to the voltage transform at another port. It is normally represented by Y (s). Subscript 1, 2 is affixed to Y to denote the port number.
Using the notation of Fig. 1 we get

$$Y_{12}(s) = \frac{I_1(s)}{V_2(s)}$$

and

$$Y_{21}(s) = \frac{I_2(s)}{V_1(s)}$$

Poles and Zeros : We will see that expression for network function N (s) may in general be put in the form of ratio of polynomials in s.

Thus,

$$N(s) = \frac{P(s)}{Q(s)}$$

$$= \frac{a_0 s^n + a_1 s^{n-1} + ...+ a_{n-1} s + a_n}{b_0 s^m + b_1 s^{m-1} + ...+ b_{m-1} s + b_m} \quad ... \text{(i)}$$

where a_n and b_m coefficients are real and positive for passive network (having no controlled sources). The equation P(s) = 0 has *n* root, let say a_1, b_2 .... a_n. Similarly, Q(s) = 0 has *m* roots, let us say b_1, b_2 ... b_m. Then both P(s) and Q(s) may be written as a product of linear factor as follows :

$$N(s) = H = \frac{(s - a_1)\,(s - a_2).....(a - a_n)}{(s - b_1)\,(s - b_2).....(s - b_m)} \quad ... \text{(ii)}$$

where $H = \frac{a_0}{b_0}$ is a constant and is often termed as scale factor.

From equation above we conclude that if the variable *s* has value equal to any of the roots, a_1, a_2 ... a_n the network function N(s) becomes zero because P(s) = 0. Hence these complex frequencies a_1, a_2.......a_n are called zeros of the network function.

Similarly when the variable *s* has any of the values, b_1, b_2.....b_m, the network function N(s) becomes infinites because Q(s) = 0. Hence these complex frequencies b_1, b_2..... b_m are called the poles of network function.

Poles and zeros are useful in describing network function. As is obvious from equations (i) and (ii), a network function is specified by its pole, zero and the scale factor.

Poles and zeros are critical frequencies. At poles the network function becomes, infinite while at zero, the network function becomes zero. At other frequencies, the function has a finite non-zero value.

The poles determine the time variation of the response while zeros determine the magnitudes of coefficients Ai and Ak in the partial fraction expansion and thus determine the magnitude of response. In each of the transfer functions described above, poles determine the time variation of the response whereas zeros determine the magnitude.

Restriction on the location of poles and zeros in the s-plane :

(i) Coefficient in the polynomials P(s) of N(s) $= \frac{P(s)}{Q(s)}$ must be real and positive. We have already stated that the coefficient 'a's and 'b's in equation (i) are real and positive. Hence when s is real P(s) and Q(s) will be real such functions and are called real functions.

(ii) Poles and zeros, if complex, must occur in conjugate pairs. Let polynomial P(s) in equation (i) be a real function. Now let one of zeros of P(s) be complex say, (–x –jy). Then its conjugate namely, –x + jy, must also form a zero of the same polynomial P(s), falling which P(s) will have same complex coefficient.

(iii) The real part of all poles and zeros must be either negative or zero. Taking the above example,

$$P(s) = (s + x + jy)\,(s + x - jy)$$
$$= (s + x)^2 + y^2$$
$$= s^2 + 2sx + (x^2 + y^2).$$

In order to make P(s) a real function, the second term, namely, 2 sx should be positive. Hence (–x) must be negative.

Thus in s-plane all poles and zeros must lie only in left half plane; of course, poles and zeros can lie on the boundary dividing the left half and the right half.

(iv) If the real part of any pole and zero is zero then that pole and zero must be simple i.e. not repeated. We have already stated above that the real part of a complex pole or zero must be either negative or zero. Therefore, the real parts of poles and zeros may be have zero value, that is, the poles and zero may lie on the boundary dividing half of the s-plane. But the restriction imposed is that such poles and zeros must be simple. This restriction crops P because if poles or zeros be of order n, then the time domain response contains terms t^{n-1} and the term becomes infinite as t increases to infinity.

(v) Polynomials P(s) and Q(s) cannot have missing terms between those of highest and lower order unless all even order terms or all odd order terms are missing.

(vi) The degree of P(s) and Q(s) may differ by zero or one only.

(vii) Lowest degree of P(s) and Q(s) may differ by almost one.

Response of Network to Step, Ramp, Impulse and Sinusoidal Inputs

Network elements and the network may be passive linear reciprocal, casual and time invariant.

A network is passive if (i) the energy delivered to the network is non-negative for any arbitrary excitation and (ii) if no voltages or currents appear between any two terminals before an excitation is applied.

A network or a system is linear if (i) the principle of proportionality end (ii) the principle of superpositions holds. Thus, if $e_1(t)$. $r_1(t)$ and $e_3(t)$, $r_2(t)$ are two excitation response pairs, then for an excitation of $k_1\, e_1(t) + k_2\, e_2(t)$, the response should be $e_1\, r_1(t) + k_2\, r_1(t)$.

A network is said to be reciprocal if when the points of excitation and measurement of response are interchanged, the relationship between excitation and response remains the same. This must be true for any choice of points of excitation and response.

A network or a system is casual if the response appears only after the application of excitation; prior to the application of excitation response is zero.

A network or a system is time invariant, if for an excitation $e(t \pm t_0)$, the response is $r(t \pm t_0)$. Thus if the excitation is delayed by t_0 the response too is delayed by t_0 and the waveshape of original response r(t) [corresponding to excitation e(t)] is preserved.

For a linear and time invariant network whose excitation response pair is e(t)–r(t) the response due to an excitation $\left(\frac{d}{dt}e(t)\right)$ is $\left(\frac{d}{dt}r(t)\right)$ and the response due to excitation $\left(\int_0^t e(t)\,(dt)\right)$ is $\left(\int_0^t r(t)\,dt\right)$.

Signals

Signals which do not exhibit random behaviour and can be explicitly characterised as functions of time are referred to as deterministic signals. Signals may be periodic or a periodic. A periodic signal can be described as

$$s(t) = s(t \pm kT);\ k = 0, 1, 2...$$

Where T is the period of signal. It can be of any waveshape. The commonly encountered periodic signals are sinusoidal, triangular, rectangular or a train of pulses. All non-sinusoidal periodic signals can be expanded into Fourier series comprising sinusoidal signals of multiple frequencies and a constant or dc signal can be found.

Signals which are not periodic are aperiodic signals. There can be almost an infinite variety of aperiodic signals. There are certain basic aperiodic signals which can be characterised in simple mathematical terms and serve as building blocks for a large number of other signals. These are step, ramp and impulse signals or functions. These signals are important because if the responses due to such signals are known, then response due to several other aperiodic signals can be determined. As such, a brief description of these functions is given below.

The Step function : The step function is a function which at a specified time jumps from zero to a finite value and remains at that value thereafter. If the specified time is the origin t = 0 and the value acquired is unity, then it is called the unit step function. If the step change occurs at some time to, then it is a shifted unit function. If the step height is k, then it is a step function of value k. Mathematically, the unit step function is defined as

$$\begin{aligned} u(t) &= 0, \quad t < 0 \\ &= 1, \quad t \geq 0 \end{aligned}$$

The shifted unit step function is expressed as $u(t - t_0)$

$$\begin{aligned} u(t - t_0) &= 0, \quad t < t_0 \\ &= 1, \quad t \geq t_0 \end{aligned}$$

A step function of value k is simply written as k u (t) and the one with shift t_0 is expressed as k u $(t - t_0)$. K and t_0 are real numbers. The step functions are shown in Fig. 2

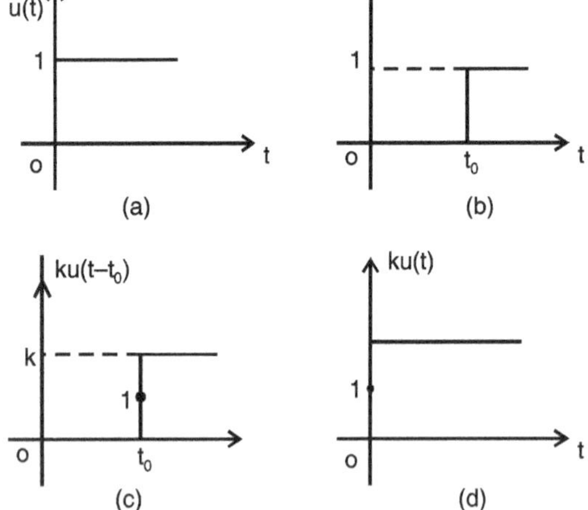

Fig. 2: *(a) Unit step function (b) Shifted Unit step function (c) Step function of magnitude k (d) Shifted function of magnitude k*

The physical analogy of excitation with k u $(t - t_o)$ corresponds to closing a switch at time $t = t_o$ which connects a dc source of k volts to a given network :

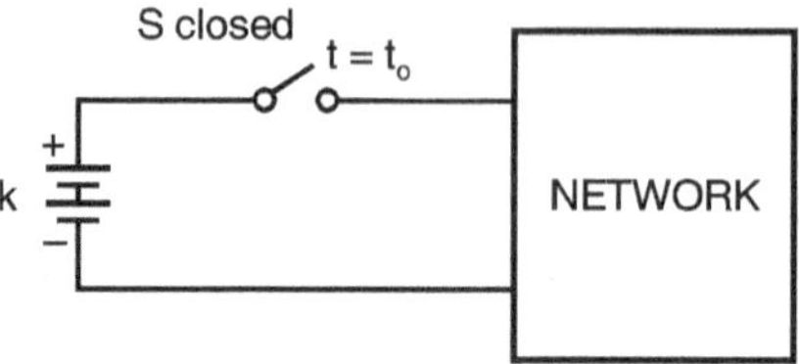

Fig. 3: *Excitation of a network with k $u(t - t_o)$ volt function*

The Ramp Function : The ramp function is a function which at a specified time starts from zero and thereafter increases linearly, with slope k. Mathematically, the unit ramp function is expressed as

$$\rho(t) = t\ u\ (t)$$

and the shifted ramp function of magnitude k is

$$k\rho(t - t_o) = k(t - t_o)\ u(t - t_o)$$

The ramp function is integral of the step function,

$$k\rho(t - t_o) = \int_{t_o}^{t} ku(t - t_o)\ dt$$

If the lower limit in the integration is changed to any value less than t_o, the value of the integral remains unaltered as the function has a value zero for $t < t_o$.

The ramp function and its derivative are shown in Fig. 4.

The Impulse Function : The impulse function is a function which at a specified time jumps to infinity and the area under the function is k. Mathematically, the unit impulse function (also known as delta function) is expressed as

$$\begin{aligned} \delta(t) &= \infty, \quad t = 0 \\ &= 0, \quad t = 0 \end{aligned}$$

$$\text{and} \quad \int_{-\infty}^{+\infty} \delta(t) dt = 1$$

and the shifted impulse function of magnitude k is

$$\begin{aligned} k\delta(t - t_o) &= \infty, \quad t = t_o \\ &= 0, \quad t = t_o \end{aligned}$$

$$\text{and} \quad \int_{-\infty}^{+\infty} k\delta(t - t_o) dt = k$$

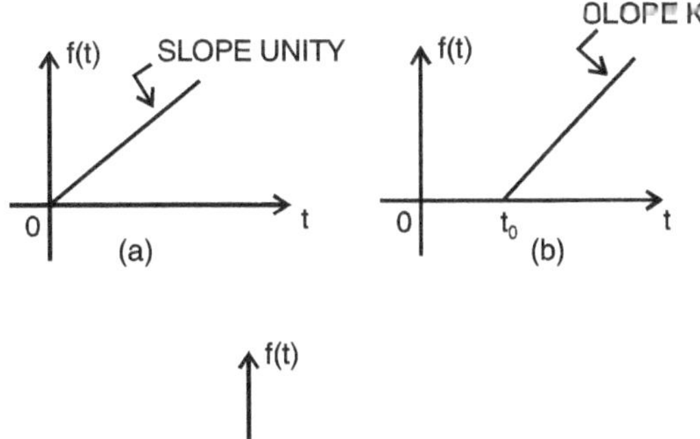

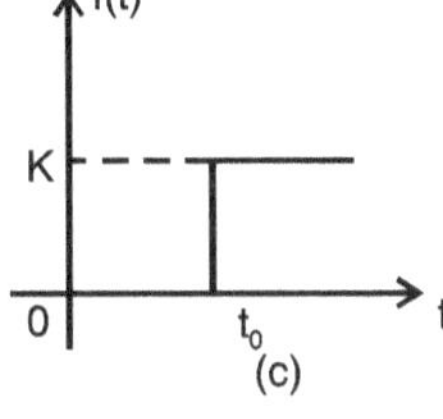

Fig. 4: *(a) Unit tramp function* $\rho(t) = t\ u(t_0)$
(b) Shifted ramp function of magnitude k,
$k\rho(t - t_o) = k(t - t_o) = k(t - t_o)u(t - t_o)$
(c) Derivative of function at (b). The derivative is a shifted step function of height k.

The impulse function is the derivative of the step function and the derivative impulse function is known as doublet (see Fig. 5). The area under the doublet is zero.

The useful mathematical properties of impulse function and doublet function are :

$$\int_{-\infty}^{+\infty} \delta(t) k\delta(t - t_o)dt = k\ f(t_o)$$

and
$$\int_{-\infty}^{+\infty} f(t) k\delta(t - t_o)dt = -k\ f'(t_o)$$

The step, ramp and impulse functions are collectively also known as singularity functions. Unless stated otherwise, the functions are intended to be of unit magnitude and without any shift.

As stated earlier, the singularity function can be used to express mathematically aperiodic signals of straight line segments. The approach is to split the given signal into several sub-signals which can be expressed in terms of singularity functions.

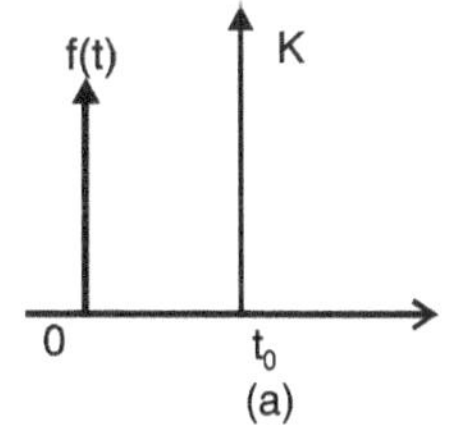

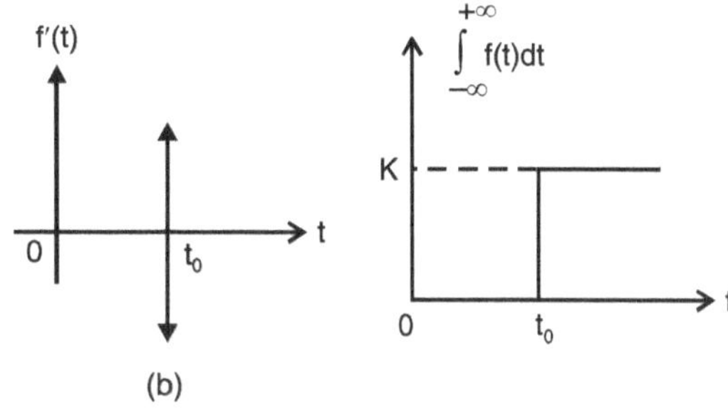

Fig. 5: *(a) Shifted impulse function of magnitude k, (b) the doublet, the derivative of function at (a) (c) Integral of the function at (a). The result is the shifted step function of value k.*

Impulse, Step and Ramp Response

As an aperiodic signal can be expressed in terms of singularity function signals and a non-sinusoidal periodic signal can be expressed in terms of sinusoidal signals. It is highly significant to know the response of the network to such excitations to be able to determine the response due to any excitation. Moreover, a reliable measure of the transient behaviour of the system is given by its step and impluse responses. Therefore, it is 'worthwhile to learn about the system of network response when excited with singularity functions.

Impulse Response : Impulse response of a network or a system is the response of the network to an impulse excitation or input. As the Laplace transform of the impulse function S(t) is unity, then R(s) is simply equal to H(s). Therefore, Laplace inverse of system function H(s) is the impulse response. r(t), denoted by h(t)

Example : *To find current i(t) in a series R-C circuit when excited by a voltage source of e(t) =* $\delta(t)$.

Solution: Since the response is the circuit current and excitation, the circuit voltage the system function is simply :

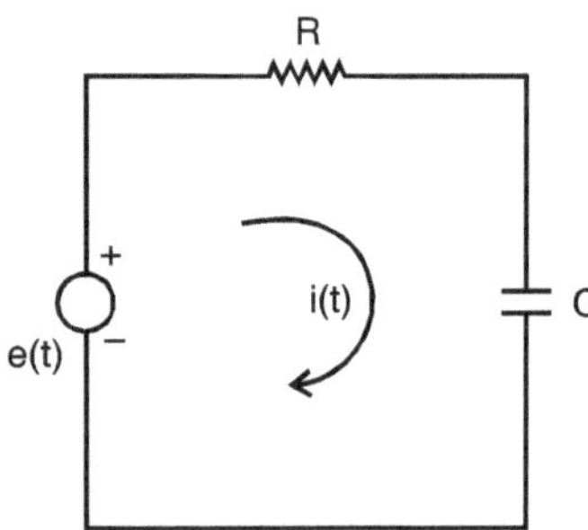

Fig. 6: *An R—C series circuit excited by an impulse voltage function.*

$$H(s) = \frac{I(s)}{E(s)} = \frac{1}{R + \frac{1}{Cs}}$$

$$= \frac{Cs}{RCs + 1} = \frac{\frac{1}{R} - s}{s + \frac{1}{RC}}$$

$$\therefore \text{ Impulse response h(t) is } £^{-1} \frac{\frac{1}{R}S}{S + \frac{1}{RC}}$$

We know Laplace inverse of $\frac{k}{s+a}$ is $ke^{at}u(t)$

$$\therefore\ £^{-1} \frac{\frac{1}{R}}{s + \frac{1}{RC}} \text{ is } \frac{1}{R} e^{-t/RC}\ u(t)$$

Further, $£^{-1}$ SF(s) is $\frac{d}{dt} f(t)$

$$\therefore\ h(t) = \frac{d}{dt}\left(\frac{1}{R} e - \frac{t}{RC} u(t)\right)$$

$$= \frac{1}{R}\left[e - \frac{t}{RC}\delta(t) + u(t)e - \frac{t}{RC}\left(-\frac{1}{RC}\right)\right]$$

$$= \frac{1}{R}\left(\delta(t) - \frac{1}{RC} e - \frac{t}{RC} ut\right)$$

This response is sketched in Fig. 7

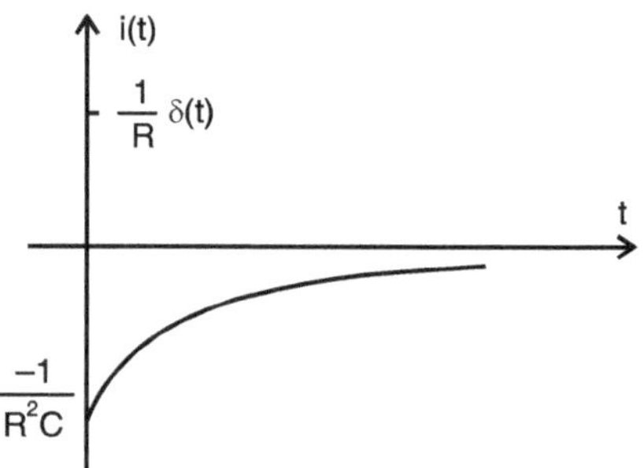

Fig. 7: *Impulse response of the circuit at Fig. 6*

Step Response : Step response of a network or a system is the response of the network to a step function input or excitation. As step function is the integral of the impulse function, the step response can be obtained simply by integrating the impulse response in time domain. Alternatively, the step response is the Laplace inverse of $\frac{H(s)}{s}$.

Example : *Considering the circuit of Fig. 6, its impulse response has already been found.*

Solution : Step response s(t) is

$$\int_0^t \left(\frac{1}{R}\delta(t) - \frac{1}{RC} e^{-\frac{t}{RC}} u(t)\right) dt$$

$$= \frac{1}{R} - \frac{1}{R^2 C}\int_0^t e^{-\frac{t}{RC}} dt$$

$$= \frac{1}{R} - \frac{1}{RC^2}\left[-RCe^{-\frac{t}{RC}}\right]_0^t$$

$$= \frac{1}{R} + \frac{1}{R} e\left[e^{-\frac{t}{RC}} u(t) - 1\right]$$

$$= \frac{1}{R} e^{-\frac{t}{RC}} u(t)$$

In fact, this has already been obtained in the previous example.

We also see that

$$£^{-1} \frac{H(s)}{s} = £^{-1} \frac{1/R}{s + 1/RC} \text{ is } \frac{1}{R} e^{-\frac{t}{RC}}$$

Ramp Response : Ramp response of a network or a system is the response of the network to a ramp function input or excitation. As the ramp function is the integral of the step function, the ramp response can be obtained simply by integrating the step response in time domain. Alternatively, the ramp response is the Laplace inverse of $\frac{H(s)}{s^2}$.

Example : *Considering the circuit of Fig. 6, its step response has already been found as* $\frac{1}{R} e^{-\frac{t}{RC}} u(t)$.

Solution : The ramp response R(t) is then

$$\int_o^t \frac{1}{R} e^{-t/RC} u(t) dt$$

$$\frac{t}{R} \int_o^t e^{-\frac{t}{RC}} dt$$

$$= \frac{1}{R} \left| \frac{e^{-\frac{t}{RC}}}{-1/RC} \right|_o^t$$

$$= C \left[1 - e^{-\frac{t}{RC}} \right] u(t)$$

Let us find the ramp response by determining

$$£^{-1} \frac{H(s)}{s^2}$$

$$\frac{H(s)}{s^2} = \frac{1}{s^2} \frac{s1/R}{(s + r/RC)} = \frac{1R}{s(s + 1/RC)}$$

$$= \frac{C}{s} - \frac{C}{s + 1/RC}$$

$$\therefore\ £^{-1} \frac{H(s)}{s^2} = cu(t) - ce^{-t/RC} u(t)$$

$$= c \left(u - (t) e^{-\frac{t}{RC}} \right) u(t)$$

Example : *If the impulse response of a time invariant system in* $h(t) = e^{-t} u(t)$, *determine the response due to an excitation* $2\delta(t-2) - 3\delta(t-3)$.

Solution : Applying the principle of linearity, superposition and time invariance, the required response is

$$r(t) = 2e^{-(t-2)} u(t-2) - 3e^{-(t-3)} u(t-3)$$

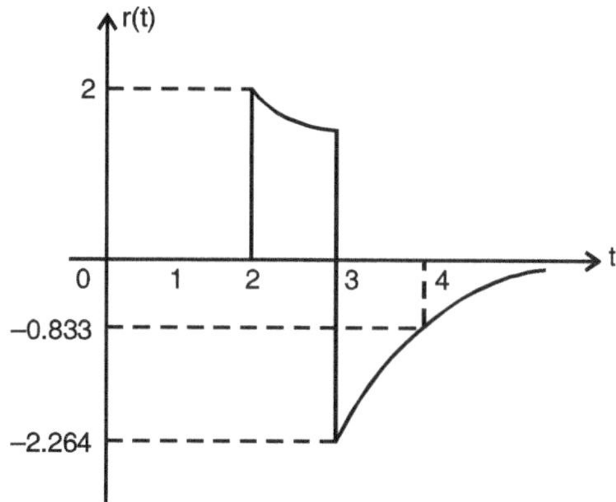

Fig. 8

Steady-State Sinusoidal Response

It has been started earlier that a non-sinusoidal periodic signal can be expanded into its Fourier series which comprises sinusoidal signals of fundamental and multiple frequencies. Therefore, if one knows the response of a network to a sinusoidal signal, its response to a non-sinusoidal periodic signal can be determined. The knowledge of steady state sinusoidal response is also valuable in studying and analysing resonance circuits, transmission circuits and filter circuits. As such, finding steady state sinusoidal response of the networks is important.

The system function H(s) of a network has been defined as the ratio of the Laplace transform R(s) of response r(t) of the network, to the Laplace transform E(s) of excitation e(t) under zero initial conditions. For a given pair of points of excitation and response; and the type of excitation and response signals (i.e., whether a voltage or a current) the system function is the same, whatever may be e(t).

It can be shown that system function is also the ratio of response r(t) to excitation e(t) when the excitation is a generalised cisoidal signal of the form A e^{st}. In a e^{st}, s is the complex frequency $(\sigma + jw)$.

Here $e\sigma^t$ is a physical signal but e^{jwt} is a non-physical signal through the real and imaginary parts of the e^{jwt} are physical signals, viz. Cos wt and Sin wt. It can further be found that the ratio of steady-state response r(t) to e(t) when e(t) is of the form A sin $(wt + \phi)$ or A cos $(wt + \phi)$ is the expression H(jw), obtained from H(s) on replacing *s* by *jw* in it. *j*A sin $(wt + \phi)$ and $\frac{1}{j}$ A sin $(wt + \phi)$ are then to be interpreted as A sin $(wt + \phi + 90°)$ and A sin $(wt + \phi - 90°)$ respectively. Similar statements apply to cosine function also. Finally, if sinusoidal excitation is expressed as a phasor E(*jw*) either in polar form or in complex form, the steady-state response expressed as a phasor R(*jw*) is H(*jw*). Therefore, it is sufficient to learn about the behaviour of the function H(*jw*) to know about steady-state sinusoidal response of a network or a system.

The function H(*jw*) can be simplified and expressed either in polar form or in complex form.

$$H(jw) = M(w)\, e^{j\phi(w)}$$

When M(*w*) is the amplitude or magnitude response function, and is an even function in *w*; and $\phi(w)$ is the phase response function and is an odd function of *w*. Alternatively.

$$H(jw) = R(w) + jI(w)$$

where R(*w*) is the real part of H(*jw*) and is an even function in *w*; and I(*w*) is the imaginary part and is an odd function in *w*.

Real part and imaginary part do not directly give information on physical quantities. As such magnitude and phase response functions are more in usage. The function M(*w*) and $\phi(w)$ are also collectively known as frequency (radian) response functions. Given a system function, H(*s*); M(*w*) and $\phi(w)$ expressions can be obtained from it and then the value of M(*w*) and $\phi(w)$ can be computed for any desired frequency. Frequency response is also commonly depicted graphically by plotting M and ϕ as functions of *w*. The w-axis scale may be a linear scale or a logarithmic scale. In the latter case, this M and ϕ plots are called Bode plots and are more commonly used in control systems studies.

Analytic Expressions

Let N(*s*) and D(*s*) be the numerator and denominator polynomials respectively of H(*s*).

$$H(s) = \frac{N(s)}{D(s)}$$

Substitute *jw* for *s* in H(*s*)

$$\text{Then} \quad H(jw) = \frac{N(jw)}{D(jw)}$$

Grouping the real and imaginary terms,

$$H(jw) = \frac{N_R(w) + jN_2(w)}{D_R(w) + jD_1(w)}, \text{ where}$$

the quantities with suffixes R and I are respectively real parts and imaginary parts of the two polynomials. The magnitudes are given by

$$N(w) = \sqrt{N_R^2(w) + N_I^2(w)} \text{ and}$$

$$D(w) = \sqrt{D_R^2(w) + D_I^2(w)}$$

$\therefore$ the magnitude response of H(*s*) is

$$M(w) = \frac{N(w)}{D(w)} \frac{\sqrt{N_R^2(w) + N_I^2(w)}}{\sqrt{D_R^2(w) + D_I^2(w)}}$$

The angles of the numerator and denominator are

$$\phi N(w) = \tan^{-1} \frac{N_I(w)}{N_R(w)} \text{ and}$$

$$\phi D(w) = \tan^{-1} \frac{D_I(w)}{D_R(w)}$$

$\therefore$ the phase response function of the network whose system function is H(*s*) is

$$\phi(w) = \phi_N(w) - \phi_D(w)$$

$$= \tan^{-1} \frac{N_I(w)}{N_R(w)} - \tan^{-1} \frac{D_I(w)}{D_R(w)}$$

Care should be taken in finding $\phi_N(w)$ or $\phi_D(w)$ when either the real part or imaginary part or both are negative.

$$\text{Let} \quad \alpha = \tan^{-1}\left|\frac{\text{Imaginary part}}{\text{Real part}}\right|$$

Real part	Imaginary part	ϕ (in degrees)
positive	positive	α
positive	negative	$-\alpha$
negative	positive	$180-\alpha$
negative	negative	$180+\alpha$

The frequency at which M(w) has the highest value is denoted by W_{max}, and the one at which the value of M(w) falls to $\frac{1}{\sqrt{2}}(=0.707)$ times its maximum value is denoted by w_e and is called the cut-off frequency, a term commonly used in filter circuits studies. This frequency is also known as the half-power frequency as the power to the circuit at this frequency becomes half of the maximum power which occurs when M(w) has maximum value.

Two-Port Network

A network is represented by a rectangular box. When a conductor is fastened to a node in the network and brought out for access, the end of the conductor is called a terminal. Terminals are required for connecting driving force to the network, for connecting load or making measurements. The minimum number of terminals that are useful are two. One pair of terminals is given the name port. Fig. 9 shows two ports viz 1-2 and 3-4, hence it is called to-port network. The port 1-2 is normally connected to the driving force or input while port 3-4 is connected to a load or the output.

The network shown has four variables identified by two voltages and two currents. We assume that

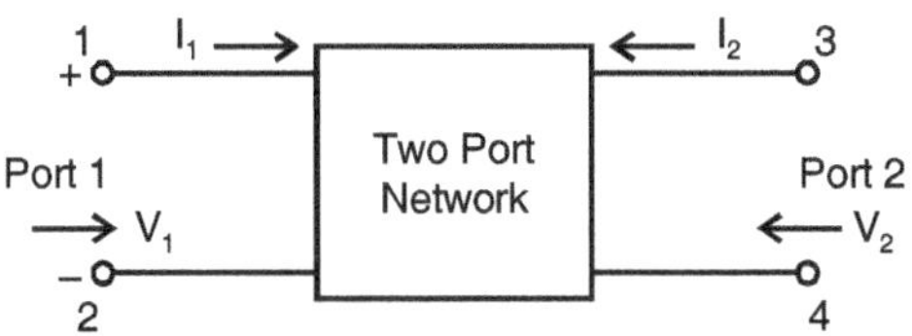

Fig. 9: *Two-port network*

the variables transform quantities and usc V_1 and I_1 as variables at the input port 1; and V_2 and I_2 as the variables at the output port 2. Now only two of the four variables are independent and the specific action of any two of then determines the remaining two.

As an example, if V_1 and V_2 are specified, then I_1 and I_2 are determined. The dependence of two of the four variables on the other two can be described in a number of ways with the help of parameters. In this chapter we will discuss four types of such parameters viz. z; y, h and ABCD. We will then discuss the relationship between these four parameters. This will help in converting one set of parameters to other.

Network parameters : Since two port networks are characterised by parameters in terms of voltage and current variables, the names of parameters are chosen to indicate dimension (impedence. z and admittance, y), lack of consistent dimensions (hybrid h) or the principal application of the parameter (transmission, ABCD).

(i) z-parameters : In this system of parameters the input and output voltages are expressed in terms of the input and output currents. Thus referring to Fig. 9, we have

$$V_1 = z_{11}I_1 + z_{12}I_2$$

and

$$V_2 = z_{21}I_1 \times v_{22}I_2$$

In the matrix from

$$\begin{bmatrix} V_1 \\ V_2 \end{bmatrix} = \begin{bmatrix} z_{11} & z_{12} \\ z_{21} & z_{22} \end{bmatrix} \begin{bmatrix} I_1 \\ I_2 \end{bmatrix}$$

where z_{11}, z_{12}, z_{21} and z_{22} are called the z-parameters of the two port network shown in Fig. 9. They are defined as follows :

z_{11} is defined as the input impedance when output is open circuited. Mathematically, it is defined as

$$z_{11} = \frac{V_1}{V_1} \text{ when } I_2 = 0$$

z_{22} is defined as the output impedance when input is open circuited. Mathematically, it is defined as

$$z_{22} = \frac{V_2}{I_2} \text{ when } I_1 = 0$$

z_{12} is called mutual or reverse transfer impedance when input is open-circuited. Mathematically, it is defined as

$$z_{12} = \frac{V_1}{I_2} \text{ when } I_1 = 0$$

z_{12} is called mutual or forward transfer impedance when output is open-circuited. Mathematically, it is defined as

$$z_2 f = \frac{V_2}{I_1} \text{ when } I_2 = 0$$

For a passive network

$z_{12} = z_{21}$

Since z-parameters are defined with either input open circuit or output open circuited, they are also called open circuited parameters.

(ii) y-parameters : In this system of parameters, the input and output current are expressed in terms of the input and output voltages. Thus referring to Fig. 9 we have

$$I_1 = y_{11}V_1 + y_2V_2$$
$$I_2 = y_{21}V_1 + y_{22}V_2$$

In the matrix form, the equations are

$$\begin{bmatrix} I_1 \\ I_2 \end{bmatrix} = \begin{bmatrix} y_{11} & y_{12} \\ y_{21} & y_{22} \end{bmatrix} \begin{bmatrix} V_1 \\ V_2 \end{bmatrix}$$

where y_{11}, y_{21}, y_{22} are called y-parameters of the four terminal network and are defined as follows:

y_{11} is defined as the output admittance when input is short-circuited. Mathematically, it is defined as

$$y_{11} = \frac{I_1}{V_1}, \text{ when } V_2 = 0$$

y_{22} is defined as the output admittance when input is short-circuited. Mathematically, it is defined as

$$y_{12} = \frac{I_2}{V_2}, \text{ when } V_1 = 0$$

y_{12} is called the mutual or reverse transfer admittance with output short-circuited. Mathematically, it is defined as

$$y_{12} = \frac{I_1}{V_2}, \text{ when } V_1 = 0$$

y_{21} is called the mutual or forward transfer admittance with output short-circuited. Mathematically, it is defined as

$$y_{12} = \frac{I_1}{V_1}, \text{ when } V_2 = 0$$

For passive network, $y_{12} = y_{21}$

Since y parameters are defined with either input or output short circuit, they are also called as short circuited parameters.

(iii) h-parameters : These parameters combine some of the properties of the z and y parameters. In this system of parameters the input voltage and output current are both expressed in terms of input current and about current voltage. Thus, referring to Fig. 9 we have

$$V_1 = I_1\ h_{11} + V_2h_{12}$$
$$I_1 = I_1\ h_{21} + V_2h_{22}$$

In the maximum

$$\begin{bmatrix} V_1 \\ I_2 \end{bmatrix} = \begin{bmatrix} h_{11} & h_{12} \\ h_{21} & h_{22} \end{bmatrix} \begin{bmatrix} I_1 \\ V_2 \end{bmatrix}$$

where h_{11}, h_{12}, h_{21} and h_{22} are called h-parameters of a four terminal network and are defined as follows :

h_{11} is defined as the input impedance with output short circuited. Mathematically, it is defined as

$$h_{11} = \frac{V_1}{I_1}, \text{when } I_1 = 0$$

h_{12} is the reverse voltage transfer ratio with output open-circuited. Mathematically, it is defined as

$$h_{12} = \frac{V_1}{V_2}, \text{ when } I_1 = 0$$

h_{21} is the forward current transfer ratio with output short-circuited. Mathematically, is it defined as

$$h_{21} = \frac{I_2}{I_1}, \text{ when } V_3 = 0$$

h_{22} is the output admittance with input open-circuited. Mathematically, it is defined as

$$h_{22} = \frac{I_2}{V_2}, \text{ when } I_1 = 0$$

Since h_{11} has the dimension of an impedance, h_{22} has the dimension of an admittance, h_{12} and h_{21} are pure numbers; these parameters are also called hybrid parameters.

(iv) ABCD parameters : In this system of parameters voltage and current at port 1 are expressed in terms of voltage and current at port 5 and defining equations are :

$$V_1 = AV_2 - BI_2$$
$$I_1 = CV_2 - DI_2$$

In the matrix form,

$$\begin{bmatrix} V_1 \\ I_1 \end{bmatrix} = \begin{bmatrix} A & B \\ C & D \end{bmatrix} \begin{bmatrix} V_2 \\ -I_2 \end{bmatrix}$$

In Eq. above, the negative sign in the second term is for I_2 and not for parameters B and D. This negative sign arises from the fact that reference direction of I_2 as in Fig. 9 is opposite to that in power transmission problems.

A is defined as the ratio of input to the output voltage when output is open-circuited. Therefore,

$$A = \frac{V_1}{V_2}, \text{ when } I_2 = 0$$

B is defined as the ratio of input voltage to output current when the output is short-circuited. Therefore,

$$B = \frac{V_1}{V_2}, \text{ when } V_2 = 0$$

C is defined as the ratio of input current to output voltage when output is open-circuited. Therefore,

$$C = \frac{I_1}{V_2}, \text{ when } I_2 = 0$$

D is defined as the ratio of input current to output current when output is short-circuited. Therefore

$$D = \frac{I_1}{I_2}, \text{ when } V_2 = 0$$

Relations between Two-port Parameters

We have studied four sets of parameters so far. Each has its own utility and is suited for certain specific applications. However, we often find it necessary to convert from on set of parameters to another. Through simple mathematical manipulation. It is possible to convert from any one set to any of the remaining sets. Table 'A' gives the relationships between different sets of parameters. In this table, matricies placed in each of the rows are equivalent. The equivalence involves determinants Δz, Δy etc. where,

$$\Delta z = z_{11}z_{22} - z_{12}z_{21}$$
$$\Delta_y = y_{11}y_{22} - y_{12}y_{21}$$

and so on.

Interconnection of Two Ports Network

Consider two ports N_1 and N_2 whose voltages and currents are as shown in Fig. 10 Suppose these two networks are connected together in a certain manner so that resulting network, say N, is also a two-port network. Then N can also be characterised by two part parameters. It is obvious that the parameters of N are related to the parameters of N_1 and N_2. We will discuss these relationships for various interconnections of N_1 and N_2.

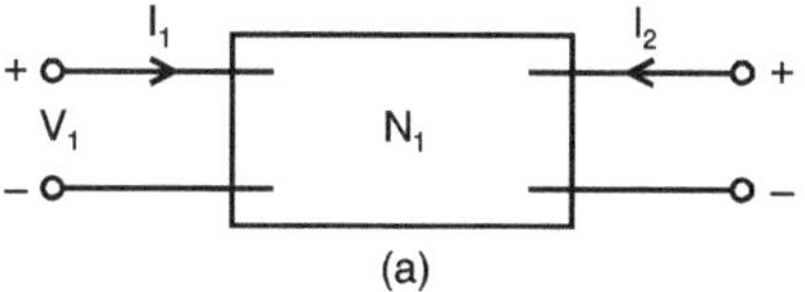

(a)

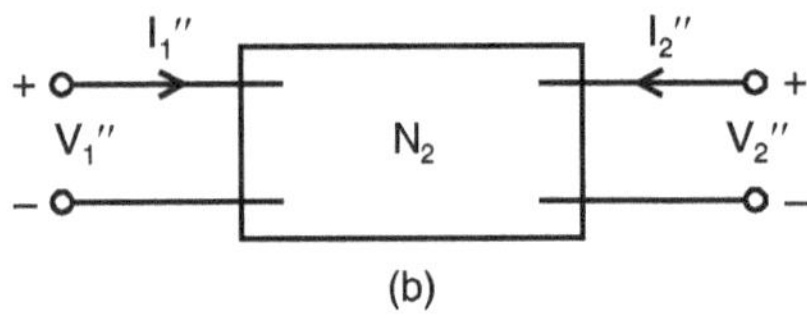

Fig. 10: *Two networks to to be interconnected.*

Transfer Function of Two Ports

For a two ports network in zero state with no internal energy source, the driving point impedance is defined as the ratio of the transform voltage at any point to the transform current at the same port. It is normally denoted by Z(s).

Therefore, refering to Fig. 11.

$$Z(s) + \frac{V_1(s)}{I_1(s)}$$

Similarly, for a two port network in zero state with no internal energy sources, driving point admittance is defined as the ratio of the transform current at any point to the transform voltage at the same point. It is normally denoted by Y (s).

Therefore, refering to Fig. 11

$$y(s) = \frac{I_1(s)}{V_1(s)}$$

Because of the similarity of impedance and admittance and to avoid writing impedance and admittance the two quantities are assigned one name immittance a combination of impedance and admittance. An immitance is thus an impedance or an admittance.

By convention transfer function is defined as the ratio of output quantity to an input quantity. These quantities can be either voltage or current or

TABLE A : Conversion of Parameters

	$[z]$		$[y]$		$[ABCD]$		$[h]$	
$[z]$	z_{11}	z_{12}	$\frac{y_{22}}{\Delta y}$	$\frac{-y_{22}}{\Delta y}$	$\frac{A}{C}$	$\frac{\Delta T}{C}$	$\frac{\Delta h}{h_{22}}$	$\frac{h_{22}}{h_{12}}$
	z_{21}	z_{22}	$\frac{-y_{21}}{\Delta y}$	$\frac{y_{11}}{\Delta y}$	$\frac{1}{C}$	$\frac{D}{C}$	$\frac{-h_{21}}{h_{22}}$	$\frac{1}{h_{22}}$
$[y]$	$\frac{z_{22}}{\Delta z}$	$\frac{-z_{12}}{\Delta z}$	y_{11}	y_{12}	$\frac{D}{B}$	$\frac{-\Delta T}{B}$	$\frac{1}{h_{11}}$	$\frac{-h_{12}}{h_{12}}$
	$\frac{-z_{21}}{\Delta z}$	$\frac{-z_{11}}{\Delta z}$	y_{21}	y_{22}	$\frac{1}{B}$	$\frac{A}{B}$	$\frac{h_{21}}{h_{11}}$	$\frac{-h_{11}}{h_{21}}$
$[ABCD]$	$\frac{z_{11}}{z_{21}}$	$\frac{\Delta z}{z_{21}}$	$\frac{-y_{22}}{y_{21}}$	$\frac{-1}{y_{21}}$	A	B	$\frac{\Delta h}{h_{21}}$	$\frac{-1}{h_{21}}$
	$\frac{1}{z_{21}}$	$\frac{z_{22}}{z_{21}}$	$\frac{-\Delta y}{y_{21}}$	$\frac{-y_{11}}{y_{21}}$	C	D	$\frac{-h_{22}}{y_{21}}$	$\frac{-1}{h_{21}}$
$[h]$	$\frac{\Delta z}{z_{22}}$	$\frac{-z_{21}}{z_{22}}$	$\frac{1}{y_{11}}$	$\frac{-y_{12}}{y_{11}}$	$\frac{B}{D}$	$\frac{\Delta T}{D}$	h_{11}	h_{22}
	$\frac{-y}{z_{22}}$	$\frac{1}{z_{22}}$	$\frac{y_{21}}{y_{11}}$	$\frac{\Delta y}{y_{11}}$	$\frac{1}{D}$	$\frac{C}{D}$	h_{21}	h_{22}

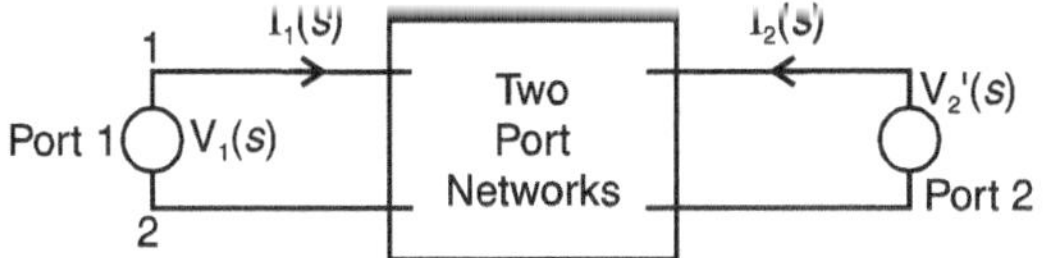

Fig. 11 : *Transform of two port network shown in Fig. 10.*

impedance or admittance. Thus transfer function can take the following four forms. Referring to the transform of the two port network shown in Fig. 11 we have

(i) **Voltage tranfer ratio :** This is defined as the ratio of voltage at one port to voltage transform at another port. It is normally denoted by G(s) Subscript. 1, 2 is affixed to G to denoted port number. Using notations of Fig. 11, we have

$$G_{12}(s) = \frac{V_1(s)}{V_2(s)}$$

$$G_{21}(s) = \frac{V_2(s)}{V_1(s)}$$

***(ii)* Transfer Impedance:** This is defined as the ratio of voltage transform at one port to the current transform at another port. It is normally denoted by Z(*s*). Subscript 1, 2 is affixed to Z to denote the port number. Using notations of Fig. 11 we get

$$Z_{12}(s) = \frac{V_1(s)}{I_2(s)}$$

$$Z_{21}(s) = \frac{V_2(s)}{I_1(s)}$$

***(iii)* Transfer admittance :** It is defined as the ratio of current transform at one port to voltage transform at another port. It is normally denoted by Y(s). Subscript 1, 2 is affixed to Y to denote the port number. Using notations of Fig. 11 we have

$$Y_{12}(s) = \frac{I_1(s)}{V_2(s)}$$

$$Y_{21}(s) = \frac{I_2(s)}{V_1(s)}$$

For finding the transfer function we start at the output and then work backward applying Kirchhoff's current law and Kirchhoff's voltage law.

SYNTHESIS OF NETWORKS

Realizability

Causality and Stability : The starting point for any synthesis problem is the system function

$$H(s) = \frac{R(s)}{E(s)}$$

where E(s) is the excitation and R(s) is the response.

Out task is the synthesize a network from a given system function.

The first step in a synthesis procedure is to determine whether H(s) can be realized as a physical passive network. There are two important considerations – causality and stability. By causality we mean that a voltage cannot appear between any pair of terminals in the network before a current is impressed, or vice versa. In other words, the impulse response of the network must be zero for $t < 0$, that is,

$$h(t) = 0 \quad t < 0$$

As an example, the impulse response

$$h(t) = e^{-at}u(t)$$

is causal, whereas $h(t) = e^{-a|t|}$ is not causal.

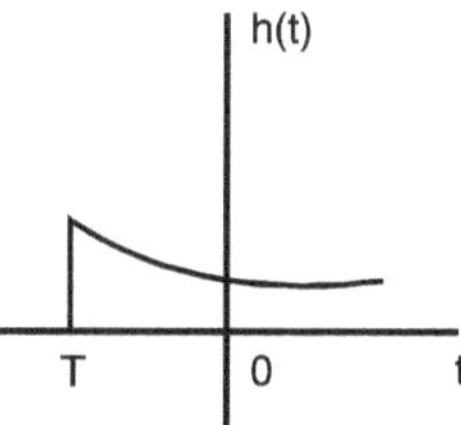

Fig. 12: *Nonrealizable impulse response*

In certain cases, the impulse response could be made realizable (causal) by delaying it appropriately. For example, the impulse response in Fig. 12 is not realizable. If we delay the response by T seconds, we find that the delayed response h(t – T) is realizable, (Fig. 13).

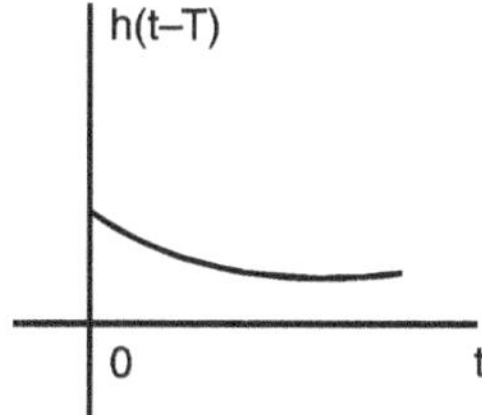

Fig. 13: *Realizable impulse response*

In the frequency domain, causality is implied when the Paley-Wiener criterion is satisfied for the amplitude function $|H(j\omega)|$. The Paley-Wiener criterion states that a necessary and sufficient condition for an amplitude function $|H(j\omega)|$ to be realizable (causal) is that

$$\int_{-\infty}^{\infty} \frac{|\log I\, H(j\omega)I|}{1+\omega^2}\, d\omega < \infty$$

The following conditions must be satisfied before the Paley-Wiener criterion is valid; h(t) must posses a Fourier transform $H(j\omega)$; the square magnitude function $|H(j\omega)|$ must be integrable, that is,

$$\int_{-\infty}^{\infty} |H(j\omega)|^2 \quad dw < \infty$$

The physical implication of the Paley-Wiener criterion is that the amplitude $|H(j\omega)|$ of a realizable network must be zero over a finite band of frequencies. Another way of looking at the Paley-Wiener criterion is that the amplitude function cannot fall off to zero faster than exponential order. For example, the ideal low pass filter Fig. 14 is not realizable because beyond ωc the amplitude is zero. The Gaussion shaped curve shown in Fig. 15

$$|H(j\omega)| = e^{-\omega^2}$$

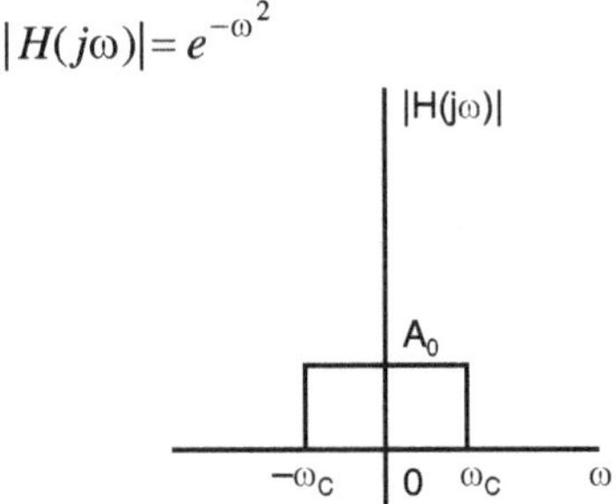

Fig. 14: *Ideal filter characteristics*

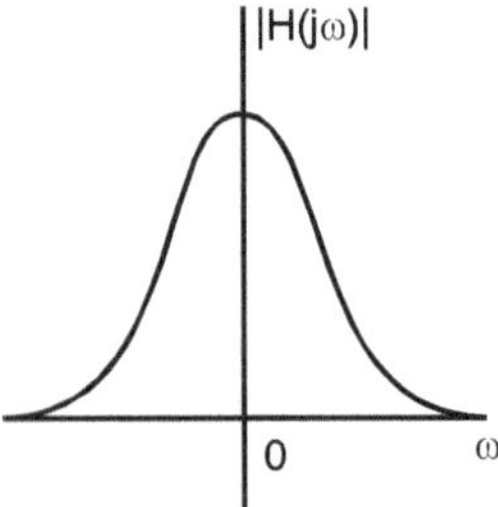

Fig. 15: *Gaussian filter characteristics*

is not realizable because

$$|\log|H(j\omega)| = \omega^2$$

so that the integral $\int_{-\infty}^{\infty} \frac{\omega^2}{1+\omega^2} d\omega$

is not finite. On the other hand, the amplitude function

$$|H(j\omega)| = \frac{1}{\sqrt{1+\omega^2}}$$

does represent a realizable network. In fact, the voltage ratio transfer function of the R-C network in Fig. 16 has an amplitude characteristic given by $|H(j\omega)|$ in eq. above.

For the ideal filter in Fig. 16 the inverse transform $h(t)$ has the form where A_0 is

$$h(t) = \frac{A_o \sin \omega_c t}{\pi t}$$

a constant, from the sin x/x curve, we see that $h(t)$ is non zero for t less than zero. In fact, in order to make $h(t)$ causal it must be delayed by an infinite amount. In practice, however, if we delay $h(t)$ by a large but finite amount t_d such that for $t < 0$ the magnitude of $h(t - t_d)$ less than a very small quantity $\in$, that is,

$$|h(t - t_d)| < \in t < 0$$

1Ω
+ +
$V_0(s)$ 1f $V_0(s)$
− −

Fig. 16

we then can approximate $h(t - t_d)$ by a causal response $h_1(t)$ which is zero for $t < 0$.

If a network is stable, then for a bounded excitation $e(t)$ the response $h(t)$ is also bounded. In other words, if

$$|e(t)| < C_1 \quad 0 \le t < \infty$$

then $|r(t)| < C_2 \quad 0 \le t < \infty$

where C_1 and C_2 are real positive, finite quantities. If a linear system is stable, then from the convolution integral we obtain

$$|r(t)| < C_1 \int_0^{\infty} |h(\tau)| d\tau < C_1$$

Equation requires that the impulse response be absolutely integrable, or

$$\int_0^{\infty} |h(\tau)| d\tau < \infty$$

One important requirement for $h(t)$ to be absolutely integral is that the impulse response approaches zero as t approaches infinity that is,

$$\lim_{t \to \infty} h(t) \to 0$$

Generally, it can be said that with exception of isolated impulse, the impulse rest must be bounded for all t, that is,

$$|h(t)| < C \quad \text{all } t$$

where C is real, positive, finite number.

Observe that our definition of stability precludes such terms as sin $\omega_0 t$ from impulse response in $\omega_0 t$ is not absolutely integrable. These undamped sinusoidal terms are associated with simple poles on the $j\omega$ axis. Since pure L-C network have system functions with simple poles on the $j\omega$ axis, and since we do not wish to call these works unstable, we say that a system is marginally stable if its impulse response is bounded according to above equation, but does not approach zero as t approaches infinity.

In the frequencies domain, the stability criterion requires that the system function possess poles in the left half plane or on the $j\omega$ axis only. Moreover the poles on the axis must be simple. As a result of the requirement of simple poles on the $j\omega$ axis H(s) is given as

$$H(s) = \frac{a_n s^n + a_{n-1} s^{n-1} + \ldots\ldots. + a_1 s + a_0}{b_m s^m + b_{m-1} s^{m-1} + \ldots\ldots. + b_1 s + b_0}$$

then the order of the numerator n cannot exceed the order of the denominator m by more that unity, that is, $n - m \leq 1$. If n exceeded m by more than unity, this would impleis that at $s = j\omega = \infty$ and there would be a multiple pole. To summarize. In order for network to be stable the following three conditions on its system function H(s) must satisfied :

1. H(s) cannot have poles in the right half plane.
2. H(s) cannot have multiple poles in the $j\omega$ axis.
3. The degree of the numerator of H(s) cannot exceed the degree of the denominator by more than unity.

Finally, it should be pointed out that a rational function H(s) with poles in the left half plane only has an inverse transform h(t), which is zero for t < 0. In this respectability implies causality. Since system functions of passive linear networks with lumpen elements are rational functions with poles in the left-half plane or $j\omega$ axis only, causality ceases to be a problem when we deal with system functions of this type. We are only concerned with the problem of causality when we have to design a filter for a given amplitude characteristic such as the ideal filter in Fig. We know we could never hope to realize exactly a filter of this type because the impulse would not be causal to this extent the Paley-Wiener criterion is helpful the limits of our capability.

Hurwitz Polynomials

We know that in order for a system function to be stable, its poles must be restricted to the left-half plane or the $j\omega$ axis. Moreover, the poles on the $j\omega$ axis must be simple. The denominator polynomial of the system function belongs to a class of polynomails known as Hurwitz polynomails. A polynomial P(s) is said to be Hurwitz if the following conditions are satisfied :

1. P(s) is real when s is real.
2. The roots of P(s) have real parts which are zero or negative. As a result of these conditions, if P(s) is a Hurwitz polynomial given by

$$P(s) = a_n s^n + a_{n-1} s^{n-1} + \ldots + a_1 s + a_0$$

then all the coefficients a_1 must be real; if $s_1 = \alpha_1 + j\beta$ is a root of P(s), then a_1 must be negative the polynomial

$$P(s) = (s+1)\,(s+1+j\sqrt{2})\,(s+1-j\sqrt{2})$$

is Hurwitz because all of its roots have negative real parts. On the other hand,

$$G(s) = (s-1)(s+2)(s+3)$$

is not Hurwitz because of the root s = 1, which has a positive real part. Hurwitz polynomials have the following properties :

1. All the coefficients a_1 are non-negative. This readily seen by examining the three types of roots that a Hurwitz polynomial might have. These are

   $S = -\gamma_1$ $\quad\quad$ γ_1 real and positive

   $S = \pm j\omega_1$ $\quad\quad$ ω_1 real

   $S = -\alpha_1 \pm j\beta_1$ $\quad\quad$ α_1 real positive

   The polynomial P(s) which contains these roots can be written as

   $$P(s) = (s+\gamma_1)(s^2+\omega_1^2)[(s+\alpha_1)^2+\beta_1^2]...$$

   Since P(s) is the product of terms with only positive coefficients, it follows that the coefficients P(s) must be positive. Corollary is that between the highest order term in s and the lowest order term, none of the coefficients may be zero unless the polynomial is even or odd. In other words, a_{n-1}, $a_{n-2}, ... a_2, a_1$ must not be zero. If the polynomial is neither even. a_{n-1}, $a_{n-2}, ... a_2, a_1$ must not be zero if the polynomial is neither even nor odd. This readily seen because the absence of a term a_1 implies cancellation brought about by a root $s-\gamma_1$ with a positive real part.

2. Both the odd and even parts of a Hurwitz polynomial P(s) have roots on the $j\omega$ axis only. If we denote the odd part of (s) as n(s) and the even part as m(s) so that

   $$P(s) = n(s) + m(s)$$

   then m(s) and n(s) both have roots on the $j\omega$ axis only.

3. As a result of properly 2 if P(s) is either even or odd, all its roots are on the $j\omega$ axis.

4. The continued fraction of the ratio of the odd to even parts or the even to odd part of a Hurwitz polynomial yields all positive quotient term. Suppose we denote the ratios as

   $\psi(s) = n(s)/m(s)$ or $\psi(s) = m(s)/n(s)$, then the continued fraction expansion of $\phi(s)$ can be written as

   $$\psi(s) = q_1 s + \cfrac{1}{q_2 s + \cfrac{1}{q_3 s + \cfrac{1}{\cdots\cdots + \cfrac{1}{q_n s}}}}$$

   where the quotients $q_1, q_2 ... q_n$ must be positive if the polynomial $P(s) = n(s)$ is Hurwitz. To obtain the continued fraction expansion, we must perform a series of long divisions. Suppose $\psi(s)$ is

   $$\psi(s) = \frac{m(s)}{n(s)}$$

   where $m(s)$ is one higher degree than $n(s)$. Then if we divide $n(s)$ into $m(s)$, we obtain a single quotient and a remainder

   $$\psi(s) = q_1 s + \frac{R_1(s)}{n(s)}$$

   The degree of the term $R_1(s)$ is one lower than the degree of $n(s)$. Therefore, if we invert the remainder term and divide, we have

   $$\frac{n(s)}{R_1(s)} = q_1 s + \frac{R_2(s)}{R_2(s)}$$

   Inverting and dividing again, we obtain

   $$\frac{R_1(s)}{R_2(s)} = q_3 s + \frac{R_3(s)}{R_2(s)}$$

We see that the process of obtaining the continued fraction expansion of $\psi(s)$ is simply involves division and inversion. At each step we obtain a quotient term $q_1 s$ and a remainder term, $R_{1+1}(s)/R_1(s)$. We then invert the remainder term and divide $R_{1+1}(s)$ into $R_1(s)$ to obtain a new quotient. There is theorem in the theory of continued fractions which states that continued fraction expansion of the even to odd or odd to even parts of a polynomial must be finite in length. Another theorem states that, if the continued fraction expansion of the odd to even or even to odd parts of a polynomial yields positive quotient term, then the polynomial must be Hurwitz to within a multiplicative factor $W(s)$. That is, if we write $F(s) = W(s)F_1(s)$ then $F(s)$ is Hurwitz, if $W(s)$ and $F_1(s)$ are Hurwitz. For example, let us test whether the polynomial

$$F(s) = s^4 + s^3 + 5s^2 + 3s + 4$$

is Hurwitz. The even and odd parts of F(s) are

$$m(s) = s^4 5s^2 + 4$$
$$n(s) = s^3 + 3s$$

we not perform a continued fraction expansion of $\psi(s) = m(s)n(s)$ by dividing $n(s)$ by $m(s)$ and then inverting and dividing again, as given by the operation.

$$s^2 + 3s)\overline{s^4 + 5s^2 + 4}(s$$
$$s^4 + 3s^2$$
$$2s^2 + 4)s^3 + 3s(s/2$$
$$s^2 + 2s$$
$$s)2s^2 + 4(2s$$
$$2s^2$$
$$4)s(s/4$$
$$s$$

so that the continued fraction expansion of $\psi(s)$ is

$$\Psi(s) = \frac{m(s)}{n(s)} = s + \cfrac{1}{\cfrac{s}{2} + \cfrac{1}{2s \cfrac{1}{s/4}}}$$

Since the quotient terms of the continued fraction expansion are positive, $F(s)$ is Hurwitz.

N.B. : Consider the case where $F(s)$ is an odd or even function. It is impossible to perform a continued fraction on the function as it stands. However, we can test the ratio of $F(s)$ to its derivative, $F'(s)$. If the ratio $F(s)$ gives a continued fraction expansion with all positive coefficients, then $F(s)$ is Hurwitz. For example, if $F(s)$ is given as

$$F(s) = s^7 + 3s^5 + 2^3 + s$$

then $F(s)$ is $F(s) = 7s^6 + 15s^4 + 6s^2 + 1$

Positive Real Functions

1. $F(s)$ is real for s; that is $F(\sigma)$ is real.
2. The real part of $F(s)$ is greater or equal to zero when the real part of s is greater than or equal to zero, that is,

$$Re\,[f(s)] \geq 0 \text{ for } Res \geq 0$$

Properties of Positive Real Functions

1. If $F(s)$ is p.r., then $1/F(s)$ is also p.r. This property implies that if a driving-point impedance is p.r., then its reciprocal, the driving-point admittance, is also p.r.
2. The sum of p.r. functions is p.r. From an impedance standpoint, we see that if two impedances are connected in series, the sum of the impedances is p.r. An analogous situation holds for to admittances in parallel. Note that the difference of two p.r. functions is not necessarily p.r.; for example, $F(s) = s - 1/s$ is not p.r.
3. The poles and zeros of a p.r., function cannot have positive real parts, i.e. they cannot be in the right half of the plane.
4. Only simple poles with real positive residues can exist on $j\omega$ axis.
5. The poles and zeros of a p.r. function are real or occur in conjugate pairs. We know that the poles and zeros of a network function are functions of the elements in the network. Since the elements themselves are real, there cannot be complex poles or zeros without conjugates because this would imply imaginary elements.
6. The highest powers of the numerator and denominator polynomials may differ at most by unity. This condition prohibits multiple poles and zeros at $s = \infty$.
7. The lowest powers of the denominator and numerator polynomials may differ by at most unity. This condition prevents the possibility of multiple poles or zeros at $s = 0$.
8. The necessary and sufficient conditions for a rational function with real coefficient $F(s)$ to be p.r. are

(a) *F(s)* must have no poles in the right-half plane.

(b) *F(s)* may have only simple poles on the $j\omega$ axis with real and positive residues.

(c) $Re\ F\ (j\omega) \geq 0,$ for all ω.

Properties of Immittance Functions

Let us consider the impedance *Z(s)* of a passive one port network. Let us represent *Z(s)* as

$$Z(s) = \frac{M_1(s)+N_1(s)}{M_2(s)+N_2(s)}$$

where M_1, M_2 are even parts of numerator and denominator, and N_1, N_2 are odd parts. The average power dissipated by the one-port is where I is input current. For a

$$\text{Average power} = \frac{1}{2} Re\left[Z(j\omega)\right]\left|1\right|^2$$

pure reactive network, we know that the power dissipated is zero. We there conclude that the real part of Z(jω) is zero that is

$$Re\,Z(j\omega) = EvZ(j\omega) = 0$$

where

$$EvZ(s) = \frac{M_1(s)M_2(s)-N_1(s)N_2(s)}{M_2^2(s)-N_2^2(s)}$$

In order for $EvZ(j\omega)=0,$ that is,

$$M_1(j\omega)M_2(j\omega)-N_1(j\omega)N_2(j\omega)=0$$

either of the following cases must hold:

(a) $M_1 = 0 = N_2$

(b) $M_2 = 0 = N_1$

In case (a), *Z(s)* is

$$Z(s) = \frac{N_1}{M_2}$$

$$Z(s) = \frac{N_1}{N_2}$$

and in case (b)

We see from this development the following two properties of L-C function :

1. $Z_{LO}(s)$ or $Y_{LO}(s)$ is the ratio of even to odd or odd to even polynomials.
2. Since both $M_1(s)$ and $N_1(s)$ are Hurwitz, they have only imaginary roots, and it follows that the poles and zero of $Z_{LO}(s)$ or $Y_{LO}(s)$ are no the imaginary axis.
3. $Z_{LO}(s)$ or $Y_{LO}(s)$ is the ratio of odd to even or even to odd polynomials.

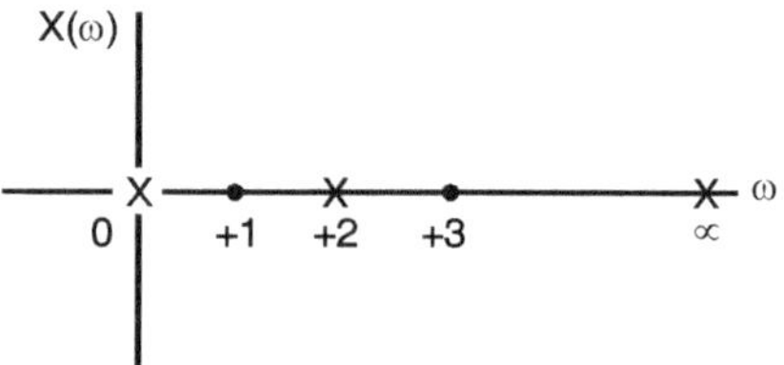

Fig. 17

4. The poles and zeros are simple and lie on the $j\omega$ axis.
5. The poles and zeros interlace on the $j\omega$ axis.
6. The highest powers of numerator and denominator must differ by unity; the lowest powers also differ by unity.
7. There must be either a zero or a pole at the origin and infinity.

The following functions are not L-C for the reasons listed at the left.

1. $Z(s) = \dfrac{Ks(s^2+4)}{(s^2+1)\,(s^2+3)}$

2. $Z(s) = \dfrac{s^5+4s^3+5s}{3s^4+6s^2}$

3. $Z(s) = \dfrac{K(s^2+1)(s^2+9)}{(s^2+2)(s^2+10)}$

On the other hand the function Z(s) below, whose pole-zero diagram is an L-C immittance.

$$Z(s) = \frac{2(s^2+1)(s^2+9)}{s(s^2+4)}$$

Properties of R-L Impedances and R-C Admittances

The immittance that represents a series Foster R-L impedance or a parallel Fost R-C admittance is given as

$$F(s) = K_\infty s + K_0 + \frac{K_1 s}{s+\sigma_1} + \ldots\ldots$$

The significant difference between an R-C impedance and an R-L impedance is that the partial fraction expansion term of the R-C "tank" circuit is $K_1/s(+ \sigma_1)$; whereas for the R-L impedance, the corresponding term must be multiplied by an s in order to give and R-L tank circuit consisting of a resistor in parallel with an inductor.

The properties of R-L impedance functions can be derived in much the same manner as the properties, of R-C impedance functions. Without going into the derivation of the properties, the more significant ones are given in the following :

1. Poles and zeros of an R-L impedance or R-C admittance are located on the negative real axis, and they alternate.
2. The singularity nearest to (or at) the origin is a zero. The singularity nearest to (or at) s = ∞ must be a pole.
3. The residues of the poles must be real and negative.

$$-\frac{K_1}{s+\sigma}$$

Properties of RC Driving-point impedances

1. Poles and zeros lie on the negative real axis and they alternate.
2. The singularity nearest to (or at) the origin must be a pole whereas the singularity nearest to (or at) $\sigma = -\infty$ must be a zero.
3. The residues of the poles must be real and positive.

An example of an R-C impedance is.

$$Z(s) = \frac{(s+1)(s+4)(s+8)}{s(s+2)(s+6)}$$

The following impedances are not R-C

$$Z(s) = \frac{(s+1)(s+8)}{(s+2)(s+4)}$$

$$Z(s) = \frac{(s+2)(s+4)}{(s+1)}$$

$$Z(s) = \frac{(s+1)(s+2)}{s\,(s+3)}$$

SUPERPOSITION THEOREM

This theorem is sometimes useful in solution of networks in which some branches may contain sources of e.m.f. It is applicable only to linear networks where current is linearly related to voltage as per Ohm's law.

The theorem may be stated as follows :

"In any network containing more than one sources of e.m.f. the current in any branch is the algebraic sum of a number of individual fictitious currents (the number being equal to the number of sources of e.m.f), each of which is due to the separate action of each source of e.m.f., taken in order, when the remaining sources of e.m.f. are replaced by conductors, the resistances of which are equal to the internal resistances of the respective sources"

The procedure of applying superposition theorem is as follows :

1. Replace all but one of the sources by their internal resistances. If the internal resistance of any source is small as compared to the other resistances present in the network, the source is replaced by a short-circuit.
2. Find the currents in different branches by using Ohm's law.
3. Repeat the process using each of the e.m.fs. as the sole e.m.f. each time.

The total current in any branch of the circuit is the algebraic sum of currents due to each source.

Example : *By using Superposition Theorem find the currents in different branches of the network shown in Fig. (18)*

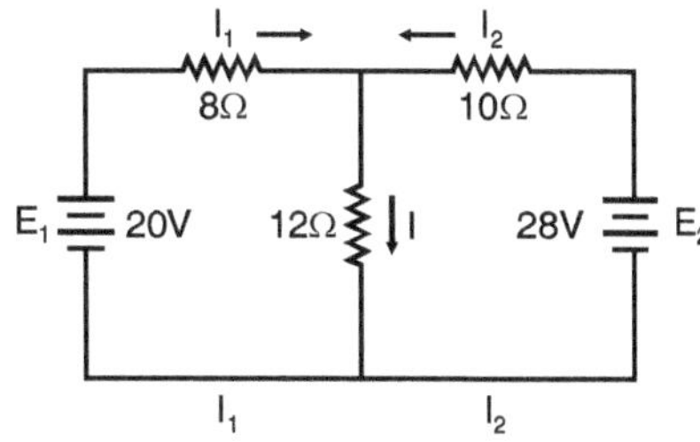

Fig. 18

Solution : $I_1 = ?, I_2 = ?, I = ?$

First Step : Take e.m.f. E_1 only and replace e.m.f. E_2 by its zero internal resistance, the circuit is shown in Fig. (19)

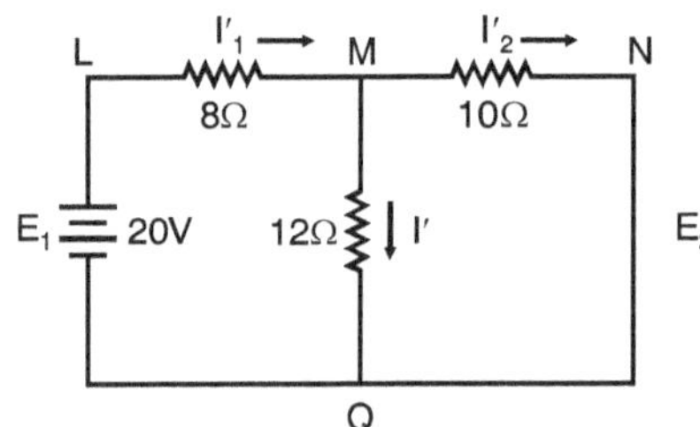

Fig. 19

$$\text{Total resistance } = 8 + \frac{10 \times 12}{10 + 12} = 13.45\ \Omega$$

Current through 8 Ω resistance,

$$I_1{}' = \frac{20}{13.45} = 1.487\text{ A}$$

Current through 10 Ω resistance,

$$I_2{}' = 1.487 \times \frac{12}{12 + 10} = 0.81\text{ A}$$

Current through 6 Ω resistance

$$I' = 1.487 \times \frac{10}{12 + 10} = 0.657\text{ A}$$

Second Step : E.m.f. E_1 is removed/short circuited and currents due to e.m.f. E_2 are found. The circuit is shown in the Fig. 20.

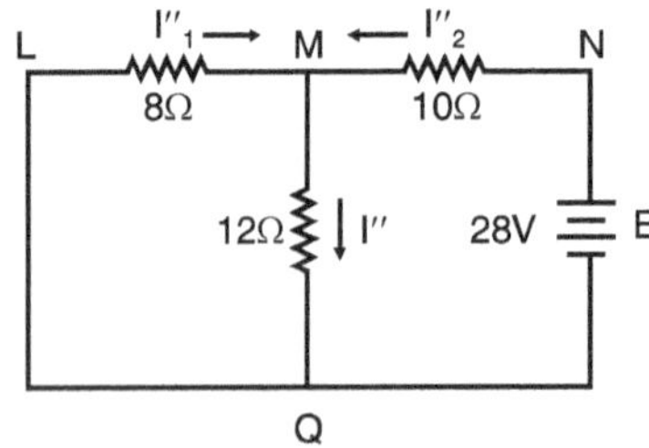

Fig. 20

$$\text{Total resistance } = 10 \times \frac{12 \times 8}{12 + 8} = 14.8\ \Omega$$

Current through 10 Ω resistance,

$$I_2{}'' = \frac{28}{14.8} = 1.892\text{ A}$$

Current through 8 Ω resistance,

$$I_1{}'' = 1.892 \times \frac{12}{12 + 8} = 1.135\text{ A}$$

Current through 12 Ω resistance,

$$I'' = 1.892 \times \frac{8}{12 + 8} = 0.757\text{ A}$$

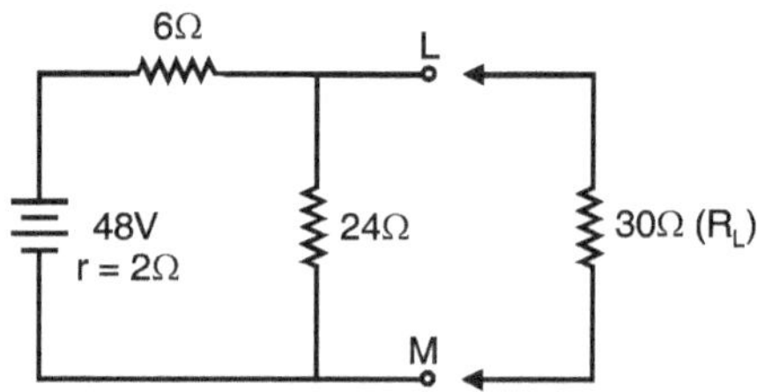

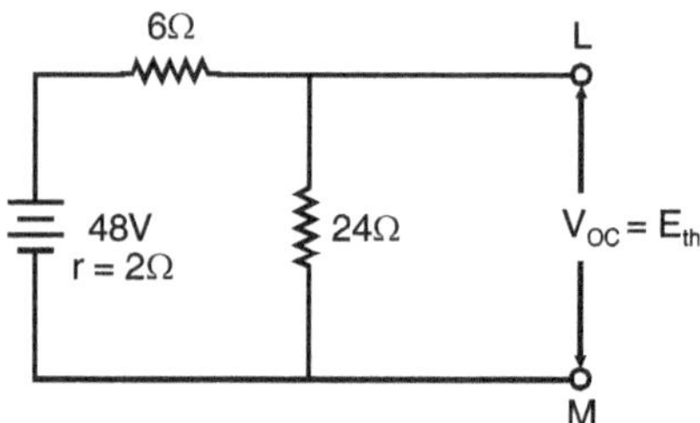

Fig. 21

Current in the network before load resistance (R_L) is connected

$$= \frac{48}{24 + 6 + 2} = 1.5\text{ A}$$

$\therefore$ Voltage across terminal LM,

$V_{oc} = E_{th} = 24 \times 1.5 = \mathbf{36\ V}$ (see Fig. 21)

Hence, so far as terminals L and M are connected, the network has an e.m.f. of 36 V (and not 48 V). **Ans.**

Equivalent Resistance of the Network = ?

There are two parallel path between points L and M. Imagine that battery of 48 V is removed but not its internal resistance. Then resistance of the circuit as looked into points *L* and *M* is (see Fig. 22)

$$R_i = R_{th} = \frac{24 \times (6+2)}{24+(6+2)} = 6\ \Omega. \text{ Ans.}$$

Current in R_L, I = ?

Refer Fig. 23

$$I = \frac{E_{th}}{R_{th} + R_L} = \frac{36}{6+30} = 1 \text{ A. Ans.}$$

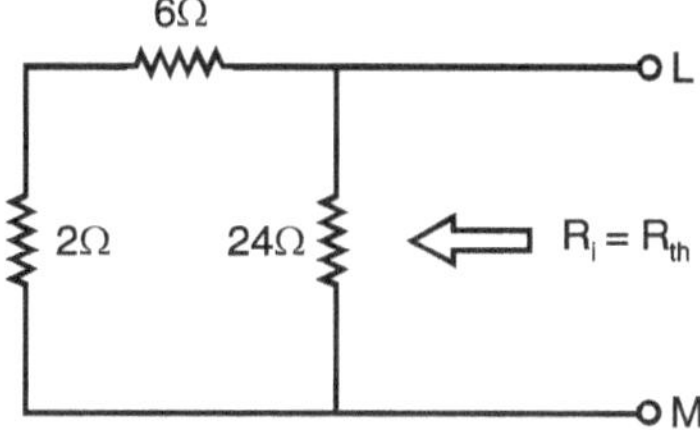

Fig 22

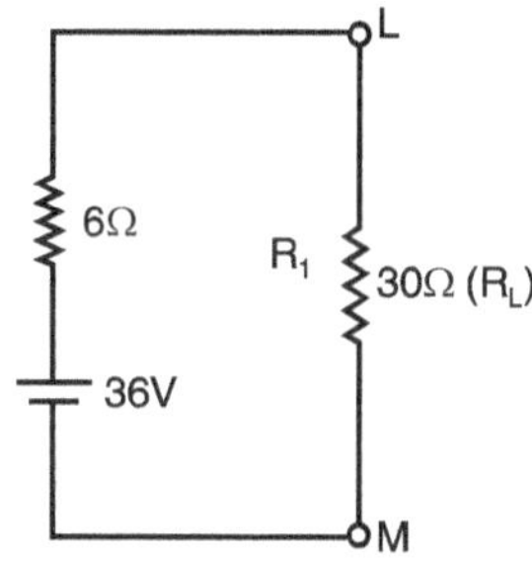

Fig. 23

NORTON'S THEOREM

Whereas Thevenin's theorem was used to simplify a network to a *constant-voltage voltage source* and a series resistance, Norton's theorem can be used to resolved a network into a constant *current current source* and a parallel resistance. The interchange of voltage sources and current sources by use of Thevenin's and Norton's theorems is sometimes useful in circuit analysis.

The theorem may be stated as follows :

''Any two-terminal linear network containing independent voltage and current sources may be replaced by an equivalent current I_N in parallel with a resistance R_N where I_N is the short-circuit current at network terminals and R_N is the equivalent resistance of network as seen from the terminals but with all voltage sources short-circuited and all current sources open-circuited''.

The following procedure may be adopted to determine the Norton's equivalent circuit :

1. Calculate the short-circuit current (I_N) at the network terminals.
2. Redraw the network with each voltage source replaced by a short-circuit in series with its internal resistance and each current source by an open circuit in parallel with its internal resistance.
3. Calculate the resistance (R_N) of the redrawn network as seen from the network terminals. (The resistance R_N is same value as used in Thevenin's equivalent circuit).

Example : *By using Norton's theorem find the current in the 12 Ω resistance of the circuit shown in Fig. 24.*

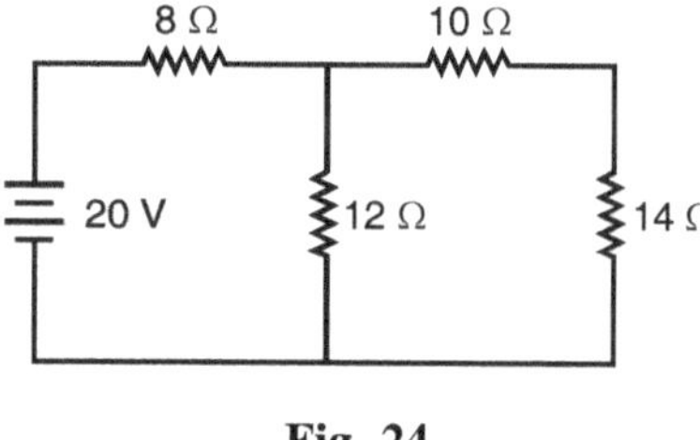

Fig. 24

Solution : With 12 Ω resistance removed and terminals L-M short-circuited, short-circuit current,

$$I_N = \frac{20}{8} = 2.5 \text{ A}$$

With 20 V battery replaced by a short-circuit, the resistance of the network as seen from terminals *L* and *M* is

$$R_N = \frac{8\ (10+14)}{8+(10+14)} = 6\ \Omega$$

The Norton's equivalent circuit is shown in figure. The current through 12 Ω resistance is

$$I = 2.5 \times \frac{6}{6+12} = 0.833\text{A. Ans}$$

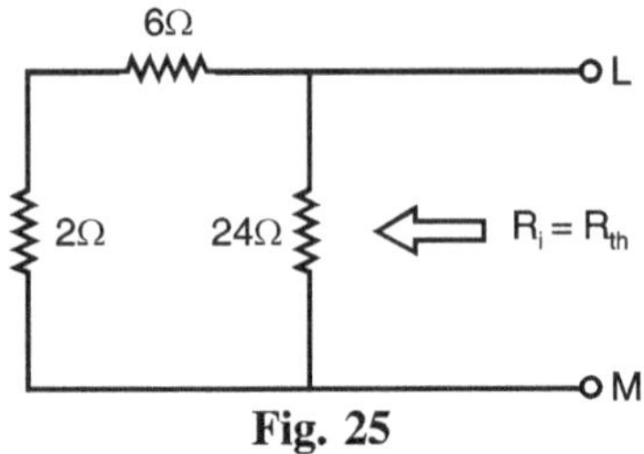

Fig. 25

Maximum Power Transfer Theorem

This theorem is particularly useful for analysing *communication networks*. It is stated as follows :

"Maximum power output is obtained from a network when the load resistance is equal to the output resistance of the network as seen from the terminals of the load."

Any network can be converted into a single voltage source by the use of Thevenin's theorem (Fig. 26). The maximum power transfer theorem aims at finding R_L such that the power dissipated in R_L is maximum

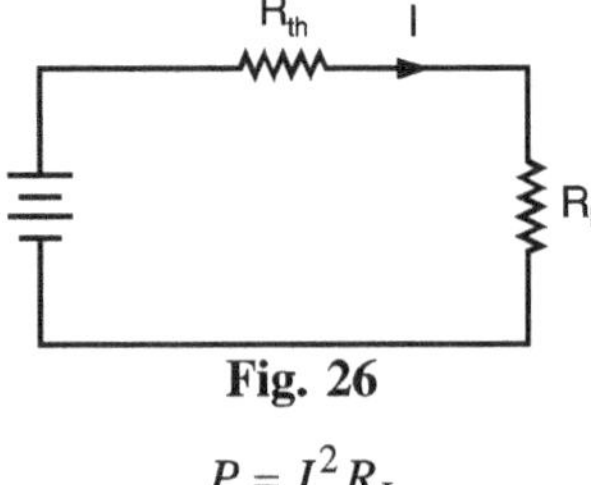

Fig. 26

$$P = I^2 R_L$$

The total current in different branches are :

Current through 8 Ω resistance,

$$I_1 = I_1' - I_1'' = 1.487 - 1.135 = 0.352 \text{ A}$$

(From L to M). **Ans.**

Current through 10 Ω resistance,

$$I_2 = I_2' - I_2'' = 1.892 - 0.81 = 1.082 \text{ A}$$

(From N to M). **Ans.**

Current through 12 Ω resistance,

$$I = I' - I'' = 0.675 + 0.757 = 1.432 \text{ A}$$

(From M to Q). **Ans.**

Thevenin's Theorem

This theorem is quite useful when the current in one branch of a network is to be determined or when the current in an added branch is to be calculated.

"It states that for the purpose of determining the current in a resistor R_L, connected across two terminals of a network which contains sources of e.m.f. and resistors, the network can be replaced by a single source of e.m.f. and a series resistor, R_{th}. This e.m.f. E_{th}, is equal to potential difference between the terminals of the network when the resistor, R, is removed : The resistance of series resistor, R_{th} is equal to the equivalent resistance of the network with the resistor, R, removed (or as it is sometimes called, "the resistance of the network when viewed from the terminals under consideration").

$$\text{Hence, } I = \frac{E}{R_L + R_{th}}$$

The procedure is outlined in Fig. 27

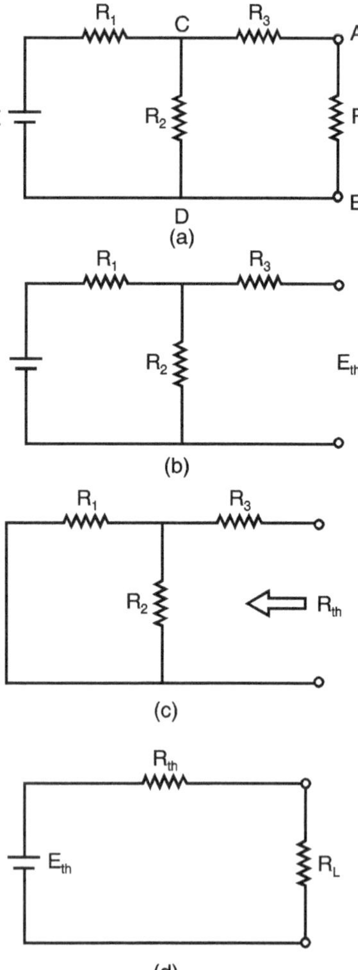

Fig. 27

Example : *For the circuit shown in Fig 28, find the current through R_L when it takes on values of 5 Ω and 25 Ω. Also, calculate the value of R_L for which the power dissipated in it would be maximum and find this power.*

Solution : The open circuit voltage V_{oc} (also called Thevenin's voltage E_{th}) which appears across terminals L and M is equal to the voltage drop across 10 Ω resistance.

Current flowing through the circuit

$$EPQ = \frac{60}{20+10} = 2A$$

∴ Voltage drop over 10 Ω resistance $= 2 \times 10 = 20$ V

Hence $V_{oc} = E_{th} = 20$ V

The resistance of the circuit as looked into the network from points L and M (when battery has been removed),

$$R = R_{th} = 30 + 10 \| 12 = 30 + \frac{10 \times 20}{10+20} = 36.67\ \Omega$$

The whole circuit upto LM can now be replaced by a single source of e.m.f. and single resistance as shown in Fig. 29.

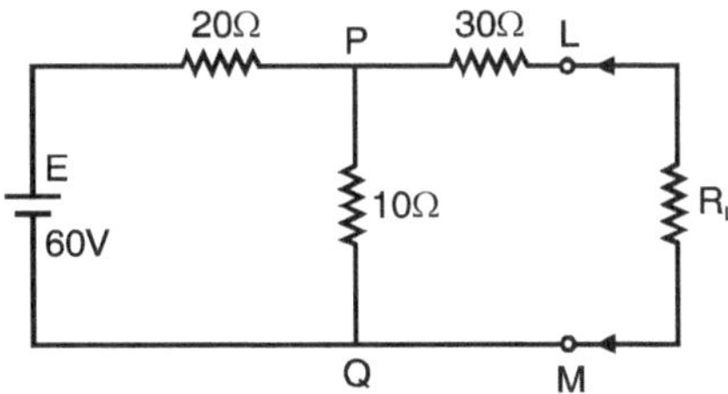

Fig. 28

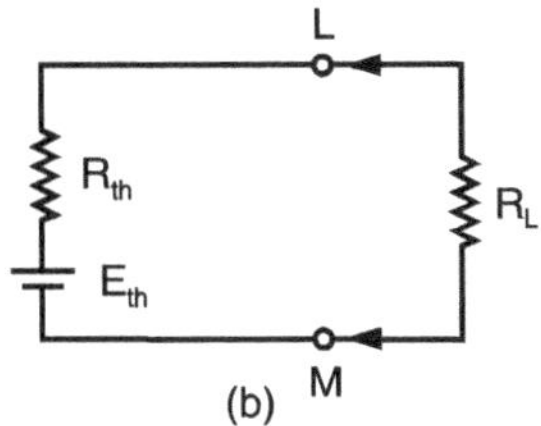

Fig. 29

(i) when $R_L = 5\ \Omega$,

$$I = \frac{E_{th}}{R_{th}+R_L} = \frac{20}{36.67+5} = 0.48\text{A. Ans.}$$

(ii) when $R_L = 5\ \Omega$,

$$I = \frac{20}{36.67+25} = 0.324\text{A. Ans.}$$

According to the Maximum Power Transfer theorem, power drawn by R_L would be maximum when $R_L = R_i$ or when $R_L = 36.67\ \Omega$

∴ Maximum power drawn by R_L

$$= I^2 R_L = \left(\frac{E_{th}}{R_{th}+R_L}\right)^2 R_L$$

$$= \left(\frac{E_{th}}{R_L+R_L}\right)^2 R_L = \frac{E_{th}^2}{4R_L} \qquad (\because R_{th} = R_L)$$

$$= \frac{20^2}{4 \times 36.67} = 2.72\ \text{W. Ans.}$$

Delta Star Transformation

When network having a large number of branches are to be solved by the use of Kirchhoff's law, a great difficulty is experienced in solving several simultaneous equations. Such complicated networks, however, can be simplified by successively replacing delta, meshes by equivalent star system and vice-versa.

Consider the two circuits shown in the Fig. 30. They will be *equivalent* if the resistance measured between any two of the terminals 1, 2 and 3 is the same in two cases.

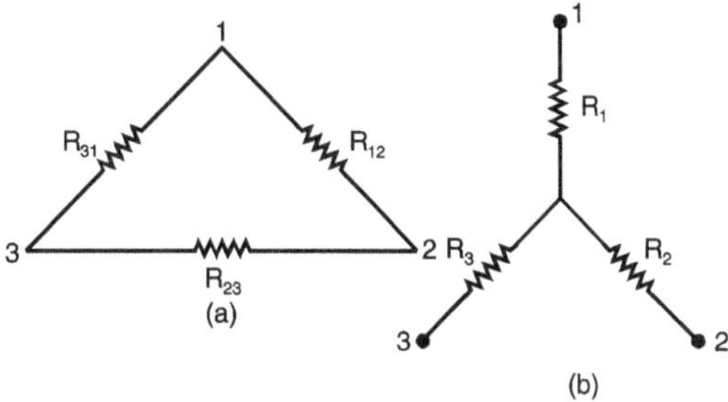

Fig. 30

$$[R_{12}]_Y = [R_{12}]_\Delta$$

$$\text{or } R_1 + R_2 = \frac{R_{12}(R_{23}+R_{31})}{R_{12}+R_{23}+R_{31}}$$

$$\text{Similarly } R_2 + R_3 = \frac{R_{23}(R_{31}+R_{12})}{R_{12}+R_{23}+R_{31}}$$

$$\text{and } R_3 + R_1 = \frac{R_{31}(R_{12}+R_{23})}{R_{12}+R_{23}+R_{31}}$$

Solving above eqns. simultaneously, we get

$$R_1 = \frac{R_{12}+R_{31}}{R_{12}+R_{23}+R_{31}}$$

$$R_2 = \frac{R_{23}R_{12}}{R_{12}+R_{23}+R_{31}}$$

$$R_3 = \frac{R_{31}R_{23}}{R_{12}+R_{23}+R_{31}}$$

From above it may be noted that resistance of each arm of the star is given by the product of the resistances of the two delta sides that meet at its end divided by the sum of the three delta resistances.

From the above eqns. for star to delta conversion can also be obtained. These are as follows:

$$R_{12} = \frac{R_1R_2 + R_2R_3 + R_3R_1}{R_3}$$

$$R_{23} = \frac{R_1R_2 + R_2R_3 + R_3R_1}{R_1}$$

$$R_{31} = \frac{R_1R_2 + R_2R_3 + R_3R_1}{R_2}$$

In electronics, star and delta circuits are generally referred to as T and π circuits respectively.

Example : *Fig. 31 shows a number of resistances connected in delta and star. Using star/delta conversion method compute the network resistance measured between (i) L and M, (ii) M and N, and (iii) N and L.*

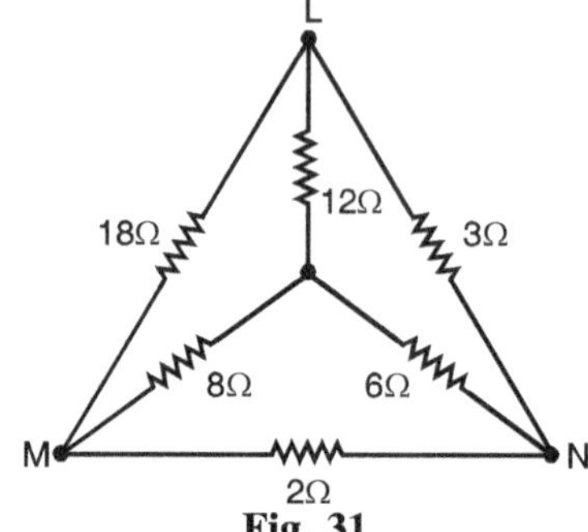

Fig. 31

Solution : Three resistance 12 Ω, 6 Ω and 8 Ω are star connected. Transform them into delta with ends at the same point as before.

Refer Fig. 32

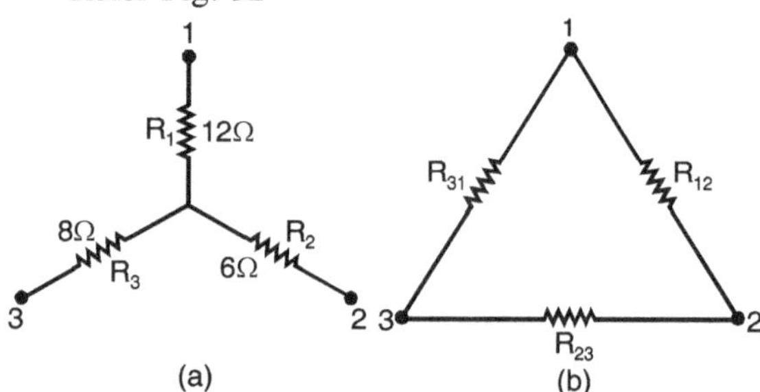

Fig. 32

$$R_{12} = \frac{R_1R_2 + R_2R_3 + R_3R_1}{R_3}$$

$$= \frac{12\times6 + 6\times8 + 8\times12}{8} = 27\ \Omega$$

$$R_{23} = \frac{R_1R_2 + R_2R_3 + R_3R_1}{R_1}$$

$$= \frac{12\times6 + 6\times8 + 8\times12}{12} = 18\ \Omega$$

$$\text{Similarly } R_{31} = \frac{12\times6 + 6\times8 + 8\times12}{6} = 36\ \Omega$$

Fig. 33 shows this transformed circuit connected to original delta connected resistances in the circuit 18 Ω, 3 Ω and 2 Ω.

Here, 18 Ω and 36 Ω are in parallel,

3 Ω and 27 Ω are in parallel,

2 Ω and 18 Ω are in parallel.

These resistances are equivalent to :

$$\frac{18 \times 36}{18+36} = 12\ \Omega,\ \frac{3 \times 27}{3+27} = 2.7\ \Omega \text{ and } \frac{2 \times 18}{2+18} = 1.8\ \Omega$$

This is shown in Fig. 34

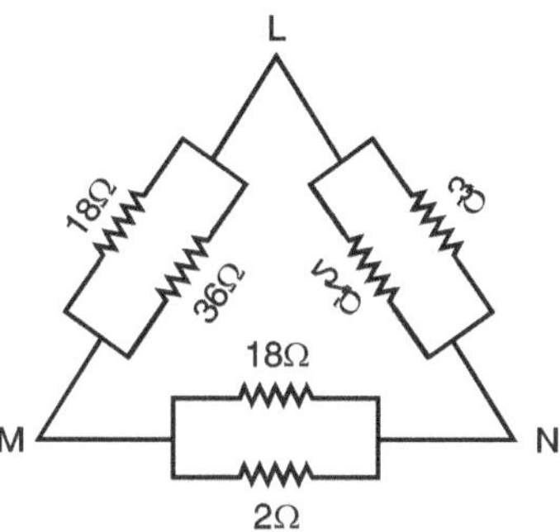

Fig. 33

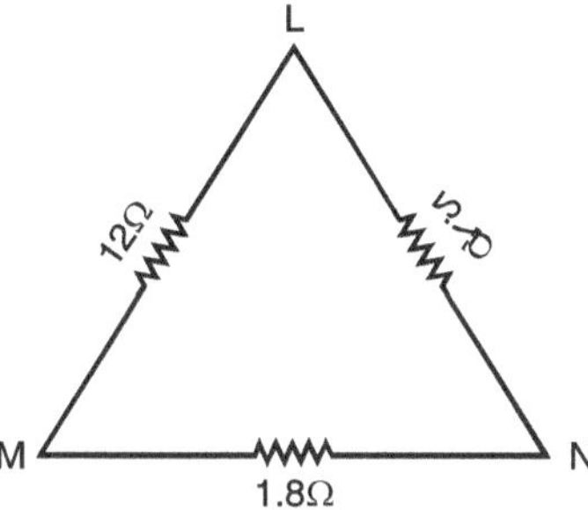

Fig. 34

(i) Resistance between L and M,

$$R_{LM} = \frac{12 \times (2.7+1.8)}{12+(2.7+1.8)} = 3.27\ \Omega \text{ Ans.}$$

(ii) Resistance between M and N,

$$R_{MN} = \frac{1.8 \times (12+2.7)}{1.8+(12+2.7)} = 1.6\ \Omega \text{ Ans.}$$

(iii) Resistance between N and L,

$$R_{NL} = \frac{2.7 \times (12+1.8)}{2.7+(12+1.8)} = 2.25\ \Omega \text{ Ans.}$$

COMPENSATION THEOREM

The compensation theorem is particularly useful for the following purposes :

(i) To calculate the sensitivity of a bridge network.

(ii) To analyse those networks where the values of the branch elements are varied and analysing the effect of tolerance on such values.

This theorem is states as follows :

"If a change, say ΔR, is made in the resistance of any branch of a network when the current was originally I, then the change of current at any other point in the network may be treated by assuming that an e.m.f. $- I\,\Delta R$ has been introduced into the changed branch of all other sources have their e.m.fs. suppressed and are represented by their internal sources only''.

RECIPROCITY THEOREM

The theorem is stated as follow :

"In any linear bilateral network, if a source of e.m.f. E in any branch produces a current I in any other branch then the same e.m.f. acting in the second branch would produce the same current I in the first branch''.

In other words, it simply means that E and I are mutually transferrable. The ratio E/I is known as the transfer resistance for impedance in A.C. systems).

MILLMAN'S THEOREM

The theorem can be applied to a network having a combination of voltage and current sources (since voltage source can be converted into a current source and vice-versa).

The theorem is applicable only to two sources connected directly in parallel. It is not applicable where there are resistance elements between the sources.

The theorem is stated in the following manner :

"Any number of current sources in parallel may be replaced by a single current source whose current is the algebraic sum of individual source currents and source resistance is the parallel combination of individual source resistances''.

The common voltage (V) across parallel branches with different V-source can be determined from the relation

$$V = \frac{(V_1/R_1)+(V_2/R_2)+(V_3/R_3)}{(1/R_1)+(1/R_2)+(1/R_3)} \ldots \text{etc.}$$

A.C. FUNDAMENTALS

Alternating Voltage and Current

Modern alternators produce an e.m.f. which is for all practical purposes sinusoidal (*i.e.*, a sine curve), the equation between the e.m.f. and time being

$$e = E_{max} \sin \omega t$$

where e = instantaneous voltage;

E_{max} = maximum voltage

ωt = angle through which the armature has turned from neutral.

Taking the frequency as f hertz (cycles per second), the value of ω will be $2\pi f$, so that the equation reads

$$e = E_{max} \sin (2\pi f)t.$$

The graph of the voltage will be as shown in Fig. 35

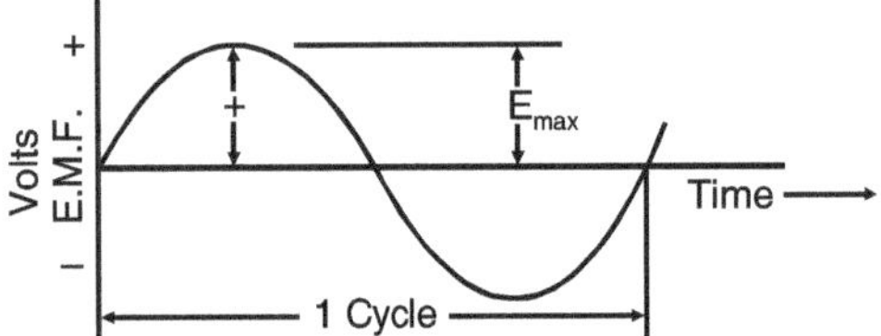

Fig. 35: *The graph of the sinusoidal voltage*

Cycle : One complete set of positive and negative values of an alternating quantity is known as *a cycle.* A cycle may also sometimes be specified in terms of angular measure. In that case, one complete cycle is said to spread over 360° or 2π radians.

Amplitude : The maximum value, positive or negative, of an alternating quantity, is known as its *amplitude*.

Frequency (f) : The number of cycles/second is called the frequency of the alternating quantity.

Its unit is hertz (Hz).

Time Period (T) : The time taken by an alternating quantity to complete the cycle is called its time period. For example, a 50 hertz (Hz) alternating current has a time period of 0.02 second.

Time period is reciprocal of frequency,

$$T = \frac{1}{f}\left(\text{or } f = \frac{1}{T}\right)$$

Root mean square (R.M.S.) value : The r.m.s. value of an alternating current is given that steady (D.C.) current *which when flowing through a given circuit for a given time produces the same heat as produced by the alternating current when flowing through the same* circuit for the same time.

R.M.S. value is the value which is taken for power purposes of any description. This value is obtained by finding the square root of the mean value of the squared ordinates for a cycle or half-cycle.

$$E_{r.m.s.} = E_{max} \times \frac{1}{\sqrt{2}}$$
$$= 0.707\ E_{max}$$

This is the value which is used for all power, lighting and heating purposes, as in these equations the power is proportional to the square of the voltage.

Average or mean value : The average value of an alternating current is expressed by *steady current which transfers across any circuit the same charge as is transferred by alternating current during the same time.*

The average value of the voltage will be found to be 0.636 of the maximum value for perfect sine wave, giving the equation

$$E_{av.} = 0.636\ E_{max.}$$

The mean value is only of use in connection with processes where the results depend on current only, irrespective of the voltage, such as electroplating or battery charging.

Form Factor and Peak Factor

Form factor : *The ratio of r.m.s. (or effective) value to average value is the form factor of the wave form.* It has use in voltage generation and instrument correction factors.

Peak factor : The ratio of maximum value to the r.m.s. value is the peak factor (K_p) value wave form.

Reasons for using alternating current (or voltage) of sinusoidal form :

An alternating current (or voltage) of sinusoidal form is normally used because of the following reasons :

1. Mathematically, it is quite simple
2. Its integrals and differentials both are sinusoidal.
3. It lends itself to vector representation.
4. A complex wave form can be analysed into a series of sine waves of various frequencies and each such component can be dial with separately.
5. This *waveform is desirable for power generation, transmission and utilisation.*

S. No.	Wave form	From factor (K_f) $= \dfrac{r.m.s\ value}{average\ value}$	Peak factor (K_p) $= \dfrac{max.value}{r.m.s.value}$
1.	**Since wave 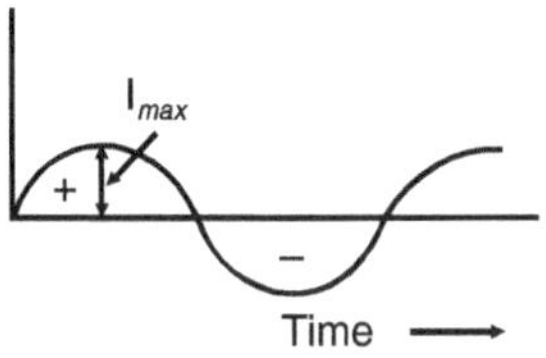 :** R.M.S. value $= \dfrac{I_{max}}{\sqrt{2}}$ $= 0.707\ I_{max}$ Average value $= \dfrac{2}{\pi} I_{max}$ $= 0.636\ I_{max}$	$K_f = \dfrac{0.707\ I_{max}}{0.636\ I_{max}} = 1.11$	$K_p = \dfrac{I_{max}}{0.707\ I_{max}} = 1.41$
2.	**Half wave rectified sine wave 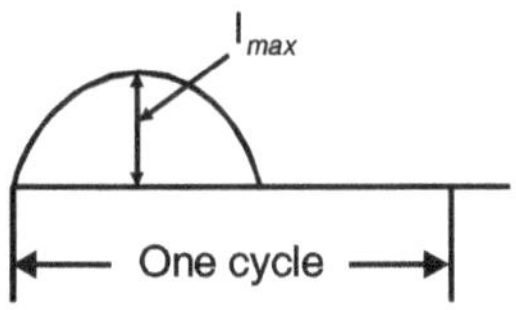 :** R.M.S. value $= \dfrac{I_{max}}{\sqrt{2}} = 0.5\ I_{max}$ Average value $= \dfrac{1}{\pi} I_{max} = 0.318\ I_{max}$	$K_f = \dfrac{0.5\ I_{max}}{0.318\ I_{max}} = 1.57$	$K_p = \dfrac{I_{max}}{0.5\ I_{max}} = 2.0$
3.	**Full wave rectifed sine wave 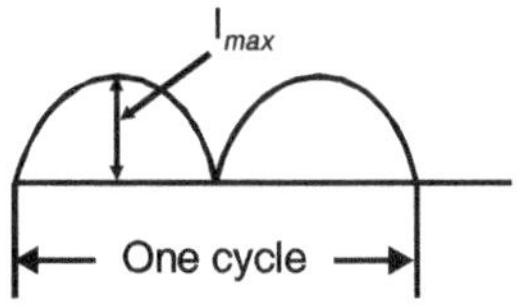 :** R.M.S. value $= \dfrac{I_{max}}{\sqrt{2}} = 0.707\ I_{max}$ Average value $= \dfrac{2}{\pi} I_{max} = 0.636\ I_{max}$	$K_f = \dfrac{0.707\ I_{max}}{0.636\ I_{max}} = 1.11$	$K_p = \dfrac{I_{max}}{0.707\ I_{max}} = 1.41$

4.	**Rectangular wave :** 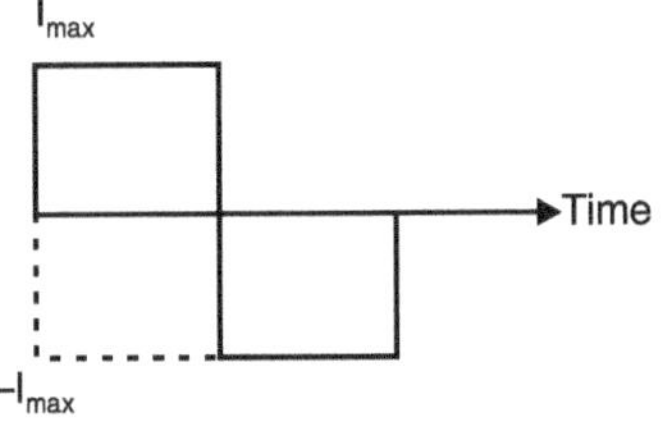 R.M.S. value $= I_{max}$ Average value $= I_{max}$	$K_f = 1$	$K_p = 1$
5.	**Triangular wave :** 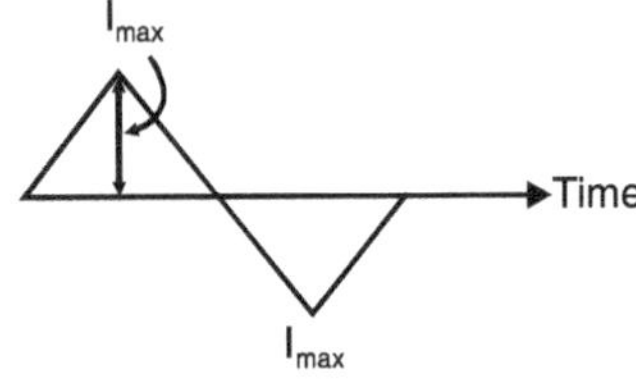 R.M.S. $= \frac{I_{max}}{\sqrt{3}} = 0.578\ I_{max}$ Average value $= \frac{I_{max}}{2} = 0.5\ I_{max}$	$K_f = \frac{0.578\ I_{max}}{0.5\ I_{max}} = 1.16$	$K_p = \frac{I_{max}}{0.578\ I_{max}} = 1.73$

A.C. Through Pure Ohmic Resistance Alone

Where a sinusoidal e.m.f. is placed across a pure resistance the current will be *in phase with e.m.f.*, and if shown graphically will be in phase with the e.m.f. curve.

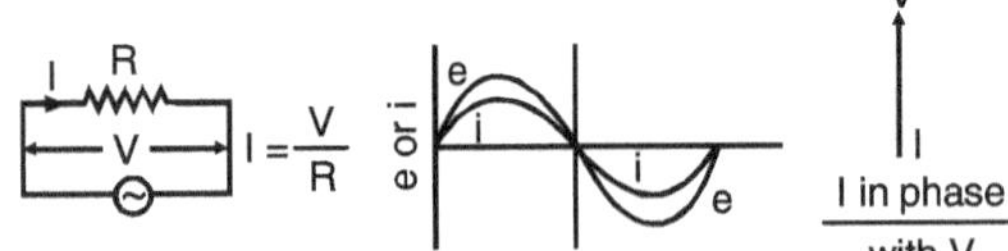

Fig. 36: *Purely resistive circuit*

The current $I = \frac{V}{R}$

where $V =$ r.m.s. value of the applied e.m.f. or voltage;

$R =$ resistance in ohms

(The value of I will be the r.m.s. value)

The power in a purely resistive circuit is given by the products of the r.m.s. voltage and the r.m.s. current *i.e.*, $P = VI$. (see Fig. 36)

A.C. Through Pure Inductance Alone

If a sinusoidal e.m.f. is placed across a pure inductance the current will be found to be,

$$I = \frac{V}{2\pi fL}$$

where $V =$ voltage (r.m.s. value);

$f =$ frequency; and

$L =$ the inductance in henries (H)

(The value of I being the r.m.s. value)

- The current will *lag behind the voltage* and the graphs will be as shown in Fig. 37, the *phase difference being 90°*.
- The expression $2\pi fL$ (or ωL) *is termed as*

inductive reactance (X_L).

- *Power consumed is zero.*

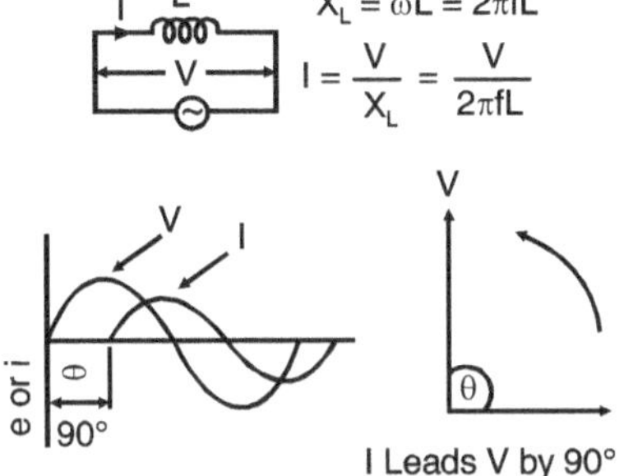

Fig. 37: *Purely inductive circuit*

A.C. Through Pure Capacitance Alone

If a sinusoidal e.m.f. is placed across a capacitor the current will be,

$$I = (2\pi f)CV$$

where $C=$ capacitance in farads (F);
$f=$ frequency; and
$V=$ voltage (r.m.s. value)

- In this case the current leads the voltage by 90°, as shown in Fig. 38

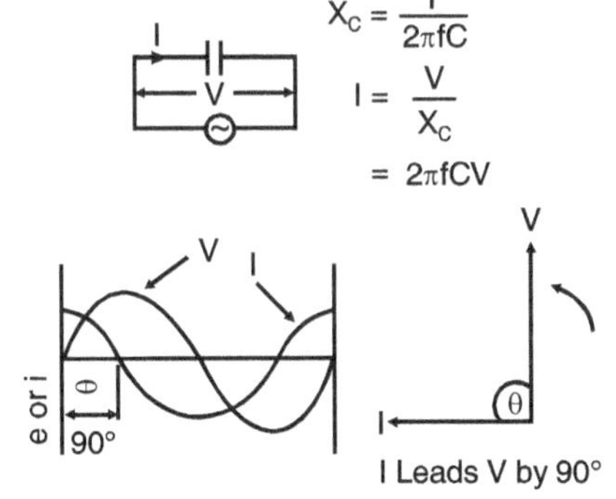

Fig. 38: *Purely capacitive circuit*

- The expression $\frac{1}{2\pi fC}\left(\text{or}\frac{1}{\omega C}\right)$ is termed the capacitive reactance (X_C) and the current is given by $I = \frac{V}{X_c}$
- *Power consumed is zero.*

PHASOR ALGEBRA

The following are the methods of representing vector quantities :

1. Symbolic notation
2. Trigonometrical form
3. Exponential form
4. Polar form

A vector as shown in Fig. 39 may be described in the above form as follows :

1. *Symbolic notation :* $E = a + jb$
2. *Trigonometrical form :*

$$E = \sqrt{a^2 + b^2}\ (\cos\theta + j\sin\theta)$$

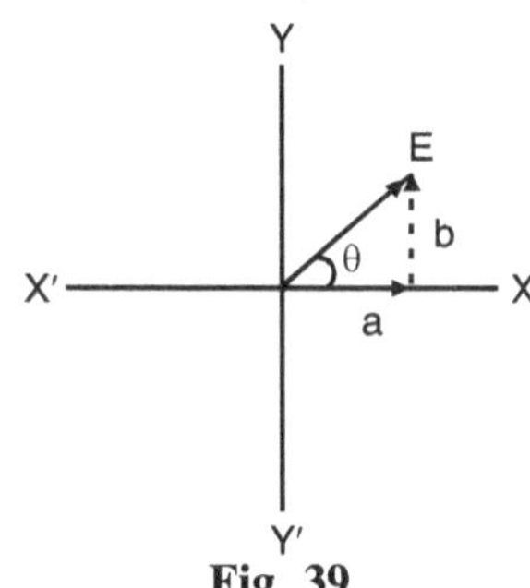

Fig. 39

$$\left[= \sqrt{a^2 + b^2}\ (\cos\theta \pm j\sin\theta)\right] \text{ in general}$$

3. *Exponential form :* $E = \sqrt{a^2 + b^2}\ e^{+j\theta}$

$$\left[= \sqrt{a^2 + b^2}\ e^{+j\theta}\right] \text{ in general}$$

4. *Polar form :* $E = \sqrt{a^2 + b^2}\angle\theta$

$$\left[= \sqrt{a^2 + b^2}\angle \pm\theta\right] \text{ in general}$$

R-L-C Circuit (Resistance induction and capacitance in series)

Fig. 40 shows a R-L-C circuit.

Important formulae :

1. Impedance, $Z = \sqrt{R_2 + (X_L - X_C)^2}$

$$\left[\begin{array}{l}\text{where } X_L = 2\pi fL, L \text{ in henries} \\ \text{and } X_C = \dfrac{1}{2\pi fC}, C \text{ in farads}\end{array}\right]$$

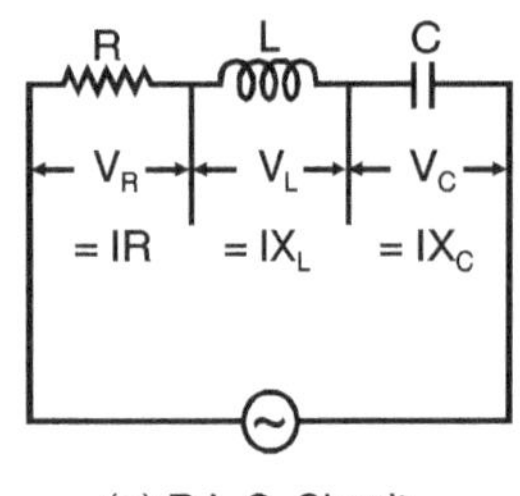

(a) R.L.C. Circuit

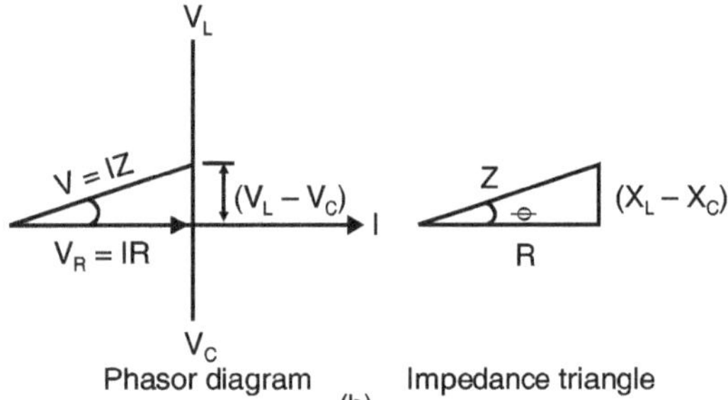

Phasor diagram (b) Impedance triangle $X_L - X_C$

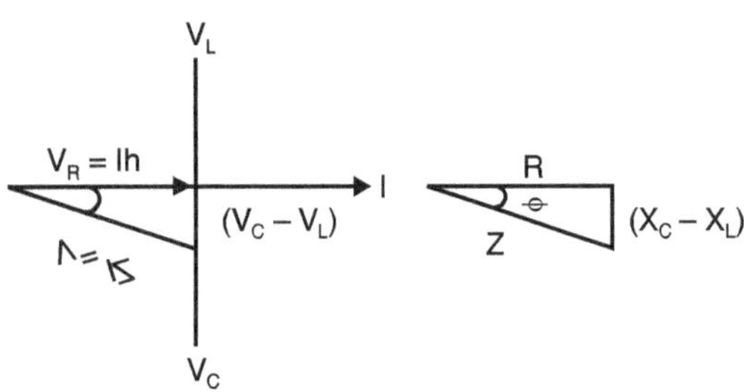

Phasor diagram (c) Impedance triangle

Fig. 40 : *Resistance, Inductance and capacitance in series*

2. Current, $I = \dfrac{V}{Z}$

3. Power factor, $\cos\phi = \dfrac{R}{Z}$

$$\left[\begin{array}{l}\text{angle of lag (when } X_L < X_C\text{) or} \\ \text{lead (when } X_C > X_L\text{), } \phi = \cos^{-1}\dfrac{R}{Z}\end{array}\right]$$

4. Power consumed = VI cos ϕ (= I^2R)

Resonance in R-L-C Circuits

The frequency of the voltage which gives the maximum value of the current in the circuit is called *resonant frequency*, and the circuit is said to the *resonant*.

At resonance, $X_L = X_C$

i.e. $$2\pi f_r L = \frac{L}{2\pi f_r C}$$

$$f_r = \frac{1}{2\pi\sqrt{LC}}$$

where f_r = resonance frequency in H_Z;
L = inductance in henries, and
C = capacitance in farads.

Fig. 41 shows variation of X_L, X_C, and X (total reactance = $X_L - X_C$) with variation of frequency f.

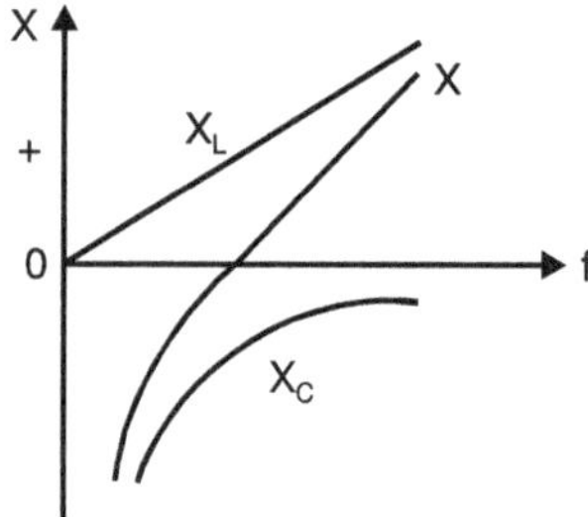

Fig. 41 : *Reactance (X) υ/s*

Fig. 42 shows the variation of current (I) with frequency (f).

At series resonance, it is seen that :

1. The impedance of the circuit is minimum and equal to the resistance (R) of the circuit $\left(i.e., I = \dfrac{V}{R}\right)$.

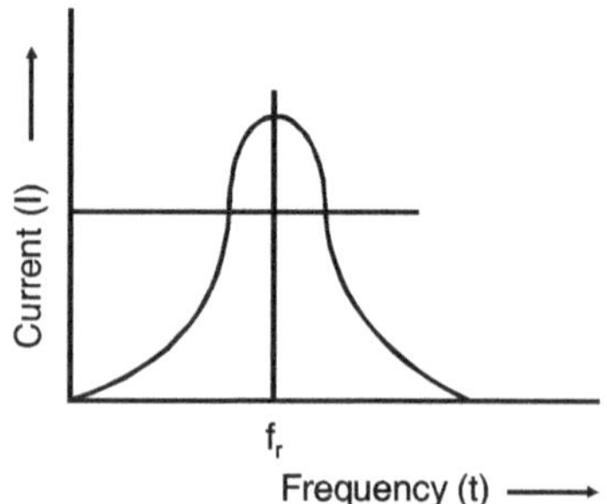

Fig. 42 : *Current in R-L-C*

2. The current drawn is maximum (*i.e.*, $I = I_{max}$).
3. The phase angle between the current and voltage is zero; the power factor is unity.
4. The resonant frequency is given by $f_r = \frac{1}{2\pi\sqrt{LC}}$, if the frequency is below the resonant frequency the net reactance in the circuit is *capacitive* and if the frequency is above the resonant frequency, the net reactance in the circuit is *inductive*.

Q-factor of a Series Circuit : In the case of a R-L-C, circuit is defined as equal to the voltage magnification in the circuit at resonance.

$$\text{Q - factor} = \frac{1}{R}\sqrt{\frac{L}{C}}$$

where R = resistance in Ω;
L = inductance in *H*; and
C = capacitance in *F*.

In the case of series resonance, the higher quality factor, i.e., Q-factor means not only higher voltage magnification but also a higher selectivity of the tuning coil.

A.C. PARALLEL CIRCUITS

In A.C. parallel circuits $\frac{1}{Z} = \frac{1}{Z_1} + \frac{1}{Z_2} + \frac{1}{Z_3}$

(in series A.C. circuits : $Z = Z_1 + Z_2 + Z_3$

The term $\frac{1}{Z}$ written as Y is called the admittance. The unit of admittance is *mho*.

Also $Y = \sqrt{G + B^2}$

where G = conductance (always positive) and
B = susceptance (+ ve for inductive reactance and negative for capacitive reactance)

The units of conductance and susceptance are *mho*.

Also power factor $= \frac{B}{G}$

Resonance in Parallel Circuits

At *parallel resonance*, it is seen that :

1. The admittance of the circuit is *minimum* and is *equal* to the conductance of the circuit.
2. The *current drawn is minimum*.
3. The phase angle between the current and voltage is zero, the *power factor is unity*.
4. The resonant frequency is given by $f_r = \frac{1}{2\pi\sqrt{LC}}$ if the resistance in the inductance and capacitance branches is negligible.

Q-Factor of a Parallel Circuit

It is defined as the ratio of the current circulating between its two branches to the line current drawn from the supply or simply, as the current magnification

Comparision of Series and Parallel Resonant Circuits

S.No.	*Aspects*	*Series Circuit (R-L-C)*	*Parallel Circuit (R-L and C)*
1.	Impedance at resonance	Minimum	Maximum
2.	Current at rersonance	Maximum $= \frac{V}{R}$	Minimum = $V/(L/CR)$
3.	Effective impedance	R	L/CR
4.	Power factor at resonance	Unity	Unity
5.	Resonant frequency	$\frac{1}{2\pi\sqrt{LC}}$	$\frac{1}{2\pi}\sqrt{\frac{1}{LC} - \frac{R^2}{L^2}}$
6.	It magnifies	Voltage	Current
7.	Magnification is	$\frac{\omega L}{R}$	$\frac{\omega L}{R}$

$$\text{Q – factor} = \frac{1}{R}\sqrt{\frac{L}{C}}$$

Half Power Frequencies

The half power frequencies are those frequencies at which the power dissipation in the circuit is half of the power dissipation at resonant frequency f_0. They are the corresponding frequencies f_1 and f_2 at the value of current $I = I_0 / \sqrt{2}$; where I_0 is the current at resounance in R-L-C series circuit (Refer Fig. 43)

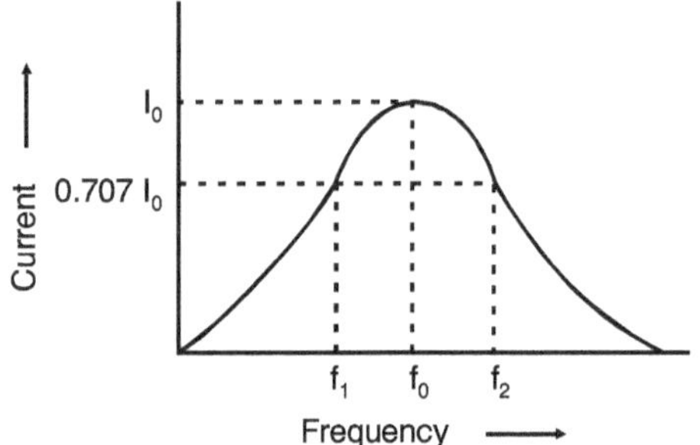

Fig. 43

Hence power P_0 drawn by the circuit at the *resonance* is

$$P_0 = I_0^2 R$$

$$\text{Power in the circuit at } f_1 = \left(\frac{I_0}{\sqrt{2}}\right)^2 R = \frac{1}{2} I_0^2 R$$

(= half the power at resonance)

$$\text{Power in the circuit at } f_2 = \left(\frac{I_0}{\sqrt{2}}\right)^2 R = \frac{1}{2} I_0^2 R$$

(= half the power at resonance)

Bandwidth and Selectivity

The difference $(f_2 - f_1)$ is called the bandwidth of the resonant network.

The ratio of the bandwidth to the resonance frequency is defined as the selectivity of the circuit.

When frequency is varied in R-L-C circuit, the selectivity becomes

$$\frac{f_2 - f_1}{f_0} = \frac{1}{Q_0}$$

where Q_0 is the quality factor of the resonant circuit.

THREE-PHASE CIRCUITS

In case of three-phase circuits, the three-phase currents are determined by considering each phase separately, and calculating the three-phase currents from the phase voltages and impedances in the same manner as for single-phase circuits. In practice, three-phase systems are *usually symmetrical*, the loads being balanced. In such cases the calculations are simple and straight forward.

Having calculated the phase currents, the line currents are obtained from the following simple rules:

1. **Star-connected system :**
   Line current (I_L) = phase current (I_{ph})
   Line voltage (E_L) = $\sqrt{3}$ phase voltage $\left(= \sqrt{3}\ E_{ph}\right)$
2. **Delta connected system** :
   Line current $\left(I_L = \sqrt{3}\right)$ phase current $\left(= \sqrt{3}\ I_{ph}\right)$
   Line voltage (E_L = phase voltage (E_{ph})

Power in a Three-phase Circuit

The total power in three-phase circuit is the sum of the power in the three-phases. The expression for power, both for *star* and *delta connections*, is given by

$$P = \sqrt{3}\ VI \cos \phi$$

where V and I are the line volts and line current and $\cos \phi$ represents the power factor.

The above expression *does not hold good for unbalanced or unsymmetrical systems.*

(Most three-phase apparatus such as motors can be assumed to form a balanced load and calculations for current etc. can be based on this assumption, using the above expression)

Measurement of Power in 3-Phase Circuit

The power in 3-phase load can be measured by using the following methods :

1. Three wattmeter method
2. Two wattmeter method
3. One wattmeter method

Two wattmeter method is generally used to measure 3-phase power. In this method current coils of the two wattmeters are connected in any two lines and their potential coils to the remaining third line. The sum of the two wattmeter readings gives the total power in the circuit. If the load is balanced, then its power factor can also be calculated from these two readings. The readings of the two wattmeters are :

(i)
$$\left.\begin{aligned} W_1 &= E_L I_L \cos(30-\phi) \\ W_2 &= E_L I_L \cos(30+\phi) \\ \tan\phi &= \frac{\sqrt{3}(W_1 - W_2)}{W_1 + W_2} \end{aligned}\right] \text{ Lagging power factor}$$

(ii)
$$\left.\begin{aligned} W_1 &= E_L I_L \cos(30+\phi) \\ W_2 &= E_L I_L \cos(30-\phi) \\ \tan\phi &= \frac{\sqrt{3}(W_1 - W_2)}{W_1 + W_2} \end{aligned}\right] \text{ Leading power factor}$$

CIRCUIT TRANSIENTS

If a circuit is switched from one condition to another either by a change in the applied voltage or change in a current parameter, there exists a transitional period during which the branch currents and voltage drops change from their former values to new ones. After transition period, the circuit becomes steady.

A.C. Transients

(i) R-L sinusoidal transient : Here the voltage function could be at any point in the period at the instant of closing the switch and therefore the phase angle ϕ can take any values from 0 to 2π red/sec.

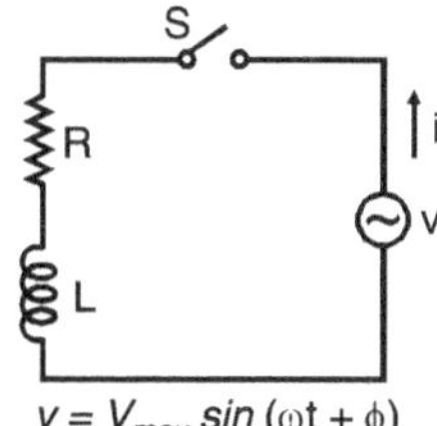

Fig. 44: *R-L Circuit*

In this case, the current (i) is given by

$$i = e^{-(R/M)}\left[\frac{-V_{\max}}{\sqrt{R^2+\omega^2 L^2}}\sin\left(\phi - \tan^{-1}\frac{\omega L}{R}\right)\right]$$

$$+\frac{V_{\max}}{\sqrt{R^2+\omega^2 L^2}}\sin\left(\omega t + \phi - \tan^{-1}\frac{\omega L}{R}\right)$$

It may be noted that :

- The first part of the above equation contains the factor $e^{-(R/L)t}$ which has a value of zero in a relatively short time.
- The second part of the above equation is the steady state current which legs the applied voltage by $\tan^{-1}\frac{\omega L}{R}$

(ii) R.C. sinusoidal transient : For R-C circuit shown in Fig. 45 the basic equation is :

$$R_i + \frac{1}{C}\int v dt = V_{\max}\sin(\omega t + \phi)$$

Here the current is given by,

$$e = e^{-(t/RC)}$$

$$\left[\frac{V_{\max}}{R}\sin\phi - \frac{V_{\max}}{\sqrt{R^2+\left(\frac{1}{\omega C}\right)^2}}\sin\left(\phi + \tan^{-1}\frac{1}{\omega CR}\right)\right]$$

$$+\frac{V_{\max}}{\sqrt{R^2+\left(\frac{1}{\omega C}\right)^2}}\sin\left(\omega t + \phi + \frac{1}{\omega CR}\right)$$

S R C V

$v = V_{max} \sin(\omega t + \phi)$

Fig. 45: *R-C Circuit*

It may be noted that :

- The first part of the above equation is the transient with decay factor $e^{-t/RC}$
- The second part is the steady current which leads the applied voltage by $\tan^{-1}\dfrac{1}{\omega CR}$

D.C. Transients

(i) R-L transients : In the *R-L* circuit shown in Fig. 46

$$t = \frac{V}{R}\left[1 - e^{-(R/L)t}\right]$$

The plot of *i* (exponential rise equation) versus time is shown in Fig. 46.

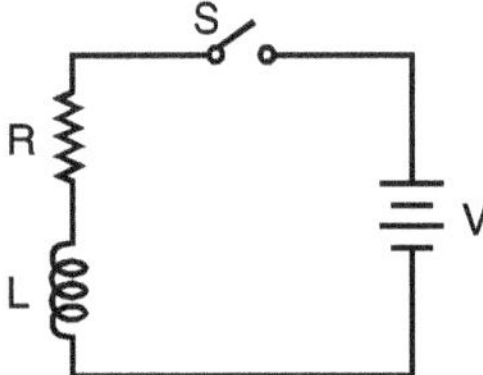

Fig. 46: *R-L Circuit*

The time constant for the above function is the time at which the exponent of e is unity. Thus in this case time constant is L/R. At one time constant, the value of i will be

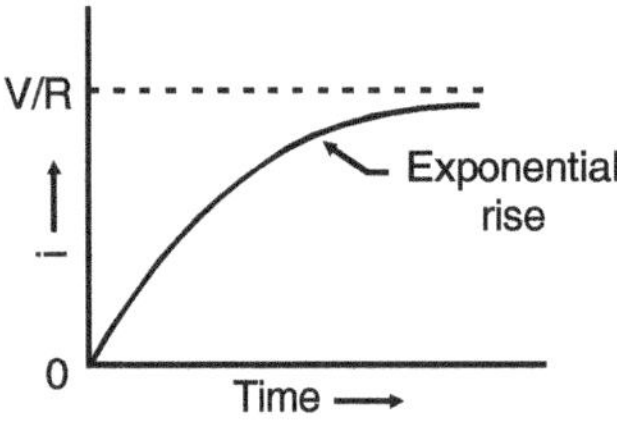

Fig. 47

$$i(1 - e^{-1}) = 1 - 0.368 = 0.632$$

At this time current will be 63.2% of its final value.

The voltage across inductance,

$$v_L = L\frac{di}{dt}$$
$$= Ve^{-(R/L)t}$$

and voltage across resistor

$$v_R = V\left[1 - e^{-(R/L)t}\right]$$

The exponential rise of resistor voltage and exponential decay in inductor voltage are shown in Fig. 48

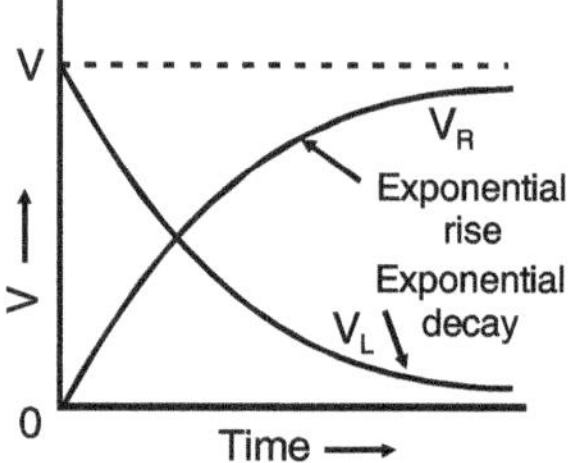

Fig. 48

Also, $v_R + V_L = V\left[1 - e^{-(R/L)t}\right] + V_e^{-(R/L)t}$
$= V$

Power in the circuit elements is given by

$$P_R = \frac{V^2}{R}\left[1 - 2e^{-(R/L)t} + e^{-2(R/L)t}\right]$$

$$P_L = \frac{V^2}{R}\left[e^{-(R/L)t} - e^{-2(R/L)t}\right]$$

Total power,

$$P = P_R + P_L$$
$$= \frac{V^2}{R}\left[1 - e^{-(R/L)t}\right]$$

(ii) R-C transients : In the R-C circuit shown in Fig. 49

$$i = \frac{V}{R}e^{-t/RC}$$

Fig. 49: *R-C Circuit*

Transients voltages across R and C are given by

$$v_R = V_e^{-t/RC}$$
$$v_C = V(1-e^{-t/RC})$$

Also, the power in circuit elements is given by

$$P_R = \frac{V^2}{R} e^{-2t/RC}$$
$$P_L = \frac{V^2}{R}\left(e^{-t/RC} - e^{-2t/RC}\right)$$

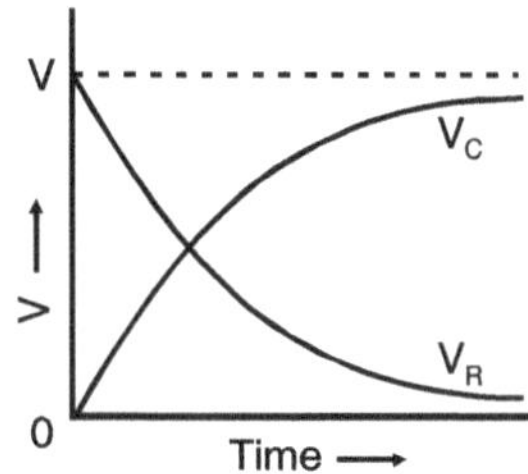

Fig. 50

(iii) R-L-C transients : For R-L-C circuit shown in Fig. 50 the following integro-differential equation can be written as follows :

$$R_i + L\frac{di}{dt} + \frac{1}{C}\int idt = V$$

While solving for i, the following three cases are considered

Case I : $\left(\frac{R}{2L}\right)^2 > \frac{1}{LC}$

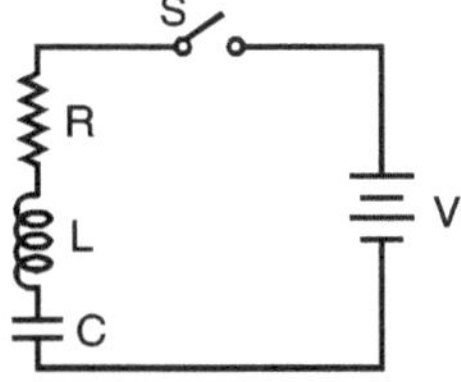

Fig. 51

In this case the current in given by

$$i = C_1e^{\alpha t} + C_1\left(e^{\beta t} + C_2e^{-\beta t}\right)$$

Case II : $\left(\frac{R}{2L}\right)^2 = \frac{1}{LC}$

Here, $i = e^{\alpha t}(C_1 + C_2t)$

Case III : $\left(\frac{R}{2L}\right)^2 < \frac{1}{LC}$

Here, $i = e^{\alpha t}(C_1 \cos \beta t + C_2 \sin \beta t)$

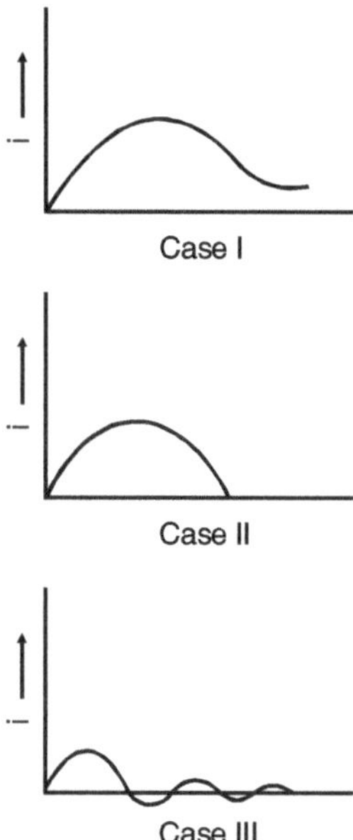

Fig. 52

NETWORK TOPOLOGY

Some Definitions

1. **Graph of network :** The diagram that gives the network configuration and uses lines with very small circles at the ends to represent a network element is called *graph of network.*
2. **Tree :** Any connected open set of branches which includes all nodes of a given graph is called *tree.*
3. **Theorem of Topology :** The basic theorem of topology is given as :

$$b = n + l - l$$

where b = number of branches,
n = number of nodes, and
l = number of independent loops.

4. Cut-set. Cut-set is that set of elements that dissociates two main portions of a network such that replacing any element will destroy this property.

Each cut-set contains one tree-branch, the remaining elements being tree links. Tree-branches connect all the nodes in the network.

LAPLACE TRANSFORMATION

Differential equations can be solved either by *classical method* or by *Laplace transform method.* The classical method is based on *time-domain analysis* and Laplace transform method is based on the *frequency-domain analysis*. The classical method for solving differential equations becomes quite cumbersome when used for network involving higher order differential equations. Therefore in such cases Laplace transform method is preferred.

Solution of differential equations by Laplace transformation involves three steps, similar to numerical calculations by logarithms.

1. Taking of the transform which automatically takes into consideration the initial condition.
2. Rearranging the algebraic equation thus obtained, using algebraic partial fraction (if necessary) to bring every term into the standard form available in the Laplace transform table.
3. Finding the desired complete solution. This table helps in finding transforms as well as

TABLE : Laplace Transform Table

S.No.	*f(t)*	*T(s)*	*S.No.*	*f(t)*	*T(s)*
1.	$C, 1$	$\frac{C}{s}, \frac{1}{s}$	12.	$t \sin \omega t$	$\frac{2\omega s}{(s^2+\omega)^2}$
2.	t	$\frac{1}{s}$	13.	$t \cos \omega t$	$\frac{s^2+\omega^2}{(s^2+\omega^2)^2}$
3.	t^n	$\frac{\angle n}{c^{n+1}}$	14.	$t \sinh t$	$\frac{2\omega s}{(s^2-\omega^2)^2}$
4.	e^{at}, e^{-at}	$\frac{1}{(s-a)}, \frac{1}{(s+a)}$	15.	$t \cosh t$	$\frac{s^2+\omega^2}{(s^2-\omega^2)^2}$
5.	te^{at}	$\frac{1}{(s-a)^2}$	16.	$z^{-at} \sinh \omega t$	$\frac{\omega}{(s+a)^2-\omega^2}$
6.	$\sin \omega t$	$\frac{\omega}{(s^2+\omega^2)}$	17.	$e^{-at} \cosh \omega t$	$\frac{(s+a)}{(s+a)^2-\omega^2}$
7.	$\cos \omega t$	$\frac{s}{(s^2-\omega^2)}$	18.	$\frac{d}{dt} f(t)$	$2F(s) - f(0+)$
8.	$\sinh \omega t$	$\frac{\omega}{(s^2-\omega^2)}$	19.	$\int f(t)dt$	$\frac{F(s)}{s} + \frac{f^{-1}(0+)}{s}$
9.	$\cosh \omega t$	$\frac{s}{(s^2-\omega^2)}$	20.	$e^{-at} t^n$	$\frac{Z_n}{(s+a)^{n+1}}$
10.	$e^{-at} \sin \omega t$	$\frac{\omega}{(s+a)^2+\omega^2}$	21.	$\delta(t)$	1
11.	$e^{-at} \cos \omega t$	$\frac{(s+a)}{(s+a)^2+\omega^2}$			

inverse transform

The Laplace transformation is denoted by the script letter *L*. The Laplace transform of any function *f(t)* is given by the expression

$$L[f(t)] = F(s) = \int_0^{\infty} f(t)e^{-st}dt$$

where f^{-1} is the integration of function.

Inverse Laplace Transformation

The mathematical process of passing from the complex variable expression to the time expression is called an *inverse transformation*. The notation for the inverse transformation is L^{-1}, so that

$$L^{-1}[F(s)] = f(t)$$

Electrical Wave Filters

Electric networks, which permit unattenuated transmission of electric signals within specified frequency ranges and produce attenuation to suppress to signals outside the specified frequency ranges are known as *Electric Wave Filters* or *Filters*. The filters may be classified as follows :

1. Low pass filter
2. High pass filter
3. Band pass filter
4. Band elimination filter.

EXERCISE

1.

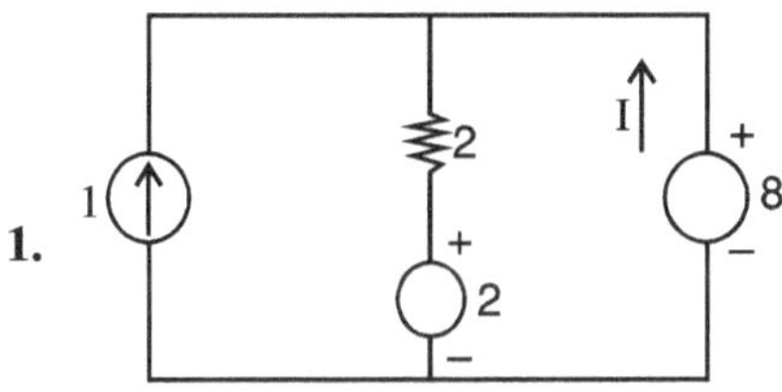

Current marked I is given by

A. 1 B. 2

C. 3 D. 4

2. By proper selection of R_f and R_1, the circuit shown can be used as a

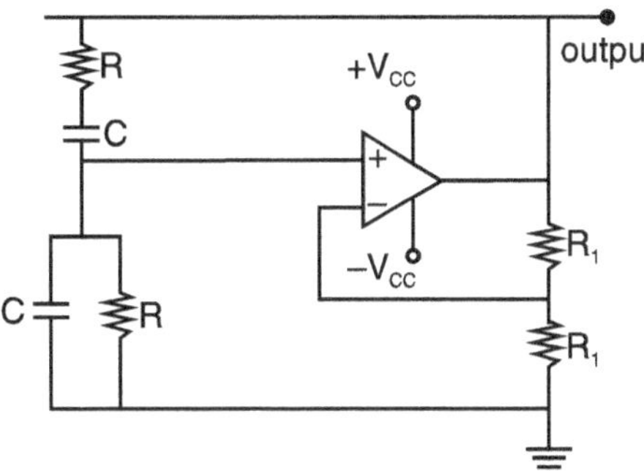

A. ramp generator

B. square wave generator

C. saw tooth generator

D. sine wave generator

3. In the circuit shown current I is

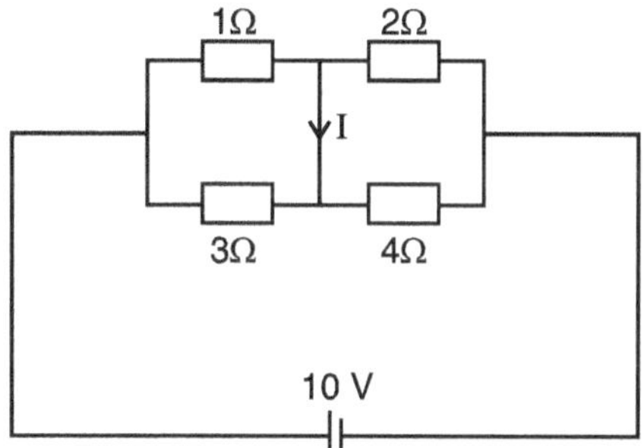

A. $-\frac{12}{5}$A B. $\frac{24}{5}$A

C. $\frac{18}{5}$A D. $\frac{12}{5}$A

4. A transfer function is characterised by

A. Zero initial conditions and dependence on input excitation

B. Given initial conditions and non-dependence on input excitation

C. Given initial conditions and dependence on input excitation

D. Zero initial conditions and non-dependence on input excitation.

5. The equivalent resistance between the terminal points X and Y of the circuit shown is

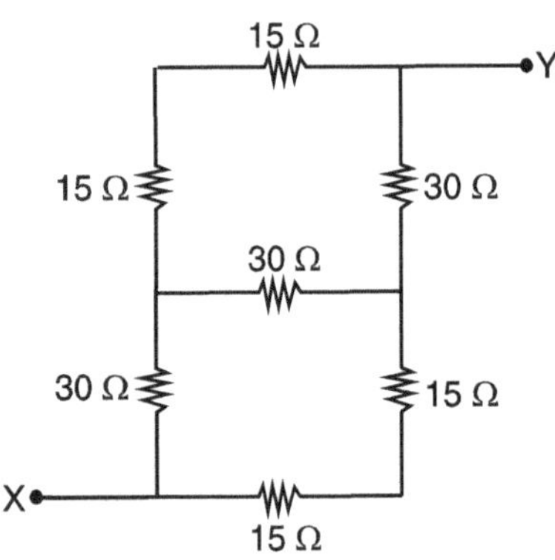

A. 15 ohms B. 45 ohms
C. 55 ohms D. 30 ohms

6. The circuit shown has been in the steady-state when the switch S is opened. The current I after the switch is opened is given by

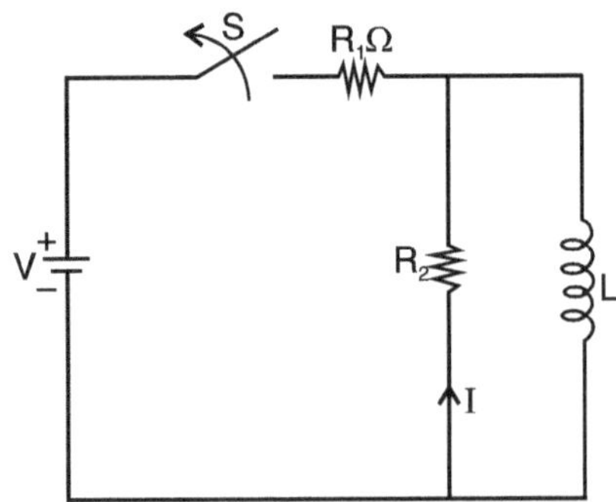

A. $\frac{V}{R_2}e^{+\frac{R_1}{L}t}$ B. $\frac{V}{R_1}e^{-\frac{R_2}{L}t}$

C. $\frac{VR_2}{R_1-R_2}e^{-\frac{L(R_1+R_2)t}{R_1R_2}}$ D. $\frac{VR_1}{R_1-R_2}e^{-\frac{L_t}{L_2}}$

7. Find z_1 and z_2.

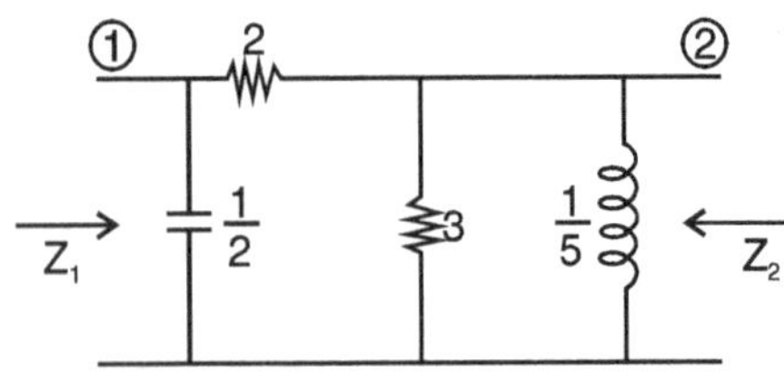

A. $s + 6 = 0$ and 1 B. $s + 6 = 0$ and $\frac{6}{5}$

C. $s + 1 = 0$ and 2 D. $s + 10 = 0$ and 0

8. The time constant associated with the capacitor charging in the circuit shown in the given figure is

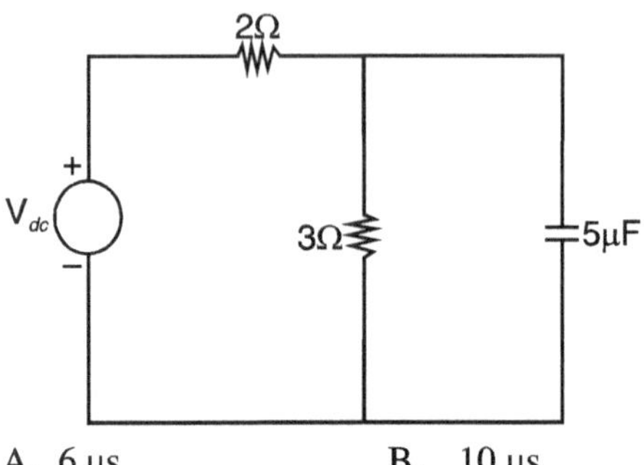

A. 6 μs B. 10 μs
C. 15 μs D. 25 μs

9. If a generator produces a current in a load which rises to a maximum value and then falls to zero periodically, it is

A. an a.c. generator
B. d.c. generator
C. a pulsating a.c. generator
D. B or C

10. The impedance traingle is similar to the _____ triangle with the resistance phasor in place of the _____

A. current, resistor current
B. current, resistor voltage
C. voltage, impedance
D. voltage, resistor voltage

11. A network containing a conventional transformer and various circuit elements has 8 elements and 5 nodes in all. The number of independent loops in the network would be

A. 3 B. 4
C. 5 D. 12

12. Two coils are wound on a common magnetic core. The sign of mutual inductance (M) between the coils in the KVL equation is positive if the

A. fluxes produced by the coils are equal
B. fluxes produced by the coils act in the same direction
C. fluxes produced by the coils are in opposition
D. two coils are wound in the same sense

13. 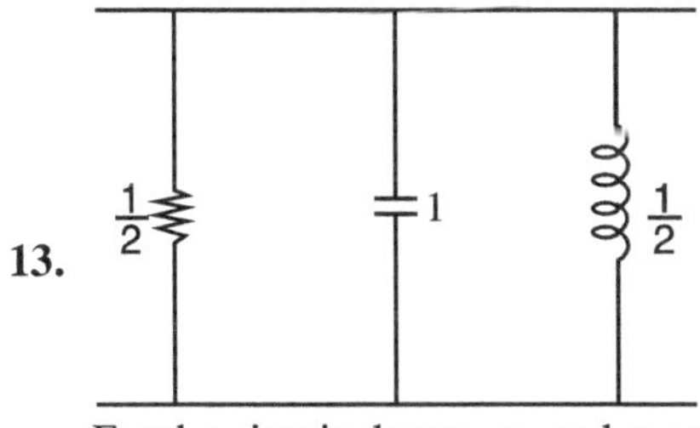

For the circuit shown, ω_0 and α are given by

A. $\sqrt{2}$ and 2

B. 1 and 2

C. 1 and 1

D. $\sqrt{2}$ and 1

14. Thevenin equivalent resistance of the circuit is

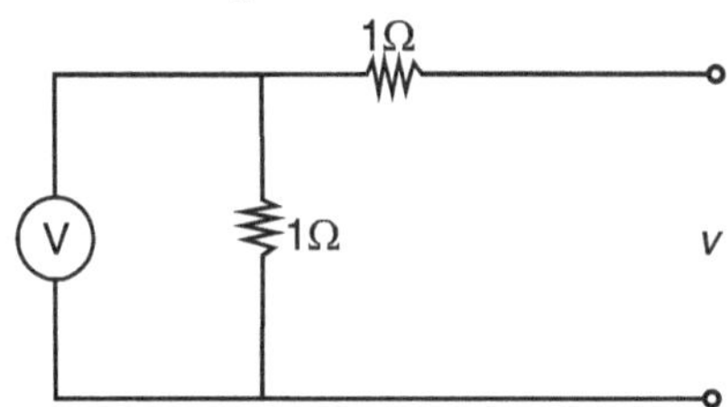

A. 1 ohm B. 0 ohm

C. 1/2 ohm D. ∞

15. 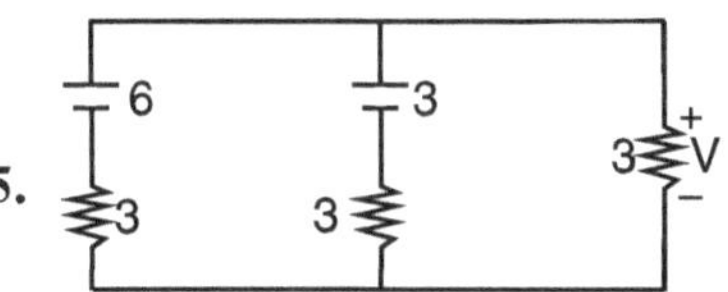

The voltage labelled V is given by

A. 2 B. 0

C. 9 D. 3

16. The reponse to a step signal $u(t)$ is given by $2 + e^{-2t}$. Then the response to the impulse signal $2\,\delta(t)$ is given by

A. e^{-2t} B. $2e^{-2t}$

C. $-2e^{-2t}$ D. $-4e^{-2t}$

17. When the network function

$Z(s) = \dfrac{(s+1)(s+3)}{(s+2)(s+4)}$ is realised, it will give the network

A. 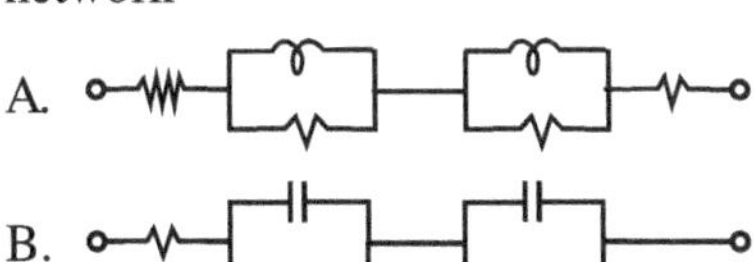

B.

C.

D.

18. A circuit has 10 V applied across a 10 Ω resistance R_2. How much resistance R_2 must be added in series with R_1 reduce the current one half

A. 5 Ω B. 10 Ω

C. 20 Ω D. 33 Ω

19. Current in parallel resonance circuit is given by

A. $\dfrac{V}{R}$ B. $\dfrac{V}{RC}$

C. $\dfrac{V}{L/CR}$ D. $\dfrac{VR}{LC}$

20. The Dimension of LC is

A. Zero B. $\left(\dfrac{\text{volt}}{\text{ampere}}\right)^2$

C. $(\text{seconds})^{-2}$ D. $(\text{second})^2$

21. Gain crossover frequency is defined as one at which $|G(jw)H(jw)|$ equals :

A. 0 B. 1

C. $\sqrt{2}$ D. $1/\sqrt{2}$

22. If all the n sources in a linear network are multiplied by a constant k, then the response gets multiplied by :

A. kn B. k

C. k D. n/k

23. In the circuit shown, $V_s = 250 \sin 500t$; $|V_2| = 200$. If R = 100, value of L is

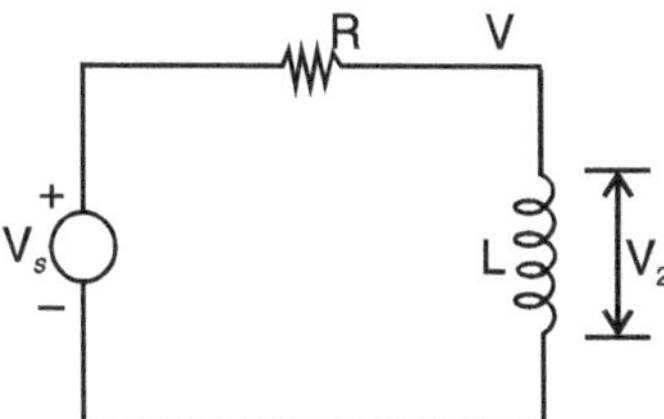

A. $\dfrac{1}{3}$ B. $\dfrac{1}{4}$

C. $\dfrac{2}{3}$ D. $\dfrac{3}{4}$

24. The equivalent inductance of two inductances with mutual coupling in between connected in parallel having the following configuration

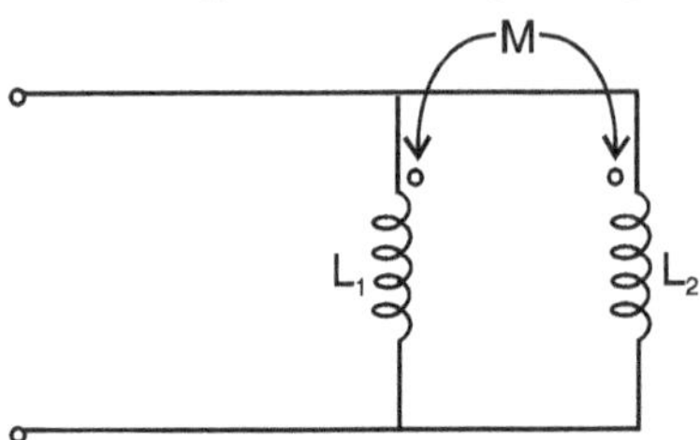

A. $\dfrac{L_1L_2+M^2}{L_1+L_2-2M}$ B. $\dfrac{L_1L_2-M^2}{L_1+L_2+2M}$

C. $\dfrac{L_1L_2-M^2}{L_1+L_2+2M}$ D. $\dfrac{L_1L_2-M^2}{L_1+L_2-2M}$

25. A real transformer has primary self-inductance $L_1 = 9$ H and secondary self inductance $L_2 = 4$ H and a mutual inductance between the primary and secondary M = 3 H. The primary and secondary leakage inductance of the transformer are

A. (3, 6) B. (6, 1)
C. (7, 22) D. (4.5, 2)

26. The passband of a typical filter network with Z_1 and Z_2 as the series and shunt arm impedances is characterised by

A. $-1<\dfrac{Z_1}{4Z_2}<0$ B. $-1<\dfrac{Z_1}{4}Z_2<0$

C. $-0<\dfrac{Z_1}{4Z_2}<1$ D. None of the above

27. The current wave form in a pure resistor of 10 Ω is shown. Power dissipated in the resistor is

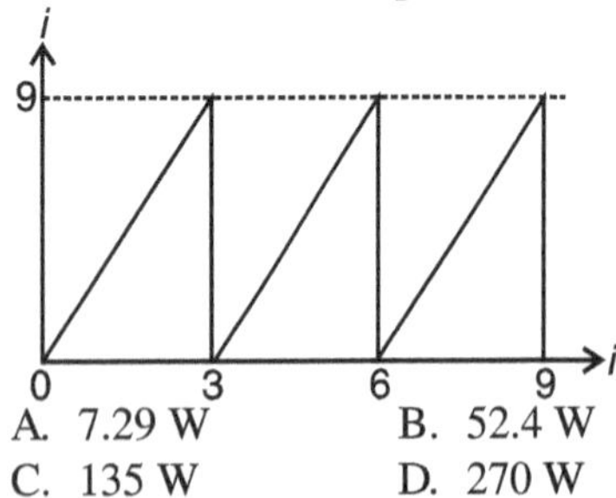

A. 7.29 W B. 52.4 W
C. 135 W D. 270 W

28. Which one of the following is the correct state model for the network shown in the given figure with

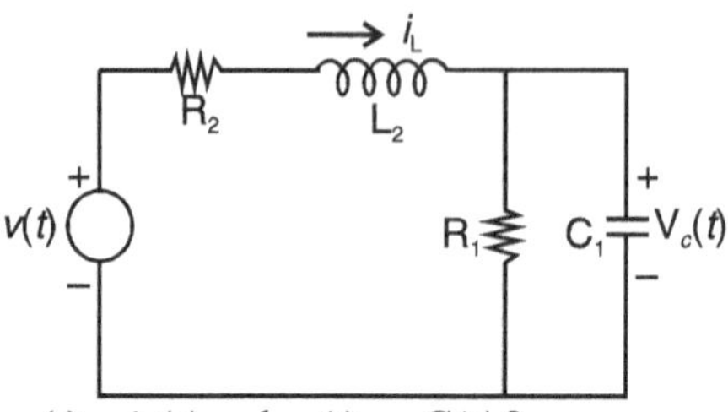

$x_2(t) = i_L(t)$ and $x_1(t) = vC(t)$?

A. $\begin{bmatrix}\dot{x}_1(t)\\ \dot{x}_2(t)\end{bmatrix}=\begin{bmatrix}-\dfrac{1}{R_1C_1} & \dfrac{1}{C_1}\\ -\dfrac{R_1}{L_2} & -\dfrac{1}{R_1C_1}\end{bmatrix}\begin{bmatrix}x_1(t)\\ x_2(t)\end{bmatrix}+\begin{bmatrix}0\\ \dfrac{1}{L_2}\end{bmatrix}v(t)$

B. $\begin{bmatrix}\dot{x}_1(t)\\ \dot{x}_2(t)\end{bmatrix}=\begin{bmatrix}-\dfrac{R_2}{L_2C_1} & \dfrac{1}{C_1}\\ -\dfrac{1}{L_2} & -\dfrac{1}{R_1C_1}\end{bmatrix}\begin{bmatrix}x_1(t)\\ x_2(t)\end{bmatrix}+\begin{bmatrix}1\\ \dfrac{1}{L_2}\end{bmatrix}v(t)$

C. $\begin{bmatrix}\dot{x}_1(t)\\ \dot{x}_2(t)\end{bmatrix}=\begin{bmatrix}-\dfrac{R_2}{L_2} & -\dfrac{1}{L_2}\\ -\dfrac{1}{L_2} & -\dfrac{1}{R_1C_1}\end{bmatrix}\begin{bmatrix}x_1(t)\\ x_2(t)\end{bmatrix}+\begin{bmatrix}\dfrac{1}{L_2}\\ 0\end{bmatrix}v(t)$

D. $\begin{bmatrix}\dot{x}_1(t)\\ \dot{x}_2(t)\end{bmatrix}=\begin{bmatrix}-\dfrac{1}{R_1C_1} & -\dfrac{1}{L_2}\\ \dfrac{1}{C_2} & -\dfrac{R_2}{L_2}\end{bmatrix}\begin{bmatrix}x_1(t)\\ x_2(t)\end{bmatrix}+\begin{bmatrix}\dfrac{1}{L_2}\\ 0\end{bmatrix}v(t)$

29. The circulating current in a parallel LC circuit at any resonant frequency is

A. directly proportional to frequency
B. inversely proportional to frequency
C. independent of frequency
D. none of the above

30. Which one of the following is the ratio V_{24}/V_{13} of the network shown in the given figure?

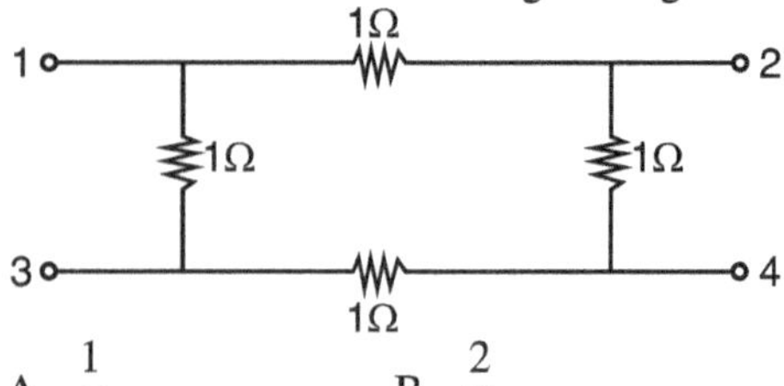

A. $\dfrac{1}{3}$ B. $\dfrac{2}{3}$

C. $\dfrac{3}{4}$ D. $\dfrac{4}{3}$

31. Which of the following expressions is/are used in computing the inductance from the terminals AD of the network

1. $L_1 \pm L_2$ 2. $L_1 + L_2 \pm M$
3. $L_1 + L_2 + 2M$ 4. $L_1 + L_2 - 2M$

Select the correct answer

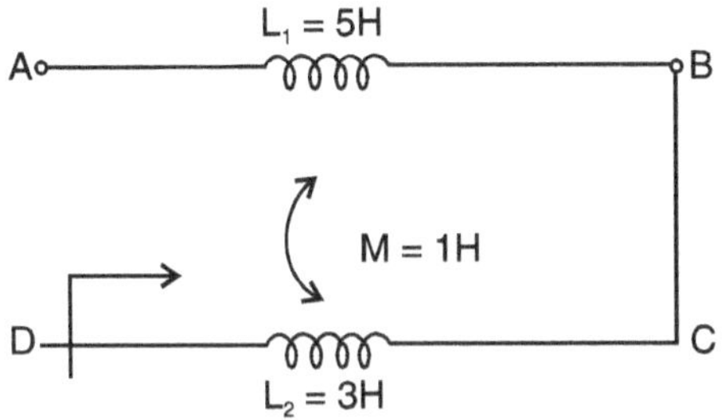

A. 1 alone B. 1 and 2
C. 3 and 4 D. 2, 3, and 4

32. The driving point impedance of the network shown in figure has

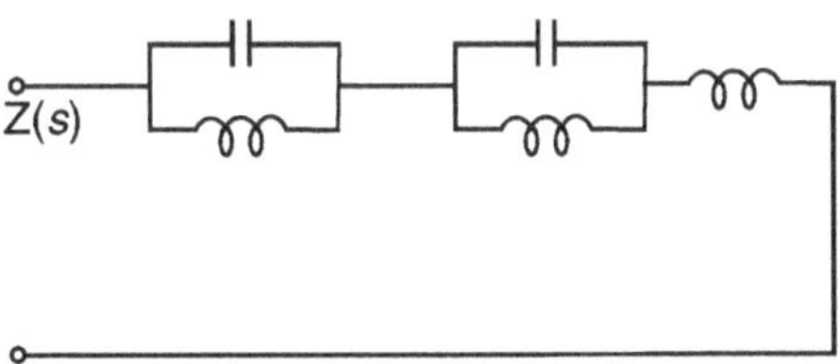

A. zeroes at 0 and infinity
B. poles at 0 and infinity
C. pole at 0 and zero at infinity
D. zero at 0 and pole at infinity

33. A $\frac{\lambda}{4}$ long high frequency transmission line is terminated into one impedance Z_R. If Z_0 is the characteristic impendance of the line, there $\frac{I}{P}$ impedance Z_{in} would be

A. $\sqrt{Z_0 Z_R}$ B. Z_{R2}/Z_0
C. Z_0^2/Z_R D. infinity

34. Two two-port networks α and β having ABCD prameters as

$A_\alpha = 4 = D_\alpha$ $A_\beta = 3 = D_\beta$
$B_\alpha = 5, C_\alpha = 3$ $B_\beta = 4$ and $C_\beta = 2$

are connected in cascade in the order of α, β. The equivalent A parameter of the combination is

A. 15 B. 17
C. 24 D. 31

35. For the circuit shown the voltage of the terminal A with respect to earth is

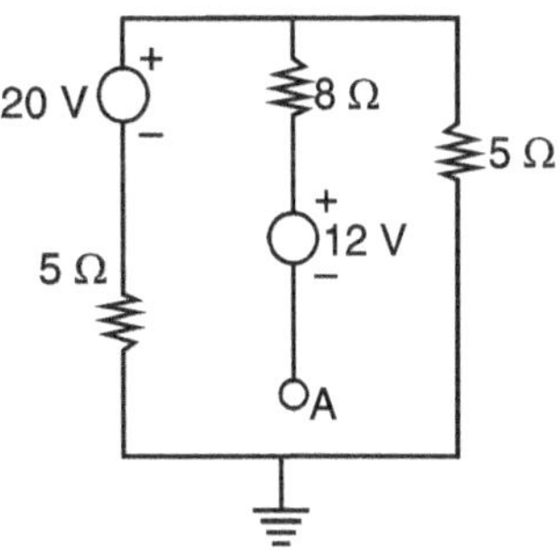

A. –2.0 V B. 2.0 V
C. –4.0 V D. +4.0 V

36. How much power is delivered by a voltage source of 10 V of *(a)* it is open circuited *(b)* if it is connected in parallel with a current source of 1 amp, and direction such that the current enters the positive terminal of the voltage source

A. infinite, – 10 W B. infinite, + 10 W
C. none, –10 W D. none, + 10 W

37. The Laplace transform of the waveform shown in the figure is

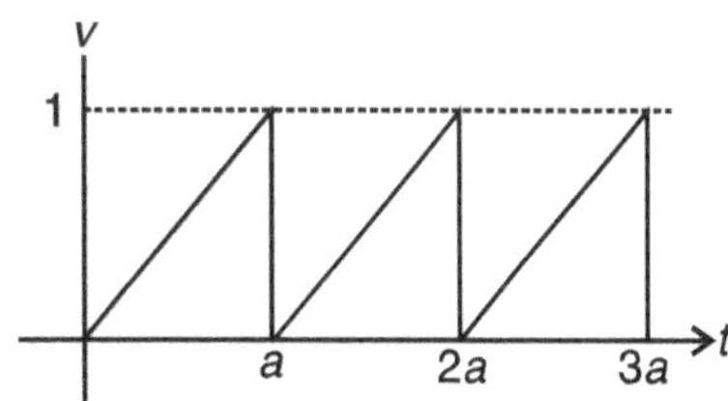

A. $V(s) = \frac{1}{(1 - e^{-as})}$

B. $V(s) = \left(\frac{1 - e^{-as}}{2s^2} - \frac{e^{-as}}{s}\right)$

C. $V(s) = \left(\frac{1 - e^{-as}}{2}\right)$

D. $V(s) = \left(\frac{1 - e^{-as}}{as^2} - \frac{e^{-as}}{s}\right)\frac{1}{(1 - e^{-as})}$

38. The value of G_{12} or $\left(\frac{V_2}{V_1}\right)$ for the circuit shown

in the figure is

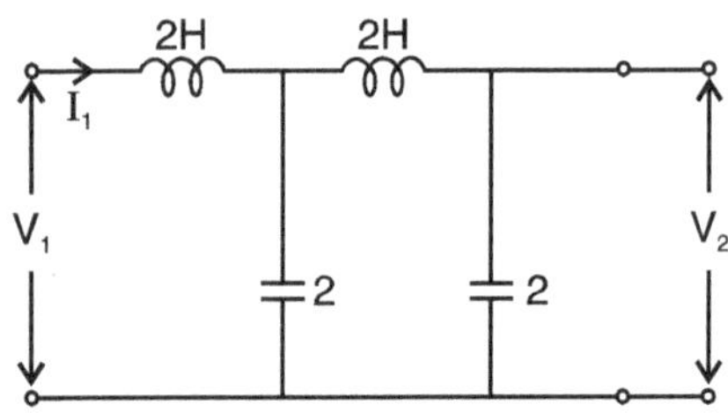

A. $\dfrac{1}{4s^4+2s^2+1}$ B. $\dfrac{s}{s^4+2s^2+1}$

C. $\dfrac{1}{s^4+1}$ D. $\dfrac{1}{16s^4+12s^2+1}$

39. The peak value of the wave form is

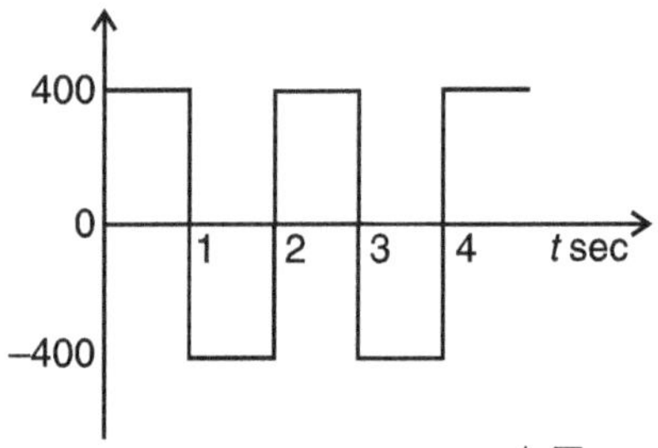

A. 400 B. $400/\sqrt{2}$

C. $-400/\sqrt{2}$ D. $800/\sqrt{2}$

40. A simple equivalent circuit of the 2-terminal network shown is

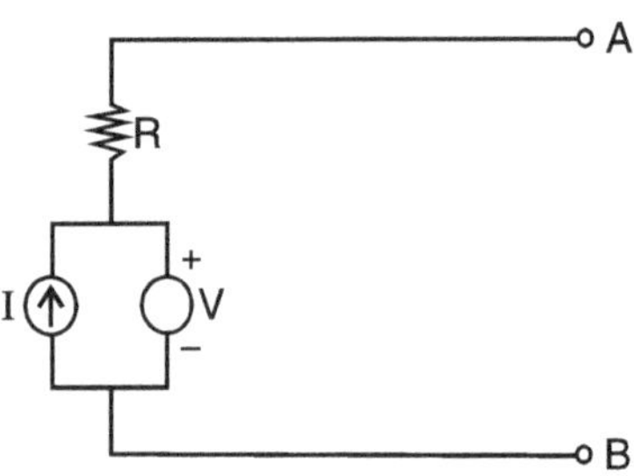

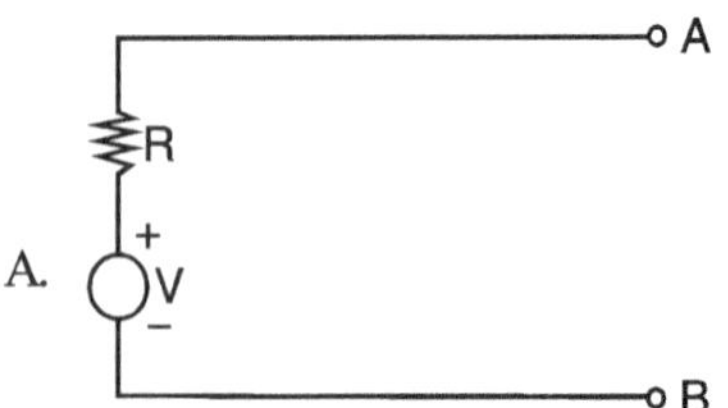

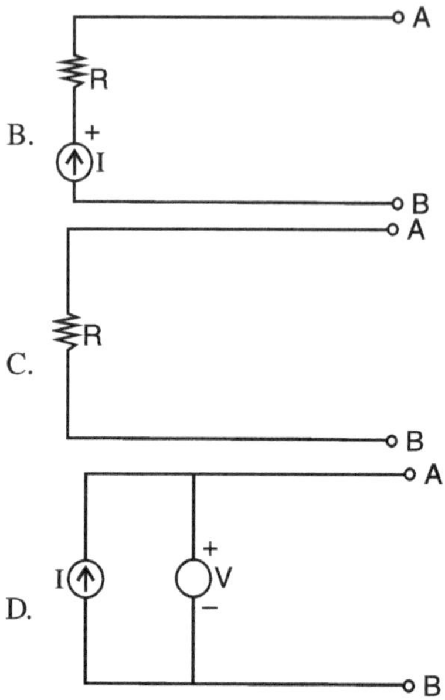

41. The network shown in the figure draw a current I. If the ends *ab* are shorted, the current drawn, would be

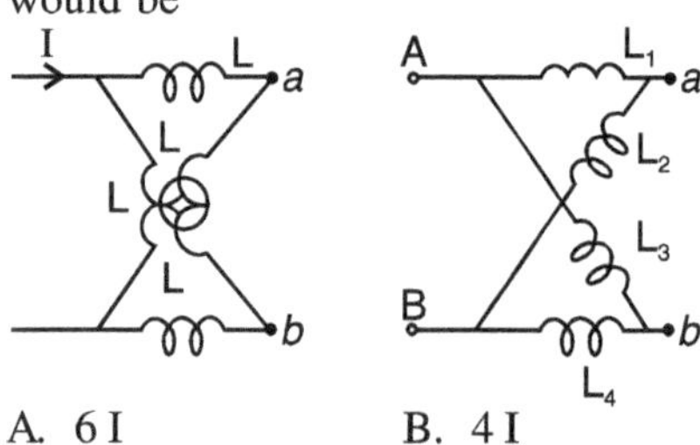

A. 6 I B. 4 I

C. 2 I D. I

42. Consider the following circuit –

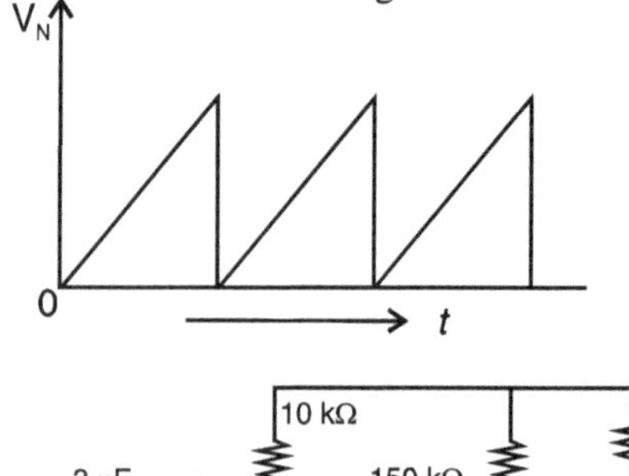

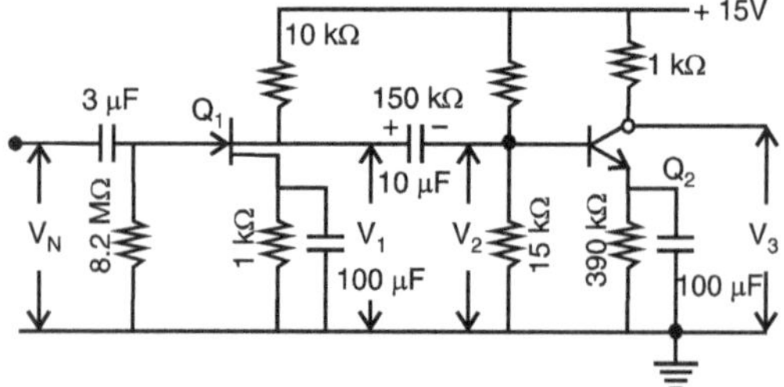

For the given input shown in the figure which of the following pair(s) of voltage and ouput waveform is/are correctly matched:

Voltage **Waveform**

1. V_1

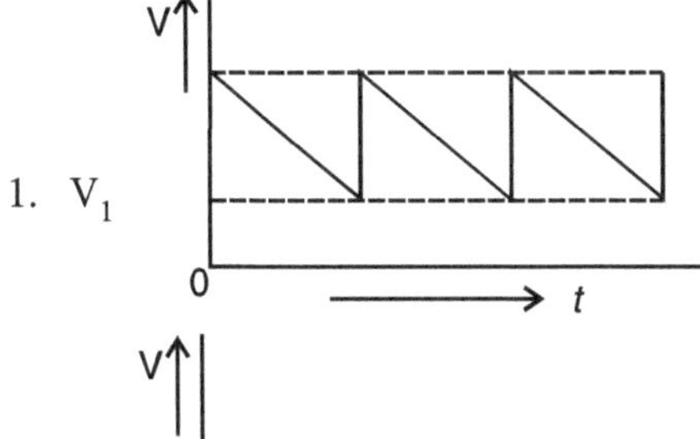

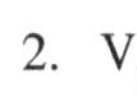

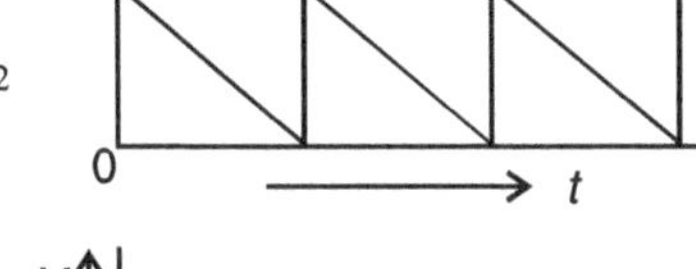

3. V_3

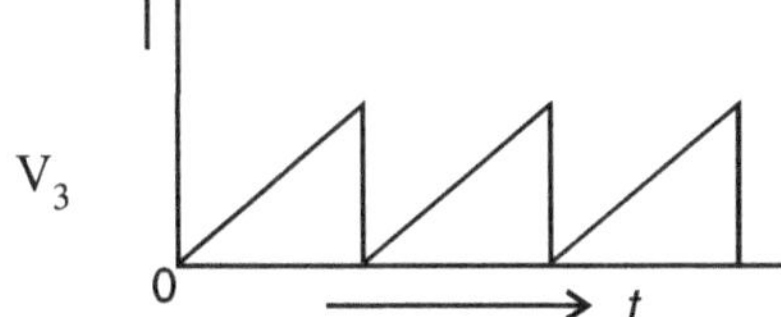

Select the correct answer

A. 2 and 3 B. 1 and 3

C. 1 alone D. 1 and 2

43.

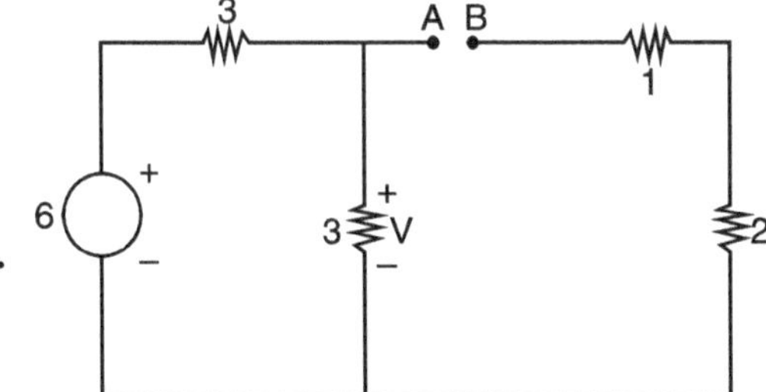

In the circuit the voltage marked V has the values V_1 when B is not connected to A and V_2 when A and B are connected together. The values of V_1 and V_2 respectively are

A. 3 and $\frac{3}{2}$ B. 2 and 1

C. 3 and $\frac{1}{2}$ D. 3 and 2

44. For the circuit shown consider the following

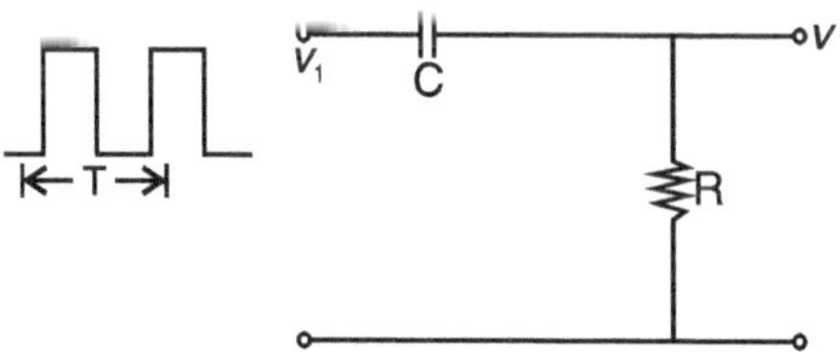

1. The output v_0 will consist of a positive at a negative spike if RC < < T.
2. The output v_0 will be similar to v_i RC > > T.
3. The output pulse will have a higher *r* time if RC is made progressively small than T.

Of these statements

A. 1, 2 and 3 are correct

B. 1 and 2 are correct

C. 2 and 3 are correct

D. 2 and 3 are correct

45. Relaxation time of the capacitor is function of

A. the dielectric medium between the plates

B. the area of the plates

C. the distance between the plates

D. all above

46. Which of the following is a dry storage cell

A. either cell B. carbon zinc cell

C. mercury cell D. nickel cadmium cell

47. Which of the following integrals represents the convolution of two functions $f_1(t)$ and $f_2(t)$?

A. $\int_0^{\tau} f_1(t) f_2(\tau - t)\, d\tau$

B. $\int_0^{t} f_1(t - \tau) f_2(\tau)\, d\tau$

C. $\int_0^{t} f_1(t - \tau) f_2(t)\, d\tau$

D. $\int_0^{t} f_1(\tau - t) f_2(\tau)\, dt$

48. If a voltmeter placed across the 3 ohm resistor in the circuit reads 45 volts, then the reading of ammeter A will be

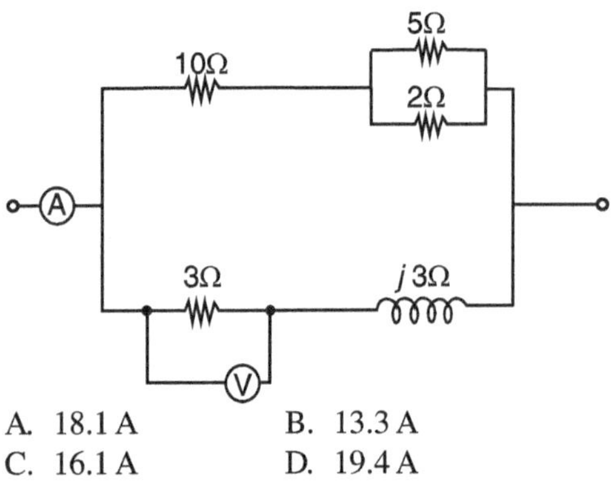

A. 18.1 A B. 13.3 A
C. 16.1 A D. 19.4 A

49. The equivalent circuit of an air-dielectric variable capacitor is as follows –

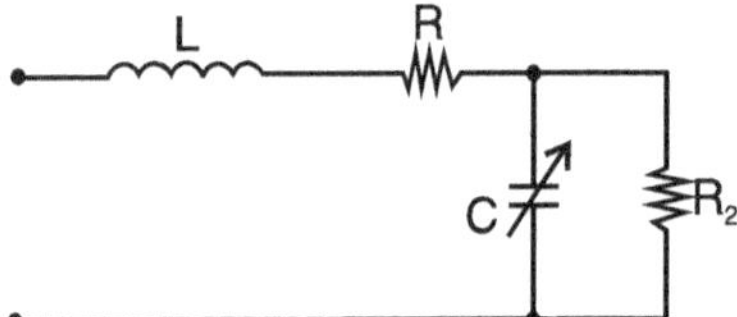

The equivalent circuit of this capacitor, at high frequencies is given by

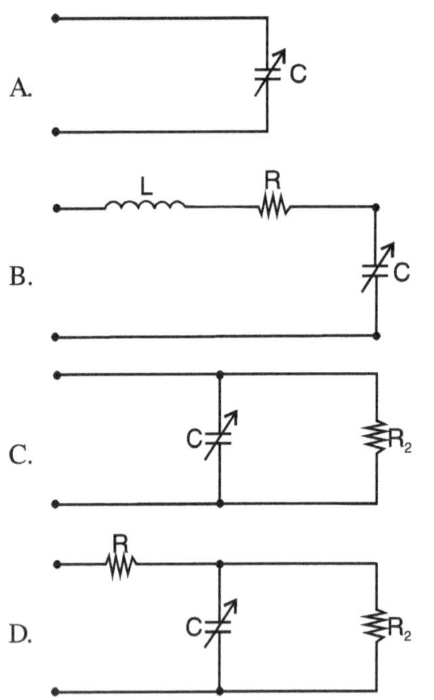

50. Match List I with List II with reference to the graph shown in the given figure and its particular tree of a circuit and select the correct answer

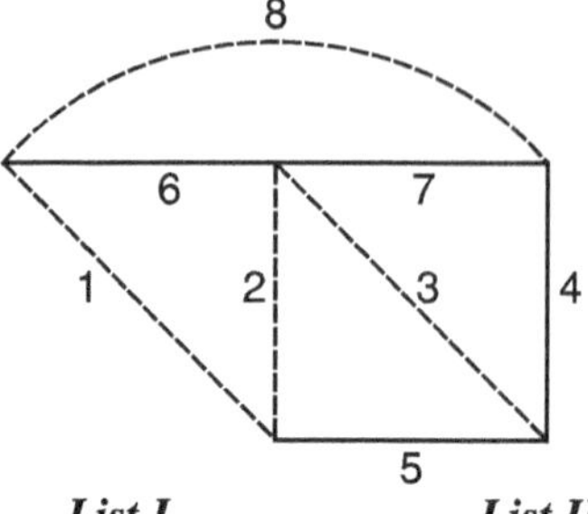

List I	*List II*
(a) 1, 2, 3, 4	1. Twigs
(b) 4, 5, 6, 7	2. Links
(c) 1, 2, 3, 8	3. Fundamental cutset
(d) 1, 4, 5, 6, 7	4. Fundamental loop

Codes :	*(a)*	*(b)*	*(c)*	*(d)*
A.	3	1	2	4
B.	3	2	4	1
C.	2	3	1	4
D.	1	4	3	2

51. Match List I and List II and select the correct answer –

List I	*List II*
(a) Poles and zeroes of the driving-point reactance function of LC network	1. lie on the real axis
(b) Canonic LC network contains	2. a zero
(c) The number of canonic networks for a given driving-point reactance function is	3. Maximum number of elements
(d) The first critical frequency nearest the origin of the complex frequency plane for an RL driving-point impedance will be	4. four
	5. minimum number of elements
	6. alternate
	7. either a pole or zero
	8. three

Codes :	*(a)*	*(b)*	*(c)*	*(d)*
A.	1	8	2	7
B.	6	5	3	2
C.	6	5	4	2
D.	1	8	2	3

52. In the circuit shown switch 'S' is closed at $t = 0$. After some time when the current in inductor was 6A, the rate of change of current through it was 4A/s. The value of the inductor is

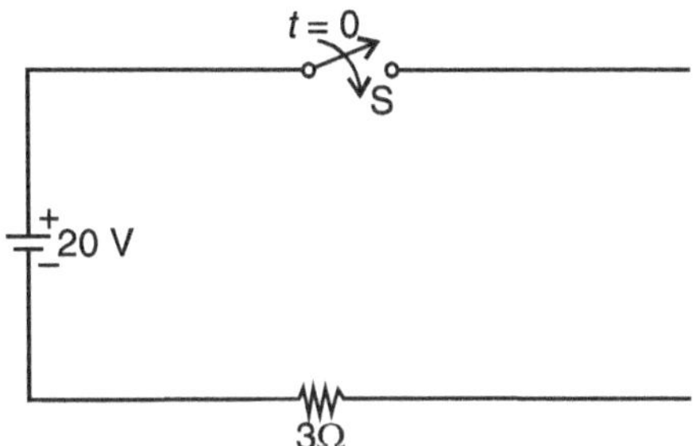

A. indeterminate B. 4.5 H
C. 1.0 H D. 0.5 H

53. There parallel branches have G_1 = 1000 mho, G_2 = 2000 mho, G_3 = 10,000 mho, I is 39 mA, I_1,I_2, I_3 are given by
A. 3, 6, 30 mA B. 6, 3, 30 mA
C. 3, 30, 6 mA D. None of these

54. The two electrical subnetworks N_1 and N_2 are connected through three resistors as shown. The voltages across 5 ohm resistor and 1 ohm resistor are given to be 10 V and 5 V, respectively. Then voltage across 15 ohm resistor is

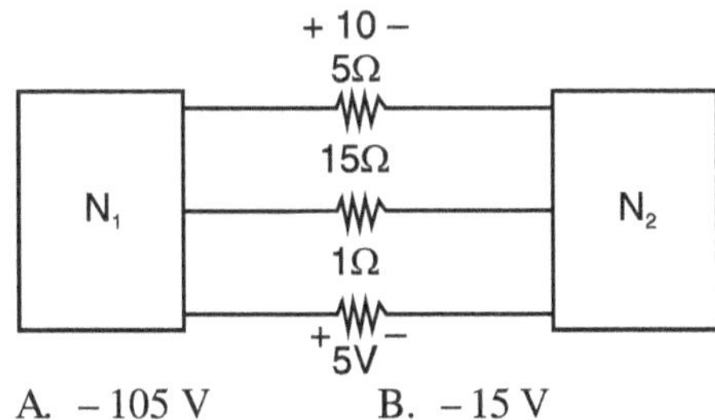

A. – 105 V B. – 15 V
C. + 105 V D. + 15 V

55. For the circuit shown in the given figure, the current through R, when V_A = 0 and V_B = 15 V is 1 amp. Now, if both V_A and V_B are increased by 15 volts, then the current through R will be

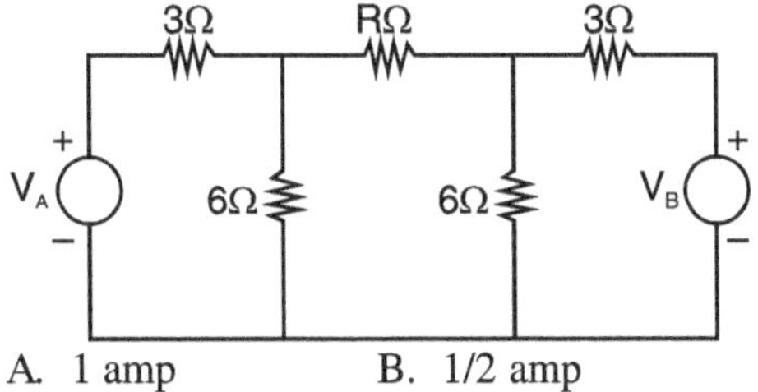

A. 1 amp B. 1/2 amp
C. 3 amp D. 1/3 amp

56. If the networks shown in figures I and II are equivalent at terminals A – B, then the values of V (in volts) and Z (in ohms), will be

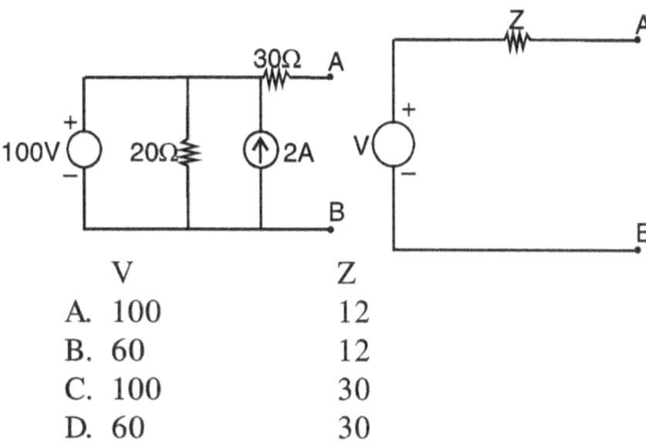

	V	Z
A.	100	12
B.	60	12
C.	100	30
D.	60	30

57. Match List I with List II and select the correct answer

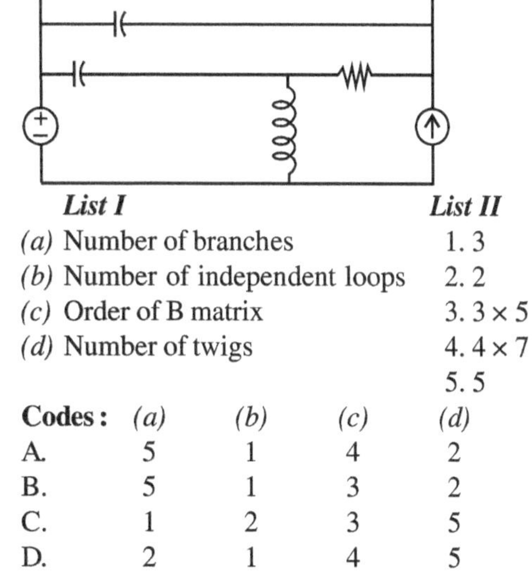

List I	*List II*
(a) Number of branches	1. 3
(b) Number of independent loops	2. 2
(c) Order of B matrix	3. 3 × 5
(d) Number of twigs	4. 4 × 7
	5. 5

Codes :	*(a)*	*(b)*	*(c)*	*(d)*
A.	5	1	4	2
B.	5	1	3	2
C.	1	2	3	5
D.	2	1	4	5

58. In series RCL circuit R = π ohms, L = 1/π henry, C = 21 π farad, and the applied voltage has an effective value of 1 volts. The resonant frequency is ______ Hz.

A. $\sqrt{\frac{2}{4}}$ B. 10
C. 5 D. $\sqrt{\pi}$

59. The circuit shown in the given figure has been in the steady-state when the switch S is opened. The current I after the switch is opened is

given by

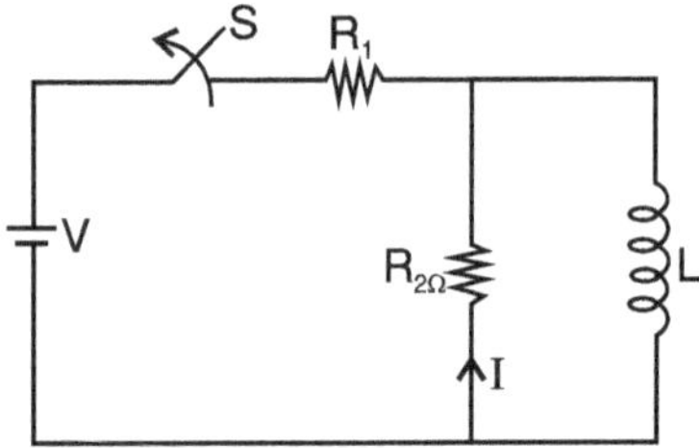

A. $\frac{V}{R_2}e^{-R_1 t/L}$

B. $\frac{V}{R_1}e^{-\frac{1}{L/R_2}}$

C. $\frac{VR_2}{R_1+R_2}e^{-L(R_1+R_2).t/R_1R_2}$

D. $\frac{V}{R_1+R_2}e^{-Lt/R_2}$

60. The transfer function, $\frac{4(s^2+25)}{s^2+2.5s+100}$ is of

A. low pass notch filter
B. low Q bandpass filter
C. high Q bandpass filter
D. high pass notch filter

61. The current through a resistance R = 10 ohms is $i = 2\cos(\omega t) + \cos(3\omega t) + \cos(5\omega t)$. The average power in the resistance is

A. 180 W B. 150 W
C. 300 W D. 1640 W

62. Thevenin's theorem can be applied to calculate the current in

A. any load B. a passive load only
C. a linear load only D. a bilateral load only

63. A current consists of a fundamental component of amplitude I_1 and a third harmonic of amplitude I_3. The rms value of current will be

A. $I_{rms} = (I_1 + I_3)/\sqrt{2}$

B. $I_{rms} = \sqrt{(I_1^2 + I_3^2)}$

C. $I_{rms} = (I_1 + I_3)/2\sqrt{2}$

D. $I_{rms} = \sqrt{(I_1^2 + I_3^2)/2}$

64. Which one of the following networks is the Y equivalent of the Δ circuit shown in Figure?

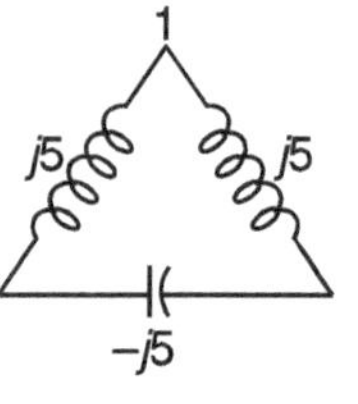

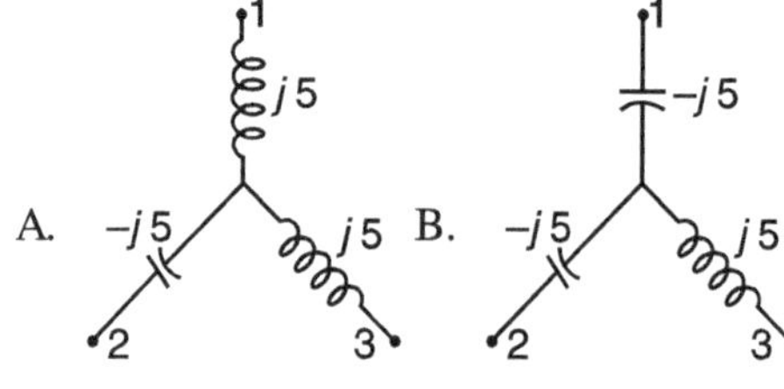

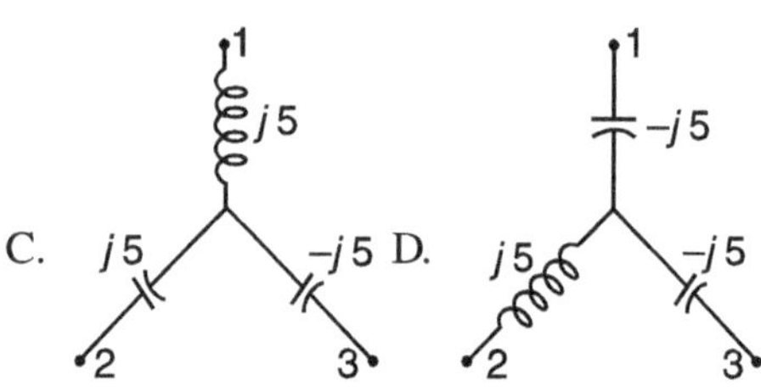

65. Match List I with List II and select the correct answer using the codes given below the lists :

List I (*Network Theorems*)	*List II* (*Most distinguished property of networks*)
(*a*) Reciprocity	1. Impedance matching
(*b*) Tellegen's	2. Bilateral
(*c*) Superposition	3. $\sum_{k=0}^{b} V_{jk}(t_1)i_{jk}(t_2) = 0$
(*d*) Maximum power Transfer	4. Linear
	5. Non linear

Codes :	(*a*)	(*b*)	(*c*)	(*d*)
A.	1	2	3	4
B.	1	2	3	5
C.	2	3	4	1
D.	2	3	5	1

66. The voltage across an impedance in a network is V(s) = Z (s) I (s), where V(s), Z (s) and I (s) are the Laplace Transforms of the corresponding time function *v(t)*, z(t) and *i(t)*. The voltage *v(t)* is

A. $v(t) = z(t).\, i(t)$

B. $v(t) = \int_0^t i(\tau) z(t-\tau) d\tau$

C. $v(t) = \int_0^t i(\tau) z(t+\tau) d\tau$

D. $v(t) = z(t) + i(t)$

67. The impedance function Z(*s*) of a one-port network is given by

$$Z(s) = \frac{3s(s^2+4)}{(s^2+1)(s^2+9)}$$

Which one of the following is correct reactance curve of the above function?

A.

B.

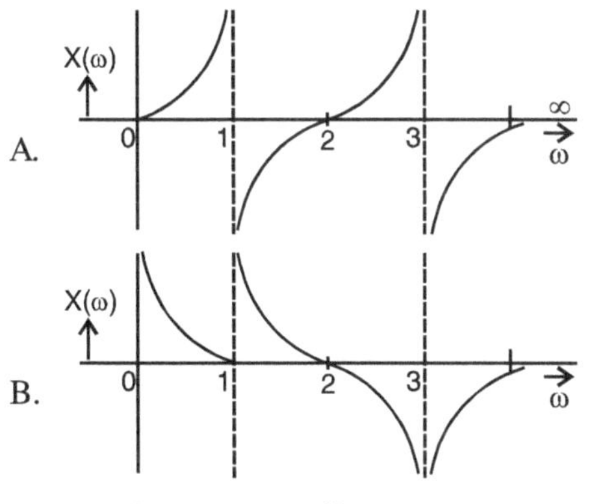

C.

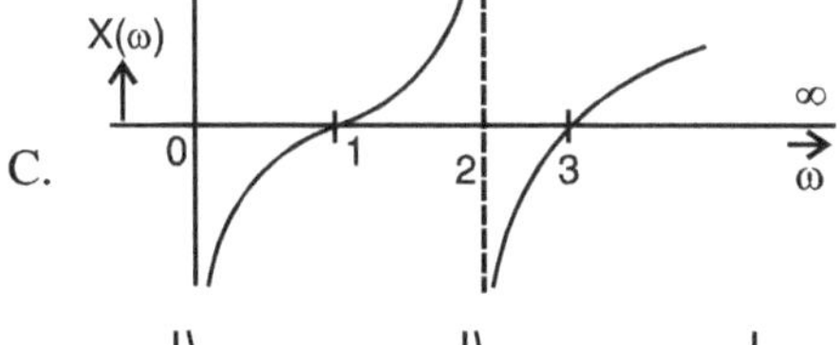

D.

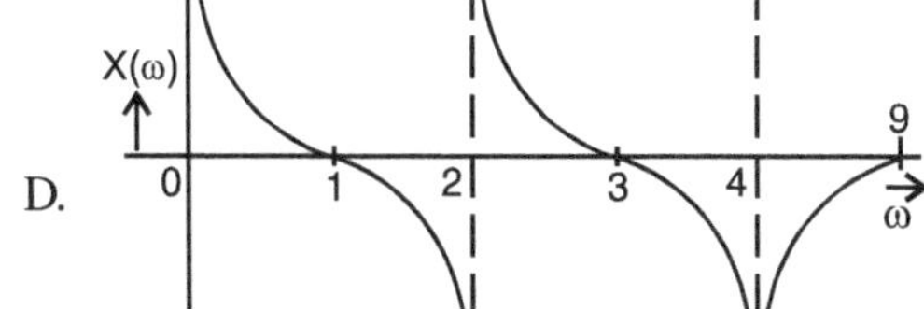

68. Consider the following statements

A unit impulse $\delta(t)$ is mathematically defined as

1. $\delta(t) = 0, t \neq = 0$
2. $\int_{0^+}^{+\infty} \delta(t) dt = 1$
3. $\int_{-\infty}^{+\infty} \delta(t) dt$

Of these statements

A. 1, 2 and 3 are correct

B. 1 and 3 are correct

C. 2 and 3 are correct

D. 1 and 2 are correct

69. A parallel RLC circuit has $\omega_0 = 10^8$ and Q = 20. Given C = 20 pF, the value of R is

A. $\frac{1}{2} \times 10^4$ B. 10^4

C. 2×10^4 D. 25

70. A unity feedback control system has a forward path transfer function equal to $\frac{42.25}{s(s+6.5)}$. The unit step response of this system starting from rest, will have its maximum value at a time equal to

A. 0 sec B. 0.56 sec

C. 5.6 sec D. infinity

71. One of the following can not be used to remove the unwanted side band in SSB. This is the

A. filter system B. phase-shift method

C. third method D. balanced modulator

72. A planar graph has 4 nodes and 5 branches. The number of meshes in the dual graph is

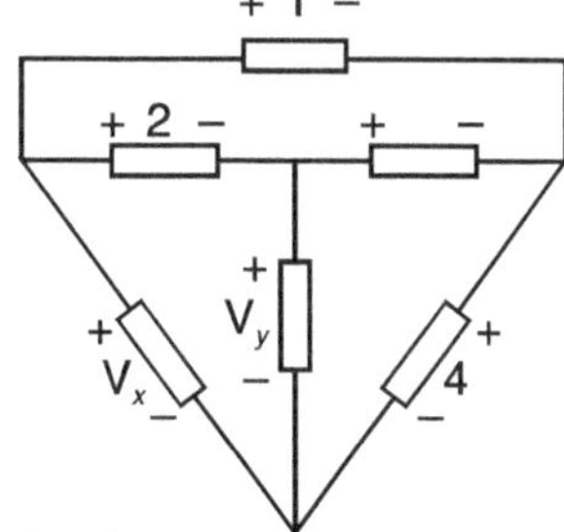

A. 5 B. 4

C. 3 D. 2

73. In the graph and the tree shown the fundamental cutset for the branch 2 is

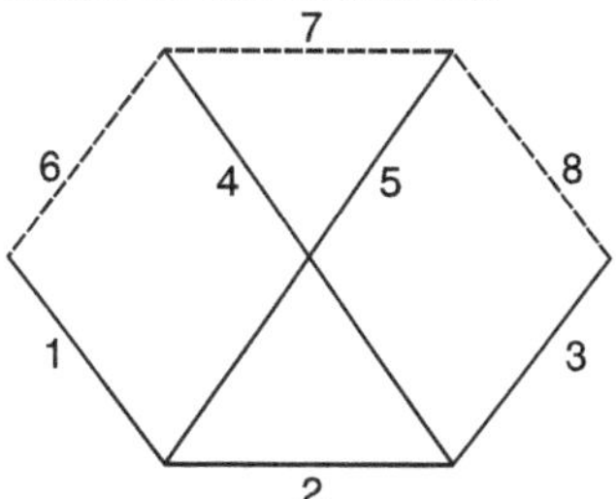

A. 2, 1, 5 B. 2, 6, 7, 8

C. 2, 1, 3, 4, 5 D. 2, 3, 4

74. Consider a system described by the following difference equation :

$y(n+3) + 6y(n+2) + 11y(n+1) + 6y(n)$
$= r(n+2) + 9r(n+1) + 20\,r(n)$

where y is the output and r is the input. The transfer function of the system will be

A. $\dfrac{2z^2 + z + 20}{3z^3 + 2z^2 + z + 6}$

B. $\dfrac{z^2 + 9z + 20}{z^3 + 6z^2 + 6z + 11}$

C. $\dfrac{z^3 + 6z^2 + 6z + 11}{z^2 + 9z + 20}$

D. none of the above

75. Match the List I with List II and select the correct answer

List I	**List II**
(a) $e^{-2(t-2)}$	1. $\dfrac{2}{s^2}$
(b) $\dfrac{d(t^2)}{dt}$	2. $\dfrac{1}{x+2}e^{-2s}$
(c) $t/2$	3. $\dfrac{1}{2s^2}$
	4. 1/2s

Codes :	*(a)*	*(b)*	*(c)*
A.	2	1	3
B.	1	2	3
C.	2	3	4
D.	3	1	4

76. Consider the following statements :

The transfer impedances and admittances of a network remain constant when the position of excitation and response are interchanged if the network

1. is linear
2. consists of bilateral elements
3. has high impedance or admittance as the case may be
4. is resonant

Of these statements

A. 1 and 2 are correct
B. 1, 3 and 4 are correct
C. 1 and 3 are correct
D. 1, 2, 3 and 4 are correct

77. Match List I with List II and select the correct answer

List I	*List II*
(a) $\dfrac{K_1 s + K_2}{K_3}$	1. P-controller
(b) $\dfrac{K_1 s^2 + K_2 s + K_3}{K_4 s}$	2. PI-controller
(c) $\dfrac{K_1 s + K_2}{K_3 s}$	3. PD-controller
(d) $\dfrac{K_1 s}{K_2 s}$	4. PID-controller

Codes :	*(a)*	*(b)*	*(c)*	*(d)*
A.	3	4	2	1
B.	3	2	4	1
C.	4	3	1	2
D.	4	1	2	3

78. In the network shown the capacitor C_1 is initially charged to a voltage V_0 before the switch S in the circuit is closed. In the steady-state

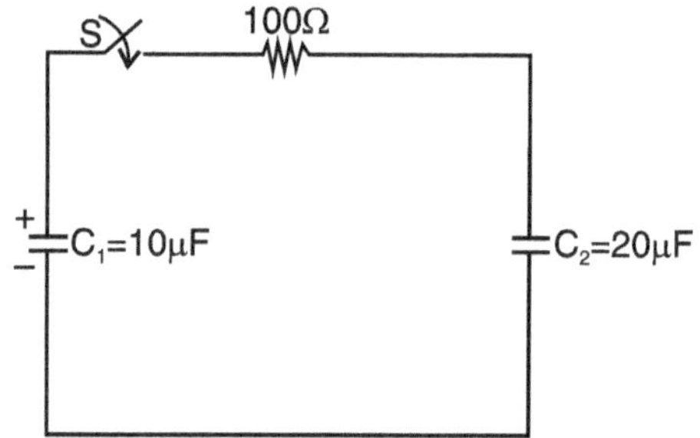

A. C_1 and C_2 are charged to equal voltages
B. C_1 and C_2 are charged with equal coulombs
C. C_1 and C_2 are discharged fully
D. C_2 alone is charged to voltage V_0

79. A resistor R is connected to a voltage source V_S having an internal resistance R_s. A voltmeter of resistance R_m is connected across the terminals of the resistor R. The voltmeter will read a voltage of

A. $\dfrac{V_S R_S R_m}{R_S R_m + R_S R + R R_m}$

B. $\dfrac{V_S R}{R + R_S}$

C. $\dfrac{V_S R_m}{R + R_S + R_m}$

D. $\dfrac{V_S R R_m}{R_S R + R_S R_m + R R_m}$

80. Which one of the following functions qualifies to be an autocorrelation function of a real-valued time signal?

A. sin (ωτ) B. exp (–τ)

C. δ (τ) D. u(τ)

81.

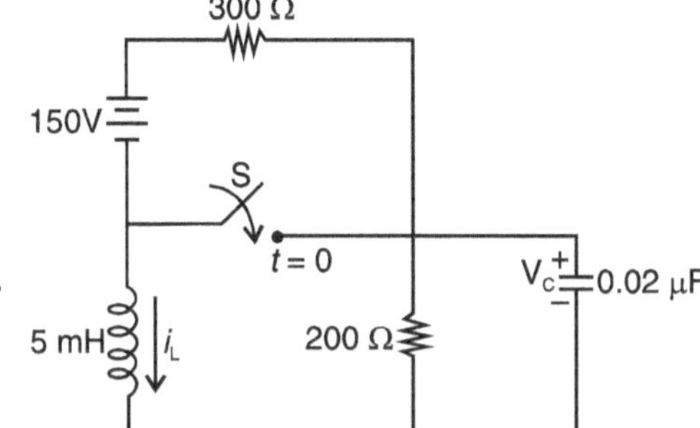

After keeping it open for a long time, the switch 'S' in the circuit shown in the given figure is closed at $t = 0$. The capacitor voltage $v_C(0^+)$ and inductor current $i_L(0^+)$ will be

A. 60 V and –0.3 A B. 150 V and zero

C. zero and 0.3 A D. 90 V and – 0.3 A

82. Laplace transform of t sin h at is

A. $\dfrac{2as}{(s^2 + a^2)^2}$ B. $\dfrac{2as}{(s^2 - a^2)}$

C. $\dfrac{2as}{(s^2 - a^2)^2}$ D. $\dfrac{2as}{(s^2 + a^2)}$

83. For the circuit shown, the current through R when $V_A = 0$ and $V_B = 15$ V is 1 amp. Now, if both V_A and V_B are increased by 15 volts, then the current through resistance R will be

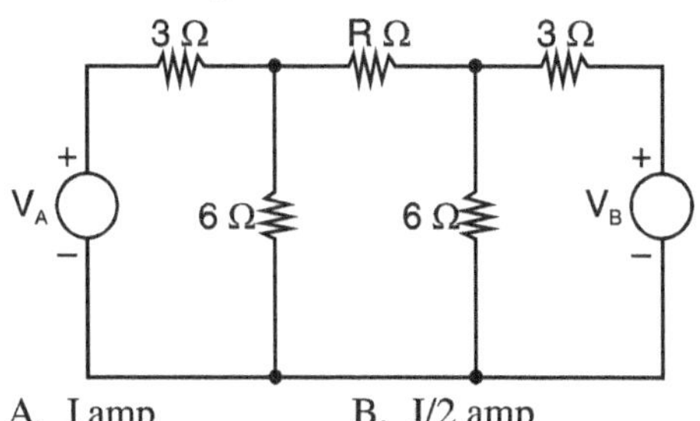

A. I amp B. I/2 amp

C. 3I amp D. I/3 amp

84. Consider the following situations/conditions

1. High loop-gain
2. Less ringing
3. Greater damping
4. Negative dB-gain margin

System stability requirements would include

A. 2 and 3 B. 1, 2 and 3

C. 1 and 4 D. 2, 3 and 4

85. An ideal transformer has turns ratio of 2 : 1. Considering high voltage side as a port 1 and low voltage side as a port 2, the transmission line parameters of the transformer will be

A. $\begin{bmatrix} 0 & -2 \\ 0.5 & 0 \end{bmatrix}$ B. $\begin{bmatrix} -2 & 0 \\ 0 & -0.5 \end{bmatrix}$

C. $\begin{bmatrix} 2 & 0 \\ 0 & 0.5 \end{bmatrix}$ D. 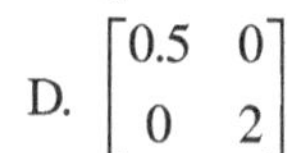 $\begin{bmatrix} 0.5 & 0 \\ 0 & 2 \end{bmatrix}$

86.

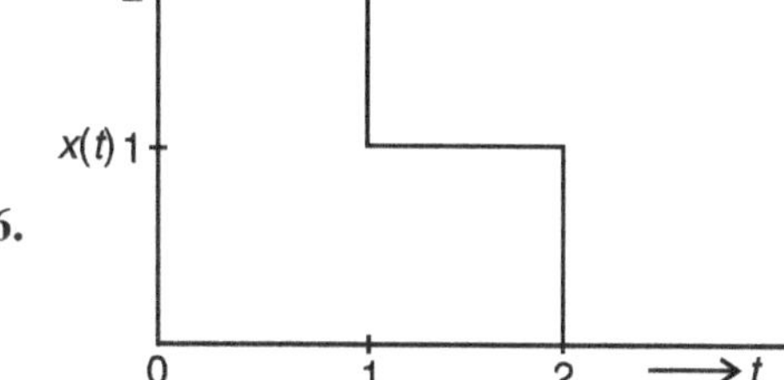

$x(t)$ shown in the figure is applied to an LTI system whose step response is $e^{-at}\, u(t)$. The response to $x(t)$ is given by

A. $2e^{-at}\, u(t) - e^{-a(t-1)}\, u(t-1) - 2e^{-a(t-2)}\, u(t-2)$

B. $2e^{-at}\, u(t) + e^{-a(t-1)}\, u(t-1) - 3e^{-a(t-2)}\, u(t-2)$

C. $2e^{-at}\, u(t) - e^{-a(t-1)}\, u(t-1)$

D. $2e^{-at}\, u(t) - e^{-a(t-1)}\, u(t-1) - e^{-a(t-2)}\, u(t-2)$

87. Any two-port network having a 6 dB loss will give

A. an output power which is one-quarter of the input power

B. an output power which is one-half of the input power

C. an output voltage which is 0.707 of the input voltage

D. an output power which is 0.707 of the input power

88. If the the charge one each of the capacitors in the given figure is 4500 μC, what is the total capacitance (in μF), assuming that the voltage distribution across C_1, C_2, C_3 is the ratio of 2 : 3: 4?

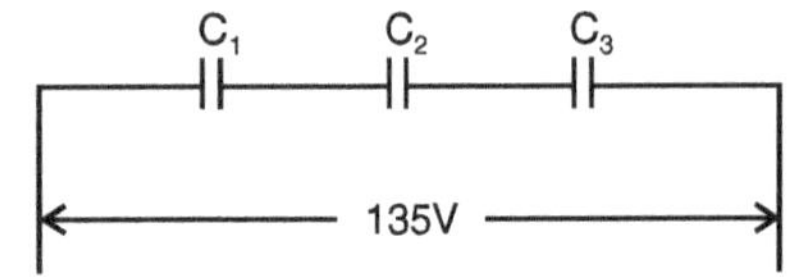

A. 325 B. 11.1
C. 22.2 D. 33.3

89. Two two-port networks are connected in cascade. The combination is to be represented as a single two-port network. The parameters of the network are obtained by multiplying the individual
A. z-parameter matrix
B. h-parameter matrix
C. y-parameter matrix
D. ABCD parameter matrix.

90. Which one of the following circuits has a driving-point impedance of

$$z(s)=\frac{2\left(s^2+s+1/2\right)}{\left(s^2+s+1\right)}$$

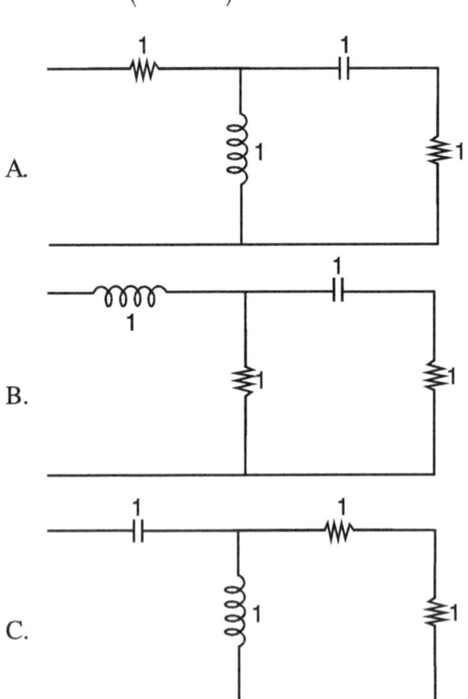

D.

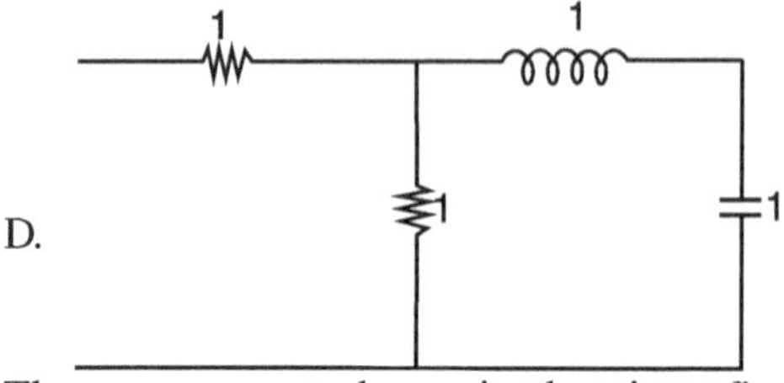

91. The arrangement shown in the given figure represents an RC potentiometer for measuring ac voltage. What should be the value of C so that V_0/V_{in} is independent of the frequency of the input signal?

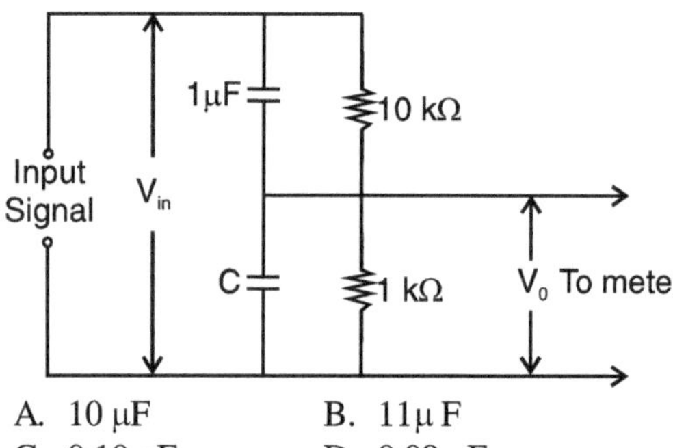

A. 10 μF B. 11μ F
C. 0.10 μF D. 0.09 μF

92. If a pulse voltage $v(t)$ of 4 V magnitude and 2 secs, duration is applied to a pure inductor of 1 H. with zero initial current, the current (in amps) drawn at $t = 3$ secs will be

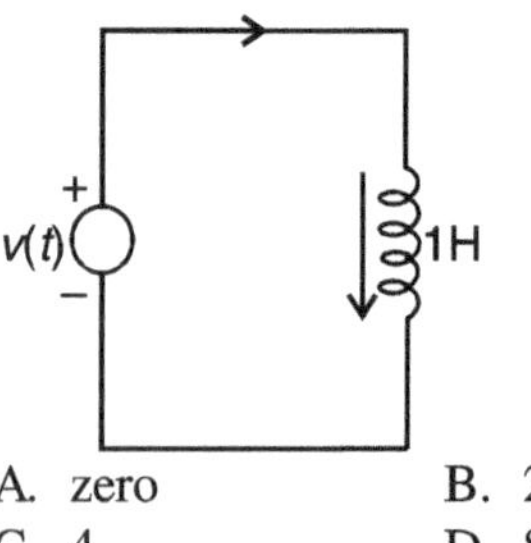

A. zero B. 2
C. 4 D. 8

93. High frequency inductors and capacitors are commonly plated with silver. The main purpose of this is to
A. reduce their dc resistances
B. reduce their ac resistances
C. increase their ac resistances
D. increase their dc resistances

94. A voltage wave form shown is applied to an ideal full-wave rectifier voltmeter. The voltmeter is calibrated to read the RMS values of sine-waves. The voltmeter will read

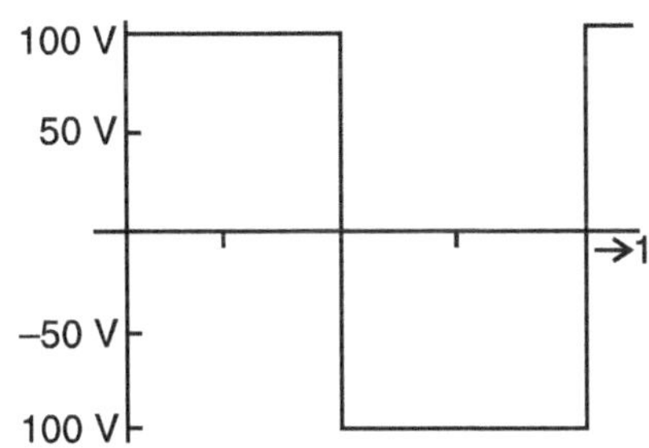

A. $\frac{50}{\sqrt{2}}$ V B. $\frac{100}{\sqrt{2}}$ V

C. 100 V D. 111 V

95. In a delta connected polyphase circuit, the line current I_L is related to I_P

A. $I_L = \sqrt{3}\, I_P$ B. $I_L = 3\, I_P$

C. $I_L = \frac{1}{\sqrt{3}} I_P$ D. $I_L = \frac{1}{3} I_P$

96. If $F(s) = L[f(t)] = \frac{K}{(s+1)(s^2+4)}$ then $\lim_{t\to\infty} f(t)$ is given by

A. K/4 B. infinite

C. zero D. undefined

97. A 200 watts electric bulb draws current of I amp from 200 V, 50 Hz source. The current I is :

A. 1.1 Amp B. $\frac{200}{220}$ Amp

C. 100 Amp D. None of these

98. The circuit diagram of an op-amp based amplifier is shown. The ratio v_{out}/v_{in} is equal to

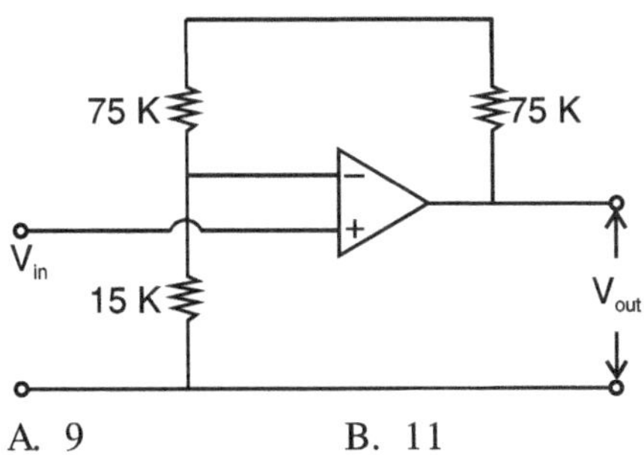

A. 9 B. 11

C. 10 D. 21

99. An iductive coil connected to a 200 V, 50 Hz ac supply with 10 A current flowing through it dissipates 1000 watts. Which of the following will have least value in ohms?

A. Resistance, R B. Reactance, X_L

C. Impedance, Z D. None of the above

100.

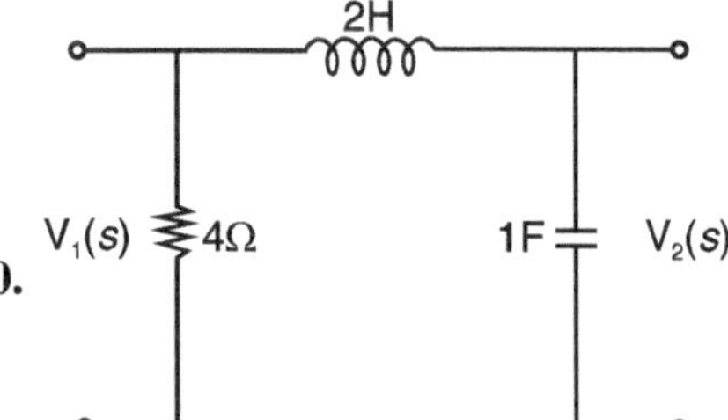

The open-circuit voltage ratio $\frac{V_2(s)}{V_1(s)}$ of the network shown is

A. $1 + 2s^2$ B. $\frac{1}{1+2s^2}$

C. $(1 + 2s)^2$ D. $\frac{1}{1+2s}$

101. System initially at rest. The response a unit step $u(t)$ is given by $\frac{1}{2}(1-e^{-2t})$. The response to a signal = $u(t) + 2\delta(t)$ (where $\delta(t)$ is the unit impulse) is given by

A. $\frac{1}{2}+\frac{3}{2}e^{-2t}$ B. $1 + e^{-2t}$

C. $1 + 3^{-2t}$ D. $\frac{1}{2}+\frac{1}{2}e^{-2t}$

102. The voltage transfer ratio of two-port networks connected in cascade may be conveniently obtained from the

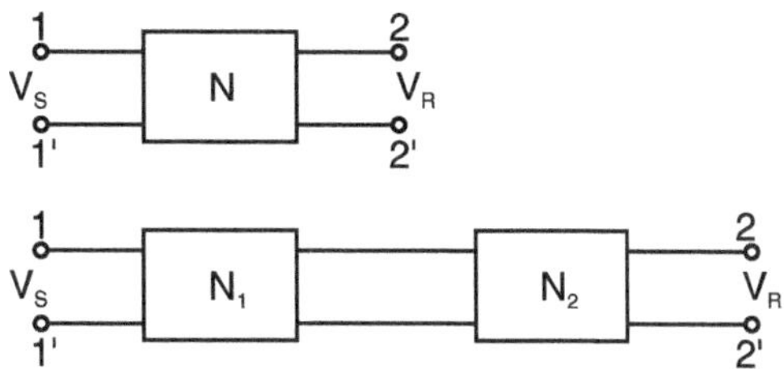

A. product of the individual ABCD matrices of the two networks

B. product of voltage transfer ratios of the two individual networks

C. sum of the Z-matrices of the two networks
D. sum of the h-matrices of the two networks

103. A planar graph has 6 branches and 3 meshes. Then the number of nodes is

A. 6 B. 4
C. 3 D. 2

104. The dissipation factor of a good dielectric is of the order of

A. 0.01 B. 0.001
C. 0.1 D. 0.0001

105. If an impedance has the pole-zero pattern shown in the given figure, it must be composed of

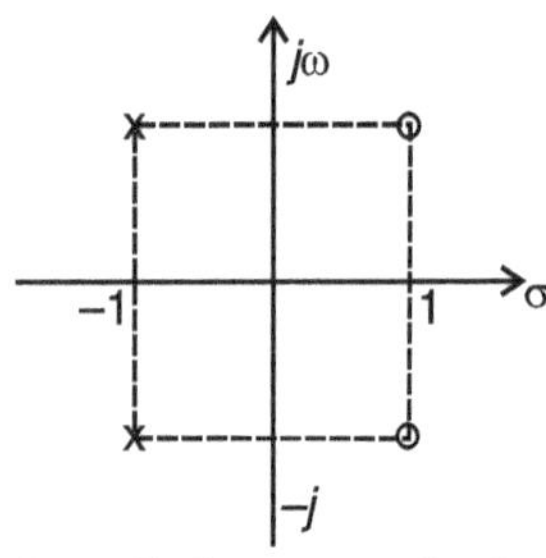

A. RC elements only B. RL elements only
C. LC elements only D. RLC elements

106. Which of the following expression is/are used in computing the inductance L as seen from the terminals AD of the network of interconnected inductances as shown in the given figure?

1. $L_1 \pm L_2$ 2. $L_1 + L_2 \pm M$
3. $L_1 + L_2 + 2M$ 4. $L_1 + L_2 - 2M$

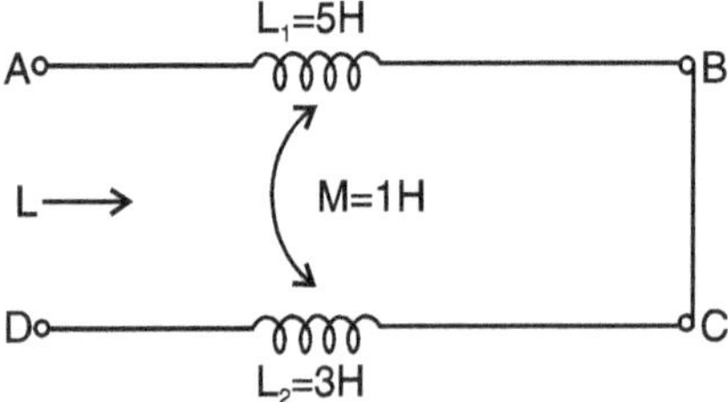

Select the correct answer using the codes given below :

Codes :

A. 1 alone B. 1 and 2
C. 3 and 4 D. 2, 3 and 4

107. If the wire in the case above is folded but not twisted the resistance will be

A. 200 Ω B. 100 Ω
C. 50 Ω D. 25 Ω

108. The two networks given below are equivalent with respect to terminals 1 and 2 at all frequencies. Find, C_A, L_B, L_C, C_C

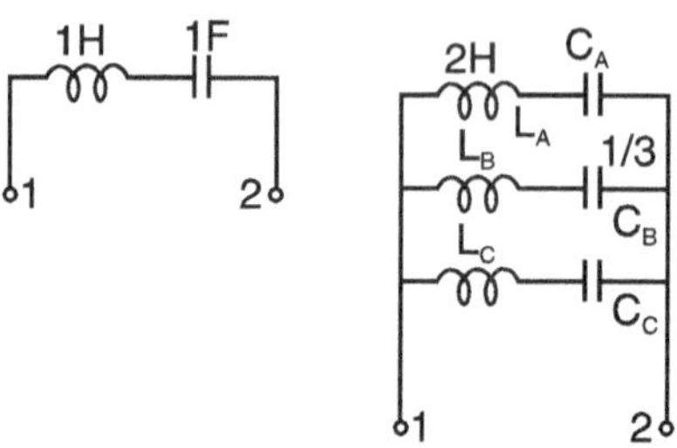

A. 0.5, 0.33, 6, 0.166 B. 0.5, 3, 6, 0.166
C. 0.5, 3, 3, 2 D. 0.5, 3, 6, 0.166

109. A series circuit consisting of a resistance, an inductance and a variable capacitance is connected to a voltage source of fixed voltage and frequency. The capacitance value required for resonance is C_r. The variation of the capacitance between half-power points is C. The value of ΔC is (Note. Q_r is Q factor at resonance).

A. $\dfrac{2C_r}{Q_r}$ B. $\dfrac{Q_r}{2C_r}$

C. $\dfrac{2Q_r}{C_r}$ D. $\dfrac{1}{2C_rQ_r}$

110. If $E_1 = 4 + j$, $Z_2 = 3 + j$, then resultant of $Z_1 + Z_2$ is given by

A. $5 \angle 45°$ B. $7 \angle 45°$
C. $1 \angle 45°$ D. none of these

111. When a voltage $V_0 \sin \omega_0 t$ is applied to the pure inductor, the ammeter shown in the figure reads I_0. If the voltage applied is :
$- V_0 \sin \omega_0 t + 2V_0 \sin 2\omega_0 t + 3 V_0 \sin 3 V\omega_0 t + 4 V_0 \sin 4 \omega_0 t$, the ammeter reading would be

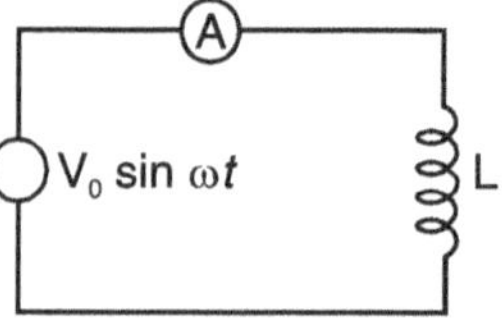

A. 0
B. $10 I_0$
C. $\sqrt{4^3 + 3^3 + 2^2 + 1}I_0$
D. $2 I_0$

112. A conductor in $(x - y)$ plane and having length 1m is moving with a volicity $V = (2i + 3j + k)$

m/sec. A magnetic induction field B = $(i + 2j)$ wb/m^2 is applied to the conductor. The potential difference between the ends of the conductor is

A. Zero B. 4.88 volt

C. $\sqrt{6}$ volt D. None of these

113. About the Fourier series expansion of a periodic function it can be said that

A. Even functions have only a constant and cosine terms in their FS expansion

B. Odd functions have only sine terms in their FS expansion

C. Functions with half-wave symmetry contain only odd harmonics

D. All the above three

114. If from the function $f(t)$ one forms the function, $\Psi(t) = f(t) + f(-t)$, then $\Psi(t)$ is

A. even

B. odd

C. neither even nor odd

D. both even and odd.

115. A capacitor charges a dc source through a series resistor R. Its charging current is:

A. an exponential decay function

B. an exponential growth function

C. a linear rise function

D. a linear decay function

116. For a high pass RC circuit, when subjected to a unit step input voltage, the voltage across the capacitor will be

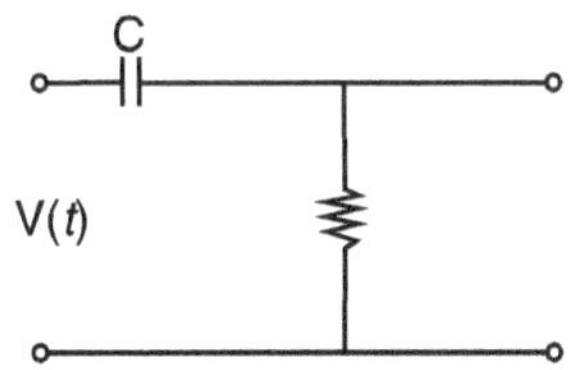

A. $1 - e^{-t/RC}$ B. $e^{-t/RC}$

C. $e^{+t/RC}$ D. 1

117. Given the Laplace transform, $v(t) = \int_{\theta}^{\infty} e^{-st}\, v(t)\, dt$.

The inverse transform $v(t)$ is

A. $\int_{\sigma-j\infty}^{\sigma+j\infty} e^{st} V(s)\, ds$ B. $\frac{1}{2\pi j}\int_{\sigma-j\infty}^{\sigma+j\infty} e^{st} V(s)\, ds$

C. $\frac{1}{2\pi j}\int_{0}^{\infty} e^{st} V(s)\, ds$ D. $\frac{1}{2\pi j}\int_{\sigma-j\infty}^{\sigma+j\infty} e^{-st} V(s)\, ds$

118. Fourier series expansion of an odd periodic function contains :

A. cosine terms only

B. sine terms only

C. both sine and consine terms

D. none of these

119. The inverse Fourier transform of F $(j\omega)$ = $\int_{-\infty}^{\infty} \exp(-j\omega t) f(t)\, dt$ is

A. $f(t) = \int_{-\infty}^{\infty} \exp(+j\omega t) f(j\omega)\, d\omega$

B. $f(t) = \frac{1}{2\pi}\int_{-\infty}^{\infty} \exp(+\varphi\omega t) F(+j\omega)\, d\omega$

C. $f(t) = \frac{1}{2\pi}\int_{-\infty}^{\infty} \exp(-j\omega t) F(+j\omega).\, d\omega$

D. $f(t) = \frac{1}{2\pi}\int_{-\infty}^{\infty} \exp(-\varphi\omega t) F(-j\omega)\, d\omega$

120. The number of 2 μF, 400 V capacitors needed to obtain a capacitance value of 1.5 μF rated for 1600 V is

A. 12 B. 8

C. 6 D. 4

121. Which of the following periodic waveforms will have only odd harmonics of sinusoidal waveforms?

1. 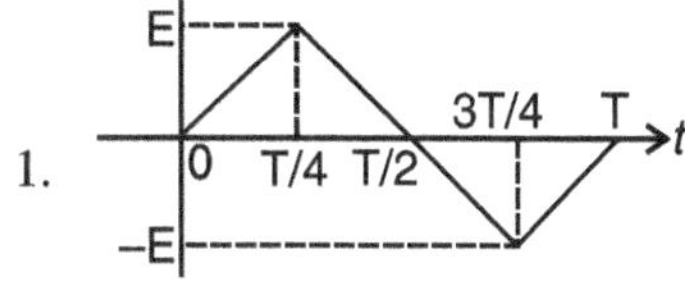

2. 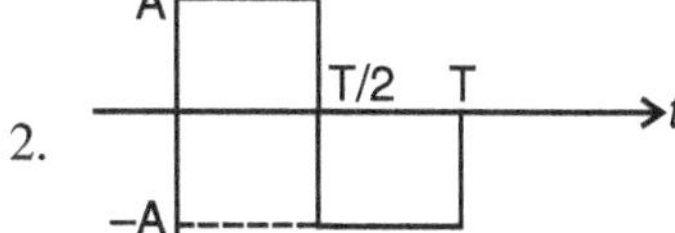

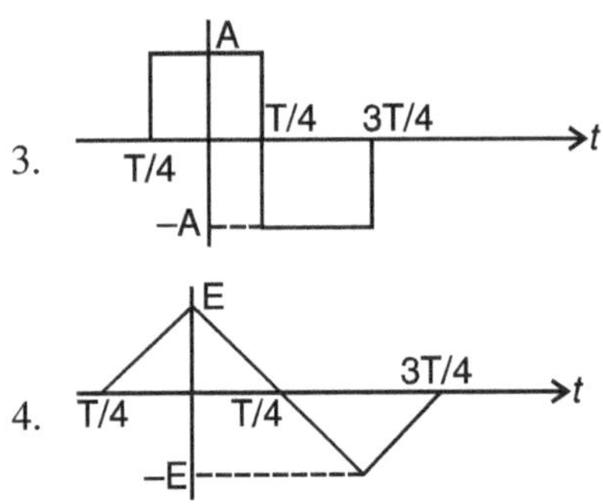

Select the correct answer using the codes given below :

Codes :

A. 1 and 2 B. 1 and 3

C. 1 and 4 D. 2 and 4

122. For a periodic function, the spectral density and the autocorrelation functions form

A. Fourier transform pair

B. Laplace transform pair

C. Hilbert's transform pair

D. Z-transform pair

123. The current through a series RL circuit $\frac{1}{4}e^{-t/2}$ when excited by a unit impulse voltage. The values of R and L are respectively

A. 8, 4 B. 4, 2

C. 2,4 D. 1, 4

124. The circuit shown in the figure is excited by a unit impulse voltage *(t)*. The current *i(t)* is given by $i(t) = \frac{\delta(t)}{2} - \frac{1}{16}e^{-t/6} + \frac{1}{4}e^{-t/2}$.

Then R_1 L, R_2 are respectively,

A. 4, 4, 2, 2 B. 2, 2, 4, 4

C. 4, 2, 2, 4 D. 2, 4, 2, 4

125. The circuit shown in the figure is excited by $V\delta(t)$. The peak voltage at the capacitor is

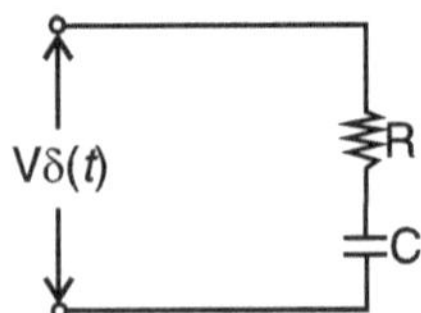

A. ∞ B. V

C. $\frac{V}{R_c}$ D. None of the above

126. The values of $i(O^+)$ and $\frac{di}{dt}(O^+)$ in the circuit shown

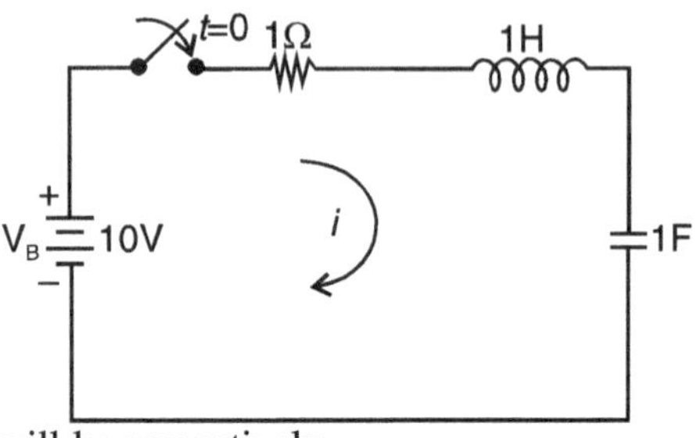

will be respectively

A. 10 A, 0 B. 0, 10 A/s

C. 0, 100 A/s D. 10A, 10 A/s

127. A four-terminal T-network is inserted between source and load resistance R as shown. The resistance seen by the source remains the same with or without the four terminal block when R is

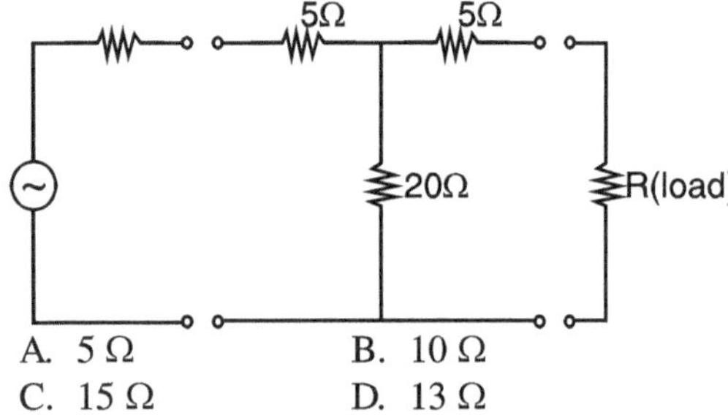

A. 5 Ω B. 10 Ω

C. 15 Ω D. 13 Ω

128. Which one of the following biasing circuits will give the best performance in preventing the variation of collector current against change in ambient temperature?

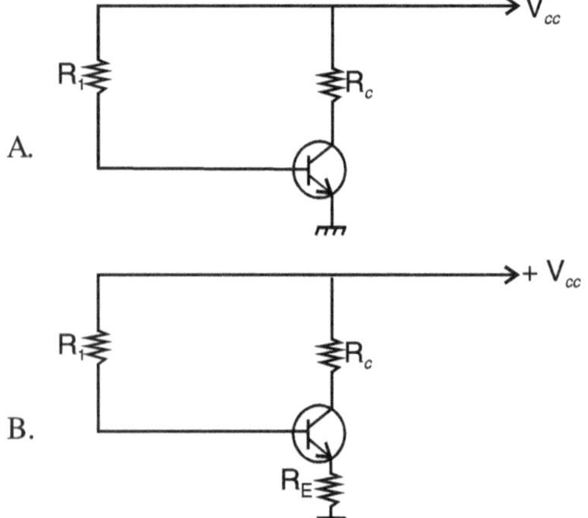

C.

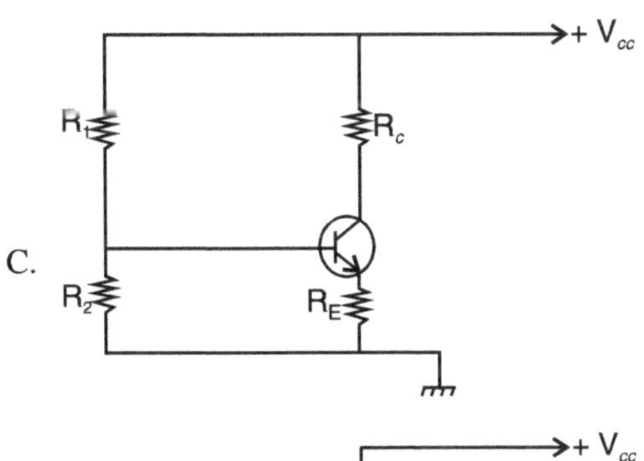

D.

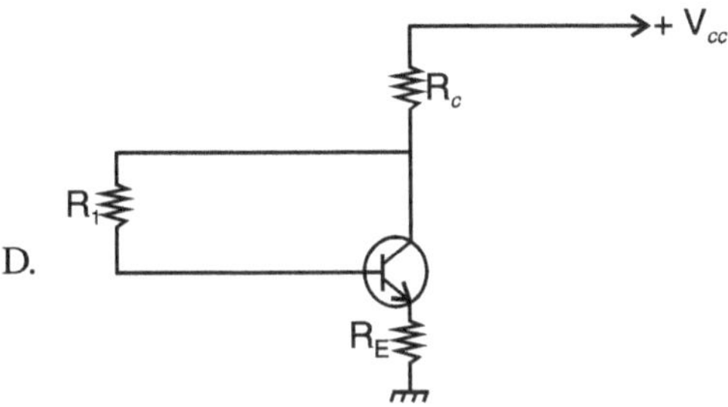

129. Which of the following oriented graphs have the same fundamental loop matrix?

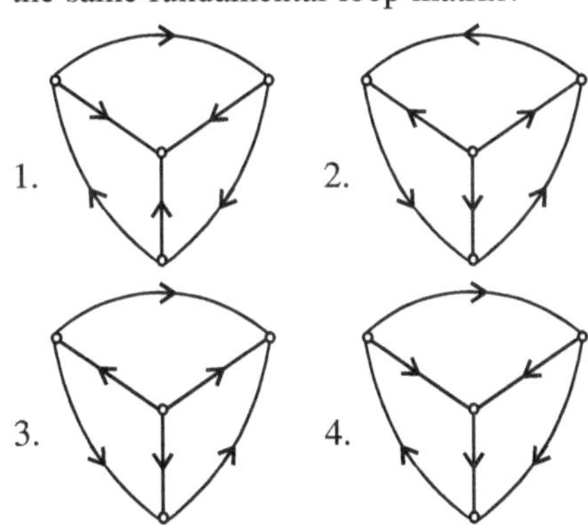

Select the correct answer :

A. 1 and 2 B. 2 and 4

C. 1, 2 and 4 D. 2, 3 and 4

130. When C_1, C_2, C_3, C = Capacitance, L = Self inductance, M_1, M_2, M = Mutual inductance, ω = angular frequency, r, R, R_1, R_2, R_3, R_4 = resistances, which of the following is dimensionally incorrect?

A. $C = \dfrac{1}{\omega^2 M}$

B. $r = \dfrac{R\omega^2 M_1 M_2}{R^2 + \omega^2 L^2}$

C. $R = \dfrac{R_2(1+\omega^2 R^2 C_3^2)}{\omega^2 + R_1 R_2 C_3^2}$

D. $I = \dfrac{V\omega M}{\left[(\omega M + R_2)^2 + \omega^2 L_1 L_2 R_1^2\right]^{1/2}}$

131. Ultra violet radiation is emitted when electron jumps from an outer stationary orbit to

A. first stationary orbit

B. second stationary orbit

C. third stationary orbit

D. fourth stationary orbit.

132. n-type semi-conductors

A. are negatively charged

B. are produced when indium is added as an impurity to Germanium

C. are produced when phosphrous is added as an impurity to silicon

D. none of the above.

133. Two resistances R_1 and R_2 are connected as shown in the given figure. If $R_1 = 528 \pm 5\ \Omega$ and $R_2 = 325 \pm 3\ \Omega$, then the total resistance R_T will be

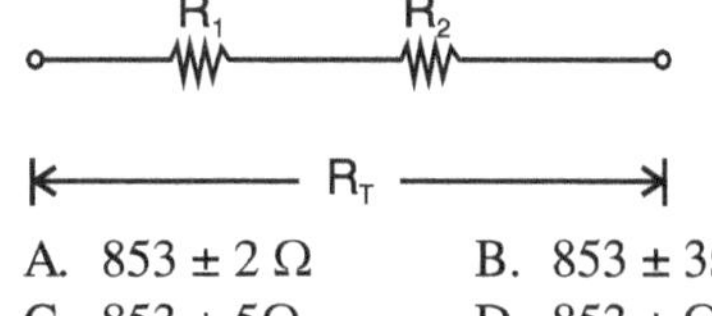

A. $853 \pm 2\ \Omega$ B. $853 \pm 3\Omega$

C. $853 \pm 5\Omega$ D. $853 \pm \Omega$

134. The dual of the network

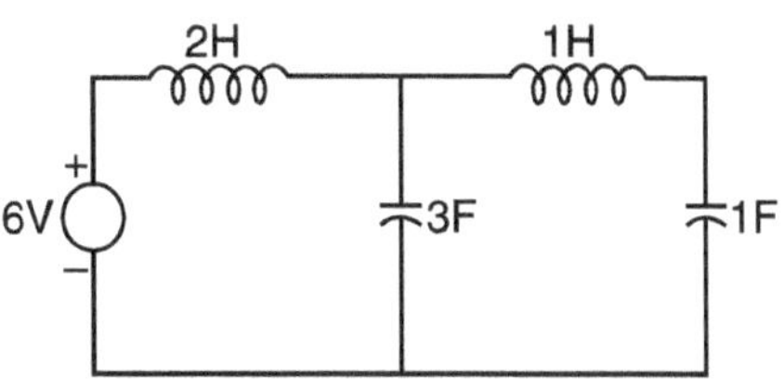

is represented by

A.

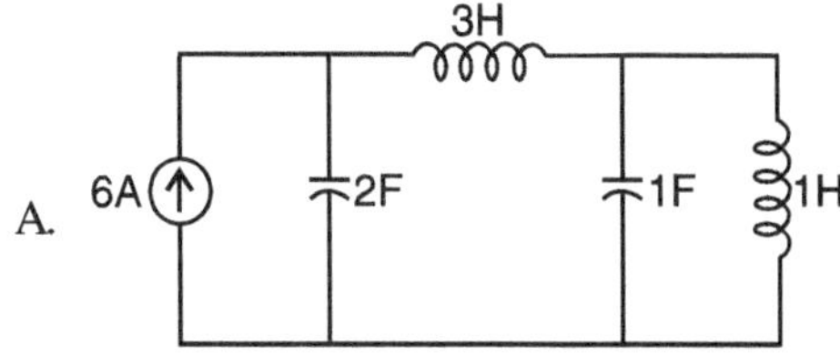

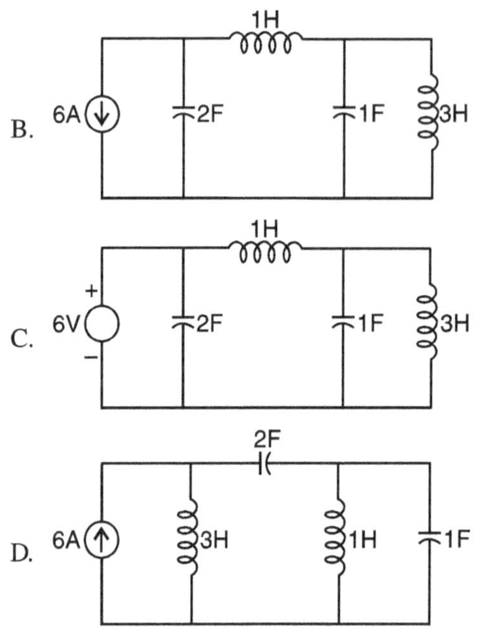

135. The voltage transfer ratio for the network shown in the figure is

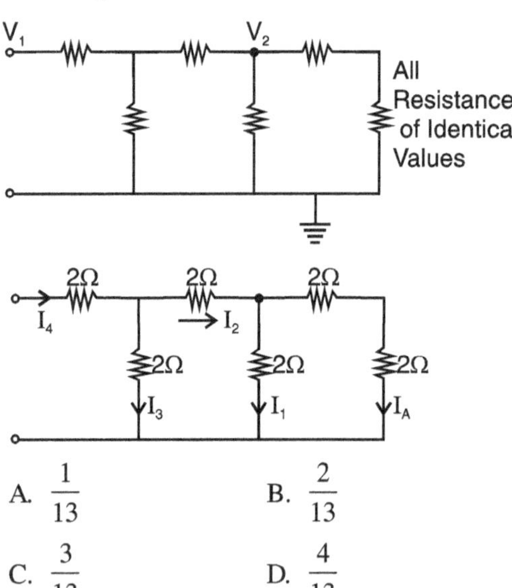

A. $\frac{1}{13}$ B. $\frac{2}{13}$

C. $\frac{3}{13}$ D. $\frac{4}{13}$

136. A step function voltage is applied to a RLC series circuit having R = 2 ohm, L = 1 H and C = 1 F. The transient current response of the circuit would be :

A. overdamped B. unerdamped

C. critically damped D. arbitary

137. Consider the vectors drawn from the poles and zeros at $j\omega = j1$ on the imaginary axis as shown. The transfer function Gain is given by

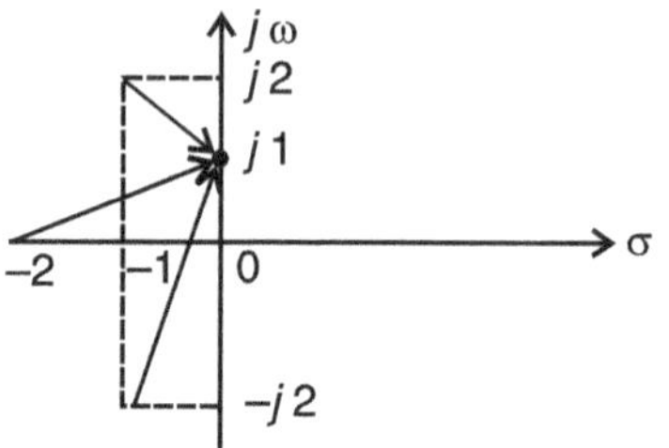

A. $2 \angle 0°$ B. $2.7 \angle -31°$

C. $2 \angle 45°$ D. $2 \angle -67.4°$

138. The 2-port network has open circuit impedance parameters given by matrix.

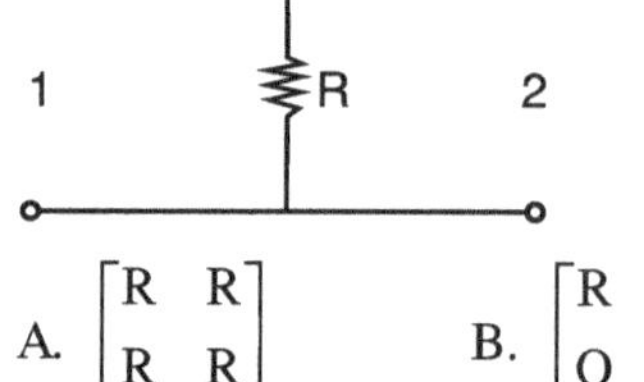

A. $\begin{bmatrix} R & R \\ R & R \end{bmatrix}$ B. $\begin{bmatrix} R & O \\ O & R \end{bmatrix}$

C. $\begin{bmatrix} \infty & R \\ R & \infty \end{bmatrix}$ D. $\begin{bmatrix} R & \infty \\ \infty & R \end{bmatrix}$

139. If the transmission parameters of network A are $\begin{bmatrix} 1 & R \\ 0 & 1 \end{bmatrix}$ and those of network B are $\begin{bmatrix} 1 & 0 \\ 1/R & 1 \end{bmatrix}$, then the transmission parameters of network are

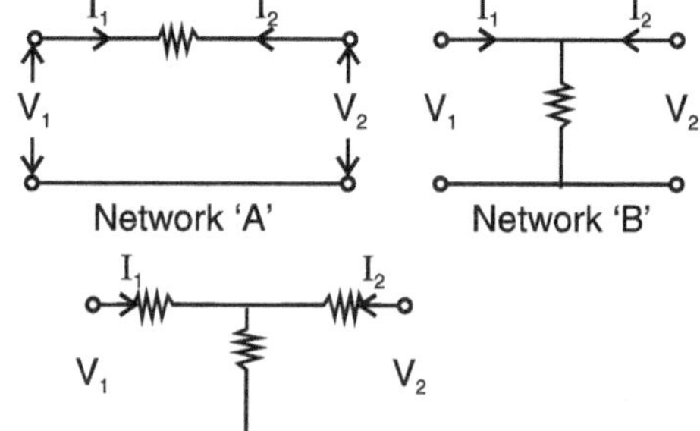

A. $\begin{bmatrix} 4 & 3R \\ 1/R & 1 \end{bmatrix}$ B. $\begin{bmatrix} 2 & 3R \\ 1/R & 2 \end{bmatrix}$

C. $\begin{bmatrix} 2 & 1/R \\ 3R & 2 \end{bmatrix}$ D. $\begin{bmatrix} 4 & 1/R \\ 3R & 1 \end{bmatrix}$

140. Which one of the following circuits is the delta equivalent of the star circuit given in figure?

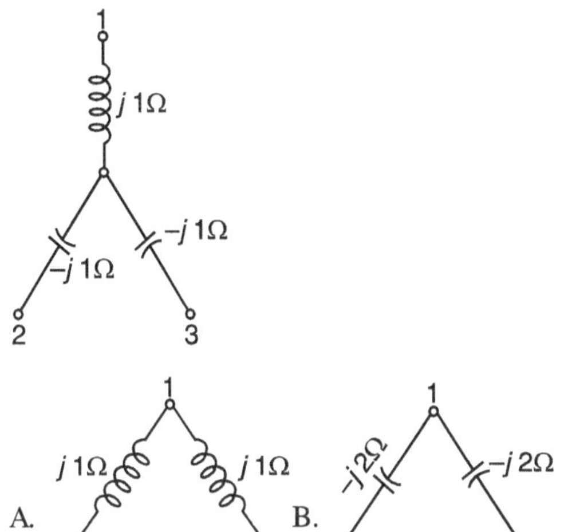

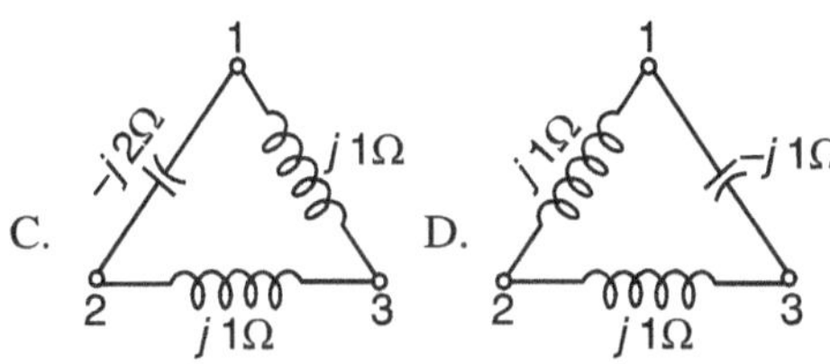

141. A system function Z(s) = V)s)/I(s) has a single pole and single zero both on the negative real axis. The pole is at $s = -\sqrt{3}$. Given that for I = cos t, the voltage leads the current by 15°. The location of the zero is at

A. $s = -\sqrt{2}$ B. $s = -1$

C. $s = -\frac{1}{\sqrt{3}}$ D. $s = 0$

142. A network having n nodes and b branches will have

A. $(b + n)$ links B. $b - n + 1$ links

C. $b + n - 1$ links D. $b + n + 1$ links

143. In the following parallel circuit, resonance will never occur, if

A. $R_1^2 = R_2^2 = \frac{L}{C}$

B. $R_1^2 < \frac{L}{C}$

C. $R_2^2 > \frac{L}{C}$ and $R_1^2 < \frac{L}{C}$

D. $R_1^2 < \frac{L}{C}$ and $R_2^2 > \frac{L}{C}$

144. Attenuator is a

A. pure resistance network producing a constant attenuation

B. pure resistance network producing a variable attenuation at variable frequencies

C. pure resistance producing a constant attenuation at all frequencies

D. pure resistance producing a constant attenuation at low frequencies

145. A two-terminal black box contains one of RLC elements. The black box is connected to 220 volts ac supply. The current through source is I. When a capacitance of 0.1 F inserted in series between the source and the box, the current through the source is 21. The element is

A. a resistance

B. an inductance

C. a capacitance of 0.5 F

D. not identifiable on the basis of the given data

146. In a low Q parallel circuit (resonant) when $X_L = X_C$

A. I_L equal I_C

B. I_L is less than I_C

C. I_L is more than I_C

D. the phase angle is 0.

147. The open circuit parameters of a 2-port network N shown in the figure, are $\begin{bmatrix} Z_{11} & Z_{12} \\ Z_{21} & Z_{22} \end{bmatrix}$. When the network is extended by a resistance R as shown in the figure, the input impedance of the network at points A, B will be

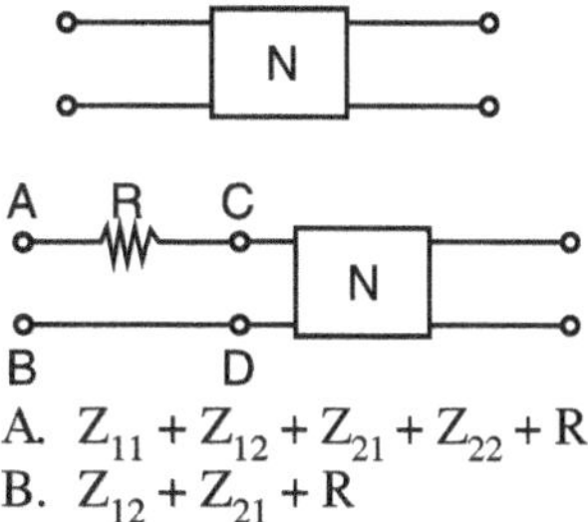

A. $Z_{11} + Z_{12} + Z_{21} + Z_{22} + R$

B. $Z_{12} + Z_{21} + R$

C. $Z_{12} + Z_{21} + R$
D. $Z_{11} + R$

148. The potential energy of a system of equal charges each of magnitude Q coulomb placed at the corners of a cube of side 1 m is

A. $8 \times \frac{1}{4\pi\varepsilon_0} Q^2$ B. $22.79 \times Q^2$

C. $\frac{6q^2}{4\pi\varepsilon_0}$ D. none

149. A transmission line with a characteristic impedance Z_1 is connected to a transmission line with a characteristic impedance Z_2. If the system is being driven by a generator connected to the first line, then the overall transmission co-efficient will be

A. $\frac{2Z_1}{Z_1 + Z_2}$ B. $\frac{Z_2}{Z_1 + Z_2}$

C. $\frac{2Z_2}{Z_1 + Z_2}$ D. $\frac{Z_1}{Z_1 + Z_2}$

150. In the given circuit, the switch S, is initially in a closed position and switch S_2 is in the open position. If S_1 is suddenly opened with simultaneous closure of S_2, then the expression for instantaneous current would be

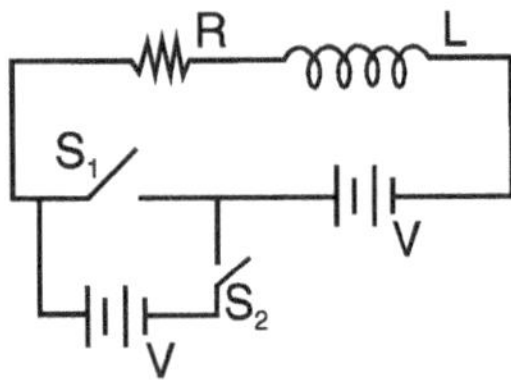

A. $2\frac{V}{R}\left\{1 - e^{-\frac{Rt}{L}}\right\}$ B. $2\frac{V}{R}\left\{1 - e^{-\frac{Rt}{2L}}\right\}$

C. $\frac{V}{R}\left\{2 - e^{-\frac{Rt}{L}}\right\}$ D. $2\frac{V}{R}\left\{2 - e^{-\frac{Rt}{2L}}\right\}$

151. The voltage applied across a capacitance is triangular in waveform. The waveform of the current is

A. triangular B. trapezoidal
C. sinusoidal D. rectangular

152 The circuit shown represents

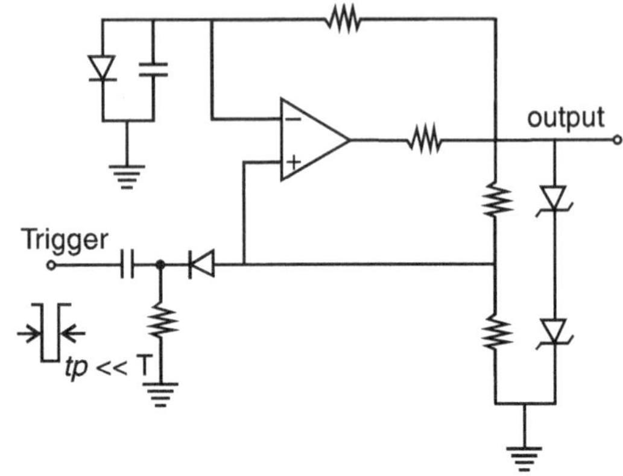

A. pulse amplifier
B. monostable multivibrator
C. astable multivibrator
D. bistable multivibrator

153. The two inputs to an analogue multiplier are $x(t)$ and $y(t)$ with Fourier transforms $X(f)$ and $Y(f)$ respectively. The output $z(t)$ will have a transform $Z(f)$ given by

A. $X(f).Y(f)$ B. $X(f) + Y(f)$

C. $\frac{X(f)}{Y(f)}$ D. $\int_{-\infty}^{\infty} X(\lambda) Y(f - \lambda)\, d\lambda$

154. The Y parameters of a four-terminal block are $\begin{bmatrix} 4 & 2 \\ 1 & 1 \end{bmatrix}$ A single element of 1 ohm is connected across as shown, the New Y parameters will be

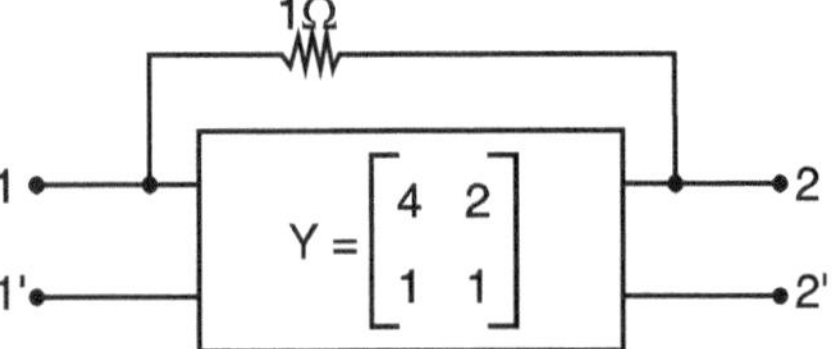

A. $\begin{bmatrix} 5 & 1 \\ 0 & 2 \end{bmatrix}$ B. $\begin{bmatrix} 4 & 3 \\ 2 & 2 \end{bmatrix}$

C. $\begin{bmatrix} 3 & 2 \\ 1 & 1 \end{bmatrix}$ D. $\begin{bmatrix} 4 & 2 \\ 1 & 1 \end{bmatrix}$

155. Obtain potential of node B with respect to node G in the network shown in the figure.

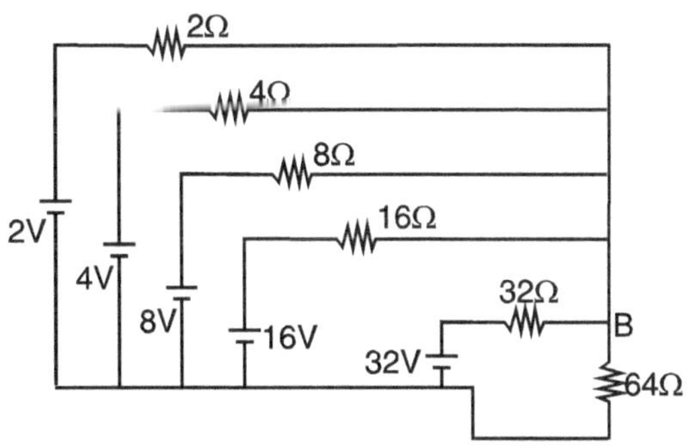

A. $\frac{64}{63}$ V B. 1 V

C. $\frac{63}{64}$ V D. $\frac{32}{63}$ V

156. Any linear active network to which a load is connected can be reduced to an equivalent current source equal to the short circuit current that flows through the load terminals when they are short circuited and having a impedance in parallel with the current source equal to the impedance of the active network. This is :

A. Thevenin's theorem
B. Millman theorem
C. Norton's theorem
D. Maximum power transfer theorem

157. In Fourier transform analysis we convert the signals

A. from time domain to frequency domain
B. from frequency domain to time domain
C. either A or B
D. neither A nor B

158. In a series circuit, the largest voltage drop is across the

A. Largest series R B. Smaller series R
C. Both A and B D. None of these

159. In two wattmeters method of measurement, if one of the wattmeters reads zero, then what will be the power factor.

A. zero B. unity
C. 0.5 D. 0.866

160. The value of resistance R in the circuit shown is varies in such a manner that the power dissipated in the 3 ohm resistor is maximum. Under this condition, the value of R will be

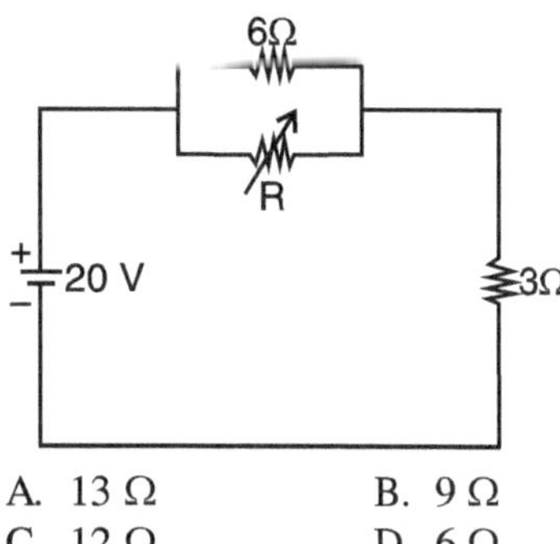

A. 13 Ω B. 9 Ω
C. 12 Ω D. 6 Ω

161. What will be the value of x_c so that the circuit is non-inductive.

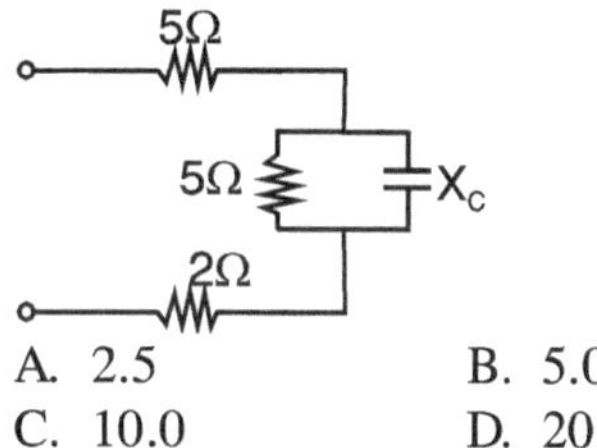

A. 2.5 B. 5.0
C. 10.0 D. 20.0

162. When a unit impulse voltage is appled to an inductor of 1 H, the energy supplied by the source is

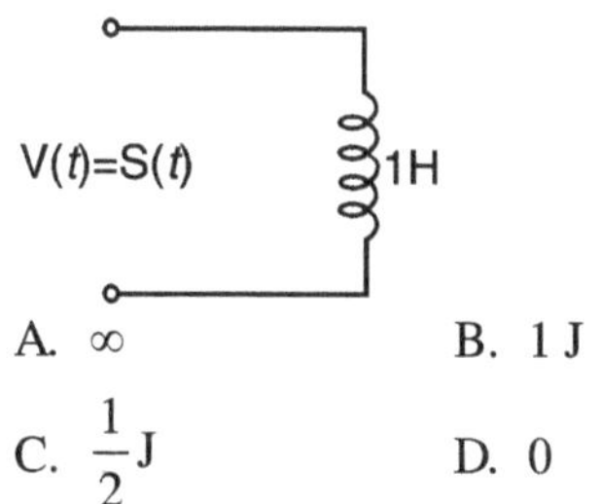

A. ∞ B. 1 J

C. $\frac{1}{2}$ J D. 0

163. What is the value of R so that i = 2A

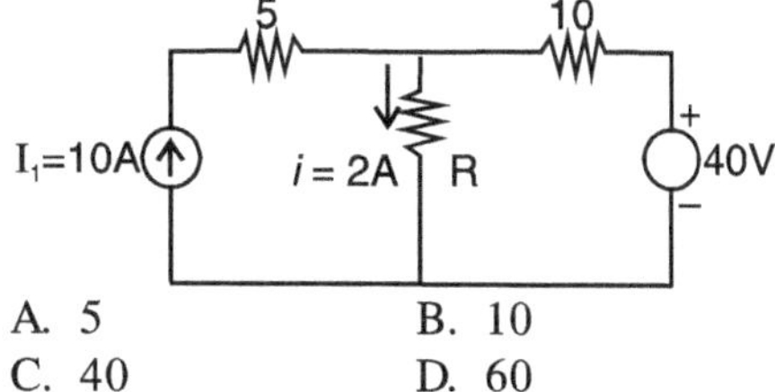

A. 5 B. 10
C. 40 D. 60

164. The dielectric-absorption loss in a capacitor is due to

A. leakage current
B. molecular distortion
C. insufficient dielectric constant

D. charging of dielectric

165. The driving point impedance function for the network shown in the figure has

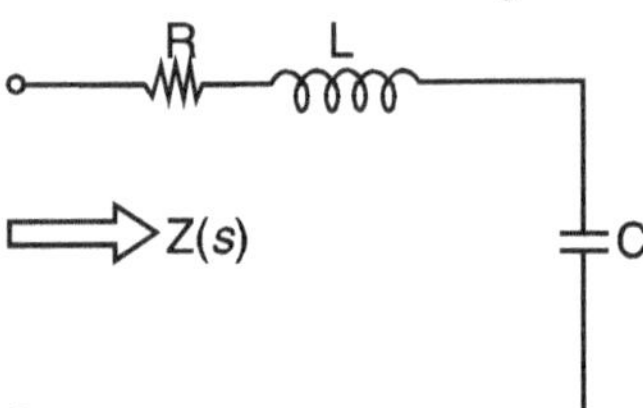

A. poles at $s = 0, s = \infty$
B. poles at $s = 0$ and zero at $s = \infty$
C. no pole at $s = 0$
D. no pole at $s = \infty$

166. The driving point admittance function of the network shown in the figure has a

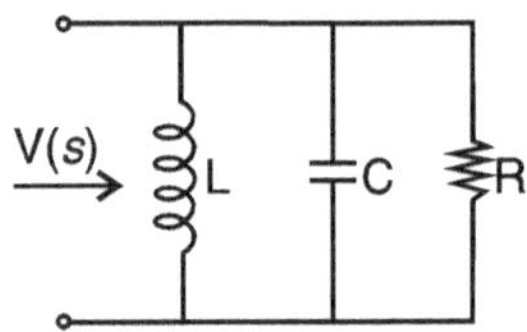

A. pole at $s = 0$ and zero at $s = \infty$
B. pole at $s = 0$ and pole at $s = \infty$
C. pole at $s = \infty$ and zero at $s = 0$
D. pole at $s = 0$ and zero at $s = \infty$

167.

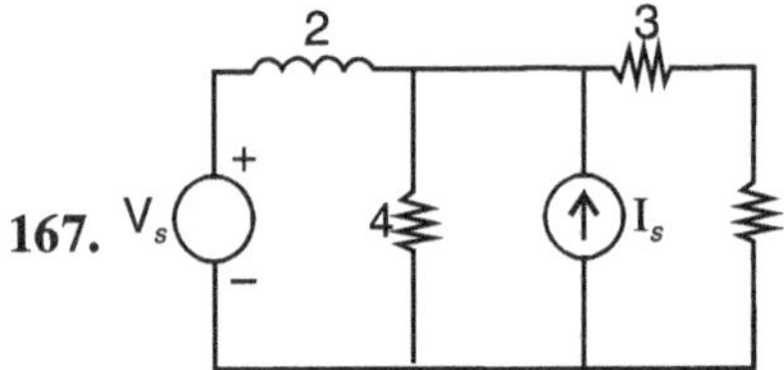

The natural frequency s_n of the given circuit is

A. 1
B. -1
C. $-\frac{1}{2}$
D. 2

168. In the circuit shown in the Figure, the switch has been in position 1 for a long time. When the switch is placed in position 2 at $t = 0$, the current $i(t)$ $(t > 0)$ can be determined from the transformed circuit given in the figure.

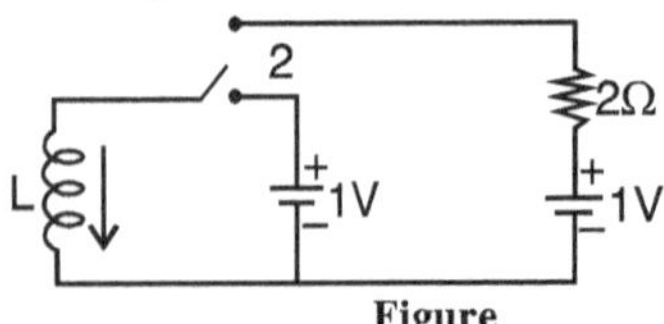

Figure

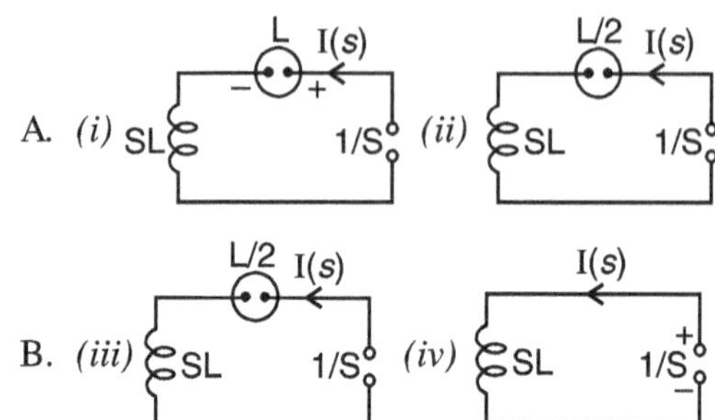

C. both A and B
D. None of these

169. In a parallel RLC circuit, at reasonant frequency:

A. line current is maximum
B. total impedance is minimum
C. inductive branch current is minimum
D. total impedance is maximum

170. Current I_0 in the given circuit will be :

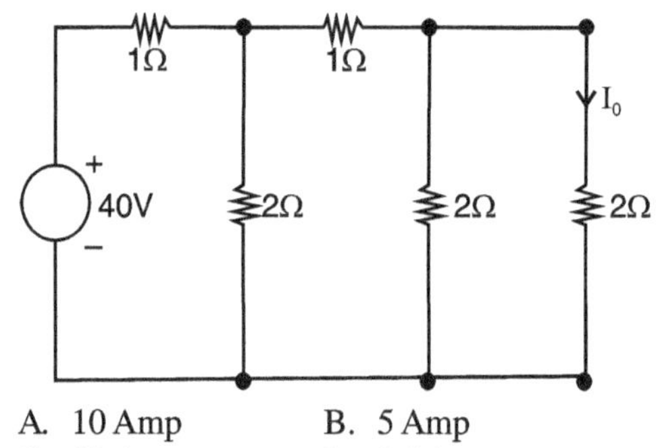

A. 10 Amp
B. 5 Amp
C. 20 Amp
D. 2.5 Amp

171. The given circuit is a

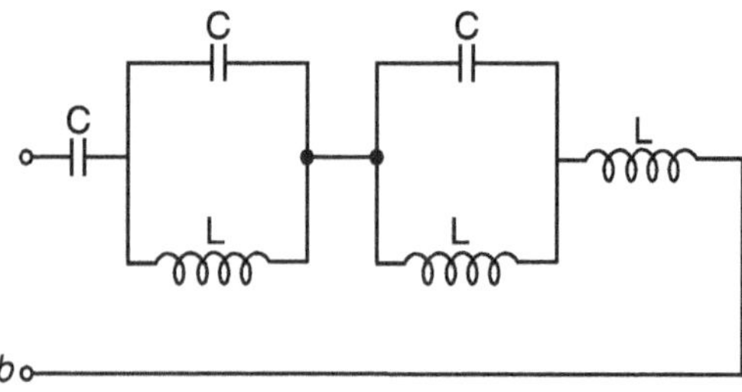

A. Foster I form
B. Foster II form
C. Cauer I form
D. Cauer II form

172. The current transfer ratio $\frac{I_2(s)}{I_1(s)}$ for the circuit shown in the figure is,

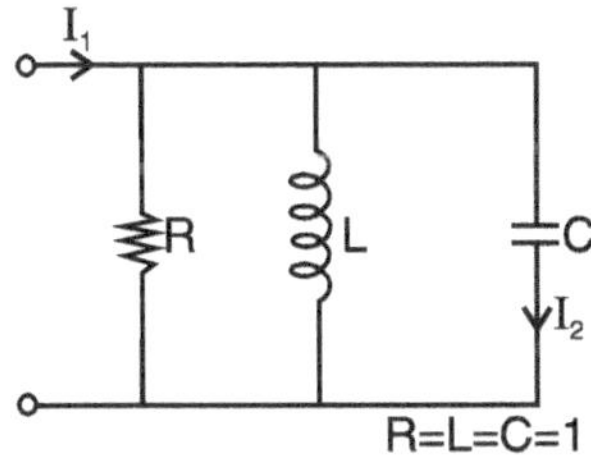

A. $\frac{s^2}{(s+1)}$ B. $\frac{s^2}{(s^2+s+1)}$

C. $\frac{s^2}{(s^3+s^2+s+1)}$ D. $\frac{s}{(s+1)}$

173. Steady state value of the current in the following circuit will be $t = 0$

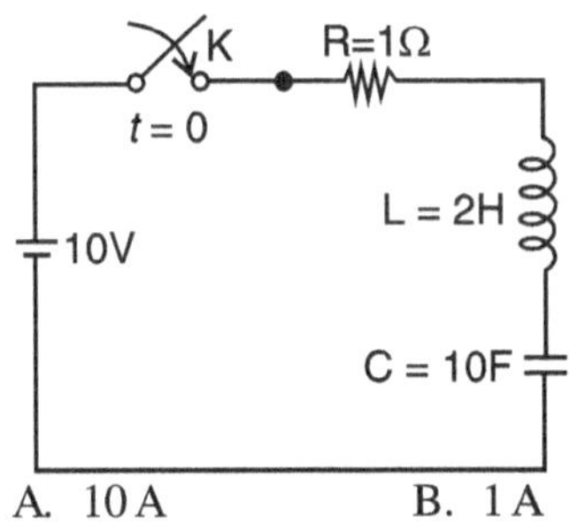

A. 10 A B. 1 A
C. 0 A D. ∞A

174. $H(s) = \frac{V(s)}{I(s)} = \frac{s+3}{(s+2)^2}$

$i(t)$ is a unit step. $v(t)$ in the steady state is given by

A. $\frac{3}{2}$ B. $\frac{3}{4}$

C. 1 D. 0

175. In the given circut, $v(t) = 200 \sin wt$. The average power in the load resistor R is :

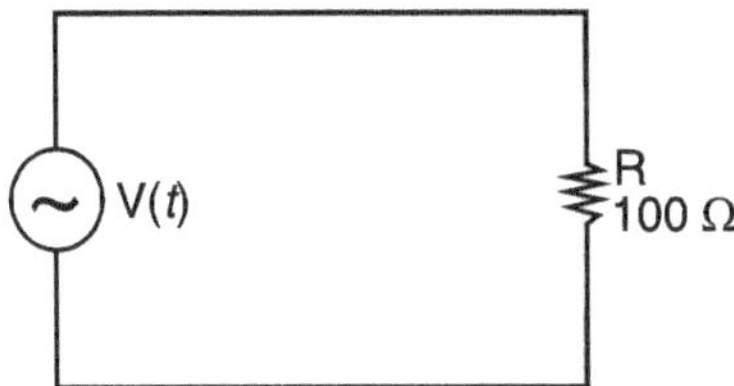

A. 100 watts B. 200 watts
C. 400 watts D. 800 watts

176. For very high frequencies the driving point impedance function, $z(s) = \frac{(s+1)(s+3)}{s(s+2)(s+4)}$

A. a resistance of $\frac{3}{2}\Omega$
B. a capacitor of 4 F
C. a capacitor of $\frac{1}{4}$ F
D. an inductance of 4 H

177. To d.c. a capacitor behaviour as
A. it behaves to A.C. B. an open circuit
C. short circuit D. None of these

178. For very high frequencies, the driving point admittance function, $y(s) = \frac{4(s+1)(s+3)}{s(s+2)(s+4)}$ behaves as

A. a resistance of $\frac{3}{2}\Omega$
B. a capacitance of 4 F
C. an inductance of $\frac{1}{4}$ H
D. an inductance of 4 H

179. The driving point impedance $z(s) = \frac{s+3}{s+4}$ behaves as
A. a resistance of 0.75 Ω at low frequencies
B. a resistance of 1 Ω at high frequencies
C. a resistance of 0.75 Ω at high frequencies
D. both (A) and (B) above.

180.

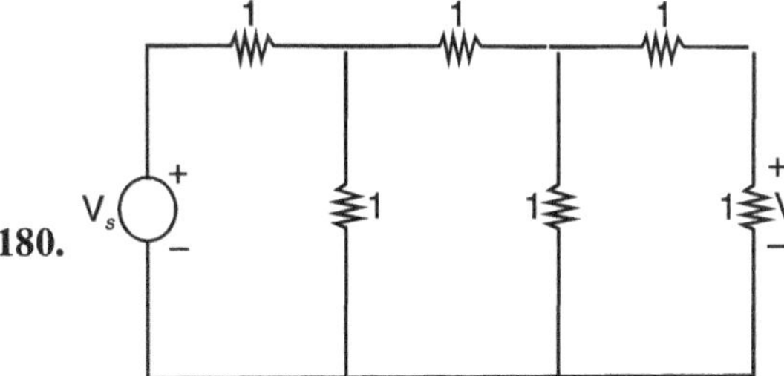

The voltage across the last resistor is V. All resistors are unity. Then V_s is given by
A. 13 V B. 8 V
C. 4 V D. none of these

181. A stepper motor
A. is a two phase induction motor
B. is a kind of rating amplifier

C. is an electromagnetic transducer commonly used to convert an angular position of a shaft into an electrical system.
D. is an electromechanical device which actuates a train of step angular (or linear) moments in response to a train of input pulses on one to one basis.

182. The transfer function of a system is $\frac{10(1+0.2s)}{1+0.5s}$

The phase shift at $\omega = 0$ and $\omega = \alpha$ will be

A. 90° and – 0° B. – 180° and 180°
C. – 90° and 90° D. none of the above

183. The transfer function $y_{12}(s) = \frac{I_2(s)}{V_1(s)}$ for the network shown in the figure, is

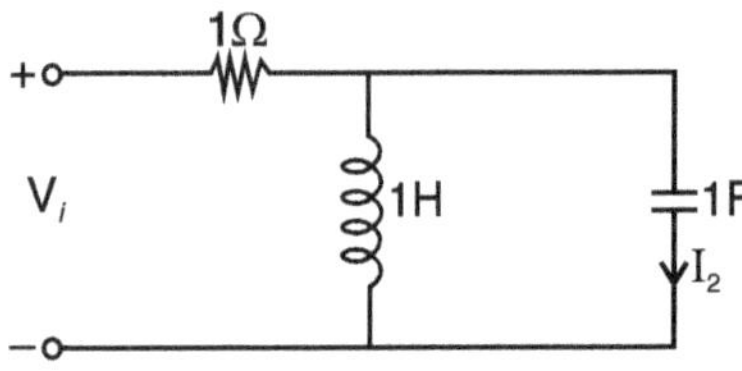

A. $\frac{s^2}{s^2+s+1}$ B. $\frac{s}{s+1}$

C. $\frac{1}{s+1}$ D. $\frac{s+1}{s^2+1}$

184. For the circuit, given the no. of current equation, voltage equations and the number of volt ampere equation

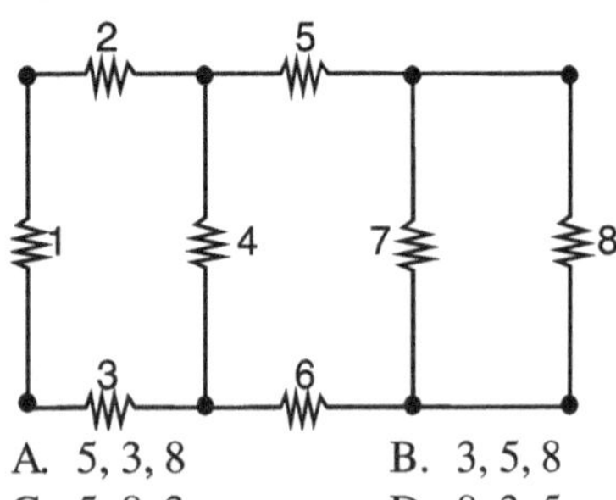

A. 5, 3, 8 B. 3, 5, 8
C. 5, 8, 3 D. 8, 3, 5

185. Damped frequency of oscillation ω_d is given by:

A. $\omega_d = \zeta\omega_n$ B. $\omega_d = \zeta^2\omega_n$

C. $\omega_d = \omega_n(1-\zeta)$ D. $\omega_d = \omega_n\sqrt{(1-\zeta^2)}$

186. Which of the following oriented graphs have the same fundamental loop matrix?

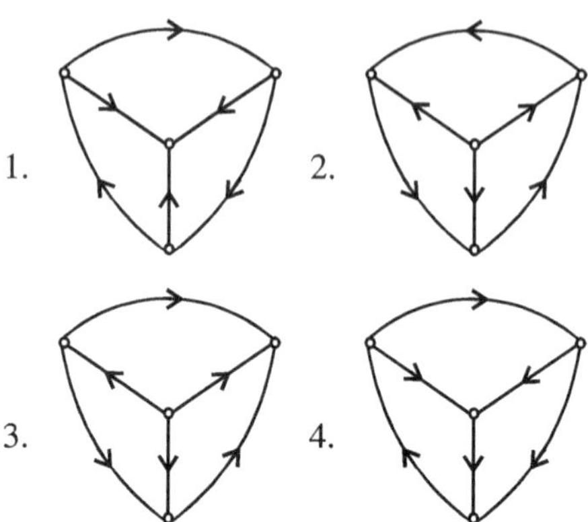

Select the correct answer using the codes given below

Codes :

A. 1 and 2 B. 2 and 3
C. 1, 3 and 4 D. 2, 3 and 4

187. The value of the parameter h_{12} for the network shown in the given figure is

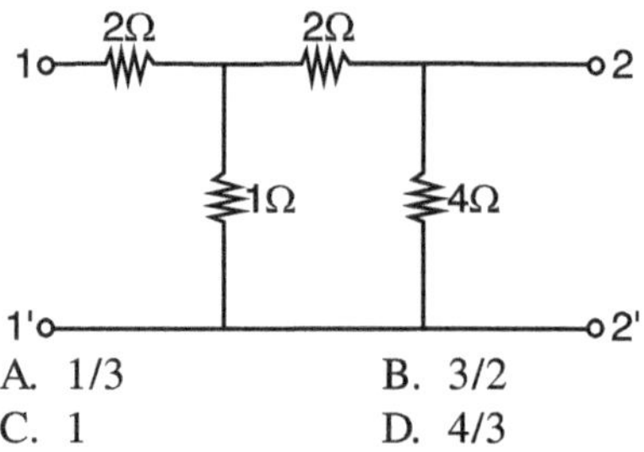

A. 1/3 B. 3/2
C. 1 D. 4/3

188. In the circuit shown, the switch closes at $t = 0$. The voltage across 4μF capacitor in ideal condition changes to

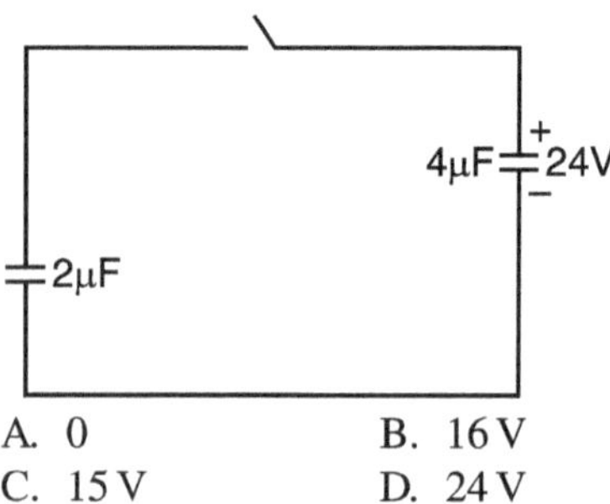

A. 0 B. 16 V
C. 15 V D. 24 V

189. In the ac circuit shown when the ammeter read 10 A, the readings on a voltmeter placed across the entire circuit and then across each element are given below. Match List I (position of the voltmeter) with List II (readings on the voltmeter) and select the correct answer

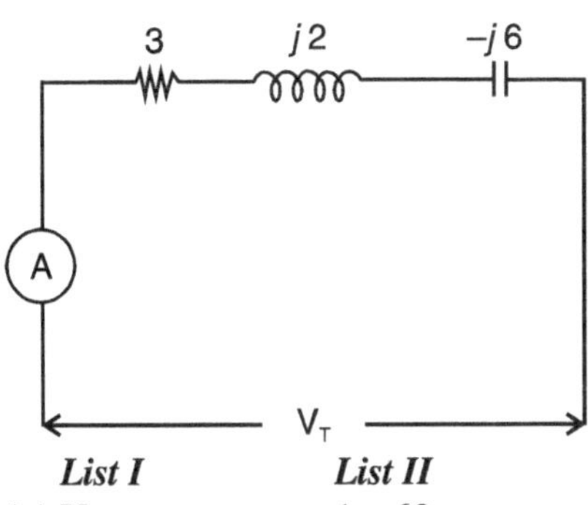

List I	*List II*
(a) V_T	1. 60
(b) V_R	2. 20
(c) V_L	3. 30
(d) V_C	4. 50
	5. 110

Codes :	*(a)*	*(b)*	*(c)*	*(d)*
A.	5	3	1	2
B.	5	4	2	1
C.	4	5	1	2
D.	4	3	2	1

190. The driving-point impedance function of a reactive network is :

$$Z_D(s) = \frac{2(s^2+1)(s^2+3)}{2(s^2+2)}$$

Which one of the following diagram realises the Cauer network for the above $Z_D(s)$?

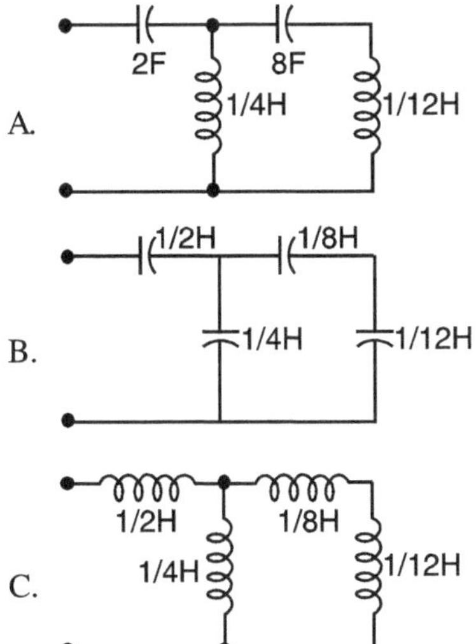

D.

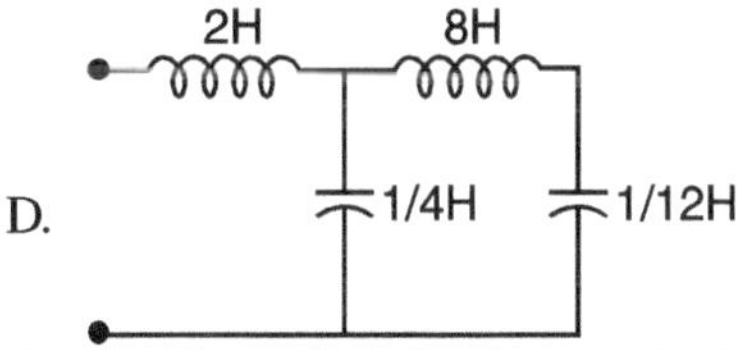

191. For the given phase lead network, the maximum possible phase lead is

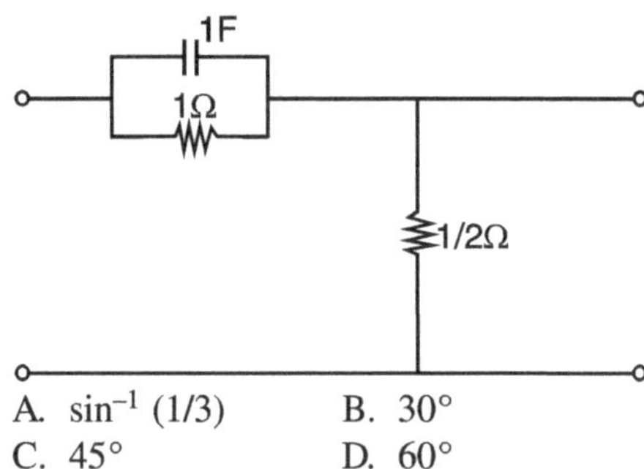

A. $\sin^{-1}(1/3)$ B. 30°
C. 45° D. 60°

192. The input impedane of a short-circuited lossless transmission line quarter wave long is :

A. purely reactive,
B. purely resistive,
C. infinite
D. dependent on the characteristic impedance of the line

193. Match List I with List II and select the correct answer using the codes given below the lists :

List I *[F(s)]*	*List II* *[Type of F(s)]*
(a) $\frac{(s^2-s+4)}{s^2+s+4}$	1. Non-positive real
(b) $\frac{(s+4)}{s^2+3s-4}$	2. Non-minimum phase
(c) $\frac{(s+4)}{s^2+6s+5}$	3. RC-impedance
(d) $\frac{s^3+3s}{s^4+2s^2+1}$	4. Unstable
	5. RL-impedance

Codes :	*(a)*	*(b)*	*(c)*	*(d)*
A.	1	2	3	4
B.	1	2	4	5
C.	2	4	3	1
D.	2	4	1	5

194. A real transformer with $L_1 = 2H$. $L_1 = 8H$ and $M = 3H$, the dots are at the top of both primary and secondary coils. The secondary leakage inductance of the transformer would be :

A. 8H B. 6H
C. 3H D. 2H

195. For the lattice type attenuator shown in the given figure, the characteristic impedance R_0 is

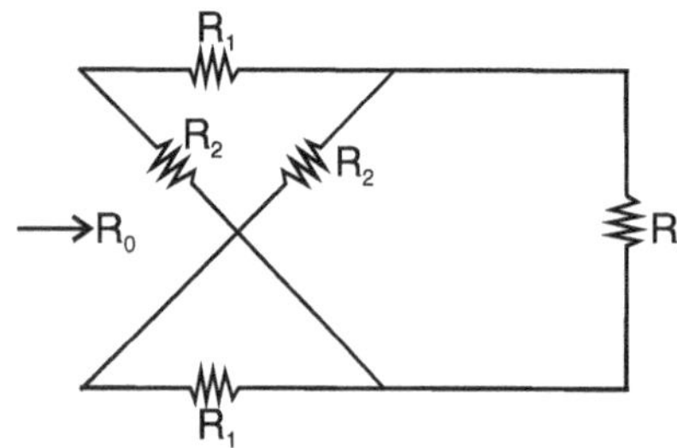

A. $\frac{R_1 + R_2}{2}$ B. $\frac{2R_1R_2}{R_1 + R_2}$

C. $\sqrt{R_1R_2}$ D. $R_1 + \frac{R_2}{2}$

196. For an RC driving-point impedance function, the poles and zeros

A. should alternate on real axis
B. should alternate only on the negative real axis
C. should alternate on the imaginary axis
D. can lie anywhere on the left half place

197. An inductor is tuned to frequency f_0 by a parallel 260 μμF capacitor. The same inductor tunes to second harmonic $2f_0$ using a capacitor 50 μμ F. The shunt self capacitance of the inductor is :

A. 10 μμ F B. 15 μμF
C. 20 μμF D. 30 μμF

198. For the design of low pass prototype filter (see figure given) with load resistance R_L ohm and angular frequency $\omega = 1$ rad/sec, the values of L and C would be

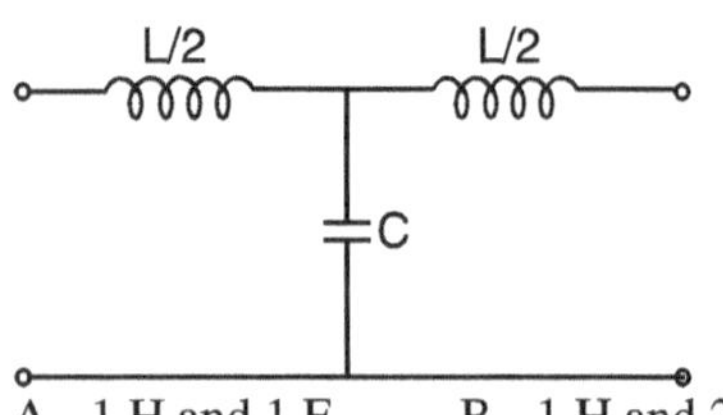

A. 1 H and 1 F B. 1 H and 2 F
C. 2 H and 1 F D. 2 H and 2 F

199. The passband of a typical filter network with Z_1 and Z_2 as the series and shunt arm impedances is characterised by

A. $-1 < \frac{Z_1}{4Z_2} < 0$ B. $-1 < \frac{Z_1}{4Z_2} < 1$

C. $0 < \frac{Z_1}{4Z_2} < 1$ D. none of the above

200. A coil has an mmf of 600 A and reluctance of 2×10^{-6} A/Wb. Total flux ϕ is given by

A. 300 μ Wb B. 3000 μ Wb
C. 3×10^{-6} Wb D. $\frac{1}{300}$ μ Wb

201. The input impedance of the circuit shown in the figure, if the respective coil impedance $Z_1 = (5 + j8)\ \Omega$ and $Z_2 = (3 + j8)\ \Omega$ is

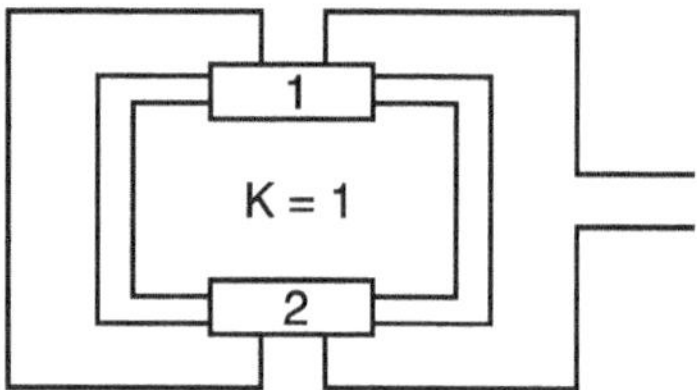

A. $(8 + 16j)\ \Omega$ B. $(2 + j0)\ \Omega$
C. $(15 + 64j)\ \Omega$ D. $(8 + 0j)\Omega$

202. Consider the following expressions which indicate the step or impulse response of an initially relaxed control system:

1. $(5 - 4e^{-2t})\,u(t)$ 2. $(e^{-2t} + 5)\,u(t)$
3. $\delta(t) + 8e^{-2t}u(t)$ 4. $\delta(t) + 4e^{-2t}\,(t)$

Those which correspond to the step and impulse response of the same system include

A. 1 and 3 B. 2 and 4
C. 2 and 3 D.1 and 4

203. Laplace transform of $t \cos at$ is :

A. $\frac{2as}{(s^2 + a^2)^2}$ B. $\frac{s^2 - a^2}{(s^2 + a^2)^2}$

C. $\frac{2as}{(s^2 - a^2)^2}$ D. $\frac{s^2 + a^2}{(s^2 - a^2)^2}$

204. In a series resonance circuit consisting of R, L and C if L_0 is the value of L at resonance, L_A is the value of L at the lower half frequency f_A, L_B is the value of L at the upper frequency f_B, Q_0 is the quality factor at resonance, then the value of

$\frac{L_O}{L_B - L_A}$ is

A. $\frac{Q_0}{2}$ B. $2Q_0$

C. $\frac{2}{Q_0}$ D. $\frac{Q_0}{4}$

205. Which one of the following is the equivalent circuit of aperture coupled lossy cavity?

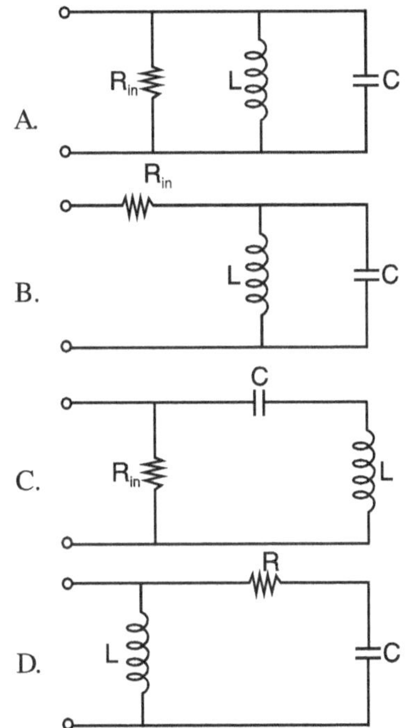

206. An electric circuit contains R, L and C is series with a voltage source. The current through the circuit is I_0. The frequencies at which the current would reduce to 0.707 I_0 is given by f_{01} and f_{02}. The resonant frequency of the circuit is the

A. geometric mean of f_{01} and f_{02}
B. arithmetic mean of f_{01} and f_{02}
C. difference of f_{01} and f_{02}
D. harmonic mean of f_{01} and f_{02}

207. In the network shown in the given figure, the

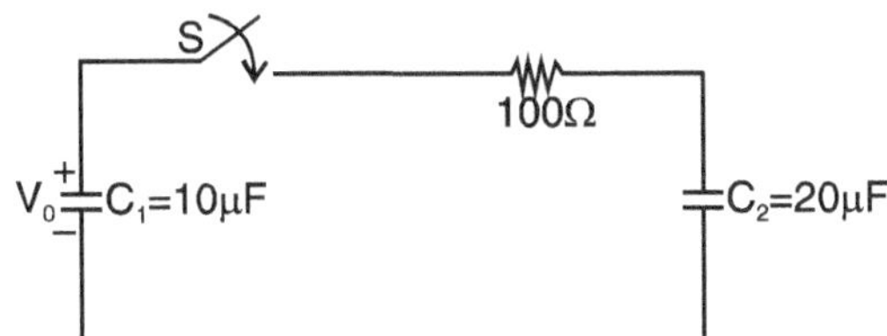

capacitor C_1 is initially charged to a voltage V_0 before the switch S is the circuit is closed. In the steady-state,

A. C_1 and C_2 are charged to equal voltages
B. C_1 and C_2 are charged with equal coulombs
C. C_1 and C_2 are discharged fully
D. C_2 alone is charged to voltage V_0

208. Which one of the following pairs is correctly matched?

A. Symmetrical two-port network : AD – BC=1
B. Reciprocal two-port network : $z_{11} = z_{22}$
C. Inverse hybrid parameter : A, B, C, D
D. Hybrid parameter : $(V_1, I_2) = f(I_1, V_2)$

209. Which one of the following is the correct Laplace transform of the signal shown

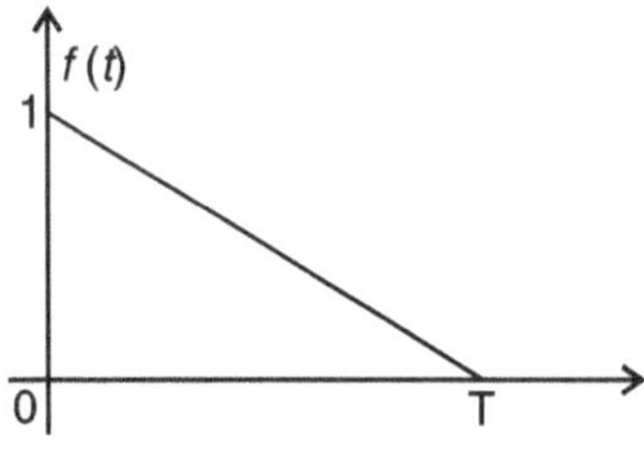

A. $\frac{1}{Ts^2\left[1 + e^{-Ts}(1 + Ts)\right]}$

B. $\frac{1}{Ts^2}\left[e^{-Ts} - 1 + Ts\right]$

C. $\frac{1}{Ts^2}\left[e^{-Ts} + 1 - Ts\right]$

D. $\frac{1}{Ts^2}\left[1 - e^{-Ts} + Ts\right]$

210. The expression for the waveform in terms of step function is given by

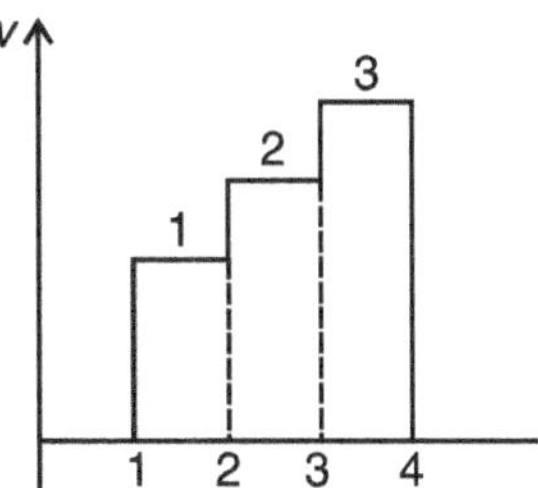

A. $v = u(t-1) - u(t-2) + u(t-3)$
B. $v = u(t-1) + u(t-2) + u(t-3)$
C. $v = u(t-1) + u(t-2) - u(t-3)$

D. $v = u(t-1) + u(t-2) + u(t-3) - 3u(t-4)$

211. A load speaker is connected across terminals A and B of the network illustrated. What should its impedance be to obtain maximum power dissipation in it?

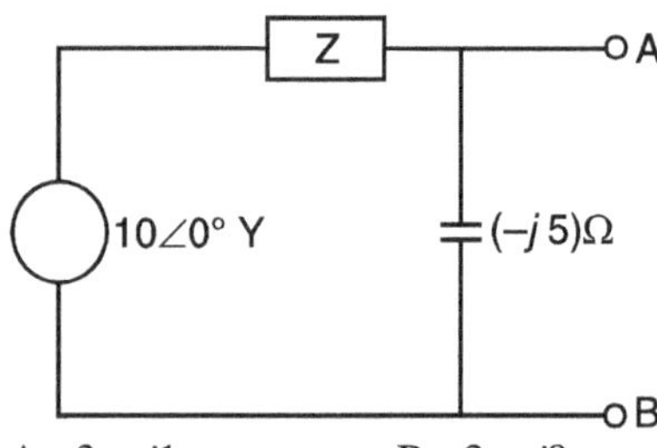

A. $3 - j1$ B. $3 + j9$
C. $7.5 + j2.5$ D. $7.5 - j2.5$

212. The apparent capacitance at a high angular frequency (ω) of an air-cored capacitor having capacitance (C) and an inductance (L) is given by

A. $C_{app} = \frac{C}{1+\omega^2 LC}$ B. $C_{app} = \frac{C}{1-\omega^2 LC}$

C. $C_{app} = \frac{L}{1+\omega^2 LC}$ D. $C_{app} = \frac{L/C}{1+\omega^2 LC}$

213. The impulse train shown in the figure represents the second derivative of a function *f(t)*. The value of *f(t)* is

A. $-tu(t-1) - tu(t-2) + tu(t-3) + tu(t-4) - tu(t-5) + 2tu(t-6) - tu(t-7)$

B. $-tu(t-1) - tu(t-2) - tu(t-3) - tu(t-4) + tu(t-5)$

C. $tu(t-3) + tu(t-4) + 2tu(t-6)$

D. $tu(t+1) + tu(t+2) + tu(t+3) + tu(t+4) + tu(t+5) + 2tu(t+6) + tu(t+7)$

214. The Fourier transform of the function Sgn *(t)* defined in the figure is

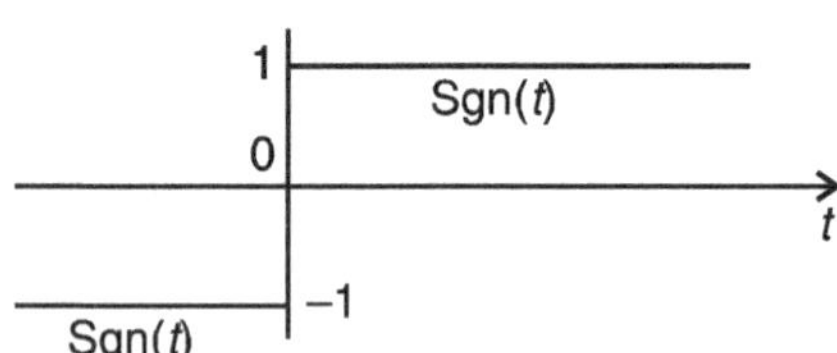

A. $\frac{-2}{j\omega}$ B. $\frac{4}{j\omega}$

C. $\frac{2}{j\omega}$ D. $\frac{1}{j\omega}+1$

215. The Thevenin resistance of the network comprising of a resistor of 1Ω in series with a parallel combination of a current source of 2I amperes and a resistor of 1Ω through which I ampere of current flows is

A. 12 B. 0.5
C. 4/3 D. 10

216. A 3-phase star connected load has one capacitor in R phase and two resistors of equal value in other two phases. If the phase sequence is RBY, then

A. $V_{BO} > V_{YO}$ B. $V_{BO} < V_{YO}$
C. $V_{BO} = V_{YO}$ D. $V_{BO} = V_{RY}$

217. Decibel is the unit adopted for the gain or attenuation of communication equipment because

A. total gain can be found by addition
B. changes on a logarithmic scale can be perceived more easily than on a linear scale
C. vision is more sensitive to the eye
D. the minimum change that can be perceived by the ear

218. For the following network input impedance $z(s)$ as $s \to 0$ and $s \to \infty$ are given by

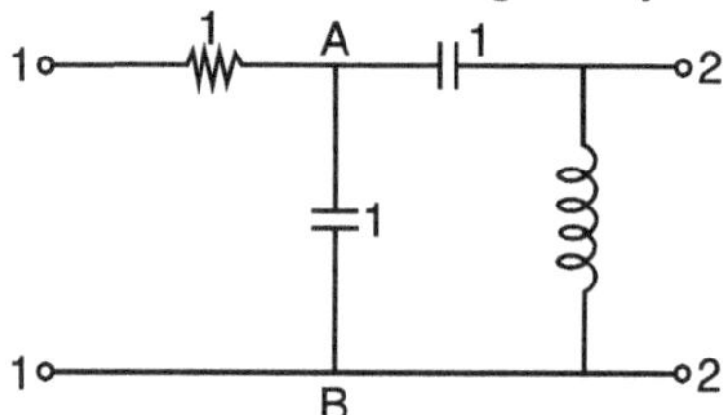

A. $\frac{1}{2s}, 1$ B. $1, \frac{2}{s}$

C. $\frac{2}{s}, 1$ D. $\frac{2}{s}, \frac{1}{s}$

219. Find the Norton's equivalent of the circuit given below :

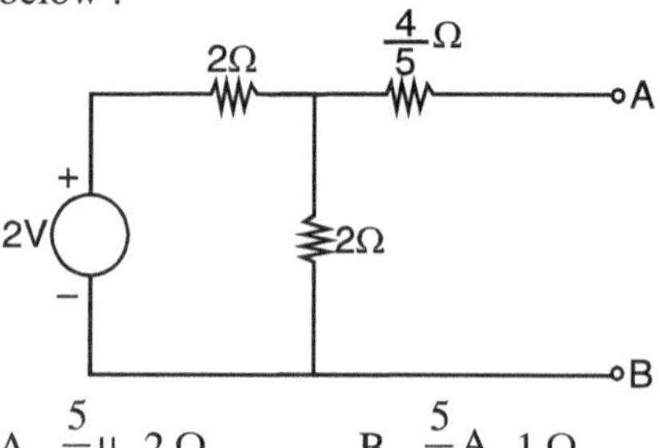

A. $\frac{5}{2}\mu, 2\,\Omega$ B. $\frac{5}{2}A, 1\,\Omega$

C. $\frac{2}{5}$ A, 1 Ω D. $\frac{2}{5}$ A, 2Ω

220. With reference to the circuit given match List I with list II and select the correct answer

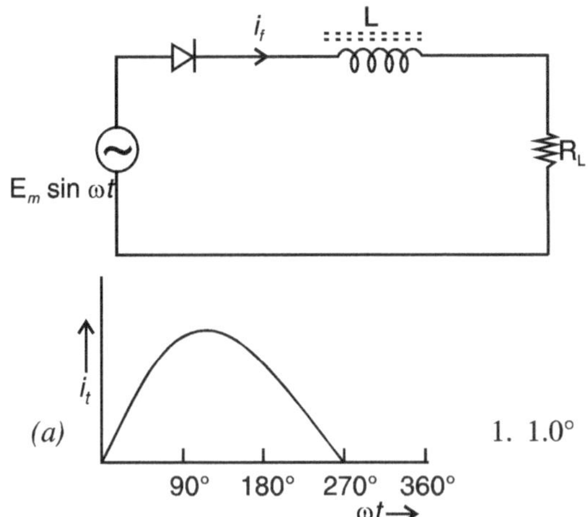

(a) 1. 1.0°

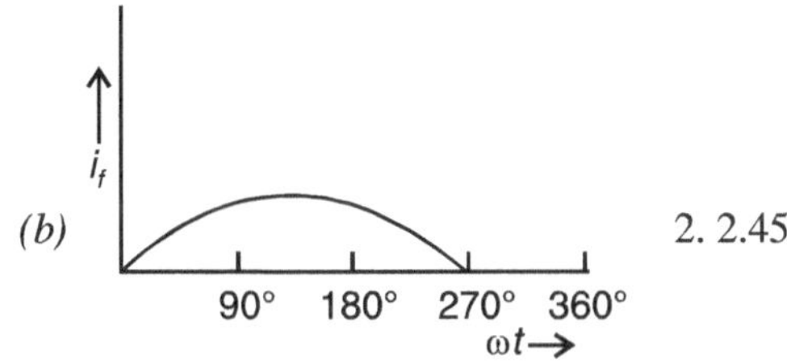

(b) 2. 2.45°

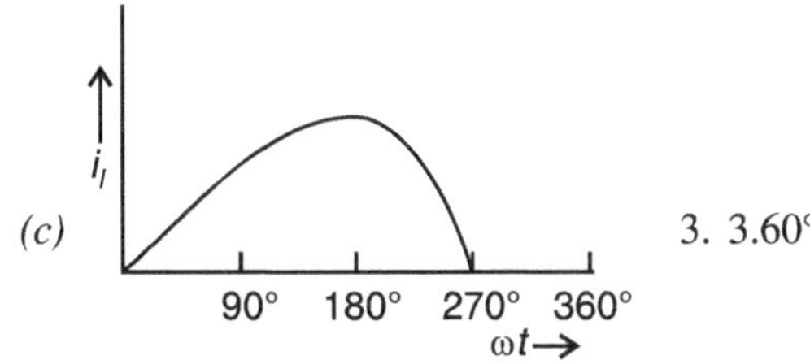

(c) 3. 3.60°

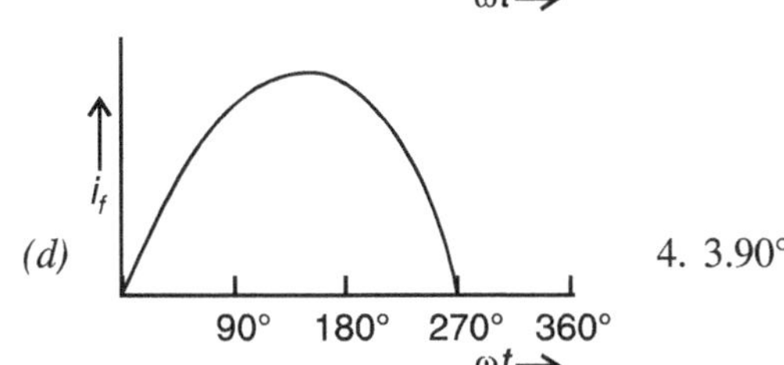

(d) 4. 3.90°

Codes:	(a)	(b)	(c)	(d)
A.	1	4	2	3
B.	1	2	3	4
C.	4	1	3	2
D.	1	3	2	4

221. A voltage source with an internal resistance R_s, supplies power to a load R_L. The power delivered to the load varies with R_L as

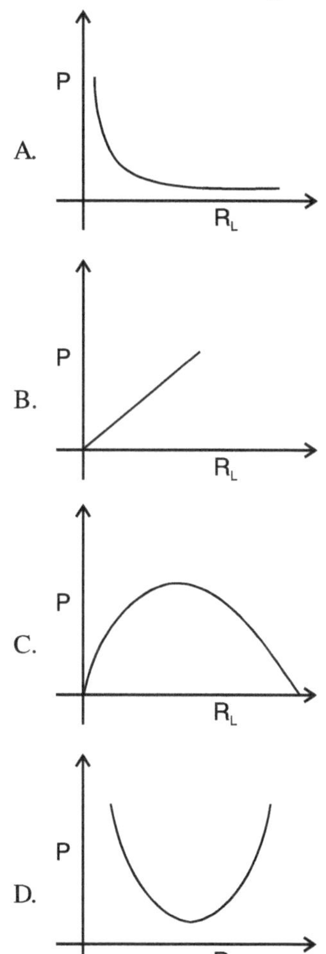

222. A Wheatstone bridge is shown with resistance thermometer connected in one arm at 20°C. At 20 °C, resistance of thermometer = 1020 Ω. The output voltage will be

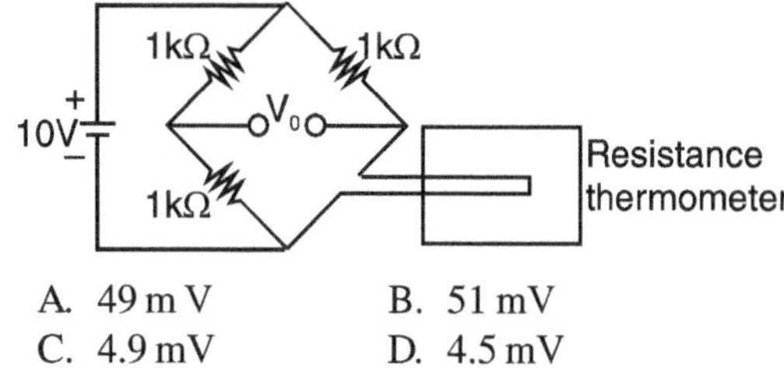

A. 49 m V B. 51 mV
C. 4.9 mV D. 4.5 mV

223. A periodic function is said to have a quarter wave symmetry, if it possesses:

A. even symmetry at an interval quarter of a wave
B. even symmetry and half-wave symmetry only
C. even or odd symmetry without the half-wave symmetry
D. even or odd symmetry with the half-wave symmetry.

224. Which one of the following star connections is equivalent of the Δ connection shown?

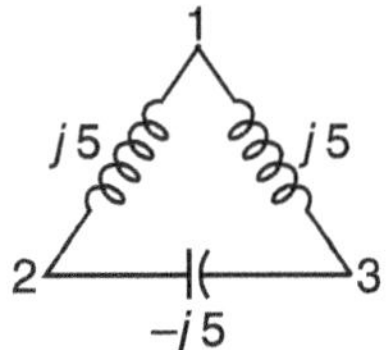

A.

B.

C.

D.

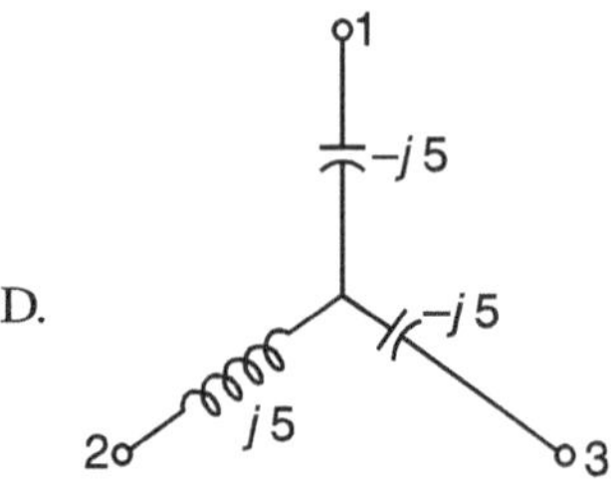

225. If the loop notation is as shown then the series executed is

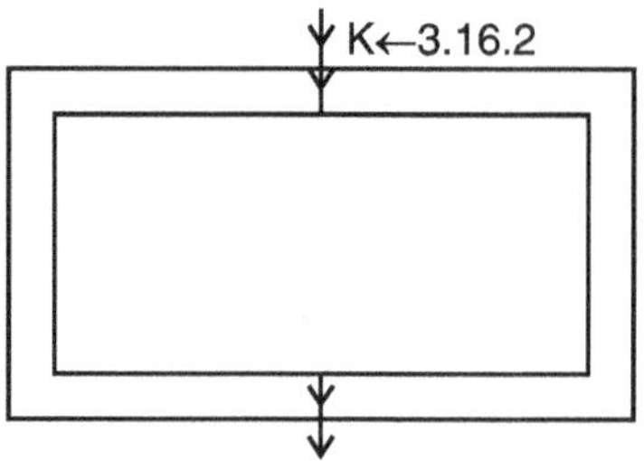

A. 3, 4, 5, 6, 7, 8
B. 3, 5, 7, 9, 11, 13, 15
C. 3, 6, 12, 8, 16
D. 3, 9, 11, 1, 3, 9

226. A network contains linear resistors and ideal voltage sources. If values of all resistors are doubled, then the voltage across each resistor is

A. halved
B. increased by four times
C. doubled
D. not changed.

227. In the Wheatstone bridge shown in the given figure, if the resistance in each arm is increased by 0.05% then the value of V_{out} will be

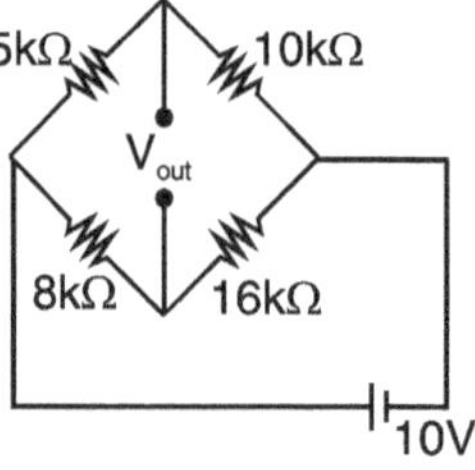

A. 50 mV
B. 5 mV
C. 0.1 V
D. zero

228. Four networks are shown below :

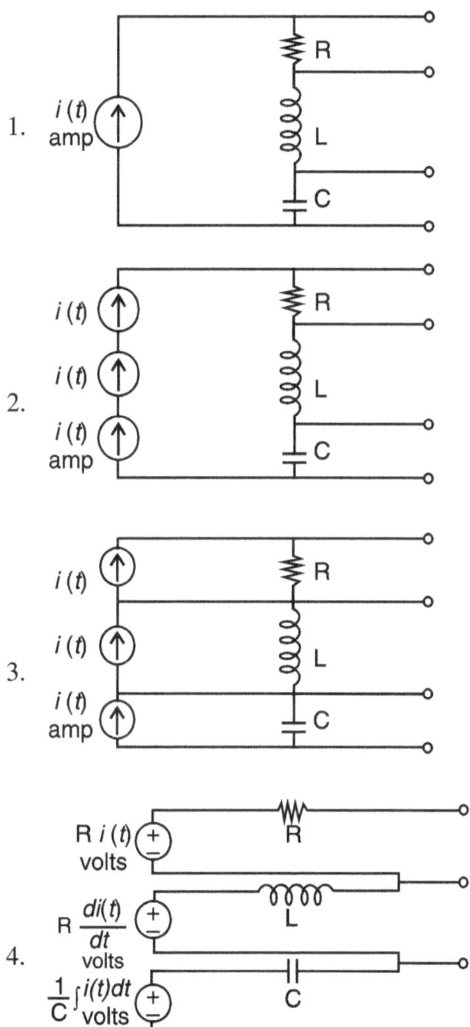

Of these networks,

A. all the four networks are equivalent
B. no two networks are equivalent
C. networks shown in Figs. (2), (3) and (4) are equivalent
D. networks shown in Figs. (3) and (4) are equivalent

229. For a given voltage, four heating coils produce maximum heat when connected

A. all in parallel
B. all in series
C. with two parallel pairs in series
D. one pair in parallel with the other two in series

230. The steady state current through the 1 H inductance in the circuit shown in the given figure is

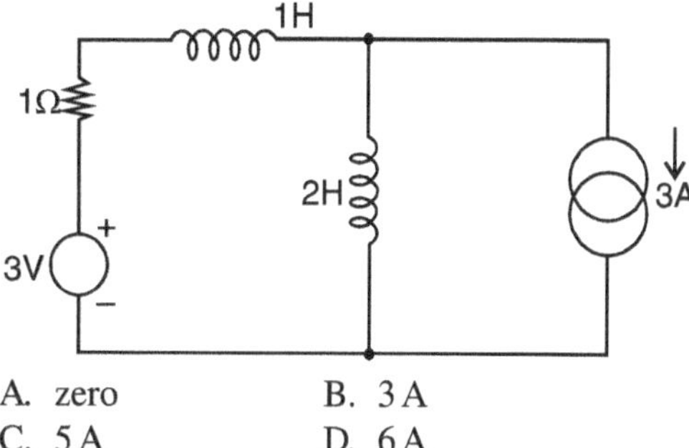

A. zero B. 3 A
C. 5 A D. 6 A

231. If $x(t)$ is the linear input to a linear network whose impulse response $h(t)$ is known, then assuming that the impulse applied at $t = \lambda$ the output response $y(t)$ will be (* = convolution)

A. $y(t) = x(t) * h(t)$
B. $y(t) = x(t - \lambda) * h(t)$
C. $y(t) = x(\lambda) * h(t - \lambda)$
D. $y(t) = x(x + \lambda) * h(t - \lambda)$

232. Consider the following statements regarding the situation at resonant frequency :

1. For a series RLC circuit, current is minimum.
2. For a series RLC circuit, voltage across C is minimum
3. For a series RLC circuit, current is maximum
4. For a parallel RLC circuit, total impedance is maximum

Of these statements

A. 3 and 4 are correct
B. 2 and 3 are correct
C. 1 and 2 are correct
D. 1 and 4 are correct

233. A network has 10 nodes and 17 branches in all. The number of node pair voltages would be

A. 7 B. 9
C. 10 D. 45

234. In the circuit shown, the switch is closed at $t = 0$. The current at $t = 0$ is (current through inductance for $t < 0$ is 0)

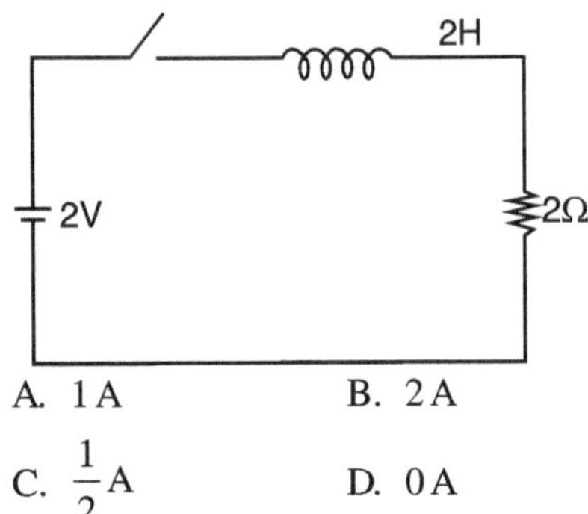

A. 1 A B. 2 A

C. $\frac{1}{2}$ A D. 0 A

235. Consider the closed-loop system shown

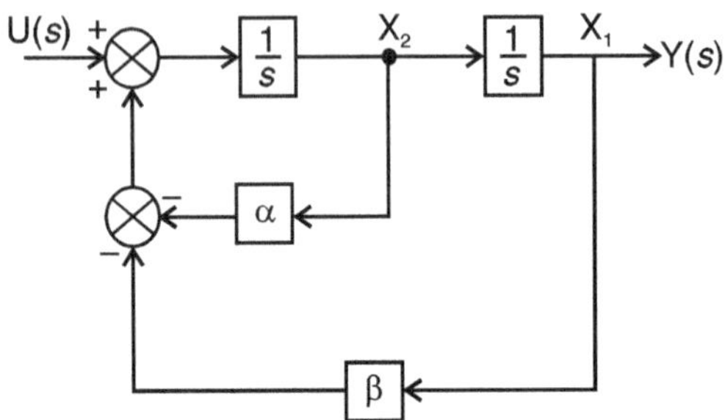

The state model of the system is

A. $\begin{bmatrix} \dot{x}_1 \\ \dot{x}_2 \end{bmatrix} = \begin{bmatrix} -1 & 0 \\ -\beta & -\alpha \end{bmatrix} \begin{bmatrix} x_1 \\ x_2 \end{bmatrix} + \begin{bmatrix} 0 \\ 1 \end{bmatrix} u$

$y = [1 \quad 0] \begin{bmatrix} x_1 \\ x_2 \end{bmatrix}$

B. $\begin{bmatrix} \dot{x}_1 \\ \dot{x}_2 \end{bmatrix} = \begin{bmatrix} 1 & -1 \\ -\beta & -\alpha \end{bmatrix} \begin{bmatrix} x_1 \\ x_2 \end{bmatrix} + \begin{bmatrix} 0 \\ 1 \end{bmatrix} u$

$y = [1 \quad 0] \begin{bmatrix} x_1 \\ x_2 \end{bmatrix}$

C. $\begin{bmatrix} \dot{x}_1 \\ \dot{x}_2 \end{bmatrix} = \begin{bmatrix} 0 & 1 \\ -\beta & -\alpha \end{bmatrix} \begin{bmatrix} x_1 \\ x_2 \end{bmatrix} + \begin{bmatrix} 0 \\ 1 \end{bmatrix} u$

$y = [1 \quad 0] \begin{bmatrix} x_1 \\ x_2 \end{bmatrix}$

D. $\begin{bmatrix} \dot{x}_1 \\ \dot{x}_2 \end{bmatrix} = \begin{bmatrix} 0 & 1 \\ -\alpha & -\beta \end{bmatrix} \begin{bmatrix} x_1 \\ x_2 \end{bmatrix} + \begin{bmatrix} 0 \\ 1 \end{bmatrix} u$

$y = [1 \quad 0] \begin{bmatrix} x_1 \\ x_2 \end{bmatrix}$

236. The network shown in the given figure, the switch K is closed at $t = 0$ with the capacitor uncharged. The value for $\frac{di(t)}{dt}$ at $t = 0^+$ will be

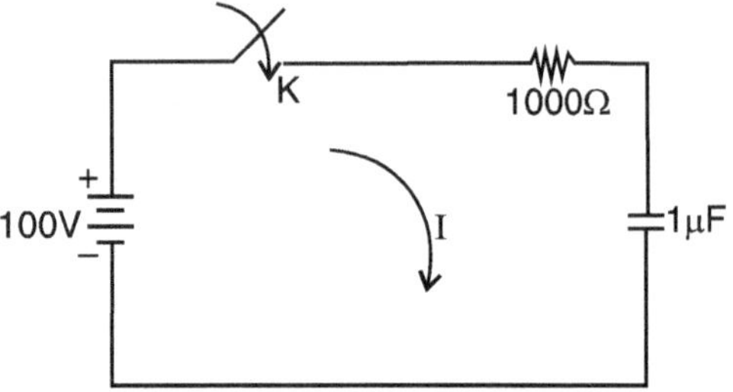

A. 100 amp/sec B. – 100 amp/sec
C. 1000 amp/sec D. – 1000 amp/sec

237. The time constant of the network shown in the given figure is given by

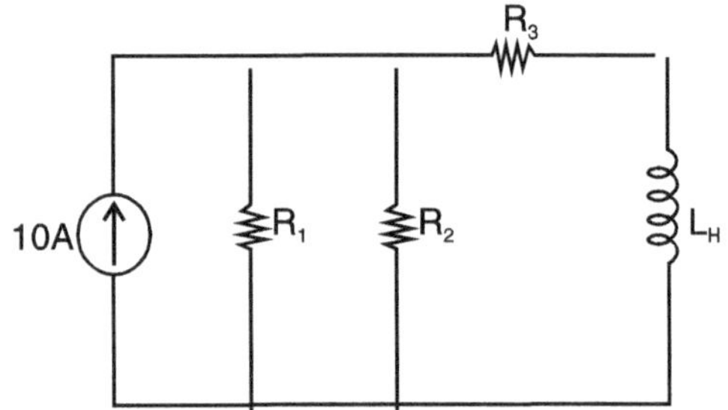

A. $\frac{L}{R_3 + \frac{R_1R_2}{R_1 \times R_2}}$ B. $\frac{L}{R_3}$

C. $\frac{L}{R_1 + R_2 + R_3}$ D. none

238. The piece-wise linear V – I characteristic of a diode is shown. When the applied voltage is 6 V, the current through the diode will be

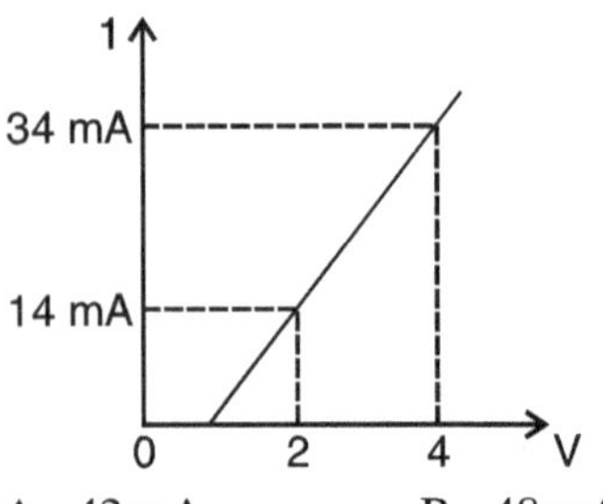

A. 42 mA B. 48 mA
C. 51 mA D. 54 mA

239. The total power in the ac circuit shown is 1100 W. Match List I with List II and select the correct answer

List I	*List II*
(a) I_1/I_2	1. 600
(b) P_1/P_2	2. 0.3
(c) P_1 in W	3. 2.0 A
(d) P_2 in W	4. 500
	5. 1.2

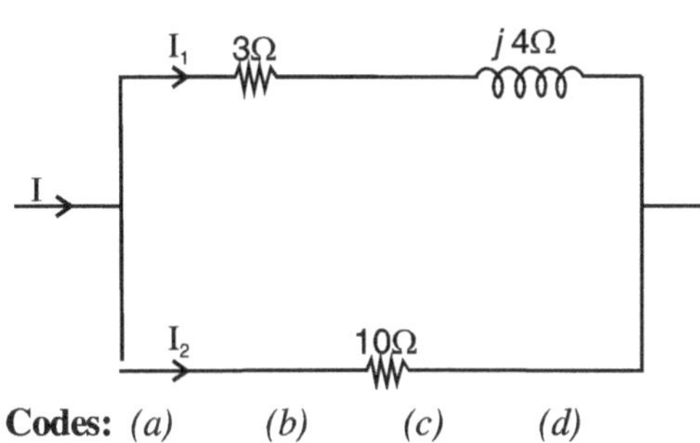

Codes:	*(a)*	*(b)*	*(c)*	*(d)*
A.	2	3	4	1
B.	3	5	4	1
C.	3	5	1	4
D.	2	3	1	4

240. Consider the following statements regarding an RC differentiating network :

1. For an applied rectangular pulse, the output is spiky in nature for RC << pulse duration
2. The output is ramp for rectangular input pulse.
3. The output has zero average value for all inputs.

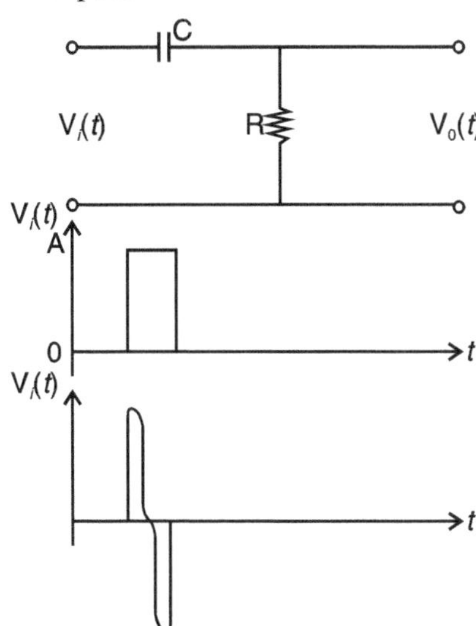

Of these statements

A. 1, 2 and 3 are correct

B. 1 and 2 are correct

C. 2 and 3 are correct

D. 1, and 2 are correct

241. A series RLC circuit is overdamped when

A. $\frac{R^2}{4L^2} > \frac{1}{LC}$ B. $\frac{R^2}{4L^2} = \frac{1}{LC}$

C. $\frac{R^2}{4L^2} < \frac{1}{LC}$ D. R = infinity

242. In the given figure, the capacitors C_1, C_3, C_4 and C_5 have a capacitance of 4 microfarads each and the capacitor C_2 has a capacitance of 10 microfarads. The effective capacitance (in microfarads) between the points X and Y will be

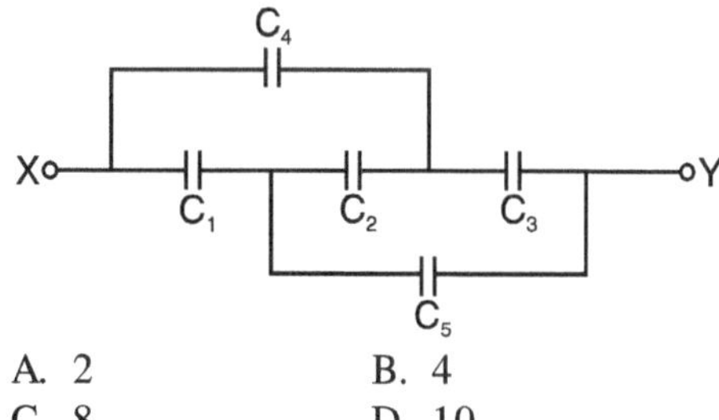

A. 2 B. 4

C. 8 D. 10

243. The amplitude and phase spectra for the few harmonics of a periodic signal of time period 1S are shown in the figure below.

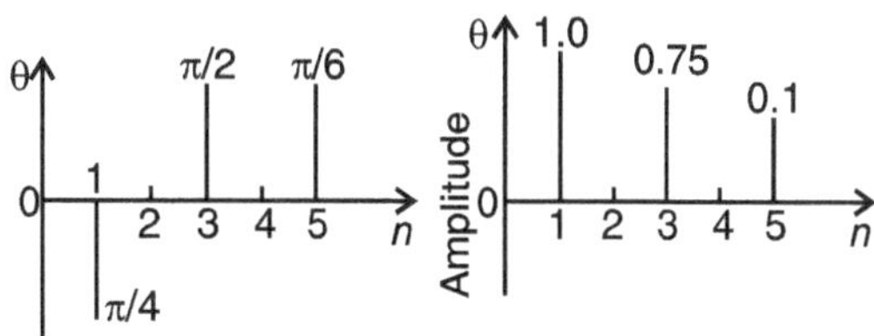

A. $\cos\left(2\pi t - \frac{\pi}{4}\right) + 0.75\cos\left(6\pi t + \frac{\pi}{2}\right) + 0.1\cos\left(10\pi t + \frac{\pi}{6}\right)$

B. $\cos\left(\frac{t}{2\pi} - \frac{\pi}{4}\right) + 0.75\cos\left(\frac{t}{6\pi} - \frac{\pi}{2}\right) + 0.1\cos\left(\frac{t}{10\pi} + \frac{\pi}{6}\right)$

C. $\cos\left(2\pi t - \frac{\pi}{4}\right) + 0.75\cos\left(6\pi t + \frac{\pi}{2}.\frac{1}{3}\right)$

$+ 0.1\cos\left(10\pi t + \frac{\pi}{6}.\frac{1}{6}\right)$

D. $-\cos\left(2\pi t - \frac{\pi}{4}\right) + 0.75\cos\left(6\pi t + \frac{\pi}{2}\right)$

$+ 0.1\cos\left(10\pi t + \frac{\pi}{6}\right)$

244. A periodic function of half-wave symmetry is necessarily

A. an even function
B. an odd function
C. neither odd nor even
D. both odd and even

245. In figure A_1, A_2 and A_3 are ideal ammeters. If A_1 reads 5A, A_2 reads 12 A, then A_3 should read

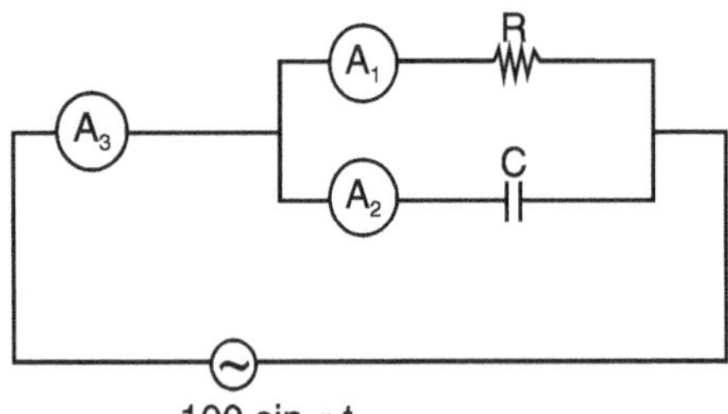

A. 7 A
B. 13 A
C. 12 A
D. 17 A

246. For the circuit shown in the figure, the current X is 3 A. The power delivered by the dependent current source D is

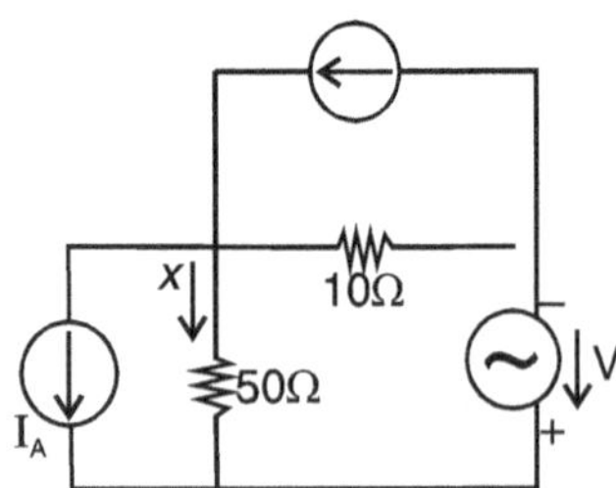

A. 50 watts
B. 2250 watts
C. 2300 watts
D. 1500 watts

247. The value of R which will enable the circuit to deliver maximum to the terminal *a* and *b* in the following circuit diagram is

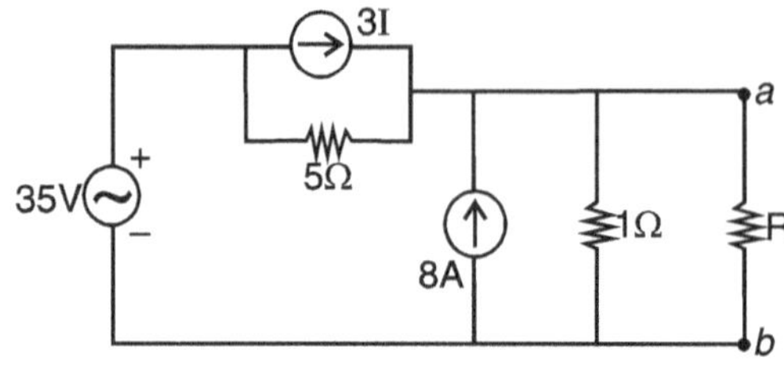

A. $\frac{5}{9}$ ohms
B. $\frac{5}{8}$ ohms
C. $\frac{5}{6}$ ohms
D. $\frac{5}{3}$ ohms

248. Fourier transform of a function *f(at)* is :

A. $af(t)$
B. $\frac{1}{a}f(t)$
C. $\frac{1}{a}F\left(\frac{\omega}{a}\right)$
D. none of these

249. The transfer function of a low pass RC network is

A. $RC_s(1 + RC_s)$
B. $1/(1 + RC_s)$
C. $RC/(1 + RC_s)$
D. $s/(1 + RC_s)$

250. A choke coil having a resistance of R ohms and an inductance of L Heneries is shunted by a capacitor of C farads. The dynamic impedance of the circuit at resonance will be

A. R/LC
B. C/RL
C. L/RC
D. 1/RLC

251. A system characteristic is defined by the equation $y = mx + c$ where m is constant and c is the intercept. The system is

A. linear
B. nonlinear
C. hyperbolic
D. none of the above

252. In the given figure, the capacitors C_1, C_3, C_4 and C_5 have a capacitance of 4 microfarads each and the capacitor C_2 has a capacitance of 10 microfarads. The effective capacitance (in microfarads) between the points X and Y will be

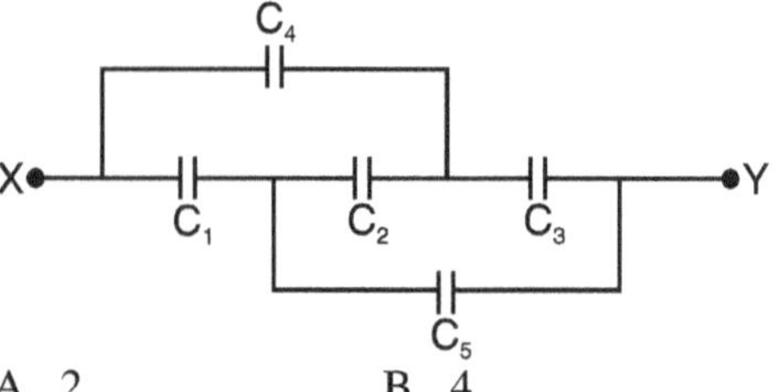

A. 2
B. 4
C. 8
D. 10

253. Match List I with List II and select the correct answer :

List I

1. 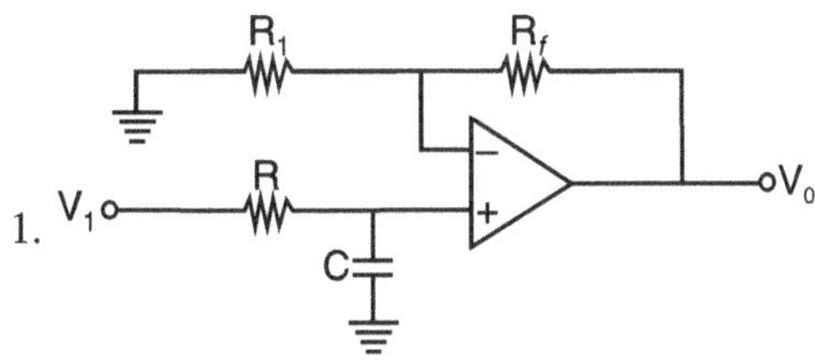

2. 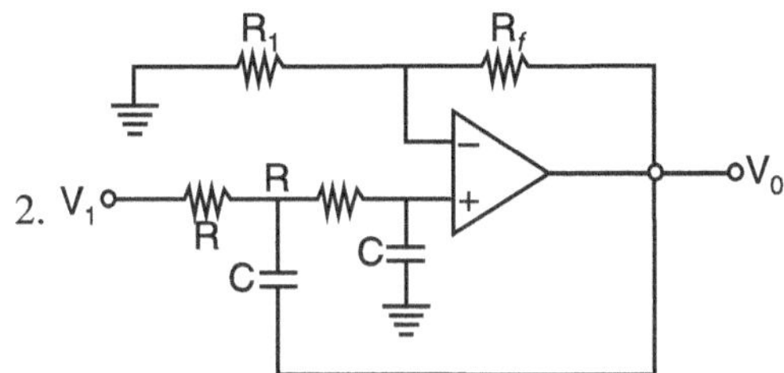

3. 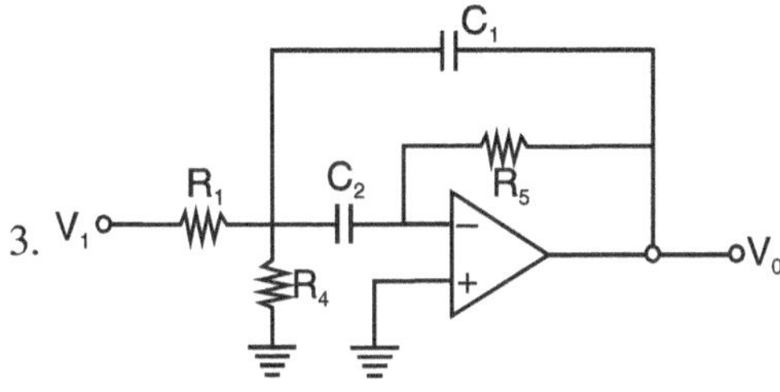

4. 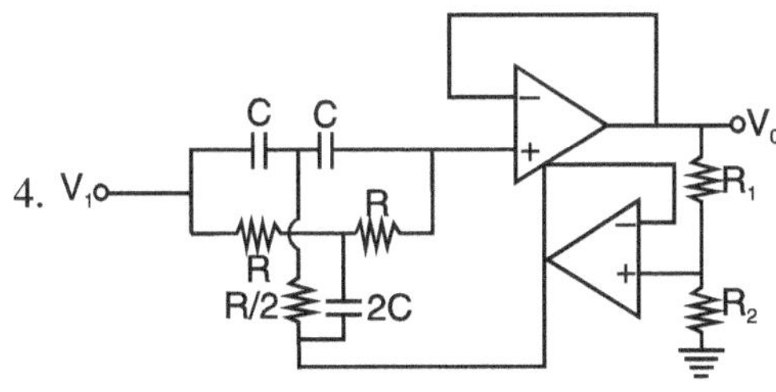

List II

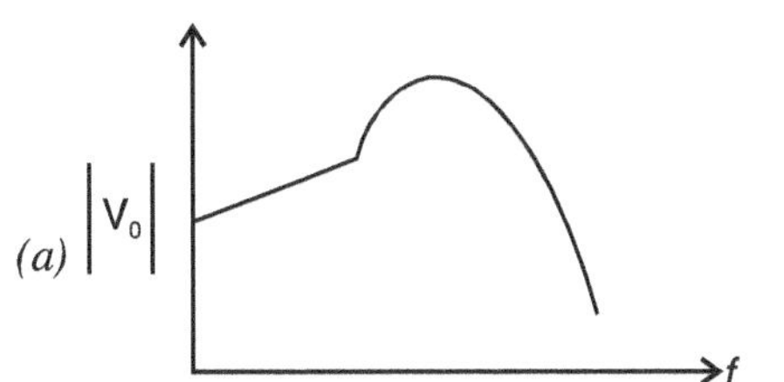

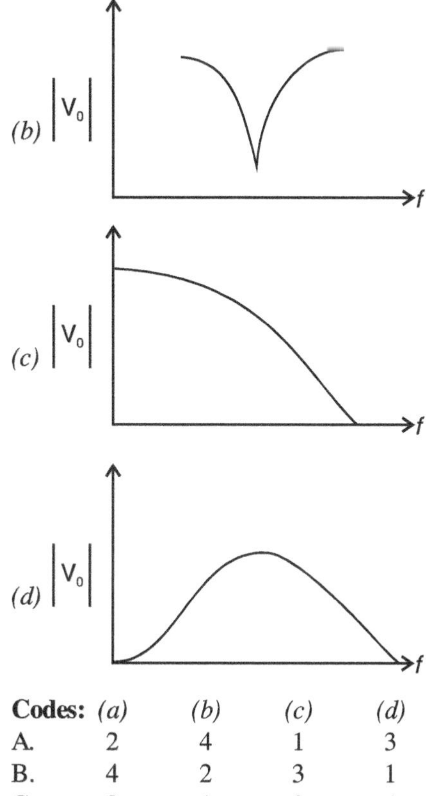

Codes:	*(a)*	*(b)*	*(c)*	*(d)*
A.	2	4	1	3
B.	4	2	3	1
C.	3	1	2	4
D.	1	2	3	4

254. A two-port network is represented by

$$V_1 = 24\,I_1 + 8I_2$$
$$V_2 = 8I_1 + 32I_2$$

Which one of the following networks is represented by these equations?

A.

B.

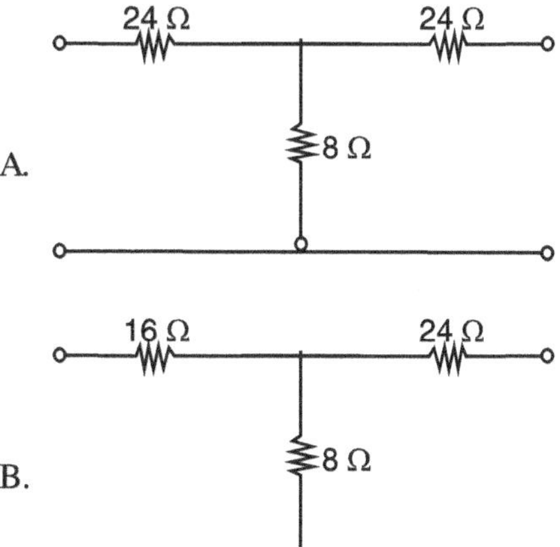

C.

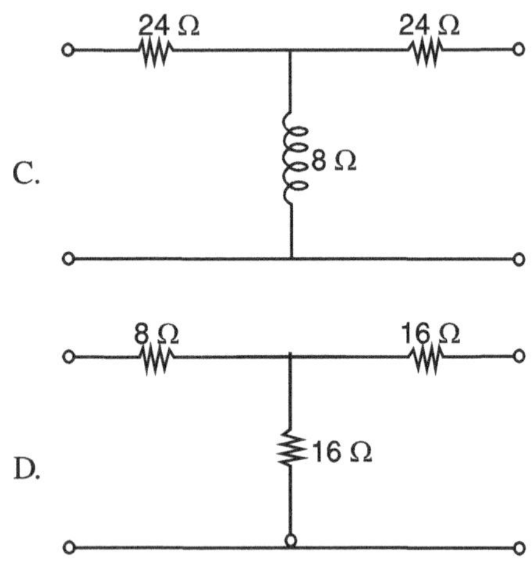

D.

255. The response to a step signal $u(t)$ is given by $2 + e^{-2t}$. Then the response to the impulse signal $2\ \delta(t)$ is given by

A. e^{-2t} B. $2e^{-2t}$

C. $-\ 2e^{-2t}$ D. $-\ 4e^{-2t}$

256. A dc circuit shown has a voltage source V, a current source I and several resistors. A particular resistor R dissipates a power of 4 Watts when V alone is active. The same resistor R dissipates a power of 9 Watts when I alone is active. The power dissipated by R when both sources are active will be

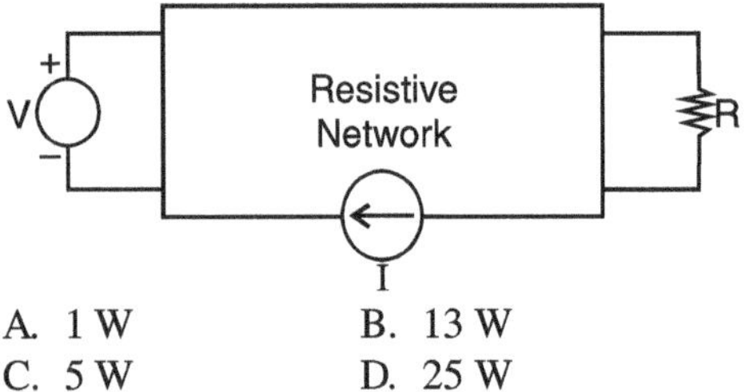

A. 1 W B. 13 W

C. 5 W D. 25 W

257. The given figure shows a computational circuit

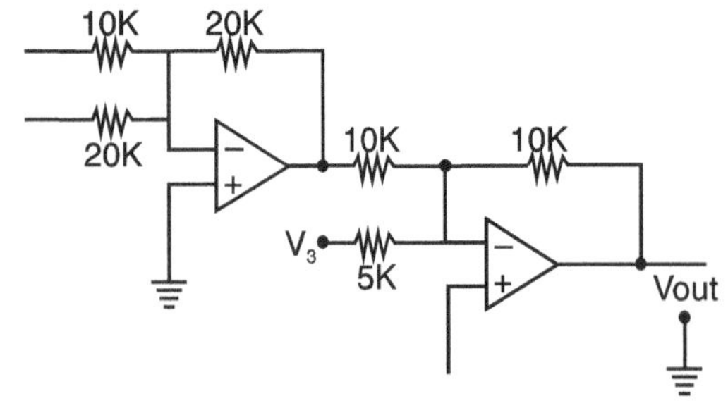

using OP-AMPS. The value of V_{out} in this circuit is

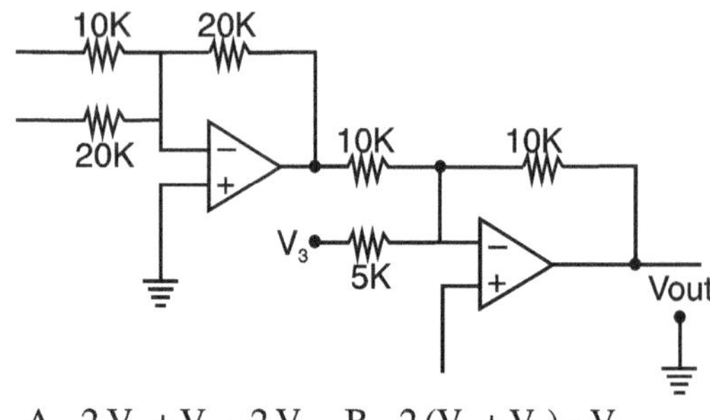

A. $2V_1 + V_2 - 2V_3$ B. $2(V_1 + V_2) - V_3$

C. $V_1 + V_2 + V_3$ D. $V_1 + 2V_2 - V_3$

258. The resolution of logic analyser is

A. the maximim number of input channels

B. the minimum duration of the glitch it can capture

C. its internal clock period

D. the minimum amplitude of input signal it can display.

259. In the series circuit shown for series resonance, the value of the coupling coefficient M will be

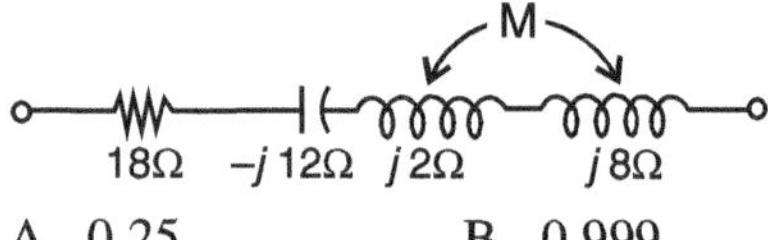

A. 0.25 B. 0.999

C. 0.5 D. 1.0

260. A two terminal black box contains one of the R.L.C elements. The black box is connected to a 220 volts A.C. supply. The current through the source is I. When a capacitance of 0.1F is inserted in series between the source and the box, the current through the source is 2I. The element is

A. a resistance

B. an inductance

C. a capacitance of 0.5 F

D. it is not possible to determine the element

261.

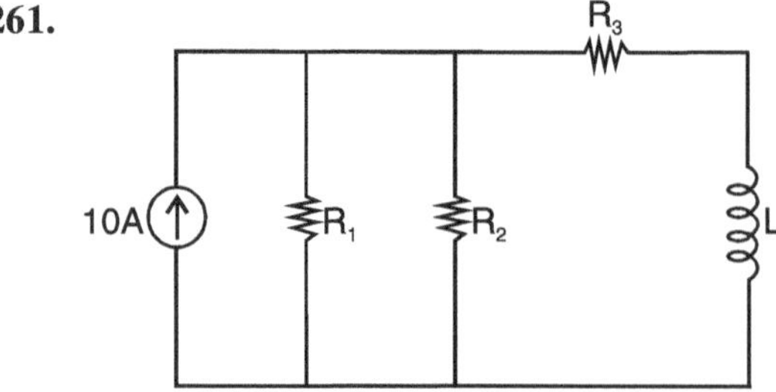

The time constant of the network shown is given by

A. $\dfrac{L}{R_3+\dfrac{R_1R_2}{R_1+R_2}}$ B. $\dfrac{L}{R_1+R_2+R_3}$

C. $\dfrac{L}{\dfrac{1}{R}+\dfrac{1}{R_2}+\dfrac{1}{R_3}}$ D. $\dfrac{L}{\dfrac{R_1R_2}{R_1+R_2}}$

262. Which factors influence the capacity of a capacitor

A. Area of plates, thickness of plates and rate of charge

B. area of the plates, dielectric and the rate of charge.

C. distance between the plates, dielectric and thickness of the plate

D. distance between the plates are of the plates and dielectric.

263. For the circuit given the output waveform v_0 is

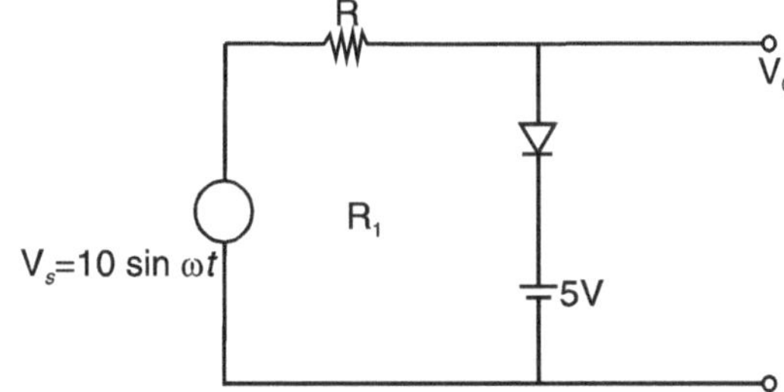

A.

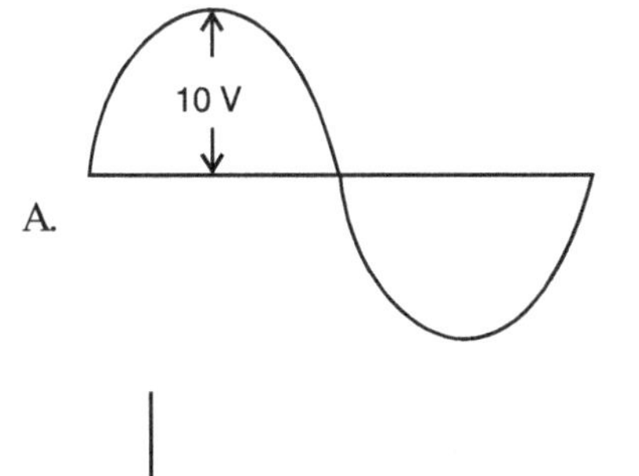

B.

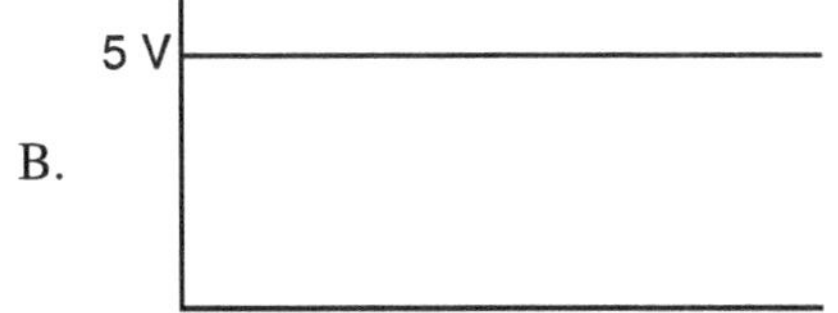

C.

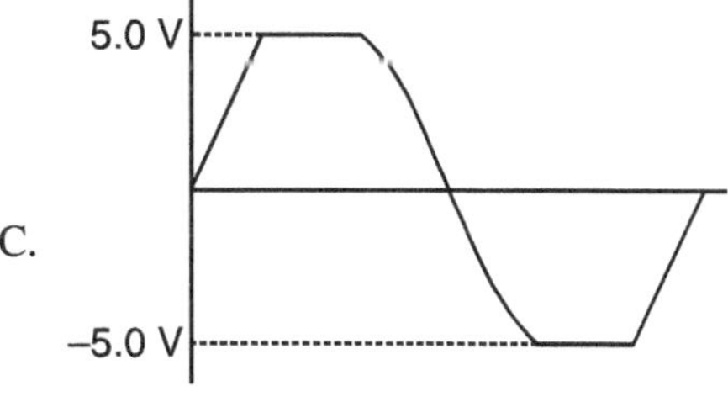

D.

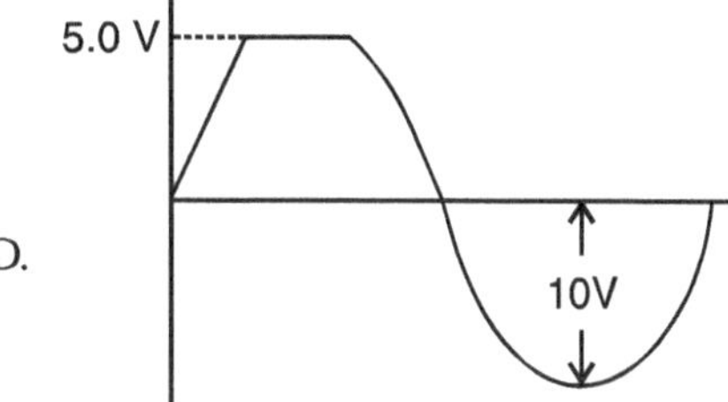

264. In the circuit shown voltmeter indicates 20 V. The reading of the ammeter will be

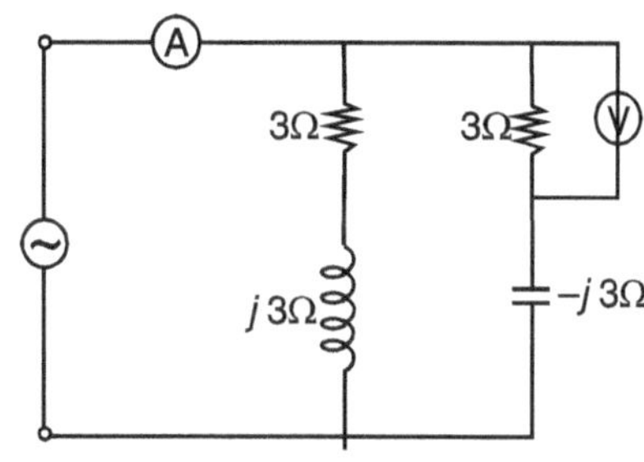

A. 20 A B. $10\sqrt{2}$A

C. 10 A D. zero

265. The approximate input impedance of the OP-AMP circuit in the given figure is

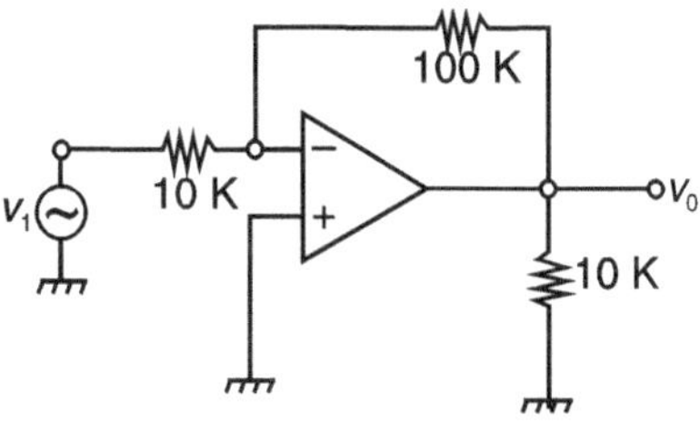

A. infinity B. 120 K

C. 110 K D. 10 K

266. If the resistance in a series RC circuit is increased the magnitude of the phase angle –

A. increase

B. remains the same
C. decreases
D. changes in an indeterminate manner

267. A series RLC circuit is overdamped when

A. $\frac{R^2}{4L^2} > \frac{1}{LC}$ B. $\frac{R^2}{4L^2} = \frac{1}{LC}$

C. $\frac{R^2}{4L^2} < \frac{1}{LC}$ D. R = infinity

268. In the circuit assume that the diodes are ideal and the meter is an average indicating ammeter. The ammeter will read

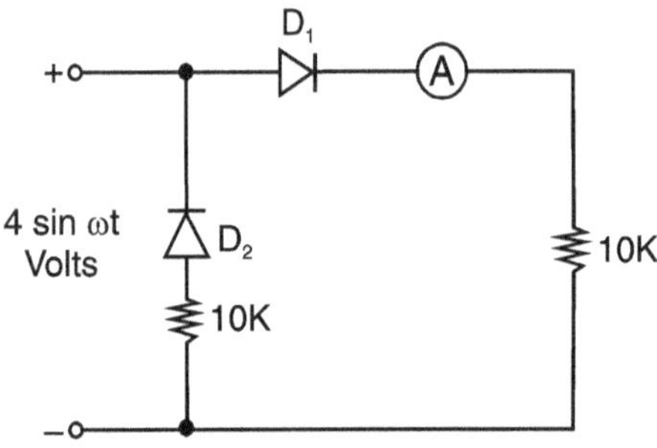

A. $0.4\sqrt{2}$A B. 0.4 A

C. $\frac{0.8}{\pi}$A D. $\frac{0.4}{\pi}$A

269. In figure shown. A_1, A_2, and A_3 are ideal ammeters. If A_2 and A_3 read 3A and 4A respectively, then A_1 should read

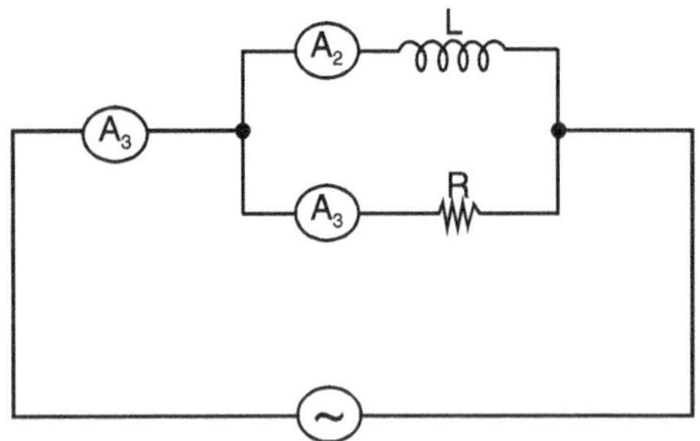

A. 1 A B. 5 A
C. 7 A D. none of the above

270. For the given parallel resonant circuit, match List I with List II and select the correct answer

List I	*List II*
(a) I at resonance	1. $\frac{\omega L}{R}$
(b) I_L	2. In phae with voltage
(c) Dynamic impedance	3. L/CR
	4. Lags the applied voltage

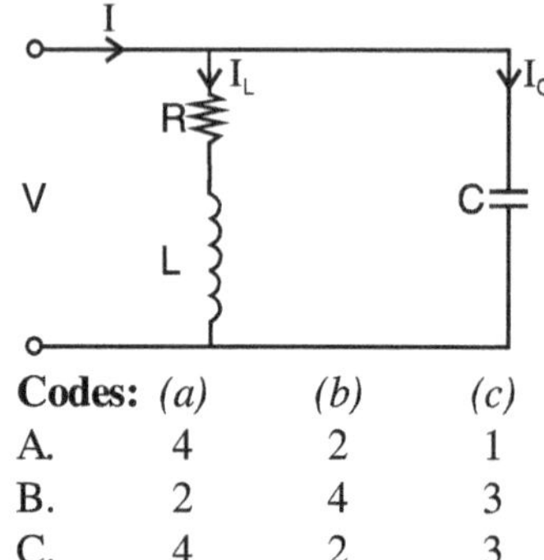

Codes:	*(a)*	*(b)*	*(c)*
A.	4	2	1
B.	2	4	3
C.	4	2	3
D.	2	4	1

271. The value of R in the oscillator circuit shown is so chosen that it just oscillates at an angular frequency of ω. The value of ω and the required of R will respectively be

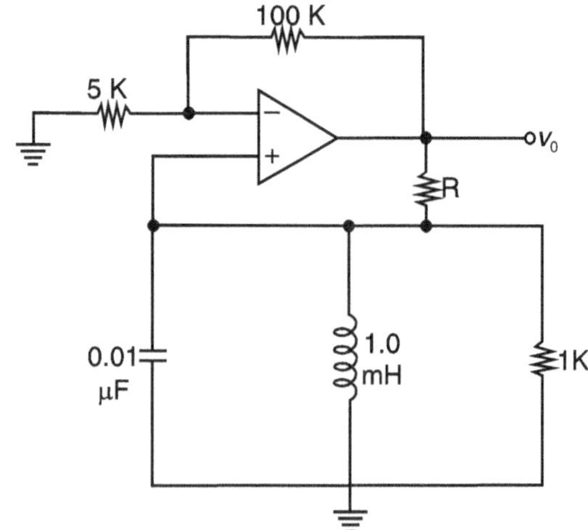

A. 10^5 rad/sec. $2 \times 10^4\ \Omega$
B. 2×10^4 rad/sec. $2 \times 10^4\ \Omega$
C. 2×10^4 rad/sec. $10^5\ \Omega$
D. 10^5 rad/sec, $10^5\ \Omega$

272. When a constant voltage source V is connected to a series R-L circuit with zero initial stored energy in the inductor, the instantaneous value of power supplied to the inductor L is given by

A. $\frac{V^2}{R}\left(e^{-Rt/L}+e^{-2Rt/L}\right)$

B. $\frac{V^2}{R}\left(1-e^{-Rt/L}\right)$

C. $\frac{V^2}{R}e^{+Rt/L}$

D. $\frac{V^2}{R}\left(1+e^{-Rt/L}\right)$

273. A driving point impedance

$$Z_d(s)=\frac{V(s)}{I(s)}=\frac{s+1}{s+2}$$

The system is initially at rest. For a voltage signal of unit step, the current $i(t)$ through Z_d is given by

A. $1-e^{-t}$ B. $2-e^{-t}$

C. $2-2e^{-t}$ D. $\frac{1}{2}\left(1-e^{-2t}\right)$

274. Match List I with List II and select the correct answer

List I ***List II***

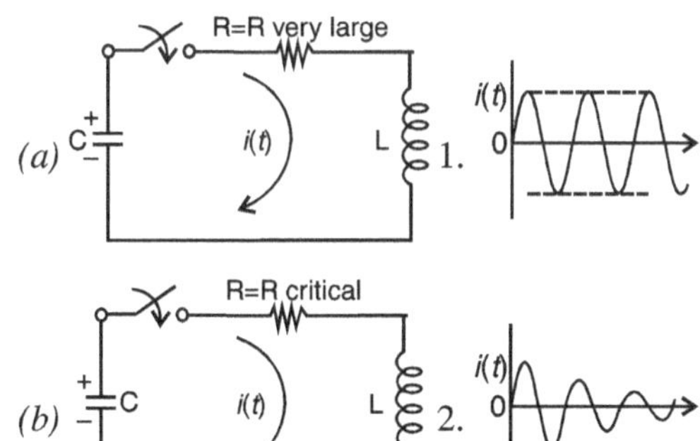

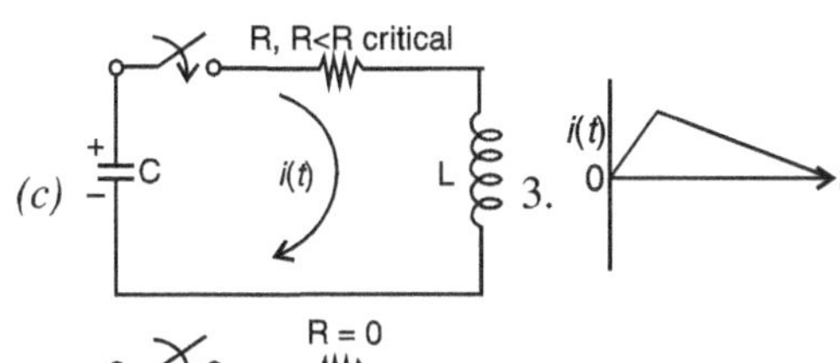

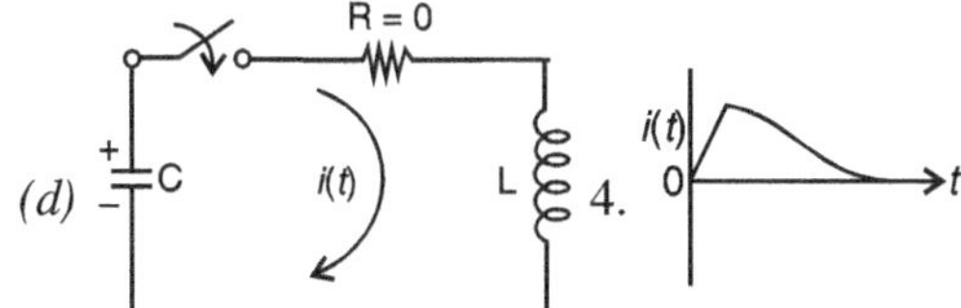

Codes:	*(a)*	*(b)*	*(c)*	*(d)*
A.	3	4	1	2
B.	1	3	1	2
C.	3	4	2	1
D.	4	3	2	1

275. For a pure capacitor, the voltage current relationship is

A. voltage lags current by 90°

B. voltage leads current by 90°

C. current lags voltage by 90°

D. current leads voltage by 180°

276. For the circuit shown below different time constants are given:

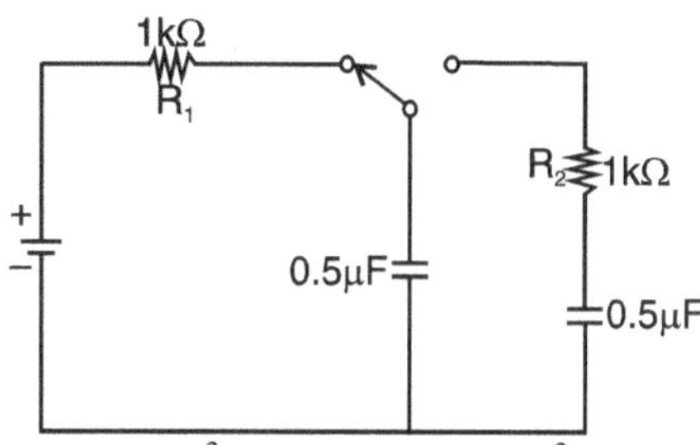

1. $0.5\times10^{-3}s$ 2. $2\times10^{-3}s$

3. $0.25\times10^{-3}s$ 4. $10^{-3}s$

What are the charging and discharging times respectively :

A. 1, 2 B. 2, 3

C. 1, 3 D. 2, 4

277. A parallel RLC circuit resonates at 100 kHz. At frequency of 110 kHz, the circuit impedance will be :

A. capacitive B. inductive

C. resistive D. none of these

278. For a series RLC circuit at resonance the current amplitude is ________ for a fixed voltage amplitude and the power factor is _______

A. minimum zero B. minimum, unity

C. maximum, zero D. maximum, unity

279. The effective value of a sinusoidal voltage is 14.14 volts and so its peak value is

A. 10 volts B. 20 volts

C. 30 volts D. 7.07 volts

280. In Fourier series expression of a function $f(t)$, the coefficient b_n is given by :

A. $b_n=\frac{2}{T}\int_{-T/2}^{T/2}f(t)\sin(n\omega t)\,dt$

B. $b_n = \frac{2}{T}\int_{-T/2}^{T/2} n\, f(t)\sin(n\omega t)\,dt$

C. $b_n = \frac{1}{2T}\int_{-T/2}^{T/2} f(t)\sin(\omega t)\,dt$

D. $b_n = \frac{2}{T}\int_{-T/2}^{T/2} f(t)\sin(\omega t)\,dt$

281. In the circuit shown, the V *(t)* and *i(t)* will be

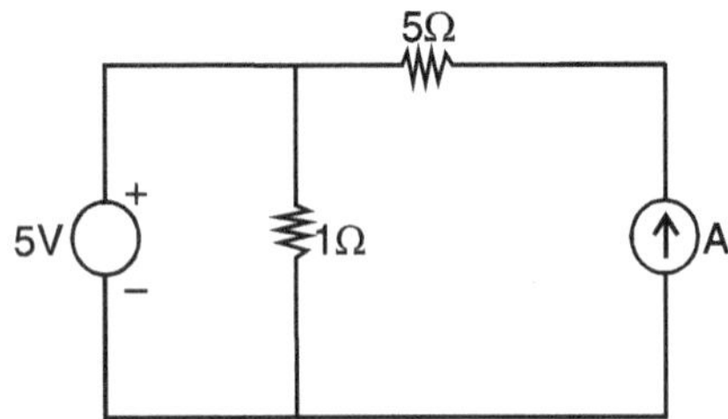

A. 1 V, 1 A B. 1 V, 6 A
C. 5 V, 5 A D. None of the above

282.

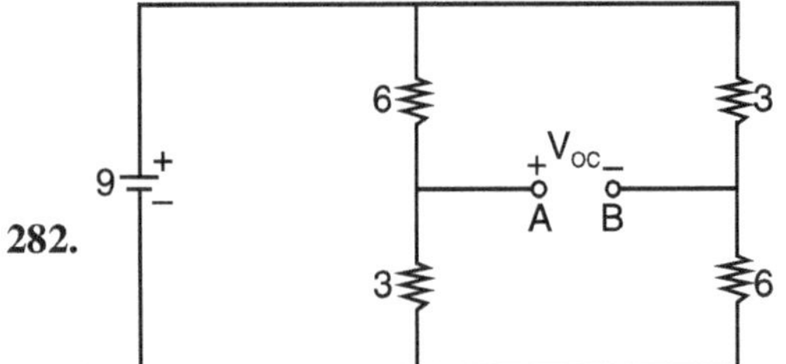

The Thevenin equivalent across points AB is given by V_{OC} and R_{eq}. Their values are

A. $V_{OC} = 3, R_{eq} = 4$
B. $V_{OC} = -3, R_{eq} = 4$
C. $V_{OC} = -3, R_{eq} = 4.5$
D. $V_{OC} = 6, R_{eq} = 4$

283. In a series RL circuit phasor diagram, total voltage may be represented by the ________ phasor and the resistor voltage may be represented by the ________ phasor

A. Current, voltage
B. Impedance, resistance
C. Current, resistance current
D. impedance, inductance

284. The poles of a certain transformed output are located as shown in figure. The responses corresponding to these poles are shown in Figs. (1, 2, 3, 4). The responses corresponding to $(p_1 p_1^*)$, P_2, (p_3, p_3^*), p_4 are given respectively by figure numbers,

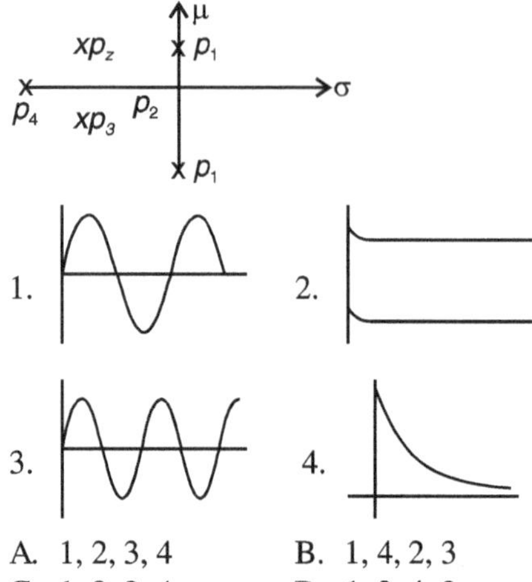

A. 1, 2, 3, 4 B. 1, 4, 2, 3
C. 1, 3, 2, 4 D. 1, 3, 4, 2

285. If the roots of the characteristic equation $s^2 + a_1 s + a_2 = 0$ of a network are negative real and unequal, then the system is :

A. oscillatory B. critically damped
C. underdamped D. overdamped

286. The line constant for the circuit shown below is:

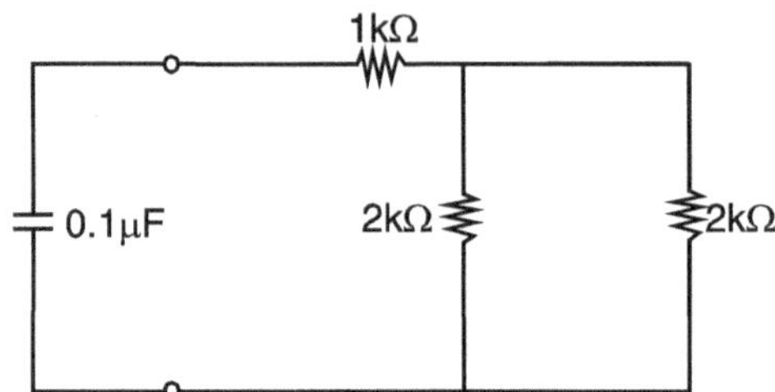

A. 0.2 microseond B. 0.8 millisecond
C. 0.4 millisecond D. 0.2 millisecond

287. An ac current of 5 A and dc current of 5 A flow simultaneously through a circuit. Which of the following statements is true?

A. An ac ammeter will read less than 10 A but more than 5 A
B. An ac ammeter will read only 5 A
C. A dc ammeter will read 10 A
D. A dc ammeter will read zero.

288. A source $V_s = 100 \cos \omega t$ is connected to an impedance which draws a power $P_{av} = 200$ watts at power factor 0.8 lagging. The current drawn from the source is $I = I_m \cos(\omega t + \theta)$. The value of I_m and θ are given by

A. 5 and $-\cos^{-1} 0.8$

B. 5 and $\cos^{-1} 0.8$

C. $\frac{5}{\sqrt{2}}$ and $\cos^{-1} 0.6$

D. 5 and 0

289. In a two port network, hybrid parameters express:

A. V_1, V_2 in terms of I_1, I_2

B. I_1, I_2 in terms of V_1, V_2

C. V_1, V_1 in terms of V_2, V_2

D. V_1, I_2 in terms of V_2, I_1

290. The driving point impedance function $\frac{s+3}{s+4}$ can be realised by

A. 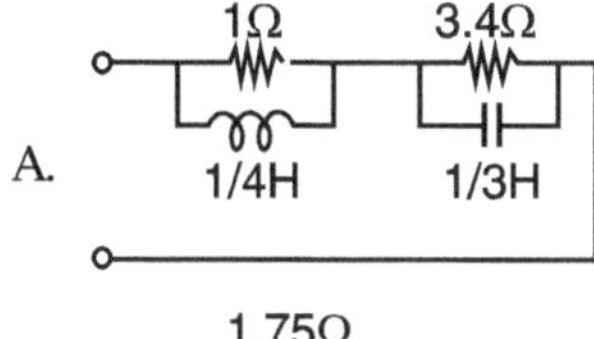

B. 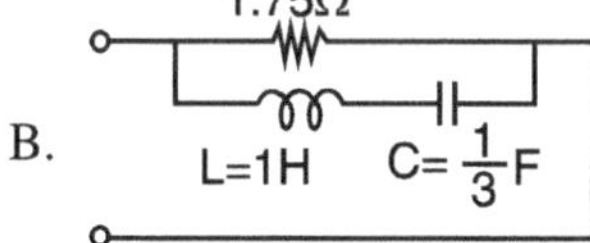

C. neither (A) nor (B) above

D. both (A) and (B) above

291. The driving point impedance of a network at a frequency of 1 Hz is $\sqrt{2j}$. The impedance can be realised as:

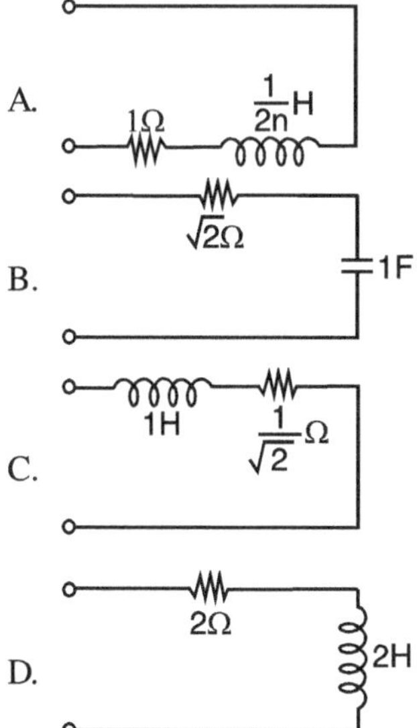

292. The equivalent inductance of the circuit is given by

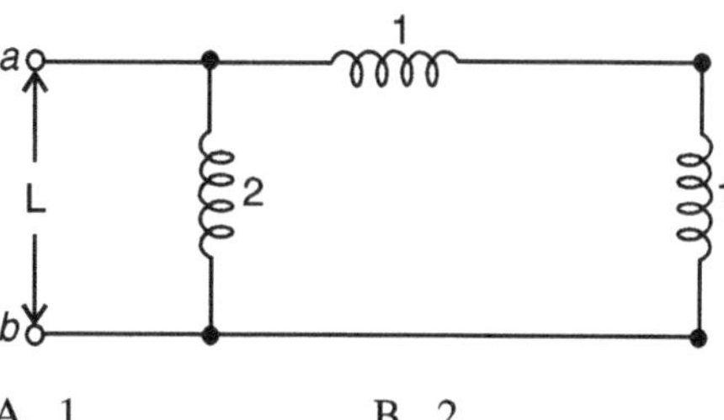

A. 1 B. 2

C. 3 D. 4

293. The roots of numerator polynomial in any network function are known as :

A. zeros

B. poles

C. either poles or zeros depending on the type of function

D. none of these

294. Voltages on R, L, C in a series circuit are shown below. Value of voltage source is

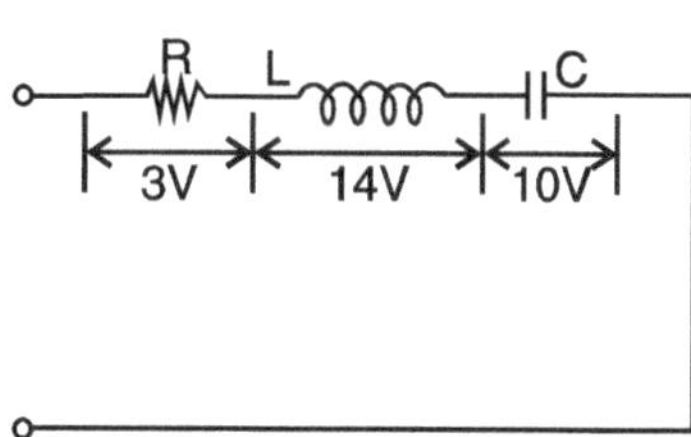

A. 10 V B. 15 V

C. 27 V D. 5 V

295. Find the current in R_L in the circuit below:

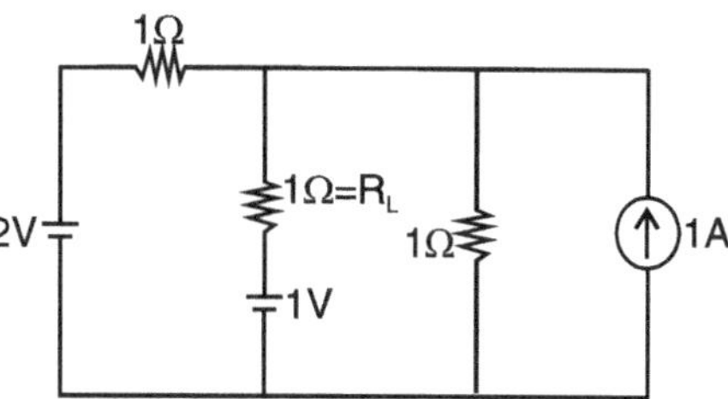

A. 0.5 A B. 1 A

C. 0.33 A D. 2 A

296. The semiconductor strain gauge has gauge factor

A. 2 B. 10

C. 100 D. 1000

297. 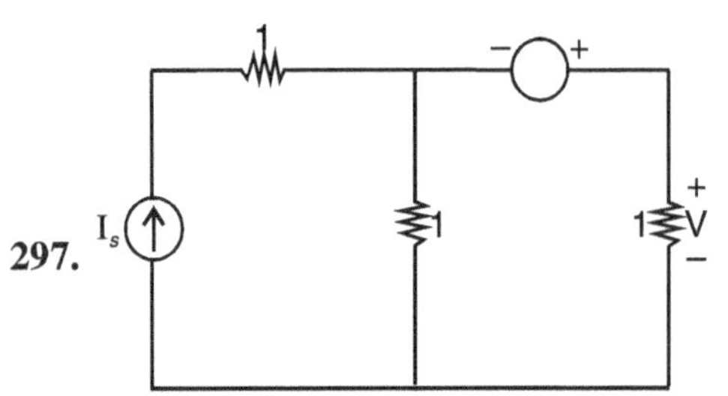

In the circuit shown $I_s = 2$, $V_s = 2$. The voltage V is given by

A. 2 B. 1
C. 0 D. −1

298. Two systems obeying equations of the same form are said to be :

A. similar systems
B. identical systems
C. equivalent systems
D. analogous systems

299. A transmission line has reflection coefficient equal to 0.5. The VSWR is :

A. 3 B. 0
C. 2 D. 1

300. The _______ is always positive while the ____- or _____ may be negative

A. abscissa, hypotenuse, ordinate
B. ordinate, hypotenuse, abscissa
C. hypotenuse, ordinate, hypotenuse
D. abscissa, ordinate, hypotenuse

301. For the network shown Z(0) = 3 Ω and Z(∞) = 2 Ω. The values of R_1 and R_2 will respectively be

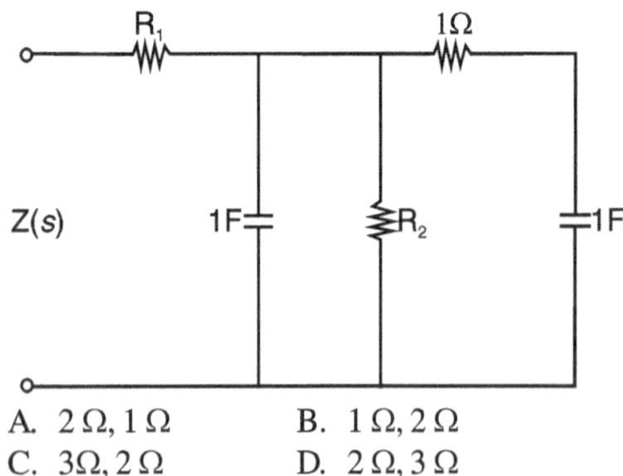

A. 2 Ω, 1 Ω B. 1 Ω, 2 Ω
C. 3Ω, 2 Ω D. 2 Ω, 3 Ω

302. Equivalent resistance of the given network is :

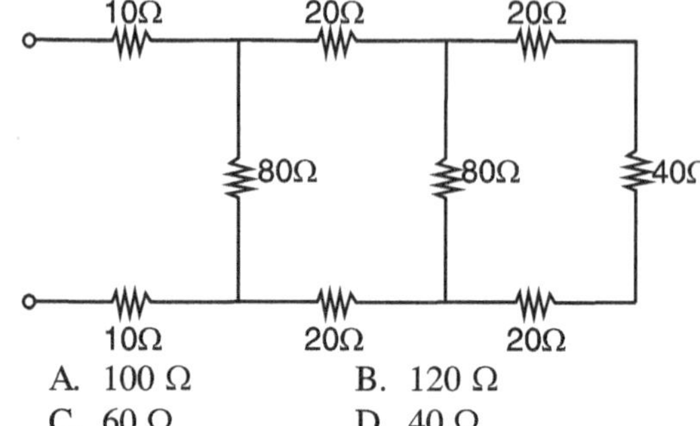

A. 100 Ω B. 120 Ω
C. 60 Ω D. 40 Ω

303. A 100 watt electric lamp draws a current I amp, from a 220 V, 50 Hz source I is

A. 100 A B. 2.2 A
C. 2 A D. none these

304. Consider the circuit shown for the measurement of resistance

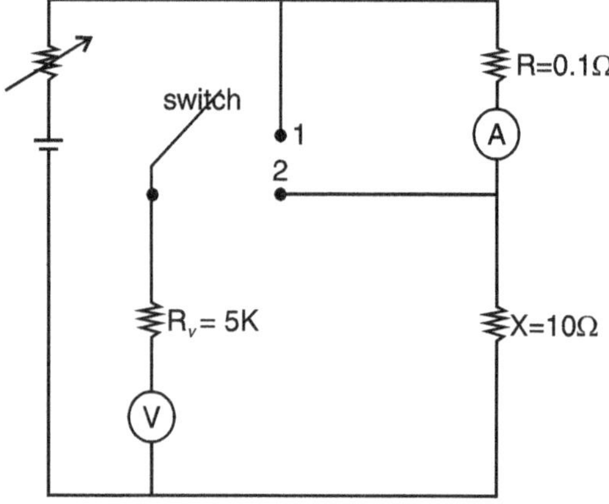

If X_1 and X_2 are the measured valued of X for switch positions 1 and 2 respectively, then the arithmetic mean of X_1 and X_2 will be

A. 9.98 Ω B. 10.00 Ω
C. 10.04 Ω D. 10.10Ω

305. The double integration of a unit step function results in

A. a unit ramp function
B. a unit impulse
C. a unit parabola
D. a unit doublet

306. The Thevenin impedance across the terminals AB of the given network is

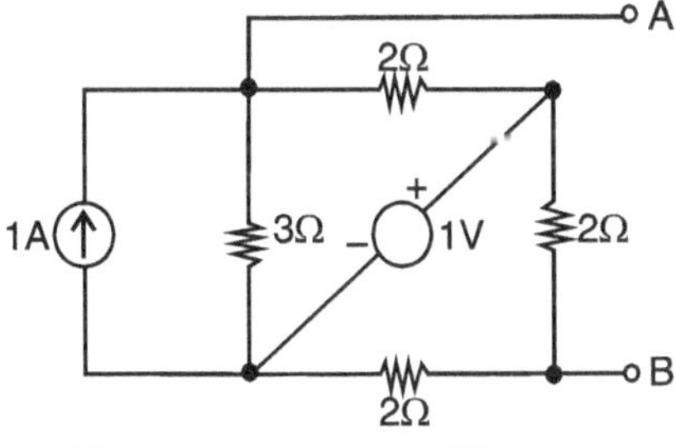

A. $\frac{10}{3}\Omega$ B. $\frac{20}{9}\Omega$

C. $\frac{13}{4}\Omega$ D. $\frac{11}{5}\Omega$

307. Consider the systems with the following open-loop transfer functions :

1. $\frac{36}{s(s+3.6)}$ 2. $\frac{100}{s(s+5)}$

3. $\frac{6.25}{s(s+4)}$

The correct sequence of these systems in increasing order of the time taken for the unit-step response to settle is

A. 3, 2, 1 B. 3, 1, 2

C. 2, 3, 1 D. 1, 2, 3

308. In synthesis of network function, partial fraction of first Foster network yields:

$$z(s) = \frac{1}{(3/4)s + (4/3)s} + \frac{1}{(4/5)s + (7/6)s}$$

In network is shown below

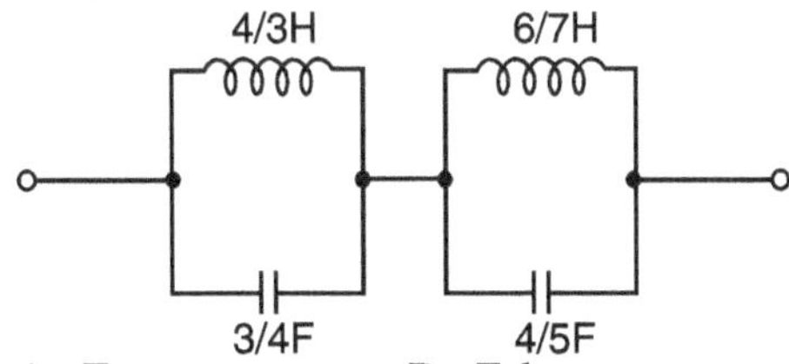

A. True B. False

C. Not known D. Incomplete Question

309.

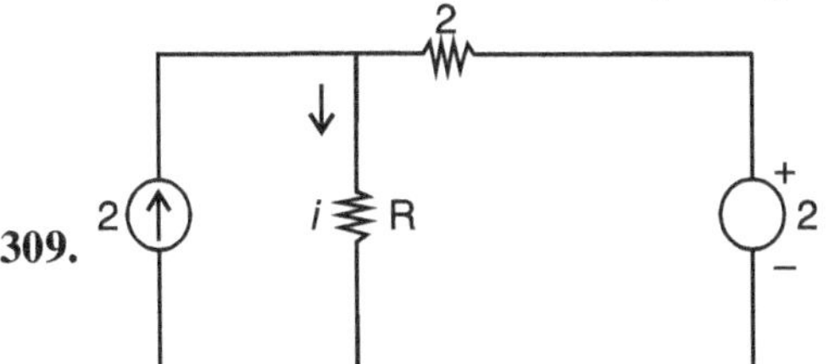

Given the current through R is $i = 1$. The value of R is

A. 0 B. 2

C. 4 D. 8

310. An RC circuit initially at rest has a step signal. The response $v(t)$ across C is

$$v(t) = 1 - e^{-at}$$

If now there is an initial voltage on C of 2 volts for the same step signal, $v(t)$ is now

A. $1 + e^{-at}$ B. $1 + 2e^{-at}$

C. $2 - e^{-at}$ D. $3 - e^{-at}$

311. A circuit consisting of a 1Ω resistor and a 2F capacitor in series is excited from a voltage source with the voltage expressed as $3e^{-t}$, as shown in the given figure. If the $i(0_-)$ and $v_c(0_-)$ are both zero, then the values of $i(0_+)$ and $i(\infty)$ will be respectively

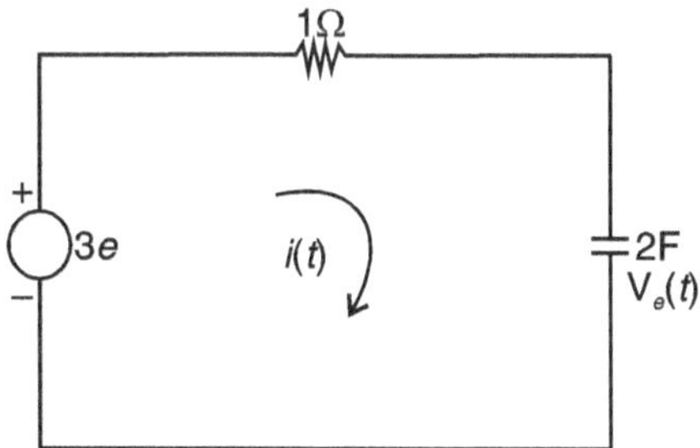

A. 3 A and 1.5 A B. 1.5 A and zero

C. 3 A and zero D. 1.5 A and 3 A

312. Impedance Z_1 = 10 ∠30° and Z_1 = 20 ∠50°. Then $Z_1 Z_2$ is :

A. 200 ∠40° B. 200 ∠50°

C. 200 ∠80° D. 200 ∠30°

313. The equivalent series resistance of a capacitor (C) in terms of its power factor and operating frequency (f) is given by

A. $\frac{\text{Power factor}}{2\pi fC}$

B. $\frac{2\pi fC}{\text{Power factor}}$

C. $2\pi fC$ (Power factor)

D. $\frac{1}{2\pi fC(\text{Power factor})}$

314. The Thevenin equivalent of the network shown in the given figure is

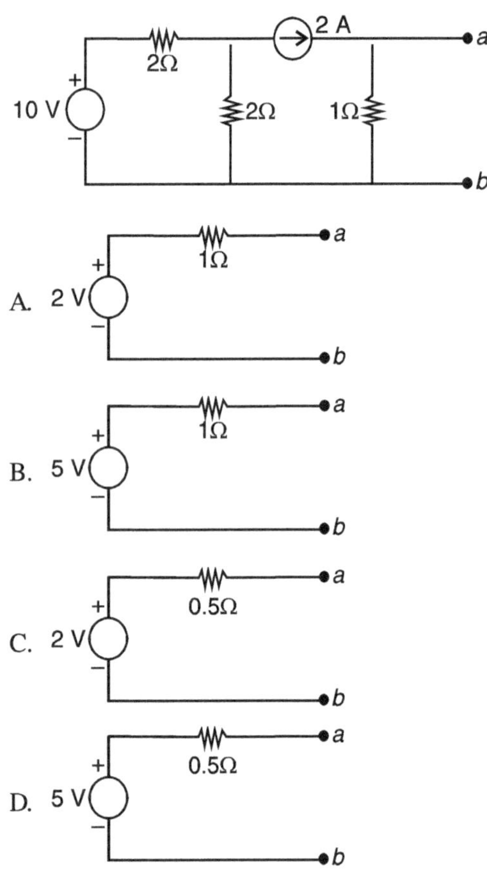

315. An inductor with inductance L and initial current I_0 is shown as

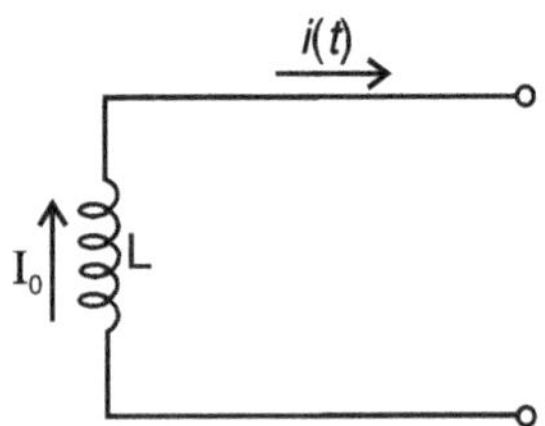

The correct admittance diagram for it is

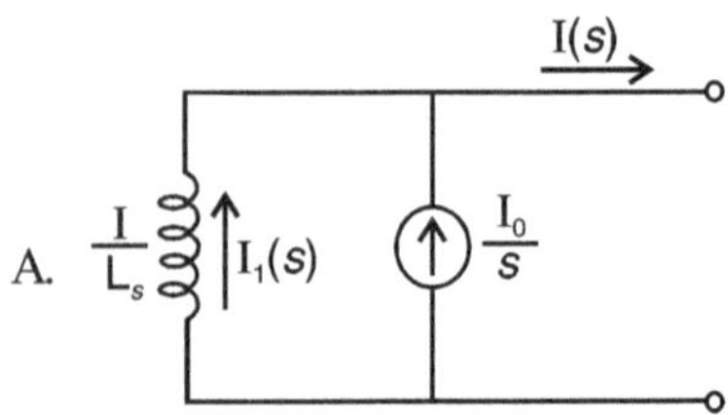

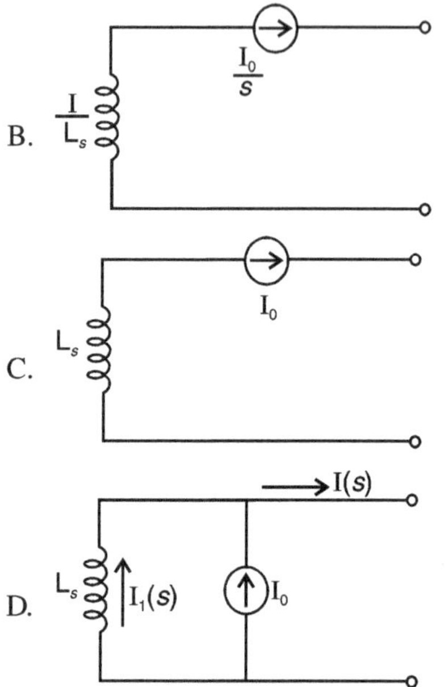

316. Laplace transform of $u(t)$ and t is given by

A. $\frac{1}{s}, \frac{1}{s^2}$ B. $\frac{1}{s^2}, \frac{1}{s}$
C. 1, s D. s, 1

317. The effect of an inductance is such that current flowing through it

A. can not change instantaneously
B. can change instantaneously
C. cannot change
D. is opposed by the inductance

318.

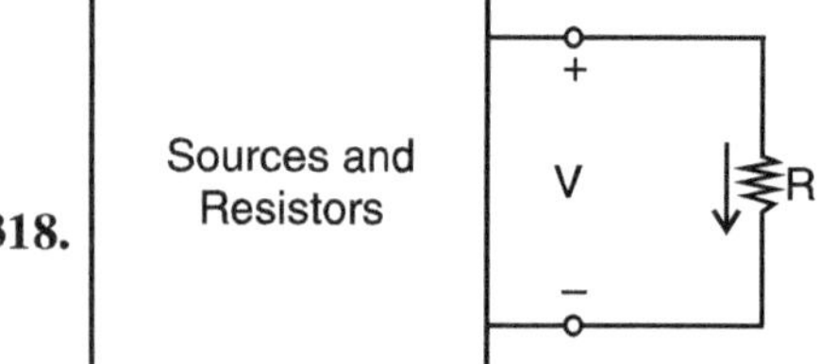

When R is infinity, V = 3. When R = 0, the current through it is 3, when R = 2, the current through it is given by

A. 1 B. $\frac{3}{2}$
C. 2 D. none of these

319. The primary purpose of the helix in a traveling-wave tube is to

A. prevent the electron beam from spreading in

the long tube

B. reduce the axial velocity of the RF field

C. ensure broadband operation

D. reduce the noise figure

320. The frequency at which the reactances of the two arms in a crystal gate are equal is the

A. pole

B. zero

C. zero attenuation frequency

D. maximum attenuation frequency

321. 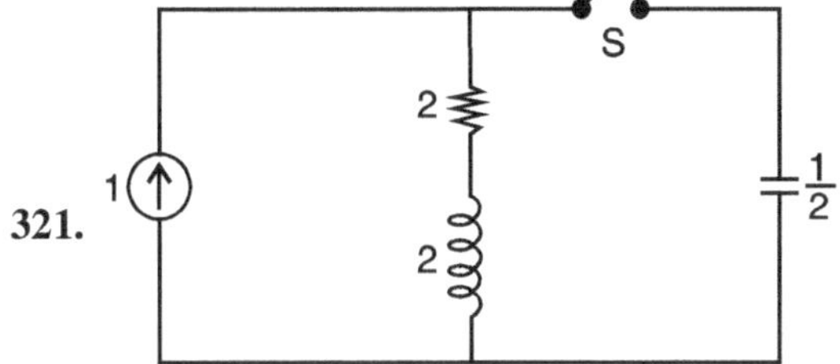

S is open for a long time and steady state is reached. S is closed at $t = 0$. Let i_L be the current in the inductor. At $t = 0^+$ $\frac{di_L}{dt}$ is given by

A. – 2 B. – 1

C. 0 D. 2

322. Thevenin's equivalent resistance of the given circuit is :

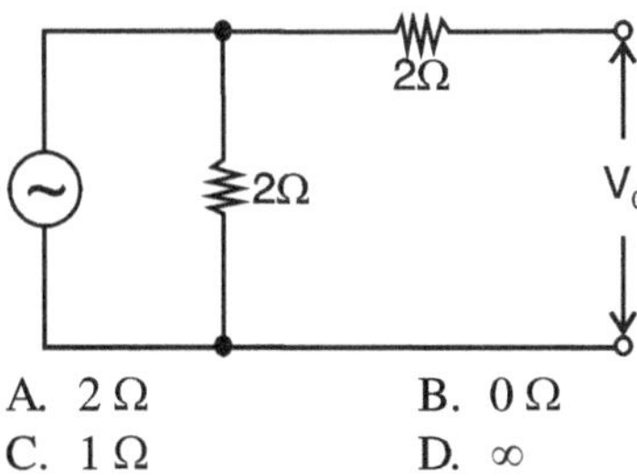

A. 2 Ω B. 0 Ω

C. 1 Ω D. ∞

323. Match List I with List II and select the correct answer

List I	*List II*
(a) Periscope antenna	1. Circular polarisation
(b) Helical antenna	2. Scanning radar
(c) Monopulse antenna	3. Communication satellite
(d) Transponder	4. Television transmission
	5. Microwave link

Codes:	*(a)*	*(b)*	*(c)*	*(d)*
A.	3	5	2	4
B.	2	3	4	5
C.	1	2	3	4
D.	5	1	2	3

324. Match List I with List II and select the correct answer

List I	*List II*
(a) Low pass filter	1. Tuned circuit
(b) High pass filter	2. Compensated attenuator
(c) Bandpass filter	3. Differentiator
(d) R-L-C circuit	4. Ringing

Codes:	*(a)*	*(b)*	*(c)*	*(d)*
A.	2	3	4	1
B.	2	3	1	4
C.	3	2	4	1
D.	2	1	3	4

325. Find the r parameter of the circuit below left

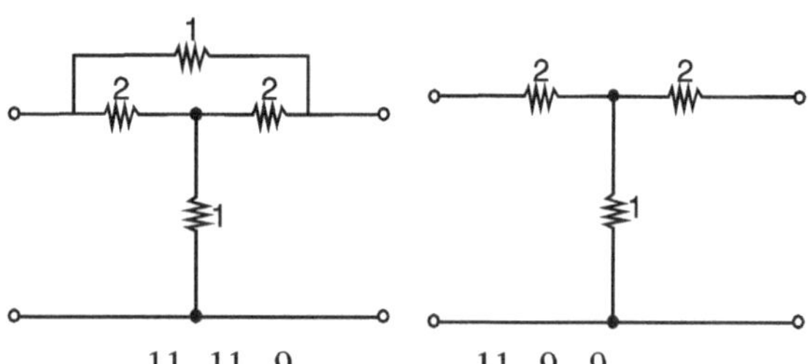

A. $\frac{11}{5}, \frac{11}{5}, \frac{9}{5}$ B. $\frac{11}{5}, \frac{9}{5}, \frac{9}{5}$

C. $\frac{9}{5}, \frac{9}{5}, \frac{11}{5}$ D. $\frac{9}{5}, \frac{11}{5}, \frac{11}{5}$

326. 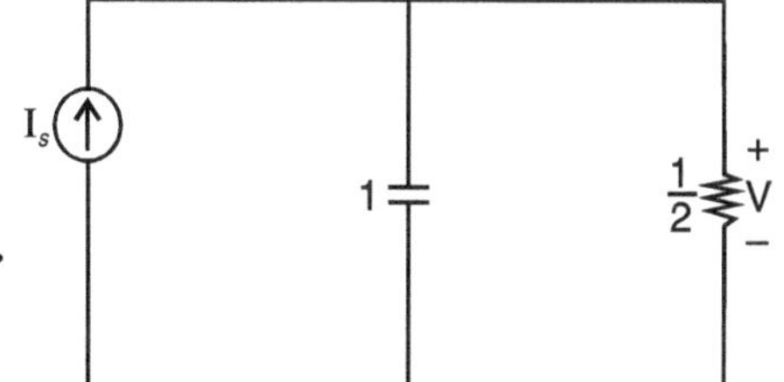

Initial voltage on capacitor $v_0 = 0.5$ volt I_s = unit step

$v(t)$ for $t > 0$ is given by

A. $1 - e^{-2t}$ B. $\frac{1}{2} - \frac{1}{2}e^{-et}$

C. $\frac{1}{2} - e^{-2t}$ D. $\frac{1}{2}$

327. In a series RLC circuit, the maximum voltage across the capacitor occurs at frequency
A. equal to resonant frequency
B. greater than resonant frequency
C. less than resonant frequency
D. $\omega = 0$

328. An RLC series circuit has f_1 and f_2 as the half power frequencies and f_0 as the resonance frequency. The Q-factor of the circuit is given by

A. $\dfrac{f_1 + f_2}{2f_0}$ B. $\dfrac{f_1 - f_2}{f_2 - f_0}$

C. $\dfrac{f_0}{f_1 - f_2}$ D. $\dfrac{f_2 - f_1}{f_0}$

329. As the modulation index of an FM signal with sinusoidal modulation is increased from zero to three, the power in the carrier component will
A. increase continuously
B. decrease continuously
C. first increase, attain a maximum and then decrease
D. first decrease, become zero and then increase

330. An LC circuit resonant at 2000 kHz has a Q of 100. Find bandwidth.
A. 10 kHz B. 20 kHz
C. 200 kHz D. 200 kHz

331. In a series RCL circuit the _______ voltage is in phase with the current
A. resistor B. inductor
C. capacitor D. total

332. Write down the voltage equation for the circuit shown above right
A. $e = \text{E}$ B. $e = i\text{R} + \text{E}$
C. $e = i\text{R} + \text{E} + \text{Ir}$ D. none of these

333. The Fourier series expansion of an odd periodic function contains only
A. Sine terms B. Cosine terms
C. Constant only D. None of above

334. Which of the following are useful in comparing the noise performance of receivers?
1. Input noise voltage
2. Equivalent noise resistance
3. Noise temperature
4. Noise figure
Select the correct answer
A. 1, 2 and 4 B. 1, 3 and 4
C. 2, 3 and 4 D. 1, 2 and 3

335. From the point of view of AGC, the main difference between tube and transistor receivers is that the latter
A. require AGC power rather than voltage
B. do not really need AGC because overloading causes limiting
C. have troubles because AGC is not very effective in them
D. must have IF stages even more stabilized than they are normally

336. Indicate the *false* statement. Flat-topped rectangular pulses must be transmitted in radar to
A. allow a good minimum range
B. make the returned echoes easier to distinguish from noise
C. prevent frequency changes in the magnetron
D. allow accurate range measurements

337. A collapsing field around a coil
A. helps the decay of coil current
B. tends to aid current flow reversal
C. tends to oppose the decay of coil current
D. does not affect coil current flow

338. Laplace transform of sin at is ;

A. $\dfrac{a}{s^2 + a^2}$ B. $\dfrac{a}{s^2 - a^2}$

C. $\dfrac{-a}{s^2 + a^2}$ D. $\dfrac{-a}{s^2 - a^2}$

339.

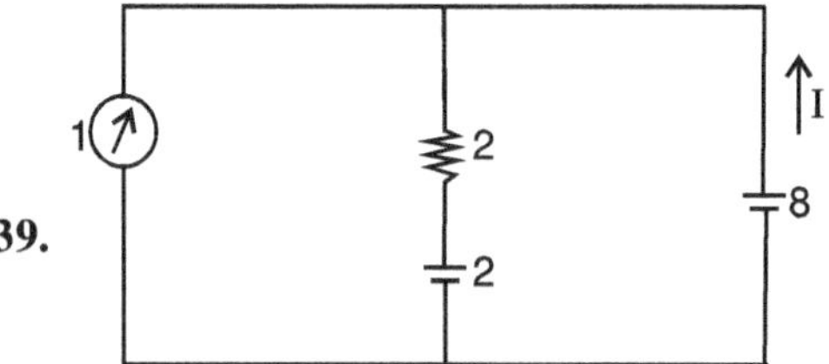

In the circuit shown, I is given by
A. -2 B. 2
C. 3 D. 4

340. Indicate the correct statement regarding satellite communications.
A. If two earth stations do not face a common satellite, they should communicate via a double-satellite hop.
B. Satellite are allocated so that it is impossible

for two earth stations not to face the same satellite.

C. Colocated earth stations are used for frequency diversity.

D. A satellite earth station must have as many receive chains as there are carriers transmitted to it.

341. The Laplace transform of a unit impulse fuction is

A. I B. I/S

C. I/S^2 D. S

342. Laplace transform of cos ωt is given by

A. $\frac{s}{s^2+\omega^2}$ B. $\frac{\omega}{s^2+\omega^2}$

C. $\frac{s^2}{s^2+\omega^2}$ D. $\frac{\omega^2}{s(s^2+\omega^2)}$

343. The Laplace transform of a unit impulse function is :

A. $\frac{1}{s}$ B. $\frac{1}{s^2}$

C. 1 D. s

344. For the graph shown incidence matrix A is given by

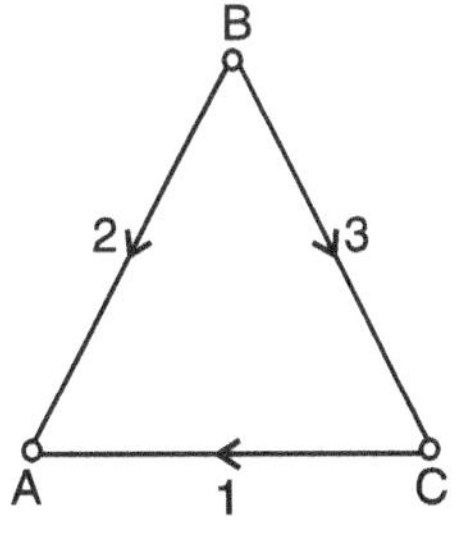

A. $\begin{bmatrix} -1 & -1 & 0 \\ 0 & -1 & 1 \\ -1 & 0 & -1 \end{bmatrix}$ B. $\begin{bmatrix} 1 & 0 & -1 \\ 1 & 1 & 0 \\ 0 & -1 & 1 \end{bmatrix}$

C. $\begin{bmatrix} -1 & -1 & 0 \\ 0 & 1 & 1 \\ 1 & 0 & -1 \end{bmatrix}$ D. $\begin{bmatrix} 1 & 0 & 1 \\ -1 & -1 & 0 \\ 0 & -1 & 1 \end{bmatrix}$

345. Consider the following statements :

A power supply uses bridge rectifier with capacitor input filter. If one of the diodes is defective, then

1. the dc load voltage will be lower than its expected value.
2. ripple frequency will be lower than the expected value.
3. the surge current will increase manifold.

Of these statements

A. 1 and 2 are correct

B. 1 and 3 are correct

C. 2 and 3 are correct

D. 1, 2 and 3 are correct

346. Consider the following statements regarding the state variable filter section given in the figure :

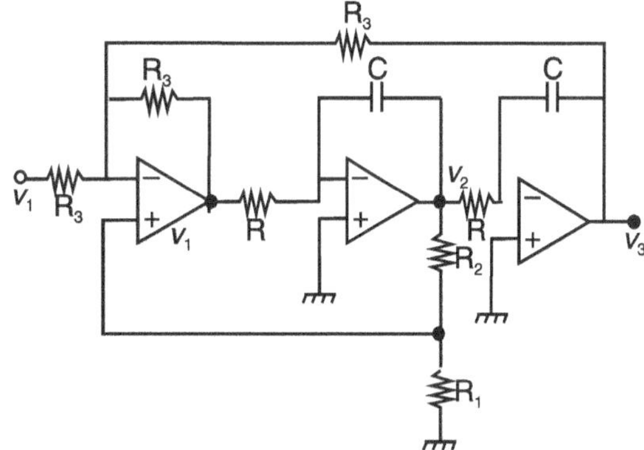

1. v_1 is the notch filtered output of v_i
2. v_3 is the low-pass filtered output of v_i
3. v_2 is the band-pass filtered output of v_i

Of these statements

A. 1, 2 and 3 are correct

B. 1 and 3 are correct

C. 2 and 3 are correct

D. 1 and 3 are correct

347. A first order system is shown. Its time reponse to a unit step input is given by

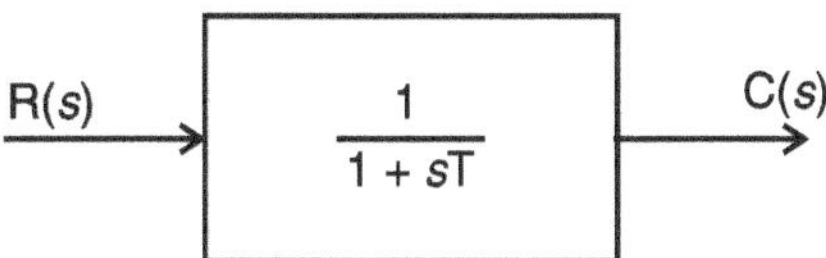

A. $C(t) = [1/T]\,[e^{-t/T}]$ B. $C(t) = T(1+e^{-t/T})$

C. $C(t) = (1-e^{-t/T})$ D. $C(t) = Te^{-t/T}$

348. For the cube the resistance between any two diagonally opposite corners of any one face of the cube will be (between A and F)

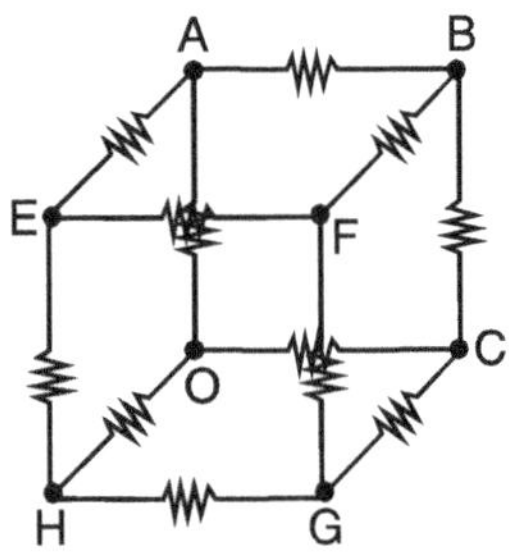

A. 2 Ω B. 3/4 Ω
C. 5/12 Ω D. 13/12 Ω

349. Match List I with List II and select the correct answer

List I	*List II*
(a) Linearity	1. Superposition
(b) Structure	2. Norton's
(c) Equivalent circuit	3. Tellegan's
(d) Bilateral	4. Reciprocity
	5. Millman's

Codes:	*(a)*	*(b)*	*(c)*	*(d)*
A.	2	3	4	5
B.	1	3	2	4
C.	1	2	3	4
D.	1	3	5	2

350. In a two-port network containing linear bilareral passive circuit elements, which one of the following conditions for *z* parameters would hold?

A. $z_{11} = z_{22}$ B. $z_{12}\, z_{21} = z_{11}\, z_{22}$
C. $z_{11}z_{12} = z_{22}z_{21}$ D. $z_{12} = z_{21}$

351. The given circuit represents a :

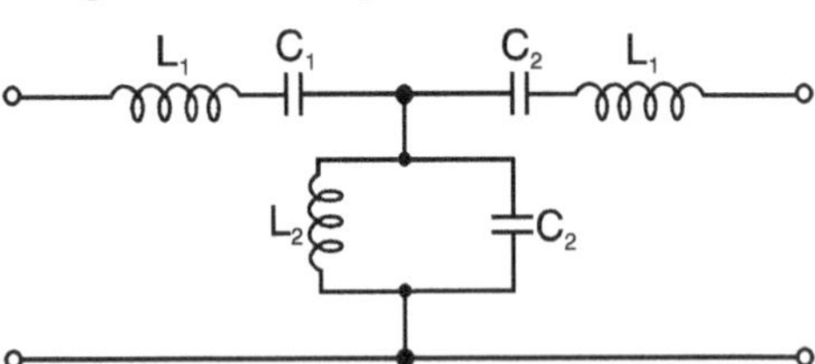

A. low pass filter B. high pass filter
C. band pass filter D. band stop filter

352. Consider the following statements :
Thyristors can be turned off by

1. reducing the current below the holding current value.
2. applying a negative voltage to the anode of the device.
3. reducing the gate current

Of these statements

A. 1 and 2 are correct
B. 1 and 3 are correct
C. 2 and 3 are correct
D. 1, 2 and 3 are correct

353. Convert the following to single voltage source in series with single resistors

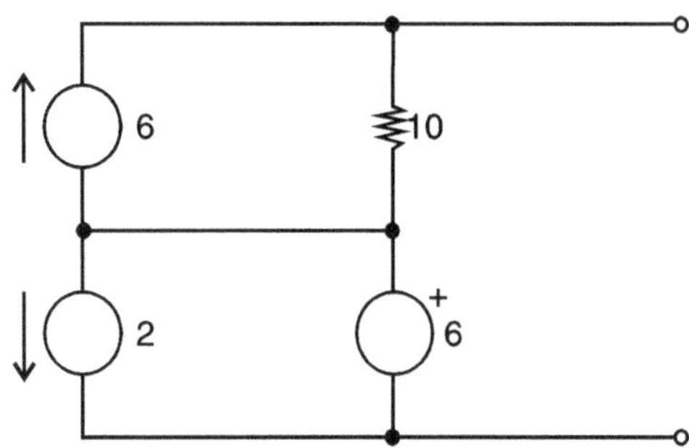

A. $5\text{V}, \frac{10}{3}\Omega$
B. $0\text{V}, 15\,\Omega$
C. $66\,\text{V}, 10\,\Omega$
D. $10\,\text{V}, 6\,\Omega$

354. In the network shown $i = 0$ when E = 4 V, and I = 2A and i = 1A when E = 8V, I = 2A. The Thevenin voltage and the resistance looking into the terminals AB are

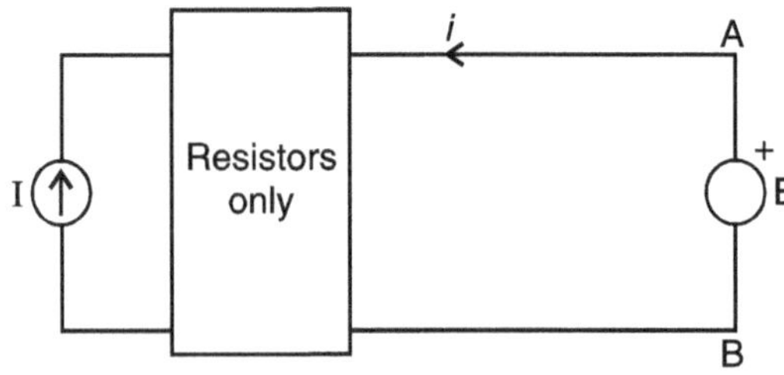

A. $4\,\text{V}, 2\,\Omega$ B. $14\,\text{V}, 4\,\Omega$
C. $18\,\text{V}, 2\,\Omega$ D. $8\,\text{V}, 4\,\Omega$

355. The equivalent circuit of a wire-wound resistor can be represented as

A. 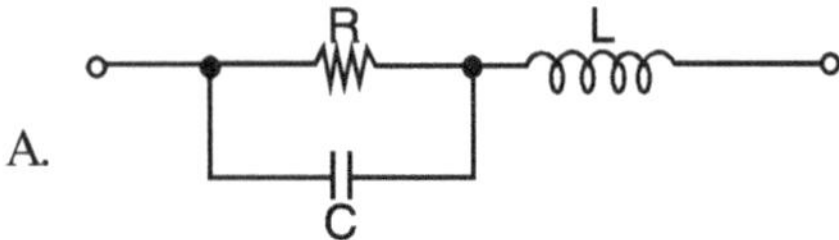

B.

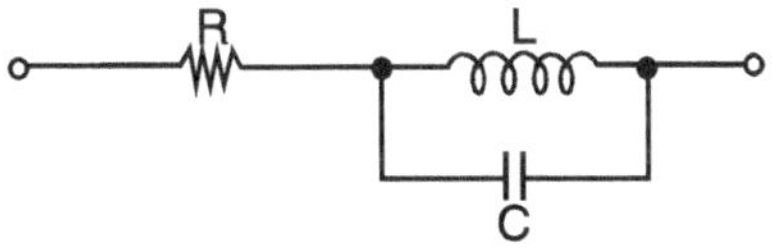

C.

D.

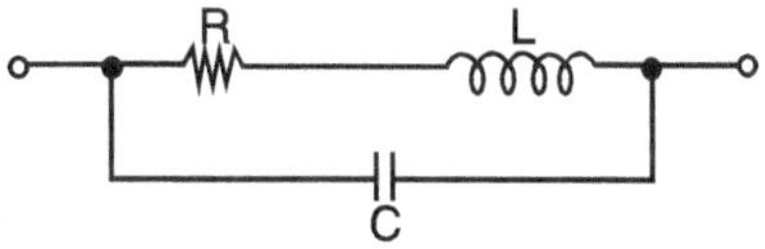

ANSWERS

1	2	3	4	5	6	7	8	9	10
B	D	C	D	D	C	B	A	B	C
11	**12**	**13**	**14**	**15**	**16**	**17**	**18**	**19**	**20**
B	B	D	A	D	D	B	C	C	D
21	**22**	**23**	**24**	**25**	**26**	**27**	**28**	**29**	**30**
B	B	A	B	D	A	D	D	C	A
31	**32**	**33**	**34**	**35**	**36**	**37**	**38**	**39**	**40**
C	C	C	B	A	C	D	D	A	C
41	**42**	**43**	**44**	**45**	**46**	**47**	**48**	**49**	**50**
D	A	D	B	D	D	B	C	B	A
51	**52**	**53**	**54**	**55**	**56**	**57**	**58**	**59**	**60**
C	D	A	D	A	C	B	A	B	D
61	**62**	**63**	**64**	**65**	**66**	**67**	**68**	**69**	**70**
B	A	D	C	C	B	A	A	B	A
71	**72**	**73**	**74**	**75**	**76**	**77**	**78**	**79**	**80**
D	C	B	A	A	A	A	B	B	D
81	**82**	**83**	**84**	**85**	**86**	**87**	**88**	**89**	**90**
D	C	A	B	D	D	A	D	D	A
91	**92**	**93**	**94**	**95**	**96**	**97**	**98**	**99**	**100**
A	B	B	B	A	C	B	B	A	B
101	**102**	**103**	**104**	**105**	**106**	**107**	**108**	**109**	**110**
A	A	B	D	D	C	C	D	A	D
111	**112**	**113**	**114**	**115**	**116**	**117**	**118**	**119**	**120**
D	C	D	A	B	A	B	B	B	C
121	**122**	**123**	**124**	**125**	**126**	**127**	**128**	**129**	**130**
A	A	C	D	C	C	C	C	A	D
131	**132**	**133**	**134**	**135**	**136**	**137**	**138**	**139**	**140**
A	C	D	A	B	C	B	A	B	A
141	**142**	**143**	**144**	**145**	**146**	**147**	**148**	**149**	**150**
B	B	D	C	B	A	D	A	C	C
151	**152**	**153**	**154**	**155**	**156**	**157**	**158**	**159**	**160**
D	C	D	D	A	C	A	B	C	D
161	**162**	**163**	**164**	**165**	**166**	**167**	**168**	**169**	**170**
A	C	A	D	A	A	B	B	D	B
171	**172**	**173**	**174**	**175**	**176**	**177**	**178**	**179**	**180**
A	B	C	B	B	C	B	C	D	A
181	**182**	**183**	**184**	**185**	**186**	**187**	**188**	**189**	**190**
B	A	A	A	D	B	A	B	D	D

191	192	193	194	195	196	197	198	199	200
B	C	C	C	C	B	B	D	A	A
201	**202**	**203**	**204**	**205**	**206**	**207**	**208**	**209**	**210**
D	A	B	A	A	B	A	D	B	D
211	**212**	**213**	**214**	**215**	**216**	**217**	**218**	**219**	**220**
C	B	A	C	C	C	D	A	D	B
221	**222**	**223**	**224**	**225**	**226**	**227**	**228**	**229**	**230**
C	A	D	C	B	D	D	A	A	B
231	**232**	**233**	**234**	**235**	**236**	**237**	**238**	**239**	**240**
C	D	A	D	B	B	C	D	C	D
241	**242**	**243**	**244**	**245**	**246**	**247**	**248**	**249**	**250**
A	B	A	C	B	D	A	C	B	C
251	**252**	**253**	**254**	**255**	**256**	**257**	**258**	**259**	**260**
B	B	A	B	D	D	A	D	B	B
261	**262**	**263**	**264**	**265**	**266**	**267**	**268**	**269**	**270**
A	D	D	A	C	C	C	B	B	B
271	**272**	**273**	**274**	**275**	**276**	**277**	**278**	**279**	**280**
D	B	B	D	A	C	A	B	B	A
281	**282**	**283**	**284**	**285**	**286**	**287**	**288**	**289**	**290**
C	B	B	C	D	D	A	A	D	A
291	**292**	**293**	**294**	**295**	**296**	**297**	**298**	**299**	**300**
A	A	A	D	C	C	A	D	A	C
301	**302**	**303**	**304**	**305**	**306**	**307**	**308**	**309**	**310**
A	C	B	B	C	D	B	A	C	A
311	**312**	**313**	**314**	**315**	**316**	**317**	**318**	**319**	**320**
C	C	C	A	A	A	A	A	B	D
321	**322**	**323**	**324**	**325**	**326**	**327**	**328**	**329**	**330**
B	A	D	B	A	D	C	C	D	A
331	**332**	**333**	**334**	**335**	**336**	**337**	**338**	**339**	**340**
A	C	A	C	A	B	C	A	B	D
341	**342**	**343**	**344**	**345**	**346**	**347**	**348**	**349**	**350**
A	A	C	C	B	C	C	B	B	D
351	**352**	**353**	**354**	**355**					
C	C	A	A	D					

EXPLANATORY ANSWERS

23. Current in the circuit,

$$I = \frac{250}{\sqrt{(\omega L)^2 + R^2}} \sin(\omega t + \alpha)$$

Voltage across the inductance,

$$V_2 = \frac{250\,L\omega \sin(\omega t + \alpha)}{\sqrt{(L\omega)^2 + R^2}}$$

$$\therefore \quad |V_2| = \frac{250\,L\omega}{\sqrt{(L\omega)^2 + R^2}},$$

$$\text{or,} \quad 200 = \frac{200 \times L \times 400}{\sqrt{(400L)^2 + (100)^2}}$$

or, $250000\,L^2 = 160000\,L^2 + (10000)$

$$\therefore \quad L = \frac{1}{3} H$$

29. Let ω be the resonant angular frequency, V be the applied voltage, I_L be the current through indicator and I_C be the current through the capacitor.

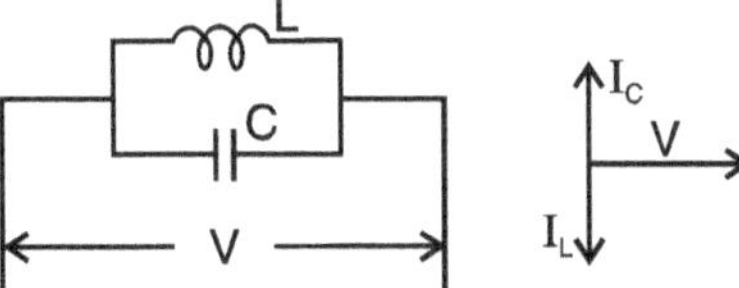

Then current taken from the supply $= I_L - I_C = 0$

circulating current, $I_L = \dfrac{V}{\omega L} = \dfrac{V\sqrt{L^{-C}}}{L}$

But, $\omega = \dfrac{1}{\sqrt{L^{-C}}}$

$\therefore$ Circulating current $= \dfrac{V}{L}\sqrt{LC} = V\sqrt{\dfrac{C}{L}}$

Hence the circulating current, $V\sqrt{\dfrac{C}{L}}$ is independent of frequency.

41. $L_1 = L_2 = L_3 = L_4$

The network can be re-drawn as :

Since all L's are equal, the points *a* and *b* are at the same potential.

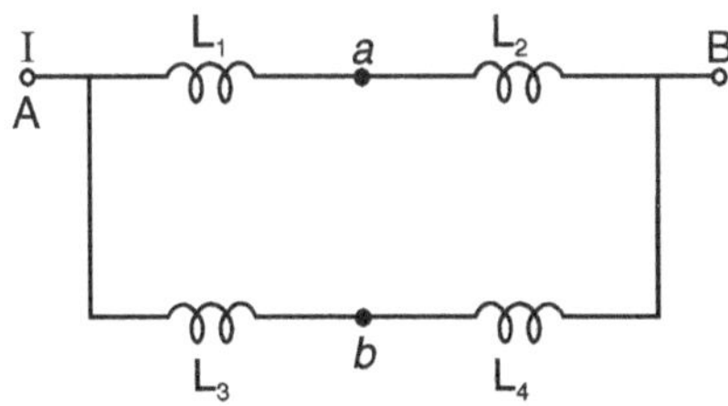

Nothing will change by going then. The current would remain I.

55. When $V_E = 30$ V and $V_A = 0$, current in R is 2A from right to left terminal of R when $V_A = 15$ V, $V_B = 0$, current in it would be 1 A from left to right terminals (the network is symmetrical).

Thus when $V_A = 15$ and $V_B = 30$ V,

Net current in R = 2 – 1 = 1 A from right to left terminal of R.

56. $E_{Thevenin}$ is found by looking into the network across A – B. It is clearly 100 V.

$R_{Thevenin}$ is found looking into A – B with voltage source replaced by a short and current source by open terminals. This resistance is easily seen to be 30 ohms.

60. $S - j\omega$

$$G(j\omega) = \frac{4(25-\omega^2)}{-\omega^2 + j25\omega + 100}$$

$$= \frac{4(25-\omega^2)}{\sqrt{(100-\omega^2)6.25\omega^2}} \angle \tan^{-1}\frac{25\omega}{100-\omega^2}$$

$= M\angle\phi$

At $\omega \to 0$, M = 1

At $\omega \to \infty$, M = 4

$$S_1\, S_2 = -2.5 \pm \sqrt{6.25 - 4.0}$$
$$= -1.25 \pm j\,9.2$$

Thus it is high pass notch filter.

64. Sum of the Δ impedances $= j5 + j5 - j5 = j5$ ohm

$$Z_1 = \frac{j5 \times j5}{j5} = j5$$

$$Z_2 = \frac{j5 \times (-j5)}{j5} = -j5$$

$$Z_3 = \frac{j5 \times (-j5)}{j5} = -j5$$

87. $\log_{10}\dfrac{P_1}{P_2} = 6$

$$\therefore \quad \frac{P_1}{P_2} = 10^{-6} \approx \Delta$$

91. For the voltage division to be independent of frequency, the impedance ratio should also be independent of frequency.

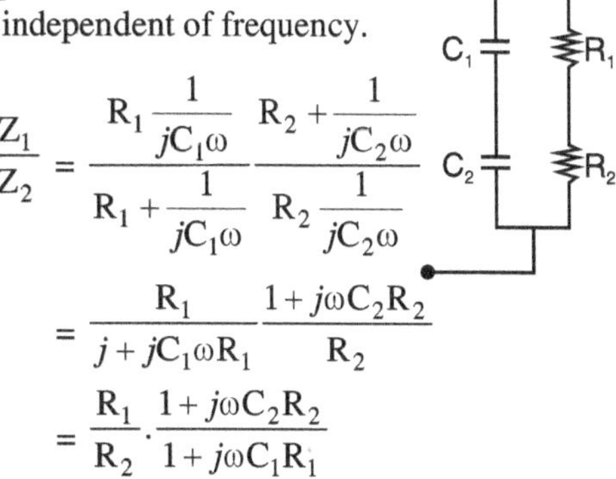

$$\frac{Z_1}{Z_2} = \frac{R_1\dfrac{1}{jC_1\omega}}{R_1 + \dfrac{1}{jC_1\omega}} \cdot \frac{R_2 + \dfrac{1}{jC_2\omega}}{R_2\dfrac{1}{jC_2\omega}}$$

$$= \frac{R_1}{j + jC_1\omega R_1} \cdot \frac{1 + j\omega C_2 R_2}{R_2}$$

$$= \frac{R_1}{R_2} \cdot \frac{1 + j\omega C_2 R_2}{1 + j\omega C_1 R_1}$$

$\therefore\ C_2R_2 = C_1R_1$

Hence $C = 10\ \mu F$

93. At high frequencies, due to skin effect, current tends to flow on the surface only and hence to reduce ac resistance, high conductivity material coating is in order.

99. Power dissipated is given by : $I^2R = W = 100$

Hence, $R = \dfrac{1000}{(10)^2} = 10$ ohms.

Impedance, $Z = \dfrac{V}{I} = \dfrac{200}{10} = 20$ ohms

Reluctance $X_L = \sqrt{Z^2 - R^2}$

$= \sqrt{(20)^2 - (10)^2} = 17.3$ ohms.

102. $V_s = AV_R + BI_R$

$I_s = CV_R + DI_R$

Voltage ratio transfer *f* is

$$\left.\frac{V_R}{V_s}\right| = \frac{1}{A}$$

$I_R = 0$

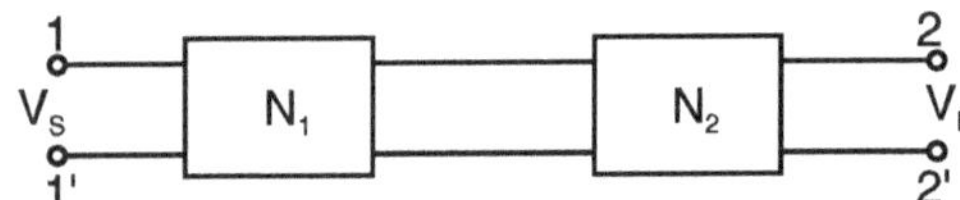

when two networks are connected in cascade, the two ABCD matrices get multiplied.

$$\begin{bmatrix} A & B \\ C & O \end{bmatrix} = \begin{bmatrix} A_1 & B_1 \\ C_1 & D_1 \end{bmatrix} \begin{bmatrix} A_1 & B_3 \\ C_2 & D_2 \end{bmatrix}$$

$$= \begin{bmatrix} A_1A_2 + B_1C_2 & A_1B_2 + B_1D_2 \\ C_1A_2 + D_1C_2 & C_1B_2 + D_1D_2 \end{bmatrix}$$

111. The impedance of the inductor for *n*th harmonic is *n* times the impedance for the fundamental. But, since the *n*th hormonic voltage is *n* times the fundamental, each harmonic gives the same current I_0

$\therefore$ Ammeter Reading $= \sqrt{I_0^2 + I_0^2 + I_0^2 + I_0^2} = 2\,I_0$

112. Electric field, $E = -(V \times B) = -(2i + 3j + k) \times (i + 2j)$ and $|\overline{E}| = (E_x^{\,2} + E_y^{\,2} + E_z^{\,2})^{1/2} = \sqrt{6}$ volt/m

$\therefore$ Potential difference across 1 m length

$= \sqrt{6}$ volt.

114. $\Psi(t) = f(t) + f(-t)$

Replace *t* by $(-t)$, $\Psi(t) = f(-t) + f(t) = \psi(t)$

Hence even.

116. $i = e^{-t/RC}$

$$V_c(t) = \frac{1}{C}\int_0^t i\,dt$$

$$= \frac{V}{RV} = \left|-RC.e^{-t/RC}\right|_0^t$$

$= V(1 - e^{t/RC})$ for $v = u(t)$

$V_c(t) = 1 - e^{-t/RC}$

119. The fourier transform pair is

$$(j\omega) = \int_{-\infty}^{\infty} f(t)e^{-j\omega t}dt$$

and $f(t) = \dfrac{1}{2\pi}\displaystyle\int_{-\infty}^{\infty} F(j\omega)e^{-j\omega t}d\omega$

123. $$1 = \frac{1}{4}.\frac{1}{s + \frac{1}{8}}(R + sL)$$

or $4s + 2 = 2 = R + sL$

$\therefore$ $R = 2$ and $L = 4$

150. After the switching action, the impressed voltage suddenly becomes

$(V + V) = 2V.$

For this new configuration, current $i = \dfrac{V}{R}$

and when steady-state is reached, current

$i = 2\dfrac{V}{R}$

Therefore, the appropriate expression for the currrent is

$$i = \frac{V}{R}\left(2 - e^{\frac{Rt}{L}}\right)$$

155. Replacing the voltage sources into current sources, and simplifying we get

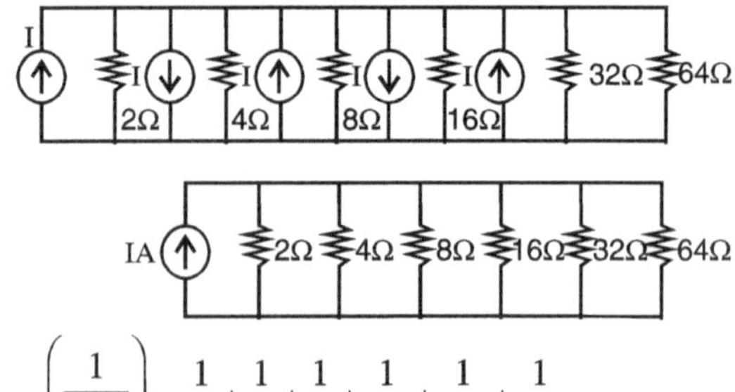

$$\left(\frac{1}{R_{eq}}\right) = \frac{1}{2} + \frac{1}{4} + \frac{1}{8} + \frac{1}{16} + \frac{1}{32} + \frac{1}{64}$$

$$= \frac{(32+16+8+4+2+1)}{64}$$

$$= \frac{63}{64}$$

$\therefore$ Equivalent resistance, $(R_{eq}) = \left(\frac{64}{63}\right)\Omega$

Voltage at B $= 1 \times \frac{64}{63} = \frac{64}{63}$ volt

159. In two wattmeter method of power measurement in balanced three-phase circular, the readings ω_1 and ω_2 are proportional to $\cos(\theta - 30°)$ and $\cos(\theta + 30°)$, where θ is the power factor angle of the load. Therefore, if θ is 60°, i.e. if load p.f. is 0.5, one of the wattmeters shall read zero.

162. The current that flows is given by

$$v(s) = \text{SL}\, I(s) = \text{S}\, I(s)$$

$$Is = \frac{1}{s} = u(t) = 1 \text{ A}$$

Energy supplied $= \frac{1}{2}\text{LI}^2$

$$= \frac{1}{2} \times 1 \times 1 = \frac{1}{2}\text{J}$$

163. The current I_1 flowing in the 10 Ω resistor and voltage source in series = 10 – 2 = 8A.

$\therefore$ Voltage across resistor, R = 40 + 10 × 8 = 120 V.

$$\therefore \quad R = \frac{120}{2} = 60\,\Omega$$

165. No continuous d.c. can flow through the circuit because of the series capacitor. Similary at infinite frequency, the inductor becomes an open circuit and no current can flow. Hence $z(s)$ has poles at 0 and ∞.

168. This question concerns initial condition on an inductive circuit. The equivalent voltage is $L(i_0)$. Now when the switch is in position 1 for a long time, the inductive current $= \frac{v}{R} = \frac{1}{2}$A. The direction must be as shown in figure.

Hence, choice is circuit (B).

172. Connect $v(s)$ to the network :

$$I_3(s) = \frac{v(s)}{R}, I_4(s) = \frac{v(s)}{sL}, I_2(s) = v(s)$$

$$I_1(s) = I_2(s) + I_3(s) + I_4(s)$$

$$= v(s)\left[s + 1 + + \frac{1}{s}\right] = \frac{v(s)}{s}\left(s^2 + s + 1\right)$$

$$\frac{I_2(s)}{I_1(s)} = \frac{s^2}{\left(s^2 + s + 1\right)}$$

183. $$V_A(s) = \frac{I_2(s)}{s}$$

$$I_1(s) = \frac{V_A(s)}{s}\frac{I_2(s)}{s^2}$$

$$I(s) = I_2(s) + I_1(s)$$

$$= I_2(s)\left(1 + \frac{1}{s^2}\right)$$

$$V_1(s) = V_A(s) + I(s) \times 1$$

$$= \frac{I_2(s)}{3} + I_2(s).\frac{1+s^2}{s^2}$$

$$\frac{I_{2(s)}}{V_{1(s)}(s)} = \frac{s^2}{s^2 + s + 1}$$

$$I_2(s) = \left[\frac{s^2 + s + 1}{s^2}\right]$$

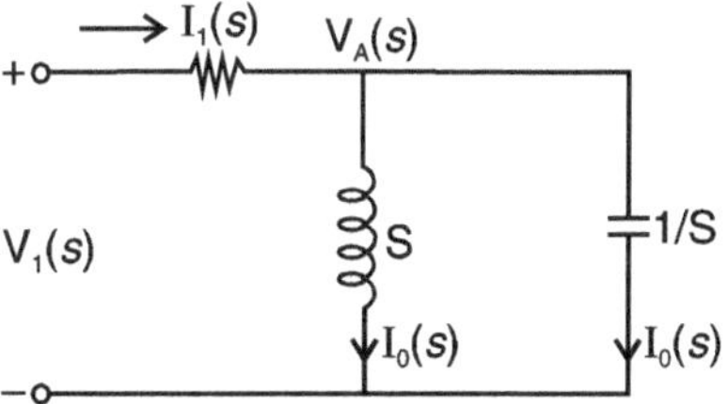

188. Before closing the switch, the charge on the capacitor is

$$Q = C \times V$$

$$= 4 \times 24 \times 10^{-6} = 96 \times 10^{-6}\,\text{C}$$

After closing the switch the charge remains same, but the total capacitance is increased to (4 + 2)

$$= 6\,\mu\text{F}$$

If V is the voltage across the combination,

$$Q = C \times V$$

or 96×10^{-6} volt $= 6 \times 10^{-6} \times V$

$$\therefore \quad V = \frac{96}{6} = 16\text{ V}$$

190. While synthesizing this function in Cauer form, the first element is seen to be an inductor of 2 H. Hence the network is at (D).

193. (A) $\dfrac{s^2 - s + 4}{s^2 + s + 4}$ has pole-zero plot as shown. It is a non-minimum phase function : (2)

(B) $\dfrac{s+4}{s^2 + 6s + 4}$ has pole in the right half plane and hence unstable : 4

(C) $\dfrac{s+4}{s^2 + 6s + 5}$ has poles and zero alternate on the negative real axis and hence an R – C impedance function : (3)

(D) $\dfrac{s^3 + 3s}{s^4 + 2s^2 + 1}$ is an L – C function : (1) Thus closest answer is (C).

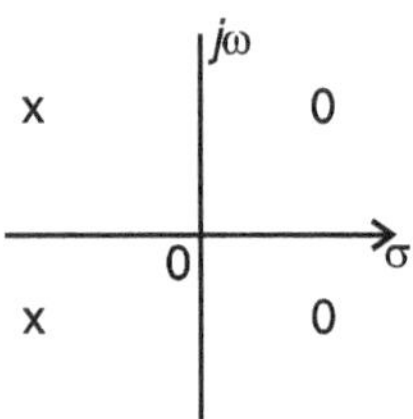

195. $$R_0 = \sqrt{R_{oc} R_{sc}}$$

where $R_{oc} = \dfrac{R_1 + R_2}{2}$ and $R_{sc} = 2 \times \dfrac{R_1 R_2}{R_1 + R_2}$

$\therefore\ R_{Oc}\, R_{sc} = R_1\, R_2$

Thus $\quad R_o = \sqrt{R_1 R_2}$

198. $$\frac{L}{C} = R_L{}^2 = 1$$

$$\omega_c = \frac{2}{\sqrt{LC}} \text{ (Assuming that given } \omega \text{ is } \omega_c)$$

Thus $\dfrac{4}{LC} = 1$

or $\quad \dfrac{4}{C^2} = 1$

or $\quad C = 2F$

$\therefore \quad L = 2H$

201. Since the coils are connected in series opposition, the equivalent impedance

$= Z_1 + Z_2 - 2\,M$

$$X_m = \sqrt{8j\,.8j} = 8j$$

or $\quad Z_1 = 5 + j8 + 3 + j8 - 2 \times j8 = 8 + j0$

204. $\omega L_A - \dfrac{1}{\omega C} = -R$ ...*(i)*

and $\omega L_B - \dfrac{1}{\omega C} = R$ ...*(ii)*

206. $$f_{01} = f_0 - \frac{1}{2} f_{bw}$$

$$f_{02} = f_0 + \frac{1}{2} f_{bw}$$

Thus, $\quad f_0 = \dfrac{f_{01} + f_{o2}}{2}$

211. Voltage source is shorted

$$Z^{th} = \frac{(3+j4)(-j5)}{3-j} = \frac{-15j + 20}{3-j} = 7.5 - j2.5$$

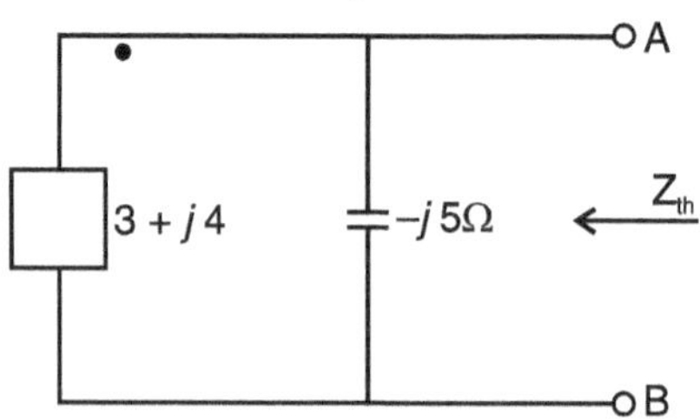

Hence load impedance should be conjugate of this for maximum power transfer.

$$Z_2 = 7.5 + 2.5\,j$$

213.

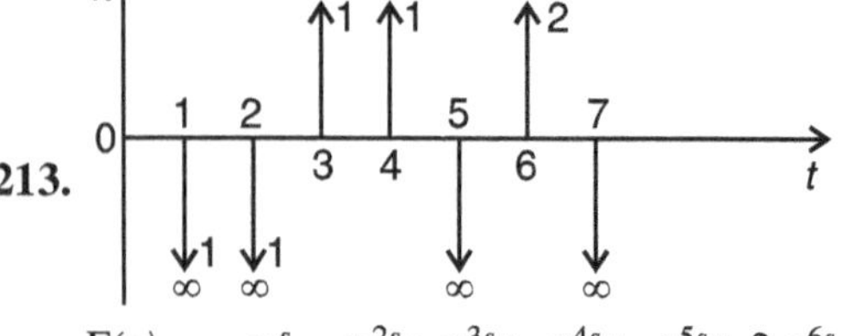

$F(s) = -\,e^{-s} - e^{-2s} + e^{-3s} + e^{-4s} + e^{-5s} + 2e^{-6s} - e^{-7s}$

$$L(f(t)) = F(s) : \int f(t)\,dt = \frac{F(s)}{s}$$

$$\int\int f(t)\,dt = \frac{F(s)}{s^2}$$

$$= \frac{-e^{-s} - e^{-2s} + e^{-3s} + e^{-4s} - e^{-5s} + 2e^{-6s} - e^{7s}}{s^2}$$

$$\Psi(t) = L^{-1}\left[\frac{-e^{-s} - e^{-2s} + e^{-3s} + e^{-4s} - e^{-5s} + 2e^{-6s} - e^{7s}}{s^2}\right]$$

$= -tu(t-1) - tu(t-2) + tu(t-3) + tu(t-4) - tu(t-5) + tu(t-6) - tu(t-7)$

219. Norton's equivalent circuit for circuit is shown in the figure :

$$I_N = \frac{2}{3 + \dfrac{2\times\left(\dfrac{4}{5}\right)}{2+\left(\dfrac{4}{5}\right)}} \times \frac{2}{\left(2+\dfrac{4}{5}\right)}$$

$$= \frac{2}{3+\dfrac{8}{14}} \times \frac{2\times 5}{14} = \frac{2}{5}\text{A}$$

For impedance the circuit is shown below :

$$Z_N = \frac{3\times 2}{3+2} + \frac{4}{5} = 2\,\Omega$$

Norton equivalent circuit is :

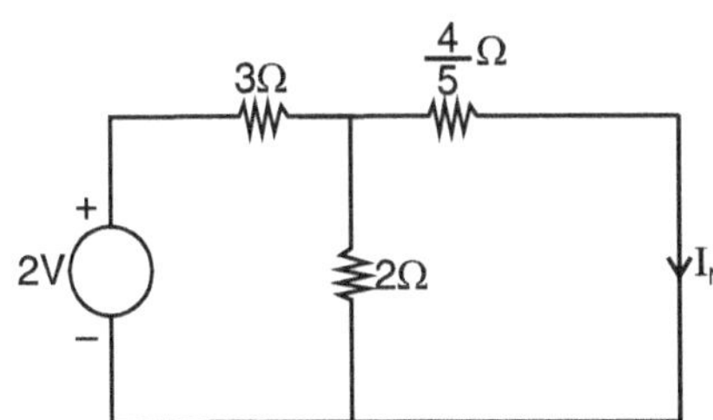

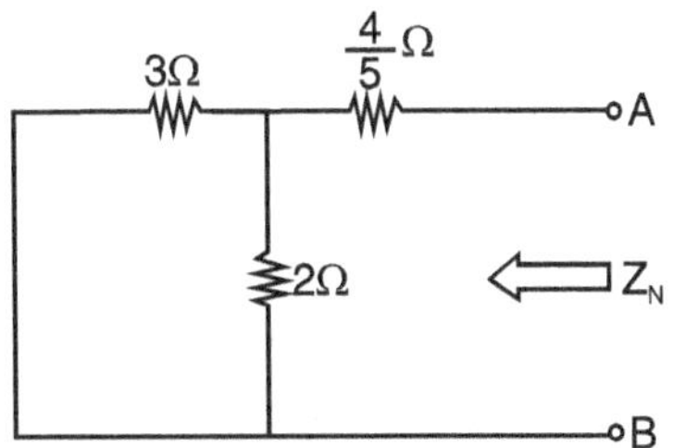

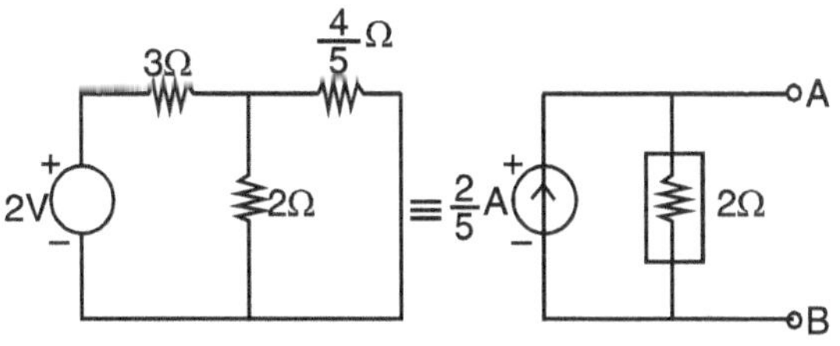

222. The equivalent circuit is shown in the figure below.

Consider section AB which consists of two 1000 ohm resistor across 10 V supply. Hence $V_1 = 5$ V.

In section CD, $V_2 = \dfrac{1020}{1000+1020} \times 10 = \dfrac{510}{101}$

$\therefore \quad V_0 = V_1 - V_2 = 5 - \dfrac{501}{101} = \dfrac{5}{101}$

$= 0.0459$ V, i.e., 49.55 mV.

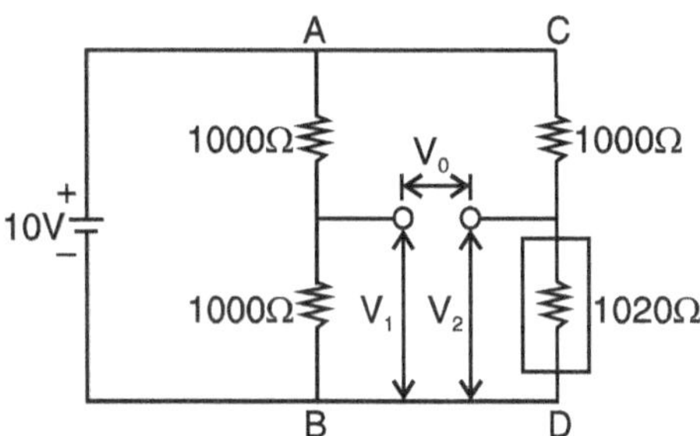

227. The bridge is in balanced condition. If all the resistances are increased by the same factor the bridge will remain balanced. Hence V_{out} is zero.

230. Assuming that every sources are dc sources, the inductors will appear as elements of zero resistance.

$\therefore$ current through 1 H inductor $= \dfrac{3}{1} = 3$ A.

231. Laplace transform of responce function (t),

$R(s) = H(s)\,E(s)$

where $H(s)$ = System function or Laplace transform of impedance responce.

$E(s)$ = Laplace transform of excitation function $e(t)$

According to convolution theorem,

if $\quad F(s) = F_1(s).F_2(s)$

$$f(t) = \int_0^t f_1(t-z)\,f_2(z)\,dz$$

$= f_1(t-z)*f_2(z)\,dz$

$$f(t) = \int_0^t f_1(z) f_2(t-z)dz$$

$= f_1(z)*f_2(t-z)$

236. Current, $i(t) = \dfrac{100}{1000}e^{-t/10^3\times10^{-6}} = 0.1\ e^{-1000t}$

$$\left.\frac{di(t)}{dt}\right|_{t=0} = -0.1\times10^3\times1$$

$= -100$ amp/sec

246. Let the current in 10 Ω by y.

Applying Kirchoff's law to the node A

$Y + 2x = 5x$

or, $Y = (5x - x - 2) = (4x - 2) = (12 - 2) = 10$

Power delivered by the dependent current source $= (10y)(5x) = 5\times3\times10\times10 = 15000$ watts.

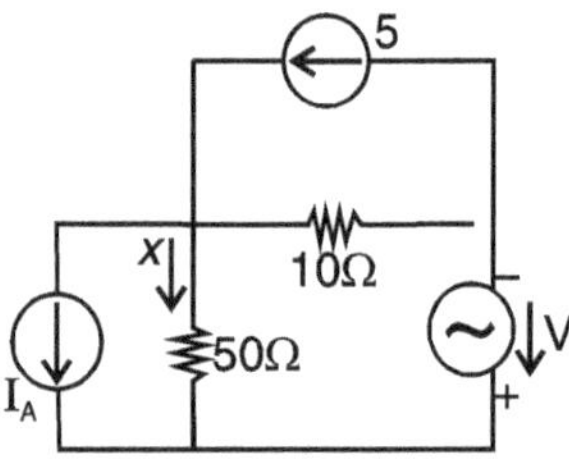

249. $$\frac{V_0(s)}{V_1(s)} = \frac{1/C_s}{R+1/C_s} = \frac{1}{1+RC_s}$$

250. In the phasor diagram shown, ϕ is the phase angle of branch 2.

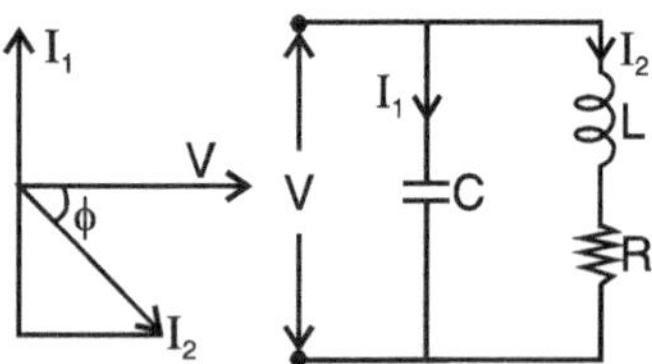

At resonance, $I_1 = I_2 \sin\phi$

$$VC\omega = \frac{V}{Z}\frac{X}{Z}$$

(where $X = L\omega$ and $Z = \sqrt{R^2 + L^2\omega^2}$

$$= \frac{VL\omega}{Z^2}$$

$\therefore \quad Z^2 = \dfrac{L}{C}$

Also, Net current $= I_2 \cos\phi$

$$= \frac{V}{Z}\frac{R}{Z} = \frac{VR}{Z^2}$$

$\therefore$ At resonance dynamic impedance,

$$Z_e = \frac{V}{I_2\cos\phi} = \frac{Z^2}{R} = \frac{L}{CR}$$

257. Output at 1st OPAM $= -\left(V_1\dfrac{20}{10} + V_2\dfrac{20}{20}\right)$

$= -(2V_1 + V_2)$

Final output $= -(2V_1 + V_2)\dfrac{10}{10} + V_3\dfrac{10}{5}$

$= 2V_1 + V_2 - 2V_3$

276. Charging time constant, $R_1 C$

$= 1000 \times 0.5 \times 10^{-6}$

$= 0.5 \times 10^{-3}s$

While discharging, the circuit consists of R_2 and two capacitors 0.5 microfarad in series.

Equivalent capacitance of these capacitors is C

$= \dfrac{0.5}{2}\ \mu F = 0.25 \times 10^{-6}$ F.

Hence discharging time constant, $R = C_2$ Ceq.

$= 1000 \times 0.25 \times 10^{-6} = 0.25 \times 10^{-3}s$.

281. The voltage across and current through the resistor of 1 ohm is constrained by the voltage source connected directly across it, i.e., 5 V and 5 A

284. Curve-1 is undamped sinusoid and therefore corresponds to poles (p, p_1^*). Curve-2 is a damped sinusoid and, therefore corresponds to poles $(p_3 = p_3^*)$

Curves-3 and 4 are exponentially decreasing curve 3 decays slowly than curve-4.

Therefore, curves 3 corresponds to p_2 and curve 4 to p_4.

Hence (1, 2, 3, 4) is correct.

286. The resistive part of the circuit is equivalent to 2 kΩ.

Time constant is given by, Req $= 2000 \times 0.1 \times 10^{-6} = 0.2$ milli second.

294. Source voltage = $\sqrt{3^2 + (14-10)^2} = 5V$

295. By superposition theorem.

Current in R_L due to 2 volt source

$$= \frac{2}{1.5} \times \frac{1}{2} = 0.67 \text{ A}$$

Current in R_L dut to 1 ampere sorce

$$= \frac{1}{2} \times 0.67 = 0.33 \text{ A}$$

Current in R_L due to 1 volt source

$= \frac{1}{1.5}$ = 0.67 A in the opposite direction.

Net current in $R_L = 0.67 + 0.33 - 0.67 = 0.33$ A.

2

DIGITAL SYSTEMS AND MICROPROCESSOR ENGINEERING

IMPORTANT TERMS

1. Each digit in the octal number system has a place value. As it is a base 8 system the place values or weights starting from ISB towards MSB will be $8^0, 8^1, 8^2, 8^3$.... for the integer part and $8^{-1}, 8^{-2}, 8^{-3}, 8^{-4}$... for the fractional part.
2. HEXADECIMAL NUMBER SYSTEM is a base-16 or radix-16 number system as it has 16 different digits. These digits are 0, 1, 2, 3, 4, 5, 6, 7, 8, 9, A, B, C, D, E, F.

The Excess-3 code, another important BCD code, is an unweighted code. The Excess-3 code is particularly significant for arithmetic operations. When the sum of two decimal digits exceeds 9, they cannot be added using 8421 BCD code, Excess-3 code has no such limitation and considerably simplifies arithmetic operations.

Gray Code

It is another important unweighted BCD code and its applications arise from the fact that each number in the gray code differs from its immediately preceding number by one bit only.

Alphanumeric Codes

The binary codes that can be used to represent all the letters of alphabet, numbers and mathematical symbols, punctuation marks are known as the ALPHANUMERIC CODES. Alphanumeric codes enable us interface the input-output devices like keyboard, punched cards, printers, video display units, etc. with the computer. One such alphanumeric code that is extensively used in microcomputers for interfacing microcomputers with keyboards, printers, video display units is the ASCII (American Standard Code for Information Intercharge) code pronounced as "ask-ee". It has become a world standard Alphanumeric Code for microcomputers. It is a 7-bit code that can be used to represent 26 upper case letters (A to Z), 26 lower case letters (a to z), 10 numerals (0 to 9), 33 special characters and symbols and 33 control characters. It comes to a total of 128 characters and that is what a 7-bit code at the most can represent $2^7 = 128$. An 8-bit version of ASCII codes known as USASC-11-8 or ASCII-8 has also been developed. The 8-bit version can represent a maximum of 256 characters.

EBCDIC (Extended Binary Coded Decimal Interchange Code) pronounced as "eb-si dik" is another widely used Alphanumeric code mainly popular with large computer systems. It is an 8 bit code and thus can accommodate upto 256 characters. Gives the listing of the characters in binary as well as Hex form in EBCDIC. The arrangement is similar to one adopted in the ASCII code.

BINARY LOGIC GATES

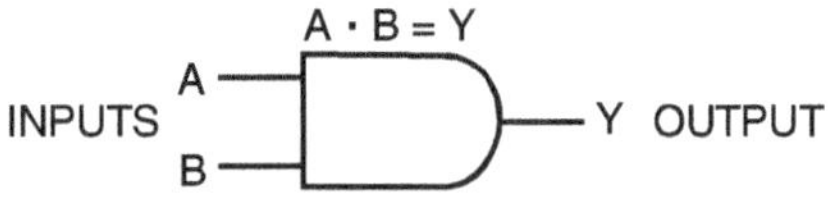

A · B = Y

INPUTS A, B — Y OUTPUT

(a)

	INPUTS		OUTPUT
	A	B	Y
	0	0	0
	0	1	0
	1	0	0
ALL →	1	1	1

(b)

The AND gate. (a) Logic symbol. (b) Truth table with binary 1 for HIGH level. All inputs must be 1 for output of 1.

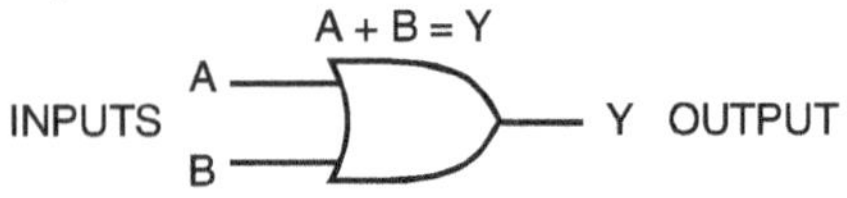

A + B = Y

INPUTS A, B — Y OUTPUT

(a)

	INPUTS		OUTPUT
	A	B	Y
	0	0	0
ANY	0	1	1
ANY	1	0	1
ANY	1	1	1

(b)

The OR gate. (a) Logic symbol. (b) Truth Any input at 1 produces output of 1.

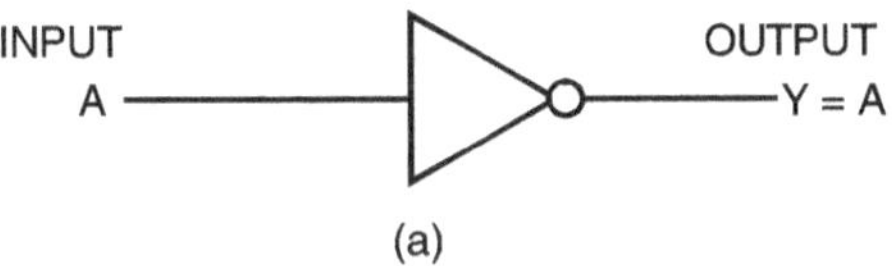

(a)

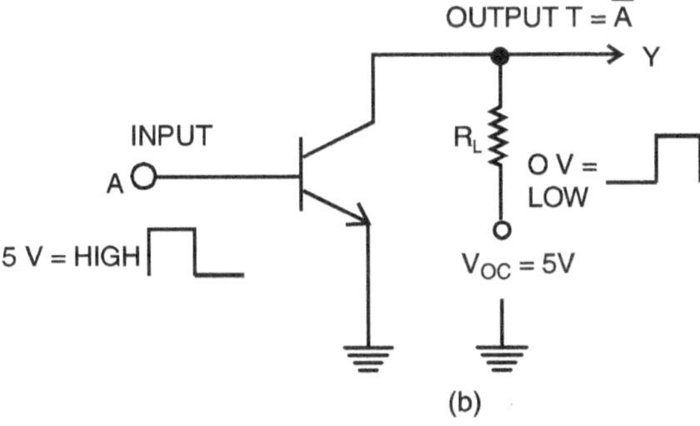

(b)

(a) Logic symbol for an inverter. Note circle or bubble at inverted output. The bar over the A indicates inverted A, NOT A, or complement of A. (b) Common-emitter circuit as an inverter.

$\overline{AB} = Y$

INPUTS A B — Y OUTPUT

(a)

	INPUTS		OUTPUT	
	A	B	AND	NAND
	0	0	0	1
	0	1	0	1
	1	0	0	1
ALL	1	1	1	0

(b)

The NAND gate. (a) Logic symbol for NAND gate with bubble at output. (b) Truth table for AND with NAND functions. Note that the NAND output is the complement of the AND output.

A + B = Y

INPUTS A B — Y OUTPUT

(a)

INPUTS		OUTPUT
A	B	Y
0	0	1
0	1	0
1	0	0
1	1	0

(b)

The NOR gate. (a) Logic symbol with bubble at output. (b) Truth table.

A + B = Y

INPUTS A B — Y OUTPUT

(a)

INPUTS		OUTPUT
A	B	Y
0	0	0
0	1	1
1	0	1
1	1	0

(b)

Exclusive-OR gate. (a) Logic symbol. (b) Truth table.

$\overline{A + B} = Y$

INPUTS A B — Y OUTPUT

(a)

INPUTS		OUTPUT
A	B	Y
0	0	1
0	1	0
1	0	0
1	1	1

(b)

Exclusive-NOR gate. (a) Logic symbol with bubble at output. (b) Truth table.

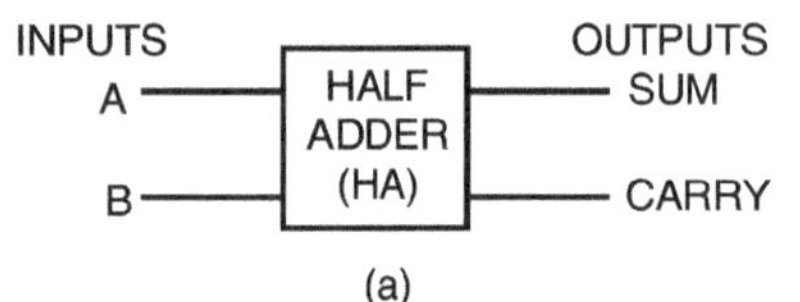

(a)

INPUTS		OUTPUTS	
B	A	SUM	CARRY
0	0	0	0
0	1	1	0
1	0	1	0
1	1	0	1
DIGITS TO BE ADDED		XOR	AND

(b)

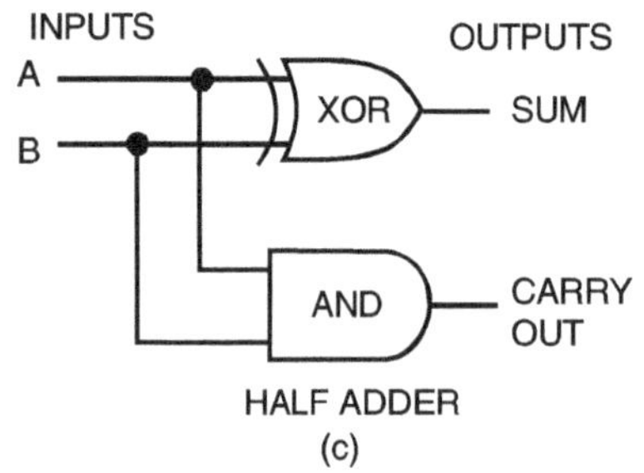

HALF ADDER
(c)

(a) Half adder,
(b) Truth table
(c) Connections of XOR and AND gates

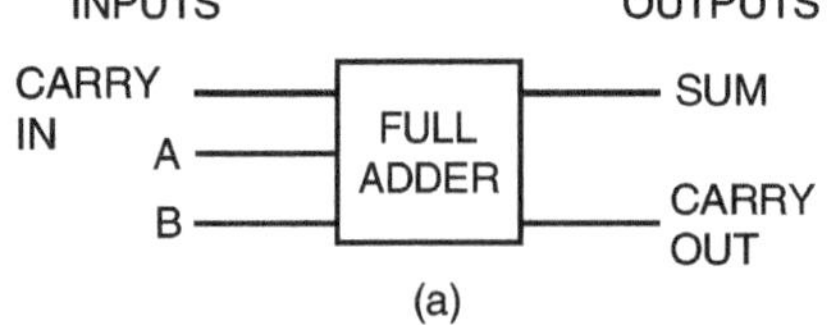

(a)

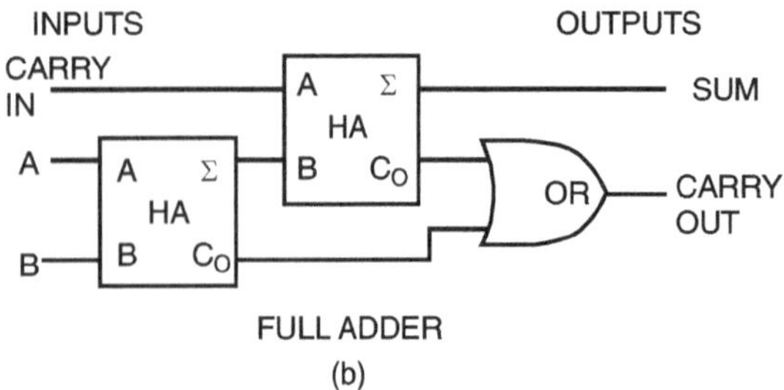

FULL ADDER
(b)

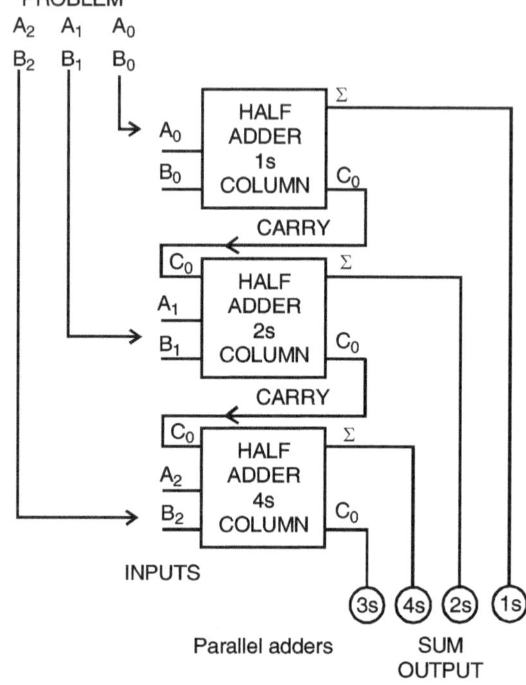

Parallel adders

BOOLEAN ALGEBRA

Boolean Algebra was developed by a ninteenth century mathematician, G. Boole. The variables used in the algebraic system occur in one of two possible which makes the system useful when applied to switching circuits such as AND and gates and other similar devices which operate with their inputs and outputs set at logical 1 or logical 0. Only these two states are possible. This system of algebra is based on three operations.

AND represented by .
OR represented by +
NOT represented by — or'

The following is least of some of the relationships in Boolean algebra.

$A.1 = A$
$A.0 = 0$
$A + 1 = 1$
$A + 0 = A$

These are the zero and unit rules. Note that the 0 is dominant in the AND operation while the 1 is dominant in the OR operation.

$A.\overline{A} = 0$

$A + \overline{A} = 1$

$\overline{\overline{A}} = A$

These are the complement relations.

$A + A = A$

$A.A = A$

This is the property of idempotence.

$A + B = B + A$

$A.B = B.A$

The system is commutative.

$A + A.B = A$

$A(A+B) = A$

$A + \overline{A}.B = A + B$

These are absorption rules.

$A(B+C) = A.B + A.C$

$A + B.C = (A+B)\,(A+C)$

The preceding two rules show that the system is distributive.

$A+B+C = (A+B) + C = A + (B+C)$

$A.B.C = A(B.C) = (A.B).C$

The system is associative.

$$\overline{A+B+C+D+E} = \overline{A}.\overline{B}.\overline{C}.\overline{D}.\overline{E}$$

$$\overline{A.B.C.D.E} = \overline{A} + \overline{B} + \overline{C} + \overline{D} + \overline{E}$$

The preceeding two relationships are known as DeMorgan's laws and show that an AND expression may be proved correct by drawing truth tables for both sides of the identities.

KARNAUGH MAP

A Karnaugh map is a pictorial device which aids in minimizing a Boolean function. Each cell on a Karnaugh map corresponds to a minterm (or a maxterm), and there exists cell for every minterm (or maxterm). Thus, the map is simply a rearranged truth table. Examine the map below :

As indicated cell, if it has a 1 in it, would indicate that the minterm $A.\overline{B}.\overline{C}.\overline{D}$ be included in the canonical minterm form of the function. (Alternatively, exterm $\overline{A}+B+C+D$ would appear in canonical maxterm form if all contained a zero).

CD \ AB	00	01	11	10	
00	0	0	0	1	
01	0	0	0	0	← $A.\overline{B}.\overline{C}.\overline{D}$
11	0	0	0	0	
10	0	0	0	0	

Notice however, that the coordinates of each cell appear in specific patterns. By Gray codes, and the binary input combinations corresponding to adjacent differ in only one bit (Recall that the defining property of a Gray code is that boring codes differ in only one bit). The adjaceny may be horizontal or vertical, see the following.

Preceding function described is $A.\overline{B}.\overline{C}.\overline{D}+A.\overline{B}.\overline{C}.D$ (This is the canonical term form).

By the rules of Boolean algebra this may be simplified to.

CD \ AB	00	01	11	10	
00	0	0	0	(1)	← $A.\overline{B}.\overline{C}.\overline{D}$
01	0	0	0	(1)	← $A.\overline{B}.\overline{C}.\overline{D}$
11	0	0	0	0	
10	0	0	0	0	

$A.\overline{B}.\overline{C}.\left(\overline{D}+D\right)$

$= A.\overline{B}.\overline{C}$ since $\left(\overline{D}+D\right) = 11$

This is the minimum sum of products form.

This function is minimized, requiring just one AND gate, rather than two AND gates and an OR gate. Furthermore, the AND gate need not have as many input parameters. Consider another example.

Minterms will combine to make a smaller but entirely equivalent function. If they are horizontally adjacent too. The function above is

$$A.\overline{B}.\overline{C}.\overline{D}+A.B.\overline{C}\overline{D}=A.\overline{C}.D.\left(\overline{B}+B\right)=A.\overline{C}.\overline{D}.$$

CD \ AB	00	01	11	10
00	0	0	1	1
01	0	0	0	0
11	0	0	0	0
10	0	0	0	0

The maps manifest wrap around, where the left hand edge is adjacent to the right-hand edge. Also the top of the map is adjacent to the bottom. See the following example :

CD \ AB	00	01	11	10
00	0	0	0	0
01	1	0	0	1
11	0	0	0	0
10	0	0	0	0

$$\overline{A}.\overline{B}.\overline{C}.D+A.\overline{B}.\overline{C}.D=\overline{B}.\overline{C}.D.$$

It is also easy to demonstrate the groups of two, four, eight and sixteen minterms will combine.

$$A.B.\overline{C}.D+A.\overline{B}.\overline{C}.D+A.B.\overline{C}.\overline{D}.+A.\overline{B}.\overline{C}.\overline{D}.$$

$$=A.\overline{C}.(B.D+\overline{B}.D+.B.\overline{D}.+\overline{B}.\overline{D})$$

$$=A.\overline{C}[B.(\overline{D}+D)+\overline{B}.(\overline{D}+D)]$$

$$=A.\overline{C}.(\overline{B}+B)$$

$$=A.\overline{C}.$$

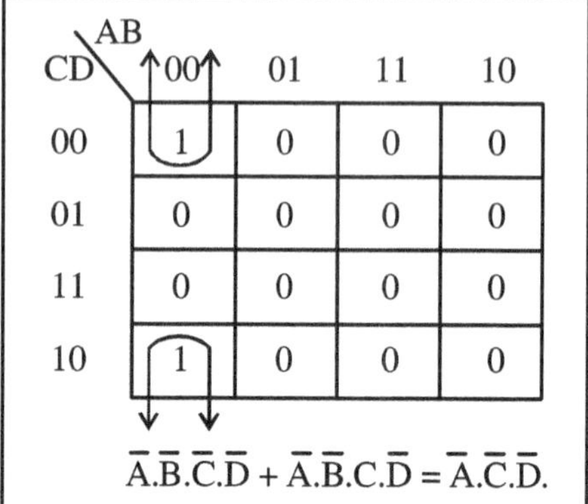

$$\overline{A}.\overline{B}.\overline{C}.\overline{D}+\overline{A}.\overline{B}.C.\overline{D}=\overline{A}.\overline{C}.\overline{D}.$$

All four corners are also adjacent.

$$\overline{A}.\overline{B}.\overline{C}.\overline{D}+A.\overline{B}.\overline{C}.\overline{D}+\overline{A}.\overline{B}.C.\overline{D}+A.\overline{B}.C.\overline{D}$$

$$=\overline{B}.\overline{D}.[\overline{C}.(\overline{A}+A)+C.(\overline{A}+A)]$$

$$=\overline{B}.\overline{D}.(\overline{C}+C)$$

$$=\overline{B}.\overline{D}$$

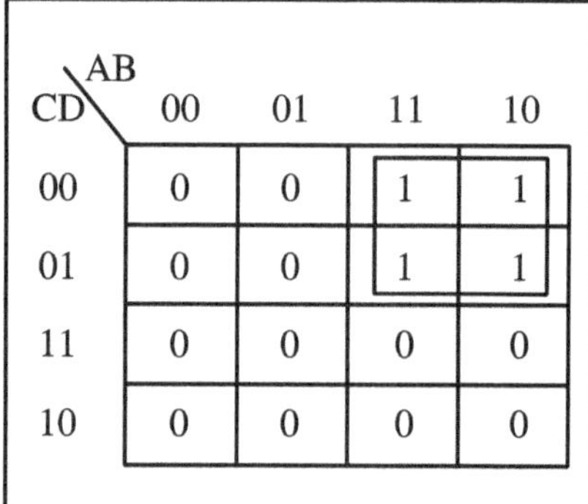

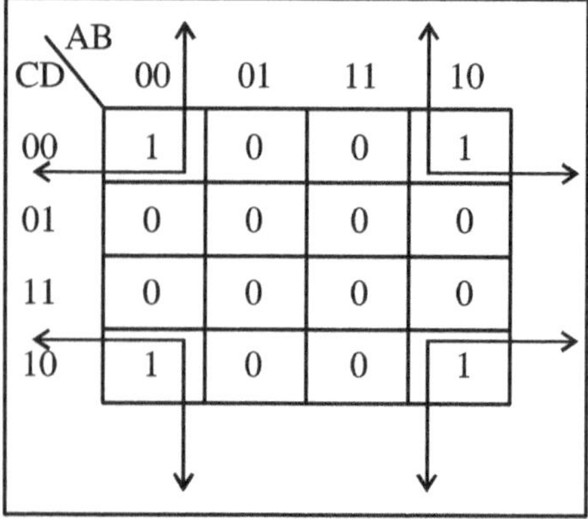

It is not necessary to perform the algebra. A Karnaugh map is read by drawing a loop around 1s, observing which variable (s) change over loop; and writing down directly the minimized Boolean sum of product expression. A minimum product of sums form can be produced by looping Os.

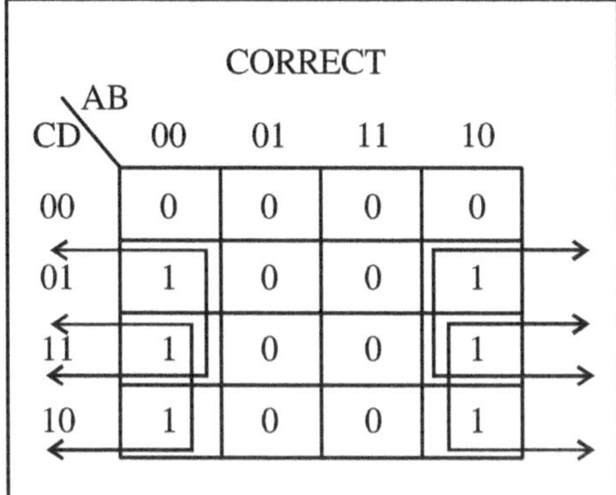

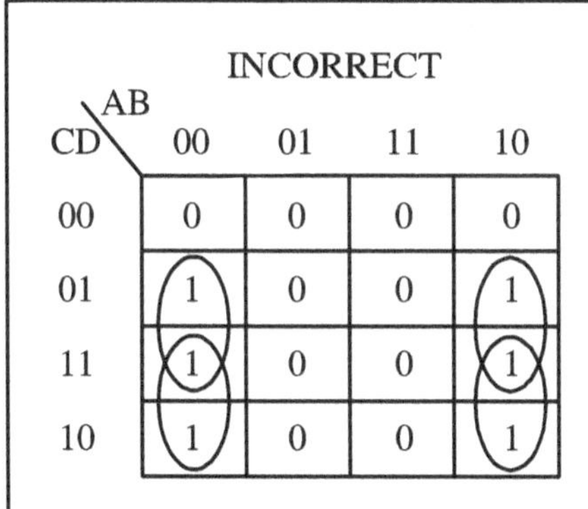

In realizing a Boolean switching function, it is advantageous for many practical reasons (such as cost space complexity of construction power dissipation, propagation delay and reliability) to minimize the function, thereby generating all its zeros and ones with the smallest number of gates each of which has the smallest number of inputs. In terms of the Karnaugh map, this corresponds to covering all the ones (zeros if working with maxterms) at least once but with the smallest possible number of loops. Since each loop generates a term the number of terms will be minimized if the number of loops is minimized. In addition, because more variables are eliminated from a term if a group is as large as possible it behaves the designer also the maximize the size of the loop. Loops may cover 1, 2, 4, 2, 16 etc. cells, but the shape is important. The dimensions of a loop must be $2^a \times 2^b$ where a and b are integers > 0. Diaganal loops or loops of 3, 5, 6, 7, 9, 11 the minterms (or maxterms) they cover will not combine into a smaller expression.

Some examples of functions and their minimizations appear in the following together with common errors.

Correct looping : Canonical minterm form

$\overline{A}.\overline{B}.\overline{C}.D.+\overline{A}.\overline{B}.C.D.+\overline{A}.\overline{B}.C.\overline{D}.+A.\overline{B}.\overline{C}.D.$

$+A.\overline{B}.C.D.+A.\overline{B}.C.\overline{D}$

Correct looping : Minimal sun of the products form

$\overline{B}.D.+\overline{B}.C.$

The incorrect looping would be read as :

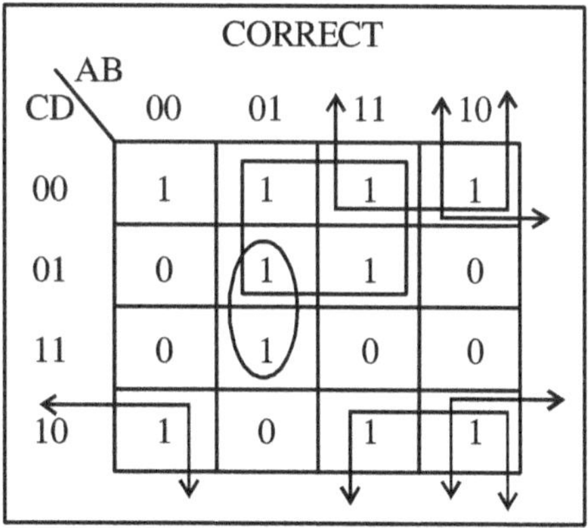

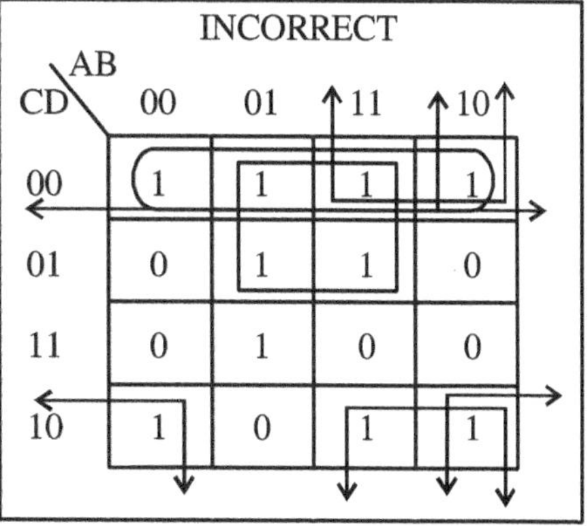

$\overline{A}.\overline{B}.C+\overline{A}.\overline{B}.D+A.\overline{B}.C+A.\overline{B}.D$ which is clearly not minimal.

Correct looping : Minimal sum of products form.

$\overline{A}.\overline{D}+B.\overline{C}+A.\overline{D}+\overline{A}.B.D$

Incorrect looping : Includes a redundant loop $(\overline{C}.\overline{D})$ resulting in an additional term.

$\overline{A}.\overline{D}+B.\overline{C}+A.\overline{D}+\overline{A}.B.D+\overline{C}.\overline{D}$ (Incorrect : expression too big).

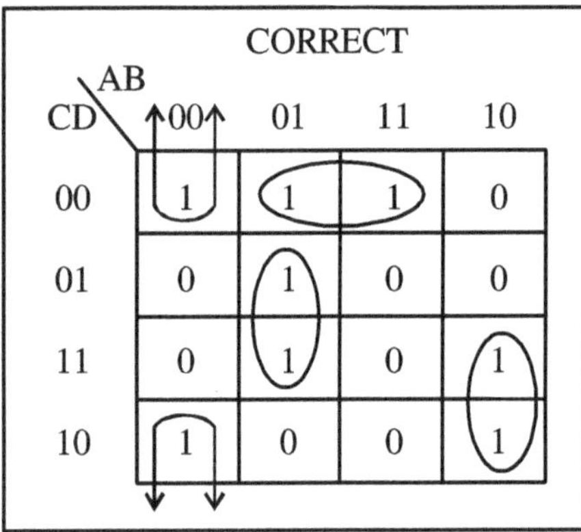

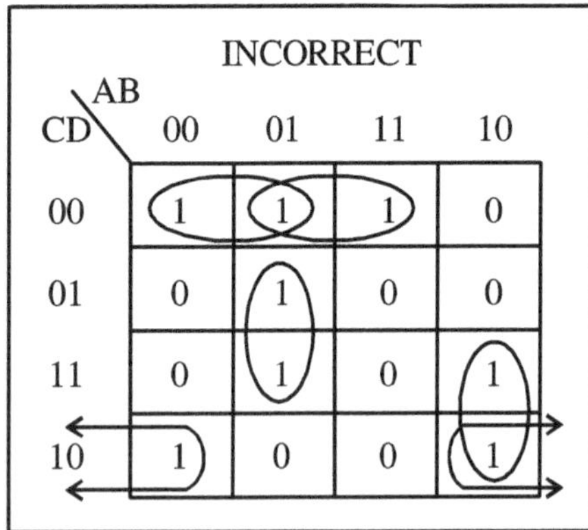

Correct : $\overline{A}.\overline{B}.\overline{D}+B.\overline{C}.\overline{D}+\overline{A}.B.D+A.\overline{B}.C$

Incorrect :

$\overline{A}.\overline{C}.\overline{D}+B.\overline{C}.\overline{D}+\overline{A}.B.D+A.\overline{B}.C+\overline{B}.C.\overline{D}$

CORRECT

CD \ AB	00	01	11	10
00	0	0	0	0
01	1	1	1	1
11	0	1	1	0
10	0	0	0	1

INCORRECT

CD \ AB	00	01	11	10
00	0	0	0	0
01	1	1	1	1
11	0	1	1	0
10	0	0	0	1

Correct : $\overline{C}D+\overline{B}.D$

Incorrect : $\overline{C}D+B.C.D$

One final maxterm example follows :

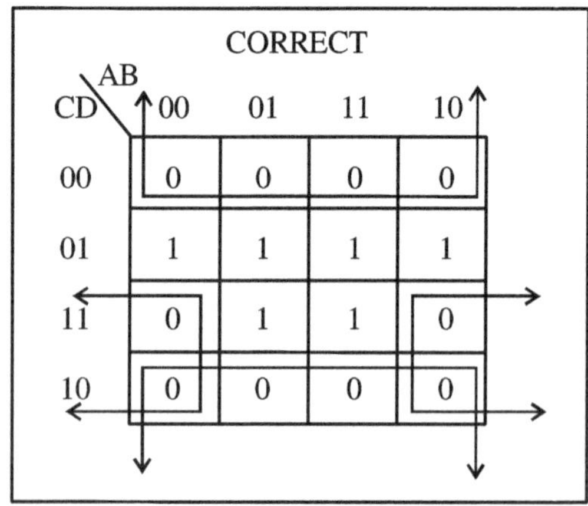

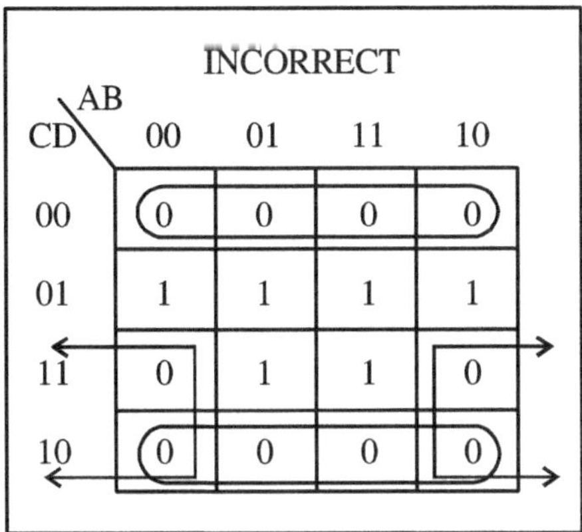

Correct : $D.(B+\overline{C})$ Minimal product of sums form :

Incorrect : $(C+D)(\overline{C}+D)(B+\overline{C})$

The incorrect version is a product of sums forms, but it is not minimal.

FAN-OUT

It is a common occurrence in logic circuits where the output of one logic gate feeds the inputs of several others. It is not that we can drive unlimited number of logic gates' inputs from the

TABLE A : LOGIC FAMILIES (LOGIC ELEMENT)

Logic Families	*Advantages*	*Disadvantages*
RTL	Low power dissipation	(1) Insufficient noise margin (2) Low speed
DTL	Low power dissipation	(1) Poor noise margin (2) Low speed
TTL	(1) Low power dissipation (2) High speed (3) High fan-out capability (4) High compatibility with existing systems (5) Low cost.	(1) Tight V_{CC} tolerances (2) Susceptible to power transients.
ECL	(1) Very high speed (2) High fan-out capability (3) Low noise generation (4) Complementary output	(1) High cost (2) Difficult to interface with other logic families
MOS	(1) Very high package density (2) Low cost (3) Fewer sub system (4) Low power drain per function	(1) Very low speed (2) More than one power source required
CMOS	(1) Very very low power consumption (2) Relatively insensitive to variations in V_{DD}. (3) High noise immunity (4) Low constant speed power product (5) Very high packing density.	(1) Low speed (2) Susceptible to static discharge damage.

TABLE B : COMPARISON OF THE MAJOR IC LOGIC FAMILIES

CHARACTERISTIC	*DTL*	*HTL*	*TTL*	*RTL*	*MOS*	*CMOS*	*ECL*
Basic gate	NAND	NAND	NAND	NOR	NAND or NOR	NAND or NOR	OR/NOR
Fan-out (recommended)	8	10	10	5	20	50	25
Propagation delay, nsec.	30	110	10	12	300	70	4
Power dissipation per gate, mW	11	44	15	25	10	1μW qu iescen 1 mW @ 1 MHz.	25
Noise immunity	Good	Excel- lent	Very good	Average	Average	Very good	Excellent for power supply noise low for signal noise.
Power supply, volts Logical unloaded	5 0.2, 5	15 0.2, 15	5 0.2, 3	3.6 0.2, 3.6	–20 0.2, –20	10 0.2, 10	–5.2 and –1.15 –1.55, –0.75

TABLE C

Logic family	*Basic gate type*	*Propogation time per gate (ns)*	*Power dissipation per gate (mW)*	*Noise margin (V)*	*Fan-in*	*Max Fan out*	*Cost (Relative)*
RTL	Positive NOR	50	10	0.2	3	4	Medium
DTL	Positive NAND	25	15	0.7	8	8	Medium
TTL	Positive NAND	10	20	0.4	8	12	Low
ECL	Positive NOR	2	50	0.4	5	25	High
MOS	Positive NAND	250	<1	2.5	10	5	Very low
CMOS	Positive NOR and positive NAND	30	0.5 μW	Depends on V_{DD}.	10	100	Low
IIL	Positive NOR	40	<1	0.35	–	8	Very low

output of a single logic gate. It is limited by the current sourcing capability of the output when it is HIGH and the current sinking capability of the output when it is LOW and the requirements of the logic gates' inputs in the two states. Number of logic gate inputs that can be driven from the output of a single logic gate with reliability (i.e. without causing any false output) is called the FAN-OUT. It is the characteristic of the family to which the IC belongs. It is 10 for the logic gates belonging to the standard TTL family and 50 for CMOS family logic gates.

INHIBIT

It is basically an AND gate with one of its inputs negated by an inverter. Negated input acts to inhibit the gate or in other words, the gate shall give a logic '1' output only when the negated input is driven to logic '0'.

A/D converter (ADC) : Short for analog-to-digital converter. A device that can accept an analog, i.e. continuous, signal whose amplitude lies within a given range, and produce an equivalent digital signal, *i.e.*, an n-bit parallel binary Word that represents this analog signal. The analog signal is "examined" at discrete fixed intervals of time by means of a "sampling process in order to produoo tho digital signal. Analog signals originating from devices such as analog sensors or tachogenerators may thus be converted into a form that can thus be processed by, say, a microprocessor.

The resolution of an A/D converter gives the smallest change in analog input that can be discriminated by the device. If the voltage range of an n-bit A/D converter is V, then its resolution is

V (2^n-1)

Since the resolution is finite, the conversion process introduces quantization noise. A/D converters are available in integrated circuit form.

EXERCISE

1. Which one of the following keeps track of intstruction execution sequence?
   A. Accumulator
   B. Program counter
   C. Stack pointer
   D. Instruction register
2. Two's complement of a given 3 or more bit binary number of non-zero magnitude is the same as the original number if all bits except the
   A. MSB are zeros B. LSB are zeros
   C. MSB are ones D. LSB are ones
3. The logic circuit shown converts $y_1 . y_2$ into :

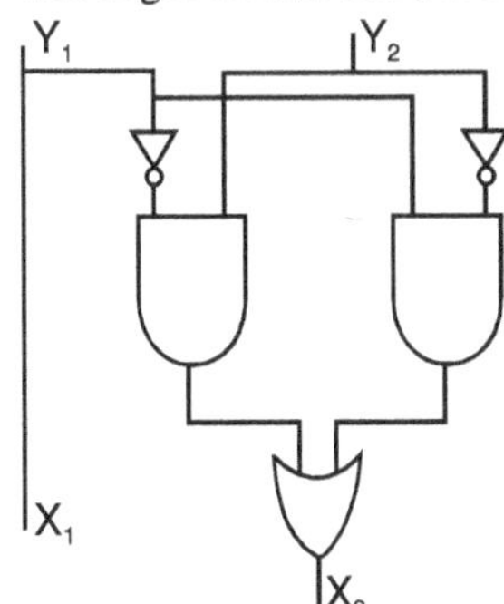

   A. gray code B. excess 3-code
   C. BCD D. error detecting code
4. One decibel represents a power ratio of
   A. 1.26 : 1 B. 2 : 1
   C. 10 : 1 D. 20 : 1
5. The output of the gate circuit shown is given by

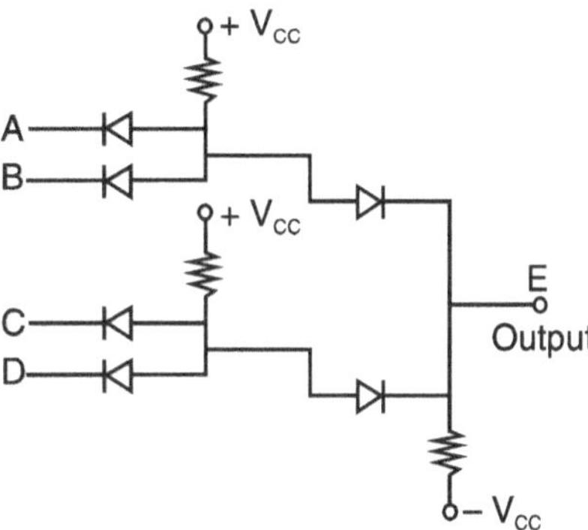

   A. $(A+B)(C+D)$ B. $AB+CD$
   C. $\overline{AB+CD}$ D. $\overline{(A+B)(C+D)}$
6. The circuit shown is

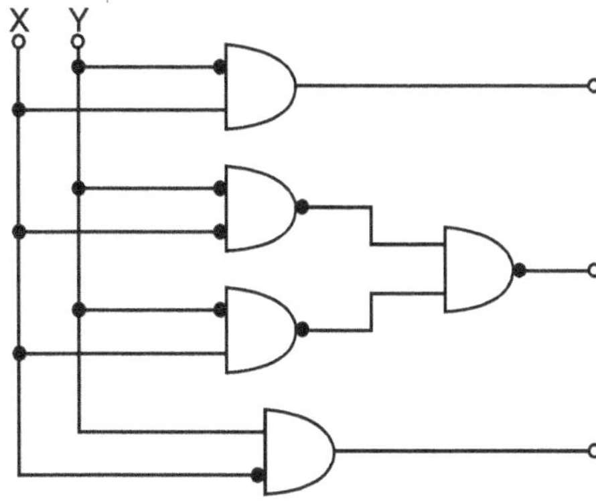

   A. an adder circuit

B. a subtractor circuit
C. a comparator circuit
D. a parity generator circuit

7. Which type of counter is shown in the figure

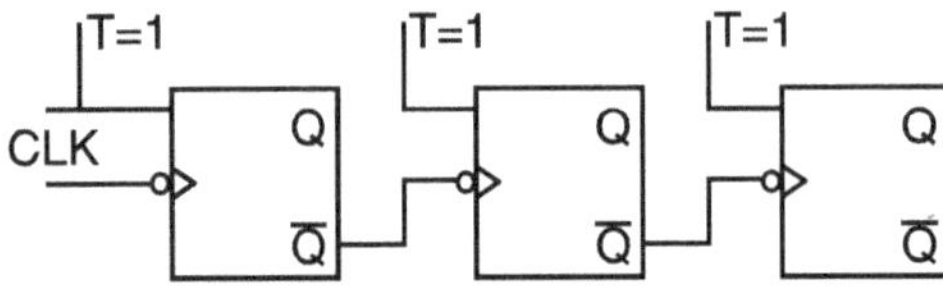

A. Synchronous B. Johnson
C. Ring D. None

8. What is dual of A + [B + (AC)] + D
A. A+[B(A+C)]+D B. A[B+AC] D
C. A+[B(A+C)]D D. A[B (A+C)] D

9. The open collector wired circuit shown below functions as :

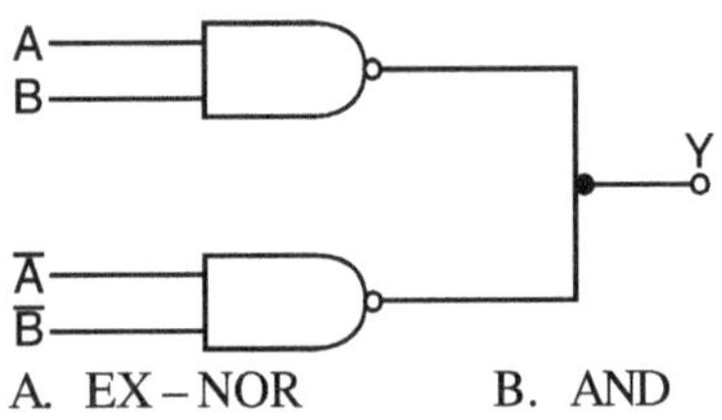

A. EX–NOR B. AND
C. EX–OR D. NOR

10. The negative logic AND gate shown is equivalent to a positive logic

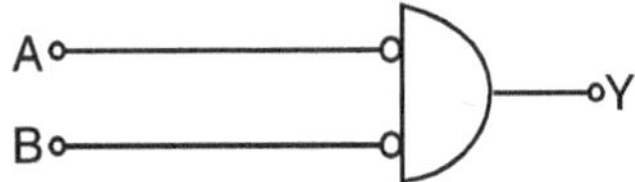

A. AND gate B. OR gate.
C. NAND gage D. NOR gate

11. Match List I with List II and select the correct answer.

List I	*List II*
(a) HTL	1. High fan-out
(b) CMOS	2. Highest speed of operation
(c) I^2L	3. High noise immunity
(d) ECL	4. Lowest product of power and delay

Codes:	*(a)*	*(b)*	*(c)*	*(d)*
A.	4	1	2	3
B.	4	3	2	1
C.	3	1	4	2
D.	3	3	1	2

12. Full adder circuit can be implemented by
A. Multiplexers B. Half adders
C. AND & OR gates D. decoders

13. In a 4-bit full adder, how half adders and OR gates are required:
A. 8 & 4 B. 7 & 4
C. 7 & 3 D. 8 & 3

14. Match List I with List II and select the correct answer

List I	*List II*
(a) Thermocouple	1. Modulated output
(b) Thermistor	2. Resistance change with pressure.
(c) Strain gauge	3. Negative temperature coefficient
(d) LVDT	4. Constant temperature at one end

Codes:	*(a)*	*(b)*	*(c)*	*(d)*
A.	2	1	4	3
B.	4	3	2	1
C.	3	2	4	1
D.	1	2	3	4

15. The *minimum* Boolean expression for the circuit is

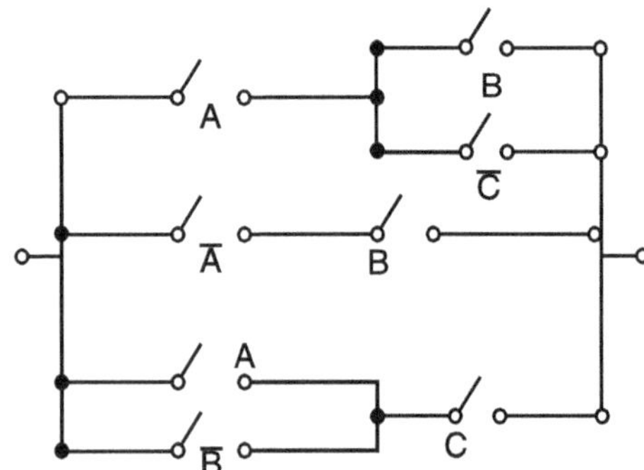

A. $AB + AC + \overline{B}C$ B. $A + BC$
C. $A + B$ D. $A + \overline{B} + C$

16. Which type of gate is shown in figure :

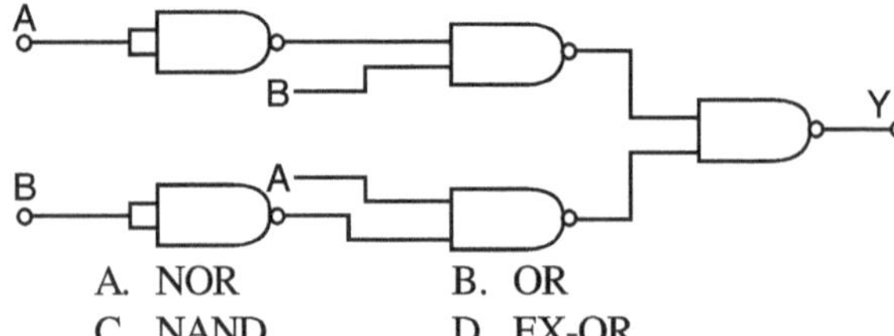

A. NOR B. OR
C. NAND D. EX-OR

17. The output (X) waveform for the combination circuit for the inputs at A and B (waveform shown in the figure) will be

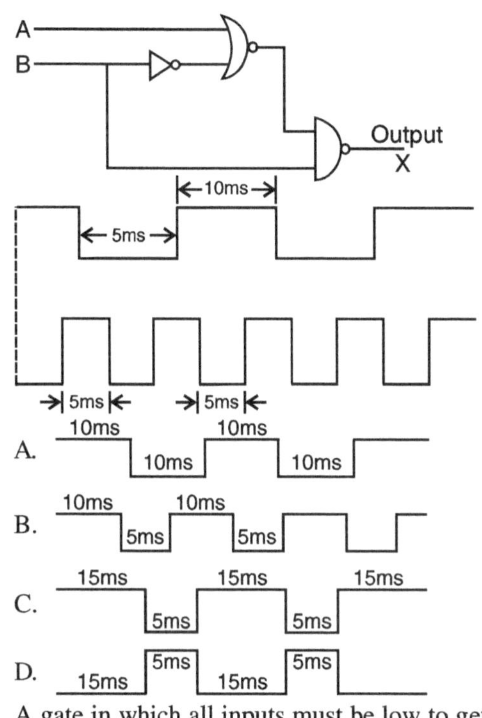

18. A gate in which all inputs must be low to get a high output is called

A. An inverter B. A NOR gate
C. An AND gate D. A NAND gate

19.

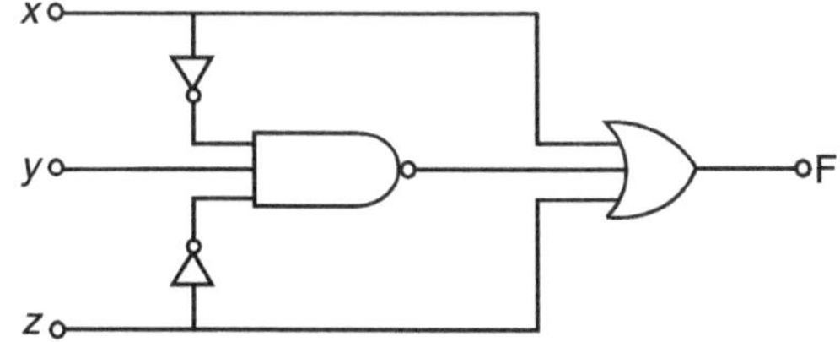

The minimized version for the logic circuit shown in the figure is :

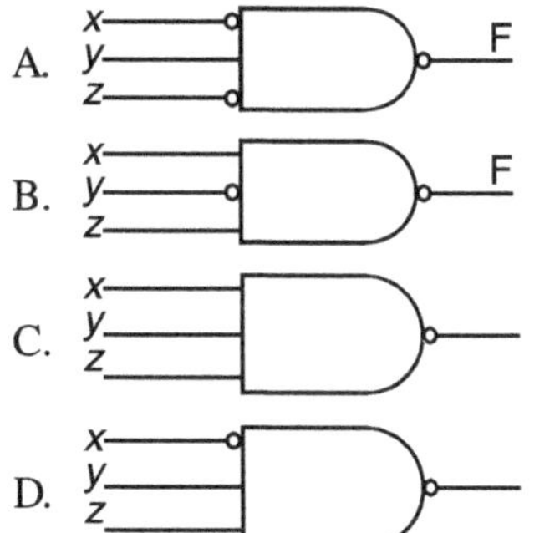

20. Sixty capsules are to be filled in bottles automatically employing an electronic counter. The number of flip-flops in such a counter will be

A. 4 B. 5
C. 6 D. 7

21. The number of bits in the binary representation of the decimal number 16 is

A. 6 B. 5
C. 4 D. 3

22. Which of the following is not a postulate of Boolean Algebra

A. X = 0 or else X = 1
B. 1.0 = 0.1 = 0
C. 1 + 0 = 0 + 1 = 1
D. None of these

23. What will be content PC after the program?

A. 00 FEH B. 0114 H
C. 0115 H D. 0119 H.

24. The output from the following wired logic will be

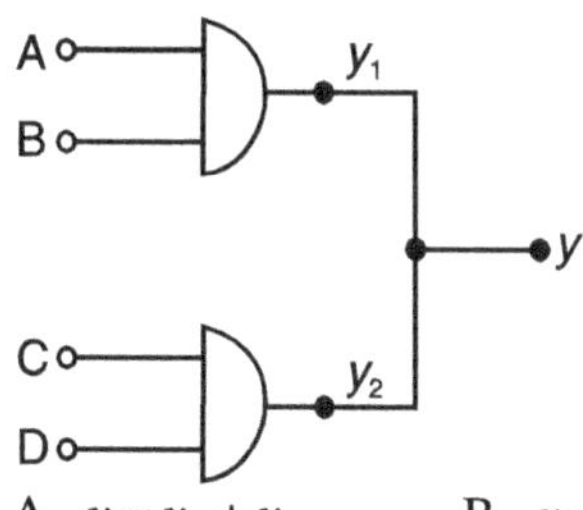

A. $y = y_1 + y_2$ B. $y = y_1 y_2$
C. $y = 2(y_1 y_2)$ D. $y = (y_1 + y_2)/2$

25. A is even or B is true

Negation of above statement will be

A. A is odd or B is false
B. A is odd and B is false
C. A is even or B is false
D. A is even and B is false

26. The circuit shown below is :

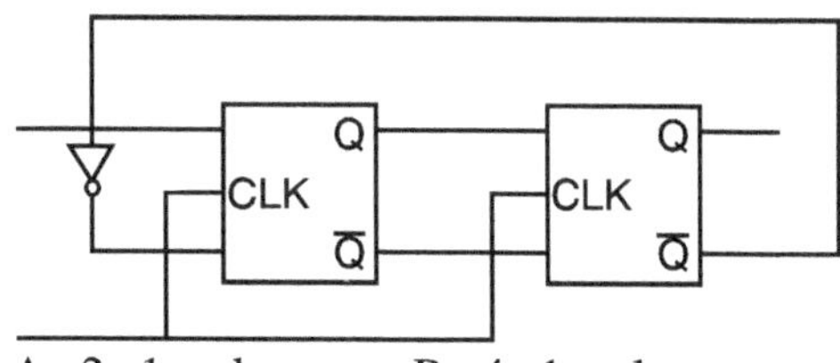

A. 2 : 1 scalar B. 4 : 1 scalar
C. up-down counter D. None

27. The difference between sequential and combinational circuit is that
A. combinational circuits store bits
B. combinational circuits have memory
C. sequential circuits store bits
D. sequential circuits have memory

28. In regard to OR gate the following statements are made
1. It has only one input
2. It has only two inputs
3. It has two or more inputs
4. If any input is high, the output is high

Of these four statements, the true statements are
A. Only 1, 2 B. Only 3, 4
C. Only 2, 3 D. Only 1, 4

29. The Boolean expression
$\overline{A}B + A\overline{B}C + \left(\overline{A+B+C}\right)$ simplifies to
A. $A\overline{B} + \overline{B}C$ B. $\overline{A}B + BC$
C. $A\overline{B} + B\overline{C}$ D. $\overline{A}B + \overline{B}C$

30. The value of 2^5 in octal system is
A. 40 B. 20
C. 400 D. 200

31. In the circuit shown below, the output X_1 and X_2 are

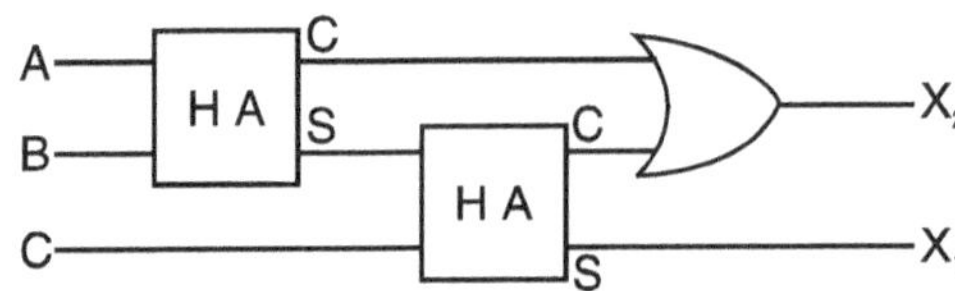

A. $X_1 = A\overline{B}\overline{C} + \overline{A}B\overline{C} + \overline{A}B\overline{C} + ABC$
$X_2 = AB + AC + BC$
B. $X_1 = (A \oplus B) \oplus C$
$X_2 = (A \oplus B) \oplus C + \overline{A}\,\overline{B}$
C. Both A and B
D. None of these

32 A 7 bit Hemming code (Even parity checker) 0001001 for a BCD digit is known to have errors. The BCD encoded digit is :
A. 9 B. 5
C. 3 D. 0

33. An DRO memory can be converted into NDRO memory by employing
A. Decade counters B. Storage registers
C. Address registers D. Inhibit lines.

34. In which of the following codes a binary number $Y_1\,Y_2\,Y_3$ is converted by the circuit given below?

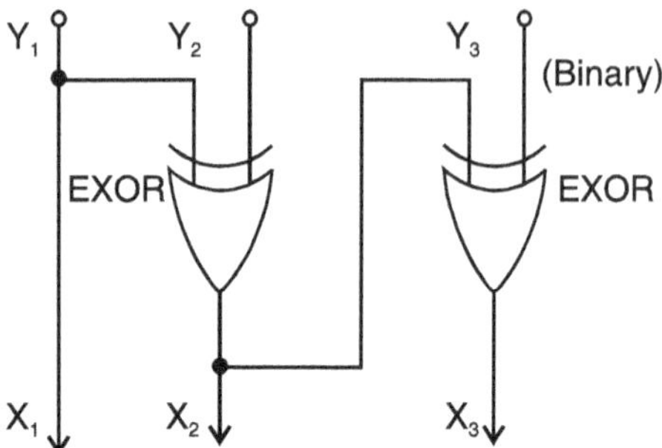

A. Excess-3 code B. Gray code
C. Decimal Code D. BCD code

35. The degree of resolution that can be obtained using an eight-bit optical encoder will be
A. 1.4° B. 2.8°
C. 3.2° D. 4.8°

36. In a five devider (ladder D/A) the value of R is 1000, the current that each input digital voltage source must be capable of supplying is given by (voltage level is + 10 V)
A. 3 mA B. 1 mA
C. $3\frac{1}{3}$ mA D. $1\frac{1}{3}$mA

37. In an all NAND gate realization input to all odd gates must be
A. A literal only B. Inverted
C. Inverted if literal D. None of the above

38. The circuit shown in figure realizes the function:

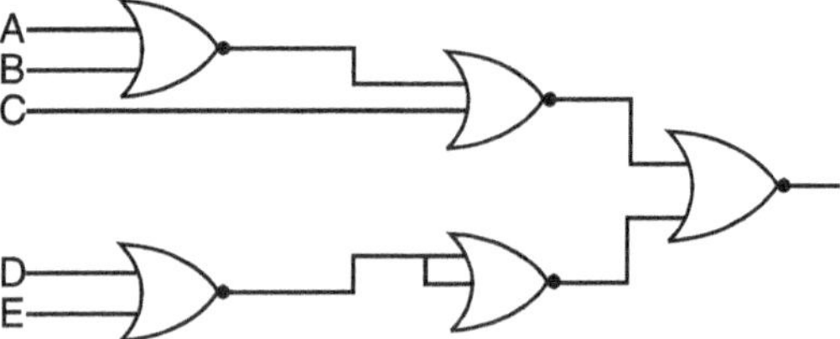

A. $(A+B+C)\left(D\overline{E}\right)$ B. $\left(A+\left(\overline{B+C}\right)\left(\overline{D}E\right)\right.$
C. $(A+B+C)\left(\overline{DE}\right)$ D. $\left(\overline{A+B}+C\right)\left(\overline{DE}\right)$

39. If two numbers in excess-3 code are added and the result is less than 9, then to get equivalent binary
A. 0011 is subtracted B. 0011 is added
C. 0110 is subtracted D. 0110 is added

40. The Truth table for half adder is :

A.

(a)	*(b)*	*(s)*	*(c)*
0	0	0	1
0	1	0	1
1	0	1	0
1	1	1	0

B.

(a)	*(b)*	*(s)*	*(c)*
0	0	1	1
0	1	1	0
1	0	0	1
1	1	0	1

C.

(a)	*(b)*	*(s)*	*(c)*
0	0	0	0
0	1	1	0
1	0	1	0
1	1	1	1

D. None

41. Which of the following code is decoded to decimal values by the diode matrix shown below?

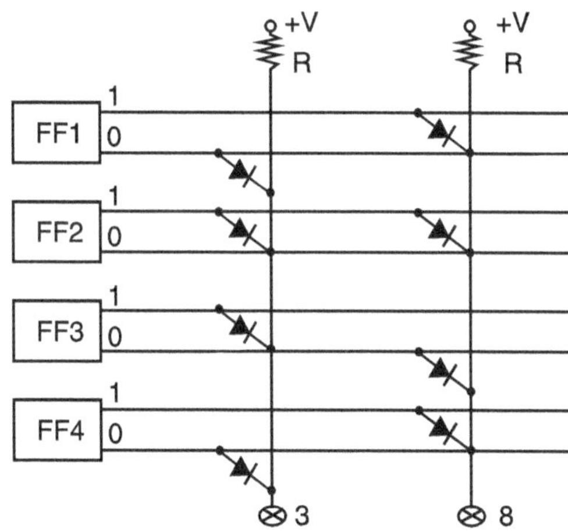

A. BCD code B. Gray code

C. Excess-3 code D. Binary code

42. A binary to decimal decoder is called 1 of 10 decoder because

A. At a time one input is high

B. Output is high only if all inputs are high

C. Only one of its output lines is high at a time

D. It has 10 input lines.

43. Which one of the following circuits converts an RS Flip to T Flip Flop?

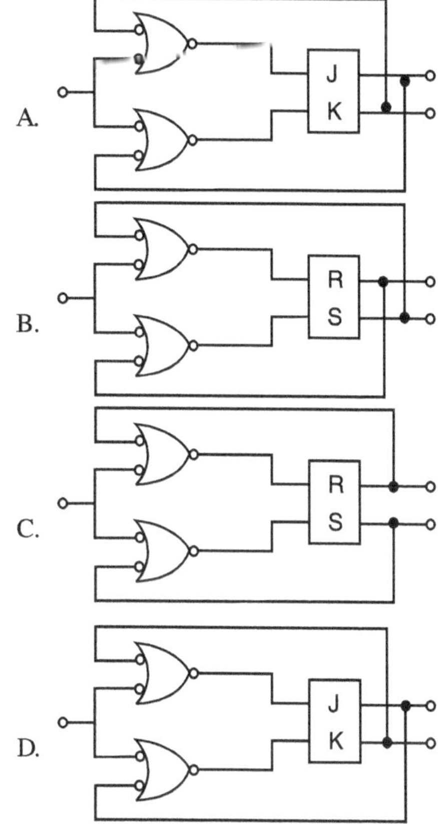

44. For counter converter, with N output bits, number of cloks cycle required for conversion operation is

A. 2^N B. N

C. $2^N - 1$ D. 2N

45. When TRAP interrupt is triggered program control is transferred to location

A. 0020 H B. 0024 H

C. 0028 H D. 002 CH

46. The binary representation of hexadecimal 'C3' is

A. 1111 B. 110011

C. 111100 D. 11000011

47. The minimum product of sums of the Boolean expression $X(Y+\overline{Z})+W$ is

A. $XY+X\overline{Z}+W$ B. $(X+YZ)W$

C. $(X+W)(Y+\overline{Z})$ D. $(X+W)(Y+\overline{Z}+W)$

48. Consider the following PASCAL declaration :

Type

Vector = array [1 .. 10] of integer;
Var p : Vector;
r, q : array [1 .. 10] of integer;
s : Vector;

Which one of the following statements is *false*?

A. p and q are structurally equivalent types
B. p and s are name equivalent types
C. q and r are structurally equivalent types
D. q and s are name equivalent types

49. In order to get correct answer when two numbers are added in Excess-3 code and sum is less than 9, it is necessary

A. To subtract 0011 from the sun
B. To add 0011 from the sum
C. To subtact 0110 from the sum
D. To add 0110 from the sum.

50. For a variable functions the number of Boxes in the K-map is

A. 8 B. 16
C. 32 D. 128

51. An AND gate has a four inputs A, B, C, D register. The input that produces an output of 1 is

A. 0000 B. 1000
C. 1111 D. 0001

52. Consider the following program segment begin

```
for    I : =1 to N – 1 do
for    J : =1 + 1 to N do
ifA(I) < A (J) THEN
begin
      T : =A (J);
    A(J) : =A [I] :
    A [I] : =T
   end
end
```

Assume that the program is syntactically correct, all variables are of type integer and N has a value greater than 1 and A is one-dimensional array. One entry to the program segment, with a value on N = 6, the values of array A are

A(1) ←	A(2) ←	A(3) ←	A(4) ←	A(5) ←	A(6)
–8	4	10	–2	7	3

After exit from the program segment, the values of A [1 ... 6], respectively, will be

A. 10, 7.4, – 2, – 8 B. 10, + 8, 4, 3, – 2
C. 4, – 8 + .10, 3, 7 D. –2, 3.4, + 8, 10

53. The flip flops used in shift register are generally

A. SR flip-flop B. JK flip-flop
C. D flip-flop D. T flip-flop

54. In NMOS acting as a switch in depletion mode

A. The device is cut off for $V_{GS} = 0$
B. The device is cut off for low V_{DS}
C. The device is cut off for large negative V_{GS}
D. None of these

55. Which one of the following represents the coincidence logic?

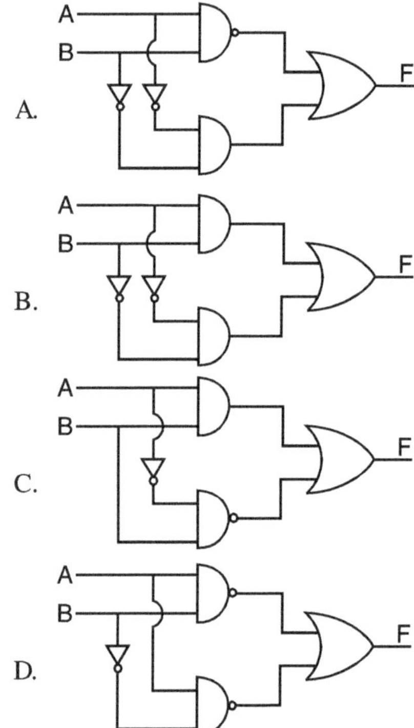

56. In the same circuit, if the AND gate is removed and output of FF_1 is directly connected to FF_2, the output of the flip-flops at 10th pulse will be

A. 0000 B. 0001
C. 0010 D. 1001

57. The astable multivibrator has :

A. Two stable states
B. One stable and one quasistable state
C. Two quasistable states
D. None of the above

58. In the sign magnitude representation, the leading bit

A. Is a part of the number itself

B. Is unity for positive numbers
C. Is always unity
D. Stands for the sign.

59. What is the length of IR (instruction register) of 8085 μP
A. 6 bits B. 8 bits
C. 12 bits D. 16 bits

60. The binary representation of the decimal number 39 is :
A. 1000111 B. 100111
C. 101011 D. 100101

61. A BYTE stands for
A. A string of two BITS
B. A string of four BITS
C. A string of eight BITS
D. None of these.

62. The simplified form of the Boolean function $F(X,Y,Z) = (\overline{X}+\overline{Y}+\overline{Z}).(X+\overline{Y}+Z).(X+Y+Z)$
A. $\overline{X}Y\overline{Z}$ B. X
C. $X+\overline{Y}Z$ D. $XY+\overline{Z}$

63. The minimised logic circuit for the circuit shown is given by

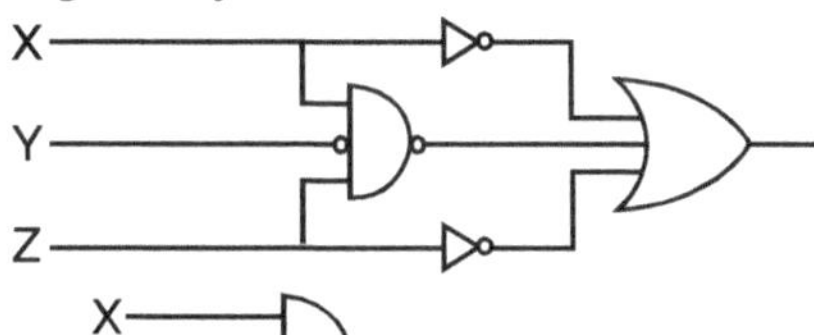

A.
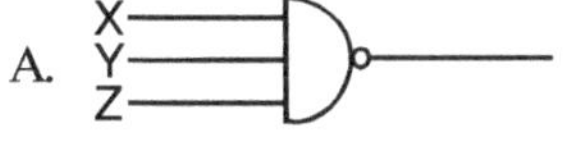

B.
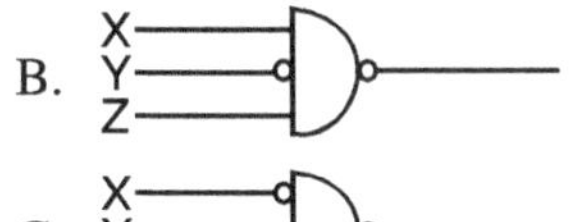

C. X Y Z

D. X Y Z

64. The commutating capacitor reduces turnoff time because
A. It acts an open circuit
B. It aids in removing the excess carriers from the base region
C. It decreases the fall time
D. All of these

65. Which one of the following is the dual-form of the Boolean identity
$AB + \overline{A}C = (A + C)(\overline{A}+B)$?
A. $AB+\overline{A}C = AC+\overline{A}B$
B. $(A+B)(\overline{A}+C) = (A+C)(\overline{A}+B)$
C. $(A+B)(\overline{A}+C) = AC+\overline{A}B$
D. $AB+\overline{A}C = AB+\overline{A}C+BC$

66. In regard to OR gate, the following statements are made :
1. It has only one input
2. It has only two inputs
3. It has two or more inputs
4. If any input is high, the output is high.
Of these four statements, the true statements are
A. Only 1, 2 B. Only 3, 4
C. Only 2, 3 D. Only 1, 4

67. The current logging takes place in
A. RTL B. HTL
C. DCTL D. TTL

68. In hexadecimal system, $77_{16} - 3B_{16}$ is equal to
A. $3D_{16}$ B. 60_{16}
C. $3C_{16}$ D. 73_{16}

69. Which of the following circuits is equivalent to the logic diagram shown below?

X Y Z

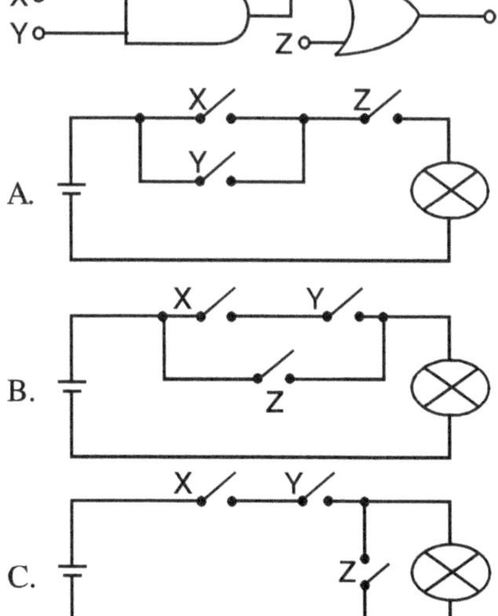

D. None of these

70. If the memory chip size is 256 × 1 bits, then the number of chips required to make up 1 k (1024) bytes of memory is

A. 32 B. 24
C. 12 D. 8

71. The aperture time of an A to D converter is given by (symbols have the usual meanings)

A. $\dfrac{E_m}{2\pi f.\Delta E}$ B. $\dfrac{2\pi f}{\Delta E E_m}$

C. $\dfrac{\Delta E}{2\pi f.E_m}$ D. $\dfrac{2\pi f.E_m}{E}$

72. For a six bit ladder D/A converter which has digital input of 101001, the analog value is (assume 0 = 0V and 1 = + 10 V)

A. 0.423 B. 0.552
C. 0.641 D. 0.923

73. In the CMOS inverter

1. Both are enhancement type
2. One transistor is N channel and the other P channel
3. Both are N channel with one enhancement and the other depletion type
4. One is enhancement and the other depletion type

Of the above 4 statements, the only true ones are

A. 4 B. 1 and 2
C. 1 and 3 D. None of these

74. In combinational circuit, the output state is a function of

A. Present input state only
B. Past output states only
C. A and B both
D. None of these

75. A microcomputer has a 64 K memory. What is the hexadecimal notation for the first memory location?

A. 0000 B. FFFF
C. 0FFF D. 3FFF

76. The most widely used universal gates are

A. OR and AND gates
B. NOR and NAND gates
C. NOR and AND gates
D. NAND and OR gates.

77. In regard to OR gate the following statements are made

1. It has only one input
2. It has only two inputs
3. It has two or more inputs
4. If any input is high, the output is high

Of these four statements, the true statements are

A. Only 1, 2 B. Only 3, 4
C. Only 2, 3 D. Only 1, 4

78. Consider the following statements regarding the circuit shown in the given figure :

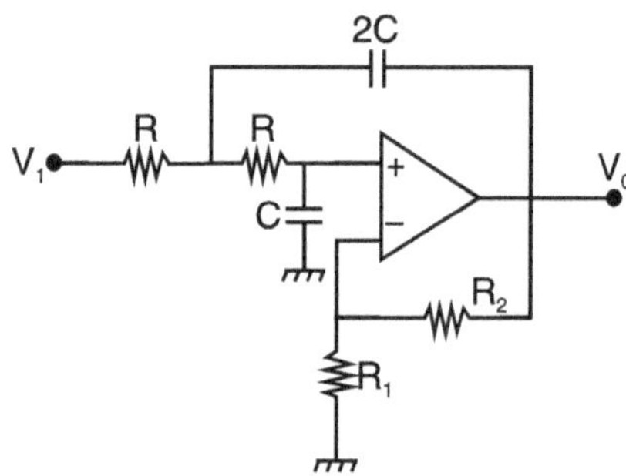

1. The circuit represents an active low pass filter
2. The circuit represents a second order active filter
3. The circuit has a roll-off rate of 40 dB/decade

Of these statements

A. 1, 2 and 3 are correct
B. 1 and 2 are correct
C. 1 and 2 are correct
D. 2 and 3 are correct

79. How is the operation dependent on A and B?

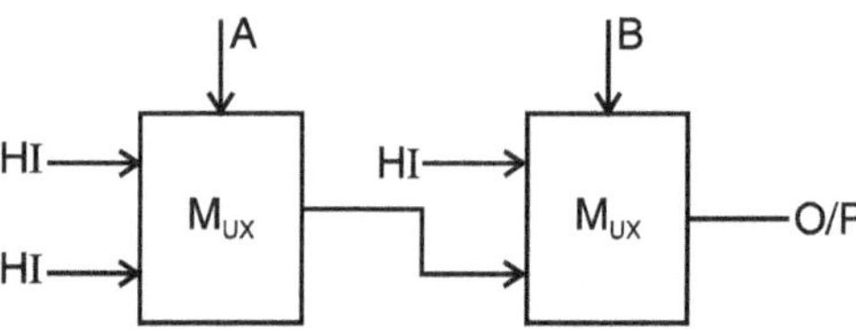

A. HI for A and B is LO
B. Independent of A and B
C. LO for A and B is HI
D. HI for A or B is LO

80. The average access time for a magnetic drum rotating at a speed of 50 revolutions per second is

A. 10 m sec B. 20 m sec
C. 30 m sec D. 50 m sec

81. In all NAND combinational circuit realization all ODD number of gates behave like

A. NAND gates B. AND gates
C. OR gates D. None of these

82. K-map method of simplification can only be applied when the given function is in
A. Sum of product form
B. Product of sum form
C. Canonical form
D. Can be applied to all form

83. The output of the circuit shown below for a digital input of 1010 would be proportional to

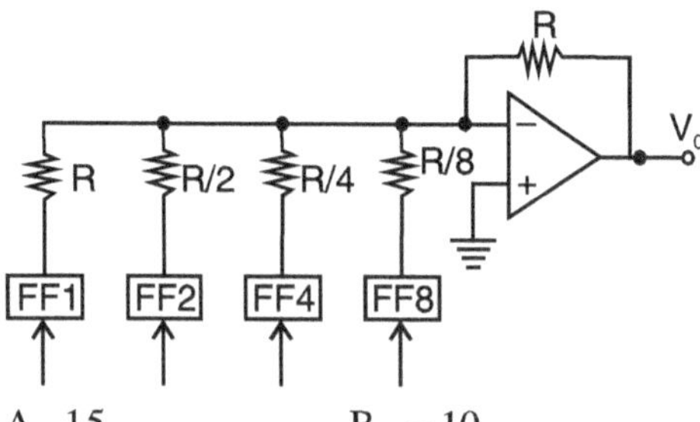

A. 15 B. – 10
C. $\frac{-10V}{R}$ D. $\frac{-15V}{R}$

84. The maximum conversion time of a 10-bit counter type converter driven by a 1 MHz clock is
A. 1.42 μ sec B. 10.54 μ sec.
C. 0.460 m sec D. 1.024 m sec.

85. The minimum Boolean for the following circuit is

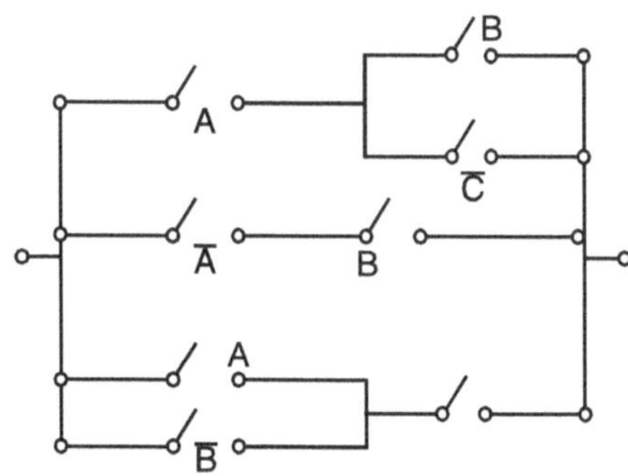

A. AB + AC + BC B. A + BC
C. A + B D. A + B + C

86. The number of switching functions of 3 variables is
A. 8 B. 64
C. 128 D. 256

87. The function shown in the fig. when simplified will yield a result with

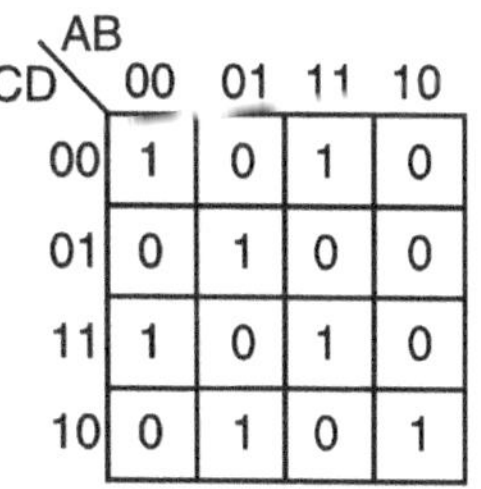

CD\AB	00	01	11	10
00	1	0	1	0
01	0	1	0	0
11	1	0	1	0
10	0	1	0	1

A. 2 terms B. 4 terms
C. 7 terms D. 16 terms

88. In FORTRAN subroutine
A. Always follows data cards
B. Always follows main program
C. Always preceeds main program
D. May be any one of (A), (B) or (C)

89. When a large number of analog signals are to be converted an analog multiplexer is used. The A.D. converter most suitable in this case will be
A. Forward counter type
B. Up-down counter type
C. Successive approximation type
D. Dual slope type

90. Letter A is punched in the punched card as follows
A. Punch in 12th row and punch in first row
B. Punch in 11th row and punch in first row
C. Punch in 10th row and punch in first row
D. Punch in first and second row.

91. A 4-line-to-1-line multiplexer shown in Figure-I (the same as per IEEE convention is shown in Figure-II) is fed with three logic inputs A, B and C as shown. The output of the multiplexer will be

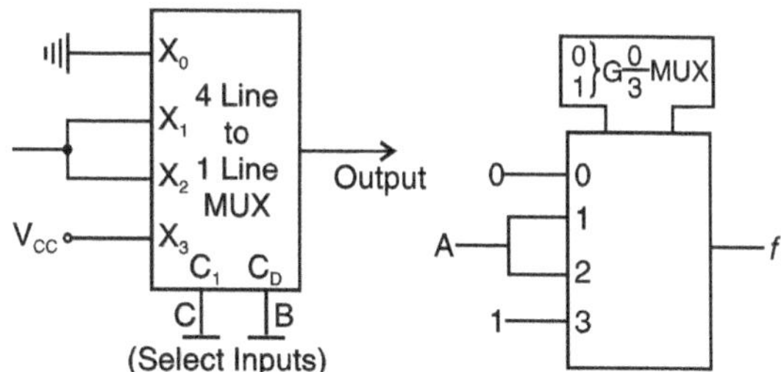

A. Σm (0, 1, 2, 4) B. Σ m(0, 3, 5, 7)
C. Σm (3, 5, 6, 7) D. Σm (1, 2, 5, 6)

92. The Boolean expression $Y = \left(A + \overline{B} + \overline{AB}\right)\overline{C}.Y$

is given by :

A. $A\overline{C}$ B. $B\overline{C}$

C. $\overline{C}$ D. None of these

93. The Boolean expression $(A+\overline{C})(\overline{B}+\overline{C})$ simplifies to

A. $\overline{C}+A\overline{B}$ B. $\overline{C}(\overline{A}+B)$

C. $\overline{B}\,\overline{C}+A\overline{B}$ D. None of these

94. The numbers of terms in a logic function of three variables X, Y and Z is

A. 3 B. 4

C. 8 D. 18

95. The standard length of magnetic tape used as digital computer memory is

A. 1200 feet B. 1600 feet

C. 2400 feet D. 3600 feet

96. The race around condition exists in J-K flip flop if

A. $J=0; K=1$ B. $J=1; K=0$

C. $J=0; K=0$ D. $J=1; K=1$

97. Which of the following is error-correcting code?

A. EBCDIC B. GRAY

C. Hamming D. ASCII

98. In a JK flip-flop, the output Q_n is 1 and it does not change when a clock pulse is applied. The possible combination of J_n and K_n could be (x denotes don't care).

A. x and 0 B. x and 1

C. 0 and x D. 1 and x

99. The Boolean expression for the lower network shown in the above question is

A. A (B + C)

B. A + BC

C. A + BC + AC + AB

D. A(A + B + C)

100. SUA 7411 is a

A. OPAMP, monolithic and short circuit protection inbuilt

B. 2 input Nand gate

C. 3 input Nand gate with open collector output

D. None of these

101. The CONTROLLED INVERTER has the property

1. When INVERT is high, the output is 1's complement of input
2. When INVERT is high, the output is 2's complement of input
3. When INVERT is low, the output word is the input word
4. When INVERT is high, the output word is the input word

Of the four statements above, the true ones are

A. Only 1, 2 B. Only 1, 3

C. Only 1, 4 D. Only 2, 4

102. The maximum time required to perform a division using 12-bit numbers and a clock of 1 MHz is

A. 12 μ sec. B. 19 μ sec

C. 31 μ sec D. 48 μ sec

103. The circuit shown below is a

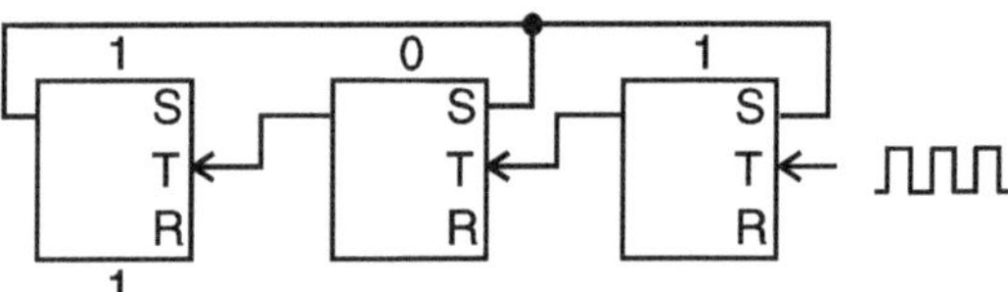

A. Mod-4 counter B. Mod-5 counter

C. Mod-6 counter D. Mod-7 counter

104. An AND gate is a

A. Sequential circuit

B. Combinational circuit

C. Memory circuit

D. Relaxation circuit

105. Which of the following interrupts is unmaskable interrupts?

A. RST 5.5 B. RST 7.5

C. TRAP D. INTR

106. The "ECL" has very high switching speed because the transistors are

A. switching between cut-off and saturation regions

B. switching between cut-off and active regions

C. switching between active and saturation regions

D. any of the above

107. Which of the following is a coincidence logic circuit :

A. 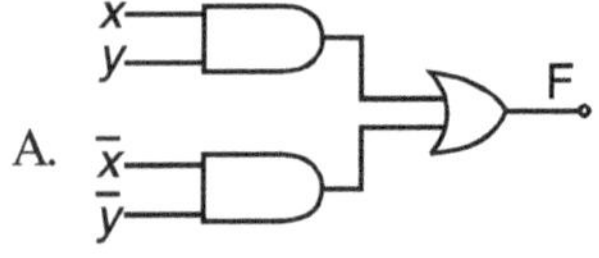

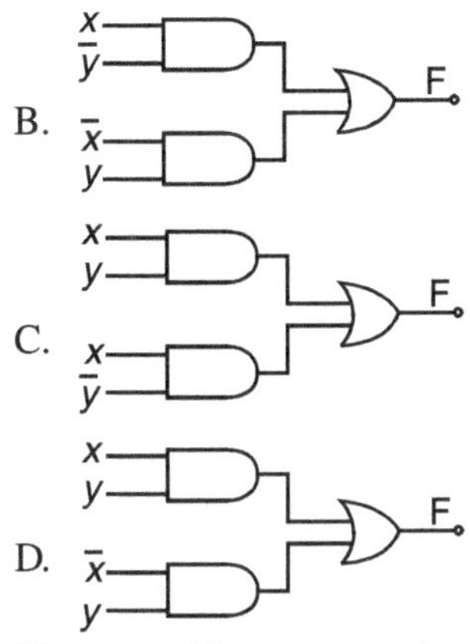

108. The true table corresponds to

I/P	O/P
An	Qn+1
1	1
0	0

A. JK flip flop B. SR flip flop
C. D flip flop D. T flip flop

109. In the Schmitt trigger shown below, the threshold voltages V_1 and V_2 are

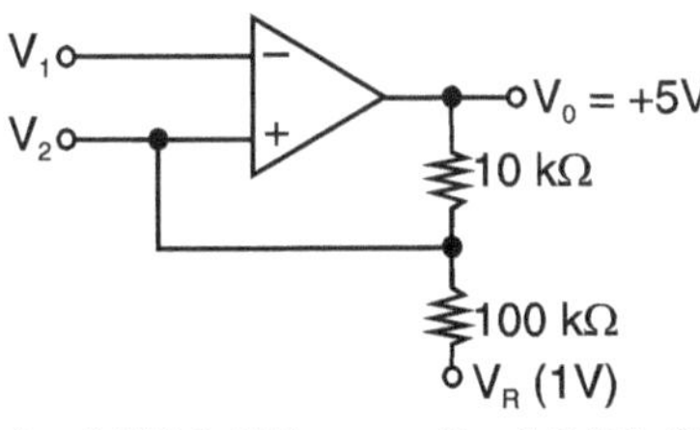

A. 2.5V, 1.5 V B. 1.04V, 094 V
C. 1.5 V, 1V D. 1.8V, 1.15 V

110. The address Bus of Intel 8085 is 16 bit wide and hence the memory which can be accessed by this address bus is

A. 1 K Byte B. 16 K Bits
C. 32 K Bits D. 64 K Bytes

111. Match List-I with List-II and select the correct answer by using the codes given below the lists:

List I	***List II***
(a) Multiplexer	1. Sequential memory
(b) De-Multiplexer	2. Converts decimal number to binary
(c) Shift-Register	3. Data selector
(d) Encoder	4. Routes out many data output with single input.

Codes :	*(a)*	*(b)*	*(c)*	*(d)*
A.	3	4	1	2
B.	4	3	1	2
C.	3	4	2	1
D.	1	2	3	4

112. A state during which nothing happens is known as

A. LDA B. Nop
C. MAR D. OP code

113. The multiple-emitter transistor-logic circuit shown below represents a

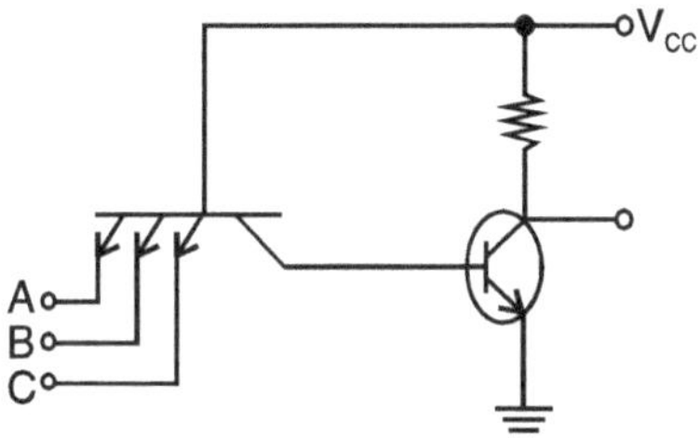

A. AND gate in positive logic
B. NOR gate in positive logic
C. NAND gate in positive logic
D. NAND gate in negative logic

114. What is the length of PC (program counter) of 8085 μP

A. 6 bits B. 8 bits
C. 12 bits D. 16 bits

115. The expression for EXCLUSIVE OR is given by:

A. $(\overline{A+B})(AB)$ B. $AB+\overline{A}\,\overline{B}$

C. $(A+B)(\overline{AB})$ D. None of these

116. The binary representation of hexadecimal 'C3' is

A. 1111 B. 110011
C. 111100 D. 11000011

117. In the binary number 110, 101 the fractional part has the value

A. 0.125 B. 0.625
C. 0.875 D. 0.5

118. Some of MOS families are PMOS, NMOS, CMOS. The family dominating the LSI field where low power consumption is necessary is

A. NMOS B. CMOS
C. PMOS D. Both NMOS and CMOS

119. In 2 out of 5 code the number 49 can be

represented as

A. 10100 10100 B. 01010 01010

C. 10100 01010 D. None of these

120. Which part of 8255 A PPI can be split into two halves?

A. Port-A B. Port-B

C. Port-C D. Port A, B

121. In positive logic, the transistor gate in the circuit below is an

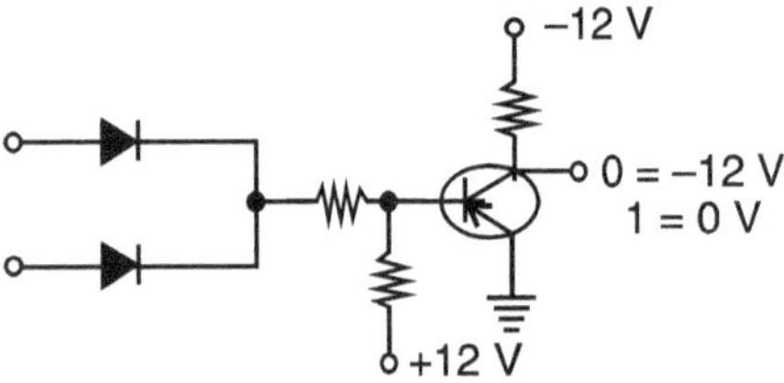

A. AND gate B. NOR gate

C. NAND gate D. OR gate

122. The open collector output of the two gates are connected together as shown in the given figure. The logic expression for Y will be

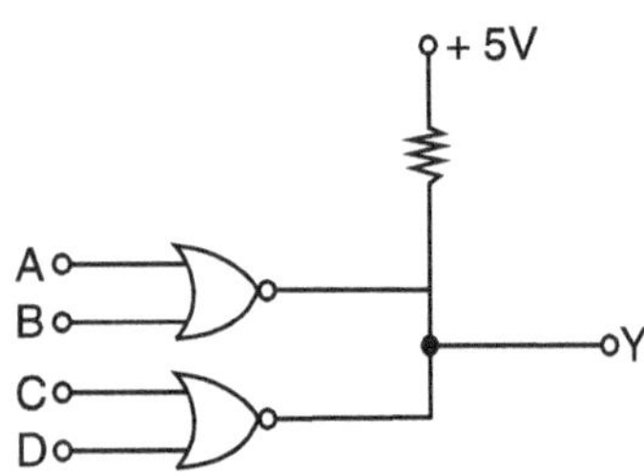

A. $\overline{A+B}+\overline{C+D}$ B. $\overline{A+B+C+D}$

C. $(A+B)(C+D)$ D. $AB+CD$

123. A dc supply of 30 V is connected across 600 Ω resistance in series with an unknown resistance. The voltmeter having a resistance of 1.2 k Ω is connected across 600 Ω resistance which reads 5 V. Then the unknown resistance should be

A. 50 Ω B. 1.7 Ω

C. 7.2 Ω D. 2.4 kΩ

124. The activity of a radioactive sample goes down to 12.57% in a time of x days. The half-life of this sample is nearly

A. $12x$ B. $8x$

C. $\frac{x}{3}$ D. $3x$

125. Match List I with List II and select the correct answer

List I	List II
(a) Static I^2L Memory	1. Erasable programable
(b) CCD memory	2. Ultra high speed
(c) ECL memory	3. Stores large volume of data
(d) GAL memory	4. Does not need refreshing
	5. Non-volatile

Codes:	(a)	(b)	(c)	(d)
A.	4	2	3	1
B.	4	3	2	1
C.	5	1	2	3
D.	3	5	2	1

126. Four the logic circuit shown the output Y is equal to

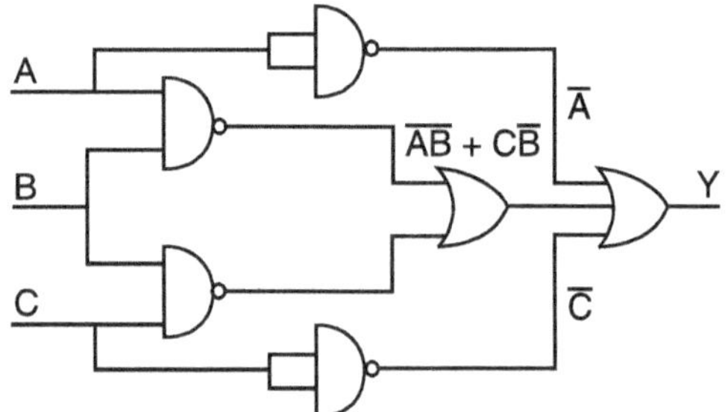

A. $\overline{ABC}$ B. $\overline{A}+\overline{B}+\overline{C}$

C. $\overline{AB}+\overline{BC}+\overline{A}+\overline{C}$ D. $\overline{AB}+\overline{BC}$

127. The logic circuit given below converts a binary code y_1, y_2, y_3 into

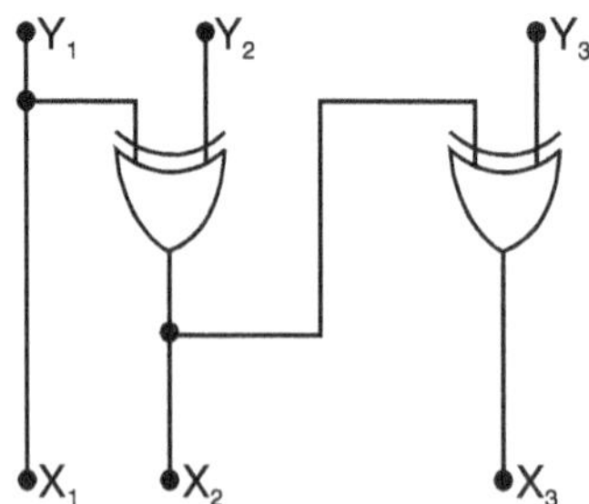

A. Excess-3 code B. Gray code

C. BCD code D. Hamming code

128. The greatest negative number which can be stroed in a computer that has 8-bit word length and uses 2's complement arithmetic is

A. − 256 B. − 255

C. − 128 D. − 127

129. A retriggerable monshot is one which
- A. can be triggered only once
- B. has two quasi stable states
- C. cannot be triggered unitl full pulse has been outputted
- D. is capable of being triggered while the output is being generated

130. The circuit shown realises the function

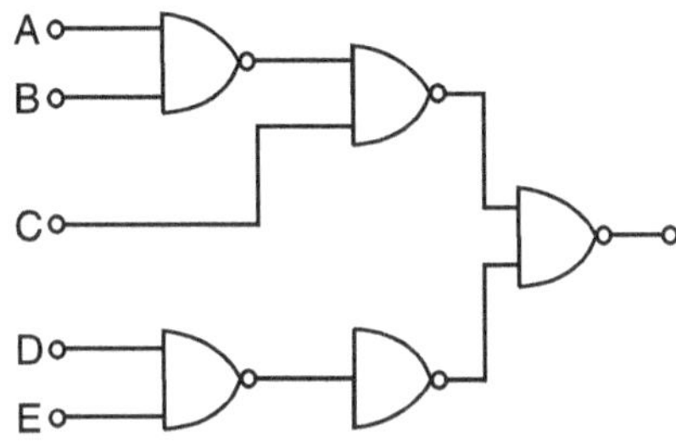

A. $\left(\overline{A}+\overline{B}\right)C+\overline{DE}$ B. $(A+B)C+D+E$
C. $AB+C+DE$ D. $AB+C(D+E)$

131. Consider the flow chart given in the figure :

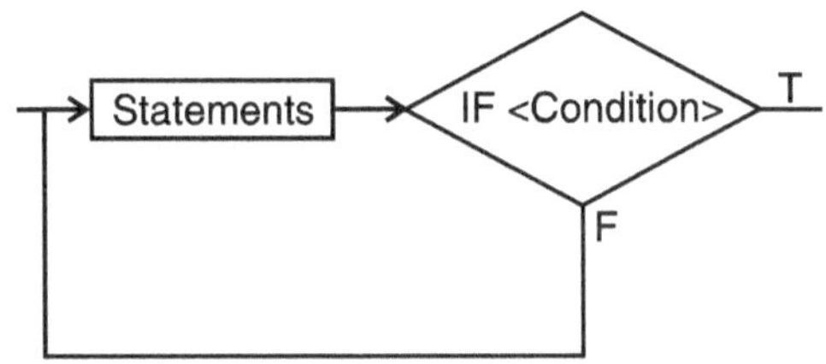

The flow chart is for
- A. IF statements
- B. while statement
- C. repeat until statement
- D. Do statment

132. The output *f(x, y)* of the multiplexer shown in Figure-I (the same as per IEEE convention is shown in Figure-II) resulting from the input logical values is

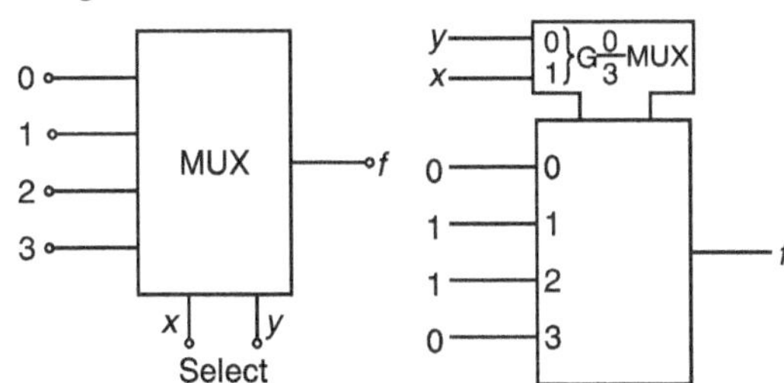

A. an EXOR gate B. a NOR gate
C. an AND gate D. a NAND gate

133. The operation $x\oplus\overline{y}$ represents

A. $x-y$ B. $\overline{x}y+\overline{xy}$
C. $xy+\overline{xy}$ D. $\overline{x}-\overline{y}$

134. The term $AB+AC+B\overline{C}$ reduce to

A. $AB+CA$ B. $AC+BC$
C. $AC+B\overline{C}$ D. $AB+\overline{B}\overline{C}$

135. In a microcomputer, wait states are used to
- A. make the processor wait during a DMA operation
- B. make the processor wait during an interrupt processing
- C. make the processor wait during a power shutdown
- D. interface slow peripherals to the processor

136. If the input and output signals of a black box are as given in the following figures :

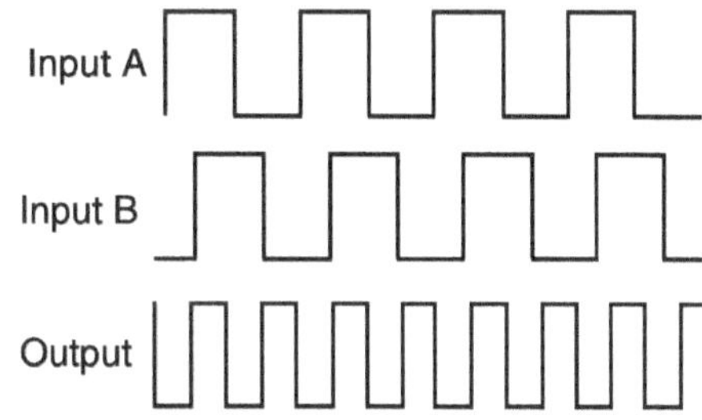

Then the black box is a/an

A. coincidence circuit B. Ex OR circuit
C. JK Flip Flop D. R-S Flip Flop

137. For a statement, DO 1001 = L, M, N, the number of iterations will be

A. $\frac{(M-L)}{N}$ B. $1+\frac{(M-L)}{N}$
C. $1-\frac{(M-L)}{N}$ D. $1+\frac{(M+L)}{N}$

138. The negative logic AND gate shown in the given figure is equivalent to a positive logic

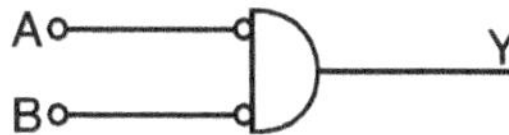

A. AND gate B. OR gate
C. NAND gate D. NOR gate

139. The Boolean expression for the shaded area in the Venn diagram is

A. $XY+Z\left(\overline{X}+\overline{Y}\right)$ B. $\overline{XY}+Z(X+Y)$

C. $XY+Z(X+Y)$ D. $\overline{X}\,\overline{Y}+Z(\overline{X}+Y)$

140. Consider the flow chart given in the figure where COND1 and COND2 are Boolean expressions, while S1 and S2 are statements:

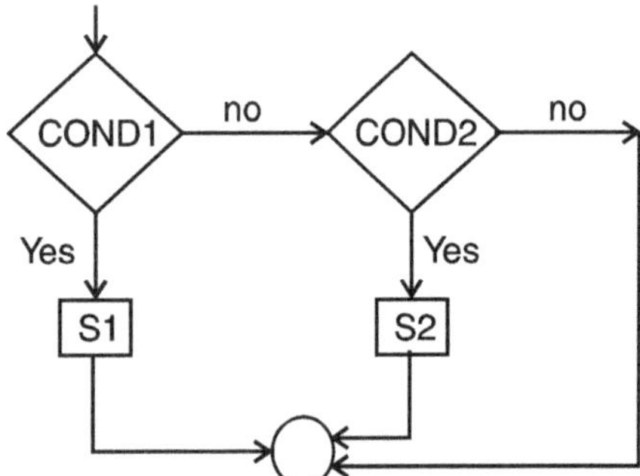

The effect of the flow chart is equivalent to :

A. if COND1 then (if COND2 then S2) else S1;
B. if (COND1 and NOT COND2) then S1 else S2;
C. if (NOT COND1) then S2 else (if COND2) then S1);
D. if COND1 then S1 else (if COND2 then S2);

141. Which one of the following logic functions is implemented by the gates when their open collector type outputs are tied together as shown?

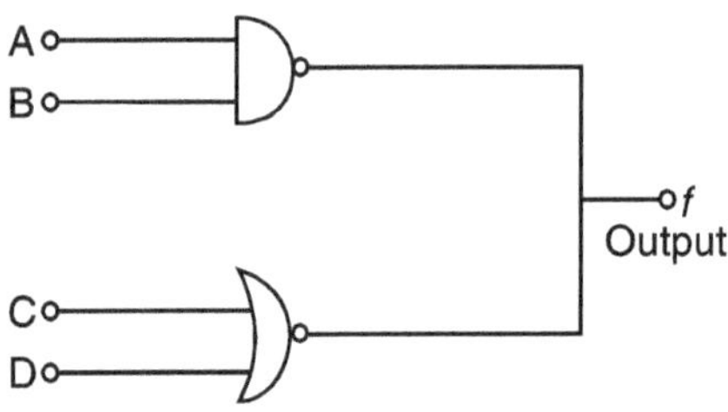

A. $f = AB + C + D$ B. $f = \overline{AB} = (\overline{C+D})$
C. $f = AB.(C+D)$ D. $f = \overline{AB + C + D}$

142. Consider the bit pattern 01010001. Which of the following has a Hamming distance of exactly 2 from this pattern?

A. 01010000 B. 010100010
C. 01010011 D. 01010110

143. The statement labels in a subroutine

A. may be the same as in the main program
B. are always the same as in the main program
C. cannot be the same as in the main program
D. canot be used

144. The given figure shows the application of 555 timer circuit as an astable multivibrator. The charging and discharging time constants are respectively

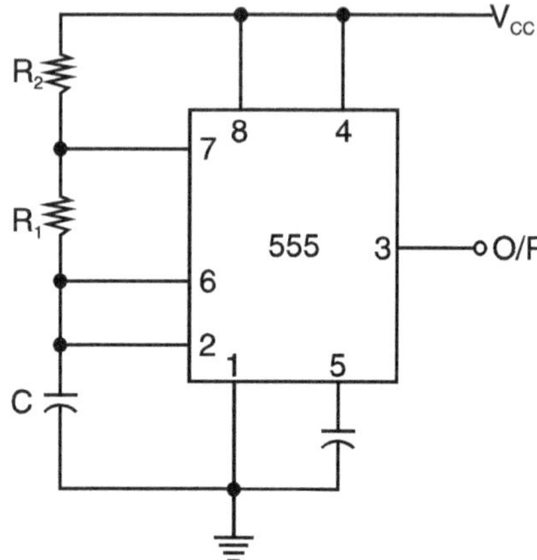

A. R_1 C and R_2 C
B. R_1 C and $(R_1 + R_2)$ C
C. $(R_1 + R_2)$ C and R_1 C
D. R_2 C and R_1 C

145. The output of a logic gate is 1 when all its inputs are at logic 0. The gate is either ________

A. a NAND or an EX-OR
B. an OR or an EX-NOR
C. an AND or an EX-OR
D. an NOR or an EX-NOR

146. Which one of the following Arithmetic IE statements is invalid?

A. IF (M – N) 50, 60, 70
B. IF (A – I) 10, 15, 20
C. IF (A – B) 20, 30, 40
D. IF (A – 4) 40, 40, 50

147. The expression, $I = 8 + \frac{7}{4}$ will be computed as

A. 3.75 B. 9.75
C. 9 D. 3

148. In the figure shown

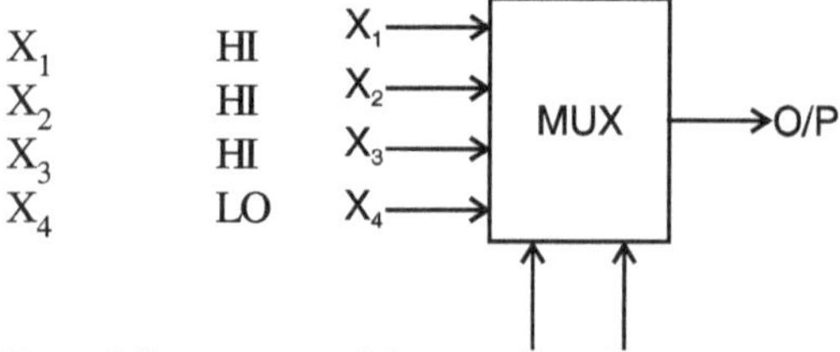

S_1 and S_0 are control inputs.

This multiplexer is equivalent to

A. NAND gate B. AND gate
C. OR gate D. EXNOR gate

149. A three-input NAND gate is to be used as an inverter. Which one of the following measures will achieve better results?

A. The two inputs not used are kept operation
B. The two inputs not used are connected to ground ("0" level)
C. The two inputs not used are connected to logic ("1" level)
D. None of the above

150. In the figure shown, X_2 X_1 X_0 will be 1's complement of A_2 A_1 A_0 if

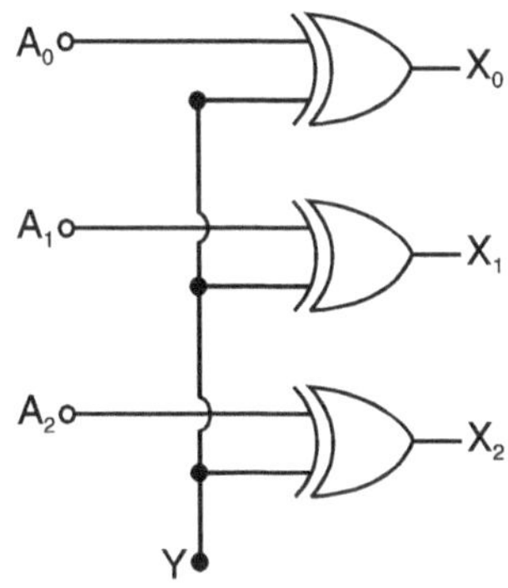

A. $Y = 0$
B. $Y = 1$
C. $Y = \overline{A}_0 = \overline{A}_1 = \overline{A}_2$
D. $Y = A_0 = A_1 = A_2$

151. Which of the following is a valid real constant

A. 579 – E03 B. 579 E
C. E – 03 D. 357 E. 03

152. A subroutine for the evaluation of the sum of numbers is to be used. The program segment given below

SUBROUTINE XYZ(N)
I = 1
$SUM_{10.0}$
26 SUM = SUM + A (I)
I = 1 + 1
If (I. LE. N) GO TO 26
C calling the subroutine with CALL XYZ (25), would compute the sum of

A. 26 elements B. 25 elements
C. 24 elements D. zero element

153. The minimised expression for the given K-map

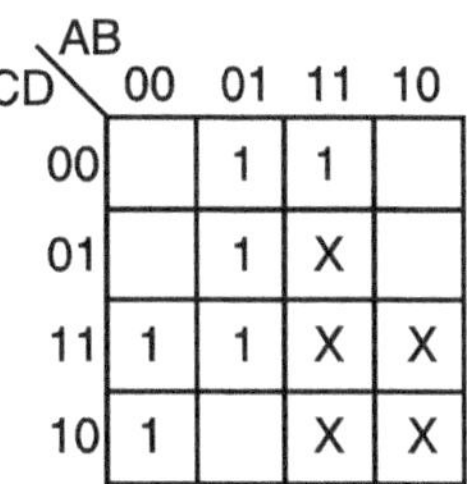

CD \ AB	00	01	11	10
00		1	1	
01		1	X	
11	1	1	X	X
10	1		X	X

A. $\overline{C}B + BD + CD$ B. $AB + C\overline{B} + B\overline{C}$
C. $C\overline{B} + AC + B\overline{C}$ D. $\overline{C}B + CD + C\overline{B}$

154. The logic circuit given below converts binary code Y_1, Y_2, Y_3 into

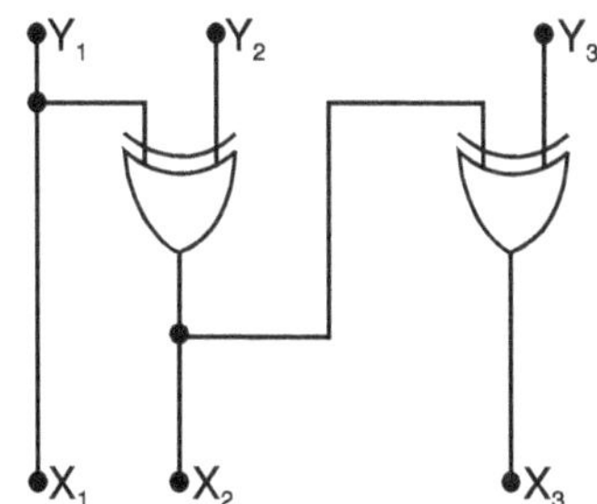

A. Excess-3 code B. Gray code
C. BCD code D. Hamming code

155. The appropriate ASSIGN statement is

A. ASSIGN 40 TO (ITEM) GO TO (ITEM) (20, 30, 40, 50)
B. ASSIGN 40 TO ITEM GO TO ITEM (20, 30, 40, 50)
C. ASSIGN 40 to (ITEM) GO TO 20
D. ITEM = 40 GO TO 20

156. Consider the function below :

function calc (x, y : integer) : integer; begin
If y =
then calc : = x
else calc : = calc (x, y – L) + x
end;

Assuming that the invocation call is "calc (a b)" and that "a" and "b" are positive integer what result does this function return?

A. a* (b – 1) B. a* b
C. a + b D. a^b

157. In a Fortran program

A. all statements must be numbered

B. the numbered statements must be referred
C. the statements referred must be numbered
D. all statements must be referred

158. Given A = 1, B = 1, $Q_n = 0$ and $P_n = 1$, what will be outputs Q_{n+1} and P_{n+1}, when the clock input (CK) is applied?

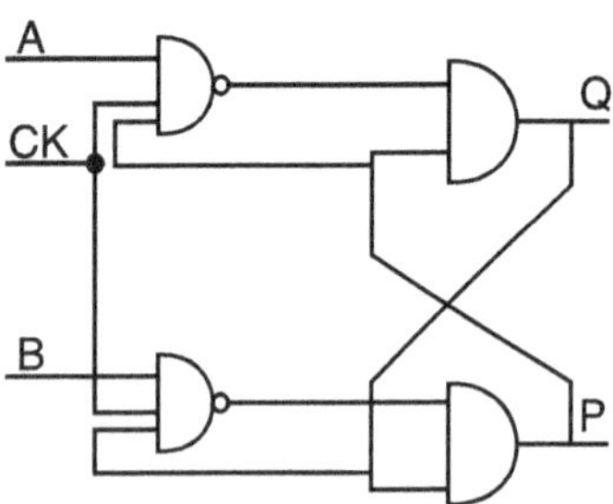

A. $Q_{n+1} = 0, P_{n+1} = 0$
B. $Q_{n+1} = 0, P_{n+1} = 1$
C. $Q_{n+1} = 1, P_{n+1} = 0$
D. $Q_{n+1} = 1, P_{n+1} = 1$

159. In the CMOS inverter
1. One transistor is N channel and the other P channel
2. Both are enhancement type
3. Both are N channel with one enhancement and the order depletion type
4. One is enhancement and the other depletion type

Of the above 4 statements, the only true ones are
A. 4 B. 1 and 2
C. 1 and 3 D. None of these

160. Which of the following flip-flop cannot be converted to D-type (delay) flip-flop
A. S – R flip flop
B. J – K flip flop
C. Master slave flip-flop
D. None of the above

161. Which of the following is not a characteristic of a flip-flop?
A. The flip-flop is a bistable device with only two stable states
B. The flip-flop has two input signals
C. The flip-flop has two output signals
D. The outputs are complement of each other

162. The expression for shaded area shown below is :

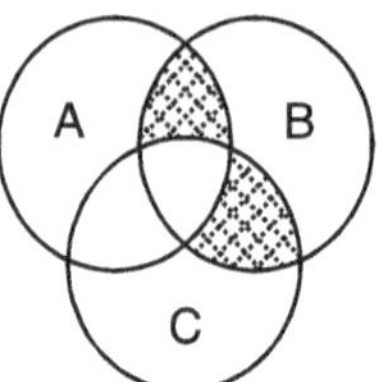

A. $AB + BC$ B. $\overline{A}\,\overline{B}C + AB\overline{C}$
C. $AB\overline{C} + \overline{A}BC$ D. None of the above

163. The minimum Boolean expression for the circuit it shown below is

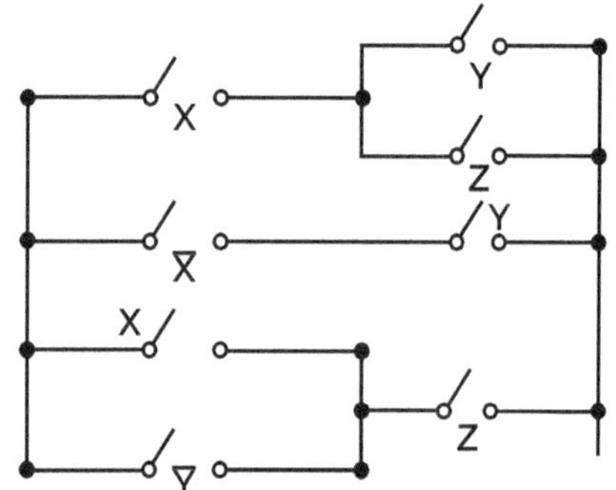

A. $XY + XZ + YZ$ B. $X + YZ$
C. $X + Y$ D. $X + Y + Z$

164. The ASCII Code is or information interchange by a binary code for
A. Numbers only
B. Alphabet only
C. Alpha numerics and other common symbols
D. None of thee

165. Given BCD number 1001 0011, its decimal equivalent is
A. 147 B. 143
C. 93 D. 39

166. Semi conductor memories are
A. Volatile
B. None volatile
C. Volatile, small size
D. Non-volatile, small size

167. How many bytes are there in
1111 1011 0111 01001 1010
A. 2 B. 3
C. $2\frac{1}{2}$ D. $3\frac{1}{2}$

168. The ripple counter functions as a
A. $2^n : 1$ B. $2 : 1$

C. $2n : 1$ D. $n : 1$ counter

169. What is dual of $X + \overline{X}Y = X + Y$

A. $X + Y = XY$ B. $\overline{X} + XY = XY$

C. $X(\overline{X} + Y) = XY$ D. $X(\overline{X} + Y) = X + Y$

170. What is boolean expression for a gating network that will have output 0 only, when $X = 1, Y = 1, Z = 1$, $X = 0, Y = 0, Z = 0$; $X = 1, Y = 0, Z = 0$?

A. $XYZ + \overline{X}\,\overline{Y}Z + X\overline{Y}Z$

B. $(XYZ)(\overline{X} + \overline{Y} + Z)(X + \overline{Y} + Z)$

C. $(\overline{X} + \overline{Y} + Z)(X + Y + Z)(\overline{X} + Y + Z)$

D. $\overline{X}\,\overline{Y}Z + XYZ + \overline{X}YZ$

171. The Address Bus of Intel 8085 is 16 bite and hence the memory which can be accessed by this address bus is

A. 1 k byte B. 16 k bits

C. 32 k bits D. 64 k bits

172. The following configuration provides

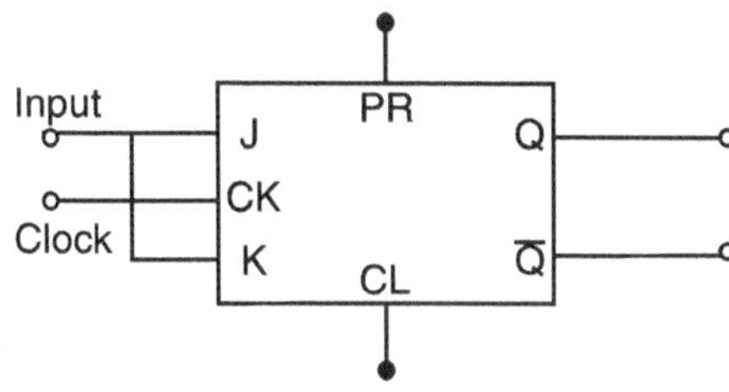

A. Master slave action

B. Delay gate

C. Toggle switch

D. None of these

173. Digital computers are more widely used as compared to analog compters, because they are

A. Less expensive

B. Always more accurate and faster

C. Useful over wider ranges of problem types

D. Easier to maintain.

174. The advantage of self-correcting code is that

A. It is a weighted code

B. It has even parity

C. It is easy to decode electronically

D. All of these

175. A NAND gate is called a universal logic element because

A. Any logic function can be realized by NAND gates alone

B. All the minimizing techniques are applicable for optimum NAND gate realization

C. $a \times b$

D. None

176. The number of Boolean functions which can be generated with four vgariables is

A. 4 B. 16

C. 256 D. 65, 536

177. Which of the following is the fastest A/D converters

A. comparater converter

B. counter type

C. successive approximation

D. dual slope converter

178. Even after a Reset operation which of the following interrupts remain enabled?

A. INTR B. RST 5.5

C. RST 7.5 D. TRAP

179. In order to enable INTR interrupt, which of the following instruction are needed?

A. EI only B. SIM only

C. EI and SIM both D. None

180. Race Condition always arises in a

A. combination circuit

B. asynchronous circuit

C. synchronous circuit

D. digital circuit

181. Given the binary 00000111, its 2's complement is obtained and expressed in hexadecimal form. The result is

A. F8 B. F9

C. – 79 D. None of these

182. Given the decimal number – 19, an eight bit 2's complement representation is given by

A. 11101110 B. 11101101

C. 11101100 D. None of these

183. Which of the following number systems has two 0's?

A. Sign plus magnitude

B. 1's complement

C. 2's complement

D. None of the above

184. Which of the following circuit represents a

subtractor to subtract A from B?

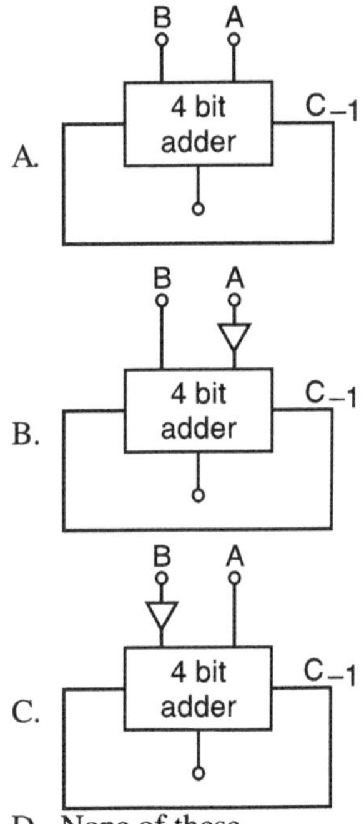

D. None of these

185. The Bipolar Technology giving the fastest logical family is

A. DTL B. TTL
C. ECL D. None of these

186. T_1 evaluate under a condition, 100 < m < 200. The computer statement shall be

A. IF (M. GE. 100. AND.M. LE. 200)
B. IF (M. GT. 100. AND M. LT. 200)
C. IF (M. LT. 100. AND. M. GT. 100)
D. IF (M. LE. 200. AND M. GT. 100)

187. Match the following :

Column I	***Column II***
(a) Digitizer	1. Terminal that can input and receive data, as well as follow users to edit and process data.
(b) Dumb terminal	2. Allows the user to touch hardware to instruct the computer to perform specialized tasks.
(c) Touch screen	3. Specialized input device often used for map making.
(d) Intelligent terminal	4. Input terminal entirely dependent for all of its capabilities on the computer system to which it is connected.

Codes:	*(a)*	*(b)*	*(c)*	*(d)*
A.	3	4	2	1
B.	3	1	2	4
C.	4	1	2	3
D.	4	2	1	3

188. Which of the following used as current source

A. BJT B. FET
C. SCR D. UJT

189. If an input 1011 drives a 74141 connected to a nixie tube, the output will be

A. 1011 B. 14
C. Zero D. a blank

190. The 2's complement representation in 8 bit format is 11010000. The equivalent hexadecimal representation is

A. – 50 B. DO
C. 30 D. – 30

191. In a full adder, there are

A. Two binary number inputs and two outputs
B. Three binary digit inputs and two binary outputs
C. Three binary digit inputs and three binary digit outputs
D. None of these

192. The decimal number 85 is encoded as 1100-0101 in

A. 8421 code B. 4421 code
C. 2421 code D. 2221 code

193. The binary fraction 0.011 in decimal form is equal to

A. 0.4375 B. 0.6225
C. 0.8325 D. 0.1105

194. Counters can be used in measurement of

A. Frequency B. Time
C. Distance D. All of these

195. For a D/A converter, the resolution required is 50 mV and the total maximum input is 10 V. The number of bits required is

A. 7 B. 8
C. 9 D. 20

196. The resolution for *n* bit system D/A converter is

A. $\frac{1}{2^N}$ B. $\frac{1}{2^N - 1}$
C. $2^N - 1$ D. 2^N

197. Which of the following circuit exhibits memory?

A. Astable multivibrator

B. Bistable multivibrator
C. NAND gate
D. Ex OR gate

198. In an stable multivibrator, the base resistors of 100 k and capacitors of 1,000 pF are used. The frequency of the output will be
A. 1.2 kHz B. 5.5 kHz
C. 6.9 kHz D. 12.8 kHz

199. The astable multivibrator has :
A. Two stable states
B. One stable and one quasi stable state
C. Two quasi-stable states
D. None of the above.

200. For the circuit shown, Y is given by

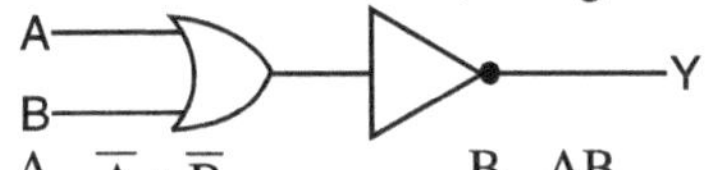

A. $\overline{A}+\overline{B}$ B. AB
C. $\overline{A}\,\overline{B}$ D. None of these

201. The n, stage register results in a delay of
A. $(n-1)$ T B. $2n$T
C. n^2 T D. $\frac{nT}{2}$
Where T is the clock period

202. Fan-out for the 74 series is
A. 4 B. 5
C. 8 D. 10

203. Which one of the following represents coincidence logic?

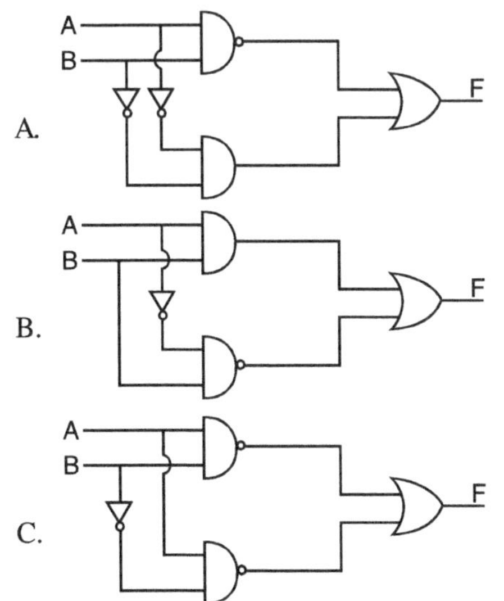

D.

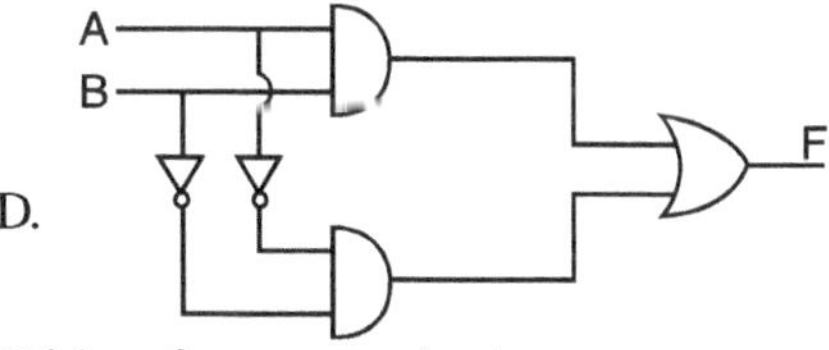

204. With reference to the inverter as a gate, the following satements are made
1. It has one or many inputs
2. It has only one input
3. The output is in a state opposite to that of the input
4. The output is low if one of the inputs is high
Of these statements, the true statements are
A. Only 1, 2 B. Only 2, 3
C. Only 3, 4 D. Only 1, 4

205. The number of 7490s to obtain positive half cycle of 10 m sec is

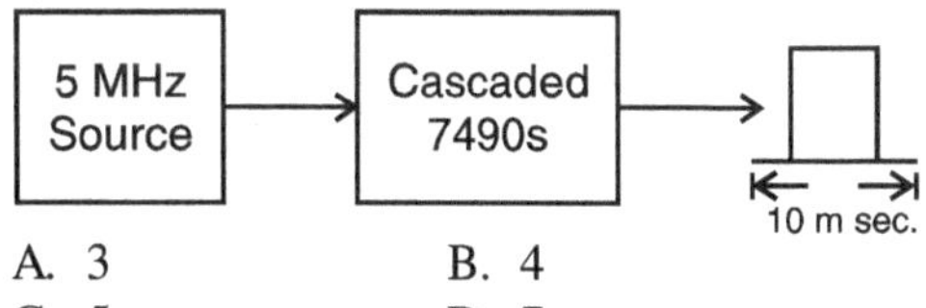

A. 3 B. 4
C. 5 D. 7

206. If we reduce CD (E + A) F + (A + E) BC to minimum number of literals the result would be
A. (A + E) C (DF + B)
B. (A + E) D (CF + B)
C. (A + E) (CD + BC)
D. (A + E) (CDF + BC)

207. Decimal number 9 in 5043210 code (biquinary code) is represented as
A. 00 01001 B. 10 10000
C. 10 01001 D. 11 00000

208. Find radix of the system shown in the figure below :

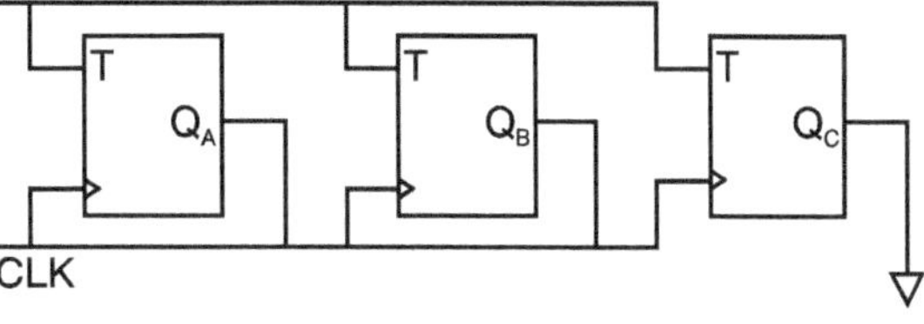

A. 2 B. 4
C. 6 D. 8

209. What will be the state of the output after the third clock cycle ?

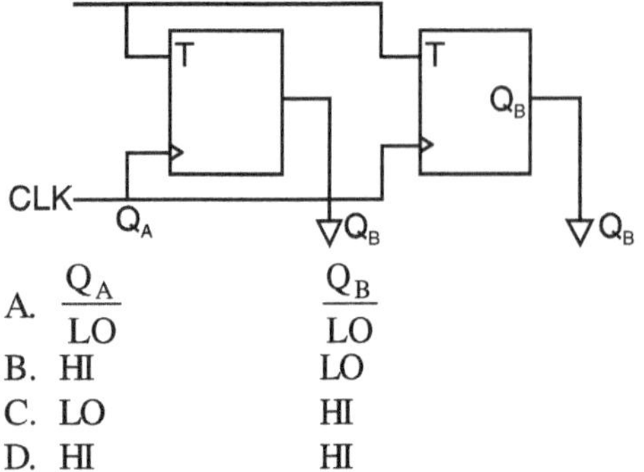

	Q_A	Q_B
A.	LO	LO
B.	HI	LO
C.	LO	HI
D.	HI	HI

210. If in an arithmetic operation carry input to the MSB is the same as carry output of MSB of the accumulator then if employs
A. An overflow
B. An underflow
C. Either A or B
D. That the result is correct

211. Which of the following codes is an unweighted code?
A. 8421 B. Excess-3 code
C. 2421 D. 63210

212. What is the maximum number (in decimal notation) that can be counted with above counter?
A. 999 B. 4096
C. 4095 D. 9999

213. SN 7401 IC is a
A. Quad 2 input Nand gate
B. Quad 2 input Nand gate with open collector output
C. Quad single input Nand gate with open collector output
D. Quad single input Nand gate

214. One of the theorem of Boolean Algebra says that $XY + \overline{X}Y + YZ$ equal to
A. $XY + YZ$ B. $X + YZ$
C. YZ D. $XY + \overline{X}Z$

215. Which of the following statement about impulse voltage is true?
A. An impulse voltage is an unidirectional voltage
B. In chopped impulse voltage, flashover does not occur
C. Time taken to rise is exactly equal to the time taken to fall
D. RMS value of impulse voltage is always less than 50% of average value

216. Match List I, List II and select the correct answer by using the Codes given below the lists :

List I	*List II*
(a) Digital Counter	1. Measurement of harmonic
(b) Scherring bridge	2. Measurement of frequency
(c) Megger	3. Measurement of oss angle in a dielectric
(d) Spectrum Analyser	4. Measurement of insulation resistance

Codes :	*(a)*	*(b)*	*(c)*	*(d)*
A.	1	2	3	4
B.	2	3	4	1
C.	4	3	2	1
D.	4	1	3	2

217. A digital voltmeter has read-out range from 0 to 9999 counts. When full scale reading is 9.999 V, the resolution of the full scale reading is
A. 0.001 B. 1000
C. 2 digit D. 1 mV

218. The input impedance of a CRO is nearly
A. zero
B. around 10 ohms
C. around 100 ohms
D. around one megaohm

219. Which of the following ICs can be used for frequency division by 5?
A. 7466 B. 7492
C. 74121 D. 7490

220. The disadvantage of magnetic drum storage is that
A. Its size is very large
B. Only read out is possible
C. Access time is high
D. Access time is low

221. How many modes of operation are there in 8255 PPI?
A. 1 B. 2
C. 3 D. 4

222. Which of the following is a 16-bit micro-processor
A. Intel 8085 B. Intel 8086
C. Zilog 80 D. Motrola 6800

223. Given two numbers A and B sign magnitude representation in an eight bit format

A = 00011110 B = 10011100

The corresponding decimal numbers respectively are

A. 30 and 156 B. 14 and – 12
C. 30 and – 100 D. – 28 and 30

224. The Gray Code for number 2 is

A. 0010 B. 0011
C. 1000 D. 0101

225. MOS family that dominates the LSI field is

A. PMOS B. NMOS
C. CMOS D. None of these

226. The boolean function F = A.B. + C.D + E can be realised as

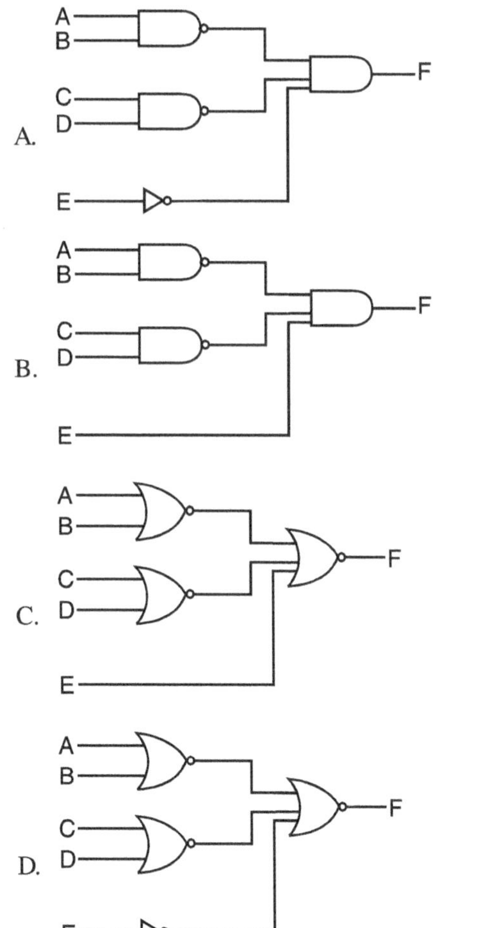

227. If m is the number of bits required to code N items in the binary form, then

A. $m = \log_{10} N$ B. $m > \log_e N$
C. $m \geq \log_2 N$ D. $m = 2 \log_{10} N$

228. The circuit diagram represents a

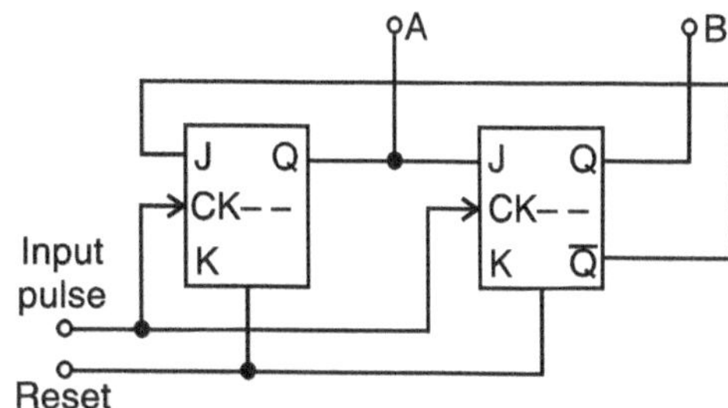

A. mod-3 counter
B. mod-5 counter
C. 3-bit shift register
D. mod-2 counter

229. The delay between successive bits for 9600 baud rate is approximately 0.1 ms

A. True B. False
C. None of these D. unit is not power

230. The shortest interval between pulse with which a binary will switch from one state to another is known as

A. Rise time B. Relaxation time
C. Resolving time D. Delay time

231. The negative decimal number – N in the 2's complement representation is 1011. Then the representation for – (N + 1) is

A. 1010 B. 1100
C. 1110 D. 1101

232. In 2's complement representation, a certain number – N is

A. 0100 B. 0101
C. 0110 D. 0011

233. The minimum number of flip-flops required in a counter to count 60 pulses is

A. 4 B. 6
C. 8 D. 10

234. The sum S of A and B is a Half Adder can be implemented by using K NAND gates. The value of k is

A. 3 B. 4
C. 5 D. None of these

235. The Josephson tunneling device illustrates principles associated with the advanced storage technique :

A. cryogenics B. CCD
C. EBAM D. holographing

236. Two variable names COLOR and COLOUR can be treated as one variable with the help of

A. COMMON statement
B. EQUIVALENCE statement
C. Declaration statement
D. Substitution statement

237. Which one of the following is an invalid integer variable?

A. INTEGER B. JSUM
C. MATRIX D. NUMBER

238. The two kinds of main memory are

A. primary and secondary
B. random and sequential
C. ROM and RAM
D. central and peripheral

239. BCD number are obtained by

A. Converting decimal number to binary
B. Converting decimal to octal numbers
C. Each-decimal digital is represented by a four bit binary
D. Converting binary to decimal

240. SN 7410 IC is a

A. Triple 3 input Nand gate
B. 2 input Nand gate
C. Triple 3 input Nand with open collector
D. None of these

241. An interrupt service request is serviced

A. Immediately on receipt of request
B. After the execution of the current instriction is completed
C. At the end of the current machine cycle
D. All of the above

242. The output of the circuit shown below is

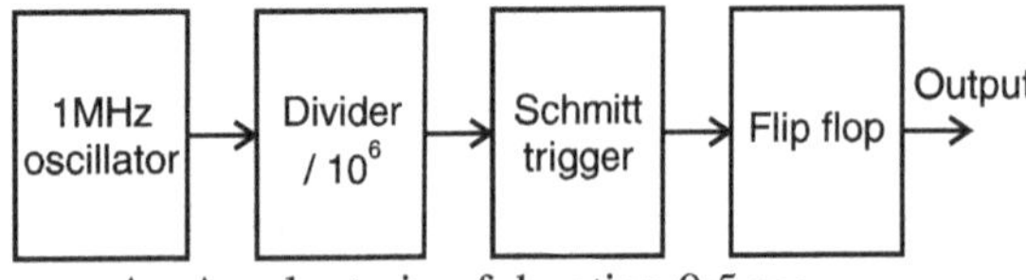

A. A pulse train of duration 0.5 sec.
B. A pulse train of duration 2 sec.
C. A pulse train of duration 1 sec.
D. a pulse train of duration 5 sec.

243. In one of the codes for transfer of numbers the code for the succeeding number differs from that of the number only in the change of a single digit. The code under consideration is

A. ASCII B. Exceess 3 CODE
C. GRAY CODE D. None of these

244. The race around condition exists in J-K flip flop if

A. J = 0; K = 1 B. J = 1; K = 0
C. J = 0; K = 0 D. J = 1; K = 1

245. Which of the following is error-correcting code?

A. EBCDIC B. GRAY
C. Hamming D. ASCII

246. In a JK flip-flop, the output Q_n is 1 and it does not change when a clock pulse is applied. The possible combination of J_n and K_n could be (x denotes don't care).

A. x and 0 B. x and 1
C. 0 and x D. 1 and x

247. If instruction RST is written in a program the problem will jump to location

A. 0020 H B. 0024 H
C. 0028 H D. 002 CH

248. The Boolean expression $(A+\overline{C})(\overline{B}+\overline{C})$ simplifies to

A. $C+AB$ B. $\overline{C}(\overline{A}+B)$
C. $\overline{B}\overline{C}+A\overline{B}$ D. None of these

249. The RST 5.5 interrupt service routine start from location

A. 0020 H B. 0024 H
C. 0028 H D. 002 CH

250. Match List I *(computer terms)* with List II *(definitions)* and select the correct answer using the codes given below the lists :

List I	***List II***
(a) Interface	1. A measure of rate of data transmission
(b) A bit	2. A binary digit
(c) Baud speed	3. The common boundary between various sections and sub sections
	4. Analog to digtial converter

Codes :	*(a)*	*(b)*	*(c)*
A.	3	2	1
B.	3	1	2
C.	2	4	1
D.	4	2	1

251. Which one of the following can be used to change data from spatial code to temporal code?
A. Shift registers
B. Counters
C. A/D converters
D. Combinational circuits

252. The number of NAND gates required for two dimensional addressing of 256 × 8 bit ROM using 8 to 1 selectors is :
A. 16 B. 48
C. 64 D. none of the above

253. In FORTRAN, the number of logical values
A. is one only B. is two only
C. is three only D. may be any number

254. If X, Y and Z are Boolean variables, then the expression $X(X+\overline{X}Y)Z(X+Y+Z)$ is equal to
A. $X+\overline{X}Y$ B. $X+Y+Z$
C. XYZ D. XZ

255. 200 numbers given in a matrix form of size (10 × 20) arc READ by the statement
A. READ (7, 25) ((N (I, J), J = 1, 30), I = 1, 10)
B. READ (7, 25) ((N (I, J), I = 1, 10), J = 1, 20)
C. READ (7, 25) ((N (I, J), J = 1, 20), I = 1, 10)
D. READ (7, 25) N (I, J), J = 1, 20), I = 1, 10)

256. Which of the following is incorrect?
A. $11100_2 - 10001_2 = 00101_2$
B. $15E_{16} = 350_{10}$
C. $81_{10} = 1010001_2$
D. $37.4_8 = 111\ 111.\ 11$

257. The FORTRAN statement, A = 2.5, J = 5*A –2** 4/2 will evaluate J as
A. 12 B. 8.5
C. 4.5 D. 4

258. ROM is composed of
A. magnetic cores
B. microprocessors
C. photoelectric cells
D. floppy disks

259. The .... is a non-impact printer that can produce very high quality, letter perfect printing
A. dot-matrix printer
B. dais-wheel printer
C. electrostatic printer
D. laser printer

260. The parity bit is added for ... purposes
A. coding B. indexing
C. error-detection D. controlling

261. In a computer program, the speed of a car is denoted by N, speed varying from 20.5 to 60.5 km/hour. This is done by using the type declaration,
A. INTEGER, N B. REAL, N
C. REAL N D. REAL (N)

262. The circuit shown below is a

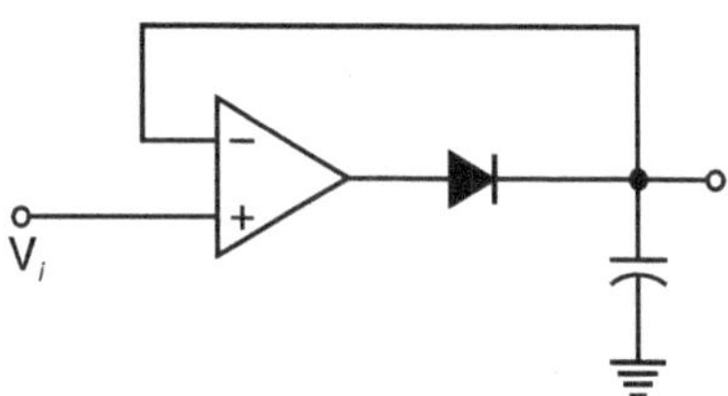

A. Half-wave rectifier
B. Average detector
C. Peak detector
D. Clamping circuit

263. For a five bit resistive divider, the weight assigned to the LSB
A. 32 B. 1/32
C. 31 D. 1/31

264. The gate whose output is LOW if and only if all the inputs are HIGH, is
A. NAND B. NOR
C. OR D. AND

265. An *n*-bit ADC using V_R as reference voltage has a resolution (in volts) of
A. $V_R/2^n$ B. V_R/n
C. $V_R/2^{n-1}$ D. $V_R/2n$

266. Exclusive OR followed by an inverter has the property
A. Output high for input word of even parity
B. Output high for input word of odd parity
C. Output is 1's complement of input word
D. Output is 2's complement of input word.

267. The significant difference between depletion type and enhancement type NMOS is
1. There is a +V threshold in depletion type
2. V_p is negative in depletion type
3. There is no channel as such in enhancement type
4. Enhancement type can be operated in both modes
5. Depletion type can be operated in both

modes.

Of these 5 statements, the only true ones are

A. 1, 2, 3 B. 3, 4, 5
C. 2, 3, 5 D. 2, 3, 4

268. The output A of the circuit shown is

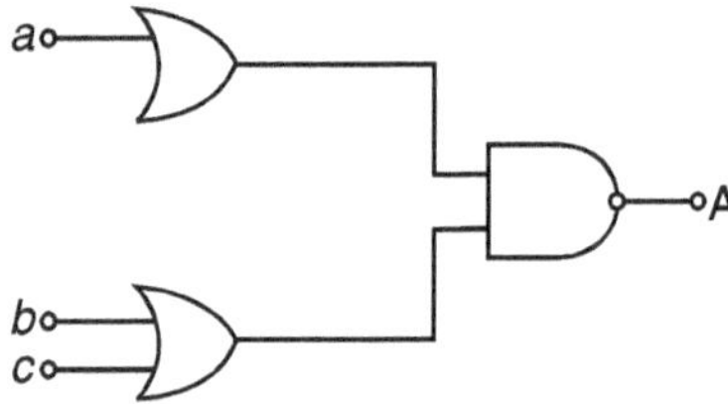

A. $\overline{a}+\overline{b}\overline{c}$ B. $a + bc$
C. $\overline{a}\overline{b}+\overline{a}\overline{c}$ D. $a+b\overline{c}$

269. Micro-program is

A. The name of the source program in microcomputers
B. The set of instructions indicating the primitive operations in a system
C. The general name of a 'MACRO's' in assembly language programming
D. The name of a program of very small size.

270. 0110-0001-1001 in standard BCD is decimal number

A. 615 B. 916
C. 619 D. 919

271. For a five bit ladder which has a digital input of 11010 (Assume 0 = 0 V and 1 = + 10 V)

A. 7.125 V B. 8.125 V
C. 26 V D. 13 V

272. Access time in Bipolar or static MOS memories are of the order of

A. 20—400 n sec
B. 20—400 μ sec
C. 200—400 milli seconds
D. 20—400 secs.

273. The expression for Y is given by :

A. AB
B. $\overline{AB}$
C. $\overline{A+B}$
D. None of these

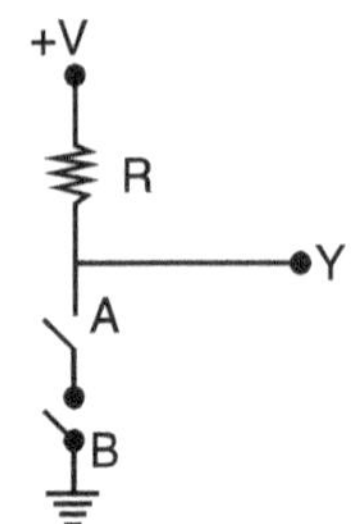

274. In order to add 1111 + 1101, we require

A. One FA, one HA
B. Three FA
C. Three FA and one HA
D. None of these

275. A Counter displays a sequence of numbers. If a reading corresponds to the hexadecimal number F 52 E, the next readings are respectively

A. F 52 F, F 520 B. F 530, F 531
C. F 52 F, F 530 D. F 52 F, F 52 G.

276. In a memory system, there are 10 bits for an address word. If first five bits are used to designate an X line and second five for an Y line, the decimal address of 1011000101 is

A. $X_{28}\,Y_4$ B. $X_{30}Y_3$
C. $X_{23}\,Y_5$ D. $X_{25}\,Y_{10}$

277. In a sequential circuit the output state depends upon

A. Inputs states only
B. Input states and output states
C. Past output states and present input states
D. None of thee

278. The bit distance between all odd numbered columns in the K-map is

A. Not fixed B. Always 1
C. Always 2 D. Not specified.

279. The circuit shown below is a

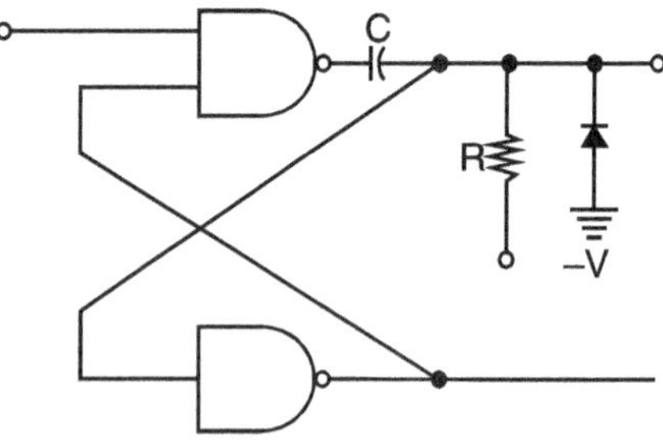

A. Bistable multivibrator
B. Astable multivibrator
C. Monostable multivibrator
D. The-delay circuit

280. NAND operation is

A. $\overline{X+Y}$ B. $\overline{X}+\overline{Y}$
C. $\overline{X}.\overline{Y}$ D. $(\overline{X}+\overline{Y})(X+Y)$

281. If the negative logic is used, the diode gate in the circuit below represents an

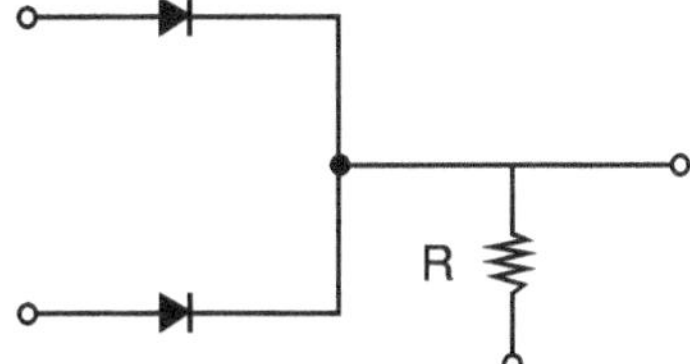

A. AND gate B. OR gate
C. NOR gate D. NAND gate

282. A six-bit alpha-numeric code is able to code
A. 36 characters B. 48 characters
C. 64 characters D. 128 characters

283. A simple flip-flop is a
A. 1 bit storage cell B. 2 bit storage cell
C. 3 bit storage cell D. 4 bit storage cell

284. The D flip flop and T flip flops are respectively used as
A. Toggle switch and delay gate
B. Delay and toggle switch
C. Both used as delay gates
D. Both used as toggle switch

285. Given the binary 00000111, its two's complement is obtained and expressed in hexadecimal form. The result is
A. F8 B. F9
C. – 79 D. None of these

286. The value of 2^5 in octal system is
A. 40 B. 20
C. 400 D. 200

287. The ASCII is an input output code
A. It is a two bit code
B. It is a four bit code
C. It is a seven bit code
D. It is an eight bit code

288. When signed numbers are used in binary arithmetic, then which one of the following notations would have unique representation for zero?
A. sign-magnitude B. 1's complement
C. 2's complement D. 9's complement

289. A carry look ahead adder is frequently used for addition because it
A. is faster B. is more accurate
C. used fewer gates D. costs less

290. Among the logical families, the lowest power dissipation is in :
A. TTL B. MOS
C. ECL D. None of these

291. The 'excess 3 code' for number 3 is
A. 0110 B. 0011
C. 0001 D. 1001

ANSWERS

1	**2**	**3**	**4**	**5**	**6**	**7**	**8**	**9**	**10**
B	B	A	A	B	C	D	D	C	D
11	**12**	**13**	**14**	**15**	**16**	**17**	**18**	**19**	**20**
C	A	C	B	B	D	C	D	A	C
21	**22**	**23**	**24**	**25**	**26**	**27**	**28**	**29**	**30**
B	D	C	A	B	B	D	B	D	A
31	**32**	**33**	**34**	**35**	**36**	**37**	**38**	**39**	**40**
A	A	B	B	A	C	C	D	C	C
41	**42**	**43**	**44**	**45**	**46**	**47**	**48**	**49**	**50**
C	C	B	A	B	D	D	C	A	C
51	**52**	**53**	**54**	**55**	**56**	**57**	**58**	**59**	**60**
C	A	B	C	B	C	C	D	B	B
61	**62**	**63**	**64**	**65**	**66**	**67**	**68**	**69**	**70**
C	C	B	B	C	B	A	C	B	A
71	**72**	**73**	**74**	**75**	**76**	**77**	**78**	**79**	**80**
C	B	B	A	A	B	B	C	B	A

81	**82**	**83**	**84**	**85**	**86**	**87**	**88**	**89**	**90**
C	A	B	D	D	A	C	B	C	A
91	**92**	**93**	**94**	**95**	**96**	**97**	**98**	**99**	**100**
B	A	A	C	C	D	C	A	B	A
101	**102**	**103**	**104**	**105**	**106**	**107**	**108**	**109**	**110**
B	B	B	B	C	B	A	C	B	D
111	**112**	**113**	**114**	**115**	**116**	**117**	**118**	**119**	**120**
A	B	C	D	C	D	B	B	C	C
121	**122**	**123**	**124**	**125**	**126**	**127**	**128**	**129**	**130**
B	C	D	C	C	C	B	C	D	A
131	**132**	**133**	**134**	**135**	**136**	**137**	**138**	**139**	**140**
D	C	C	C	A	C	B	D	C	D
141	**142**	**143**	**144**	**145**	**146**	**147**	**148**	**149**	**150**
D	B	A	B	D	D	C	A	C	B
151	**152**	**153**	**154**	**155**	**156**	**157**	**158**	**159**	**160**
D	C	A	B	B	B	C	A	B	D
161	**162**	**163**	**164**	**165**	**166**	**167**	**168**	**169**	**170**
B	C	D	C	C	C	C	A	D	C
171	**172**	**173**	**174**	**175**	**176**	**177**	**178**	**179**	**180**
D	C	C	C	B	B	A	D	A	B
181	**182**	**183**	**184**	**185**	**186**	**187**	**188**	**189**	**190**
B	B	B	B	C	B	A	A	D	B
191	**192**	**193**	**194**	**195**	**196**	**197**	**198**	**199**	**200**
B	B	A	D	B	B	B	C	C	C
201	**202**	**203**	**204**	**205**	**206**	**207**	**208**	**209**	**210**
A	D	D	B	C	A	B	A	D	D
211	**212**	**213**	**214**	**215**	**216**	**217**	**218**	**219**	**220**
B	C	B	D	A	B	D	D	D	C
221	**222**	**223**	**224**	**225**	**226**	**227**	**228**	**229**	**230**
C	B	B	B	B	A	C	A	A	C
231	**232**	**233**	**234**	**235**	**236**	**237**	**238**	**239**	**240**
A	B	B	B	A	B	A	C	C	A
241	**242**	**243**	**244**	**245**	**246**	**247**	**248**	**249**	**250**
B	B	C	D	C	A	C	A	D	A
251	**252**	**253**	**254**	**255**	**256**	**257**	**258**	**259**	**260**
A	D	B	D	B	D	D	A	D	C
261	**262**	**263**	**264**	**265**	**266**	**267**	**268**	**269**	**270**
C	C	D	A	A	A	C	A	B	C
271	**272**	**273**	**274**	**275**	**276**	**277**	**278**	**279**	**280**
B	A	B	C	C	C	C	C	C	B
281	**282**	**283**	**284**	**285**	**286**	**287**	**288**	**289**	**290**
A	C	A	B	B	A	C	C	A	**B**
291									
A									

EXPLANATORY ANSWERS

3. The truth table of circuit is

y_1	y_2	x_1	x_2
0	0	0	0
0	1	0	1
1	0	1	1
1	1	1	0

Equations for x_1 and x_2 are : $x_1 = y_1$, $x_2 = \bar{y}_1 y_2 + y_1 \bar{y}_2$. From the truth table, we see that it is a gray code

4. $1 = 10 \log_{10} N$

Hence $N = 10^{0.1} = 1.26$

8. To obtain dual of any boolean expression, the rule is

(i) Change + t_0 * and vice versa

(ii) Change + t_0* and vice-versa

Hence, dual of A + [B + (AC)] + D is

A [B (A + C)] D

12. Multiplexers can implementful adder by using proper control signals. Neither half adder nor decoders can alone do this. Similarly AND and OR gates require NOT gate for implementation.

13. The circuit of a 4-bit full adder using half adders and OR gates has been shown below. It is clear that it requires 7 full adders and 3 OR gates.

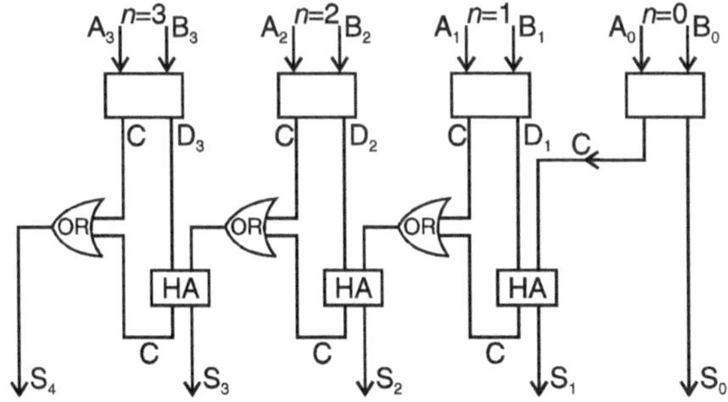

16. The expression for Y can be solved in terms of A and B as

$Y = \overline{A}\overline{B} + \overline{A}B$

$Y = \overline{A}B + A\overline{B}$

Hence, it is EX-OR gate.

19. F $= x + \bar{x}y\bar{z} + z$

$= x + (x + \bar{y} + z) + z$

$= x + \bar{y} + z$

$= \bar{x}\ \bar{y}\ \bar{z} = \bar{x}y\bar{z}$

Hence, Figure (A) shows correct minimized version.

26. The circuit is of 4 : 1 scalar states will be :

Initially	$Q_0 = 0$	$Q_1 = 0$
After first pulse	$Q_0 = 1$	$Q_1 = 0$
After second pulse	$Q_0 = 1$	$Q_1 = 1$
After third pulse	$Q_0 = 0$	$Q_1 = 1$
After four pulse	$Q_0 = 0$	$Q_1 = 0$

32. Given code is 0001001

Even parity at positions 1, 3, 5, 7 fails; $C_1 = 1$

Even parity at positions 2, 3, 6, 7 fails; $C_2 = 1$

Even parity at positions 4, 5, 6, 7 passes; $C_4 = C$

∴ Error is at position 011 = 3

Hence correct code is 0011001 and the corresponding BCD digit is 1001 = 9

70. For 1 byte memory, number of chips required

$$= \frac{1024 \times 8}{256} = 32$$

79. As both lines of A are high so its output irrespective of control input A and now second multiplexer has both high inputs so its output is independent of control input B. Hence, output is independent of both A and B.

85. The Boolean expression is

$A(B + \overline{C}) + \overline{A}B + (A + \overline{B})C$

$= AB + A\overline{C} + \overline{A}B + \overline{A}C + \overline{B}C$

$= (AB + \overline{A}B) + (A\overline{C} + AC) + \overline{B}C$

$= B + A + \overline{B}C$

$= A + B + C$

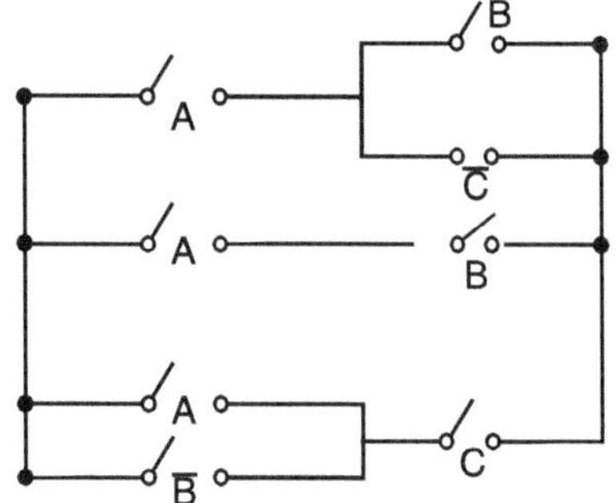

91.

C	B	1/P	f
0	0	0	x_0
0	1	A	x_1
1	0	A	x_2
1	1	1	x_3

$f = \overline{A}\,\overline{B}\,\overline{C} + AB\overline{C} + ABC + ABC$

96. When J = 1 and K = 1, the result is

$$\overline{Q}_{n+1} = \overline{Q}_n$$

In such a case, the output oscillates between 0 and 1. At the end of pulse, the value of Q_{n+1} is ambiguous.

98. If we have J = 0 and K = 0, then there will be no change in state so $Q_{n+1} = 1$. Similarly if J = 1 and K = 0, then also we will have $Q_{n+1} = 1$. So possible combination of J and K is $J = x$, K = 0.

107. For coincidence logic we must have

$$F = Z\overline{Y} + \overline{X}Y$$
$$= (\overline{X} + Y)(X + \overline{Y})$$
$$= XY + \overline{XY}$$

123. Since the combination of voltmeter resistance and 600 Ω is 400 Ω and voltage across it is 5 V, the unknown resistance across which the voltage is 35 – 5 = 30 V must be $\frac{30}{5} \times 400 = 2.4\ k\Omega$.

133. The operation A ⊕ B, i.e., exclusive OR is represented as $A\overline{B} + \overline{A}\,\overline{B}$. Hence $x \oplus \overline{y}$ is equal to $xy + \overline{x}.\overline{y}$

134. $AB + AC + B\overline{C} = AB(C + \overline{C}) + AC + AC + B\overline{C}$ .... $(\text{because } C + \overline{C} = 1)$

$$= ABC + AB\overline{C} + AC + B\overline{C}$$
$$= AC(B + 1) + B\overline{C}(A + 1)$$
$$= AC + B\overline{C}$$

137. Here L represents the initial value and M represents the greatest value which I can have and N is the increment given to I every time.

142. The number of bits in which two code words differ is known as the Hamming distance. In this case, only alternative with a Hamming distance of exactly 2 is (B).

143. The statement labels in a subroutine may be the same as in the main program because the Fortan compiler deals with one module at a time.

146. We cannot use real values in the arithmetic IF statement. Hence (D) is invalid.

148. According to the MUX's working principle, its truth table is, O – LO, 1 – HI

S_R	S_0	O/P	as per given condition	
0	0	X_1	=	HI i.e. 1
0	1	X_2	=	HI i.e. 1
1	0	X_3	=	HI i.e. 1
1	1	X_4	=	LO i.e. 0

At this matches the truth table of a NAND gate, so it is equivalent to NAND

A	B	NAND
0	0	1
0	1	1
1	0	1
1	1	0

151.

(A) 579. – E03	Minus sign cannot preceede E, hence incorrect
(B) 579. E	E must be followed by some two digit integer value
(C) E – 03	Some real value must preceede E, hence incorrect

only statement (D) is correct

152. Since after calculating the sum I is incremented by 1 and then compared with N and N is 25 here, so the sum of 24 elements will be computed.

156. The functional reflects a recursive definition of multiplication by repeated addition. The product $a^*\ b$, where a and b are positive integers, may be defined as a added to itself b times.

In other words $a * b = a$ if $b = 1$

$= a^*\ (b - 1) + a$ if $b > 1$

160. In D-type flip-flops the only inputs are either 1 and ϕ or ϕ and 1, which is a valid condition in all flip-flops. So any of the given flip-flops may be converted to a D-type flip-flop.

170. About O outputs, one can easily write product of sums form. If the input variable has a 1, then complement it otherwise not. So the product-of sums form would be

$$(\overline{X} + \overline{Y} + \overline{Z})(X + Y + Z)(\overline{X} + Y + Z)$$

176. Number of Boolean functions is 2^{2n} though the number of combination for 4 variables is

$$2^4 = 16.$$

For instance the 16 functions for 2 variables are:

f	*Binary row number*	*Name*
f_0	0000	Zero
f_1	0001	AND
f_2	0010	INHIBIT
f_3	0011	IDENTITY X
f_4	0100	INHIBIT
f_5	0101	IDENTITY Y
f_6	0110	EXCLUSIVER-OR
f_7	0111	OR
f_8	1000	NOR
f_9	1001	EQUALITY
f_{10}	1010	NOT Y
f_{11}	1011	IMPLICATION
f_{12}	1100	NOT X
f_{14}	1110	NAND
f_{15}	1111	ONE

183. 1's complement system has two O's which represent + O and —O example in 4 bit system the two O's are 0000 and 1111.

195. As minimum resolution required is 50 mV for 10 V maximum input, the minimum number states required $= \dfrac{10\text{V}}{50\text{mV}} = 200$

$\therefore$ Minimum number of bits required is given by, $200 \leq 2^N - 1$

Solving we get, N = 8

196. For an N-bit D/A converter, total number of non-zero states are $2^N - 1$.

$\therefore$ Resolution is $\dfrac{1}{2^N - 1}$ or % resolution is $\dfrac{100}{2^N - 1}$.

203. Coincidence logic means the condition when inputs are same, i.e., all are 0 or all are 1

To detect this condition, the boolean expression is

$AB + \overline{A}\,\overline{B}$ which has been implemented in circuit (D).

208. As each T – FF receives its independent T and clock input, it is a simple binary counter.

215. An impulse voltage rises rapidly to its maximum value and falls rapidly to zero. If an impulse voltage develops without causing a flash over or puncture, it is called *chopped impulse voltage*.

217. Resolution $= \dfrac{1}{9999}$ count

$= \dfrac{1}{9999} \times 9.999$ V

$= 10^{-3}\text{V} = 1\text{m V}$

226. The output of (A) will be A.B + C.D + E

The output of (B) will be $(A+B).(C+D).\overline{E}$

The output of (C) will be $A.B + C.D + \overline{E}$

The output of (D) will be (A + B) . (C + D) . E

236. The EQUIVALENCE statement in FORTRAN is used to treat two variable names as same.

237. Since INTEGER is a reserved word in FORTRAN, we can't use it as integer variable.

244. When J = 1 and K = 1, the result is $\overline{Q}_{n+1} = \overline{Q}_n$

In such case, the output oscillates between 0 and 1. At the end of pulse, the value of Q_{n+1} is ambiguous.

246. If we have J = 0 and K = 0, then there will be no change in state so $Q_{n+1} = 1$. Similarly if J = 1 and K = 0 then also we will have $Q_{n+1} = 1$. So possible combination of J and K is $J = x$, K = 0.

252. 256 × 8 = 2048 bit ROM

M = $8(2^8 = 256)$,

N = 8

Normally 256 NAND gates are required. But by 2 dimensional addressing, the number of gates may be reduced.

For 8 output, eitht 8 × 1 MUX_s are required.

This will give 8 × 8 = 64 lines (Y address)

Number of columns (X address) $= \dfrac{2048}{64} = 32$

Now 32 NAND gates required.

254. $X(X+\overline{X}Y)Z(X+Y+Z) = XZ(X+Y+Z)$

$= XZ + XYZ$

$= XZ(1+Y) = XZ$

256. Choice (A) is : $11100_2 - 10001_2 = 28_{10} - 17_{10}$

$= 01011_2$

Choice (B) : $15\text{E}_{16} = (1 \times 16^2) + (5 \times 16) + (14 \times 16^0)$

$= 256_{10} + 80_{10} + 14_{10} = 350_{10}$

Choice (C) : $81_{10} = (1\times 2^6) + (0\times 2^5) + (0\times 2^4) + (0\times 2^3) + (0\times 2^2) + (0\times 2^1) + (0+2^0) = 1010001_2$

Choice (D) Octal : 3 7 4

Binary : 011 111 100

257. Given, J = 5 × A – 2** 4/2

Substituting the value of A in this expression, J

$= 5 \times 2.5 - 2^{**} 4/2$

$= 12.5 - \underline{16}$

$= 4.5 - \overline{2}$

(because J is an integer variable so the fractional part of the result will not be stored).

261. Since N has to store real values, it must be declared as REAL N

264. Output of the four gates has been summarised in the following table :

A	B	NAND	NOR	OR	AND
0	0	1	1	0	0
0	1	1	0	1	0
1	0	1	0	1	0
1	1	0	0	1	1

So the gate whose output is low if and only if all imparts are high is NAND gate.

288. In sign-magnitude, 1's complement and 9's complement, + 0 and – 0 are represented differently but in 2's complement there is single notation, for zero.

3

ELECTROMAGNETIC THEORY

FORCE BETWEEN POINT CHARGES—COULOMB'S LAW

Coulomb's law, states that the force F between two stationary point charges Q_1 and Q_2 is proportional to the product of the charges and inversely proportional to square of the distance R between them; i.e.,

$$F = k\frac{Q_1 Q_2}{R^2} \qquad (N) \qquad ...(1)$$

The force F is along the line connecting the two charges. It is a force of repulsion if the charges are of like sign and a force of attraction if the charges are of unlike sign. K in (1) is a constant of proportionality whose value depends on the system of units. In vector notation form, and using the International System of units, abbreviated SI, (1) be

$$\bar{F}_2 = \left(\frac{1}{4\pi \in_0}\right)\frac{Q_1 Q_2}{R_{12}^2} a_{R12} \qquad (N) \qquad ...(2)$$

where $\bar{F}_2$ = force on point charge Q_2 due to point charge Q_1 (N) ;

Q_1 = charge at location (1) (C)

Q_2 = charge at location (2) (C)

R_{12} = distance between the location of Q_1 and Q_2 (m)

a_{R12} = unit vector along a straight line from Q_1 to Q_2 locations (from Q_1, the location of the force, to the location of the force $\bar{F}_2$ at Q_2)

$\in_0$ = permittivity of free space (F m^{-1} farads per meter)

One can readily show that (2), the vector form of Coulomb's law (1), gives the correct magnitude and direction for like or unlike point charges. It should be noted that Q_1 and Q_2 take on the sign of the charge. In units, the constant of proportionality in (1) becomes

$$k = \frac{1}{4\pi \in_0}$$

where $\in_0$, the permittivity of free space, is

$$\in_0 = 8.854 \times 10^{-12} \cong \frac{1}{30\pi}10^{-9} \qquad \left(F\ m^{-1}\right)$$

Coulomb's law is based on point charges and linearity exists; thus, when Q_1 is increased by a factor m, the force $\bar{F}_2$ is also increased by the same factor m. More than two point charges are present, the force on one of these charges is equal to the vector sum of the forces on that charge due to each of the other charges acting. Thus, we can say that superposition can be extended to the summation of vector due to point charges. Through the use of (2) we will find the force between two charges of 1 coulomb, separated 1 meter, to equal approximately 9×10^9 N or approximately 9 million tons. This shows that the coulomb is a large unit of charge.

Electric Field of many charges

If there are several point charges q_1, q_2, q_3..........q_n located at different point of electric field intensity at point p is.

$$\bar{E} = \frac{\bar{F}}{q} = \frac{q_1}{4\pi \in_0 R_1^2}\bar{i}_{R1} + \frac{q_2}{4\pi \in_0 R_1^2}\bar{i}_{R2} + ... + \frac{q_n}{4\pi \in_0 R_n^2}\bar{i}_{Rn}$$

$$= \sum_{j=i}^{n} \frac{q_j}{4\pi \in_0 R_i^2}\bar{i}_{Rj}$$

Electric Flux

1. Electric flux begins on positive charges and terminates on negative charges.
2. Flux is in the same direction as the electric field $\overline{E}$.
3. Flux density is proportional to the magnitude of E.
4. In the SI system of units, the total flux emanating from a charge of Q (C) is N-m^2/ coulomb

   A single line will emanate from 1 C of charge.

Electric Flux Density Vector

In free space the electric flux density vector $\overline{D}$ is defined to be in, the same direction as the electric flux lines (same as $\overline{E}$) and to have magnitude.

$$|\overline{D}| = \lim_{\Delta s \to 0}\left[\frac{\Delta\Psi_E}{\Delta_s}\right] \quad \text{(lines m}^{-2}\text{ (aap)}$$

where $\Delta\Psi_E$ equals the number of electric lines of flux that are perpendicular to the suface Δs.

GAUSS'S LAW

The total electric flux passing through any closed imaginary surface, enclosing the charge Q (C), is equal to Q (C) in the SI units. Note that Q is enclosed by the closed surface. For this reason we will refer to Q as Q enclosed, or just Q_{en}. A graphical representation of this last statement, $\Psi_E = Q_{en}$, is shown in Fig.1. The total flux Ψ_E is thus equal to

$$\Psi_E = \oint_B d\Psi_E = \oint \overline{D}s.\overline{ds.} = Q_{en} \qquad (C)$$

where $\oint_s$ indicates a double integral over the closed surface s. The mathematical form, from above equation

$$\oint_s \overline{D}s.\overline{ds} = Q_{en} \qquad (C)$$

is named Gauss's law. The Q_{en} enclosed by surface s, due to a. p_v distribution, becomes

$$Q_{en} = \int_v pv \ du \qquad (C)$$

Divergence

The divergence of $\overline{D}$ equals the net flux of the vector $\overline{D}$ that flows outwardly through a closed surface s per unit volume (enclosed by $\oint s$) as the volume goes to zero.

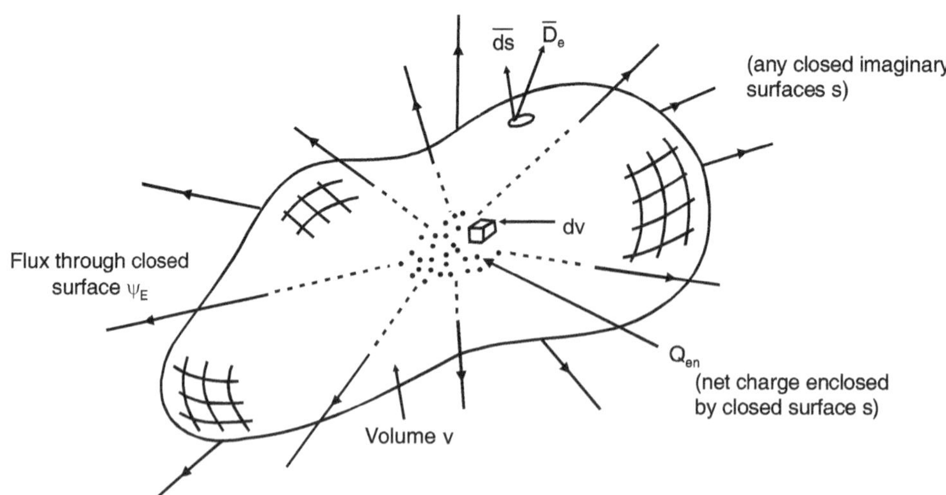

Fig. 1: *The total electric lines of flux Ψ_E passing through any closed imaginary surface enclosing the charge Q_{en} (C) is equal to Q_{en} (lines)*

$$\underbrace{\text{Div } \overline{D}}_{\text{Concept}} = \underbrace{\nabla \cdot \overline{D}}_{\substack{\text{Vector analysis}\\\text{compact}\\\text{symboli-}\\\text{cation}}} \overset{\Delta}{=} \underbrace{\lim_{\Delta h - 0} \left[\frac{\oint_s \overline{D} \cdot \overline{ds}}{\Delta u} \right]}_{\text{Defining equation}}$$

$$= \underbrace{\left(\frac{dD}{dx} + \frac{dD}{ds} + \frac{dD}{dz} \right)}_{\substack{\text{Mathematical relationship}\\\text{resulting from application}\\\text{of defining equation in}\\\text{the rectangular coordinate}\\\text{system}}} = p \quad \underbrace{(\text{Cm i aap})}_{\substack{\text{Physical quantity}\\\text{of the point p}}}$$

Thus, in rectangular coordinates

$$\text{Div. } \overline{D} = \nabla. \overline{D} = \frac{\partial D_x}{\partial x} + \frac{\partial D_y}{\partial y} + \frac{\partial D_z}{\partial z} \ (\text{cm}^{-3})(\text{aap}).$$

From the last two terms of eq. in box the divergence concept gives us a partial differential equation that relates rates of change of the $\overline{D}$ component to the charge density p_v at a point. This equation is expressed in compact form as

$$\Delta.\overline{D} = p_v \qquad \left(\text{Cm}^{-8}\right) \qquad (\text{aap})$$

Divergence Theorem

$$\oint s\overline{D}.\overline{ds} = \int_v \nabla.\overline{D}dv$$

The divergence theorem relates a closed surface integral to a volume integral involving the same vector. It should be noted that the closed surface s encloses the volume v.

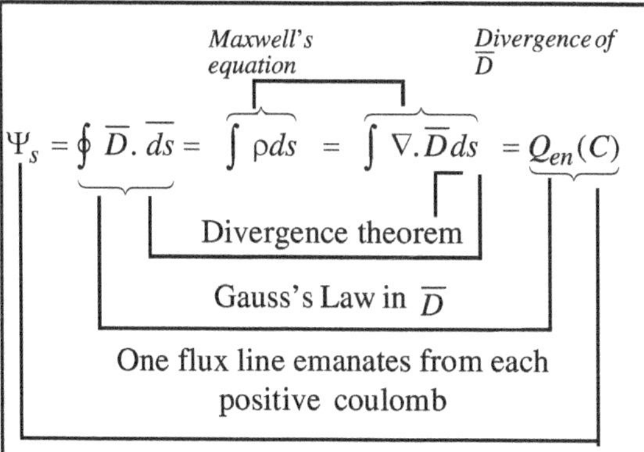

Potential Difference (Potential) About A Point Charge

The potential difference between points a and b, V_{ab} is defined as the work in joules that we must expend in moving a unit positive point charge from point b to a in an electric field $\overline{E}$.

CURRENT

Current is defined as the movement of charge through a given surface and is equal to the coulombs per second through that surface.

DIPOLE

The electric dipole, or simply a dipole, is the name given to two point charges of equal magnitude and opposite sign, separated by a distance which is small compared to the distance to the point P at which we want to know the electric and potential fields.

CONDUCTORS AND CONDUCTIVITY

A conducting material (medium) has electric charge conduction as its prominent characteristic. A metal conductor has enormous electric charge conduction through the presence of a large number of outer-orbit electrons that are free to move about in the lattice structures of the material. These free electrons, with zero external E applied within the conductor, move in random directions and with varying velocities to produce zero net current through any surface in the conductor. The direction of the electron motion is changed in the process of collision with the thermally excited lattices. Now, if we apply an electric field, the electrons will still have motion in random directions but will drift slowly in the–E direction with an average velocity called drift velocity $\overline{U}_d$. The motion of electrons in the $-\overline{E}$ direction with a velocity $\overline{U}_d$ gives rise to a conduction current in the conductor. The drift velocity $\overline{U}_d$ of the electrons is related to the externally applied $\overline{E}$ field through

$$\overline{U}_d = -\mu\overline{E}$$

where μ is a positive quantity called the electron mobility

Point form of Ohm's law

$$\overline{J} = \sigma\overline{E} \qquad (\text{A m}^{-2})$$

where σ is the conductivity.

The conductivity σ is thus a measure of the free electron conductive properties of a conductor. In good conductors its value decreases with temperature since increased thermal energy of the lattice structure increases lattice vibration, thus increasing the collision possibility with the moving free electrons. Material possessing low values of conductivity are called insulators.

The reciprocal of σ is called resistivity and is generally symbolized by ρ with no subscript and has the units of (Ω.m). Care should be taken not to confuse this with the symbols used for charge density ρ_e, ρ_s, and ρ_u.

Also

$$R = \frac{L}{\sigma s} \quad (\Omega)$$

where R is the resistance in ohms to free-electron flow or to the current I of a conductor of uniform cross section s length L and uniform $\bar{E}$ (thus uniform $\bar{J}$).

SEMICONDUCTORS

All materials are classified on the basis of their value of conductivity σ as conductors, semi-conductors, or insulators. σ varies over a range of about 25 orders of magnitude from that of silver, a good conductor, to fused quartz, a good insulator. A display of this overall range and defined subranges of σ for conductors, semiconductors and insulators can be found in Fig. 2.

A middle range of conductivity, from about 10^{-8} to about 1S m^{-1}, exists where the materials are called intrinsic (pure) semiconductors, for they are neither conductors nor insulators. Two intrinsic semiconductors of importance in electrical engineering are intrinsic (pure) germanium and intrinsic (pure) silicon. Their importance is due to their property of increase in conductivity with temperature and through addition of impurities in a process called doping. The increase in conductivity with temperature property is found in a device called a thermistor while the controlled increase in conductivity through doping property is found in semiconductor diodes and transistors.

In a good conductor, the free electrons are the carriers while in an intrinsic (pure) semiconductor two current carriers are present–free electrons and bound electrons bound positive holes). The lattice structure of intrinsic germanium or silicon has four covalent bonds between each of its atoms. At temperature near absolute zero, the outer shell electrons are held tightly within the covalent bonds and thermal vibration of the lattice structure is zero. Under this condition, there are no free carriers and the material is an insulator. As the temperature is increased, thermal vibration of the lattice structure is increased with the result that some electrons gain sufficient thermal energy to break away from the covalent bond and become free electrons in random motion similar to free electrons in a conductor. Each electron that leaves a covalent bond creates a bond

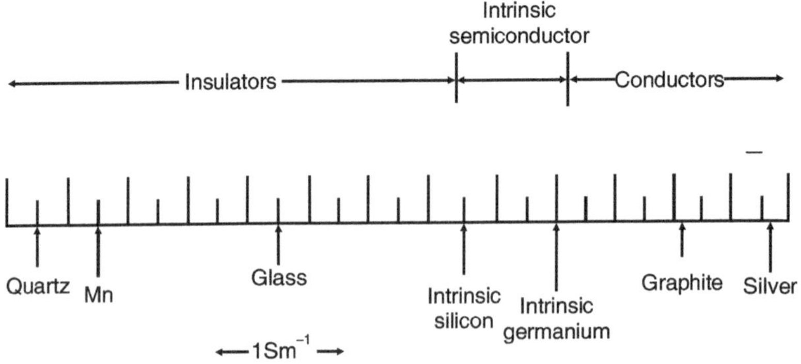

Fig. 2: *Graphical display of range of conductivity from conductors to good insulators and defining sub-ranges for conductors, semiconductors; and insulators*

vacancy called a hole. These holes also move randomly throughout the lattice since electrons from other covalent bonds can make a transition into a hole position, and thus the hole replaces the initial position of the covalent bond electron. It should be noted that the covalent bond electron is not a free electron but a bound electron that has insufficient thermal energy to become a free electron. With zero applied $\overline{E}$ within the semiconductor, the net current due to free electrons and holes (bound electrons) is zero since the drift velocity of each is zero. Now, with an applied $\overline{E}$ the free electrons will drift with a $\overline{U}_c$ in the $-\overline{E}$ direction while the holes drift with $\overline{U}_h$ in the $\overline{E}$ direction. Thus, the total conduction current is due to two carriers : free electrons and covalent-bound electrons that can be visualized as covalent-bound holes. Since each free electron creates a covalent-bound hole, the free electron density is equal to the covalent-bound hole density.

Many semiconductor for properties can be explained by treating the hole as if it had positive charge equal to $|q_e|$ and a mobility of μh.

With two current carriers, the expression for the conduction current density becomes

$$\overline{J} = \left(-\rho ve\mu_c{}^{-1}\rho vh\mu h\right)\overline{E} = \sigma\overline{E} \qquad ;(A\ m^{-2})$$

From (A), σ becomes

$$\sigma = \left(-\rho ve\mu_e + \rho uh\mu h\right) \qquad (S\ m^{-1})$$

Table 1, lists some of the important parameters of germanium and silicon

With increasing temperature, the mobilities μ_e and μh decrease, but the free-electron density increases more rapidly to increase the conductivity.

DIELECTRICS (INSULATORS)

Insulators are commonly referred to as dielectrics. The prominent characteristic of dielectrics materials is polarization; i.e., the formation of electric dipoles.

Polarization in Dielectrics

These electrics dipoles have an effect on the electric field in which the dielectric material is placed.

There are three basic polarization mechanisms that produce polarization; they are :

1. Electronic polarization exists in an atom when, upon the application of an electric field, the center of the cloud of electrons is displaced relative to the center of the nucleus. (see Fig. 3.)
2. Ionic polarization exists in a molecule having ionic bonds that can be viewed as a build up of positive and negative ions. An applied electric field will displace the positive ions relative to the negative ions, Sodium chloride (NaCl) is a good example of this type of polarization.
3. Orientational polarization exists in a material whose molecules, because of structure, possess permanent dipole moments that are randomly oriented in the absence of an applied electric field but tend to orient themselves in the direction of an applied electric field. This type of material is also called polar material, water, (H_2O) being a good example.

TABLE 1: Some Important Parameters of Germanium and Silicon at 300°K

Parameter	*Symbol units*	*Intrinsic germanium*	*Intrinsic silicon*
Free-electron mobility	$\mu_e\left(m^2V^{-1}s^{-1}\right)$	0.39	0.135
Hole mobility	$\mu_h\left(m^2V^{-1}s^{-1}\right)$	0.19	0.048
Free-electron density	$\eta_e\left(m^{-3}\right)$	2.4×10^{19}	1.5×10^{16}
Hole density	$\eta_h\left(m^{-3}\right)$	$2.4 + 10^{19}$	1.5×10^{16}
Conductivity	$\sigma\left(Sm^{-1}\right)$	2.2	0.44×10^{-3}

Dielectric material polzarized through orientational polarization is called polar material while material polarized through electronic or ionic polarization is called non-polar meterial. It should be noted that in the process of polarization the charges forming the electric dipoles are bound charges.

Linearity, Homogeneity, and Isotropy

A dielecttric material is considered linear when, at a given point, $\in$ is a constant. Thus, a plot of $|\overline{D}|$ versus $|\overline{E}|$ will be a straight line. If the permittivity, at a given point is a function of $|\overline{E}|$. $\in = \in(|\overline{E}|)$. The material is non-linear and a plot of $|\overline{D}|$ versues $|\overline{E}|$ will not be a straight line.

A dielectric material is homogeneous when $\in$ does not vary from point to point; thus, $\in$ is not a function of position. If the permittivity varies with position, $\in = \in$ (X, Y, Z), then the material is non-homogeneous.

The permittivity of a dielectric can also vary with frequency and temperature. The dependency on frequency becomes an important factor at high frequencies.

CAPACITANCE

The capacitane of a two-conductor capacitor is defined as the ratio of the magnitude of charge on one of the conductors to the magnitude of the potential difference between the two conductors.

Thus, the mathematical definition of capacitance becomes

$$C = \frac{|Q|}{|V_{ab}|} \quad (C\,V^{-1} \text{ or } F)$$

where V_{ab} is the potential difference between any point on conductor *b* and any point on conductor *a*. It should be noted that under static conditions each conductor is at a constant potential.

Laplace and Poisson's Equations

The equation

$$\nabla^2 V = -\frac{\rho v}{\in} (V\,m^{-2}) \qquad ...(i)$$

is called Poisson's equation and applies to a homogeneous media. When the free charge density $\rho v = 0$, (i) will reduce to

$$\nabla^2 V = 0 \,(V\,m^{-2})$$

and is called Laplace's equation for homogeneous media. Here ∇^2 is called the Laplacian operator and $\nabla^2 V$ is the Laplacian of V.

The expanded expressions for Laplace's equation in rectangular, cylindrical, and spherical coordinate systems became

Rectangular

$$\nabla^2 V = \frac{\partial^2 V}{\partial x^2} + \frac{\partial^2 V}{\partial y^2} + \frac{\partial^2 V}{\partial z^2} = 0$$

Cylindrical :

$$\nabla^2 V = \frac{1}{r_c}\frac{\partial}{\partial r_c}\left(r_c \frac{\partial V}{\partial r_c}\right) + \frac{1}{r_c^2}\frac{\partial^2 V}{\partial \phi^2} + \frac{\partial^2 H}{\partial z^2} = 0$$

ELECTRIC FLUX DENSITY (D)

Electric flux density or electric displacement *(D)* is given by the normal flux per unit area.

Mathematically, $$D = \frac{\psi}{A} C/m^2$$

It is related to electric field intensity by the relation

$$D = \varepsilon_0 \varepsilon_r E$$

GAUSS'S THEOREM

It states as follows :

"If charge Q is enclosed within a surface, the total flux crossing the surface is Q whatever may be distribution of the charge enclosed." It is assumed that medium is homogenous.

Also $\psi = \sum Q$

"The electric flux through any closed surface is equal to the charge enclosed"

$$\oiint \vec{D} \cdot \vec{ds} = Q = \iiint_V$$

In the point or differential form it is given by

$$\Delta . \vec{D} = \rho$$

Above equations are also known as Maxwell's equations in integral and differential forms.

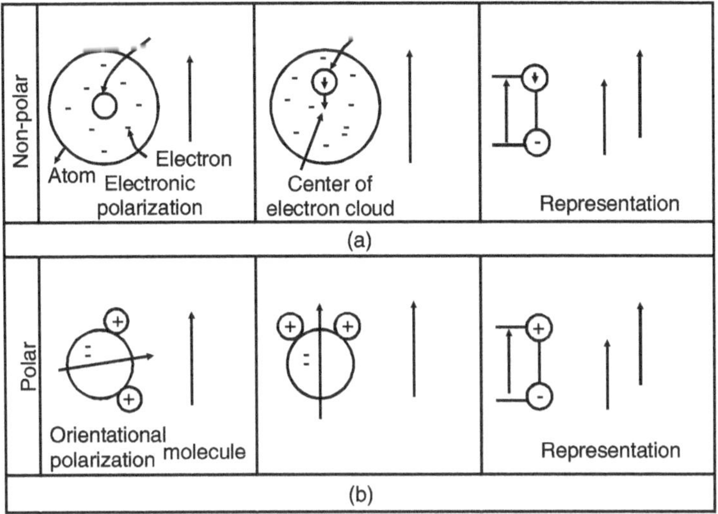

Fig. 3: *Graphical representation of a polarization mechanism on a microscopic base: (a) electronic (non-polar) polarization; (b) orientational (polar) polarization*

Electrical Potential and Potential Difference

"The eletric potential at any point is given by the work which would have to be done in bringing a unit positive charge from an infinite distance to that point, or alternatively from a place of zero potential to the point".

The difference of electric potential between two points is the work done in moving a unit charge from one point to another.

Equipotential surface : An *"equipotential surface"* is one having all point at the same potential.

Potential gradient : The potential gradient is the rate of change of potential measured in the direction of electric force.

The *potential gradient*, in any electrostatic field, *is equal to the field strength.*

CAPACITANCE

Capacitor

A capacitor is a device *capable of storing electric charge*. It consists of *two conducting surfaces* (may be in form of either circular or rectangular plates or of spherical or cylindrical shapes) separated by an insulating material called a *dielectric*.

Capacitance (C)

Capacitance is a measure of ability of a capacitor to store an electric charge. It is the ratio of the charge (Q) that can be stored to the voltage applied (V) across the plates.

Mathematically, $C = \frac{Q}{V}$.

The formulae of capacitance for various capacitors are given below:

1. Parallel plates capacitor :

(i) *Uniform dielectric medium :*

$$C = \frac{\varepsilon_0 \varepsilon_r A}{t} \text{..... in a medium}$$

$$= \frac{\varepsilon_0 A}{t} \text{..... in air}$$

[where t = thickness of the dielectric, and

ε_r = relative permittivity].

(ii) *Medium partly air :*

$$C = \frac{\varepsilon_0 A}{\left[d - \left(t - \frac{t}{\varepsilon_r}\right)\right]}$$

(iii) *Composite medium :*

$$C = \frac{\varepsilon_0 A}{\frac{t_1}{\varepsilon_{r_1}} + \frac{t_2}{\varepsilon_{r_2}} + \frac{t_3}{\varepsilon_{r_3}}}$$

(iv) *Multi-plate capacitor* :

$$C = \frac{(n-1)\varepsilon_0\varepsilon_r A}{t}$$

(where n = number of plates)

2. Cylindrical capacitor :

$$C = \frac{2\pi\varepsilon_0\varepsilon_r}{\log_e r_2 / r_1} F/m$$

Potential gradient in a cylindrical capacitor (g)

$$g_{max} = \frac{V}{r_1 \log_e r_2 / r_1}$$

$$g_{min} = \frac{V}{r_2 \log_e r_2 / r_1}$$

Capacitors in Series

In series combination, charge on all capacitors is same but p.d. across each is different.

$$\frac{1}{C} = \frac{1}{C_1} + \frac{1}{C_2} + \frac{1}{C_3}$$

$$\left[\text{or } C = \frac{C_1C_2C_3}{C_1C_2 + C_2C_3 + C_3C_1}\right]$$

Capacitors in Parallel : In this case, *p.d. across each capacitor is same but charge on each is different.*

$$C = C_1 + C_2 + C_3$$

Energy Stored in a Capacitor : The energy stored in a capacitor equals the work done to put the charge on the capacitor in opposition to the voltage on the capacitor.

Electric energy, $W = \frac{1}{2}CV^2$

Dielectric Strength : The dielectric strength *refers to the breakdown voltage of an insulator.* It is given by the maximum potential difference which a unit thickness of the medium can withstand without breaking down.

Insulating material	*Dielectric strength (kV/mm)*
Air	3.2
Glass	12-20
Mica	20-60
Paper (varnished)	18
Porcelain	15

Charging and Discharging of a Capacitor

Charging :

(i) $v = V\left(1 - e^{-t/\lambda}\right)$

(ii) $q = Q\left(1 - e^{-t/\lambda}\right)$

(iii) $i = I_m e^{-t/\lambda}$

Time constant (λ)**.** "The time constant of R-C circuit is the time during which voltage across capacitor would have reached its maximum value V had it maintained its initial rate of rise"

$$\lambda = CR$$

$$\begin{bmatrix} v = 0.632V \\ i = 0.37\ I_m \end{bmatrix}$$

Discharging :

(i) $v = Ve^{-t/\lambda}$

(ii) $q = Qe^{-t/\lambda}$

(iii) $i = -I_m e^{-t/\lambda}$

Types of Capacitors :

1. Air capacitors
2. Paper capacitors
3. Plastic film capacitors
4. Mica capacitors
5. Ceramic capacitors
6. Electrolytic capacitors.

Classification of Magnetic Materials

In accordance with the value of *relative permeability the magnetic materials may be classified in the following three ways* :

1. **Ferromagnetic materials :** The relative permeabilities of these materials are *much greater than unity* and are dependent on the field strengths.
   Examples : *Iron, cobalt and nickel.*
   Gadolinium however also comes under this classification. These materials have *have high susceptibility.*
2. **Paramagnetic materials :** These have relative permeabilities *slightly greater than unity* and are *magnetised slightly.*
   Examples : *Aluminium, platinum* and *oxygen.*
3. **Diamagnetic materials :** The relative permeabilities of these materials are *slightly less*

than unity. They *repel* the lines of force *slightly*. **Examples** : *Bismuth, silver, copper* and *hydrogen*.

Magnetic Soft Materials

The magnetically soft materials (*suitable for making electromagnets*) are characterised as follows:

1. They have high permeability.
2. The magnetic energy stored is not high.
3. They have negligible co-service force (due to which these are not suitable for making permanent magnets).
4. They have low remanance.

Examples : *Pure or ingot iron, manganese and nickel steels, cast iron, silicon steels, carbon steels, mumetal, perminvar, permalloy.*

Magnetically Hard Materials

These are suitable for making *permanent magnets* and have the following characteristics.

1. They possess high value of *BH* product.
2. High retentivity.
3. High coercivity.
4. Strong magnetic reluctance.
5. Hysteresis loop is more rectangular in shape.

Examples : *Tungsten steel, cobalt steel, chromium steel, alnico, cunife, hypernic.*

LAWS OF MAGNETIC FORCE

Coulomb, through his experiments, found that the force between two magnetic poles placed in a medium is

(i) directly proportional to their pole strengths (m_1, m_2),

(ii) inversely proportional to the square of the distance (d) between them, and

(iii) inversely proportional to the absolute permeability (μ) of the surrounding medium.

i.e., $$F \propto \frac{m_1 m_2}{d^2}$$

or $$F = k\frac{m_1 m_2}{\mu d^2}$$

(where k = constant)

In the S.I. system, the value of $k = \dfrac{1}{4\pi}$

$$\therefore \quad F = \frac{m_1 m_2}{4\pi\mu d^2} = \frac{m_1 m_2}{4\pi\mu_0\mu_r d^2}$$

$(\because \mu = \mu_0\mu_r)$

Magnetic Field due to a Current Carrying Conductor

1. When an electric current flows through a wire, a magnetic field is built up around the wire itself. This can be seen using a cardboard, iron filings, and a current-carrying wire. When the wire passed through the cardboard and the current flowing, iron filing are sprinkled on the cardboard. They can be seen arranging themselves in a magnetic field (Fig. 4) The magnetic lines of force are referred to as flux. Just as in a natural magnet, the field is strongest near the wire and diminishes as the distance from the wire increases.
2. Flux around a wire does have direction. Flux direction is determined by the direction of electron flow within the wire. As shown in Fig. 5 the North pole of the compass needle indicates the direction of flux or magnetic field around the wire. The dot in the centre of the wire on the left indicates the point of the current-direction arrow coming toward the observe; the X at the right represents the tail of the current arrow pointing away from the observer. If the direction of electron flow within the wire is reversed, the compass needles will reverse themselves, indicating a change in flux direction.

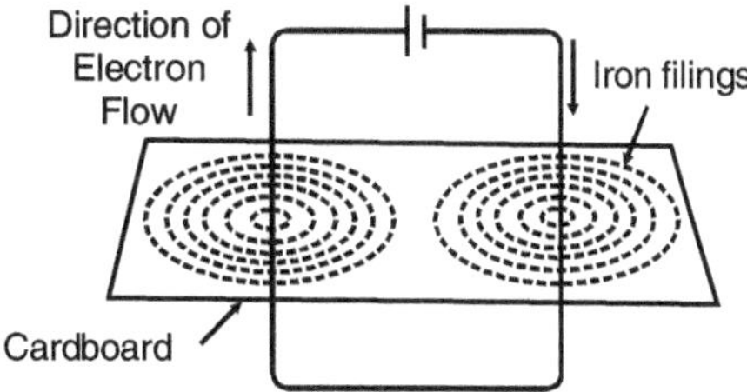

Fig. 4: *Magnetic field around a wire that is carrying electric current*

Right hand rule (or right hand screw rule)

The direction of the magnetic field can be found by using *right hand rule or the right hand screw*

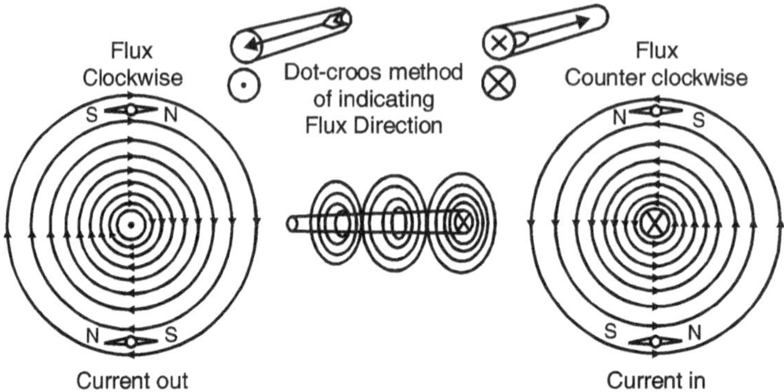

Fig. 5: *Compasses indicate the direction of flux around a wire*

rule. The right hand rule states as follows :

"Grasp the wire in the right hand with the thumb pointing in the direction of the current. The fingers will curl around the wire in the direction of the magnetic field".

Fig. 6 illustrates this rule.

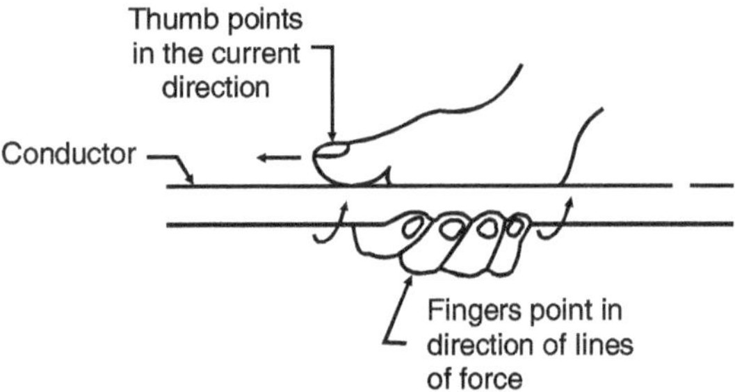

Fig. 6: *Right hand rule (or right hand screw rule)*

The right hand screw rule can be explained as follows:

As a wood screw is turned clockwise it moves (or progresses) into the wood. The horizontal direction of the screw is analogous to the direction of current in a conductor. The circular motion of the screw shows the direction of the magnetic flux around the conductor.

Force on A Current-Carrying Conductor Lying in A Magnetic Field

Refer Fig. 7 It has been found that whenever a current-carrying conductor is placed in a magnetic field, it experiences a force which acts in a direction perpendicular both to the direction of the current and the field.

The force developed in the conductor is given by the relation :

$$F = BIl \text{ newtons}$$

$$(= \mu_0 \mu_r HIl \text{ newtons})$$

where F = force developed in the conductor

B = flux density, T (Wb/m^2)

I = current in the conductor, A

l = exposed length of the conductor, m

$$\begin{bmatrix} \mu_0 = \text{absolute permeability} \\ \mu_r = \text{relative permeability} \\ H = \text{magnetising force} \end{bmatrix}$$

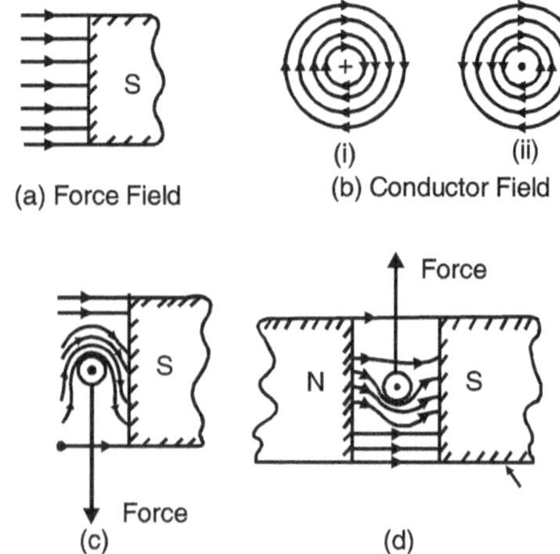

Fig. 7: *Force on a current-carrying conductor applying a magnetic field*

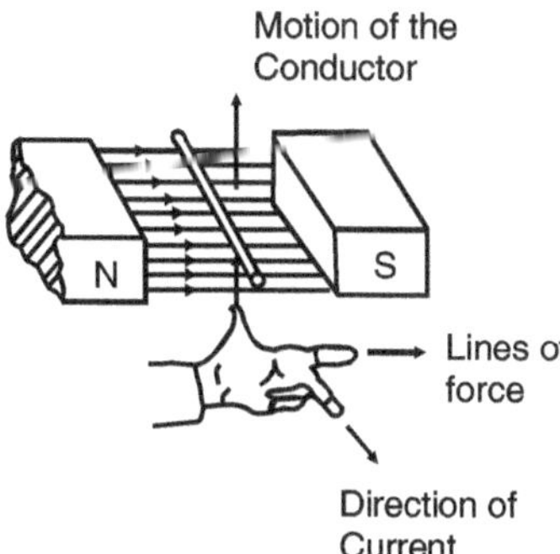

Fig. 8: *Fleming's left hand rule*

The Direction of this force may be easily found by Fleming's left hand rule which states that,

Place your left hand with index finger, middle finger and thumb at right angles. If the index finger points in the direction of the flux from north to south and middle finger points in the direction of the imposed voltage and its resulting conventional current flow, then the thumb points in the direction of the force that is developed.

Magnetising Force (H) of A Long Straight Conductor and A Solenoid

Straight Conductor :

$$H = \frac{NI}{2\pi r} \text{AT/m}$$

$$B = \frac{\mu_0 \mu_r NI}{2\pi r} \text{Wb/m}^2 \ (or\ T) \text{ .... in a medium}$$

$$= \frac{\mu_0 NI}{2\pi r} \text{Wb/m}^2 \ (or\ T) \text{ .... in air}$$

Distance of the point from the centre of the conductor.

Solenoid :

$$H = \frac{NI}{l} \text{AT/m}$$

$$B = \frac{\mu_0 \mu_r NI}{l} \text{ ... in a medium}$$

$$= \frac{\mu_0 NI}{l} \text{ ... in air}$$

FORCE BETWEEN PARALLEL CONDUCTORS - AMPERE'S LAW

$$F = \frac{\mu_0 I_1 I_2 l}{2\pi d} \text{ newtons} \qquad \text{...(1)}$$

where, $F =$ force between two parallel conductors,

$I_1, I_2 =$ current flowing through two parallel conductors,

$l =$ length of each conductor, and

$d =$ distance between the conductors.

Eqn. 1 is known as Ampere's law and is used to define the ampere in S.I. units.

If $l = d = 1$ m ; $I_1 = I_2 = 1$ A, then $F = 2 \times 10^{-7}$ N

Hence, one amperes is defined as follows :

"An ampere is that current when flowing in each of the two infinitely long parallel conductors situated in vacuum and separated 1 metre between centres, produces on each conductor a force of 2 × 10^{-7} per metre length".

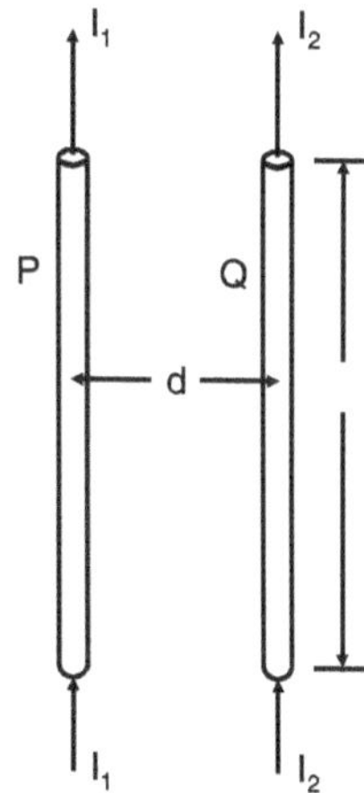

Fig. 9: *Force between two parallel conductors*

Ampere's Circuital Law

It states that the *line integral of magnetic field intensity $\vec{H}$ around a single closed path is equal to the current enclosed.*

$$\text{Thus, } \oint_c \vec{H} \cdot \vec{dl} = I = \iint_s \vec{J} \cdot \vec{ds}$$

In differential or point form, it is given by

$$\Delta \times \vec{H} = \vec{J}$$

Comparison of Electric and Magnetic Circuits

The analogy between electric and magnetic circuits is given in the Table.

Differences

1. The *resistivity* of conductors is more or less *constant* but the *permeability* of the ferro-magnetic materials *varies* greatly with magnetic field strength.
2. Strictly speaking, flux does not actually 'flow' in the sense in which an electric current flows.
3. In an electric circuit energy must be supplied to *maintain* the flow of electricity in the circuit, whereas (in a magnetic circuit) the magnetic flux once, it is set up, does not require any further supply of energy.
   (Example : Once the flux produced by a current in the solenoid has attained it steady value, the energy subsequently absorbed by that solenoid is all dissipated as heat due to the resistance of the winding).
4. In ferromagnetic materials if the magnetic field strength is increased the flux density increases till a state of saturation is reached after which there is no increase in flux density for further increase in field strength. There is no such phenomenon in electric circuit.

Ohm's Law :

In differential form Ohm's law may be written as

$$\vec{J} = \sigma \vec{E}$$

Thus according to Ohm's law the *field strength within a conductor is proportional to the current density* (σ is the conductivity of the material).

Continuity Equation

It states "*The net flow of current out of the volume must equal the negative rate of change of charge with time*",

i.e. $$\oiint_s \vec{J}\cdot\vec{ds} = -\iiint_p \frac{\partial\rho}{\partial t}dv$$

In the differential form it is given by

$$\Delta\cdot\vec{J} = -\frac{\partial\rho}{\partial t}$$

For steady currents, $\frac{\partial\rho}{\partial t} = 0$

$$\Delta\cdot\vec{J} = 0$$

Rise of Current in an Inductive Circuit

Let i = current at any instant,

I_m = maximum steady value of current $\left(=\frac{V}{R}\right)$,

R = resistance,

L = inductance, and

λ = time constant of the circuit $\left(=\frac{L}{R}\right)$.

Then, important relations are :

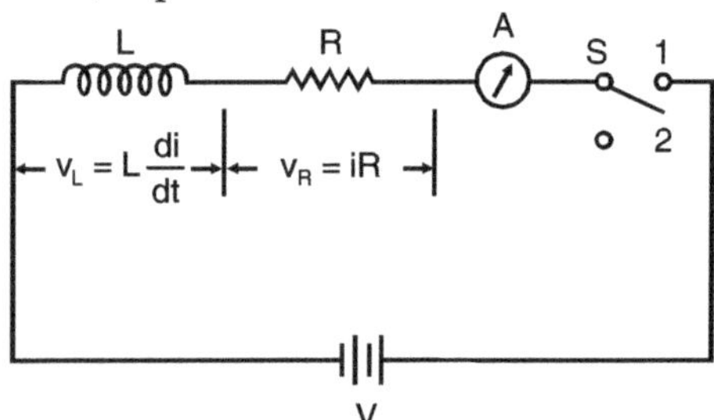

Fig. 10 : *Inductive circuit*

(a) **Rise of Current :**

$$i = \frac{V}{R}\left(1-e^{-t/\lambda}\right) = I_m\left(1-e^{-t/\lambda}\right)$$

$$\left(\text{where } I_m = \frac{V}{R}\right)$$

Putting $t = \frac{L}{R} = \lambda$ in eqn. we get

$i = 0.632\ I_m$.

Hence the time constant λ of an R-L circuit may be defined as the *time during which the current actually rises to 0.632 of its maximum steady value* (Fig. 11)

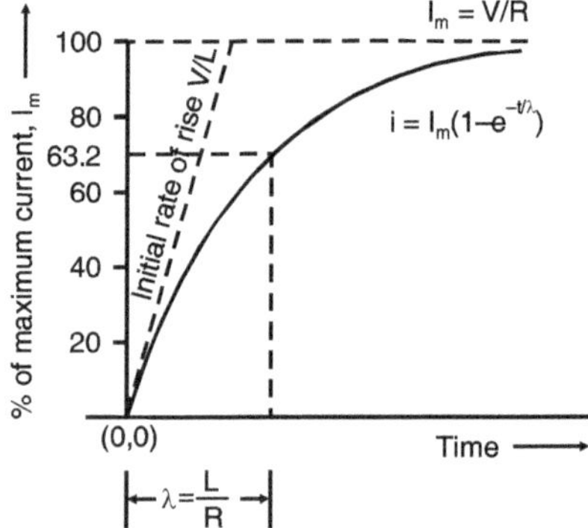

Fig. 11 : *Rise of current in an R-L circuit*

TABLE 2 : Analogy Between Electric and Magnetic Circuits (Similarities)

Aspects	*Electric Circuit*	*Magnetic Circuit*
1. Equivalent circuit	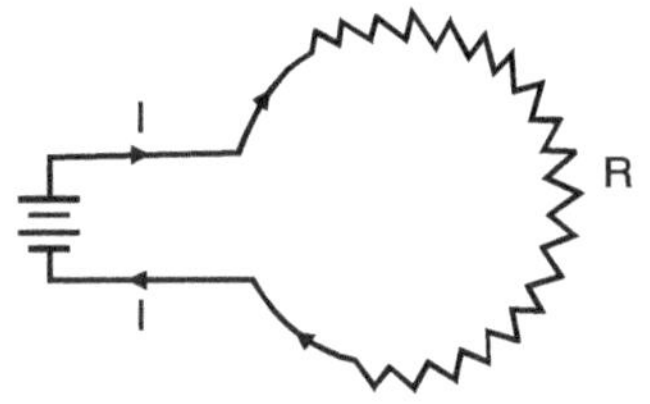 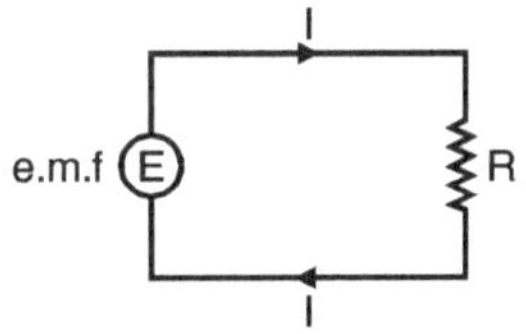 I, R, e.m.f (E)	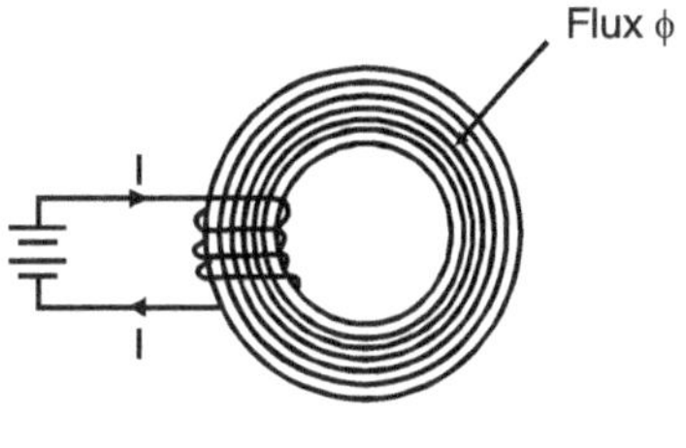 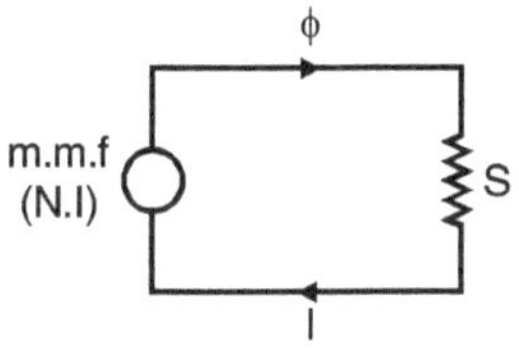 Flux ϕ, I, m.m.f (N.I), ϕ, S
2. Exciting force	Battery voltage (E)	Ampere-turns (AT)
3. Response	Current (I)	Flux (ϕ)
4. Ohm's law	$I = \frac{E}{R}$	$\phi = \frac{\text{m.m.f}}{\text{reluetance}} = \frac{\text{NI}}{\text{S}}$
5. By dimensions	$R = \rho \frac{l}{a}$ (conductance = 1/R)	$S = \frac{l}{\mu_0 \mu_r A}$ (permeance = 1/S)
6. Proportionality	ρ	$\frac{1}{\mu}\left(=\frac{1}{\mu_0 \mu_r}\right)$
7. Field intensity	Electric field intensity $= \frac{\text{E}}{l} \text{V/m}$	Magnetic field intensity $= \frac{\text{NI}}{l} \text{AT/m}$
8. Density	Current density (A/m^2)	Flux density (Wb/m^2).

The delayed rise of current in an inductive circuit is utilized in providing time lag in operation of electric relays and trip coils etc.

(b) **Decay of Current :**

$$t = I_m e^{-t/\lambda}$$

Putting $t = \lambda$ in above eqn. we get

$$i = 0.37 \text{ I}_{\text{m}}.$$

Hence time constant (λ) of an R-L circuit may be defined as the time during which current falls 0.37 or 37% of its maximum steady value while decaying (Fig. 12).

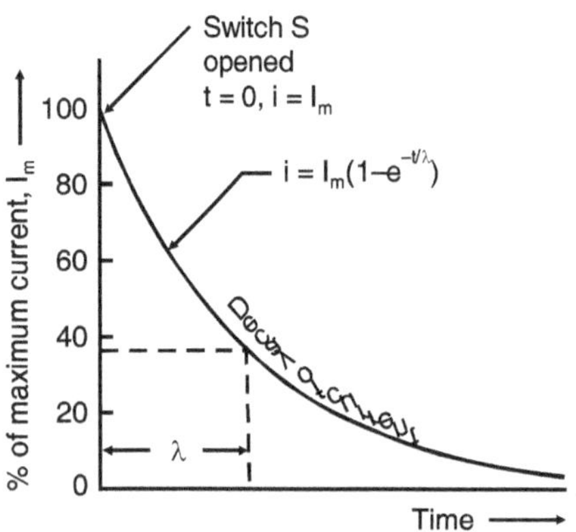

Fig. 12 : *Decay of current in an R-L circuit*

Faraday's Laws of Electromagnetic Induction

"The phenomenon whereby an e.m.f. and hence current is induced in any conductor which cut across or is cut by a magnetic flux is known as electromagnetic induction".

First Law : It states as follows :

"Whenever a magnetic flux linked with a circuit changes, an e.m.f. is always induced in

Or

"Whenever conductor cuts magnetic flux, an e.m.f. is induced in that conductor".

Second Law : It states as follows:

"The magnitude of induced e.m.f. is equal to the rate of change of flux-linkages".

Mathematically, $e = -\dfrac{Nd\phi}{dt}$ volts.

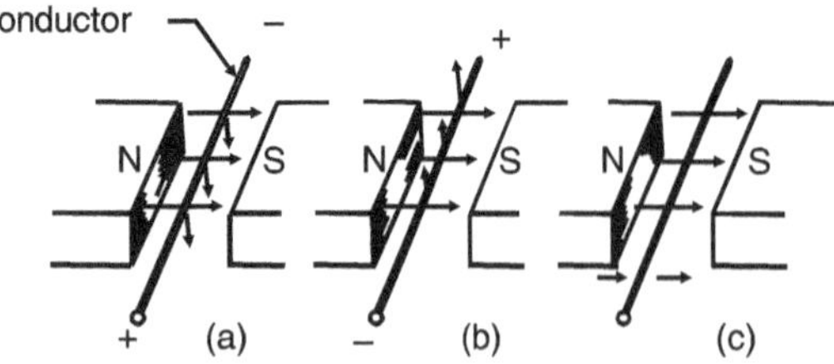

Fig. 13: *When a conductor is moved across a magnetic field a voltage is induced in the conductor. (a) Voltage induced across a wire moving downward. (b) Voltage induced across a wire moving upward. (c) No voltage is induced in a wire moving parallel to the field*

where, e = induced e.m.f.

$\dfrac{d\phi}{dt}$ = rate of change of flux

N = number of turns of the coil

[Usually, a minus sign is given to the right-hand side expression to signify the fact that the induced e.m.f. sets up current in such a direction that magnetic effect produced by it *opposes* the very cause producing it.]

Direction of induced e.m.f. and current

The direction of the induced, current may be found easily by applying either Fleming's Right-hand Rule (Fig.14) or Lenz's Law. *Fleming's rule is used where induced e.m.f. is due to flux-cutting (i.e.,*

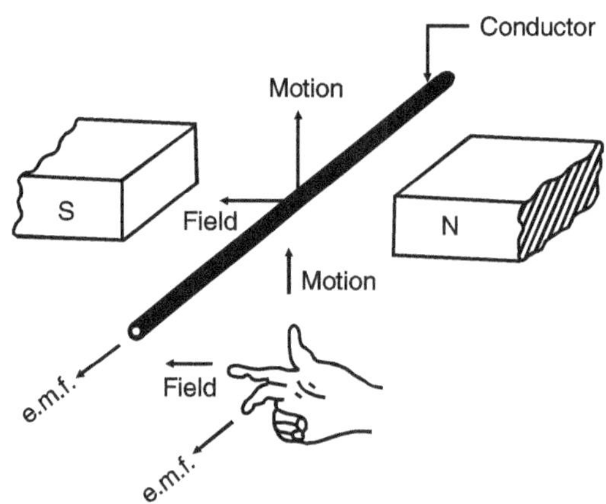

Fig. 14: *Fleming's right hand rule*

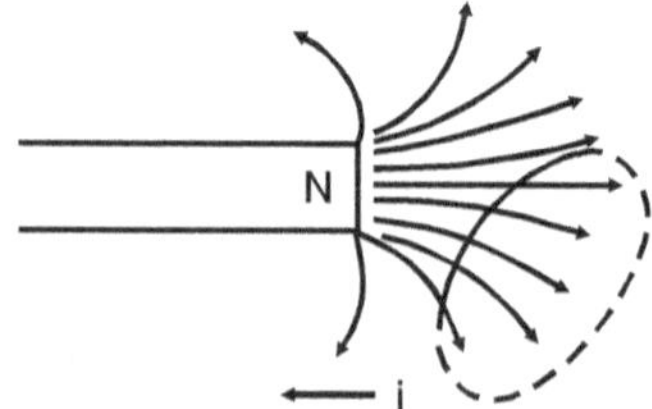

Fig. 15: *Induction of e.m.f. in a simple circuit*

dynamically induced e.m.f.) and Lenz's law when it is due to change by flux-linkages (*i.e.* statically induced e.m.f.).

Lenz's law. Fig. 15 shows induction of an e.m.f. in a simple circuit. The direction of the induced e.m.f. is determined by Lenz's law, which states that the current produced by the induced e.m.f. opposes the change in flux.

Lenz's law may also be stated as follows :

"In all cases of electromagnetic induction, an induced voltage will cause a current to flow in closed circuit in such a direction that the magnetic field which is caused by that current will oppose the change that produced the current".

INDUCED E.M.F.

Induced e.m.f. may be of the following two types.

1. Dynamically induced e.m.f.
2. Statically induced e.m.f.

1. Dynamically induced e.m.f.

Refer Fig.16. The e.m.f. induced (e) in the conductor is given by :

$$e = Blv \text{ volts}$$

where B = flux density of the magnetic field in *tesla*.

l = length of the conductor in *metres*, and

v = velocity of the conductor in *m/s*.

If the conductor moves at an angle θ with the direction of flux then the induced e.m.f.

$$e = Blv \sin\theta \text{ volts}$$

The direction of the induced e.m.f. is given by Fleming's Right Hand Rule.

2. Statically induced e.m.f.

The e.m.f. induced by *variation of flux is termed as "statically induced e.m.f."*.

Statically induced e.m.f. can be further subdivided as follows :

(i) Self-induced e.m.f.

(ii) Mutually induced e.m.f.

Self-induced e.m.f. : Self-induced e.m.f. is the e.m.f. induced in a coil due to the change of its own flux linked with it. If the current through the coil (Fig.15) is changed then the flux linked with its own turns will also change which will produce in it, what its called self-induced e.m.f.

$\left(e = -N\dfrac{d\phi}{dt}\right)$ The direction of the e.m.f. is given by Lenz's law (and would be such as to oppose any change of flux which is, in fact the very cause of its production.

Self-inductance : The property of the coil due to which it opposes any increase or decrease of current or flux through it, is known as self-inductance. It is measured in terms of self-induction L (in henry).

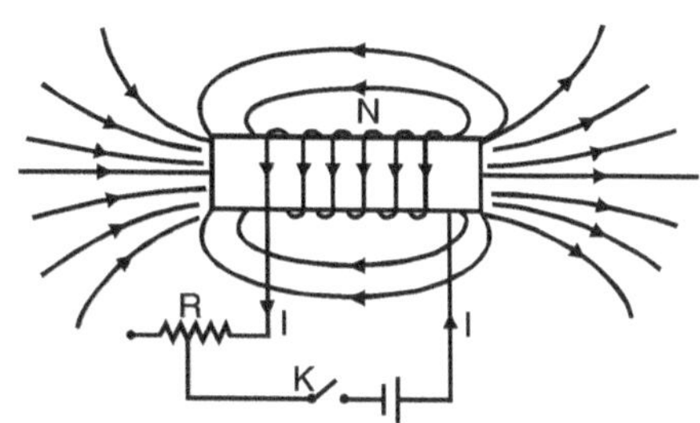

Fig. 16: *Self-induced e.m.f. and self-inductance*

Self-induction is sometimes analogously called *electromagnetic or electrical inertia.*

Co-efficient of self-induction (L) may be found by the following relations :

1. $L = \dfrac{N\phi}{I}$ henry

2. $L = \dfrac{\mu_0\mu_r AN^2}{l}$ henry

3. $L = \dfrac{e_L}{\frac{dI}{dt}}$ henry

where N = number of turns of the solenoid,

A = area of cross-section,

e_L = induced e.m.f. and

$\dfrac{dI}{dt}$ = rate of change of current.

Energy in electromagnetic field : The energy in electromagnetic field is given by:

$$W = \frac{1}{2}LI^2 \text{ joules}$$

The above eqn. gives an expression for stored energy in the magnetic field when the current is increased from zero and the same amount of energy is released when the current is reduced to zero.

Mutually induced e.m.f. : Refer Fig.17. Production of e.m.f. in coil B due to change in current in coil A is called mutually induced e.m.f.

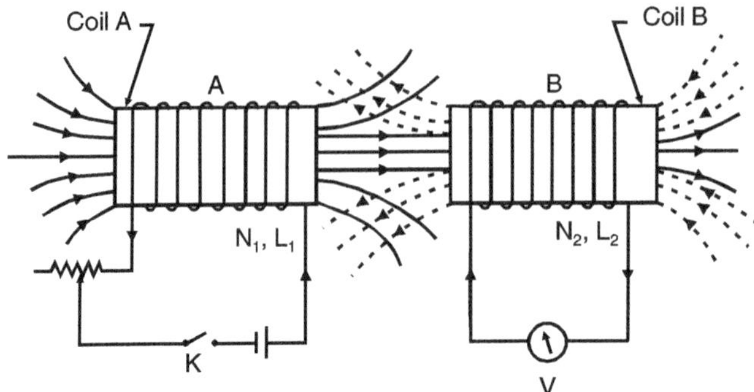

Fig. 17: *Mutually induced e.m.f. and mutual inductance*

Mutual inductance: It is defined as *"The phenomenon by which one circuit causes an e.m.f. induced in the adjacent circuit by induction when flux produced by it is changed".*

Co-efficient of mutual inductance (M) may be found by the following relations :

1. $M = \dfrac{N_2\phi_1}{I_1}$ henry

2. $M = \dfrac{\mu_0\mu_r A N_1 N_2}{l}$ henry

3. $M = -\dfrac{e_M}{\dfrac{dI_1}{dt}}$

Co-efficient of coupling (k) : It is defined as the *ratio of mutual inductance between the coils and the square root of product of self-inductance of each coil.*

In other words, $k = \dfrac{M}{\sqrt{L_1 L_2}}$

INDUCTANCES IN SERIES

In gereral, we have,

$L = L_1 + L_2 + 2M$.....if *m.m.fs.* are additive

$L = L_1 + L_2 - 2M$....if *m.m.fs.* are subtractive

INDUCTANCES IN PARALLEL

In general, we have,

$L = \dfrac{L_1 L_2 - M^2}{L_1 + L_2 - 2M}$ when mutual field *assists* the separate fields

$L = \dfrac{L_1 L_2 - M^2}{L_1 + L_2 + 2M}$ when two fields *oppose* each other.

Magnetic hysteresis : A phenomenon observed in ferromagnetic materials below the Curie point where the magnetization of the material varies nonlinearly with the magnetic field strength and also lags behind it (see Fig. 18). The magnetic susceptibility of such materials is large and positive, a large value of magnetization (M) being produced for comparatively small fields. A characteristic plot of either magnetization, M, or magnetic flux density, B, against magnetic field strength, H, demonstrates the hysteresis effect and is termed a hysteresis loop (see diagram). If an initially unmagnetized sample of iron is subjected to an increasing magnetic field the magnetization follows the curve shown by the dotted line OAS. This is known as the magnetization curve. If the specimen is then subjected to a complete magnetizing cycle with magnetic field varying symmetrically between +H and –H the curve shown by the solid line is followed.

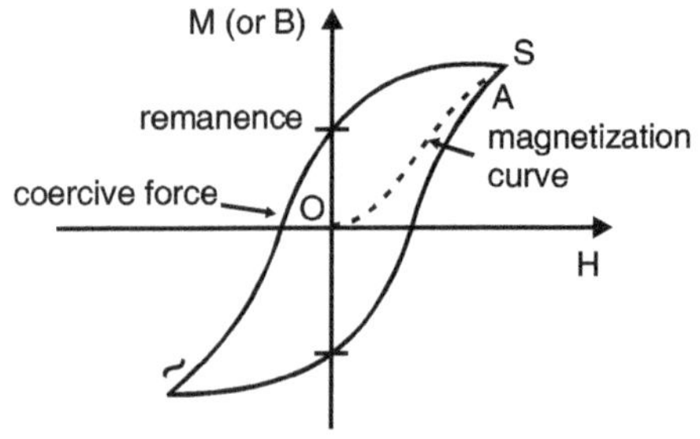

Fig. 18: *Magnetic hysteresis loop*

Table A: Maxwell's equations in integral form

	From Ampere	From Faraday	From Gauss	
Dimensions and SI units → *Case ↓*	*mmf, A*	*emf, V*	*Electric flux, C*	*Magnetic flux, Wb*
General	$F=\oint H.dl=\int_s\left(J+\frac{\partial D}{\partial t}\right).ds=I_{total}$	$V=\oint E.dl=-\int_s\frac{\partial B}{\partial t}.ds$	$\psi=\oint_s D.ds=\int_v \rho dv$	$\psi_m=\oint_s B.ds=0$
Free space	$F=\oint H.dl=\int_s\frac{\partial D}{\partial t}.ds=I_{disp}$	$V=\oint E.dl=-\int_s\frac{\partial B}{\partial t}.ds$	$\psi=\oint_s D.ds=0$	$\psi_m=\oint_s B.ds=0$
Harmonic variation	$F=\oint H.dl=(\sigma+j\omega\in)\int_s E.ds=I_{total}$	$V=\oint E.dl=-j\omega\mu\int_s H.ds$	$\psi=\oint_s D.ds=\int_v \rho dv$	$\psi_m=\oint_s B.ds=0$
Steady	$F=\oint H.dl=\int_s J.ds=I_{cond}$	$V=\oint E.dl=0$	$\psi=\oint_s D.ds=\int_v \rho dv$	$\psi_m=\oint_s B.ds=0$
Static	$U=\oint H.dl=0$	$V=\oint E.dl=0$	$\psi=\oint_s D.ds=\int_v \rho dv$	$\psi_m=\oint_s B.ds=0$

Table B: Maxwell's equations in differential form

	From Ampere	*From Faraday*	*From Gauss*	
Dimensions → *case ↓*	*Electric current area*	*Electric potential area*	*Electric flux volume*	*Magnetic flux volume*
General	$\nabla\times H=J+\frac{\partial D}{\partial t}$	$\nabla\times E=-\frac{\partial B}{\partial B}$	$\nabla.D=\rho$	$\nabla.B=0$
Free space	$\nabla\times H=\frac{\partial D}{\partial t}$	$\nabla\times E=-\frac{\partial B}{\partial t}$	$\nabla.D=0$	$\nabla.B=0$
Harmonic variation	$\nabla\times H=(\sigma+j\omega\varepsilon)E$	$\nabla\times E=-j\omega\mu H$	$\nabla.D=\rho$	$\nabla.B=0$
Steady	$\nabla\times H=J$	$\nabla\times E=0$	$\nabla.D=\rho$	$\nabla.B=0$
Static	$\nabla\times H=0$	$\nabla\times E=0$	$\nabla.D=\rho$	$\nabla.B=0$.

The value of M at zero field is termed the remanence, the corresponding value of B being known as the residual induction. The value of H at the point at which M (or B) falls to zero is known as the coercive force and is the reverse field required to demagnetize the sample. The area enclosed by the

Table C : Resistance, capacitance and inductance in terms of circuit and field quantities.

	Circuit definition	*Physical definition*	*Field cell*	*Energy definition*
Resistance R	$\frac{\int E.dl}{\oint H.dl}$	$\frac{1}{\sigma A}$	$\frac{1}{\sigma d}$	$\frac{\iiint \frac{J^2}{\sigma} dv}{I^2}$
Capacitance C	$\frac{\iint D.ds}{\oint E.dl}$	$\frac{\varepsilon A}{l}$	εd	$\frac{\iiint \varepsilon E^2 dv}{V^2}$
Inductance L	$\frac{\iint B.ds}{\oint H.dl}$	$\frac{\mu A}{l}$	μd	$\frac{\iiint \mu H^2 dv}{I^2}$.

M–H curve represents the work done in taking the sample through a complete cycle and is therefore the energy, dissipated in each complete cycle when ferromagnetic material is subjected to an alternating magnetic field. This is known as the hysteresis loss. The incremental hysteresis loss is the total energy dissipated in the sample when subjected to an alternating magnetic field and is the product of the hysteresis loss and the number of cycles.

The effect of hysteresis in a magnetic core is to cause an increase in the effective resistance of the coil surrounding the core; the hysteresis factor is the increase in the effective resistance of a coil carrying a current of one ampere at a specified frequency.

The general form of the hysteresis curve is shown in the diagram. The area enclosed by the curve depends on the nature of the ferromagnetic sample : minimum area (and minimum coercive force) occur with soft iron, rising to a value some twenty times greater with tungsten steel. Any complete magnetizing cycle (say between the values of H+h and H–h) will give rise to a hysteresis loop.

Magnetostriction is mechanical deformation of a ferromagnetic material when subjected to a magnetic field. The effect arises because of internal stresses in the material due to the anisotropy energy required to magnetize it in certain directions relative to the crystal axes (see ferromagnetism). Conversely, when subjected to mechanical stress a change in the magnetization of the material is observed.

Joule magnetostriction (or positive magnetostriction) is an increase in the length of a rod or tube of ferromagnetic material when an axial magnetic flux is applied to it. Negative magnetostriction occurs in materials, such a nickel, that suffer a decrease in length with an increase in magnetic flux density. Magnetostriction also affects deformed ferromagnetic samples. A bent bar tends to straighten under the influence of a magnetic flux applied along its length (Guillemin effect); a twisted bar also tends to straighten under similar conditions (Wiedemann effect).

High-frequency alternating fields set up longitudinal vibrations in the sample, which thus acts as a source of sound waves ; if the frequency of these vibrations corresponds to the natural frequency of the sample large values of amplitude occur. Magnetostrictive vibrations have many applications. At ultrasonic frequencies they form a source of ultrasonic energy that can be used commercially, for example for ultrasonic cleaning or for breaking the oxide film on aluminium for soldering. Magnetostriction oscillators utilize the magnetostrictive effect of alternating currents to produce frequency controlled oscillations at frequencies in the range 25-000 hertz downwards.

Magnetostriction is also used to provide the vibrating part of a loudspeaker; conversely audio vibrations applied to a magnetostictive rod produce corresponding flux changes, as in a magnetostriction microphone.

EXERCISE

1. The instrument landing system has which of the following components?
   1. Goniometer
   2. Marker beacons
   3. Ad cock antenna
   4. Localizer
   5. Glide slope

   Select the correct answer

   A. 2, 3 and 5 B. 1, 2, 4 and 5
   C. 1, 2, 3 and 5 D. 2, 3, 4 and 5
2. Spot the odd one out?
   A. Bismuth B. Silver
   C. Cobalt D. Copper
3. $-\oint \vec{E} \times \vec{H} . \vec{ds}$ in the Polynting's theorem indicates
   A. electric and magnetic fields are perpendicular to each other
   B. energy is transported in space by E and H fields
   C. energy is decreasing while advancing in space
   D. the loss of energy is being made up by energy flowing into the volume
4. A dominant wave is characterized by
   A. highest cutoff wavelength
   B. lowest cutoff wavelength
   C. no attenuation
   D. All of the above
5. By inserting a plate of dielectric material between the plates of a parallel plate capacitor, the energy stored in the capacitor is increased five times. The dielectric constant of the material is
   A. 1/25 B. 1/5
   C. 15 D. 25
6. A parallel plate capacitor of plate area A and plate separation 't' has a capacity C. If a metallic plate P of area A and of negligible thickness is introduced in the capacitor at a distance $t/2$ from either of the two plates, as shown in the given figure, then the capacity of the capacitor will become

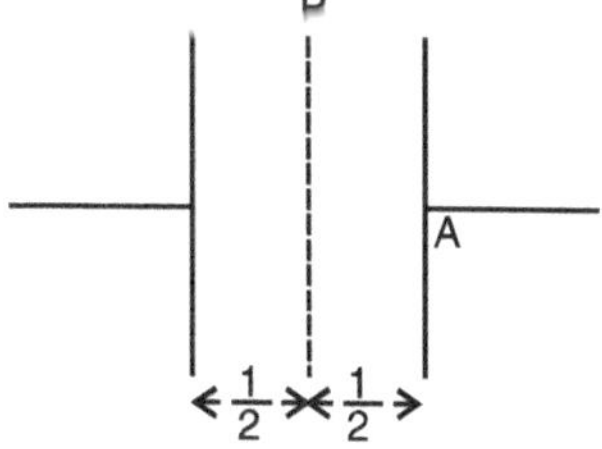

   A. $\frac{C}{2}$ B. C
   C. 2C D. 4C
7. The field configuration in two different views of waves propagating in a rectangular wave

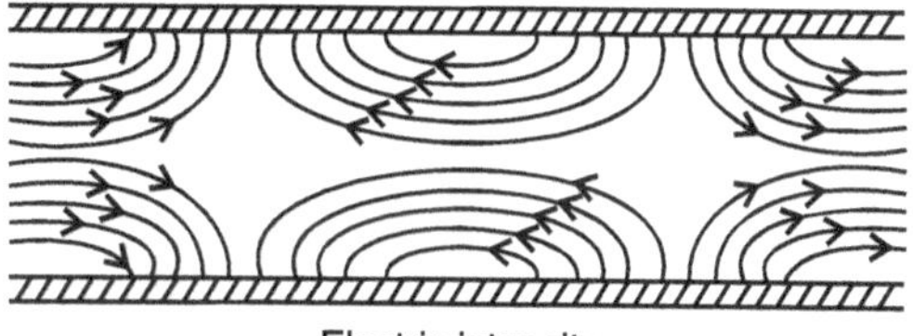

Electric intensity

Magnetic intensity

   A. TM_{11} B. TE_{11}
   C. TE_{21} D. TE_{10}
8. F_{12} and F_{13} are two forces of 2 and 3 newtons. F_{13} makes an angle 30° with normal. The force acting in x-direction is
   A. 5 nt B. 3.5 nt
   C. 1 nt D. – 1 nt
9. Which of the following are wave equations
   A. $\nabla^2 \vec{E} = \mu \varepsilon \vec{E}''$ B. $\nabla^2 \vec{E} = -\mu \varepsilon \vec{E}'$
   C. A, B both D. none of the above
10. Pick up the correct answer :
    A. curl grad V = $\nabla \times (\nabla V) = 0$
    B. curl grad V = $\nabla \times (\nabla V) = 1$
    C. curl grad V = $\nabla \times (\nabla V) = -1$
    D. curl grad V = $\nabla \times (\nabla V) = \infty$

11. A prominent wave should have
A. no phase shift
B. no attenuation
C. highest cutoff frequency
D. lowest cutoff frequency

12. The function $f(x - v_0 t)$ represents
A. a wave motion in forward direction
B. a wave motion in reverse direction
C. a stationary wave
D. None of these

13. The total current in a conductor of radius R is I and is uniformly distributed. The axis of the conductor is in the z-direction and $\overline{\text{H}}$ is given by

$$\overline{\text{H}} = \bar{a}_\theta \cdot \frac{1}{2\pi r} \text{ for } \text{R} \le r \le \infty$$

what is $\overline{\nabla} \times \overline{\text{H}}$ outside the conductor?

A. $\bar{a}_z \cdot \frac{1}{2\pi r}$
B. $\bar{a}_r \cdot \frac{1}{2\pi r}$
C. $\bar{a}_\theta \cdot \frac{1}{2\pi r}$
D. Zero

14. Consider the following statements pertaining to a uniform plane wave travelling in the x-direction :
1. $\overline{\text{E}}$ and $\overline{\text{H}}$ are both independent of y and z
2. $\overline{\text{E}}$ and $\overline{\text{H}}$ have no x-component
3. It has the same phase for all points and its amplitude is constant over a plane surface.
4. If both Ey and Ez are present, the direction of resultant vector is constant with time.

Of these statements
A. 1 and 3 are correct
B. 1 and 4 are correct
C. 1, 2 and 3 are correct
D. 1, 3 and 4 are correct

15. Maxwell's wave equations are

A. $\nabla \times \vec{\text{E}} = \frac{-\partial \vec{\text{B}}}{\partial t}$

B. $\nabla \times \vec{\text{E}} = \frac{-\partial \vec{\text{B}}}{\partial t} . \nabla \times \vec{\text{H}} = \frac{\partial \vec{\text{D}}}{\partial t}$

C. $\nabla \times \vec{\text{D}} = 0, \nabla^2 \times \vec{\text{E}} = \mu_0 \varepsilon_0 \frac{\partial^2 \vec{\text{E}}}{\partial}$

D. $\nabla^2 \times \vec{\text{E}} = \mu_0 \varepsilon_0 \frac{\partial^2 \vec{\text{E}}}{\partial t^2}, \nabla^2 \vec{\text{H}} = \mu_0 \varepsilon_0 \frac{\partial^2 \vec{\text{H}}}{\partial t^2}$

16. Poynting vector can be represented by the equation
A. $\vec{\text{E}} \times \vec{\text{B}} / \mu_0$
B. $\vec{\text{E}} \vec{\text{B}} / \mu_0$
C. $\vec{\text{E}} . \vec{\text{B}} / \mu_0$
D. none of these

17. As an electromagnetic wave travels in free space, the only one of the following can hapen to them
A. Reflection
B. Absorption
C. Attenuation
D. Refraction

18. The impulse response of a first order system is $\text{K}e^{-2t}$. If the input signal is sin $2t$, then the steady state response will be given by

A. $\frac{-1}{2\sqrt{2}} \sin\left(2t + \frac{\pi}{4}\right)$

B. $\frac{1}{4} \sin 2t$

C. $\frac{1}{2\sqrt{2}} \sin\left(2t - \frac{\pi}{4}\right)$

D. $\frac{1}{2\sqrt{2}} \sin\left(2t + \frac{\pi}{4}\right) + \text{K}e^{-2t}$

19. A plane electromagnetic wave of the form $\vec{\text{E}}$ = $\hat{y}\text{E}_0 \cos\left[2\pi\left\{\left(5 \times 10^{14} \text{ sec}^{-1}\right)t - \left(2.5 \times 10^6 \text{ m}^{-1}\right)x\right\}\right]$ represents a wave propagating along
A. + x direction
B. – x direction
C. + y direction
D. – y direction

20. The polarization of a dielectric material is given by.
A. $\vec{\text{P}} = \varepsilon_r \vec{\text{E}}$
B. $\vec{\text{P}} = (\varepsilon_r - 1).\vec{\text{E}}$
C. $\vec{\text{P}} = \vec{\text{E}} \varepsilon_0 (\varepsilon_r - 1)$
D. $\vec{\text{P}} = (\varepsilon_r - 1)\varepsilon_0$

21. The maximum effective aperture of a $\frac{\lambda}{2}$ (half-wave) antenna is
A. 0.013 λ^2
B. 0.13 λ^2
C. 1.3 λ^2
D. 13 λ^2

22. In case of concentric conductors (of radius a and b).
A. $\text{C} = \frac{2\pi\varepsilon}{\ln b/a}$
B. $\text{C} = \frac{\ln b/a}{2\pi\varepsilon}$

C. C = 0 D. none of these

23. Laplacian operation of a scalar function V is :

A. ∇V B. $\nabla^2 V$

C. $\nabla \times V$ D. $\nabla \times \nabla V$

24. Two straight parallel conductors carry equal currents in opposite directions. The force between them is

A. repulsive B. attractive

C. zero D. infinite

25. Which of the following material will have the least value of relative permeability?

A. Silver B. Cobalt

C. Iron D. Palladium

26. One Newton metre equals :

A. one watt B. one joule-second

C. one joule D. one volt-ampere

27. Consider the following statements regarding an electromagnetic flowmeter :

1. A.C. excitation is used to avoid polarisation of the fluid.
2. The meter calibration is changed as the viscosity of the fluid changes
3. Stainless steel pipes can be used for measuring the flow of fluids of high conductivity.

Of these statements

A. 1, 2 and 3 are correct

B. 1 and 2 are correct

C. 1 and 3 are correct

D. 2 and 3 are correct

28. Circular waveguide may be preferred to rectangular one for same application because of

A. Rotation

B. The smaller cross-section

C. Freedom from spurious modes

D. None of the above

29. In the Fermi-Dirac statistics, the probability of electron occupation of an energy level equal to the Fermi level is

A. 0 B. 0.25

C. 0.5 D. 1.0

30. The most important primary force of attraction in the formation of solids is

A. electromagnetic

B. electrostatic

C. quantum-electronic

D. quantum-magnetic

31. In order to reduce cross-sectional dimension, which of the following wave guide is preferred?

A. Flexible B. Rectangular

C. Circular D. Ridged

32. The following device is suitable for very low power oscillator circuit only :

A. Gunn diode B. Tunnel diode

C. IMPATT diode D. TRAPATT diode.

33. A pin diode is

A. A metal semi-conductor point contact diode

B. A microwave mixer diode

C. Often used as a microwave detector

D. Suitable for use as a microwave switch

34. Wave guides are generally not used for frequencies

A. Below 1 GHz B. Below 10 GHz

C. Below 50 GHz D. Below 150 GHz

35. As compared to stripline a microstrip offers certain disadvantages. One of the disadvantages is that microstrips

A. Are heavier

B. Are expensive

C. Are more likely to radiate

D. Do not lend themselves to printed circuit techniques.

36. A cavity may be considered as :

A. Low pass filter B. High pass filter

C. Band pass filter D. Band reject filter

37. In a transmission line terminated with a load equal to the characteristics impedance, the reflection coefficient is

A. zero B. + 1

C. – 1 D. infinity

38. The signal to quantization noise ratio in a PCM system depends upon

1. Sampling rate
2. Number of quantization levels
3. Message signal bandwidth

A. 1, 2 and 3 B. 2 and 3 only

C. 2 only D. 3 only

39. The skin effect causes current to flow

A. In the centre of the conductor

B. Near the surface of the conductor

C. Through central core of the conductor

D. Uniformly through the conductor.

40. Klystron operates on the principle of :
A. Amplitude modulation
B. Frequency modulation
C. Pulse modulation
D. velocity modulation

41. Match List I with List II and select the correct answer using the codes given below the lists :

List I (Electron spin) ***List II (Phenomenon)***

(a) ↑↑↑↑↑↑ 1. Antiferromagnetism
(b) ↑↓↑↓↑↓ 2. Ferrimagnetism
(c) ↑↓↑↓↑↓ 3. Ferromagnetism

Codes :	*(a)*	*(b)*	*(c)*
A.	1	2	3
B.	3	2	1
C.	2	3	1
D.	3	1	2

42. Two positive charges 10 μ coulomb and 15 μ coulomb are separated by a distance of 10 cms with a dielectric of alcohol. Find the force between them, ε_r of alcohol is 20.
A. 180 nt B. 135 nt
C. 9 nt D. 6.75 nt

43. For different loads the range of the values for VSWR is
A. 0 to 0.5 B. 0.5 to 1
C. 0 to 1 D. 1 to ∞

44. Side lobe of an antenna causes
A. Reduction in the gain of antenna
B. Reduces beam width of antenna
C. Ambiguity in direction finding
D. None of these

45. When electromagnetic waves are propagated in a waveguide :
A. They travel along the walls of the waveguide
B. They travel through the dielectric without touching the walls
C. They are reflected from the walls but do not travel along the walls
D. None of these

46. Which one of the following modes has the characteristic of attenuation becoming less as the frequency is increased and is attractive at microwave frequencies of circular cylindrical waveguides?
A. TE_1 mode B. TM_{01} mode
C. TE_{01} mode D. Higher order mode

47. A signal propagated in a waveguide has a full wave of electric intensity change between the two farther walls and no component of the electric field in the direction of propagation. The mode is
A. TE_{11} mode. B. TM_{10}
C. TM_{22} D. TE_{20}

48. When a particular mode is excited in a waveguide there appears an extra electric component in the direction of propagation. The resulting mode is
A. longitudinal electric
B. transverse electromagnetic
C. transverse magnetic
D. transverse electric

49. For a cavity resonator with, Q_0 = unloaded Q and ρ = standing wave ratio, Q_e the loaded Q is given by
A. $Q_e = \rho Q_o$ B. $Q_e = \frac{\rho+1}{\rho} Q_o$
C. $Q_e = \frac{\rho}{1+\rho} Q_o$ D. $Q_e = \frac{1}{\rho} Q_o$

50. An electron is assumed to be a sphere of radius R on the surface of which the charge Q is distributed. The total electrical energy from the entire surface of the electron is given by
A. $\frac{e^2}{\varepsilon_0 R}$ B. $\frac{e}{8\pi\varepsilon_0 R}$
C. $\frac{e}{8\pi\varepsilon_0 R}$ D. None of the above

51. The peak power of pulsed microwave system is while the average power is 400 watts. The duty cycle is :
A. 0.02 percent B. 0.2 percent
C. 2 percent D. 20 percent

52. The power handling capacity of a ferrite component depends upon its
A. Curve temperature
B. Gyrometric resonance
C. Line width
D. All of the above.

53. When a particular mode is excited in a waveguide, there appears an extra electric current, in the direction of propagation. The resulting mode is
A. Longitudinal
B. Transverse-electric
C. Transverse-magnetic
D. Transverse-electromagnetic

54. Choke flange may be used to couple two waveguides
A. Top help in the alignment of the waveguides
B. Because it is simpler than any other joint
C. To compensate for discontinuties at the joint
D. To increase the band width of the system.

55. Match List I with List II and select the correct answer by using the codes given below the lists:

List I	*List II*
(a) T (Reflection coefficient) = 0	1. $\dfrac{Z_L - Z_0}{Z_L + Z_0}$
(b) T = – 1	2. $Z_L = Z_0$
(c) T = + 1	3. $Z_L = 0$
(d) $-1 < 0 < T < +1$	4. $Z_L = \infty$

Codes :	*(a)*	*(b)*	*(c)*	*(d)*
A.	2	3	4	1
B.	1	2	3	4
C.	2	4	1	3
D.	3	2	4	1

56. A TM wave in general is characterized by
A. The electric field component wholly transverse
B. The absence of magnetic field component in the direction of propagation
C. The absence of electric field component in the direction of propagation
D. None of the above.

57. To protect the replacer of a reflex Klystron (indicate the false statement)
A. A permanent negative voltage is conneted to the cathode
B. The repeller voltage is applied before the anode voltage
C. A diode is sometimes connected across the klystron
D. A cathode resistor is often used.

58. If the peak power is 10^4 watt and the average power is 800 watts then the duty cycle is
A. 0.08 per cent
B. 0.8 per cent
C. 8.0 per cent
D. 0.04 per cent

59. The glass of a TWT may be coated with aquadag to
A. Help focussing
B. Provide attenuation
C. Improve bunching
D. Increase gain.

60. The magnitude and direction of the electric field balancing the weight of alpha particle is
A. 20.36×10^{-8} nt/C downward
B. 20.36×10^{-8} nt/C upward
C. 20.36×10^{-8} nt/C at $\angle 45°$
D. 2.036×10^{-8} nt/C

61. When a wave travelling in air enters into a wave guide
A. the phase velocity will increase
B. the group velocity will increase
C. the phase velocity will decrease
D. none of these

62. Distilled water is a dielectric having $\epsilon_r = 81$ and $\mu_r = 1$. If a wave is incident from water into a water air interface, the critical angle will be nearly
A. 6°
B. 12°
C. 15°
D. 22°

63. A short vertical grounded antenna is required to radiate at 1 MHz. The effective height of the antenna is 30 m. The calculated value of radiation resistance is
A. 1.58 Ω
B. 158 Ω
C. 15.8 Ω
D. none of these

64. Which of the following statements are correct :
A. In case of an antenna, the radiation resistance and loss resistance are not two different quantities
B. The loss resistance includes loss by eddy currents, improper earth connections, insulation leakage etc, but not I^2R losses
C. Radiation resistance varies directly as square-root of the frequency
D. none of the above.

65. Standing in a room against the wall, I fire an electron beam horizontally which is deflected to the right.

A. The magnetic field of induction is along the beam
B. The direction of magnetic field is from left to right in a clockwise direction
C. The direction of magnetic field is towards bottom of room
D. The direction of magnetic field is from left to right in an anticlockwise direction

66. Propagation of frequencies in UHF range takes place by means of
A. surface waves B. space waves
C. ground waves D. sky waves

67. E = 0 between any two charges on the line joining them but not between. The magnitude and sign of the charges and location of the null point are
A. the same
B. opposite sign
C. opposite sign and beyond the line depending on the magnitude of the charges
D. None of the above

68. For a spherical surface which of the following relation is valid?
A. $D = \frac{Q}{4\pi r}$ B. $D = \frac{Q}{4\pi r^2}$
C. $D = \frac{Q}{4\pi r \varepsilon}$ D. None of the above

69. Polarization of a ratio wave is taken as the :
A. direction of the magnetic field
B. direction of the electric field
C. normal to the direction of magnetic field
D. none of these

70. A charge + 2 × 10^{-6} C is situated in air at the origin of a rectangular coordinate system. E at a point 4 m from the origin on Z-axis is
A. 1/1125 V/m
B. 1125 nt/coul along Z-axis
C. 1125 V/m
D. 112.5 nt/coul

71. A 12 volt battery sets up a current of 5 amp. for 15 min. The chemical energy of the battery is reduced by
A. 60 watts B. 900 watts
C. 16 J D. 5.4 × 10^4J

72. Match List I with List II and select the correct answer by using the codes given below the lists:

List I	*List II*
(a) Dirichlet problem	1. The potential V is unique if it satisfies Laplace's equation, the known V_b on part of surface and the known $\partial V/\partial n/b$ on the remaining closed surfaces.
(b) Neumann problem	2. The potential solution V is unique within a constant, if it satisfies Laplace's equation and the known $\partial V/\partial n/b$ on the closed surfaces
(c) Mixed Boundary value problem	3. The potential solution V is unique if it satisfies Laplace's equation and the known v_b on the closed surfaces.

Codes :	*(a)*	*(b)*	*(c)*
A.	1	2	3
B.	2	1	3
C.	3	2	1
D.	1	3	2

73. When electromagnetic waves are propagated in a waveguide,
A. they travel along the broader walls of the guide
B. they are reflected from the walls but don't travel along them
C. they travel through the dielectric without touching the walls
D. they travel along all four walls of the waveguide

74. The dimensions of relation permittivity are
A. Farads B. Farads/metres
C. meters/Farads D. it is dimensionless

75. SHF range is
A. 30 – 300 MHz B. 300 – 3000 MHz
C. 3 – 30 GHz D. 30 – 300 GHz

76. Which of the following relation is valid for a perfect conductor?
A. $D_n = \rho_s$ B. $D_n = 0$
C. $D_n = 0$ D. $D_n = \infty$

77. Polynting vector P = E × H has the dimensions :
A. watts/metre^2 B. watts/metre
C. watts-metre D. watts-metre^2

78. Figures (A), (B), (C) and (D) given below

represent patterns of electric field of TE waves in circular waveguides. Which one of them represents TE_{02} mode?

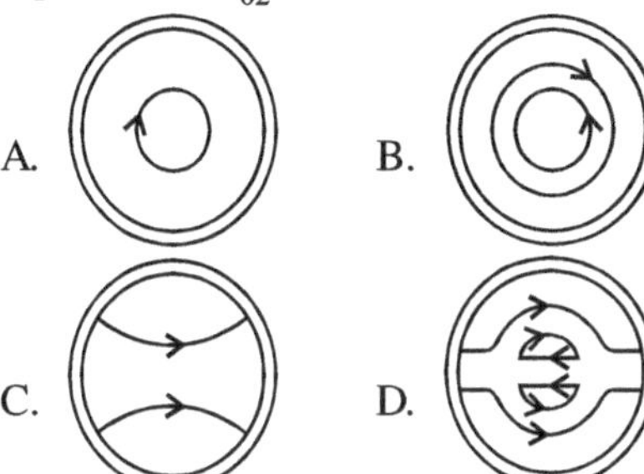

79. The dominat mode in a waveguide is characterized by
A. longest cutoff wavelength
B. shortest cutoff wavelength
C. infinite attenuation
D. zero attenuation

80. The frequency of resonance of water vapour in the atmosphere is
A. 10.225 GHz B. 12.850 GHz
C. 17.500 GHz D. 60.000 GHz

81. Mark the false statement regarding the boundary conditions between different media
A. The tangential component of E is continuous at the surface. That is, it is the same just outside the surface as it is just inside the surface.
B. The tangential component of H in continuous across a surface except at the surface of a perfector conductor.
C. The normal component of D in continuous if there is no surface charge density otherwise D is discontinuous by an amount equal to the surface charge density
D. The normal component of B is zero at the surface of discontinuity

82 The ground wave coverage of the medium wave transmitter is 100 km and in the night the first reflected ray is at 800 km. The skip distance is
A. 900 km B. 700 km
C. 100 km D. 800 km

83. In a 1 mfd capacitor an instantaneous displacement current of 1 amp is to be established in the space between the plates. It is possible by
A. 10^6 amp B. 10^6 amp/sec
C. 10^6 volts sec D. 10^6 volts/sec

84. A wire 12 cm long cuts directly across a flux of 14,000 gauss at a speed of 2 m/s. The volts generated are
A. 336×10^3 B. 3.36×10^3
C. 336 D. 336×10^{-3}

85. Flux/current could be the dimension of
A. Inductance B. Charge
C. Capacitance D. none of the above

86. An observer is an inertial frame of reference measures surface charge density σ of a sheet of + ve charges. The magnitude of the surface charge density as measured by another observer in another inertial frame F' which is moving towards left with respect to F, with a velocity v, is given by
A. σ
B. zero
C. $> \sigma$ and $= \dfrac{\sigma}{\sqrt{1-\dfrac{v^2}{c^2}}}$
D. $< \sigma$ and $= \sigma\sqrt{1-\dfrac{v^2}{c^2}}$

87. In broad side array
A. the principal direction radiation is perpendicular to the array axis and also to the plane containing the array elements.
B. the principal direction radiation is parallel to the array axis and also to the plane containing the array elements
C. the principal direction radiation is parallel to the array axis and perpendicular to the plane containing the array elements
D. the principal direction radiation is perpendicular to the array axis and parallel to the plane containing the array elements

88. The difference between conductors and insulators is :
A. metals are conductors and insulators are non-conductors
B. conductors have free electrons to move about whereas in insulators they are bound
C. decided by frequency

D. $\frac{\sigma}{\omega\varepsilon}$ is the dividing line between conductor and insulator

89. When the peak power is 10,00,000 watts and the average power is 800 watts then the duty cycle is

A. 0.008 per cent B. 0.08 per cent
C. 0.8 per cent D. 8 per cent

90. One of the following is not used as a microwave mixer or detector

A. Crystal diode
B. Schottky barrier diode
C. Backward diode
D. PIN diode

91. Biot-Savart's law states the relation between magnetic intensity and

A. filament current only
B. surface current only
C. volume current only
D. (A), (B) and (C) are all correct

92. A piston attenuator is :

A. Mode filter
B. Flap attenuator
C. Waveguide below cutoff
D. Vane attenuator.

93. TEM mode in co-axial line has

A. No cut off wavelength
B. Direct current can't pass
C. A, B is true for TEM mode
D. None of these

94. Helical antenna is used for satellite tracking because of its :

A. Circular polarization
B. Broad bandwidth
C. Good front-to-back ratio
D. Small size

95. A 'rodome' is a

A. Protective cover for antenna
B. Radar housed in a dome
C. Dome shaped radar antenna
D. none of the above

96. In electromagnetic waves, polarization

A. is caused by reflection
B. is due to the transverse nature of the waves
C. results from the longitudinal nature of the waves
D. always vertical in an isotropic medium

97. To overcome difficulties with strapping high frequencies, the type of cavity structure desired for magnetron is :

A. Hole and slot B. Slot
C. Rising sun D. Vane

98. Which of the following is valid :

A. 1 GHz = 1 kHz × 10 KHz
B. $1\text{ MHz} = \frac{1\text{GHz}}{1\text{MHz}}$
C. 1 GHz = 100 MHz
D. $\frac{1\text{GHz}}{1\text{MHz}} = 1000.$

99. Component given by

$H_z = 3x \cos \beta + 6y \sin \gamma$

Also the field is invariant with time. The current...

A. $J = 6k \sin \gamma$ B. $J = -6k \sin \gamma$
C. $J = -6k \cos \gamma$ D. $J = 6k \cos \gamma$.

100. For a parabolic dish antenna, if λ be the free space wavelength and D be the diameter, then bandwidth is given by

A. $70\,\lambda\,D$ B. $\frac{70D}{\lambda}$
C. $\frac{70\lambda}{D}$ D. $\frac{D}{70\lambda}$

101. Which of the following is easily adjustable tuning component in a wave guide?

A. Screw
B. Plunger
C. Plunger and stub
D. Plunger, stub and screw

102. In order to couple two generators to a waveguide system without coupling them to each other, one could not use a

A. rat-race B. E-plane T
C. hybrid ring D. magic T

103. Which of the following statements are true of piezoelectric crystals?

1. Their crystal structure possesses a centre of symmetry.
2. They always exhibit ferroelectricity
3. They are necessarily poor electrical conductors.
4. They exhibit a relative permittivity less than unity.

Select correct answer

A. 1, 2 and 3 B. 2, 3 and 4

C. 1 and 3 D. 1, 2 and 4

104. At microwave frequencies, the circuit elements are taken as :

A. Lumped elements

B. Distributed elements

C. Both (A) and (B)

D. None of these

105. The electric field vector $\overline{E}$ of wave in free space (ε_0, μ_0) is given by

$$\overline{E} = \hat{y}\left[A\cos\omega\left(t - \frac{z}{c}\right)\right]$$

Its magnetic vector $\overline{H}$ will be given by

A. $\overline{H} = \hat{y}\left[A\sin\omega\left(t + \frac{z}{c}\right)\right]$

B. $\overline{H} = \hat{z}\left[A\cos\omega\left(t - \frac{z}{c}\right)\right]$

C. $\overline{H} = \hat{x}\left[-j\frac{\sqrt{\varepsilon_0}}{\mu_0}A\cos\omega\left(t - \frac{z}{c}\right)\right]$

D. $\overline{H} = \hat{x}\left[-j\frac{\sqrt{\varepsilon_0}}{\mu_0}A\sin\omega\left(t + \frac{z}{c}\right)\right]$

106. A magnetron has an average power output of 50 W and a duty cycle of 2 per cent, its peak output power is

A. 25 W B. 100 W

C. 2500 W D. 10,000 W

107. Cylindrical cavity resonators are not used with klystrons because they have

A. Q that is too low

B. Too heavy losses

C. Harmonically related resonant frequencies

D. A shape whose resonant frequency is too difficult to calculate.

108. In which of the following two ground planes are used?

A. Stripline B. Coaxial line

C. Parallel wire line D. Microstrip

109. The current in a conductor when connected to a dc source is established

A. instantaneously

B. with a very small (of the order of the fraction of the millionth of a nanosecond) delay

C. with a small (of the order of a nanosecond) delay

D. with a delay of the order of milliseconds.

110. Waveguides are used mainly for microwave signals because

A. they depend on straight-line propagation which applies to microwaves

B. losses would be too heavy at lower frequencies

C. there are no generators powerful enough to excite them at lower frequencies

D. they would be too bulky at lower frequencies

111. A helical antenna is used for satellite tracking because of its

A. Broad bandwidth

B. Easy manoeuverability

C. Circular polarization

D. Good front to back ratio

112. Two particles of equal mass *m* charged with + Q_1, + Q_2 are suspended from a point by weightless strings of the same length. The charges and point of support will form

A. isosceles triangle

B. irregular triangle with the string suspending the larger charge making a larger angle with the horizontal plane

C. irregular triangle with the string suspending the larger charge making a larger angle with the vertical plane

D. equilateral traingle.

113. The coupling between the tunnel diode and its cavity is kept loose in order to :

A. Increase the frequency stability

B. Increase the available negative resistance

C. Allow operation at high frequencies

D. Provide a larger voltage swing

114. Which amplifier has the least nosie?

A. Tunnel diode amplifier

B. Multicavity klystron

C. Parametric amplifier

D. All have almost same noise

115. The advantage of microwave power frequency signals are

A. Increased BW

B. Ability to use high gain, directive antenna
C. Greater secrecy
D. A, B, C combined

116. In radar engineering in order to make use of narrow beam antennas for achieving accuracy in target direction indication, it is preferable to use frequencies in the range of
A. lower VHF B. upper VHF
C. lower UHF D. upper UHF

117. If a vertical dipole antenna is used in conjunction with a loop antenna for direction finding, then the field pattern obtained will be

A.  B. 

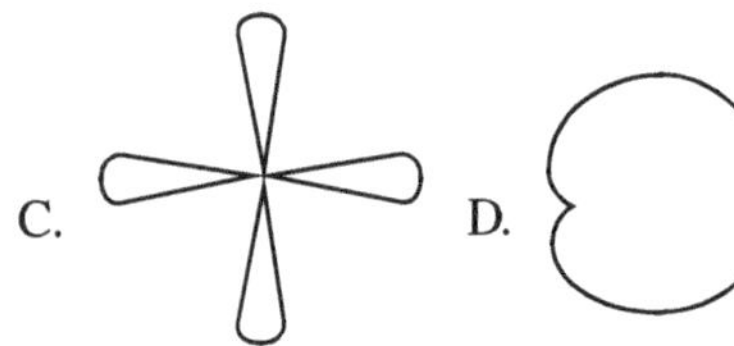

118. Scalar or dot product of two vectors A and B can be expressed as :
A. A. B = AB sin α B. A. B = BA sin α
C. A. B = AB tan α D. A. B = AB cos α

119. Very low frequency waves are used for some types of services because
A. They penetrate the ionosphere easily
B. They are very reliable
C. Of the low powers required
D. All of the above

120. Which of the following is an Maxwell Equation
A. $\nabla.\vec{E} = \rho$ B. $\nabla.\vec{B} = 0$
C. $\nabla \times \vec{E} = B$ D. $\nabla \times \vec{H} = -\vec{B}$

121. The positive point charges Q_1 and Q_2 are moving with velocities V_x and V_y as shown in the figure. Which of the statement/statements are correct.

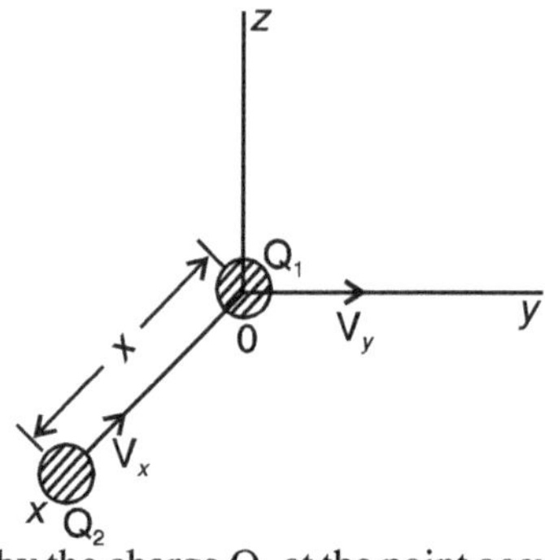

1. There will be magnetic induction set up by the charge Q_1 at the point occupied by Q_2 and its magnitude will be $\frac{\mu_0}{4\pi}.\frac{Q_1 V_y}{x^2}$
2. There will be no magnetic induction set up by the charge Q_1 at the point occupied by Q_2
3. There will be a force acting on Q_2 and its magnitude will be
$$F_y = \frac{\mu_0}{4\pi}\left(\frac{Q_1 Q_2 V_x V_y}{x^2}\right)$$
4. There will be a force of repulsion acting on Q_2 and its magnitude will be, $\frac{\mu_0}{4\pi}\frac{Q_1 Q_2}{x^2}$ along x direction.

A. 1, 2 and 3 B. 1 and 2
C. 3 and 4 D. 2, 3 and 4

122. The important postulate arising out of Maxwell's modification of ampere's law is
A. A steady current produces a magnetic field
B. A changing magnetic field produces an electric field
C. A changing electric field produces a magnetic field
D. A motional emf is produced by a coil moving in a magnetic field

123. Magnetic flux density B and vector magnetic potential A are related by :
A. $\vec{B} = \nabla.\vec{A}$ B. $\vec{B} = \nabla \times \vec{A}$
C. $\vec{A} = \nabla \times \vec{B}$ D. $A = \nabla.\vec{B}$

124. The characteristic impedance of a lossless transmission line is given by
A. $Z = \sqrt{LC}$ B. $Z = \sqrt{C/L}$
C. $Z = \sqrt{L/C}$ D. $Z = LC$

125. The absolute permeability value is (μ_r)
A. $4\pi \times 10^{-7}$ Henry/meter
B. $4\pi \times 10^{-12}$ H/m
C. 8.854×10^{-12} H/m
D. 8.854×10^{-7} H/m

126. The numerical value of directive gain may lie

between

A. 1 and + 1 B. 0 and 1

C. 0 and ∞ D. 1 and ∞

127. Diffraction of electromagnetic waves

A. Is caused by reflections from the ground

B. Arises only with spherical wave fronts

C. Will ocur when the waves pass through large slot

D. May occur around the edge of a sharp obstacle

128. In case of surface wave, to prevent short-circuiting of the electric component, the wave should be

A. vertical polarised

B. horizontally polarised

C. circularly polarised

D. of high power

129. Consider a time varying magnetic induction fields B→ increasing at a rate of $\left(\frac{dB}{dt}\right)$ within a cylindrical region of radius *r*. This changing magnetic induction would induce electric field E→ at any point at the loop with radius *x*. Select the corrrect statement/statements :

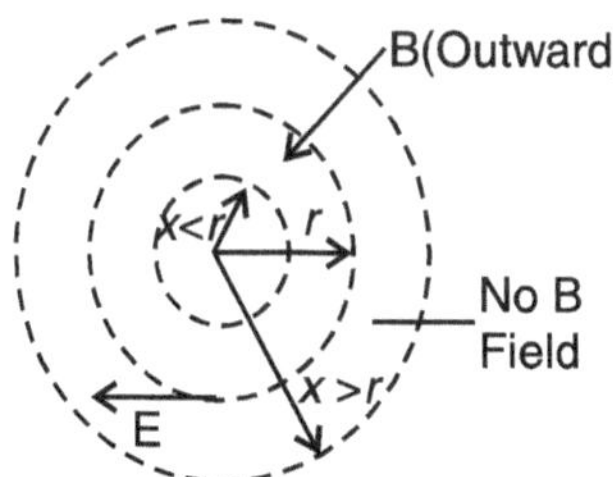

1. $E = -\left(\frac{dB}{dt}\right).x$ for $x < r$

2. $E = -\left(\frac{dB}{dt}\right).x$ for $x > r$

3. $E = \left(\frac{dB}{dt}\right)\cdot\frac{x}{2}$ for $x < r$

4. $E = \left(\frac{dB}{dt}\right)\cdot\frac{x}{2}$ for $x < r$

5. $E = \left(\frac{dB}{dt}\right)\cdot\frac{r^2}{2x}$ for $x > r$

A. 1, 2 and 3 B. 2, 3 and 4

C. 2 and 3 D. 4 and 5

130. In a waveguide, the attenuation near the cutoff frequency is :

A. low B. high

C. very high D. zero

131. The characteristic impedance of a lossless transmission line is given by

A. $Z_0 = \sqrt{LC}$ B. $Z_0 = \sqrt{L/C}$

C. $Z_0 = \sqrt{C/L}$ D. $Z_0 = \frac{1}{\sqrt{LC}}$

132. A wave incident normal on a perfect conductor is

A. partially transmitted and partially reflected

B. full transmitted

C. entirely reflected

D. none of these

133. An antenna coupler is used due to all of the following reasons EXCEPT.

A. to prevent reradiation of the local oscillator

B. to discriminate against harmonics

C. to make the antenna look resistive

D. to provide the output amplifier with the correct load impedance

134. The equation of continuity for current is

A. $\text{Div.}\vec{J} + \frac{\partial q}{\partial t} = Q$ B. $\text{Div.}\vec{J} + \frac{\partial \rho}{\partial t} = 0$

C. $\text{Div.}\vec{J} = \frac{\sigma}{\vec{F}}$ D. none of these

135. The refractive index of the ionosphere for radio wave is

A. greater than 1 B. equal to 1

C. less than 1 D. equal to 0

136. The energy of electric field due to a spherical charge distribution of radius *r* and uniform charge density σ in vacuum is

A. $1.8 \times 10^9.\frac{Q^2}{r}$ where $Q = \frac{4}{3}\pi r^3\sigma$

B. $5.4 \times 10^9.\frac{Q^2}{r}$ where $Q = 4\pi r^2\sigma$

C. $5.4 \times 10^9.\frac{Q^2}{r}$ where $Q = \frac{4}{3}\pi r^3\sigma$

D. $\frac{1}{4\pi\varepsilon_0} \times 0.6\left(\frac{4}{3}\pi R^3\sigma\right)\frac{1}{R}$

137. The disturbance from adjacent telephone line is called :

A. cross fire
B. cross talk
C. inductive disturbance
D. none of these

138. Permittivity is given by

A. $\varepsilon_0 = 8.854 \times 10^{-19}$ F/m

B. $\varepsilon_0 = \dfrac{1}{36\pi \times 10^{-19}}$ F/m

C. $\varepsilon_0 = 8.854 \times 10^{-12}$ F/m

D. $\varepsilon_0 = \dfrac{1}{36\pi \times 10^{-10}}$ F/m

139. The important postulate arising out of Maxwell's modification of Ampere's law is

A. A steady current produces a magnetic field
B. A changing magnetic field produces an electric field
C. A changing electric field produces a magnetic field
D. None of these

140. A rectangular loop of resistance R, length l and breadth b is pulled at constant speed u through a thickness of region $x(a < x < 2a)$ θ in which a uniform field of induction B is present. The plot between induced e.m.f. e at various positions of x is as shown below :

A. 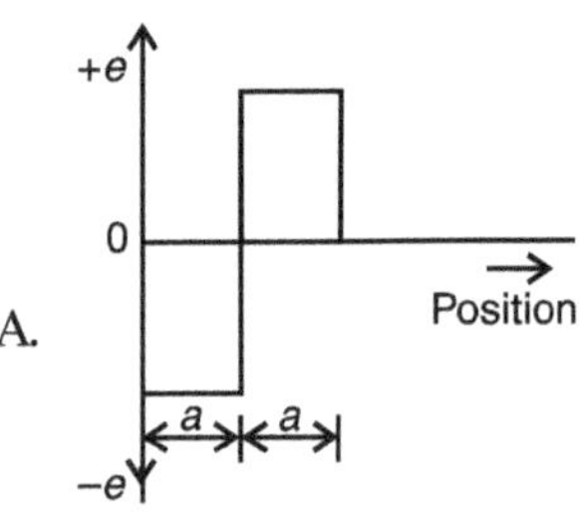

B. 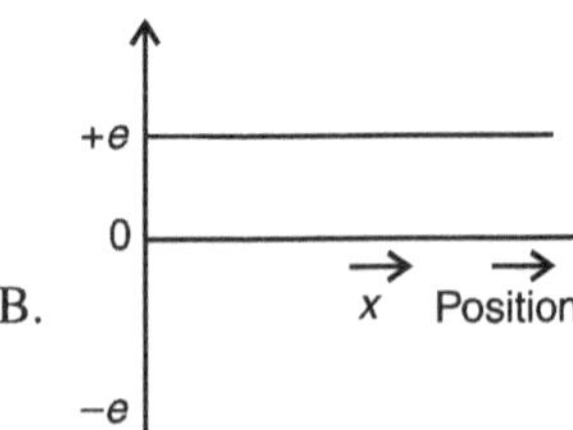

C. 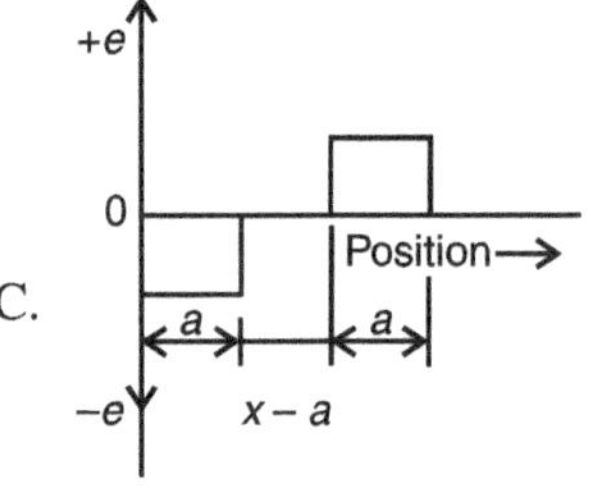

D. 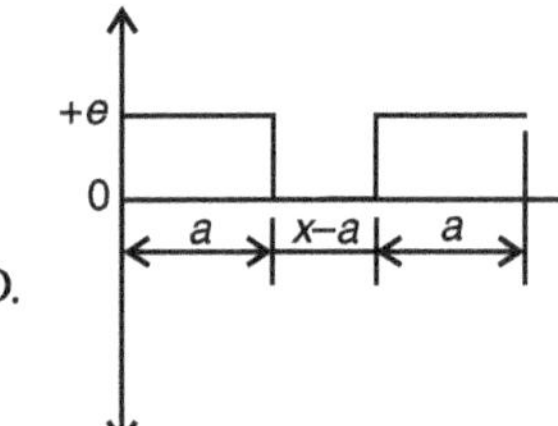

141. A parallel plate air capacitor has a capacitance of 100 pfd. A p.d 50 V is applied. The stored energy is

A. 1.25×10^{-7} J
B. 2.50×10^{-7} J
C. 40×10^{-9} J
D. 20×10^{-19} J

142. The equations $\nabla^2 \vec{E}\ \mu_0\varepsilon_0 \dfrac{\partial^2 \vec{E}}{\partial t^2}$ and $\nabla^2 \vec{H} = \mu_0\varepsilon_0 \dfrac{\partial^2 \vec{H}}{\partial t^2}$ are called wave equations because

A. they are second order differential equations
B. they contain the term $\mu_0\, \varepsilon_0$ which is equal to velocity of light
C. they are 2nd order differential equations for which the solution is a sine or cosine term
D. all the above 3 are true

143. A series combination of L and R is connected to a battery by a switch and the switch is suddenly closed. The time constant for reaching the maximum value and the percentage of the maximum value are

A. 100 sec, 100%
B. $\dfrac{R}{L}$ sec, 100%
C. $\dfrac{L}{R}$ sec, 37%
D. L/R sec, 63%

144. Which of the following statement is false ? Compared with equivalent transmission lines, 3 GHz wave guides

A. Are less lossy

B. Are less bulky
C. Have lower attenuation
D. Can carry higher powers.

145. Gallium antimonide is preferred to germanium for use in tunnel diode since it permits :
A. Lower noise
B. Larger voltage swing
C. Better frequency stability
D. Simpler fabrication process.

146. The most fundamental method of frequency measurement is
A. With an absorption wavemeter
B. With a slotted line
C. With a transmission wavemeter
D. By comparison

147. A slot line is etched on the ground plane of a microstrip. Coupling between the slot line and the microstrip, on the two sides of the same substrate, takes place through which one of the following mechanisms?
A. The coupling is magnetic and is the minimum when they are at right angles
B. The coupling is electric and is the maximum when they are parallel
C. The coupling is magnetic and is the maximum when they are at right angles
D. The coupling is electric and is the minimum when they are at right angles

148. At the interface of two conducting surfaces.
A. the tangential component's of both electric intensity and current density are continuous
B. the tangential component of electric intensity and normal component of current density are continuous
C. the normal component of both electric intensity and current density are continuous
D. none of the above is correct.

149. The following microwave diode is suitable for very low power oscillators only :
A. Gunn diode B. Tunnel diode
C. LSA diode D. IMPATT diode

150. Indicate the false statement.
Transit time in microwave tubes will be reduced if
A. The electrodes are brought closer together
B. A higher anode current is used
C. Multiple or coaxial leads are used
D. The anode voltage is made larger.

151. Cut off wavelength rectangular waveguide for $TE_{0.1}$ is (a is broader dimension, b is the narrow dimension)
A. $2a$ B. $2b$
C. a D. b

152. A disadvantage of microstrip with respect strip line circuits is that the former
A. Don't lend themselves to printed circuit techniques
B. Are more likely to radiate
C. Are bulkier
D. Are more than expensive and complex to manufacture

153. The field configuration in two different views of waves propagating in a rectangular waveguide as shown in the given figure represents

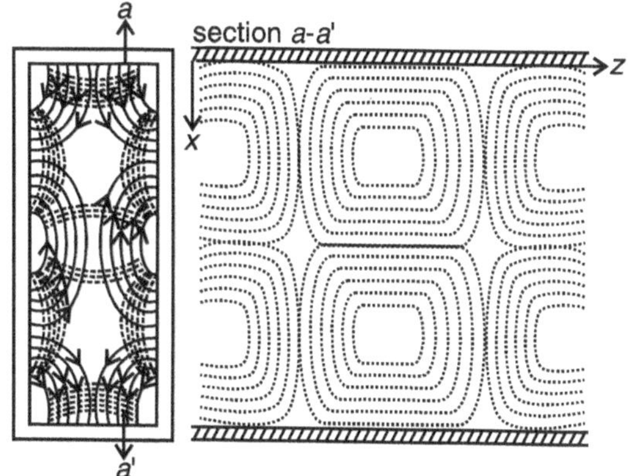

_______________ electric intensity
.... magnetic intensity

A. $TE_{1.0}$ mode B. $TM_{1.1}$ mode
C. $TE_{1.1}$ mode D. $TE_{2.1}$ mode

154. The quantity which is directly measured in sweep reflectometry is the
A. Impedance B. SWR
C. Returnloss D. Reflection coefficient

155. Which of the following is unlikely to be used as pulsed device?
A. TWT B. CFA
C. BOW D. Multicavity klystron

156. The attenuator is used in TWT to
A. Help bunching B. Prevent oscillation
C. Prevent saturation D. Increase gain

157. For a rectangular waveguide, having a broader dimension and b as narrow dimension the cut off wavelength for TE_{10} will be

A. a B. b

C. $2a$ D. $2b$

158. Indicate the false statement. Fourier analysis shows that a sawtooth wave consists of

A. fundamental and sub harmonic sine waves

B. a fundamental sine wave and an infinite number of harmonics

C. fundamental and harmonic sine waves whose amplitude decreases with the harmonic number

D. sinusoidal voltages, some of which are small enough to ignore in practice

159. A magnetron whose oscillating frequency is electrically adjustable over a wide range is called a

A. Co-axial magnetron

B. Frequency agile magnetron

C. VTM

D. None of these

160. A wave guide can be considered as

A. Band pass filter B. Band stop filter

C. Low pass filter D. High pass filter

161. Match List I with List II and select the correct answer

List I	***List II***
(a) Evanescent wave	1. Pattern multiplication
(b) Skip distance	2. Mismatched transmission line
(c) Return loss	3. Propagation in lossy medium
(d) Antenna Array	4. Short-wave propagation
	5. Propagation below cut-off

Codes:	*(a)*	*(b)*	*(c)*	*(d)*
A.	2	4	1	5
B.	5	2	4	3
C.	5	4	2	1
D.	1	3	4	5

162. A directional coupler with three or more hole is sometimes used in preference to the two holes coupler

A. To increase the bandwidth of the system

B. To reduce spurious mode generation

C. To increase coupling of the signal

D. As it is more efficient

163. When the free space wavelength of a signal equals the cutoff wavelength of the waveguides the characteristic impedance of the wave guide becomes:

A. Zero B. 377 Ω

C. 3.7 MΩ D. Infinite

164. For best lowlevel noise performance in the X-band an amplifier should use

A. a biopolar transistor

B. A Gunn diode

C. A field effect transistor

D. An IMPATT diode

165. Bolometers are used in measurement of

A. RF power

B. Audio power

C. Microwave power

D. None of the above

166. One of the following is very useful as a multiband HF receiving antenna. This is the :

A. conical horn B. folded dipole

C. log-periodic D. square loop

167. If $s^3 + 3s^2 + 4s + A = 0$, then all the roots of this equation are in the left half plane provided that

A. $A > 12$ B. $-3 < A < 5$

C. $0 < A < 12$ D. $5 < A < 12$

168. Figures (A), (B), (C) and (D) given below represent patterns of electric field of TE waves in circular waveguides. Which one of them represents TE_{02} mode?

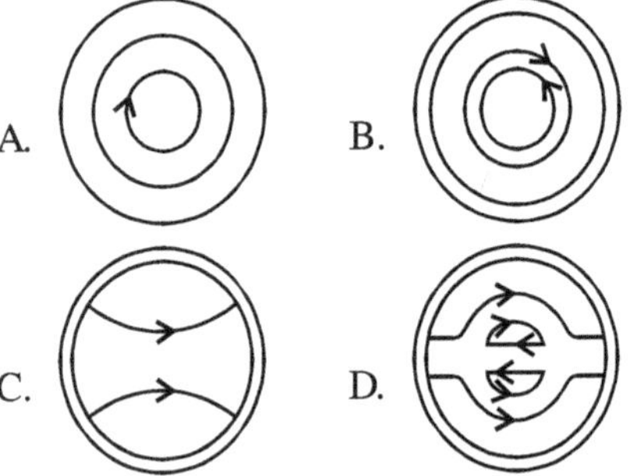

169. In a homogeneous non-magnetic dielectric medium of dielectric constant 9.0, the propagation delay per unit length is

A. 5 ps/mm B. 10 ps/mm

C. 1 ns/mm D. 5 ns/mm

170. Which of the following statement is not valid? A varactor diode may be useful at microwave frequencies

A. For electronic tuning
B. As an oscillator
C. As a parametric amplifier
D. For frequency multiplication

171. The negative resistance in Gunn diode is due to:

A. Electron transfer to a less mobile energy level
B. High reverse bias
C. Electron domain formation at the junction
D. Tunnelling across the junction.

172. Noise figure is used as figure of merit of

A. Oscillator B. Amplifier
C. Moderator D. Isolator

173. A magnetic circuit is analogous to an electrical cirruit because

A. as electrons flow in an electrical circuit flux flows in magnetic circuit
B. as we can open and short-circuit an electrical circuit, so can we open and close a magnetic circuit
C. like leakage flux, there is always a leakage current
D. only the sets of equations of electrical and magnetic circuits are similar

174. Which of the following is not an advantage of stripline over microstrip?

A. Higher Q
B. Lower tendency to radiate
C. Higher isolation between adjacent circuit
D. Easier integration with semiconductor devices.

175. Which of the following properties pertain to a circularly polarized wave having no component of electric field in the *x*-direction but having components E_y and E_z?

1. E_y and E_z are equal in magnitude
2. Direction of resultant electric vector is constant with time.
3. Direction of resultant electric vector varies with time.
4. E_y and E_z have a time phase difference of 90°.

Select the correct answer

A. 2 and 4 B. 2 and 3
C. 1, 3 and 4 D. 1 and 3

176. A carry look ahead adder is frequently used for addition because, it

A. is faster B. is more accurate
C. uses fewer gates D. costs less

177. The retarded vector potential (at P) is given by

$$\vec{A} = \frac{\mu_0}{4\pi}\int_{\text{volume}} \frac{\vec{i}\left(t-\frac{r}{v}\right)}{r}\,dV$$

The retarded scalar potential V (at P) can be expresed as

A. $V = \dfrac{1}{4\pi\varepsilon_0}\displaystyle\int_{\text{vol}} \frac{\rho\left(t-\frac{r}{v}\right)}{r}\,dV$

B. $V = \displaystyle\int_{\text{vol}} \frac{\rho\left(t+\frac{r}{v}\right)}{r}\,dV$

C. $V = \dfrac{\varepsilon_0}{4\pi}\displaystyle\int_{\text{vol}} \frac{\rho\left(t-\frac{r}{v}\right)}{r}\,dV$

D. $V = \dfrac{1}{4\pi}\displaystyle\int_{\text{vol}} \frac{\rho\left(t+\frac{r}{v}\right)}{r}\,dV$

178. The figure shows a string of positive and negative ions of dielectric (List II) in the presence of a field. Match the type of polarization (List I) with that described by the figure (List II).

Slect the correct answer :

List I

(*a*) Orientational
(*b*) Ionic
(*c*) Electronic
(*d*) Both ionic and electronic

List II

1. ⊕ ⊕ ⊕ ⊕
2. ⊕ ⊕ ⊕ ⊕
3. ⊕ ⊕ ⊕ ⊕

Codes:	(*a*)	(*b*)	(*c*)
A.	2	3	1
B.	1	2	3
C.	3	1	2
D.	2	1	3

179. When electromagnetic waves are reflected at an angle from a wall, their wavelength along the wall is

A. The same as in free space
B. The same in the wavelength and perpedicular to the wall
C. Shortend because of the doppler effect
D. Greater than in the actual direction of propagation

180. The electric field component of a uniform plane electromagnetic wave propagating in the Y-direction in a lossless medium will satisfy the equation

A. $\frac{\partial^2 E_y}{\partial y^2} = \mu\varepsilon \frac{\partial^2 E_y}{\partial t^2}$

B. $\frac{\partial^2 E_y}{\partial x^2} = \mu\varepsilon \frac{\partial^2 E_y}{\partial t^2}$

C. $\frac{\partial^2 E_x}{\partial y^2} = \mu\varepsilon \frac{\partial^2 E_x}{\partial t^2}$

D. $\frac{\sqrt{E_x^2 + E_z^2}}{\sqrt{H_x^2 + H_z^2}} = \sqrt{\mu/\varepsilon}$

181. A material is described by the following electrical parameters at a frequency of 10 GHz : $\sigma = 10^6$ mho/m, $\mu = \mu_0$, and $\varepsilon/\varepsilon_0 = 10$. The material at this frequency is considered to be $\left(\varepsilon_0 = \frac{1}{36\pi} \times 10^{-9} \text{ F/m}\right)$

A. a good conductor
B. a good dielectic
C. neither a good conductor, nor a good dielectric
D. a good magnetic material

182. A TWT operates with accelerating potential of 1600 V. For helix structure, the ratio of wire length to helix length is approximately :

A. 20 B. 15
C. 10 D. 5

183. Which of the following frequency bands fall under microwave frequency?

A. UHF and SHF
B. SHF and EHF
C. UHF, SHF and EHF
D. VLF, LF and MF

184. A plane wave is incident normally on a perfect conductor as shwon E_x^i, H_y^i and $\vec{P}_i$ are electric field, magnetic field and Poynting vector, respectively, for the incident wave. The reflected wave should have

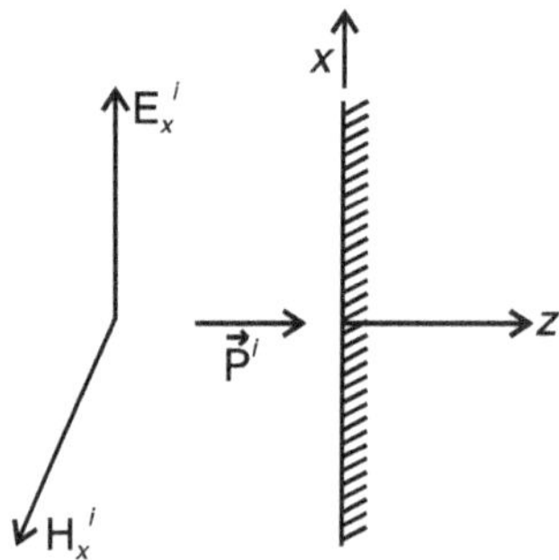

A. $E_x^r = -E_x^i$ B. $H_y^r = -H_y^i$
C. $\vec{P}^r = -\vec{P}^r$ D. $E_x^r = E_x^i$

185. A diplexer is an equipment which represents

A. Two antennas
B. Two transmission lines
C. Two signals by frequency
D. None of these

186. Magnetic field is used in a ruby maser to :

A. Provide sharp focussing of the electron beam
B. Allow room temperature operation
C. Provide frequency adjustments
D. Increase population inversion.

187. Two circular rings carry equal and opposite change as shown. The dipole moment of the charge distribution is

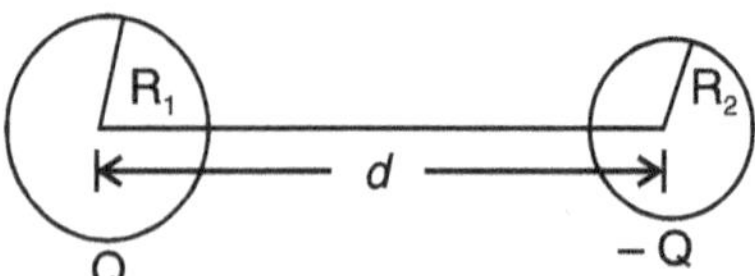

A. $q\frac{R_2}{R_1}d$ B. $Q\frac{R_1}{R_2}d$

C. $Q\frac{R_1^2 - R_2^2}{d}$ D. Qd

188. If the load impedance terminating a transmission line is complex, then the reflection coefficient is complex. This would mean that the reflected

A. voltage and reflected current at the load are out of phase with the incident voltage and incident current respectively

B. voltage and reflected current at the load are in-phase with the incident voltage and incident current respectively

C. current at the load is out of phase with the incident voltage and the reflected voltage at the load is in-phase with the incident voltage

D. voltage is out of phase with the incident voltage and the reflected current is in phase with the incident current

189. In an m-derived low pass filter, the value of m is [f_c = cut-off frequency, f_∞ = series resonant frequency (series arm)]

A. $\sqrt{1-\left(\frac{f_\infty}{f_c}\right)^2}$ B. $\sqrt{1-\left(\frac{f_c}{f_\infty}\right)^2}$

C. $\sqrt{1+\left(\frac{f_\infty}{f_c}\right)^2}$ D. $\sqrt{1+\left(\frac{f_c}{f_\infty}\right)^2}$

190. In microwave we take the elements as

A. Lumped circuit element

B. Distributed circuit element

C. A, B both

D. None of these

191. A reflex klystron

A. Only one cavity as both the buncher and catcher

B. Two cavity one for buncher, second for catcher

C. Either of the above

D. None of the above

192. Indicate the false statement. An advantage of stripline over waveguide is its

A. Smaller bulk

B. Greater bandwidth

C. Higher power handling capacity

D. Greater campatibility with semi-conductor devices.

193. Solid state maser is an amplifier of the following type :

A. Diamagnetic B. Paramagnetic

C. Ferromagnetic D. Electromagnetic

194. The power 'P_t' transmitted by an antenna is being received by another antenna at a distance 'r' and at a power level 'P_r'. With reference to this, which one of the following statements is true? (D_t and D_r represent the gains of the transmitting and the receiving antennas, respectively) P_r/P_t is

A. directly proportional to $D_t D_r$

B. directly proportional to D_r/D_t

C. directly proportional to $(D_t/D_r)^2$

D. inversely proportional to $\left(\frac{\lambda}{4\pi r}\right)^2$

195. An arbitrarily eliptically polarised wave can be broken up into

A. two circularly polarised components rotating in the same direction

B. two oppositively rotating circularly polarised components

C. two stationary circularly polarised components

D. None of thse

196. What is the numerical aperture of an optical fibre when its critical angle is 30°?

A. 0.5 B. 0.704

C. 0.866 D. 0.2

197. If the antenna diameter in a radar is increased by a factor of 4, the maximum range will be increased by a factor of

A. $\sqrt{2}$ B. 2

C. 4 D. 8

198. The utility of VLF waves in communication is due to their

A. low power requirements

B. high reliability

C. simpler antennas

D. these waves are not ued at all

199. When a wave is propagated in a good dielectric where $\sigma/\omega\varepsilon << 1$, then the attenuation factor and the phase shift factor are given by

A. $\alpha = \frac{\sigma}{2}\sqrt{\frac{\mu}{\varepsilon}}, \beta = \omega\sqrt{\mu\varepsilon}$

B. $\alpha = \frac{\sigma}{2}, \beta = \omega\sqrt{\mu\varepsilon}$

C. $\alpha = \sigma\sqrt{\frac{\mu}{2\varepsilon}}, \beta = \sqrt{\frac{\omega\mu\varepsilon}{2}}$

D. $\alpha = 1, \beta = 45°$

200. A transmission line of characteristic impedance 300 ohms is terminated by a load of (300 – j 300) ohm. The transmission coefficient is

A. 1.265 $\angle -18.43°$ B. 1.01 $\angle -10°$
C. 1.14 $\angle 66.68°$ D. 1.09 $\angle 66.68°$

201.

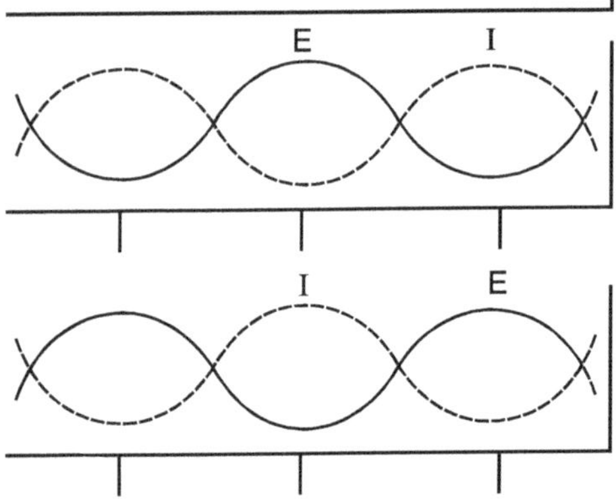

Indentify the correct load termination

A. $Z_L > Z_C$, $Z_L = Z_C$, $Z_L < Z_C$
B. $Z_L = Z_C$, $Z_L < Z_C$, $Z_L > Z_C$
C. $Z_L < Z_C$, $Z_L = Z_C$, $Z_L > Z_C$
D. $Z_L = Z_C$, $Z_L > Z_C$, $Z_L < Z_C$

202. For a lossy transmission line short-circuited at the receiving end, the input impedance is given by (Z_0 is the characteristic impedance, γ is the propagation constant and 1 is the length of the line)

A. $Z_0 \coth \gamma l$ B. $Z_0 \cot \gamma l$
C. $Z_0 \tanh \gamma l$ D. $Z_0 \tan \gamma l$

203. A short grounded vertical antenna has a length L which is 0.05 λ at frequency f. If its radiation resistance at f is R ohms, then its radiation resistance at a frequency $2f$ will be

A. R/2 ohms B. R ohms
C. 2 R ohms D. 4R ohms

204. Which of the following equations is correct?

A. $\vec{A} \times \left(\vec{B} \times \vec{C}\right) = \left(\vec{A} \times \vec{B}\right) - \left(\vec{A} \times \vec{C}\right)$

B. $\vec{A} \times \left(\vec{B} \times \vec{C}\right) = \left(\vec{A}.\vec{C}\right)\vec{B} - \left(\vec{A} \times \vec{B}\right)\vec{C}$

C. $\vec{A} \times \left(\vec{B} \times \vec{C}\right) = \left(\vec{A}.\vec{C}\right)\vec{B} - \left(\vec{A}.\vec{C}\right)\vec{B}$

D. $\vec{A} \times \left(\vec{B} \times \vec{C}\right) = \left(\vec{A}.\vec{C}\right)\vec{B} - \left(\vec{A}.\vec{B}\right)\vec{C}$

205. In a static magnetic field, the energy density W_m is given by

A. $W_m = mH^2$ B. $W_m = \frac{\mu H^2}{2}$

C. $W_m = \frac{\mu^2 H^2}{2}$ D. $W_m = \frac{\mu^2 H^2}{4}$

206. Polarisation of electromagnetic waves is due to

A. transverse nature of the waves
B. longitudinal nature of the waves
C. reflection from another medium
D. none of these

207. When a charge q moves with velocity v in an electric field E and magnetic field B, the Lorentz force F is given by :

A. $F = qE + vqB$ B. $F = vBq$
C. $F = qE$ D. $F = qE - vqB$

208. A two wire transmission line of characteristic impedance Z_0 is connected to a load of impedance Z_L ($Z_L \neq Z_0$). Impedance matching cannot be achieved with

A. a quarter wavelength transformer
B. a half-wavelength transformer
C. an open circuited parallel stub
D. a short circuited parallel stub

209. A p.d. of 300 V is applied to a series combination of 3μ*fd* and 9 μ*fd* capacitors. The charge on each capacitor is

A. 3600 μ C B. 40×10^{-3} C
C. 7.5×10^{-3} C D. 675 μC

210. The type of time variation $\left(\frac{\partial}{\partial t}\right)$ implied with reference to the field quantities in Maxwell's equations is

A. sinusoidal
B. non-sinusoidal
C. both sinusoidal and non-sinusoidal
D. exponential

211. Lorentz' force law for a point charge q is given by

A. $\vec{F} = q\left(\vec{E} + \vec{v}.\vec{B}\right)$

B. $\vec{F} = q\left(\vec{E} + \vec{v} \times \vec{B}\right)$

C. $\vec{F} = q\left(\vec{B} + \vec{v} \times \vec{E}\right)$

D. $\vec{F} = q\left(\vec{B} + \vec{v}.\vec{E}\right)$

212. The equation $\Delta \times \overline{E} = \frac{\partial B}{\partial t}$ is the generalisation of

A. Amere's law B. Gauss's law
C. Faraday's law D. Biot-Savart's law

213. The skin depth (δ) due to RF current is equal to (ω = angular frequency, μ = permeability and σ = conductivity).

A. $\frac{2}{\omega \mu \sigma}$ B. $\frac{2}{\sqrt{\omega \mu \sigma}}$

C. $\left(\frac{2}{\omega \mu \sigma}\right)^{1/2}$ D. $\frac{\sqrt{2}}{\omega \mu \sigma}$

214. The relation for electric field intensity is

A. $E = \frac{Q}{4\pi r}$ B. $E = \frac{Q}{4\pi r^2}$

C. $E = \frac{Q}{4\pi \varepsilon r}$ D. $E = \frac{Q}{4\pi \varepsilon r^2}$

215. A uniform plane wave is one in which

A. $\vec{E}$ and $\vec{H}$ are perpendicular

B. $\vec{E}$ and $\vec{H}$ lie in a plane

C. $\vec{E}.\vec{H} = 0$

D. $\vec{E} \times \vec{H} = 0$

216. Arrange the following forces in descending order of their strength

1. Electromagnetic
2. Nuclear
3. Gravitational
4. Radioactive decay

A. 1, 2, 3, 4 B. 1, 3, 2, 4
C. 1, 4, 3, 2 D. 2, 1, 4, 3

217. A loop produces a maximum emf of 150 volts when rotated at 60 rev/sec in a field of magnetic induction 5000 gauss. The dimensions of the loop are

A. 1.8 or 2 turns B. $\frac{5}{2\pi}$ turns metre^2

C. $\frac{5}{2\pi}$ turn/metre^2 D. 2 turn metre^2

218. 1. If there is no net charge inside a Gaussian surface, E is zero all around the surface
2. If E is zero, everywhere on the surface, there is no net charge

A. 1, 2 are true B. 1 true, 2 false
C. 1, 2 false D. 1 false, 2 true

219. Fibres have numerical aperture (N. A.) in the region of 0.15 to 0.4. Fibres with higher N. A. values generally exhibit

A. reduced losses
B. high bandwidth
C. reduced losses and low bandwidth
D. greater losses and low bandwidth

220. Microwave link repeaters are typically 50 km apart

A. because of atmospheric attenuation
B. because of output tube power limitations
C. because of earth's curvature
D. to ensure that the applied dc voltage is not excessive.

221. The product of RC and L/R is

A. ohms farad and henry/ohm
B. ohms farad and seconds
C. both seconds
D. coulomb and ampere

222. A series combination of resistance and condenser is charged by a battery. They charging is done

1. linearly
2. exponentially
3. exponentially with time for build up of charge
4. current decreases exponentially with time

A. 2 and 4 B. 1 only
C. 1 and 4 D. none

223. In a vertical earthed antenna the base input impedance will be inductive for a height

A. less than $\frac{\lambda}{4}$

B. between $\frac{\lambda}{4}$ and $\frac{\lambda}{2}$

C. between $\frac{\lambda}{2}$ and $\frac{3\lambda}{4}$

D. more than $\frac{3\lambda}{4}$

224. The relative permeability of a paramagnetic substance will be
A. Less than 1
B. 1
C. Slightly more than 1
D. Around 0.5

225. Which of the following statements is true for isotopes?
A. They have different atomic numbers and the same atomic masses
B. Their electronic configurations differ
C. Slight differences in chemical behaviour of isotopes (called isotopic effects) influence rather the kind of reaction itself
D. The first ones discovered were those of neon.

226. Assuming that the energy of an electron is quantized, which of the following statements is not true?
A. The centrifugal force created by the electron's motion about nucleus is counter balanced by the electrostatic attraction.
B. The radius (R) of an electron's orbit about the nucleus is equal to $R = \frac{n^2 a_0}{z}$ (where z = atomic number, a_0 = size of the orbit and n = quantum number).
C. If the principal quantum number *(n)* of an electron is known, only the size of its orbit and the energy of the electron can be calculated.
D. As the principal quantum number increases, the difference in energy between the levels decreases.

227. In case of antenna the ratio of the power radiated in the desired direction to the power radiated in the opposite direction is known as
A. transmission efficiency
B. front to back ratio
C. loos coefficient
D. none of the above

228. The effective length of a tower radiator is
A. $l_e = \frac{l}{2\pi}$ B. $l_e = \frac{2\pi}{l}$
C. $l_e = \frac{2l}{\pi}$ D. none of the above

229. In a waveguide, the suffix mn of the modes TE/TM denote
A. Half wavelength of E field and full wavelength of H field
B. Half waelengths of E and H fields in directions other than guide axis
C. Full wavelength of E field and half wavelength of H field
D. Half wavelength of E and H fields

230. In a circularly polarized wave travelling in Z direction
A. E_x and E_y are in phase
B. E_x and E_y are 90° out of phase and unequal in magnitude
C. E_x and E_y are 90° out of phase and equal in magnitude
D. $E_x = -E_y$

231. Yagi antenna has :
A. only driven antenna
B. driven antenna and a reflector
C. driven antenna and directors
D. driven antenna, reflector and directors

232. Directivity depends on
A. the distribution of radiated power in space
B. solid angle of the radiated field pattern
C. both (A) and (B) above
D. none of the above

233. The emission of an alpha particle from the nuclear of $^{226}_{88}Ra$ will yield
A. $^{223}_{86}Rn$ B. $^{222}_{86}Rn$
C. $^{223}_{88}Rn$ D. $^{222}_{88}Rn$

234. Which of the following is not a polymer?
A. Plastic B. Petroleum
C. Glycogen D. Starch

235. The electric field intensity at a distance '*d*' from a infinitely long charged line will vary as
A. d^2 B. d^4

C. $\frac{1}{d}$ D. $\frac{1}{d^2}$

236. The ground waves eventually disappear as one moves away from the transmitter because of
A. tilting
B. maximum single-hop distance limitation
C. interference from the sky waves
D. loss of line-of sight condition

237. An arrangement of several individual antennas so spaced and phased that their individual contributions combine in one preferred direction and cancel in all other directions, to get greater directive gain or directivity, is known as
A. directional radiation of antennas
B. anetnna gain stability
C. array of antennas
D. none of the above

238. The maximum usable frequency for sky wave propagation lies in the range
A. 10 – 500 kHz B. 8 – 35 MHz
C. 30 – 300 MHz D. 100 – 500 MHz

239. A typical radar antenna has a bandwidth of
A. 5° B. 3°
C. 2° D. 1°

240. Laplacian of scalar function V is :
A. gradient of V
B. divergence of V
C. gradient of V
D. divergence of the gradient of V

241. The approximate input impedance of the OPAMP circuit shown in given figure is

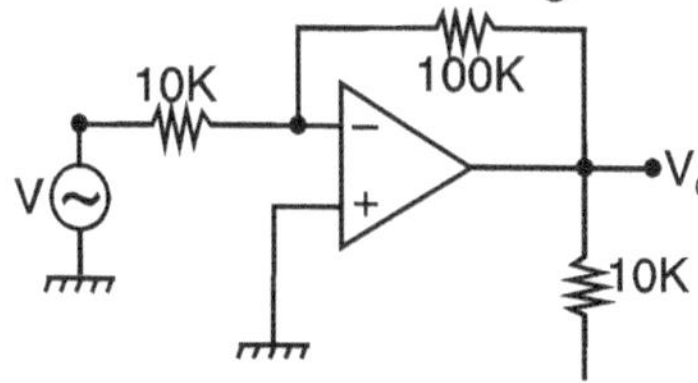

A. infinity B. 120 K
C. 110 K D. 10 K

242. Refraction of electromagnetic waves occurs when they
A. pass through a small slot in a conducting plane
B. pass into a medium of different dielectric constant
C. are polarized at right angles to the direction of propagation
D. encounter a perfectly conducting surface

243. The circuit shown in the given figure is a

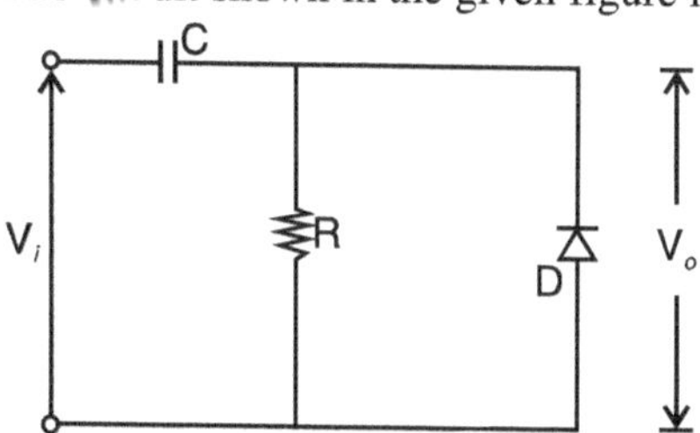

A. positive clamper B. negative clamper
C. differentiator D. positive peak clipper

244. Two parallel plates, having a separation x ($x <<$ lateral dimensions of the plates) are connected by a wire. A point charge q is located between the plates at a distance x_1 from one plate ($x_1 < x$). The magnitude of the total surface charge on the inner surface of each plate is
A. Zero, because two plates are connected
B. $+\frac{q}{2}$
C. $+q$
D. $-q$

245. Pick out the wrong statement.
The electric and magnetic fields travel through a medium of perfect insulator. The modified Maxwell's equations are
A. $\nabla.\vec{D} = 0$, rest 3 not changed
B. $\nabla.\vec{D} = 0$, $\nabla \times \vec{E} = 0$, rest 2 not changed
C. $\nabla.\vec{D} = 0$, $\nabla \times \vec{H} = \frac{\partial \vec{D}}{\partial t}$, rest 2 not changed
D. none changed

246. In a horizontally polarized wave
A. electric component is perpendicular to the ground
B. electric component is parallel to the ground
C. magnetic component is parallel to the ground
D. none of these

247. The effective height of an antenna is slightly greater than physical height becaue
A. wave velocity in conductor is less than its velocity in free space
B. wave velocity in conductor is more than its

velocity in free space
C. resistance of conductor is less than that of free space
D. none of the above

248. Directive gain G_d of an antenna in a given direction is the ratio of :
A. radiation intensity in that direction to the average power radiated
B. radiation intensity in that direction to the total input radiated power
C. average power to maximum power
D. none of these

249. Which of the following is true for isotopes of an element?
A. they are atoms of the same atomic number with different masses.
B. the only difference in composition between isotopes of the same elements is in the number of neutrons in the nucleus
C. the atomic weight of an element is an average of the weight of the isotopes of the element in the proportions in which they normally occur in nature
D. all of the above

250. Which of the following best characterizes electron affinity?
A. Neutral atoms with unfilled orbitals having repulsion for electrons
B. Neutral atoms of noble gases having attraction for electrons
C. Neutral atoms with unfilled orbitals having attraction for electrons
D. Atoms that lie to the upper left of the periodic table

251. Between a hollow and solid metal sphere charges reside
A. on the outer surface in both
B. on outer surface in hollow and throughout in solid
C. throughout in both
D. none of the above

252. A plane monochromatic electro-magnetic wave travels in a perfect conducting medium, which is charge-free and external current-free. Then,
A. $\vec{E}$ field lags $\vec{B}$ field by π
B. $\vec{E}$ field leads $\vec{B}$ field by π
C. $\vec{E}$ and $\vec{B}$ field are co-phasal
D. $\vec{E}$ and $\vec{B}$ fields differ in phase by $\frac{\pi}{2}$

253. Impedance inversion is obtained by use of :
A. a quarter wave line
B. Balun transformer
C. halfwave line
D. fullwave line

254. The VHF band is
A. 0.3 – 3 MHz B. 3 – 30 MHz
C. 30 – 300 MHz D. 0.3 – 3 GHz

255. Plasma signifies :
A. electronic medium
B. solid dielectric
C. liquid dielectric
D. gaseous medium

256. The inconsistency of continuity equation for time varying fields was corrected by Maxwell and the correction applied was
A. Ampere's law, $\frac{\partial D}{\partial t}$
B. Gauss's law, J
C. Faraday's law, $\frac{\partial B}{\partial t}$
D. Ampere's law, $\frac{\partial \rho}{\partial t}$

257. There exists a +ve charge in an inertial frame of reference F. The electric field as viewed from another inertial frame of reference F' moving uniformly in the –ve x direction with speed v, can be represented by

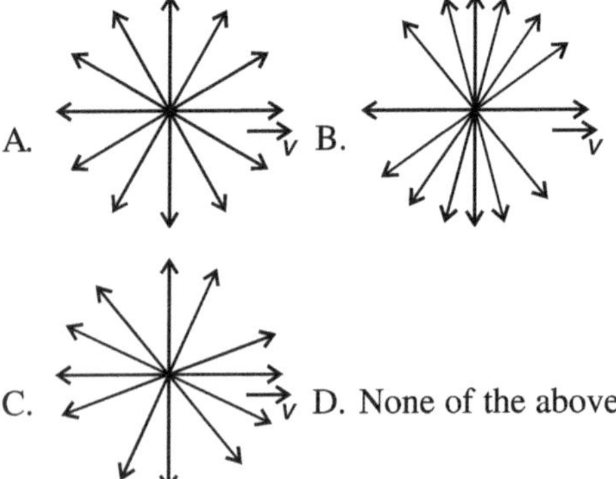

258. In Faraday's induction phenomenon, a changing

magnetic field is accompanied by an electric field. Which of the following equation or equations represents/represent it

A. $\nabla.\overline{E} = \frac{\partial}{\varepsilon}$ B. $\nabla.\overline{B} = 0$

C. $\nabla \times \overline{E} = \frac{\partial B}{\partial t}$ D. $\nabla \times \overline{B} = \mu\varepsilon \frac{\partial E}{\partial t}$

259. As a result of reflection from a plane conducting wall, the electromagnetic waves acquire an apparent velocity greater than the velocity of light in free space. This velocity is called :

A. group velocity
B. phase velocity
C. normal velocity
D. velocity of propagation

260. Divergence theorem relates surface integration with volume integration as :

A. $\int_v \left(\nabla.\vec{J}\right) dv = \vec{J}\, ds$

B. $\int_v \left(\nabla.\vec{J}\right) dv = \oint_v \oint_s \vec{J}.\vec{ds}$

C. $\int_v \left(\nabla.\vec{J}\right) dv = \oint_v \vec{J}.\vec{ds}$

D. none of these

261. Pick out the wrong answer
Two point charges of $+ 2.0 \times 10^{-7}$ coul and $+ 8.5 \times 10^{-8}$ coul are 12 cm apart. The electric field each produces at the site of the other is

A. 1250×10^2 nt/coul, 530×10^2 nt/coul
B. 1.25×10^3 nt/coul, 5.3×10^2 nt/coul
C. 1250 nt coul, 530 nt coul
D. 530×10^2 nt/coul, 1250×10^2 nt/coul

262. A pith ball of 10 g carries a charge of 10 μc. What must be a charge on a ball placed 10 cm directly above which will hold the pith ball in equilibrium?

A. 1 μc B. .011 μc
C. – 0.011 μc D. .1 μc

263. There is an infinite linear array of alternately placed +ve and – ve charge each of magnitude Q and separation *r*. The potential energy of a charge is

A. + ve and equal to ln $(2)^2$. $\frac{Q^2}{4\pi\varepsilon_0 r}$

B. – ve and equal to – ln $(2)^2$. $\frac{Q^2}{4\pi\varepsilon_0 r}$

C. + ve and equal to ln (2). $\frac{Q}{4\pi\varepsilon_0 r}$

D. + ve and equal to

$$\frac{Q}{4\pi\varepsilon_0 r}\left[1-\frac{1}{2}+\frac{1}{3}-\frac{1}{4}+\frac{1}{5}\ldots\right]$$

264. Consider the following diagrams showing vector fields

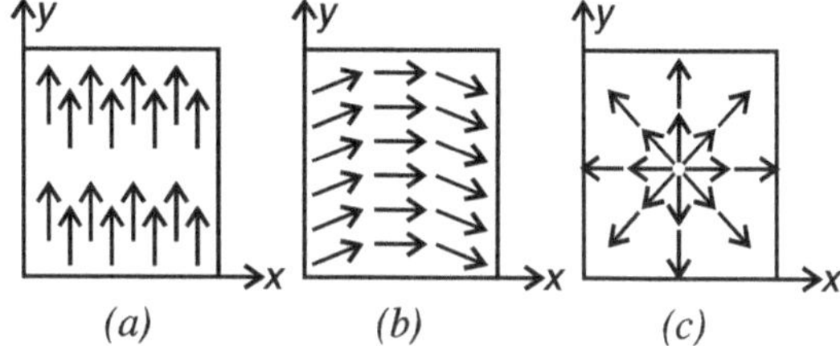

Select the correct statements :

A. For Fig. *(a)*, $\nabla.\overline{F} = 0$, $\nabla \times \overline{H} = 0$
B. For Fig. *(a)*, $\nabla.\overline{F} = 0$, $\nabla \times \overline{H} \neq 0$
C. For Fig. *(b)*, $\nabla.\overline{F} = 0$, $\nabla \times \overline{F} \neq 0$,
D. For Fig. *(c)*, $\nabla.\overline{F} = 0$, $\nabla \times \overline{F} \neq 0$

265. An antenna cut of a resonant frequency of *f* and having a Q factor of 20 can handle a bandwidth of 10 MHz Then *f* is

A. 10 MHz B. 20 MHz
C. 200 MHz D. none of these

266. If the diameter of the parabolic reflector of a microwave antenna is doubled, its gain will increase by

A. 0 *d*B B. 2 *d*B
C. 4 *d*B D. 6 *d*B

267. Inductive reactances applies to sine waves because it

A. increases with lower frequency
B. increases with lower inductance
C. increases with higher frequency
D. decreases with higher frequencies

268. Between the plate of a capacitor, an aluminium foil is introduced midway and connected to one plate. The capacitance

A. increases B. decreases
C. remains same D. increases twice

269. An infinite number of concentric rings carry a charge Q each alternately positive and negative. Their radii are 1, 2, 4, 8 ... meters in geometric progression as shown. The potential at the centre will be

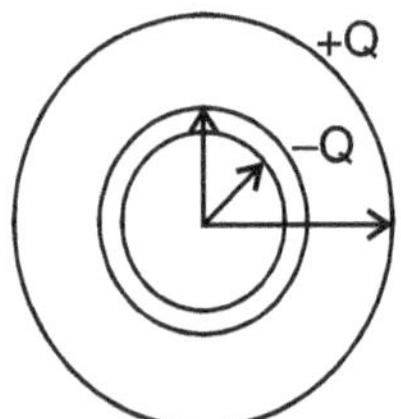

A. zero B. $\frac{Q}{12\pi\varepsilon_0}$

C. $\frac{Q}{8\pi\varepsilon_0}$ D. $\frac{Q}{6\pi\varepsilon_0}$

270. Indicate which one of the following type of noise does not ocur in transistors?

A. Shot noise B. Partition noise

C. Resistance noise D. None of these

271. Which of the following is not a wide band antenna?

A. Marconi B. Discone

C. Helical D. Folded dipole

272. The magnetic field intensity emanating from a closed surface

A. depends on dipole moment

B. depends on volumetric compaction factor

C. is infinite

D. is zero

273. $J = \sigma E$ is known as

A. Biot Savart's law B. Ampere's law

C. Coulomb's law D. Ohm's law

274. The polarization of electro-magnetic wave is

A. the direction of electric and magnetic field

B. the direction of electric field

C. the direction of magnetic field

D. none of the above

ANSWERS

1	2	3	4	5	6	7	8	9	10
D	C	D	B	B	B	A	B	A	A
11	12	13	14	15	16	17	18	19	20
D	A	D	A	D	A	C	C	A	C
21	22	23	24	25	26	27	28	29	30
B	A	B	A	A	C	B	D	C	B
31	32	33	34	35	36	37	38	39	40
D	B	D	A	C	C	A	D	B	D
41	42	43	44	45	46	47	48	49	50
D	D	D	A	C	C	D	C	B	B
51	52	53	54	55	56	57	58	59	60
A	A	C	C	A	B	B	C	B	B
61	62	63	64	65	66	67	68	69	70
A	A	C	D	C	B	C	B	B	B
71	72	73	74	75	76	77	78	79	80
D	B	B	D	C	A	A	B	A	C
81	82	83	84	85	86	87	88	89	90
B	B	D	B	B	C	A	C	B	D
91	92	93	94	95	96	97	98	99	100
D	C	A	A	A	B	B	D	B	D
101	102	103	104	105	106	107	108	109	110
D	A	C	B	C	C	C	A	B	D
111	112	113	114	115	116	117	118	119	120
C	A	A	C	D	C	C	D	B	B

121	122	123	124	125	126	127	128	129	130
B	C	B	C	D	C	D	A	D	D
131	132	133	134	135	136	137	138	139	140
B	C	A	B	C	C	B	C	C	C
141	142	143	144	145	146	147	148	149	150
A	D	D	B	B	C	B	A	B	C
151	152	153	154	155	156	157	158	159	160
B	B	B	C	C	B	D	A	C	D
161	162	163	164	165	166	167	168	169	170
C	A	D	C	C	C	C	D	A	B
171	172	173	174	175	176	177	178	179	180
A	B	D	D	C	A	A	D	D	C
181	182	183	184	185	186	187	188	189	190
B	A	C	C	C	C	D	A	B	B
191	192	193	194	195	196	197	198	199	200
A	C	B	A	B	A	C	B	A	A
201	202	203	204	205	206	207	208	209	210
D	C	D	D	B	A	A	B	D	C
211	212	213	214	215	216	217	218	219	220
B	C	C	D	B	D	B	D	D	C
221	222	223	224	225	226	227	228	229	230
D	A	B	C	D	C	B	C	B	C
231	232	233	234	235	236	237	238	239	240
D	A	B	B	D	A	C	B	C	D
241	242	243	244	245	246	247	248	249	250
D	A	A	D	C	B	A	C	D	C
251	252	253	254	255	256	257	258	259	260
A	D	A	A	D	A	B	C	B	C
261	262	263	264	265	266	267	268	269	270
D	C	B	B	C	D	C	D	B	D
271	272	273	274						
C	D	D	B						

EXPLANATORY ANSWERS

37. Reflection coefficient is given by,

$$K_r = \left|\frac{Z_L - Z_0}{Z_L + Z_0}\right|$$

when, $Z_L = Z_0$, $K_r = 0$

38. Quantization noise can be reduced by increasing the number of standard levels. But more levels require more bits to send them. Obviously, it needs more bandwidth because bandwidth required is proportional to the number of bits sent per second.

64. In case of an antenna, the radiation resistance (R) and loss rsistance (R_L) are two different things. R_L includes I^2R losses also and R varies directly as square of the frequency.

The efficiency of an antenna $= \dfrac{R}{(R + R_L)}$

86. Since to an observer in F', the dimensions of the sheet of charge are no longer same, they are contracted by a factor $\sqrt{1 - v^2/c^2}$, but total charge is invariant and independent of the frame of reference, hence, the apparent charge density would now be more than > σ and equal to $\frac{\sigma}{\sqrt{1-v^2/c^2}}$,

where c is the velocity of light.

136. The spherical charge distribution may be considered to be equivalent to a combination of many charged spherical shells. Consider any such shell of radius x and thickness dx. Surrounding the charged spherical core of radius x. If σ be the surface charge density, charge on spherical core of radius $x = \frac{4}{3}\pi x^3 \sigma$

Charge on the shell = $4\ \sigma x^2.\ dx.\ \sigma$,

Electric potential energy of the core and shell

$$= \frac{1}{4\pi\varepsilon_0}\left(\frac{\text{charge on core} \times \text{charge on shell}}{x}\right)$$

$$= \frac{1}{4\pi\varepsilon_0}.\frac{16}{3}\pi^2\sigma^2 x^4.dx$$

∴ Total potential energy

$$= \int_0^r \frac{1}{4\pi\varepsilon_0}\frac{16}{3}\pi^2\sigma^2 x^4 dx$$

$$= \frac{1}{4\pi\varepsilon_0}\frac{16}{15}\pi^2\sigma^2 r^5 = 5.4\times10^9\frac{Q^2}{r}$$

But $Q = \frac{4}{3}\pi r^3 \sigma$

∴ Total potential energy = $5.4\times10^9\frac{Q^2}{r}$

200. $T = \frac{2Z_L}{Z_L + Z_C}$

$$= \frac{2(300 - j300)}{(300 - j300) + 300}$$

$= 1.265\ \angle -18.43°$

203. Radiation resistance, $R_R = 80\ \pi^2 = \left(\frac{dl}{\lambda}\right)^2$

219. By Snell's law,

$N.A = \sin\theta\sqrt{(\mu_1^2 - \mu_2^2)}$

Where μ_1 is the refractive index of the core material and μ_2 is the refractive index of the cladding material.

225. Isotopes of a particular element are characterized by the same atomic number and different atomic masses. The only difference in composition between them is in the number of neutrons in the nucleus.

226. The energy of an electron in an atom cannot vary continuously, rather it is limited to a specific number of discrete values, thus the electron's energy is said to be quantized. If the principle quantum number n is known, the energy, the size of orbit and angular momentum of the electron can be determined.

233. An alpha particle is a helium nucleus having atomic number 2 and mass number 4.

$${}^{226}_{88}\text{Ra} \rightarrow {}^{4}_{2}\text{He} + {}^{222}_{86}\text{Rn}$$

241. From the equivalent circuit and Miller's Theorem Assuming R_1

$A_v \rightarrow \infty$

$R_1 = 10$ K

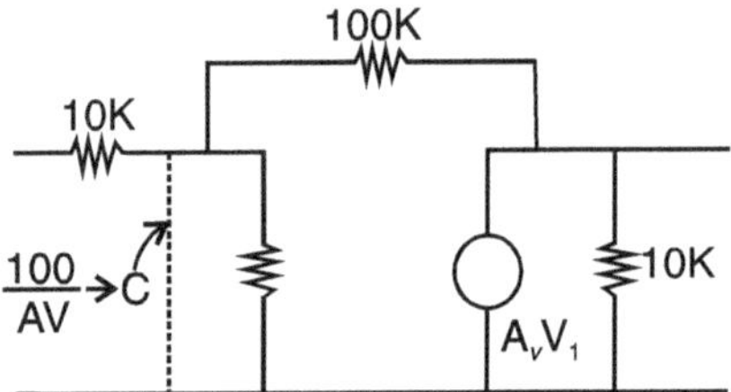

244. Referring the adjacent diagram.

The charge + q would induce charges of opposite nature at the inner surfaces of the parallel plates, and a larger amount would be produced on the nearer plate. If it is too close to A, most of the charges would be induced on A and presence of B would be insignificant. However, the total induced charge must be equal and opposite to the inducing charge + q and hence answer is – q.

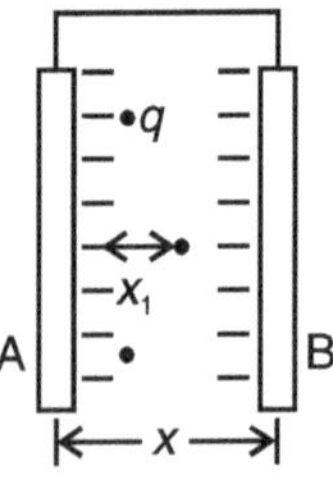

257. A charge in motion would produce a unique

type of spherically asymmetrical electric field because of a preferred direction of motion of the charge. In the course of time, the electric at any point in the frame F would change as the charge moves.

264. In figure-(a), $\frac{\partial F}{\partial Y} = 0$, $F_x = 0$ hence div F = $\nabla.\overline{F}$ = 0, but curl $\overline{F} = \nabla \times \overline{F} \neq 0$. In figure-(b), there is no change in the magnitude of F and hence $\nabla.\overline{F}$ = 0 and also it is possible that $\nabla \times \overline{F} = 0$, as there may be an electrostatic field in which $\overline{F}_x \times \frac{1}{\text{distance}}$ and circulation around any closed path in it may be equal to zero.

In Figure-(c), indicating a Central field (radial) $\nabla \times \overline{F} = 0$ and $\nabla.\overline{F} \neq 0$.

270. Potential at the centre due to a ring of charge Q is, $V = \frac{Q}{4\pi\varepsilon_0 R}$

$\therefore$ Net potential,

$$V = \frac{Q}{4\pi\varepsilon_0}\left(\frac{1}{1} - \frac{1}{2} + \frac{1}{4} - \frac{1}{8} + \frac{1}{16}\ldots\right)$$

$$= \frac{Q}{4\pi\varepsilon_0 R}\left[\left(\frac{1}{1} + \frac{1}{4} + \frac{1}{16}\ldots\right) - \left(\frac{1}{2} + \frac{1}{8} + \frac{1}{32}\ldots\right)\right]$$

Sum of an infinite geometric progression series when common ratio $(r < 1)$ $S = \frac{a}{1-r}$, where a is the first term.

$$\therefore \; S_1 = \frac{1}{1 - \frac{1}{4}} = \frac{1}{2},$$

$$S_2 = \frac{1/2}{1 - \frac{1}{4}} = \frac{2}{8}$$

$$\therefore \; V = \frac{Q}{4\pi\varepsilon_0}\left(\frac{4}{3} - \frac{2}{3}\right) = \frac{Q}{6\pi\varepsilon_0}$$

4

COMMUNICATION SYSTEM

Modulation Index and Number of Sidebands

It is found that the number of sidebands

1. depends *directly* on the amplitude of the modulating signal
2. depends *inversely* on the frequency of the modulating signal

Since frequency deviation is directly related to the amplitude of the modulating signal, the above two factors can be combined in one factor called **modulation index**.

Hence, number of sidebands depends on $m_f = \Delta f / f_m$.

Obviously, the number of pairs of sidebands.

(i) **increases** as frequency deviation (or amplitude of modulating signal) **increases**

(ii) **increases** as the modulating signal frequency **decreases**

TABLE 1

m_f	0.5	1	2	3	4	5	6	7
n	2	3	4	6	7	8	9	11
$BW = 2nf_m$	$4f_m$	$6f_m$	$8f_m$	$12f_m$	$14f_m$	$16f_m$	$18f_m$	$22f_m$
m_f	8	9	10	11	12	13	14	15
n	12	13	14	15	16	17	18	19
$BW = 2nf_m$	$24f_m$	$26f_m$	$28f_m$	$30f_m$	$32f_m$	$34f_m$	$36f_m$	$38f_m$

Table 1 summarizes the relation between the number (n) of sidebands on either side of the carrier and the corresponding bandwidth of the spectrum.

Example : *A 5-kHz audio signal is used to frequency-modulate a 100 MHz carrier causing a frequency deviation of 20 kHz. Determine*

(i) modulation index

(ii) bandwidth of the FM signal.

Solution : (i) $m_f = \dfrac{\Delta f}{f_m} = \dfrac{20}{5} = \mathbf{4}$

As seen from above Table

$$BW = 14f_m = 14 \times 5 = \mathbf{70\ kHz}$$

Note : We cannot use the alternate expression of BW given in above because $m_f > 6$.

Example : *In an FM circuit, the modulation index is 10 and the highest modulation frequency is 20 kHz. What is the approximate bandwidth of the resultant FM signal?*

Solution : Since the value of m_f is more than 6, we will use the expression

$$BW = 2(\Delta f + f_m)$$

Now, $m_f = \dfrac{\Delta f}{f_m}$ or $10 = \dfrac{\Delta f}{20}$

$\therefore \quad \Delta f = 200\,\text{kHz}$

$\therefore \quad \text{BW} = 2(200 + 20) = \mathbf{440\ kHz.}$

Mathematical Expression of FM Wave

The unmodulated carrier is given by $e_c = A \sin 2\pi f_o t$

The modulating signal frequency is given by $e_m = B \sin 2\pi f_m t$

The modulated carrier frequency of swings around the resting frequency f_o, thus

$$f = f_o + \Delta f.\sin 2\pi f_m t$$

Hence, equation for the frequency-modulated wave becomes

$$e = A \sin 2\pi ft = A \sin [2\pi(f_o + \Delta f.\sin 2\pi f_m t)\, t]$$

$$= A \sin (2\pi f_o t + \frac{\Delta f}{f_m} \cos 2\pi f_m t)$$

$$= A \sin \left(2\pi f_o t + m_f \cos 2\pi f_m t\right)$$

Demodulation or Detection

When the RF modulated waves, radiated out from the transmitter antenna, after travelling through space, strike the receiving aerials, they induce very weak RF currents and voltages in them. If these high-frequency currents are passed through headphones or loudspeakers, they produce no effect on them because all such sound-producing devices are unable to respond to such high frequencies due to large inertia of their vibrating discs etc. Neither will such RF currents produce any effect on human ear because their frequencies are much beyond the audible frequencies (20 to 20,000 Hz approximately). Hence, it is necessary to demodulate them first in order that the sound-producing devices may be actuated by audio-frequency current similar to that used for modulating the carrier wave at the broadcasting station.

This process of *recovering AF signal from the modulated carrier wave is known as demodulation or detection.*

The demodulation of an AM wave involved to operations :

(i) rectification of the modulated wave and

(ii) elemination of the RF component of the modulated wave.

However, domodulation of an FM wave involves three operations (i) conversion of frequency changes produced by modulating signal into corresponding amplitude changes (ii) rectification of the modulating signal and (iii) elimination of RF component of the modulated wave.

Essential of AM Detection

For recovering the AF waveform from modulated wave (a mixture of AF wave and RF carrier), it is essential to find some way of reducing (or better, eliminating) one half of the modulated wave. The result of this elimination (or rectification) would be that the average value of the wave would not be zero because, now, the impulse would be all in one direction as shown in Fig. 1. If this new wave is now passed through a headphone shunted by a suitable capacitor, then AF wave will pass through the headphone whereas the RF wave will be bypassed by the capacitor (because the high inductance of magnet coils of the headphones will offer tremendous impedance to RF currents). Hence, two will become separated.

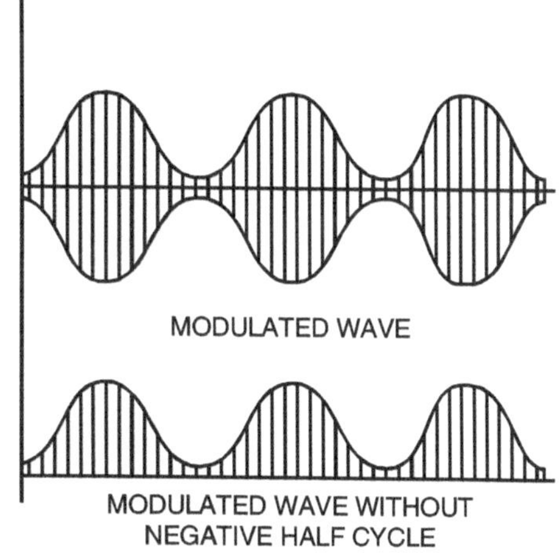

Fig. 1

Diode Detector for AM Signals

Diode detection is also known as **envelope-detection** or **linear detection**. In appearance, it looks like an ordinary half-wave rectifier circuit with capacitor input as shown in Fig. 2. It is called **envelope detection** because *it recovers the AF signal envelope* from the composite signal. Similarly, diode detector is called **linear detector** because its output is *proportional to the voltage of the input signal.*

Circuit Action : Of the various *RF* voltages induced in the receiver aerial, only those having the *same* frequency as the resonant frequency *LC* circuit

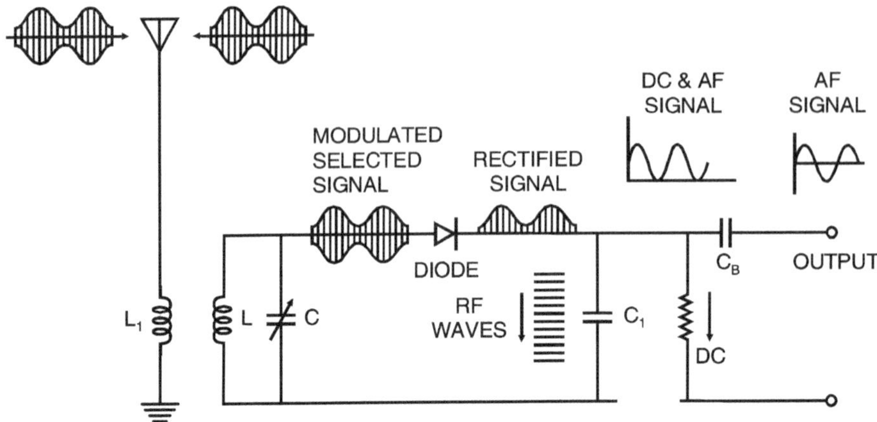

Fig. 2

are tuned in due to electromagnetic induction between coils L_1 and L. By varying C, the resonant frequency of the LC circuit can be varied and hence RF signal of any desired frequency can be tuned in. This input signal is rectified by the diode and passed on to the low-pass filter RC_1.

The RF carrier wave is filtered out by low-reactance capacitor C_1 which is called RF filter capacitor or RF by-pass capacitor. The dc component of the remaining signal is shunted out through R because it cannot pass through blocking capacitor C_B. But the low frequency signal AF can easily get through C_B and becomes available across the output. When passed through a stuiable device, say, a headphone, the original sound can be heard.

Advantages : Diode detectors are extensively used in AM broadcast receivers because they have the following advantages :

1. they can handle comparatively large input signals
2. they can be operated as linear or power detectors
3. they rectify with negligible distortion and, hence, have good linearity.
4. they are well-adopted for use in simple automatic gain control circuits.

Disadvantages : However, the disadvantages are that

1. they do not have the ability to amplify the rectified signal by themselves as is done by a transistor detector. However, it is not a very serious drawback since signal amplification can be affected both before and after rectification.
2. while conducting, the diode consumes some power which reduces the Q of its tuned circuit as well as its gain and selectivity.

Transistor Detectors for AM Signals

Transistors can be used as detector amplifiers *i.e.*, both for rectification and amplification. As shown in Fig. 3, the RF signal is applied at the base-emitter junction where rectification takes place. The amplification of the recovered signal takes place in the emitter-collector circuit.

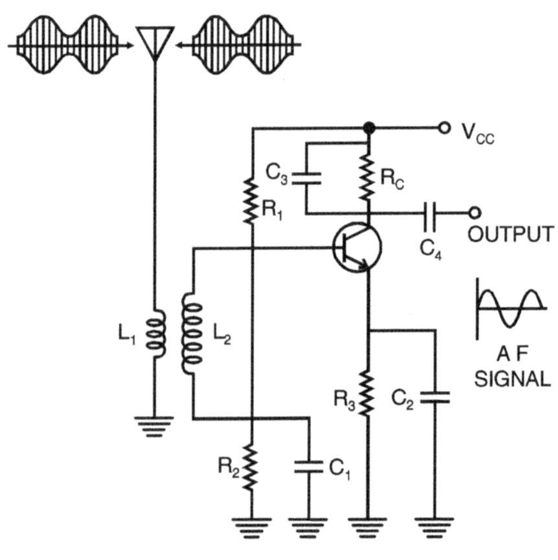

Fig. 3

It should be noted that when used as a detector, the transistor has to be biased at cut-off or operated as class-B. As is usual, C_2 ac grounds R_3 whereas voltage divider R_1—R_2 establishes proper bias which holds the transistor just at cut-off. Obviously, only positive half-cycles of the modulated RF input signal will drive the transistor into conduction whereas negative half-cycles would be cut-off. Thus, rectification of the input signal *takes place in the base-emitter circuit.* The small variations in base current produced by the positive half-cycle of the input signal produced large (β times) variations in the collector current. Hence, an amplified version of the rectified signal appears across RC-C_3 combination. C_3 eliminates RF component so that only AF signal voltage drops across R_C. The capacitor C_4 permits AF signal to pass through but blocks its dc component.

FM Detection

As discussed earlier, an FM carrier signal contains information (or intelligence we wish to convey) in the form of frequency variations above and below the centre frequency of the carrier. For recovering the information, we must first convert the FM signal in such of a way that it appears as a modulated RF voltage across the diode. A simple method of converting frequency variations into voltage variations is to make use of the principle that reactance (of coil or capacitor) varies with frequency. When an FM signal is applied to an inductor, the current flowing through it varies in amplitude according to the changes in frequency of the applied signal. Now, changes in frequency of the FM signal depend on the amplitude of the modulating AF signal. Hence, the current in the inductor varies as per the amplitude of the original modulating signal. In this way, frequency chanes in FM signal are converted into amplitude changes in current. These changes in current when passed through a resistor produce corresponding changes in voltage.

Hence, we find that, ultimately, frequency variations in FM signal are converted into voltage changes. Also, there exists a linear relation between the two—something essential for distortion-less demodulation.

FM demodulation may be carried out with the help of (i) ratio detector and (ii) quadrature detector.

Quadrature Detector

This detector depends on the frequency/phase relationship of a tuned circuit. It uses only one tuned circuit and is becoming increasingly popular in the integrated FM strips.

Theory : Let us first consider the general principle. A sinusoidal current is given by the equation

$$i = I_m \sin \theta = I_m \sin \omega t$$

Suppose, it flows through a circuit shown in Fig. 5(a). The voltage V_L across the inductor (assumed pure) leads the current I by 90°.

$$\therefore\ v_L = V_L \cos \omega t$$

The voltage, v_z across the parallel tuned circuit will be in phase with I at resonance. However, at frequencies slightly different (± 1%) from the resonant frequency, the phase angle ϕ will be given by

$$\tan \phi = \frac{yQR}{R} = yQ$$

assuming $f_o / f \cong 1$. (Fig. 4)

$$\text{Here, } y = \frac{\omega}{\omega_o} = \frac{\omega_0}{\omega}$$

$$= \frac{f}{f_o} - \frac{f_o}{f} \cong \frac{2.\Delta f}{f_o}$$

where $\Delta f = f - f_o$

f_o = resonant frequency

f = slightly off-resonance frequency

Hence, equation for v_z is given by

$$v_z = V_z \sin(\omega t - \phi)$$

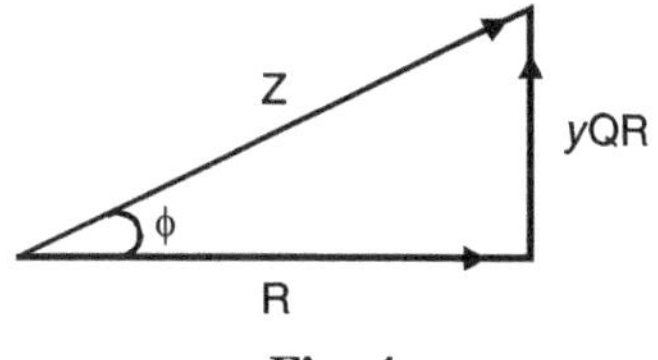

Fig. 4

When the two voltage v_L and v_Z are applied as inputs to a multiplier, the output voltage v_o is found to be proportional to their product as shown in Fig. 5(b).

$$v_o \propto v_L, v_z \propto \cos \omega t.\sin (\omega t.\phi) \propto \sin(2\omega t + \phi) + \sin \phi$$

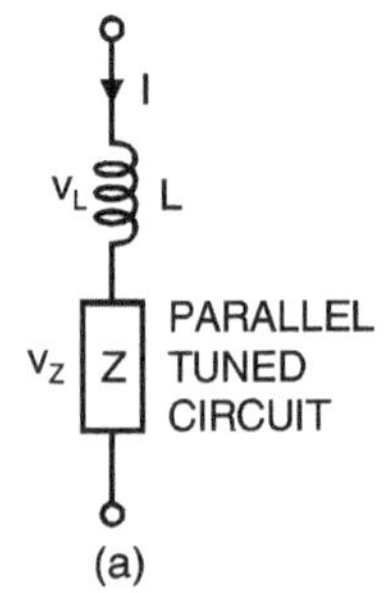

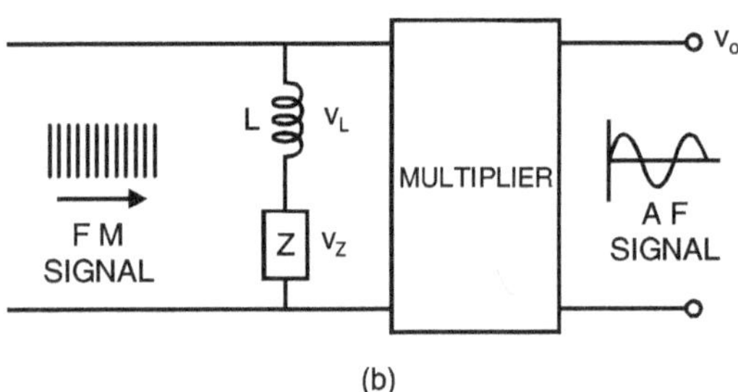

Fig. 5

A low-pass filter is used to reject the double frequency component sin $(2\omega t + \phi)$ and select only the low-frequency sin ϕ component.

$$\therefore\ v_o \propto \sin\phi \tan\phi \qquad \text{since } \phi \text{ is very small}$$

$$\propto y\phi \propto y \propto \frac{2.\Delta f}{f_o}$$

where v_m is the modulating AF voltage.

Now, $\Delta f = f - f_o \propto kv_m$

$$\therefore \quad v_o \propto v_m$$

It shows that output voltage v_o is proportional to the original modulating signal voltage.

Frequency Conversion

Let us first consider the necessity for radio broadcasting purposes and the basic principle of heterodyning action.

1. **Need :** It is very difficult to design amplifiers which give uniformly high gain over a wide range of radio frequencies used in commercial broadcast stations. However, it is possible to design amplifiers which can provide high-gain uniform amplification over a narrow band of comparatively lower frequencies called intermediate frequencies (IF). Hence, it is necessary to convert the modulated RF carrier into modulated IF carrier by using a frequency converter.
   This IF signal is then amplified by narrow-band IF amplifiers and passed on the AM detectors.
2. **Basic Principle :** The frequency conversion can be achieved by utilizing the heterodyne principle. For this purpose, the modulated RF signal is mixed (in a mixer) with an unmodulated RF signal produced by local oscillator as shown in Fig. 6.

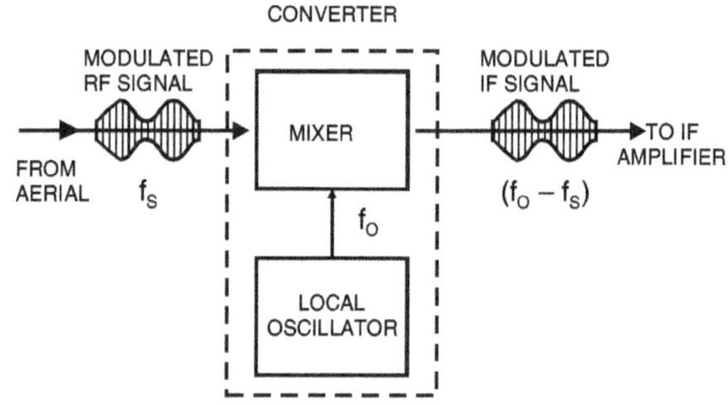

Fig. 6

The oscillator and the mixer may be either two separate devices or may be combined into one device called **converter**. The process of combining two ac signals of different frequencies in order to obtain a signal of new frequency is called **heterodyning action**.

3. **Heterodyning Action :** Suppose the carrier signal of frequency f_s is heterodyned with another signal of frequency f_o, then two additional signals are produced whose frequencies are :

(i) $f_o + f_s$ — the sum component
(ii) $f_o - f_s$ — the difference component

Usually, the sum frequency is removed by bandpass filtering. The difference frequency (also called *beat frequency*) is retained and forms the IF frequency in AM receivers.

Suppose, an AM carrier signal of frequency 1500 kHz is mixed with an unmodulated carrier (produced by local oscillator of frequency 1955 kHz. Then, following two frequencies are produced.

(i) 3455 kHz and
(ii) 455 kHz

The **higher frequency** is generally filtered out leaving behind the difference frequency of 455 kHz which forms the IF frequency.

Superhetrodyne AM Receiver

It is also referred to as superhet and is extensively used in modern AM receivers. Its block diagram with signals is shown in Fig. 7. The operation of this receiver is as under :

(i) Let us assume that the incoming signal frequency is 1500 kHz. It is first amplified by the R.F. amplifier.
(ii) Next, it enters a mixer circuit which is so designed that it can conveniently combine two radio frequencies–one fed into it by the R.F. amplifier and the other by a local oscillator.
(iii) The local oscillator is an RF oscillator whose frequency of oscillation can be controlled by varying the capacitance of its capacitor. In fact, the tuning capacitor of the oscillator is ganged with the capacitor of the input circuit so that the difference in the frequency of the selected signal and oscillator frequency is always constant. Usually, the difference is maintained at 455 kHz. If signal frequency is 1,500 kHz, then oscillator frequency can be either 1,955 or 1,445 kHz. Let us suppose that it is 1,955 kHz. In fact, local oscillator frequency is always higher than the frequency of the incoming signal.
(iv) When two alternating currents of these two different frequencies are combined in the mixer transistors, then phenomenon of beats is produced. In the present case, the beat frequency is 1955–1500 = 455 kHz. Since this frequency is lower than the signal frequency but still above the range of audio-frequencies, it is called *intermediate frequency* (IF).
(v) The 455 kHz output of the mixer is then passed on the IF amplifier which is *fixed-tuned* to 455 kHz frequency. In practice, one or more stages of IF amplification may be used.

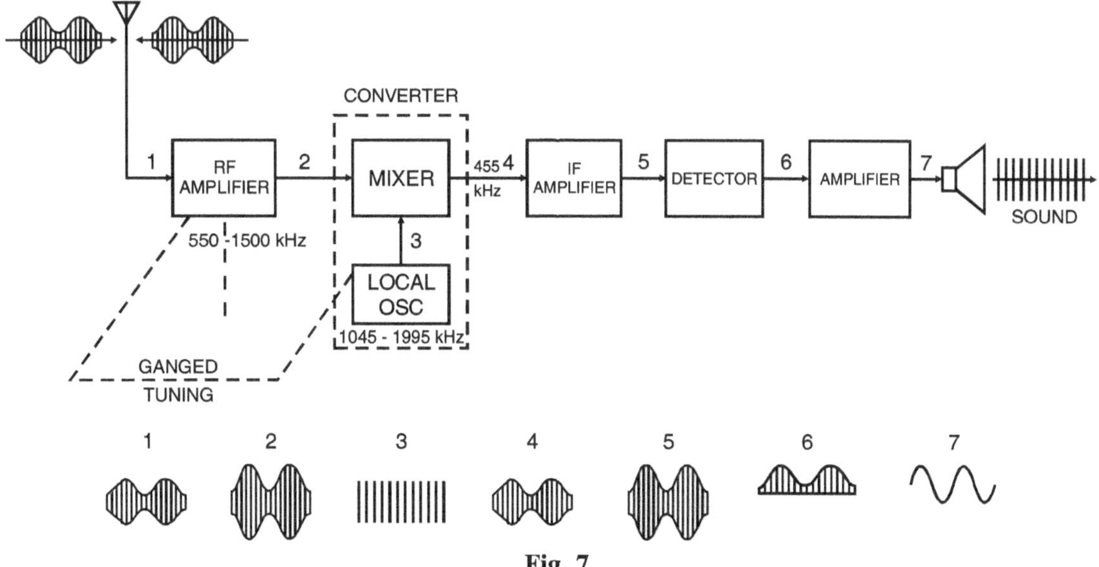

Fig. 7

(vi) The output of IF amplifier is demodulated by a detector which provides the audio signal.

(vii) This audio signal is amplified by the audio-frequency (AF) amplifier whose output is fed to a loud-speaker which reproduces the original sound.

Standard Superhet AM Receiver

Fig. 8 shows the block diagram of a standard superhet. AM receiver alongwith its seven-transistor circuit diagram. Usually, transistor radios are all-wave radios having two or more bands. With the help of a band switch, the coils and trimmers of any one band can be connected to the circuit at will. For the sake of simplicity, the coils and trimmers of one band only have been shown in Fig. 8. The operation of various stages and the functions of various radio components used in the radio circuit of Fig. 8 described below.

1. **Frequency Changer or Converter :** It is the first stage in a transistor radio and has two sections – oscillator and mixer. A single silicon transistor Q_1 performs both the above functions. L_4 is the oscillator coil whereas L_2 is the signal input coil. Resistors R_1 and R_2 form a voltage divider for giving base bias to Q_1.
   The signal intercepted by the aerial are picked up the coil L_1 which is tuned to the transmission frequency of the desired station with the help of ganged capacitor of C_1 and its fine-tuning trimmer C_2. This signal is then induced into the secondary winding L_2 which passes it on to the base of Q_1 via C_3. The local oscillations produced by L_4 and ganged capacitor C_7 are given to the emitter of Q_1 via C_4.
   As seen, Q_1 has been used in *CE* configuration for producing oscillations. The positive feedback required for producing oscillations is given from the collector of Q_1 to its emitter through L_3, L_4 and C_4.
   The transistor Q_1 mixes the incoming signal frequency f_s and oscillation frequency f_o thereby producing the sum and difference frequencies of $(f_l + f_s)$ and $(f_o - f_s)$ respectively. The difference frequency called IF frequency is the passed on to the IF transformer IFT-1 connected in the collector of Q_1.
2. **IF Amplifier:** As the name indicates, its function is to amplify the IF signal obtained from the frequency changer. Usually, two stages are used to obtain sufficient gain. The gain of the first stage is controlled by the automatic gain control circuit (AGC) with the help of dc voltage drop across R_{10}. This dc voltage is obtained from the detector diode. Three fixed-tuned IF transformers are used–one at the collector of the frequency changer and the other two for the two IF amplifier stages. These transformers are also used as collector load as well as for inter-stage coupling.
   Resistor R_5 gives base bias to Q_2 whereas capacitor C_8 decouples the AGC voltage. As usual, R_8 and R_9 form the voltage divider circuit for giving base bias to Q_3.
   The amplified IF signal obtained from IFT-2 is further amplified by IFT-3 of the second IF amplifier stage before it is passed on to the detector diode.
3. **Detector Circuit :** It consists of a diode which rectifies the IF signal. Resistor R_{13}, capacitors C_{10} and C_{11} together filter out RF. The audio signal is given to the following audio section through potentiometer R_{14} which is also used as a volume control. Capacitor C_{12} blocks the dc component in the rectified signal. This dc voltage is used for AGC via R_{10}. The main purpose of using AGC is to minimize the variations in sound due to changes in signal strength (because of fading) and to prevent overloading of second IF stage because of a very strong signal a close-by transmitter.
4. **Audio Frequency (AF) Amplifier :** The audio section consists of three stage *i.e.* audio amplifier, driver and output which feeds a loud speaker. The audio-amplifiers amplify the audio signal given to Q_4 through volume control and C_{12}, R_{15} and R_{16} are biasing resistors whereas R_{19} is the emitter resistor with a by-pass capacitor C_{13}. The amplified signal is developed over collector load resistor R_{17}. This amplified signal is capacitively-coupled to the driver stage through C_{14}.

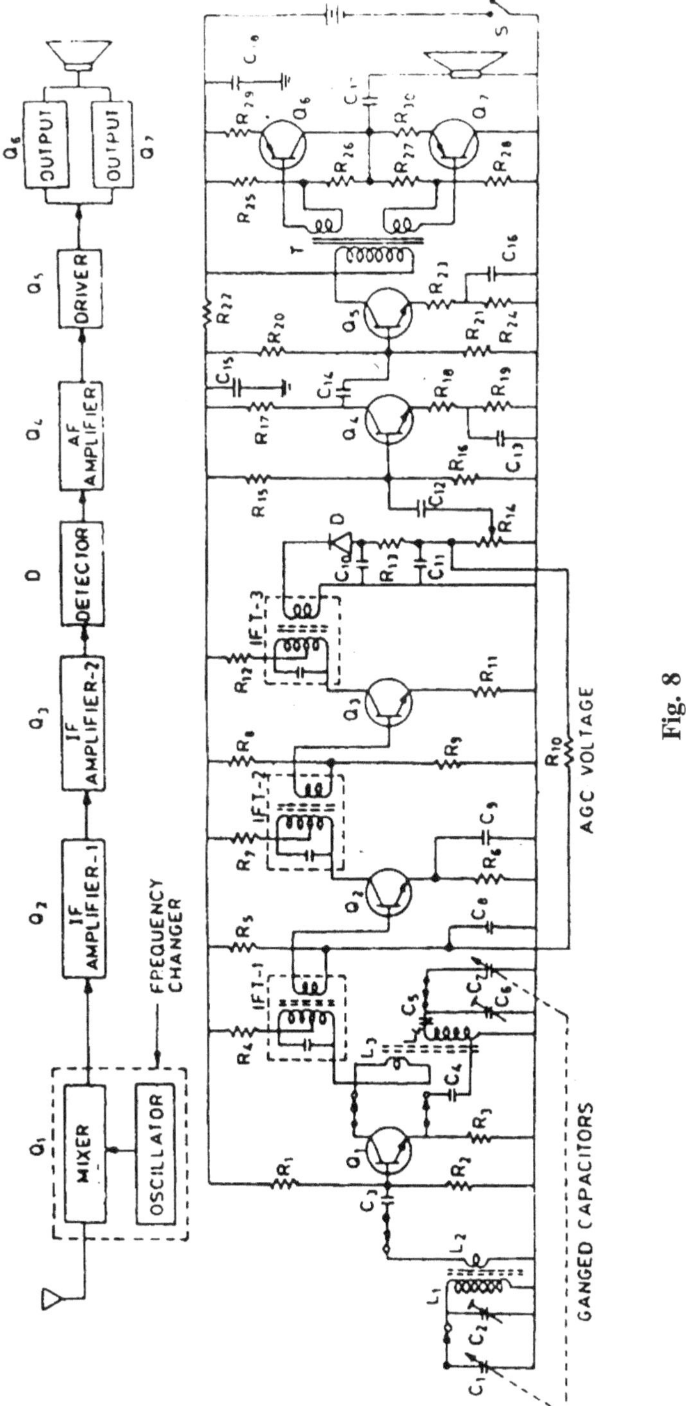

MIXER
OSCILLATOR
FREQUENCY CHANGER
IF AMPLIFIER-1
IF AMPLIFIER-2
DETECTOR
AF AMPLIFIER
DRIVER
OUTPUT
OUTPUT
IFT-1
IFT-2
IFT-3
AGC VOLTAGE
GANGED CAPACITORS

Fig. 8

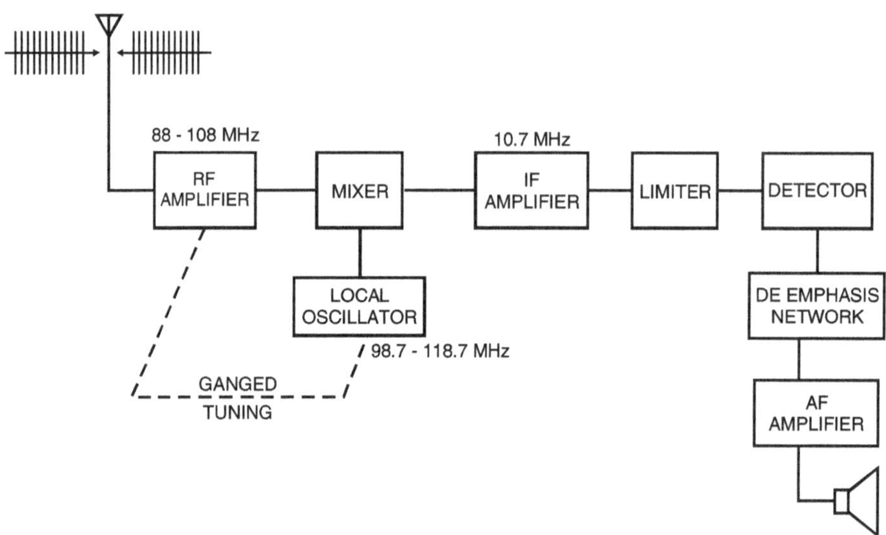

Fig. 9

5. **Driver :** The driver stage further amplifies the audio signal obtained from the previous AF amplifier till it is large enough to derive the output stage. Transformer T (having two secondary windings) is used for coupling to the output stage. The emitter resistor of Q_5 is split into two parts (R_{23} and R_{24}) with only R_{24} being by-passed by C_{16}. This provides some negative feedback and reduces distortion.
6. **Output Stage :** The function of this stage is to amplify the audio signal received from the driver stage and hence provide sufficient power to drive the loudspeaker. Class-B push-pull amplifier employing transistors Q_6 and Q_7 is used for economizing battery consumption. Resistors R_{25} and R_{26} provide base bias to Q_6 and R_{27} and R_{28} to Q_7 whereas R_{29} and R_{30} are the emitter resistors of the two respective transistors. The audio signal is given via C_{17} to the loudspeaker which converts it into sound similar to the one transmitted by the broadcasting transmitter.

FM Receiver

The block diagram for such a receiver is shown in Fig. 9. These receivers operate in the VHF band of 88–108 MHz and have an IF of 10.7 MHz with a bandpass of 2 × 75 kHz = 150 kHz.

Like AM receivers, such receivers also have a frequency converter and IF amplifiers.

The function of the **limiter** is to remove all *amplitude various* (caused by noise) from IF signal which might have crept into the FM signal. This removal of amplitude variations is necessary for distortion less demodulation. **Limiter is a sort of clipping circuit.**

The de-emphasis network *reduces the amplitude of high frequencies in the audio signal* which was earlier increased by the pre-emphasis network at the transmitting station. It serves to re-establish the **tonal balance** of the speech or music etc. lost in the pre-emphasis. Without it, the sound signal would have a heavy treble effect. It is, in fact, a low-pass filter having a time constant of 75 μs.

Comparison Between AM and FM

Frequency modulation (FM) has the following advantages as compared to amplitude modulation (AM) :

1. All transmitted power in FM is useful whereas in AM most of it is in carrier which *serves no useful purpose*.
2. It has high signal-to-noise (*S/N*) ratio. It is due two reasons : firstly, there happens to be less noise at VHF band and secondly, FM receivers

are fitted with amplitude limiters which remove amplitude variations caused by noise.

3. Due to 'guard-band' there is hardly any adjacent-channel interference.
4. Since only transmitter *frequency* is modulated in FM, only fraction of a watt of audio power is required to produce 100% modulation as compared to high power required in A_m.

However, FM has the following disadvantages :

1. It requires **much wider channel**–almost 7 to 15 times as large as needed by A.M.
2. It requires complex and expensive transmitting and receiving equipment.
3. Since FM reception is limited to only line of sight, area of reception for FM is much smaller than for AM.

The Four Fields of FM

There are four major areas of application for FM transmission :

1. First use is in FM broadcast band 88-108 MHz with 200 kHz channels in which commercial FM stations broadcast programmes to their listeners.
2. Second use is in T.V. Though video signal is amplitude-modulated, sound is transmitted by a separate transmitter which is frequency-modulated.
3. Third use in the mobile or emergency services which transmit voice frequencies (20-4000 Hz) only.
4. Fourth use is in the amateur bands where again only voice frequencies are transmitted.

In FM demodulation, the FM demodulator circuit is such that it generates voltages in proportion to the frequency changes. The voltage waveform thus generated duplicates the original intelligence.

Pulse Modulation

Here, it is assumed that the time function is uniquely determined by sampling every 1/2 f_m sec and that the sampling is periodic. Thus the sampling frequency f_s of a periodic pulse train is kept in such a way that the sampling theorem is always satisfied, *i.e.* $f_s \geq 2\ f_m$.

Once the information source has been sampled, there are four methods used for transmitting the amplitude of the sampled value. All these methods utilize modulating the carrier with a pulse whose parameters are made to vary.

1. Pulse amplitude modulation (PAM), whereby the height of a pulse is made proportional to the sampled value.
2. Pulse position modulation (PPM), whereby the position of a pulse is changed as a function of the sampled value with reference to standard position.
3. Pulse duration modulation (PDM), whereby the width or duration of the pulse is varied as a function of the sampled value.
4. Pulse code modulation (PCM), whereby the sampled value is first coded, usually into a digital code and then transmitted.

Delta modulation

DM is a technique by which an analog signal can be encoded into binary digits. Hence, DM is a differential PCM system. DM has the merit that the electronic circuitry required for modulation at the transmitter and particularly for demodulation at the receiver is substantially simpler than the corresponding hardware needed for PCM.

Slope-overload noise in DM systems occurs when the modulating signal changes by an amount greater than the size of the step. To avoid slope-overload, we must have some of the modulating signal A as

$$A = \frac{Sf_s\Delta}{2\pi\, f_m}$$

If above condition is not met, variable step size of the DM is used.

For transmission of binary signal through antenna, it becomes necessary to superimpose a binary signal (output of PCM) on a carrier, then amplitude modulation, phase modulation or frequency modulation may be used. Such resulting signals are then called ASK, PSK and FSK signals.

COMPANDING (with respect to PCM)

As we know, in pulse-code modulation, the analog signal to be transmitted is sampled. More the number of sampling intervals, lesser is the quantizing noise. 128 level have been observed to be satisfactory. With 128 levels, maximum error that

Comparison of various communication systems :

System	*Relationship between input and output S/N*	*Bandwidth requirement*
AM–DSB	$\frac{S_0}{N_0}=\frac{2S_i}{N_i}$	W = 2 fm
AM–SSB	$\frac{S_0}{N_0}=\frac{S_i}{N_i}$	W = fm
NBFM	$\frac{S_0}{N_0}=3\,m^2 t\frac{S_i}{N_i}$	W ≅ 2 fm $m_f \le 0.6$
WBFM	$\frac{S_0}{N_0}=3\,m^2 t\frac{S_i}{N_i}$	W ≅ 2 fm.m_f $m_f > 6$
PAM/AM	$\frac{S_0}{N_0}=\frac{2t_0^2W^3}{f_m}\frac{S_i}{N_i}$	W = 2m fm m = no. of channels
PPM/AM	$\frac{S_0}{N_0}=\frac{2S_i}{N_0}$	W = 1/t_r
PCM/AM	$\frac{S_0}{N_0}=2^{2n}-1$	W = nm fm n = no. of pulses per code

can occur is 1/256 of the total amplitude range. This holds good as long as the amplitude is near to the maximum. If the signal happens to be a small one in 128 level system, this error will form quite a large portion of this small singal which immensely increases the quantizing noise. COMPANDING which means compressing the signal at the transmitter and expanding it at the receiver, is done to overcome the above mentioned defect. The signal is modified at the transmitter by artificially making the smaller signals immune to quantizing noise by increasing the amplitudes of smaller signals. The original amplitudes are restored at the receiver.

Transmission of Signals and Spectral Densities

If f(t) is the input signal, for distortionless transmission we need the response r(t)

$$r(t) = kf(t-T)$$

For distortionless transmission, the transfer function of the communication system is $H(\omega) = ke^{-j\omega T}$

Hence for distortionless transmission, the magnitude of the transfer function is k and is constant for all frequencies. The phase shift, on the other hand is proportional to frequency

The energy E of a signal f(t) is

$$E = \int_{-\infty}^{\infty} f^2(t)dt = \int_{-\infty}^{\infty} \frac{1}{2\pi}\, IF(\omega)I^2 d\omega$$

Where entropy spectral density is given by

$$G_E(\omega) = |F(\omega)|^2$$

Power contained trained in a signal f (t) is

$$P = \lim_{T\to\infty} \frac{1}{T}\int_{-T/2}^{T/2} f^2(t)dt$$

and the power spectral density is

$$Gp(\omega) = \lim_{T\to\infty} \frac{|F_T(\omega)|^2}{T}$$

Power spectral density of a periodic signal is

$$G_p(\omega) = 2\pi \sum_{n=-\infty}^{\infty} |F_n|^2 \delta(\omega - n\omega_0)$$

where $$F_n = \frac{1}{T}\int_{-T/2}^{T/2} f(t)e^{jn\omega_0 t}\,dt$$

The matched filter

A matches filter represents a communication system which maximizes signal-to-noise ratio (SNR) in the presence of white noise. This filter will also have least probability of error in making a decision. For a matched filter

$$H(\omega) = S(-\omega)e^{-j\omega T}$$

$$h(t) = s(T-t)$$

$$(SNR)_{\max} = \frac{1}{\pi\eta}\int_{-\infty}^{\infty} |S(\omega)|^2\, d\omega$$

where $S(\omega)$ is the spectrum of input signal and $G(\omega) = \eta/2$ represents spectral density of white noise else

$$(SNR)_{\max} = \frac{E}{\eta/2}$$

The maximum amplitude of the signal component at the output occurs at t = T and has the magnitude E, the energy of the signal s(t). The maximum amplitude is independent of the waveform s(t) and depends only upon its energy.

The matched filter detection is essentially a synchronous detection.

The minimum probability of error for such filter is

$$P_{e(\min)} = erfc\left(\sqrt{\frac{E}{2\eta}}\right)$$

For binary PCM :

$$h(t) = -s(t)$$

$$P_{e(\min)} = erfc\left(\sqrt{\frac{K^2}{4}}\right) = erfc\left(\frac{K}{2}\right)$$

$$(SNR)_{\max} = \frac{2E}{\eta} = \frac{2A^2T}{\eta}$$

where C is the ratio of the pulse amplitude A to the root mean square value of noise.

For amplitude shift keying (ASK) :

$$h(t) = -s(t)$$

$$P_{e(\min)} = erfc\left(\frac{A}{2}\sqrt{\frac{T}{\eta}}\right) = erfc\left(\sqrt{\frac{P_sT}{\eta}}\right)$$

$$(SNR)_{\max} = \frac{A^2T}{\eta} = \frac{2E}{\eta}$$

where P_s represents total signal power

For phase shift keying (PSK) :

$$h(t) = s(T-t) = s_1(T-t) = -s_2(T-1)$$

$$P_{e(\min)} = erfc\left(A\sqrt{\frac{T}{\eta}}\right) = erfc\left(\sqrt{\frac{2P_sT}{\eta}}\right)$$

$$(SNR)_{\max} = \frac{A^2T}{\eta}$$

For frequency shift keying :

$$h(t) = s_1(T-t) = s_2(T-t)$$

$$P_{e(\min)} = erfc\left(\sqrt{\frac{E}{\eta}}\right) = erfc\left(\sqrt{\frac{P_sT}{\eta}}\right)$$

$$(SNR)_{\max} = \frac{2A^2T}{\eta}$$

Mathematical model of noise :

The voltage spectral density due to thermal noise or Johnson's noise is

$G_v(f) = 2kTR$; volts^2/Hz

Mean square noise voltage

$\overline{v_n}^2 = 4k\ TR\ \Delta f$; volts^2

where k = Boltzman constant = 1.38×10^{-23} J°/K

Δf = band of frequency between f_1 and f_2

mean square voltage is

$$v_{rms} = \sqrt{\overline{v_n}^2} = \sqrt{4kTR\,\Delta f}\ ;\ \text{volts}$$

Current spectral density

$$G_1(f) = \frac{G_v(f)}{R^2};\ \text{Amp}^2/\text{Hz}$$

Two resistors in series

$$G_v(f) = 2kT(R_1 + R_2)$$

Two resistors in parallel :

$$G_v(f) = \frac{2kTR_1R_2}{R_1 + R_2}$$

Series combination of R and C :

$$G_v(f) = 2kT\ \text{Re}(Z) = \frac{2kTR}{1+(\omega RC)^2}$$

$$\bar{v}n^2 = \frac{kT}{C}$$

Noise figure is given by

$$F = \frac{S_1/N_1}{S_0/N_0} = \frac{N_0}{AN_i} = \frac{dN_0}{AdN_i}$$

where A is the power gain of the system. Other expression for noise figure are

$$F = \frac{G_0(f)}{AG_i(f)} = 1 + \frac{R_{eq}}{R_s} = 1 + \frac{N_s}{AN_i}$$

where N_s is the system noise power.

$$F = \frac{dN_{oa}}{A_a dN_{ia}} = \frac{dN_{oa}}{A_s KT/2} = \frac{N_{oa}}{A_a\, N_{ia}}$$

where A_a is the available power gain of the system and N_{oa} is the available output noise power when the output is matched.

For cascaded communication system :

$$F = F_1 + \frac{F_2 - 1}{A_{a1}}$$

Random from signal theory

General relations in discrete probabilty theory :

$P(A + B) = P(A) + P(B) - P(AB)$

$P(AB) = P(B/A) = P(A.B)\,P(B)$

For statistically independent events,

$P(AB) = P(A).\,P(B)$

$P(A/B) = P(A)$ and $P(B/A) = P(A/B).P(B)$

Probability distribution functions :

$$P(X \le x_1) = \sum_{k=-\infty}^{j} P(x_k)$$

$$P(X \le xj;\, Y \le y_t) = \sum_{k=-\infty}^{j} \sum_{i=-\infty}^{j} P(x_k, y_t)$$

$$\sum_{k=-\infty}^{\infty} \sum_{i=-\infty}^{\infty} P(x_k, y_t) = 1$$

and $P(x_k) = \sum_{i=-\infty}^{\infty} P(x_k, y_t)$

Statistical averages :

$$\overline{X} = E(X) = m = \sum_{k=1}^{\infty} P_k X_k$$

$$\overline{X}^2 = E(X^2) = \sum_{k=1}^{\infty} X_k^{\,2} P_k$$

$$X^n = E(X^2) = \sum_{k=1}^{\infty} X^n{}_k\, P_k$$

$$\text{variance}\ \sigma^2 = E(X^2) - [E(X)]^2$$

$$= \overline{X}^2 - (\overline{X})^2 = \overline{X}^2 - m^2$$

General relations for continuous probability theory:

Cumulative distribution function

$$P(x_1 \le X \le x_2) = \int_{x_1}^{x_2} P(x)\,dx$$

and $\int_{-\infty}^{\infty} P(x)\,dx = 1$

$$P(x) = \int_{-\infty}^{x_1} P(x)\,dx\,;\ P(x) = \frac{dP(x)}{dx}$$

Joint cumulative distribution is

$$P(x_1 \le X \le x_2\,;\, y_1 \le Y \le y_2) = \int_{x_1}^{x_2} \int_{Y_2}^{Y_1} P(x, y)\,dx\,dy$$

$$\int_{-\infty}^{\infty} \int_{-\infty}^{\infty} P(x, y)\,dx\,dy = 1$$

Statistically independent events :

$p(x, y) = p(x).\,p(y)$

In general

$p(x, y) = p(x).\,p(y/x) = p((y).\,p(x/y)$

$$E(X) = \overline{X} = m = \int_{-\infty}^{\infty} x\,p(x)\,dx$$

$$E(X^2) = \overline{X}^2 = \int_{-\infty}^{\infty} x^2 p(x)\,dx$$

$$E(X^n) = \overline{X}^n = \int_{-\infty}^{\infty} x^n p(x)\,dx$$

$$\sigma^2 = E(X^2) - [E(X)]^2 = \overline{X}^2 - \left(\overline{X}\right)^2$$

$$= \overline{X}^2 - m^2$$

probability density function of sum of random varaiable is

$$Z = X+Y$$

then $p(z) = \int_{-\infty}^{\infty} p(x).p(z-x)\,dx = p(x) * p(y)$

Also, if $F_1(\omega) = FT\,[p(x)]$, $F_2(\omega) = FT\,[p(y)]$ and $F_3(\omega) = FT\,[p(z)]$, then

$$F_3(\omega) = F_1(\omega), F_2(\omega)$$

The probability density function $p(z)$ of the random variable Z when

$$Z = f(Y)$$

Basic Information Theory

Basic to the concept of the measure of information is the notion of uncertainty, i.e., the more uncertain we are about the outcome of a message state, the greater will be the amount of information associated with the outcome. The measure of information is given by

$I = -\log_2 P(x_1)$; bits

$= -\log_e P(x_1)$; nits

$= -\log_{10} P(x_1)$; Hartley

Average information of entropy is given by

$$H(x) = -\sum_{i-1}^{m} P(i) \log P(i) \quad ; \text{bits/symbol}$$

Properties of H(x)

1. If all the P(i)s except one are zero, H(x) = 0 since there is no uncertainty in the outcome of the message state.
2. If P(i)s are equal for all m symbols, then H(x) is a maximum and is given by

$$H(x) = -\sum_{i-1}^{m} \frac{1}{m} \log \frac{1}{m} = \log m$$

3. Joint entropy of the events is

$$H(x, y) = \sum_{i=1}^{m}\sum_{j=1}^{n} P(i, j) \log P(i/j)$$

4. The conditional entropy is given by

$$H(y/x) = \sum_{i=1}^{m}\sum_{j=1}^{n} P(i, j) \log P(i/j)$$

5. Joint entropy may be expressed as

$H(x, y) = H(x) + H(y/x)$

If independence between symbols is assumed, then we have

$H(x, y) = H(x) + H(y)$

Redundancy

A sequence of symbols which are dependent upon one another are termed redundant. We measure the redundancy of a sequence of symbols by nothing the amount by which the entropy has been reduced. The redundancy is defined as

$$E = 1 - \frac{H(y/x)}{H(x)}.$$

TELEVISION

Colour Television

Transmission and reception of transient visual images in full colour. The technique used for colour television broadcasting in India compatible with monochrome, or black-and-white, television. That is, colour television signals are sufficiently similar to monochrome signals so that they produce acceptable monochrome pictures on black-and-white receivers while producing full-colour picture on colour receivers. Conversely, a standard monochrome signal produces a satisfactory monochrome image on a colour receiver. Compatible colour television is based on a combination of the primary-colour principle, conventional television technology, and special encoding techniques for combining primary-colour television signals in such a way that they can be transmited through a single broadcast channel.

Use of primary colours : Colour television images are formed by combining three primary-colour images. Although many sets of primary-colours are theoretically possible (provided only that no combination of any two primaries in a set be capable of matching the standard set which has proved to be most practical for colour television consists of highly saturated, red, green and blue colours. The colour kinescope, or picture tube, used to display the picture in a colour television receiver can produce independent red, green, and blue images, usually in the form of tiny dots so closely intermingled that they cannot be resolved individually at the normal viewing distance. Three images are thus effectively fused by the visual system of the observer so that only the combined, full-colour image is seen.

Encoding and decoding : Because of the great demand for television broadcast channels in India it is not practical to utilize three separate channels for transmitting the red, green, and blue signals required to control a colour kinescope. It is necessary, therefore, to encode the primary colour signals at the transmitter in such a way that they can be transmitted through a single television channel. The encoded colour video signal contains three independent components, and is enough like a conventional monochrome signal to produce good images on monochrome receiver. Each colour receiver contains decoder circuits (between the second detector and kinescope) to retrieve from the colour signal its red, green and blue components.

Colour encoder : In a compatible colour encoder the red (R) green (G) and blue (B) signals are first applied to a matrix, where they are cross-mixed in accordance with eqs. (1)-(3). The signal designated M conveys luminance information, it is nominally

$$M = 0.30\,R + 0.59\,G + 0.11\,B \quad \text{...(1)}$$

$$I = 0.60\,R - 0.28\,G - 0.32\,G \quad \text{...(2)}$$

$$Q = 0.21\,R - 0.52\,G + 0.33\,B \quad \text{...(3)}$$

equivalent to the signal that would be generated by a monochrome camera viewing the same scene as the colour camera. That is, the M or luminance signal controls the sensation of brightness. The I and Q signals conveys chrominance information, which controls hue and saturation in the final image. The I and Q signals are reduced in bandwidth to about 1.5 and 0.5 MHz, respectively, in recognition of the fact that the human eye has limited resolving power for chrominance difference between small picture areas.

The I and Q signals are converted to modulated waves by means of a pair of balanced modulators. These modulated waves (including the necessary side bands) are designed to fall within the 0–4.2 MHz video passband of a conventional television channel. The subcarrier inputs for the two modulators are both at the same frequency (3.5795 45 MHz), but are 90° apart in phase. When the modulated I and Q signals are added together, they form a two-phase-modulated wave (variable in both amplitude and phase known as the chrominance subcarrier).

The complete colour video signal is formed as the sum of (1) the chrominance subcarrier, (2) the monochrome or M signal, (3) conventional synchronizing pulses, and (4) a special colour synthronizing signal consisting of a brief sample (or burst) of the subcarrier frequency timed (by means of burst flag pulses) to occur shortly after each horizontal synchpulse. The purpose of the burst flag is to establish a frequency and phase reference for the receiver circuits that demodulate the chrominance subcarrier.

Performance standards and requirements : A simplified block diagram for a colour television receiver is shown in Fig. 10. Many of the circuits in a colour receiver are the same in principle as the corresponding circuits in a monochrome receiver, but all circuits handling the complete colour signal must be designed for high performance standards. Because the chrominance information is received in the form of sidebands occupying the upper portion of the video spectrum (centered on approximately 3.6 MHz) it is necessary that the antenna, tuner, intermediate-frequency amplifier, and video detector be designed to handle the full 4 MHz bahdwidth provided in the broadcast transmission standards if degradation on the colour information is to be avoided. Because the colour subcarrier signal is simply added to the normal monochrome signal before transmission, it is necessary that all stages handling the complete signal be linear, so as to avoid intermodulation or distortion of the various signal components. The deflection circuits for a colour receiver are similar in principle

TABLE A : Principle Television Systems

Particulars	*Western Europe Middle East India and most Asian countries*	*North and South America including US, Canada Mexico and Japan*	*England*	*Russia**	*France***
Line per frame	625	525	625	625	625
Frames per second	25	30	25	25	25
Field frequency (Hz)	50	60	50	50	50
Line frequency (Hz)	15,625	15,750	15,625	15,625	15,625
Video bandwidth (MHz)	5 or 8	4.2	5.5	6	6
Channel bandwidth (MHz)	7 to 8	6	8	8	8
Video modulation	Negative	Negative	Negative	Negative	Negative
Picure modulation	AM	AM	AM	AM	FM
Sound signal modulation	FM	FM	FM	FM	AM
Colour system	PAL	NTSC	PAL	SECAM	SECAM

* England earlier used 405 lines system in the 5 MHz channel.

** France also uses 819 line system with a channel bandwidth of 14 MHz.

to those used in monochrome receivers although the output stages are normally designed for a higher power level because of the greater deflection requirements for colour cathode-ray tubes. Colour displays use 22 to 30 kV for the high-voltage supply.

Sound reproduction : The typical circuit for recovery of the sound from the aural transmitter consists of a band-pass filter to extract the frequency-modulated 4.5 MHz carrier from the video detector amplification and amplitude limiting, frequency capacitance (RC) low filter having a time constant 75 microseconds, followed by volume control, a power amplifier, and a loudspeaker. The result is single-channel monophonic sound.

RF Tuner

The various functions performed by a television receiver tuner are :

1. It selects the desired station and rejects others.
2. It converts all incoming channel frequencies to a common IF band.
3. It provides gain to the weak input signal picked up by the antenna. This improves S/N ratio and reduces snow on the picture.
4. It isolates the local oscillator from the feeder circuit to prevent undesired radiation through the antenna.
5. It improves image rejection ratio and thus reduces interference from sources operating at frequencies close to the image frequencies of various channels.
6. It prevents spurious pick-ups from sources which operate in the IF band of the receiver.
7. It provides matching between the antenna and input circuits of the receiver, thus eliminating the appearance of ghost images on the screen.
8. It rejects any pick-up from stations operating in the FM band.
9. It has provision to select any channel out of the various allotted in the VHF and UHF bands for TV transmission.
10. It has a fine tuning control for accurate setting of the selected station.

Factors Affecting Tuner Design

1. Choice of IF and Local Oscillator frequencies.
2. Need for an RF Amplifier stage
3. Coupling Networks.

Various Sections of a VHF Tuner

As shown in Fig. 11 the tuner consists of the following sub-sections.

(a) Balun-input Transformer

(b) HP Filter and Trap Circuits

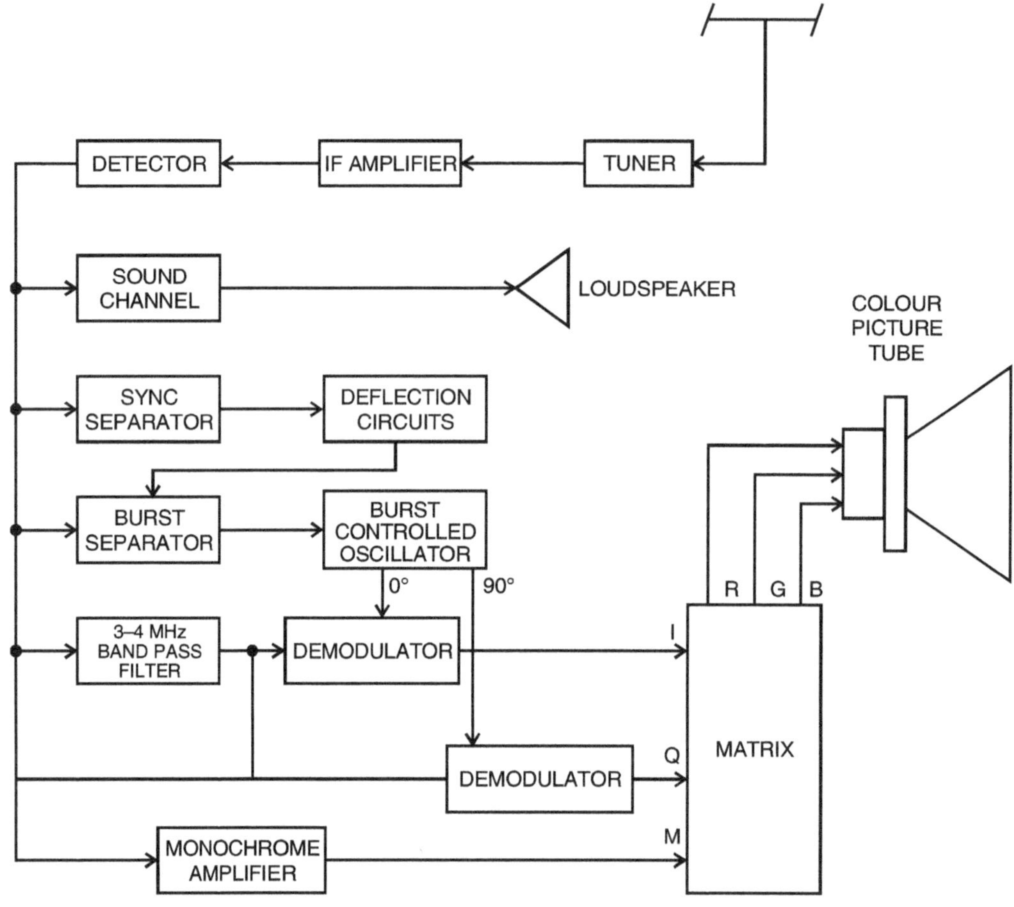

Fig. 10: *Simplified block diagram of a colour television receiver*

(c) RF Amplifier
(d) Local Oscillator
(e) Mixer Stage

UHF Tuners

There are two types of UHF Tuners. In one type the converter section of the UHF tuner converts the incoming signals from the UHF band (470 to 890 MHz) to the VHF band. Usually the UHF channels are converted to VHF channel numbers 5 or 6 and the VHF tuner, in turn, processes these signals in the usual way. Since the relative positions of video and sound IFS are specified, the converter oscillator of the UHF tuner is tuned to a frequency lower than the incoming UHF signal to take into account the double-heterodyning action involved.

In the second type, the UHF signal is heterodyned with local oscillator output to translate the incoming channel directly to the IF band. The output thus obtained from the UHF tuner is coupled to the VHF tuner, where the RF amplifier and mixer stages operate as IF amplifiers, when the VH station selector is set to the UHF position. In this position the VHF oscillator is disabled by cutting off its dc supply. The RF tuned circuits are also changed to IF tuned circuits.

In present day tuners, this type is preferred. Fig. 12, illustrates the block schematic of this circuit. Separate antennas are used for VHF, and UHF channels. The UHF tuner does not employ an RF

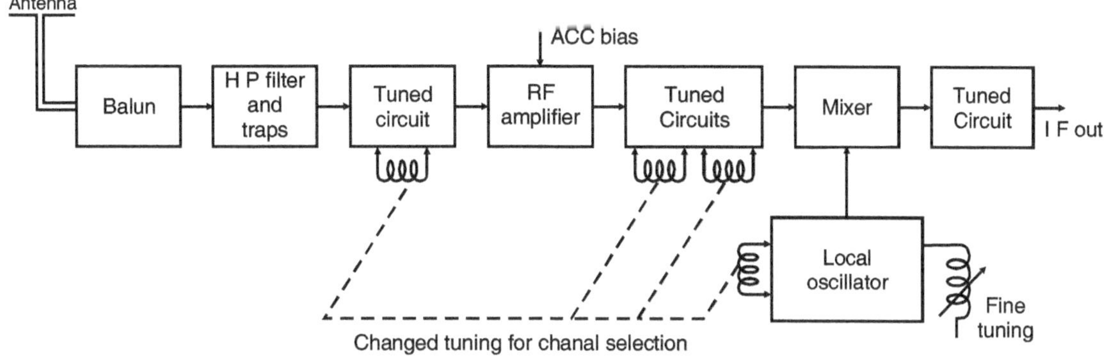

Fig. 11: *Sub-sections of a VHF tuner in a monochrome receiver*

amplifier because the input signal strength being very low is comparable to the noise generated in the RF amplifier. If used it lowers the signal-to-noise ratio and fails to boost the signal level above the noise level. Therefore, the incoming UHF signal is directly coupled to the mixer where a diode is used for heterodyning instead of a transistor or a special tube. The reason for using a diode is two-fold. Firstly, the local oscillator output at ultra high frequencies, is too low to cause effective mixing with an active device whereas it's adequate when a diode is employed. Secondly, the noise level with a diode is lower than when an active device is used. Since the diode is a non-linear device it can produce beating of the incoming channel frequencies with the local oscillator frequency to generate side-band frequencies.

The weak output from the UHF tuner is coupled to the VHF tuner where both the RF amplifier and mixer stages acts as IF amplifiers to boost the signal to the level normally obtained when a VHF channel is tuned.

Electronic Tuning

Varactors or varicaps are used for electronic tuning in tuner circuits. Varactor is a special silicon diode, the junction capacitance of which is used for tuning. This capacitance varies inversly with the amount of reverse bias applied across the diode. The resonant frequency of the tuned circuits in which they are connected, is controlled merely by changing the reverse bias across the varactor. Figure 13 (a) illustrates how a varactor diode D is used in a resonant circuit. The capacitor C_1 is quite large and thus has negligible reactance at resonant frequency of the tuned circuit, but is needed to block the dc supply voltage. The dc supply through R_1 supplies reverse

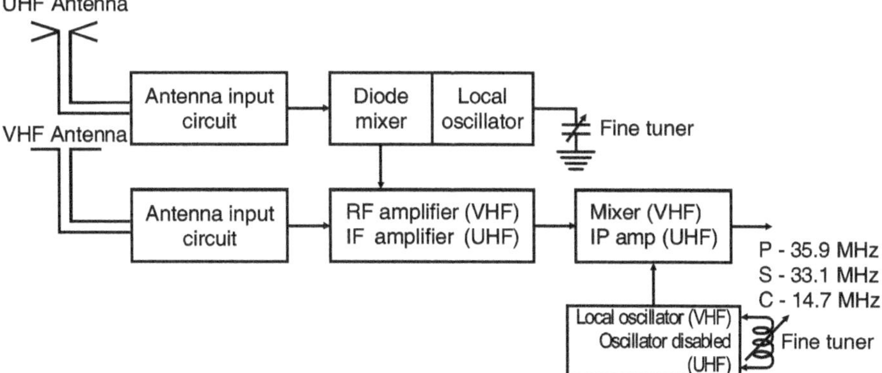

Fig. 12: *Block diagram of a UHF-VHR tuner*

bias to the diode R_2 is the isolating resistance and R_3 can be varied to change the reverse bias voltage.

Since C_1 is a short for ac, it puts the capacitance of the varactor in parallel with C_2. At higher channels, capacitor C_2 is not necessary and only the varicap is enough to provide necessary capacitance.

Channel Selection : A simplified circuit arrangement of a common method by which various channels can be selected by electronic tuning is shown in Fig. 13. In this circuit selection of dc voltage for reverse biasing the varactor diode in each tuned circuit is done by push buttons. The amount of reverse bias determines the diode capacitance and thereby the resonant frequency of each tuned circuit. All the tuned circuits of a channel are adjusted simultaneously to ensure proper tracking. A separate potentiometer is provided for each channel. Each potentiometer acts as a fine tuning control for the associated channel tuned circuits. The dc tuning voltage is obtained through a voltage regular for stable operation.

In Fig. 13 (b) the resistors R_1, R_2, R_3 and R_4 couple dc voltage to the varactors from potentiometers and isolate one varactor from the other. The capacitors C_1, C_2, C_3 and C_4 are large and provide a short to the RF signals. Therefore, from the ac point of view the varactors, in effect, appear in parallel with the tuned circuits.

Principle of Colour mixing

Mixing of colours can take place in two ways– subtractive mixing and additive mixing. In subtractive mixing, reflecting properties of pigments are used which absorb all wavelengths but for their characteristics colour wavelengths. When pigments of two or more colours are mixed, they reflect wavelengths which are common to both. Since the pigments are not quite saturated (pure in colour) they reflect a fairly wide band wavelength. This type of mixing takes place in painting and colour printing.

In additive mixing which forms the basis of colour television, light from two more colours

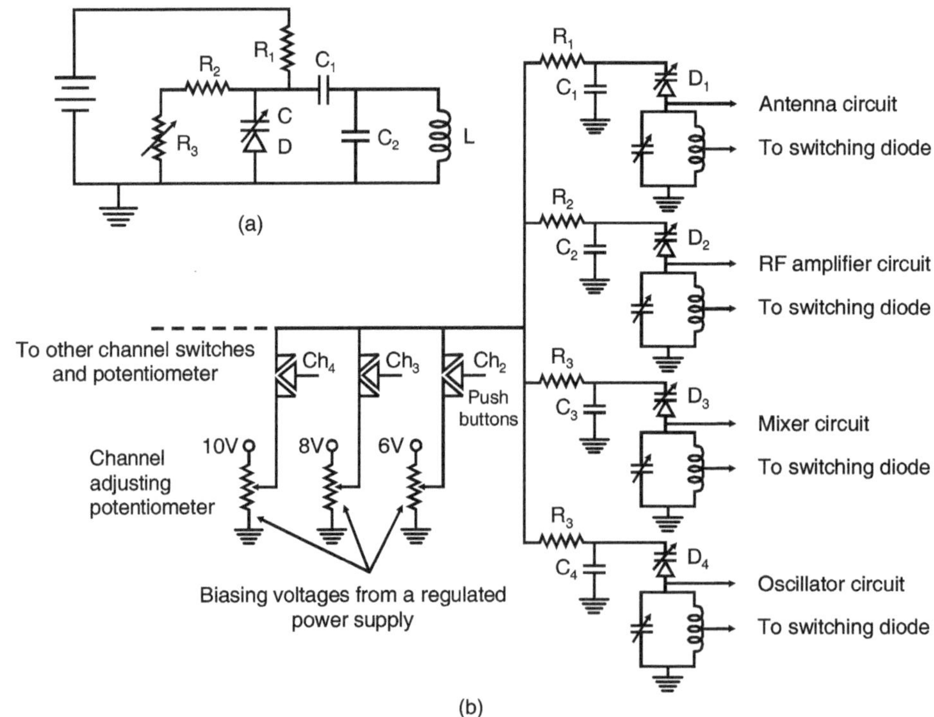

Fig. 13: *Channel selection by electronic tuning*

obtained either from independent sources or through filters can create combined sensation of a different colour. Thus different colours are created by mixing pure colours and not by subtracting parts from white.

The additive mixing of three primary colours– red, green and blue in adjustable intensities can create most of the colours encountered in everyday life. The impression in white light can also be created by choosing suitable intensities of these colours.

Red, green and blue are called primary colours. These are used as basic colours in television. By pairwise additive mixing of the primary colours the following complementary colours are produced :

Red + Green = Yellow

Red + Blue = Magenta (purplish red shade)

Blue + Green = Cyan (greenish blue shade)

If a complementary is added in appropriate proportion to the primary which it itself does not contain white is produced. This is illustrated in Fig. 14 where each circle corresponds to one primary colour. Smilarly Fig. 15 illustrates the process of subtractive mixing. Note that as additive mixing of the three primary colours produces white, their subtractive mixing results in black.

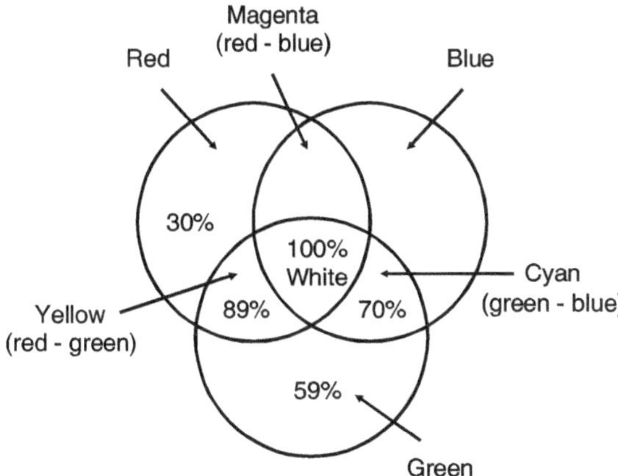

Fig. 14 : *Additive colour mixing. The diagram shows the effect of projecting green, red and blue beams on a white screen in such a way that they overlap*

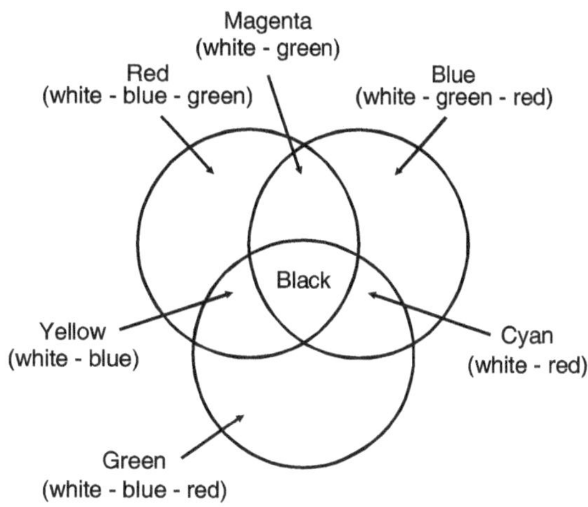

Fig. 15

PAL Colour Television System

The PAL system which is a variant of the NTSC system, was developed at the Telefunken Laboratories in the Federal Republic of Germany. In this system, the phase error susceptibility of the NTSC system has been largely eliminated. The main features of the PAL system are :

(i) The weighted (B–Y) and (R–Y) signals are modulated without being given a phase shift of 33° as is done in the NTSC system.

(ii) On modulation both colour difference quadrature signals are allowed the same bandwidth of about 1.3 MHz. This results in better colour reproduction. However the chroma signal is of vestigial sideband type. The upper sideband attenuation slope starts at 0.57 MHz, *i.e.* (5–4.43 = 0.57 MHz) but the lower, sideband extends to 1.3 MHz before attenuation begins.

(iii) The colour subcarrier frequency is chosen to be 4,43261875 MHz. It is an odd multiple of one-quarter of the line frequency instead of the half-line offset as used in NTSC system. This results in some what better cancellation of the dot pattern interference

EXERCISE

1. In a collector modulated AM amplifier the component that adds the modulating signal to the carrier is
   A. Tank circuit
   B. Modulation choke
   C. Condenser
   D. Modulation transformer.
2. In an unlaminated as well as conducting magnetic circuit excited under ac conditions, the flux is proportional to
   A. the permeability and the inverse of the frequency
   B. The permeability, the inverse of the frequency and the inverse of the conductivity
   C. the square root of option (B)
   D. none of the above
3. The VSWR of a transmission line is :
   A. Directly proportional to characteristic impedance and inversely to load impedance
   B. Directly proportional to load impedance and inversely to characteristic impedance
   C. Proportional to load and characteristic impedances
   D. Directly to voltage minimum and inversely to voltage maximum.
4. Which of the following statement (s) is/are correct?
   1. Ferromagnetic substances have very large negative values of susceptibility.
   2. Diamagnetic substances have very small negative values of susceptibility
   3. Paramagnetic substances have very small positive values of susceptibility
   4. The permeability of diamagnetic substance is slightly greater than unity.

   Select the correct answer
   A. 3 and 4 B. 2 and 3
   C. 1 and 4 D. 4 alone
5. A SSB system is not used for broadcasting because
   A. There will be poor fidelity as only one sideband is transmitted
   B. There is more power is sidebands
   C. Transmitters and receivers are complicated
   D. None of the above
6. A geostationary satellite completes one orbit in:
   A. One hour B. 5 hours
   C. 24 hours D. 28 days.
7. Which one of the following is the correct angular aperture for a paraboloidal reflector antenna for which the aperture number is 0.25?
   A. 45° B. 90°
   C. 120° D. 180°
8. The following is an indirect method of generating FM:
   A. Reactive FM modulator
   B. Armstrong modulation
   C. Varactor diode modulation
   D. Reactance BJT modulator
9. For satellite communication the frequency should be
   A. Less than the critical frequency of ionosphere
   B. Equal to the critical frequency of ionosphere
   C. More than the critical frequency of ionosphere
   D. None of the above
10. A RF carrier of 10 kV at 1 MHz is amplitude modulated and the modulation percentage is 50. The peak voltage of the signal is
    A. 500 kV B. 5 kV
    C. 1200 kV D. 10 kV
11. Channel capacity is basically a measure of:
    A. Entropy rate
    B. Maximum rate of information
    C. Amount of information a channel can take
    D. None of these.
12. A circular metallic disc rotates along its axis in a magnetic field. This would be termed Faraday's homopolar generator where one can collect a potential difference between the centre and any point of the disc.
    The emf is maximum when
    A. the field is parallel to the plane of the disc and the emf is picked up between the centre and edge

B. the field is parallel, to the plane of the disc and the emf is picked up between the centre and another point midway between the centre and the edge

C. the field is perpendicular to the plane of the disc, and the emf is picked up between the center and the edge

D. none of the above

13. At the air-dielectric interface :

A. the normal component of displacement vector and tangential component of electric field are continuous

B. the circulation of the field is constant

C. the normal component of electric field and tangential component of displacement vector are continuous

D. none of the above

14. A source generates four messages. The entropy of the source will be maximum, when

A. all probabilities are equal

B. one of the the probabilities equal 1.0 and others are zero

C. the probabilities are 1/2, 1/4, 1/8 and 1/8

D. two of the probabilities are 1/2 each and others are zero

15. The maximum possible value of entropy is

A. 1 B. 4

C. 8 D. infinity

16. The image orthicon uses silicon photodiodes for target plate.

A. Statement is true B. False

C. None of these D. It uses Ge photodiodes

17. In case atoms of a material have a net magnetic moments of zero, the external magnetic field

A. reduces magnetic flux inside

B. causes no change in inside flux

C. increases magnetic flux inside

D. destroys the material

18. Which stage is not necessary for producing horizontal output:

A. Horizontal oscillaor

B. Damper

C. Horizontal amplifier

D. Horizontal AFC

19. In satellite transmission, analog signals may be converted into digital form using

A. Modem B. Transponder

C. Compander D. Codec

20. A square loop carrying current is introduced in a uniform magnetic field inclined at angle α with perpendicular to the plane. The coil experiences a torque proportional to

A. α B. $\sin \alpha$

C. $\cos \alpha$ D. none of the above

21. A charged particle with an initial uniform velocity along the x-direction is introduced in a uniform magnetic field acting in the same direction. The particle experiences

A. a force in the same direction

B. a force in the z-direction

C. a force in the y-direction

D. no force

22. A straight conductor of length l moving with a velocity v in the presence of a magnetic field of flux density B directed at an angle θ with the direction of v experiences a force. Which of the following statement (s) is/are true for the magnitude of the force?

1. It is independent of θ
2. It is proportional to l^2
3. It is proportional to B
4. It is independent of v

Select the correct answer

A. 1, 2 and 4 B. 4 alone

C. 3 alone D. 2, 3 and 4

23. Which of the following lines has the greatest losses

A. Twin lead

B. Open wire

C. Coaxial cable – RG – 59U

D. From RG – 11 U coaxial cable

24. The Doppler shift is given by

A. $\Delta_f = \frac{V_r}{\lambda}$ B. $\Delta_f = \frac{2\sqrt{V_r}}{\lambda}$

C. $\Delta_f = \frac{V_r}{2\lambda}$ D. $\Delta_f = \frac{2V_r}{\lambda}$

25. An FM signal with frequency deviation δ is passed through a mixer and has its frequency reduced three fold. The frequency deviation in the output of the mixer is :

A. 3δ B. $\delta/3$

C. δ D. none of these

26. D layer roughly extends from :
A. 20 to 40 km B. 55 to 90 km
C. 90 to 110 km D. Beyond 110 km

27. In a telephone exchange serving 6400 subscribers, the number of calls originating per hour during peak is 8315. The calling rate is
A. 8315 B. 1.3
C. 1.77 D. 6400

28. In communication receivers the fidelity is provided by
A. Detector range
B. Audio stage
C. Mixer stage
D. All the three (A), (B) and (C) above

29. VLF waves are used for some types for services because
A. of the low powers required
B. the transmitting antennas are of convenient size
C. they are very reliable
D. they penetrate the ionosphere easily

30. Companding is used:
A. To protect small signals in PCM from quantizing distortion
B. To overcome quantized noise in PCM
C. To overcome impulse noise
D. None of these

31. Indicate the false statements. Flat topped rectangular pulses must be transmitted in radar to
A. Allow a good minimum range
B. Make the returned echoes easier to distinguish from noise
C. Prevent frequency changes in the magnetron
D. Allow accurate range transmission

32. Frequency shift keying is used mostly in
A. Radio transmission
B. Telegraphy
C. Telephony
D. None of these

33. In PCM, the biggest advantage as compared to AM is
A. larger bandwidth
B. larger noise
C. inability to handle analog signals
D. incompatibility with time division multiplex system.

34. MODEM stands for
A. Modulator at transmitting side and detector at the receiving side
B. which deals with analog signals and shows digital information
C. analog to digital at transmitting side and digital to analog at a receiving side
D. a device which deals with digital signals only

35. Consider the following statements regarding an electromagnetic flowmeter :
1. A.C. excitation is used to avoid polarisation of the fluid.
2. The meter calibration is changed as the viscosity of the fluid changes
3. Stainless steel pipes can be used for measuring the flow of fluid of high conductivity.

Of these statements
A. 1, 2 and 3 are correct
B. 1 and 2 are correct
C. 2 and 3 are correct
D. 1 and 3 are correct

36. A rectangular waveguide is designed to propagate the dominant mode TE_{10} at a frequency of 5 GHz. The cut off frequency is 0.8 of signal frequency. The ratio of the guide height to width is 2. The dimensions of the guide are
A. 3.75 cm, 1.875 cm
B. 4.4 cm, 2 cm
C. 8.8 cm, 4 cm
D. 2.54 cm, 1.27 cm

37. A. loading is done only on cables
B. Loading can be done on open wire transmission lines and cables
C. Loading is done only on cables carrying voice frequency and not for carrier communication
D. Loading is done on any cable or line for any purpose.

38. In ionosphere, the incidence angle is 60. The maximum usable frequency will be
A. $2f_c$ B. f_c
C. $\frac{2}{\sqrt{3}}f_c$ D. $\sqrt{3}\,f_c$

39. A 3 A modulation is sometimes used to
A. Allow the receiver to have a frequency synthesizer
B. Simplify the frequency stability problem in reception
C. Reduce the power that must be transmitted
D. Reduce the bandwidth required

40. Probability density function of thermal noise is:
A. Binomial B. Gaussian
C. Poisson D. None of these

41. One of the main functions of the RF amplifier in a superhetrodyne receiver is to:
A. Provide improved tracking
B. Permit better adjacent channel rejection
C. Increase the tuning range of the receiver
D. Improve the rejection of the image frequency

42. If the critical frequency is 30 MHz angle of departure is 60° then maximum usable frequency is given by
A. 30 MHz B. 15 MHz
C. 60 MHz D. 10 MHz

43. The saving in power in DSBSC system modulated at 80% is:
A. Nil B. 80%
C. 88.3% D. 50%

44. Linear velocity of the satellite in a circular orbit is
A. Directly proportional to its mass.
B. Directly proportional to square root of its mass
C. Directly proportional to square of its mass
D. Independent of its mass

45. The secondary constants of a transmission line are
A. α, β, Z_C B. R, G, γ
C. γ, v_p, β D. α, β

46. Vidicon and plumbicon operate on the principle of
A. photo conduction B. photo emission
C. photo voltaic D. none of these

47. To eliminate ghosts in the picture
A. use a longer transmission line
B. connect a booster
C. change the antenna orientation of location
D. twist the transmission line

48. Leak type bias is used in a plate modulated class C amplifier to
A. Increase the bandwidth
B. Prevent excessive grid current
C. Prevent tuned circuit damping
D. None of the above

49. The visual omni range (VOR) navigation system gives
A. Height of aeroplane
B. Position of ship or location of plane
C. Range of target
D. None of these

50. Troposcatter antenna generally used is:
A. Horn antenna B. Parabolic antenna
C. Yagi antenna D. Helical antenna

51. A solution of the blind speed problem is to
A. Change the Doppler frequency
B. Vary the PRF
C. Use monopulse
D. Use MTI

52. In instrument landing system, we have
A. Master and slave station
B. Glidepath localizer
C. MTI
D. None of these

53. Pilot carrier in SSB is provided for having
A. Lower power consumption
B. Better noise immunity
C. Frequency stability purpose
D. None of these

54. In radio receivers tube EM 84 is used as
A. Magic eye B. Full wave rectifier
C. Audio amplifier D. None of the above

55. High IF in superhet receiver
A. Improves selectivity
B. Increases tracking problems
C. Reduces adjacent channel rejection
D. None of the above

56. Signals in UHF range use:
A. Space wave propagation
B. Sky wave propagation
C. Surface wave propagation
D. None of the above

57. The damper tube is over heated, the result is
A. Wide white bar at the left
B. Thin white bars at the left
C. Folder at the right

D. No brightness

58. How long does it take to scan across two elements if 400 are scanned in 50 μ s?

A. 0.50 μs B. 0.25 μ s

C. 1.0 μ s D. None of these

59. Broadband long-distance communications were originally made possible by the advent of

A. telegraph cables

B. repeater amplifiers

C. HF radio

D. geostationary satellites

60. An FM signal is being broadcast in the 88-108 MHz band having a carrier swing of 125 kHz. The modulation index is

A. 100% B. 83%

C. 67% D. 50%

61. In an operational amplifier circuit, the maximum frequency for the undistorted output is 100 kHz for 10 volts p-p output because of the finite slew rate of the OPAMP. The maximum frequency for undistorted output of 5 volts p-p

A. will be approximately 100 kHz

B. will be less than 100 kHz

C. will be more than 100 kHz

D. cannot be determined from the given data

62. The highest modulation frequency typically used in FM broadcast is:

A. 5 kHz B. 10 kHz

C. 15 kHz D. 25 kHz

63. During hetrodyne process in a receiver the modulation of the signal

A. Remains unaffected

B. Increases

C. Decreases

D. Is totally eliminated

64. Pre-emphasis is used to amplify:

A. Low frequencies

B. High frequencies

C. Both low and high frequencies

D. None of these

65. Short noise current in a diode is proportional to:

A. $\sqrt{I_p}$ B. I_p^2

C. I_p D. $1/I_p^2$

Where I_p is the d.c. diode current

66. The primary constants of a transmission line are

A. α, β, γ B. R, G

C. R, L, C D. R, G, L, C

67. The following circuit can not demodulate SSB:

A. Balanced modulator

B. BFO

C. Phase discriminator

D. Product detector

68. The autocorrelation function of an ergodic random process is given by:

A. $\underset{T\to\infty}{\text{Lt}} \frac{1}{2T} \int_{-T}^{T} x(t)x(t-\tau)\,dt$

B. $\underset{T\to\infty}{\text{Lt}} \frac{2}{T} \int_{0}^{T} x(t)x(t-\tau)\,dt$

C. $\underset{T\to\infty}{\text{Lt}} \frac{1}{2T} \int_{0}^{T} x(t)x(t-\tau)\,dt$

D. $\underset{T\to\infty}{\text{Lt}} \frac{2}{T} \int_{-T}^{T} x(t)x(t-\tau)\,dt$

69. A short transmission line is terminated by its characteristic impedance. (1) Its input impedance is zero (2) It behaves as an infinite line (3) It has no reflection (4) It has input impedance equal to characteristic impedance. The wrong statement is

A. 1 B. 2, 4

C. 2, 3 D. 1, 3, 4

70. MW links are generally preferred to coaxial cable for TV transmission because:

A. They have less overall phase distortion

B. They are cheaper

C. Of their greater bandwidth

D. Of their relative immunity to impulse noise

71. One of the following is not a reason for the use of an antenna coupler :

A. To make the antenna look resistive

B. To provide the output amplifier with the correct load impedance

C. To discriminate against harmonics

D. To prevent reradiation of the local oscillator

72. A low ratio of the a.c. to the d.c. load impedance

of a diode detector results in
A. Diagonal clipping
B. Poor AGC operation
C. Negative peak clipping
D. Poor AF response

73. Typical gain of an amplifier for the UHF channels is
A. 0 dB B. 2 dB
C. –5 dB D. 40 dB

74. An air filled rectangular waveguide has dimensions 6 cm, 4 cm. The cut off frequency for TE_{10} is :
A. 2.5 GHz B. 25 GHz
C. 5.25 MHz D. 4.5 GHz

75. The intrinsic impedance of free space is known to be 377 Ω. The intrinsic impedance of a medium with relative permittivity and permeability of 4 and 1 responsibility of 4 and 1 respectively will be
A. 75 Ω B. 94 Ω
C. 188 Ω D. 377 Ω

76. The difference between a DSB and a SSB transmission is
A. DSB has a carrier and two sidebands and SSB, a carrier and a sideband
B. DSB has two sidebands and SSB one sideband
C. DSB may or may not have a carrier with two sidebands and SSB either of the two sidebands without carrier
D. DSB has a carrier and two sidebands and SSB without carrier and two different sidebands.

77. High frequency inductors and capacitors are commonly plated with silver. The main purpose of this is to
A. reduce their dc resistances
B. reduce their ac resistances
C. increase their ac resistances
D. increase their dc resistances

78. The relation between electric polarization and susceptibility indicates that electric polarization is
A. independent of susceptibility
B. inversely proportional to susceptibility
C. proportional to square root of susceptibility
D. proportional to susceptibility

79. The pilot carrier in SSB is used for:
A. Providing better noise immunity
B. Frequency stability purpose
C. Lower power consumption
D. None of these

80. The characteristic impedance of a line in terms of its primary line constants are
A. $\sqrt{Z_{SC}\,Z_{OC}}$ B. $\sqrt{Z_1 Z_2 + \frac{Z_1^2}{4}}$
C. $\sqrt{\frac{R}{G}}\cdot\sqrt{\frac{L}{C}}$ D. $\sqrt{\frac{R+jwL}{G+jwC}}$

81. Cassegrain feed is used with a parabolic reflector to
A. increase the gain of the system
B. increase the bandwidth of the system
C. reduce the size of the main reflector
D. allow the feed to be placed at a convenient point

82. The shadow mask in a colour picture tube is used to:
A. Reduce X-ray transmission
B. increase screen brightness
C. Ensure that each beam hits only its own dots
D. None of these

83. The rms antenna current of an AM transmitter increases by 20% over the unmodulated value. The modulation percentage is:
A. 20 B. 120
C. 94 D. 50

84. When a dielectric is placed in an electric field, the electric flux density $\overline{D}$, electric field intensity $\overline{E}$ and polarization $\overline{P}$ are related as
A. $\overline{D} = \varepsilon_0 \overline{E} \times \overline{P}$ B. $\overline{D} = \varepsilon_0 \overline{E} - \overline{P}$
C. $\overline{D} = \varepsilon_0 (\overline{E} + \overline{P})$ D. $\overline{D} = \varepsilon_0 \overline{E} + \overline{P}$

85. A magnetic circuit carriers an alternating flux in the *z*-direction, the length being L and the cross-section being $b \times d$, in the x and y-directions where $d < b$. N-laminations are used to make up the dimensions L, b and d as
A. *b*, *d* and L/N B. *b*, L and *d*/N
C. L, *d* and *b*/N D. *d*, L and *d*/N

86. If 4 V peak audio amplitude changes the RF

signal frequency from 200 to 210 kHz, the frequency deviation

A. 5 B. 10
C. 15 D. 100

87. Indicate which of following signal is not transmitted in colour TV

A. Y B. Q
C. R D. I

88. Sky wave propagation is used for

A. Mobile transmission
B. Short distance transmission
C. Medium distance transmission
D. Long distance transmission

89. In Troposcatter the antenna used is:

A. Parabolic antenna
B. Helical antenna
C. Conical antenna
D. Lens antenna

90. Ground controlled approach uses

A. Slave and master station
B. Two radars
C. Localizer and glider path
D. None of these

91. The unit of magnetic permeability is

A. weber
B. weber per square metre
C. henry metre
D. henry per metre

92. The distance between the transmitting antenna and the point where the sky wave is first received after returning to the earth is known as:

A. Line-of-sight distance
B. Skip distance
C. radial distance
D. Effective distance

93. The shape factor of a filter-response

A. It the same as the peak-to-valley ratio
B. is equal to the 60 dB bandwidth of a mechanical filter
C. Cannot be applied to crystal filters
D. Determines the adjacent-channel selectivity of the filter.

94. A pre-emphasis circuit provides extra noise immunity by

A. Boosting the bass frequencies
B. Amplifying the higher audio frequencies
C. Pre-amplifying the whole audio bound
D. Converting the phase modulation to FM

95. The critical frequency of an ionospheric layer during the day time is

A. More than that for night time
B. Less than that for night time
C. Less than that for night time
D. None of the above

96. In a coaxial cable, braided copper is used as a

A. conductor B. shield
C. dielectric D. jacket

97. The orientational polarization is

1. directly proportional to temperature
2. inversely proportional to temperature
3. inversely proportional to square of the permanent dipole moment
4. directly proportional to square of the permanent dipole moment

A. 1 and 3 only B. 2 and 3 only
C. 1 and 4 only D. 2 and 4 only

98. 4 Volt peak audio modulating signal changes the carrier frequency from 200 kHz to 210 kHz. The frequency deviation is:

A. 5 kHz B. 10 kHz
C. 15 kHz D. 20 kHz

99. The mean free path for electron drift increases with

A. purity B. strain hardening
C. elastic modulus D. none of the above

100. Which law is synonymous to the occurrence of diamagnetism?

A. Ampere's law B. Maxwell's law
C. Coulomb's law D. Lenz's law

101. The number of repeaters along a coaxial cable link depends on:

A. number of coaxial cables in the tube
B. the bandwidth of the system
C. separation between the equalizers
D. None of these

102. Pulse code modulation (PCM) employing 4-bit code is used to transmit a data signal having frequency components from dc to 2 kHz. The minimum bandwidth of the carrier channel should be

A. 2 kHz B. 4 kHz
C. 8 kHz D. 16 kHz

103. The bandwidth required for base-band signals in Hi-Fi radio is :
A. 0.3 to 3 kHz B. 3 to 5 kHz
C. Upto 10 kHz D. Upto 20 kHz

104. EHF range is between (GHz)
A. 3-300 B. 30-3030
C. 3-3000 D. 300-3000

105. A 10 kW transmitter is modulated to 80%. The average sideband power will be
A. 1.80 kW B. 8 kW
C. 3.2 kW D. 4.6 kW

106. The effective height of a vertical radiator can be increased by
A. Loading with lumped inductance
B. By capacity hat
C. Supporting on an insulator
D. Pumping more power

107. Cassegrain feed is used with a parabolic reflector to
A. increase the gain of the system
B. increase the beamwidth of the system
C. reduce the size of the main reflector
D. allow the feed to be placed at a convenient point

108. Frequencies in UHF range propagate by means of
A. ground waves B. sky waves
C. space waves D. surface waves

109. A transmission line has a VSWR of 2, the reflection coefficient is
A. $\frac{1}{3}$ B. 0
C. $\frac{1}{4}$ D. $\frac{1}{2}$

110. When there is snowy picture the trouble lies in:
A. R.F. amplifier
B. too much AGC
C. Antenna
D. Weak vertical amplifier

111. The maximum usable frequency (MUF) equals:
A. $f_c \cos\theta$ B. $\sqrt{f_c}\cos\theta$
C. $f_c \sec\theta$ D. $\sqrt{f_c}\sec\theta$

112. 24 hour satellite (synchronous satellite) should be placed in an equatorial orbit with a approximate height of
A. 1500 kilometer B. 2000 kilometer
C. 5000 kilometer D. 36000 kilometer

113. A carrier voltage is simultaneously modulated by two sine waves causing modulation indices of 0.4 and 0.3. The overall modulation index is:
A. 0.35
B. 0.7
C. 0.5
D. Can not be calculated unless the phase relations are known.

114. In a communication system, noise is most likely to affect the signal
A. In the transmitter
B. In the channel
C. In the destination
D. At the destination

115. Diffraction of electro magnetic waves
A. is caused by reflection from the ground
B. rises only with spherical wave fronts
C. will occur when the waves pass through a large slot
D. may occur around the edge of a sharp obstacle

116. Considering the use in a microwave spectrum analyser, match List-I with List-II and select the correct answer using the codes given below the lists:

List I	*List II*
(a) Crystal diode	1. Local oscillator
(b) Reflex Klystron	2. Narrow-band signal
(c) Saw-tooth generator	3. Varying repellar voltage
(d) IF amplifier	4. Mixing

Codes:	*(a)*	*(b)*	*(c)*	*(d)*
A.	1	2	3	4
B.	2	3	4	1
C.	4	1	3	2
D.	2	3	1	4

117. For microwave propagation through microstrip lines, there will be
A. radiation, eddy current and dielectric losses
B. dielectric, ohmic and hysteresis losses
C. dielectric, ohmic and radiation losses
D. ohmic, hysteresis and eddy current losses

118. De-emphasis circuit is used:
A. After modulation B. Before modulation
C. Before detection D. After detection

119. Which of the following crystal filter is used only at the higher frequencies?
A. Ladder B. Full lattice
C. Half lattice D. Crystal gate.

120. Indicate which of the following circuit could not demodulate SSB:
A. Balanced modulator
B. Product detector
C. BFO
D. Phase discriminator

121. Which of the following statement is incorrect? Nothing that no carrier is transmitted with A 3 J we see that
A. The receiver cannot use a phase comparator for AFC
B. Adjacent-channel rejection is more difficult
C. Production of AGC is rather complicated process
D. The transmission is not compatible with A 3.

122. Which of the following is true?
1. series resistance is zero
2. shunt conductance is zero
3. shunt conductance is infinite
4. series resistance is infinite
A. 1 & 2 B. 2 & 3
C. 2 & 4 D. 3 & 4

123. $\frac{\lambda}{4}$ a transformer is used for
A. light loads
B. high frequency load
C. connecting high impedance loads to low impedance loads
D. reducing distortion in transmission lines.

124. A RF carrier of 8 kV is amplitude modulated by an audio signal of 7 kV. The percentage modulation is:
A. 87.5% B. 187.5%
C. 114% D. 0%

125. In T.V. system, indicate the true statement:
A. Picture is AM, sound is FM
B. Picture is FM, Sound is AM
C. Picture is AM, sound is AM
D. Both are frequency modualted

126. Long distance communication system via satellites should perferably use frequencies from
A. 1 to 13 GHz B. 10 to 20 GHz
C. 20 to 30 GHz D. 3 to 6 GHz

127. After a target has been acquired, the best scanning system for tracking is
A. Nodding B. Spiral
C. Conical D. Helical

128. In frequency modulation the modulation index is proportional to:
A. ω_m B. $1/\omega_m$
C. ω_m^2 D. $1/\omega_m^2$

129. Ionosphere roughly extends from:
A. 50 km to 110 km
B. 50 km to 200 km
C. 50 km to 500 km
D. 50 km to several earth radii

130. If a sample of germanium and a sample of silicon have the same impurity density and are kept at room temperature
A. both will have equal value of resistivity
B. both will have equal negative resistivity
C. resistivity of germanium will be higher than that of silicon
D. resistivity of silicon will be higher than that of germanium

131. Which control performs the function; D.C. bias for picture tube
A. Volume B. Tuning
C. Brightness D. None

132. If R be the resistance, then noise power generated in resistor will be proportional to:
A. $R^{3/2}$ B. R^2
C. R D. $\sqrt{R}$

133. A passive satellite
A. Only reflects back signals
B. Only generates signals
C. Only absorbs signals
D. Receives, modulates and then reflects.

134. By inserting a plate of a dielectric between the plates of a parallel plate capacitor, the energy stored in the capacitor is increased five times. The dielectric constant of the material is
A. 1/25 B. 1/5
C. 5 D. 25

135. For small size, high frequency coils, the most common core material is

A. air B. ferrite
C. powdered iron D. steel

136. The modulation index of an AM signal is increased from 0 to 1. The carrier power:

A. Remains unchanged
B. Gets doubled
C. Increases by 50%
D. Gets quadrupled

137. In a low level AM system, amplifiers following the modulation stage must be:

A. Harmonic device
B. Class C amplifier
C. Non-linear device
D. Linear device

138. In a trapezoidal display of modulation, the ratio of long side to short side is 65. The modulation percentage is

A. .65 B. 35
C. 21 D. 50

139. Trigonometric function, exponential functions are example of

A. Singular function
B. Orthogonal
C. Gaussian
D. None of these

140. The purpose of buffer amplifier in transmitter is to:

A. Function as harmonic generator
B. Amplify RF signals
C. Restrict over modulation
D. Isolate master oscillator from harmonic generator

141. Four charges are arranged at the corners of a square of each side $\sqrt{2}$ m, as shown in the figure. The electric field at point P, at distance R is given by

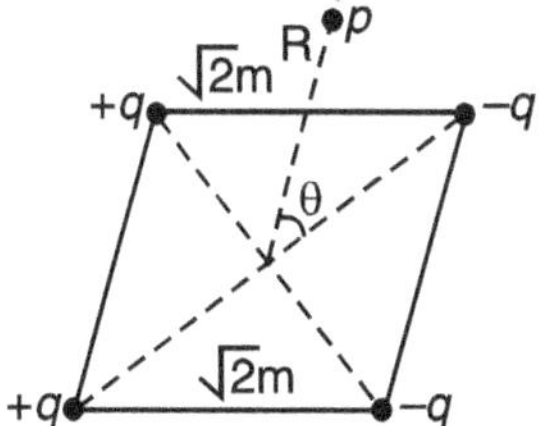

A. $\frac{9q(2\sin^2\theta-1)}{4\pi\varepsilon_0 R^3}$ B. $\frac{3q(2\sin^2\theta-1)}{4\pi\varepsilon_0 R^3}$

C. $\frac{3q(2\sin^2\theta-1)}{4\pi\varepsilon_0 R^4}$ D. $\frac{9q(2\sin^2\theta-1)}{4\pi\varepsilon_0 R^4}$

142. A proton is released from rest at a distance of 10^{-10}m from another proton (charge on proton $= 1.6 \times 10^{-19}$ Coul.). Select the correct statement.

A. The velocity of the released proton would go on decreasing and would be brought to rest after a certain distance since the force of repulsion rapidly decreases
B. The velocity of the released proton would go on increasing and would be finally equal to 50 km/sec, at infinite distance apart
C. The velocity of the released proton would first increase and then acquire constant velocity even when it approaches infinite separation
D. None of the above.

143. Indicate which of the following is not a binary code

A. Morse B. Baudot
C. CCITT-2 D. ARQ

144. Image signal frequency in a superhet receiver is given by :

A. $f_s + f_i$ B. $f_s + 2f_i$
C. $f_s - f_i$ D. $f_s - 2f_i$

145. Medium wave broadcast is used for:

A. Regional broadcast
B. National broadcast
C. International broadcast
D. None of these

146. Equalising pulses in T.V. are sent during

A. Horizontal blanking
B. Vertical blanking
C. Serrations
D. Horizontal retrace

147. From the point of view AGC, the main difference between tube and transistor receivers is that the latter:

A. Require AGC power rather than voltage
B. Do not really need AGC because over loading causes limiting

C. Have troubles because AGC is not very effective in them
D. Must have IF stages even more stabilized than they are normally.

148. Frequency shift keying is basically a method involving:
A. Amplitude modulation
B. Frquency modulation
C. Phase modulation
D. None of these

149. In a radio receiver padders are provided
A. For fine tuning
B. For reducing noise
C. To improve tracking
D. To improve sensitivity

150. Maxwell's equations establish the relationship between $\overline{E}$, $\overline{D}$, $\overline{H}$ and $\overline{B}$ at any point in a continuous medium. When we move from one medium to another, then which of the following statement/statements is/are correct?
A. At the boundary separating the two media, the tangential components of $\overline{E}$ and $\overline{H}$ are continuous
B. At the boundary separating the two media, the tangential components of only $\overline{E}$ are continuous
C. If there are surface charges present on the boundary, then normal component of $\overline{D}$ is continuous
D. If there are no surface charges then normal component of $\overline{D}$ is discontinuous

151. Which of the following equations satisfies Maxwell's equations.
A. $E_x = E_y = 0; E_z = \cos(y - ct)$
$B_x = \cos(y - ct)$; $B_y = B_z = 0$
B. $E_x = E_y = 0; E_z = \cos(y - ct)$
$B_y = \cos(y - ct)$; $B_x = B_z = 0$
C. $E_y = \cos(y - ct); E_x = E_z = 0$
$B_x = \cos(y - ct)$; $B_y = B_z = 0$
D. None of the above

152. A receiver is tuned to carrier frequency of 1000 kHz. The image frequency is:
A. 1000 kHz B. 1455 kHz
C. 1910 kHz D. 545 kHz

153. 'COHO' stands for
A. Coherent output
B. Counter housed oscillator
C. Coherent local oscillator
D. None of the above

154. In a receiver which stage rectifies the IF signal?
A. Mixer B. Frequency converter
C. Detector D. Loud speaker

155. In a short electric doublet, the radiation properties are so that
A. the induction field diminishes as the square-root of the distance and is only appreciable in the vicinity of the conductor.
B. in the radiation field, magnetic field is minimum when the current is maximum.
C. the radiation resistance of a short doublet antenna is extremely high
D. mean rate of power through a unit area of spherical sphere surrounding this doublet is proportional to square of the elemental length, other factors remaining constant.

156. Shannon's law relates
A. antenna gain to bandwidth
B. frequency to antenna gain
C. antenna gain to transmission losses
D. information carrying capacity to S/N ratio

157. Which of the following noise has a source different from that of the remaining?
A. Galactic noise B. Cosmic noise
C. Solar noise D. Atmospheric noise

158. The channel required for FM telemetry is
A. the same as that required for AM telemetry
B. smaller than that required for AM telemetry
C. 100 times that required for AM telemetry
D. 10 times that required for AM telemetry

159. We have the following six sets of electro-magnetic waves with the frequency ranges as follows:
30 – 300 kHz
10 – 30 kHz
3 – 30 MHz
300 – 3000 kHz
30 – 300 MHz
> 300 MHz
These are designated in the following orders:
A. VLF, LF, MF, HF, VHF and UHF

B. LF, VLF, MF, HF, VHF and UHF
C. LF, VLF, HF, MF, VHF and UHF
D. VHF, VLF, HF, MF, LF and UHF

160. The traffic flow in a system with 5000 calls/ busy hour if the average conversation time is 3 minutes the setting up time is 24 secs is:
A. 4562.5 TUs B. 1620 TUs
C. 4700 TUs D. 283.33 TUs

161. A transmission line of characteristic impedance 300 Ω is terminated by a load of 300 – *j* 300 Ω. The reflection coefficient at the load is
A. .88 ∠ 76.68° B. .447 ∠ – 63.43°
C. .75 ∠ 30° D. .447 ∠ – 90°

162. One of the following can't be used to remove the unwanted side band in SSB. This is the
A. Filter system B. Phase shift method
C. Third method D. Balanced modulator

163. The HV anode supply for the picture tube of TV receiver is generated in the
A. mains transformer
B. vertical output stage
C. horizontal output stage
D. horizontal deflection oscillator

164. In an ADM transmission system, the output signal amplitudes for 1's and 0's are
A. fixed and the repetition rate is also fixed
B. fixed but the repetition rate is variable
C. variale and the repetition rate is also variable
D. variable but the repetition rate is fixed

165. If an FET is used as the first AF amlifier in a transistor receiver, this will have the effect of
A. Improving the selectivity of the receiver
B. Reducing the effect of noise at low modulation depths
C. Improving the effectiveness of the AGC
D. Reducing the effect of negative peak clipping

166. Which stage contains the primary of the first IF transformer?
A. Mixer B. Detector
C. Oscillator D. None of the above

167. The biggest advantage of the TRAPATT diode over the IMPATT diode is its
A. lower noise
B. higher efficiency
C. ability to operate at higher frequencies
D. lesser sensitivity to harmonies

168. Match List I (computer terms) with List II (definitions) and select the correct answer

List I	*List II*
(a) Interface	1. A measure of rate of data transmission
(b) A bit	2. A binary digit
(c) Baud speed	3. The common boundary between various sections and sub-sections
	4. Analog to digital converter

Codes:	*(a)*	*(b)*	*(c)*
A.	3	2	1
B.	2	4	1
C.	3	1	2
D.	4	2	3

169. A (75 – *j* 50) Ω load is connected to a coaxial transmission line of Z_0 = 75 Ω, at 10 GHz. The best method of matching consists in connecting
A. a short-circuited stub at the load
B. an inductance at the load
C. a capacitance at some specific distance from the load
D. a short-circuited sub at some specific distance from the load

170. If an FET is used as the First AF amplifier is a transistor receiver, this will have the effect of
A. Improving the effectiveness of the AGC
B. Reducing the effect of negative peak clipping
C. Reducing the effect of noise at low modulation depths
D. Improving the selectivity of the receiver.

171. Noise can be measured by
A. AC voltmeter
B. DC voltmeter
C. Either a.c. or d.c. voltmeter
D. None of these

172 In a low level AM transmitter, amplifier following the modulated stage must be:
A. Class A B. Linear Class B
C. Class C D. Class AB

173. In case IF is very high, then which of the following statement is incorrect?
A. Tracking will be improved
B. The selectivity will be poor

C. Image frequency rejection will be good
D. The local oscillator need not be extremely stable

174. Since noise phase-modulates the FM wave, as the noise sideband frequency approaches the carrier frequency, the noise amplitude
A. remains constant B. is decreased
C. is increased D. is equalized

175. The number of lines per second in the United States monochrome TV system is
A. 31,500 B. 15,750
C. 63.5 D. 525

176. Match List I with List II and select the correct answer

List I	*List II*
(a) Gonimeter	1. Based on phase difference between signals of two stations
(b) Radio Compass	2. Based on time interval between signals of two stations
(c) LORAN-A	3. Provides three dimensional data in azimuth, elevation and range
(d) DECCA	4. Automatic direction finder
	5. Couples fixed loop antennas and the receiver

Codes:	*(a)*	*(b)*	*(c)*	*(d)*
A.	5	4	2	1
B.	1	5	3	2
C.	5	3	2	4
D.	2	4	5	1

177. The modulation index of an AM wave is changed from 0 to 1. The transmitted power is:
A. Unchanged B. Halved
C. Doubled D. Increased by 50%

178. At very high frequencies, earth acts as a:
A. Dielectric B. Conductor
C. Perfect conductor D. Perfect dielectric

179. A hetrodyne frequency changer is called a:
A. Modulator
B. Mixer
C. Frequency translator
D. None of the above

180. The autocorrelation function $x(t) = \text{V} \sin wt$ is given by :
A. $\frac{1}{2}\text{V}^2 \cos wt$ B. $\text{V}^2 \cos wt$
C. $2\,\text{V}^2 \cos wt$ D. $\text{V}^2 \cos^2 wt$

181. Match List I with List II and select the correct answer

List I	*List II*
(a) Helical antenna	1. Circular polarisation
(b) Horn antenna	2. Simplicity and compactness
(c) Hoghorn antenna	3. Low noise
(d) Parabolic dish with horn feed	4. High directivity

Codes:	*(a)*	*(b)*	*(c)*	*(d)*
A.	1	2	3	4
B.	2	1	3	4
C.	4	3	2	1
D.	1	2	4	3

182. The energy density spectrum $|G(f)|^2$ of a rectangular pulse shown is:

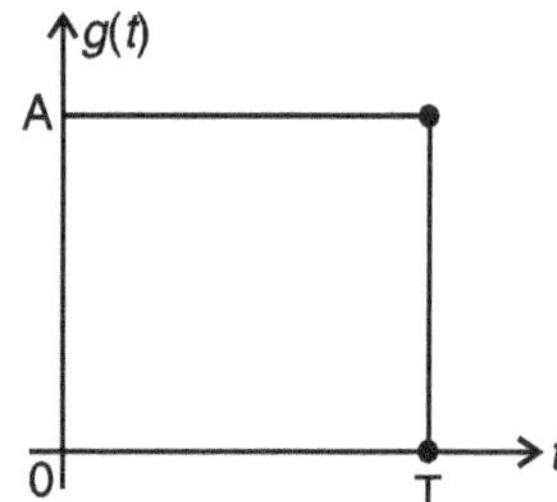

A. $\text{AT}\left(\frac{\sin \pi f \text{T}}{\pi f \text{T}}\right)$

B. $(\text{AT})^2\left(\frac{\sin \pi f \text{T}}{\pi f \text{T}}\right)^2$

C. $(\text{AT})^2\left(\frac{\sin \pi f \text{T}}{\pi f \text{T}}\right)$

D. $\text{A}^2\left(\frac{\sin \pi f \text{T}}{\pi f \text{T}}\right)$

183. The COHO in MTI radar operates as the
A. Intermediate frequency
B. Transmitted frequency

C. Received frequency
D. Pulse repetition frequency

184. The channel capacity of a band limited Gaussian channel is given by:

A. $C = B\log_2\left(1+\frac{S}{N}\right)$

B. $C = B\log_2\left(\frac{S}{N}\right)$

C. $C = \frac{1}{B}\log_2\left(\frac{S}{N}\right)$

D. $C = \frac{1}{B}\log_2\left(1+\frac{S}{N}\right)$

185. Pre-emphasis circuit is used:
A. After modulation B. Before modulation
C. Before detection D. After detection

186. The radiation pattern of the antenna used in VHF omnidirectional range (VOR) is
A. pencil beam B. omnidirectional
C. cardioid D. elliptical

187. The controlled oscillator synthesizer is sometimes preferred to the direct one because
A. it is a simpler piece of equipment
B. its frequency stability is better
C. it does not require crystal oscillators
D. it is relatively free of spurious frequencies

188. In order to reduce quantizing noise, one must
A. increase the number of standard amplitudes
B. send pulses whose sides are more near vertical
C. use an RF amplifier in the receiver
D. increase the number of samples per second

189. The anode voltage of TV picture tube is of the order of:
A. 6 kV B. 8 kV
C. 12 kV D. 20 kV

190. Blind speed causes target appear
A. Moving
B. Stationary
C. Distinguish correctly
D. None of the above

191. The minimum value of m_f for an FM system required to produce a noticeable improvement in $\frac{S}{N}$ ratio over a comparable AM system with $m = 1$ is
A. $\sqrt{3}$ B. $\sqrt{2}$
C. $1\frac{1}{\sqrt{2}}$ D. $\frac{1}{\sqrt{3}}$

192. One of the following microwave diodes is suitable for very low power oscillator only
A. Tunnel B. Gunn
C. IMPATT D. LSA

193. In order to get the original signal from the sampled signal, it is necessary to use:
A. Low pass filter B. High pass filter
C. Band pass filter D. Band stop filter

194. In photo emission; when the intensity of light from the scene to be televised is increased
A. The number of electrons emitted from the target decreases
B. The number of electrons emitted from the target increases
C. The number of electrons emitted from the target remains unchanged
D. None of these

195. Two sine waves of amplitude A_1 and A_2 are multiplied; the amplitude of the spectral components is
A. $2A_1A_2$ B. $\frac{A_1A_2}{4}$
C. $\frac{A_1A_2}{2}$ D. A_1A_2

196. For a Gaussian process, autocorrelation also implies:
A. Statistical dependence
B. Statistical independence
C. Ergodic process
D. Stationary dependence

197. Linear diode detector utilizes:
A. Linear portion of the dynamic characteristic of diode
B. Square law portion of the dynamic characteristic of a diode
C. Rectification property of a diode
D. None of these

198. If the probability of occurrence of a message is $\frac{1}{16}$, then the amount of information is given by
A. 2 bit B. 4 bit

C. 1 bit D. 16 bit

199. The equation of the amplitude modulated wave is

A. $E_c\left(1+\frac{E_m}{E_c}\cos w_m t\right)\cos w_c t$

B. $E_c \cos w_c t + E_c E_m \cos w_m t \sin w_c t$

C. $E_c \sin w_c t + \frac{E_c}{E_m}\sin w_c t \sin w_m t$

D. $E_c \cos w_c t + E_c E_m \cos^2 (w_m + w_c) t$

200. Which of the following will serve as a donar impurity in silicon?

A. Boron B. Indium

C. Germanium D. Antimony

201. Two concentric rings 1 and 2 carrying equal and uniform charge densities revolve at the same angular speed ω about their common axis passing through O as shown. The ratio of flux densities due to the two rings at O will be

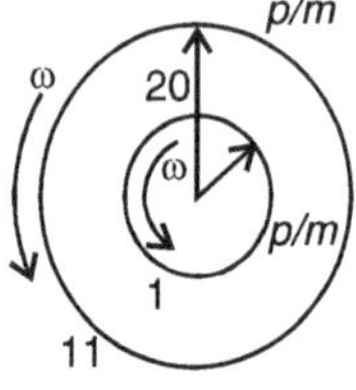

A. 1 : 1 B. 1 : 2

C. 1 : 4 D. 2 : 1

202. The Gaussian probability density is defined:

A. $f(x)=\frac{1}{\sqrt{2\pi\sigma^2}}$

B. $f(x)=\sqrt{2\pi\sigma^2}\,e^{-(x-m)^2/2\sigma^2}$

C. $f(x)=\frac{1}{2\pi\sigma^2}e^{-(x-m)^2/2\sigma^2}$

D. $f(x)=2\pi\sigma^2 e^{-(x-m)^2/2\sigma^2}$

203. With 0.5 MHz video signal modulating the picture carrier:

A. Both upper and lower side frequency is transmitted

B. Only the upper side frequency is transmitted

C. Only the lower side band frequency is transmitted

D. No side frequency is transmitted.

204. Indicate which of the following statements above the advantages of the phase discriminator over the slope detector is false.

A. Much easier alignment

B. Better linearity

C. Greater limiting

D. Power tuned circuit

205. The auto-correlation function of a function *x(t)* is defined as:

A. $R(t_0)=\frac{1}{T}\int_0^T x(t)x(t+t_0)\,dt$

B. $R(t_0)=\frac{1}{T}\int_0^T x(t-t_0)x(t+t_0)\,dt$

C. $R(t_0)=\frac{1}{T}\int_0^T x(t)x(t-t_0)\,dt$

D. $R(t_0)=\frac{1}{T}\int_{-T}^T x(t-t_0)x(t+t_0)\,dt$

206. Match List-I with List-II and select the correct answer by using the codes given below the lists:

List I	*List II*
(a) Isotopes	1. Same A, same Z
(b) Isotones	2. Same A, Different Z
(c) Isobar	3. Different A, Different Z
	4. Same (A – Z), Different Z
	5. Different N, same Z,

Codes:	*(a)*	*(b)*	*(c)*
A.	5	4	2
B.	5	2	3
C.	5	3	2
D.	4	5	3

207. The value or R_x so that power dissipated in it is maximum is

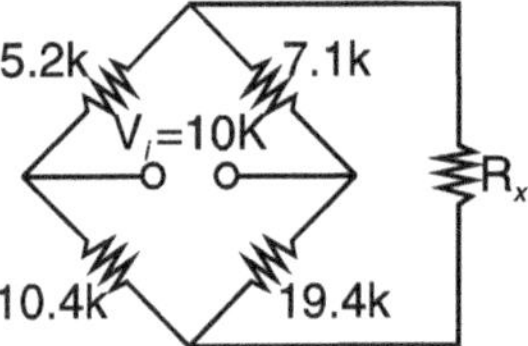

A. 33.4 K B. 17.6 K

C. 10 K D. 5 K

208. The number of spectral components when two sine waves are multiplied are

A. One B. Two

C. Eight D. Four

209. Three point tracking is achieved with:
A. The padder capacitor
B. Variable selectivity
C. Double frequency conversion
D. None of these

210. One of the following is not a useful quantity for comparing the noise performance of receivers.
A. Noise figure
B. Noise temperature
C. Input noise voltage
D. Equivalent noise resistance

211. The highest modulation frequency typically used in AM broadcast is:
A. 5 kHz B. 10 kHz
C. 15 kHz D. 25 kHz

212. A 1000 kHz carrier is modulated with 3 kHz audio. The upper side frequency in the RF signal will be
A. 1003 kHz B. 1000 kHz
C. 997 kHz D. 455 kHz

213. Brass is an alloy of
A. copper and tin B. copper and zinc
C. copper and lead D. lead and tin

214. Nonmetals when compared to metals
A. are less electronegative and have smaller atomic radii
B. have greater ionization energies and larger atomic radii
C. are more electronegative and have smaller atomic radii
D. have greater atomic radii and are more electronegative

215. Show which of the following statements about the amplitude limiter is untrue
A. The circuit is always biased in class C, by virtue of the leak type bias
B. When the input increases past then there should be of limiting the gain decreases to keep output constant
C. The output must be tuned
D. Leak type bias must be used.

216. The following filter is used only at higher frequencies:
A. Crystal gate B. ladder
C. Full lattice D. Half lattice

217. The autocorrelation function of the white noise is:
A. Constant B. Step function
C. Impulse function D. None of these

218. There point tracking is achieved with
A. Variable selectivity
B. The padder capacitor
C. Double spotting
D. Double conversion

219. The following is untrue regarding autocorrelation function:
A. It is an average function
B. It is defined only for periodic signals
C. It is a measure of both time variation and statistical dependence
D. None of these

220. A moving coil microphone has a coil length of 0.3 m placed in a magnetic field of 2 Wb/m^2. The velocity of the diaphragm is .06 m/sec. The voltage generated is:
A. .036 V B. .02 V
C. 10 V D. .4 V

221. Match List I with List II and select the correct answer by using the codes given below the lists:

List I	*List II*
(a) Television	1. 1 kHz
(b) Radio	2. 3 kHz
(c) Telephony	3. 9 kHz
(d) Telegraphy	4. 5 MHz

Codes:	*(a)*	*(b)*	*(c)*	*(d)*
A.	4	1	2	3
B.	2	1	3	4
C.	4	3	2	1
D.	1	2	3	4

222. An engineer uses an ohmmeter to measure the direct current resistance of a transformer winding. The engineer notices that when one of the ohm meter lead is disconnected from the transformer, a spark results. What accounts for this phenomenon?
A. Ohm's law B. Faraday's Law
C. Thomson effect D. Seeback effect

223. Which of the following statement about the amplitude limiter is not valid?
A. The output must be tuned
B. Leak type bias must be used
C. By virtue of the leak-type bias, the circuit is

always biased in clas C

D. when the input increases past the threshold of limiting, the gain decreases to keep the output constant.

224. Cosmic noise is caused by:

A. Lighting discharge

B. Solar erruptions

C. Distant stars

D. None of these

225. The shadow mask in a colour picture tube is used to

A. Reduce X-ray transmission

B. Insure that each beam hits only its own dots

C. Increase screen brightness

D. None of these

226. The saving in power in a SSB system is

A. 100% B. 75%

C. ≥ 75% D. 50%

227. Assuming R = G = B = 1 V, fully conducted red is

A. 0.3 B. 0.59

C. 0.11 D. 1.0

228. During daytime ionosphere is composed of:

A. E, E, F_2 layers B. D, F_1, F layers

C. D, E layers D. D, E, F_1, $F_{\backslash 2}$ layers

229. The following can not be used to remove the unwanted side band is SSB:

A. Filter system

B. Balanced modulator

C. Phase shift method

D. The third method (Weaver's method)

230. A moving coil loudspeaker has a coil of 3000 turns of means diameter 2.5 cam and placed in a radial magnetic field of 0.5 Wb/m^2. The force on the diaphragm attached to the coil per ampere is

A. .0125 N B. .0375 N

C. 118.5 N D. 75 N

Question 231 to 233 refer to data below :

Figure shows the frequency multiplier and heterodyne portion of an FM transmitter

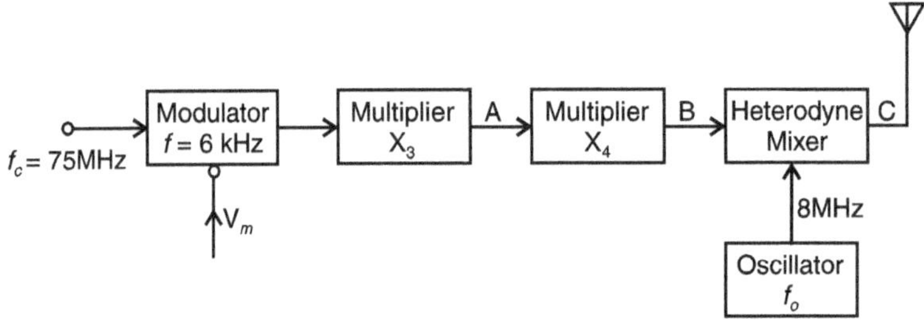

231. Carrier frequency at B is

A. 22.5 MHz B. 72 MHz

C. 90 MHz D. 180 MHz

232. Carrier frequency at C is

A. 108 MHz B. 98 MHz

C. 90 MHz D. 80 MHz

233. The frequency deviation at C is

A. 7.5 kHz B. 30 kHz

C. 60 kHz D. 72 kHz

234. Which is the correct statements?

A. Impedance inversion is obtained by shorting a quarter wave line

B. Impedance inversion is obtained by shorting or open circuiting a quarter wave line

C. Impedance inversion is obtained by both quarter and half wave lines.

D. Impedance inversion is obtained by both quarter and half wave lines

235. The primary constants (which may be assumed independent of frequency) for a transmission line are R = 0.5 Ω/m, L = 250 nH/m, C = 100 pF/m, and G = 10^{-6} S/m. The characteristic impedance is

A. 2.5 m ohm B. 2.5 K ohm

C. 50 ohm D. .4 ohm

236. Given that L = 1 μ H/m. C = 11.11 p F/m group delay time is:

A. Nil B. 3.33 n sec

C. 3.3 μ sec D. 33 sec

237. The channel capacity of a band limited Gaussian

channel is (B is the channel BW):

A. $C = B \log_2\left(2 + \frac{S}{N}\right)$ bits/sec

B. $C = \frac{1}{B}\log_2 S/N$

C. $C = B \log_2 S/N$

D. $C = B \log_2 (1 + S/N)$

238. A receiver producing an accurate reproduction of the modulating signal is said to have:

A. Good fidelity B. Wide bandwidth

C. High sensitivity D. Better selectivity

239. Neutralization is generally done in

A. Diode B. Triode

C. Tetrode D. Pentode

Questions 240 to 241 refer the data below:

A binary system linking teletypewriters to a computer operates at a mean frequency of 1080 Hz shifting to 980 Hz when a one is transmitted and to 1180 Hz for a zero.

240. The maximum permitted data rate for this system is

A. 200 bits/seconds B. 200 baud

C. 100 baud D. 1080 bits/second

241. In case a zero-crossing detector used in this system indicates a time of 510 microseconds, the state received would be

A. 1 B. 0

C. 1/0 D. 0/1

242. A 60 Hz carrier is amplitude modulated by the speech band of 300 to 3000 Hz. The range of upper side bands will be

A. 60.3 to 63 kHz B. 60 to 59.7 kHz

C. 57 to 59.7 kHz D. 56.7 to 56.3 kHz

243. The voltage waveshape in the output of the vertical amplifier is a:

A. Trapezoid B. Sawtooth

C. Rectangle D. Square

244. Indicate which one of the following application or advantages of radar become false:

A. Target identification

B. Navigation

C. Very significant extension of the maximum range

D. none

245. Bridge duplex works on the principle

A. Resistance balance

B. Current balance

C. Potential balance

D. None of the above

246. EM 84 tube is used in radio receiver as

A. Full wave rectifier

B. audio amplifier

C. Magic eye

D. Live output pentode

247. If P_k is the probability of a message (being received or transmitted), then the amount of information I_k associated with it in bits is given by:

A. $I_k = \log_2 P_k$ B. $I_k = 10 \log_2 P_k$

C. $I_k = \log_2 \frac{1}{P_k}$ D. $I_k = 2\log_2 \frac{1}{P_k}$

248. Consider the following statements:

SWR for a transmission line is infinity when the load is

1. a short-circuit
2. a complex impedance
3. an open-circuit
4. a pure reactance

Of these statement

A. 1, 3 and 4 are correct

B. 1, 2 and 3 are correct

C. 2, 3 and 4 are correct

D. 1, 2 and 4 are correct

249. Microwave link repeaters are typically 50 km apart in TV transmission because

A. of atmospheric attenuation

B. of output power tube limitations

C. microwave transmission is through surface wave which attenuates faster

D. of earth's curvature

ANSWERS

1	2	3	4	5	6	7	8	9	10
D	C	C	B	C	C	B	B	C	B

11	**12**	**13**	**14**	**15**	**16**	**17**	**18**	**19**	**20**
B	C	A	A	A	B	A	D	D	C
21	**22**	**23**	**24**	**25**	**26**	**27**	**28**	**29**	**30**
D	C	B	D	C	B	B	B	C	A
31	**32**	**33**	**34**	**35**	**36**	**37**	**38**	**39**	**40**
B	B	A	B	D	A	C	A	D	B
41	**42**	**43**	**44**	**45**	**46**	**47**	**48**	**49**	**50**
D	C	C	D	A	A	C	B	B	B
51	**52**	**53**	**54**	**55**	**56**	**57**	**58**	**59**	**60**
B	B	C	A	B	A	D	B	C	B
61	**62**	**63**	**64**	**65**	**66**	**67**	**68**	**69**	**70**
C	C	A	B	A	D	C	B	A	A
71	**72**	**73**	**74**	**75**	**76**	**77**	**78**	**79**	**80**
D	C	D	A	C	C	B	B	B	D
81	**82**	**83**	**84**	**85**	**86**	**87**	**88**	**89**	**90**
D	C	B	D	B	B	C	D	A	B
91	**92**	**93**	**94**	**95**	**96**	**97**	**98**	**99**	**100**
D	B	D	B	A	B	D	B	A	D
101	**102**	**103**	**104**	**105**	**106**	**107**	**108**	**109**	**110**
B	B	A	A	C	B	D	C	A	A
111	**112**	**113**	**114**	**115**	**116**	**117**	**118**	**119**	**120**
C	D	A	B	D	D	C	D	A	D
121	**122**	**123**	**124**	**125**	**126**	**127**	**128**	**129**	**130**
B	A	C	A	A	D	C	B	D	D
131	**132**	**133**	**134**	**135**	**136**	**137**	**138**	**139**	**140**
C	D	A	C	A	C	D	C	D	D
141	**142**	**143**	**144**	**145**	**146**	**147**	**148**	**149**	**150**
D	B	A	B	A	B	A	B	C	A
151	**152**	**153**	**154**	**155**	**156**	**157**	**158**	**159**	**160**
A	C	C	C	D	D	D	D	C	D
161	**162**	**163**	**164**	**165**	**166**	**167**	**168**	**169**	**170**
B	D	C	C	D	A	B	A	D	B
171	**172**	**173**	**174**	**175**	**176**	**177**	**178**	**179**	**180**
A	B	A	B	B	A	D	A	B	A
181	**182**	**183**	**184**	**185**	**186**	**187**	**188**	**189**	**190**
A	B	B	A	B	B	D	A	C	B
191	**192**	**193**	**194**	**195**	**196**	**197**	**198**	**199**	**200**
D	A	A	A	B	B	C	B	A	D
201	**202**	**203**	**204**	**205**	**206**	**207**	**208**	**209**	**210**
A	A	C	C	A	A	C	D	A	C
211	**212**	**213**	**214**	**215**	**216**	**217**	**218**	**219**	**220**
A	A	B	C	A	B	C	B	B	A
221	**222**	**223**	**224**	**225**	**226**	**227**	**228**	**229**	**230**
C	B	C	C	B	C	A	D	B	C
231	**232**	**233**	**234**	**235**	**236**	**237**	**238**	**239**	**240**
C	B	D	B	C	B	D	A	B	B
241	**242**	**243**	**244**	**245**	**246**	**247**	**248**	**249**	
A	A	A	A	C	C	C	C	D	

EXPLANATORY ANSWERS

35. Statement 1 is correct as many hydrogen bearing or aqueous solutions exhibit polarization effects when the excitation is dc. The positive ions migrate to the negative electrode and dissociate, forming an insulation pocket of gaseous hydrogen. There is no such action when ac is used.

Statement 2 is incorrect. In electromagnetic flow meters, the output is unaffected by changes in characteristics of liquid such as viscosity, pressure and temperature.

In electromagnetic flowmeter non-conducting, non-magnetic pipe is used for carrying the liquid whose flow is to be measured. Non-conducting pipe has to be used as the output voltage gets short circuited, if metallic pipes are used. This is true when liquids of low conductivity are being measured. But when flow at liquids of high conductivity is to be measured, the short circuiting has no effect. Stainless steel pipes can then be used. Thus statement 3 is correct.

38. From $\sin^2 \phi = 1 - \frac{f_c^2}{f_m^2}$

$$\sin^2 60° = 1 - \left(\frac{f_c}{f_m}\right)^2$$

or, $\left(\frac{f_c}{f_m}\right)^2 = 1 - \frac{3}{4} = \frac{1}{4}$

$\therefore \quad f_m = 2f_c$

60. $\Delta f = \frac{\text{carrier swing}}{2} = \frac{125 \times 10^3}{2} = 62.5 \text{ kHz}$

Maximum frequency deviation for the FM broadcast band is 75 kHz.

$\therefore$ Modulation index $= \frac{62.5 \times 10^3}{75 \times 10^3} = 0.83$ or 83%

61. Frequency for undistorted output is 100 kHz, 10 V peak to peak.

Slew rate $= \frac{10}{100 \times 1000} = 10$ microseconds.

Hence, for 5 volt peak to peak input, the maximum frequency will be more than 100 kHz.

96. A coaxial cable has a central conductor covered with a dielectric material, a metallic shield and a jacket conductor, the shield is generally made of braided copper, metal tape or solid shield.

109. VSWR $= \frac{K_r + 1}{K_r - 1}$

where, K_r is the reflection coefficient.

Since, VSWR = 2, therefore, $2 = \frac{K_r + 1}{K_r}$

or $\quad K_r = \frac{1}{3}$

123. When a transmission line of high characteristic impedance is to be connected to a low impedance load, or vice-versa, the $\frac{\lambda}{4}$ transformer of a suitable Z_0 can act as a connecting link for impedance matching. The $\lambda/4$ transformer is inherently a narrow band device, for it can be exctly a quarter wavelength at one frequency only.

130. Since the mobility of charge carriers is better in germanium than in silicon; for the same concentration of charge carriers (impurity density) the conductivity of germanium is higher than that of silicon.

151. This shows a time varying electric field travelling with speed c along z axis in a plane perpendicular to the plane containing time varying magnetic field (B_x), so that the magnitudes of both are same. This is the property of an electromagnetic wave as established by Maxwell's equations.

155. In a short electric doublet antenna of length dl, the induction field diminishes as the square of the distance r, whereas radiation field is inversely proportional to distance. The radiation field is given by

$$dH = \frac{I_0 . dl . \sin\theta}{2\lambda r} \cos\theta\left(t - \frac{r}{c}\right) \text{ amp. turn/m.}$$

Mean rate of power, $dp = \frac{80\pi^2 l^2 r.m.s. dl^2}{2}$ watt

Radiation resistance, $dR = 80\pi^2\left(\frac{dl}{\lambda}\right)^2$ ohm

Since $\lambda >> dl, \frac{dl}{\lambda} = 10^{-2}$ $dR = 0.88$ ohm. (hence very small).

190. Gun diode can replace reflex klystron in receiver local oscillators, parametric amplifier pumps, signal generators, frequency modulated power oscillators, etc.

191. $\left(\frac{S_0}{N_0}\right)_{F.M} = 3(m_f)^2\left(\frac{S_0}{N_0}\right)_{A.M}$

For improvement to be noticeable, $3\,(m_f)^2 = 1$

or, $m_f = \frac{1}{\sqrt{3}}$

201. Equivalent currents of the two rings are :

$I_1 = \frac{2\pi a\rho}{2\pi/\omega} = a\,\rho\omega$, and $I_2 = \frac{2\pi 2a/\rho}{2\pi/\omega} 2\rho\omega$

The magnetic flux density produced at O shall be inversely proportional to radii and directly proportional to equivalent current.

$\therefore B_1 = \frac{Ka\pi\omega}{a} = K.\,\rho\omega$ and $B_2 = \frac{K2a\rho\omega}{2a} = K.\,\rho\omega$

$\therefore B_1 : B_2 :: 1 : 1$

207. R_x should have a value equal to the resistance seen by it as looking in the network, replacing the sources by their accompanying resistances.

$$\therefore \quad R_x = \frac{10.9\times 29.4}{10.9+19.4} + \frac{5.2\times 7.1}{5.2+7.1} = 10\text{ K}$$

222. Faraday's Law. Whenever the magnetic flux linked with an electric circuit is altered, an e.m.f. is induced in the circuit. By disconnecting the instrument, the flux gets altered and hence the spark.

231. $f_c = 7.5$ MHz and $\Delta f = 6$ kHz

Carrier frequency at A,

$f_{cA} = 3\times f_c = 3\times 7.5 = 22.5$ MHz,

$\Delta f_A = 3\,\Delta_f = 3\times 6 = 18$ kHz.

$\therefore$ Carrier frequency at B,

$f_{cB} = 4f_{cA} = 4\times 22.5\text{ MHz} = 90$ MHz

232. The mixer has the effect of raising or lowering the carrier frequency by an amount equal to oscillator frequency. Assuming a rise in frequency,

$f_{cC} = f_{cB} + f_0 = 90 + 8 = 98$ MHz

241. 510 micro-seconds per half cycle = 2 × 510 = 1020 μs/cycle = 1.02 ms/cycle

$\therefore$ Signal frequency, $f = \left(\frac{1}{1.02}\times 10^{-3}\right) = 980$ Hz

Hence, binary 1 is received on channel 1.

248. Standing wave ratio $S = \frac{1+|k|}{1-|k|}$

where k = reflection coefficient $= \frac{Z_R - Z_o}{Z_R + Z_o}$

Here Z_o = circuit impedance, and

Z_R = receiving end or load impedance

$S = \infty$ when $K = 1$

$|k| = 1$ when $Z_R = 0$

or $Z_R = \infty$

Thus statements 1 and 3 are correct.

249. Microwave Link repeaters are used to compensate for the transmission loss in long-haul data communication. The repeating amplifiers are inserted at regular intervals in order to maintain sufficient signal levels. Their gain is just sufficient to compensate for the loss of signal levels between stations.

5

MICROWAVES

MICROWAVE BAND AND SOURCES

The microwave frequencies span three major bands at the highest end of RF spectrum. They are the ultra-high frequency (UHF) band of 0.3 to 3.0 GHz, the superhigh frequency (SHF) band of 3 to 30 GHz, and the extra-high frequency (EHF) band of 30 to 300 GHz.

Microwave Letter Designations

The letter designation in Table-1 are commonly used for specific parts of the microwave band. They do not have any official international standing, but they are convenient for specifying microwave applications. Most commercial microwave equipment is made for specific band from 1 to 40 GHz. Note that the new letter designations C to K are in ascending order of frequencies. The old S band of 2 to 4 GHz is split into E and F bands which are popular microwave frequencies.

TABLE 1 : IEEE Microwave Frequency Bands

Designation	*Frequency range in gigahertz*	
HF	0.003 –	0.030
VHF	0.030 –	0.300
UHF	0.300 –	1.000
L band	1.000 –	2.000
S band	2.000 –	4.000
C band	4.000 –	8.000
X band	8.000 –	12.000
Ku band	12.000 –	18.000
K band	18.000 –	27.000
Ka band	27.000 –	40.000
Millimeter	40.000 –	300.000
Submillimeter	▷	300.000

Microwave Applications

Probably the most common use of microwaves is in the field of telecommunications; the transmission of information, including analog and digital signals, from one point to another. The 5.925– to 6.425-GHz band is used for microwave relay stations in telecommunications.

Microwaves are also used for distance–measuring equipment (DME) in air navigation. The DME system indicates kilometers to the airport. The frequencies used are 962 to 1213 MHz.

Another application is microwave links used in television for relaying signals from a remote location to the studio, or from the studio to the transmitter, and for intercity networks. The studio to transmitter microwave links use 947 to 952 MHz.

Some of the industrial applications of microwaves are RF heating, bonding and gluing. A popular consumer application is the microwaves oven in which microwaves are used for very fast cooking. Microwave ovens operate at 2.45 GHz

With the advent of communication via satellites in space, the application of microwaves has become very important.

Fibre Optics

As the frequency of microwaves is increased the radiation approaches the short wavelengths of visible light. Actually, in a communication system with fibre optics, visible light is used as the carrier of modulated signals. One application is in, long distance telephone transmission lines. Instead of using radio, cables containing optical fibres are used. They transmit light by internal reflections along the entire length. The advantage is that the losses are very small.

At the transmitting end a laser beam is modulated to produce pulses of light. At the receiving end, special photodiodes convert the light back into electrical signals.

WAVEGUIDE

A device which constrains or guides the propagation of electromagnetic waves along a path defined by the physical construction of the guide. In a broad sense, devices such as a pair of parallel wires and a coaxial cable can certainly be called waveguides. When used in a more restricted sense, however, the term waveguide usually means, a metallic tube which can confine and guide the propagation of electromagnetic waves in the hollow space along the lengthwise direction of the tube. Hollow waveguides of convenient sizes are best adopted to the transmission of microwaves.

1. The properties of media affecting transmission are permittivity, permeability, and conductivity
2. At the boundary between media of different electrical properties, the electromagnetic wave will, in the general case, be partially transmitted and partially reflected.
   (a) The tangential components of the electric and magnetic fields are equal on the two sides of any boundary between physically real media.
   (b) The tangential component of electric field vanishes at the surface of a perfect conductor.
   (c) The normal component of a time-varying magnetic field vanishes at the surface of a perfect conductor.
3. Electric field lines may begin and end on charges. If an electric field ends on a conductor. It must represent a charge induced on the conductor.
4. Magnetic field lines can never end since magnetic charges are not known physically. Magnetic fields must always form continuous closed loops surrounding either a conduction current or a charging electric field (displacement current).
5. Electric field lines may form continuous closed path, surrounding a changing magnetic field.
6. Electromagnetic waves are classified according their equiphase surfaces or wavefronts, and the polarization is defined by the direction of the electric field.

   Rays : Lines indicating the direction of propagation.

   Plane waves : Wavefronts of the electric and magnetic are parllel to the plane of the source.

   Transverse waves : Electric and/or magnetic fields lie in a plane perpendicular to the direction of propagation.
   1. TEM (transverse-electromagnetic) – Both the electric and magnetic fields are transverse to the direction of propagation.
   2. TE (transverse-electric) – The electric field is transverse to the direction of propagation.
   3. TM (transverse-magnetic) – The magnetic field is transverse to the direction of propagation.
7. **Linear polarized waves :** The components of the electric and magnetic fields in the transverse plane do not change in direction from instant to instant or point to point.

   Vertical polarization : The electric field is in the vertical direction.

   Horizontal polarization : The electric field is in the vertical direction.

   Horizontal polarization : Resultant electric field lines in the horizontal plane.

   Circular polarization : Resultant electric field when equal amplitude horizontal and vertical polarized waves 90° out of phase are combined. The direction of this wave varies with time and distance. The wave is elliptically polarized if the horizontal and vertical waves are not equal in amplitude.

Advantages of Hollow Waveguides

There is no power loss by radiation from metal pipes of any type, including coaxial lines, if the ends are closed. Hollow waveguides are superior to coaxial lines in their ability to handle large concentrations of charges since these concentrations can be kept farther apart so that the electric field is less intense.

The construction of the waveguide is more simple than that of the coaxial cable since the inner conductor and its supports are eliminated. It is also more rugged and less susceptible to vibration and shock. Elimination of the insulating supports also results in a decrease of attenuation. Waveguides are usually air-filled, and for practical purposes they are considered to have no dielectric loss.

The current-carrying capacity is greater since, in practice, the waveguide is likely to have much greater conducting surface than a coaxial line. The overall power lost as heat in the walls of the waveguide is lower than the heat dissipated in the conductors of conventional size coaxial lines.

Although hollow waveguides and coaxial cables are the commonest in application, some other types of waveguides are also occasionally used. A single conductor (called a G string) is sometimes used as a waveguide. Another waveguide takes the form of a flat conduction strip having a certain spacing from a ground plane, known as microwave strip. Still another example is found in a dielectric rod.

Transmission Modes

Consider a hollow waveguide with a given cross section which is uniform throughout its entire length. As a result of the application of these boundary conditions to the wave equation, it can be shown that only certain unique patterns for the distribution of E and H (taken together) can exist in the waveguide. Each unique pattern of the field distribution is called a mode. There are two types of modes possible in a hollow waveguide. One type is called the transverse elecetric (TE) mode, in which E has only a component transverse (that is perpendicular) to the direction of propagation, whereas the magnetic intensity H has both transversal and longitudinal components. The other type is called the transverse magnetic (TM) mode in which the magnetic intensity has only a transverse component and the electric intensity has both components. Each type (TE or TM) of mode has an infinite number of submodes which have the common characteristic of the type to which they belong but differ among themselves in the details of field distribution. Since it is known that the transverse electric and magnetic (TEM) mode is not possible in a hollow waveguide any arbitrary electromagnetic wave inside such a waveguide can be considered as a linear superposition of all possible modest of both the TE and TM types.

Rectangular Waveguides

The type of waveguide with a rectangular cross section is not only the commonest in use but also the simplest in theoretical analysis. It is used here as a concrete example to illustrate various common properties of a wave guide. Consider a rectangular waveguide as shown in Fig. 1. The wave propagates along the z-axis. The simplest and also the most commonly used mode is called the $_{TE01}$ wave; its electric and magnetic intensities can be described by the expressions satisfying the boundary conditions of Eqs. (1) where A = arbitrary constant depending.

$$'E_x = A\sin\left(\frac{\pi}{b}y\right)\sin\omega\left(t-\frac{z}{v}\right)$$

$$H_y = \frac{1}{\mu v}A\sin\left(\frac{\pi}{b}y\right)\sin\omega\left(t-\frac{z}{v}\right)$$

$$H_z = -\frac{\pi}{\mu\omega b}A\cos\left(\frac{\pi}{b}y\right)\cos\omega\left(t-\frac{z}{v}\right)$$

$$H_x = E_y = E_z = 0.$$

upon the strength of wave excitation, w = $2\pi\times$ frequency, t = time, v = velocity of propagation of the wave and μ = permeability of the dielectric filling the waveguide. It is particularly important to note that the velocity of propagation v, in a waveguide is different from that in an infinite space filled with the same dielectric material v_0. These MO quantities are related for the TE_{01} wave by the Eq. (2), in which f = frequency, By

$$v = \frac{v_o}{\sqrt{1-(v_o/2fb)^2}} \quad \text{... (2)}$$

applying the simple formula v = $f\lambda$, it is seen that the wavelength in the waveguide is similar to the corresponding quantity (for the TE_{01} wave) in an infinite dielectric λ_0) as in Eq. (3).

$$\lambda = \frac{\lambda_0}{\sqrt{1-(\lambda_0/2b)^2}} \quad \text{... (3)}$$

The expressions in Eq. (1) reveal the wave nature of the field quantities through the sinusoidal functions of $\omega(t-z/v)$, which is description of wave motion with velocity v along the z-axis. On any cross section (constant z) these sinusoidal functions depend

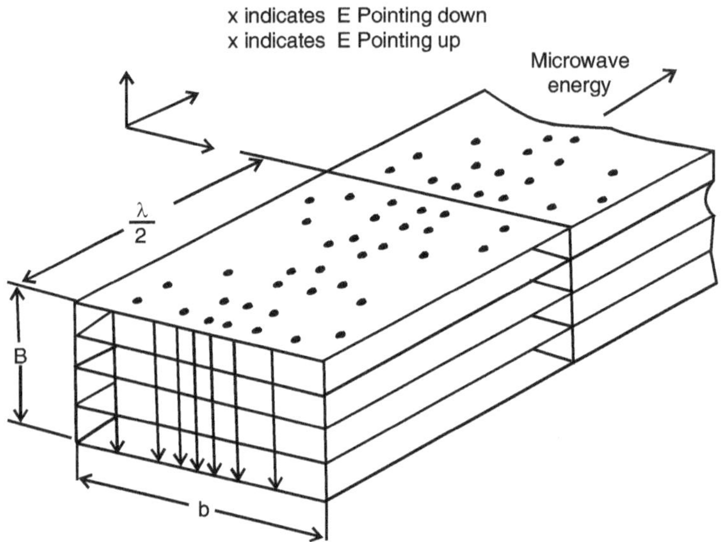

Fig. 1 : *Instantaneous field pattern for the TE_{01} wave in a rectangular waveguide. Solid lines indicate the electric intensity E, and broken lines the intensity H.*

simply on time. Hence, each of the field quantities oscillates at the common frequency f and with an amplitude which varies with the field point (y in this particular case).

Phase Velocity

Equation (2) represents the phase velocity of the guided wave because the quantity v is contained in w(t–z), which is called the phase of wave propagation. If $\lambda_0 < 2b$ the phase velocity in a waveguide is larger than that in an open space filled with the same dielectric material. Correspondingly, the wavelength in a waveguide is longer than that in an infinite dielectric medium, as indicated by Eq. (3). All waves with $\lambda_0 < 2b$ belong to the transmission region of the waveguide because only the waves in this region are allowed to pass through. When $\lambda_0 = 2b$, both v and λ are infinite and when $\lambda_0 > 2b$, they are imaginary. It can be shown that waves are not allowed to propagate in a waveguide when $\lambda_0 \geq 2b$. The critical value of $\lambda = 2b$ is known as the cutoff, or critical wavelength.

Generalization to other TE Waves

The preceding discussion can be carried through in a parallel fashion for the general case of any TE wave designated as TE_{nm}, where n and m are integers. Each TE_{nm} wave has its characteristic field distribution, velocity, and wavelength. The expression for the wavelength in the waveguide, for

$$\lambda = \frac{\lambda_0}{\sqrt{1-\left(\frac{n\lambda_0}{2a}\right)^2-\left(\frac{m\lambda_n}{2b}\right)^2}} \quad \text{... (4)}$$

example, is given by Eq. (4). The corresponding cutoff wavelength is given by Eq. (5) below

$$\lambda_0 = \frac{2}{\sqrt{\left(\frac{n}{a}\right)^2+\left(\frac{m}{b}\right)^2}} \quad \text{... (5)}$$

which reduces to $\lambda_0 = 2b$ for the special case of TE_{01} where n = 0 and m = 1, A TE_{oo} wave (n = 0 and m = 0) would have an infinite cutoff wavelength which is characteristic of a principal mode; if such were possible. Actually the solution or the field equation shows that the principal mode cannot exist in a hollow metallic waveguide. Equation (5) further shows that any other values of n and m in TE_{nm} would lead to a cutoff wavelength shorter than that of TE_{01}, which is called the dominant mode for this reason.

TM Waves

The TM waves can be treated in almost exactly the same manner as the TE waves. Although the field distribution in the two cases are completely different, the wavelengths (similarly, the velocities of propagation) in the guide for both cases obey Eqs. (4) and (5), except that neither n nor m can become zero for a TM_{nm} wave.

Designation of Rectangular Waveguides

It is highly desirable and customary for practical use to excite only the dominant mode (TE_{01}) in a rectangular waveguide. This means that a waveguide of a given size is usefull for this one-mode operation only for a certain range of free-space wavelengths (or frequencies). Four waveguides of certain dimensions and their corresponding wavelength bands have been conventionally designated by the letters S.C.X and K. as shown in the Table 2.

TABLE 2 : Band Designation for Rectangular Waveguides

Band	*Dimension cm*^{2}	λ_0 *cm*
S	7.62 × 2.54	8.9–10.5
C	3.41 × 1.58	3.7–5.1
X	2.54 × 1.27	3.0–3.5
K	1.06 × 0.43	1.2–1.5

Higher Order Modes

A double infinity of higher order modes is possible in the rectangular guide indicated by the integral number of half waves that may exist between the a and b dimensions. Practical use of wave guides is centered around the dominant mode and these higher order modes are mainly of academic interest in general these waves are more highly attenuated and more difficult to recover from the waveguide system than the corresponding dominant wave.

If the source frequency is increased, so that the 'a' dimension is greater than one wavelength the TE_{20} mode can exist. Also, half-wave distributions can be obtained in the 'b' dimension resulting in for example, the TE_{11} mode. The wave tends to remain in the dominant mode so that the even though the source frequency may be high enough for higher modes to exist they do not necessarily exist unless through deliberate or accidental distortion of the fields.

Higher order modes are present in the vicinity of a discontinuity where the fields are distorted. Since the waveguide dimensions, for the particular operating frequency are such that the higher order mode can not be propagated these waves exist as fringing effects which become negligible at relatively short distances from their source.

Higher order modes are frequently given relative cutoff frequency values. These numbers are obtained by normalizing the cutoff frequency of the mode in question to the cutoff frequency of the TE_{10} mode in a rectangular waveguide which has 'a' diameter equal to the a dimension of the rectangular waveguide. An example for the TE_{11} mode in a circular waveguide.

$$\text{Relative cutoff frequency} = \frac{f_c(TE_{11})}{f_c(TE_{10})}$$

$$= \frac{7{,}700}{6{,}557} = 1.17$$

several of the higher order modes are shown in Fig. 2

Practical Operating Range

The dominant mode propagation without the presence of higher order modes is usually desirable since for a given excitation frequency, it has lower dissipation, requires smaller, lighter and cheaper guiding structures, and requires simpler associated components.

The practical operating range of the TE_{10} mode in a rectangular waveguide where b/a = 0.5 is from 62 to 95 percent of cutoff of the TE_{20} mode. Thus the 'a' dimension must be greater than $\lambda/2$ and less than λ .

The region of practical operation is shown in Fig. 3 which shows the relative phase velocity and group velocity for various condition of waveguide operation.

Reviewing again the previous analysis, we find that below cutoff the waveguide acts as a nondissipative attenuater, the group velocity is zero, and the phase velocity is infinite. As the operating

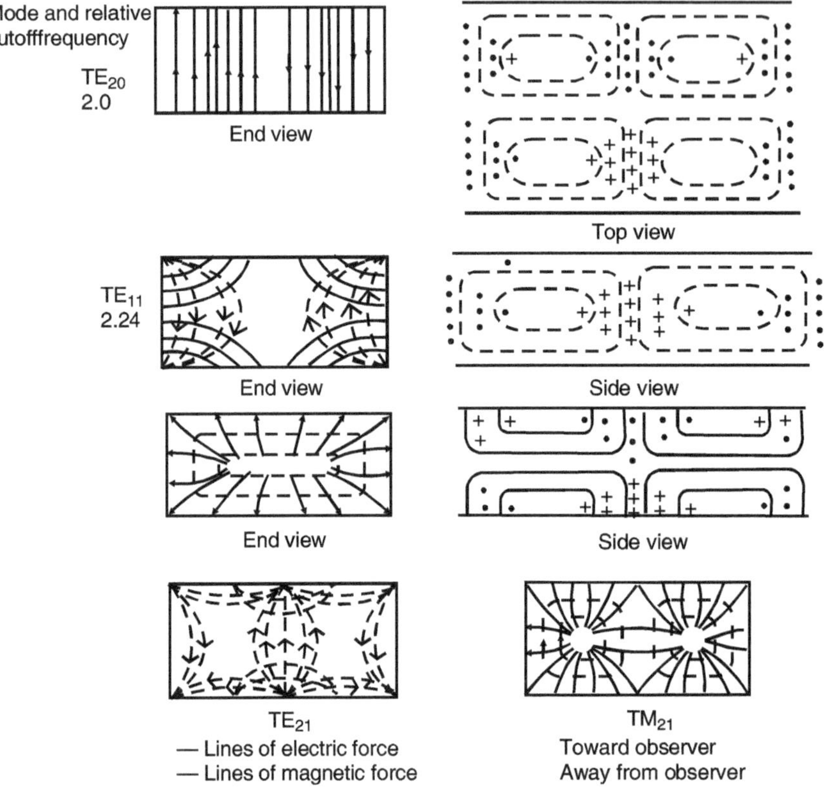

Fig. 2: *Field configuration of higher order modes in the rectangular waveguide*

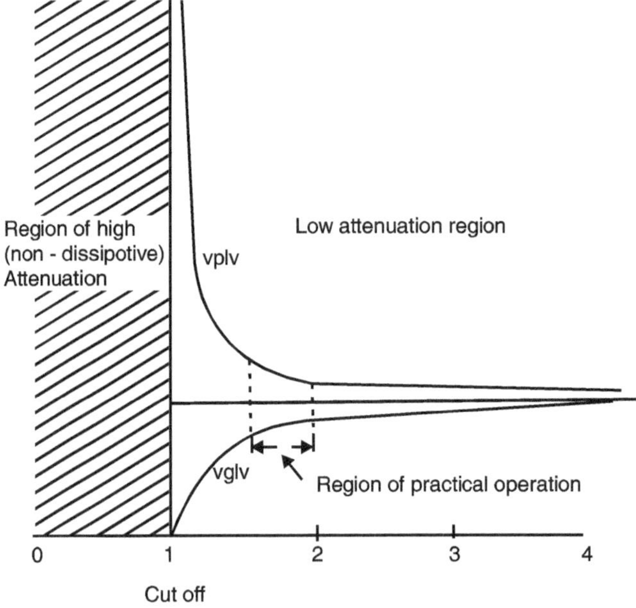

Fig. 3: *Variation of phase velocity and group velocity with frequency*

frequency is increased far above the cutoff frequency, the phase and group velocities approach the free-space velocity.

Waveguide Dimensions

Rectangular waveguides are usually chosen so that only the dominant mode exists over a certain frequency range. The above consideration determinas the 'a' dimension and is important because of the following considerations :

(a) The attenuation loss is greater as the 'b' dimension is made smaller.

(b) There 'b' dimension determines the voltage breakdown characteristics and therefore determines the maximum power capacity.

The power transmitted by a waveguide has been determined from the Poynting vector evaluated at points on a cross section of guide. The total power transmitted is P.

$$P = \frac{1}{4} E_m H_{xm} ab$$

and $$H_{xm} = \frac{E_m}{Z_1}\frac{\lambda}{\lambda_g} = \frac{E_m}{Z_1}\sqrt{1-\left(\frac{\lambda}{\lambda_o}\right)^2}$$

Therefore, $$P = \frac{1}{4}\frac{E_m^2}{Z_1}\sqrt{1-\left(\frac{\lambda}{\lambda_o}\right)^2}\, ab$$

The power capacity depends on the maximum electric field intensity, a, and b. Therefore, a and b should be made as large as possible for higher power capacity. In practice, the dimension b is usually chosen to be about one-half of a. The ratio of b/a is $\frac{1}{2}$.

Waveguide Wave Impedance

The characteristic wave impedance is analogous to the characteristic impedance of the two-wire and coaxial lines. The wave impedance represents the ratio of the electric to the magnetic fields, in which case, the electric fields is the analogue of voltage and the magnetic field is the analogue of current.

The actual wave impedance is little used and is seldom determined. There is no advantage in extending calculations beyond the determination of the per-unit impedance since the actual measurement of impedance in a waveguide results in a value of normalized impedance. The wave impedance for TE and TM modes are

$$Z_{TE} = \frac{\eta}{\sqrt{1-(\lambda/\lambda_o)^2}} = \frac{\eta}{\sqrt{1-(f_o/f)^2}}$$

$$Z_{TM} = \eta\sqrt{1-\left(\frac{\lambda}{\lambda_o}\right)^2} = \eta\sqrt{1-\left(\frac{f_o}{f}\right)^2}$$

where η is the intrinsic impedance of the medium (377 ohms for free space).

Comparison and summary of modes of propagation

1. The characteristic impedance and phase velocity are essentially independent of frequency in the TEM mode of propagation. Lossless transmission lines carrying TEM waves are nondispersive.
2. Propagation occurs in a waveguide only above a critical frequency called the cutoff frequency. Thus a waveguide carrying a TE or TM wave may be considered to be a high pass filter as for as the particular mode is concerned. At frequencies far above cutoff, the propagation characteristics approach that of free space, *i.e.*, the waveguide wavelength approaches the free unbounded wavelength in the medium contained inside the waveguide.
3. In the waveguide the phase and group velocities and the wave impedance of TE and TM waves are functions of frequency. Therefore, the waveguide is a dispersive medium in that it tends to spread out waves of different frequencies which may be applied simultaneously at a common point, This dispersion phenomenon would cause distortion in waves which have modulation components extending over a wide frequency range.
4. TEM waves cannot exist in closed waveguides without inner conductors, but TE and TM wave may exist at higher order modes on two conductor or coaxial lines if the operating frequency is sufficiently high.

Circular Waveguides

The circular waveguide is used in many special applications in microwave techniques. The circular guide has the advantage of greater power-handling capacity and lower attenuation for a given cutoff wavelength, but it has the disadvantage of somewhat greater size and weight. Also the polarization of the transmitted wave can be altered due to minor irregularities of the wall surface of the circular guide, whereas the rectangular cross section definitely fixes the polarization.

The wave of lowest frequency or the dominant mode in the circular waveguide is the TE_{11} mode. The subscripts which describe the modes in the circular waveguide are different than for the rectangular waveguide. For the circular waveguide, the first subscript m indicates the number of full-wave variations of the radial component of the electric field around the circumference of the guide. The second subscript n indicate the number of half-wave variations across a diameter. Also, the second subscript indicates the number of diameters that can be drawn perpendicular to all electric field lines; and in the case of TE_{01} waves it indicates the half-wave variations of the electric field across a radius of the guide. Illustrations of various circular waveguide modes at shown in Fig. 4

The higher order modes which have properties of importance are the TM_{01} and TE_{01} modes. The TM_{01} mode has a circular symmetry which makes it adaptable for use in rotating joints. The TE_{01} mode has the property of decreasing attenuation as the frequency of the source is increased relative to cutoff. This mode is used for high-Q frequency meters.

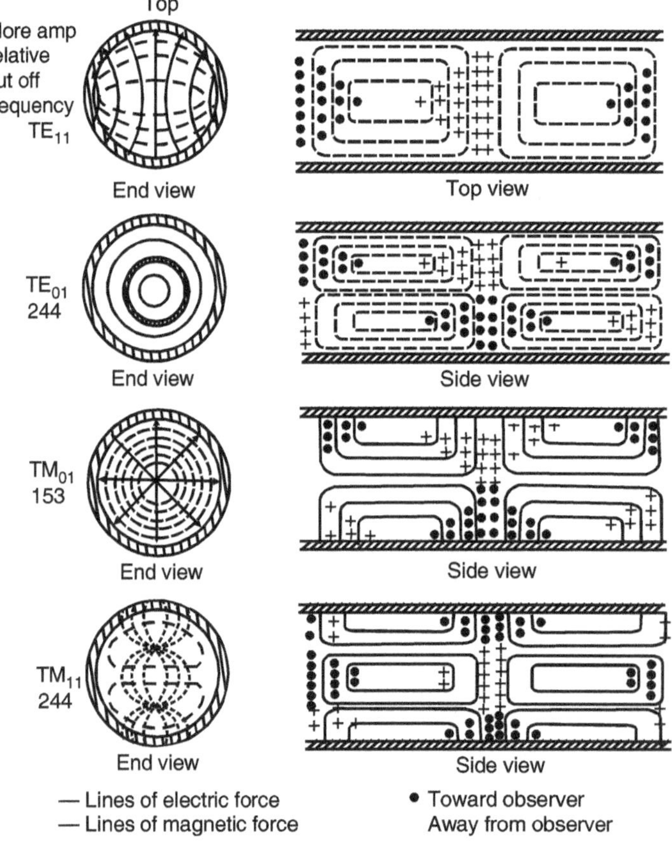

Fig. 4: *Field configurations of modes in the circular waveguides*

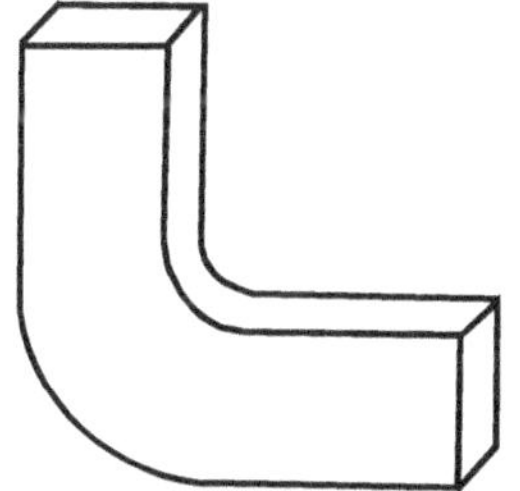

Fig. 5: *H-plane 90° bend for a rectangular waveguide*

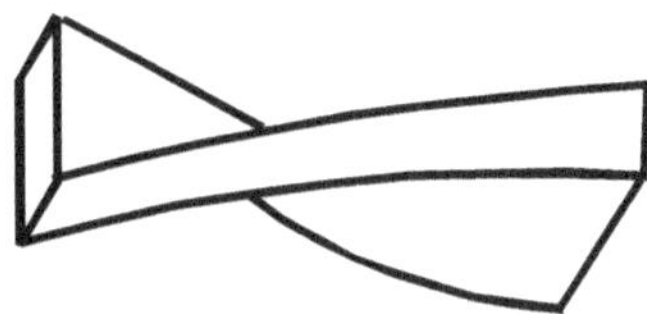

Fig. 6: *A 90° twist for rectangular waveguide*

Joints, Bends and Junctions

To achieve certain effects, waveguides are frequently joined together, bent or twisted, or formed into networks known as junctions.

Joints : Waveguides are often joined together under various conditions. Identical waveguide terminals may be connected from end to end for extension or for inserting circuit elements. This can be done by providing a flange at each end of the waveguide and then butting the two flat surfaces of the flanges together so that the waveguide ends form a contact joint smoothly. Sometimes, to alleviate the difficulty of making good contact, a choke joint is operated. This is based on the principle that an artificially created current nodal line at the joint will make the physical contact of the waveguide ends unnecessary.

Another type of waveguide joint is required when a coaxial cable is to be joined to a hollow waveguide. Here, a transformation of the wave is to take place between the coaxial mode and , for instance; the TE_{01} mode of a rectangular wave guide. In this case the coaxial cable is usually led into the blocked end (or into a tunable short-circuiting plunger) of the regular waveguide, and the centre conductor is extended to touch the opposite face so that the extended wire will act as an antenna for the excitation of the wave in the rectangular guide (Fig. 7)

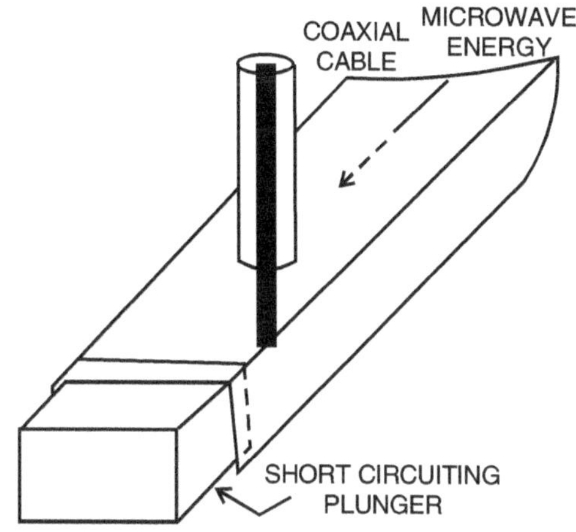

Fig. 7 *Coupling between waveguide and coaxial cable*

Bends and twists : Waveguide bends are often used to change the direction of the waveguide by a desired angle. There are two types of 90° band—one in the H plane and the other in the E plane. A 90° bend in the H plane is shown in Fig. 5 Twists are used to change the plane of polarization by a desired angle while maintaining the direction of the waveguide. A 90° twist is shown in Fig. 6

Waveguide Tees

Waveguide junctions play an important part in waveguide techniques; two of the common forms are shown in Fig.8 The interesting properties and characteristics of these junctions can be examined by considering the behaviour of the electromagnetic fields in passing through the junctions. These junctions are referred to as wave guide tees, and they differ with regard to the plane in which the branch lies. The tee in Fig. 8(a) is called an E plane or series tee because the axis of the side arm is parallel to the E field of the main transmission line. The tee at Fig. 8(b) is called an H plane or shunt tee because tee axis of the side arm is parallel to the planes of the magnetic field of the main transmission line.

The progress of the representative line of electric force of the wavefront is shown in the series tee. The line of electric force bends as it leaves the side arm and causes the electric fields in the side arms to be of opposite polarity (out of phase). These waves set up in the side arms are equal in magnitude and the junction is completely symmetrical.

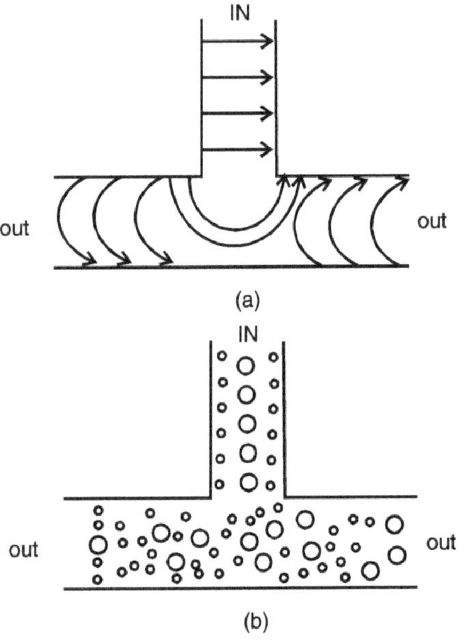

Fig. 8

RADAR

Basic Functions

The basic functions of radar are in the word, which stands for RAdio Detection And Ranging. Measurement of target angles has been included as a basic function of most radars, and doppler velocity is often measured directly as a fourth basic quantity. Resolution of the desired target from background noise and clutter is a prerequisite to detection and measurement, and resolution of surface features is essential to mapping or imaging radar. The radar resolution cell is a four-dimensional volume, bounded by antenna beamwidths, width of the processed pulse, and bandwidth of the receiving filter. Within each such resolution cell, a decision may be made as to presence or absence of a target and if a target, is present , its position may be interpolated to some fraction of the cell dimensions.

The block diagram of a typical pulsed radar is shown in Fig. 9. The equipment has been divided arbitrarily into seven subsystems corresponding to the organisation of the paragraphs on radar

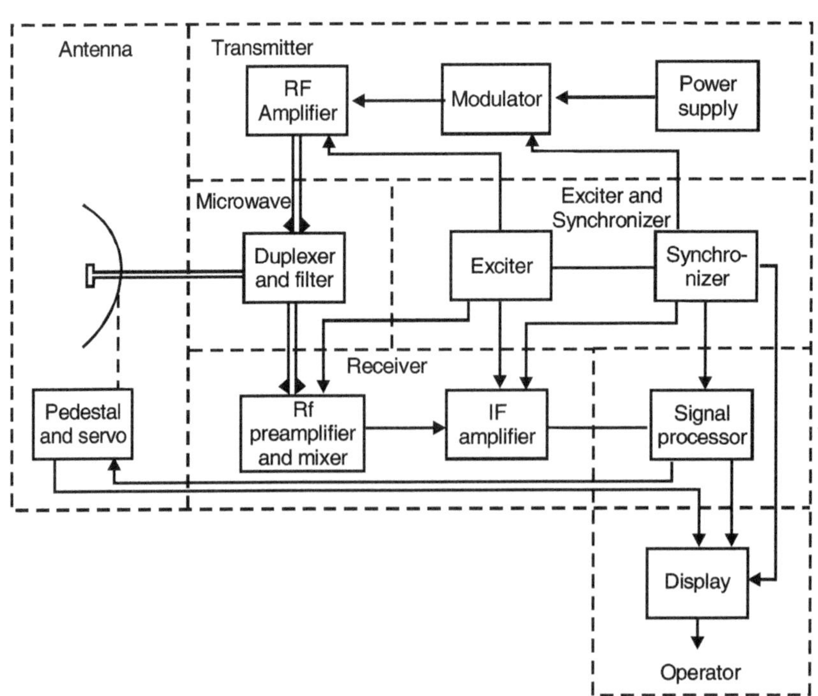

Fig. 9 : *Block diagram of typical radar*

technology. Radar operation is initially by a synchronizer which controls the time sequence of transmissions, receiver gates and gain settings, signal processing and display. When called for by the synchronizer, the modulator applies a pulse of high voltage to the rf amplifier simultaneously with an rf drive signal from the exciter. The resulting highpower rf pulse is passed through a transmission line or waveguide to the duplexer, which connects it to the antenna for radiation into space. The antenna shown is of the reflector type, steered mechanically by a servo-diver pedestal. A stationary array may also be used, with electrical steering of the radiated beam.

After reflection from a target, the echo signal re-enters the antenna, which has been connected to the receiver preamplifier or mixer by a duplexer. A local-oscillator signal furnished by the exciter translates the echo frequency to i.f., which can be amplified and filtered in the receiver prior to more refined signal processing. The processed i.f. signal is passed through an envelope detector and displayed, with or without video processing. Data to control the antenna steering and to provide outputs to an associated computer are extracted from the time delay and modulation on the video signal.

Target Characteristics

Given a set of radar parameters, the detection and location performance will depend also or target characteristics : (a) target cross-section (average or median) and fluctuation with time and frequency, (b) target size, shape, distribution of scatters, aspect angle, and rate of turn; (c) target position in range, azimuth elevation and its velocity and acceleration in three coordinates.

Radar-Range Equations : (Signal to Noise Ratio).

The signal power S received by the radar antenna is calculated from transmitted power P_t and a series

TABLE 3: Radar Parameters

Sub-system	*Parameters*
Transmitter	Frequency wavelength, peak and average power pulse width, pulse energy pulse repetition frequency, duty ratio, signal band-width.
Antenna	Size, effective aperture area, beamwidths, scan rate and coverage.
Receiver and signal processor	Effective input temperature, bandwidth, input noise power, output pulsewidth, MTI improvement factor, video integration gain.
System performance	System losses, received power and energy on standard target time on target-to-noise ratios in power and energy, measurement slope factors, data rate, instrument errors.

TABLE 4: Radar Applications

All surveillance	Long-range early warning ground-controlled intercept acquisition for weapon system, height finding and three-dimensional radar-airport and air-route surveillance.
Space and missile surveillance	Ballistic missile warning, missile acquisition, satellite surveillance
Surface search and battlefield survillance	Sea search and navigation ground mapping, mortar and artillery location, airport taxiway control.
Weather radar	Observation and prediction, weather avoidance (aircraft), cloud-visibility indicators.
Tracking and guidance	Antiaircraft fire control, surface fire control, missile guidance, range instrumentation, satellite instrumentation, precision-approach and landing.
Astronomy and geodesy	Planetary observation earth survey, ionospheric sounding.

TABLE 5 : Radar Frequency Band (IEEE Standard 521–1976)

Name	*Frequency range*	*Radiolocation bands based on ITU assignments in region II*
VHF	30-300 MHz	137-144 MHz; 216-225 MHz
UHF	300-1000 MHz	216-225 MHz; 890-940 *MHz
P band	230-1000 MHz	
L band	1000-2000 MHz	1215-1400 MHz
S band	2000-4000 MHz	2300-2550 MHz 2700-3700 MHz
C band	4000-8000 MHz	5255-5925 MHz
X band	8000-12500 MHz	8500-10700 MHz
Ku band	12.5-18 GHz	13.4-14.4 GHz 15.7-17.7 GHz
K band	18.5-26.5 GHz	23-24.25 GHz
Ka band	26.5-40 GHz	33.4-36 GHz
Millimeter	40 GHz	

*Sometimes included in L band.
Seldom used nomenclature

of factors that describe basically the geometry of the radar beam and target.

$$S = P_t G_t \sigma / (4\pi)^2 \; R^4 \qquad \text{... (a)}$$

where,

G_t = transmitting antenna gain,

A_t = effective receiving aperture area,

σ = target cross-section, and

R = target range. Since the receiving aperture and gain are related by

$$A_t = G_t \lambda^2 / 4\pi \qquad \text{... (b)}$$

the signal power can also be given as

$$S = P_t G_t G_t \lambda^2 \sigma / (4\pi)^3 R^1 \qquad \text{... (c)}$$

In pulsed radar, P_t and S are the so called peak power levels (actually the mean power over rf cycle at the peak of pulse envelope).

Radar noise is a combination of receiver-generated and environmental noise, extending over a broadband with a power density No.

$$N_o = kT_s = kT_o F_{no} \qquad \text{... (d)}$$

where

k = Boltzmann's constant = 1.38×10^{-23} W/(Hz K),

T_s = system noise temperature referred to the antenna terminals,

T_o = standard temperature (290 K) used in measuring noise factor, and

F_{no} = operating noise factor ($F_{no} = T_s/To$).

The system noise temperature is calculated from receiver noise factor F_n, line losses Lt. and antenna temperature Ta.

$$T_s = T_a + T_t + L_t T_e \qquad \text{... (e)}$$

$$T_a = (0.876\ a—254)/L_a + 290 \qquad \text{... (f)}$$

$$T_t = T_{tr}\ (L_t—1) \qquad \text{... (g)}$$

$$T_e = To\ (F_n—1) \qquad \text{... (h)}$$

where

T_t = temperature contribution of loss L_t,

$T'a$ = sky temperature,

L_a = antenna ohmic loss,

T_{tr} = physical temperature of the receiving line (290 K) and

T_e = temperature contribution of the receiver.

For T_a = 290 K, the operating noise factor

F_{no} = F_a and

T_a = 290 F_n. For antennas directed into space F_{no} can be much less than F_n, and T_s much less than 290 K.

The noise power at the i.f. output of the receiver will depend upon receiver band width and gain, but this noise power is equivalent to an input power at the antenna terminals of

$$N = No\ B = kT_E B_n$$

where

B_n = noise bandwidth of i.f filter for a wide-band filter $(B_n \tau >\!\!> 1)$ the signal peak is not affected by the filter and

S/N = $S/k\ T_s B_n$. In general, however, the S/N ratio at the receiver i.f. output is calculated from the ratio of received pulse energy 'S_τ' to noise density :

Ideal energy ratio for single pulse,

$$E_1 / No = P_1 \tau G_t G_t \lambda^2 \sigma / (4\pi)^3 R^4 kT_E \quad \text{... (i)}$$

Intermediate-frequency power ratio for single pulse:

$$S/N = E_1 / No L_m = P_t G_t G_t \lambda^2 \sigma / (4)^3 R^4 kT_s L_m \quad \text{... (j)}$$

where

L_m = i.f. filter matching loss and

τ = pulse width for a cw or coherent pulse radar integrates over an observation interval to in a predetection filter, the i.f. output S/N ratio is :

Intermediate frequency power ratio over interval to:

$$S/N = E/N_o L_m = P_{av} T_o G_t G_t \lambda^2 \sigma / (4\pi)^3 R^4 kT_a L_m$$

where

P_{av} = average transmitter power $(P_t \tau\ f_t$ for pulsed radar$)$ and

L_m = matching loss of the filter to the entire waveform over t_o S.

Moving Target Indicator

A moving target indicator (MTI) is a device which limits the display of radar information primarily to moving targets. The sensitivity to moving target is provided primarily through their doppler shifts, although area MTI systems have been built which cancel targets on the basis of overlap of their signal envelopes in both range and angle.

In the usual pulsed amplifier coherent MTI system, two cw oscillators in the radar are used to produce a phase and frequency reference for both transmitting and receiving so that echoes received over the train of pulses from fixed targets have a constant phase at the detector. These echoes will be cancelled, leaving in the output only those signals whose phase varies from pulse to pulse. Coherent MTI can be implemented with a pulsed oscillator transmitter (Fig. 10) in which the coherent oscillator is locked in phase to remember the random phase with which each successive pulse is transmitted. Although the transmitted signal is noncoherent the i.f. signal is coherent and has the same line structure as in the pulsed amplifier system. Both systems attenuate targets in a band centered at zero radial velocity the depth and width of the rejection notch depending on design of the canceller and stability of the received signals.

Two variations on the coherent MTI are available for rejection of clutter with non-zero radial velocity. In the clutter-locked MTI, the average doppler shift of a given volume of clutter is measured and used to control an offset frequency oscillator in the receiver, shifting the clutter into the rejection notch. Short or long-term averages can be used to obtain rapid adaptation to varying clutter velocity (as in weather clutter) or better rejection of selected parts of the complex clutter background. The alternative in noncoherent MTI, which the clutter surrounding a target provides the phase reference with which the target signal is mixed to produce a doppler signal. Although simpler to implement, noncoherent MTI does not cancel as completely and may lose target signals when the clutter is too small (as well as when it is too large).

Performance of MTI

The basic measure of MTI performance is the MTI improvement factor l. This is equal to the clutter attenuation when the canceller gain is set to unity on noise or targets uniformly distributed in velocity. The basic relationships between l and radar or clutter parameters can be expressed in terms of the ratio of rms clutter spread to repetition rate of blind speed :

where

σ_f = standard deviation of clutter power spectrum (H_z),

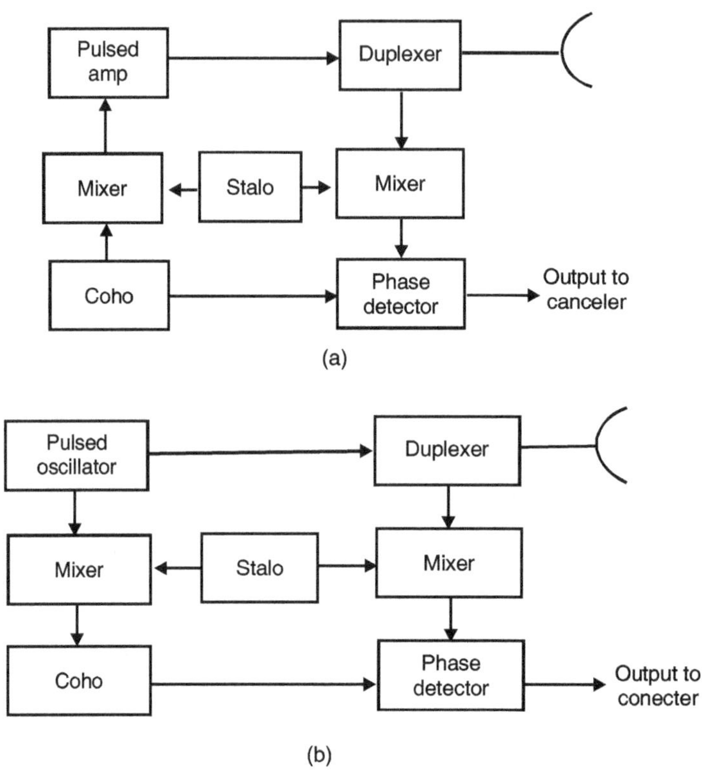

Fig. 10 : *Block diagrams of MTI radars :*
(a) simplified block diagram for power amplifier
(b) pulsed oscillator

f_t = repetition rate,

σ_V = standard deviation in m/s and

V_{bl} = first blind speed from Eq. (1). For a scanning gaussian beam

$$Z = 2\sqrt{\ln 2}\left(w / f_t \theta_3\right) = 1.665 / n$$

for a (sin x)/x beam, the constant 1.665 should be changed to 1.760, giving slightly. Poorer MTI performance.

In general, the rms speed should be calculated using the sum of the componants σs due to scanning, σ_1 due to internal motion of the clutter , σ_m due to relative motion between the radar and the clutter and σ_x due to low frequency instabilities in the radar.

CW Radar

A cw transmission has no velocity ambiguity, and so cw radar equipment can easily be designed to provide 80 to 100 dB rejection of fixed or moving clutter. Coherent integration of target signals in selected doppler bands is also provided by narrow-band filters or relatively simple construction. Three problems, however, restrict the performance and usefulness of cw radar for search :

(a) **Isolation of Receiver from Transmitter :** Direct feedthrough of transmitter power to the receiver must be minimized, requiring separate antennas in high power systems and careful design in all systems to avoid receiver saturation.

(b) **Magnitude of Short-Range Clutter Echo :** The echo power received from clutter in a cw radar is the integrated product of the reflectivity $(\text{range})^{-4}$ and antenna gain factors of the radar over the common volume of the transmitting and receiving beams. In both volume and surface clutter the echo power is controlled by the clutter at the shortest range in the common volume and the effective clutter cross section is a function of beamwidth and range R_c to the point where the beams substantially overlap. The required clutter improvement for a cw radar may therefore be very high because of the $(R/R_c)^4$ term.

(c) **Transmitter Noise :** Both the direct feed-through from transmitter to receiver and the echoes from short-range clutter will contain random-noise component from the transmitter. Special circuits may be designated to cancel the direct feed-through and low frequency components of reflected noise, but the higher-frequency components will appear with phase shift from the range delay and cannot be cancelled completely. Subclutter visibility in cw systems is generally controlled by these noise points.

Displays

Radar indicators are the coupling link between radar information and a human operator. Radar information is generally a time-dependent function of range of distance. Thus the display format generally uses a displacement of signal proportional to time of range. The signal may be displaced as an orthogonal displacement (such on an oscilloscope) or as an intensity modulation of brightening of the display. Most radar signal presentations have the intensity-modulated type of display where additional information such as azimuth, elevation, angle, or height can be presented.

A common format is a polar-coordinate, or plan-position indicator (PPI) which results in a maplike display. Radar azimuth is presented on the PPl angle and range as radial distance. In a cariesian coordinate display, one coordinate may represent range and the other may represent azimuth elevation or height. Variations may use one coordinate for azimuth and the other cordinate for elevation and gate a selected time period or range for to the display. The increasing use of processed radar data provide further variations. In each case, the display technique is optimized to improve the information transfer to the operator.

EXERCISE

1. The bandwidth of an RC-coupled amplifier is limited by

A. coupling capacitors at the low-frequency end and bypass capacitors at the high-frequency end

B. coupling capacitors at the high frequency end and bypass capacitors at the low-frequency end

C. bypass and coupling capacitors at the low frequency end and shunt capacitors at the high frequency end

D. shunt capacitors at the low frequency end and bypass as well as coupling capacitors at the high frequency end

2. MW signals follow the curvature of the earth. This is known as :

A. Faraday effect

B. Ducting

C. Tropospheric scatter

D. Ionospheric reflection

3. The anode voltage for the picture tubes is of the order of:

A. 4 kV B. 8 kV

C. 12 kV D. 18 kV

4. A line with lumped parameters of 10 T sections, series resistance of 10 ohms per section and 200 ohms shunt resistance, is matched to a generator with emf of 3 volts and teminated by its characteristic impedance. The sending end current is

A. 1.667 A B. 2.133

C. 6.67 A D. .033 A

5. For a transfer function H (s) = P (s)/Q(s) where P(s) and Q(s) are polynomials in *s*:

A. the degree of P(s) is always greater than the degree of Q(s)

B. the degree of P (s) and Q (s) are same

C. the degree of P(s) is independent of the degree of Q(s)

D. the maximum degree of P(s) and Q(s) differ at most by one.

6. For the circuit shown in the given figure, the quiescent point is

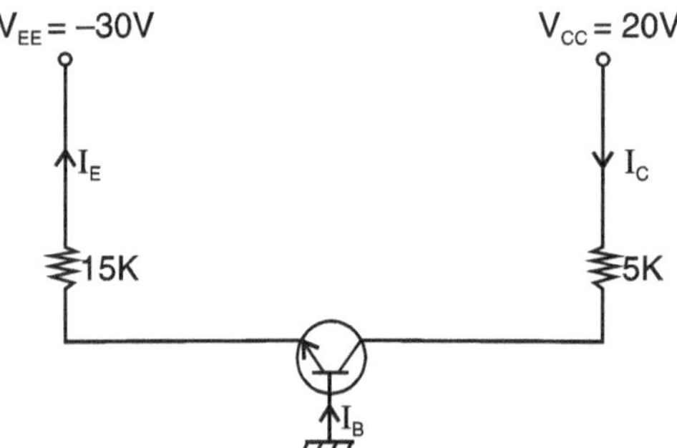

A. 12 V, 50 mA B. 12 V, 2 mA
C. 10 V, 20 mA D. 10 V, 5 mA

7. A charge is introduced instantly in the deep core of a solid conductor (you could assume the conductor to be built around the charge instantly). The charge will
A. remain at the same point where introduced
B. settle on the approximate central
C. be distributed as a uniform volume charge
D. distributed as a surface charge.

8. At low and medium frequencies.
A. Horizontal waves suffer more attenuation
B. Vertical polarized waves suffer more attenuation
C. Both suffer equal attenuation
D. None of the above

9. In a radio receiver with a simple AGC
A. An increase in signal strength produces more AGC
B. The audio stage gain is normally controlled by the AGC
C. The faster the AGC time constant the more accurate the output
D. The highest AGC voltage is produced between stations.

10. Gain or level for video signal in TV receiver is provided by:
A. Volume control B. Control
C. Brightness control D. None of these

11. Hight frequency is
A. Absorbed by F2 layer
B. Capable of use for long distance common the moon
C. Affected by the solar cycle
D. None of the above

12. In ergodic process
A. Ensemble and time averages are different
B. Ensemble and time averages are identical
C. Ensemble and time averages are opposite to each other
D. None of these

13. If negative feedback is introduced in an amplifier in a shunt current configuration, then the input resistance will
A. increase
B. remain the same
C. decrease
D. increase, decrease or remain the same depending upon input

14. In a stabilized reactance modulator AFC system
A. The discriminator must have a full time constant to prevent demodulation
B. The higher the discriminator frequency, the better the oscillator frequency stability
C. The discriminator frequency must not be too low, or the system will fail
D. Phase modulation is converted into FM by the equalizer circuit.

15. A wave is frequency modulated with an index of 1 and its frequency is multiplied 8 times. The modulation index is
A. Not changed
B. Increased eight times
C. Decreased eight times
D. Increased four times.

16. Consider a single crystal of an intrinsic semiconductor. The number of free carriers at the Fermi level at room temperature is
A. half the total number of electrons in the crystal
B. half the number of free electrons in the crystal
C. half the number of atoms in the crystal
D. zero

17. The local oscillator of a broadcast receiver is tuned to a frequency higher than the incoming frequency:
A. To permit easier racking

B. To allow adequate frequency coverage without switching
C. To help the image frequency rejection
D. Because otherwise an intermediate frequency could not be produced.

18. For a given ionospheric layer, the highest frequency that will be reflected back for vertical incidence is given by:
A. $f_c = 81\ N_{max}$ B. $f_c = 81\ N^2_{max}$
C. $f_c = \sqrt{81\ N_{max}}$ D. $f_c = 81\sqrt{N_{max}}$
where N_{max} is the maximum ionisation density.

19. The peak-responding ac voltmeter employing a half-wave precision rectifier has been calibrated to give a reading of 1 V for 1 V rms sinusoidal input. For 1 V dc input, the meter reading will be
A. 0.707 V   B. 0.707 V or 0 V
C. 1.414 V   D. 1.414 V or 0 V

20. For the design of low pass prototype filter with load resistance R_L = 1 ohm and angular frequency ω = 1 rad/sec the values of L and C would be

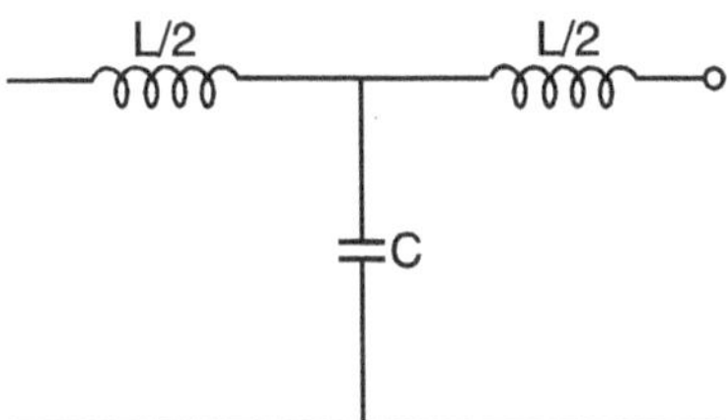

A. 1 H and 1 F   B. 1 H and 2 F
C. 2 H and 1 F   D. 2 H and 2 F

21. Fading can be avoided or improved by
A. Providing space diversity reception
B. Providing AVC in a receiver
C. Providing frequency diversity reception
D. All are possible

22. A submarine cable repeater contains among other equipment
A. A d.c. power supply
B. Filter for the two direction of transmission
C. Multiplexing and demultiplxing equipment
D. pilot inject and pilot extract equipment

23. The effect of transit-time noise is signficant in
A. LF   B. MF
C. HF   D. All of the above

24. To prevent loading in the last I. F. amplifier in a receiver, we should use:
A. Squelch circuit
B. Variable-selectivity
C. Variable sensitivity
D. Double conversion

25. A helical antenna is used for satellite tracking because of its
A. Broad bandwidth
B. Circular polarization
C. Manoeuverability
D. Good front-to-back ratio

26. In a transistor push-pull amplifier
A. there is no d.c. present in the output
B. there is no distortion in the output
C. there are no even harmonics in the output
D. there are no odd harmonics in the output

27. Match List I (Waveguide discontinuities) with List II (Equivalent circuits) and select the correct answer

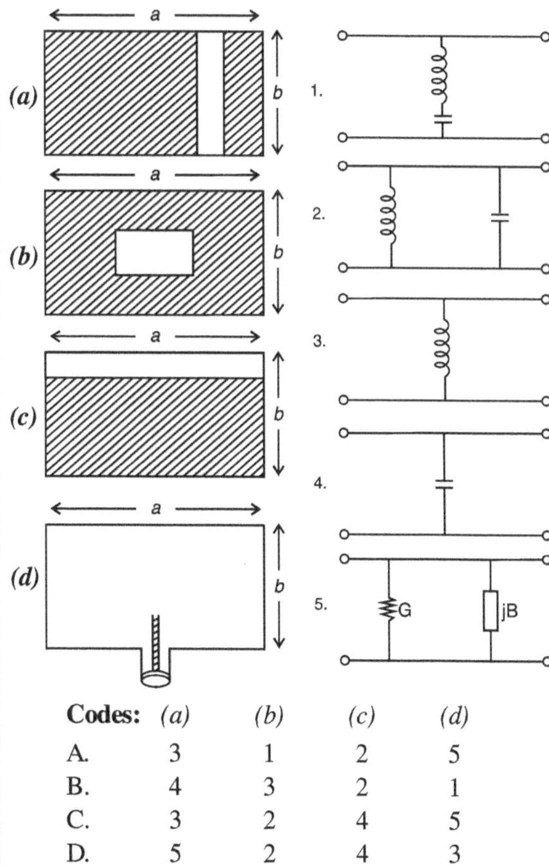

Codes:	*(a)*	*(b)*	*(c)*	*(d)*
A.	3	1	2	5
B.	4	3	2	1
C.	3	2	4	5
D.	5	2	4	3

28. Silicon is not used for the construction of tunnel

diode due to all of the following reasons except

A. High noise
B. High forbidden gap
C. Low ion mobility
D. Difficult to work at

29. Which of the following is a false statement? When the free-space wavelength of a signal equals the cut off wavelength of guide

A. The group velocity of the signal becomes zero
B. The phase velocity of the signal becomes infinite
C. The characteristic impedance of the guide becomes infinite
D. The wavelength within the waveguide becomes infinite.

30. In laboratory experiments the output from reflex Klystrons are modulated by square waves because

A. it is easy to generate a square wave
B. crystal diode operates in the square law region of the I -V characteristics
C. it prevents frequency modulation
D. detector circuit is less complicated

31. The guide wavelength is measured by short-circuiting a waveguide and shifting the tunable probe along the slotted line to locate the voltage minima. If the shorting plate is replaced by a matched load, then

A. it would improve the accuracy of the measurement
B. the guide wavelength cannot be measured
C. it will give less accurate result, as no reflected wave is present
D. it will change the value of the guide wavelength

32. Which of the following condition should obeyed by every Laser

A. rate of stimulated emission should be greater than the rate of spontaneous emission.
B. there should be no spontaneous emission
C. rate of stimulated emission should be equal to the rate of spontaneous emission
D. there should be no stimulated emission

33. Which of the following fields are involved in a travelling wave magnetron?

A. The axial magnetic field only
B. The axial magnetic field and axial electric field
C. The axial magnetic field and the gap RF field
D. The axial magnetic field, direct electric field and gap RF field.

34. Beam load is high, if the velocity of electron is:

A. Very low B. Low
C. Medium D. High

35. 6 dB pad introduces

A. 6 dB attenuation B. 6 dB gain
C. 4 dB attenuation D. 4 dB gain

36. Which one of the following waveguide tuning components is not easily adjustable?

A. Screw B. Stub
C. Iris D. Plunger

37. When a desired mode is excited in a waveguide, there appears an extra component is the direction of propagation. The resulting mode is:

A. Transverse electric
B. Longitudinal
C. Transverse magnetic
D. Transverse electromagnetic

38. In a peak power measurement with the arrangement given in the figure, the oscilloscope measures the PRF to be 1 kHz and the pulse width as 2 μs. If the peak power to be measured is 50 kW and the power sensor has a rating of 10 mW average, then for protecting the sensor the minimum value of the attenuator R should be

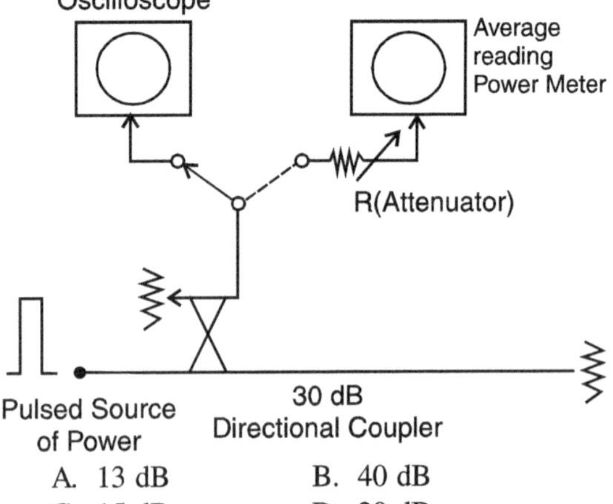

A. 13 dB B. 40 dB
C. 15 dB D. 20 dB

39. The parametric amplifier has ... gain and low noise figure

A. Medium B. Negligible
C. Low D. High

40. For some application, circular waveguide may be preferred to rectangular ones because of

A. The smaller cross-section
B. Lower attenuation
C. Freedom from spurious modes
D. Rotation

41. The following device uses an axial magnetic field and a radial electric field:

A. Coaxial magnetron
B. CFA
C. Reflex klyston
D. Travelling wave magnetron

42. Tropospheric scatter is used with frequencies in the following range:

A. HF B. VHF
C. UHF D. VLF

43. The flux density $\overline{B}$ in a ferromagnetic medium of large extent and permeability μ is as shown. Two cavities, cavity I shaped like a thin disc and cavity II long and needle like are cut. The ratio of H_1 to H_2 for air filled cavities is

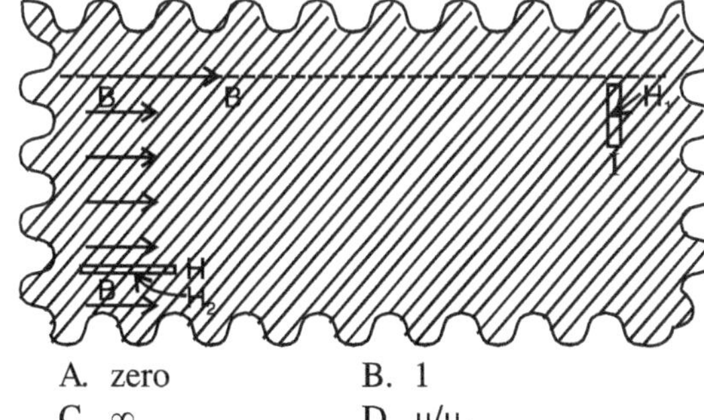

A. zero B. 1
C. ∞ D. μ/μ_0

44. The given figure shows an ideal 4-port circulator. The connections of

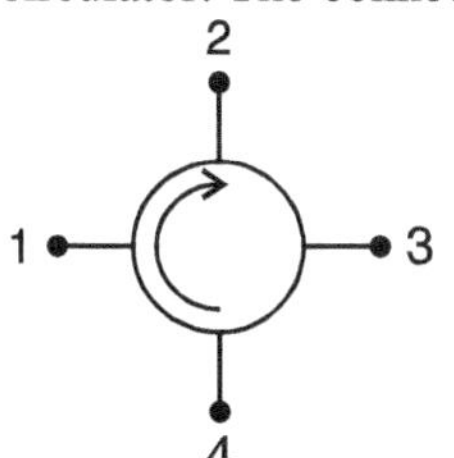

(a) microwave generator.
(b) synchronising master source for injection locking
(c) antenna, and
(d) receiver are

Codes:	*(a)*	*(b)*	*(c)*	*(d)*
A.	1	4	2	3
B.	1	2	3	4
C.	4	3	2	1
D.	2	4	1	3

45. The maximum unambiguous range in a radar system depends on

A. maximum power of the transmitter
B. pulse repetition frequency
C. width of the transmitted pulse
D. sensitivity of the radar receiver

46. What type of a device does the given figure microstrip pattern represent?

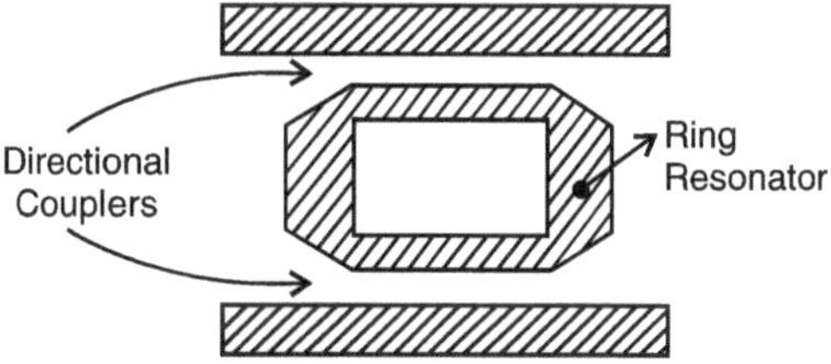

A. Balanced mixer
B. Hybrid ring
C. Channel dropping filter
D. Low-pass filter

47. One of the following microwave diodes is suitable for very low power oscillators only

A. TUNNEL B. LSA
C. GUNN D. IMPATT

48. The wavelength of a waveguide

A. Is greater than in free space
B. Depends only on the waveguide dimensions and the free space wavelength
C. Is inversely proportional to the phase velocity
D. Is directly proportional to the phase velocity.

49. TWT is characterized by :

A. Low noise figure, wide bandwidth and average gain
B. Gain exceeding 40 dBs, wide bandwidth and low noise figure
C. More noise and wide bandwidth
D. None of these

50. Cassegrain feed is used with a parabolic reflector to
- A. Increase the gain of the system
- B. Increase the beam width of the system
- C. Reduce the size of the main reflector
- D. Allow the feed to be placed at a convenient point.

51. Microwave links are preferred for TV transmission because
- A. They are free from impulse noise
- B. They produce less phase distortion
- C. They are cheaper
- D. They have large bandwidth

52. The following device is not used for pulse operation:
- A. CFT B. BWO
- C. TWT D. Multicavity klystron

53. If the molecular structure and density of the material are known, then which of the following material parameters may be determined from the ε_r versus $\frac{1}{T}$ plot?
- A. Sum of electronic and ionic polarizabilities
- B. Lattice Constant
- C. Breakdown voltage
- D. Loss tangent

54. When electromagnetic waves are reflected at an angle from a wall, their wavelength along the wall is ... than in the actual direction of propagation.
- A. Same B. Less
- C. Greater D. None of the above

55. An attenuator is used with TWT to :
- A. Prevent oscillations
- B. Increase gain
- C. Prevent saturation
- D. Help bunching

56. Waveguides are pressurised above normal atmospheric pressure for
- A. increasing their power handling capacity
- B. improving the conductivity of their walls
- C. preventing higher order modes from propagating
- D. varying the wave inpedance

57. A duplexer is used to
- A. couple two antennas to a transmitter without interference
- B. isolate the antenna from the local oscillator
- C. prevent interference between two antennas connected to a receiver
- D. use an antenna for receiption or transmission without interference

58. Klystron operate on the principle of
- A. Velocity modulation
- B. Amplitude modulation
- C. Frequency modulation
- D. Pulse modulation

59. Standard mismatching in microwave circuits have SWR from
- A. 0.5 : 1 to 2 : 1 B. 1 : 0 to 2 : 1
- C. 1.2 : 1 to 2 : 1 D. 1.33 : 1 to 2 : 1

60. A uniform plane wave travelling in a perfect dielectric is incident normally on the surface of a perfect conductor. Then
- A. the wave is transmitted into the conductor without attenuation
- B. 50% of the incident wave is transmitted and 50% is reflected
- C. a standing wave is set up in the conducting medium
- D. a standing wave is set up in the dielectric

61. Baretters and Bolometer are used in
- A. Measurement of microwave power
- B. VSWR
- C. Transmission losses
- D. None of these

62. In a low-level AM system, amplifiers following the modulated stage must be
- A. linear devices
- B. harmonic devices
- C. class C amplifiers
- D. nonlinear devices

63. Consider the following statements associated with microstrip patch antennas :
1. The microstrip patch behaves more like a leaky cavity rather than like a radiator and this is not a highly efficient antenna.
2. They are broadband antennas and are well-suited for mm-waves.
3. They can be adaped for radiation of circularly polarised waves.

Of these statements
- A. 1, 2 and 3 are correct

B. 1 and 2 are correct
C. 1 and 3 are correct
D. 2 and 3 are correct

64. A signal propagated in a waveguide has a full wave of electric intensity change between the two further walls, and no component of the electric field in the direction of propagation. The mode is

A. $TE_{1.1}$ B. $TE_{1.0}$
C. $TM_{2.2}$ D. $TE_{2.0}$

65. Match List I with List II and select the correct answer

List I	*List II*
(a) Propagation ______	1. $\left(\frac{2}{\omega \mu_0 \sigma}\right)^{1/2}$
(b) Skin depth	2. $\left(\frac{\omega \mu_0}{2\sigma}\right)^{1/2}$
(c) Sheet resistance	3. $(j \omega \mu_0 \sigma)^{1/2}$
	4. $\left(\frac{\omega \mu_0 \sigma}{2}\right)$

Codes:	*(a)*	*(b)*	*(c)*
A.	2	4	3
B.	3	1	2
C.	2	3	4
D.	3	2	1

66. When the modulation index of AM wave is doubled, antenna current is also doubled. The AM system being used is

A. A3H B. A5C
C. A3J D. A3

67. A magnetron whose oscillating frequency is electronically adjustable over a wide range is called a

A. coaxial magnetron
B. dither-tuned magnetron
C. frequency-agile magnetron
D. VTM

68. In a broadcast superheterodyne receiver the

A. local oscillator operates below the signal frequency
B. mixer input must be tuned to the signal frequency
C. local oscillator frequency is normally double the IF
D. RF amplifier normally works at 455 kHz above the carrier frequency

69. Match the following

Column I	*Column II*
(a) Ampere's Law	*(i)* $\nabla \times H = \frac{\partial D}{\partial t}$
(b) Faraday's Law	*(ii)* $\nabla . D = \rho$
(c) Gauss's Law	*(iii)* $\nabla \times E = -\frac{\partial B}{\partial t}$
(d) Free space	*(iv)* $\nabla \times H = J + \frac{\partial D}{\partial t}$

Codes:	*(a)*	*(b)*	*(c)*	*(d)*
A.	*(i)*	*(iv)*	*(ii)*	*(iii)*
B.	*(ii)*	*(iv)*	*(iii)*	*(i)*
C.	*(iii)*	*(i)*	*(ii)*	*(iv)*
D.	*(iv)*	*(iii)*	*(ii)*	*(i)*

70. Indicate which of the following diodes will produce the highest pulsed power output:

A. Varactor B. Gunn
C. Schottky barrier D. RIMPATT

71. Which one of the following quantities has the same dimension in both electromagnetic and electrostatic systems?

1. Current
2. Electric energy
3. Electric power

Select the correct answer using the codes given below:

Codes :

A. 1, 2 and 3 B. 1, and 2
C. 1, and 3 D. 2 and 3

72. Microwave antenna aperture efficiency depends on

A. feed pattern B. antenna aperture
C. surface losses D. low side lobe level

73. The positions of the probe for half power points in the slotted line of a microwave bench are interpreted to measure

A. guide wavelength
B. load impedance
C. source frequency
D. high SWR

74. The following figure shows the conductor pattern of a microstrip or strip-line version of a

microwave filter. This represents a

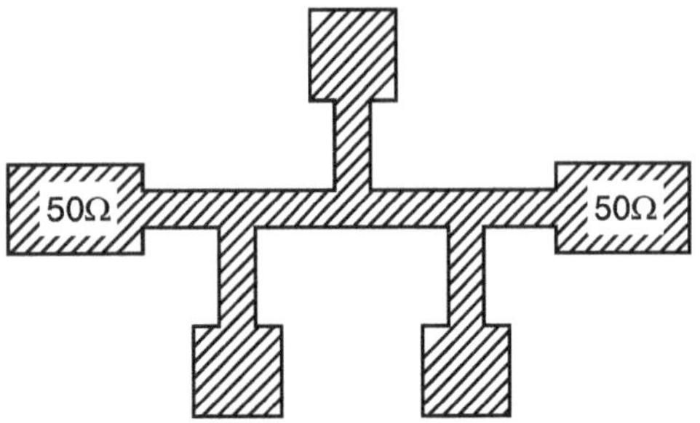

A. band pass filter
B. high pass filter
C. low pass filter
D. channel dropping filter

75. A microwave junction is supposed to be matched at all ports if in the S matrix
A. all the diagonal elements are zero
B. all the diagonal elements are equal but not zero
C. all the diagonal elements are complex
D. is Hermitian

76. The electron relaxation time of metal A is 2.7×10^{-4} secs., that of B is 1.37×10^{-4} secs. The ratio of resistivity of B to resistivity of A will be
A. 4.05 B. 2.0
C. 0.55 D. 0.25

77. Indicate which of three is different from the other
A. Schottky barrier diode
B. Backward diode
C. ESBAR diode
D. Hot carrier diode

78. A duplexer is used
A. To couple two different antennas to transmitter without mutual interference
B. To allow the one antenna to be used for reception or transmission without mutual interference
C. Both (A) and (B)
D. None of these

79. The dominant mode in a rectangular waveguide is
A. TE_{01} B. TM_{01}
C. TE_{10} D. TM_{11}

80. Match List-I with List-II and select the correct answer by using the codes given below the lists:

List-I	*List-II*
(a) Reflex klystron	1. High efficiency
(b) Magnetron	2. Slow wave structure
(c) TWT	3. Cross-field device
(d) Multi-cavity klystron	4. Oscillator

Codes:	*(a)*	*(b)*	*(c)*	*(d)*
A.	1	2	3	4
B.	4	3	2	1
C.	2	4	1	3
D.	3	1	4	2

81. In Gunn diodes, electrons are transferred from
A. high to low mobility energy bands
B. low to high mobility energy bands
C. valley to domain formation
D. domain to valley formation

82. For a noise figure of 25 and 2 and gain 20 the total noise figure is
A. 3.2 B. 6.4
C. 1.6 D. None of these

83. MASER finds application in
A. Opto electronics B. Fibre optics
C. Telephones D. All of the above

84. A helical antenna is used for satellite tracking because of its
A. Circular polarization
B. Broad BW
C. Good front to back ratio
D. None of these

85. The electric field lines and equipotential lines
A. are parallel to each other
B. are one and the same
C. cut each other orthogonally
D. can be inclined to each other at any angle

86. Two electrons moving parallel to each other with the same velocity will
A. attract each other if they move in the same direction
B. repel each other more strongly when moving in same direction than when moving in opposite directions.
C. repel each other less strongly when moving in same direction than when moving in opposite directions.
D. repel each other with the same force when moving in the same or opposite direction.

87. The term RADAR stands for
A. Radio Direction and Reflection
B. Radio Detection and Ranging
C. Radio waves Despatching and Receiving
D. Random Detection and Reradiation

88. Tees and junction constitute :
A. Single port network
B. Two port network
C. Three port network
D. Three or more port network

89. If D, G and η denote the directivity, power gain and efficiency of an antenna, then the following relation holds
A. $D = \eta G$ B. $D = G/\eta$
C. $\eta = DG$ D. $\eta^2 = DG$

90. In order to reduce cross-sectional dimension, the waveguide to use is
A. Circular B. Ridged
C. Rectangular D. Flexible

91. VOR stands for
A. VHF omni range
B. Visually operated radar
C. Voltage output of regulator
D. VLF omni range

92. Which microwave tube uses buncher and catcher cavities
A. Magnetron B. Klystron
C. Reflex klystron D. Travelling-wave tube

93. Indicate the noise whose source is in a category different from that of the other three?
A. Solar noise
B. Cosmic noise
C. Atmospheric noise
D. Galactic noise

94. A cavity wavemeter measures microwave frequencies because, the cavity
A. offers high impedance to microwaves
B. has more volume for microwaves to interact
C. will have a resonance with one frequency of the microwaves signal
D. offers a resistive load.

95. In the slotted line measurement for Standing Wave Radio (SWR), the distance between the two 3-dB points is Δd and λ' is the wavelength along the transmission line. SWR of the transmission system is given by
A. $\frac{\pi \Delta d}{\lambda'}$ B. $\frac{\pi \lambda'}{\Delta d}$
C. $\frac{\lambda'}{\pi \Delta d}$ D. $\frac{\Delta d}{\pi \lambda'}$

96. The wavelength for microwave frequencies are
A. More than 1 m
B. Less than 1 mm
C. From 1 mm to 1 m
D. 0.3 to 300 m

97. Magnetostriction is
A. induced magnetism
B. volume changes accompanying magnetization
C. rate of magnetisation
D. rate of loss of magnetism

98. When n_e and n_h are the electron and hole densities, and μ_e and μ_h are the carrier mobilities, the Hall coefficient is positive when
A. $n_h \mu_h > n_e \mu_e$ B. $n_h \mu^2_h > n_e \mu^2_e$
C. $n_h \mu_h < n_e \mu_e$ D. $n_h \mu^2_h < n_e \mu^2_e$

99. Joule/Coulomb is the unit of
A. Electric field potential
B. Potential
C. Charge
D. None of the above

100. The following diode does not use negative resistance in its operation:
A. Gunn diode B. Tunnel diode
C. Backward diode D. IMPATT diode

101. A reflex klystron is capable of operating at such high frequencies as:
A. 1 GHz B. 10 GHz
C. 80 GHz D. 5 GHz

102. In the figure below, A, B and C are equal charges. When a unit charge moves from 'C' to 'O', the force on the charge when it reaches will be

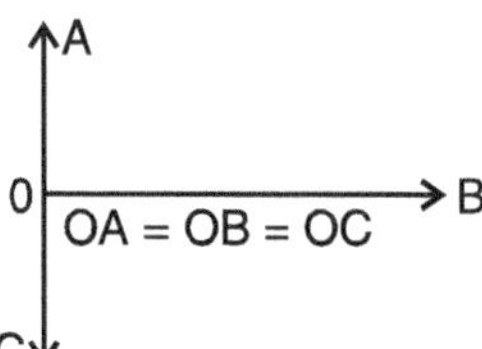

A. zero
B. a pull in the direction of A

C. a pull in the direction of B

D. none of the above

103. Which of the following antenna is best excited from a waveguide?

A. Horn B. Discone

C. Helical D. Biconical

104. The ratio of the charges stored by two metallic spheres raised to the same potential is 6. The ratio of the surface areas of the spheres is

A. 6 B. $\frac{1}{6}$

C. 36 D. $\frac{1}{\sqrt{6}}$

105. A helical antenna is used for satellite tracking because of its

A. Broad bandwidth

B. Manoeuverability

C. Circular polarization

D. Good front-to-back ratio

106. The following is not used as a microwave mixer or detector:

A. Crystal diode

B. PIN code

C. Schottkty barrrier diode

D. Backward diode

107. The idler frequency of a nondegenerate parametric amplifier with input frequency f_i and pump frequency f_p is:

A. $f_p - f_i$ B. $f_i - f_p$

C. $f_i + f_p$ D. $(f_i + f_p)/2$

108. The reflex klystron is often preferred to the two-cavity klystron oscillator because the former

A. Is more efficient

B. Has higher number of modes

C. Is easier to tune

D. Produces a higher output power

109. One of the following microwave diodes is suitable for very low-power oscillator only

A. Tunnel B. Gunn

C. IMPATT D. LSA

110. Microwave links are preferred for TV transmission because:

A. They produce less phase distortion

B. They are cheaper

C. They have no impulse noise

D. Of small S/N ratio.

111. Match List I with List II and select the correct answer using the codes given below the lists:

List I	***List II***
(a) Resistance	1. $M^{-1} L^{-2} T^2 I^{-2}$
(b) Inductance	2. $ML^2 T^{-2} I^2$
(c) Capacitance	3. $M^{-1} L^{-2} T^4 I^2$
(d) Reluctance	4. $ML^2 T^{-2} I^{-1}$

Codes :	*(a)*	*(b)*	*(c)*	*(d)*
A.	1	2	3	4
B.	4	2	3	1
C.	2	1	3	4
D.	4	2	3	1

112. If the transient during switching of a power supply are to be studied, which oscilloscope will be preferred?

A. An ordinary oscilloscope with high frequency sweep generator

B. Dual beam oscilloscope

C. Dual trace oscilloscope

D. Storage oscilloscope

113. A magic tee is nothing but a modification of

A. H-plane tee

B. E-plane tee

C. Two E-plane tees

D. A combination of E-plane and H-plane tee.

114. TWT is basically:

A. An oscillator

B. Tuned amplifier

C. Wideband amplifier

D. None of these

115. Which of the following does not have the same units as the others? The symbols have their usual meanings.

A. $\frac{L}{R}$ B. RC

C. $\sqrt{LC}$ D. $\frac{1}{\sqrt{LC}}$

116. Maxwell's divergence equation for the magnetic field is given by

A. $\overline{\nabla} \times \overline{B} = 0$ B. $\overline{\nabla}.\overline{B} = 0$

C. $\overline{\nabla} \times \overline{B} = P$ D. $\overline{\nabla}.\overline{B} = P$

117. The primary purpose of the helix in a travelling wave tube is to

A. Reduce the noise figure

B. Ensure broadened operation

C. Reduce the axial velocity of the RF field
D. Prevent the electron beam from spreading in the long tube

118. The Lorentz force law is given by
A. $\vec{F} = Q\vec{E}$
B. $\vec{F} = \vec{V} \times \vec{B}$
C. $\vec{F} = Q\left(\vec{E} + \vec{V} \times \vec{B}\right)$
D. $\vec{F} = Q\left(\vec{V} \times \vec{B}\right)$

119. If the transmission bandwidth is doubled in FM, then the SNR is
A. also doubled
B. improved four-fold
C. decreased by one fourth
D. unaffected

120. Zoning is used with a dielectric antenna in order to
A. reduce the bulk of the lens
B. increase the bandwidth of the lens
C. permit pin-point focussing
D. correct the curvature of the wavefront from a horn that is too short

121. When the polarisation of the receiving antenna is unknown, to ensure that it receives atleast half the power (except in one particular situation) the transmitted wave should be
A. horizontally polarised
B. vertically polarised
C. circularly polarised
D. elliptically polarised

122. The demarcation between Fresnel region and Fraunhoffer region fairly approximated by minimum distance between antennas for pattern measurement is (given that R = distance between antennas, D = largest aperture dimension, and λ = wavelength).
A. $R = \lambda/2D^2$ B. $R = \lambda^2/2D^2$
C. $R = 2D^2/\lambda$ D. $R = 2D^2/\lambda^2$

123. One of the following is unlikely to be used as a pulsed device, it is the
A. multicavity klystron
B. BWO
C. CFA
D. TWT

124. Identity the waveguide tuning component which is not easily adjustable?
A. Iris B. Plunger
C. Screw D. Stub

125. A reflex Klystron function as:
A. Microwave amplifier
B. Microwave oscillator
C. A high gain cavity
D. Both as amplifier and as oscillator

126. The envelope detector is a/an
A. synchronous detector
B. asynchronous detector
C. product demodulator
D. coherent detector

127. Which one of the following statements regarding the following signal
$x(t) = 5 \sin(2\pi \times 10^3 t) \sin(2\pi \times 10^6 t)$ is incorrect?
A. The upper side band frequency is 1001000
B. The lower side band frequency is 999000
C. The carrier amplitude is 5
D. $x(t)$ is a DSB-SC signal

128. A 75 ohm transmission line is first short terminated and the minima location are noted. When the short is replaced by a resistive load R_L, the minima locations are not altered and the VSWR is measured to be 3. The value of R_L is given by
A. 75 ohms B. 500 ohms
C. 225 ohms D. 250 ohms

129. Microwave link repeaters are typically 50 km apart
A. Because of earth's curvature
B. Because of atmospheric attenuation
C. Because of output tube power limitations
D. To ensure that the applied dc voltage is not excessive.

130. Tees and junctions are
A. Three port network
B. Two port network
C. Single port network
D. Three port or more

131. Standard mismatches are built with standing wave ratios from
A. 1.2 : 1 to 2 : 1 B. 0.5 : 1 to 2 : 1

C. 1.0 : 1.5 to 2 : 1 D. None of these

132. Delay distortion is transmission lines exists when

A. different frequency components have different attenuation
B. alpha is a constant
C. beta is proportional to frequency
D. beta is a constant

133. A screw protruding into a waveguide constitutes:

A. Capacitive discontinuity
B. Inductive discontinuity
C. Capacitive or inductive discontinuity depending on its position inside the waveguide
D. Makes no difference in the waveguide.

134. Match List I (type of antenna) with List II (communication system using such an antenna) and select the correct answer using the codes given below the lists :

List I	*List II*
(a) Turnstile	1. RADAR
(b) Yagi	2. TV transmission
(c) Dish	3. MW broadcasting
(d) Vertical mast	4. TV reception

Codes :	*(a)*	*(b)*	*(c)*	*(d)*
A.	4	2	3	1
B.	2	4	1	3
C.	2	4	3	1
D.	4	2	1	3

135. The shadow mask in a colour picture tube is used to

A. reduce *x*-ray emission
B. ensure that each beam hits only its own dots
C. increase screen brightness
D. provide degaussing for the screen

136. A radio wave is incident on a layer of ionosphere at an angle of 30 degrees with the vertical. If the vertical frequency is 1.2 MHz, the maximum usable frequency is

A. 1.2 MHz B. 2.4 MHz
C. 0.6 MHz D. 1.386 MHz

137. For a transmission line terminated by a load, the reflection co-efficient magnitude $|\Gamma|$ and the voltage standing wave ratio S are related as

A. $S = 1/(1+|\Gamma|)$
B. $S = 1/(1-|\Gamma|)$
C. $S = (1-|\Gamma|)/(1+|\Gamma|)$
D. $S = (1+|\Gamma|)/(1-|\Gamma|)$

138. The Hysteresis phenomenon in Schmitt Trigger has, which of the following features :

1. It is due to negative feedback.
2. In the inverted mode, circuit triggers at a higher voltage for increasing signals than for decreasing signals.
3. Noise signal voltages greater than V_n are cut-off, where V_n is the Hysteresis voltages.
4. Slowly varying signals are converted to output waveform with abrupt changes at two specified levels of input signals.

Of these four statements, the true statements are:

A. 2 and 3 only B. 2 and 4 only
C. 2, 3 and 4 only D. 3 and 4 only

139. Which of the following are useful in comparing the noise performance of receivers?

1. Input noise voltage
2. Equivalent noise resistance
3. Noise temperature
4. Noise figure

Select the correct answer using the codes given below

Codes :

A. 1, 2 and 3 B. 1, 3 and 4
C. 2, 3 and 4 D. 1, 2 and 4

140. In order to couple two generators to a waveguide system without coupling them to each other, one could not use a

A. Rat race B. E plane T
C. Hybrid ring D. Magic T.

141. The skin depth (δ) due to RF current is equal to (ω = angular frequency, μ = permeability and σ = conductivity)

A. $\left(\frac{2}{\omega\mu\sigma}\right)^2$ B. $\frac{2}{\sqrt{\omega\mu\sigma}}$
C. $\left(\frac{2}{\omega\mu\sigma}\right)^{1/2}$ D. $\left(\frac{\sqrt{2}}{\omega\mu\sigma}\right)^{1/3}$

142. In any electromagnetic region,

A. $\vec{F} = q\left(\vec{E} + \vec{V}.\vec{B}\right)$

B. $\vec{F} = q\left(\vec{E} + \vec{V} \times \vec{B}\right)$

C. $\vec{F} = q\left(\vec{B} + \vec{V} \times \vec{E}\right)$

D. $\vec{F} = q\left(\vec{B} + \vec{V} . \vec{E}\right)$

143. The following figure represents the basic arrangement of a/an

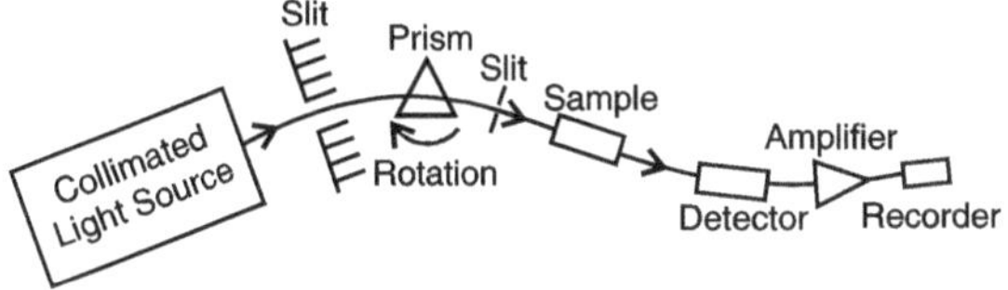

A. interferometer
B. absorption spectrometer
C. refractometer
D. photometer

144. Periodic permanent magnet focussing is used with TWT to:

A. Avoid bulky electromagnet
B. Allow pulsed operation
C. Improve electron bunching
D. Allow coupled cavity operation at the higher frequencies.

145. Baretters and bolometers are used for measurement of:

A. VSWR
B. Transmission losses
C. Microwave power
D. None of these

146. Which of the following is not a TWT slow-wave structure?

A. Ring bar
B. Coupled cavity
C. Periodic permanent magnet
D. Helix.

147. The TWT is sometimes preferred to the magnetron as a radar transmitter output tube because it is

A. Less noisy
B. More broad band
C. Capable of a longer duty cycle
D. A more efficient cycle.

148. Match List I with List II and select the correct answer

List I	***List II***
(a) Reflex klystron	1. Wide band amplification
(b) TWT	2. Very low level
(c) 2-cavity klystron	3. Low power FM wave generation
(d) MASER	4. Frequency

Codes :	*(a)*	*(b)*	*(c)*	*(d)*
A.	4	3	2	1
B.	1	2	3	4
C.	3	4	1	2
D.	3	1	4	2

149. With 100 mW input power, the power measured at port 4 is 0.1 mW and 10 mW when port is terminated in a matched load and a short, respectively. The directivity of the coupler shown

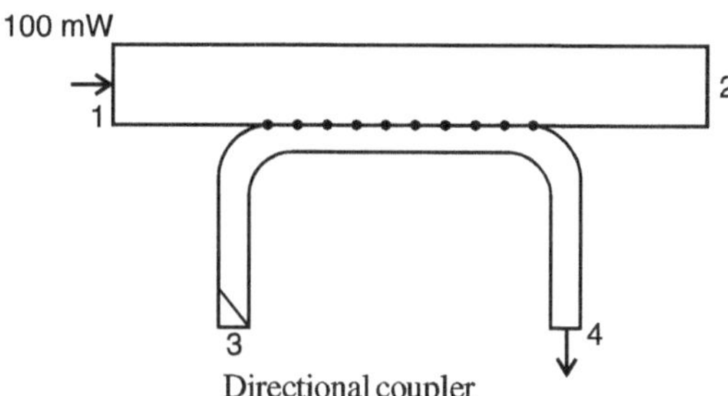

Directional coupler

A. 15 dB
B. 40 dB
C. 20 dB
D. 20 dB

150. The advantage of microwave over lower frequency signal is

A. Increased bandwidth
B. Ability to use high gain
C. Increased secrecy
D. All of the above

151. Non-degenerate one port parametric amplifier should have a high ratio of pump to signal frequency because this:

A. Yields a low noise figure
B. Reduces the pump power requirement
C. Permits satisfactory low power operation
D. Permits satisfactory high power operation

152. The principle of bridge duplex working involves

A. potential balance
B. current balance
C. resistance balance

D. impedance balance

153. A master RF amplifier is not really suitable for

A. Radioastronomy
B. Satellite communication
C. Radar
D. Traposcatter receivers.

154. The attenuator is used in the travelling wave tube to

A. Increase gain
B. Help bunching
C. Prevent saturation
D. Prevent oscillations

155. If the receiving antenna is polarized at 90° with respect of transmitting antenna it will receive

A. No signal B. Maximum signal
C. Minimum signal D. None of these

156. Which of the following antennas is best excited from waveguide?

A. Biconical B. Horn
C. Helical D. Discone

157. A magnetron for which the oscillating frequency is electronically adjustable over a wide range is known as

A. VTM
B. Coaxial magnetron
C. Either tuned magnetron
D. Frequency-agile magnetron

158. To avoid difficulties with strapping high frequencies, the type of cavity-structure in the Magnetron is the

A. Hole and slot B. Slot
C. Vane D. Rising sun

159. Indicate the false statement. Klystron amplifiers may use intermediate cavities to

A. prevent the oscillations that occur in two-cavity klystrons.
B. increase the bandwidth of the device
C. improve the power gain
D. increase the efficiency of the klystron

160. A backward wave oscillator is based on the

A. Coaxial magnetron
B. Travelling wave tube
C. Rising sun magnetron
D. Crossed field amplifier

161. A small current carrying circular loop lies on a vertical plane. P is a point of observation on the orthogonal vertical plane passing through the centre of the loop and θ is the angle between a horizontal line through the centre of the loop O and OP as shown. The θ-component of the H-field will vary as

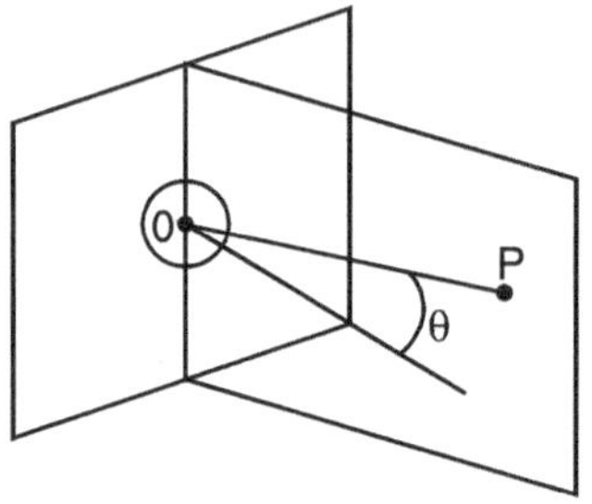

A. $\cos\theta$ B. $\cot\theta$
C. $\cos^2\theta$ D. $\sin^2\theta$

162. The scattering matrix of a magic-tee, shown on the given figure is

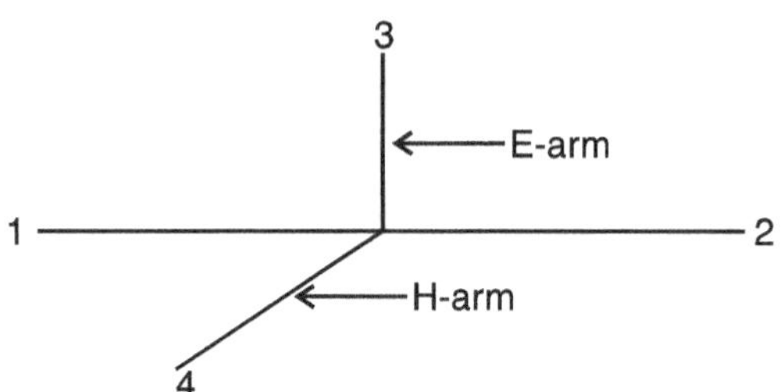

A. $\frac{1}{\sqrt{2}}\begin{bmatrix} 0 & 0 & -1 & 1 \\ 0 & 0 & -1 & 1 \\ 1 & -1 & 0 & 0 \\ 1 & -1 & 0 & 0 \end{bmatrix}$

B. $\frac{1}{\sqrt{2}}\begin{bmatrix} 0 & 0 & 1 & -1 \\ 0 & 0 & -1 & 1 \\ 1 & -1 & 0 & 0 \\ -1 & 1 & 0 & 0 \end{bmatrix}$

C. $\frac{1}{2}\begin{bmatrix} 0 & 0 & 1 & 1 \\ 0 & 0 & -1 & 1 \\ 1 & -1 & 0 & 0 \\ 1 & 1 & 0 & 0 \end{bmatrix}$

D. $\frac{1}{2}\begin{bmatrix} 0 & 0 & 1 & -1 \\ 0 & 0 & 1 & 1 \\ 1 & 1 & 0 & 0 \\ -1 & 1 & 0 & 0 \end{bmatrix}$

163. Waveguide may be considered as :
A. High pass filter B. Low pass filter
C. Band pass filter D. Band reject filter

164. Klystron as compared to magnetron
A. Low pulling figure
B. More pulling figure
C. Equal pulling figure
D. Can't be compared

165. A MASER R.F. amplifier is not truly suited for use in:
A. Radar
B. Radio astronomy
C. Satellite communication
D. Troposcatter receiver.

166. In loop coupling the plane of the loop is ... to the flux lines
A. Vertical B. Parallel
C. Normal D. Inverse

167. For a microwave transistor to operate at the highest frequencies, the (indicate the false answer)
A. collector voltage must be large
B. collector current must be high
C. base should be thin
D. emitter are must be large

168. Varactor diode is useful at microwave frequencies:
A. For frequency multiplication
B. As an oscillator
C. As a parametric amplifier
D. For electronic tuning

169. A reflex Klystron
A. Only one cavity as both the buncher and catcher
B. Two cavity one for buncher, second for catcher
C. Three cavity, two for buncher, one for catcher
D. No cavity at all

170. Which of the following is a unit of loudness?
A. mel B. decibel
C. sone D. phon

171. Match the following:

Column I (Navigational Aid)	*Column II (Frequency)*
(*a*) Loran – C	1. 100 kHz continuous wave
(*b*) Omega	2. 300 kHz
(*c*) Consol	3. 100 kHz pulsed
(*d*) Decca navigator	4. 10.2 kHz
	5. 400 MHz

Codes :	(*a*)	(*b*)	(*c*)	(*d*)
A.	1	2	4	5
B.	5	1	3	4
C.	4	1	5	4
D.	3	4	2	1

172. A carbon microphone is a variable
A. resistance B. inductance
C. capacitance D. all of the above

173. The ruby maser has been preferred to the ammonia maser for microwave amplification because the former has
A. A much greater bandwidth
B. A better frequency stability
C. A lower noise figure
D. No need for a circulator

174. A ruby maser is preferred to ammonia masser for microwave amplification because it has:
A. Much greater bandwidth
B. Lower noise figure
C. Better frequency stability
D. No need for a circulator

175. The multicavity klystron
A. Is not a good low-level amplifier because of noise
B. Has high repeller voltage to ensure a rapid transit time
C. Is not suitale for pulsed operation
D. Needs a long transit time through the buncher cavity to ensure current modulation

176. A varactor diode may be useful at microwave frequencies (Indicate the false answer)
A. For electronic tuning
B. For frequency multiplication
C. As an oscillator
D. As a (parametric) amplifier

177. Magic tee is a combination of :
A. One H-plane tee and one E-plane tee

B. 2 H-plane tees and one E-plane tees
C. One H-plane tee and two E-plane tees
D. None of these

178. When microwave signals follow the curvature of the earth, this is known as
A. The Faraday effect
B. Ducting
C. Tropospheric scatter
D. Ionospheric reflection

179. Which of the following is the typical range for multicavity klystron?
A. 250 MHz to 50 GHz
B. 100 MHz to 250 MHz
C. 100 kHz to 250 kHz
D. 50 kHz to 50 MHz

180. Fig. shows the resistance versus power characteristics of a

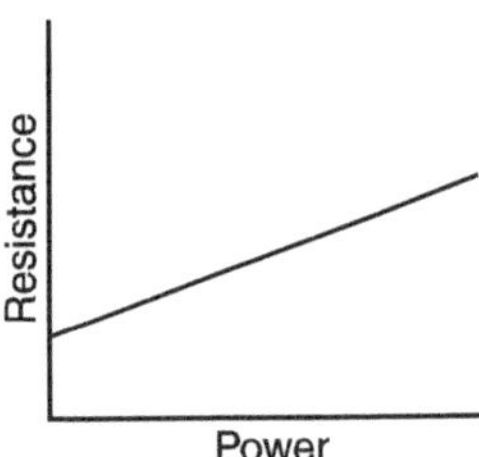

A. thermister B. LDR
C. baratter D. photodiode

181. A magnetic field is used in the cavity magnetron to
A. Ensure that the oscillations are pulsed
B. Prevent anode current in the absence of oscillations
C. Ensure that the electrons will orbit around the cathode
D. Help in focusing the electron beam, thus preventing spreading.

182. Which of the following statement regarding electric flux is true?
1. Electric flux begins on positive charges and terminates on negative charges
2. Flux is in the same direction as the electric field $\overline{E}$
3. Flux density is proportional to the magnitude of $\overline{E}$
4. In the SI system of units, total flux emanating from a charge of Q(C) is Q(C). A single line will emanate from 1 C of charge.

A. 1 only B. 1 & 2 only
C. 1, 2 & 3 only D. 1, 2, 3 & 4

183. Cassegrain feed is used with a parabolic reflector to:
A. Increase the beam width of the system
B. Increase the gain of the system
C. Allow the feed to be placed at a convenient point
D. Reduce the size of the main reflector

184. Gallium arsenide is preferred to silicon for use in Gunn diode because it has:
A. Lower noise at high frequencies
B. High ion mobility
C. Suitable empty energy band which silicon does not have
D. Better frequency stability

185. A Gunn diode can replace reflex klystron in
I. Signal generators
II. Frequency modulated power oscillators
III. Parametric amplifier pumps

A. 1 and 2 B. 2 and 3
C. 1 and 3 D. 1, 2 and 3

186. The transferred electron bulk effect occurs in
A. Germanium
B. Gallium arsenide
C. Silicon
D. Metal semiconductor function

187. Match the following:

***Column I* (*Frequency Band*)**	***Column II* (*Frequencies*)**
(*a*) UHF	1. 500 MHz
(*b*) SHF	2. 200 MHz
(*c*) VHF	3. 20 GHz
(*d*) VLF	4. 20 kHz
	5. 10 MHz
	6. 500 kHz

Codes :	(*a*)	(*b*)	(*c*)	(*d*)
A.	1	3	4	6
B.	1	3	2	4
C.	1	4	6	5
D.	6	4	5	3

188. Which of the following word cannot be followed by the word "wave guide"?
A. Ridged B. Coaxial
C. Flexible D. Elliptical

189. Match the following:

Column I	*Column II*
(Display)	*(Applications)*
(a) Type – A	1. For elevation of a target
(b) Type – B	2. Range and azimuth bearing
(c) Type – C	3. Giving information about other ships around
(d) Type – D	4. Range estimation

Codes:	*(a)*	*(b)*	*(c)*	*(d)*
A.	1	2	3	4
B.	4	3	2	1
C.	4	1	3	2
D.	2	4	1	3

190. For a given average power the peak output power of a ruby laser may be increased by

A. Using cooling
B. Using Q-spoiling
C. Increasing the magnetic field
D. Dispensing with the Fabry Perot resonator

191. Travelling wave parametric amplifier is used to:

A. Avoid need for cooling
B. Provide greater gain
C. Prevent noise feedback
D. Provide greater bandwidth

192. A quarter-wave stub shorted at the end has high impedance

A. At shorted end
B. $\frac{\lambda}{4}$ back from the short
C. $\frac{\lambda}{2}$ back from the short
D. At the centre

193. For measurement of high values of VSWR, the method to be used is:

A. Single minima method
B. Double minima method
C. Either *(a)* or *(b)*
D. None of these

194. Indicate the false statement
Klystron amplifiers may use intermediate cavities to

A. Prevent the oscillations that occur in two-cavity klystron
B. Improve the power gain
C. Increase the efficiency of the klystron
D. Increase the bandwidth of the device

195. Helix is used in TWT

A. reduces the bandwidth
B. acts as a coaxial transmission line
C. reduce inductance/unit length
D. increase phase velocity

196. The glass tube of a TWT may be coated with a quadag to

A. Help focussing
B. Provide attenuation
C. Improve bunching
D. Increase gain.

197. Which of the following microwave tube is based on the principle of velocity modulation?

A. TWT B. Pencil triode
C. Magnetron D. Klystron

198. A 100 watt incandescent lamp radiates all of its energy in the form of $E = E_0 \sin \frac{2\pi}{\lambda}(vt - x)$;

$$H = H_0 \sin \frac{2\pi}{\lambda}(vt - x),$$

where, E, E_0, v, H, H_0 and λ have usual meanings. The energy is emitted uniformly in all directions. The maximum electric and magnetic intensities at distance of 200 cm from the lamp are

A. 38.6 V/m, Zero
B. 38.6 V/m, 0.103 amp/m
C. Zero, 0.103 amp/m
D. Zero, Zero

199. The transmission loss for a 3 GHz microwave system over a certain distance is 130 dB. If the frequency is now doubled, then the transmission loss will be

A. 136 dB B. 133 dB
C. 127 dB D. 139 dB

200. The precision of an instrument indicates its ability to reproduce a certain reading with a given

A. drift B. resolution
C. shift D. consistency

201. Which one of the following quantities is analogous to permeability in magnetic circuits?

A. Current sensitivity
B. Conductivity in electrical circuits
C. Resistance in electrical circuits
D. Consistency

202. The main advantage of the two hole directional

coupler is

A. Low directional coupling

B. Poor directivity

C. High SWR

D. Narrow bandwidth

203. A signal propagated in a waveguide has a full wave of electric intensity change between the two further walls and no component of the electric field in the direction of propagation. The mode is

A. $TE_{1.1}$ B. $TE_{1.0}$

C. $TM_{2.3}$ D. $TE_{2.0}$

204. In a rectangular cavity resonator having *a* = width, *b* = height and *d* = length for *a* > b < *d*, the dominant mode is

A. TE_{101} B. EM_{101}

C. TE_{110} D. TM_{110}

205. Two linearly polarised plane waves A and B travelling in free space in the positive X direction meet at a point. The wave A is polarised in the Y direction and the wave B is polarised in the Z direction. The resulting wave is linearly polarised. Therefore, A and B.

A. must have equal amplitudes and must also be in time phase with each other

B. must have equal amplitudes but need not be in time phase

C. may have different amplitudes but must be in phase with each other in time

D. may have different amplitudes and may also differ in time phase

206. The biggest disadvantage of the IMPATT diode is its

A. Low efficiency than the other microwave diodes

B. Low power handling ability

C. High noise

D. Inability to provide pulsed operation

207. Transferred electron bulk effect occurs in:

A. Gallium assenide

B. Germanium

C. Silicon

D. None of these

208. Two isotropic antennas are separated by a distance of two wavelengths. If both the antennas are with currents of equal phase and magnitude, the number of lobes in the radiation pattern in the horizontal plane is

A. 2 B. 4

C. 6 D. 8

209. Backward wave oscillator is based on:

A. Crossed field amplifier

B. Coaxial magnetron

C. Travelling wave tube

D. Rising sun magnetron

210. The largest application of microwaves is in

A. Industrial heating

B. Research

C. Cooking

D. Communications

211. Wave guide can be considered as

A. High pass filter B. Low pass filter

C. Band pass filter D. None of these

212. If all television satellites use the same 3.7 to 4.2 GHz band for downlink signals, how can an earth station select one satellite?

A. Tuning the waveguides

B. Narrow-beam receiving antenna

C. Both A and B

D. None of above

213. Which of the following is not used as a pulsed device?

A. TWT B. CFA

C. BWO D. Multicavity klystron

214. The behaviour of which of the following device is governed by bulk effect?

A. PIN code B. Gunn diode

C. Tunnel diode D. IMPATT diode

215. The cavity structure in magnetron in order to avoid difficulties with stopping high frequencies, is

A. Vane B. Slot

C. Hole and slot D. Rising sun

216. Cathode material used for magnetron is:

A. Tungsten

B. Thoriated tungsten

C. Thorium oxide

D. Barium strontium oxide.

217. Which one of the following wave guide tuning components is not easily adjustable

A. Screw B. Stub

C. Iris D. Plunger

218. For a two open-wire transmission line excited

by a harmonically oscillating source with 3x p (– jωt) as the time factor, the voltage on the transmission line will satisfy which one of the following relations? (L and C are the inductance and capacitance per unit length of the line respectively).

A. $\frac{dV(x)}{dx} = -j\,\omega LI(x)$

B. $\frac{dV(x)}{dx} = j\,\omega LI(x)$

C. $\frac{dV(x)}{dx} = j\,\omega CI(x)$

D. $\frac{dV(x)}{dx} = -j\,\omega CI(x)$

219. Consider the following statements about a transmission line having non-zero parameters R, L, G, and C:

1. By proper choice of the parameters, the line can be made lossless even though R and G are non-zero.
2. The line will have both amplitude and phase distortion in general.

Of these statements

A. 1 alone is true
B. 2 alone is true
C. both 1 and 2 are true
D. both 1 and 2 are false

220. An earth station receiver has a signal of 4×10^{-12} W across 50 Ω input. What is the signal level, in microvolts?

A. 1.6 B. 14.14
C. 338 D. 1000

221. TEM mode in a coaxial line:

A. Has no cutoff wavelength
B. Disallows d.c. current
C. both A and B
D. None of these

222. Materials generally preferred for wave guides are

A. Brass and aluminium
B. Cast iron and steel
C. Non-metallic solids including plastics
D. High carbon steel and Vanadium steel.

223. Which of the following is directly measured in sweep reflectometry?

A. SWR
D. Impedance
C. Reflection coefficient
D. Return loss

224. For a directional coupler, the power is in the ratio of

A. 40 dB B. 30 dB
C. 20 dB D. 10 dB

225. In an optical fibre, the refractive index of the cladding material should be

A. nearly unity
B. very low
C. less than that of the core
D. more than that of the core

226. A screw projecting in a wave guide is

A. Inductive discontinuity
B. Capacitive discontinuity
C. Either of above
D. None of the above

227. For matching purpose, waveguide tapers are preferred to transformers because of :

A. Low cost B. Ease of fabrication
C. Durability D. None of these

228. Which of the following uses an axial magnetic field and a radial electric field?

A. CFA B. BWO
C. Reflex klystron D. Coaxial magnetron

229. To couple two generators to a waveguide without coupling them to each other, we can use a:

A. E plane T B. Hybrid ring
C. Magic T D. Rat race.

230. The standard reference antenna for the directive gain is:

A. Halfwave dipole
B. Infinite single dipole
C. Elementary dipole
D. Isotropic antenna

231. Which of the following diode is different from the remaining?

A. Backward diode
B. Hot carrier diode
C. ESBAR diode
D. Schottky barrier diode

232. If 1 watt of RF power is fed to a directional

coupler having 30 dB coupling, the power available at the coupled port is

A. $\frac{1}{30}$W B. $\frac{1}{10}$W

C. $\frac{1}{100}$W D. $\frac{1}{1000}$W

233. A ruby maser must be cooled:

A. Because it generates a lot of heat

B. Because it cannot operate at room temperature

C. To improve bandwidth

D. To improve noise figure

234. The mouth diameter of two parabolic reflectors are in the ratio of 2 : 1. Their power gains with respect, to a resonant half-wave dipole will be in the ratio of

A. 2 :1 B. 1 : 4

C. 4 : 1 D. 1 : 2

235. The beamwidth between nulls of 140 cm parabolic reflector used at 6 GHz is

A. 2.5 degrees B. 5 degrees

C. 10 degrees D. 7.5 degrees

236. In the *x*-band, for best low level noise performance, an amplifier should use:

A. Bipolar junction transistor

B. FET

C. IMPATT diode

D. Gunn diode

237. Large MW (microwave) power may be measured with a

A. Thermistor B. Baretter

C. Calorimeter D. Bolometer

238. A magnetron is virtually a

A. High power amplifier

B. Low power amplifier

C. High power oscillator

D. Low power oscillator

239. The TWT is sometimes preferred to the multicavity klystron amplifier, because the former

A. Is more efficient

B. Has a greater bandwidth

C. Has a higher number of modes

D. Produces a higher output power

240. The ratio arm is used in bridge method of measurement basically

A. To speed up the response time for balancing

B. To reduce the error due to the errors in known resistances

C. To simplify the calculations

D. To enhance the range of measurement

241. Find the shunt needed to extend the range of a 50 Ω 1 mA movement to *(i)* 2 mA, *(ii)* 10 mA

A. 50, 5.5 Ω B. 50, 5 Ω

C. 50, 40 Ω D. 5.5, 5.5 Ω

242. A meter has a full-scale angle of 90° at a current of 1 A. This meter has perfect square-law response. What is the current when the deflection angle is 45°?

A. 0.5 A B. 0.25 A

C. 0.707 A D. 0.67 A

243. Consider a parallel plate capacitor being charged. Identify the incorrect statements

A. Energy stored in the capacitor does not enter it through the connecting wires but through the space around the wires and plates of capacitor.

B. Rate at which energy flows into this volume is equal to the integration of the Poynting vector over the boundary of the volume between the plates

C. The Poynting vector points everywhere radially outward of the volume between the plates.

D. The Poynting vector points everywhere radially into the volume between the plates

244. As a result of reflections from a plane conducting wall, electromagnetic waves acquire an apparent velocity greater than the velocity of light in space. This is known as

A. Phase velocity

B. Group velocity

C. Normal velocity

D. Velocity of propagation

245. A PIN diode is

A. A metal semiconductor point-contact diode

B. A microwave mixer diode

C. Often used as a microwave detector

D. Suitable for use as a microwave switch

246. Conisder the following statements regarding a phase-sensitive detector:

1. The detector reads zero when the frequencies of input signal and reference signal are different.
2. The detector acts as a linear rectifier
3. The detector can be used as a vector voltmeter.

Of these statements

A. 1, 2 and 3 are correct
B. 1 and 3 are correct
C. 2 and 3 are correct
D. 1 and 2 are correct

247. The wavelength of microwaves at 100 GHz will be

A. 3 cm B. 0.3 cm
C. 0.03 m D. 0.3 m

248. The local action occurs in the primary cell

A. When the cell is delivering a small current for a long duration
B. When the cell is delivering a large current even for a small duration
C. When the cell is delivering a large current for long duration
D. Wheather or not the cell is delivering a current

249. Various fields involved in TWT operation are

A. An axial magnetic field and a beam of electrons
B. An axial magnetic field, a beam of electrons and RF signal
C. A beam of electron and the RF gap signal
D. None of the above

250. Five isotropic elements placed along the x-axis are fed by in-phase currents of equal amplitudes. The spacing between successive elements is $\lambda/4$. Consider the following statements about the directivity pattern of the array :

1. The array will have a maximum in the direction of Y-axis.
2. The field strength in the + X-axis direction will be Zero.
3. The directivity pattern will have minor lobes.

Of the statements given above

A. 1, 2 and 3 are correct
B. 1 and 2 are correct
C. 2 and 3 are correct
D. 1 and 3 are correct

251. Two isotropic radiators A and B are placed along the X-axis at the positions ($-\lambda/8$, 0) and ($+\lambda/8$, 0) respectively. The elements are fed by currents of equal amplitude with the phase of B leading that of A by 90°. The array pattern will have maxima

A. in the direction of the + X axis
B. at an angle of + 45° with respect to the Y axis
C. in the direction of + Y axis
D. in the direction of – X axis

252. Two ground planes are used in:

A. Microstrip B. Parallel wire line
C. Coaxial line D. Strip line

253. The output from a laser is monochromatic, this means that it is

A. Infrared B. Polarized
C. Narrow beam D. Single frequency

254. The deflection of hot wire instruments depends on

A. Instantaneous value of alternating current
B. Average value of current
C. r.m.s. value of alternating current
D. None of the above

255. For measurement of high values of VSWR, we should use

A. Single minima method
B. Double minima method
C. Double maxima method
D. None of these

256. The radiation pattern of the antenna used in VHF omnidirectional range (VOR) is

A. Pencil beam B. omnidirectional
C. cardioid D. elliptical

257. Earth-coverage satellite antennas from synchronous altitudes have a beamwidth of

A. 1.73° B. 7.3°
C. 17.3° D. 34.6°

258. The full-duplex rond-trip delay through a synchronous satellite is approximately

A. 300 m sec B. 550 m sec
C. 600 m sec D. 800 m sec

259. Indicate the false statement.

The parabolic reflector is

A. Omnidirectional antenna
B. pencil beam antenna
C. Popularly used antenna is terrestrial line-of-

sight communication

D. Not a broad cast antenna.

260. Noise figure of a system is defined as:

A. $\frac{S_i/N_i}{S_0/N_0}$ B. $\frac{S_0/N_0}{S_i/N_i}$

C. $\frac{\sqrt{S_0/N_0}}{S_i/N_i}$ D. $\frac{\sqrt{S_i/N_i}}{S_0/N_0}$

261. The primary purpose of the helix in a travelling-wave tube is to

A. Prevent the electron beam from spreading in the long tube
B. Reduce the axial velocity of the RF field
C. Ensure broadband operation
D. Reduce the noise figure

262. The greater the coupling mechanism

A. Wider the BW B. Greater the directivity
C. Lesser the BW D. A and B both

263. Two linearly polarised plane waves A and B travelling in free space in the +ve X-direction meet at a point. Wave A is polarised in the Y direction and the wave B is polarised in the Z direction. If the two waves are of equal amplitude and are in phase quadrature, the resultant wave will be

A. circularly polarised
B. elliptically polarised
C. linearly polarised
D. vertically polarised

264. Permeance is analogous to

A. resistance B. conductance
C. reluctance D. capacitance

265. With double the number of turns but the same length and area, the inductance is

A. the same B. double
C. quadruple D. one quarter

266. The energy stored in a magnetic field is given by

A. $W = \frac{1}{2}\sqrt{\frac{\mu H}{4}}$ B. $W = \mu\sqrt{\frac{H^2}{2}}$

C. $W = \frac{\mu H}{2}$ D. $W = \frac{\mu H^2}{2}$

267. The given figure shows five thin slots on the walls of a rectangular waveguide carrying TE_{10} mode. Which of these are radiating (R) and which are non-radiating (NR)?

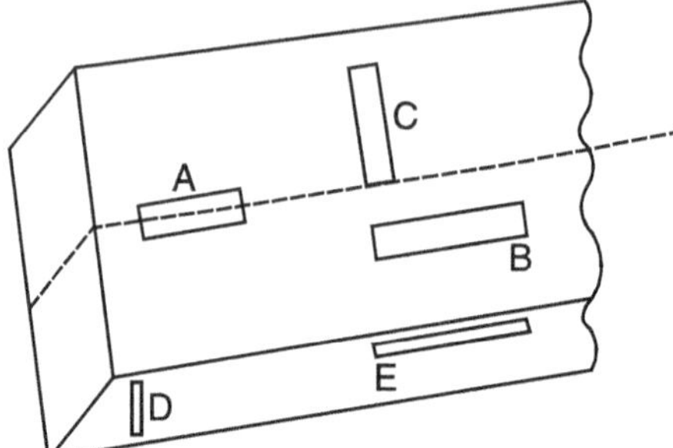

	R	NR
A.	D, E	A, B, C
B.	A, B, C	D, E
C.	C, E	A, B, D
D.	B, C, E	A, D

268. Which of the following represents the Maxwell's continuity equation for time varying electro-magnetic fields.

A. $\left(\Delta.\vec{J} + \frac{\partial p}{\partial t}\right) = 0$ B. $\Delta \times \vec{J} = \frac{\partial p}{\partial t}$

C. $\Delta.\vec{J} = \frac{\partial p}{\partial t}$ D. none of the above

269. A. C. currents can induce voltage because it has:

A. constant magnetic field
B. varying magnetic field
C. stronger magnetic field
D. high peak value

270. For a two open-wire transmission line excited by a harmonically oscillating source with exp $(-j\omega t)$ as the time factor, the voltage on the transmission line will satisfy which one of the following relations? (L and C are the inductance and capacitance per unit length of the line respectively)

A. $\frac{dV(x)}{dx} = -j\omega LI(x)$

B. $\frac{dV(x)}{dx} = j\omega LI(x)$

C. $\frac{dV(x)}{dx} = j\omega CI(x)$

D. $\frac{dV(x)}{dx} = -j\,\omega\, CI\,(x)$

271. On equipotential surface
A. potential is every where same
B. potential is every where different
C. potential varies radially only
D. none of the above

272. Displacement density D and electric field strength E are related by:
A. $D = \varepsilon E$ B. $D = E/\varepsilon$
C. $D = \varepsilon E^2$ D. none of these

273. Which of the following relation is valid for Ampere's law of force?

A. $F = \frac{\mu}{4\pi}\frac{I_1 I_2}{r^3}$ Newton

B. $F = \frac{\mu\varepsilon}{4\pi}\frac{I_1 I_2}{r^2}$ Newton

C. $F = \frac{\mu}{4\pi}\frac{I_1 I_2}{r^2}$

D. $F = \frac{\mu\,\varepsilon}{4\pi}\frac{I_1 I_2}{r}$

274. Heat is to be generated in 0.1 ohm resistor at the rate of 10 Watts from a battery of 1.5 V emf. The internal resistance of the battery is
A. .05 ohm B. .15 ohm
C. .2 ohm D. .225 ohm

275. The belt of an electrostatic generator is 60 cm wide and travels at 30 metres/sec carrying charge equivalent to 10^{-3} amp. The surface charge density is
A. 5.56×10^{-5} coul/m^2
B. 5.56×10^{-5} m/coul^2
C. 5.56×10^{-5} m^2/coul
D. 5.56×10^{-5} coul/metre^2

276. A permanent magent moving coil meter measures the
A. true rms value B. dc value
C. peak value D. rms plus dc value

277. The impedance of a 10 μF capacitor at 50 Hz is
A. 314 B. 318
C. 3180 D. 31

278. The impedance of a 1 henry inductor at 50 Hz is
A. 314 B. 50
C. 31.4 D. 3.14

279. The electromagnetic waves are refracted when they
A. Pass into a medium of different dielectric constant
B. Are polarized at right angles to the direction of propagation
C. Encounter a perfectly conducting medium
D. Pass through a small slot in a conducting plane.

280. The following is the condition for field $\vec{E}$ to be conservative:

A. $\oint_c \vec{E}.\vec{dl} = 0$ B. $\oint_c \vec{E}.\vec{ds} = q/\varepsilon_0$

C. $\nabla \times \vec{E} = 0$ D. $\oint_c \vec{E}.\vec{dl} = \delta$

281. A tube of force is
A. the force exerted by a steel tube
B. the fictitious tube of electric lines of force over a surface
C. force exerted by a tube of water
D. None of the above

282. An air-filled parallel-plate capacitor has a stored energy of W_0 and a charge Q_0 when connected to a source of voltage V_0. If the air-space is now filled completely by a dielectric having a dielectric constant of 2 and connected to a voltage source of 2 V_0, the stored energy will be:
A. 4 W_0 B. 2 W_0
C. W_0 D. $W_0/4$

283. In a semiconductor J, J_p and J_n indicate total diffusion current density, hole current density and electron current density respectively $\frac{\partial n}{\partial x}$ and $\frac{\partial P}{\partial x}$ are the electron and hole concentration gradients respectively in x-direction and D_f and D_n are the hole and electron diffusion constants respectively. Which one of the following equations is correct?

A. $J_n = -qD_n \frac{\partial n}{\partial x}$ for electrons

B. $J_p = -qD_p \frac{\partial P}{\partial x}$ for holes

C. $J = -qD_p \frac{\partial P}{\partial x} - qD_n \frac{\partial n}{\partial x}$

D. $J = qD_p \frac{\partial P}{\partial x} + qD_n \frac{\partial n}{\partial x}$

284. If a charge density σ exists within a conductor at certain time, it would be reduced to $1/e$ times this value in

A. σ/e seconds B. σ/ε second

C. ε/σ seconds D. ρ/e seconds

285. $\nabla.\left(\nabla \times \vec{A}\right)$ is equal to

A. 0 B. 1

C. ∞ D. none of these

286. When the distance between two charges is doubled, the force between them will be

A. Four times B. Double

C. Half D. One fourth

ANSWERS

1	**2**	**3**	**4**	**5**	**6**	**7**	**8**	**9**	**10**
A	B	C	D	C	B	D	D	A	B
11	**12**	**13**	**14**	**15**	**16**	**17**	**18**	**19**	**20**
A	A	C	C	B	A	B	C	D	D
21	**22**	**23**	**24**	**25**	**26**	**27**	**28**	**29**	**30**
D	B	C	C	D	C	C	D	C	B
31	**32**	**33**	**34**	**35**	**36**	**37**	**38**	**39**	**40**
B	A	C	D	A	C	C	C	D	B
41	**42**	**43**	**44**	**45**	**46**	**47**	**48**	**49**	**50**
B	C	C	B	A	D	A	A	B	D
51	**52**	**53**	**54**	**55**	**56**	**57**	**58**	**59**	**60**
B	B	A	C	A	C	D	A	C	C
61	**62**	**63**	**64**	**65**	**66**	**67**	**68**	**69**	**70**
A	A	B	D	B	C	D	B	D	D
71	**72**	**73**	**74**	**75**	**76**	**77**	**78**	**79**	**80**
D	B	D	D	A	B	B	B	C	B
81	**82**	**83**	**84**	**85**	**86**	**87**	**88**	**89**	**90**
A	A	A	A	C	A	B	D	B	B
91	**92**	**93**	**94**	**95**	**96**	**97**	**98**	**99**	**100**
A	B	C	C	C	C	B	B	B	C
101	**102**	**103**	**104**	**105**	**106**	**107**	**108**	**109**	**110**
B	D	A	C	C	B	A	C	A	A
111	**112**	**113**	**114**	**115**	**116**	**117**	**118**	**119**	**120**
D	D	B	C	D	B	C	C	C	A
121	**122**	**123**	**124**	**125**	**126**	**127**	**128**	**129**	**130**
B	C	B	A	B	B	C	C	A	D
131	**132**	**133**	**134**	**135**	**136**	**137**	**138**	**139**	**140**
A	A	C	C	B	D	D	B	D	B
141	**142**	**143**	**144**	**145**	**146**	**147**	**148**	**149**	**150**
C	B	C	A	C	C	C	C	D	B
151	**152**	**153**	**154**	**155**	**156**	**157**	**158**	**159**	**160**
A	B	C	D	A	B	A	D	A	B
161	**162**	**163**	**164**	**165**	**166**	**167**	**168**	**169**	**170**
A	C	A	A	B	B	D	D	A	C

171	172	173	174	175	176	177	178	179	180
D	A	A	A	A	A	B	D	A	C
181	182	183	184	185	186	187	188	189	190
C	D	C	C	D	B	A	B	C	B
191	192	193	194	195	196	197	198	199	200
D	B	B	A	B	B	D	B	B	D
201	202	203	204	205	206	207	208	209	210
B	D	D	A	C	C	A	D	C	D
211	212	213	214	215	216	217	218	219	220
A	B	C	B	D	D	C	A	B	B
221	222	223	224	225	226	227	228	229	230
A	A	D	C	C	C	B	A	A	D
231	232	233	234	235	236	237	238	239	240
A	D	B	A	B	B	C	C	B	B
241	242	243	244	245	246	247	248	249	250
A	C	C	A	D	C	B	D	A	D
251	252	253	254	255	256	257	258	259	260
D	D	D	C	B	A	C	A	A	A
261	262	263	264	265	266	267	268	269	270
B	D	A	B	C	D	B	A	B	A
271	272	273	274	275	276	277	278	279	280
A	A	C	A	A	B	B	A	A	A
281	282	283	284	285	286				
B	A	B	C	A	D				

EXPLANATORY ANSWERS

57. A duplexer is a circuit designed to allow the use of same antenna for both transmission and reception, with minimal interference between the transmitter and the receiver.

95. The shape of the voltage standing wave on line or waveguide is parabolic near a minimum, when measured with a probe whose response is a square load. Sharper the parabola, higher the SWR.

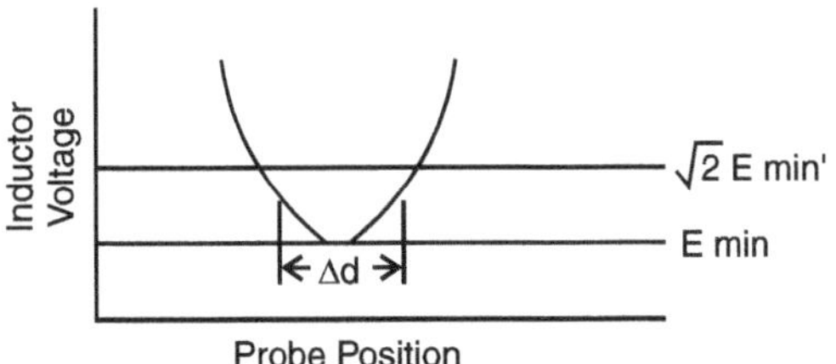

104. For a sphere, $V = \dfrac{Q}{4\pi\varepsilon_0 R}$

$$\therefore \frac{Q_1}{4\pi\varepsilon_0 R_1} = \frac{Q_2}{4\pi\varepsilon_0 R_2} \text{ or, } \frac{Q_1}{Q_2} = \frac{R_1}{R_2} = 6$$

Hence $\dfrac{A_1}{A_2} = \dfrac{R_1^2}{R_2^2} = 36$

120. Zoning is often used to cure the problem of great thickness required of lenses used at lower microwave frequencies or for strongly curved wavefronts. Not only would the lens be thick and heavy without zoning, but it would also absorb a large proportion of the radiation passing through it. This is because any dielectric with a large enough refractive index must, for that very reason absorb a lot of power.

126. An envelope detector is a diode half wave rectifier followed by an RC - low pass detector. The output of the detector represents the envelope of the incoming high frequency signal. It is used for FSK signals.
However, PSK needs synchronous detection.

138. The hysteresis phenomenon in Schmitt Trigger is

(i) In the inverted mode, the circuit triggers at a higher voltage for increasing signals than for decreasing signals.

(ii) Slowly varying signals are converted to output waveform with abrupt changes at two specified levels of input signals.

139. Noise figure is defined for the transmission system as the noise power output which would be obtained in the same bandwidth provided the transmission system were completely noise free and the only source of noise were the thermal noise generated at the input in the internal resistance of the signal source. Noise figure is used as an index of the relative "noisiness" of amplifiers and other transmission systems, particularly radio receivers.

195. The helix is thought of as a special coaxial transmission line with large inductance per unit length. The inductance provides the coaxial circuit with a phase velocity much less than that in free space. In fact, the input signal is slowed down to match the velocity of the electron beam heading toward the collector. As the beam and the input signal move together, their fields interact to produce velocity modulation of the electrons.

204. Cut off frequency, $f_r = \dfrac{2}{\sqrt{\left(\dfrac{m}{a}\right)^2 + \left(\dfrac{n}{b}\right)^2 + \left(\dfrac{p}{d}\right)^2}}$

TE is the transverse electric mode for $a > b < d$, and is the dominant mode.

208. Radiation field in the horizontal plane is given by

$E(\phi) = I_1 e^{-jk_d \cos\phi} + I_1 e^{-jk_d \cos\phi} \qquad 0 \le \phi \le 2\pi$

where k_d is the separation between the two antennas

Since $\quad I_1 = I_2 = I$

$$\therefore \quad E(\phi) = I\left(e^{-jk_d \cos\phi} + e^{-jk_d \cos\phi}\right) = 2\, I \cos(k_d \cos\phi)$$

$$\text{or,} \quad k_d = \frac{2\pi}{\lambda}, \lambda = 2\pi$$

Now plot this function with respect to ϕ for $0 \le \phi \le 2\pi$

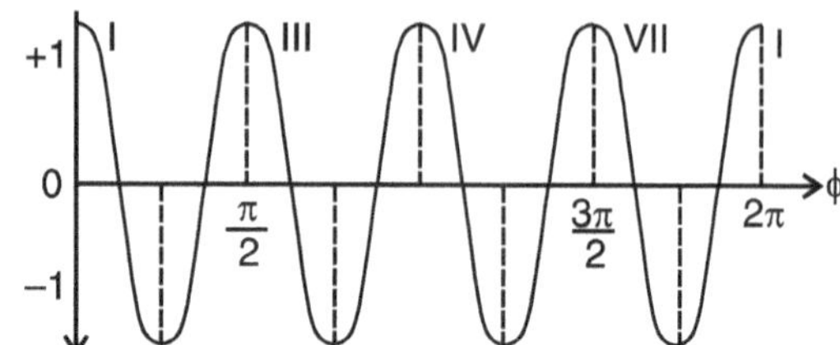

Hence number of lobes is 8 since 2 π is again 0.

224. Coupling Factor $= 10\log_{10}\left(\dfrac{P_B}{P_A}\right)$

$= 10 \log (100) = 20$ dB

225. In an optical fibre, the core material is either plastic or glass. The cladding is a material whose refractive index is less than that of the core. Total internal reflection at the core/cladding interface confines the light to travel within the core.

235. Given, $\quad f = 6$ GHz, hence $\lambda = 5$ cm

$\therefore \quad 140 \text{ cm} = 28\lambda$

$\therefore \quad \text{HPBW} = \dfrac{5.73}{2\pi} \approx 2.05 \text{ rad}$

$\therefore \quad \text{BWFM} = 2 \times \text{HBW} = 4.1 \approx 5.$

243. During charging, a current I is passing through the connectors. $\vec{S}$ is Poynting vector which is pointing every where radially into the volume and is defined as rate of energy flow per unit area in a plane electromagnetic wave $\left(\vec{S} = \dfrac{1}{\mu_0}\vec{E} \times \vec{B}\right)$ and direction of $\vec{S}$ gives the direction in which energy moves.

The rate at which energy flows into the volume between the plates

$$= \int \vec{S}.d\vec{A} = Ad.\frac{d}{dt}\left(\frac{1}{2}\varepsilon_0 E^2\right)$$

where E is instantaneous value of electric field. Thus, energy stored in the capaciter does not enter through the connecting wires but through the space surrounding the wires and plates of capacitor.

251. $E_{TR} = \cos\dfrac{\Psi}{2} = \cos\dfrac{\pi}{2}\cos(\phi + 45°)$

$\Psi = \beta d \cos(\phi + \alpha)$

$= \beta d \cos(\phi + 90°)$

$= \dfrac{2\pi}{\lambda}.\dfrac{\lambda}{4}\cos(\phi + 90°) = \dfrac{\pi}{2}\cos(\phi + 90°)$

6

CONTROL SYSTEMS

SYSTEM

A system is an arrangement, set or collection of things connected or related in such a manner as to form an entirety or whole. A system is also defined as an arrangement of physical components connected or related in such a manner as to form and/or act as an entire unit.

Control System

The word control is usually taken to mean **regulate, direct** or command. Accordingly a control system is an arrangement of physical components connected or related in such a manner as to command, direct or regulate itself or another system.

Classification

Control systems are classified into two general categories; open-loop and closed-loop systems. The direction is determined by the control action, which is that quantity responsible for activating the system to produce the output.

An open-loop control system is one in which the control action is independent of the output.

A closed-loop control system is one in which the control action is somehow dependent on the output.

Two outstanding features of open-loop control systems are :

1. Their ability to perform accurately is determined by their calibration. To calibrate means to establish or re-establish the input-output relation to obtain a desired system accuracy.
2. They are not generally troubled with problems of instability, a concept to be subsequently discussed in detail.

Closed-loop control systems are more commonly called feedback control system.

Feedback

Feedback is that characteristic of closed-loop control systems which distinguished them from open-loop systems.

Feedback is that property of a closed-loop system which permits the output (or some other controlled variable of the system) to be compared with the input to the system (or an input to some other internally situated component or subsystem) so that the appropriate control action may be formed as some function of the output and input.

1. More generally, feedback is said to exist in a system when a closed sequence of cause and-effect relations exists between system variables.
2. In essence, every passive system (one containing no energy sources) may be viewed as a feedback system. We shall consider only those closed-loop control systems where the existence and purpose of feedback are readily identified.

Characteristics of Feedback

The most important features, the presence of feedback imparts to a system, are the following :

1. Increased accuracy. For example, the ability to faithfully reproduce the input.
2. Reduced sensitivity of the ratio of output to input to variations in system characteristics.
3. Reduced effects of nonlinearities and distortion.
4. Increased bandwidth. The bandwidth of a system is that range of frequencies (of the input) over which the system will respond satisfactorily.
5. Tendency toward oscillation or instability.

Analogous Systems : For mathematical relations analogies are drawn between features of a system and features of some known elements or properties. Some analogous systems are given on next page.

Block Diagrams

A block diagram is a shorthand, pictorial representation or the cause and effect relationship between the input of a physical system. It provides a

Comparison of Open-Loop and Closed-Loop Control Systems

Open Loop	*Closed-Loop*
1. The accuracy of an open-loop system depends on the calibration of the input. Any departure from pre-determined calibration affects the output.	1. As the error between the reference input and the output is continuously measured through feedback the closed-loop system works more accurately.
2. The open-loop system is simple to construct and cheap.	2. The closed-loop systems is complicated to construct and costly.
3. The open-loop systems are generally stable.	3. The closed-loop systems can become unstable under certain conditions.
4. The operation of open-loop system is affected due to the presence of non-linearities in its elements.	4. In terms of the performance the closed-loop system adjusts to the effects of non-linearities present in its elements.

TABLE 1 : Translational and Rotational Mechanical Systems

Translational	*Rotational*
Force, f	Torque, T
Acceleration, a	Angular acceleration, α
Velocity, u	Angular velocity, ω
Displacement x	Angular displacement, x
Mass, M	Moment of inertia, Iθ
Damping co-efficient, D	Rotational damping co-efficient, Dθ
Compliance, K $\left(\frac{1}{\text{stiffness}}\right)$	Torsional compliance, Kθ

TABLE 2 : Force-Voltage Analogy

Mechanical System	*Electrical System*
Force, f	Voltage, v
Velocity, u	Current, i
Displacement x	Charge, q
Mass, M	Inductance, L
Damping co-efficient, D	Resistance, R
Compliance, K	Capacitance, C

TABLE 3 : Force-Current Analogy

Mechanical System	*Electrical System*
Force, f	Current, i
Velocity, u	Voltage, v
Displacement, x	Flux linkage, ϕ
Mass, M	Capacitance, C
Damping co-efficient, D	Conductance, G
Compliance, K	Inductance, L

TABLE 4

Electrical Systems	*Thermal systems*	*Liquid-level Systems*	*Pneumatic systems*
Charge, q	Heat flow, J	Liquid flow, m^3	Air flow, m^3
Current, A	Heat flow rate, J./s	Liquid flow rate, m^3/s	Air flow rate, m^3/s
Voltage, V	Temperature, °C	Head, m	Pressure, N/m^2
Resistance, Ω	Resistance, $°CsJ^{-1}$	Resistance, $m^{-2}s$	Resistance, $N\text{-}ms^{-1}$
Capacitance, F	Capacitance	Capacitance	Capacitance

convenient and useful method for characterizing the functional relationships among the various components of a control system. System components are alternatively called elements of the system.

Block Diagram of a Feedback Control System :

The block diagram representing the various components of a control system are connected in a fashion which characterizes their functional relationship within the system. The basic configuration of simple closed loop (feedback) control system is illustrated in the block diagram below, Fig. 1. It is emphasized that the arrows of the closed loop, connecting one block with another, represent the direction of flow of control energy or information, and not the main source of energy for the system. For example, the major source of energy for the thermostatically controlled furnance may be chemical, from burning fuel oil or coal. But this energy source would not appear in the closed control loop of the system :

The generalized feedback control system is given by Fig. 1.

Terminology of the Closed-Loop Block Diagram :

It is important that terms used in the closed-loop block diagram be clearly understood and remembered.

Lower case letters are used to represent the input the output variables of each element as well as the symbols for the blocks g_1, g_2, and h. These quantities represent functions of time, unless otherwise specified.

Example : r = r(t)

Capital letters denote Laplace transformed quantities as functions of the complex variables, or Fourier transformed quantities (frequency functions) as functions of the pure imaginary variable $j\omega$. Functions of s are usually abbreviated to the capital letter appearing alone. Frequency functions are never abbreviated.

Example : R(s) is abbreviated to R. R($j\omega$) is never abbreviated.

The letters r, c, m etc. were chosen to preserve the generic nature of the block diagram.

The plant g_2, also called the controlled system, is the body, process, or machine, of which a particular quantity or condition is to be controlled.

The control elements g_1, also called the controller, are the components required to generate the appropriate control signal m applied to the plant.

The feedback elements h are the components required to establish the functional relationship between the primary feedback signal b and the controlled output c.

The reference input r is an external signal applied to a feedback control system in order to command a specified action of the plant. It often represents ideal plant output behaviour.

The controlled output c is that quantity or condition of the plant which is controlled.

The primary feedback signal b is a signal which is a function of the controlled output c, and which is algebraically summed with the reference input r to obtain the actuating signal e.

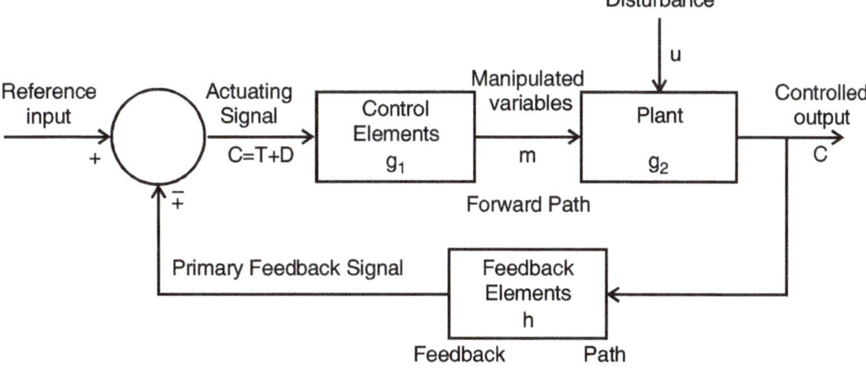

Fig. 1

TABLE 5 : Laplace and z-Transforms

Time function	$f(t)$	$F(s) = £\,[f(t)]$	$F^*(z) = (x)$
Unit ramp function	t	$\frac{1}{s^2}$	$\frac{Tz}{(z-1)^3}$
Unit acceleration function	$\frac{t^2}{2}$	$\frac{1}{s^3}$	$V_2 T^2 \frac{z(z+1)}{(z-1)^3}$
Exponential function	e^{-at}	$\frac{1}{s+a}$	$\frac{z}{z-t^{-at}}$
Sinusoidal	$\sin \beta t$	$\frac{\beta}{s^2+\beta^2}$	$\frac{z \sin \beta T}{z^2 - 2z \cos \beta T + 1}$
Cosinusoidal function	$\cos \beta t$	$\frac{s}{s^2+\beta^2}$	$\frac{z\,(z-\cos \beta T)}{z^2 - 2z \cos \beta T + 1}$

The actuating signal e, also called the error or **control action**, is the algebraic sum consisting of the reference input r plus or minus (usually minus) the primary feedback b.

The manipulated variable m (control signal) is that quantity or condition which the control elements g_1 apply to the plant g_2.

The disturbance u is an undesired input signal which affects the value of the controlled output c. It may enter the plant by summation with m, or via an intermediate point, as shown in the block diagram of Fig. 1.

The forward path is the transmission path from the actuation signal e to the controlled output c.

The feedback path is the transmission path from the controlled output c to the primary feedback signal b.

Block Diagram Transformation Theorems

Block diagrams of complicated control systems may be simplified using easily deriviable transformations. The letter P is used to represent any transfer function, and W, X, Y, Z denote any s-domain signals.

Stability

The stability of a system is determined by its response to inputs or disturbances. Intuitively, a stable system is one that will remain at rest unless excited by an external source and will return to rest if all excitations are removed. Stability can be precisely defined in terms of the impulse response of a system as follows :

A system is stable if its impulse response approaches zero as time approaches infinity.

A system is stable is every bounded input produces a bounded output.

Consideration of the degree stability of a system often provides valuable information about its behaviour. That is, if it is stable, how close is it to being unstable? This is the concept of relative stability. Usually, relative stability is expressed in terms of some allowable variation of a particular system parameter for which the system will remain stable.

Marginally Stable : If the system has some roots with real parts equal to zero, but none with positive area parts, the system is said to be marginally stable which is unstable.

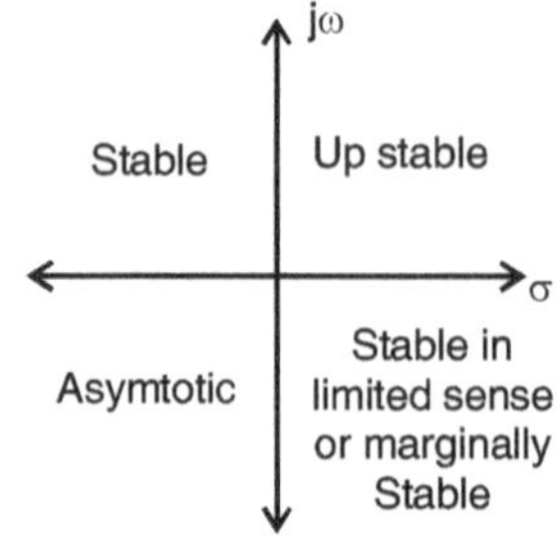

Transformation	*Equation*	*Bloack Diagram*	*Equivalent Block Diagram*
1. Combining Blocks in Cascade	$Y = (P_1P_2)X$	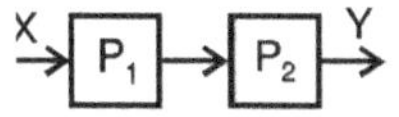	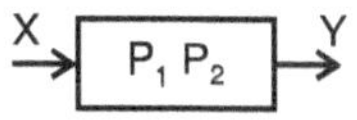
2. Combining Blocks in Parallel; or Eliminating a Forward Loop	$Y = P_1X \pm P_2X$	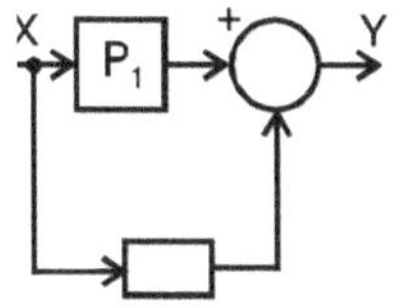	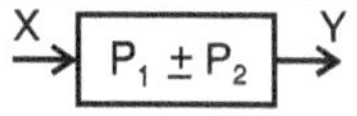
3. Removing a Block from a Forward Path	$Y = P_1X \pm P_2X$		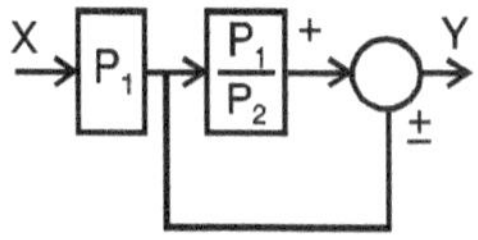
4. Eliminating a Feedback Loop	$Y = P_1X \pm P_2X$	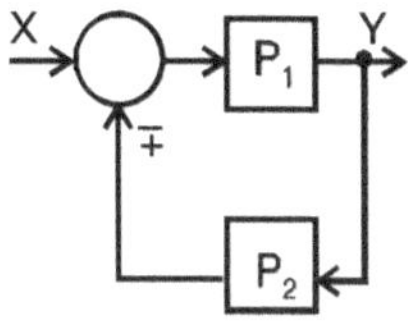	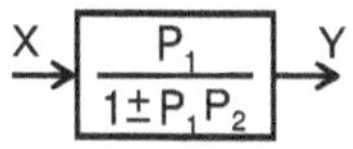
5. Removing Block from a Feedback Loop	$Y = P_1(X \mp P_2Y)$		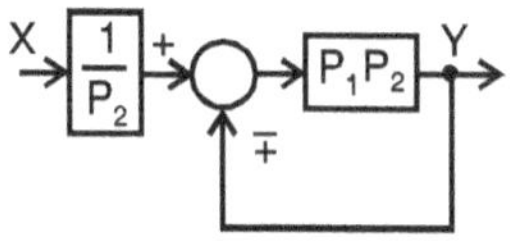
6a. Rearranging Summing Points	$Z = W \pm X \pm Y$	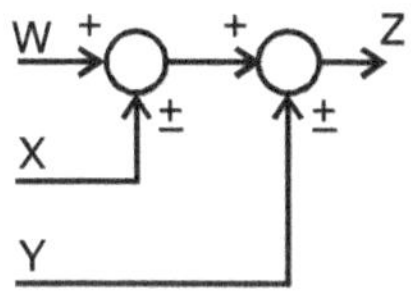	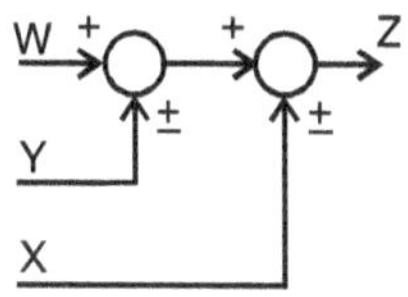
6b. Rearranging Summing Points	$Z = W \pm X \pm Y$	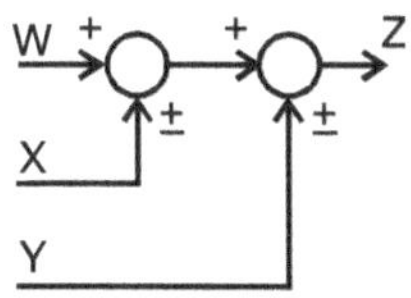	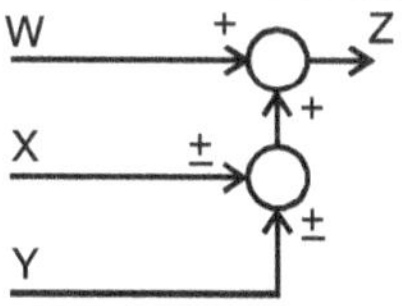
7. Moving a Summing Point Ahead of a Block	$Z = PX \pm Y$	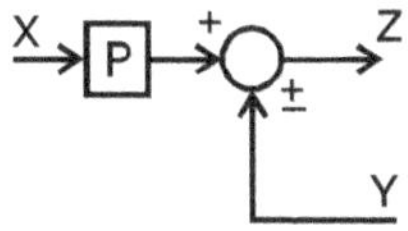	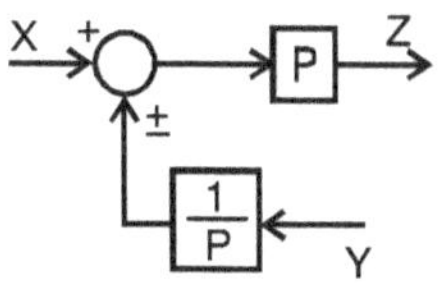
8. Moving a Summing Point Beyond a Block	$Z = P[X \pm Y]$	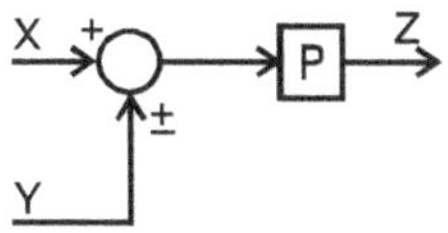	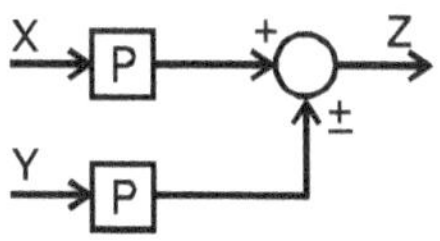
9. Moving a Takeoff Point Ahead of a Block	$Y = PX$	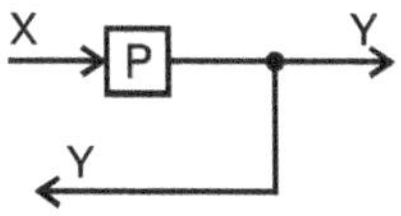	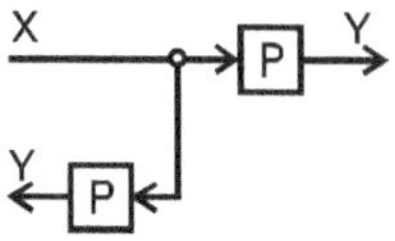

10. Moving a Takeoff Point Beyond a Block	Y = PX		
11. Moving a Takeoff Point Ahead of a Summing Point	Z = X ± Y		
12. Moving a Takeoff Point Beyond a Summing Point	Z = X ± Y		

Routh Stability Criterion

The Routh stability criterion is a method for determining system stability that can be applied to an nth order characteristic equation of the form

$$a_n s^n + a_{n-1} s^{n-1} + \ldots\ldots + a_1 s + a_0 = 0$$

The criterion is applied through the use of a Routh table defined as follows :

$$\begin{array}{c|cccc}
s^n & a_n & a_{n-2} & a_{n-4} & \ldots\ldots \\
s^{n-1} & a_{n-1} & a_{n-4} & a_{n-5} & \ldots\ldots \\
\ldots & b_1 & b_2 & b_3 & \ldots\ldots \\
\ldots & c_1 & c_2 & c_3 & \ldots\ldots \\
\ldots & \ldots\ldots & & &
\end{array}$$

where a_n, a_{n-1}, ... a_o are the coefficients of the characteristic equation and

$$b_1 \equiv \frac{a_{n-1}\, a_{n-2} - a_n\, a_{n-s}}{a_{n-1}},$$

$$b_2 \equiv \frac{a_{n-1}\, a_{n-4} - a_n\, a_{n-s}}{a_{n-1}}, \text{ etc.}$$

$$c_1 \equiv \frac{b_1\, a_{n-s} - a_{n-1}\, b_2}{b_1},$$

$$c_2 \equiv \frac{b_1\, a_{n-s} - a_{n-1}\, b_s}{b_1}, \text{ etc.}$$

The table is continued horizontally and vertically until only zeros are obtained. Any row can be multiplied by a constant before the next row is computed without disturbing the properties of the table. All the roots of this characteristic equation have negative real parts if and only if the elements of the first column of the Routh table have the same sign. Otherwise, the number of roots with positive real parts is equal to the number of changes of sign.

A row of zeros for the s^1 row of the Routh table indicates that the polynomial has a pair of roots which satisfy the axuxiliary equation formed as follows :

$$As^2 + B = 0$$

where A and B are the first and second elements of the s^2 row.

Hurwitz Stability Criterion

The Hurwitz stability criterion is another method for determining whether or not all the roots of a characteristic equation have negative real parts. This criterion is applied through the use of determinants formed from the coefficients of the characteristic equation. It is assumed that the first coefficient, a_n is positive. The determinants Δ_1 for

$i = 1, 2 \ldots\ldots n-1$ are formed as the principal minor determinants of the determinant.

$$\Delta_n = \begin{vmatrix} a_{n-1} & a_{n-1} & \cdots & \begin{bmatrix} a_0 \text{ if } n \text{ odd} \\ a_1 \text{ if } n \text{ even} \end{bmatrix} & 0 & \ldots & 0 \\ a_n & a_{n-2} & \cdots & \begin{bmatrix} a_1 \text{ if } n \text{ odd} \\ a_0 \text{ if } n \text{ even} \end{bmatrix} & 0 & \ldots & 0 \\ 0 \; a_{n-1} & a_{n-3} & \cdots & \cdots & & & 0 \\ 0 & a_n & a_{n-2} & \cdots & & & 0 \\ \cdots & \cdots & \cdots & \cdots & \cdots & \cdots & \cdots \\ 0 & \cdots & \cdots & \cdots & & & a_0 \end{vmatrix}$$

The determinants are thus formed as follows :

$$\Delta_1 = a_{n-1}$$

$$\Delta_2 = \begin{vmatrix} a_{n-1} & a_{n-3} \\ a_n & a_{n-2} \end{vmatrix} = a_{n-1}\, a_{n-2} - a_n\, a_{n-3}$$

$$\Delta_3 = \begin{vmatrix} a_{n-1} & a_{n-3} & a_{n-5} \\ a_n & a_{n-2} & a_{n-4} \\ 0 & a_{n-1} & a_{n-3} \end{vmatrix}$$

$$= a_{n-1}\, a_{n-2}\, a_{n-3} + a_n a_{n-1} a_{n-5} - a_n a^2{}_{n-3} - a_{n-4} a^2{}_{n-1}$$

and so on up to Δ_{n-1}.

All the roots of the characteristic equation have negative real parts if and only if $\Delta_1 > 0$ for i = 1, 2, ... n.

Continued Fraction Stability Criterion

The continued fraction stability criterion can be applied to the characteristic equation by forming a continued fraction from the odd and even portions of the equation in the following a continued fraction from the odd and even portions of the equation in the following manner. Let

$$Q(s) \equiv a_n s^n + a_{n-1} s^{n-1} + \ldots\ldots + a_1 s + a_0$$
$$Q_1(s) \equiv a_n s^n + a_{n-2} s^{n-2} + \ldots\ldots$$
$$Q_2(s) \equiv a_{n-1} s^{n-1} + a_{n-3} s^{n-3} + \ldots\ldots$$

Form the fraction Q_1 / Q_2, and then divide the denominator into the numerator and Invert the remainder to form a continued fraction as follows :

$$\frac{Q_1(s)}{Q_2(s)} = \frac{a_n s}{a_{n-1}} + \frac{\left(a_{n-2} - \dfrac{a_n a_{n-3}}{a_{n-1}}\right) s^{n-2} + \left(a_{n-4} - \dfrac{a_n a_{n-5}}{a_{n-1}}\right) s^{n-1} + \ldots}{Q_2}$$

$$= h_2 s + \cfrac{1}{h_2 s + \cfrac{1}{h_3 s + \cfrac{1}{h_4 s \cfrac{1}{\cdots\cdots \cfrac{1}{h_n s}}}}}$$

If $h_1, h_2 \ldots, h_n$ are all positive, then all the roots of Q(s) have negative real parts.

Transfer Functions

The response of a time-invarient linear system can be separated into two parts : the forced response and the free response. The forced response includes terms due to initial values x_0^k of the input, and the free response depends only on initial conditions y_0^k on the output. If terms due to all initial values, *i.e.* x_0^k and y_0^k, are lamped together; equation can be written as

$$y(t) = \alpha^{-1}\left[\left(\sum_{i=0}^{m} b_1 s^1 \Big/ \sum_{i=0}^{n} a_1 s^1\right) X(s) + (\text{terms due to all initial values } x_0^k, y_0^k)\right]$$

$$Y(s) = \left(\sum_{i=0}^{m} b_1 s^1 \Big/ \sum_{i=0}^{n} a_1 s^1\right) \times (s) + (\text{terms due to all initial values } x_0^k, y_0^k)$$

The Transfer function P(s) of a system is defined as that factor in the equation for Y(s) multiplying the transform of the input X(s). For the system described above, the transfer function is

$$P(s) = \sum_{i=0}^{m} b_i s^i / \sum_{i=0}^{n} a_i s^i$$
$$= \frac{b_m s^m + b_{m-1} s^{m-1} + ... + b_0}{a_n s^n + a_{n-1} s^{n-1} + ... + a_0}$$

and the transfer of the response may be written as

Y(s) = P(s) X(s) + (terms due to all initial values x_0^k, y_0^k)

If the quantity (terms due to all initial values x_0^k, y_0^k) is zero, the Laplace transform of the output (s) in response to an input X(s) is given by

Y(s) = P(s)X(s)

If the system is at rest prior to application of the input, *i.e.* $d^k y / dt^k = 0, k = 3, 1, ..., n-1$ for $t < 0$, then

(terms due to all initial values x_0^k, y_0^k) = 0

and the output as a function of time y(t) is simply the inverse transform of P(s)X(s).

It is emphasized that not all transfer functions are rational algebraic expressions. The transfer function of a system including time delays contains terms of the form e^{-sT}. In fact, the transfer function of an element representing a pure time delay is P(s) = e^{-sT} where T is the time delay in units of time.

Since, the formation of the output transform Y(s) is purely an algebraic multiplication of P(s) and X(s) when (terms due to all initial values x_0^k, y_0^k) = 0, the multiplication is commutative, that is

Y(s) = X(s) P(s) = P(s) X(s)

Properties of Transfer Function

The transfer function of a system has several useful properties :

1. The transfer function of a system is the Laplace transform of its impulse response. That is, if the input to a system with transfer function P(s) is an impulse and all initial values are zero, the transform of the output is P(s).
2. The system transfer function can be determined from the system differential equation by taking the Laplace transform and ignoring all terms arising from initial values. The transfer function P(s) is then given by
$$P(s) = \frac{Y(s)}{X(s)}$$
3. The system differential equation can be obtained from the transfer function by replacing the s variable with the differential operator D defined D ≡ d/dt.
4. The stability a time invarient linear system can be determined from the characteristic equation. The denominator of the system transfer function set equal to zero is the characteristic equation. Consequently if all the roots of the denominator have negative real parts, the system is stable.
5. The roots of the denominator are the system poles and the roots of the numerator are the system zeros. The system transfer function can then be specified to within a constant by specifying the system poles and zeros. This constant, usually denoted by K, is the system gain factor. Poles and zeros can be represented schematically by a pole-zero map in s-plane.

Signal Flow Graph

A signal flow graph is pictorial representation of the simultaneous describing a system, shown in Fig. 2

An input node or source is a node with only outgoing branches.

An output node or sink is a node with only incoming branches.

A forward path is a path from the input node to the output node.

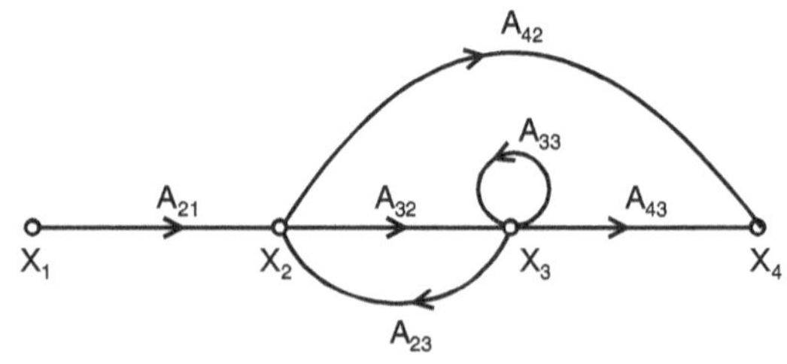

Fig. 2

A feedback path is a path which originates and terminates on the same node.

A self-loop is a feedback loop consisting of a single branch (A_{33} is a self-loop).

The gain of a branch is the transmission function of that branch when the transmission function is a multiplicative operator. For example, A_{33} is the gain of self loop It A_{33} is a constant or transfer function.

The path gain is the product of the branch gains encountered in traversing a path.

For example, the path gain of the forward path from X_1 to X_2 to X_3 to X_4 is $A_{21}\, A_{32}\, A_{42}$.

The loop gain is the product of the branch gains of the loop. For example, the loop gain on the feedback loop from X_2 to X_3 and back to X_2 is A_{32}, A_{23}.

Mason's gain formula for signal flow graph

The transmittance between an input node and an output node is the overall transmittance. Between these two nodes, Mason's gain formula, which is applicable to the overall gain, is

$$P = \frac{1}{\Delta}\sum_k P_k \Delta_k$$

where

P_k = path gain transmittance of kth forward path

Δ = Determinant of the graph

= 1–(sum of all different loop gains) + (sum of gain products of all possible combinations two non-touching loop) –(sum of gain products of all possible combination of three non-touching loop)

$$= 1 - \sum_a L_a + \sum_{b,\,c} L_b L_c - \sum_{d,\,e,\,f} L_d L_e L_f$$

Δ_k = Cofactor of the k_{th} path determinant of this graph with the loops touching the k_{th} forward path removed.

Frequency Response

In time domain analysis, the nature of input are either step, ramp or parabolic inputs. But, sometimes, system has inputs other than the above such as sinusoidal. Inputs whose amplitude and frequency may vary. The analysis of the system whose frequency and amplitude are discussed in frequency response. The system is actuated by a sinusoidal input and allowed to settle. The output amplitude and its phase with respect to input are measured. The phase difference and amplitude the nature of the system.

Frequency Response of a Closed-loop System

Let the forward path function be G(s) and the feedback path transfer function is H(s). Closed loop transfer function will be

$$\mathrm{C(s)} = \frac{\mathrm{G(s)}}{1+\mathrm{G(s)H(s)}} = \mathrm{R(s)}$$

Since the input is sinusoidal having amplitude and variable frequency ω

$$C(s) = \frac{G(s)}{1+G(s)H(s)} \cdot \frac{\omega}{\omega^2 + s^2}$$

$$\text{Let } T(s) = \frac{G(s)}{1+G(s)H(s)}$$

$$\text{Thus the output } c(t) = £^{-1}\left[\frac{\omega}{s^2+\omega^2} . T(s)\right]$$

Since R(s) may be assumed to be stable system the steady state response $y_{ss}(t)$ is that component of the response y(t) which is due to the input poles $j\omega$ and $-j\omega$ thus

$$y_{ss}(t) = \alpha \sin(\omega t + \theta)$$

These can be plotted in any one of the following ways :

1. $|G(j\omega)|$ vs frequency and phase angle vs frequency.
2. Magnitude $|G(j\omega)|$ and phase angle in polar plot
3. Magnitude vs phase shift.

The three predominant objectives of feedback control systems analysis are the determination of the following system characteristics.

1. The degree or extent of system stability.
2. The steady state performance
3. The transient response.

There are primarily four graphical methods available to control systems analysis which are simpler and more direct than the time remain method for practical linear models of feedback control systems. These are

1. Bode's-Plot representation
2. Nyquist Diagrams
3. Nichols Charts
4. The Root Locus method.

The first three are frequency-domain techniques.

Bode's Plot

A sinusoidal function may be represented by two separate plots, one giving the magnitude versus frequency and the other phase angle versus frequency. A logarithmic plot of Bode's plot of logarithm of the magnitude of a sinusoidal transfer function and the other is a plot of the phase angle both are plotted against the frequency in logarithmic.

The main advantages of Bode's plot is that multiplication of magnitudes can be converted into addition. It is simplest method. Transfer function can be determined easily by Bode's plot, the low frequency and high frequency characteristics of the transfer function can be determined.

Gain Margin : It is a measure of relative stability, is defined as the magnitude of the reciprocal of the open-loop transfer function evaluated at the frequency $\omega\pi$ at which the phase angle is –180 degrees

$$\text{i.e., Gain Margin } = \frac{1}{|GH(j\omega)|}$$

where $\arg GH(j\omega\pi) = -180$ degrees

$\omega\pi$ is called phase cross over frequency

Phase Margin : It is a measure of relative stability, is defined as 180 degrees plus the phase angle ϕ_1 of the open-loop transfer function at unity gain that is

$$\phi_{PM} = [180 + \arg GH(j\omega_1) \text{ degrees}$$

where $|GH(j\omega_1)| = 1$ and ω_1 is called gain crossover frequency.

Bandwidth (BW) of a system is defined as that range of frequencies over which the system will respond satisfactorily.

Cut-off Rate is the frequency rate at which the magnitude ratio decreases beyond the cut-off frequency ω_c.

Resonance Peak M_p is a measure of relative stability, is the maximum value of the magnitude of the closed-loop frequency response. That is

$$M_p = \max\left|\frac{C}{R}(j\omega)\right|$$

Resonant frequency ω_p is the frequency at which M_p occurs.

Polar Plots

The polar plot of a sinusoidal transfer function $G(j\omega)$ is a plot of the magnitude of $G(j\omega)$ vs phase angle of $G(j\omega)$ on polar coordinates as ω is varied from 0 to infinity. Thus, the polar plot is the locus of vectors $|G(j\omega)|\angle G(j\omega)$ as ω is varied from zero to infinity. A positive angle is measured from counter clockwise from positive rear axis. Polar plot is called Nyquist plot. Both magnitude $|G(j\omega)|$ and phase angle $\angle\,(G(j\omega))$ must be calculated directly for each frequency ω in order to construct polar plot.

Properties of Polar Plots

The following are several useful properties of Polar Plots.

1. The Polar Plot for $P(j\omega) + a$ where a is any complex constant, is identical to the plot for $P(j\omega)$ with the origin of coordinates shifted to the point $-a = -(\text{Re } a + j \text{ lm a})$.
2. The Polar Plot of the transfer function of a time-invarient, constant coefficient, linear system exhibits conjugate symmetry. That is, the graph for $-\infty < \omega < 0$ is the mirror image about the horizontal axis of the graph for $0 \le \omega < \infty$.
3. The Polar Plot may be constructed directed from a Bode Plot if one is available. Values of magnitude and phase angle at various frequencies ω on the Bode Plot represent points along the locus of the Polar Plot.
4. Constant increments of frequency are not generally separated by equal intervals along the Polar Plot.

The Nyquist Method

The Nyquist method handles systems with time delay without the necessary of approximations and hence yields exact results about both absolute and relative stability of the system. This is also useful for obtaining information about transfer functions of components or systems from experimental frequency response data.

The Nyquist Stability Criterion : A closed loop control system is absolute stable if the roots of the characteristic equation have negative real parts. Equivalent, the poles or the closed-loop transfer function of the zeros of the denominator

$$1 + GH(s)$$

of the closed-loop transfer function must lie in L.H.P.

Relative Stability : The relative stability of a feedback control system is determined from the polar of Nyquist stability plot.

The phase crossover frequency ω_π is that frequency at which the phase angle of GH(Jω) is –180°, that is the frequency at which the polar plot crosses real axis, the gain margin is given by

$$\text{gain margin} = \frac{1}{|GH(j\omega_\pi)|}$$

Gain crossover frequency ω is that frequency at which |GH| = 1

The phase margin ϕ_{PM} is the angle by which polar must be rotated to cause it to pass through the (–1, 0) point. It is given by

$$\phi_{PM} = [180 = \arg GH(j\omega)] \text{ degrees}$$

Root-Locus Method

The closed-loop transfer function of this system is

$$\frac{C}{R} = \frac{G}{1+GH}$$

centre of asymptotes σ_c is given by

$$\sigma_1 = \frac{-1}{n-m}\left[\sum_{i=1}^{n} p_1 - \sum_{i=1}^{m} z_1\right]$$

where p_1 are the poles z_1 are the zeros, n is the number of poles and m the number of zeros of GH Angle between the asymptotes and real axis are given by

$$\beta = \begin{bmatrix} \dfrac{(2L+1)180}{n-m} & \text{degree for } k < 0 \\ (2L)180/(n-m) & \text{degree for } k < 0 \end{bmatrix}$$

for i = 0, 1, 2, ... n – m – 1

Breakaway point can be determined by solving the following equation

$$\sum_{i=1}^{n} \frac{1}{\sigma_k + p_i} = \sum_{i=1}^{m} \frac{1}{\sigma_b + z_i}$$

where $-p_i$ and $-z_i$ are the poles and zero of GH respectively.

Time Response of Control System

For a unit step input, the transient response of a second older system is shown in Fig. 3

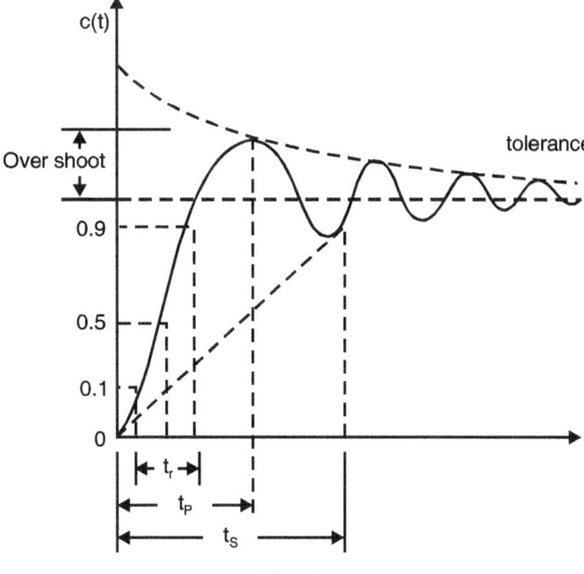

Fig. 3

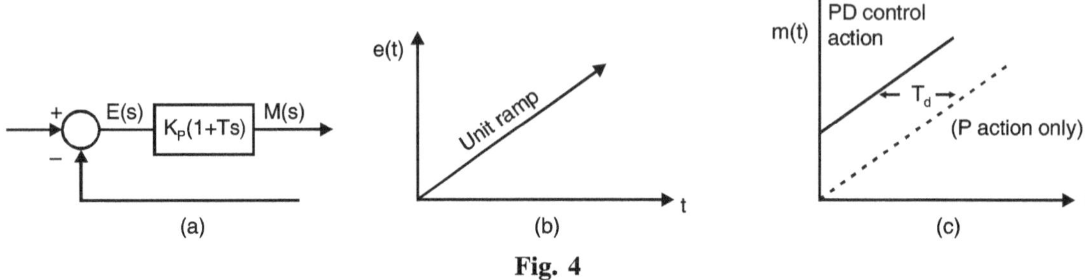

Fig. 4

1. Delay time t_d is the time required for the response to reach line half the final value.
2. Rise time t_r, is the time required for the response to rise from 10 to 90% of its final value.
3. Peak time t_p, is the time required for the response to reach the first peak of the over shoot.
4. Maximum overshoot M_p is the max. value of the response curve measured from unity.
5. Overshoot is the maximum difference between the transient and steady state solutions for a unit step function input. It is measure of relative stability and is represented as percentage of final value of the output.
6. Settling time t_s, is defined as the time required for the response to unit step function input to reach and remain within a specified percentage of its final value.

Test Function : Typical Test Signals Commonly used test signals, are step fraction, ramp function, impulse function and sinusoidal functions.

Proportional Plus-Derivative Control Action (PD Action) : The control action of a proportional plus derivative controller is defined by the equation.

$$m(t) = K_p e(t) + K_p T_d \frac{de(t)}{dt}$$

$$\frac{M(s)}{E(s)} = K_p(1 + T_d s)$$

where K_p is proportional sensitivity and T_d derivative time. Fig. 4 shows the proportional plus derivative controller.

Proportional Plus Derivative Plus-Integral Control Action (PID Action) : It is the combination of proportional control action, derivative control action and integral control action. The equation of controllers with this combined action is given by

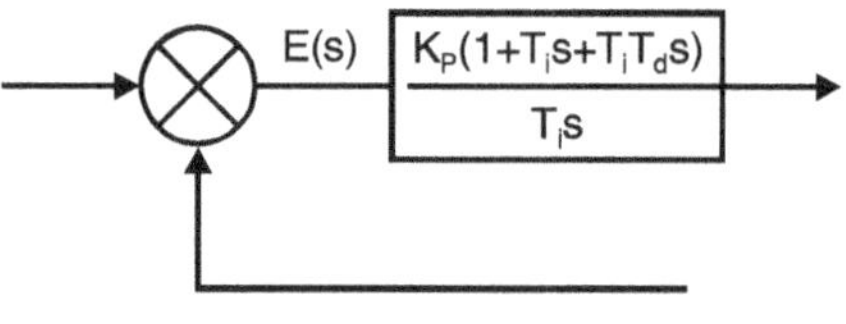

Fig. 5

$$m(t) = K_p e(t) + k_p T_d \frac{de(t)}{dt} + \frac{K_p}{T_i} \int_0^t e(t)dt$$

$$\frac{M(s)}{E(s)} = K_p\left(1 + T_d s + \frac{1}{T_i s}\right)$$

It is shown in block diagram in Fig. 5.

Generation of Proportional Plus Integral Action : A proportional plus integral controller is required to give to the correcting unit, an output proportional to the magnitude and to the duration of the deviation. The latter requires that the output should increase at a rate proportional to the deviation.

Electric Controllers

Fig. 6 shows a three term controller. The proportional signal is obtained from the resistors R_1, R_2 and R_3. Derivative control can be added by including a conductor in parallel with R_1 so that the transfer function becomes

$$\frac{V_0(s)}{V_1(s)} = \frac{Ts + 1}{Ts + \frac{R_1}{R_2} + 1}$$

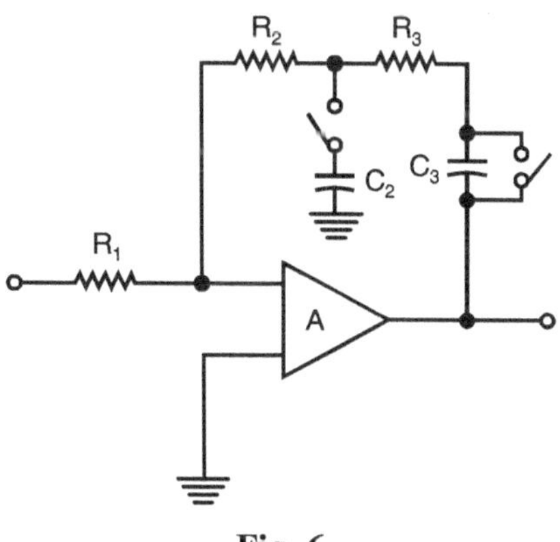

Fig. 6

Error Constants

Position Error Constant is a measure of steady state error between the input and output when the input is a unit step function.

Velocity Error Constant is a measure of the steady-state error if the system when the input is until parabolic function, that is $r = \frac{t^2}{2}$ and $R = \frac{1}{s^3}$

The error constants and steady errors for type 0, type 1 and 2 unity feedback systems are tabulated below in Table A.

Potentiometers : Since the output voltage of a potentiometer is proportional to the shaft displacement, when a voltage is applied across its fixed terminals, the device can be used to compare two shaft positions. In this case, one shaft may be fastened to the potentiometer can and other to shaft of the potentiometer. When a constant voltage is applied to the fixed terminals of the potentiometer, the voltage across the variable and the reference terminals will be proportional to the difference between the two shaft positions.

DC Servomotors : Among various dc servomotors are series motor, split phase motor, shunt control motor and permanent magnet shunt motor. These units develop high output power in a given size and in the case of field controlled shunt motor require little control power, the problems are radio interference, brush wear and dc amplifier drift.

The series motor has high starting torque and draws large current and has poor speed regulation. Reversal can be obtained by reversing field voltage polarity with split series winding. A typical speed torque curve shows a high starting torque and rapid reduction of torque with increasing speed. This results is good damping and large velocity error.

The shunt motor has good starting torque and good speed regulation. Reversibility is possible by reversing armature of field polarity. Performance is similar to shunt motor with armature control and fixed field.

Armature controlled DC Motor : A dc motor is often employed in a control system where an appreciable amount of shaft power is required. The dc motors are much more efficient than two phase ac servomotor.

The dc motors have separately excited fields. They are armature controlled with fixed armature current.

Table A

Input	*Unit Step*		*Unit Ramp*		*Unit Parabola*	
System type	K_p	*Steady state error*	K_v	*Steady state error*	K_a	*Steady state error*
Type 0	$\frac{C\beta_1(0)}{\beta_2(0)}$	$\frac{1}{1+K_p}$	0	∞	0	∞
Type 1	∞	0	$\frac{K\beta_2(0)}{\beta_2(0)}$	$\frac{1}{K_v}$	0	∞
Type 2	∞	0	∞	0	$\frac{K\beta_2(0)}{\beta_2(0)}$	$\frac{1}{K_a}$

Two Phase Servo Motors : These motor transform electric signals into equivalent speed, angle or torque qualitities. Accuracy, speed and reliability of signal transmission are more important than the efficiency. Thus two phase servo motors can be identified by their operating characteristics from the normal fractional h.p. two phase capacitor motor etc. The typical characteristics of servo motors are the decreasing speed torque characteristics. Rotors are specially designed to have low inertia. That is why they are usually long and have small diameters.

Pneumatic Control System

The closed-loop of the automatic control system consists of

(i) Measuring unit (ii) Controlling unit
(iii) Correcting unit and (iv) Process

Proportional Control Action (P-Action) : It is defined as the action of a controller in which the output signal *m(t)* is proportional to the measured actuating error signal *e(t)* as shown in Fig. 7

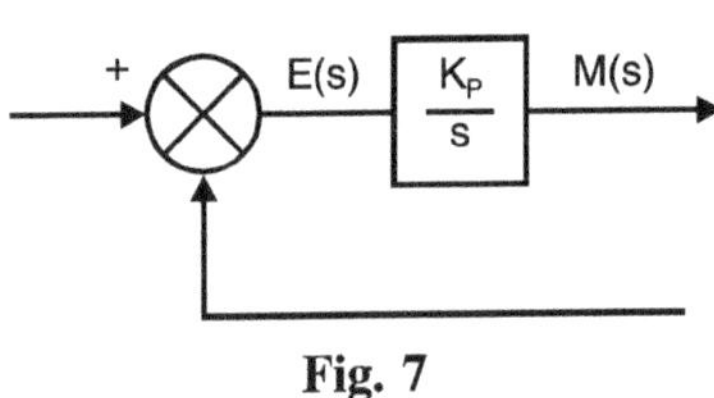

Fig. 7

$$m(t) = k_p a e(t)$$

$$\frac{M(s)}{E(s)} = K_p$$

where K_p is termed the proportional sensitivity or the gain. The proportional controller is essentially an amplifier with an adjustable gain A.

Integral Action : In a controller with integral control action, the value of the output $m(t)$ is proportional to the measured actuating error signal $e(t)$ as shown in Fig. 8

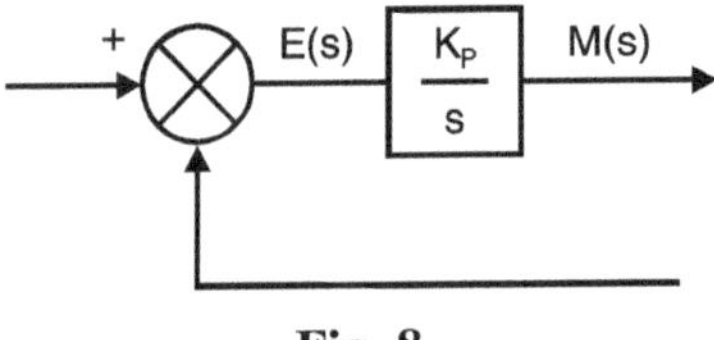

Fig. 8

$$\frac{dm(t)}{dt} = K_i e(t) \quad \text{or} \quad m(t) = K_i \int_0^t e(t)\, dt$$

where K_i is an adjustable constant, the transfer function of the integral controller is

$$\frac{M(s)}{E(s)} = \frac{K_t}{s}$$

For zero actuating error, the value *m(t)* remains stationary. The integral control action is called reset control.

Fig. 7 shows a block diagram of such a controller.

Proportional-plus Integral Control Action (PI Action) : The control action of a proportional plus integral controller is defined by the following equation.

$$m(t) = K_p e(t) + \frac{K_p}{T_i} \int_0^t e(t)dt$$

$$\frac{M(s)}{E(s)} = K_p \left(1 + \frac{1}{T_i s}\right)$$

where K_p and T_i represents the proportional sensitivity and integral time. Both are adjustable. If actuating error signal is step input the controller output is shown in Fig. 9

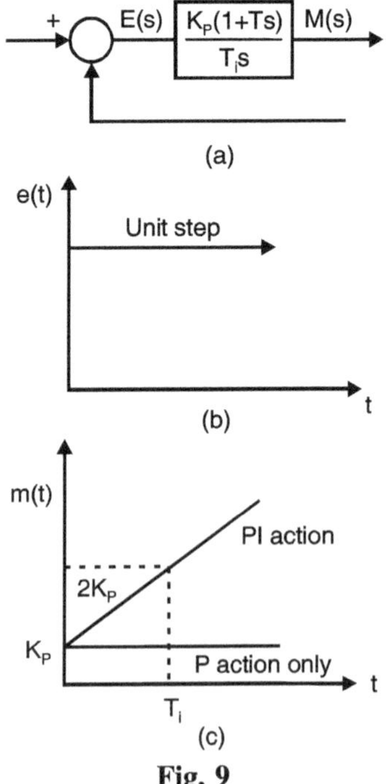

Fig. 9

EXERCISE

1. The end points of root locii are:
   A. Open loop poles
   B. Open loop zeros
   C. Closed loop poles
   D. Closed loop zeros
2. In n are the no. of poles and m are the no. of zeros, then how many root locus branches tend to infinity
   A. $n+m$ B. $\frac{n+m}{2}$
   C. $n-m$ D. $\frac{n-m}{2}$
3. A second order system exhibits 100% overshoot. Its damping coefficient is
   A. equal to 0 B. equal to 1
   C. less than 1 D. greater than 1
4. A stepper motor
   A. is a two phase induction motor
   B. is a king of rotating amplifier
   C. is an electromagnetic transducer commonly used to convert an angular position of a shaft into an electrical system
   D. is an electromechanical device which-actuates a train of step angular (or linear) movements in response to a train of input pulses on one to one basis.
5. The Nyquist plot for the function
   $G(s)\,H(s) = \frac{K}{s(1+T_1 s)(1+T_2 s)}$
   is shown in the figure. The condition for stability is given as

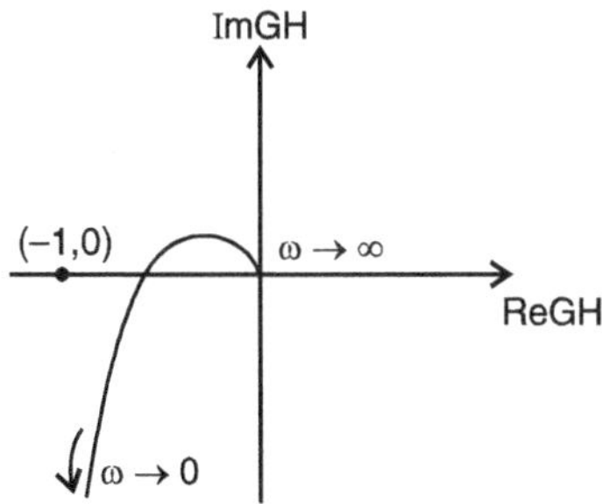

   A. $\frac{K(T_1+T_2)}{T_1T_2(T_1-T_2)} < 1$
   B. $\frac{K(T_1+T_2)}{T_1T_2} < 1$
   C. $\frac{T_1+T_2}{K(T_1+T_2)} < 1$
   D. None of these
6. For all frequencies a unit circle in the Nyquist plot transform into
   A. 10 dB line of amplitude plot in Bode diagram
   B. 0 dB line of amplitude plot in Bode diagram
   C. 1 dB line of amplitude plot in Bode diagram
   D. B or C
7. For the transfer function
   $G(s)H(s) = \frac{1}{s(s+1)(s+0.5)}$ the phase cross-over frequency is
   A. 0.5 rad/sec B. 0.707 rad/sec
   C. 1.732 rad/sec D. 2 rad/sec
8. In the root locus analysis, the angle between adjacent asymptotes is:
   A. $\frac{360°}{n+m}$ B. $\frac{360°}{n-m}$
   C. $\frac{180°}{n+m}$ D. $\frac{180°}{n-m}$
   Where n is the number of poles and m is the number of zeros.
9. The maximum overshoot is a function of:
   A. Damping ratio
   B. Damped frequency of oscillation
   C. Natural frequency of oscillation
   D. None of these
10. The system shown in the figure:

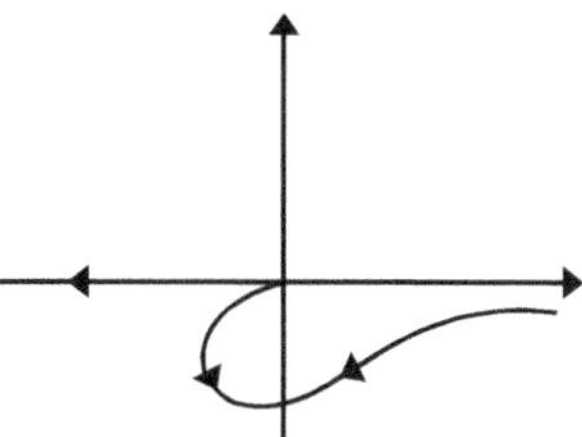

   A. Stable B. Conditionally stable

C. Unstable D. None of these

11. In force current analogy indicate the correct combination

(*i*) Current in analogous to force

(*ii*) Capacitance is analogous to displacement

A. True, true B. False, false

C. True, false D. False, true

12. The polar plot of a system shows

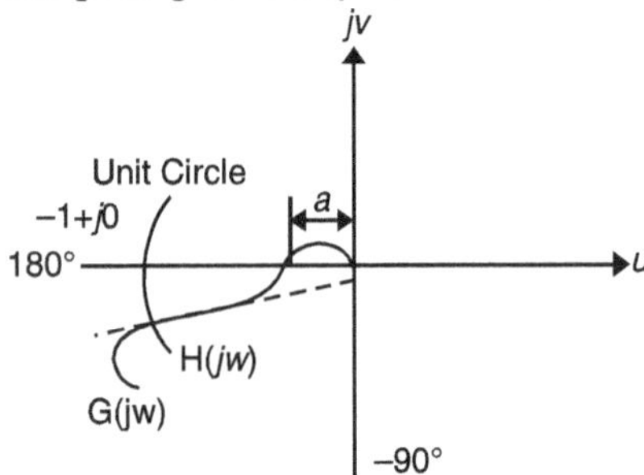

A. Good GM and Poor PM

B. Good GM and Good PM

C. Poor GM and Poor PM

D. Poor GM and Good GM

13. A linear stable time-invariant system is forced with an input

$x(t) = \text{A} \sin \omega t$

Under steady-state conditions, the output Y(t) of the system will be

$x(t)$ —— G(s) —— $y(t)$

A. A sin $(\omega t + \phi)$, where $\phi = \tan^{-1} |G(j\omega)|$

B. $|G(j\omega)|\, A \sin [\omega t + \angle G(j\omega)]$

C. $|G(j\omega)|\, A \sin [2\,\omega t + \angle G(j\omega)]$

D. $AG(j\omega) \sin [\omega t + \angle G(j\omega)]$

14. The transfer function of system is $\dfrac{5(1+0.1s)}{(1+0.5s)}$.

The corner frequencies will be

A. 5 and 2 B. – 02 and 05

C. 10 and 2 D. 0.1 and 5.

15. If the gain of a critically damped system is increased, the system will behave as:

A. Underdamped B. Overdamped

C. Oscillatory D. None of these

16. The algebraic sum of the angles of the vectors from all poles and zeros to the point on any root locus segment is

A. 0°

B. 180°

C. 180° or odd multiple thereof

D. 90° or odd multiple thereof

17. The first derivative output when fed back negatively will:

A. Decrease the damping of the system

B. Increase the damping of the system

C. Decrease the velocity-lag error

D. Decrease the steady state error.

18. The state representation of a second order system is

$x_1 = -x_1 + u,\ x_2 = x_1 - 2x_2 + u$

Consider the following statements regarding the above system:

1. The system is completely state controllable.
2. If x_1 is the output, then the system is completely output controllable.
3. If x_2 is the output, then the system is comletely output controllable

Of these statements

A. 1, 2 and 3 are correct

B. 1 and 2 are correct

C. 2 and 3 are correct

D. 1 and 3 are correct

19. Integral error compensation in a control system

A. Minimizes steady state error

B. Increases steady state error

C. Does not have any effect on steady state error

D. None of these

20. Which of the following is not a method for describing linear systems

A. Differential equations

B. Dynamic equations

C. State transition equations

D. None of these

21. The open-loop transfer function of a unity feedback control system is

$$G(s) = \frac{1}{(s+2)^2}$$

The closed-loop transfer function will have poles at

A. – 2, – 2 B. – 2, – 1

C. $-2 \pm j$ D. – 2, 2

22. For the transfer function

$$G(s)\,H(s) = \frac{1}{s(s+1)(s+0.5)}$$

the phase cross-over frequency is

A. 0.5 rad/sec B. 0.707 rad/sec
C. 1.732 rad/sec D. 2 rad/sec

23. The maximum phase shift that can be provided by a lead compensator with transfer function,

$$G_c(s) = \frac{1+6s}{1+2s}$$

A. 15° B. 30°
C. 45° D. 60°

24. The radius and the centre of M circles are given respectively by

A. $\frac{M}{M^2-1} \cdot \left(\frac{-M^2}{M^2-1}, 0\right)$

B. $\frac{M^2}{M^2-1} \cdot \left(\frac{+M^2}{M^2-1}, 0\right)$

C. $\frac{M^2}{M-1} \cdot \left(\frac{M^2}{M-1}, 0\right)$

D. $\frac{M^2}{M^2-1} \cdot \left(\frac{-M^2}{M^2-1}, 0\right)$

25. The loop transfer function of a feedback control system is given by

$$G(s)\,H(s) = \frac{K}{s(s+2)\left(s^2+2s+2\right)}$$

The number of asymptotes of its root loci is

A. 1 B. 2
C. 3 D. 4

26. In a feedback system

$G = \frac{4}{s(s+3)}$ and $H = 1/s$. The system is of:

A. Type 2 B. Type 0
C. Type 1 D. None of the above

27. A plant has the following transfer function

$$G(s) = \frac{1}{\left(s^2+0.2s+1\right)}$$

For a step input it is required that the response settles to within 2% of its final value. The plant setting time is

A. 20 sec B. 40 sec
C. 35 sec D. 45 sec

28. Match List I with List II and select the correct answer using the codes given below the lists :

List I (Transfer function)	*List II (Controller)*
(a) $\frac{K_1 s + K_2}{K_3}$	1. P-controller
(b) $\frac{K_1 s^2 + K_2 s + K_3}{K_4 s}$	2. PI-controller
(c) $\frac{K_1 s + K_2}{K_3 s}$	3. PD-controller
(d) $\frac{K_1 s}{K_2 s}$	4. PID-controller

Codes:	*(a)*	*(b)*	*(c)*	*(d)*
A.	2	1	3	4
B.	3	4	2	1
C.	2	3	4	1
D.	3	2	4	1

29. Well-designed first-order control systems have

A. A negative time constant
B. A small bandwidth
C. A small negative even value
D. A large negative transfer function pole.

30. If one wants to keep the basic form of the transfer function of a system unchanged and change its time constant and gain, he must use

A. Resistance feedback
B. Unity feedback
C. Potentiometer feedback
D. Tachometer feedback

31. Match the system open-loop transfer functions given in List I with the steady-state errors produced for a unit ramp input. Select the correct answer using the code given below the lists:

List I	**List II**
(a) $\frac{30}{s^2+6s+9}$	1. Zero
(b) $\frac{30}{s^2+6s}$	2. 0.2
(c) $\frac{30}{s^2+9s}$	3. 0.3
(d) $\frac{s+1}{s^2}$	4. Infinity

Codes :

	(a)	(b)	(c)	(d)
A.	1	2	3	4
B.	4	3	2	1
C.	1	3	2	4
D.	4	2	3	1

32. For the asymptotic Bode plot (dB versus log ω) of the transfer function with a factor of the form

$$\frac{1}{\left(s^2 + 2\zeta\omega_n s + w_n^2\right)}$$

the corner frequency (I), slope in dB/decade (II) and error at the corner frequency (III) will be

	I	II	III
A.	$\zeta\omega_n$	+ 40	Error is ζ dependent
B.	ω_n	– 40	Error is ζ dependent
C.	$\zeta\omega_n$	+ 20	– 3 dB
D.	ω_n	+ 20	+ 3 dB

33. A system has a transfer function

$$\frac{C(s)}{R(s)} = \frac{4}{s^2 + 1.6s + 4}$$

For the unit step response, the settling time (in seconds) for 2% tolerance band is

A. 1.6 B. 2.5

C. 4 D. 5

34. In the root-locus for open-loop transfer function

$$G(s)\,H(s) = \frac{K(s+6)}{(s+3)(s+5)},$$ the break away and 'break in' points are located respectively at

A. – 2 and – 1

B. – 2.47 and – 3.77

C. – 4.27 and – 7.73

D. – 7.73 and – 4.27

35. For a closed-loop transfer function given below

$$\frac{C(s)}{R(s)} = \frac{2600\,K(s+25)}{s^4 + 125s^3 + 5100s^2 + 65000s + 65000\,K}$$

The imaginary axis intercepts of the root loci will be

A. $\pm j\,22.80$ B. $\pm j\,2.28$

C. $\pm j\,1.14$ D. $\pm j\,11.48$

36. A transfer function G(s) has type pole-zero plot as shown in the given figure. Given that the steady-state gain in 2, the transfer function G(s) will be given by

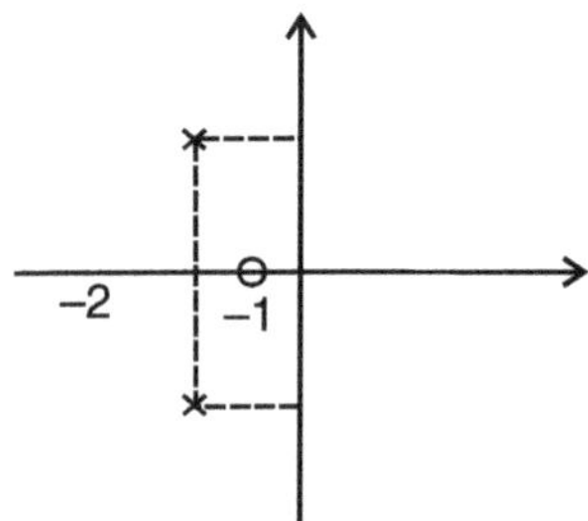

A. $\frac{2(s+1)}{s^2+4s+5}$ B. $\frac{5(s+1)}{s^2+4s+4}$

C. $\frac{10(s+1)}{s^2+4s+5}$ D. $\frac{10(s+1)}{(s+2)^2}$

37. The transfer function $H(s) = \frac{s^2 - 5s + 100}{s^2 + 5s + 100}$ represents

A. a high-pass filter

B. a band elimination filter

C. a resonator

D. an all pass filter

38. The characteristic equation $s^3 + 3s^2 + 3s + k = 0$ is stable for which value of k?

A. – 6 B. 0

C. 5 D. 12

39. In the circuit shown in the given figure, R_f provides

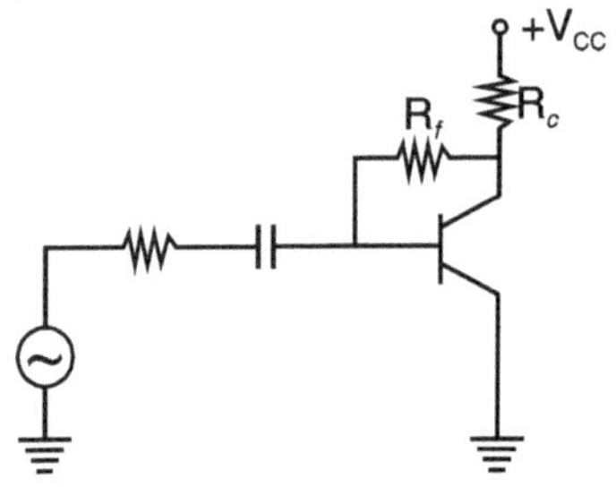

A. current series feedback

B. current shunt feedback

C. voltage series feedback

D. voltage shunt feedback

40. A lag compensator is basically a

A. high pass filter

B. band pass filter

C. low pass filter

D. band elimination filter

41. A characteristic equation of a system has roots

such that the response is a damped sinusoid as shown below. IF the roots are made more negative, the response would be:

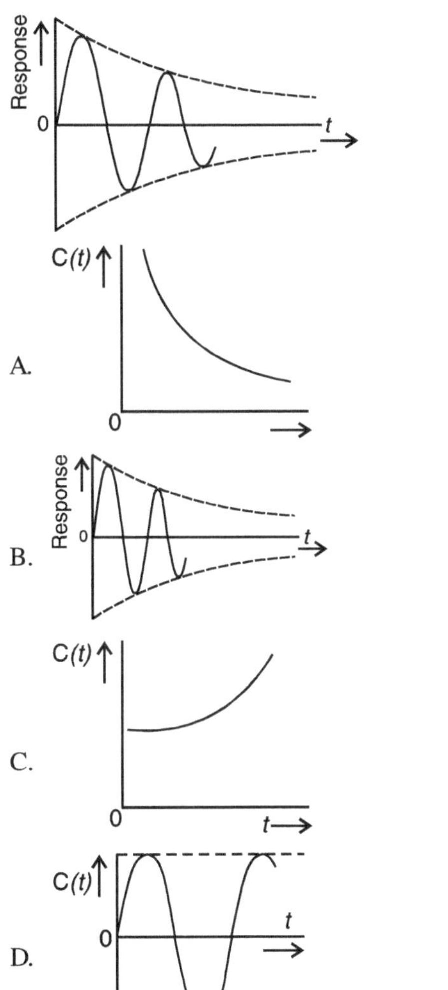

42. Typical current-voltage characteristic of a solar cell is given in the following figure by

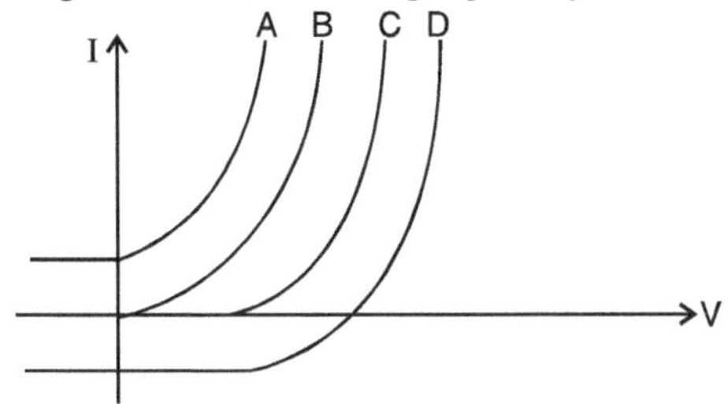

A. curve A B. curve B

C. curve C D. curve D

43. The Inverse Laplace transform of the function $\dfrac{s+5}{(s+1)(s+3)}$ is

A. $2e^{-t} - e^{-3t}$ B. $2e^{-t} + e^{+3t}$

C. $e^{-t} - 2e^{-3t}$ D. $e^{+t} + 2e^{-3t}$

44. Match List I with List II and select the correct answer

List I	***List II***
(a) Emitter	1. Moderate resistivity
(b) Base	2. Very high resistivity
(c) Collector	3. Large size
(d) Substrate	4. Very high conductivity

Codes:	*(a)*	*(b)*	*(c)*	*(d)*
A.	4	1	2	3
B.	1	4	2	3
C.	4	1	2	3
D.	1	4	3	2

45. The following system conveniently displays the impact of poles and zeros on phase margin and gain margin:

A. Root locus

B. Nyquist plot

C. Bode plot

D. Routh-Hurwitz criterion

46. The final value of $\dfrac{2s+1}{s^4+8s^2+16s^2+s}$ is

A. infinity B. 20

C. zero D. 1

47. In hydraulic control system, power is transmitted through:

A. Mechanical gears

B. Shafts

C. Fluid under pressure

D. None of these

48. The given figure shows the root-focus of open-loop transfer function of a control system.

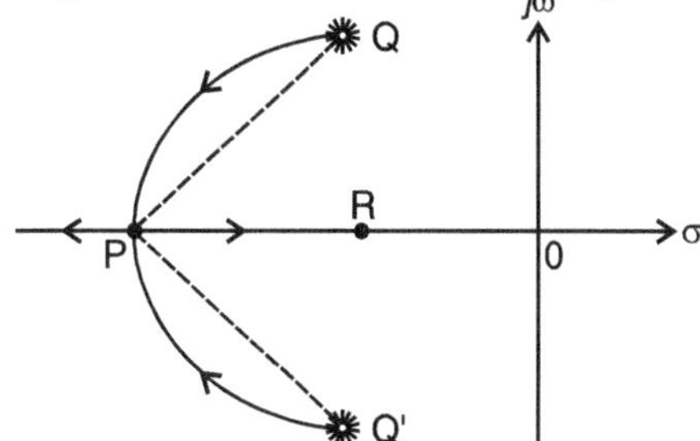

PQ = 2.6 = PQ'
PR = 1.4
OR = 2.0
OQ = 1.4 = OQ'

The value of the forward path gain K at the point P is

A. 0.2 B. 1.4
C. 3.4 D. 4.8

49. The number of times the point A and B are encircled is

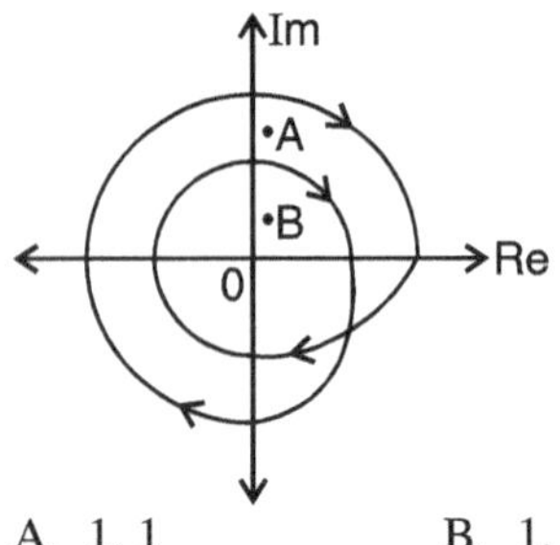

A. 1, 1 B. 1, 2
C. 2, 1 D. 2, 2

50. A.C. systems are usually preferred to the D.C. systems in control applications because:

A. A.C. systems are cheaper
B. A.C. system are more stable
C. A.C. systems have better performance characteristics and are smaller in size
D. All of these

51. At the point where 180° locus crosses the *j*-axis the system is:

A. Absolutely stable
B. Absolutely unstable
C. Conditionally stable
D. None of these

52. An impedance has the pole-zero pattern shown in Fig. It must be composed of

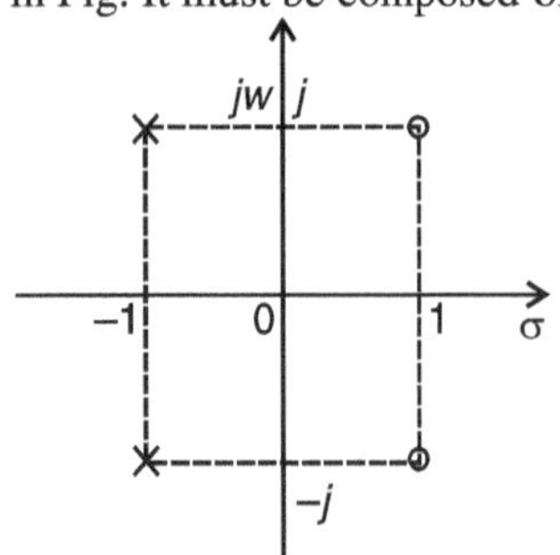

A. RC elements only
B. RL elements only
C. LC elements only
D. RLC elements

53. A system has characteristic equation as $s^2 + 2s + 8$. The damping ratio and natural frequency of oscillation of the system respectively are:

A. $2\sqrt{2}, 0.5$ B. $0.5, 2\sqrt{2}$
C. $0.353, 2\sqrt{2}$ D. 2, 0.353

54. For a given G(s) H(s) the complete polar plot is shown in the given figure. If all T's are positive then how many zeros of the closed-loop system will be there in the RHP?

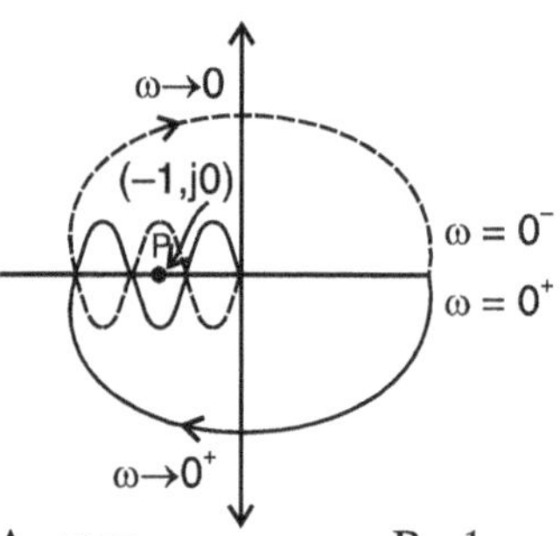

A. zero B. 1
C. 2 D. 3

55. If the poles of system are lying on the imaginary axis in *s*-plane. The system will be

A. Conditionally stable
B. Stable
C. Marginally stable
D. Unstable

56. The magnitude plot for a minimum phase function is:

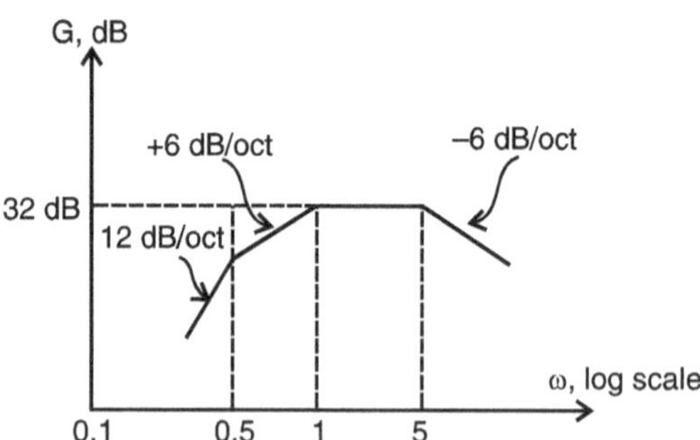

The phase plot for this function

A. cannot be uniquely determined
B. will be monotonically increasing from 0° to 180°

C. will be monotonically decreasing from 180° to 0°

D. will be monotonically decreasing from 180° to – 90°

57. For all frequencies, a unit circle in the Nyquist plot transforms into:

A. 0 dB line of amplitude plot in Bode diagram

B. 1 dB line of amplitude plot in Bode diagram

C. (B) or (A)

D. None of these

58. For the control system in the given figure to be critically damped the value of the gain K required is :

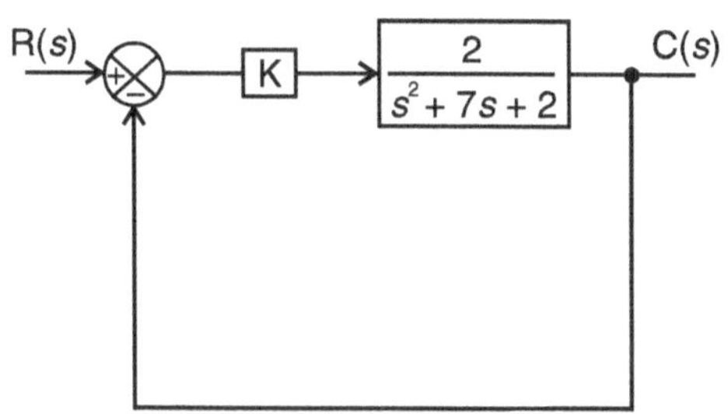

A. 1.4 B. 5.125

C. 6.831 D. 10.35

59. The voltage applied to the rotor of a synchro is always

A. D.C. – ve B. D.C. + ve

C. A.C. D. May be A.C. or D.C.

60. 1. Nyquist criterion is in frequency domain

2. Bode Plot is in frequency domain

3. Root locus plot is in time domain

4. Routh Hermitz's criterion is in time domain.

A. 1, 2, & 3 are correct

B. 2, 3 & 4 are correct

C. 1 & 2 are correct

D. all four are correct

61. In the derivative error compensation

A. damping decreases and settling time decreases

B. damping increases and settling time increases

C. damping decreases and settling time increases

D. damping increases and settling time decreases

62. When all the roots of the characteristic equation are found in the left half of s-plane, the system response due to initial condition will

A. increase to infinity as time approaches infinity

B. decreases to zero as time approaches infinity

C. remain constant for all time

D. be oscillating

63. Match List I with List II and select the correct answer

List I **List II**

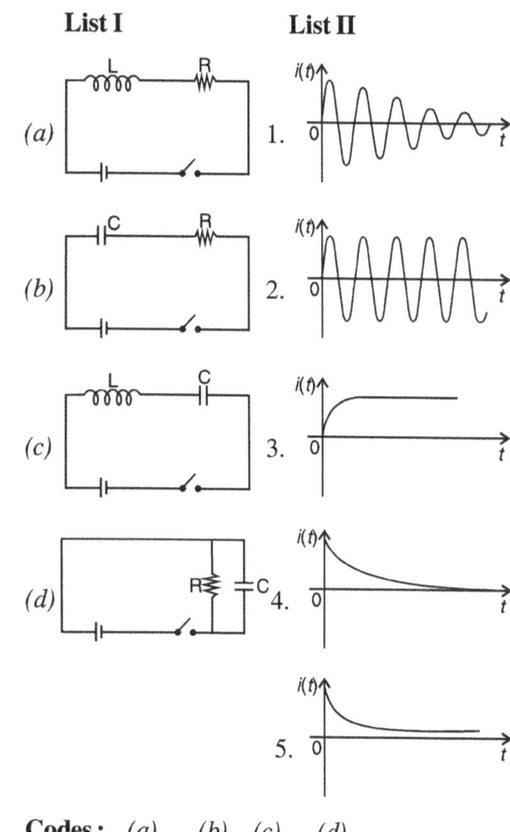

Codes :	*(a)*	*(b)*	*(c)*	*(d)*
A.	5	2	3	1
B.	2	4	3	5
C.	5	3	2	1
D.	2	3	1	4

64. Two RC circuits each haing transfer function $\left(\frac{1}{1+s\tau}\right)$ are connected in cascade. The overall transfer function of the combination is:

A. $\frac{1}{(1+s\tau)^2}$ B. $\frac{1}{1+3s\tau+s^2\tau^2}$

C. $\dfrac{2}{1+s\tau}$ D. $\dfrac{1}{1+(1+s\tau)^2}$

65. The first column of a Routh array is

s^5	1
s^4	2
s^3	3/2
s^2	– 1/3
s^1	10
s^0	2

How many roots of the corresponding characteristic equation are there in the left-half of the s-plane?

A. 4 B. 3
C. 1 D. 2

66. Match List I with List II and select the correct answer by using the codes given below the lists:

List I *(Characteristic Root Location)*	**List II** *(System characteristic)*
(a) $(-1+j), (-1-j)$	1. Marginally stable
(b) $(-2+j), (-2-j), (2j), (-2j)$	2. Unstable
(c) $-j, j, -1, 1$	3. Stable

Codes:	*(a)*	*(b)*	*(c)*
A.	1	2	3
B.	2	3	1
C.	3	1	2
D.	1	3	2

67. The signal flow diagram of a system is shown in the figure. The number of forward paths and the number of pairs of non-touching loops are respectively,

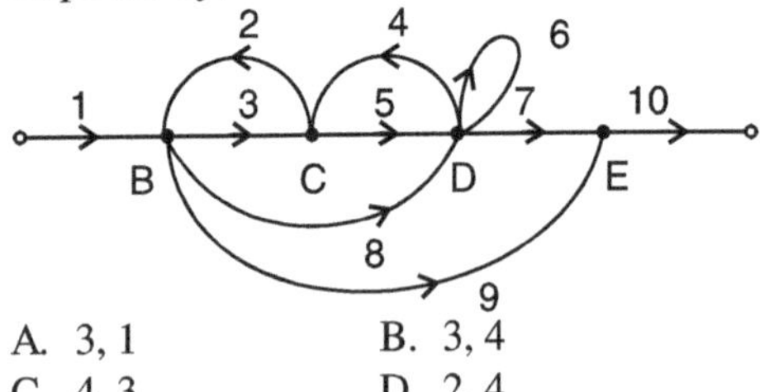

A. 3, 1 B. 3, 4
C. 4, 3 D. 2, 4

68. The number of integrators in the transfer function $G(s) = \dfrac{(s+3)}{s^2(s+5)}$ are:

A. One B. Zero
C. Two D. Three

69. What is the slope change at ω = 10 of the magnitude *v/s* frequency characteristic of a unity feedback system with the following open-loop transfer function?

$$G(jw) = \frac{5(1+j0.1\,\omega)}{j\omega(1+j0.5\omega)\left[1+j0.6(\omega/50)+(j\omega/50)^2\right]}$$

A. – 40 dB/dec to – 20 dB/dec
B. 40 dB/dec to 20 dB/dec
C. – 20 dB/dec to – 40 dB/dec
D. 40 dB/dec to – 20dB/dec

70. Match List I with List II and select the correct answer by using the codes given below the lists:

List I	**List II**
(a) Synchro	1. Amplifier
(b) Amplidyne	2. Actuator
(c) Servo	3. Compensator
(d) RC Network	4. Transducer

Codes :	*(a)*	*(b)*	*(c)*	*(d)*
A.	1	2	3	4
B.	4	3	2	1
C.	3	2	4	1
D.	4	1	2	3

71. In tachogenerators used for measurement of speed of a shaft, frequent calibration has to be done because

A. the contacts wear off
B. the strength of permanent magnet decreases with age
C. the armature current produces heating effect
D. there is back emf

72. System function V(s)/ I (s) has a pole at $s = -2$ and a zero at $s = 1$. For sinusoidal current excitation, the voltage response:

A. Is zero
B. Leads with reaspect to the current
C. Lags with respect to the current
D. In phase with the current

73. For a second order type 1 system, the resonance peak will occur when the system gain is at the:

A. Critical damping value
B. Under damping value
C. Over damping value
D. None of these

74. For a second order system $2\dfrac{d^2y}{dt^2}+4\dfrac{dy}{dt}+8y$

$= 8x$

The damping ratio is

A. 0.1 B. 0.25
C. 0.333 D. 0.5

75. What is transfer function of the circuit shown in the figure?

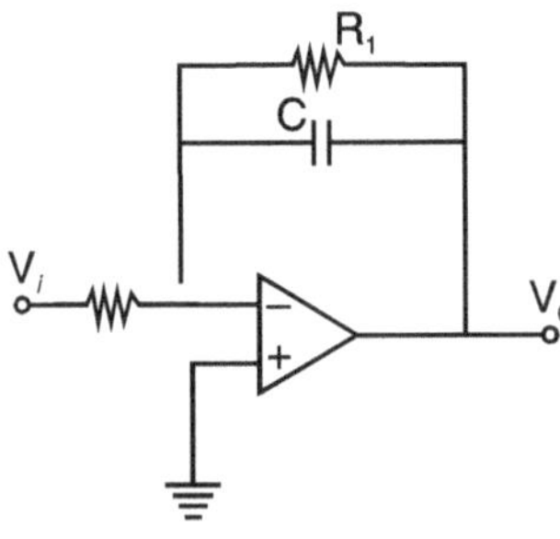

A. $\frac{V_0}{V_i} = +\frac{1}{R_2(1 + j\omega\, C_1R_1)}$

B. $\frac{V_0}{V_i} = -\frac{R_1}{R_2(1 + j\omega\, C_1R_1)}$

C. $\frac{V_0}{V_i} = +\frac{1}{R_1(1 + j\omega\, C_1R_1)}$

D. $\frac{V_0}{V_i} = -\frac{1}{R_2(1 + j\omega\, C_1R_1)}$

76. A transfer function with one or more zeros in the RHS plane is called:

A. Minimum phase transfer function
B. All pass transfer function
C. Non-minimum phase transfer function
D. None of these

77. Find the number of times points A and B have been enclosed by the curve

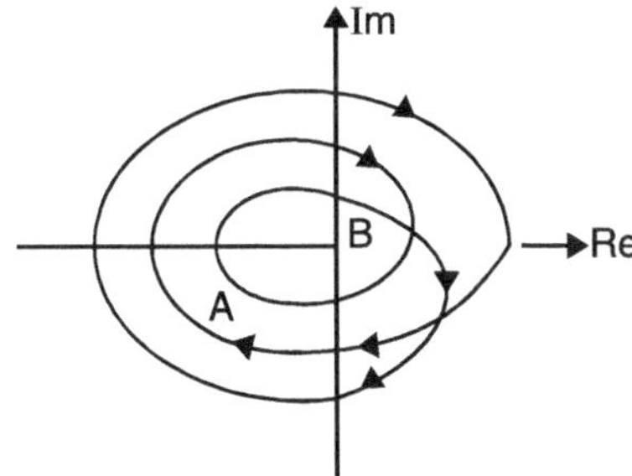

A. 2, 3 B. 3, 2
C. 1, 0 D. 2, 0

78. A Phase lag-lead network shifts the phase of a control signal in order that the phase of the output

A. Lags at low frequencies and leads at high frequencies relative to input
B. Leads at low frequencies and lags at high frequencies relative to input
C. Lags at all frequencies relative to input
D. Leads at all frequencies relative to input

79. The block diagram shown below is equivalent to:

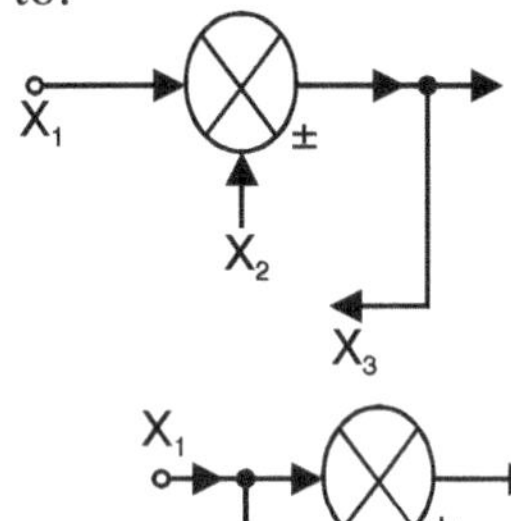

A.

B.

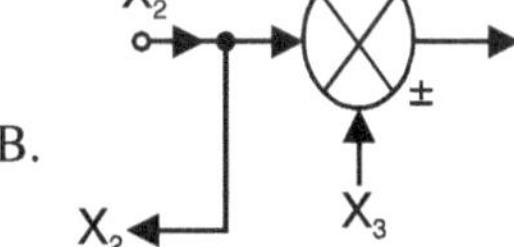

C.

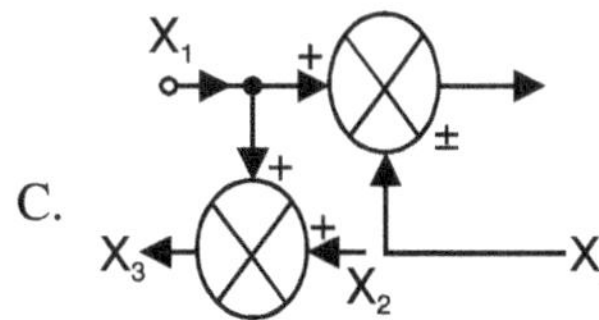

D. None of the above

80. In control systems, we have

I. Nyquist criterion
II. Bode plot
III. Root Locus plot
IV. Routh Hurwitz criterion

Which of the above are in time domain?

A. I and II only B. II and IV only
C. I and III only D. II and IV only

81. Octave frequency range is given by:

A. $\frac{\omega_1}{\omega_2} = 2$ B. $\frac{\omega_1}{\omega_2} = 4$

C. $\frac{\omega_1}{\omega_2} = 8$ D. $\frac{\omega_1}{\omega_2} = 10$

82. The transfer function for the system given below is:

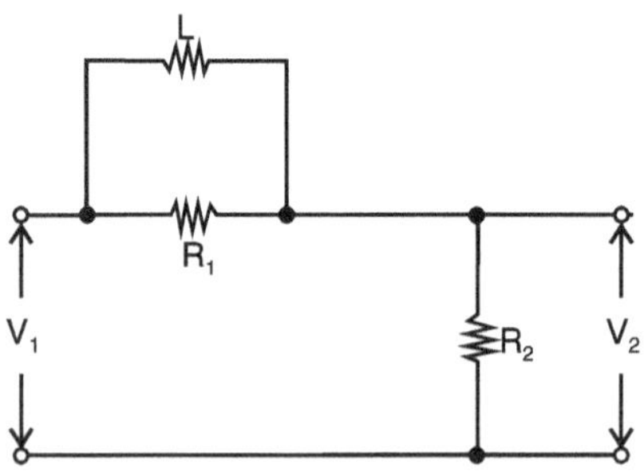

A. $\frac{R_2}{(1+R_1Ls)/R_1 + Ls}$

B. $\frac{R_1 + sL}{sL\left(\frac{R_1+R_2}{R_2}\right)+R_1}$

C. $s\frac{R_2}{R_1+R_2}+\frac{R_1}{L}$

D. None of these

83. The Bode plot of the transfer function G(s) is

A. constant magnitude and constant phase shift angle

B. – 20 dB/decade and constant phase shift angle.

C. 20 dB/decade and phase shift of $\pi/2$

D. zero magnitude and phase shift.

84. The system response can be tested better with

A. Sinusoidal input signal

B. Unit impulse input signal

C. Ramp input signal

D. Exponentially decaying signal

85. Under which of the following conditions, the transfer function technique is considered as inadequate?

A. Systems having stability problems

B. Systems having multiple input disturbances

C. Systems having complexities and non-linearities

D. All of the above

86. Consider the vectors drawn from the poles and zero at $j\omega = j1$ on the imaginary axis as shown in the given figure. The transfer function $G(j1)$ is given by

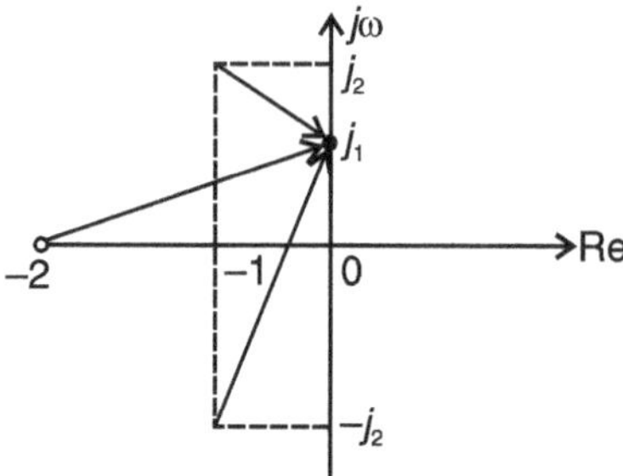

A. $\frac{1}{2}\angle 0°$ B. $2.7\angle -31°$

C. $2\angle 45°$ D. $2\angle -67.4°$

87. A device used for the conversion of ordinates is

A. Selsyn

B. Miscrosyn

C. Synchro transformer

D. Synchro resolver

88. The transfer function C/R of the system block shown below is given by:

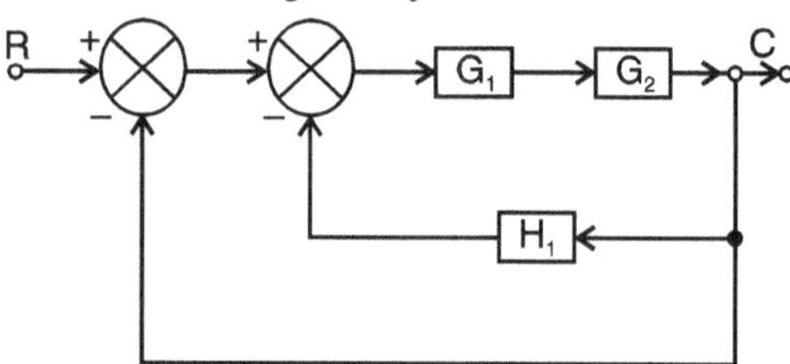

A. $\frac{G_1G_2}{1+H_1G_1G_2}$

B. $\frac{G_1+G_2}{1+(G_1+G_2)H_1+G_1G_2}$

C. $\frac{G_1G_2}{1+G_1G_2H_1+G_1G_2}$

D. $\frac{G_1G_2(1+H_1+G_1G_2)}{1+G_1G_2H_1+G_1G_2}$

89. Thr Routh criterion tells us the number of roots lying:

A. In the left hand side half s-plane

B. In the right half s-plane

C. On the orgin of s-plane

D. None of these

90. The open-loop transfer function of a unity feedback control system is

$$G(s)=\frac{1}{(s+2)^2}$$

The closed-loop transfer function will have poles at

A. $-2, -2$ B. $-2, -1$

C. $-2, \pm j1$ D. $-2, 2$

91. For a linear transfer function of the variable, the transfer function of the system

A. Changes

B. Does not change

C. Has no relation at all

D. Depends on other factors also

92. Loop transfer function zeros are the same as:

A. Poles of G(s) H (s)

B. Zeros of G(s) H(s)

C. Poles of 1 + G(s) H(s)

D. Zeros of 1+ G(s) H(s)

93. The form of the transfer function used in the Bode plot

A. $G(j\omega)$ B. G(s)

C. (A) or (B) D. None of the above

94. When the synchro transmitter power is made on its shaft should

A. Be at no load

B. Be at full load

C. Either of A or B

D. Should be engaged with its driving gear

95. Which one of the following is a D to A conversion technique?

A. Successive approximation

B. Weighted resistor technique

C. Dual slope technique

D. Single slope technique

96. The feedback control system shown in the given figure represents a

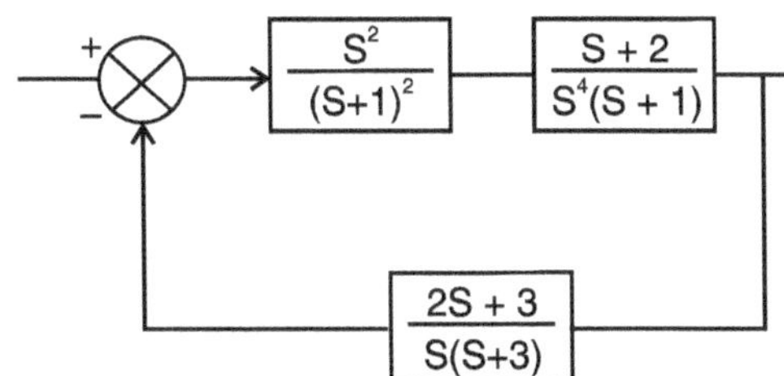

A. Type 0 system B. Type 1 system

C. Type 2 system D. Type 3 system

97. The Bode plot of a constant is a

A. Line with a slope of – 20 dB/decade

B. Line with a slope of – 6 dB/decade

C. Straight line parallel to frequency axis and corresponding phase is 0° or 180° depending on whether constant is positive or negative

D. None of the above

98. The reference phase and control phase in the A.C. servomotor are

A. Always out of phase by 90°

B. Always in phase with each other

C. Always at a phase difference of 120°

D. It can be any value depending upon the type reference voltage used.

99. In system using relays, the signals in the part of the system are in the form of

A. Sinusoidal wave

B. Modulated wave with frequency modulation

C. Demodulated wave with amplitude modulation

D. Rectangular wave with polarity depending upon that of the actuating signal.

100. The following device is based on the law of conservation of momentum:

A. Synchros B. Gyroscope

C. Hydraulic lifter D. None of these

101. In pneumatic control systems, the medium is:

A. Air B. Liquid

C. Oil D. Hydrogen

102. The system matrix of a continuous-time system, described in the state variable form is

$$A=\begin{bmatrix} x & 0 & 0 \\ 0 & y & -1 \\ 0 & 1 & -2 \end{bmatrix}$$

The system is stable for all values of x and y satisfying

A. $x < 1/2, y < 1/2$ B. $x > 1/2, y > 0$

C. $x < 0, y < 2$ D. $x < 0, y < 1/2$

103. Consider the following overall transfer function for a unity of feedback system:

$$\frac{4}{s^2+4s+4}$$

Which of the following statements regarding this

system are correct?

1. Position error constant K_p for the system is 4.
2. The system is type one.
3. The velocity error constant K_v for the system is finite

Select the correct answer

A. 1, 2 and 3 B. 1 and 3
C. 2 and 3 D. 1 and 2

104. The signal flow diagram of a system is shown in the given figure. The number of forward paths and the number of pairs of non-touching loops are respectively

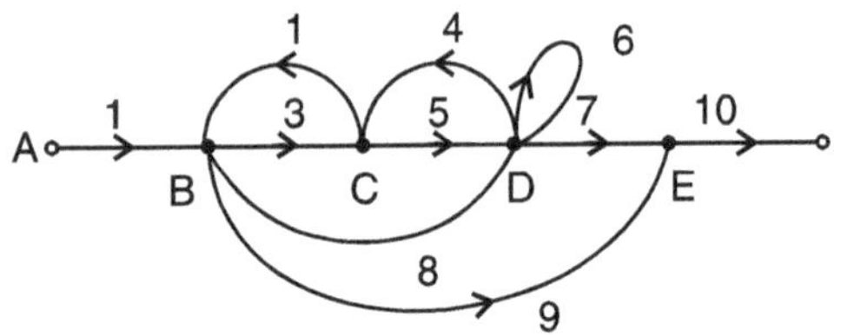

A. 3, 1 B. 3, 2
C. 4, 2 D. 2, 4

105. In a root locus plot, increase in K will:

A. Decrease the overshoot of the response
B. Incrase the overshoot of the response
C. Not alter the overshoot of the response
D. None of these

106. The error signal produced in a control system is a constant. The output of D action will be:

A. Zero B. Linear
C. Constant D. Infinity

107. An amplidyne can give the characteristics as

A. Constant voltage
B. Constant current
C. Both (A) and (B) above
D. Constant voltage, constant current and constant power

108. The stator of synchro is made up of

A. Cast iron
B. High speed steel
C. High carbon steel
D. Silicon steel

109. The step size of a stepper motor with 12 rotor tooth is

A. 10° B. 30°
C. 60° D. 120°

110. The following Bode plot represents :

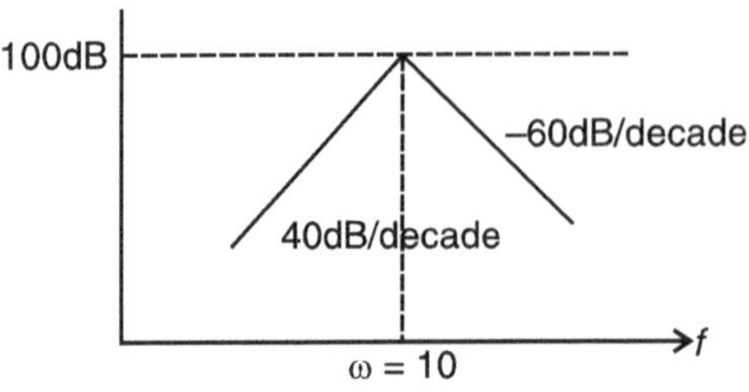

A. $\frac{100\,s^2}{(.1+0.1s)^3}$ B. $\frac{1000\,s^2}{(1+0.1s)^3}$

C. $\frac{100\,s^2}{(1+0.1s)^5}$ D. $\frac{1000\,s^2}{(1+0.1s)^5}$

111. The circuit given represents as:

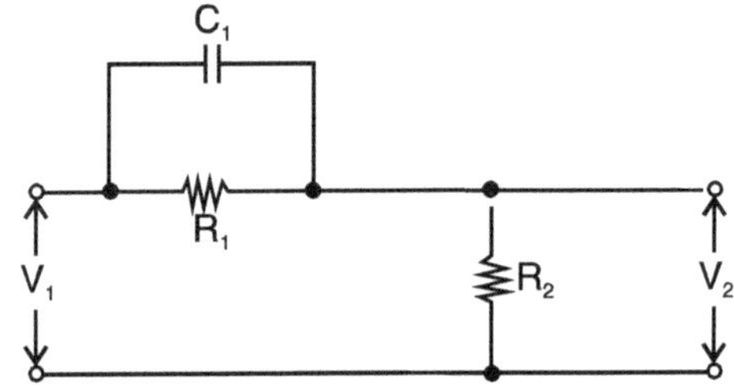

A. Lag network
B. Lead network
C. Lag-lead network
D. None of these

112. The following device works as a error detector:

A. Synchro-transmitter
B. Control transformer
C. Both the above two
D. None of these

113. The characteristic represented in options relay with dead zero and ideal relay respectively are

A. A and B B. B and C
C. A and C D. B and A

114. Which of the following is an electromechanical device

A. Induction relay B. D.C. motor
C. LVDT D. All of these

115. Non-linearity in the servosystem due to saturation is caused by:

A. Servo motor B. Relays
C. Gear trains D. All of these

116. The ac motor used in servo application is a

A. single-phase induction motor
B. two-phase induction motor

C. three-phase induction motor
D. synchronous motor

117. In position control systems, the device used for providing rate-feedback voltage is called
A. potentiometer
B. synchro transmitter
C. synchro transformer
D. technogenerator

118. A synchro transmitter-receiver unit is a
A. two-phase ac device
B. 3-phase ac device
C. dc device
D. single-phase ac device

119. Match the control system components in List I with their functions in List II and select the correct answer using the codes given below the lists:

List I	List II
(a) Servo motor	1. Error detector
(b) amplidyne	2. Transducer
(c) Potentiometer	3. Actuator
(d) Flapper value	4. Power amplifier

Codes:	*(a)*	*(b)*	*(c)*	*(d)*
A.	3	4	2	1
B.	4	1	2	3
C.	4	1	3	2
D.	3	2	4	1

120. Which of the following is not true for signal Flow Graph
A. A signal flow graph applies only to linear systems
B. Nodes are used to represent variables
C. Branch directing from node y_k to y_j represents the dependence of variable y_j upon k and vice versa
D. A signal y_k travelling along a branch between nodes y_k and y_j is multiplied by the gain of branch a_k, so that a signal $a_{kj}\, y_k$ is delivered at node y_j.

121. Consider the following statements:
The gain margin and phase margin of an unstable system may respectively be
1. positive, positive
2. positive, negative
3. negative, positive
4. negative, negative

Of these statements
A. 2 and 4 are correct
B. 1 and 3 are correct
C. 1, 2 and 3 are correct
D. 2, 3 and 4 are correct

122. The transfer function of a compensating network is of the form $\frac{1+\alpha Ts}{(1+Ts)}$.
If this is phase-lag network the value of α should be
A. exactly equal to 0
B. between 0 and 1
C. exactly equal to 1
D. greater than 1

123. Which differential equation correctly represents the mathematical model of the mechanical system shown?

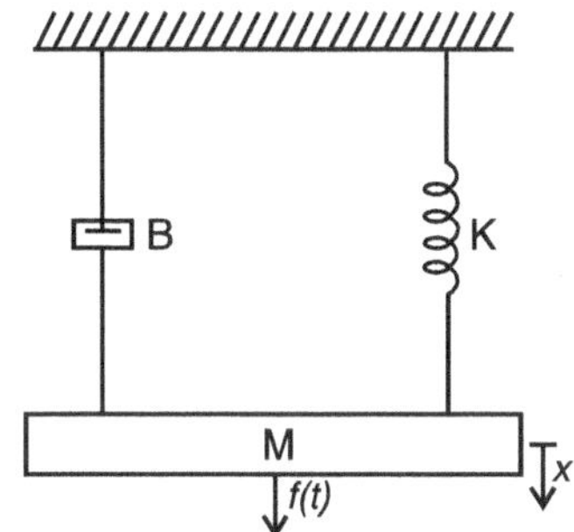

x = displacement in M
M = mass in kg
B = damping constant, $\frac{\text{N}-\text{sec}}{m}$
K = spring constant, N/m
f = force, Newton

A. $f(t)-\text{M}\frac{d^2x(t)}{dt^2}-\text{B}\frac{dx(t)}{dt}-\text{K}x(t)=0$

B. $f(t)+\text{M}\frac{dx(t)}{dt}-\text{B}x(t)-\text{K}\int x(t)\,dt=0$

C. $f(t)+\text{M}\frac{dx(t)}{dt}+\text{B}x(t)+\text{K}\int x(t)\,dt=0$

D. $f(t)-\text{M}\frac{d^2x(t)}{dt^2}+\frac{1}{\text{B}}\frac{dx(t)}{dt}+\frac{1}{\text{K}}x(t)=0$

124. The algebraic sum of the angles of the vectors from all poles and zeros to the point on any root

locus segment is:
A. 45°
B. 90°
C. 180°
D. 110° or its odd multiple

125. A unity feedback control system has a forward path transfer function equal to

$$\frac{42.25}{s(s+6.5)}$$

The unit step response of this system starting from rest, will have its maximum value at a time equal to

A. 0 sec B. 0.56 sec
C. 5.6 sec D. infinity

126. The gain-phase plots of open-loop transfer functions of four different systems labelled A, B, C and D are shown. The correct sequence of the increasing order of stability of these four systems will be

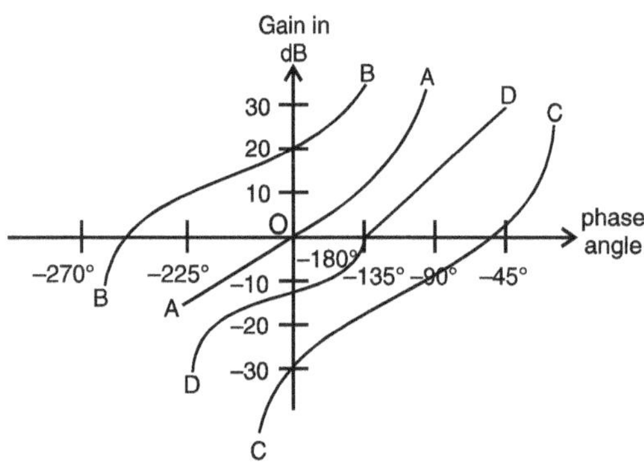

A. A, B, C, D B. D, C, B, A
C. B, A, D, C D. B, C, D, A

127. If the open-loop transfer function of the system is $G(s)\,H(s) = \dfrac{K(s+10)}{s(s+8)(s+16)(s+72)}$, then a closed-loop pole will be located at $s = -12$, when the value of K is

A. 4355 B. 5760
C. 9690 D. 9862

128. The following device is usually used for transmission of data:
A. Rotary LVDT
B. Synchros
C. Both the above two
D. None of these

129. For the signal-flow graph shown

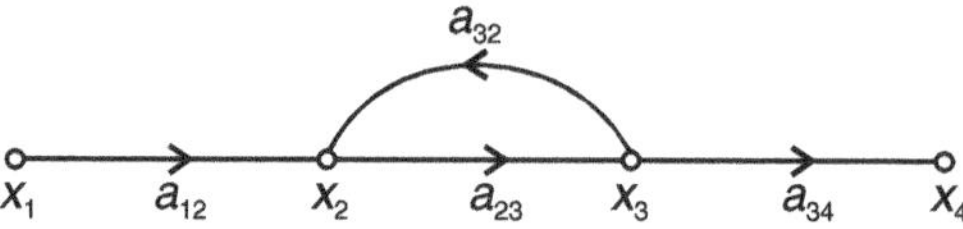

the value of x_2 is
A. $a_{12}\, x_1$
B. $a_{12}\, x_1 + a_{32}\, x_3$
C. $a_{12}\, x_1 + a_{23}\, x_3 + a_{32}\, x_3$
D. $-\, a_{23}\, x_3 + a_{34}\, x_4$

130. A transfer function having open right-half plane zeros is a/an
A. minimum phase function
B. non-minimum phase function
C. unstable function
D. constant phase function

131. From the signal flow graph shown in the figure, the value of x_6 is:

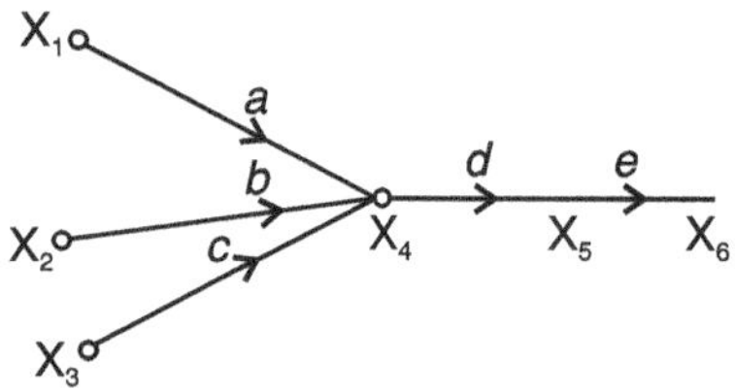

A. de $(ax_1 + bx_2 + cx_3)$
B. $(a + b + c)\,(x_1 + x_2 + x_3)\,(d + e)$
C. $(ax_1 + bx_2 + cx_3)\,(d + e)$
D. $abcde\,(x_1 + x_2 + x_3)$

132. From the figure shown the transfer function of the signal flow graph is

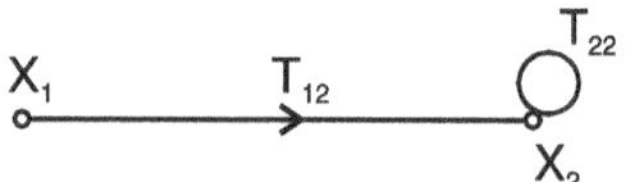

A. $\dfrac{T_{12}}{1-T_{22}}$ B. $\dfrac{T_{22}}{1-T_{12}}$

C. $\dfrac{T_{12}}{1+T_{22}}$ D. $\dfrac{T_{22}}{1+T_{12}}$

133. With variable reluctance tachometer having 60 rotor teeth, the counter records 3600 counts per minute. The device speed, in rpm, would be
A. 36 B. 60
C. 360 D. 3600

134. Consider a negative feedback system where

$$G(s) = \frac{1}{(s+1)},\ H(s) = \frac{K}{s(s+2)}$$

The closed-loop system is stable for

A. $K > 20$ B. $15 < K < 29$

C. $8 \le K \le 14$ D. $K < 6$

135. For the linear feedback system shown in the given figure, the transfer function and states are respectively

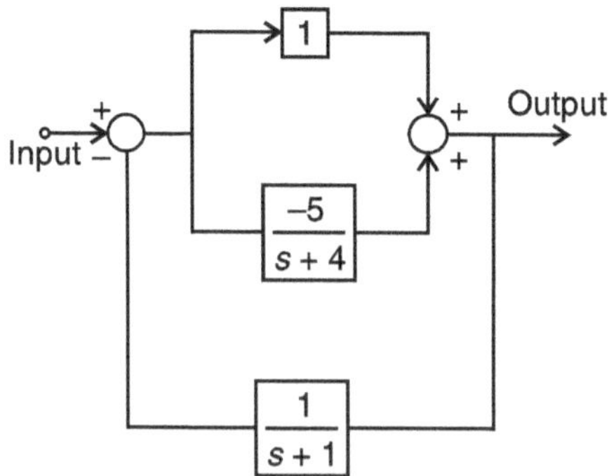

A. strictly stable and not stable

B. strictly stable and stable

C. not strictly stable and stable

D. not stable and not stable

136. Consider the Root Locus Diagram of a system and the following statements:

1. The open loop system is a second order system
2. The system is overdamped for $K > 1$
3. The system is absolutely stable for all values of R

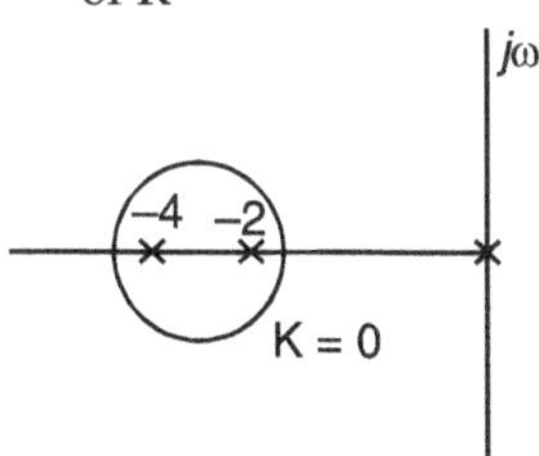

Which of these statements are correct?

A. 1, 2 & 3 B. 1 & 3

C. 2 & 3 D. 1 & 2

137. A unit impulse function on differentiation results in

A. unit doublet

B. unit triplet

C. unit parabolic function

D. unit ramp function

138. The pole zero pattern of a certain filter is shown. The filter is of the following type.

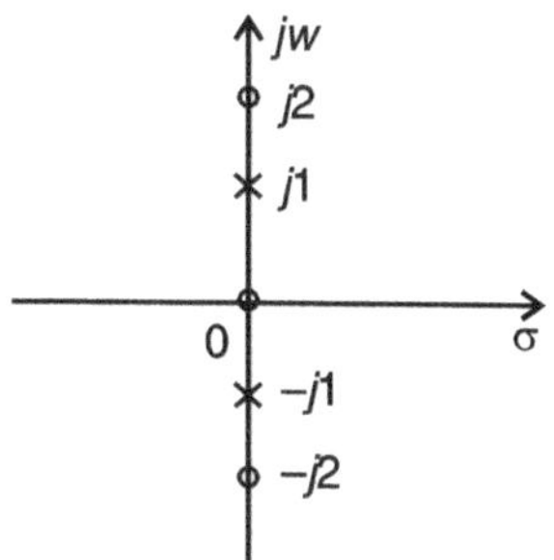

A. low-pass B. high-pass

C. band-pass D. all-pass

139. The impedances of starter and rotor windings of the control transformer are:

A. Considerably lower than those of equivalent sized synchros

B. Considerably higher than those of equivalent sized synchros

C. Equal to those of equivalent sized synchros

D. None of these

140. A system has 12 poles and 2 zeros. Its high frequency asymptote in its magnitude plot has a slope of

A. – 200 dB/decade B. – 240 dB/decade

C. – 280 dB/decade D. – 320 dB/decade

141. The effect of adding poles and zeros can be determined quickly by

A. Nicholas chart B. Nyquist plot

C. Bode plot D. Root locus

142. Which one of the following is a positive real function?

A. $\frac{s(s^2+4)}{(s^2+1)(s^2+6)}$

B. $\frac{s(s^2-4)}{(s^2+1)(s^2+6)}$

C. $\frac{s^3+3s^2+2s+1}{4s}$

D. $\frac{s(s^4+3s^2+1)}{(s+1)(s+2)(s+3)(s+4)}$

143. Consider the following statements regarding the number of sign changes in the first column of Routh table in respect of the characteristic equation

$s^2 + 2as + 4$:

1. If $a = +\varepsilon$, where ε = near zero, the number of sign change will be equal to zero
2. If $a = 0$, the nmber of sign changes will be equal to one.
3. If $a = -\varepsilon$, where ε = near zero, the number of sign changes will be equal to two.

Of these statements

A. 1, 2 and 3 are correct
B. 2 and 3 are correct
C. 1 and 2 are correct
D. 1 and 3 are correct

144. Which of the following terms is not related to potentiometers

A. Least count B. Reference voltage
C. Confirmly D. All are related

145. When a current source of value I is suddenly connected across a two-terminal relaxed RC network at time $t = 0$, the observed nature of the voltage across the current source is shown

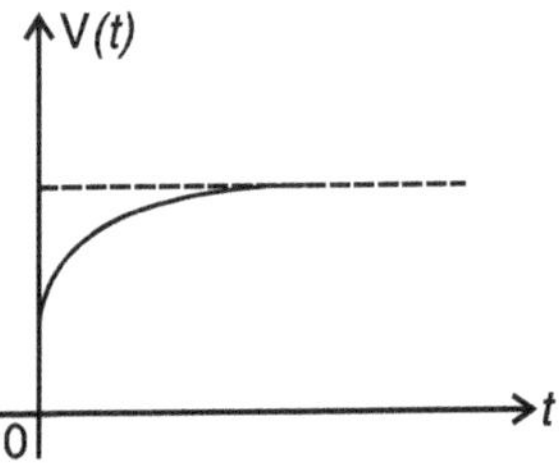

The RC network is

A. a series combination of R and C
B. a parallel combination of R and C
C. a series combination of R_1 and parallel combination of R and C
D. a pure capacitor

146. By adding a pole at $s = 0$ the Nyquist plot of a system

A. Will remain unchanged
B. Will rotate by 90° in clockwise direction
C. Will rotate by 90° in anticlockwise direction
D. Will rotate by 180°

147. In a root locus plot, increase in K will

A. Decrease the natural frequency
B. Increase the natural frequency
C. Not alter the natural frequency
D. None of these

148. Consider the following statements for a variable reluctance stepper motor used in control systems:

1. The static torque acting on the rotor is a function of angular misalignment between stator and rotor teeth.
2. There are two positions of zero torque: $\theta = 0°$, 180°/T (T = number of rotor teeth)
3. Both torqe zero positions are stable.
4. As the stator is excited, the rotor is pulled into the nearest minimum reluctance position.

Of these statements

A. 1, 3 and 4 are correct
B. 1, 2 and 3 are correct
C. 1, 2 and 4 are correct
D. 2, 3 and 4 are correct

149. The transfer function T(s) of the system shown is given by

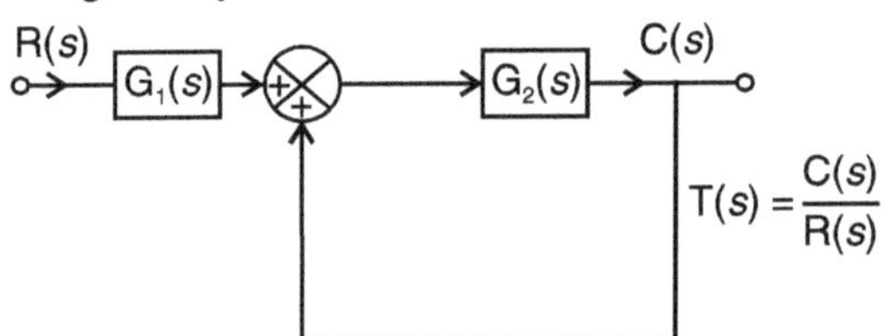

A. $T(s) = \dfrac{G_1(s)G_2(s)}{1 - G_2(s)}$

B. $T(s) = \dfrac{G_1(s)}{1 + G_1(s)G_2(s)}$

C. $T(s) = \dfrac{G_2(s)}{1 + G_1(s)G_2(s)}$

D. $T(s) = \dfrac{1}{1 + G_1(s)G_2(s)}$

150. The necessary and sufficient condition for a rational function of s, T(s) to be driving point impedence of an RC network is that all poles and zeros should be

A. Simple and lie on the negative real axis in the s-plane
B. Complex and lie in the left half of the s-plane
C. Complex and lie in the right half of the s-

plane

D. Simple and lie on the positive real axis of the s-plane

151. Consider the following statements:
Monolithic regulators can be used to
1. realize a constant voltage regulator
2. realize a constant current regulator
3. introduce flexibility in setting output voltage level

Of these statements
A. 1 and 2 are correct
B. 2, and 3 are correct
C. 1 and 3 are correct
D. 1, 2 and 3 are correct

152. Consider a control system shown in the given figure:

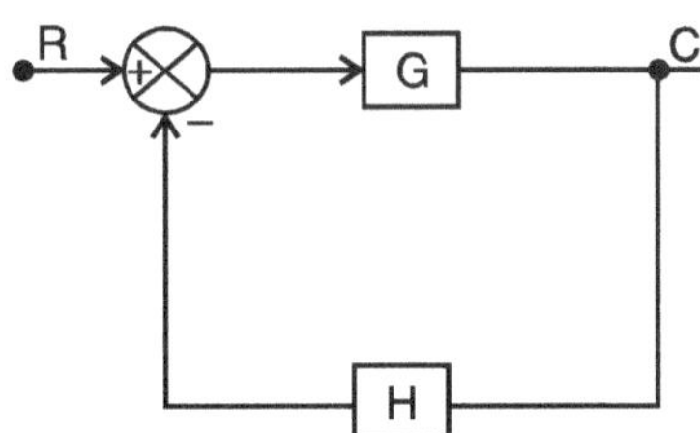

For a slight variation in G, the ratio of open-loop-sensitivity to closed-loop-sensitivity will be given by
A. $1:(1+GH)$ B. $1:(1+GH)^{-1}$
C. $1:(1-GH)$ D. $1:(1-GH)^{-1}$

153. Consider the control system shown and the statements given

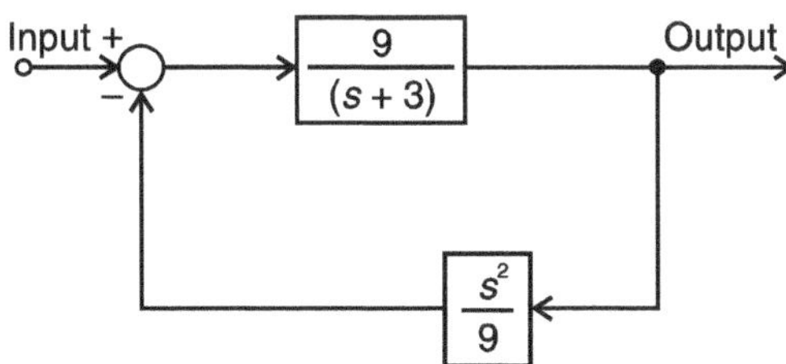

1. The system is of second order
2. Basically, the system is having positive feedback.
3. The system is of type 1.
4. The dimension of the output is not the same as that of the input.

Of these statements
A. 1 and 4 are correct
B. 2 and 4 are correct
C. 2, 3 and 4 are correct
D. 1, 2 and 3 are correct

154. A conductor oriented along the Z-axis moves along a path so as to describe a right circular cylinder with the Z-axis coinciding with its longitudinal axis. There is a uniform magnetic field in the region. In order to generate alternating emf in the conductor, the magnetic field must be along the:
A. Z-axis B. X-axis
C. Y-axis D. X and Y-axis

155. In a two-phase ac servomotor, the rotor has a resistance R and a reactance X. The torque-speed characteristic of the servomotor will be linear provided that

A. $\frac{X}{R} << 1$ B. $\frac{X}{R} >> 1$

C. $\frac{X}{R} = 1$ D. $X^2 = R$

156. $G_A(j\omega)$, $G_B((j\omega)$ and $G_C(j\omega)$ are three transfer functions whose phase variations are shown. The magnitude variation with respect to frequency is the same for both $G_A(j\omega)$ and $G_C(j\omega)$, but their phase variation is as indicated by the curves labelled A and C in the given figure. The transfer function $G_B(j\omega)$ is such that.
$\phi_C(j\omega) = \phi_A(j\omega) + \phi_B(j\omega)$.

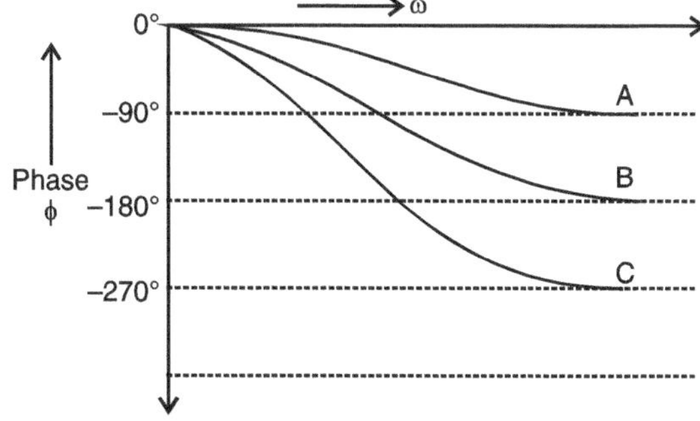

G_A, G_B and G_C are respectively
A. minimum phase, all pass and non-minimum phase functions
B. minimum phase, non-minimum phase and all pass functions
C. all pass, minimum phase and non-minimum

phase functions

D. all pass, non-minimum phase and minimum phase functions

157. Given the unity feedback system with $G(s) = \dfrac{K}{s(s+1)(s+2)}$, the root-locus plot of the system is given by

A. 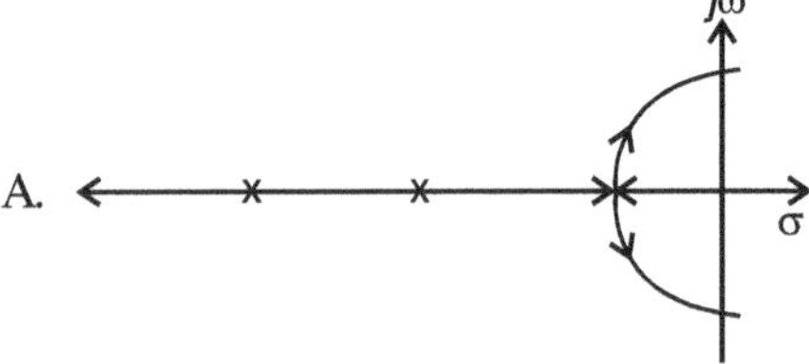

B. 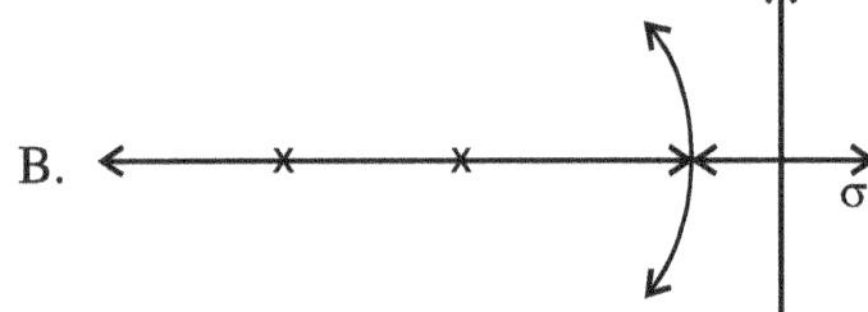

C. 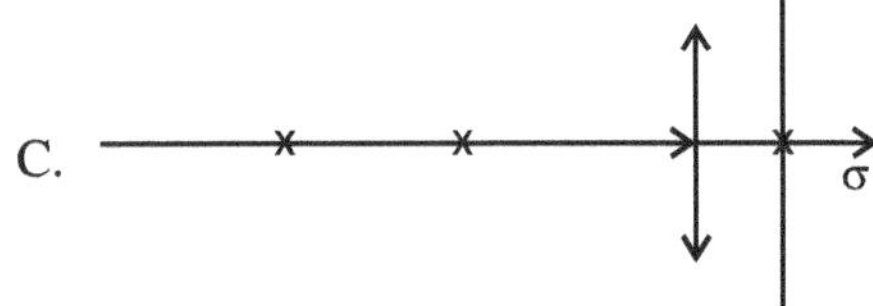

D. 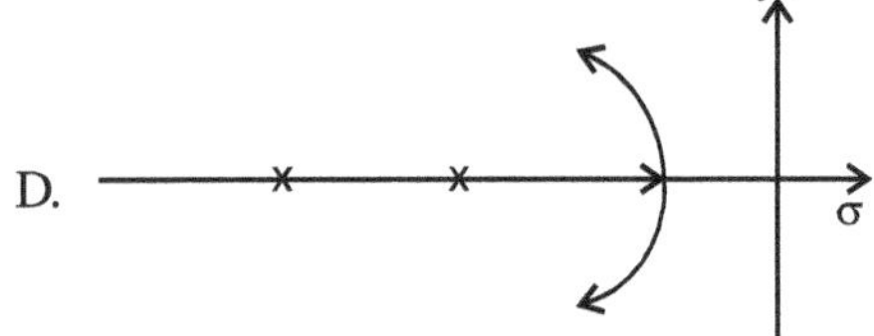

158. The number of operational amplifiers required to design an electronic PID controller is:

A. 1 B. 2
C. 3 D. 4

159. For a second order system with unit step excitation, the steady state output

$$C_{ss}(t) = \lim_{t \to \infty} [c(t)]$$

A. 1 B. Zero
C. Infinity D. 1/2

160. The polar plot of $G(s) = \dfrac{10}{s(s+1)^2}$ intercepts real axis at $\omega = \omega_0$. Then, the real part and ω_0 are respectively given by:

A. – 5, 1 B. – 2.5, 1
C. – 5, 0.5 D. – 5, 2

161. For changing the time constant and providing sufficient damping (supressing the oscillations) the factor responsible is

A. Input signal
B. First derivative of the input signal
C. Second derivative of the input signal
D. None of these

162. Pick up the correct statement:

A. An inductor may be used to form a lag-lead network.
B. A capacitor may be used to form a lag-lead network.
C. Both inductor and capacitor together may be used to form a lag-lead compensator
D. None of these

163. In Nyquist criterion, the roots of the characteristic equation are given by:

A. Zeros of the closed loop transfer function
B. Poles of closed loop transfer function
C. Zeros of the open loop transfer function
D. Poles of the open loop transfer function

164. A restriction-volume combination is basically a pneumatic:

A. Rectifier B. RL circuit
C. RC circuit D. RLC circuit

165. Root loci starts from

A. poles and ends on zeros
B. zeros and ends on poles
C. zeros and ends on infinity
D. poles and ends on infinity

166. The impedance, $Z(s) = \dfrac{s^2+1}{2s}$ is a

A. inductance in series with resistor
B. inductance in series with capacitor and resistor
C. inductance in series with capacitor
D. capacitor in series with resistor.

167. Which of the following statement about the equation below, for Routh Hurwitz criterion is true?

$$2s^4 + s^3 + 3s^2 + 5s + 10 = 0$$

A. It has only one root on the imaginary axis

B. It has two roots in the right half of the s-plane
C. The system is unstable
D. The system is stable

168. What will happen if the value of K (gain of the system) is zero
A. The roots move away from the poles
B. The roots coincide with the poles
C. The roots move away from the zeros
D. None of the above

169. Which feedback which of the following reduces?
A. System gain
B. System stability
C. Both (A) and (B) above
D. None of the above

170. A network has its response with time as shown. Which one of the following diagrams represents the location of the poles of this network?

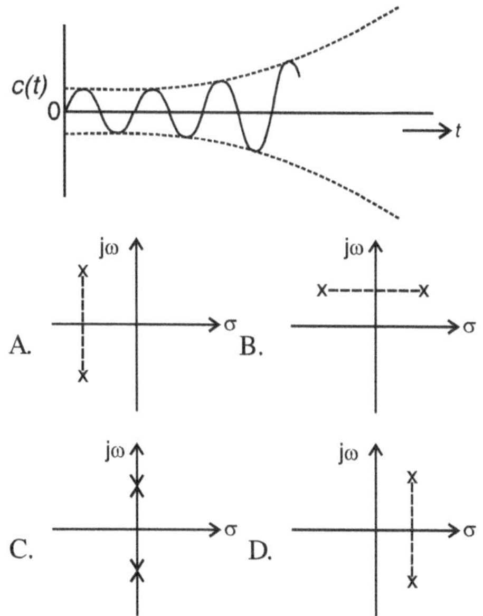

171. In the type 1 system, the velocity log error is
A. inversely proportional to bandwidth of the system
B. directly proportional to gain constant
C. inversely proportional or gain constant
D. independent of gain constant

172. Dither phenomenon can be used to overcome the
A. large steady state error
B. Coulombic friction
C. unstability of a system
D. viscous friction

173. The main cause of absolute instability is
A. parameters of the controlled system
B. Parameters of the controlling system
C. Parameters of the feedback system
D. Error detector where the two signals are compared.

174. When the X/R in a two phase induction motor is decreased, the maximum torque occurs at:
A. Rated speed
B. Constant speed
C. Maximum voltage
D. Minimum voltage

175. To achieve a linear torque-speed characteristics in servomotors the X/R ratio is
A. Kept large
B. Kept small
C. Kept Medium
D. Speed is independent of the X/R ratio

176. The purpose of the series quadrature windings in an amplidyne is to
A. neutralise the effect of armature reaction
B. reduce commutation difficulties
C. increase the gain
D. increase the response time

177. A control system is as shown in the given figure. The maximum value of gain K for which the system is stable is

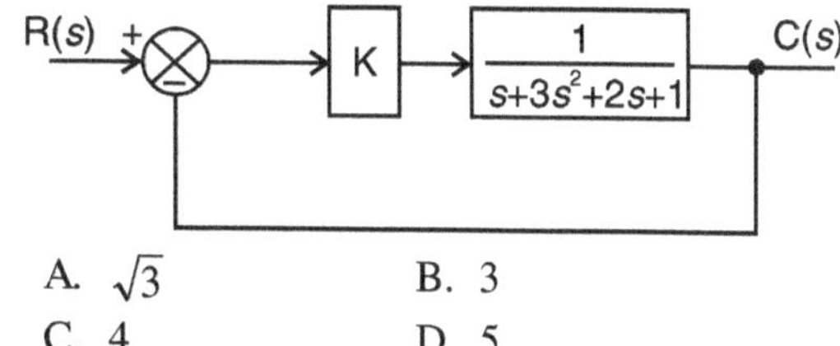

A. $\sqrt{3}$ B. 3
C. 4 D. 5

178. Addition of poles in transfer function causes:
A. Lead compensation
B. Lag compensation
C. Lag-lead compensation
D. None of these

179. The root locii of a system has 4 separate locii The system can have:
A. Four poles and four zeros
B. Four poles or four zeros
C. Two poles and two zeros

D. Six poles and two zeros

180. The open-loop transfer function of a system is given by

$$G(s)\,H(s) = \frac{K(s+1)(s+3)}{s^2+4s+8}$$

Its root locus diagram is

A.

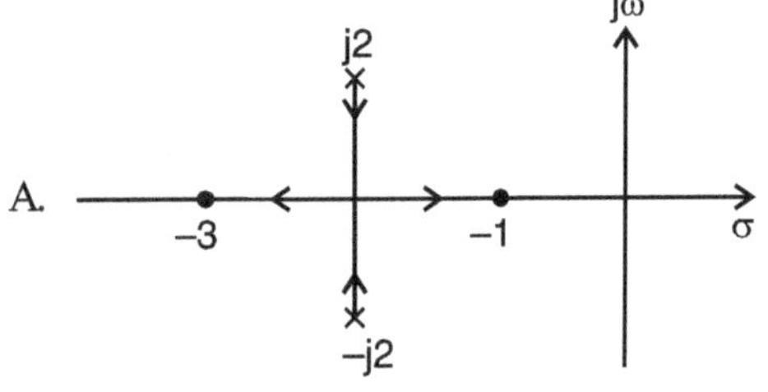

B.

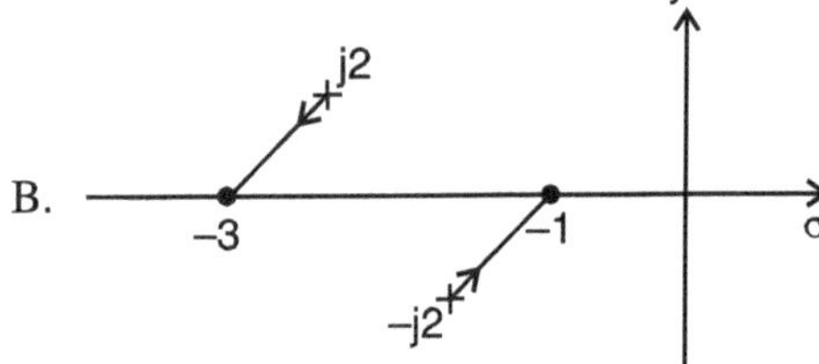

C.

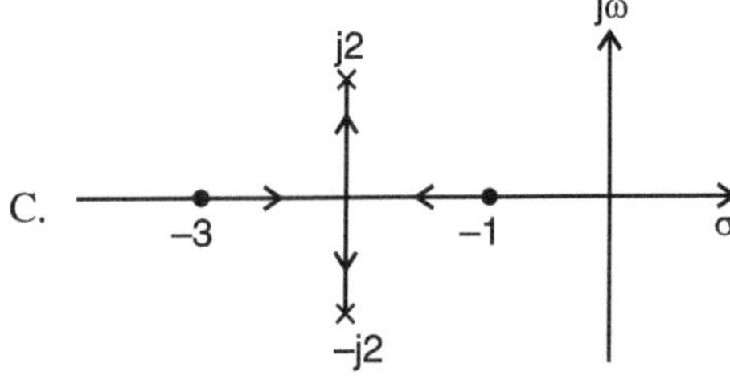

D.

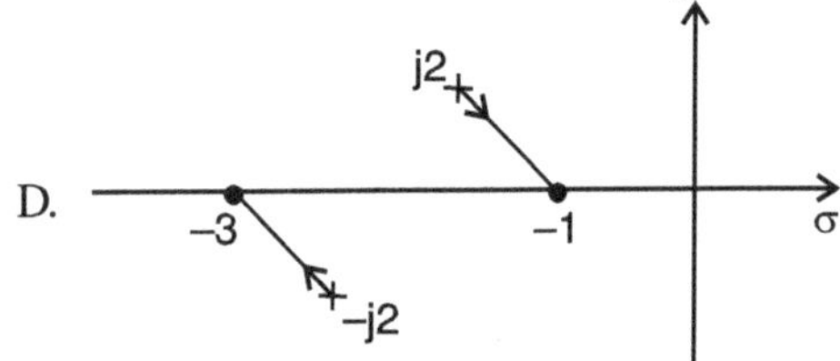

181. Dither phenomenon can be used to overcome the

A. Large steady state error

B. Coulombic friction

C. Unstability of a system

D. Viscous friction

182. A bimetallic thermostat is a:

A. Zero term controller

B. One term controller

C. Two term controller

D. On-off controller

183. The slope in Bode plot is expressed as:

A. – 6 dB/octave B. – dB/decade

C. – 20 dB/octave D. – 20 dB/decade

184. Match List I with List II and select the correct answer:

List I	**List II**
(a) $s^2 + 15s + 56.25$	1. Undamped
(b) $s^2 + 5s + 6$	2. Underdamped
(c) $s^2 + 20.25$	3. Critically damped
(d) $S^2 + 4.5s + 42.25$	4. Overdamped

Codes:	*(a)*	*(b)*	*(c)*	*(d)*
A.	3	4	2	1
B.	3	4	3	2
C.	4	3	2	1
D.	4	3	3	2

185. Match List I with List II and select the correct answer

List I	***List II***
(a) Lag Compensation	1. Exterior of unit circle $\lvert Z \rvert = 1$
(b) Lead Compensation	2. PD control
(c) Stability	3. Non-minimum phase
(d) Instability	4. Left half of *s*-plane
	5. PI control

Codes:	*(a)*	*(b)*	*(c)*	*(d)*
A.	3	1	2	5
B.	2	5	1	3
C.	1	2	3	4
D.	5	2	4	1

186. Consider the following statements:

1. The derivative control improves the overshoot of a given system.
2. The drivative control reduces steady-state error
3. Integral control reduces steady-state error.
4. Integral control does not affect stability of the system.
5. Integral control improves the overshoot of the system.

Of these statements

A. 2 and 3 are correct

B. 1, 2 and 5 are correct

C. 2, 4 and 5 are correct

D. 1, 3 and 5 are correct

187. When phase lag compensation is used in a system, gain crossover frequency, band width and undamped frequency are respectively

A. increased, increased, increased
B. increased, increased, decreased
C. increased, decreased, decreased
D. decreased, decreased, decreased

188. The magnitude plot for a transfer function is:

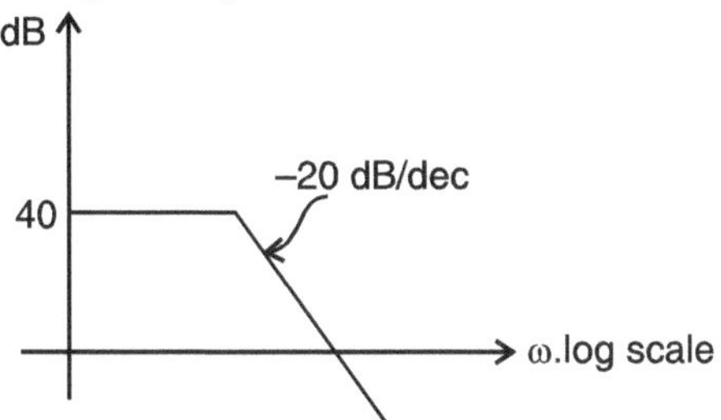

What is the steady-state error corresponding to a unit step input?

A. 1/101 B. 11/100
C. 1/410 D. 1/40

189. Match List I with List II and select the correct answer

List I	List II
(a) On-off	1. Steam kettle with different time settings
(b) Proportional	2. Elevator
(c) Cascade	3. Robot positioning
(d) Digital	4. Domestic refrigerator

Codes:	*(a)*	*(b)*	*(c)*	*(d)*
A.	2	4	1	3
B.	4	2	1	3
C.	3	1	4	2
D.	4	2	1	1

190. Gain crossover frequency is the one at which $|G(j\omega)H(j\omega)|$ is:

A. Equal to 1 B. < 1
C. Equal to – 1 D. > – 1

191. A linear stable time-invariant system is forced with an input

$x(t) = A \sin \omega t$

x(t) —— G(s) —— y(t)

Under steady-state conditions, the output *y(t)* of the system the system will be

A. $A \sin(\omega t + \phi)$, where $\phi = \tan^{-1}|G(j\omega)|$
B. $|G(j\omega)| A \sin[\omega t + \angle G(j\omega)]$
C. $|G(j\omega)| A \sin[2\omega t + \angle G(j\omega)]$
D. $AG(j\omega) \sin[\omega t + G(j\omega)]$

192. Consider the frequency response (amplitude and phase) of two control system A and B shown

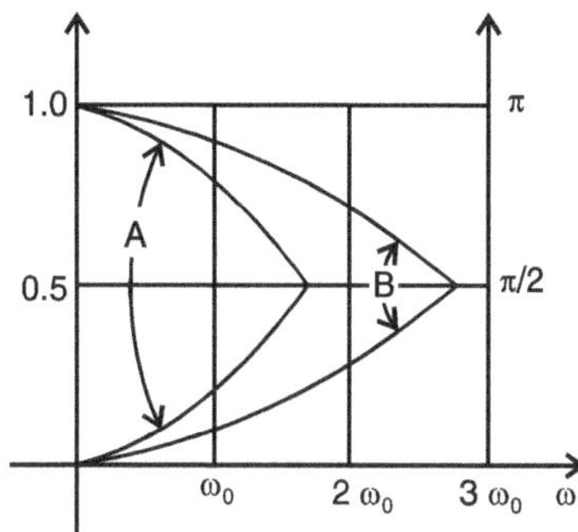

Which one of the following is true for the above system?

A. Transient process will be faster in A
B. Transient process will be faster in B
C. Transient process will be faster in A or B depending upon the nature of input
D. No inference can be drawn about the fastness of transient process

193. A system is described by:

$$\begin{bmatrix} \dot{x}_1 \\ \dot{x}_2 \end{bmatrix} = \begin{bmatrix} 0 & 1 \\ -1 & -1 \end{bmatrix} \begin{bmatrix} x_1 \\ x_2 \end{bmatrix}$$

To test its stability by Lyapunov's method, the following V-functions are considered:

$V_1 = 2x_1^2 + x_2^2$
$V_2 = x_1^2 + x_2^2$

Which of these V-functions are suitable in this case?

A. Only V_1 B. Only V_2
C. Both V_1 and V_2 D. Neither V_1 nor V_2

194. The phase margin of system specifies:

A. Relative stability
B. Absolute stability
C. Frequency response
D. none of the above

195. For a unity feedback system, the open-loop transfer function is

$$G(s) = \frac{16(s+2)}{s^2(s+1)(s+4)}$$

What is the steady-state error if the input is

$r(t) = (2 + 3t + 4t^2)\, u(t)$

A. 0 B. 1

C. 4 D. 8

196. Match List I with List II and select the correct answer

List I	List II
(a) $\frac{s^2 - s + 1}{s^2 + s + 1}$	1. RL admittance
(b) $\frac{s^2 + s + 1}{s^2 - s + 1}$	2. RL impedance
(c) $\frac{s^2 + 4s + 3}{s^2 + 6s + 8}$	3. Unstable
	4. Non-minimum phase

Codes:	(a)	(b)	(c)
A.	1	4	2
B.	1	2	3
C.	4	3	2
D.	4	3	1

197. Match the control system components in List I with their function in List II and select the correct answer

List I	List II
(a) Servo motor	1. Error detector
(b) Amplidyne	2. Transducer
(c) Potentiometer	3. Actuator
(d) Flapper value	4. Power amplifier

Codes:	(a)	(b)	(c)	(d)
A.	2	4	1	3
B.	3	1	2	4
C.	4	3	2	1
D.	3	4	1	2

198. In the feedback system shown the noise component of output is given by (assume high loop gain at frequencies of interest)

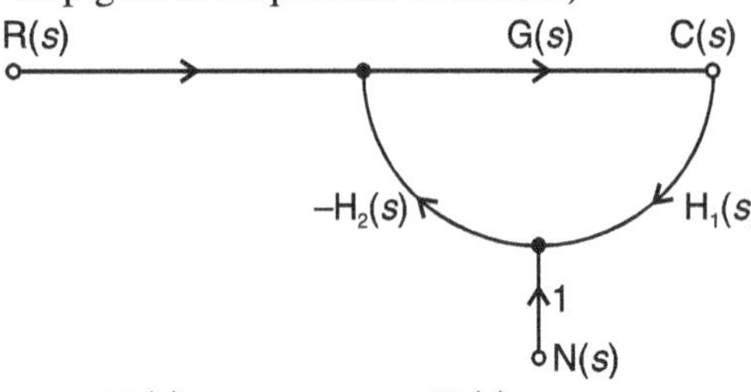

A. $\frac{-N(s)}{H_1(s)}$ B. $\frac{N(s)}{H_1(s)}$

C. $\frac{N(s)}{H_1(s)H_2(s)}$ D. $\frac{-N(s)}{H_1(s)H_2(s)}$

199. The Nyquist plot of an open-loop system with the transfer function

$$G(s)H(s) = \frac{s+2}{(s+1)(s-1)}$$

will be the form

A. 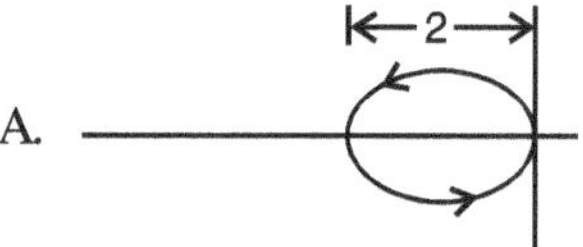

B. 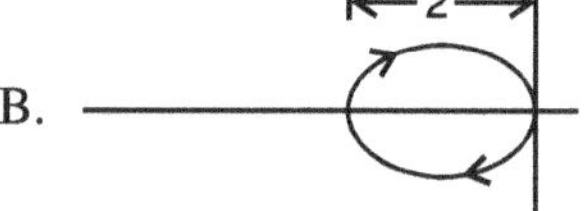

C. 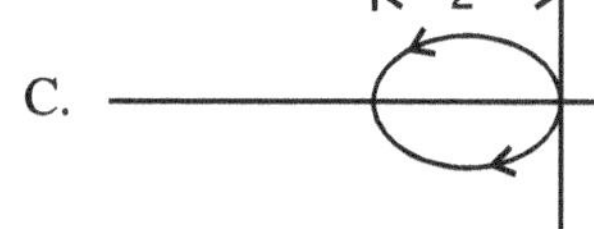

D. 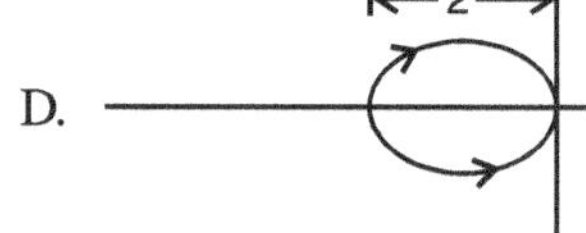

200. Laplace transform and Fourier integrals are related through:

A. Frequency domain

B. Time domain

C. Both frequency and time domains

D. None of these

201. The advantage of Nyquist stability test is

A. Its applicability to experimental results of frequency response open loop system

B. Its ability to predict its closed loop stability from open loop results

C. Its ability to indicate how to stabilise an unstable system

D. All of the above

202. Consider the following units:

1. sec^{-1} 2. $rad^2\ sec^2$

3. sec 4. ohm

The units of $\frac{R}{L}$, $\frac{1}{LC}$, CR and $\sqrt{\frac{L}{C}}$ are respectively

A. 1, 3, 4 and 2 B. 3, 2, 1 and 4

C. 2, 4, 1 and 3 D. 1, 2, 3 and 4

203. Match List I with List II and select the correct answer

	List I	List II
(a)	$\frac{(s^2 - s + 4)}{s^2 + s + 4}$	1. Non-positive real
(b)	$\frac{(s+4)}{s^2 + 3s - 4}$	2. Non-minimum phase
(c)	$\frac{(s+4)}{s^2 + 6s + 5}$	3. RC-impedance
(d)	$\frac{s^3 + 3s}{s^4 + 2s^2 + 1}$	4. Unstable
		5. RL-impedance

Codes:	(a)	(b)	(c)	(d)
A.	1	2	4	5
B.	1	2	3	4
C.	2	4	3	1
D.	2	4	1	5

204. For the following signal flow graph, feedback path can be

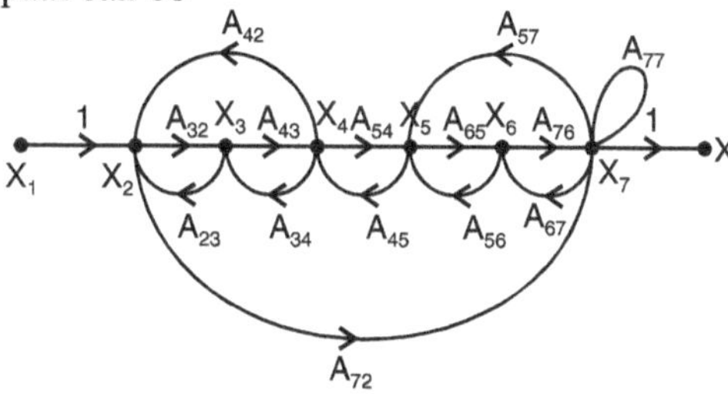

A. Six B. Eight

C. Ten D. Twelve

205. The transfer function, when the Bode diagram is plotted should be of the form

A. $\left(\frac{1}{s+1}\right)$ B. $(s\text{T} + 1)$

C. $\left(\frac{s+1}{\text{T}}\right)$ D. $(s + \text{T})$

206. Match the system open-loop transfer functions given in List I with the steady-state errors produced for a unit ramp input. Select the correct answer

	List I	List II
(a)	$\frac{30}{a^2 + 6s + 9}$	1. Zero
(b)	$\frac{30}{s^2 + 6s}$	2. 0.2
(c)	$\frac{30}{s^2 + 9s}$	3. 0.3
(d)	$\frac{s+1}{s^2}$	4. Infinity

Codes:	(a)	(b)	(c)	(d)
A.	4	3	2	1
B.	4	2	3	4
C.	1	3	2	4
D.	4	2	3	1

207. Assuming the transient response of a second-order system to be given by

$$c(t) = 1 - \frac{e^{-4t}}{\sqrt{s - \delta^2}} \sin\left(\omega_n \sqrt{1 - \delta^2} + \theta\right)$$

the settling time for the 5% criterion will be

A. $\frac{1}{4}$ sec B. $\frac{3}{4}$ sec

C. $\frac{5}{4}$ sec D. 4 sec

208. The peak overshoot in a control system is a function of:

A. Natural frequency of oscillation w_n only

B. Damped frequency of oscillation w_d only

C. Damping ratio ζ only

D. None of these

209. The number of separate parts of a root locus is equal to the:

A. Number of zeros in open loop transfer function

B. Number of poles in the closed loop transfer function

C. The difference between (A) and (B)

D. The order of the characteristic equation of the system.

210. In the root locus analysis the angle between the adjacent asymptotes is

A. $360°/(n - m)$ B. $180°/(n - m)$

C. $90°/(n + m)$ D. $360°/(n + m)$

211. Match List I (transfer functions) with List II (impulse response) and select the correct answer

List I | **List II**

(a) $\frac{1}{s(s+1)}$ 1.

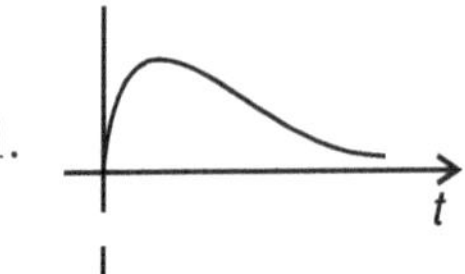

(b) $\frac{1}{(s+1)^2}$ 2.

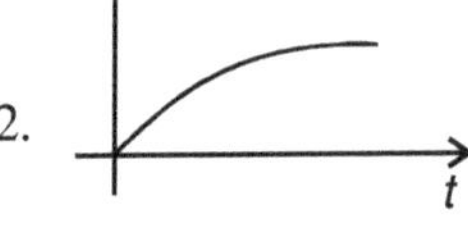

(c) $\frac{1}{s(s+1)+1}$ 3.

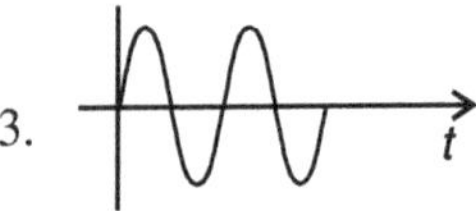

(d) $\frac{1}{s^2+1}$ 4.

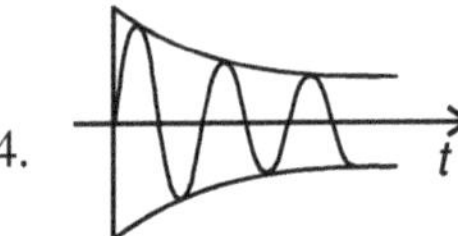

Codes:	*(a)*	*(b)*	*(c)*	*(d)*
A.	2	1	3	4
B.	1	2	3	4
C.	2	1	4	3
D.	1	2	3	4

212. When all the roots of the characteristic equation are found in the left half of *s*-plane the system response due to initial condition will

A. Increase to infinity as time approaches infinitely

B. Decrease to zero as time approaches infinity

C. Remain constant for all time

D. Be oscillating

213. A decade frequency range is specified by

A. $\frac{w_2}{w_1} = 2$ B. $\frac{w_2}{w_1} = 10$

C. $\frac{w_2}{w_1} = 8$ D. None of the above

214. The force required to initiate motion between two contacting surfaces is known as:

A. Stiction

B. Coulomb friction force

C. Viscous friction force

D. None of these

215. The transfer function of a system is defined as the ratio of output to input in:

A. Z-transform

B. Fourier transform

C. Laplace transform

D. Single algebraic equation

216. In control system compensation methods, lead compensation network

A. achieves the desired results through the merits of its attenuation properties at high frequencies

B. in the frequency domain decreases the bandwidth but increases phase margin

C. in the frequency domain increases the bandwidth and phase margin

D. improves steady-state accuracy but reduces bandwidth

ANSWERS

1	**2**	**3**	**4**	**5**	**6**	**7**	**8**	**9**	**10**
B	C	A	D	B	D	B	B	C	A
11	**12**	**13**	**14**	**15**	**16**	**17**	**18**	**19**	**20**
C	A	B	C	A	C	B	C	A	D
21	**22**	**23**	**24**	**25**	**26**	**27**	**28**	**29**	**30**
C	A	B	A	D	A	B	B	D	D
31	**32**	**33**	**34**	**35**	**36**	**37**	**38**	**39**	**40**
D	B	A	C	A	A	D	C	D	C
41	**42**	**43**	**44**	**45**	**46**	**47**	**48**	**49**	**50**
B	A	A	C	B	D	C	D	B	D
51	**52**	**53**	**54**	**55**	**56**	**57**	**58**	**59**	**60**
B	C	C	A	C	B	C	B	A	D

61	62	63	64	65	66	67	68	69	70
D	A	C	C	B	C	A	C	A	D
71	**72**	**73**	**74**	**75**	**76**	**77**	**78**	**79**	**80**
D	B	A	D	C	C	B	A	C	D
81	**82**	**83**	**84**	**85**	**86**	**87**	**88**	**89**	**90**
A	B	D	B	D	A	D	C	B	C
91	**92**	**93**	**94**	**95**	**96**	**97**	**98**	**99**	**100**
B	B	B	D	B	D	C	A	D	B
101	**102**	**103**	**104**	**105**	**106**	**107**	**108**	**109**	**110**
A	C	C	A	B	A	D	D	B	D
111	**112**	**113**	**114**	**115**	**116**	**117**	**118**	**119**	**120**
B	B	D	C	A	B	D	D	A	C
121	**122**	**123**	**124**	**125**	**126**	**127**	**128**	**129**	**130**
D	B	A	D	D	A	B	B	B	B
131	**132**	**133**	**134**	**135**	**136**	**137**	**138**	**139**	**140**
A	A	B	D	A	B	A	C	B	A
141	**142**	**143**	**144**	**145**	**146**	**147**	**148**	**149**	**150**
C	A	D	D	A	B	C	B	A	A
151	**152**	**153**	**154**	**155**	**156**	**157**	**158**	**159**	**160**
A	B	B	B	A	A	A	C	A	A
161	**162**	**163**	**164**	**165**	**166**	**167**	**168**	**169**	**170**
C	B	B	C	A	C	C	B	A	D
171	**172**	**173**	**174**	**175**	**176**	**177**	**178**	**179**	**180**
C	A	D	A	B	C	D	A	B	A
181	**182**	**183**	**184**	**185**	**186**	**187**	**188**	**189**	**190**
A	D	A	B	D	D	C	A	D	A
191	**192**	**193**	**194**	**195**	**196**	**197**	**198**	**199**	**200**
B	A	B	A	A	D	D	A	A	C
201	**202**	**203**	**204**	**205**	**206**	**207**	**208**	**209**	**210**
D	D	C	D	B	D	B	C	D	A
211	**212**	**213**	**214**	**215**	**216**				
B	A	B	A	C	B				

EXPLANATORY ANSWERS

3. Maximum overshoot = $e^{-\xi\pi}\sqrt{1-\xi}$

7.
$$G(s)\,H(s) = \frac{1}{s(s+0.5)(s+1)}$$

$$G(j\omega)\,H(j\omega) = \frac{1}{j\omega\,(j\omega+0.5)(j\omega+1)}$$

$$\phi = \angle G(j\omega)\,H(j\omega)$$

$$= -90^\circ - \tan^{-1} 2\omega - \tan^{-1}\omega$$

At phase crossover point $\phi = -180^\circ$

$\therefore\ \tan^{-1}\omega + \tan^{-1} 2\omega = 90^\circ$

or, $\alpha_1 + \alpha_2 = 90^\circ$

$$\frac{\tan\alpha_1 + \tan\alpha_2}{1-\tan\alpha_1\,\tan\alpha_2} = \tan 90^\circ = \infty$$

$\therefore\ \tan\alpha_1 \tan\alpha_2 = 1$

or, $\omega \times 2\omega = 1$

$\therefore\ \omega = 0.707$ rad/s

13. Response $Y(s) = G(s) \times (s)$

Steady state response is $Y(j\omega) = G(j\omega) \times (j\omega)$

Above equation is terms of complexors and the complex function $G(j\omega)$ which may be expressed

as $|G(j\omega)| \angle G(j\omega)$
As a function of time,

18. $$x_1 = -x_1 + U$$
$$x_2 = x_1 - 2 \times 2 + U$$

$\therefore$ $$A = \begin{bmatrix} -1 & 0 \\ 1 & 2 \end{bmatrix}, B = \begin{bmatrix} 1 \\ 1 \end{bmatrix}$$

For the system to be state controllable [B A B] should be non-singular, i.e. determinant of [B A B] should be non-zero

$$[B\,A\,B] = \begin{bmatrix} 1 & -1 \\ 1 & -1 \end{bmatrix} = 0$$

Therefore, the system is not state controllable. Hence all the responses containing statement 1 are incorrect.

23. $$G_c(s) = \frac{1+6s}{1+2s}$$

Phase shift is $\phi = \tan^{-1} 6\omega - \tan^{-1} 2\omega$

$$\frac{d\phi}{d\omega} = \frac{6}{1+36+\omega^2} - \frac{2}{1+4\omega^4}$$

For maximum value of ϕ, $\frac{d\phi}{d\omega} = 0$

$\therefore$ $3(1 + 4\omega^2) = 1 + 36\,\omega^2$

or, $$\omega = \frac{1}{\sqrt{12}}$$

$\therefore$ $$\phi = \tan^{-1}\sqrt{\frac{36}{12}} - \tan^{-1}\sqrt{\frac{4}{12}}$$

$$= \tan^{-1}\sqrt{3} - \tan^{-1}\frac{1}{\sqrt{3}} = 30°$$

27. The characteristic equation is
$s^2 + 0.2\,s + 1 = 0$
When expressed in the form $s^2 + 2\delta\,\omega_n + \omega_n^2$, we have

$$\omega_n = 1 \text{ and } \delta = 0.1$$

Settling time, $$t_s = \frac{n\ pu \text{ difference}}{\delta\,\omega_n}$$

$$= \frac{\ln 0.02}{0.1 \times 1}$$

$$= \frac{3.91}{0.1} = 40 \text{ s}$$

31. Assuming unity feedback, system steady state error for ramp input is given by

$$\in_{ss} = \lim_{s\to 0} \frac{1}{sG(s)}$$

A. $\lim_{s\to 0} sG(s) = 0 \quad \therefore \in_{ss} = \infty$

B. $\lim_{s\to 0} sG(s) = \frac{30}{6} = 5 \quad \therefore \quad \in_{ss} = 0.2$

36. $$G(s) = \frac{2(s+1)}{(s+2+j1)(s+2-j1)}$$

$$= \frac{2(s+1)}{(s+2)^2+1} = \frac{2(s+1)}{s^2+4s+5}$$

39. As the voltage is fed from the collector to base terminating through R_f, it is a case of voltage shunt feedback.

74. Dividing both sides of the equation by 2

$$\frac{d^2y}{dt^2} + 2\frac{dy}{dt} + 4y = 4x$$

Hence damping ratio, $\xi = \frac{1}{2} = 0.5$

86. $|G(j1)| =$

$$\frac{\text{product of lengths of vectors drawn zero } j1}{\text{product of lengths of vectors drawn from poles to } j1}$$

$$= \frac{\sqrt{5}}{\sqrt{10}\sqrt{2}} = \frac{1}{2}$$

$$\angle G(j1) = 26.56 - (71.56 - 45°)$$

$\therefore$ $$G(j1) = \frac{1}{2}\angle 0°$$

None of the given answers is correct.

90. $$G(s) = \frac{1}{(s+2)^2}$$

$$H(s) = 1$$

$\therefore$ $$M(s) = \frac{G(s)}{1+G(s)}$$

$$= \frac{\frac{1}{(s+2)^2}}{1+\frac{1}{(s+2)^2}} = \frac{1}{(s+2)^2+1}$$

$\therefore$ The poles are at $-2 \pm j1$

96. For the given system

$$G(s) = \frac{s^2(s+2)}{(s+1)^2 s^4 (s+1)} = \frac{(s+2)}{s^2(s+1)^3}$$

$$H(s) = \frac{(2s+3)}{s(s+3)}$$

$$G(s)H(s) = \frac{(s+2)(2s+3)}{s^3(s+1)^3(s+3)}$$

$$k_p = \lim_{s\to 0} G(s)H(s)$$

$$k_p = \lim_{s\to 0} S^2 G(s)H(s)$$

For a type '0' system, $k_p = k, k_v = 0, k_a = 0$
For a type '1' system, $k_p = \infty, k_v = k, k_a = 0$
For a type '2' system, $k_p = \infty, k_v = \infty, k_a = k$
For a type '3' system, $k_p = \infty, k_v = \infty, k_a = \infty$
It is therefore type 3 system

109. Step size $= \frac{360}{12} = 30°$

110. Here 1000 contributes 60 dB and s^2 contributes 40 dB/decade with 0 dB at $\omega = 1$. At $\omega = 10$, this will contribute additional 40 dB so that net gain will become 100 dB $(1 + 0.1s)$ will contribute negative slope of – 100 dB/decade starting at $\omega = 10$.

For ω greater than 10, slope deviates to s^2 and contributes to + 40 dB/decade and due to fifth order pole at $\omega = 10$, slope is – 100 dB/decade. Therefore, net slope is – 60 dB/decade as given.

122. Output leads the input by an angle,
$\phi = \tan^{-1} \alpha T - \tan^{-1} T$
If $\alpha > 1$; ϕ is positive, and
if $0 < \alpha < 1$, then ϕ becomes negative meaning that output will lag the input by ϕ.
Since the given function is to represent a phase lag network α should be between 0 and 1.

125. Step response transform is

$$\frac{42.25}{s^2(s+6.5)} = \frac{1}{s+6.5} + \frac{6.5}{s^2} - \frac{1}{s}$$

$$w = e^{-6.5t} = 6.5\, t(t) - u(t)$$

At $\quad t = \infty, c(t) \to \infty$

131. $\quad x_4 = ax_1 + bx_2 + cx_3$

$\therefore \quad x_6 = de\, x_4$

$$= de\,(ax_1 + dx_2 + cx_3)$$

132. $\quad x_2 = T_{12}\, x_1 + T_{22}\, x_2$

$$\therefore \quad x_2 = \frac{T_{12}}{1 - T_{22}} . x_1$$

140. The system has 12 poles and 2 zeros.
A pole gives a – 20 dB/decay and a zero gives a + 20 dB/decade decay.
Hence slope of high frequency asymptote
= – 20 (12) + 2(20) = – 200 dB/decade

160. $$G(s) = \frac{10}{s(s+1)^2}$$

Putting $\quad s = j\omega$,

$$G(j\omega) = \frac{10}{j\omega(j\omega+1)^2}$$

$$= \frac{10}{j\omega(-j\omega^2 + 2j\omega + 1)^2}$$

$$= \frac{10}{-j\omega^2 + j(\omega - \omega^2)}$$

This can be divided into the real and imaginary parts as shown below:

$$G(j\omega) = \frac{10\left[72\omega^2 - j(\omega - \omega^2)\right]}{\left[-2\omega^2 + (\omega - \omega^3)\right]\left[-2\omega^2 - j(\omega - \omega^3)\right]}$$

$$= \frac{-20\omega^2}{4\omega^2 + (j\omega - \omega^3)^2} = \frac{10j(\omega - \omega^3)}{4\omega^4 + (\omega - \omega^3)^2}$$

In the real axis at $\omega = \omega_0$, the imaginary portion goes to zero. Therefore

$$\frac{(\omega_0 - \omega_0^2)}{4\omega_0^4 + (\omega_0 - \omega_0^3)^2} = 0$$

or, $\quad \omega_0 - \omega_0^3 = 0$
or, $\quad \omega_0(1 - \omega_0^2) = 0$
which gives the possible solution
$\omega_0 = 0, |G(j\omega)| = \alpha$

$$\text{Real part of } G(j\omega) = \frac{-20\omega_0^2}{4\omega^4 + (\omega_0 - \omega_0^3)^2}$$

$$\omega_0 = +1,$$

$$G_R(j\omega) = \frac{-20}{4+0} = -5$$

and $\omega_0 = -1,$

$$G_R(j\omega) = \frac{-20}{4+4} = 2.5$$

Hence the possible solutions are – 5, 1.

177. The characteristic equation is

$$1 + GH = 0$$

or, $$1 + \frac{k}{s^3 + 3s^2 + 2s + 1} = 0$$

or, $$\frac{s^3 + 3s^2 + 2s + 1 + k}{s^3 + 3s^2 + 2s + 1} = 0$$

or, $s^3 + 3s^2 + 2s + 1 + k = 0$

Hurwitz criterion can be applied

s^3	1	2	$\frac{5-k}{3}$ and $(1 + k)$ both should be positive.
s^2	3	$1 + k$	
s^1	$\frac{6-(1+k)}{3}$	0	$\therefore$ k should be more than – 1 and less than 5.
s^0	$1 + k$		

7

INSTRUMENT AND MEASUREMENTS

ELECTRICAL INSTRUMENT

Electrical instruments may be divided into two classes :

1. **Absolute Instruments :** Those which measure electrical quantities in terms of the constants of instrument and its deflection.
2. **Secondary Instruments :** Those in which the value of electrical quantity to be measured can be determined from the deflection of the instruments only when they have been pre-calibrated by comparison with an absolute instrument. Of course without calibration, the deflection of such instruments is meaningless.

   Instruments may also be classified as

   (i) **Indicating Instruments :** Those which indicate the instantaneous values of electrical quantity being measured. A pointer moving over a calibrated scale gives the reading. Ordinary voltmeters, ammeters etc., are indicating instruments.

   (ii) **Recording Instruments :** Those which give a continuous record of the variations of an electrical quantity with respect to time usually a graph paper is used in these instruments and a pointer registers the reading on the graph paper.

   (iii) **Integrating Instruments :** Those which measure and register by a set of dials and pointers. A energy meter installed at the residence is an integrating instrument.

Effects Utilised in Measurements : All instruments operate utilizing one more of the following effects :

Effect utilised	*Category of instruments*
1. Magnetic effect	Voltmeters, ammeters
2. Thermal effect	Voltmeters, ammeters
3. Electrodynamic effect	Voltmeters, ammeters, wattmeters
4. Electromagnetic effect	Voltmeters, ammeters, wattmeters, watthour meters
5. Chemical effect	DC ampere hour meters
6. Electrostatic effect	Voltmeters only

Torques on Moving Systems

The moving system of an instrument is subjected to following torques :

1. **Deflecting Torque :** Produced due to one or more effects, mentioned above. This torque causes the moving system and hence pointer to move from initial zero condition.
2. **Controlling Torque :** Which opposes the deflecting torque and increases with deflection of the moving system. The pointer comes to rest at a position where these torques are equal (and opposite). Controlling torque may be provided by
   (i) Spring control
   (ii) Gravity control
3. **Damping Torque :** Which acts only when the instrument system is moving. This torque tries to stabilize the moving system quickly so that pointer comes to rest at the point of reading without oscillating around. Damping force can be produced by
   (i) Air friction
   (ii) Eddy currents
   (iii) Fluid friction

Comparison Between Meter Movements

Moving coil meters have a linear scale and require a minute current (of the order of a few microamperes) to operate. This current varies from about 25 μA to 100 μA. These meters have an accuracy of ±2% and are delicate instruments in comparison with other types of meter movements. More than one quantity can be measured with moving-coil meters by suitably modifying the circuit connections. These instruments are suitable for directly measuring direct currents and voltages. Being delicate and sensitive these instruments are costly.

Moving iron meters can be used for measuring d.c. as well as a.c. These meters are less sensitive as compared to moving coil meters and have a rugged construction. The accuracy of these meters is $\pm 2\%$ or even higher. The scale of these meters is nonlinear. Moving iron meters are cheaper than their counterts, moving-coil meters.

Hot wire meter can be used for measuring both d.c. and a.c. They consume more power. The scale of these meters is **nonlinear**. Hot-wire meters are a bit sluggish (take some time to operate) and should be protected from mechanical shocks which can cause permanent damage to these instruments.

Thermocouple meters are most suited to measuring radio frequency currents. An accuracy of $\pm 1\%$ can be achieved even at frequencies as high as 50 MHz. Of all the different types of meter movements thermocouple meters, in particular, are delicate, sensitive, and costly. These meters have a nonlinear scale. Thermocouple meters should be protected from overloads.

N.B.

1. Gravity controlled instruments have scales which are not uniform, but cramped or crowded at their lower end.
2. Gravity controlled instruments must be used in a vertical position.
3. Damping force can be produced by air friction, fluid friction and eddy currents.
4. Mechanical damping is caused mainly by the motion of the moving coil through the air surrounding it.
5. In eddy current damping, the rotation of the coil in the magnetic field sets up eddy currents in the conductive metal frame, causing a retarding torque that opposes the motion of the coil.
6. Eddy current damping is the most efficient.
7. When used in conjunction with rectifiers or thermo-junctions, moving-coil instruments can also be used for a.c. measurement over a wide range of frequencies.
8. Since the jewel-and bearing construction is replaced by a band that twists in accordance with the amount of current through it, friction is practically eliminated.
9. Moving iron instruments have a non-linear scale.
10. The electrodynamometer movement is less sensitive than the PMMC movement.
11. Crossed-coil movements have two moving coils mounted at 90° to each other.
12. Because of the square law response, electro-dynamometer instruments can be used on both d.c. and a.c.
13. Thermocouple meters are suitable for measuring both d.c. and a.c.
14. Electrostatic instruments cannot be adopted to measure current or resistance.
15. Electrostatic instruments have a nonlinear scale.

Advantages of Instrument Transformers

Keeping in view the disadvantages of shunts and multipliers practical use for measuring high a.c. currents and voltages is limited. Current transformers and potential transformers are used respectively in place of shunts and multipliers. Instrument transformers have the following advantages :

(i) The scale of meters using instrument transformer does not depend on the resistance or reactance of transformers. If different meters are used with the same instrument transformers, the scale of these meters will not be affected in any way.

(ii) The current flowing in the secondary of instruments is standardised at 5 A. Similarly the voltage across the secondary of instrument transformers is standardised at 100 to 150 V. Meters used with instrument transformers are selected accordingly. If it is required to measure 1000 A of a.c. current, then a 1000/5A or 200 : 1 current transformer will be used. Similarly if it is required to measure 66 kV a.c. then a 66000/220 or 300 V potential transformer will be used.

(iii) Because of standardisation of instrument transformers the measuring instruments used with these transformers are also standardised.

In this way replacement of these transformers posses no problems.

(iv) The primary of instrument transformers is connected to the circuit under test. This provides electrical isolation to the meter movement which is connected to the secondary winding. It becomes easier to provide good electrical insulation to these transformers during manufacture. It also provides physical protection to the user of instrument.

(v) Measuring instruments using instrument transformers consume less power.

(vi) The same instrument transformer can be used with more than one measuring instrument.

Difference Between Current Transformer and Potential Transformer

Based upon their working, the following differences exist between current transformers and instrument transformers.

(i) The primary winding of current transformers is connected in series whereas the primary winding of potential transformers is connected in parallel.

(ii) The current flowing through the primary winding of current transformers is independent of load conditions in the secondary. The current flowing through the primary winding of potential transformers depends on the load conditions in the secondary.

(iii) The whole of the current to be measured flows through the primary winding of the current transformers. The whole of the unknown voltage to be measured is impressed upon the primary winding of potential transformers.

(iv) The primary flux of current transformers depends on the current flowing through their primary winding. The flux density of potential transformers is almost constant.

N.B.

1. Ammeters are always connected in series.
2. Voltmeters are always connected in parallel.
3. The resistance of the rectifier changes with the current passing through it.
4. The current transformer enables the various current ranges to be scaled to an appropriate magnitude for the rectifier characteristic.
4. Instrument transformers are used as attachments to measuring instruments.
5. If different meters are used with the same instrument transformer, the scale of these meters will not be affected in any way.

POWER MEASUREMENTS

The electrodynamometer movement is used extensively in measuring power. It may be used to indicate both dc power for any waveform of voltage and current and it is not restricted to sinusoidal waveforms.

Polyphase Wattmeter

Power measurements in a polyphase system require the use of two or more wattmeters. The total real power is then found by algebraically adding the readings of the individual wattmeters. If the neutral wire of the three phase system is also present as in case of four wire-star connected load, three wattmeters would be needed to make the total real power measurement.

Three Wattmeter Method : In this method, three wattmeters are connected, one in each phase. The three wattmeter readings add up to give the total power supplied if the potential coils measure V_r and the current coil I_P. The total power is $P_P = 3\ V_{P/P}$ cos θ.

Two Wattmeter Method

The two wattmeter method can be used in a three wire system delta or star connected load.

$$\therefore\ P_T = \sqrt{3} V_L I V_L \cos\theta$$

This is the total power in a three phase system. This proves that the power in a three wire system is the sum of readings of the two wattmeters. This method is also used to calculate the phase angle θ.

$$\tan\theta = \sqrt{3}\frac{P_2 - P_1}{P_2 + P_1} \text{ for balanced load}$$

Reactive Power Measurements . The reactive power is given by

$$P_T = VI\sin\theta$$

It indicates the exchange of power between the

source and the load due to the reactance of the circuit. Reactive power can be measured by an instrument called the varmeter.

A varmeter is in effect a wattmeter in which the current in its potential coil has been shifted to lag the applied potential by 90°. The current in the potential moving coil is shifted 90° by means of a tapped inductor and a shunt resistor. The meter is now sensitive to reactive power and is calibrated in vars.

In three-phase three-wire system the reactive power is given by

$$P_T = \sqrt{3}\, V_L \, / \, V_L \sin\theta$$

Power Measurement at High Frequency : For the measurement of power at high frequency (in the microwave region) absorption meters are used. The two types of absorption meters which are extensively used in the band of frequencies from 500 MHz to 40 GHz are : 1. Calorimeter power meter, 2. Bolometer power meter.

Calorimeter type Power Meter : The basic principle of operation of the calorimeter type power meter is quite simple as illustrated in Fig. 1. A resistor is heated by the signal to be measured. The resistor is in a completely enclosed, well insulated fluid bath. The insulation of the walls of the bath are such that very little heat is allowed to radiate from the bath. If the bath temperature is measured before and after the signal is applied the difference in temperature is a measure of the heat developed by the resistor. Knowing the volume, the specific heat of the fluid and the characteristic of the bath, the power can be calculated. The need to know the exact mass and specific heat of calorimeter bath material restricts the usage of the basic calorimeter power meter. Some of these difficulties are avoided in an instrument called the substitution calorimeter. Another calorimeter type instrument is the comparison flow calorimeter. In this instrument, the fluid is made to pass in separate streams by the input load and the comparison load. The power to the comparison load is a function of the difference in temperature between the loads.

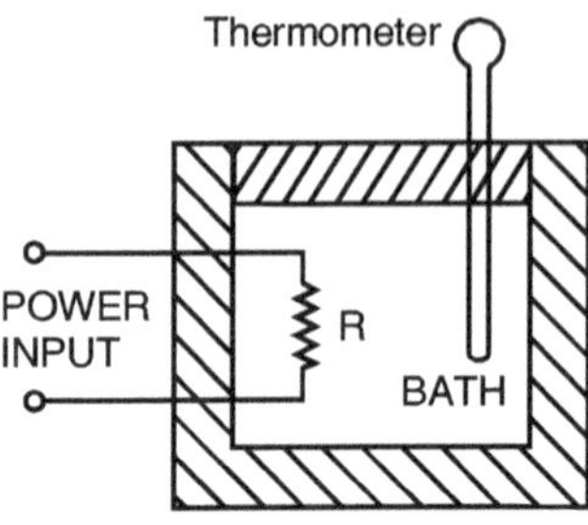

Fig. 1: *Calorimeter type Power meter*

BOLOMETER

The bolometer is basically a bridge circuit where one of the arms contains a temperature sensitive resistor. The temperature-sensitive resistor is balanced in the field of microwave signal whose power is to be measured. This is shown in Fig. 2. The power is absorbed by the resistor and the heat generated causes a change in resistance. This change of resistance is measured with a bridge circuit. It consists of differential amplifier and a bridge. Oscillates at a particular amplitude when the bridge is unbalanced. The sensitive resistance elements R_T absorbs the power and heats itself until the bridge is very nearly balanced. The d.c. bias is than adjusted until the exact balance is achieved. Now the element is placed in the microwave field. The element absorbs the power and is heated, which causes a change in its resistance, thereby unbalancing the bridge. The unbalance is in a direction opposite to that of initial cold resistance. This action causes the output from the oscillator to decreases so as to effect the balance. The EVM circuit measures the amount of this power-decrease from the oscillator and displays the measurement as a power which is supplied by the microwave field.

Bolometer bridges use two types of element; barretters and thermistors. Barretters are metallic wires with positive temperature coefficient of resistance, and thermistor are resistors made of semiconductors material which has a negative temperature coefficient of resistance. The bolometer is customarily used for power measurements in the range 0.01-10mW. In the range of 10 mW to 10 W, a comparison-type calorimeter is suitable. Above 10 W, power measurements can be made by using attenuators and a low-power bolometer.

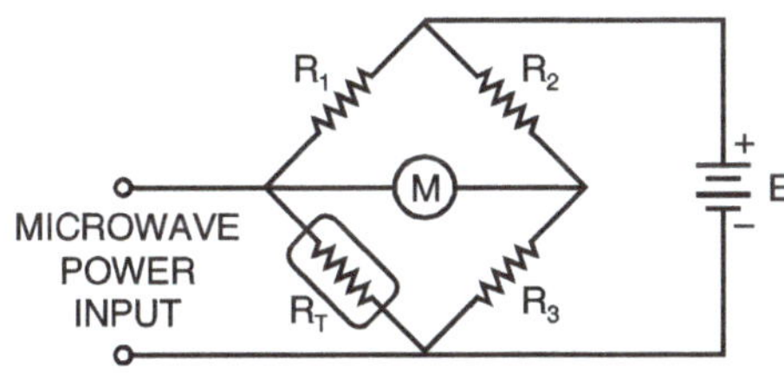

Fig. 2: *Bolometer*

Power-Factor Meter : It essentially measures the cosine of phase angle between voltage and current. The most common type is crossed coil power factor meter. This instrument is basically an electrodynamometer movement where the moving element consists of two coils, mounted on the same shaft but at right angles to each other. The moving coils rotate in the magnetic field provided by the field coil that carries the line current.

The connections for this meter in a single-phase circuit are shown in Fig. 3. The field coil is connected as usual in series with the line and carries the line current. One coil of the movable element is connected in series with an inductor across the lines. Since no control springs are used, the balance position of the movable element depends on the resulting torque developed by the two crossed coils. When the movable element is in a balanced position, the contribution to the total torque by each elements must be equal but of opposite sign.

The developed torque in each coil is a function of the current through the coil and therefore depends on the impedance of that coil circuit. The torque is also proportional to the mutual inductance between each part of the crossed coil and the stationary field coil. This mutual inductance depends on the angular position of the crossed coil elements with respect to the position of the stationary field coil. When the movable element is at balance, it can be shown that its angular displacement is a function of the phase angle between line current (field coil) and line voltage (crossed coil). The indication of the pointer which is connected to the movable element, is calibrated directly is terms of the phase-angle or power-factor.

Frequency Meters . Indicating frequency-meters use the effect of frequency upon such factors as mutual inductance, resonance of a tuned circuit, and mechanic a resonance. Three important types of frequency meters are described below.

Electrodynamo-meter type Frequency Meter. It is shown in Fig. 4. Here the field coil forms part of two separated resonant circuits. Field coil 1 is in series wit inductor L_1 and capacitor C_2, and forms a resonant circuit which is tuned to a frequency slightly below the lower and of the instrument scale. Field coil 2 is in series with inductance L_2 and capacitor C_2, and form a resonant circuit which is tuned to a frequency slightly higher than the high end of the instrument scale. The torque on the movable element proportional to the current through the moving coil.

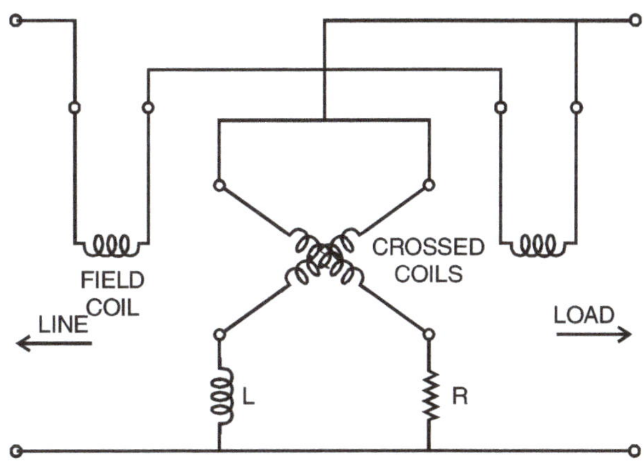

Fig. 3: *Circuit for Power Factor meter*

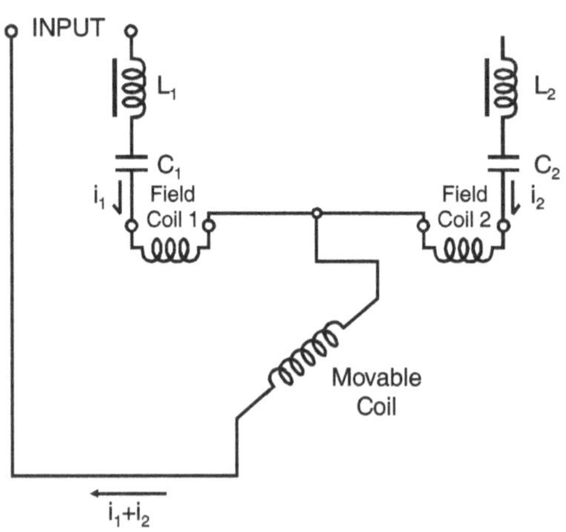

Fig. 4: *Frequency Meter*

The Tuned-Reed Frequency Meter : It operates on the principle of mechan resonance. A series of reeds is fastened to a flexible common base which is mounted the armature of an electromagnet. The coil of the electromagnet is energized form a.c. power line whose frequency is to be determined. The reeds are tuned to exact natural frequency by careful selection of their length and mass. The reed, which has a nat frequency equal to the frequency with which the electromagnet is energized, builds vibration. the vibration of the reed is visible at the front of the meter where the vibration tip of the reed is seen through a window. If the frequency to be measured is intermediate between the natural frequencies of two adjacent reeds, both reeds will vibrate and line frequency will be closest to the reed with the largest vibration. Interpole between the natural frequencies of the reeds can be made very easily and accurate since the reed frequencies are exact.

Instrument Transformers

The transformer used for measurement of current is called a current transformer that for the measurement of voltage is known as potential transformer.

Nominal Rated is the ratio of primary rated current to secondary rated co For CT,

$$K_n = \frac{\text{Rated primary current}}{\text{Rated secondary current}}$$

Turns Ratio

$$N = \frac{\text{Number of turns of secondary bridge}}{\text{Number of turns of primary bridge}}$$

The rated burden is the volt-ampere loading which is permissible without exceeding the limits for particular class of accuracy.

Transformation Ratio for current transformer is given by

$$R = n + \frac{I_0}{I_s}\sin(\delta + \alpha)$$

where

$n =$ the turns ratio

$I_0 =$ exciting current,

$I_s =$ secondary current,

$\delta =$ angle between secondary induced emf and secondary current and

$\alpha =$ angle between exciting current I_0 and working flux.

Phase Angle for current transformer is the angle by which the secondary current phasor, when reversed, differs in phase from the primary current.

Phase Angle Error for current transformer is given by

$$\theta = \frac{180}{\pi}\left[\frac{I_m \cos\delta - I_e \sin\delta}{nI_s}\right] \text{degrees}$$

Ratio Error is usually as percentage ratio error. That is,

$$\text{Ratio error} = \left(\frac{\text{nominal ratio} - \text{actual ratio}}{\text{actual ratio}}\right) \times 100\%$$

The Bridge Circuits

These are extensively used for measuring component values, such as resistance, inductance and capacitance, etc. The bridge circuit compares the values of an unknown component to that of an accurately known component. Its measurement accuracy is very high.

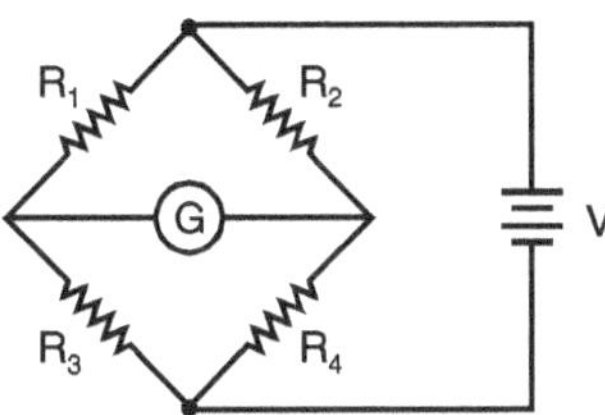

Fig. 5: *Wheatstone Bridge*

Wheatstone Bridge : It is used for the measurement of **dc resistance** and a simple version is shown in Fig. 5. The bridge has four resistor arms together with a source of emf and a null detector (usually a galvanometer). the bridge is said to be balanced when the potential difference across the galvanometer is zero volts so that there is no current through the galvanometer. Hence the bridge is balanced, when

$$R_1R_4 = R_2R_3$$

The wheatstone bridge is widely used for precision measurement of resistance from approximately 1 ohm to the low mega ohm range.

Kelvin Bridge : The Kelvin double bridge is a modification of the Wheatstone bridge so as to improve the accuracy in the measurement of low resistance. It consists of double set of ratio arms. There are many difficulties that arise in the Wheatstone bridge when measuring resistance that are so small that the resistance of the leads and contacts become appreciable, Here

$$R_a = R_s \frac{R_1}{R_2}$$

The Kelvin bridge is used for measuring resistance from 1 ohm to 0.0001 ohm

A.C. Bridges : AC bridge shown in Fig. 6 consists of four bridge arms, a source of excitation and a null detector. The power source supplies and ac voltage to the bridge at the desired frequency. At higher frequencies, an oscillator generally supplies the excitation voltage. The null detector must respond to ac unbalance currents and in its cheapest form consists of a pair of headphones. Other null detector may be ac amplifier with an output meter or an electron ray tube indicator.

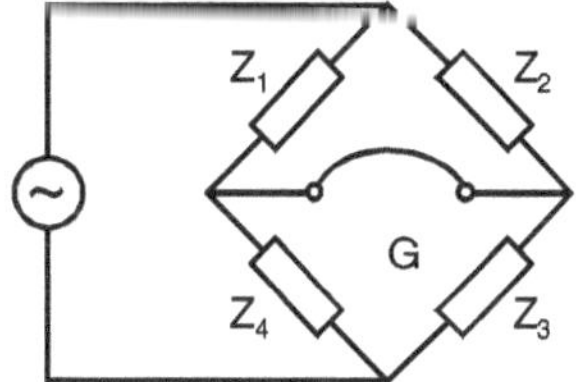

Fig. 6: *Kelvin Bridge*

The condition for bridge balance requires that the potential difference from A to C be zero. Then.

$$Z_1Z_4 = Z_2Z_3$$

Maxwell Bridge : It is shown in Fig. 7 and it measures an unknown inductance in terms of known capacitance. One, of the ratio arms has a resistance and capacitance in parallel.

$$R_x = \frac{R_2R_3}{R_1} \qquad L_x = R_2R_3C_1$$

Maxwell bridge is limited to the measurement of medium Q-coils (1 < Q < 10). It is unsuited for the measurement of coils with very low Q value (Q < 1).

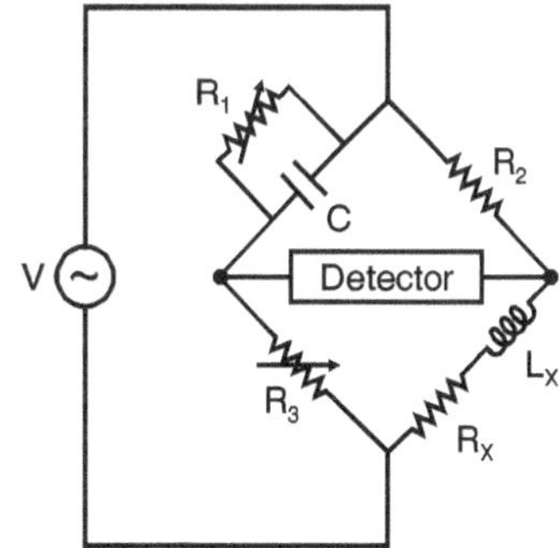

Fig. 7: *Maxwell Bridge*

Hay Bridge : Hay bridge shown in Fig. 8 differs from the Maxwell bridge in having resistor R_1 in series with standard capacitor C_1 instead of being in parallel.

The value of R_1 and L_x are given by

$$R_x = \frac{\omega^2 C_1^2 R_1 R_2 R_3}{1+\omega^2 C_1^2 R_1^2} \qquad L_x = \frac{R_2 R_3 R_1}{1+\omega^2 C_1^2 R_1^2}$$

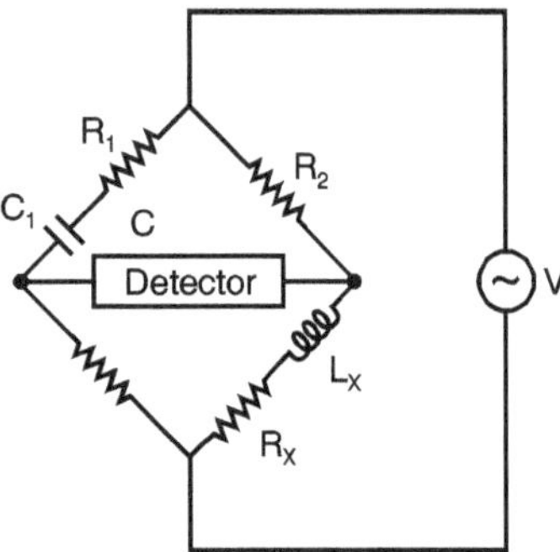

Fig. 8: *Hay Bridge*

Hay bridge is suited for the measurement of high-Q inductors having Q more tan 10.

Schering Bridge : It is used for the measurement of capacitor, particularly, for measuring insulating properties, i.e., for phase angles very nearly 90°. The basic circuit arrangement is shown in Fig. 9. Arm has a parallel combination of resistor and capacitor and standard arm contains only a capacitor.

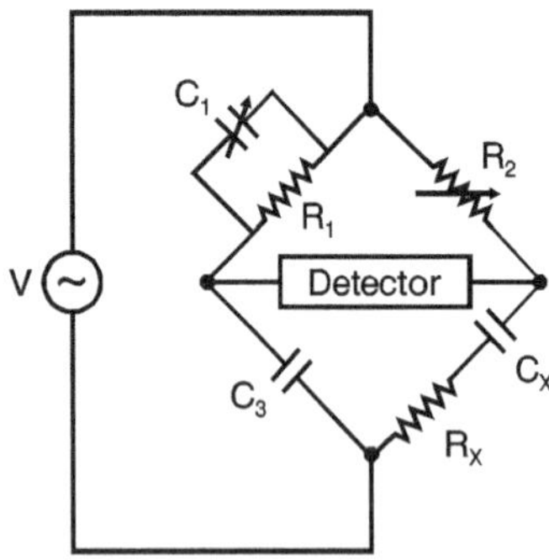

Fig. 9: *Schering Bridge*

The value of R_x and C_x are given by

$$R_x = \frac{R_2 C_1}{C_3} \qquad C_x = C_3 \frac{R_1}{R_2}$$

$$\text{Power factor} = \frac{R_x}{X_x} = \omega C_x R_x$$

Wien Bridge : The wein bridge has a series RC combination in one arm and a parallel RC combination in the adjoining arms as shown in Fig. 10. The frequency of the applied voltage is

$$f = \frac{1}{2\pi RC}$$

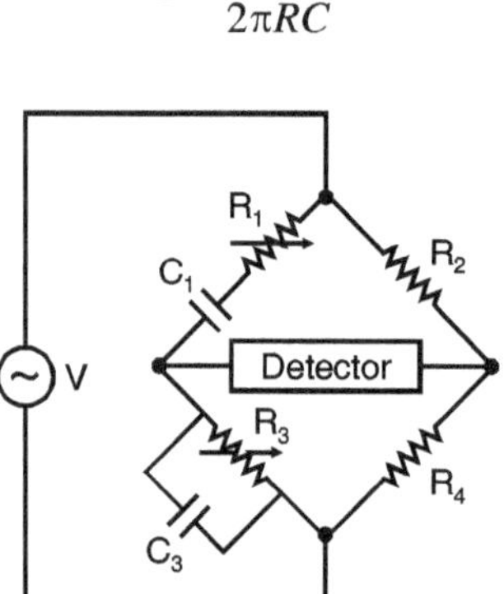

Fig. 10: *Wein Bridge*

Wein bridge is used as a notch filter in harmonic distortion analyzer. It is used in audio and HF oscillators as frequency-determining elements.

Null Detector

There are basically three types of null detectors :

1. The pointer type galvanometer having from 1 μA to 0.1 μA per scale division.
2. The reflecting galvanometer with enclosed lamp and scale. It has high sensitivity range of 0.1 μA to 0.01 μA per scale division.
3. Electronic null detector having excellent sensitivity at high input impedance, capable of measuring a wide variety of rapidly changing electrical signal. The signals may be repetitive or may occur only once and last a fraction of a microsecond. It presents a picture of the exact signal as a function of time. The oscilloscope graphs the changes in signal amplitude on its vertical axis and time (how long the event lasts) on its horizontal axis.

Waveform Measurements

Methods . The instantaneous variations of current and voltage in a circuit can be measured by oscillographs, whose basic operating principle may be either that of a D Arsonval galvanometer whose inertia is low enough to permit it to follow the variations or that of an electron beam which has no sensible inertia and whose deflection is governed by electric or magnetic fields. In addition to tracing waveforms, oscillographs are used for measurements of transient phenomena, such as those which occur in switching operations or in the impulse voltage testing of insulating structures and disturbances resulting from lightning discharges. Transient phenomena may also be captured using digitizing oscilloscopes and transient digitizers (waveform recorder).

Cathode-ray-oscillograph : Tube element are shown in Fig. 11. Electrons emitted by heated cathode are accelerated towards an anode which has a small central aperture. A grid system serves to concentrate the electrons into a narrow beam and focus it into a small spot on a fluorescent screen at the far end of the tube. The grid nearest the cathode controls the beam current and the brightness of the spot on the fluorescent screen. After going through the focussing electrode system, the beam passes between two plate pairs set at right angles to one another and is deflected by the electric field impressed on the plates. If a sawtooth voltage is impressed on the plate pair controlling horizontal deflection and is synchronized with an ac signal impressed on the vertical deflection plate pair, the waveshape of the signal is traced out on the fluorescent screen shown in Fig. 12 i.e. Lessajous pattern of different frequency ratios and phase differences. Since the deflection sensitivity of this system is relatively low, the plate pairs are preceded by amplifiers which control the actual magnitude of deflections the screen. A variety of circuits is required to control the screen displays for epetitive signals, to initials form and synchronize a variety of sweepsignal voltages, etc.

Digitizing oscilloscopes and transient digitizers (waveform recorders) represent a new class of instrument responding to the need for measurements fast enough to characterize integrated circuit parameters (where speeds are high because of extremely small dimensions of circuit paths, i.e. involvement of small capacitances and inductances) at high accuracy for quality-control purpose. The waveform recorder typically consists of a wide band amplifier: high-speed sample and hold and analog-to-digital conversion circuits; temporary (or buffer) memory; display; and digital outputs, all under the control of internal digital logic circuits. The display is generally a CRO fed with an analog signal reconstructed from the digital information, using operator-selected time scales. Better waveform recorders are stated by their manufacturers to have 20 to 100-MHz sampling rates and 0.25 to 0.1% amplitude resolution. As of 1983, neither performance standards nor means of verifying performance (physical standards, measurement

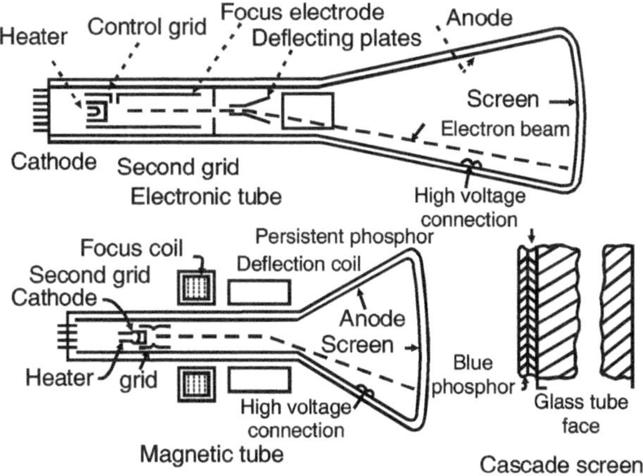

Fig. 11: *Elements of cathode ray oscilloscope*

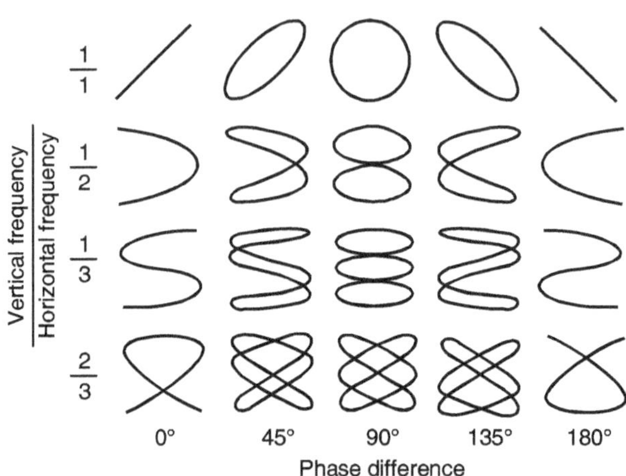

Fig. 12: *Lessajous pattern for different frequency ratios and phase differences*

techniques etc.) have been formulated for this class of instruments.

Oscilloscope Probes : Important types of oscilloscope probes are :

Compensated Attenuator Probe : The input impedance of general purpose oscilloscope is 1 MΩ in shunt with a 10 to 80 pf capacitance. A wire connected form the scope vertical input to the circuit under test is adequate for low frequency measurements but suffers from stray signal and operating frequency pick-up (50 Hz). If a coaxial cable is used to shield the probe from stray signals, it will significantly increase the input shunt capacitance, loading down the circuit. To overcome this, a compensated (attenuator) probe is used. This probe presents a higher input resistance and lower shunt capacitance, attenuating all frequencies equally.

Active Probe : The active probe, consisting of an FET input, is used for high frequency measurements (very small rise time) of extremely low level pulse signals. It has a lower shunt capacitance than compensated attenuator, with the added feature of zero signal attenuation.

Demodulator Probe : The demodulator probe is used for converting an RF signal into a dc voltage with its output voltage directly proportional to the peak RF signal.

Q-Meter

The Q meter is an instrument designed to measure some of the electrical properties of coils and capacitors. Its operation is based on the familiar characteristics of a series resonant circuit that the voltage across the coil or capacitor is equal to the applied voltage times the Q of the circuit.

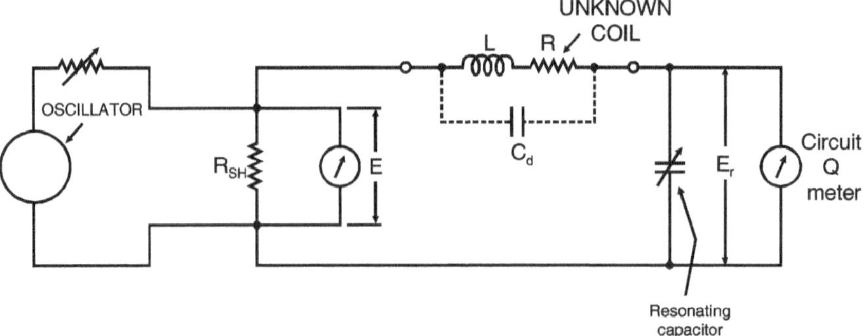

Fig. 13: *Q-meter*

A typical Q-meter circuit is shown in Fig. 13

To make a measurement the unknown coil is connected to the test terminals of the instrument and the circuit is tuned to resonance either by setting the oscillator to a given frequency and varying the internal resonating capacitor. The reading Q on the output of the meter must be multiplied by the index setting of the Q-meter to obtain the actual Q value.

The inductance of the coil can be calculated from the known values of frequency f and resonating capacitance C. Since

$$X_L = X_C \text{ and } L = \frac{1}{(2\pi f)^2 C} \text{Henry.}$$

Digital Voltmeters : The digital voltmeter (DVM) displays measurements of d.c. or a.c. voltages or discrete numerals instead of pointer deflection on a continuous scale as in analog devices.

Advantages of digital voltmeters are :

1. It reduces human reading and interpolation errors
2. It eliminates parallax error
3. It increases reading speed
4. It provides output in digital form suitable for reading

Digital voltmeters can be classified as

(a) Ramp-type DVM

(b) Integrating DVM

(c) Continuous balance DVM

(d) Successive approvmation DVM

Ramp-type DVM : The operating principle of the ramp-type DVM is based on the measurement of the time it takes for a linear ramp voltage to raise from zero to the level of the input voltage or to decrease from the level of the input voltage to zero. This time interval t is measured with an electronic time-interval counter and the count is displayed as a number of digits on electronic indicating devices.

Conversion from a voltage to a time interval is illustrated in Fig. 14.

At the start of the measurement cycle, a ramp voltage is initiated. This voltage can be positive-going or negative going. The negative-going ramp is continuously compared with the unknown input voltage. At the instant the ramp voltage equals the unknown voltage, a coincidence circuit generates a pulse, which opens a gate as shown in Fig. 14.

The ramp voltage continues to decrease with time until finally it reaches zero voltage and a second comparator generates clock pulse which closes the gate. An oscillator generates clock pulses which are allowed to pass through the gate to number of decade counting units which totalize the number of pulses passed through the gate. The decimal number displayed by the display system associated with the counter is a measure of magnitude of input voltage.

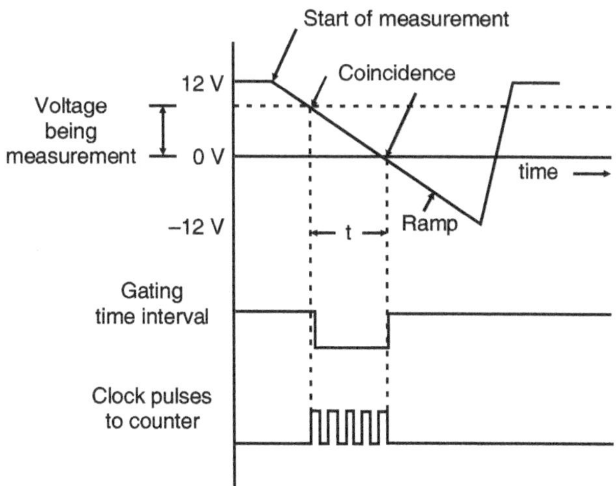

Fig. 14

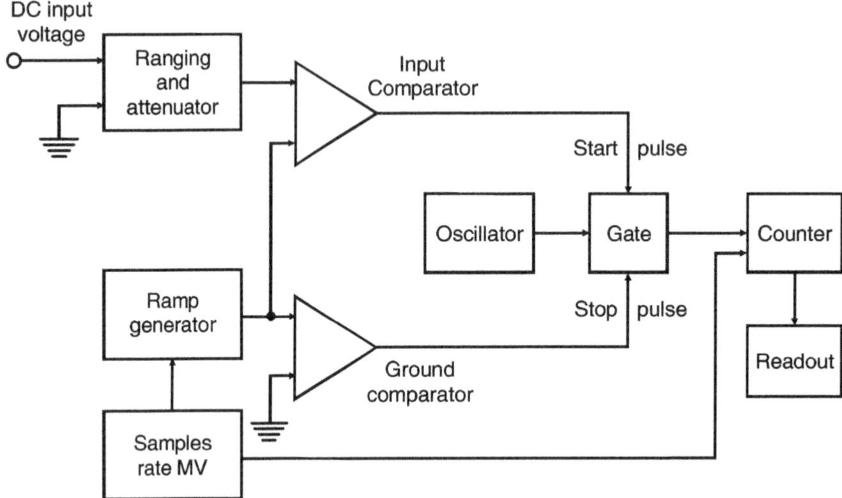

Fig. 15

Integrating DVM : The integrating DVM measures the true average of the input voltage over a fixed measuring period, in contrast to the ramp-type DVM which samples the voltage at the end of measuring cycle. A widely used technique to accomplish integration employs a voltage-to-frequency converter. The converter functions as a feedback control system that governs the rate of pulse generation in proportion to the magnitude of input voltage.

The major advantage of this system of A/D conversion is its ability to measure accurately in the presence of large amounts of superimposed noise since the input in integrated.

Strobing in Digital Voltmeters : It is important to note that the ramp repetition rates and the clock frequency are quite high, to minimise the time required to display readings (about 1 m sec) and to permit strobing or multiplexing of the displays. When strobing is present each of the four digits to lighted up separately, in turn. However this is done at such a high rate they all appear to be stationary. It saves an appreciable amount of battery power.

The multiplexing required is fed to logic circuit and its affects both the operation of the digital drivers and the segment drivers. Thus as one particular digit is caused to the energised at the same exact time the segment drivers cause the proper combination of segment for that digit to appear. Since all corresponding segment drivers in this way, it can b seen that in order to have only one digit at a time light up, showing the correct number, the strobing scheme is necessary.

The $\frac{1}{2}$ digit : For relatively little additional expense and circuitry, it is possible to provide an additional digit in front of digit displays. This added digit can only be programmed to be either unlighted or to show a 1. Thus, if energised it can make the reading from 9999 to 19999 or practically double the original reading. When the $\frac{1}{2}$ digit is added to a four-digit display it is called a $4\frac{1}{2}$ digit display. When the $\frac{1}{2}$ digit is switched on, he meter is said to the over-ranging.

1. In the Kelvin bridge for low resistance measurement, the effects of contact and load resistance are eliminated and the resistance of the unknown may be measured at its rated or working current.
2. The measurement of high resistance is prone to errors because parallel leakage paths are difficult to eliminate and the values obtained are affected by the magnitude.

Recording and Display

Recording Instruments : Recording instruments are designed to record the time variation of electrical quantities. A recording instrument usually consists of the following parts :

(a) the electrical measuring instrument
(b) the writing mechanism
(c) the moving chart (moving surface) for receiving the record
(d) the chart-driving mechanism.

Recording is made on the chart :

(A) Continuous-line recording :

1. By a pointer in permanent contact with the chart;
   (a) by ink, directly applied by a pen on the pointer to the chart
   (b) by ink pressed by a point from an inked ribbon on to the chart
   (c) by the tip of the pointer engraving the trace on a suitably prepared surface (waxed paper, film etc.)
   (d) by low-voltage d.c led by the pointer on a dark paper coated with a clear metal (metallic paper), by melting the covering metal film.
2. By a coloured liquid, squirted from the end of pointer of the recording instrument (without contact with the chart).
3. By the heated tip of the pointer (not touching the chart) which melts and removes a fine white coating from coloured base;
4. By a light beam reflected from the moving mirror of the instrument onto a photosensitive surface (paper or film).

(B) Dotted-line recording :

5. By rapid intermittent pressure (e.g. of 0.5 sec duration) of a stylus upon ordinary paper, by interposition of an inked ribbon, or upon chemcially prepared paper at longer intervals (e.g. 10 to 15 sec);
6. By a high-voltage spark from the pointer, without contact with the chart.
7. By a stylus made of some special material, pressed for equal intervals of time to a chemically prepared paper, giving a row of dots or dashes.

The chart may be driven at constant speed by :

(a) a built-in spring driven clockwork
(b) a built-in spring-driven clockwork, pulse controlled
(c) a built-in synchronous motor
(d) a built-in synchronous motor plus spring driven clockwork.

Graphical Recorders

Hard copy data from a process or system can be displayed in a graphical form either as the relationship between one or more variables and time, using an x–t recorder or as the relationship between two variables using an x–y recorder. x–t recorders can be classified as either strip chart recorders, in which data are recorded on a continuous roll of chart paper, or circular chart recorders, which as their name implies, record the data on a circular chart.

Strip Chart Recorders

Most strip chart recorders employ a servo-feedback system (as shown in Fig. 16) to ensure that the displacement of the writing head across the paper tracks the input voltage over the required frequency range. The position of the writing head is generally measured by a potentiometer system. The error signal between the demanded position and the actual position of the writing head is amplified using an a.c. or d.c. amplified and the output drives either an a.c. or d.c. motor. Movement to the writing head is effected by a mechanical linkage between the output of the motor and writing head. The chart paper movement is generally controlled by a stepping motor.

The methods used for recording the data onto the paper include :

1. **Pen and Ink :** In the past these have used pens having ink supplied from a refillable resevoir. Increasingly, such system use disposable fibre-tipped pens. Multi-channel operation can be achieved using up to six pens. For full-width recording using multiple pens, staggering of the pens is necessary to avoid mechanical interference.
2. **Impact Printing :** The 'ink' for the original impact systems was provided by a carbon ribbon placed between the pointer mechanism and the paper. A mark was made on the paper by

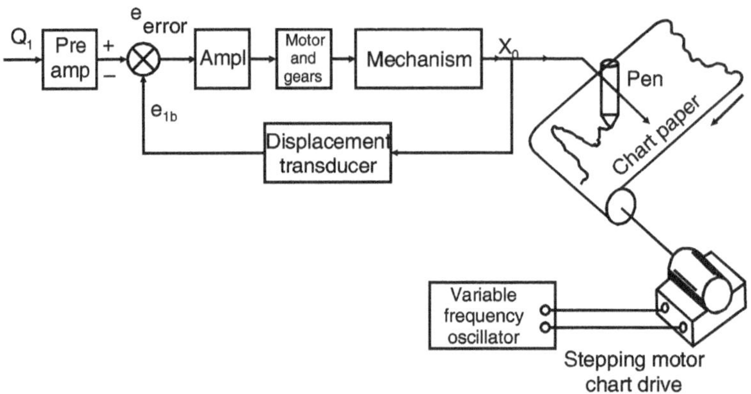

Fig. 16: *Strip Chart Recorder*

pressing the paper. Newer methods can simultaneously record the data from up to twenty variables. This is achieved by having a wheel with an associated ink pad which provides the ink for the symbols on the wheel. By rotating the wheel different symbols can be printed on the paper for each of the variables. The wheel is moved across the paper is response to the variable being recorded.

3. **Thermal Writing :** These systems employ thermally sensitive paper which changes colour on the application of heat. They can use a moving writing head which is heated by an electric current or a fixed printing head having a large number of printing elements (up to several hundred for 100 mm wide paper). The particular printing element is selected according to the magnitude of the input signal. Multichannel operation is possible using such systems and time and time and date information can be provided in alphanumeric form.
4. **Optical Writing :** This technique is commonly used in galvanometer system. The source of light is generally ultraviolet to reduce unwanted effects from ambient light. The photographic paper used is sensitive to ultra-violet light. This paper develops in daylight or artificial light without the need for special chemicals. Fixing of the image is necessary to prevent long-term degradation.
5. **Electric Writing :** The chart paper used in this technique consists of a paper base coated with a layer of coloured dye–black, blue, or red–which in turn is coated with a very thin surface coating of aluminium. The recording is effected by a tungsten wire stylus moving over the aluminium surface. When a potential of approximately 35 V is applied to this stylus an electric discharge occurs which removes the aluminium, revealing the dye. In multichannel recording the different channels are distinguished by the use of different line configurations (for example, solid, dashed, dotted). Alphanumeric information can also be provided in these systems.

Circular Chart Recorders

These recorders, as shown in Fig. 17, generally use a 300 mm diameter chart. They are particularly

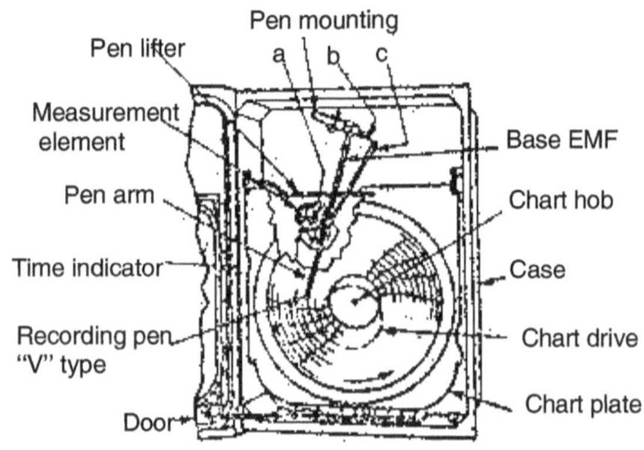

Fig. 17: *Circular Chart Recoders*

well suited for direct actuation by number of mechanical sensors without the need for a transducer to convert the measured quantity into an electrical one. Thus they can be configured for a temperature recorder using a filled bulb thermometer system or an absolute, differential, or gauge pressure recorder employing a belows or a burden tube. Up to four variables can be recorded on a single chart. The rotation of the circular scale can be provided by a clockwork motor, which makes the recorder ideally in remote location having no source of power.

X-Y Recorders

In an analogue x–y recorder (Fig. 18) both the x and y deflections of the writing head are controlled by servo feedback systems. The paper, which is usually either A3 or A4 size, is held in position by electrostatic attraction or vaccum. These recorders can be provided with either one or two pens. Plug-in timebases are also generally available to enable them to function as x-t recorders. A typical device may provide x and y input ranges which are continuously variables between 0.25 mV/cm and 10/cm with an accuracy of $\pm$ 0.1 percent of full scale and a deadband of 0.1 percent of full scale. Zero offset adjustments are also provided. A typical timebase can be adjusted between 0.25 and 50s/cm with an accuracy of 1 percent.

The dynamic performance of x-y recorders is specified by their slowing rate and acceleration. A very high-speed x-y recorder capable of recording a single up to 10 Hz at an amplitude of 2cm peak to peak will have a slewing rate of 97 cm/s and a peak acceleration of 7620 cm/s^2. Remote control of such functions as sweep start and reset, pen lift and chart hold can be provided.

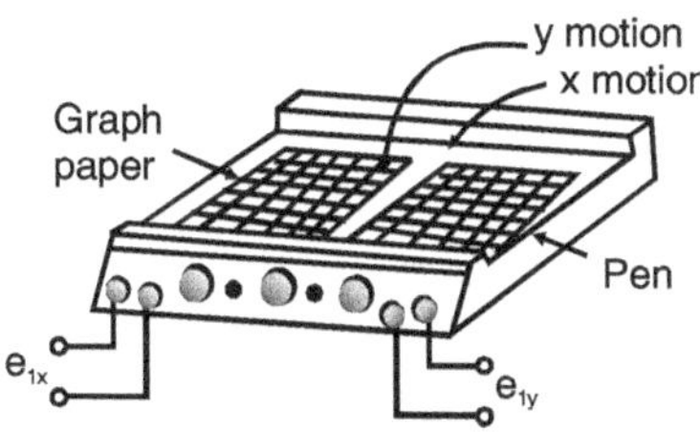

Fig. 18: *X-Y Recorder*

Digital x-y plotters replace the servo feedback with an open-loop steeping motor drive. These can replace traditional analogue recoders and provide increased measurement and graphics capabilities. A digital measurement and plotting system can provide, for example, simultaneous sampling and storage or a number of input channels; a variety of trigger modes, including the ability to display pre trigger data; multi pen plotting of the data annotation of the record with data, time and set-up conditions; and an ability to draw grids and axes. Communication with such devices can be means of the IEEE-488 of RS-232 interfaces.

For obtaining hard copy from digital data input, graphics plotters are available. With appropriate hardware and software these devices can draw grids, annotate chart, and differentiate data by the use of different colours and line types. They are specified by their line quality, plotting speed, and paper size. Intelligence built into the plotter will free the system's CPU for other tasks. The availability of a graphics language to control the plotter functions and graphics software packages also simplifies the user's programming tasks.

Data Loggers

Data loggers can be broadly defined as data-acquisition systems whose functions are programmed from the front panel. They take inputs from a mixture of analogue and digital signals, perform limited mathematical and logical operations on the inputs, and provide storage either in the form of semiconductor memory or magnetic tape or disc systems. Many data loggers are provided with an integral alphanumeric display and printer. Fig. 19 shows the elements of a typical data-logging system. These systems can be characterized by:

1. **The number and type of analogue or digital inputs which can be accepted .** The analogue input may be a.c. or d.c. current, d.c. current or from thermocouples or RTDs.
2. **The scanning rates and modes of scanning available .** These can include single-scan, monitor, interval scan, and continuous scan.
3. **The method of programming .** This may be one button per function on the front panel; by

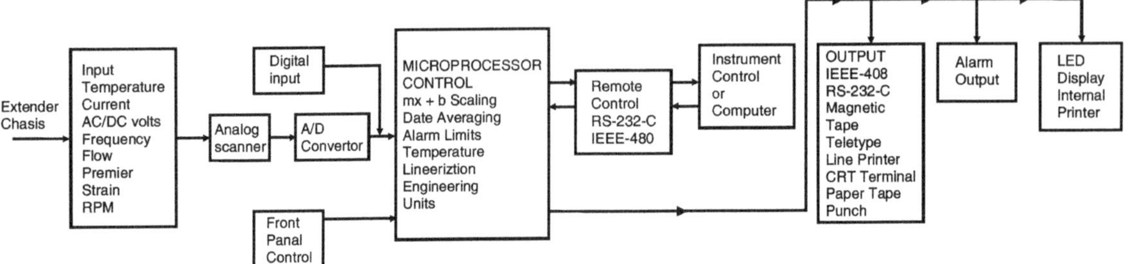

Fig. 19: *Data Logger*

use the of a menu-driven approach by presenting the information to user on a built-in CRT screen; or by using a high-level language such as BASIC.

4. **The methematical function provided by the logger .** These may include linearization for thermocouple inputs scaling to provide the user with output in engineering units and data averaging over a selectable number of scans.
5. **The nature and capacity of the internal memory .** If the memory is semiconductor memory, is it volatile ? If so, is it provided with battery back-up in case of power failure ?
6. **The printer technique and the width of the Print output .** Data loggers typically use dot matrix ink printers or thermal or electric writing technique with continuous strip paper.
7. **The display .** Data loggers generally provide display using LEDs or LCDs, although some incorporate CRTs.
8. **Communication with other systems .** This is generally provide by Rs. 232/422/423 interfaces. Some data loggers are capable of being interrogated remotely using the public.

Commonly used Display Techniques

Display technique	*Mechanism*	*Configurations*
Indicating devices		
Moving pointer	Mechanical/electromechanical movement of pointer over a fixed scale	Horizontal/vertical straight, art, cicrular or segment scales with edgewise strip, hairline, or arrow-shaped pointers
Moving scale	Mechanical/electromechanical movement of scale. Indication given by position of scale with respect to fixed pointer	Moving dial or moving drum analogue indicators, digital drum indicators.
Bar graph	Indication given by height or length of vertical or horizontal column	Moving column provided by mechanical driven ribbon or LED or LCD elements
Recording	*Technique*	*Configurations*
Illuminative displays		
Light emitting diodes	Light output provided by recombination electroluminescence in a forward biased semiconductor diode	Red, yellow, green, displays configured as lamps, bar graphs, 7 and 16-segment alphanumeric displays, dot matrix displays
Liquid crystal displays	The modulation of intensity of transmitted-reflected light by the application of an electric field to a liquid crystal cell	Reflective or transmissive displays bar graph, I-segment, dot matrix displays alphanumeric panels

Recording	*Technique*	*Configurations*
Plasma displays discharge	Cathode glow of a neon gas	Nixie tubes, 7-segment displays, plasma panels
CRT displays	Conversion into light of the energy scanning electron beam by phosphor	Monochrome and colour tubes, storage tubes, configured as analogue, storage, sampling, or digitizing oscilloscopes, VDUs, graphic displays.

Commonly used Recording Techniques

Recording	*Technique*	*Configurations*
Graphical records	Provide hard copy of data in graphical form using a variety of writing tecniques, including pen-ink, impact printing, thermal, optical, and electric writing	Single/multichannel x–t strip chart and circular chart recorders, galvanometer recorders, analogue and digital x-y recorders, digital, digital plotters
Printers	Provide hard copy of data in alphanumeric form using impact and non-impact printing techniques	Serial impact printers using cyllinder, golf ball, daisywheel or dot matrix heads. Line printes using drum, chain-belt, oscillating bar, comb, and needle printing heads. Non impact printers using thermal, electrical electrostatic, magnetic, ink-jet, electrophotographic, and lasser printing techniques
Magnetic recording	Use the magnetization of magnetic particles on a substrate to store information	Magnetic type recorders using direct, frequency modulation or pulse code modulation tecnique for the storage of analogue or digital data. Spool to spool or cassette recorders. Floppy or hard discs for the storage of digital data
Transient recorders	Use semiconductor memory to store high-speed transient waveforms	Single/multichannel devices using analogue to-digital conversion techniques. High speed transient recorders using optical scanning techniques to capture data before transfer to semiconducor memory.
Data loggers	Data-acquistion system having functions programmed from the front panel	Configured for a range of analogue or digital inputs with limited logical or mathematical functions, Internal display using LED; LCD, CRT. Hard copy provided by dot matrix ink or thermal or electrical writing technique. Data storage provided by semiconductor or megnetic storage using tape or disc.

EXERCISE

1. For measuring high value of capacitor and low value of inductor one will use
   A. Series connection Q-meter
   B. Parallel connection Q-meter
   C. Direct connection Q-meter
   D. All the above
2. At a frequency below the resonant frequency ... circuit is capacitive and ... circuit is inductive.
   A. series, parallel B. parallel, series
   C. parallel, parallel D. series, series
3. Reactive current through the capacitive road produces ... field and reactive current through the inductive load process ... field:
   A. magnetic, electric
   B. electric, magnetic
   C. magnetic, magnetic
   D. electric, electric
4. Which class of amplifier has the least distortion?
   A. A B. B
   C. C D. AB
5. A moving coil ammeter has a fixed shunt of 0.02 ohm resistance. If the coil resistance of the meter is 1000 ohms, a potential difference of 500 mV is required across it for full-scale deflection. Under this condition, the current in the shunt would be
   A. 2.5 A B. 25 A
   C. 0.25 A D. 0.025 A
6. The Main application of a self balancing potentiometer is
   A. In calibrating instruments
   B. In control systems
   C. In power systems
   D. It can be used any where
7. With a sweep time 10 m-seconds across the screen the approximate horizontal sawtooth frequency will be
   A. 50 HZ B. 100 Hz
   C. 1 Hz D. 500 Hz
8. Which of the following is a null defection device
   A. Ballistic galvanometer
   B. D'Arsonval galvanometer
   C. Potentiometer
   D. Ammeter
9. T = K*i* is the torque equation of a
   A. PMMC instrument
   B. Rectifier instrument
   C. Moving iron instrument
   D. D'Arsonval instrument.
10. The current capacity of the battery depends upon
    A. The kind of electrode metals used
    B. The size of the electrodes
    C. Volume of the electrolyte container
    D. All the above
11. A 200 V, 5A wattmeter is tested on full load using the circuit shown in the figure. The power consumed for testing is

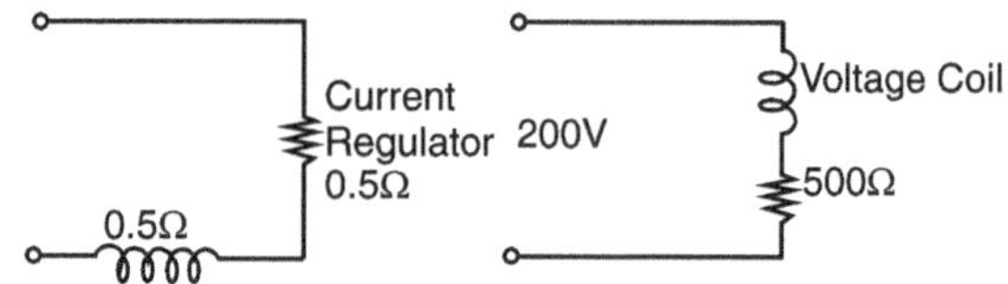

    A. 8 W B. 25 W
    C. 33 W D. 1000 W
12. A galvanometer can't be used for the measurement of
    A. Measuring current and voltage of small magnitude
    B. Determining the equality of two currents
    C. Measuring the quantity of electricity or current impulse
    D. Measuring the voltage impulse
13. Current passing through a resistor of 100 ± 0.2 Ω is 2.00 ± 0.01 A. Computed value of power will be
    A. 400 ± 0.21 W B. 400 ± 0.002 W
    C. 400 ± 5.2 W D. 400 ± 4.8 W.
14. Consider the circuit shown in the given figure for the measurement of resistance X using voltmeter-ammeter method.
    Ammeter resistance, $R_a = 0.1\ \Omega$
    Voltmeter resistane, $R_v = 5000\ \Omega$
    If X_1 and X_2 are the measured values of X for switch positions 1 and 2 respectively, then the

arithmetic mean of X_1 and X_2 will be

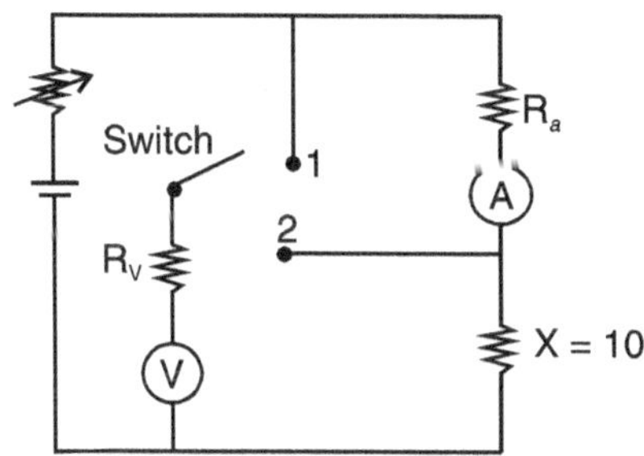

A. 9.98 Ω B. 10.00 Ω
C. 10.04 Ω D. 10.10 Ω

15. Schering bridge is used to
A. determine dielectric loss
B. Determine the inductance
C. Measure mutual inductance
D. Measure low resistance

16. Frequency compensation in a moving iron instrument is achieved by connecting a
A. High resistance in series with the coil
B. Low resistance in series with the coil
C. A capacitor in series with the fixed coil
D. A capacitor across the fixed coil

17. '50 MHz' on the specification plate of a CRO indicates that
A. Sweep signal is of 50 MHz
B. Horizontal oscillator frequency is 50 MHz
C. Vertical oscillator frequency is 50 MHz
D. Input signal frequency should be less than 50 MHz

18. The price of capacitor will increase as
A. Voltage rating of the capacitor increases
B. The size of the capacitor increases
C. Microfarad rating of the capacitor increases
D. None of the above

19. Wiens bridge is used for measurement of frequency in the applied voltage waveform is
A. sinusoidal B. square
C. rectangular D. triangular

20. Which one of the following types of indicating instruments is an electrometer
A. Electrodynamometer
B. PMMC
C. Electrostatic
D. Moving iron

21. The Reed frequency meter is essentially a:
A. Recording measuring system
B. Deflection measuring system
C. Vibrational measuring system
D. Oscillator measuring system

22. Total flux of 15000 lines equals
A. 22.26 Gilberts B. 15,000 gauss
C. 15,000 maxwells D. 15,000 kilolines.

23. It is desired to convet a 0–1000 μA meter moment, with an internal resistance of 100 ohms, into a 0 – 100 mA meter. The required value of shunt resistance is about
A. 1 ohm B. 10 ohms
C. 99 ohm D. 100 ohms

24. A moving iron ammeter may be compensated for frequency errors by
A. A series inductance
B. A shunt resistance
C. Shunt capacitor
D. Series resistance

25. Generally the number of turns on the primary winding of a CT are
A. 1 to 5 B. 5 to 10
C. 10 to 50 D. 10 to 100

26. Two milliammeters, with a full-scale current of 1 mA and 10 mA, are conneted in parallel and they read 0.5 mA and 2.5 mA respectively. Their internal resistances are in the ratio of
A. 1 : 10 B. 10 : 1
C. 1 : 5 D. 5 : 1

27. Consider the following statements about temperature measurement:
1. A thermistor is much more sensitive than a platinum thermometer.
2. The resistance of a thermistor is solely a function of its absolute temperature, whether the source of heat is external, internal or both.
3. Highly linear temperature-resistance relationship is a great advantage of the thermistor
4. Thermistor has negative temperature coefficient.

Of these statements
A. 1 and 2 are correct
B. 3 and 4 are correct

C. 1, 2, 3 and 4 are correct
D. 1, 2 and 4 are correct

28. Shaft encoder is used for the measurement of
A. angular position B. linear position
C. linear velocity D. linear acceleration

29. The instrument whose deflection is given by the expression $\theta \propto I^2 \frac{dM}{d\theta}$ is known as:
A. Electrodynamic type
B. Repulsion type
C. Electrostatic type
D. Attraction type

30. When using ohm meter, applied voltage is to be disconnected from the circuit because
A. Voltage source will increase resistance
B. Current will decrease resistance
C. The ohmmeter has its own internal battery
D. None of the above

31. Which of the following bridges can be used for the measurement of dielectric loss of a capacitor?
A. Schering bridge
B. Heaviside Campbell equal ratio voltage
C. Owen bridge
D. Anderson bridge

32. The thermocouple instrument will measure
A. Current B. Voltage
C. Flux D. None of the above

33. Which of the following instruments are not a current operated instrument
A. Ferrodynamic inruments
B. Electrostatic instruments
C. Electrodynamometer type instruments
D. Hot wire instruments

34. By increasing the concentration of electrolyte in a battery
A. The output voltage increases
B. The current given by the battery increases
C. The capacity of the battery increases
D. The internal resistance of the battery decreases

35. The instrument transformers are normally used to extend the range of
A. Moving iron instruments
B. Electrostatic instruments
C. Induction instruments
D. Can be used with any of the three.

36. A trimmer is
A. Variable capacitor
B. Another name for a gang
C. Another name for preset capacitor
D. None of these

37. Schering bridge measures
A. Capacitance dielectric loss
B. Inductance
C. Resistance
D. Mutual inductance

38. Burden is a term associated with
A. Only CT
B. Only PT
C. CT or PT
D. Any electrical measuring instrument

39. A dual-beam CRO has
A. Two horizontal amplifiers
B. Two trigger circuits
C. Two vertical amplifiers
D. All the above

40. Which of the following methods can be used for measuring power without using wattmeter.
(i) One voltmeter, one ammeter
(ii) Two voltmeters, two ammeters
(iii) Three voltmeters
(iv) Three ammeters
A. I and II only B. II and IV only
C. I and III only D. IV only

41. The maximum power transferred to the load R_L, P_L max is
A. $2V_{CC}^2/R_L$ B. $V_{CC}^2/4R_L$
C. V_{CC}^2/R_L D. $V_{CC}^2/2R_L$

42. Which of the following contributes to harmonics distortion in amplifiers?
A. Positive feedback
B. Presence of noise
C. Defective active device
D. Non-linearity in active device.

43. A parabolic dish has a diameter of 10 m. The maximum possible (ideal) gain of the antenna at $\lambda = 314$ cm will be
A. 30 dB B. 40 dB
C. 50 dB D. 60 dB

44. The thermal emf's in the bridge measurement
A. Does not affect the measurement

B. Causes a small inaccuracy in the measurement
C. Causes considerable trouble in balancing the bridge
D. None of these

45. Equivalent circuit for capacitor is

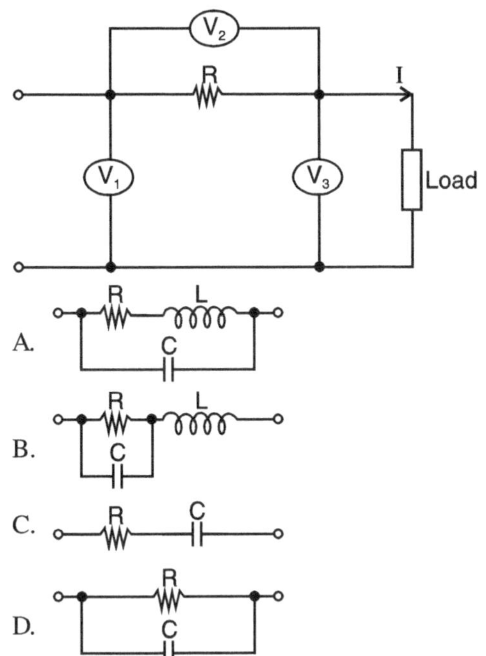

46. Which of the following is a new material emerging from advances in biotechnology?
A. PTEE
B. Nylon
C. Kevlar
D. Poly 3-hydroxybutyrate

47. Which of the following gives piezoelectric effect?
A. Mu metal B. PVDF
C. Sapphire D. Ferrites

48. If the scale of an instrument is uniform it generally means that the instrument is meant for reading
A. RMS values
B. Average values
C. Peak values
D. Nothing can be said with certainty

49. The defection factor G of a CRT is
A. The same as the deflection sensitivity
B. The reciprocal of defection sensitivity
C. Independent of the accelerating potential
D. Dependent on the deflection voltage.

50. The design of capacitor is desired ideally for
A. Low time constant
B. Zero time constant
C. High time constant
D. Infinite time constant

51. For a power factor meter, mark the incorrect statement:
A. The deflection is proportional to the phase angle between field coil and crossed coil
B. The restoring torque is provided by a control spring
C. The balance position depends upon the resulting torque developed by the two coils
D. It consists of two coils mounted at right angles to each other.

52. An instrumemt has sensitivity of 1000 ohm/volt. On the 100 volt scale, this instrument will have internal resistance of:
A. 10 ohms B. 10,000 ohms
C. 100,000 ohms D. 1000 ohms

53. Electrical conductivity of metals expressed in $ohm^{-1}\ m^{-1}$ is of the order of
A. 10^{10} B. 10^{5}
C. 10^{-4} D. 10^{-6}

54. Induction method of testing is used for
A. Short ckt fault in the cables
B. Open ckt fault in the cables
C. Short ckt and ground fault in the cables
D. Open ckt and ground fault in the cables

55. X and Y plates of a CRO are connected to unequal voltages of equal frequency with phase shift of 90°. The Lissajous figure on the screen will be
A. Circle B. Straight line
C. Ellipse D. Figure of eight

56. Two 300 V d.c. voltmeters having resistances of 15000 Ω and 5000 Ω are connected in series. The safe voltage which can be measured by this combination is
A. 600 V B. 400 V
C. 300 V D. 200 V

57. Match List I with List II and select the corrrect answer by using the Codes given below the lists:

	List I		*List II*
(a)	Mu metal	1.	Very high magnetic permeability
(b)	Samurium/ cobalt	2.	Very high magnetic remanence
(c)	Sapphire	3.	Very high thermal conductivity
(d)	Ferrites	4.	Very stable magnetic permeability

Codes:	(a)	(b)	(c)	(d)
A.	1	2	3	4
B.	2	3	4	1
C.	3	4	1	2
D.	4	1	2	3

58. The bandwidth of an audio amplifier extends from 20 Hz to

A. 2000 Hz B. 10,000 Hz

C. 20,000 Hz D. 100,000 Hz

59. Which of the following is a low pass filter

A. L type with series C and shunt L

B. T type with series C and shunt L

C. T type with series L and shunt C

D. L type with series L and shunt C.

60. In an experiment with a thermocouple with one junction at 0°C, the following data were obtained at θ° C:

θ°C	20°C	100°C	300°C
e(in millivolts)	0.194	0.850	1.650

Assuming parabolic relation, the calculated neutral temperature for this thermocouple will be about

A. 300°C B. 333°C

C. 366°C D. 399°C

61. Thermocouple transducer is used for:

A. Temperature measurement

B. Velocity and vibration measurement

C. Pressure measurement

D. Gas flow measurement

62. Which of the following instruments will be used to measure 1500 kV a.c. voltage

A. Moving coil voltmeter

B. Electrostatic voltmeter

C. Moving iron voltmeter

D. Hot wire instrument

63. $_{92}U^{235}$ nucleus absorbs a slow neutron and undergoes fission into $_{50}X^{140}$ and $_{38}Sr^{94}$ nuclei. The other particles produced in this fission process are

A. 1 β and 1 neutron

B. 2 β and 1 neutron

C. 2 neutrons

D. 3 neutrons

64. De Brogli wavelength λ of the electron is given by the relation

A. $\lambda = \frac{h}{mv}$ B. $\lambda = \frac{hv}{m}$

C. $\lambda = \frac{nh}{2\pi}$ D. $\lambda = \frac{2\pi}{h}$

65. The 45° – 45° system refers to stereo for

A. Magnetic tape recorders

B. Phonograph records

C. Crystal microphones

D. Loudspeakers

66. In the context of ac bridge measurements, the term "Wagner Ground" means

A. a special RC connection that eliminates stray magnetic effects

B. any unwanted and unintended ground connection

C. a large metal plate buried in the ground and connected to one corner of the bridge

D. a special RC connection that eliminates stray capacitance effects

67. The poor overload capacity is the main disadvantage of

A. Permanent moving coil instruments

B. Electronic instruments

C. Hotwire instruments

D. Electrostatic instruments

68. Induction wattmeters can be used for measuring:

A. D.C. power only

B. A.C. power only

C. Both d.c. and a.c. powers

D. A.C. power of single phase only

69. What will be the power consumed by the voltage source, current source and resistance respectively?

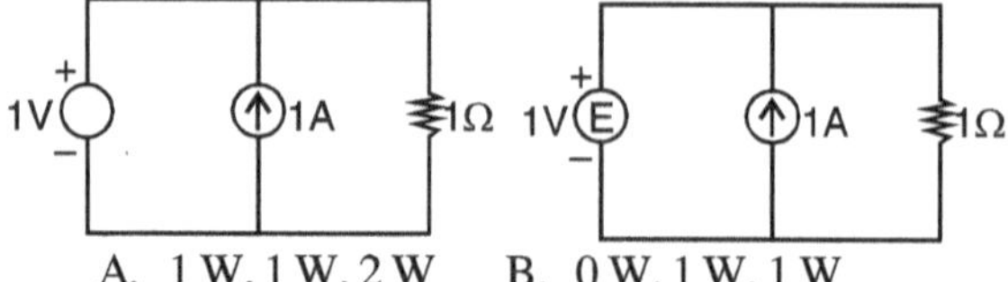

A. 1 W, 1 W, 2 W B. 0 W, 1 W, 1 W

C. 1 W, 0 W, 1 W   D. 0 W, 0 W, 0 W

70. A non-sinusoidal voltage is fed to both a moving iron voltmeter and a rectifier-type instrument. The readings are 12.1 V and 11.1 V respectively The harmonic content as a percentage of average value is

A. 63.8%   B. 100%
C. 43.4%   D. 10%

71. If retrace is visible on the CRT screen, the trouble may be that

A. There is loss of SYNC signal
B. Blanking control is not set properly
C. Sweep signal is too lengthy
D. All of these

72. EPROM is

A. ultraviolet-light erasable and electrically programmable
B. infrared-light erasable and magnetically programmable
C. electrostatically erasable and magnetically programmable
D. magnetically erasable and electrically programmable

73. The energy required to change the speed of one electron from rest to 0.6 c is nearly

A. 0.085 Me V   B. 0.13 Me V
C. 0.26 Me V   D. 0.37 Me V

74. For the same ac voltage and load impedance which of the following statements about rectifiers are correct?

A. The average load current in a FW rectifier is twice that in HW rectifier
B. The average load current in a FW rectifier is π times that in HW rectifier
C. HW rectifier will have a bigger transformer compared to FW rectifier
D. HW rectifier will have a larger sized transformer compared to a FW rectifier.

75. A 150μ- amp meter movement has 1000 ohm resistance, what shunt resistance is required to extend the range to 750μ-amp?

A. 222 ohms   B. 225 ohms
C. 250 ohms   D. 150 ohms

76. The advantage of the LVDT over the inductive bridge-type transducer is that

A. it produces high output voltage for small changes in core position
B. It can be used for operation at temperatures as low as 265°C and as high as + 600°C.
C. it can measure displacement of a moving object
D. the error due to stray magnetic field is less

77. Thermostat used for temperature control is:

A. An absolute instrument
B. An indicating instrument
C. A controlling instrument
D. A recording instrument

78. Which of the following statement is true

A. Eddy current loss is determined from the area of B-H curve
B. Hysteresis and eddy current losses can be determined from the area of the B-H loop
C. Hysteresis loss can be determined from the area of B-H curve
D. None of the above losses is determined from the B-H loop.

79. The resistance of 125 ohm strain gauge changes by one ohm for 4000 microstrain. The gauge factor of the strain gauge is

A. 1.5   B. 2
C. 2.5   D. 3

80. A vacuum tube has electrodes confined inside an evacuated

A. Glass envelop only
B. Metal envelop only
C. Either of (A) and (B) above
D. None of these

81. The output of a piezoelectric crystal has

A. low amplitude and low impedance
B. high amplitude and high impedance
C. low amplitude and high impedance
D. high amplitude and low impedance

82. Three phase four wire energy meter is used to measure:

A. Single phase energy
B. Two phase energy
C. Three phase balanced energy
D. Three phase unbalanced energy

83. Solar cell is an example of

A. Photo voltaic cell
B. Photo conductive cell
C. Photo emissive cell

D. None of these

84. In the Miller sweep circuit

A. Constant charging current is maintained by keeping the voltage across the charging resistor and hence the current through it
B. A step input is converted into a linear ramp by using an operational integrator
C. The timing capacitor is charged linearly from a constant current source
D. The timing capacitor is charged from a constant voltage source.

85. Which of the following relation is vaild?

A. MTBF = MTTF + MTTR
B. MTTR + MTTF + MTBT = 1
C. $\frac{1}{MTTR} + \frac{1}{MTTF} = \frac{1}{MTBF}$
D. MTBF.MTTF.MTTR = 1

where, MTNF = Mean Time Between Failures
MTTF = Mean Time To Failure
MTTR = Mean Time To Repair

86. Much of initial body resistance to electricity is in

A. skin B. bones
C. blood D. hair

87. Ballistic galvanometer can be used to measure

A. Current B. Flux
C. A and B D. None of the above

88. Low resistance is measured by

A. De Sauty's bridge
B. Maxwell's bridge
C. Kelvin's double bridge
D. Wien bridge

89. Match the following:

Column I	*Column II*
(a) Conducting material	I. HF_6
(b) Semi-conducting material	II. Platinum
(c) Super-conducting material	III. Silicon
(d) Insulating material	IV. Cadmium

Codes :	*(a)*	*(b)*	*(c)*	*(d)*
A.	4	1	3	2
B.	1	3	2	4
C.	3	4	1	2
D.	2	3	1	4

90. Two voltmeters one with a full scale reading of 100 V and another with a full scale reading of 200 V, are connected in series across a 100 V supply. The internal resistance of both meters is the same. What are the readings?

A. 100 V and 0 V B. 75 V and 25 V
C. 67 V and 33 V D. 50 V and 50 V

91. If the current sensitivity of a galvanometer is very high then its critical resistances are

A. High and low respectively
B. Low and high respectively
C. High and high respectively
D. Low and low respectively

92. In a set-up for the determination of self-capacitance of a coil, the standard variable capacitor has a value of 460 pF at 2 MHz and a value of 100 pF at 4 MHz. The coil self-capacitance is

A. 2pF B. 20 pF
C. 60 pF D. 200 pF

93. The coil of a moving iron voltmeter has a resistance of 20 ohms and an inductance of 0.3 H. It is connected in series with a swamping resistance of 2000 ohms. The capacitor connected in shunt with the swamping resistance for making the meter read correctly at dc as well as at 50 Hz ac will have a value of

A. 0.0125 μF B. 0.0311 μF
C. 0.0521 μF D. 0.0622 μF

94. In a resistance loaded amplifier, the gain decreases at high frequencies because

A. of decreasing reactance of the stray C_r
B. increasing reactance of the stray C_r
C. increasing reactance of C_C
D. decreasing reactance of C_C

95. The combination of two 0.02 μF 500 V capacitors in series has a capacitance and breakdown ratings of

A. 0.01 μF, 500 V B. .01 μF, 1000 V
C. 0.02 μF, 500 V D. 0.04 μF, 500 V.

96. Which of the following statements is not true for complementary push-pull amplifier?

A. Its efficiency is same as that of class B push-pull amplifier
B. It requires one power supply
C. It does not require any transformer
D. It employs PNP and NPN transistor

97. LBDT is used as a

A. Displacement transducer

B. Pressure transducer
C. Temeperature transducer
D. Any of the above

98. Which of the following is an active transducer?
A. Photoelectric
B. Photovoltaic
C. Photo-conductive
D. Photoemission

99. The flux density of 12000 gauss equals
A. 12,000 lines/in^2
B. 12,000 lines/cm^2
C. 12,000 kilolines/cm^2
D. 14400,000 lines/cm^2

100. The X and Y inputs to an oscilloscope have same frequency but differ in phase. List I gives some of the patterns on the oscilloscope screen whilst List II gives various values of phase differences:

List I	**List II**
(a) Ellipse	1. 0 degrees
(b) Circle	2. 30 degrees
(c) Straight line with positive slope	3. 60 degrees
(d) Straight line with negative slope	4. 90 degrees
	5. 120 degrees
	6. 180 degrees

Match the two lists by choosing the correct answer

Codes:	*(a)*	*(b)*	*(c)*	*(d)*
A.	2	4	1	6
B.	5	1	4	6
C.	4	2	3	1
D.	1	2	3	6

101. OP-AMP is stabilised using dominant pole compensation. The following statements are made
1. The compensating circuit contains one C and one R
2. The compensating network has a zero which cancels one pole of the open loop transfer function
3. The closed loop gain intersects the open loop gain where the fall off is 12 dB/octave
4. Since the capacitor introduces a phase shift less than 90° the circuit is stable

Of these four statements the true statements are
A. Only 2 B. Only 2 and 3
C. Only 1 D. Only 1 and 4

102. The arrangement for calibrating a phase shifter is shown in the following diagram:

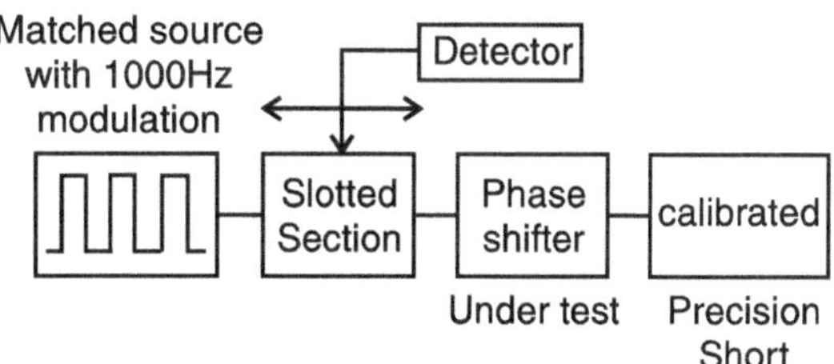

The calibrated short is adjusted while phase shift is introduced by the phase shifter under test to maintain a null at the probe fixed in location. In one such measurement, the displacement of the short is 5 mm while the guide wavelength in it is 5 cm. At this setting of the phase shifter, the value of the phase shift (in degrees) will be
A. 90 B. 72
C. 45 D. 36

103. Hall effect can be used to measure
A. Electric field intensity
B. Magnetic field intensity
C. Carrier concentration
D. None of these

104. In a RC coupled amplifier, which of the following component is mainly responsible for harmonic distortion of the signal?
A. Transistor
B. Biasing resistors
C. Coupling capacitor
D. power supply

105. Which of the following periodic wave forms will have only odd harmonics of sinusoidal wave forms?

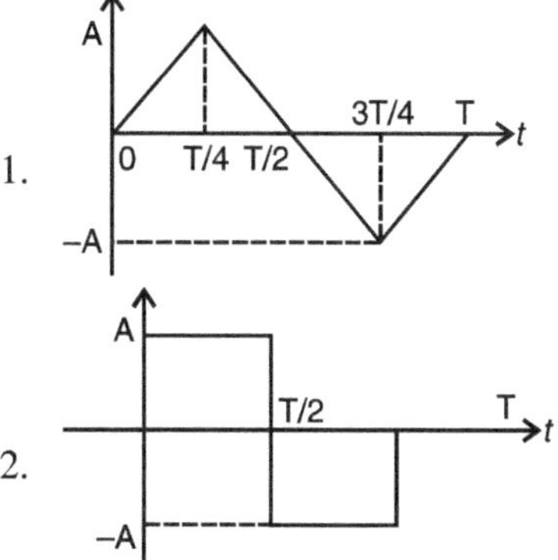

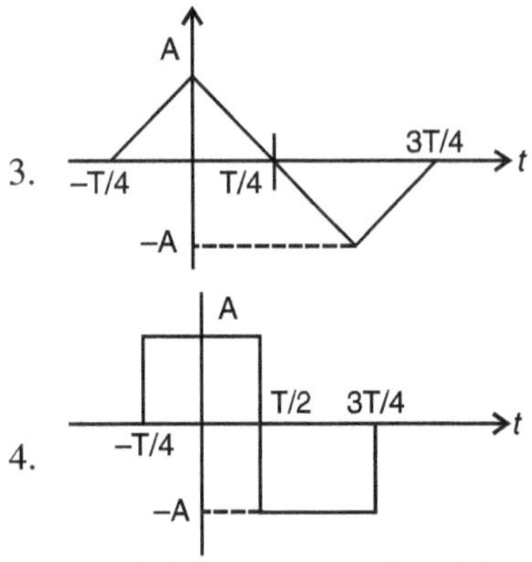

A. 1 and 2 B. 1 and 4
C. 2 and 4 D. 2 and 3

106. If the voltmeter resistance is increased the error in the reading given by this voltmeter will
A. Increase
B. Decrease
C. Increase or decrease depending upon the value of the quantity being measured
D. Be independent of the voltmeter resistance

107. The VA rating of instrument transformers is normally
A. 0.1 VA B. 10 VA
C. 100 VA D. kVA

108. Flash ADC is
A. a serial ADC
B. a parallel ADC
C. partly serial and partly parallel ADC
D. successive approximation ADC

109. Which of the following positions of holding a power transistor to identify its terminals is correct?

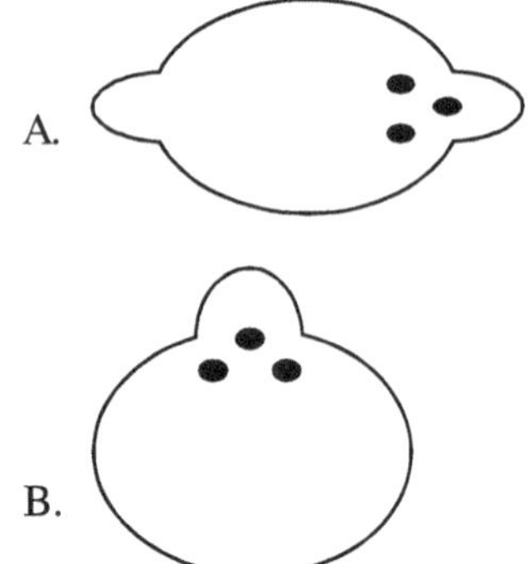

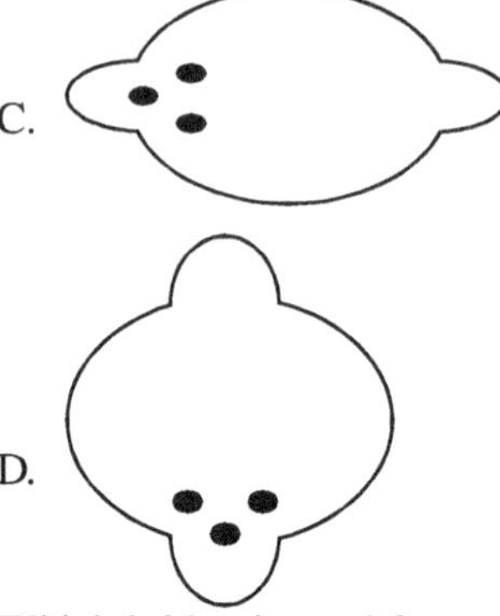

110. Which bridge is used for measuring audio and H.F. oscillations
A. Wien bridge
B. Maxwell bridge
C. Schering bridge
D. None of the above

111. The waveshape of V_0 for the circuit shown is

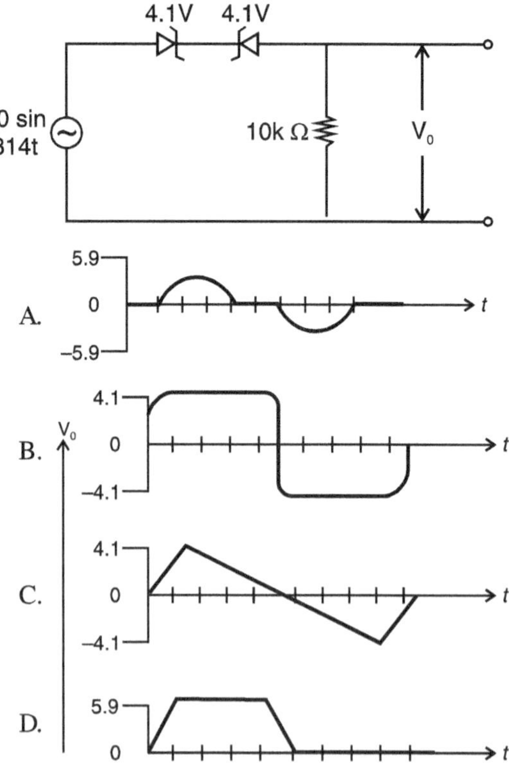

112. If a capacitor C having selectivity factor Q is scaled down in all its dimensions by a factor K without in any way altering the dielectric material used, then the new capacitor will have its value (= C) and selectivity factor (= Q) respectively as

A. $C, \frac{Q}{K}$ B. $\frac{C}{K}, Q$

C. $\frac{C}{K}, \frac{Q}{K}$ D. C, Q

113. The stored energy associated with inductance is 1/2 LI^2, whereas the power in the resistance is RI^2. This apparent discrpency of a factor of 1/2 is due to:

A. electromagnetic radiation where 1/2 LI^2, is radiated

B. a relation between RMS and maximum of $1/\sqrt{2}$

C. the power and hence energy balance up to the steady state

D. None of the above

114. The Q of a radio coil

A. is indepedent of frequency

B. increases monotonically as frequency increases

C. decreases monotonically as frequency increases

D. increases upto a certain frequency and then decreases beyond that frequency

115. In the circuit of an electrostatic wattmeter given below, the instantaneous torque is proportional to

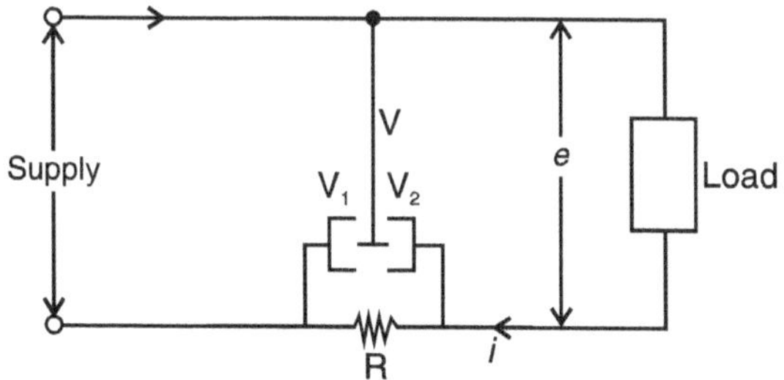

A. $ei+\frac{i^2 R}{2}$ B. $ei-\frac{i^2 R}{2}$

C. $e i + i^2 R$ D. $e i - i^2 R$

116. If the voltage across the plates of a capacitor is held constant and the area of overlap between the plates is increased, the energy:

A. increases B. decreases

C. remains constant D. becomes zero

117. The diffusion capacitance of the PN junction

A. Decreases with increasing current and increasing temperature

B. Decreases with decreasing current and increasing temperature

C. Increases with increasing current and increasing temperature

D. Does not depend on current and temperature.

118. Match the following :

Column I (Colour)	*Column II (Colour Code value for resistor)*
(a) Black	1. 3
(b) Brown	2. 0
(c) Red	3. 2
(d) Orange	4. 1

Codes:	*(a)*	*(b)*	*(c)*	*(d)*
A.	2	4	3	1
B.	1	3	2	4
C.	3	1	2	4
D.	4	1	2	3

119. Match the following:

Column I	*Column II*
(a) Mass spectrum	1. Wave function
(b) X-ray spectrum	2. Unpaired electron
(c) Paramagnetism	3. Atomic number
(d) Orbitals	4. Isotopes

Codes:	*(a)*	*(b)*	*(c)*	*(d)*
A.	2	4	3	1
B.	4	3	2	1
C.	3	4	2	1
D.	1	4	2	3

120. The flux of 2000 lines through 0.25 cm^2 equal a flux density of

A. 1,000 gauss B. 2,000 gauss

C. 4,000 gauss D. 8,000 gauss.

121. The dielectric loss of capacitance can be measured by

A. Hay bridge B. Schering bridge

C. Maxwell bridge D. Anderson bridge

122. With a variable reluctance tachometer having 60 rotor teeth, the counter 3600 counts per minute. The device speed, in rpm, would be

A. 36 B. 60

C. 360 D. 3600

123. The gauge factor of strain gauge is 2, stress is 1050 kg/cm², Y = 2.1 × 10⁶ kg/cm² and R is 1000 ohms. The value of Δ R will be

A. 2 Ω B. 3 Ω

C. 4 Ω D. 1 Ω

124. The barrier capacitance C_T

A. Increases with the width of the space charge layer.

B. Increases with increasing reverse voltage

C. Is due to the immobile charges at the junction varying with the applied voltage

D. Can be defined as $\frac{Q}{V}$

125. Which of the following compensation methods in amplifiers leads to reduction in bandwidth?

A. Lead compensation

B. Pole-zero compensation

C. Miller Effect compensation

D. Dominant-pole cancellation

126. A capacitance transducer has two plates of area 5 cm² each, separated by an air gap of 2 mm. Displacement sensitivity (in pF/cm) due to gap change would be

A. 11.1 B. 44.2

C. 52.3 D. 66.3

127. In regard to class A single ended transformer coupled amplifier the following statements are made.

1. In the presence of a signal, the transistor dissipation is less than in the case of direct coupled load
2. The output a.c. power obtainable is the same as in the series fed directly coupled load for a given transistor dissipation capability
3. The d.c. current flowing through primary causes considerable dissipation of power
4. The drain on the power supply is ideally one half of what it is in direct coupled amplifier.

Of these four statements, the true one are 1 and 2

A. 1 and 2 B. 2 and 3

C. 3 and 4 D. 2 and 4

128. In a JFET, if the gate voltage V_{gs} is made more negative, then

A. Channel conductivity increases

B. Depletion region decreases

C. Channel conductivity decreases

D. Channel current increases

129. A semiconductor strain gauge has the advantage over the normal strain gauge that

A. it is more sensitive

B. it is more linear

C. it is less temperature dependent

D. its cost is low

130. For the transistor amplifier, the component values and high frequency specifications are R_s = 10 K, R_b = 2 K, r_{bb} = 20 ohms; r_{bc} = 150 ohms, C_{bc} = 2pF, C_{be} = 200 pF, g_m = 0.5 ohm and R_L = 200 ohms ($R_c >> R_L$)

The mid-band gain will be

A. + 120 B. – 62

C. + 90 D. – 75

131. In Hall effect, the output voltage produced across the crystal is due to

A. Induced voltage by the applied magnetic field

B. Drop across the crystal due to the current passed through it

C. Movement of charge carries towards one end

D. None of these

132. The percentage limiting error, in the case of an instrument reading of 8.3 V with a 0 to 150 voltmeter having a guaranteed accuracy of 1% full-scale reading is

A. 1.81% B. 0.181%

C. 18.10% D. 0.0018%

133. For good stabilised biasing of the transistor of the CE amplifier we should have

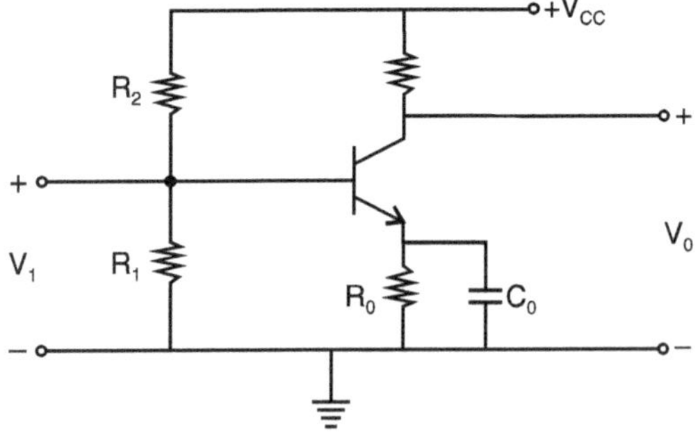

A. $\frac{R_e}{R_b} << 1$ B. $\frac{R_e}{R_b} >> 1$

C. $\frac{R_b}{R_e} << h_{FE}$ D. $\frac{R_b}{R_e} >> h_{FE}$

where $R_b = R_1 \parallel R_2$

134. Tunnel diodes
  A. Are mainly operated with reverse bias
  B. Have a negative resistance region with reverse bias
  C. Has very high impurity concentration thus increasing the barrier width
  D. Electrons have a finite probability of penetrating the potential barrier.

135. When the biasing of the base-emitter junction of a transistor is decreased, then
  A. Only emitter current decreases
  B. Only collector current decreases
  C. Emitter current decreases and current increases
  D. Both emitter current and collector current decrease

136. The actual equivalent ciruit of an air-dielectric variable capacitor is as follows:

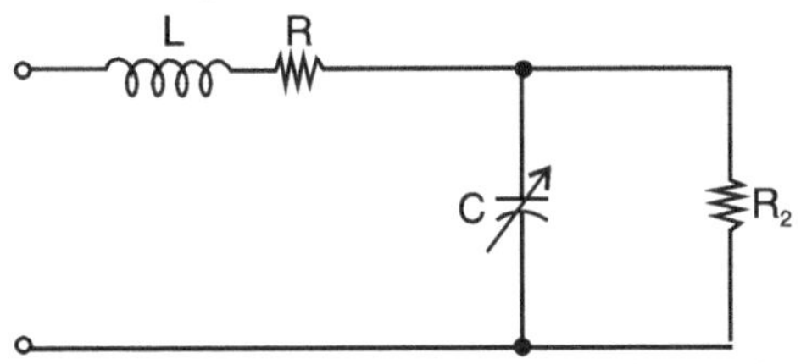

The equivalent circuit of this capacitor, valid at high frequencies only, is given by

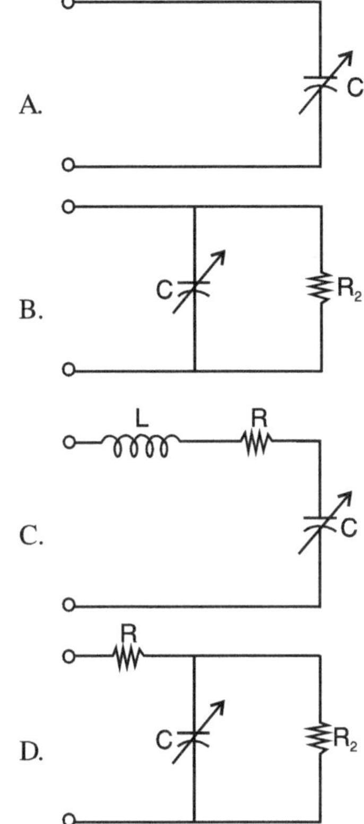

137. If the reference sound presure level is 2 × 10 N/m², then a sound pressure of 90 dB will be equal to

A. 0.632 N/m² B. 0.707 N/m²
C. 0.835 N/m² D. 0.925 N/m²

ANSWERS

1	2	3	4	5	6	7	8	9	10
A	A	B	A	B	B	B	B	D	B
11	**12**	**13**	**14**	**15**	**16**	**17**	**18**	**19**	**20**
C	A	D	C	A	D	A	A	A	C
21	**22**	**23**	**24**	**25**	**26**	**27**	**28**	**29**	**30**
C	B	A	C	A	D	D	A	A	C
31	**32**	**33**	**34**	**35**	**36**	**37**	**38**	**39**	**40**
A	A	B	D	C	C	A	D	B	A
41	**42**	**43**	**44**	**45**	**46**	**47**	**48**	**49**	**50**
B	D	D	C	D	D	B	D	B	B
51	**52**	**53**	**54**	**55**	**56**	**57**	**58**	**59**	**60**
C	C	B	B	A	B	A	C	B	B

61	62	63	64	65	66	67	68	69	70
A	B	B	A	B	D	C	B	B	C
71	**72**	**73**	**74**	**75**	**76**	**77**	**78**	**79**	**80**
B	A	B	A	C	C	C	C	B	C
81	**82**	**83**	**84**	**85**	**86**	**87**	**88**	**89**	**90**
C	D	A	B	A	A	C	C	D	D
91	**92**	**93**	**94**	**95**	**96**	**97**	**98**	**99**	**100**
C	B	D	C	B	(A, C, D)	A	D	C	A
101	**102**	**103**	**104**	**105**	**106**	**107**	**108**	**109**	**110**
B	C	B	A	A	B	B	B	D	A
111	**112**	**113**	**114**	**115**	**116**	**117**	**118**	**119**	**120**
B	B	C	D	A	A	C	A	B	D
121	**122**	**123**	**124**	**125**	**126**	**127**	**128**	**129**	**130**
B	B	D	C	D	A	D	C	A	D
131	**132**	**133**	**134**	**135**	**136**	**137**			
C	C	C	D	D	B	A			

EXPLANATORY ANSWERS

2. A series resonant circuit is capacitive below the resonant frequency and a parallel resonant circuit is inductive.

3. A capacitor stores energy in the form of electric field and an inductor stores energy in the form of magnetic field.

5. Current in shunt

$$= \frac{\text{potential difference across the coil}}{\text{shunt resistance}}$$

$$= \frac{500 \times 10^{-3}}{20 \times 10^{-3}} = 25 \text{ A}$$

11. Watt-meter current coil circuit resistance

$$= (0.5 + 0.5) = 1\ \Omega$$

Wattmeter voltage coil circuit resistance

$$= 5000\ \Omega$$

The watt-meter is tested for full load, i.e. current in the current coil is 5 A and the voltage across voltage coil is 200 V.

The associated powers consumed are :

$$5^2 \times 1 \text{ and } \frac{200^2}{5000}$$

∴ Power consumed in the wattmeter

$$= 25 + 8 = 33 \text{ W.}$$

14. $$X_1 = \frac{(10 + 0.1)\,1}{1} = 10.1$$

$$X_2 = \left(5000 \times \frac{1 \times 10}{(50 + 5000)}\right)\frac{1}{1}$$

$$= 9.98004$$

$$\text{Arithmatic mean} = \frac{10.1 + 9.98004}{2}$$

$$= 10.04002 \text{ ohms}$$

19. The bridge can be balanced at one frequency only. If the applied signal is of a waveshape other than sinusoidal, it will contain harmonics and the bridge cannot be balanced perfectly even if the harmonic strength is small. As such bridge can be used to measure frequency of a sinusoidal signal only.

23. Shunt resistance, $S = \frac{r}{m - 1}$

where r = movement resistance 100 ohms

m = multiplying factor

$= \frac{I}{i}$

I = total current

i = current through the movement

$$\therefore \quad m = \frac{100}{1} = 100$$

$$\text{Hence} \quad S = \frac{100}{100 - 1} \approx 1 \text{ ohm}$$

31. Only Schering bridge can be used for the measurement of dielectric loss of a capacitor. Heaviside Campbell equal ratio bridge is used for the measurement of self inductance. In Owen's bridge the value of inductance is determined in terms of resistance and capacitance.

Anderson bridge is also used to measure unnknown inductance is terms of a known capacitance and resistance.

37. Schering bridge is widely used for the measurement of capacitance and dissipation factor. Thus it is widely used in the measurement of the properties of insulators, consider bushings, insulating all and other insulating materials.

40. Since power = V × I, first two methods can be used for measuring power, Voltmeters or ammeters alone cannot be used for measurement of power.

43. Antenna gain, $G = \frac{4\pi A}{\lambda^2}$

where $A = \frac{\pi D^2}{4}$

D = diameter of dish

$$\therefore \quad G = \left(\frac{\pi D}{\lambda}\right)^2$$

53. Electrical conductivity is measured in $\text{ohm}^{-1}\ \text{m}^{-1}$. Electrical conductivity of metals is 10^7 to 10^5 of semiconductors, from 1 to 10^{-6} and of insulators 10^{-9} onwards. However, electrical resistivity is measured in terms of ohm – m.

69. Since the voltage across the resistor of 1 Ω is also to be 1 V, the current from the current source flows in the resistor only (= 1 A). The voltage source is carrying no current. Hence the corresponding powers are 0, 1 W, 1 W

79. Gauge factor $K = \frac{\Delta R/R}{\Delta L/L} = \frac{\Delta R/R}{\text{Strain}}$

$$= \frac{1/125}{4000\times10^{-6}} = 2$$

98. An active transducer is also known as self-generating type transducer. These transducers generate their own voltage or current. Energy required for production of an output signal is obtained from physical phenomenon being measured.

Passive transducers are externally powered transducers. These transducers derive the power required for energy conversion form an external power source. However, they may absorb some energy from physical phenomenon under study.

112. $C_1 = \frac{A_1 \in}{d_1}, A_2 = \frac{A_1}{K^2}, d_2 = \frac{d}{K}$

$$\therefore C_2 = \frac{A_2 \in}{d_2} = \frac{A_1 K \in}{K^2 d} = \frac{1}{K}\frac{A_1 \in}{d_1} = \frac{C_1}{K}$$

$$R_1 = \frac{d_1 \rho}{A_1}$$

$$R_2 = \frac{d_2 \rho}{A_2} = \frac{d_1}{K}\rho\frac{K^2}{A_1} = K\frac{d_1 \rho}{A_1} = KR_1$$

$$Q_1 = \frac{1}{C_1 \omega R_1}$$

$$Q_2 = \frac{1}{C_2 \omega R_2} = \frac{1}{\frac{C_1}{K}\omega K R_1}$$

$$= \frac{1}{C_1 \omega R_1} = Q$$

123. Change in resistance R, DR = GER

where, G = gauge factor, R = 1000 ohms

$$E = \text{strain} = \frac{\text{Stress}}{\text{Youngs modulus}}$$

$$= \frac{1050}{2.1\times10^{-6}}$$

$$\therefore \quad \Delta R = 2\times\frac{1050}{2.1\times10^{6}}\times1000 = 1 \text{ ohm}$$

126. $C = \frac{A \in}{g}$

or, $\frac{dC}{dg} = -\frac{A\in}{g^2}$

$$\therefore \quad \left|\frac{dC}{dg}\right| = \frac{5\times10^{-4}\times8.85\times10^{-12}}{\left(2\times10^{-3}\right)^2}$$

= 11.06 pF/cm.

129. The advantages of semiconductor strain gauges are:

1.High gauge factor, i.e., a relatively higher change in resistance which can be easily measured with a good degree of accuracy. This results in higher sensitivity of the gauge.

2. High fatigue life.

3. Excellent hysteresis characteristics.

4. Smaller size.

132. Maximum possible error at any reading

$$= \frac{150 \times 1}{100} = 1.5 \text{ V}$$

$\therefore$ Percentage error at a reading of 8.3 V

$$= \frac{15}{8.3} \times 100 = 18.1\%$$

135. At high frequencies, R_1 will be comparatively much larger than the capacitive reactance. Hence R_2, which appears across *c* can be omitted.

137. Sound pressure

$$= 20 \log_{10} \frac{\text{Given pressure in N/m}^2}{\text{Reference pressure in N/m}^2} \text{ db}$$

$$\therefore \quad 20 \log_{10} \frac{P}{P_0} = 90$$

$$\text{or,} \quad \log_{10} \frac{P}{P_0} = 4.5$$

$$\text{or,} \quad \frac{P}{P_0} = 10^{4.5} = 31622.777$$

$$\therefore \quad P = P_0 \times 31622.777$$

$$= 2 \times 10^{-5} \times 31622.77$$

$$= 0.632 \text{ N/m}^2.$$

8

ELECTRONIC DEVICES

TRANSISTORS

TABLE 1 : Characteristics of Silicon and Germanium Transistors

	Transistor type		
	Silicon grown	*Germanium*	
		Grown	*Alloy*
Collector :			
Voltage (maximum), V	40	40	25
Dissipation (maximum), mW	150	50	50
Cutoff current, μA	0.02	2	10
Capacitance, pF	7	14	40
Conductance parallel, μS†	0.3	0.2	1.0
Emitter :			
Current (minimum usable), mA	1	0.01	0.1
Reverse voltage (maximum), V	2	10	5
Bias voltage, mV	500	160	160
Resistance, Ω	100	25	25
Base, resistance, Ω	500	150	300
Gain :			
Power, dB	35	47	40
Current	26	35	40

† S is the abbreviation for siemens, the SI unit of conductance. It is equivalent to mho.

Parameters Variations

The general rules for transistor parameters variations with temperature are

(a) I_{CBO} increases by 8% per °C increases in Ge transistor.

(b) I_{CBO} increases by 4% per °C in Si transistor.

(c) $V_{B\pi}$ increases by 2 mV per °C increase.

(d) β doubles every 75°C increases.

At room temperature I_{CBO} for silicon transistor ≤1 μA and, for germanium ≤1 μA. Therefore, effect of increase in I_{CBO} with temperature is more profound in germanium transistor as compared to silicon.

Variations due to β. The β variations of a transistor of the same types may vary typically as 3 : 1. Therefore, Q point may vary over a wide range so that transistor may be in saturation if β is high, and in cut-off if β is low.

An ideal amplifier has an infinite input resistance and zero output resistance.

AMPLIFIERS

Harmonic distortion is caused by the nonlinearity of an active device's characteristics and may be observed in an amplifier's dynamic transfer curve, which is a plot of output voltage of current against input excitation.

TABLE 2 : h-Parameters

h-Parameter	*CE-Configuration*	*CB-Configuration*	*CC-configuration*
h_{fe}	—	$\frac{h_{fb}}{(1+h_{fb})}$	h_{fc}
h_{re}	—	$\frac{h_{fb}\,h_{ob}}{1+h_{fb}} - h_{rb}$	$(1-h_{rc})$
h_{fe}	—	$\frac{h_{fb}}{(1+h_{fb})}$	$-(1+h_{fe})$
h_{oc}	—	$\frac{h_{ob}}{(1+h_{fb})}$	h_{oe}
h_{fd}	$\frac{h_{fe}}{(1+h_{fe})}$	—	$\frac{h_{fc}}{h_{fe}}$
h_{rb}	$\frac{h_{fe}\,h_{oe}}{(1+h_{fe})} - h_{re}$	—	$-\frac{h_{fe}\,h_{oe}}{h_{fe}}$
h_{fb}	$\frac{-h_{fe}}{(1+h_{fe})}$	—	—
h_{ob}	$\frac{h_{oe}}{(1+h_{fe})}$	—	—
h_{fc}	h_{fe}	$\frac{h_{fb}}{(1+h_{fb})}$	—
h_{re}	$(1-h_{re})=1$	≈ 1	—
h_{fc}	$-(1+f_e)$	$\frac{1}{(1+h_{fb})}$	—
h_{oe}	h_{oe}	$-\frac{h_{ob}}{(1+h_{fb})}$	—

TABLE 3 : Parameter

Parameter	*Common Emmiter (CE)*	*Common Base (CB)*	*Common Collector (CC)*
Gurrent Gain (A_f)	High (Typ. 50)	Less than unity (Typ. 0.98)	Highest (Typ. 100)
Voltage Gain (A_v)	High (Typ. 50)	Highest (Typ. 100)	Less than unity (Typ. 0.98)
Input impedence (Z_f)	Several Kilo-ohms	Lowest (Several ohm)	Highest (hundreds of K)
Output impedence (Z_0)	Several Kilo-ohms	Highest (Mega ohms)	Lowest (Several ohms)
Phase diff. between input and output	180°	0°	0°

TABLE 4 : The characteristics of three major classes of operation.

Class of operation	*Current flows through active device for a period which is*	*Conversion Efficiency* (η)	*Distortion*	*Typical applications*
Class-A	Full 360° of the input signal period	Least	Least	Where intelligence is to be amplified; like audio and video amplifiers for example
Class-B	180° of the input signal period. Two transistors are needed at the output to reconstruct signal	Intermediate	Intermediate	Has replaced class-A in most of the applications
Class-C	Less than 180° of the input signal period. Load is a resonant circuit.	Maximum	Maximum	Where pure sinusoidal signals are to be amplified

1. A square wave is rich in odd harmonics and may be used to quickly show the frequency response of an amplifier by observing the output for squareness.
2. The ear responds differently to different frequencies, depending upon the loudness level; consequently bass and treble tone controls are often used to boost an amplifier's response at low and high frequencies.
3. The CE ac current gain $\beta_{ac} = h_{fe} = \frac{\Delta I_C}{\Delta I_B}$ (V_{CE} = constant), while $\beta_{dc} = h_{FB} = \frac{I_C}{I_B}$.
4. The CB ac current gain $\alpha_{ac} = h_{fb} = \frac{\Delta I_C}{\Delta I_E}$ (V_{CB} = constant), while $\alpha_{dc} = h_{FB} = \frac{I_C}{I_E}$, and is less than unit.
5. $\alpha = \frac{\beta}{1+\beta}$ and $\beta = \frac{\alpha}{1-\alpha}$.
6. In either CB or CE action, majority carriers are swept across the narrow, light doped base to the collected. This current is controlled by a relatively small base input current that is the result of the forward-baised base-emitter junction, which accounts for the amplifying action in a transistor.
7. The estability factor S is defined as the rate of change of collector current I_C with respect to the leakage current I_{CO}, with both β and I_B held constant. This instability factor indicates how much change there is I_C for a given change in I_{CO} and depends upon the bias circuit used.
8. The stability factor S for a CE circuit using fixed bias is $\beta+1$ for either a silicon or germanium transistor.

Tuned Amplifiers

Ideal response of a tuned amplifier provided constant and over the complete pass bound $(\omega_h - \omega_L)$ centred around a resonant frequency ω_O. If rejects all frequencies $<\omega_L$ and ω_h. The actual response is shown in Fig. 1 where ω_h and ω_L are defined in terms of 3 dB frequencies.

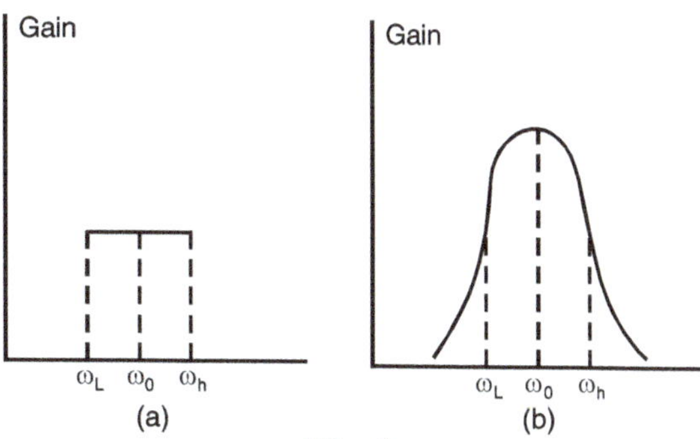

Fig. 1

Single-Tuned Amplifier : It consists of a CE amplifier in which a parallel-tuned circuit has been included.

The gain is maximum at $\omega = \omega_0$ i.e.,

$$BW = f_h - f_L = \frac{\omega_0}{2\pi Q_1} = \frac{1}{2\pi RC}$$

Gain Bandwidth Product GBW $= \dfrac{g_m}{2\pi C}$

Double-Tuned Amplifier : The frequency response of the double tuned amplifier shown in Fig. 2 is far flatter within the bandpass region than that of the single tuned amplifier with the result that all signals within this range receive approximately equal amplification. Secondly, the slope of the sides of the curve for the single turned amplifier is less sharp than that of the double tuned amplifier which indicates a very rapid attenuation of the signals immediately adjacent to the bandpass region ensuring good selectivity.

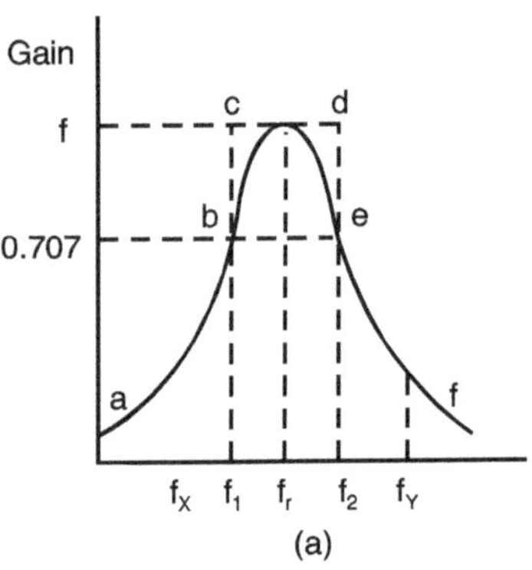

(a)

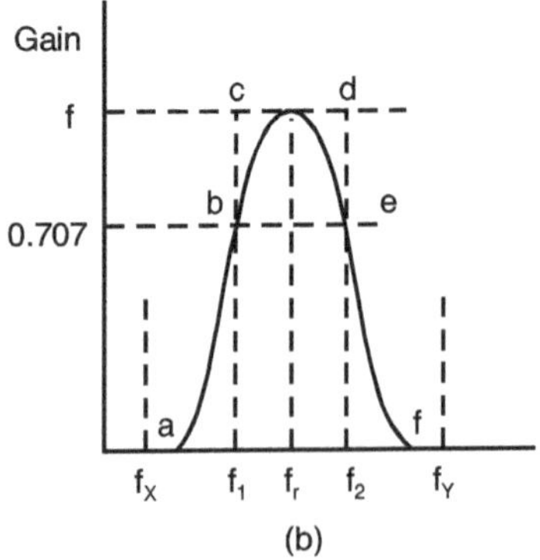

(b)

Fig. 2

Synchronously-Tuned Amplifier : It consists of cascaded tuned amplifiers in which all stages are identical and tuned to the same frequency ω_0. Such an amplifiers has increased gain and a bandwidth which is narrower than the bandwidth of each of the stages. The gain of the n-stage amplifier will be

$$A = \frac{(-g_m R)^n}{[1 + jQ_1(\omega/\omega_0 - \omega_0/\omega)]^n}$$

Again BW can be obtained by equating $|A| = \dfrac{A_m}{\sqrt{2}}$ and this leads to

$$BW = \frac{\omega_0}{2\pi Q_1}(2^{1/n} - 1)^{1/2}$$

Difference Amplifier

It amplifies the difference of two input signals V_{11}, and V_{12}. It provides very high gain and very high input impedance.

$$V_0 = A(V_{12} - V_{11})$$

The basic circuit is shown in Fig. 3. Its operation can be understood in terms of common-mode operation, and differential mode operation.

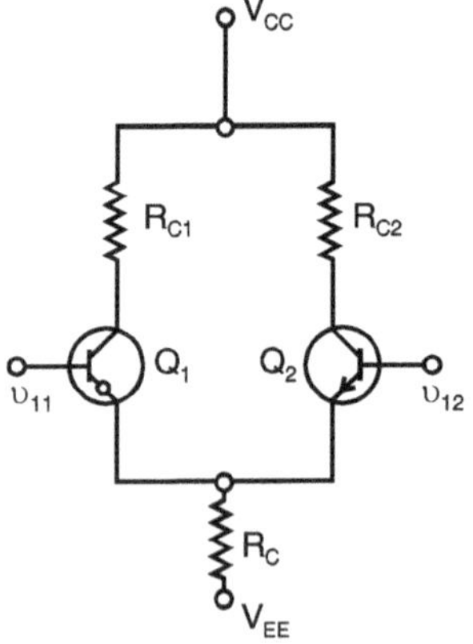

Fig. 3

Common-Mode Operation : If both the input voltages are zero, the circuit is completely symmetrical and thus any variation in temperature which effects the two transistors identically will cause no variation in the output measured between the two collectors. Similarly, supply voltage variations will have same effect on both transistors. Moreover, if a common-mode signal is introduced e.g., a noise a signal, it will be amplified and the circuit will remain balanced. The current will increase and drop across R_e will also increase correspondingly.

Differential Mode Operation : When a differential input signal is applied, the input voltages V_{11} and V_{12} will be equal in amplitude and opposite in phase. Therefore, if I_{c1} increases, I_{c2} will decrease by an equal amount. Since V_{c1} decreases and V_{c2} increases simultaneously, the output voltage $(=V_{c1} - V_{c2})$ and hence the gain will be very high as compared to a circuit using single transistor. Differential mode signals are not affected by R_e because as I_{c1} increases, I_{c2} decreases, and as a result the total emitter current $(= I_{c1} + I_{c2})$ remains constant.

In order to amplify differential signals and suppress common-mode signals, R_e should be as large as possible so that a small change in emitter voltage does not lead to any change in the emitter current.

A difference amplifier is characterised by a parameter called Common-Mode Rejection Ratio (CMRR) which is defined as

$$\text{CMRR} = \frac{\text{Differential-mode gain}}{\text{Common-mode gain}} = \frac{A_d}{A_c}$$

Obviously, CMRR should be as high as possible. Typical values of CMRR is 40 – 120 dB.

Direct-Coupled Amplifiers

If the amplifier is to respond to low frequency alternating signals, a direct connection is made from the collector, Q_1 to the base of in the two stage direct-coupled amplifier shown in Fig. 4. If a small increase in signal raises the base of Q_1, this increases the collectors current of Q_1 causing larger voltage to appear across R_{c1}. This lowers the collector of Q_1, decreasing the base current of Q_2, and thus decreasing its collector current. Due to this, the output voltage at the collector of Q_2 increases.

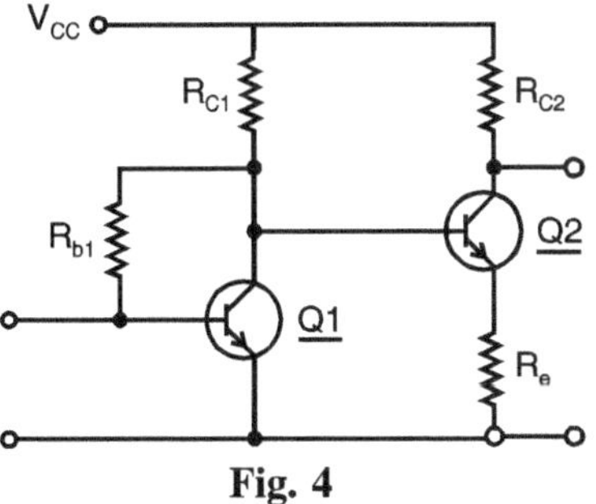

Fig. 4

These amplifiers are very important in integrated circuits as the coupling capacitors which require lot of space proportional to their values and take much larger area that needed to form a transistors are not present. Moreover in designs where size is a crucial factor (e.g., in case of hearing aids), dc coupled amplifiers are very common.

The Cascode Amplifier

Here, a CE amplifier is followed by CB stage. The circuit and its equivalent are shown in Fig. 5.

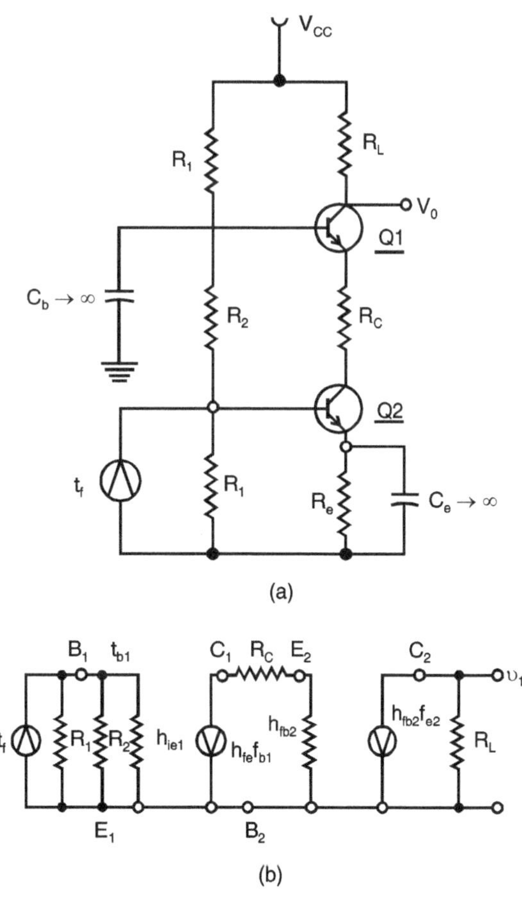

Fig. 5

The transfer gain $A_t \cong -h_{fb2}R_L$.

Therefore, the gain is nearly the same as that of a CE configuration. But its advantage is obtained at high frequencies as it provides higher gain than provided by CE amplifier alone. This is because at high frequencies, the feedback provided by the internal capacitor C_{bc} reduces the gain of CE amplifier.

In the cascode configuration, the load on its CE stage is a CB stage presenting a very low Z_f. The miller capacitance (C_m) is considerably reduced and thus there is no need of neutralisation ef K_{bc}. Therefore, this amplifier is commonly used in ICs. The same effect can be obtained by using a CB amplifier, but it does not provide any current gain.

Darlington Pair : A compound connection of two transistors that operates as if it were a single transistor with an extremely high forward-current transfer ratio. The input signal applied to the base of transistor T_1 and the emitter current supplies the input to transistor T_2 (as shown in Fig. 6). Since in normal transistor operation.

$$I_c = \alpha I_e$$

where α (the alpha current factor) is near unity, then the collector current I_{c1} is approximately equal to the emitter current I_{c1} and

$$I_{c1} = \beta_1 I_{b1}$$

β_1 is the beta current gain factor and I_{b1} the base current. In transistor T_2 the collector current I_{c2} given by,

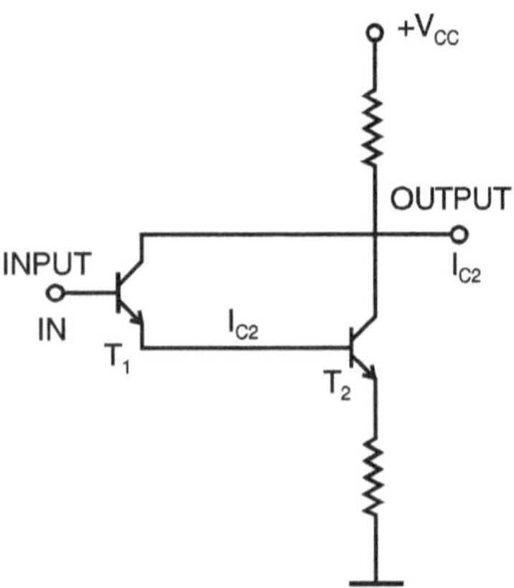

Fig. 6: *Darlington pair*

$$I_{c2} = \beta_2 I_{b2} = \beta_1\beta_2 I_{b1}$$

If the transistors are matched, i.e.

$$\beta_1 = \beta_2$$

the overall beta current gain of the combination is given by the square of the value for a single transistor.

Push-pull Amplifier

The use of two matched devices in such a way that they operate with a 180° phase difference. The output circuits combine the separate outputs in phase (Fig.7). One common means of achieving the desired 180° phase shift in the inputs is a transformer-coupled input circuit (Fig.8). Complementary transistors may also be used (Fig.9), in which case no phase shift is required in the inputs.

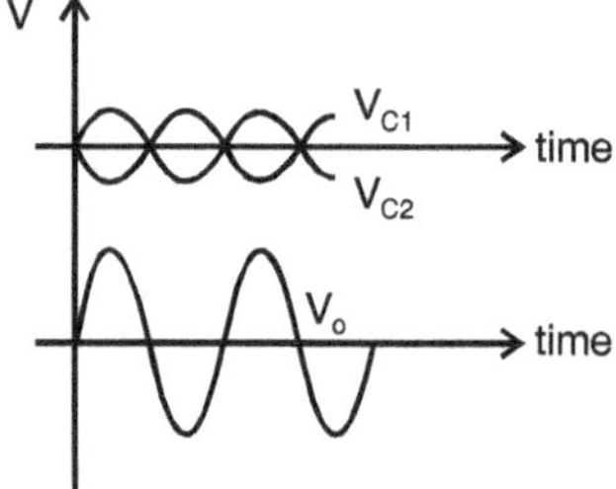

Fig. 7: *Output of transformer-coupled push-pull operation*

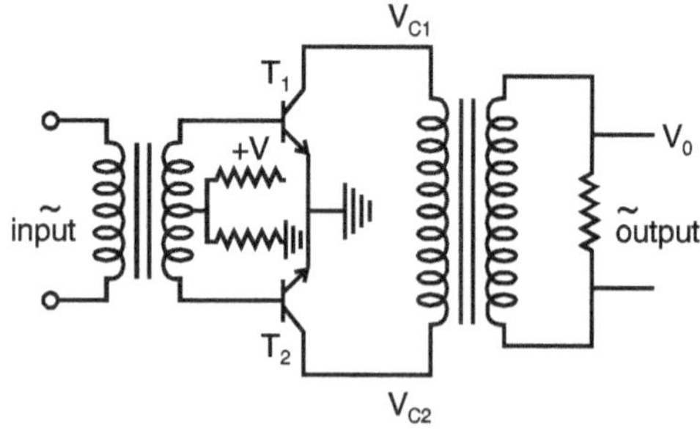

Fig. 8: *Transformer coupled class A push-pull operation*

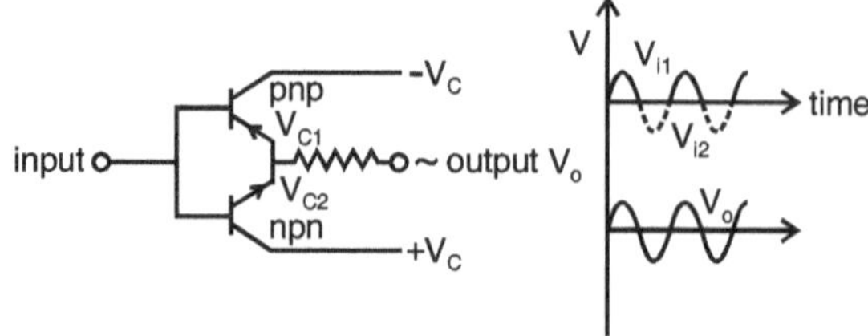

Fig. 9: *Complementary transistor push-pull operation*

Push-pull circuits are frequently used for class A and class B amplification and are then termed push-pull amplifiers. A push-pull amplifier that is suitably biased to give negligible output current when no input signal is present is termed a quiescent push-pull amplifier.

Feedback Amplifiers

Fig. 10 shows, general flow diagram of a feedback amplifier. The gain of the amplifier without feedback is A and feedback from output to input occurs through the feedback network F.

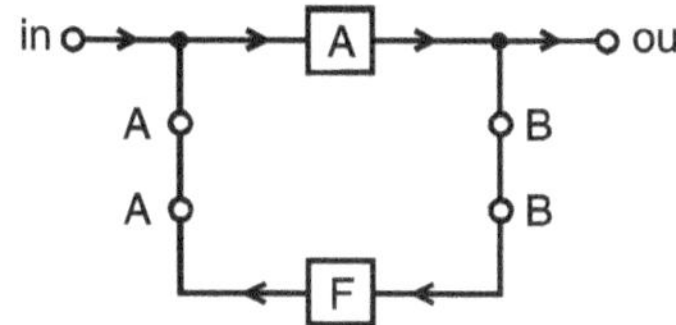

Fig. 10

1. If an amplifier having a voltage A_v has some of the output voltage feedback in series with the input, the voltage gain with feedback is

   $A_{vf} = \dfrac{A_v}{1 - kA_v}$, where k is the feedback factor,

   negative for negative feedback, and positive for positive feedback.
2. Although negative feedback reduces voltage gain, the result is a stabilization of gain, since

   $A_{vt} \approx \dfrac{1}{k}$.
3. Advantages of negative feedback include extended bandwidth, reduced harmonic distortion, and possibly a reduction in noise.
4. The effect of negative feedback on input and output resistance depends upon the type of negative feedback. The four types are current and voltage-series, and current and voltage-shunt.
5. Loop gain is the product, kA_v and is negative for negative feedback.
6. Feedback in dB $= 20 \log \dfrac{A_{vt}}{A_v} = 20 \log \dfrac{1}{1 - kA_v}$
7. Desensitivity. Consider the feedback formula,

   $$A_f(s) = \frac{A(s)}{1 - A(s)\,F(s)}$$

   Differentiating this equation with respect to A,

   $$\frac{dA_f}{dA} = \frac{1}{(1-AF)^2} = \frac{A_f}{A(1-AF)}$$

   $$\text{or} \quad \frac{dA_f}{A_f} = \frac{1}{(1-AF)} \cdot \frac{dA}{A}.$$

   The term (1 – AF) = (1 – T) is called desensitivity factor. Thus, larger the loop gain, the more insensitive or stable the amplifier is to parameter variations.
8. **Noise Reduction :** Let a noise voltage V_n appears at the output of the amplifier A due to noise, distortion or any other nonlinear, disturbance is generated. If V_{nf} is noise voltage present at the output of the complete amplifier, then a voltage of AFV_{nf} will be present at the output. Thus, the total amount of noise voltage will be

   $$V_{nf} = V_n + AFV_{nf}$$

TABLE 5 : Effect of negative feedback on amplifier characteristics

Characteristic	*Type of Feedback*			
	Current-Series	*Voltage Series*	*Voltage Shunt*	*Current-Shunt*
Voltage gain	Decreases	Decreases	Decreases	Decreases
Bandwidth	Increases	Increases	Increases	Increases
Harmonic distortion	Decreases	Decreases	Decreases	Decreases
Noise*	Decreases	Decreases	Decreases	Decreases
Input resistance	Increases	Increases	Decreases	Decreases
Output resistance	Increases	Decreases	Decreases	Increases

or $V_{nf} = \frac{V_n}{1-AF} = \frac{V_n}{1-T}$

Thus, any internally generated disturbance will be reduced by the desensitivity factor (1 – T). But it must be noticed that feedback cannot improve the signal-to-noise ratio if the noise signal enters at the amplifier's input.

9. **Increased Bandwidth :** Let the high frequency gain A(w) of the basic amplifier be

$$A(w) = \frac{A_m}{(1 + jw / w_h)}$$

$$A_f(w) = \frac{A_0 / (1 + jw / w_h)}{1 - A_0F / (1 + jw / w_h)}$$

$$= \frac{A_0}{(1 + jw / w_h) - A_0F}$$

$$= \frac{A_0}{(1 - A_0F)} \; \frac{1}{[1 + jw / (1 - A_0F)w_h]}$$

The new mid band gain A_{mf} of the amplifier is

$$A_{mf} = \frac{A_0}{1 - A_0F}$$

And its high cut-off frequency w_{hf} or bandwidth is $w_{hf} = w_h(-A_0F)$.

Thus, the gain decreases, but bandwidth of the amplifier increases by the same factors.

MILLER EFFECT

According to Miller's theorem, ''if an impedance (Z) is connected between input and terminals of a network and if there is voltage gain factor (K) between input and output, then the impedance (Z) can be replaced by two impedances, one from input to ground equal in magnitude to (Z/1-K) and the other from output to ground equal in magnitude to (ZK/K-1). If there is a phase reversal between input and output, (K) will have negative sign.

What is an OP-AMP ?

It is a very high-gain, high-r_{in} directly-coupled negative-feedback amplifier which can amplify signals having frequency ranging from **0 Hz to a little beyond 1 MHz.** They are made with different internal configurations in linear ICs. An OP-AMP is so named because it was originally designed to perform mathematical operations like summation, subtraction, multiplication, differentiation and integration etc. in analog computers. Present day usage is much wider in scope but the popular name OP-AMP continues.

Typical uses of OP-AMP are : scale changing, analog computer operations, in instrumentation and control system and a great variety of phase-shift and oscillator circuits. The OP-AMP is available in three different packages (i) standard dual-in-line package (DIL) (ii) TO-5 case and (iii) the flat-pack.

Although an OP-AMP is a complete amplifier, it is so designed that external components (resistors, capacitors etc.) can be connected to its terminals to change its external characteristics. Hence, it is relatively easy to tailor this amplifier to fit a particular application and it is, in fact, due to this versatility that OP-AMPs have become so popular in industry.

An OP-AMP IC may contain two dozen transistors, a dozen resistors and one or two capacitors.

Examples of OP-AMPs

1. μA 709—is a high-gain operational amplifier constructed on a single silicon chip using planar epitaxial process.
   It is intended for use in dc servo systems, high-impedance analog computers and in low-level instrumentation applications.
   It is manufactured by Semiconductors Limited, Pune

2. LM 108, LM 208 – Manufactured by Semiconductor Ltd. Bombay.

3. CA 741 CT and CA 741 T —these are high-gain operational amplifiers which are intended for use as (i) comparator (ii) integrator (iii) differentiator (iv) summer (v) dc amplifier (vi) multivibrator and (vii) bandpass filter.

Manufactured by Bharat Electronics Ltd (BEL), Bangalore.

OP-AMP Symbol

Standard triangular symbol for an OP-AMP is shown in Fig. 27 (a) though the one shown in Fig.

27 (b) is also used often. In Fig. 27 (b), the common ground line has been omitted. It also does not show other necessary connections such as for dc power and feedback etc.

The OP-AMP's input can be single-ended or double-ended (or differential input) depending on whether input voltage is applied to one input terminal only or to both. Similarly, amplifier's output can also be either single-ended or double-ended. The most common configuration is *two input terminals and a single output.*

All OP-AMPs have a minimum of five terminals

1. inverting input terminal
2. non-inverting input terminal
3. output terminal
4. positive bias supply terminal
5. negative bias supply terminal.

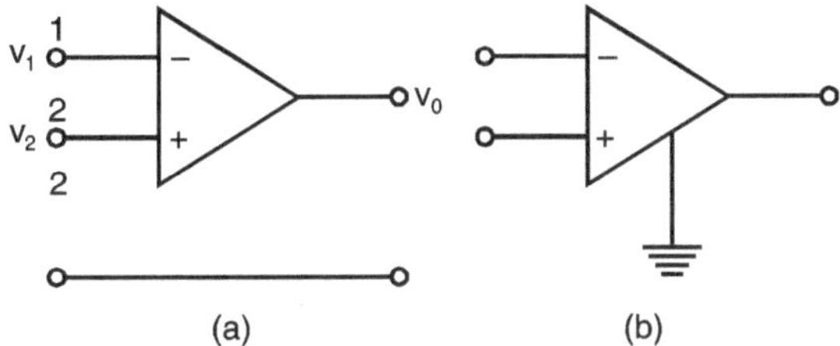

Fig. 27

Polarity Conventions

In Fig. 11(b), the input terminals have been marked with minus (–) and plus (+) signs. These are meant to indicate the inverting and non-inverting terminals only [Fig. 11(a)]. It simply means that a signal applied at negative input terminal will appear amplified but phase inverted at the output terminal as shown in Fig. 11(b). Similarly, signal applied at the positive input terminal will appear amplified and inphase at the output. Obviously, these plus and minus polarities indicate phase reversal only. It does not mean that voltages v_1 and v_2 in Fig. 11(a) are negative and positive respectively. Additionally, it also does not imply that a positive input voltage has to be connected to the plus-marked non-inverting terminal 2 and negative input voltage to the negative-marked inverting terminal 1. In fact, the amplifier can be used 'either way up' so to speak. It may also be noted that all input and output voltages are

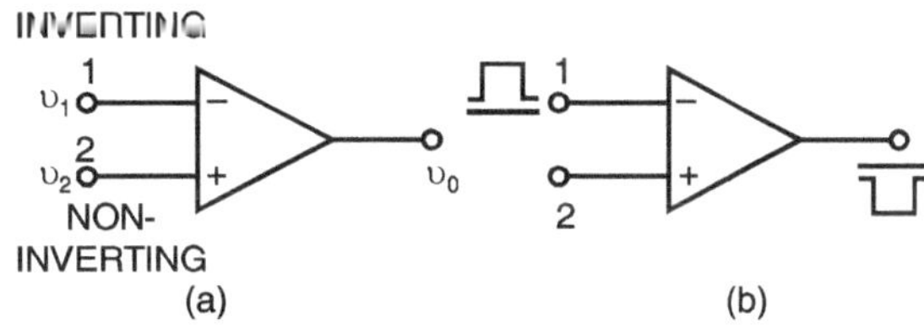

Fig. 11

referred to a common reference usually the ground shown in Fig. 11(a).

Ideal Operational Amplifier

When an OP-AMP is operated without connecting any resistor or capacitor from its output to any one of its inputs (i.e., without feedback), it is said to be in the **open-loop condition.** The word 'open loop' means that **feedback path or loop is open.** The specification of an OP-AMP under such condition are called open-loop specifications.

An ideal OP-AMP (Fig. 12) has the following characteristics :

1. its open-loop gain A_v is infinite *i.e.* $A_v = -\infty$
2. its input resistance R_i (measured between inverting and non-inverting terminals) is infinite i.e. $R_i = \infty$ ohm
3. its output resistance R_0 (seen looking back into output terminals) is *zero i.e.* $R_0 = 0\ \Omega$
4. it has infinite bandwidth i.e. it has flat frequency response from dc to infinity.

Though these characteristics cannot be achieved in practice, yet an ideal OP-AMP serves as a convenient reference against which real OP-AMPs may evaluated.

Following additional points are worth noting :

1. infinite input resistance means that input current $i = 0$ as indicated in Fig. 12.
   It means that an ideal OP-AMP is a voltage-controlled device.
2. $R_o = 0$ means that v_o is not dependent on the load resistance connected across the output.
3. though for an ideal OP-AMP $A_v = \infty$, for an actual one, it is extremely high *i.e.* about 10^6. However, it does not mean that 1 V signal will be amplified to 10^6 V at the output. Actually, the maximum value of v_o is limited by the bias supply voltage, typically ± 15 V. With $A_v = 10^6$

and v_o = 15 V, the maximum value of input voltage is limited to $15/10^6$ = 15 μV. Though 1 V cannot become 1 million volt in the OP-AMP, J μV can certainly become 1V.

Virtual Ground and Summing Point

In Fig. 13 (a) is shown an OP-AMP which *employs negative feedback* with the help of resistor R_f which feeds a portion of the output to the input.

Since input and feedback currents are algebraically added at point A, it is called the **summing point**.

The concept of **virtual** ground arises from the fact that input voltage r_i at the inverting terminal of the OP-AMP is forced to such a small value that, for all practical purposes, it may be assumed to be zero. Hence, point *A* is essentially at ground voltage and is referred to as *virtual ground*. Obviously, *it is not the actual ground*, which as seen from Fig. 13(a), is situated below.

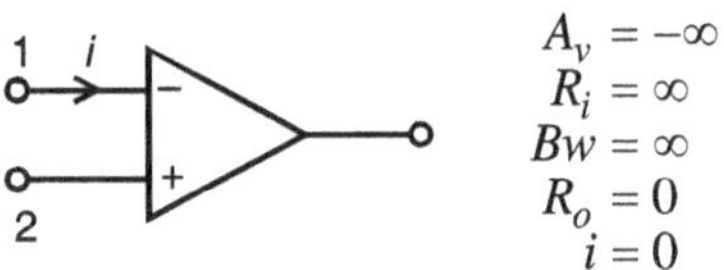

Fig. 12

Why V_i is Reduced to Almost Zero?

When v_1 is applied, point *A* attains some positive potential and at the same time v_o is brought into existence. Due to negative feedback, some fraction of the output voltage is feedback to point *A* antiphase with the voltage already existing three (due to v_1).

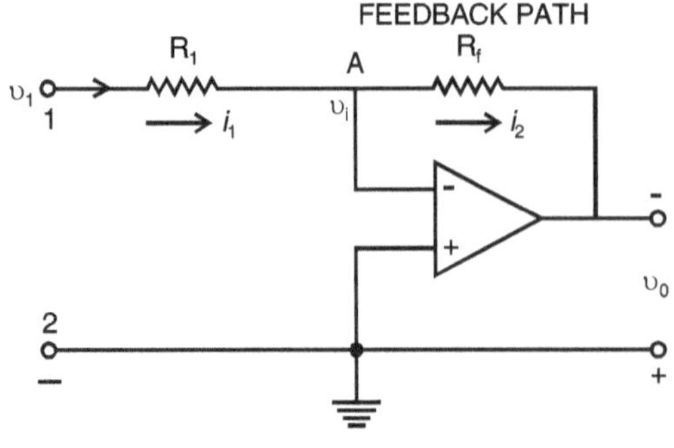

Fig. 13 (a)

The algebraic sum of the two voltages is almost zero so that $v_i \cong 0$. Obviously, v_i will become exactly zero when *negative feedback voltage at A is exactly equal to the positive voltage produced by v_1 and A.*

Another point worth considering is that there exists a virtual short between the two terminals of the OP-AMP because $v_i = 0$. It is virtual because no current flows (remember $i = 0$) despite the existence of this short.

OP-AMP Applications

We will consider the following applications :

1. as scaler or linear (i.e. small-signal) constant-gain amplifier; both inverting and non-inverting
2. as unity follower
3. Adder or Summer
4. Subtractor
5. Integrator
6. Differentiator
7. Comparator

Now, we will discuss the above circuits one by one assuming an ideal OP-AMP.

Linear Amplifier

We will consider the functioning of an OP-AMP as a constant-gain amplifier both in the inverting and non-inverting configurations.

(a) Inverting Amplifier or Negative Scalar : As shown in Fig. 13(b), non-inverting terminal has been grounded whereas R_1 connects the input signal v_1 to the inverting input. A feedback resistor R_f has been connected from the output to the inverting input.

Gain : Since point *A* is at ground potential,

$$i_1 = \frac{v_{in}}{R_1} = \frac{v_1}{R_1}$$

$$i_2 = \frac{-v_o}{R_f}$$ —please note –ve sign

Using KCL for point *A*.

$$i_1 + (-i_2) = 0 \text{ or } \frac{v_1}{R_1} + \frac{v_o}{R_f} = 0$$

$$\text{or } \frac{r_o}{R_f} = -\frac{v_1}{R_1} \text{ or } \frac{v_o}{v_1} = -\frac{R_f}{R_1}$$

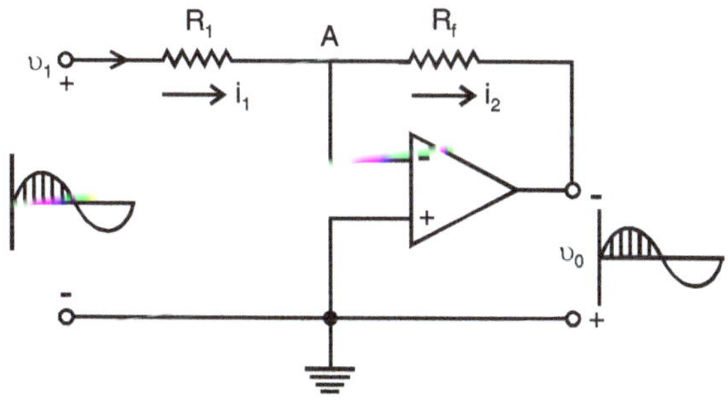

Fig. 13 (b)

$$\therefore \ A_v = -\frac{R_f}{R_1} \quad \text{or} \quad A_r = -K$$

Also, $v_o = -K\ v_{in}$

It is seen from above, that closed-loop gain of the inverting amplifier depends on the ratio of the two external resistors R_1 and R_f and is independent of the amplifier parameters.

It is also seen that the OP-AMP works as a negative scaler. It scales the input *i.e.* it multiplies the input by a minus constant factor *K*.

(b) Non-inverting Amplifier or Positive Scaler : This circuit is used when there is need for an output which is equal to the input multiplied by a positive constant. Such a positive scaler circuit which uses negative feedback but provides an output that equals the input multiplied by a positive constant is shown in in Fig. 13 (c).

Since input voltage v_2 is applied to the non-inverting terminal, the circuit is also called **non-inverting amplifier.**

Here, polarity of v_o is the same as that v_2 *i.e.* both are positive.

Gain : Because of virtual short between the two OP-AMP terminals, voltage across R_1 is the input voltage v_2. Also, v_o is applied across the series

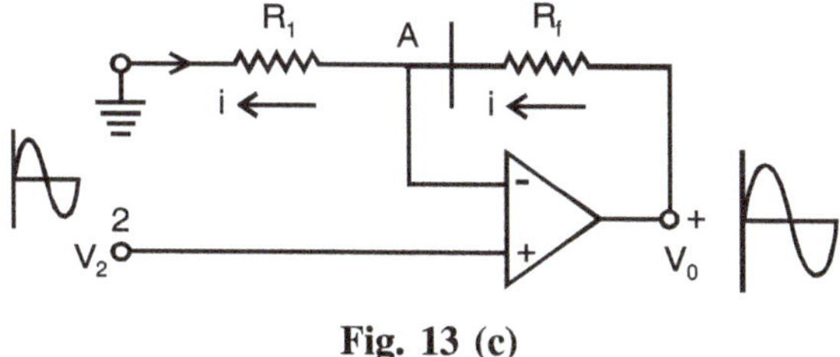

Fig. 13 (c)

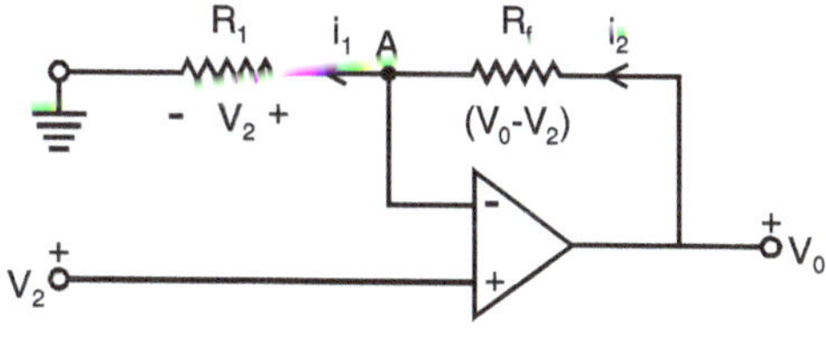

Fig. 13 (d)

combination of R_1 and R_f.

$$\therefore \quad v_{in} = v_2 = iR_1, \quad v_o = i(R_1 + R_f)$$

$$\therefore \quad A_v = \frac{v_o}{v_{in}} = \frac{i(R_1 + R_f)}{iR_1} \quad \text{or}$$

$$A_v = \frac{R_1 + R_f}{R_1} = \left(1 + \frac{R_f}{R_1}\right)$$

Alternative Derivation : As shown in Fig. 13 (d) let the currents through the two resistors be i_1 and i_2.

The voltage across R_1 is v_2 and that across R_f is $(v_o - v_2)$.

$$\therefore \quad i_1 = \frac{v_2}{R_1} \quad \text{and} \quad i_2 = \frac{v_o - v_2}{R_f}$$

Applying KCL to junction *A*, we have

$$(-i_1) + i_2 = 0 \quad \text{or} \quad \frac{-v_2}{R_1} + \frac{(v_o - v_2)}{R_f} = 0$$

$$\therefore \quad \frac{v_o}{R_f} = v_2\left(\frac{1}{R_1} + \frac{1}{R_f}\right) = v_2\frac{R_1 + R_f}{R_1 R_f}$$

$$\therefore \quad \frac{v_o}{v_2} = \frac{R_1 + R_f}{R_1} \quad \text{or} \quad A_v = 1 + \frac{R_f}{R_1} \text{ as before}$$

Example : *For the inverting amplifier of Fig. 13 (b), R_1 = 1K and R_f = 1 M. Assuming an ideal OP-AMP amplifier, determine the following circuit values :*

(a) *voltage gain (b) input resistance (c) output resistance*

Solution : It should be noted that we will be calculating values of the circuit and not for the OP-AMP proper.

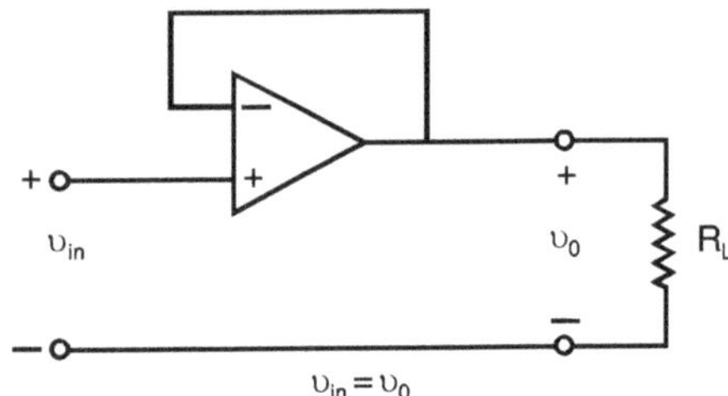

Fig. 13 (e)

(a) $A_v = -\frac{R_f}{R_1} = -\frac{1000\text{ K}}{1\text{K}} = \mathbf{-1000}$

(b) Because of virtual ground at A, $R_{in} = R_1 = \mathbf{1K}$

(c) Output resistance of the circuit equals the output resistance of the OP-AMP *i.e.*, **zero ohm**.

Unity Follower

It provides a gain of unity without any phase reversal. It is very much similar to the emitter follower except that its gain is very much closer to being exactly unity.

This circuit (Fig. 13 (e)) is useful as a buffer or isolation amplifier because it allows input voltage v_{in} to be transferred as output voltage v_o while at the same time preventing load resistance R_L from loading down the input source. It is due to the fact that its $R_i = \infty$ and $R_e = 0$.

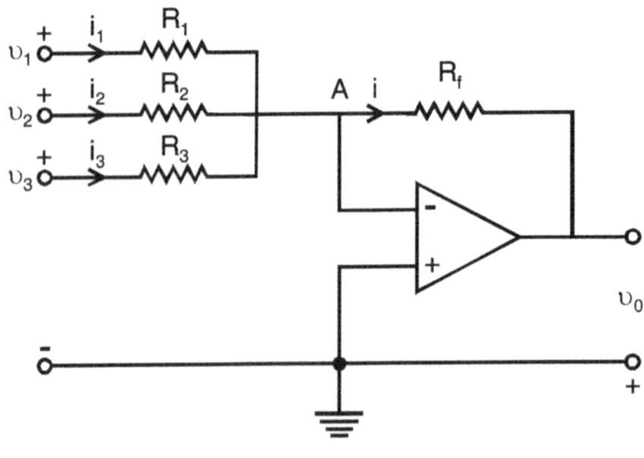

Fig. 13 (f)

Adder or Summer

The adder circuit provides an output voltage proportional to or equal to the algebraic sum of two or more input voltages each multiplied by a constant gain factor. It is basically similar to a scaler except that it has more than one input. Fig. 13 (f) shows a three-input inverting adder circuit. As seen, the output voltage is phase-inverted.

Calculations

As before, we will treat point A as virtual ground.

$$i_1 = \frac{v_1}{R_1} \text{ and } i_2 = \frac{v_2}{R_2}$$

$$i_3 = \frac{v_3}{R_3} \text{ and } i = \frac{v_o}{R_f}$$

Applying KCl to point A, we have

$i_1 + i_2 + i_3 + (-i) = 0$ or

$$\frac{v_1}{R_1} + \frac{v_2}{R_2} + \frac{v_3}{R_3} - \left(\frac{-v_o}{R_f}\right) = 0$$

$$\therefore \quad v_o = -\left(\frac{R_f}{R_1}v_1 + \frac{R_f}{R_2}v_2 + \frac{R_f}{R_3}v_3\right)$$

or $v_o = -(K_1v_1 + K_2v_2 + K_3v_3)$

The overall negative sign is unavoidable because we are using the inverting input terminal.

If $R_1 = R_2 = R_3 = R$, then

$$v_o = -\frac{R_f}{R}(v_1 + v_2 + v_3) = -K(v_1 + v_2 + v_3)$$

Hence, output voltage is proportional to (not equal to) the **algebraic sum** of the three input voltages.

If $R_f = R_3$ then output exactly equals the sum of inputs. However, if $R_f = R/3$, then

$$v_o = -\frac{R/3}{R}(v_1 + v_2 + v_3) = -\frac{1}{3}(v_1 + v_2 + v_3)$$

Obviously, the output is equal to *the average of the three inputs.*

Subtractor

The function of a subtractor is to provide an

output proportional to or equal to the difference of two input signals. As shown in Fig. 13 (g) we have to apply the inputs at the inverting as well as non-inverting terminals.

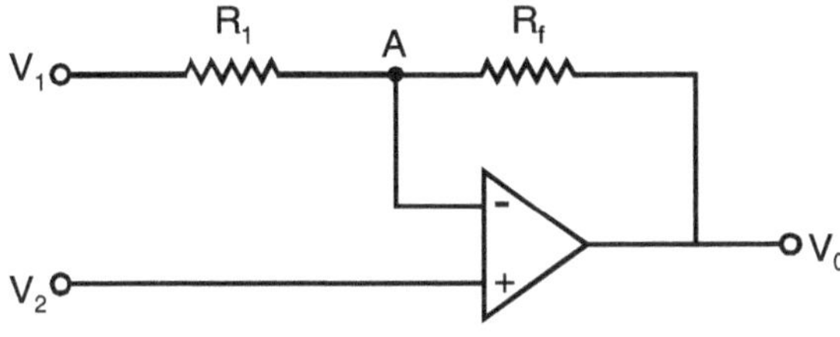

Fig. 13 (g)

Calculations

According to Superposition Theorem

$v_o = v_o' + v_o''$

where v_o' is the output produced by v_1 and v_o'' is that produced by v_2.

Now, $v_o' = -\frac{R_f}{R_1} . v_1$ — Fig. 13 (a)

$v_o'' = \left(1 + \frac{R_f}{R_1}\right) v_2$ — Fig. 13 (b)

$$\therefore \quad v_o = \left(1 + \frac{R_f}{R_1}\right) v_2 - \frac{R_f}{R_1}$$

Since $R_f \geq R_1$ and $R_f / R_1 > 1$, hence

$$v_o \cong \frac{R_f}{R_1}(v_2 - v_1) = K(v_2 - v_1)$$

Further, if $R_f = R_1$, then

$v_o = (v_2 - v_1) =$ difference of the two input voltages

Obviously, if $R_f \neq R_1$, then a scale factor is introduced.

Example : *Find the output voltage of an OP-AMP inverting adder for the following sets of input voltages and resistors. In all cases, $R_f = 1$ M.*

$v_1 = -3V,\ v_2 = +3V,\ v_3 = +2V;$
$R_1 = 250\ K,\ R_2 = 500\ K,\ R_3 = 1M$

Solution : $v_o = -(K_1 v_1 + K_2 v_2 + K_3 v_3)$

$$K_1 = \frac{R_f}{R_1} = \frac{1000\ K}{250\ K} = 4,$$

$$K_2 = \frac{1000}{500} = 2, K_3 = \frac{1\ M}{1\ M} = 1$$

$$\therefore\ v_o = -[(4 \times -3) + (2 \times 3) + (1 \times 2)] = +\mathbf{4V}$$

Example : In the subtractor circuit of Fig. 13 (g), *$R_1 = 5$ K, $R_f = 10$ K, $v_1 = 4$ V and $v_2 = 5$V. Find the value of output voltage.*

Solution :

$$v_o = \left(1 + \frac{R_f}{R_1}\right) v_1 - \frac{R_f}{R_1} v_2 = \left(1 + \frac{10}{5}\right) 4 - \frac{10}{5} \times 5 = +2\mathbf{V}$$

Integrator

The function of an integrator is to provide an output voltage which is proportional to the integral of the input voltage.

A simple example of the integration is shown in Fig. 13 (h) where input in de level and its integral is *a* a linearly-increasing ramp output.

The actual integrating circuit is shown in Fig. 13(h). This circuit is similar to the scalar circuit except that the feedback component is a capacitor C instead of a resistor R_f.

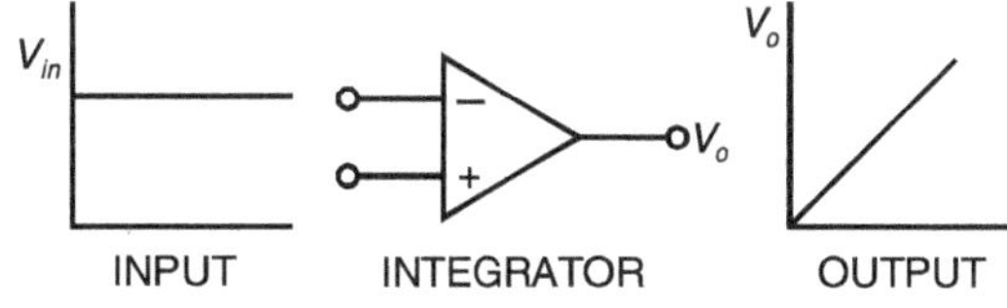

Fig. 13 (h)

Calculations

As before, point *A* will be treated as virtual ground.

$$i_1 = \frac{v_1}{R};\ i_2 = -\frac{v_o}{X_C} = -\frac{v_o}{1 / j\omega C}$$

$$= -\frac{v_o}{1 / sC} = -sCv_o$$

where $s = j\omega$ is the Laplace notation.

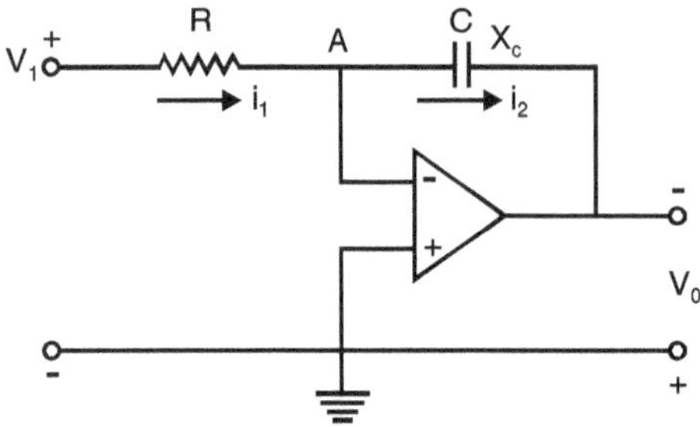

Fig. 13 (i)

Now $i_1 = i_2$

$$\therefore \quad \frac{v_1}{R} = -sCv_o$$

$$\therefore \quad \frac{v_o}{v_{in}} = \frac{v_o}{v_1} = -\frac{1}{sCR} \qquad \ldots (i)$$

$$\therefore \quad A_v = -\frac{1}{sCR}$$

Now, the expression of Eq. (i) can be written in time domain as

$$v_o(t) = -\frac{1}{CR}\int v_1(t) \cdot dt$$

It is seen from above that output (right-hand side expression) is an integral of the input, with an inversion and a scale factor of $1/CR$.

This ability to integrate a given signal enables an analog computer to solve differential equations and to set up a wide variety of electrical circuit analogs of physical system operations. For example, let

$$R = 1\,M$$

and $C = 1\mu\text{F}$. Then scale factor $= -\dfrac{1}{CR}$

$$= -\frac{1}{10^6 \times 10^{-6}} = -1$$

As shown in Fig. 13 (j) the input is a step voltage whereas output is a ramp (or linearly-changing voltage) with a scale multiplier of –1.

However, when $R = 100$ K, then

$$\text{scale factor} = -\frac{1}{10^6 \times 10^{-6}} = 10$$

$$\therefore \quad v_o(t) = -10\int v_1(t).dt$$

It is also shown in Fig. 13 (j). Of course, we can integrate more than one input as shown below in Fig. 13 (d). With multiple inputs, the output is given by

$$v_o(t) = -[K_1 \int v_1(t)dt + K_2 \int v_2(t) + K_3 \int v_3(t)dt]$$

where $K_1 = \dfrac{1}{CR_1}$, $K_2 = \dfrac{1}{CR_2}$ and $K_3 = \dfrac{1}{CR_3}$

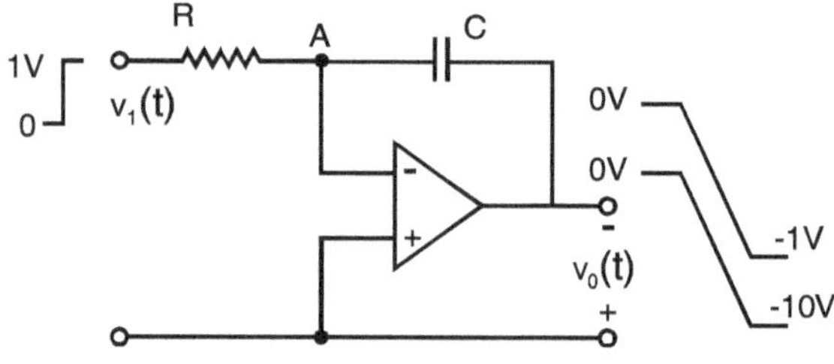

Fig. 13 (j)

Differentiator

Its function is to provide an output voltage which is *proportional to the rate of change of the input voltage*. It is an inverse mathematical operation to that of an integrator. As shown in Fig. 14 (a) when we feed a differentiator with linearly-increasing ramp input, we get a constant dc output.

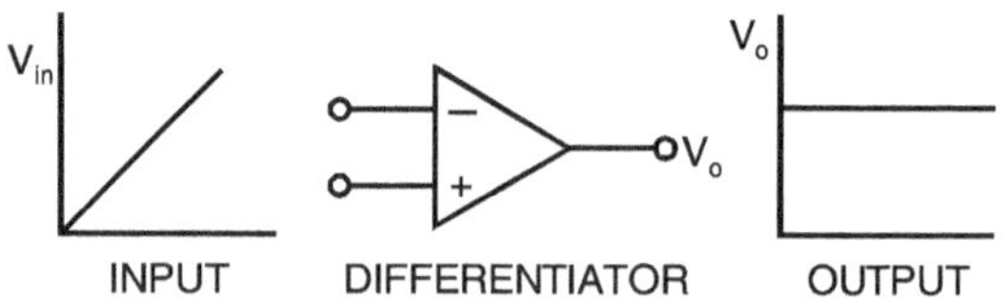

Fig. 14(a)

Circuit

Differentiator circuit can be obtained by interchanging the resistor and capacitor of the integrator circuit of Fig. 14 (b).

Let i = rate of change of charge

$$= \frac{dq}{dt}$$

Now, $q = Cv_c$

$$\therefore \quad i = \frac{d}{dt}(Cv_c) = C\frac{dv_c}{dt}$$

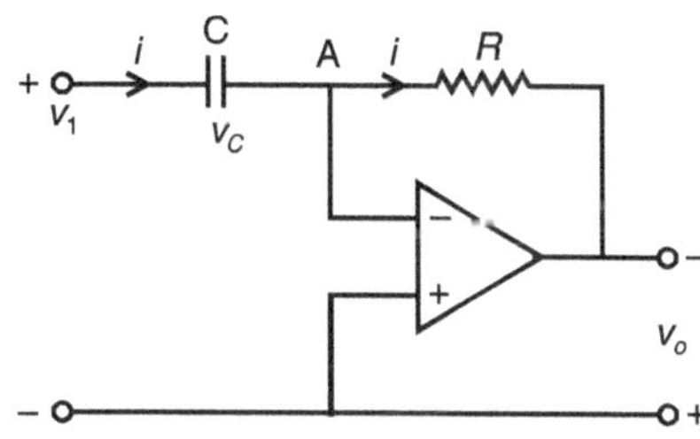

Fig. 14(b)

Taking point A as virtual ground

$$v_o = -iR = -\left(C.\frac{dv_c}{dt}\right)R = -CR.\frac{dv_c}{dt}$$

As seen, output voltage is proportional to the derivate of the input voltage, the constant of proportionality (*i.e.* scale factor) being ($-CR$).

Example : *The input to the differentiator circuit of Fig. 14 (b) is a sinusoidal voltage of peak value 5 mV and frequency 1 kHz. Find out the output if R = 100 K and C = 1 μ F.*

Solution : The equation of the input voltage is

$$v_1 = 5\sin 2\pi \times 1000t = 5\sin 2000\pi t \text{ mV}$$

scale factor $= CR = 10^{-6} \times 10^5 = 0.1$

$$v_o = 0.1\frac{d}{dt}(5\sin\ 2000\pi t) = (0.5 \times 2000\pi)$$

$$\cos 2000\,\pi l = 1000\pi\ \cos\ 2000\pi t \text{ mV}$$

As seen, output is a cosinusoidal voltage of frequency 1 kHz and peak value 1000π mV.

Comparator

It is a circuit which compares two signals or voltage levels. The circuit is shown in Fig. 14 (c) and (like that of the unity follower) is the simplest because it needs no additional external components.

If v_1 and v_2 are equal, then v_o should ideally be zero. Even if v_1 differs from v_2 by a very small amount, v_o is large because of amplifier's high gain. Hence, circuit of Fig. 14 (c) can detect very small level changes which is another way of saying that it compares two signals.

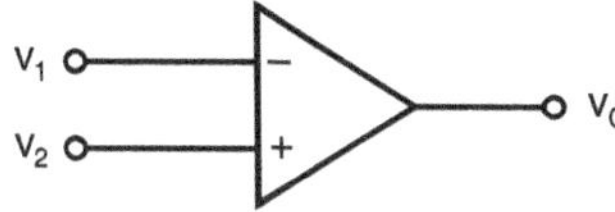

Fig. 14(c)

Oscillator

An electronic oscillator may be defined in any one of the following four way :

1. it is a circuit which converts dc energy into ac energy at a very high frequency.
2. it is an electronic source of alternating current or voltage having sine, square or sawtooth or pulse shapes
3. it is a circuit which generates an ac output signal without requiring any externally-applied input signal
4. it is an unstable amplifier.

These definitions exclude electromechanical alternators producing 50 Hz ac power or other devices which convert mechanical or heat energy into electric energy.

Comparison Between an Amplifier and an Oscillator

An amplifier produces an output signal whose waveform is similar to the input signal but whose power level is generally high. This additional power is supplied by the external dc source. Hence, an amplifier is essentially an energy convertor *i.e.* it takes energy from the dc power source and converts it into ac energy at signal frequency. The process of energy conversion is controlled by the input signal. If there is no input signal, there is no energy conversion and hence there is no output signal.

An oscillator differs from an amplifier in one basic respect : the oscillator *does not require an external signal* either to start or maintain energy conversion process (Fig. 15). It keeps producing an output signal so long as the dc power source is connected.

Moreover, the frequency of the output signal is determined by the passive components used in the oscillator and can be varied at will.

Classification of Oscillators

Electronic oscillators may be broadly divided into following two groups :

(i) sinusoidal (or harmonic) oscillators — which produce an output having sine waveform.

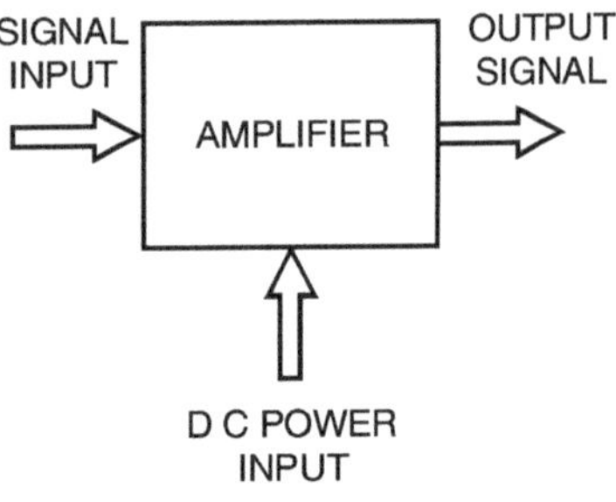

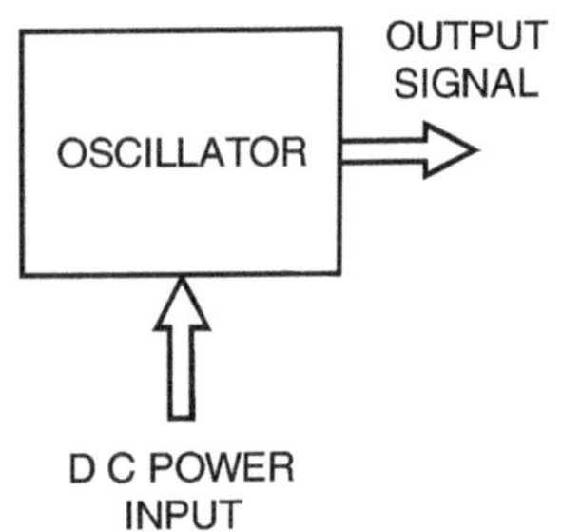

Fig. 15

(ii) non-sinusoidal (or relaxation) oscillators — they produce an output which has square, rectangular or sawtooth waveform or is of pulse shape.

Sinusoidal oscillators may be further subdivided into :

(a) tuned-circuits or LC feedback oscillators such as Hartley, Colpitts and Clapp etc.

(b) RC phase-shift oscillators such as Wein-bridge oscillator

(c) negative-resistance oscillators such as tunnel diode oscillator

(d) crystal oscillators such as Pierce oscillator

(e) heterodyne or beat-frequency oscillator (BFO)

The active devices (bipolars, FETs or unjunction transistors) in the above mentioned circuits may be biased class-A, B or C. Class-A operation is used in high-quality audio frequency oscillators. However, radio frequency oscillators are usually operated class-C.

Damped and Undamped Oscillations

Sinusoidal oscillations produced by oscillators may be (i) damped or (ii) undamped

(i) Damped Oscillations : Oscillations whose

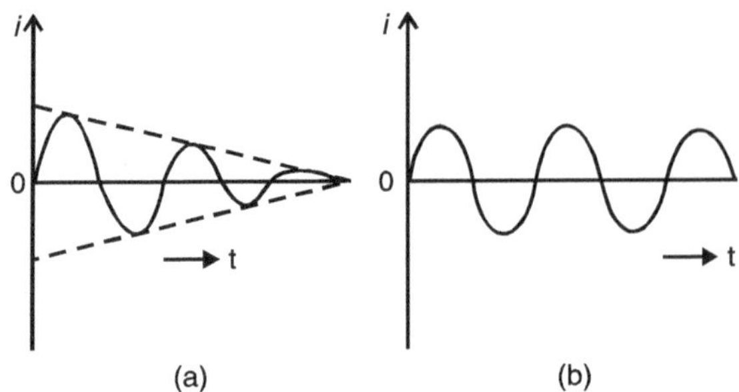

Fig. 16

amplitude keeps decreasing (or decaying) with time are called **damped** or **decaying** oscillations. The waveform of such oscillations is shown in Fig. 16 (a). These are produced by those oscillator circuits in which I^2R losses take place continuously during each oscillation without any arrangement for arrangement for compensating the same. Ultimately, the amplitude of the oscillations decays to zero when there is not enough energy to supply circuit losses. However, the frequency or time-period remains constant because *it is determined by the circuit parameters.*

Sinusoidal oscillators serve a variety of functions in telecommunications and in electronics. The most important application in telecommunication is the use of sine waves as carrier signal in both radio and cable transmissions.

Sine wave signals are also used in frequency response testing of various types of systems and equipment including analogue communication channels, amplifiers and filters and closed-loop control systems.

(ii) Undamped Oscillations : Oscillations whose amplitude remains constant *i.e.* does not change with time are called **undamped oscillations**. These are produced by those oscillator circuits which have no losses or if they have, there is provision for compensating them. The constant-amplitude and constant-frequency sinusoidal waves shown in Fig. 16 (b) are called *carrier waves* and are used in communication transmitters for transmitting low-frequency audio information to far off places.

The Oscillatory Circuit

It is also called LC circuit or tank circuit. The oscillatory circuit consists of two reactive elements *i.e.* an inductor and a capacitor. Both are capable of storing energy. The capacitor stores energy in its electric field whenever there is potential difference across its plates. Similarly, a coil or an inductor stores energy in its magnetic field whenever current flows through it. Both L and C are supposed to be loss-free (i.e. their Q-factors are infinite)

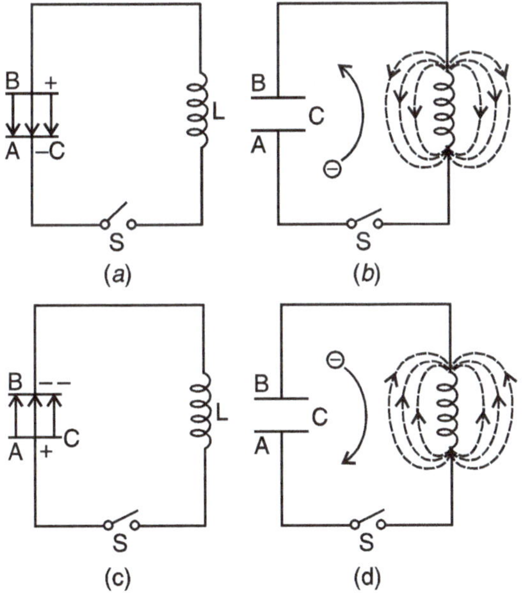

Fig. 14

As shown in Fig. 14 (*a*), suppose the capacitor has been fully-charged from a dc source. Since S is open, it cannot discharge through L. Now, let us see what happens when S is closed.

1. When S closed [Fig 14 (*b*)] electrons move from plate A to plate B through coil L as shown by the arrow (or conventional current flows from B to A). This electron flow reduces the strength of the electric field and hence the amount of energy stored in it.
2. As electronic current starts flowing, the self-induced emf in the coil opposes the current flow Hence, rate of discharge of electrons is somewhat slowed down.
3. Due to the flow of current, magnetic field is set up which stores the energy given out by the electric field [Fig. 14 (*b*)].
4. As plate A loses its electrons by discharge, the electron current has a tendency to die down and will actually reduce to zero when all excess electrons on A are driven over to plate B so that both plates are reduced to the same potential. At that time, there is no electric field but the magnetic field has maximum value.
5. However, due to self-induction (or electrical inertia) of the coil, more electrons are transferred to plate B than are necessary to make up the electron deficiency there. It means that now plate B has more electrons than A. Hence, capacitor becomes charged again though in opposite direction as shown in Fig. 14 (*c*).
6. The magnetic field L collapses and the energy given out by it is stored in the electric field of the capacitor.
7. After this, the capacitor starts discharging in the opposite direction so that, now, the electrons move from plate B to plate A [Fig. 14 (*d*)]. The electric field starts collapsing whereas magnetic field starts building up again though in the opposite direction. Fig. 14 (*d*) shows the condition when the capacitor becomes fully discharged once again.
8. However, these discharging electrons overshoot and again an excess amount of electrons flows to plate A, thereby charging the capacitor once more.
9. This sequence of charging and discharging continues. The to and fro motion of electrons between the two plates of the capacitor constitutes an oscillatory current.

It may be also noted that during this process, the electric energy of the capacitor is converted into magnetic energy of the coil and vice versa.

These oscillations of the capacitor discharge are damped because energy is dissipated away gradually so that their amplitude becomes zero after sometime. There are two reasons for the loss of this energy:

(a) some energy is lost in the form of heat produced in the resistance of the coil and connecting wires.
(b) and some energy is lost *in the form of electromagnetic (EM) waves* that are radiated out from the circuit through which an oscillatory current is passing.

Both these losses subtract energy from the circuit with the result that circuit current decreases gradually till it becomes zero. The waveform of the oscillatory discharge is similar to that shown in Fig. 16 (a).

Frequency of Oscillatory Current

The frequency or time-period of the oscillatory current depends on two factors :

(a) **Capacitance of the Capacitor :** Larger the capacitor, greater the time required for the reversal of the discharge current *i.e.* lower its frequency.
(b) **Self-inductance of the Coil :** Larger the self-inductance, greater the inertial effect and hence longer the time required by the current to stop flowing during discharge of the capacitor.

The frequency of this oscillatory discharge current is given by

$$f = \frac{1}{2\pi\sqrt{LC}} = \frac{159}{\sqrt{LC}} \text{kHz}$$

where $L =$ self-inductance in μH and

$C =$ capacitance in μF

It may, however, be pointed out here that damped oscillations so produced are no good for radio transmission purpose because of their limited range and excessive distortion. For good radio transmission, we need undamped oscillations which can be produced if some additional energy is supplied in correct phase and correct direction to the LC circuit for making up the I^2R losses continually occurring in the circuit.

Frequency Stability on an Oscillator

The ability of an oscillator to maintain a constant frequency of oscillation is called its frequency stability. Following factors affect the frequency stability :

1. **Operating Point of the Active Device :** The Q-point of the active device (*i.e.* transistor) is so chosen as to confine the circuit operation on the linear portion of its characteristics. Operation on non-linear portion varies the parameters of the transistor which, in turn, affects the frequency stability of the oscillator.
2. **Inter-element Capacitances :** Any changes in the inter-element capacitances of a transistor particularly the collector-to-emitter capacitance cause changes in the oscillator output frequency, thus affecting its frequency stability. The effect of changes in inter-element capacitances, can be neutralized by adding a swamping capacitor across the offending elements—the added capacitance being made part of the tank circuit.
3. **Power Supply :** Changes in the dc operating voltages applied to the active device shift the oscillator frequency. This problem can be avoided by using regulated power supply.
4. **Temperature Variations :** Variations in temperature cause changes in transistor parameters and also change the values of resistors, capacitors and inductors used in the circuit. Since such changes take place slowly, they cause a slow change (called drift) in the oscillator output frequency.
5. **Output Load :** A change in the output load may cause a change in the Q-factor of the LC tuned circuit thereby affecting the oscillator frequency.
6. **Mechanical Vibrations :** Since such vibrations changes the values of circuit elements, they result in changes of oscillator frequency. This instability factor can be eliminated by isolating the oscillator from the source of mechanical vibrations.

Essentials of a Feedback LC Oscillator

The essential components of a feedback LC oscillator shown in Fig. 17 are :

1. a resonator which consists of an LC circuit. It is also known as frequency-determining network (*FDN*) or tank circuit.
2. an amplifier whose function is to amplify the oscillations produced by the resonator.
3. a positive feedback network (*PFN*) whose

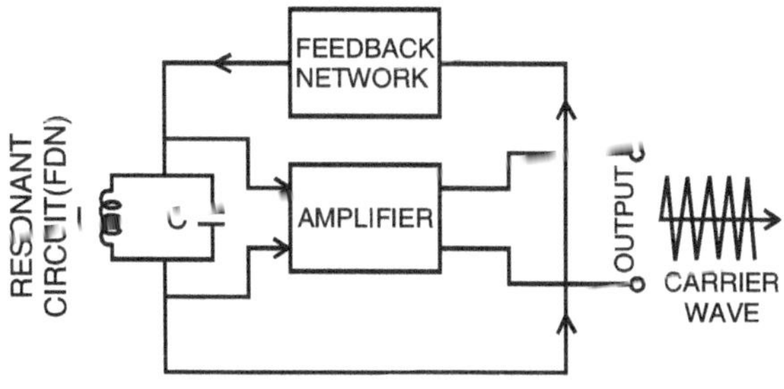

Fig. 17

function is to transfer part of the output energy to the resonant *LC* circuit *in proper phase*. The amount of energy fed back is sufficient to meet I^2R losses in the *LC* circuit.

The essential condition for maintaining oscillations and for finding the value of frequency is

$$\beta A = 1 + j0 \quad \text{or} \quad \beta A \angle \phi = 1 \angle 0$$

It means that

(i) the feedback factor or loop gain $|\beta A| = 1$.

(ii) the net phase shift around the loop is 0° (or an integral multiple of 360°). In other words, *feedback should be positive*.

The above conditions for Barkhausen criterion for maintaining a steady level of oscillation at a specific frequency.

Majority of the oscillators used in radio receivers and transmitters use tuned circuits with positive feedback. Variations in oscillator circuits are due to the different way in which this feedback is applied. Some of the basic circuits are :

1. *Armstrong or Tickler or Tuned-base Oscillator*—it employs inductive feedback from collector to the tuned *LC* circuit in the base of a transistor.
2. *Tuned Collector Oscillator*—it also employs inductive coupling but the *LC* tuned circuit is in the collector circuit.
3. *Hartley Oscillator*—here feedback is supplied inductively.
4. *Colpitts Oscillator*—here feedback is supplied capacitively.
5. *Clapp Oscillator*—it is a slight modification of the Colpitts oscillator.

Tuned Based Oscillator

Such an oscillator using a transistor in *CE* configuration is shown in Fig. 18. Resistors R_1, R_2 and R_3 determine the dc bias of the circuit. The parallel R_3–C_2 network in the emitter circuit is a stabilizing circuit to prevent signal degeneration. As usual, C_1 is the dc blocking capacitor. The mutually-coupled coils L_1 and *L* forming primary and secondary of an *RF* transformer provide the required feedback between the collector and base circuits. The amount of feedback depends on the coefficient of coupling between the two coils. The *CE* connected transistor itself provides a phase shift of 180° between its input and output circuits. The transformer provides another 180° phase shift thus producing a total phase shift of 360° which is an essential condition for producing oscillations.

The parallel-tuned *LC* circuit connected between base and emitter is the frequency determining network (*FDN*) i.e. it generates the oscillations at its resonant frequency.

Circuit Action

The moment switch *S* is closed, collector current is set up which tends to rise to its quiescent value. This increases in I_C is accompained by :

1. an expanding magnetic field through L_1 which links with *L* and

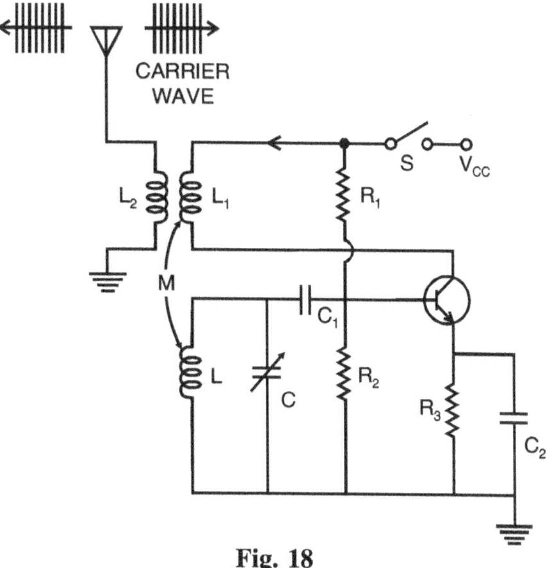

Fig. 18

2. an induced e.m.f. called feedback voltage in L

Two immediate reactions of this feedback voltage are

(i) increase in emitter-base voltage (and base current) and

(ii) a further increase in collector current I_C

It is followed by a succession of cycles of

1. an increase in feedback voltage
2. an increase in emitter-base voltage and
3. an increase in I_C until saturation is reached.

Meanwhile, C gets charged. As soon as I_C ceases to increase, magnetic field of L_1 ceases to expand and thus no longer induces feedback voltage in L. Having been charged to maximum value, C starts to discharge through L. However, decrease in voltage across C causes the following sequence of reactions :

1. a decrease in emitter-base bias and hence in I_B
2. a decrease in I_C
3. a collapsing magnetic field in L_1
4. an induced feedback voltage in L though, this time, in opposite direction
5. further decrease in emitter-base bias and so on till I_C reaches its cut-off value.

During this time, the capacitor having lost its original charge, again becomes fully charged though with opposite polarity. Transistor being in cut-off, the capacitor will again begin to discharge through L. Since polarity of capacitor charge is opposite to that when transistor was in saturation, the sequence of reactions now will be

1. an increase in emitter-base bias
2. an increase in I_C
3. an expanding magnetic field in L_1
4. an induced feedback voltage in L
5. a further increase in emitter-base bias and
6. so on till I_C increases to its saturation value.

This cycle of operation keeps repeating so long as enough energy is supplied to meet losses in the LC circuit.

The output can be taken out by means of a third winding L_2 magnetically coupled to L_1. It has approximately the same waveform as collector current.

The frequency of oscillation is equal to the resonant frequency of the LC circuit.

Tuned Collector Oscillator

Such an oscillator using a transistor in CE configuration is shown in Fig. 16

***(i)* Frequency Determining Network (FDN)**

It is made up of a variable capacitor C and a coil L which forms primary winding of a step-down transformer. The combination of L and C forms an oscillatory tank circuit to set the frequency of oscillation.

Resistors R_1, R_2 and R_3 are used to dc bias the transistor. Capacitors C_1 and C_2 act to bypass R_3 and R_2 respectively so that they have no efect on the ac operation of the ciruit. Moreover, C_2 provides ac ground for transformer secondary L_1.

***(ii)* Positive Feedback**

Feedback between the collector-emitter circuit and base-emitter circuit is provided by the transformer secondary winding L_1 which is mutually-coupled to L. As far as ac signals are concerned, L_1 is connected to emitter via low-reactance capacitors C_2 and C_1.

Since transistor is connected in CE configuration, it provides a phase shift of 180° between its input and output circuits. Another phase shift of 180° is provided by the transformer thus producing a total phase shift of 360° between the output and input voltages resulting is positive feedback between the two.

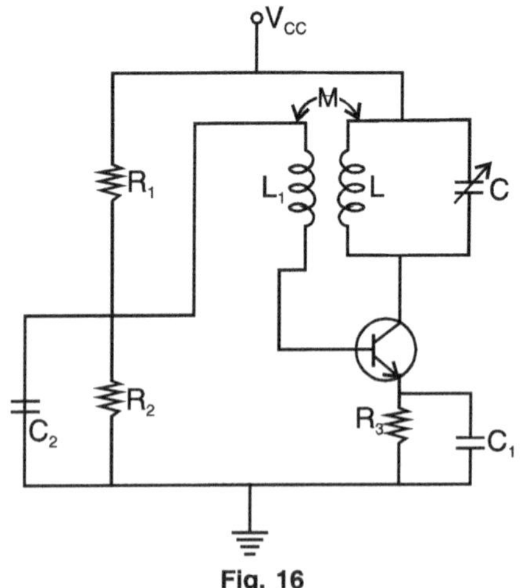

Fig. 16

***(iii)* Amplifying Action**

The transistor amplifier provides sufficient gain

for oscillator action to take place.

(iv) **Working**

When the supply is first switched on, a transient current is developed in the tuned LC circuit as the collector current rises to its quiescent value. This transient current initiates natural oscillations in the tank circuit. These natural oscillations induce a small emf into L_1 by mutual induction which causes corresponding variations in base current. These variations in I_B are amplified β times and appear in the collector circuit Part of this amplified energy is used to meet losses taking place in the oscillatory circuit and the balance is radiated out in the form of electromagnetic waves.

The frequency of oscillatory current is almost equal to the resonant frequency of the tuned circuit.

$$\therefore \quad f_0 = \frac{1}{2\pi\sqrt{LC}}$$

Tuned Drain Oscillator (FET)

The basic circuit is illustrated in Fig. 17. It is similar to the tuned collector oscillator of Fig. 16. Because of its high input impedance and high voltage amplification, a FET can be used to construct very simple and efficient oscillator circuits. This frequency of oscillation is given by

$$f_0 = \frac{1}{2\pi\sqrt{LC}}\sqrt{\left(1+\frac{R}{r_d}\right)}$$

where r_d = ac drain resistance

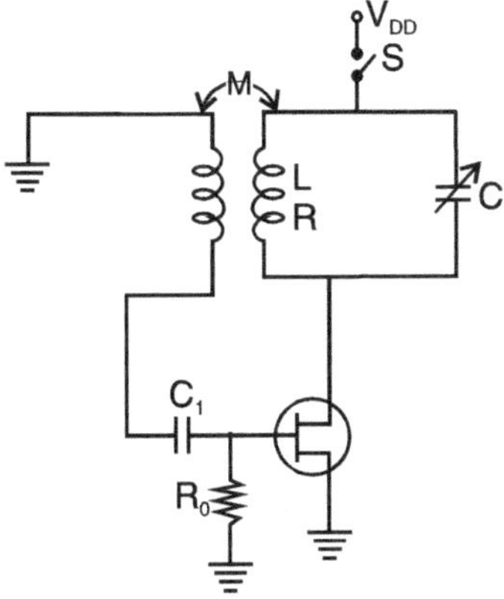

Fig. 17

The value of mutual inductance required for maintaining oscillations is

$$M = \frac{r_d RC + L}{\mu}$$

(a) **Piezoelectric Effect**

The quartz crystal described above has peculiar properties. When mechanical stress in applied across its two opposite faces, a potential difference is developed across them. It is called *piezo-electric effect*. Conversely, when a potential difference is applied across its two opposite faces, it causes the crystal to *either expand or contract.* If an alternating voltage is applied, the crystal water is set into vibration. The frequency of vibration is equal to the resonant frequency of the crystal as determined by its structural characteristics. When the frequency of the applied ac voltage equals the natural resonant frequency of the crystal, the amplitude of vibration will be maximum. As a general rule, thinner the crystal, *higher its frequency of vibration.*

(b) **Equivalent Electrical Circuit**

The electrical equivalent circuit of the crystal is shown in Fig. 17. It consists of a series RLC_1 circuit in parallel with a capacitor C_2.

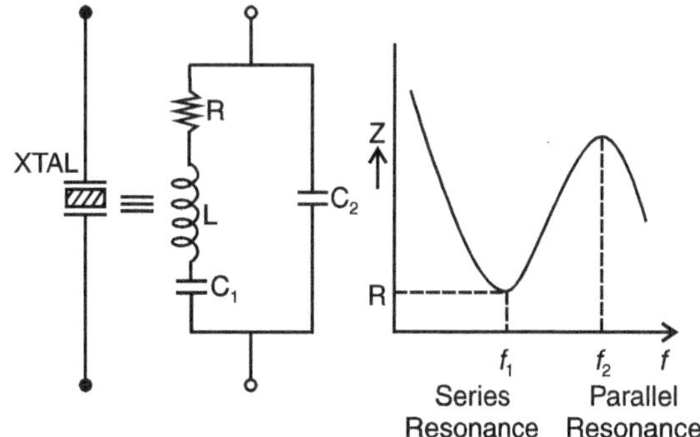

Fig. 18

The circuit has two resonant frequencies:

(i) one is the lower series resonance frequency f_1 which occurs when $X_L = X_{C1}$. In that case, Z =

$$f_1 = \frac{1}{2\pi\sqrt{CL_1}}$$

(ii) the other is the parallel resonance frequency f_2 which occurs when reactance of the series leg equals the reactances of C_2. At this frequency, the crystal offers very high impedance to the external circuit.

$$f_0 = \frac{1}{2\pi\sqrt{CL}} \text{ where } C = \frac{C_1 C_2}{C_1 + C_2}$$

The impedance versus frequency graph of the crystal is shown in figure. Crystals are available at frequency of 15 kHz and above. However, at frequencies above 100 MHz, they become so small that handling them becomes a problem.

(c) **Q-factor**

The equivalent inducttance of a crystal is very high as compared to either its equivalent capacitance or equivalent resistance. Because of high L/R ratio, the Q-factor of a crystal circuit is 20,000 as compared to a maximum of about 1000 for high-quality LC circuits. Consequently, greater frequency stability and frequency discrimination are obtained because of extremely high Q (upto 10^6) and high L/R ratio of the series RLC_1 circuit.

Example : *A tuned-collector oscillator has a fixed inductance of 100 μH and has to be tunable over the frequency band of 500 kHz to 1500 kHz. Find the range of variable capacitor to be used.*

Solution : Resonant frequency is given by

$$f_o = 1/2\pi\sqrt{LC} \quad or \quad C = 1/4\pi^2 f_o^2 L$$

where L and C refer to the tank circuit.

When f_o = 500 kHz

$$C = 1/4\pi^2 \times (500 \times 10^3)^2 \times 100 \times 10^{-6} = 1015 \text{ pF}$$

When f_o = 1500 kHz

$$C = 1015/(1500/500)^2 = 113 \text{ pF}$$

Hence, capacitor range required is 113–1015 pF.

Example : *The resonant circuit of a tuned-collector transistor oscillator has a resonant frequency of 5 MHz. If value of capacitance is increased by 50%, calculate the new resonant frequency.*

Solution : Using the equation for resonant frequency, we have

$$5 \times 10^6 = 1/2\pi\sqrt{LC} \qquad \text{— 1st case}$$

$$f_o = 1/2\pi\sqrt{L \times 1.5C} \qquad \text{— 2nd case}$$

$$\therefore \frac{f_o}{5 \times 10^6} = \frac{1}{\sqrt{1.5}} \text{ or } f_o = \mathbf{4.08\ MHz}$$

HARTLEY OSCILLATOR

In Fig. 19 (a) is shown a transistor Hartley oscillator using *CE* configuration. Its general principle of operation is similar to the tuned-collector oscillator discussed in Fig. 20.

It uses a single tapped-coil having two parts marked L_1 and L_2 instead of two separate coils. So far as ac signals are concerned, one side of L_2 is connected to base via C_1 and the other to emitter via ground and C_3. Similarly, one end of L_1 is connected to collector via C_2 and the other to common emitter

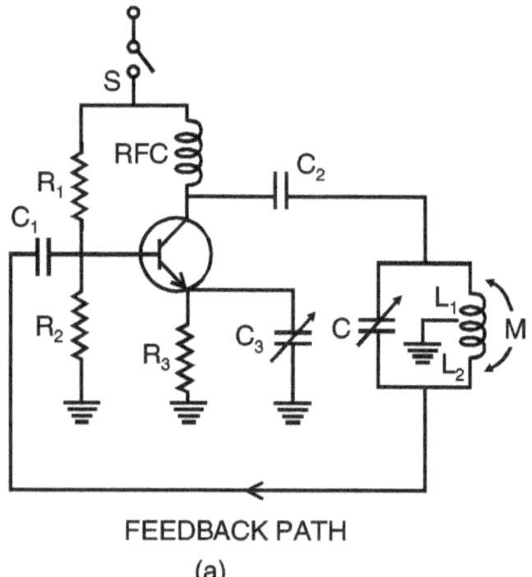

Fig. 19(a)

A C EQUIVALENT CIRCUIT

(b)

Fig. 19(b)

terminal via C_3. In other words, L_1 is in the output circuit *i.e.* collector-emitter circuit whereas L_2 is in the base-emitter circuit *i.e.* input circuit. These two parts are inductively-coupled and form an auto-transform, or a split-tank inductor. Feedback between the output and input circuits is accomplished through autotransformer action which also introduces a phase reversal of 180°. This phase reversal between two voltages occurs because they are taken from opposite ends of an inductor (L_1–L_2 combination) with respect to the tap which is tied to common transistor terminal *i.e.* emitter which is ac grounded via C_3. Since transistor itself introduces a phase shift of 180°, the total phase shift becomes 360° thereby making the *feedback positive* or *regenerative* which is essential for oscillations. As seen, positive feedback is obtained from the tank circuit and is coupled to the base via C_1. The feedback factor is given by the ratio of turns in L_2 and L_1 *i.e.* by N_2/N_1 and its value ranges from 0.1 to 0.5 Fig. 19 (b) shows the equivalent circuit of Hartely oscillator.

Resistors R_1 and R_2 form a voltage divider for providing the base bias and R_2 is an emitter swamping resistor to add stability to the circuit. Capacitor C_3 provides ac ground thereby preventing any signal degeneration while still providing temperature stabilisation. Radio-frequency choke (RFC) provides dc load for the collector and also keeps as currents out of the dc supply V_{CC}.

When V_{CC} is first switched on through S, an initial bias is established by R_1–R_2 and oscillations are produced because of positive feedback from the LC tank circuit (L_1 and L_2 constitute L). The frequency of oscillation is given by

$$f_o = \frac{1}{2\pi\sqrt{LC}} \quad \text{where } L = L_1 + L_2 + 2M$$

The output from the tank may be taken out by means of another coil coupled either to L_1 or L_2.

FET Hartley Oscillator

The basic circuit is shown in Fig. 20. R_G is the gate biasing resistor. There is mutual induction between the two parts L_1 and L_2 of the coil.

$$f_o = \frac{1}{2\pi\sqrt{LC}}$$

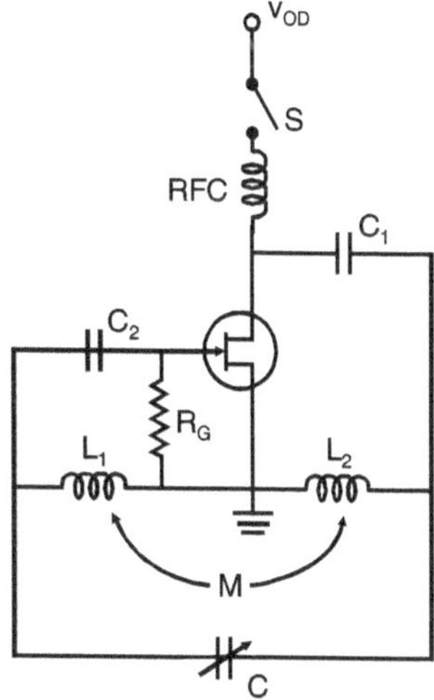

Fig. 20

where $L = L_1 + L_2 + 2M$

Colpitts Oscillator

This oscillator is essentially the same as Hartley oscillator except for one difference. Colpitts oscillator uses **tapped capacitance** whereas Hartley oscillator uses **tapped inductance**. Fig. 21 (a) shows the complete circuit with its power source and dc biasing circuit whereas Fig. 21 (b) shows its ac equivalent circuit. The two series capacitors C_1 and C_2 form the voltage divider used for providing the feedback voltage (the voltage drop across C_2 constitutes the feedback voltage). The feedback factor is C_1/C_2. The minimum value of amplifier gain for maintaining oscillations is

$$A_{v(min)} = \frac{1}{C_1/C_2} = \frac{C_2}{C_1}$$

The tank circuit consists of two ganged capacitors C_1 and C_2 and a single fixed coil. The frequency of oscillation (which does not depend on mutual inductance) is given by

$$f_o = \frac{1}{2\pi\sqrt{LC}} \quad \text{where } C = \frac{C_1 C_2}{C_1 + C_2}$$

Transistor itself produces a phase shift of 180°. Another phase shift of 180° is provided by the capacitive feedback thus giving a total phase shift of

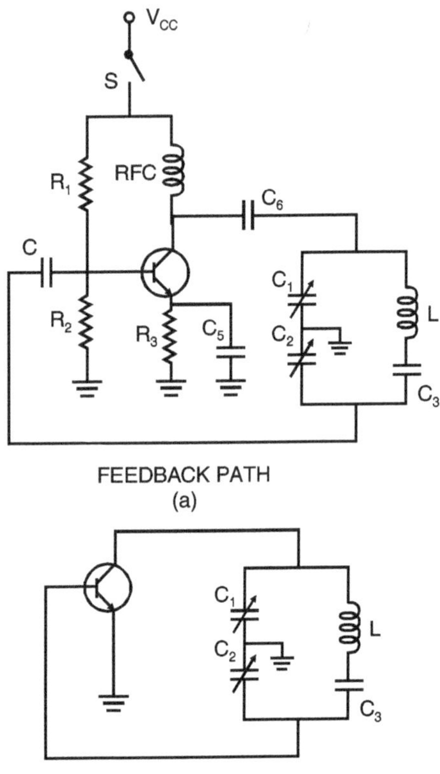

Fig. 21

360° between the emitter-base and collector-base circuits.

Resistors R_1 and R_2 form a voltage divider across V_{CC} for providing base bias, R_3 is for emitter stabilisation and RFC provides the necessary dc load resistance R_C for amplifier action. It also prevents as signal from entering supply dc V_{CC}. Capacitor C_5 is a by-pass capacitor whereas C_4 conveys feedback from the collector to-base circuit.

When S is closed, a sudden surge of collector current **shock-excites** the tank circuit into oscillations which are sustained by the feedback and the amplifying action of the transistor.

Colpitts oscillator is widely used in commercial signal generators upto 1 MHz. Frequency of oscillation is varied by gang-tuning the two capacitors C_1 and C_2.

Example : *For the Colpitts oscillator circuit shown in Fig. 22 find the values of*

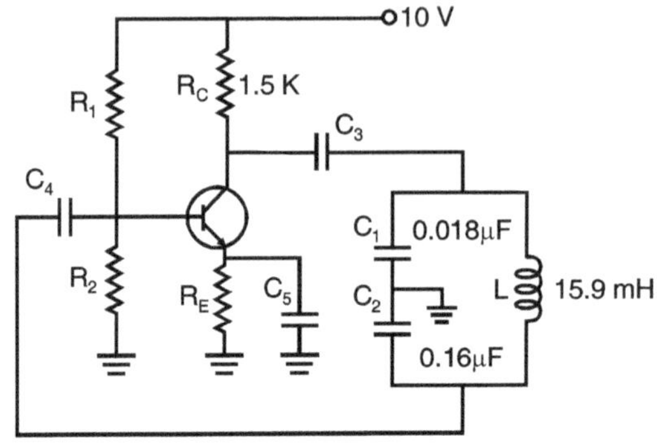

Fig. 22

(a) feedback fraction

(b) minimum gain to sustain oscillation

(c) emitter resistor R_E.

Solution : (a) The voltage drop across C_2 is fed back to the input circuit.

$$\text{feedback fraction} = \frac{\text{drop across } C_2}{\text{drop across } C_1} = \frac{IX_2}{IX_1} = \frac{C_1}{C_2} = \frac{0.018}{0.16} = \mathbf{0.11}$$

(b) $$A_{v(min)} = \frac{1}{\text{feedback fraction}} = \frac{1}{0.11} = \mathbf{9}$$

(c) Now, $A_v \cong \dfrac{R_C}{R_E}$

$$\therefore\ R_E = \frac{R_C}{A_v} = \frac{1500}{9} = \mathbf{167\ \Omega}$$

It should be noted that the above calculations do not take into account the losses in the coil and the loading effect of the amplifier input impedance.

Clapp Oscillator

It is a variation of Colpitts oscillator. It differs from Colpitts oscillator in respect of capacitor C_3 only which is joined in series with the tank inductor (Fig. 23).

Addition of C_3 (i) improves frequency stability and (ii) eliminates the effect of transistor's parameters on the operation of the circuit.

The operation of this circuit is the same as that of the Colpitts oscillator.

The frequency of oscillations is given by

$$f_o = \frac{1}{2\pi\sqrt{LC}} \text{ where } \frac{1}{C} = \frac{1}{C_1} + \frac{1}{C_2} + \frac{1}{C_3}$$

FET Colpitts Oscillator

The circuit is shown in Fig. 24. It is similar to the transistor circuit of Fig. 21. Here, R_G is the gate biasing resistor. The radio-frequency coil choke (RFC) performs two functions :

(i) it keeps ac current out of the dc drain supply and

(ii) it provides drain load.

As seen, C_1 is in the input circuit whereas C_2 is in the output circuit.

The frequency of oscillation is given by

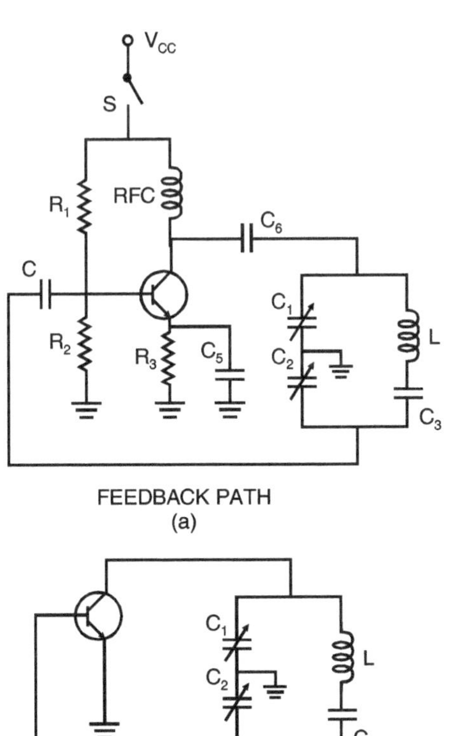

Fig. 23

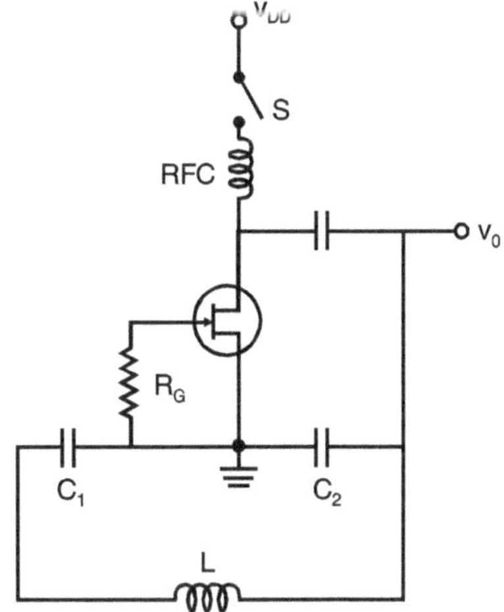

Fig. 24

$$f_o = \frac{1}{2\pi\sqrt{LC}}$$

$$\text{where } C = \frac{C_1C_2}{C_1 + C_2}$$

Crystals

For an exceptionally high degree of frequency stability, use of crystal oscillators is essential. The crystal generally used is a finely-ground wafer of transluscent quartz (or tourmaline) stone held between two metal plates and housed in a package about the size of a postage stamp. The crystal wafers are cut from the crude quartz in two different ways. The method of 'cutting' determines the crystal's natural resonant frequency and its temperature coefficient. When the wafer is cut so that its flat surfaces are perpendicular to its **electrical axis**, it is called an **X-cut** crystal (Fig. 25). But if the wafer is so cut that its flat surfaces are perpendicular to its **mechanical axis**, it is called **Y-cut** crystal.

Temperature Coefficient

Temperature variations affect the resonant frequency of a crystal. The number of cycles change per million cycles for a 1°C change in temperature is called the **temperature coefficient** (TC) of the crystal. It is usually expressed in parts per million (ppm) per °C. For example, a TC of 10 ppm per °C

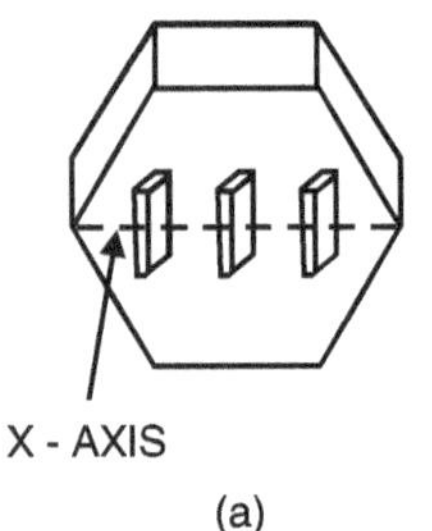

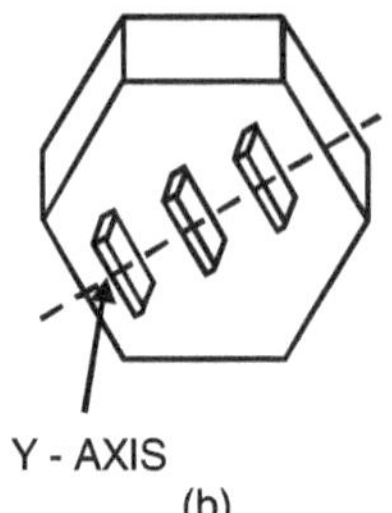

Fig. 25

means that frequency variation is 0.001 percent* per °C change in temperature. It can also be expressed as 10 Hz/MHz/°C. When kept in temperature-controlled ovens, crystal oscillators have frequency stability of about ± 1 ppm.

Usually, X-cut crystals have negative TC whereas Y-cut crystals have positive TC.

Example : *A 600-kHz X-cut crystal when calibrated at 50°C has a TC of – 20 ppm/°C. What will be its resonant frequency when its temperature is raised to 60°C?*

Solution : A TC of 20 ppm/°C is the same thing as 20 Hz/MHz/°C. Since temperature rise is 10°C and original frequency is 600 kHz = 0.6 MHz, the total change as calculated by ratio proportion is = 20 × 10 × 0.6 = 120 Hz.

Since TC is negative, the new resonance frequency of the crystal is

= 600, 000 – 120 = 599,880 = **599.88 kHz.**

Example : *A certain X-cut quartz crystal resonates at 450 kHz. It has an equivalent inductance of 4.2 H and an equivalent capacitance of 0.0297 PF. If its equivalent resistance is 60 Ω, calculate its Q-factor.*

Solution :

$$Q = \frac{\omega L}{R} = \frac{2\pi f L}{R} = \frac{2\pi \times 450 \times 10^3 \times 4.2}{600} = \mathbf{19{,}790}$$

Crystal Controlled Oscillator

Fig. 26 shows the use of a crystal to stabilise the frequency of a tuned-collector oscillator which has a crystal (usually quartz) in the feedback circuit.

The *LC* tank circuit has a frequency of oscillation $f_o = 1/2\pi\sqrt{LC}$

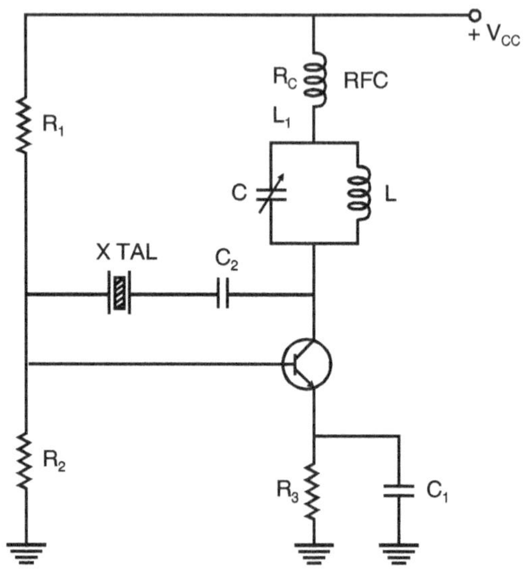

Fig. 26

The circuit is adjusted to have a frequency near about the desired operating frequency but the exact frequency is set by the crystal and stabilized by the crystal. For example, if natural frequency of vibration of the crystal is 27 MHz, the *LC* circuit is made to resonate at this frequency.

As usual, resistors R_1, R_2, and R_3 provide a voltage-divider stabilised dc bias circuit. Capacitor C_1 by-passes R_3 in order to maintain large gain. RFC coil L_1 prevents ac signals from entering dc line whereas R_C is the required dc load of the collector. The coupling capacitor C_2 has negligible impedance at the operating frequency but prevents any dc link between collector and base. Due to extreme stability of crystal oscillations, such oscillators are widely used in communication transmitters and receivers where frequency stability is of prime importance.

Spacing Between Energy Levels of a Band

It will be quite interesting as well as instructive to calculate the spacing between different energy levels in an energy band. A crystal weighing one milligram contains about 10^{19} atoms. If we assume the valency band to be an *s*-band, it will contain 2×10^{19} levels. Suppose the width of energy band is 2 eV. Then, it is obvious that 2×10^{19} levels per milligram are spread over an energy band width of 2 eV. Hence, spacing between different levels = 2/2 × $10^{19} = 10^{-19}$ eV. It will be appreciated that even though energy levels are discrete, the picture of a band as a continuum of energy levels is a very good approximation. This splitting of the single energy level of an *isolated* atom into a band of energy in the case of a *solid* is responsible for most of the electrical, magnetic and optical properties of that solid. It is worth pointing out here that in gases under normal conditions of temperature and pressure, the atomic spacing is so great that there is no splitting of energy levels and hence no band formation.

Energy Bands in Lithium and their Occupancy

Consider the case of lithium metal—the simplest atom which forms a solid at ordinary temperature. Its atom has three electrons, two of which have the same energy and the third one has higher value of energy. In an isolated single atom, two electrons move round the electron orbit with $n = 1$ whereas the third occupies the orbit with $n = 2$ as shown in Fig. 28 (a). Now consider a piece of lithium metal containing 100 atoms. It will be found that the lower level (with $n = 1$) forms a band of 200 electrons occupying 100 different energy states. The higher level (with $n = 2$) forms a wider band of 100 energy states which could, as before, accommodate 200 electrons. But as there are only 100 electrons available (one from each atom), this energy band remains half-filled [Fig. 28 (b)].

Valence and Conduction of Bands

The outermost electrons of an atom *i.e.* those in the shell furthermost from the centre are called *valence electrons* and have the *highest energy* or least binding energy. It is these electrons which are most affected when a number of atoms are brought very close together as during the formation of solid. The states of lower-energy electrons orbiting in shells nearer to the nucleus are little, if at all, affected by this atomic proximity.

The band of energy occupied by the valence electrons is called the *valence band* and is, obviously, the *highest occupied band*. It may be completely filled or partilly filled with electrons but never empty.

The next higher permitted energy band is called the *conduction* band and may either be *empty* or *partially filled* with electrons. In fact, it may be defined as the lowest unfilled energy band. (Fig. 29)

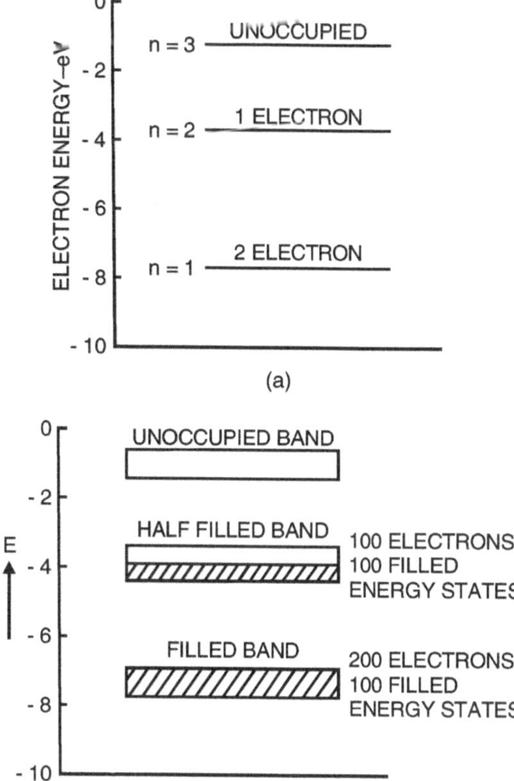

Fig. 28

In conduction band, electrons can move freely and hence are known as *conduction* electrons. The gap between these two bands is known as the *forbidden energy gap*.

It may be noted that the covalent forces of the crystal lattice have their source in the valence band. If a valence electron happens to absorbs enough energy, it jumps across the forbidden energy gap and enters the conduction band (Fig. 30). An electron in the conduction band can jump to an adjacent conduction band more readily than it can jump back to the valence band from where it had come earlier. However, if a conduction electron happens to radiate too much energy, it will suddenly reappear in the valence band once again.

When an electron is ejected from the valence

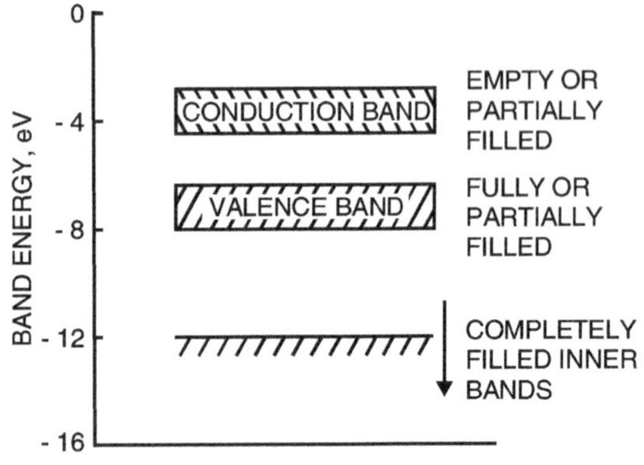

Fig. 29

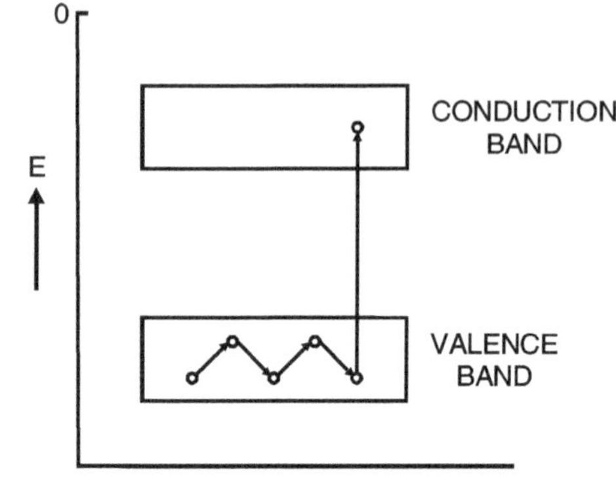

Fig. 30

band, a covalent bond is broken and a positively-charged hole is left behind. This hole can travel to an adjacent atom by acquiring an electron from that atom which involves breaking an existing covalent bond and then re-establishing a covalent bond by filling up the hole. It is to be noted carefully that holes are filled by electrons which move from adjacent atoms without passing through the forbidden energy gap.

It is simply another way of saying that conditions in the conduction band have nothing to do with the hole flow. It points to a very important distinction between the hole current electron—although holes flow with ease, they experience more opposition than electron flow in the conduction band.

To summarize the above, it may be repeated that :

1. conduction electrons are found in and freely flow in the *conduction* band
2. holes exist in and flow in the *valence* band
3. conduction electrons move almost twice as fast as the holes.

Fig. 31 (a) shows the energy band and diagram of an unexcited silicon atom (Z = 14) with its electronic distribution. When silicon crystal is given thermal or light energy from outside [Fig. 31 (b)]., some electrons gain sufficient energy to jump the gap from the valence band into the conduction band thereby becoming free electrons [Fig. 31 (c)]. For every electrons which jumps to conduction band, a hole is created in the valence band. In this way, an electron-hole pairs is created.

INSULATORS, CONDUCTORS AND SEMICONDUCTORS

The electrical conduction properties of different elements and compounds can be explained in terms of the electrons having energies in the valence and conduction bands. The electrons lying in the lower energy bands, which are normally filled, play no part in the conduction process.

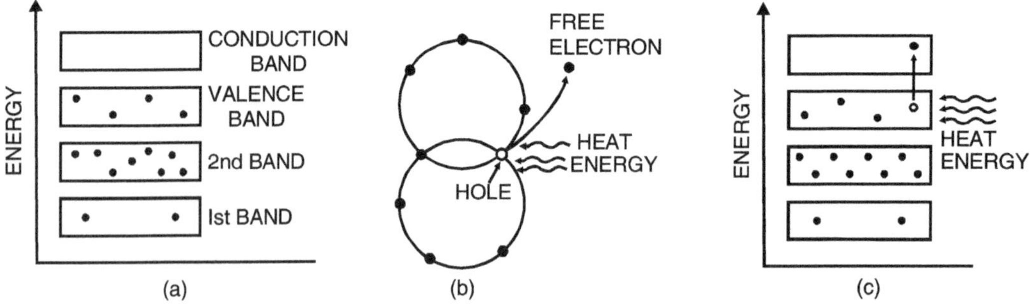

Fig. 31

(i) **Insulators :** Stated simply, insulators are those materials in which valence electrons are bound very tightly to their parent atoms, thus requiring very large electric field to remove them from the attraction of their nuclei. In other words, insulators have no free charge carriers available with them under normal conditions.

In terms of energy bands, it means that insulators

(a) have a full valence band

(b) have an empty conduction band

(c) have a large energy gap (of several eV) between them and

(d) at ordinary temperatures, the probability of electrons from full valence band gaining sufficient energy so as to surmount energy gap and thus become available for conduction in the conduction band, is slight.

This shown in Fig. 32 (a). For conduction to take place, electrons must be given sufficient energy to jump from the valence band to the conduction band. Increase in temperature enables some electrons to go to the conduction band which fact accounts for the negative resistance-temperature coefficient of insulators.

(ii) **Conductors :** Put in a simple way, conducting materials are those in which plenty of free electrons are available for electric conduction.

In terms of energy bands, it means that electrical conductors are those which have overlapping valence and conduction bands as shown in Fig. 32 (b).

In fact, there is no physical distinction *between the two bands*. Hence, the availability of a large number of conduction electrons.

Another point worth nothing is that in the absence of forbidden energy gap in good conductors, there is no structure to establish holes. The total current in such conductors is simply a flow of electrons. It is exactly for this reason that the existence of holes was not discovered until semi-conductors were studied thoroughly.

(iii) **Semiconductors :** A semiconductor material is one whose electrical properties lie in between those of insulators and good conductors. Examples are : germanium and silicon.

In terms of energy bands, semiconductors can be defined as those materials which have almost an empty conduction band and almost filled valence band with a very narrow energy gap (of the order of 1 eV) separating the two.

At 0°K, there are no electrons in the conduction band and the valence band is completely filled. However, with increases in temperatures, width of the forbidden energy band is decreased so that some of the electrons are liberated into the conduction band. In other words, conductivity of semiconductors increases with temperature. Morever, such departing electrons leave behind positive holes in the valence band (Fig. 30). Hence, semiconductor current is the sum of electron and hole current flowing in opposite directions.

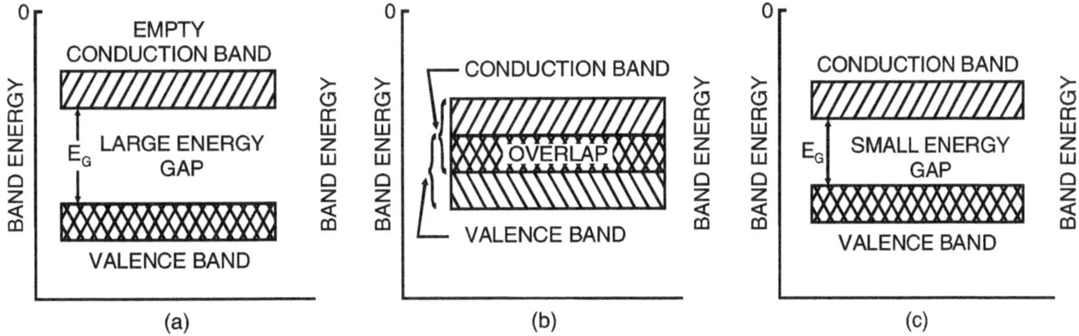

Fig. 32

Atomic Binding in Semiconductors

Semiconductors like germanium and silicon, have crystalline structure. Their atoms are arranged in an ordered array known as crystal lattice. Both these materials are tetravalent *i.e.* each has four valence electrons in its outermost shell. The neighbouring atoms form *covalent* bonds by sharing four electrons with each other so as to achieve inert gas structure (*i.e.* electrons in the outermost orbit). A two-dimensional view of the germanium crystal lattice is shown in Fig. 33 (b) in which circles represent atom cores consisting of the nuclei and inner 28 electrons. Each *pair* of lines represents a covalent bond. The dots represent the valence electrons. It is seen that each atom has 8 electrons under its influence.

A 3-dimensional view of germanium crystal lattice is shown in Fig. 33 (a) where each atom is surrounded symmetrically by four other atoms forming a tetrahedral crystal. Each atom shares a valence electron with each of its four neighbours, thereby forming a stable structure.

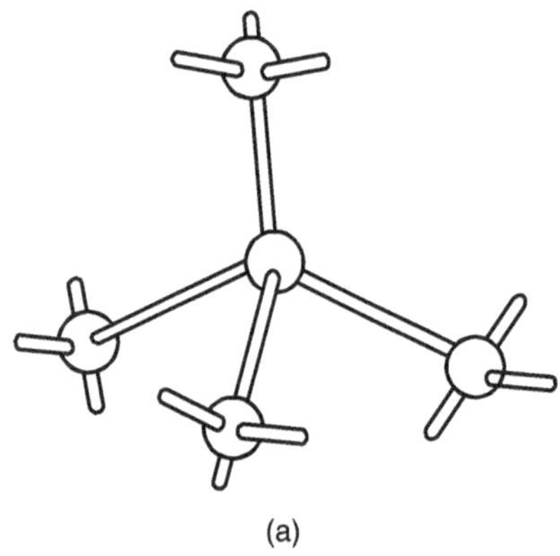

Fig. 33 (a)

In the case of pure (*i.e.* intrinsic) germanium, the covalent bonds have to be broken to provide electrons for conduction. There are many ways of rupturing the covalent bond and thereby setting the electrons free. One way is to increase the crystal temperature above 0°K.

It may be noted that covalent crystals are characterised by their hardness and brittleness. Their brittleness is due to the fact that in such crystals,

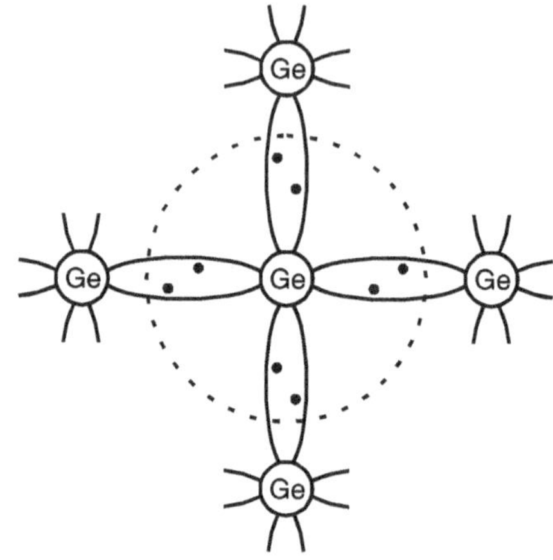

Fig. 33 (b)

adjacent atoms must remain in accurate a alignment since the bond is strongly directional and formed along a line joining the atoms. The hardness is due to the great strength of the covalent bond itself.

TYPES OF SEMICONDUCTORS

Semiconductors can be classified as shown below :

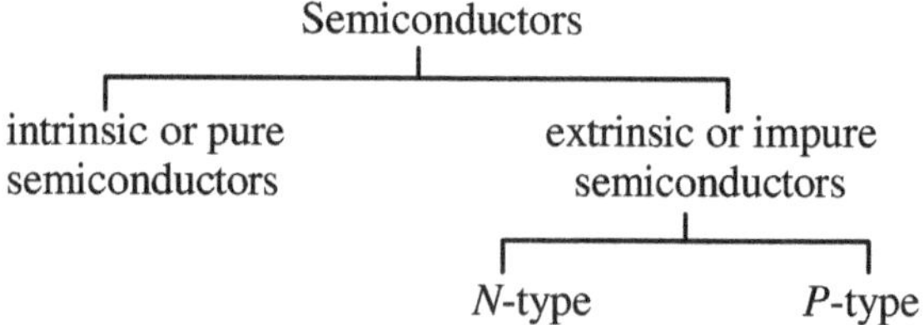

Intrinsic Semiconductors

An intrinsic semiconductor is one which is made of the semiconductor material in its extremely pure form.

Common examples of such semiconductors are : pure germanium and silicon which have forbidden energy gaps of 0.7 eV and 1.1 eV respectively. The energy gap is so small that even at ordinary room temperature, there are many electrons which possess sufficient energy to jump across the small energy gap between the valence and the conduction bands. However, it is worth noting that for each electron liberated into the conduction band, a positively-charged hole is created in the valence band (Fig. 34). When an electric field is applied to an intrinsic semiconductor at a temperature greater than 0°K, conduction electrons move to the anode and the holes in the valence band move to the cathode. Hence, semiconductor current consists of movement of electrons and holes in opposite directions. Electron current is due to movement of electrons in the conduction band whereas hole current is within the valence band as result of holes 'jumping' from one atom to another.

Hole Formation in Semiconductors

The formation of a hole which is positive charge carrier is explained below :

As shown in Fig. 35, suppose the covalent bond is broken at A and the electron has moved through the crystal lattice leaving behind a hole in the covalent bond. An electron B may jump into the vacant hole at A and later, an electron at C may jump into the hole at B and so on. In this way, by succession of electron movements, a hole will appear at G and a negative charge would have moved from G to A. It would, however be more convenient to regard positive charge to have moved from A to G and this conception gives rise to a hole as a positive charge carrier as if it were an electron with a positive charge. It should be clearly understood that these holes are due to the movement of electrons in the valence band and that each electron movement corresponds to a collision. The drift velocity of holes is, obviously, much less than the drift velocity of electrons.

Alternatively, an intrinsic semiconductor may be defined as one in which the number of conduction electrons is equal to the number of holes.

Schematic energy band diagram of an intrinsic

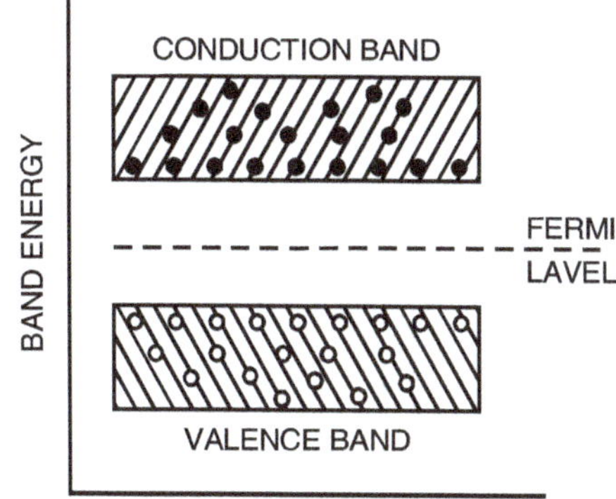

Fig. 34

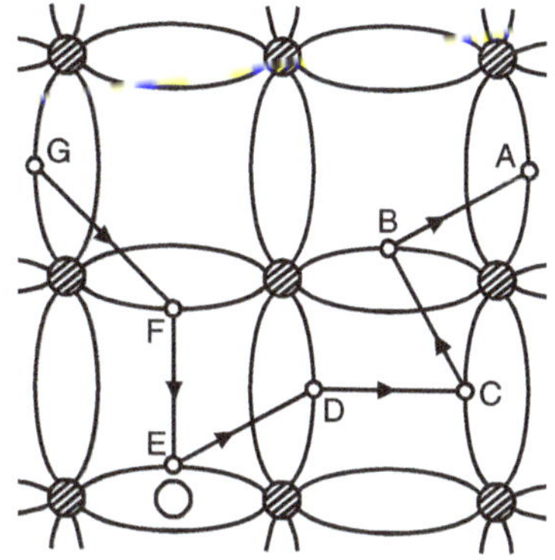

Fig. 35

semiconductor at room temperature is shown in Fig. 34. Only two bands *i.e.* valence and conduction bonds have been shown since lower filled bands are not of any consequence. Here, Fermi level lies exactly in the middle of the forbidden gap.

Fermi Level in an Intrinsic Semiconductor

It can be proved that in an intrinsic semiconductor, Fermi level E_F lies in the middle of the energy gap *i.e.* midway between the conduction and valence bands.

Let, at any temperature $T°K$

$n_c =$ No. of electrons in the conduction band

$n_v =$ No. of electrons in the valence band

$N = n_c + n_v$

$=$ No. of electrons in both bands.

Further, let us make the following simplifying assumptions :

1. widths of energy bands are small as compared to forbidden energy gap between them :
2. since band widths are small, all levels in a band have the same energy.
3. energies of all levels in valence band are zero as shown in Fig. 36.
4. energies of all levels in the conduction band are $= E_g$.

In Fig. 36, the zero-energy reference level has been arbitrarily taken at the top of valence band.

Now, number of electrons in the conduction band is

$$n_c = N.P(E_g)$$

where $P(E_g)$ represents the probability of an electron having energy E_g. Its value can be found from Fermi-

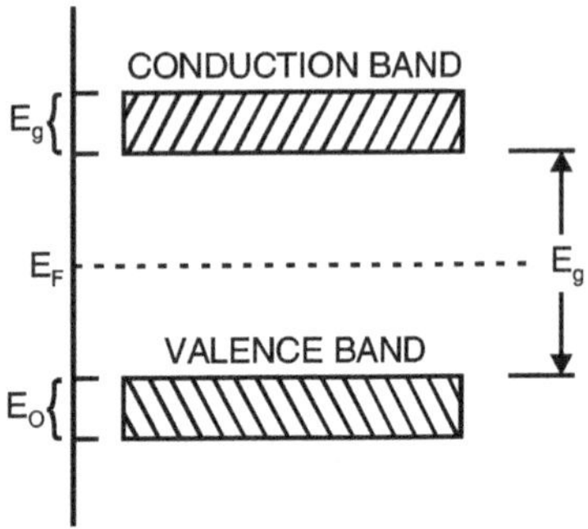

Fig. 36

Dirac probability distribution function given by

$$P(E) = \frac{1}{1 + e^{(E - E_g)/kT}}$$

where $P(E)$ is the probability of finding an electron having any particular value of energy E and EF is Fermi level.

$$\therefore\ P(E_g) = \frac{1}{1 + e^{(E_g - E_F)/kT}}$$

$$\therefore\ n_c = \frac{1}{1 + e^{(E_g - E_F)/kT}}.$$

Now, number of electrons in the valence band is $n_v = N.\,P(0)$

The probability $P(0)$ of an electron being found in the valence band with zero energy can again be calculated by putting $E = 0$ in the Femi-Dirac probability distribution function.

$$P(0) = \frac{1}{1 + e^{(0 - E_F)/kT}} = \frac{1}{1 + e^{-E_F/kT}}.$$

Hence, $n_v = \dfrac{N}{1 + e^{-E_F/kT}}$

Now $N = n_c + n_v$

or $N = \dfrac{N}{1 + e^{(E_g - E_F)/kT}} + \dfrac{N}{1 + e^{-E_F/kT}}$

$$\therefore\ 1 - \frac{1}{1 + e^{-E_F/kT}} = \frac{1}{1 + e^{(E_g - E_F)/kT}}$$

which on simplification gives $E_F = E_g / 2$.

Hence, it shows that in an intrinsic semiconductor, the Fermi level lies midway between the conduction and valence bands. That this conclusion is physically plausible can be best seen with the help of Fig. 37.

Extrinsic Semiconductors

Those intrinsic semiconductors to which some suitable impurity or doping agent or dopant has been added in extremely small amounts (about 1 part in 10^8) are called *extrinsic* or impurity semiconductors. The usual doping agents are :

1. *pentavalent* atoms having five valence electrons (arsenic, antimony, phosphorus) or
2. *trivalent* atoms having three valence electrons (gallium, indium, aluminium, boron).

Pentavalent doping atom is known as *donor* atom because it donates or contributes one electron to the conduction band of pure germanium. The trivalent atom, on the other hand, is called *acceptor*

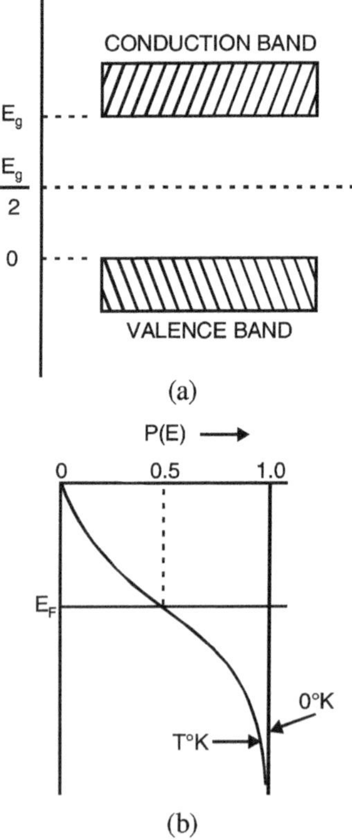

Fig. 37

atom because it accepts one electron from the germanium atom.

Depending on the type of doping material used, extrinsic semiconductors can be sub-divided into two classes :

(a) *N*-type semiconductors and

(b) *P*-type semiconductors.

(a) N-type Extrinsic Semiconductor : This type of semiconductor is obtained when a pentavalent material like antimony (Sb) is added to pure germanium crystal. As shown in Fig. 38 (a), each antimony atom forms covalent bonds with the surrounding four germanium atoms with the help of four of its five electrons. The fifth electron is superfluous and is loosely bound to the antimony atom. Hence, it can be easily excited from the valence band to the conduction band by the application of electric field or increase in thermal energy. Thus, practically every antimony atom introduced into the germanium lattice, contributes one conduction electron into the germanium lattice *without creating a positive hole*. Antimony is called *donor* impurity and makes the pure germanium an *N* type (*N* for negative) extrinsic semi-conductor. As an aid to memory, the student should associate the *N* in donor with the *N* in the *N*-type material and *N* in Negative charge carrier.

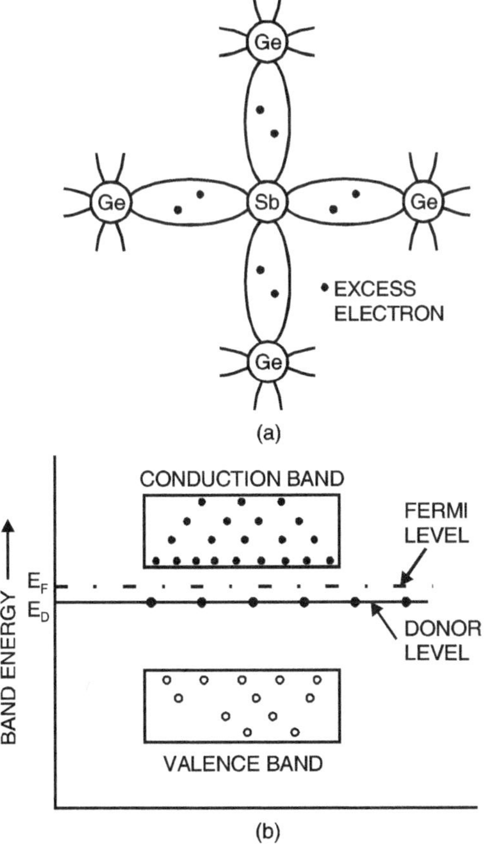

Fig. 38

It may be noted that by giving away its one valence electron, the donor atom becomes a positively-charged *ion*. But it cannot take part in conduction because it is firmly fixed or tied into the crystal lattice. It will be send that apart from electrons and holes *intrinsically available in germanium*, the addition of antimony greatly increases the number of conduction electrons. Hence, concentration of electrons in the conduction band is increased and exceeds the concentration of holes in the valence band. Because of this, Fermi level shifts upwards towards the bottom of the conduction band as shown in Fig. 38 (b), because the number of charge carriers has become more in conduction band than in valence band.

In terms of energy levels, the fifth antimony electron has an energy level (called donor level) just below the conduction band. Usually, the donor level is 0.01 eV below conduction band for germanium and 0.054 eV for silicon.

It is seen from the above description that in *N*-type semiconductors, electrons *are the majority carriers* while holes constitute the minority carriers. Hence, *N*-type semiconductor conducts principally by electrons in the nearly empty conduction band and the process is called '*excess*' conduction.

Another point worth noting is that even though *N*-type semiconductor has excess of electrons, still it is electrically neutral. It is so because by the addition of donor impurity, number of electrons available for conduction purposes becomes more than the number of holes available intrinsically. But the total charge of the semiconductor does not change because the donor impurity brings in as much negative charge (by way of electrons) as positive charge (by way of protons in its nucleus).

(b) P-type Extrinsic Semiconductor : This type of semiconductor is obtained when traces of a trivalent impurity like boron (B) are added to a pure germanium crystal.

In this case, the three valence electrons of boron atom form covalent bonds with four surrounding germanium atoms but one bond is left incomplete and gives rise to a hole as shown in Fig. 39 (a).

Thus, boron which is called an acceptor impurity causes as many positive holes in a germanium crystal as there are boron atoms thereby producing a *P*-type (*P* for positive) extrinsic semiconductor. As an aid to memory, the student should associate the **P** in acceptor with **P** in *P*-type material and **P** with Positive charge carrier.

In this type of semiconductor, conduction is by the movement of holes in the valence band.

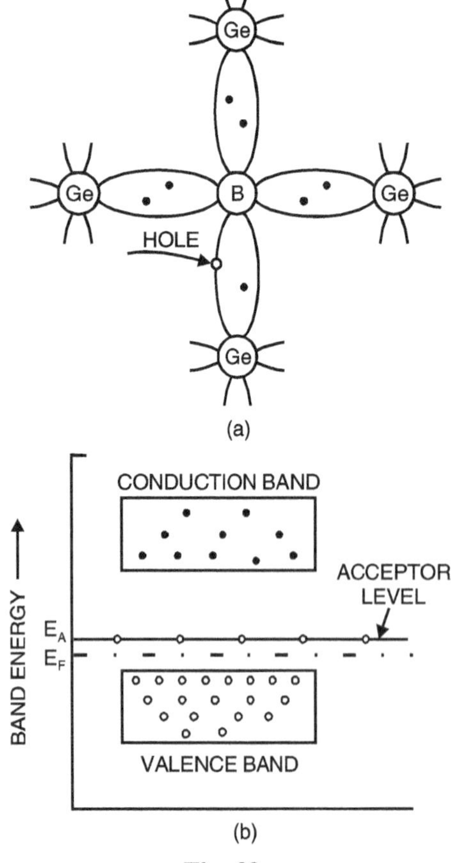

Fig. 39

Accordingly, holes form the majority carriers *whereas electrons constitute minority carriers*. The process of conduction is called 'deficit' conduction.

Since concentration of holes in the valence band is more than concentration of electrons in the conduction band, Fermi level shifts nearer to the valence band [Fig. 39 (b)]. The acceptor level lies immediately above the Fermi level. Conduction is by means of hole movement at the top of valence band, the acceptor level readily accepting electrons from the valence band.

Again, it may be noted that even though *P*-type semiconductor has excess of holes for conduction purposes, on the whole it is electrically neutral for the same reasons as given above.

Majority and Minority Carriers

In a piece of *pure* germanium or silicon, no free charge carriers are available at 0°K. However, as its temperature is raised to room temperature, some of the covalent bonds are broken by heat energy and as a result, electron-hole pairs are produced. These are called thermally generated charge carriers. They are also known as intrinsically available charge carriers. Ordinarily, their number is quite small.

An intrinsic or pure germanium can be converted into a *P*-type semiconductor by the addition of an acceptor impurity which adds a large number of holes to it. Hence, a *P*-type material contains following charge carriers :

(a) large number of positive holes—most of them being the added impurity holes with only a very small number of thermally-generated ones.

(b) a very small number of thermally-generated electrons (the companions of the thermally-generated holes mentioned above).

Obviously, in a *P*-type material, the number of holes (both added and thermally-generated) is much more than that of electrons. Hence in such a material, holes constitute *majority* carriers and electrons form carries as shown in Fig. 40 (a)

Similarly, in an *N*-type material, the number of electrons (both added and thermally-generated) is much larger than the number of thermally-generated holes. Hence, in such a material, electrons are majority carriers whereas holes are minority carriers as shown in Fig. 40 (b).

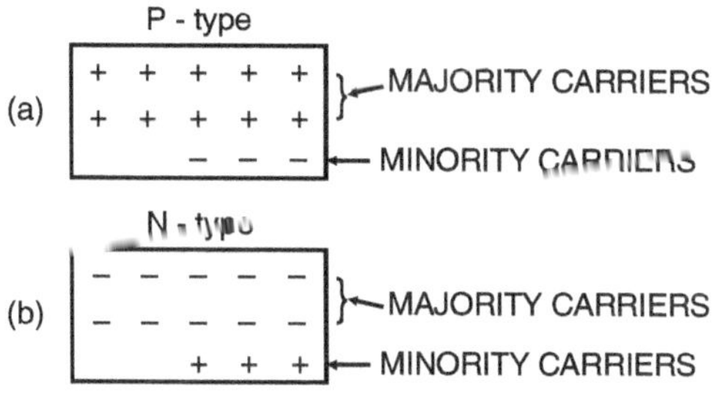

Fig. 40

Mobile Charge Carriers and Immobile Ions

P-type material is formed by the addition of acceptor impurity atoms like boron to the pure Ge or Si crystals. The number of holes added is equal to the number of boron atoms because each such atom contributes one hole. Now, when a hole moves *away* from its parent atom, the remaining atom becomes a *negative ion*. Unlike the mobile and free-moving hole, this ion cannot take part in conduction because it is fixed in the crystal lattice. In Fig. 41 (a), these immobile ions are shown by *circled* minus signs whereas free and mobile holes are shown by *uncircled* plus signs. Thermally-generated electrons (which form minority carriers) are shown by *uncircled* minus signs.

Similarly, addition of pentavalent atoms like antimony to pure Ge or Si crystal produces an *N*-type material. The number of free and mobile electrons which are added equals the number of donor Sb atoms. Again, when an electron moves away from its parent atom, it leaves behind a positive ion. This ion, being fixed in crystal structure, cannot take part in conduction. As shown in Fig. 41 (b), these immobile ions are represented by *circled* plus signs whereas free and mobile electrons are represented by *uncircled* minus signs. The thermally-generated

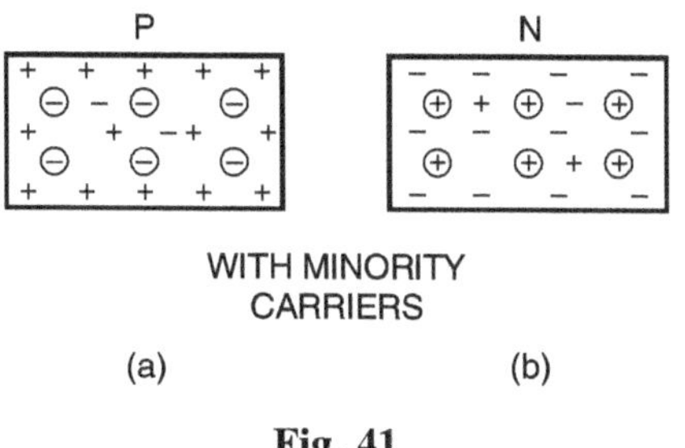

Fig. 41

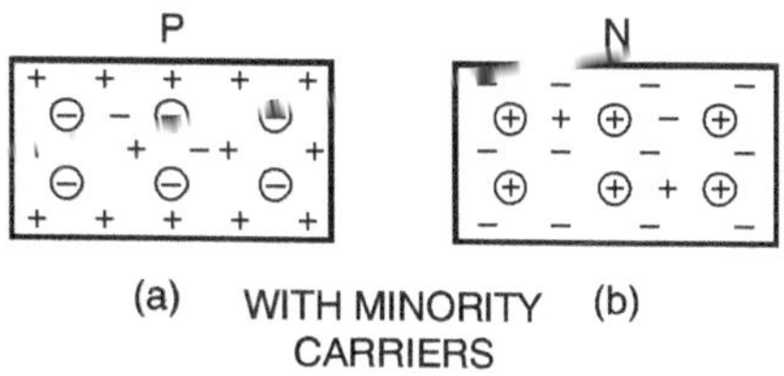

Fig. 42

holes (which form minority carriers in this case) are shown by *uncircled* plus signs. In Fig. 42 minority carriers of both types have been neglected. Hence, the figure does not show the small number of free electrons in the *P*-type material or the small number of holes in the *N*-type material.

Electron Conductivity of a Metal

According to free electron model of an atom, the valence electrons are not attached to individual atoms but are free to move about in all directions among the atoms. These electrons are called *conduction* electrons and are said to form 'free electron cloud' or free electron 'gas' or the Fermi gas. For example, in copper there is one such free electron per atom, the other 28 electrons remaining bound to the copper nuclei to form positive ion cores.

When no external field is applied to the metal, the free electrons move randomly in all directions as shown in Fig. 43 (a). However, when an external

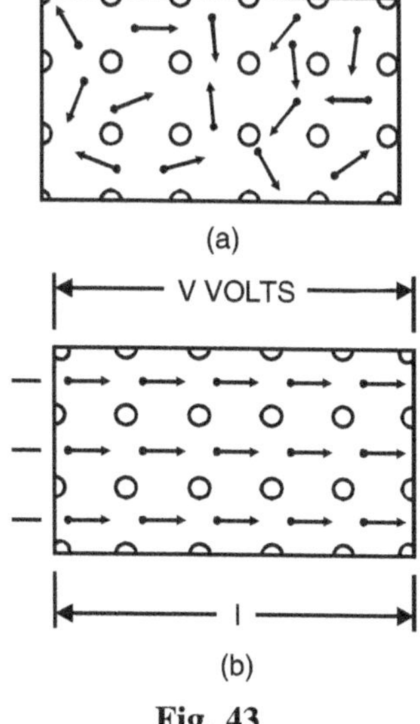

Fig. 43

electric field is applied to the metal, the free electron motion becomes directed. This directed flow of electrons results in a net charge displacement in a definite direction. This type of motion is known as *drift* and the phenomenon is referred to as process of conduction by drift charge. The drift velocity (v) of the electrons is dependent upon the electron mobility (μ_e) and the applied electric field E. The actual relation is $v = \mu_e E$.

Let, e = electron charge (coulomb)

v = electron drift velocity (m/s)

A = conductor cross-section (m^2)

n = number of free electrons per unit volume of the conductor *i.e.* electron density (per m^3)

l = length of the conductor (m)

E = V/l — applied electric field (V/m)

Now, electric current flowing in any conductor is given by the amount of charge which flows in one second across any plane of the conductor (Fig. 44) The total number of electrons which cross the plane P of cross-section A is one second = $n \times (v \times A)$

Charge carried by them per second is = $envA$. Hence, $I = venA$.

Substituting the value v, we get

$$I = neA\mu_e E = nAe\mu_e V/l$$

$$\therefore\ R = \frac{V}{I} = \frac{l}{A}\left(\frac{1}{ne\mu_e}\right) = \pi\frac{l}{A}$$

$\therefore$ resistivity $\rho = 1/ne\mu_e$ ohm-m and conductivity $\sigma = ne\mu_e$ Siemens/m

Incidentaly, it may be noted that conductivity of a semiconductor differs from that of a metal in one important respect *i.e.* in a semiconductor, charge carriers are both holes as well as electrons whereas in metals, electrons are the only charge carriers.

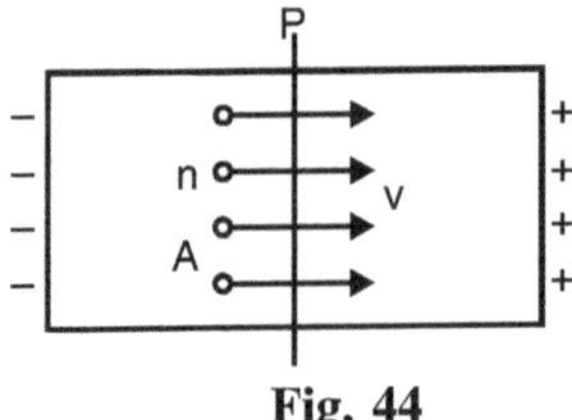

Fig. 44

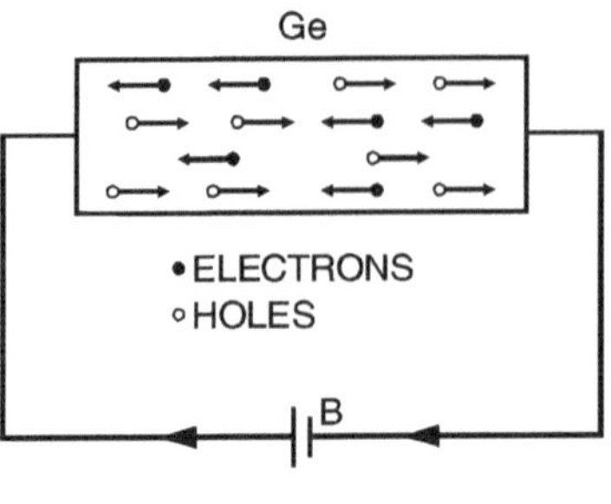

Fig. 45

Conductivity of Intrinsic Semiconductors

In their case, current flow is due to the movement of electrons and holes in opposite directions. However, since their charges are of opposite sign, the current due to each is in the same direction. Even though the number of electrons equals the number of holes, hole mobility μ_h is practically half of electron mobility μ_e.

As shown in Fig. 45 the total current flow which is due to the sum of electron flow and hole flow, is given by

$$I = I_e + I_h$$

Let v_e = drift velocity of electrons (m/s)

v_h = drift velocity of holes (m/s)

n_I = density of free electrons in an intrinsic semiconductor (per m^3)

p_I = density of holes in an intrinsic semiconductor (per m^3)

e = electron charge (coulomb)

A = cross-section of the semiconductor (m^2)

Since in an intrinsic semiconductor $n_1 = p_1$

$$\therefore\ I = n_I e(v_e + v_h)A = n_I e(\mu_h + \mu_e)EA$$

where μ_e = electron mobility = v_e/E

μ_h = hole mobility = v_h/E

Since $E = V/l$ where l is the length of the intrinsic semiconductor,

$$\therefore\ I = n_I e(\mu_e + \mu_h)AV/l$$

$$\therefore\ \frac{V}{I} = \frac{l}{A}\cdot\frac{1}{n_I e(\mu_e + \mu_h)} = \rho\frac{l}{A}$$

where ρ is the resistivity of the semiconductor. It is given by

$$\rho_I = \frac{1}{n_I e(\mu_e + \mu_h)} \text{Ohm} - \text{m}$$

The electrical conductivity which is the reciprocal of resistivity is given by

$\sigma_I = n_I e(\mu_e + \mu_h) \text{S/m}$

Now, current density, $J = I/A$

$\therefore\ J = n_I e(\mu_e + \mu_h)E = \sigma_I E$

$\therefore\ \sigma_I = J/E$

Obviously, conductivity of semiconductors depends on two factors (i) number of current carriers present per unit volume and (ii) the mobility of the current carriers. It is found that with increase in temperature, n_I as well p_I increase and correspondingly, the conductivity of intrinsic semiconductors increases *i.e.* resistivity decreases.

Conductivity of Extrinsic Semiconductors

The general expression for current density (derived above) in the case of an extrinsic semiconductor when an electric field is employed is

$$J = (ne\mu_e + pe\mu_h)E \qquad \text{... (i)}$$

(a) If it is an *N*-type semiconductor, then the above expression becomes

$$J_n = e(n_n\mu_e + p_n\mu_h)E$$

where n_n and p_n represent the electron and hole densities in the *N*-type semiconductor after doping.

(b) If it is a *P*-type semiconductor, then

$$J_p = e(n_p\mu_e + p_p\mu_h)E$$

where n_p and p_p represent similar quantities in a *P*-type semiconductor *after doping.*

The conductivity is given by $\sigma = J/E$

$\therefore\ \sigma = ne\mu_e + pe\mu n$

or $\sigma_n = e(n_n\mu_e + p_n\mu n)$ — for *N*-type

and $\sigma_p = e(n_p\mu_e + p_p\mu n)$ — for *P*-type

(i) In *N*-type semiconductors, electrons form the majority carriers although holes are also available as minority carriers.
The current density in such a semiconductor is given by Eq. (i) above. However, since electron density in such extrinsic semiconductors is much more than hole density *i.e.* $n_n > p_n$, the above expressions are simplified to

$$J_n = n_n e\mu_e\ E \text{ and } \sigma_n = n_n e\mu_e \qquad \text{... (ii)}$$

(ii) In *P*-type semiconductors, conduction is by means of holes in the valence band which form majority carriers in this case although electrons are available as minority carriers.
Since in such extrinsic semiconductors, $n_p < p_p$, the above expressions become

$$J_p = p_p e\mu h\ E \text{ and } \sigma_p = p_p\ e\ \mu h \qquad \text{... (iii)}$$

Conductivity when Intrinsic Charge Carrier Densities are Neglected

In case density of charge carriers available intrinsically is negligible as compared to the added impurity atoms (whether of donor or acceptor type), then the formulae for conductivity given by E.q. (ii) and (iii) above will be changed as follows :

(a) For N-type Semiconductor : As seen from Eq. (ii) above, the conductivity is given by $\sigma_n = n_n e\ \mu_e$ where n_n is electron density after doping.

In this relation, intrinsic hole density has already been neglected. The remaining electron density is also made up of the following two components :

1. intrinsic electron density due to electrons available in a *pure* semiconductor.
2. electron density N_d contributed by the added donor impurity.

However, if we further neglect the intrinsic electron density, then

$$\sigma_n = N_d e\mu_e$$

(b) For P-type Semiconductor : A seen from Eq. (iii) above, the conductivity is given by $\sigma_p = p_p e\mu_h$ were p_h represents hole density. Again, in this relation, intrinsic electron density has been already neglected. This hole density further consists of the following two components :

1. intrinsic hole density due to holes available in a pure semiconductor
2. hole density (N_a) contributed by added acceptor impurity.

However, if we further neglect the intrinsic hole density, then $\sigma_p = Nae\mu_h$

Conductivity of Pure and P-type-Germanium

The conductivity of pure germanium is given by

$$\sigma = n_I e(\mu_e + \mu_h) = p_I e(\mu_e + \mu_h)$$

When germanium is doped by a trivalent impurity like indium, it becomes a P-type semiconductor. After doping, its conductivity depends on the number of charge carriers available in it. The Law of Mass Action can be used for finding this number. For acceptor impurity, the law may be stated as follows :

$$n_p p_p = n_1 p_1 = n_1^2 = p_1^2$$

where n_p and p_p represent the 'free' electron and hole densities respectively in the semiconductor *after* doping and n_1 and p_1, the electron and hole densities *before* doping *i.e.* in an intrinsic semiconductor. Put in another way, it simply means that, at constant temperature, the product of the number of electron carriers and the number of hole carriers is independent of the density of acceptor atoms. In physical terms, it means that the introduction of *P*-type impurity fills some of the electron levels produced by thermal action.

By calculating n_p and p_p from above and knowing μ_e and μ_h, conductivity after doping can be found out as illustrated by the following example.

(a) Pure Germanium

$$\sigma = n_1 e(\mu + \mu_h) = p_1 e(\mu_e + \mu_h)$$

Let,

$$n_1 = p_1 = 2 \times 10^{19} \text{ per m}^3, \ \mu_e = 0.36 \text{ m}^2/\text{V}-\text{s}$$

$$\mu_h = 0.17 \text{ m}^2/\text{V}-\text{s} \text{ and } \quad e = 1.6 \times 10^{-19}\text{C}$$

$$\sigma = 2 \times 10^{-19} \times 1.6 \times 10^{-19}(0.36 + 0.17) = 1.69 \text{ S/m}$$

(b) P-type Germanium

Here, $\sigma_p = e(n_p\mu_e + p_p\mu_h)$

Suppose, we add 10^{22} atoms/m^3 of indium and that $n_1 = 2 \times 10^{19}$ charge carriers (either electrons or holes) per m^3.

Then, $n_n p_p = n_1^2 = (2 \times 10^{19})^2 = 4 \times 10^{38}$

and $p_p - n_p = 10^{20}$

$$\therefore \ p_p - 4 \times 10^{38}/p_p = 10^{20}$$

or $p_p^2 - 10^{20} p_p - 4 \times 10^{38} = 0$

Solving the above quadratic equation and taking positive value only,

$$p_p = 1.04 \times 10^{20} \text{ and } n_p = 0.04 \times 10^{20}$$

$$\sigma_p = 1.6 \times 10^{-19}(0.04 \times 10^{20} \times 0.36 + 1.04 \times 10^{20} \times 0.17) = 3.1 \text{ S/m}$$

It is seen that conductivity is almost doubled.

Example : *Mobilities of electrons and holes in a sample of intrinsic germanium at room temperature are 0.36 m²/ volt-second and 0.17 m²/volt-second respectively. If the electron and hole densities are each equal to 2.5 × 10¹⁹ per m³, calculate the conductivity.*

Solution : The conductivity of an intrinsic conductor is

$$\sigma_1 = n_1 e(\mu_e + \mu_h)\text{S/m}$$

$$n_1 = 2.5 \times 10^{19}/m^3; \ e = 1.6 \times 10^{-19}\text{C},$$

$$\mu_e = 0.36 \ m^2/\text{V-s}; \ \mu_h = 0.17 \text{ m}^2/\text{V-s}.$$

$$\therefore \quad \sigma_1 = 2.5 \times 10^{19} \times 1.6 \times 10^{-19}(0.36 + 0.17)$$

= 2.12 S/m.

Example : *Calculate donor concentration in N-type germanium having resistivity of 100 Ω-m. Derive the formula you use. Take e = 1.6 × 10⁻¹⁹ C,* $\mu_e = 0.36 \ m^2V^{-1}s^{-1}$.

Solution : As we know that

$$\rho_n = 1/N_d e\mu_e$$

$\therefore$ $100 = 1/N_d \times 1.6 \times 10^{-19} \times 0.36$;
$N_d = 1.74 \times 10^{17}$ atoms/m^3

Example : *A p.d. of 10 V is applied longitudinally to a rectangular specimen of intrinsic germanium of length 2.5 cm, width 0.4 cm and thickness 0.15 cm. Calculate at room temperature*

(i) electron and hole drift velocities

(ii) the conductivity of intrinsic Ge if intrinsic carrier density is 2.5 × 10¹⁹/m³.

(iii) the total current.

Given,

$\mu_e = 0.38 m^2V^{-1}s^{-1}$, $\mu_h = 0.18\ m^2V^{-1}s^{-1}$,

$k = 1.38 \times 10^{-23} J\ \deg^{-1}$, $e = 1.6 \times 10^{-19} C$

Solution :

(i) $v_e = \mu_e E = 0.38 \times (10/2.5\times10^{-2}) = \mathbf{152\,m/s}$

$v_h = \mu_h E = 0.18 \times (10/2.5\times10^{-2}) = \mathbf{72\ m/s}$

(ii) $\sigma_1 = n_1 e(\mu_e + \mu_h) = 2.5\times10^{19} \times 1.6\times10^{-19}$ $(0.38+0.18) = \mathbf{2.24\ S/m}$

(iii) $I = \sigma_1 EA = 2.24 \times (10/2.5\times10^{-2})$ $(0.4\times0.15\times10^{-4}) = \mathbf{5.38\ mA.}$

Example : *A specimen of pure germanium at 300°K has a density of charge carriers of 2.5 × 10*^{19}*/m*^{3}*. It is doped with donor impurity atoms at the rate of one impurity atom for every 10*^{6} *atoms of germanium. All impurity atoms may be supposed to be ionized. The density of germanium atoms is 4.2 × 10*^{28} *atoms/m*^{3}*.*

Find the resistivity of the doped germanium if electron mobility is 0.36 m^{2}*/V-s.*

Solution : Density of added impurity atoms is $N_d = 4.2 \times 10^{28}/10^6 = 4.2 \times 10^{12}$ atoms/m^3

As seen, it is very large as compared to the intrinsic charge carrier density of 2.5×10^{19}/m^3 which will, therefore, be neglected.

$$\sigma_n = N_d e\mu_e = 4.2\times10^{21} \times 1.6\times10^{-19} \times 0.36 = 2.24\times10^3 \text{S/m}$$

$$\sigma_n = I/\sigma_n = 1/2.42\times10^3 = \mathbf{0.413\times10^{-3}\ \Omega\text{-m.}}$$

DRIFT

Directed motion of charge carriers in semiconductors occurs through two mechanisms :

(i) charge drift under the influence of applied electric field and

(ii) diffusion of charge from a region of high charge density to one of low charge density.

Consider the drift phenomenon first. When no electric field is applied to the semiconductor which is above 0°K, the conduction electrons (as well as holes) move within the crystal with random motion and repeatedly collide with each other and the fixed ions. Due to randomness of their motion, the net average velocity of these charge carriers in any given direction is zero. Hence, no current exists in the crystal under this condition of no field.

Now, consider the case when an electric field is applied to the crystal. Under the influence of this field, the charge carriers attain a directed motion which is superimposed on their random thermal motion. This results in a net average velocity called *drift* velocity in the direction of the applied electric field. Of course, electrons and holes move in opposite directions but because of their opposite charges, both produce current in the same direction. In extrinsic semiconductors, this current is essentially a majority carrier flow.

The drift velocity is proportional to electric field strength E, the constant of proportionality being called mobility μ. The exact relation between the two is $v = \mu E$.

Let us find the value of drift current in a semiconductor.

(i) current density due to electron drift is

$J_e = e\mu_e nE$ where μ_e is electron mobility, n is electron density and E is the electric field strength.

(ii) current density due to hole drift is

$J_h = e\mu_h pE$ where p is hole density.

Total current density due to electron and hole drift is

$$J = J_e + J_h = e\mu_e nE + e\mu_e npE = e(n\mu_e + p\mu_h)E$$
$$= e\mu_h(p+bn)E \quad \text{where } b = \mu_e/\mu_h$$

DIFFUSION

It is gradual flow of charge from a region of high density to a region of low density. It is a force-free process based on non-uniform distribution of charge carriers in a semiconductor crystal. It leads to an electric current without the benefit of an applied field. This flow or diffusion or carriers is proportional to the carrier density gradient, the constant of proportionality being called diffusion constant or diffusion coefficient D which as a unit of m^2/s.

Current density due to hole diffusion is $J_h = -eD_h\, dp/dx$. Similarly, current density due to electron diffusion is $J_e = eD_e\, dn/dx$.

where

D_e, D_h = electron and hole diffusion constants respectively

dn/dx = density gradient of electrons

dp/dx = density gradient of holes.

It is obvious that diffusion depends on charge inhomogeneity or on the presence of a space gradient of charge density. It can occur in regions free of electric field. On the other hand, drift current is a function of both electric field and charge density.

Incidentally, it may be noted that, generally, diffusion leads to redistribution of charges which further results in the development of potential difference between different parts of the semiconductor. The electric field due to this potential difference between different parts of the semiconductor. The electric field due to this potential difference sets up drift current in opposition to diffusion current. Find equilibrium is achieved when the potential difference developed becomes sufficiently large so as to create a drift current equal and opposite to the diffusion current thus resulting in zero net flow of current.

Combined Drift and Diffusion Currents

In semiconductors, drift and diffusion processes may be present simultaneously. The expressions for total electron and hole densities become,

$$J_e = e\mu_e nE + eD_e\, dn/dx \text{ A/m}^2$$

$$\text{and } J_h = e\mu_h pE - eD_h\, dp/dx \text{ A/m}^2$$

Relation Between D and μ

Both diffusion constant and mobility are statistical thermodynamic phenomena and are related to each other by the following equation

$$\mu_\varepsilon = \frac{e}{kT} D_e \text{ and } \mu_h = \frac{e}{kT} D_h$$

$$\text{or } \frac{D_e}{\mu_e} = \frac{D_h}{\mu_h} = \frac{kT}{e} = \frac{T}{11{,}600}$$

This relationship is known as *Einstein's equation*

At $t = 23°C$, $T = 300°K$, hence $D/\mu = 300/11{,}600$

or $\mu = 39$ D

Recombination

Apart from drift and diffusion, a third phenomenon which occurs in semiconductors is called recombination that results from the collision of an electron with a hole.

This process is essentially the return of a free conduction electron to the valence band and is accompanied by the emission of energy. Obviously, the recombination rate is directly proportional to the carrier concentration for the simple reason that larger the number of carriers, the more likely is the occurrence of electron-hole recombination. This phenomenon is important in describing minority carrier flow.

As is well-known, in a semiconductor, thermal generation of electron-hole pairs also takes place continuously. Hence, there is net *recombination rate* given by the difference between the recombination and generation rates.

Carrier Life Time

It is defined as the time for which, on an average, a charge carrier will exist before recombination with a carrier of opposite charge. Its value varies from nanoseconds (10^{-9}) to hundreds of microseconds (μs) and depends on temperature and impurity concentration in the semiconductor material.

Total Carrier Flow

The total carrier flow in a semiconductor is the sum of the three flow phenomena discussed above. Each type of carrier has to be treated separately and the number of electrons or holes leaving the sample being accounted for by drift or diffusion or net recombination. The current in the semiconductor is then the sum of the electron and hole currents.

P-N Junction

It is possible to manufacture a single piece of a semiconductor material half of which is doped by P-type impurity and the other half by N-type impurity as shown in Fig. 46. The plane dividing the two zones is called *junction*. Theoretically, junction plane is assumed to lie where the density of donors and acceptors is equal. The P-N junction is fundamental to the operation of diodes, transistors and other solid-state devices.

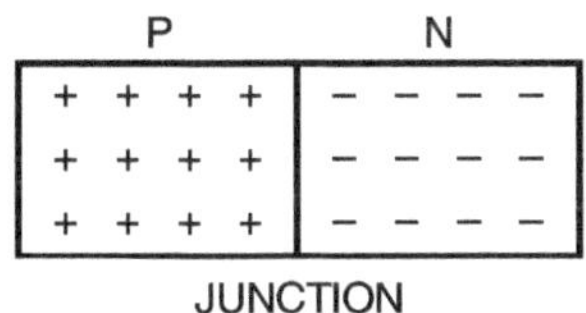

Fig. 46

Let us see if anything unusual happens at the junction. It is found that following three phenomena take place :

1. a thin *depletion layer* or region (also called space-charge region or transition region) is established on both sides of the junction and is so called because it is depleted of *free charge carriers*. Its thickness is about 10^{-6} m.
2. a barrier potential or junction potential is developed across the junction.
3. the presence of depletion layer gives rise to junction and diffusion capacitances.

Formation of Depletion Layer

Suppose that a junction has just been formed. At that instant belonging with the P-region and electrons in the N-region. However, there is greater concentration of holes in P-region than in N-region (where they exist as minority carriers). Similarly, concentration of electrons is greater in N-region than in P-region (where they exist as minority carriers). This concentration difference establishes density gradient across the junction resulting in carrier diffusion. Holes diffuse from P- to N-region and electrons from N- to P-region and terminate their existence by recombination [Fig. 47 (a)]. This recombination of free and mobile electrons and holes produces the narrow region at the junction called depletion layer. It is so named because this region is devoid of (or depleted of) *free and mobile charge carriers like electrons and holes*—there being present only positive ions which are not free to move.

It might seem from above that eventually all the holes from the P-side would diffuse to the N-side and all the electrons from the N-side would diffuse to the P-side but this does not occur due to the formation of ions on the two sides of the junction. The impurity atoms which provide these migratory electrons and holes are left behind in an ionized state bearing a charge which is opposite to that of the departed carrier. Also, these impurity ions, just like germanium atoms, are fixed in their positions in the crystal lattice in the P- and N-regions of the diode. Hence, as shown in Fig. 47 (b), they from parallel rows or 'plates' of opposite charges facing each other across the depletion layer. Obviously, row of *fixed* positive ions in the N-region is produced by the migration of electrons from the N- to P-region. Similarly, the row of *fixed* negative ions in the P-region is produced by the migration of holes from the P- to N-region.

If a majority carrier (either an electron or a hole) tries to cross into depletion layer, it can meet either of the following two fates :

(i) either it can be trapped or captured by the row of fixed impurity ions of opposite sign which guard its own region. For example, a hole trying to approach the depletion layer may be neutralized by the row of fixed negative ions situated in the P-region itself at the edge of the depletion layer. So will be the case with the electron trying to approach the depletion layer from N-region [Fig. 47 (c)]. or

(ii) it may succeed in entering the depletion layer where it will be repelled by the row of similarly-

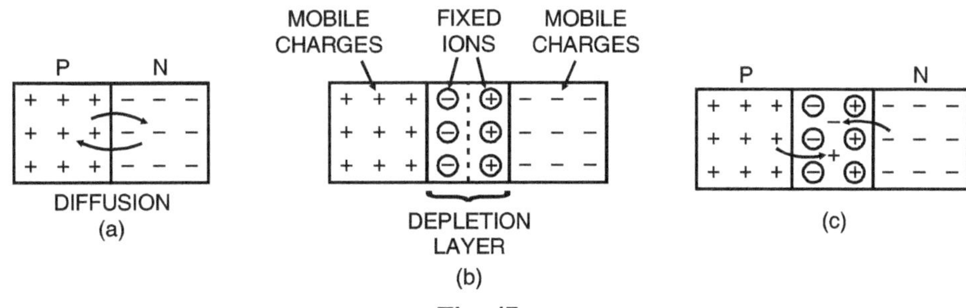

Fig. 47

charged impurity ions guarding the other region. But its life will be cut short by recombination with a majority carrier of opposite sign which has similarly entered the depletion layer from the other half of the diode.

Ultimately, an equilibrium condition is reached when depletion layer has widened to such an extent that no electrons or holes can cross the P-N junction.

Junction of Barrier Voltage

Even though depletion layer is cleared of charge carriers, it has oppositely charged fixed rows of ions on its two sides. Because of this charge separation, an electric potential difference V_B is established across the junction even when the *junction is externally isolated* (Fig. 48). It is known as *junction* or *barrier potential.* It stops further flow of carrier across the junction unless supplied by energy from an external source. At room temperature of 300°K, V_B is about 0.3 V for Ge and 0.7 V for Si.

The value of barrier voltage is given by $V_B = V_T \log_e N_a N_d/n_1^2$ where N_a, N_d, n_1^2 and V_T have the meanings explained. The value of V_T at room temperature of 300°K is given by

$$V_T = V_{300} = \frac{kT}{e} = \frac{1.38 \times 10^{-23} \times 300}{1.6 \times 10^{-10}} = 26 \text{ mV}$$

$$\therefore \quad V_B = 26 \log_e(N_a N_d / n_1^2)\text{mV}$$

Barrier voltage depends on doping density, electronic charge and temperature. For a given junction, the first two factors are constant, thus making V_B dependent on temperature. With increase in temperature, more minority charge carriers are produced, leading to their increased drift across the junction. As a result, equilibrium occurs at a lower barrier potential. It is found that both for Ge and Si, V_B decreases by about 2 mV/°C.

$$\therefore \quad \Delta V_B = -0.002 \, \Delta t$$

where Δt is the rise in temperature in °C.

The strong field set up by V_B causes drift of carriers through depletion layer. As seen from Fig. 49 (b), under the influence of this field, *holes drift from N-to-P-region and electron from P- to N-region.* This drift current must be equal and opposite to the diffusion current [Fig. 49 (a)] because under condition of equilibrium and with no external supply, net current through the crystal is zero.

To summarize the main points, we have :

1. as soon as *P-N* junction is formed, free electrons and holes start diffusion across the junction and recombining.
2. their recombination leads to the appearance of a depletion layer across the junction which contains no *mobile* carriers but only immobile ions.
3. these immobile ions set up a barrier potential and hence an electric field which sets up drift current that is equal and opposite to the diffusion current when final equilibrium is reached.

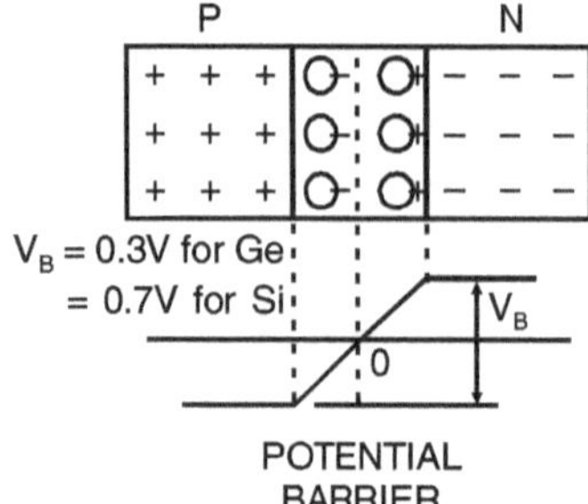

Fig. 48

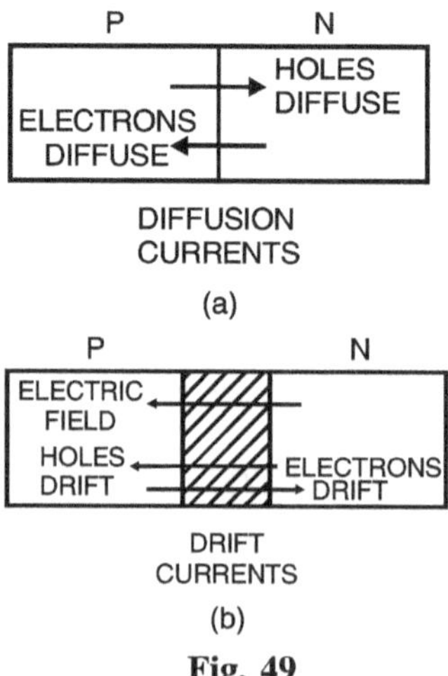

Fig. 49

Reverse V/I Characteristic

As said earlier, the reverse saturation current is also referred to as *leakage current* of the P-N junction. Fig. 51 shows V/I characteristics of a reverse-biased P-N junction. It is seen that as reverse voltage is increased from zero, the reverse current quickly rises

Energy Band Diagram of a P-N Junction

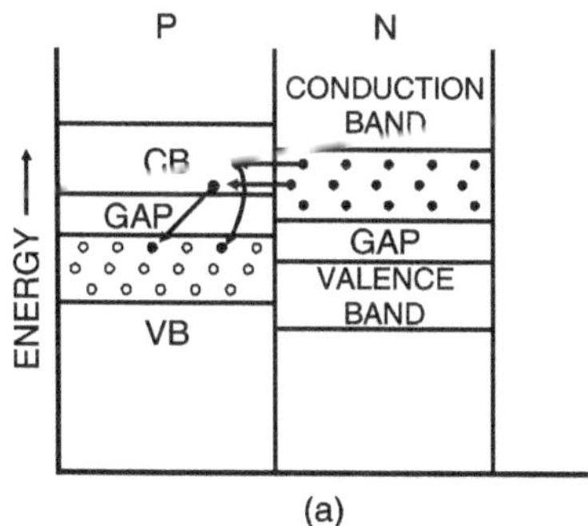

(a)

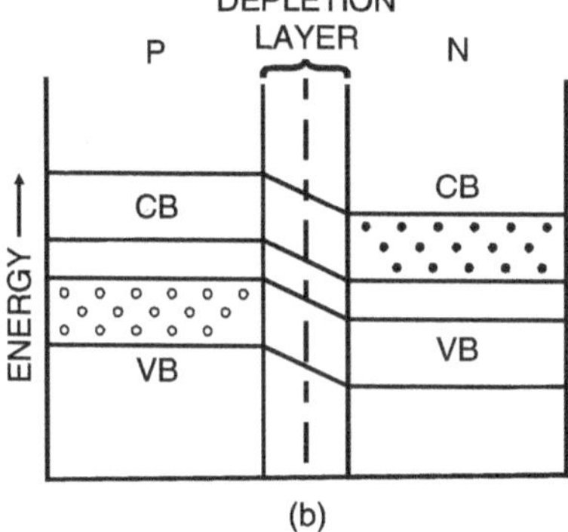

(b)

Fig. 50

to its maximum or saturation value. Keeping temperature constant, as the reverse voltage is increased, I_0 is found to increase only slightly. This slight increase is due to the impurities on the surface of the semiconductor which behaves as a resistor and hence obeys Ohm's law. This gives rise to a very small current called *surface leakage current.* Unlike

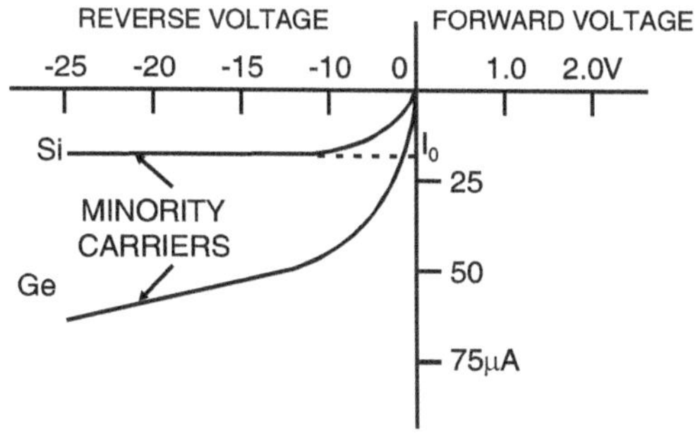

Fig. 51

the main leakage (or saturation) current, this surface leakage current is independent of temperature but depends on the magnitude of the reverse voltage.

A reverse-biased junction can be represented by a very large resistance. As seen from Fig. 51 in the case of Si, for a reverse voltage of about 15V, I_o = 10 μA. Hence, reverse resistance is R_R = 15 V/10 μA = 1.5 M

Hall Effect

If a specimen (whether of a metal or a semiconductor) carrying a current I is placed in a *transverse* magnetic field of flux density B, an electric field is developed along a direction perpendicular to both B and I. This phenomenon is known as *Hall effect* and is used for the following purposes :

1. to determine whether a semiconductor is of *N*-type or *P*-type
2. to find carrier concentration
3. to measure the conductivity of the material
4. to find carrier mobility
5. to detect and measure magnetic fields one million times smaller than that of earth with the help of Hall-effect magnetometers.

As shown in Fig. 52 (a), a current I is flowing through the semiconductor in the direction MN under the influence of an external applied electric field E_L. Obviously, electrons comprising this current move along NM with a velocity of v_e. Consider one such electron shown in the figure. The direction of the force exerted on it by the magnetic field B can be found by using Fleming's left-hand rule and is as shown. The magnitude of force is $Be\ v_e$.

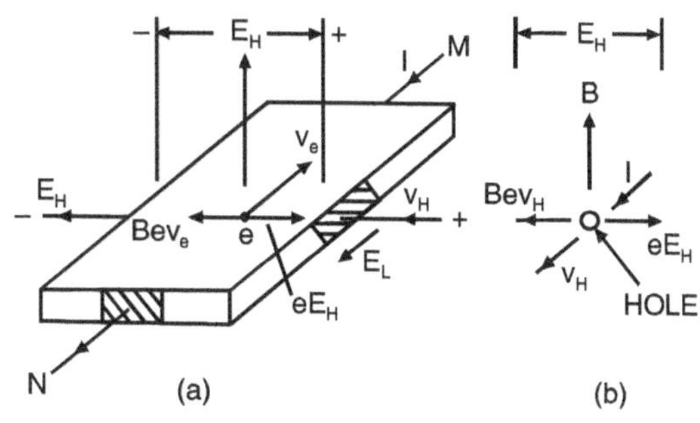

Fig. 52

Under the influence of this force, electrons will tend to crowd towards the left side of semiconductor. This collection of electrons to one side gives rise to electric potential difference V_H (called Hall voltage) and hence to an electric field E_H. This field begins to prevent additional electrons from arriving there. Force exerted by this field on the electrons in eE_H. Equilibrium is established when the two oppositely-directed magnetic and electric forces acting on the electron become equal in magnitude.

i.e. $Bev_e = eE_H$ or $Bv_e = E_H$ ... (i)

Also $v_e = \mu_e E_L$

$\therefore E_L = E_H$

or $\mu_e B = E_H / E_L$

Hence, knowing E_L and B and measuring E_H, we can find electron mobility.

Now, current density $J = I/A = nev_e A/A = nev_e$,

Substituting the value of v_e from (i) above, we get

$J = neE_H/H$

or $n = JB/eE_H$

This equation may be used to find electron density.

Now, Hall coefficient, R_H = 1/charge density = 1/ ne or $R_H = E_H/JB$

As shown in previous

$$\mu_e = 1/ne\rho = R_H/\rho$$

where ρ is the resistivity of the semiconductor.

It may be noted that the above treatment is equally applicable to *P*-type semiconductors where current flow is made up of hole movement. If holes were to move in the same direction as that of the electrons in Fig. 52 (a), then polarity of Hall voltage E_H is same as above in Fig. 52 (b) in fact it was the observation of the polarity.

EXERCISE

1. In a push-pull amplifier, a speaker can be put in place of load resistance because
   A. DC current will flow through it
   B. No dc current will flow through it
   C. Speaker is also a resistance
   D. None of these
2. A class A amplifier is the one in which
   A. Base is biased to cut-off
   B. I_C flows most of the time
   C. I_E flows all the time
   D. None of the above
3. The combination of two 0.02 μF 500 V capacitors in parallel has capacitance and breakdown ratings of
   A. 0.01 μF, 1000 V B. 0.02 μF, 500 V
   C. 0.04 μF, 500 V D. 0.04 μF, 1000 V.
4. Oscillators have
   A. No feedback
   B. Negative feedback
   C. Positive feedback
   D. Either positive or negative feedback.
5. If the residual resistivity is high the temperature coefficient of the metal will be
   A. High
   B. Small
   C. Medium
   D. Temperature coefficient has no relation with thermal resistivity.
6. The voltage gain of an amplifier is 100 and the current gain is 2. The power gain of the amplifier will be
   A. 40000 B. 400
   C. 200 D. none
7. Given that ρ = resistivity, l = length of wire, a = cross-sectional area of the wire, ω = operating frequency, F = form factor, n = number of turns, d = coil diameter and r = resistance value, match List I with List II and select the correct answer

List I		List II
(a) Value of inductor	1.	$\rho . l/a$
(b) Power factor	2.	$\omega L/r$
(c) Q-factor	3.	$r/\omega L$
(d) Value of resistance	4.	$F . n^2 . d$

Codes:	*(a)*	*(b)*	*(c)*	*(d)*
A.	4	2	3	1
B.	4	3	2	1
C.	1	2	3	4
D.	2	3	2	4

8. A 10-bit D/A converter provides an analog output which has a maximum value of 10.23 volts. The resolution is

A. 40 mV B. 20 mV
C. 15 mV D. 25 mV

9. The electron and hole concentration in an N-type semiconductor is $10^{16}/cm^3$ and $10^{16}/cm^3$, if the mobilities of electrons and holes are 400 and 600 cm^2/V-sec respectively, the conductivity of the material approximately is

A. $10^{22} \times e$ B. $10^{19} \times e$
C. $4 \times 10^{18} \times e$ D. $6 \times 10^{8} \times e$

10. A 0 to 200 V dc moving coil voltmeter has a guaranteed accuracy of 0.75% of full-scale reading. The voltage measured by the instrument is 100 V. The limiting error is

A. 3% B. 2%
C. 1.5% D. 0.75%

11. The signal handling capacity of an amplifier will be high when the operating point is selected

A. Close to saturation region
B. Close to cut-off region
C. At the extremities of the active region
D. In the middle of the active region

12. Which of the following methods cannot be used to increase the bandwidth for cascaded RF amplifier stages?

A. Wave traps in each stage
B. Stagger tuning
C. A shunt damping resistor across each tuned circuit
D. Double tuned transformers with tighter coupling.

13. The cut in voltage $V\gamma$.

A. Is the same as the barrier potential
B. It is the same for germanium and silicon
C. Is the forward voltage below which the current is very small
D. Is in the region of reverse bias.

14. The wavelength of light emitted by a GaAs lase is 8670×10^{-10} m. Given Planck's constant = 6.626×10^{-34} Js velocity of light = 2.998×10^{8} ms^{-1} and 1 eV = 1.602×10^{-19} J, the energy gap in GaAs is

A. 0.18 eV B. 0.7 eV
C. 1.43 eV D. 2.289 eV

15. A high fidelity audio amplifier has a passband as

A. 20 Hz to 20 kHz B. 0 Hz to 10 kHz
C. 1 kHz to 40 kHz D. 50 Hz to 10 kHz

16. The equation that does not give the power dissipated in a resistor is :

A. $P = I^2 R$ B. $P = V^2/R$
C. $P = VI$ D. $P = R/V^2$

17. No eddy current and hysteresis losses occur in

A. Electrostatic instruments
B. PMMC type instruments
C. Moving iron instruments
D. Electrodynamometer instruments

18. In a measurement system the transducer is the

A. Input element
B. Output element
C. Processing device
D. Signal-conditioning device

19. Which of the instrument is used both for a.c. and d.c.

A. moving iron (attraction type) ammeter
B. moving iron (repulsion type) voltmeter
C. permanent magnet type ammeter
D. induction type ammeter

20. In regard to Wien bridge oscillator

A. The bridge is unbalanced so that the amplitude of oscillations can be stabilised
B. At the frequency of oscillations the bridge is exactly balanced.
C. In the bridge arms containing only resistors, one of the resistors is temperature sensitive to stabilise the amplitude of oscillations
D. At the frequency of oscillations the gain is exactly 3.

21. Electronic null detectors have

A. Low sensitivity
B. High sensitivity
C. Excellent sensitivity, rugged and more expensive
D. Excellent sensitive, rugged and fairly expensive

22. The output voltage V out for the circuit shown

is given by

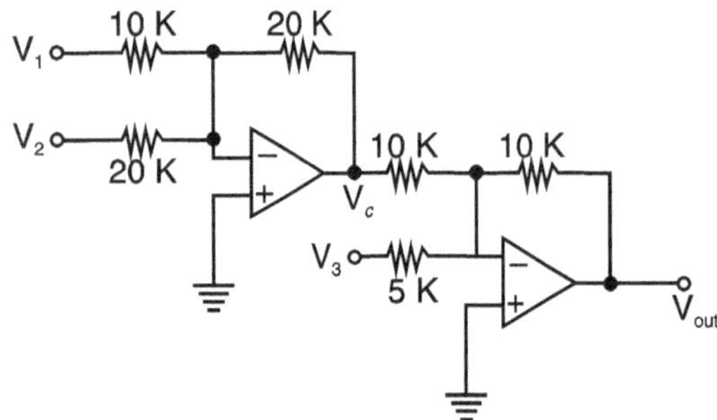

A. $2v_1 + v_2 - 2v_3$ B. $2(v_1 + v_2) + v_3$
C. $2v_1 + v_2 + v_3$ D. $2v_1 + 2v_2 - v_3$

23. The power loss in wattmeter and energy meter instruments is more than the power loss in all ammeter and voltmeter instruments and is of the order of

A. 1 to 2 W B. 0.5 to 3 W
C. 2 to 5 W D. 4 to 10 W

24. Match List I with List II and select the correct answer

List I	*List II*
(a) Wien Bridge	1. RF oscillator : two inductances and one capacitance on the reactance network
(b) Colpitts	2. LC oscillator for RF frequency : three capacitances and one inductance in the reactance network
(c) Hartley	3. RC oscillator for audio frequency applications
(d) Clapp	4. RF oscillator, two capacitances and one inductance as the reactance network

Codes:	*(a)*	*(b)*	*(c)*	*(d)*
A.	2	4	1	3
B.	2	1	4	3
C.	3	4	1	2
D.	3	1	4	2

25. A saw-tooth generator is built using a UJT with a constant-current-charged capacitor C of 1 μF driving the emitter. If the voltage V_{BB} is 15V and the capacitor charging current is 1 mA, then the period of the saw-tooth is approximately

A. 10 ms B. 1 ms
C. 0.1 ms D. 0.01 ms

26. Usually the instruments are designed (with out range extending arrangement) for a voltage of the order of

A. 45 to 150 μV B. 45 to 150 mV
C. 45 to 250 V D. None of these.

27. If *m* identical coils of e.m.f. V and internal resistance *l* are connected in series, they produce a current through an external resistance R equal to

A. m V/R + l B. m V/R
C. V/R + l D. m V/R + $m.l.$

28. For the measurement of very high resistance (insulation resistance) the instrument used is

A. An avometer B. A multimeter
C. A megger D. An ohm meter

29. The temperature coefficient of resistance for thermistors is

A. Low and negative B. Low and positive
C. High and negative D. High and positive

30. Treble tone of loud speakers corresponds to

A. High frequencies B. Low frequencies
C. High amplitudes D. Low amplitudes.

31. The signal(s) applied to the yoke of a picture tube from the vertical amplifier output include(s)

A. a saw-tooth voltage
B. a saw-tooth current
C. amplified vertical sync pulses
D. blanking pulses

32. The drain-source output characteristics of an N-channel depletion FET has

A. $I_{DS} = 0$ at $V_{GS} = 0$
B. I_{DS} = positive maximum at $V_{GS} = 0$
C. I_{DS} = negative maximum at $V_{Gs} = 0$
D. I_{Ds} independent of V_{GS}.

33. Which of the following is more sensitive

A. PIN diode B. neither (A) or (C)
C. APD D. either (A) or (C)

34. The amplifiers which are inserted at intervals to amplify the signal and compensate for transmission loss on the cable are called

A. line amplifiers B. equalizing amplifiers
C. compandors D. repeaters

35. A class-B push-pull power amplifier uses two transistors each having a power dissipation capacity of 2 watts. The maximum power that the circuit can deliver to the load without extending the power dissipation capacity of the transistors is

A. 1 watt B. 2 watts

C. 5 watts D. 10 watts

36. Which of the semiconductor can be used for fabricate a LED

A. Si B. Ge

C. GaAs D. none of these

37. Consider the following statements : Drift in direct coupled amplifiers may be reduced effectively by use of

1. a diode as a thermal compensating element
2. another active device, making it a differential amplifier
3. bypass capacitors
4. choppers

Of these statements

A. 1 and 3 are correct

B. 1, 3 and 4 are correct

C. 1 and 4 are correct

D. 1, 2 and 4 are correct

38. Schottky barrier diode can be used as

A. low noise amplifier

B. variable capacitance device

C. power supply rectifier

D. low level detector

39. The filter shown uses an ideal OP-AMP. It represents a

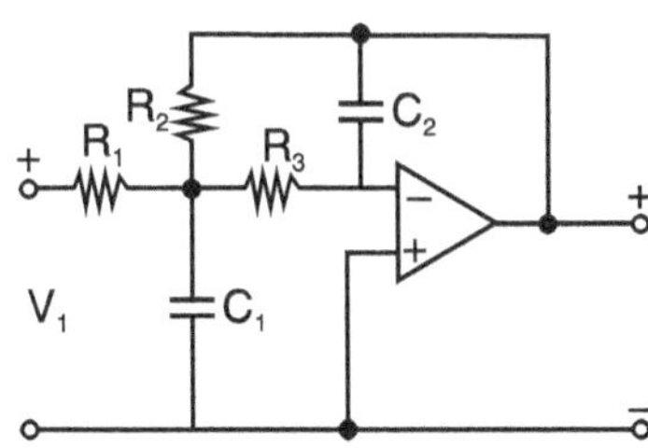

A. low pass filter B. high pass filter

C. band pass filter D. band stop filter

40. If a transistor is operating with both of its junctions forward biased, but with the collector-base forward bias greater than the emitter-base forward bias, then it is operating in the

A. forward active mode

B. reverse saturation mode

C. reverse active mode

D. forward saturation mode

41. For a high pass RC circuit, when subjected to a unit step input voltage, the voltage across the capacitor will be

A. $1 - e^{-t/RC}$ B. $e^{-t/RC}$

C. $e^{+t/RC}$ D. 1

42. For an abrupt junction varactor diode, the dependence of device capacitance (C) on applied reverse bias (V) given by

A. $C \propto V^{1/3}$ B. $C \propto V^{-1/3}$

C. $C \propto V^{1/2}$ D. $C \propto V^{-1/2}$

43. The characteristic equation of an armature controlled d.c. motor is of :

A. Zero order equation

B. First order equation

C. Second order equation

D. Third order equation

44. Typical values of h parameters, at about 1 m A collector current, for small signal audio transistors in the CE configuration

	h_{ie}	h_{re}	h_{fe}	h_{ve}
A.	100 Ω	10^{-1}	50	1 mΩ
B.	5 kΩ	10^{-4}	200	20 μΩ
C.	5 kΩ	0	20	3 mΩ
D.	100 kΩ	10^{2}	100	10 mΩ

45. Two non-inverting amplifiers, one having a unity gain and the other having a gain of twenty, are made using identical operational amplifiers. As compared to the unity gain amplifier, the amplifier with gain twenty has

A. less negative feedback

B. greater input impedance

C. less bandwidth

D. all of the above

46. A silicon sample is uniformly doped with 10^{16} phosphorous cm^3 and 2×10^{16} boron atoms/cm^3. If all the dopants are fully ionised, the material is

A. N-type with carrier concentration of $10^{16}/cm^3$

B. P-type with carrier concentration of $10^{16}/cm^3$

C. P-type with carrier concentration of $2 \times 10^{16}/cm^3$

D. N-type with a carrier concentration of $2 \times 10^{16}/cm^3$

47. Match List I with List II and select the correct

answer

List I	List II
(a) LED	1. Optical scattering
(b) LCD	2. Glow discharge
(c) NIXIE	3. Electromagnetic radiation emitter

Codes:	(a)	(b)	(c)
A.	2	3	1
B.	3	1	2
C.	1	3	2
D.	3	2	1

48. A signaling system in which each letter of the alphabet is represented by a different symbol is not used because
  A. it would be too difficult for an operator to memorize
  B. it is redundant
  C. noise would introduce too many errors
  D. too many pulses per letter are required

49. In the wavelength range (11-16) μm, the principal optical source material is
  A. In GaAsBr B. In GaAsF
  C. In GaAsP D. None of these

50. In an intrinsic semiconductor, the mobility of electrons in the conduction band is
  A. zero
  B. less than the mobility of holes in the valence band
  C. equal to the mobility of holes in the valence band
  D. greater than mobility of holes in the valence band

51. Source of light for optical fibre is
  A. PIN diode B. Photodiode
  C. LED D. None of these

52. Which one of the following is true with regard to photoemission?
  A. Velocity of emitted electrons is dependent on light intensity
  B. Rate of photo emission is inversely proportional to light intensity
  C. Maximum velocity of eletrons increases with decreasing wavelength
  D. Both electrons and holes are produced

53. Which one of the following is correct of the biasing of a bipolar junction transistor in its active region?

	biasing	
	Emitter junction	Collector junction
A.	Forward	Forward
B.	Forward	Reverse
C.	Reverse	Forward
D.	Reverse	Reverse

54. A NPN transistor can be considered to be equivalent to two diodes as shown in

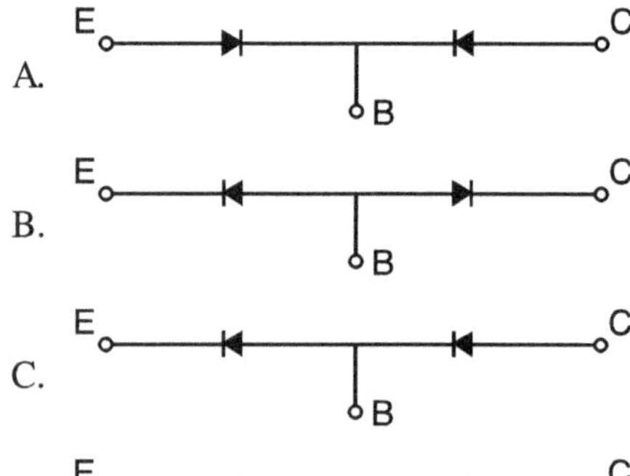

55. In a triode, tetrode or pentode, the plate current increases when the
  A. Control grid voltage is made more negative
  B. Control grid voltage is made less negative
  C. Plate voltage is made less positive
  D. Screen grid voltage is made less positive

56. With a 50 μA Movement a VOM has an input resistance of 6 MΩ on the d.c. voltage range of
  A. 3 B. 12
  C. 60 D. 300

57. Which of the given frequency meters is suitable to measure radio frequency
  A. Weston frequency meter
  B. Electrical resonance frequency meter
  C. Hetrodyne frequency meter
  D. Either (B) or (C)

58. Consider the following statements regarding the circuit given in the figure, where the output voltage is constant :
  1. v_{in} > the voltage at which the zener breaks down
  2. I_L < the difference between I and I_Z, the current at which the zener breaks down
  3. R_S < the zener nominal resistance

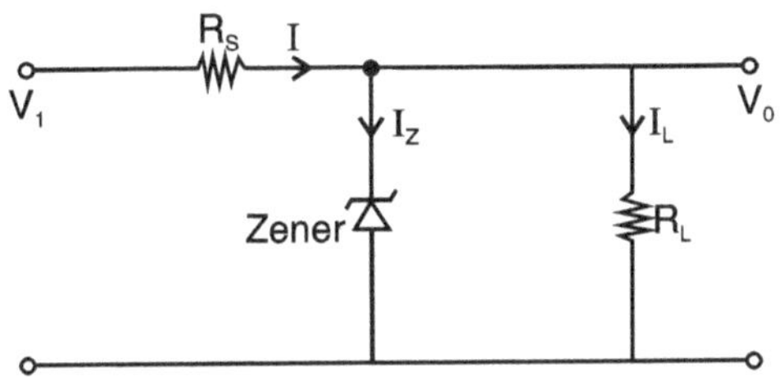

Of these statements

A. 1, 2 and 3 are correct B. 1 and 2 are correct

C. 1 and 3 are correct D. 2 and 3 are correct

59. Consider the circuit shown in Figure, diode used here has the V – I characteristic as shown, then the output wave-form v_0 is

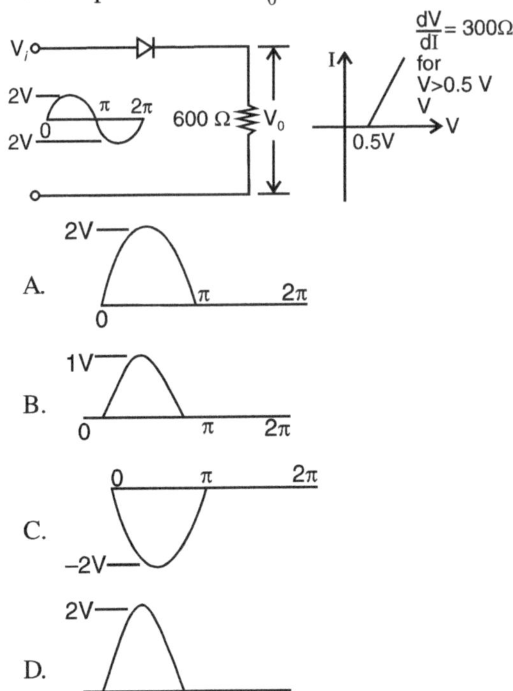

60. The multirange instruments generally have

A. Multicoil arrangement inside the meter

B. Multiple shunt or series resistances in side the meter

C. A dimension setting feature with the coil of the system

D. There is only one coil with a variable number of turns arrangement with it.

61. Consider the following statement regarding the circuit shown in Figure :

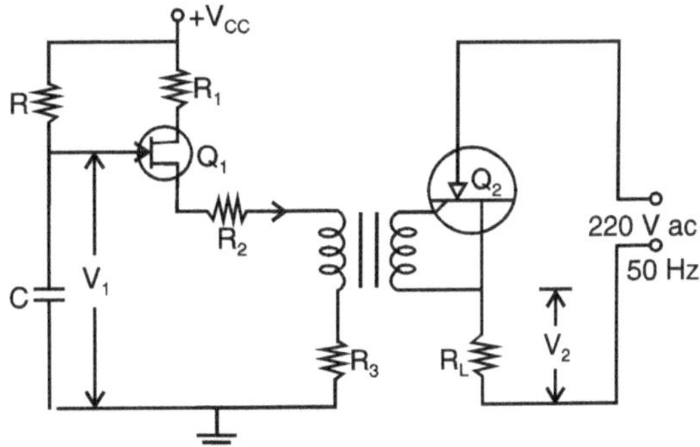

1. The waveform of V_1 is

2. The waveform of 'I' is

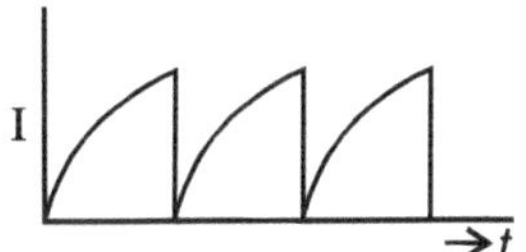

3. The waveform of V_2 is

Of these statements

A. 1 and 3 are correct

B. 1 and 2 are correct

C. 1 alone is correct

D. 2 and 3 are correct

62. The circuit shown is a

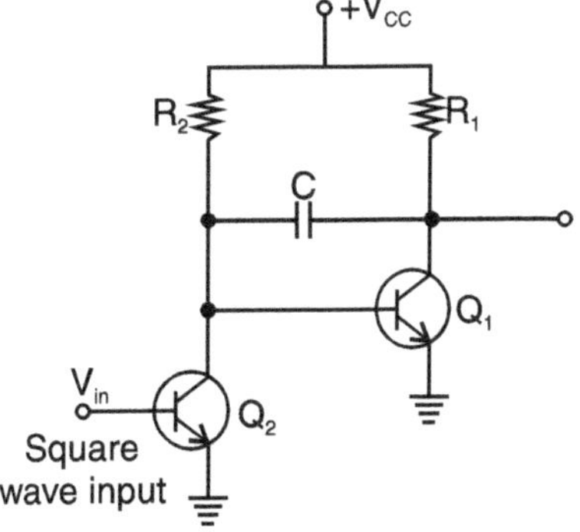

A. monostable multivibrator

B. frequency division circuit
C. Miller sweep circuit
D. bootstrap sweep circuit

63. For the circuit shown assuming ideal diodes, the output waveform V_0 will be

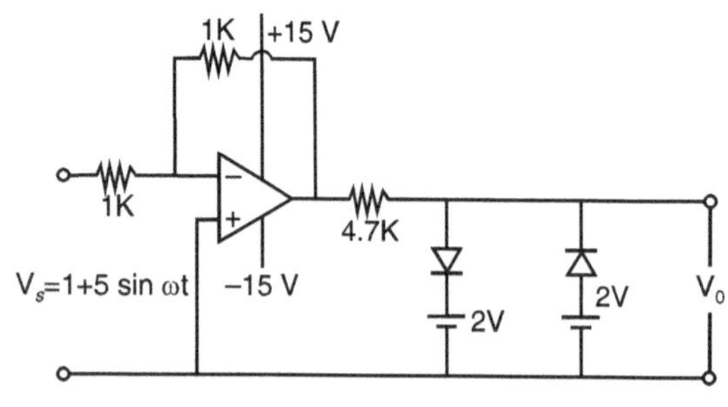

A. 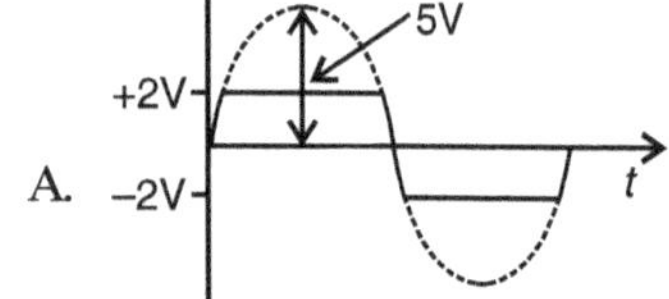

B. 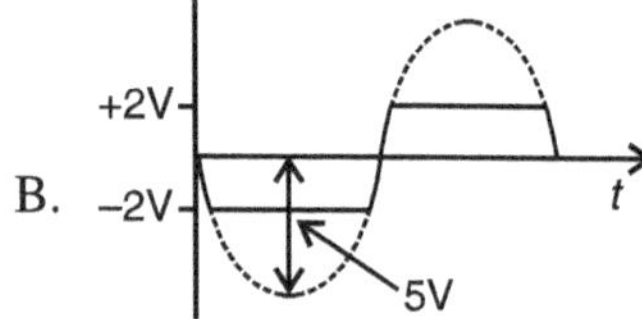

C. 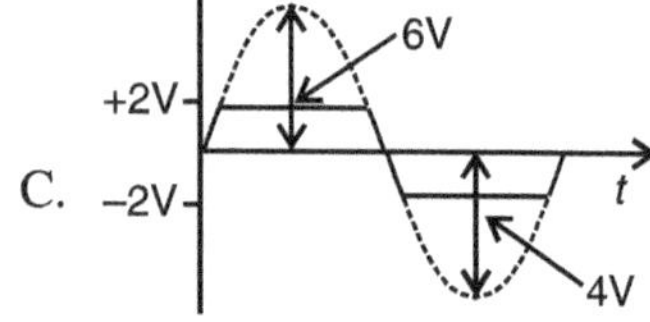

D. 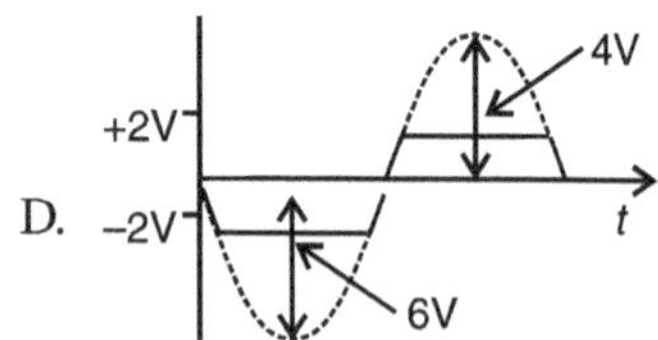

64. One of the following microwave diodes is suitable for very low-power oscillators only :

A. Tunnel B. LSA
C. Gunn D. IMPATT

65. The dielectric power factor of mica capacitor is

A. 0.001 B. 0.0001
C. 0.00001 D. 0.000001

66. Electrostatic instruments are normally used for

A. Low current measurments
B. High current measurements
C. Low voltage measurements
D. High voltage measurements

67. A second-order low-pass Butterworth active filter is shown in the given figure. What is the 3-dB cutoff frequency (ω_0) and the damping factor (K) of the filter?

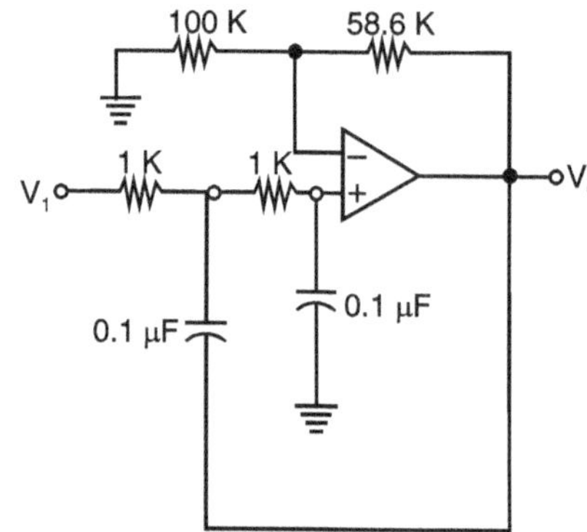

A. 10.06 kHz and 0.586
B. 1.6 kHz and 0.707
C. 16 kHz and 0.586
D. 10 kHz and 0.707

68. A Darlington stage is shown. If the trans-conductance of Q_1 is g_{m1} and that Q_2 is g_{m2}, then the overall transconductance g_m is given by

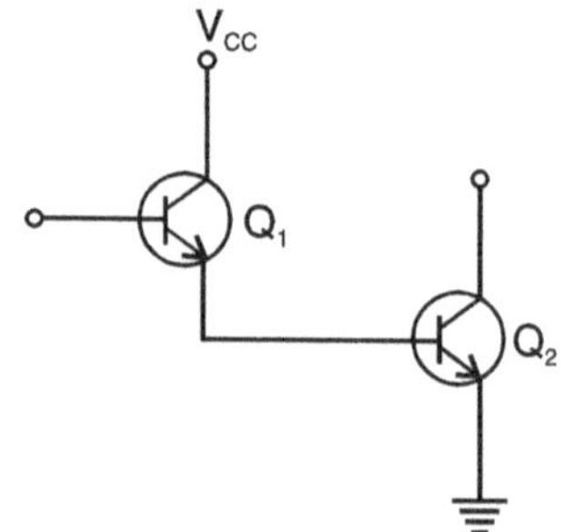

A. g_{ml} B. $0.5\ g_{ml}$
C. g_{m1} D. $0.5\ g_{m2}$

69. Which of the following statement is true

A. Alternating current and voltage wave forms do not in general, contain even harmonics
B. Alternating current and voltage waveforms, in general, do not contains odd harmonics
C. Alternating current and voltage waveforms, in general, contain even harmonics

D. Alternating current and voltage wave forms contain even and odd harmonics in general.

70. The output characteristic of FET is given in the figure.

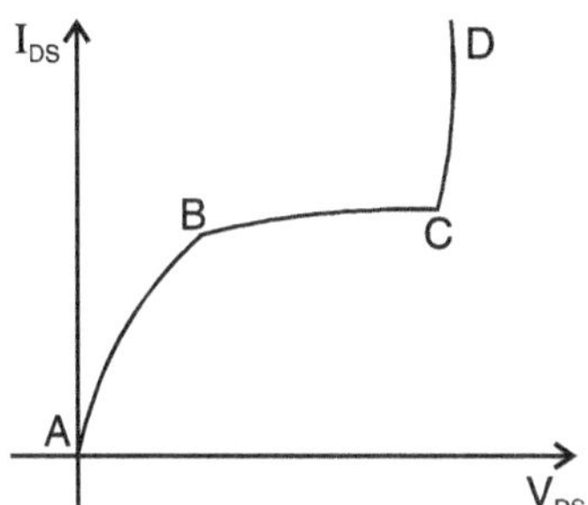

In which region is the device biased for small signal amplification?

A. AB B. BC
C. CD D. BD

71. Gold is often diffused into silicon PN junction devices to

A. increase the recombination rate
B. reduce the recombination rate
C. make silicon a direct gap semiconductor
D. make silicon a semi-metal

72. In the circuit shown N is a finite gain amplifier with a gain of k, a very large input impedance, and a very low output impedance. The input impedance of the feedback amplifier with the feedback impedance Z connected as shown will be

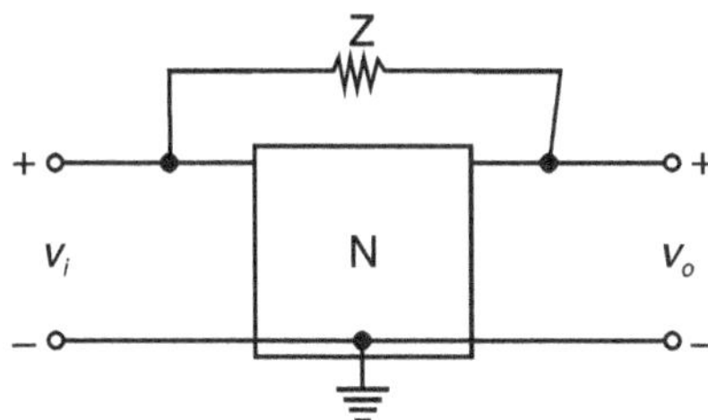

A. $Z(1 - 1/k)$ B. $Z(1 + k)$
C. $Z/(k - 1)$ D. $Z/(1 - k)$

73. The speed of a cassette tape for tap recorder is

A. 1 inch/sec B. $1\frac{7}{8}$ inch/sec
C. 5 cm/sec D. 10 cm/sec.

74. Match List I with List II and select the correct answer

List I	List II
(a) Tunnel diode oscillator	1. Produces high current pulses of short duration
(b) UJT oscillator	2. An LC oscillator used for generation of sine wave at RF
(c) Hartley oscillator	3. A negative resistance oscillator for MW frequency
(d) Blocking oscillator	4. Uses negative resistance property for the generation of sawtooth waveform

Codes:	*(a)*	*(b)*	*(c)*	*(d)*
A.	1	2	4	3
B.	3	2	1	4
C.	3	4	2	1
D.	4	3	1	2

75. A zener diode in the circuit shown has a knee current of 5 mA, and a maximum allowed power dissipation of 300 mW. What are the minimum and maximum load currents that can be drawn safely from the circuit, keeping the output voltage V_n constant at 6 V?

A. 10 mA, 180 mA B. 15 mA, 140 mA
C. 10 mA, 55 mA D. 60 mA, 180 mA

76. In a power supply a shorted input capacitor is likely to result in

A. Excessive hum
B. Reduced voltage output
C. No voltage output
D. None of the avove

77. The output voltage (v_0) of the circuit shown is given by

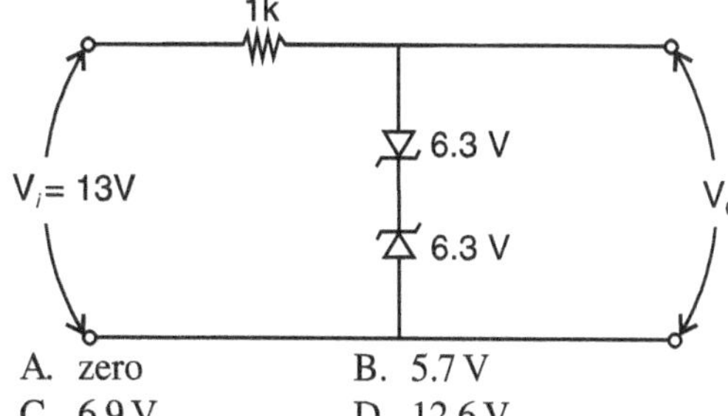

A. zero B. 5.7 V
C. 6.9 V D. 12.6 V

78. An amplifier has $R_i = 2$ K and $R_0 = 40$ K, $A_v = 90$, the amplifier is modified to provide 10%

negative voltage feedback in series with input. The value of R_{if} and R_{of} (i.e. with feedback) respectively are (in ohms)

A. 2 K and 40 K B. 20 K and 4 K
C. 0.2 K and 400 K D. 16 K and 5 K.

79. A cylindrical rod of magnetic material fits tightly when it is inserted into a long solenoid, half way through. The nature of the force on the rod and the type of material of the rod and listed below. Match List I with List II and select the correct answer

List I (Material)	**List II (Force)**
(a) Diamagnetic	1. Nil
(b) Paramagnetic	2. Repulsive
(c) Ferromagnetic	3. Attractive

Codes:	*(a)*	*(b)*	*(c)*
A.	1	3	3
B.	2	2	3
C.	2	3	3
D.	3	2	2

80. The OP-AMP cricuit shown can be used for

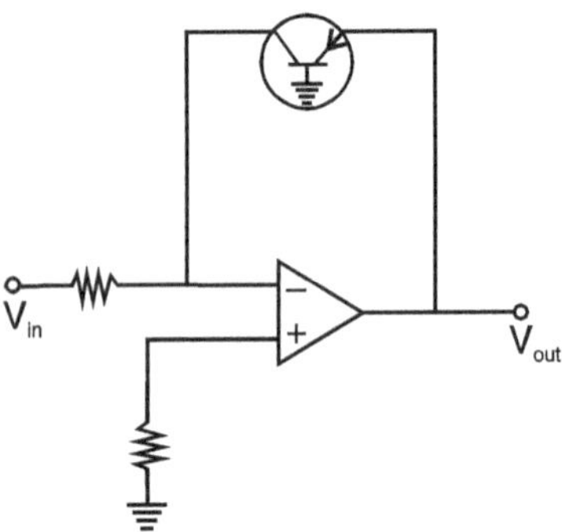

A. addition
B. subtraction
C. both addition and subtraction
D. multiplication

81. A 120 volt source connected across a resistor produces a 2 amp current. The power dissipated in the resistor is :

A. 240 watts B. 60 watts
C. 120 watts D. 30 watts

82. The Q factor of an inductor would be higher if it is made up to

A. Thinner wire B. Long wire
C. Small wire D. Thicker wire

83. Some commonly used power devices are given in List I and the approximate maximum frequencies to which they can be presently used for switching large currents is given in List II. Match List I with List II and select the correct answer

List I	**List II**
(a) Power Bipolar	1. 1 MHz
(b) power MOSFET	2. 100 kHz
(c) Silicon controlled rectifier	3. 10 kHz
(d) Insulated gate bipolar transistor	4. 1 kHz

Codes :	*(a)*	*(b)*	*(c)*	*(d)*
A.	3	1	4	2
B.	3	1	2	4
C.	2	3	4	1
D.	2	3	1	4

84. The purpose of the delay line is to retard the arrival of the input waveform at the

A. Vertical deflection plates until the sweep is started
B. Vertical deflection plates until the sweep is completed
C. Horizontal deflection plates until the sweep is started
D. Horizontal deflection plates until the sweep is completed.

85. For full scale deflection ($\theta = \pi/2$) the ratio of the width of the control spring to the thickness of the control spring strip should not be less than

A. 1000 B. 2000
C. 3000 D. 4000

86. The output off-set voltage is zero for the circuit given, when

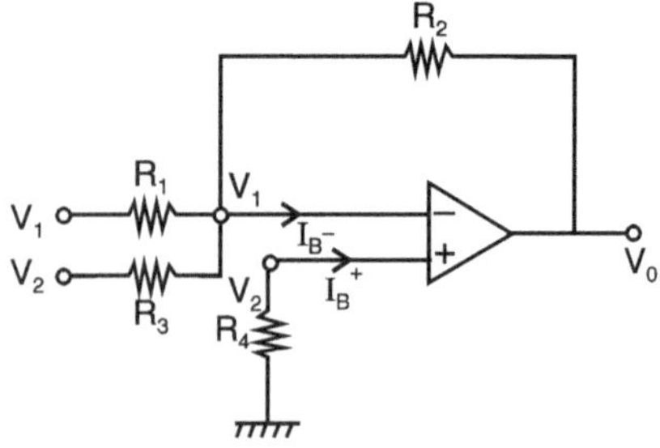

A. $V_1 = 0, I_{B-} = I_{B+}$ and $V_1 = V_2$
B. $V_1 = 0$, and $I_{B-} = I_{B+}$

C. $I_{B-} = I_{B+}$ and $V_1 = V_2$
D. $V_1 = 0$ and $V_1 = V_2$

87. The value of R_b required to drive the transistor in the circuit below into saturation is

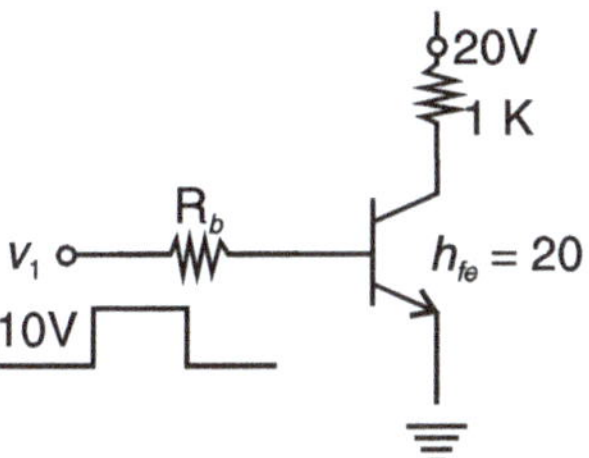

A. 1 K B. 10 K
C. 20 K D. 30 K

88. In a junction diode
A. The depletion capacitance increases with increase in the reverse bias
B. The depletion capacitance decreases with increase in the reverse bias
C. The diffusion capacitance increases with increase in the forward bias
D. The diffusion capacitance is much higher than the depletion capacitance when it is reverse biased.

89. A copper metal ring is spinning at a constant very high speed around an axis passing through centre of the ring and perpendicular to its plane. The ring is suddenly stopped. Which of the following statements is the most correct.
A. On stopping the ring there would be a definite amount of charge to flow in the ring and hence an electric current would be produced.
B. Since the copper metal is electrically neutral, there would not be any net flow of charges and hence no current.
C. Since the free electrons tend to keep moving due to spinning of the ring, there must exist a simple relation between the electron's motion and an electric field which would cause the same electrons movement in the stationary ring, stopping the ring causes a definite amount of charge to flow and hence current is produced.
D. This is never possible that a current is produced as there is no external magnetic field which would charge the magnetic flux across the spinning ring, as per Faraday's law of electromagnetic induction.

90. Fig. represents a

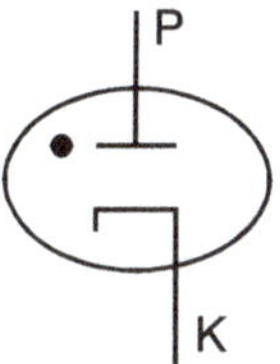

A. Zener diode B. Crystal diode
C. Vacuum diode D. Gas diode

91. The circuit shown is a

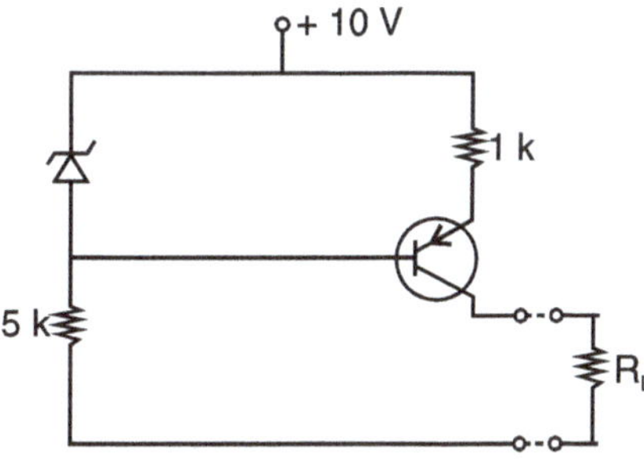

A. constant voltage source
B. constant current source
C. common emitter amplifier
D. common base amplifier

92. It is not advisable to use solid conductors with large diameter for ac circuits since
A. the hysteresis loss is more
B. the current is not uniformly distributed and effective resistance is more
C. it is difficult to fabricate large cross-section conductions
D. none of the above

93. A solid conducting slab falls under gravity. However, after some time, it tends to acquire constant speed. This due to
A. air friction B. conduction current
C. both the above D. none of the above

94. Which of the following are piezo-electric substances
1. Barium titanate 2. Lead titanate
3. Lead zirconate 4. Cadmium sulphate
A. 1, 2 and 4 B. 1, 3 and 4
C. 1, 2 and 3 D. 2, 3 and 4

95. The equivalent capacitance across '*ab*' will be

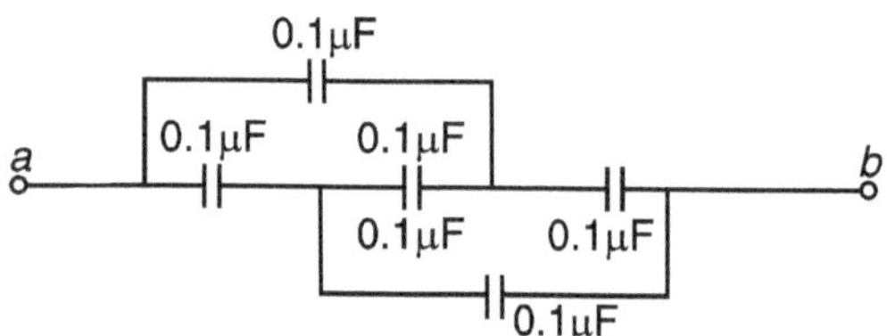

A. 0.2 μF B. 0.1 μF
C. 0.5 μF D. 0

96. The equivalent circuit of the following circuit :

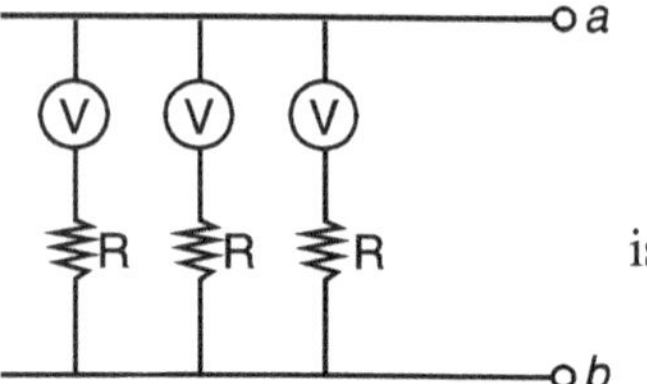

is

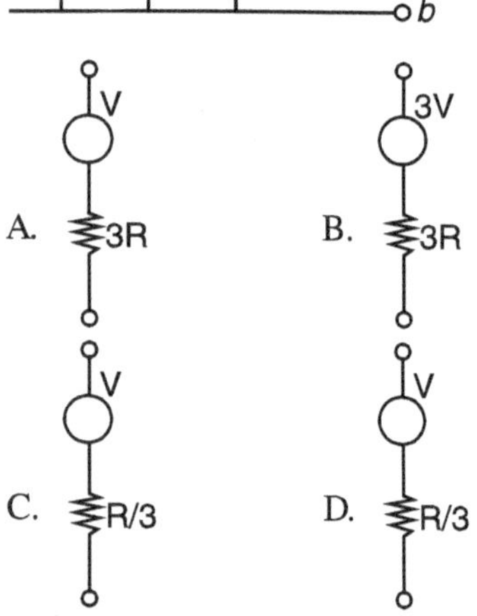

97. In CRT aquadag carries
A. aqueous solution of graphite
B. sweep voltage
C. secondary emission electrons
D. none of the above

98. The upper 3 dB frequency in a common emitter amplifier is given in terms of the hybrid π parameters as

A. $\dfrac{g_{b'e}}{2\pi C}$ B. $\dfrac{g_{b'e}}{C}$

C. $\dfrac{h_{fe}}{g_{b'e}}$ D. $\dfrac{g_m}{h_{fe}}$

99. Match List I with List II and select the correct answer

List I	List II
(a) Class A single ended	1. Requires PNP & NPN transistor pairs
(b) Class B push-pull transformer coupled	2. Prevents cross-over distortion
(c) Class AB push-pull	3. Has maximum efficiency amongst these amplifiers
(d) Class AB complementary symmetry	4. Has minimum efficiency amongst these amplifiers

Codes :	*(a)*	*(b)*	*(c)*	*(d)*
A.	4	3	2	1
B.	4	2	3	1
C.	1	2	2	4
D.	3	2	1	4

100. Match List I with List II and select the correct answer by using the codes given below :

List I ***(Material)***	***List II*** ***(Electrical Conductivity*** $ohm^{-1}\ m^{-1}$***)***
(a) Gold	1. 10^{-14}
(b) Aluminium	2. 10^{5}
(c) Graphite	3. 3.5×10^{7}
(d) Diamond	4. 6.3×10^{7}

Codes :	*(a)*	*(b)*	*(c)*	*(d)*
A.	1	2	3	4
B.	2	3	4	1
C.	3	4	1	2
D.	4	3	2	1

101. Which of the following closely represents the temperature (T), Resistance (R) characteristics of a thermistor.
A. $R_T = a\,R_o e^{b/T}$ B. $R_T = a\,R_o e^{-b/T}$
C. $R_T = -a\,R_o e^{-b/T}$ D. $R_T = -a\,R_o e^{b/T}$

102. Consider the following semiconductor diodes.
1. Germanium diode 2. Silicon diode
3. Tunnel diode 4. Schottky diode
The correct increasing order of forward voltage drop of these diodes is
A. 1, 3, 4, 2 B. 1, 2, 3, 4
C. 3, 4, 2, 1 D. 3, 1, 4, 2

103. In a difference amplifier, a large R_e leads to decrease in
A. CMRR of the amplifier

B. Differential-mode gain
C. Common-mode gain
D. All of these.

104. Match List I with List II and select correct answer

List I	*List II*
(a) Darlington	1. A circuit using *pnp* and *npn* transistors; used in power amplifier
(b) Common Base amplifier	2. A common-source amplifier driving a common-gate amplifier
(c) Complementary symmetry amplifier	3. A circuit with overall voltage gain close to 1 and very large input impedance
(d) Cascade amplifier	4. A circuit with low input impedance; mainly used in high frequency applications

Codes :	*(a)*	*(b)*	*(c)*	*(d)*
A.	1	2	3	4
B.	2	1	3	4
C.	3	4	1	2
D.	4	3	2	1

105. The temperature below which certain materials are antiferromagnetic and above which they are paramagnetic is called
A. Weiss temperature
B. Curie temperature
C. Neel temperature
D. Double poing temperature

106. Which of the following is not an advantage of semiconductor strain gauges as compared to conventional strain gauges?
A. Excellent hysteresis characteristics
B. Least sensitive to temperature changes
C. High fatigue life
D. Smaller size

107. In the self-bias CE transistor amplifier, a single battery V_{CC} is used because
A. Of the voltage divider action of biasing resistances
B. The emitter base junction is of low resistance
C. The collector base junction is reverse biased by V_{CC}
D. All of these

108. The frequency modulated (FM) radio frequency range is nearly
A. 250-300 MHz B. 150-200 MHz
C. 90-105 MHz D. 30-70 MHz

109. Which of the following is not electromagnetic in nature?
1. alpha rays 2. X-rays
3. gamma rays 4. cathode rays
A. 1 and 2 only B. 2 and 3 only
C. 3 and 4 only D. 1 and 4 only

110. The nature of crystal binding in germanium is
A. ionic B. metallic
C. covalent D. Van der wall type

111. The transition capacitance C_T is related to the reverse bias $|V_r|$ of a step graded PN junction as

A. $C_T \alpha |V_r|$ B. $C_T \alpha |V_r|^{\frac{1}{2}}$

C. $C_T \alpha |V_r|^{-\frac{1}{2}}$ D. $C_T \alpha |V_r|^2$

112. A CE amplifier has an unbypassed R_e. Given, $R_C = 10$ K ohms, $R_e = 1$ K ohm, $r_\pi = 500$ ohms, $h_{fe} = 100$, the voltage gain $|A_v|$ is approximated by
A. 200 B. 1000
C. 20 D. 10

113. For large power applications
A. D.C. systems are ideal choice
B. A.C. systems are ideal choice
C. Both of these can be used efficiently
D. None of the two of preferred.

114. For the circuit shown in given figure, the quiescent point is

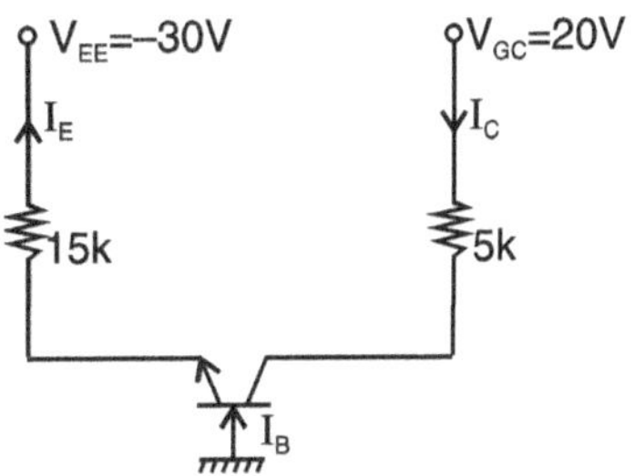

A. 12 V, 5 mA B. 12 V, 2 mA
C. 10 V, 2 mA D. 10 V, 5 mA

115. The circuit shown is modified according to List I. Match List I with List II and select the correct answer

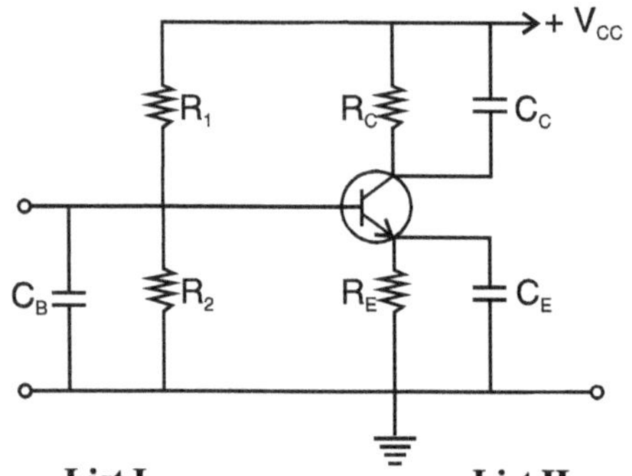

	List I		**List II**
(a)	$C_B = 0, C_C = 0, C_E = \infty$. Input at base point and output from collector point	1.	amplifier in CC mode with feedback
(b)	$C_B = 0, C_E = 0, C_C = \infty$. Input at base point and output from emitter point	2.	amplifier is CB mode with no feedback
(c)	$C_B = 0, C_C = 0, C_E = 0$. Input at emitter point and output from collector point	3.	Amplifier in CE mode with feedback
(d)	$C_B = 0, C_E = 0, C_C = 0$. Input at base point and output from collector point	4.	Amplifier in CE mode with no feedback

Codes :	*(a)*	*(b)*	*(c)*	*(d)*
A.	4	3	2	1
B.	1	2	3	4
C.	4	1	2	3
D.	2	3	4	1

116. In transformer-coupled amplifiers,
   A. DC load line is nearly vertical
   B. AC load line is nearly vertical
   C. Both dc as well as ac load line are nearly vertical
   D. Both load lines are nearly horizontal.

117. When both the primary and secondary resonant circuits of a double tuned transformer have a Q of 50, the value of critical coupling is
   A. 0.2 B. 0.5
   C. 1 D. 2

118. The drain-source output V-I characteristics of an *n*-channel depletion FET has
   A. $I_{DS} = 0$ at $V_{GS} = 0$
   B. I_{DS} = positive maximum at $V_{GS} = 0$
   C. I_{DS} = negative maximum at $V_{GS} = 0$
   D. I_{DS} independent of V_{GS}

119. The equivalent circuit of a CE transistor is represented by

A.
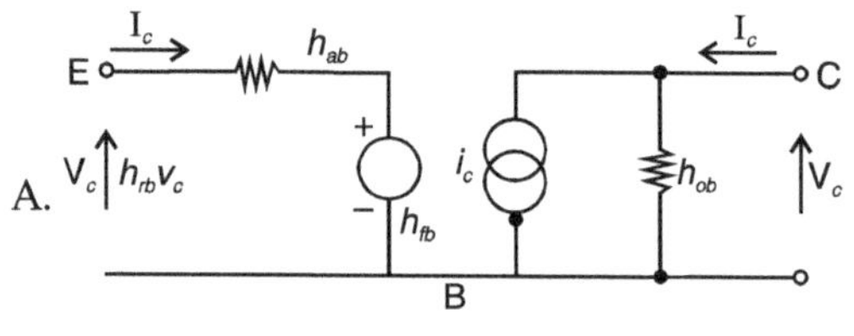

B.
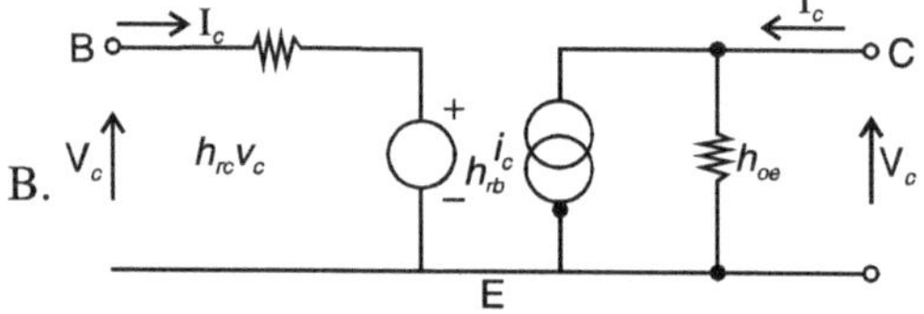

C.

D.
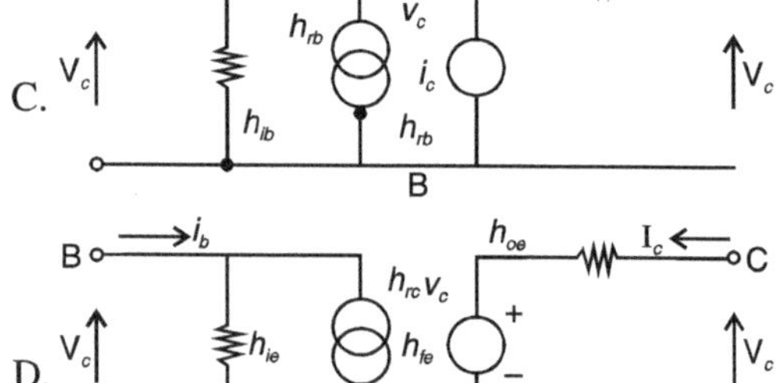

120. The suppressor grid in pentode
   A. Reduces the grid-plate capacitance, as compared with a triode
   B. Eliminates the problem of secondary emission from the plate
   C. Usually has a positive dc potential less than the plate voltage
   D. Is usually connected internally to the control grid.

121. What value of L is needed with a C of 20 pF for an oscillatory frequency of 110.7 MHz?
   A. 0.001 μH B. 0.01 μH
   C. 0.1 μH D. 1 μH.

122. A.C.T. is overloaded when
   A. Its terminals are short-circuited
   B. The load connected to it is less than its burden

C. When its terminals are open-circuited
D. When the primary current is 110% of its rated value

123. The staggered tuned amplifier generally employs
A. Common emitter configuration
B. Common collector configuration
C. Common base configuration
D. FET

124. The following system provides excellent transient as well as steady state response :
A. Proportional action
B. proportional + Integral action
C. proportional + differential action
D. Proportional + Integral + differential action.

125. When a square wave pulse is fed to the input of the circuit shown in the given figure, then V_C the voltage across C will have the waveform

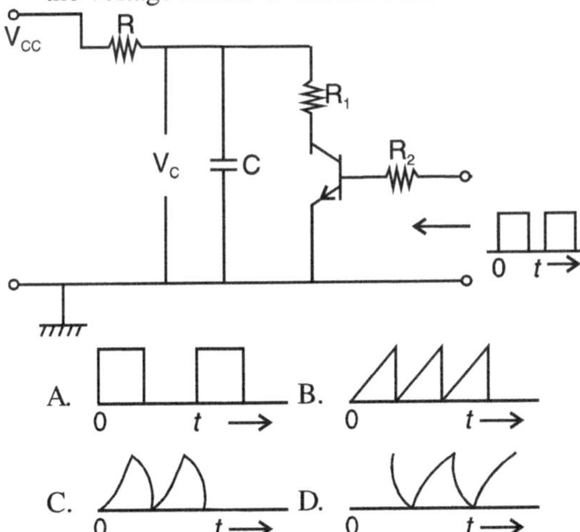

126. A Darlington pair is used for
A. low distortion B. high frequency range
C. high power gain D. high current gain

127. Which of the following configurations has low thermal stability?
A. Common emitter
B. Common base
C. Common collector
D. All of these

128. The forward current amplification factor α of a transistor is the ratio of output current to input current with
A. The input short-circuit
B. The output short-circuit
C. The input open-circuit
D. The output open-circuit

129. The circuit shown in figure below represents

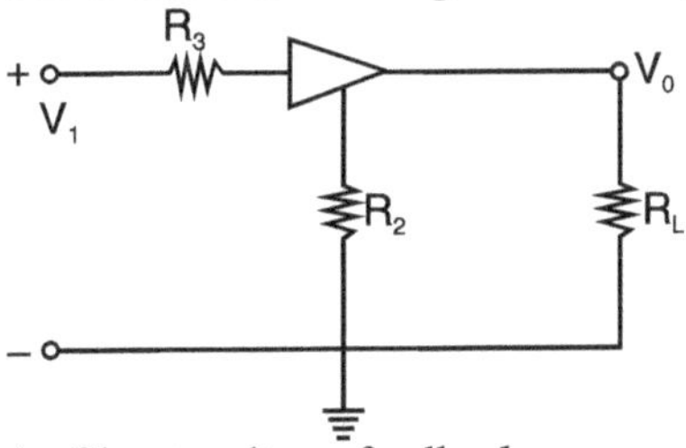

A. Shunt-voltage feedback
B. Series-voltage feedback
C. Series-voltage feedback
D. Shunt-current feedback

130. The phenomenon of secondary emission is common in
A. Diodes B. Triodes
C. Tetrodes D. None of the above

131. Which of the following pairs is/are correctly matched?

Waveform	*Circuitry and input signal*
1. Triangular wave	Integrating circuit and square wave
2. Impulsive wave	Differentiating circuit and step signal
3. Saw-tooth wave	Differentiating circuit and triangular wave

Select the correct answer using the codes given below :
A. 2 and 3 B. 1 and 2
C. 1 alone D. 1 and 3

132. Which of the following represents ferromagnetic and ferrimagnetic substances?

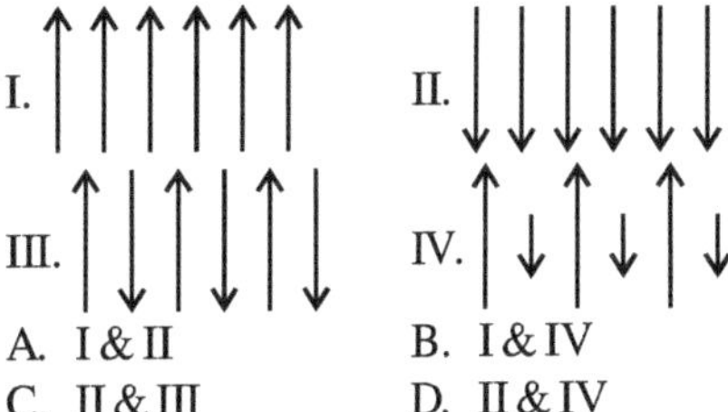

A. I & II B. I & IV
C. II & III D. II & IV

133. As the temperature goes higher the vibrational frequency of atoms in a solid
A. Increases more rapidly

B. Decreases more rapidly
C. May increase or decrease
D. Tends to become constant.

134. For higher frequency response of a transistor amplifier, suitable model to use is

A. Kelvin network B. Lattice network
C. Hybrid π D. T-network

135. The screen grid in a pentode

A. Makes the plate current more dependent on the plate voltage
B. Has the same potential as the cathode
C. Decreases the grid plate capacitance as compared with a triode
D. Eliminates the problem of secondary emission from the plate.

136. Match the following :

Column-I	***Column-II***
(a) Invar	*(i)* 90% Cu, 10% Zn
(b) Gun metal	*(ii)* 90% Cu, 10% Sn
(c) Constantan	*(iii)* 64% Cu, 36% Ni
(d) Bronze	*(iv)* 50% Cu, 50% Ni

Codes :	*(a)*	*(b)*	*(c)*	*(d)*
A.	*(i)*	*(ii)*	*(iii)*	*(iv)*
B.	*(ii)*	*(iv)*	*(iii)*	*(i)*
C.	*(iii)*	*(i)*	*(iv)*	*(ii)*
D.	*(iii)*	*(iv)*	*(i)*	*(ii)*

137. Which of the following can be used as a jewel bearing material?

A. Corundum B. agate
C. diamond D. any of the above

138. The small-signal model of common-base transistor configuration is shown in the figure below. The current source (I) and the resistor (R) in the output side are (α is the CB current gain of the transistor)

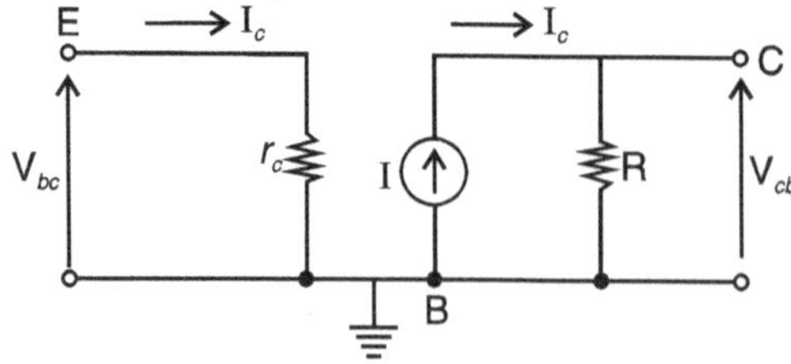

A. $\frac{\alpha}{1-\alpha} i_e, r_e$

B. $\frac{\alpha}{1-\alpha} i_e, (1-\alpha) r_e$

C. $(1-\alpha)\, i_e, (1-\alpha)\, r_e$

D. $\alpha\, i_e, r_e$

139. Of the target cross-section is changing, the best system for accurate tacking is

A. lobe switching B. sequential lobing
C. conical scanning D. mono-pulse

140. Differential amplifier has a differential gain of 20,000. CMRR = 80 dB. The common mode gain is given by

A. 2 B. 1
C. 1/2 D. 0

141. Two transistors have the same value of α but different current gain-bandwidth products. One of them is a germanium transistor and the other is a silicon transistor. Both the transistors have similar geometries and the value of base width. The transistor with lower GB product

A. is the germanium
B. is the silicon
C. may be either of two
D. cannot be identified unless more information is available.

142. Match List I with List II and select the correct answer using the codes given below the lists

List-I (Maximum Efficiency in Per cent)	***List-II (Amplifiers)***
(a) 25	1. Class–B transformer-coupled
(b) 78.5	2. Class – A RC-coupled
(c) 100	3. Class–A transformer-coupled
(d) 50	4. Class–D switching mode

Codes :	*(a)*	*(b)*	*(c)*	*(d)*
A.	1	2	3	4
B.	2	3	4	1
C.	2	1	4	3
D.	3	4	2	1

143. The overall voltage gain in the amplifier shown in the figure is :

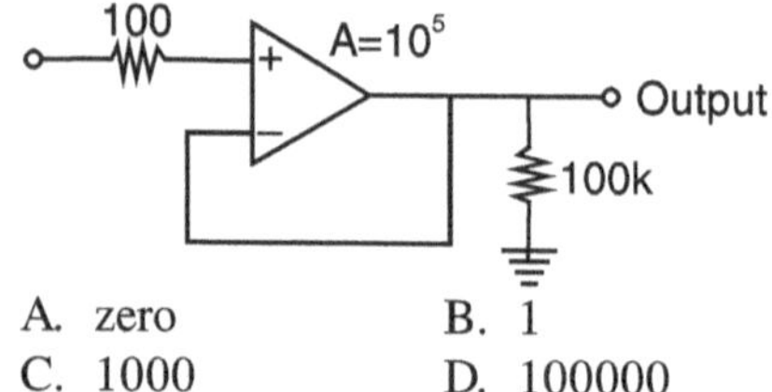

A. zero B. 1
C. 1000 D. 100000

144. If V_m is the peak value of an applied voltage in a half wave rectifier with a large capacitor across the load, then the peak inverse voltage will be

A. $V_m/2$ B. V_m

C. $\sqrt{2}V_m$ D. $2V_m$

145. An amplifier has an input power of 2 microwatts. The power gain of the amplifier is 60 dB. The output power will be

A. 6 microwatts B. 120 microwatts

C. 2 milliwatts D. 2 watts

146. In which one of the following transistor configuration, is the input impedance LEAST dependent on the load resistance?

A. CE

B. CB

C. CE

D. CE with unbypassed emitter resistance

147. In the OPAMP circuit shown in the figure, the output would be

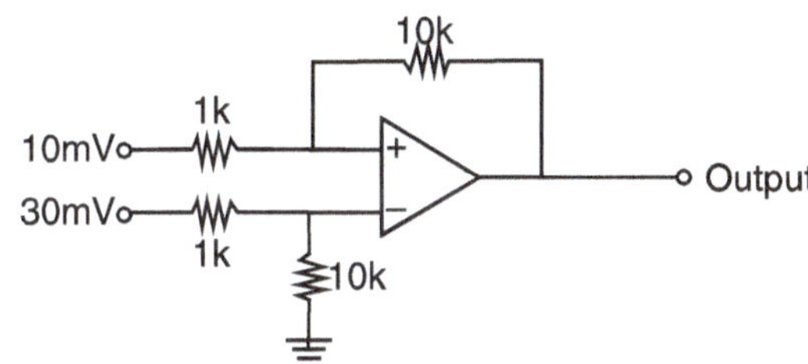

A. 400 mV B. 300 mV

C. 200 mV D. 100 mV

148. Given that the open-loop of the OPAMP shown in the figure is 100000, the approximate output voltage (V_0) will be

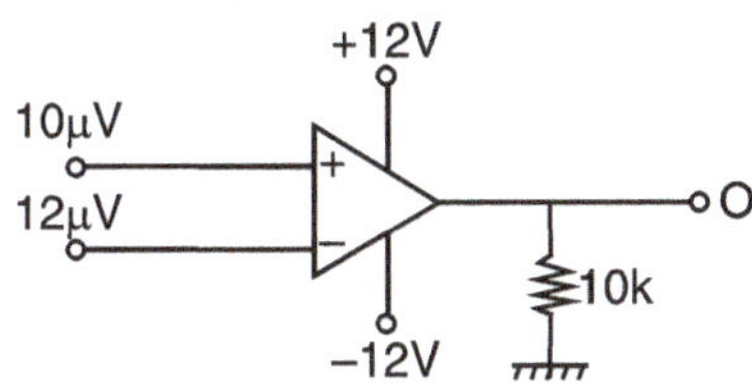

A. + 12 V B. − 12 V

C. − 2 microvolt D. − 0.2 V

149. A half-effect transducer with Hall coefficient $K_H = -1 \times 10^{-8}$ is required to measure a magnetic field of 10,000 gauss. A 2 mm bismuth slab is used as the transducer with a current of 3 A. The output voltage of the transducer will be

A. -7.5×10^{-6} V B. -15×10^{-6} V

C. -20×10^{-4} V D. -22.5×10^{-4} V

150. Which of the following instruments may be used to measure the optical activity of compounds?

A. Infrared spectrometer

B. Atomic absorption spectrometer

C. Polarimeter

D. Flouroscope

151. The excitation of outer electrons in atoms and molecules is associated with which of the following bands of radiation?

A. X-rays B. Gamma rays

C. Ultra violet D. Infrared

152. An operational amplifier has a slew rate of 100 V/microsecond. For a frequency of 10 MHz, the maximum (peak) value of the sine-wave output voltage will be

A. 100 V B. $\frac{50}{\pi}$ V

C. 10 V D. 5 V

153. When a junction transistor is operated under saturated conditions

A. both the CB and EB junctions are forward biased

B. the CB junction is forward biased but the EB junction is reverse biased

C. the CB junction is reverse biased but the EB junction is forward biased

D. both the CB and EB junctions are reverse biased

154. If P_r is the rate at which carriers are produced by light in a photo-diode, τ is the average life time for the newly created carriers, T_t is the average transit time for carriers to reach the ohmic contacts of the diode and 'e' is the electronic charge, then the steady-state photo-current will be given by

A. $i = \frac{2eP_r\tau}{T_t}$ B. $i = \frac{e.P_r.\tau}{2T_t}$

C. $i = \frac{eP_r\tau}{T_t}$ D. $i = \frac{e.P_r.\tau}{3T_t}$

155. In an ideal junction transistor, the impurity concentration in the emitter (E), base (B) and collector (C) are such that

A. E > B > C B. B > C > E

C. C > E > B　　D. C = E = B

156. A wide pulse is given as input to the monostable multi-vibrator shown in the figure. The collector outputs of Q_1 and Q_2 will be pulses duration.

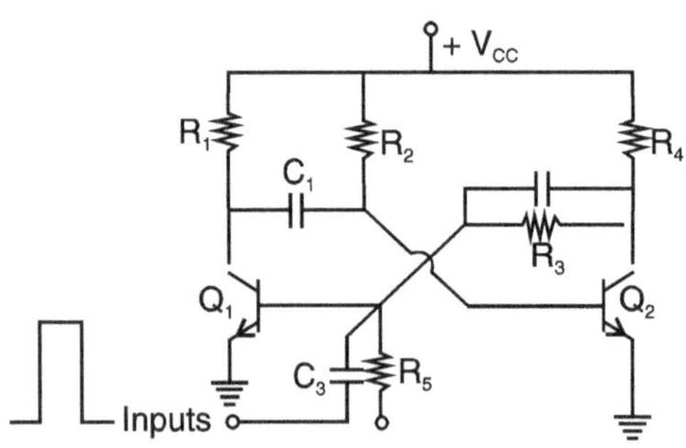

A. $0.69\ R_1\ C_1$　　B. $0.69\ R_2\ C_1$

C. $0.69\ R_3\ C_2$　　D. $0.69\ R_5\ C_3$

157. A BJT has CE parameters as given below

$h_{ie} = 5$ k ohms,　　$h_{re} = 10^{-5}$

$h_{fe} = 100$,　　$h_{oe} = 1.5 \times 10^{-5}$ mho

The value of the h_{ib} for CB configuration is

A. 10^{-5} ohm　　B. 49.5 ohms

C. 5000 ohms　　D. 10^5 ohms

158. "In any network of linear impedances, the current flowing at any point is equal to the algebraic sum of the currents caused to flow at that point by each of the sources of emf taken separately with all other emf's reduced to zero.

This statement represents

A. Kirchhoff's law

B. Norton's theorem

C. Thevenin's theorem

D. Superposition theorem

159. Which of the following theorems is a manifestation of the Law of conservation of energy?

A. Tellegen's theorem

B. Reciprocity theorem

C. Thevenin's theorem

D. Norton's theorem

160. Which of the following statements is correct?

A. Copper has partially filled conduction band

B. Diamond has completely filled conduction band but empty valence band

C. Silicon has a partially field conduction band and an empty valence band

D. The energy gap between conduction and valence bands in diamond is smaller than in silicon.

161. Polarization is a measure of

A. dielectric constant per unit volume

B. voltage gradient to produce electrical breakdown

C. product of charge and distance

D. excess charge density

162. AE 139 is a

A. Tunnel diode

B. Germanium power transistor

C. Photoconductive cell

D. Silicon diode

163. A temperature sensitive transducer is subjected to a sudden temperature change. It takes 10 seconds for the transducer to reach equilibrium condition (five time constants). The time taken by the transducer to read half of the temperature difference will be nearly

A. 1.38 second　　B. 5.00 second

C. 8.62 second　　D. 9.8 second

164. For the junction transistor, which of the following relation is true?

A. $\alpha = \frac{\beta}{1-\beta}$　　B. $\alpha = \beta(1+\beta)$

C. $\beta = \frac{\alpha}{1-\alpha}$　　D. $\beta = \frac{\alpha}{1+\alpha}$

165. Skin depth depends on

A. f　　B. f^2

C. $\sqrt{f}$　　D. $\frac{1}{\sqrt{f}}$

166. One of the advantages of Ebers-Moll equations for transistor is that they provide

A. True terminal voltages dependent of junction biases

B. True terminal currents regardless of junction biases

C. Separate input and output circuits

D. All of these

167. Which one of the following is correct?

A. The amplitude as well as the frequency of the waveform will get doubled

B. The amplitude will get doubled but the frequency will reduce to half its value

C. The amplitude will get doubled but the

frequency will remain unchanged

D. The amplitude will remain unchanged but the frequency will get doubled

168. In the circuit of the relaxation oscillator shown what will be the change in the voltage waveform across capacitor, if the voltage V is doubled?

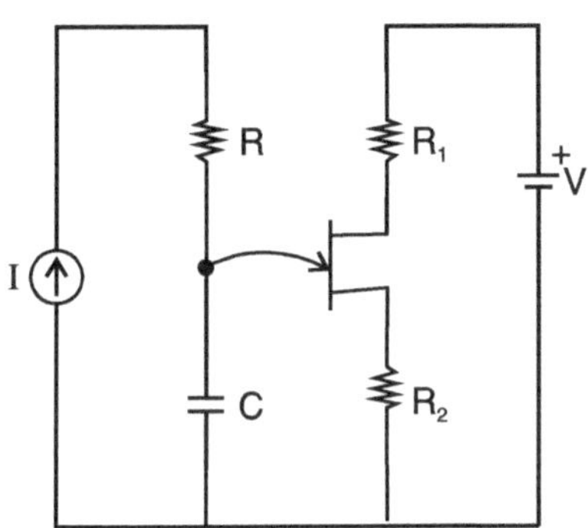

A. RC B. 1.39 RC

C. 0.69 RC D. none

169. In a CE amplifier, the effect of an unbypassed Re is to increase the input resistance.
This happens because

1. The system tends towards an ideal CVS
2. Both input and output currents flow through Re.
3. The fed back quantity is a voltage
4. The fed back quantity is a current.

The true statements are

A. Only 1, 2 B. Only 3, 4

C. Only 2, 3 D. Only 1, 4.

170. In the circuit shown below, if temperature decreases, then the d.c. voltage across R_c

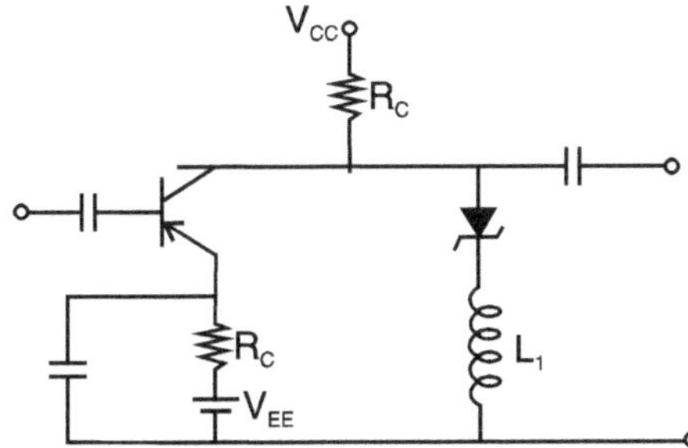

A. Decreases

B. Increases

C. Remains constant

D. May increase or decrease.

171. The given feedback amplifier circuit is modified according to List I and the possible feedback is as per List II. Match the two lists and select the correct answer

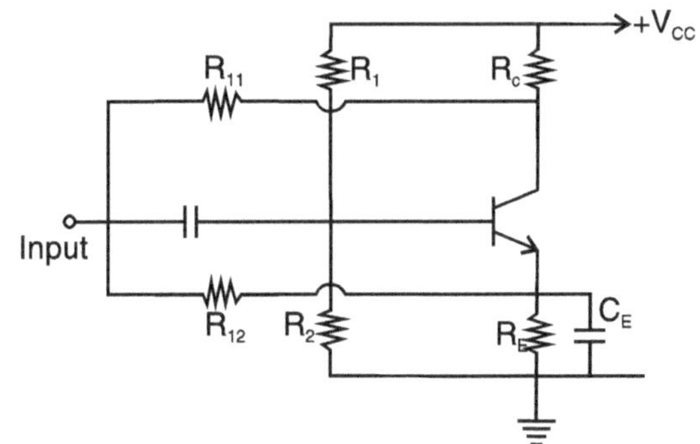

List I	*List II*
(a) $R_C = 0, C_E = 0.$ $R_{F1} = \infty. R_{F2} = \infty,$ and output is taken from emitter point	1. Current shunt positive feedback
(b) $C_E = 0, R_{F1} = \infty$ $R_{F2} = \infty$ and output is taken from collector point	2. Voltage series negative feedback
(c) $R_{F2} = \infty$ and output is taken from collector point	3. Current series negative feedback
(d) $C_E = 0, R_{F1} = \infty$ and output is taken from collector point	4. Voltage shunt negative feedback

Codes :	*(a)*	*(b)*	*(c)*	*(d)*
A.	3	2	1	4
B.	2	3	4	1
C.	1	3	4	2
D.	4	3	2	1

172. If a pentode, the suppressor grid is placed between

A. Plate and screen grid

B. Cathode and heater

C. Screen grid and control grid

D. Control grid and cathode.

173. P and Si have valencies of

A. 5 and 4 B. 4 and 5

C. 3 and 5 D. 3 and 4

174. Match List I with List II and select the correct

answer

List I	*List II*
(a) Voltage controlled device	1. Bipolar junction Transistor
(b) Current controlled device	2. Uni-Junction Transistor
(c) Conductivity modulation device	3. Field Effect Transistor
(d) Negative conductance device	4. IMPATT Diode

Codes :	*(a)*	*(b)*	*(c)*	*(d)*
A.	2	3	4	1
B.	2	3	1	4
C.	3	1	2	4
D.	3	1	4	2

175. If both the emitter-base and the collector-base junctions of a bipolar transistor are forward biased the transistor is in the

A. active region B. saturated region
C. cut-off region D. inverse mode

176. In the circuit given, in order to get an ac current through the load R_L, it is necessary that input 1 and input 2 should be

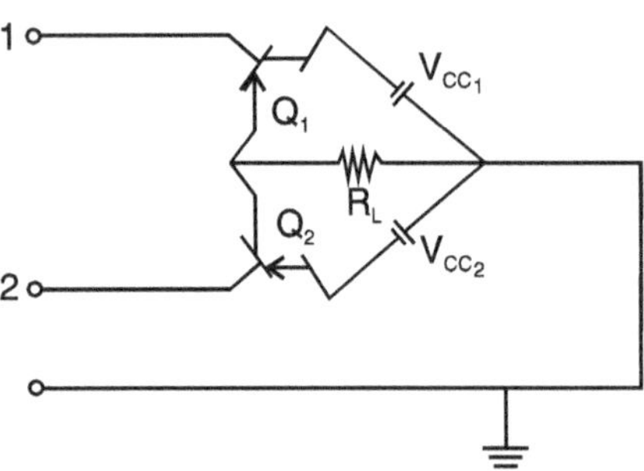

A. 180° out of phase
B. In phase
C. Of opposite polarity
D. 90° out of phase.

177. In a PNP transistor with V_{EB} set to 0 and V_{CE} reverse bias there is a current flow. The following statements refer to this current.

1. It is dependent on V_{CB}
2. The current is due to diffusion
3. It is due to drift in the electric field
4. It is due to thermally generated minority carriers

Of these the true statements are

A. Only 1, 2 B. Only 2, 3
C. Only 3, 4 D. Only 4, 1

178. In an unbiased *p-n* junction the current in equilibrium is

A. Zero because no charges cross the junction
B. Zero because equal number of charges cross the junction
C. Due to diffusion of minority carriers
D. Due to diffusion of majority carriers.

179. With reference to the equivalent circuit of a transistor shown in the given figure, match List I with List II and select the correct answer

List I	**List II**
(a) h_{11}	1. $(r_e + r_b) - \frac{\alpha r_e + r_b}{r_e + r_b} r_b$
(b) h_{12}	2. $\frac{r_b + \alpha r_e}{r_e + r_b}$
(c) h_{21}	3. $\frac{1}{r_e + r_b}$
(d) h_{22}	4. $\frac{r_b}{r_c + r_b}$

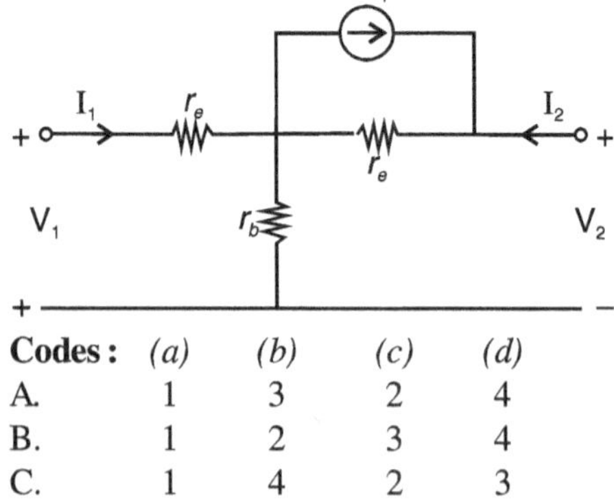

Codes :	*(a)*	*(b)*	*(c)*	*(d)*
A.	1	3	2	4
B.	1	2	3	4
C.	1	4	2	3
D.	2	3	4	1

180. The additivity rule of Matteissen describes that

A. $\rho_T = \rho + \rho_r$ B. $\rho = \rho_T + \rho_r$
C. $\rho_T = \rho - \rho_r$ D. None of these

Where ρ_T = thermal part of resistivity
ρ_r = residual resistivity
ρ = resistivity coverages.

181. In the circuit shown transistor parameter changes on variation in temperature which in turn causes change in pulse width. The variation in pluse width would be the *least* if A is connected

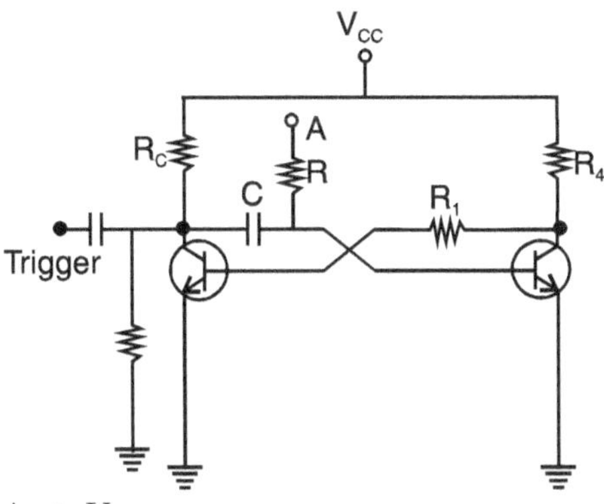

A. to V_C
B. to a voltage greater than V_C
C. to a voltage less than V_C
D. through a thermistor to V_C

182. A complementary symmetry amplifier has
A. 1 PNP and 1 NPN transistor
B. 2 PNP transistors
C. 2 NPN transistors
D. 2 P-channel FETs.

183. In a pentode the suppressor grid is used to
A. Limit anode voltage
B. Limit anode current
C. Dissipate heat
D. Suppress secondary emission

184. In the circuit shown, assume that the capacitor C is almost shorted for the frequency range of interest of the input signal. Under this condition, the voltage gain of the amplifier will be approximately

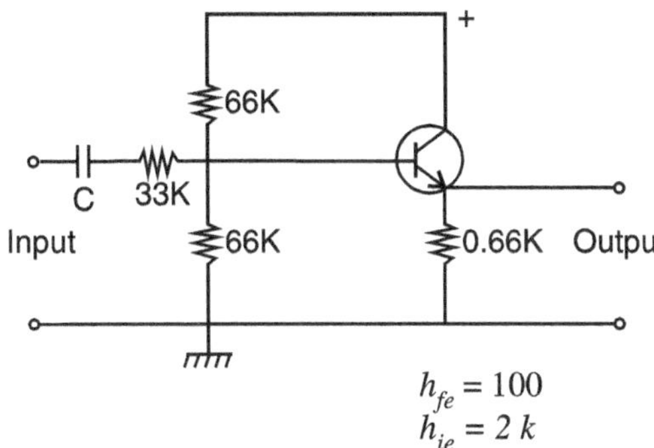

A. 0.335 B. 0.55
C. 0.66 D. 1

185. Match List I with List II and select the correct answer

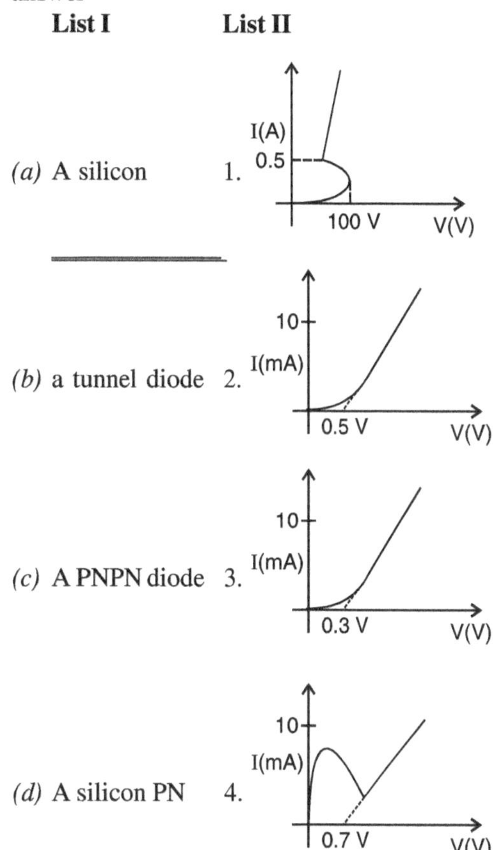

doped to $10^{17}/cm^3$)

Codes :	*(a)*	*(b)*	*(c)*	*(d)*
A.	2	4	1	3
B.	4	2	3	1
C.	1	2	4	3
D.	3	4	1	2

186. With reference to idealised class A BJT amplifiers, the following statements are made about their power relations. Given maximum collector dissipation = 20 Watts.
1. In series-fed load $P_{ac} = 10$ watts
2. In transformer coupled load $P_{ac} = 20$ watts
3. With series-fed load, DC power $P_{dc} = 40$ watts.
4. With transformer coupled load, $P_{dc} = 40$ watts.

The true statements are
A. Only 1, 2 B. Only 1, 3
C. Only 2, 3 D. Only 2, 4

187. Match List I and List II and select the correct answer

List I	List II
(a) Polarization	1. Mechanical vibration
(b) Magnetostriction	2. Heat dissipation
(c) Hall effect	3. Voltage generation
(d) Grain orientation	4. Magnetic field improvement
	5. Charge storage

Codes :	(a)	(b)	(c)	(d)
A.	5	1	3	4
B.	5	2	1	3
C.	2	5	3	1
D.	2	1	4	5

188. Consider the following statements : The functions of an oxide layer in an IC device is to

1. mask against diffusion or ion-implant
2. insulate the surface electrically
3. increase the melting point of silicon
4. produce a chemically stable protective layer.

Of these statements

A. 1, 2 and 3 are correct
B. 1, 3 and 4 are correct
C. 2, 3 and 4 are correct
D. 1, 2 and 4 are correct

189. The approximate values of input and output impedance are

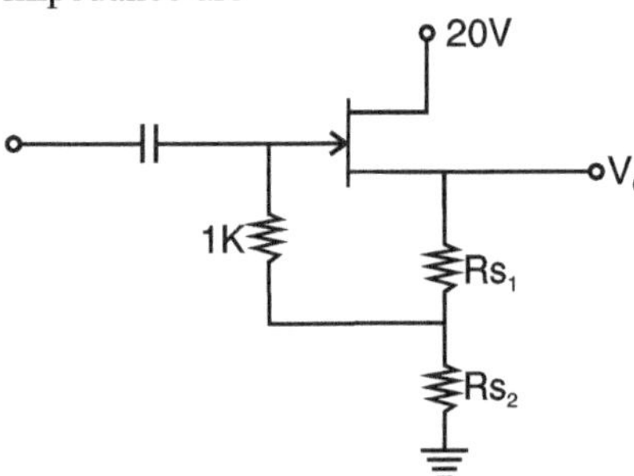

A. 30 K, 390 ohms B. 151 K, 1 K
C. 167 K, 500 ohms D. 200 K, 1 K.

190. If P_r is the rate at which carries are produced by light in a photo-diode. τ is the average life time for the newly created carriers. T_t is average transit time for carriers to reach the ohmic contacts of the diode and 'τ' is the electronic change, then the steady-state photo-current will be given by

A. $i = 2e\,P_r\tau/T_t$ B. $i = e\,P_r\tau/2T_t$
C. $i = e\,P_r\tau/T_t$ D. $i = e\,P_r\tau/3T_t$

191. The current gain of the amplifier shown with output short-circuited is

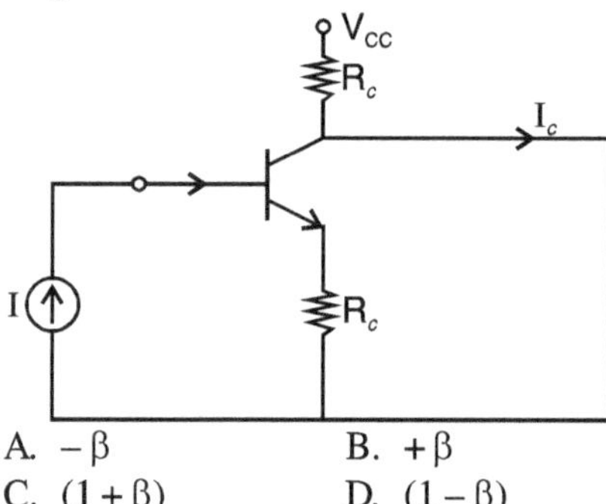

A. $-\beta$ B. $+\beta$
C. $(1+\beta)$ D. $(1-\beta)$

192. Transformer coupling in transistor amplifier circuits provides high efficiency, because

A. DC resistance is low
B. Collector voltage is stepped up
C. Flux linkages are incomplete
D. All of the above

193. The circuit shown employs a silicon transistor with $\beta = 60$. The value of the biasing resistance R_b should be

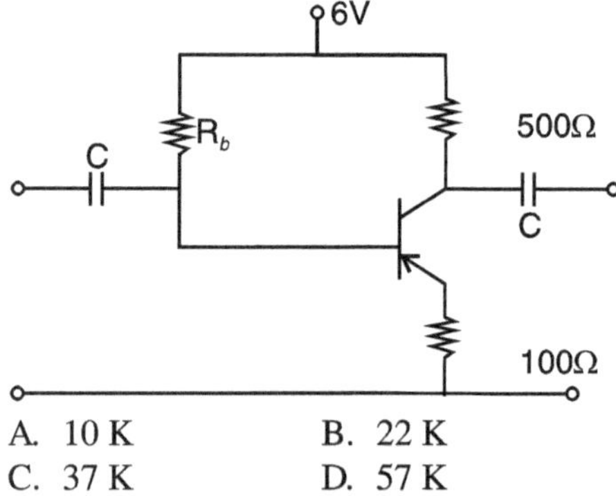

A. 10 K B. 22 K
C. 37 K D. 57 K

194. In circuit shown the feedback causes

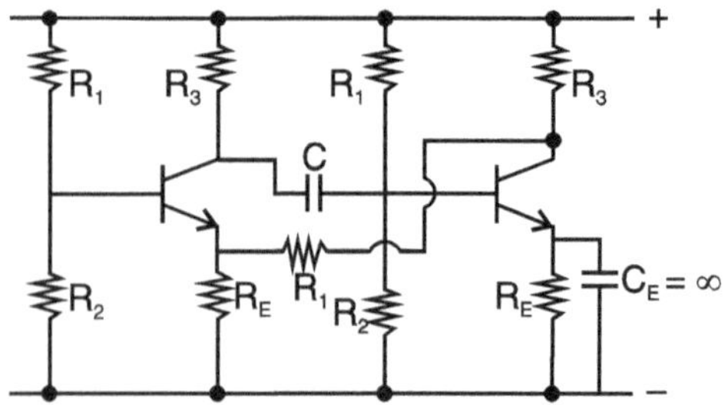

A. an increase in both the input and output impedances

B. an increase in the input impedance but a decrease in the output impedance
C. a decrease in both the input and output impedances
D. a decrease in the input impedance but an increase in the output impedance

195. Match List I with List II and select the correct answer

List I	List II
(a) Darlington amplifier	1. Low input impedance
(b) Cascade amplifier (FET)	2. Low output impedance
(c) Common gate amplifier	3. Low input capacitance but high R_{in}
(d) Differential amplifier	4. Large common mode rejection

Codes:	*(a)*	*(b)*	*(c)*	*(d)*
A.	1	2	4	2
B.	1	2	3	4
C.	2	1	3	4
D.	2	3	1	4

196. The value of parasitic capacitances for a triode may range from
A. 2 μF to 12 μF B. 20 μF to 120 μF
C. 2 *p*F to 12 *p*F D. 20 *p*F to 120 *p*F

197. The density of the copper is 8.96 gm/cc. The number of free electrons per m^3 in the metal will be
A. 8.5×10^{23} B. 7.7×10^{27}
C. 4.5×10^{29} D. None of these

198. When a square wave pulse is fed to the input of the circuit shown then V_C the voltage across C will have the waveform

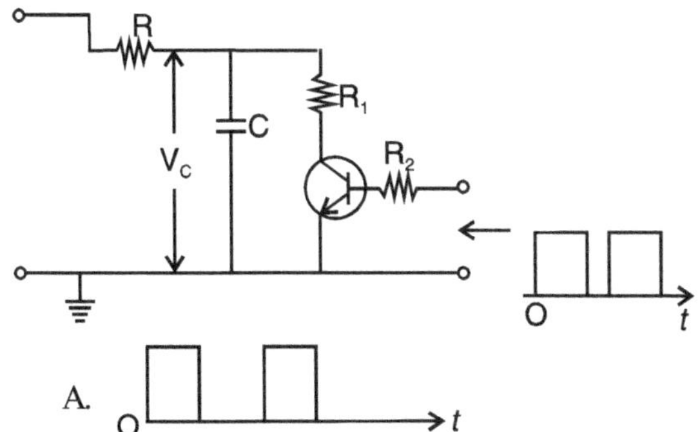

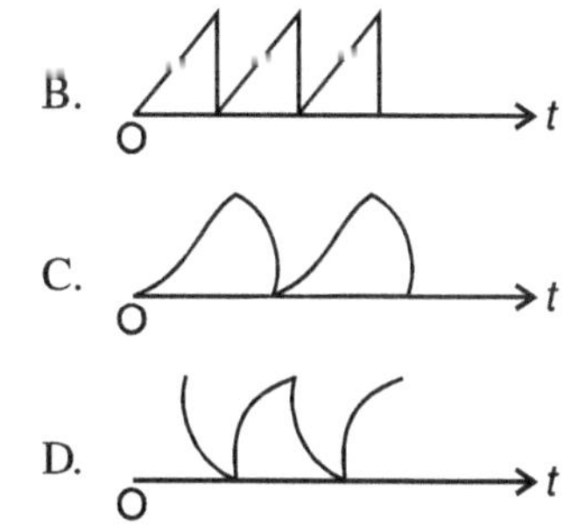

199. The circuit shown is that of

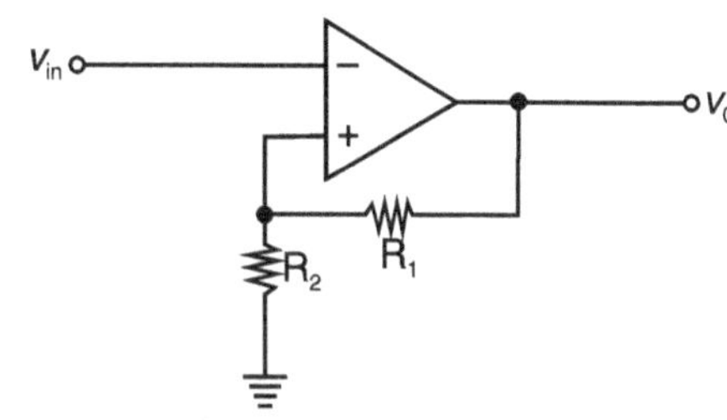

A. a non inverting amplifier
B. an inverting amplifier
C. an oscillator
D. a Schmitt trigger

200. The push-pull amplifier shown below has reduced power efficiency because

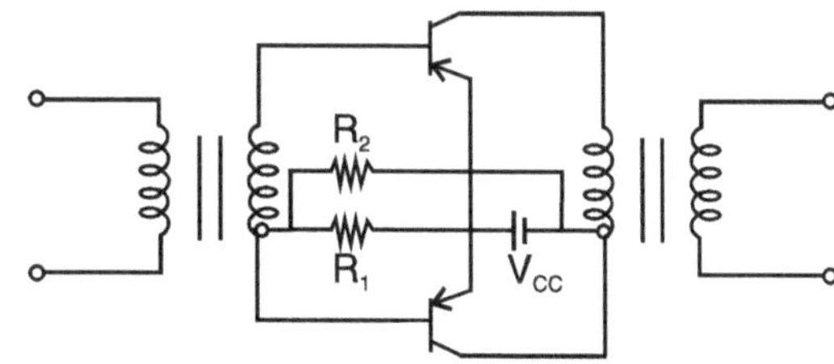

A. There are voltage drops across R_1 and R_2
B. Collector voltage is reduced due to voltage R_1 and R_2
C. There will be collector current in both the transistor for zero input
D. There will be collector current in one of the transistor for zero input.

201. In switching a junction diode from forward to reverse bias, the transition interval is the period
A. From the time the diode voltage begins to reverse to the time when the diode has nominally recovered
B. From the time of forward bias to the time diode voltage has attained zero value

C. From the time when the diode is forward biased to the time when it is reverse biased

D. None of these

202. Following statements about conditions that make a metal-semiconductor contact rectifying are given as :

1. N-type semiconductor with its work function ϕs, greater than the work function ϕ_M of the metal.
2. N-type semiconductor with its work function ϕs, smaller than the work function ϕ_M of the metal.
3. P-type semiconductor with its work function ϕs, greater than the work function ϕ_M of the metal.
4. P-type semiconductor with its work function ϕs, smaller than the work function ϕ_M of the metal.

Of these statments

A. 1 and 3 are correct

B. 2 and 4 are correct

C. 1 and 4 are correct

D. 2 and 3 are correct

203. When Z' << Z, the OP-AMP circuit shown in the given figure behaves as a/an

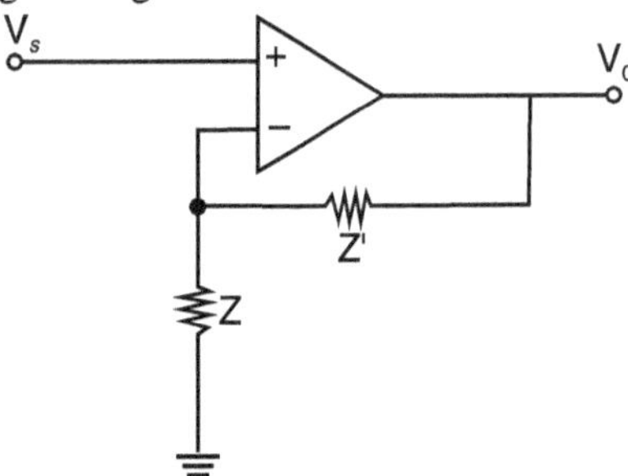

A. inverting amplifier

B. Schmitt trigger

C. non-inverting amplifier

D. voltage follower

204. The capacitor filter are characterised by

A. Small ripple and high voltage at high load

B. Large ripple and low voltage

C. Small ripple and high voltage at light load

D. None of these

205. The diffusion capacitance of a foward biased *p-n* junction diode varies

A. Linearly with the current

B. Inversely with the current

C. As the square of the current

D. As the square root of the current

206. Match List I with List II and select the correct answer

List I	List II
(a) Zener diode	1. High speed switching
(b) Tunnel diode	2. Multivibrator circuits
(c) Gun diode	3. Voltage stabilizer
(d) PIN diode	4. Microwave oscillator

Codes :	*(a)*	*(b)*	*(c)*	*(d)*
A.	4	(1, 4)	1	3
B.	4	(2, 4)	4	1
C.	3	(1, 2)	4	1
D.	3	(1, 2, 4)	4	1

207. Match List I with List II and select the correct answer

List I	List II
(a) Resistance	1. Current Noise
(b) Diode	2. Partition Noise
(c) Triode	3. Shot Noise
(d) *pn* Junction	4. Atmospheric Noise
	5. Johnson Noise

Codes :	*(a)*	*(b)*	*(c)*	*(d)*
A.	5	3	2	4
B.	5	3	2	1
C.	1	3	4	5
D.	1	5	3	1

208. The cut-off frequency of the amplifier is

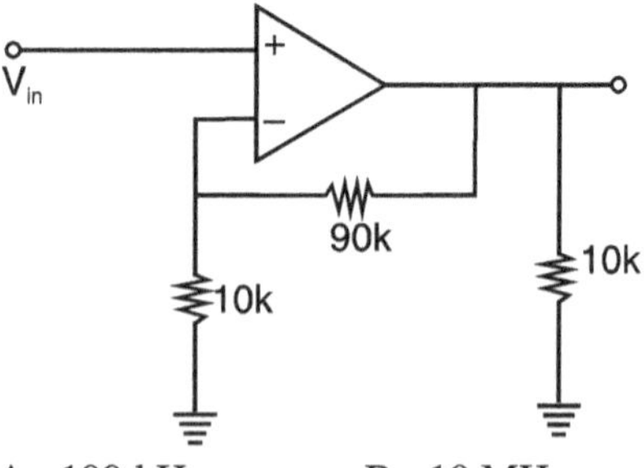

A. 100 kHz
B. 10 MHz
C. 10.10 MHz
D. 90 MHz

209. The following statements refer to an N channel FET operated in the active region.

1. The gate voltage V_{GS} reverse biases the

junction.

2. The drain voltage V_{DD} is negative with respect to the source
3. The current in the N channel is due to electrons
4. increasing the reverse bias V_{GS} increases the cross-section for conduction.

Of these the true statements are

A. 1, 2 B. 1, 3
C. 1,4 D. 2, 4

210. Match List I and List II and select the correct answer

List I	List II
(a) PIN diode	1. Microwave amplification
(b) GaAs MOSFET	2. Low noise microwave generation
(c) Transferred electron devices	3. Electronic tuning of MW oscillator
(d) Varactor diode	4. Light wave detection

Codes:	*(a)*	*(b)*	*(c)*	*(d)*
A.	2	3	1	4
B.	2	4	3	1
C.	4	2	3	1
D.	4	3	2	1

211. For the transistor amplifier in the given figure, the voltage amplification is approximately

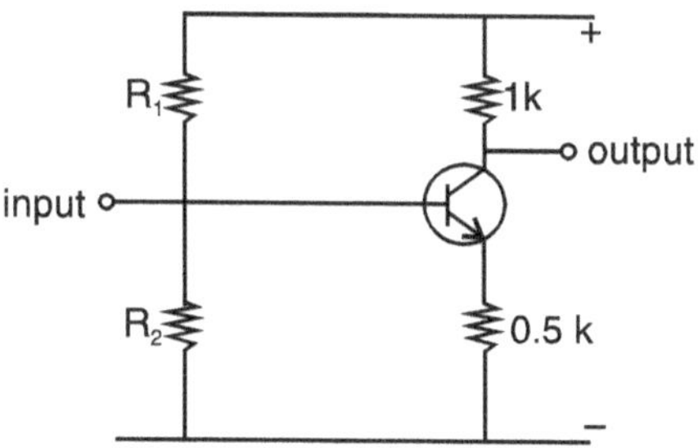

A. 100 B. 840
C. 0.723 D. – 20

212. Which is single 8-input NAND gate

A. SN 7430 B. SN 7410
C. SN 7440 D. SN 7432

213. For the model of a transistor in the common emitter connection shown match List I with List II and select the correct answer

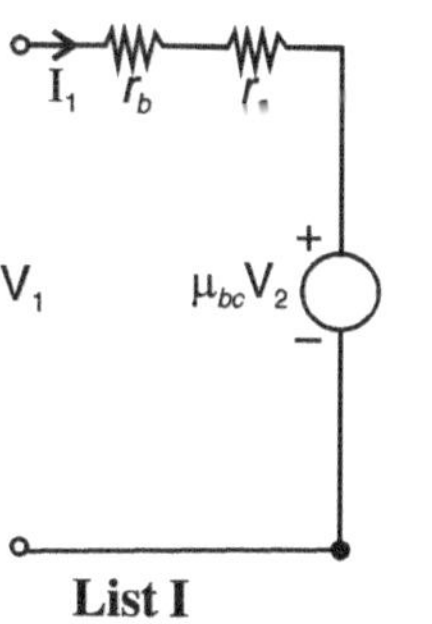

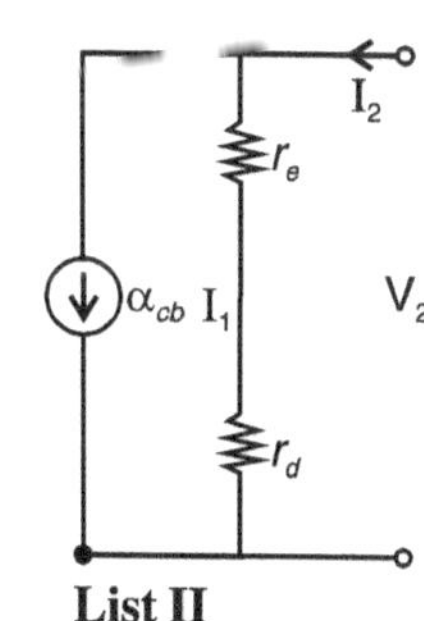

List I	List II
(a) h_{22}	1. $r_b + r_e$
(b) h_{11}	2. α_{cb}
(c) h_{21}	3. $\dfrac{1}{r_e + r_d}$

Codes:	*(a)*	*(b)*	*(c)*
A.	3	2	1
B.	1	3	2
C.	3	2	1
D.	3	1	2

214. The circuit given below employs a silicon transistor with $\beta = 50$. The value of biasing resistance R_b for maximum symmetrical swing at the output should be

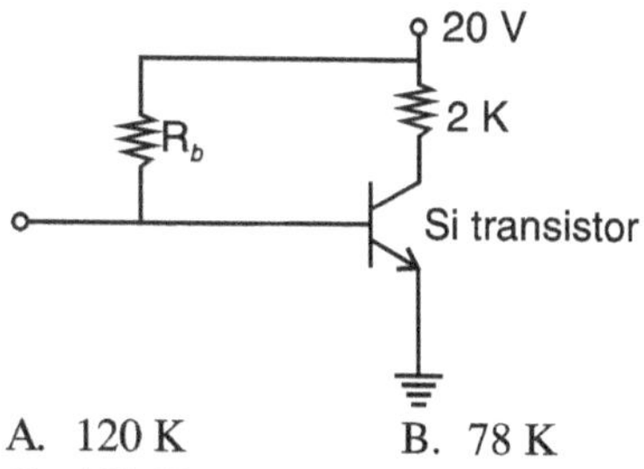

A. 120 K B. 78 K
C. 193 K D. 122 K.

215. Match List I with List II and select the correct answer

List I	List II
(a) BJT	1. Pinch-off effect
(b) FET	2. Controlled, rectification
(c) SCR	3. negative resistance characteristics
(d) Tunnel diode	4. Punch-through effect

Codes:	*(a)*	*(b)*	*(c)*	*(d)*
A.	1	2	3	4
B.	1	3	2	4
C.	4	1	2	3
D.	1	4	3	2

216. A semi-conductor material (Germanium) is doped with 10^{15} donor type (Phosphorus) atoms per cm^3. The intrinsic charge carrier concentration in the semi-conductor is 10^{10} per cm^3. The approximate number of holes in the doped semi-conductor under equilibrium condition is

A. 10^5 B. 0
C. 10^{10} D. 10^{15}

217. Which of the following circuit has the greatest bandwidth?

A. f_r = 1 MHz; Q = 10
B. f_r = 455 kHz; Q = 100
C. f_r = MHz; Q = 100
D. f_r = 50 MHz; Q = 50

218. In the circuit shown in the given figure, the frequency of oscillation depends mainly upon

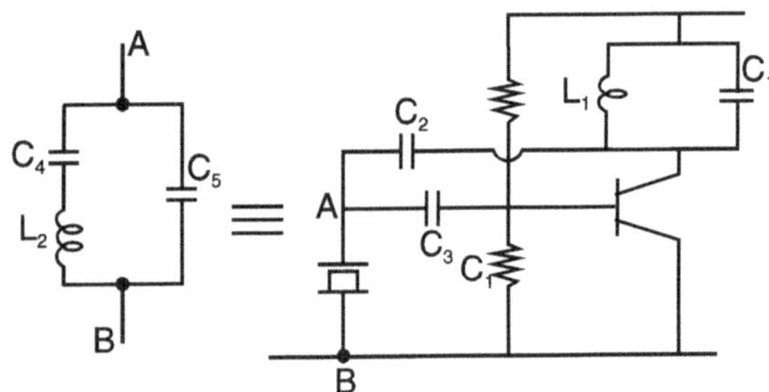

A. L_1, C_1
B. L_1, C_1 and C_2
C. L_1, C_1, C_4, C_5 and L_2
D. L_2, C_4 and C_5

219. For the transistor BFY 51, emitter, base and collector terminals are

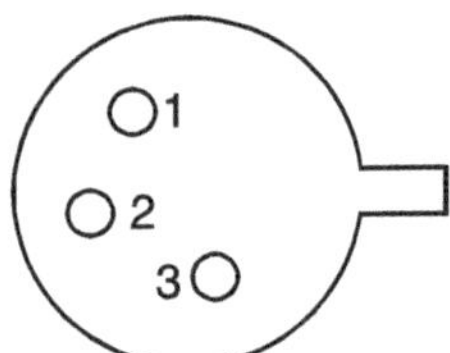

A. 1, 2, 3 B. 3, 2, 1
C. 2, 3, 1 D. 3, 1, 2.

220. Differential amplifier can be used as a limiter

A. As the linear region of transfer characteristics abruptly ends for a specific value of input
B. As g_m is proportional to I_E
C. As CMRR can be made very large
D. As I_E can be made constant.

221. Each day there comes to Earth from the sun energy in the form of direct solar radiation. Summer radiation intensity is of the order of

A. 1 megawatt per square meter
B. 1 kilowatt per square metre
C. One watt per square metre
D. None of these.

222. Minority carrier injection in an NPN transistor biased in the active region means

A. Predominantly holes injected into the base region
B. Predominantly electrons injected into the emitter region
C. Predominantly electron injected into the base region
D. Predominantly holes injected into the emitter region.

223. Which of the following statements are correct for basic transistor amplifier configuration.

A. CB amplifier has low input impedance and a low current gain
B. CC amplifier has low output impedance and a low current gain
C. CE amplifier has very poor voltage gain but very high input impedance
D. The current gain of CB amplifier is higher than the current gain of CC amplifier.

224. A two-stage amplifier is required to have an upper cut-off frequency of 2 MHz and a lower cut-off frequency of 30 Hz. The upper and lower cut-off frequencies of individual stage are respectively approximately

A. 4 MHz, 60 Hz B. 3 MHz, 20 Hz
C. 3 MHz, 60 Hz D. 4 MHz, 20 Hz

225. Consider the following statements : A clamper circuit

1. adds or subtracts a dc voltage to or from a waveform
2. does not change the shape of the waveform
3. amplifies the waveform

Of these statements

A. 1 and 2 are correct
B. 2 and 3 are correct
C. 1 and 3 are correct
D. 1, 2 and 3 are correct

226. A bypass capacitor C_b across R_b provides low pass filtering because

A. Current in C_b branch is maximum for low frequencies
B. Voltage across C_b is minimum for high frequencies
C. Voltage across C_b is minimum for low frequencies
D. Voltage across R_b is minimum for low frequencies.

227. The gain of an amplifier with feedback is given by the relation

A. $\frac{A}{1+\beta A}$ B. $\frac{\beta}{1+\beta A}$

C. $\frac{\beta}{1-\beta A}$ D. $\frac{A}{1-\beta A}$

228. In a CE configuration transistor amplifier, there is

A. No phase change between input and output voltages
B. 180° phase change between input and output currents
C. No phase change between input and output currents
D. None of these

229. For a transistor the normal collector voltage is 12 V. If actually it is found to be 28^+ V, the trouble could be

A. R_L is open B. R_E is open
C. C_E is shorted D. C_E is open

230. Which one of the following conditions would give $V_e = 0$ in the circuit shown in the given figure?

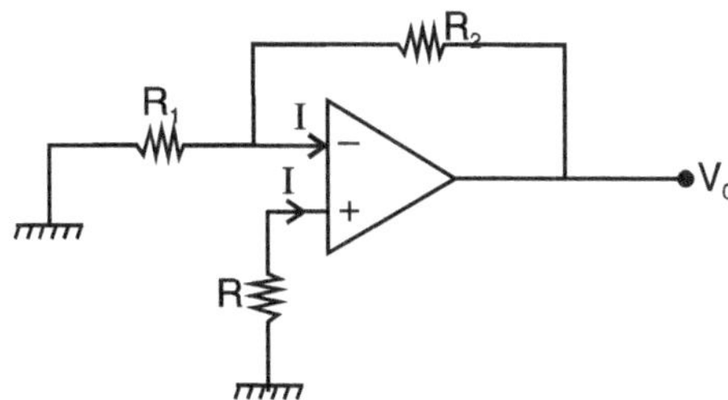

A. $R = R_1 + R_2$ B. $R = R_2/R_1$
C. $R = R_2 - R_1$ D. $R = R_1 \| R_2$

231. In intrinisic semi-conductor the free electron concentration depends on

A. Effective mass of electrons only
B. Effective mass of holes only
C. Temperature of the semi-conductor
D. Width of the forbiddeen energy band of the conductor.

232. The turn off time of a SCR will be normally of the order of

A. 1 microsecond B. 2 microsecond
C. 50 microsecond D. 1 second

233. In common source configuration the input resistance of a JFET will be approximately

A. 10^{15} ohms B. 500 ohms
C. 100 K D. 3 megaohms

234. The given feedback amplifier circuit is modified according to List I and the possible feedback is as per List II. Match the two lists and select the correct answer using the codes given below the lists

List I	List II
(a) $R_c = 0, C_E = 0, R_{r1} = \infty, RF_2 = \infty$ and output is taken from emitter point	1. Current shunt positive feedback
(b) $C_E = 0, R_{F1} = \infty, RF_2 = \infty$ and output is taken from collector point	2. Voltage series negative feedback
(c) $R_{F2} = \infty$ and output is taken from collector point	3. Current series negative feedback
(d) $C_E = 0, R_{F1} = \infty$ and output is taken from collector point	4. Voltage shunt negative feedback

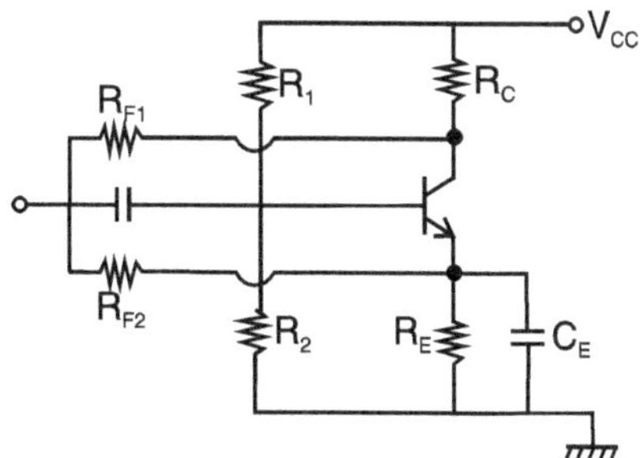

Codes :	*(a)*	*(b)*	*(c)*	*(d)*
A.	1	3	4	2
B.	2	3	4	1
C.	3	2	1	4
D.	4	3	2	1

235. In a P type semiconductor
A. The hole concentration is less than the electron concentration
B. Electrons are the majority carriers
C. At room temperature the hole concentration equals the acceptor concentration
D. Electron density does not change with temperature.

236. The circuit shown represents

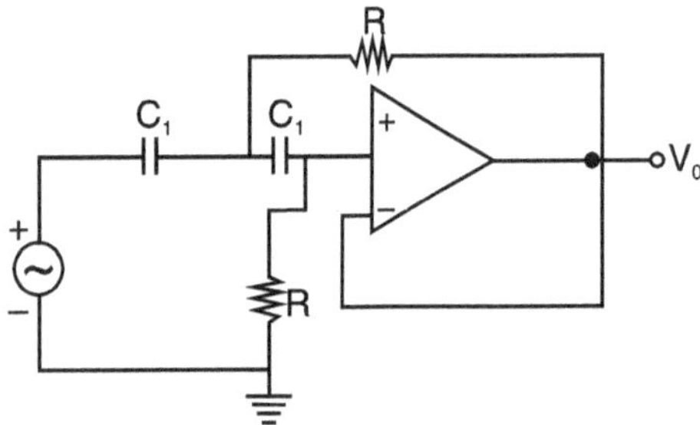

A. an amplifier
B. an oscillator
C. a high pass filter
D. a saw tooth generator

237. In which of the following types of operation, base-emitter junction of the transistor moves alternately from reverse to forward bias and vice-versa for the sinusoidal base input voltage?
A. Class A B. Class B
C. Class C D. None of these

238. In regard to improving the perfomance of a BJT differential amplifier, consider the following statements.
1. The value of A_{DM} is increased by increasing R_e
2. The value of CMRR is increased by increasing R_e
3. Increasing R_e increases A_{CM}
4. Increasing β_o and R_C increases A_{DM}.

The true statements are
A. Only 1, 2 B. Only 2, 3
C. Only 3, 4 D. Only 2, 4

239. The kinetic energy of a bounded electron is .... the kinetic energy of a free electron
A. Equal to
B. Less than
C. Greater than
D. Kinetic energies not defined for a free electron.

240. In a tetrode tube, secondary emission means the emission of ...
A. High velocity electrons from the cathode
B. Electrons form the plate due to bombardment of the fast moving electrons emited from the cathode
C. Electrons from the filament due to heat energy
D. None of these

241. Resistivity of semiconductors lies between the range of
A. 10^{-3} to 10^{3} ohm-m
B. 10^{-5} to 10^{5} ohm-m
C. 10^{-5} to 10^{3} ohm-m
D. 10^{-3} to 10^{5} ohm-m

242. In a PNP transistor in the active region of operation the base current is due to
A. Holes entering the base from the terminal for recombination
B. Holes injected into the emitter from base
C. Only saturation current due to reverse bias of collector junction
D. Electrons recombining with a part of the holes diffusing in the base region.

243. A push-pull amplifier balances out
A. Odd harmonics
B. Even harmonics
C. Both odd as well as even harmonics
D. Neither odd nor even harmonics.

244. The circuit shown is of

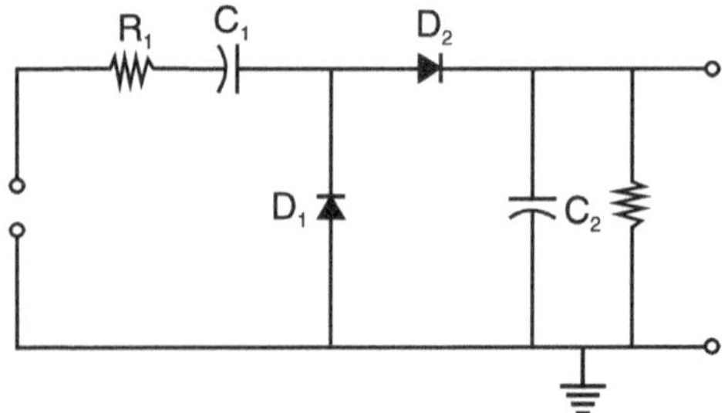

A. Bridge rectifier
B. Low frequency oscillator

C. Half wave voltage doubler

D. Full wave voltage doubler.

245. In a CE amplifier with R_C as collector load and R_e as emitter to ground resistor, the voltage gain A_v can be approximated by

A. $g_m R_c$ provided $h_{fe} R_e >> h_{ie}$

B. $\frac{R_e}{R_c}$ provided h_{fe} is large and $h_{ie} << h_{fe} R_e$

C. $\frac{R_c}{R_e}$ provided h_{fe} is large and $h_{ie} << h_{fe} R_c$

D. $\frac{R_c}{R_e}$ provided h_{fe} is large and $h_{ie} << h_{fe} R_e$.

246. For an amplifier, $R_i = 1$ K and $R_L = 100$ ohms. If the input voltage of 1 m V peak leads to a load voltage of 10 V peak, the voltage gain in decibels is

A. 70 dB　　B. 80 dB

C. 90 dB　　D. 100 dB

247. Among the three configurations the highest output impedance is obtained by

A. CE　　B. CC

C. CB　　D. CE and CB

248. In a resistance-load amplifier, the gain decreases at low frequencies because of

A. Increasing resistance of C_C

B. Decreasing resistance of C_C

C. Decreasing reactance of the stray C_S

D. All of the above

249. The work function in case of thoriated tungsten is

A. 1.1 eV　　B. 2.63 eV

C. 3.98 eV　　D. 4.52 eV

250. The output characteristics of a transistor are to be observed on oscilloscope. Which of the following is the correct procedure to achieve this?

A. Apply I_C and V_{CE} to the inputs of a dual beam CRO

B. Apply I_C and V_{CE} to as single beam CRO by disconnecting the internal sweep

C. Apply voltage across collector resistance to the Y input and V_{CE} to X input by disconnecting the internal sweep

D. Apply V_{CE} to Y input and voltage across collector resistance to X input by disconnecting the internal sweep.

251. Which of the following types of amplifier can correspond to a half-wave rectifier with reference to output waveforms?

A. Class A amplifier B. Class B amplifier

C. Class C amplifier D. None of these

252. In case of amplifiers which coupling gives the highest gain

A. Transformer coupling

B. Resistance coupling

C. Impedance coupling

D. Capacitance coupling

253. In a PNP transistor with V_{EB} set to 0 and V_{CB} bias there is a current flow. The following statements refer to this current.

1. The current is due to diffusion
2 It is dependent on V_{CB}
3. It is due to drift in the electric field
4. It is due to thermally generated minority carriers

Of these the true statements are

A. Only 1, 2　　B. Only 2, 3

C. Only 3, 4　　D. Only 4, 1

254. The high and low frequency response of an RC coupled circuit can be increased by

A. Increasing load resistance

B. Decreasing load resistance

C. Increasing coupling capacitor

D. Decreasing coupling capacitor.

255. For a p-n junction, match the type of breakdown with phenomenon

(a) Avalanche breakdown	1. Collision of carriers with crystal ions
(b) Zener breakdown	2. Early effect
(c) Punch through	3. Rupture of covalent bond due to strong electric field

Codes :	*(a)*	*(b)*	*(c)*
A.	2	1	3
B.	2	3	1
C.	1	2	3
D.	1	3	2

256. Figure shows a PNP transistor carrying an emitter current of 2 mA. The collector base junction of

the transistor has

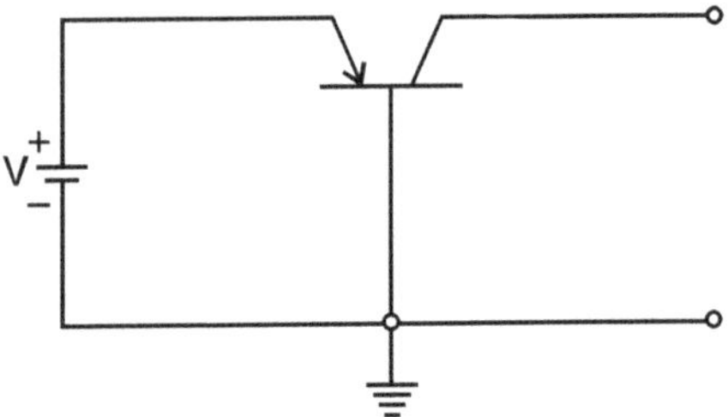

A. Forward bias
B. A reverse bias
C. No bias
D. The same bias as the emitter-base junction.

257. In a PNP transistor, hole emitter current is due to flow of 100 holes. Four of these holes recombine in the base region. The alpha and beta of the transistor would be

A. 0.95, 20 B. 0.90, 48
C. 0.96, 48 D. 0.96, 24

258. In a regulated power supply using a Zener diode the unregulated input voltage as compared to the regulated output voltage must be

A. Same
B. Smaller
C. Larger
D. Larger with opposite polarity

259. In regard to I_s the saturation current of a pn diode,

A. I_s strongly depends on the reverse voltage bias
B. I_s decreases with increasing temperature
C. I_s is due to thermally generated minority carriers
D. I_s is mainly a deffusion current at the junction.

260. For the FET amplifier shown below, the values of drain resistance R_d and source resistance R_S for a desired Q-point of (4.5 mA, 15 V), gate voltage $V_{GSQ} = 2$ V and $V_{PO} = 1$ V are given by

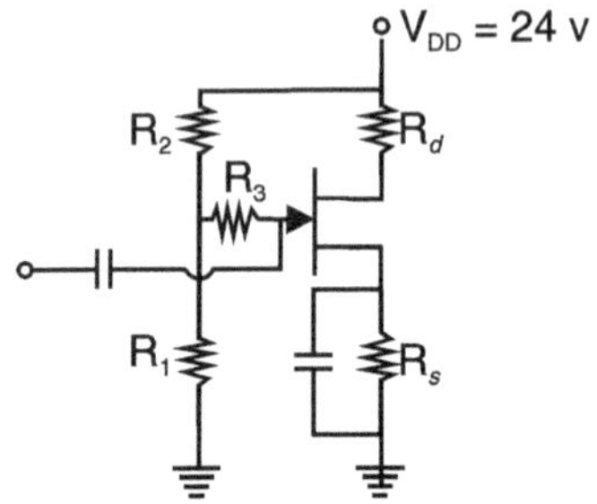

A. 10 K, 1 K B. 1.3 K, 680 ohms
C. 15 K, 1 K D. 4 K, 3000 ohms.

261. Among the three configurations the lowest input impedance is obtained by

A. CE B. CC
C. CB D. Both CB and CC.

262. If we increase the impurity concentration in the metal the residual part of the resistivity

A. Always decreases
B. Remains constant
C. Always increases
D. May increase or decrease

263. RF feedback oscillators are usually tuned by varying the

A. Bias B. Supply voltage
C. L or C D. Load impedance.

264. The normal value of relaxation time is of the order of

A. Nano seconds B. Micro seconds
C. Mili seconds D. Seconds.

ANSWERS

1	2	3	4	5	6	7	8	9	10
B	C	C	C	B	C	B	B	C	C
11	**12**	**13**	**14**	**15**	**16**	**17**	**18**	**19**	**20**
D	A	C	C	A	D	A	A	A	C
21	**22**	**23**	**24**	**25**	**26**	**27**	**28**	**29**	**30**
D	A	D	C	D	C	D	C	C	A
31	**32**	**33**	**34**	**35**	**36**	**37**	**38**	**39**	**40**
A	B	B	D	C	C	A	A	D	B

41	**42**	**43**	**44**	**45**	**46**	**47**	**48**	**49**	**50**
A	B	D	B	C	A	B	C	C	D
51	**52**	**53**	**54**	**55**	**56**	**57**	**58**	**59**	**60**
C	B	B	B	B	D	C	B	B	B
61	**62**	**63**	**64**	**65**	**66**	**67**	**68**	**69**	**70**
C	C	B	A	C	D	B	A	A	B
71	**72**	**73**	**74**	**75**	**76**	**77**	**78**	**79**	**80**
A	D	D	C	C	C	B	B	C	D
81	**82**	**83**	**84**	**85**	**86**	**87**	**88**	**89**	**90**
A	D	A	A	C	C	B	B	B	D
91	**92**	**93**	**94**	**95**	**96**	**97**	**98**	**99**	**100**
B	B	C	C	B	C	A	A	B	D
101	**102**	**103**	**104**	**105**	**106**	**107**	**108**	**109**	**110**
A	C	C	C	C	B	B	C	D	C
111	**112**	**113**	**114**	**115**	**116**	**117**	**118**	**119**	**120**
C	D	A	C	C	B	A	B	B	B
121	**122**	**123**	**124**	**125**	**126**	**127**	**128**	**129**	**130**
C	C	C	D	D	D	A	B	C	C
131	**132**	**133**	**134**	**135**	**136**	**137**	**138**	**139**	**140**
B	B	A	C	C	C	D	D	D	A
141	**142**	**143**	**144**	**145**	**146**	**147**	**148**	**149**	**150**
B	C	B	D	D	D	C	B	B	C
151	**152**	**153**	**154**	**155**	**156**	**157**	**158**	**159**	**160**
C	D	A	A	A	A	B	D	A	A
161	**162**	**163**	**164**	**165**	**166**	**167**	**168**	**169**	**170**
A	B	A	C	D	B	B	B	C	C
171	**172**	**173**	**174**	**175**	**176**	**177**	**178**	**179**	**180**
B	C	A	C	B	A	C	B	C	C
181	**182**	**183**	**184**	**185**	**186**	**187**	**188**	**189**	**190**
A	A	D	D	D	B	A	D	C	C
191	**192**	**193**	**194**	**195**	**196**	**197**	**198**	**199**	**200**
A	A	D	B	C	A	D	B	D	C
201	**202**	**203**	**204**	**205**	**206**	**207**	**208**	**209**	**210**
A	A	B	C	A	D	D	A	B	A
211	**212**	**213**	**214**	**215**	**216**	**217**	**218**	**219**	**220**
A	C	A	C	C	A	D	D	A	A
221	**222**	**223**	**224**	**225**	**226**	**227**	**228**	**229**	**230**
B	C	A	B	A	A	D	B	B	D
231	**232**	**233**	**234**	**235**	**236**	**237**	**238**	**239**	**240**
C	A	D	B	C	C	C	D	B	B
241	**242**	**243**	**244**	**245**	**246**	**247**	**248**	**249**	**250**
A	D	B	B	D	B	D	A	B	C
251	**252**	**253**	**254**	**255**	**256**	**257**	**258**	**259**	**260**
B	A	C	B	C	A	D	D	C	B

EXPLANATORY ANSWERS

95. The circuit can be redrawn to see that it is a bridge circuit.
Since the bridge is balanced, the capacitance across *a* and *b* irrespective of the value of capacitance between '*c*' and '*d*' is given by
$C = 0.1\ \mu F$

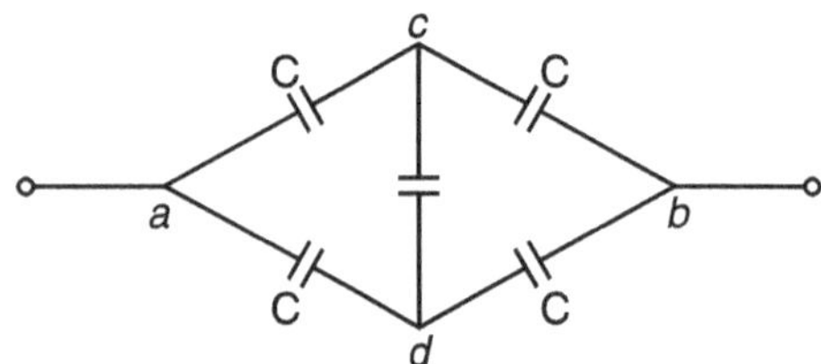

97. The bombarding electrons, striking the screen release secondary emission electrons. The secondary emission electrons are collected by an aqueous solution of graphite called *aquadag* which is collected at the second electrode. Collector of secondary electrons is necessary to keep the screen in a state of electrical equilibrium.

105. The Neel temperature plays a similar role in antiferromagnetic materials as does the Curie temperature in ferromagnetic materials. Thus above the Neel temperature, the susceptibility is observed to follow the equation, $x = \frac{C}{T+\theta}$.
Where C is the Curie constant of θ is the paramagnetic Curie temperature. Below the Neel temperature, the spin system tends to be 'ordered' in a way similar to the spm system in a ferromagnetic material, except that at $T = \theta$, half the spins are oriented in one direction and the other half in the opposite direction.

Para Ferro Antiferro

$x = \frac{C}{T}, x = \frac{C}{T}(t-\theta)$ for $T > \theta$ $X = \frac{C}{(T+\theta)}$ for $T > T_N$

The difference between the three groups of materials is illustrated in figure shown in terms of a plot of $\frac{1}{x}$ versus-temperature.

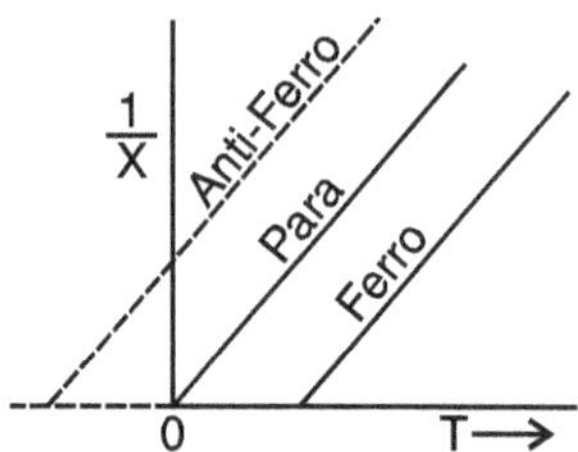

114. Transistor in active region
Assume, $V_{BE} = 0.8$ V

$$I_E = \frac{29.2}{15} = 1.95 \text{ mA} \approx 2\text{mA}$$

$V_{CE} = 0.2$ volts, and
Potential difference = 30 – 20 = 10 volts.

118. $I_{DS} = L_{DSS}\left(1 - \frac{V_{GS}}{V_r}\right)$ i.e. at $V_{GS} = 0$, I_{DS} is maximum possible.

125. A square wave when imparted to an integrating circuit causes a triangular wave output. A unit step function derivative is an impulse function. A triangular wave when differentiated will lead to a square wave. Thus only 1 and 2 are correct.

126. A Darlington pair circuit, has two emitter followers in cascade. An emitter follower is the common collector circuit. In this case current gain is high as the overall gain is the product of the individual stages.

137. Corundum, agate, diamond have been used as jewels. The oriental ruby and the sapphire are still the best stones for jewel cups.

138. i_e is input current, α is current gain, therefore
$\alpha\, i_e = i_c$

140. $20 \log . \frac{A_{DM}}{A_{CM}} = 80$

or, $\log \frac{A_{DM}}{A_{CM}} = 4$

or, $\frac{A_{DM}}{A_{CM}} = 10^4$

$\therefore$ $A_{CM} = \frac{2 \times 10^4}{10^4} = 2$

141. As α is same for both Si and Ge, germanium because of greater mobility of electrons can operate at much higher frequency than silicon.

144. When the diode conducts, the voltage across the load and hence across the capacitor increases to V_m.

During remaining part of the cycle (3/4th), on conduction occurs.

145. Given, $10 \log \frac{P_o}{P_i} = 60,$

$$\therefore \quad \frac{P_o}{P_i} = 10^6$$

$$\therefore \quad P_o = (10)^6 \times 2 \times 10^{-6} = 2\text{ W}$$

146. For CE configuration the input resistance of the circuit, is given by,

$$R_i = R_{ie} + (1 + h_{fe}) R_e$$

This does not depend on R_L.

149. Half coefficient, $K_H = \frac{V_H . w}{BI}$

where, V_H = Hall voltage, (V)

w = width of the slab in the direction of magnetic field, *(m)*

B = magnetic flux density, (wb/m^2), and

I = current (A)

$$V_H = \frac{K_H BI}{w}$$

$$= \frac{1 \times 10^{-8} \times 1 \times 3}{2 \times 10^{-3}} = 1.5 \times 10^{-5}\text{V}$$

152. Slew rate limit indicates how far the output voltage can change with time. As a large signal parameter, it defines the maximum rate at which the output can change from its most positive to its most negative values.

Slew rate = 100 V/microsecond, frequency = 10 MHz

In a cycle, the voltage varies by 10 V *p* to *p*.

For a sinewave output, the voltage will be 5 V.

224.
$$\frac{f_L^*}{f_L} = \frac{1}{\sqrt{2^{1/n} - 1}}$$

and
$$\frac{f_H^*}{f_H} = \sqrt{2^{1/n} - 1}$$

For $n = 2$

$f_H^* = 2$ MHz and $f_L^* = 30$ Hz.

$f_H \approx 3.1$ MHz; $f_L \approx 19.3$ Hz

230. Taking $R_1 \to \infty$ and

for $V_2 = 0$

$$R = R_1 \parallel R_2$$

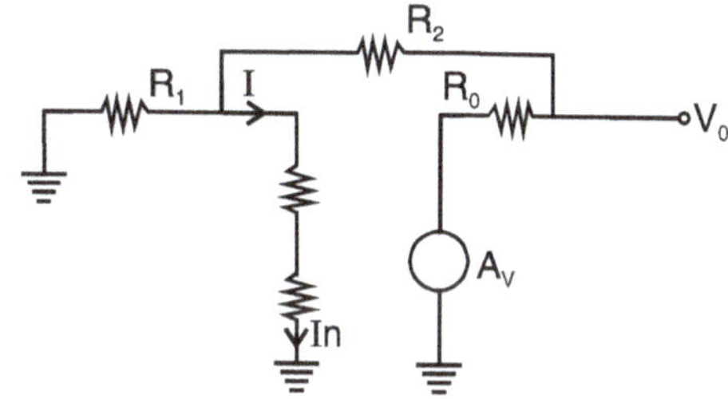

234.

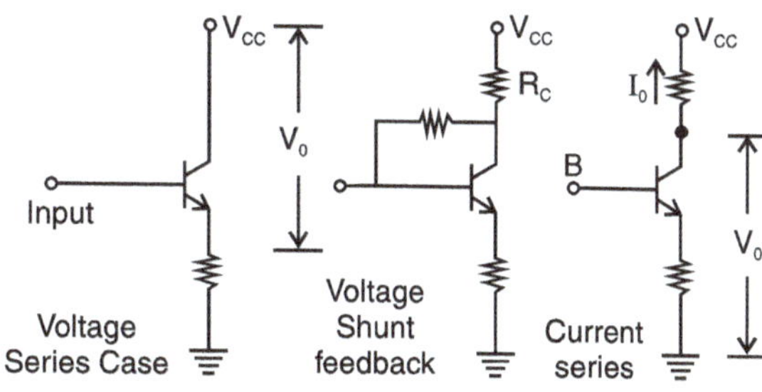

YOUR SPACE

ENGINEERING STREAM

MATERIALS AND COMPONENTS

MULTIPLE CHOICE QUESTIONS

1. A capacitor that stores 0.5 C at 10 volts has a capacitance of .... farad:
   A. 5
   B. 20
   C. 10
   D. 0.05
2. The main function of a capacitor is to:
   A. block current flow
   B. help current flow
   C. store energy
   D. dissipate heat
3. The reactance of a coil at 100 Hz is 20 Ω. Its reactance at 1 kHz would be .... ohm.
   A. 2
   B. 10
   C. 40
   D. 200
4. A coil of 2 H develops a counter emf of 4 V in it. The rate of change of current through it must be .... amperes/second.
   A. 2
   B. 0.5
   C. 8
   D. 1
5. A coil has an inductance of 1 H if current changing at the rating of 2 A/s induces:
   A. one volt in it
   B. two volts in another coil
   C. two volts in it
   D. half a volt in it
6. Point out the incorrect answer. Inductance of a coil depends:
   A. directly on the square of the number of its turns
   B. directly on the cross-sectional area of the core
   C. directly on the thickness of the wire used
   D. inversely as the length of the core
7. Ferrite cores commonly used at high frequencies:
   A. increase core losses
   B. decrease core losses
   C. decrease inductance
   D. increase resistance
8. While measuring resistance with an ohm-meter, one should not touch test leads because:
   A. of the possibility of electric shock
   B. it may lower the ohm-meter reading
   C. it may ground the resistor
   D. it may 'short' the resistor being checked
9. When checked with an ohm-meter, an 'open' resistor reads:
   A. infinite
   B. zero
   C. low but not zero
   D. high but not infinite
10. The combined resistance of the two resistors connected as in figure below is about:

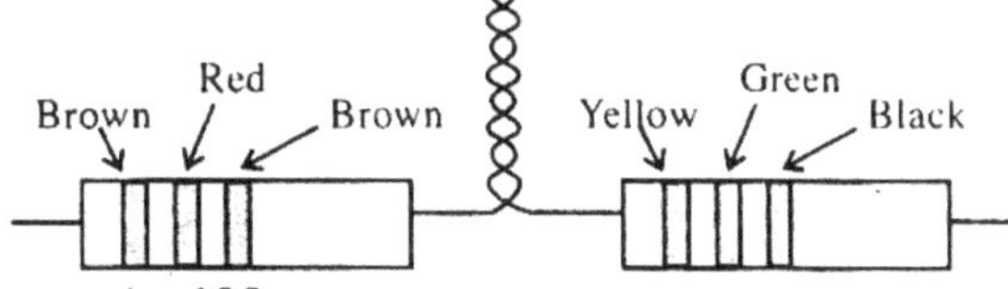

   A. 120
   B. 45
   C. 165
   D. 35
11. The colour code for the resistor shown below is

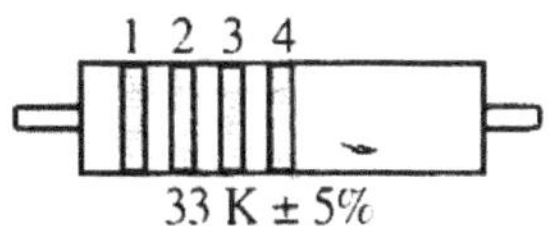

   A. red, organge, red, gold
   B. orange, orange, red, gold

C. orange, orange, orange, gold

D. red, red, violet, silver

12. The carbon-composition resistor shown in the figure below has a resistance of:

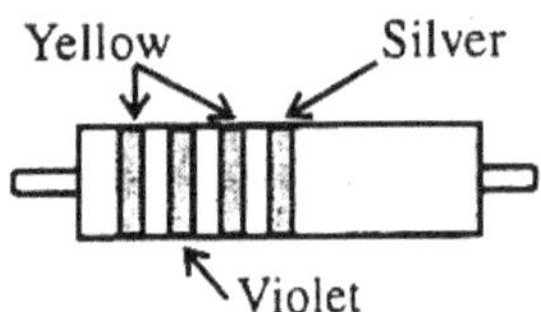

A. 270 K ± 27Ω

B. 274 ± 5%

C. 470 K ± 47K

D. 274 Ω ± 10%

13. If current through a resistor is halved, wattage developed by it would be:

A. halved

B. quadrupled

C. cut to one-fourth

D. doubled

14. The main cause of a avelanche breakdown is:

A. high doping

B. recombination of electrons and holes

C. collision ionization

D. shifting of electrons from conduction band to valence band

15. The height of potential barrier in PN junction is:

A. $\propto T^2$

B. $\propto T$

C. $\propto \frac{1}{T^2}$

D. $\propto T^3$

16. Two similar PN junctions can be connected in three different ways as shown in the figure. The two connections across which the potential difference is same are:

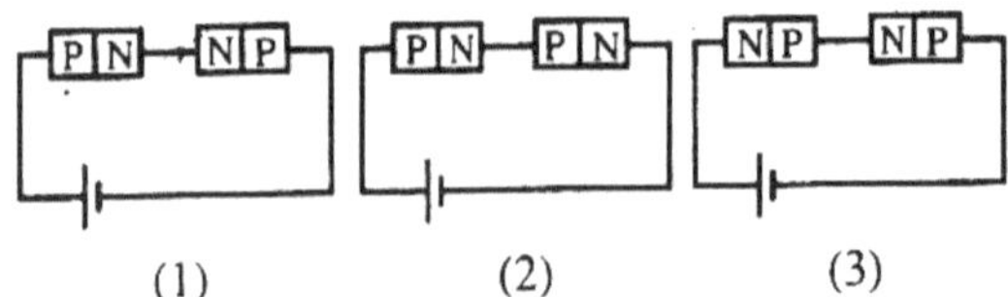

A. only circuit 1

B. circuits 1 and 2

C. circuits 2 and 3

D. circuits 3 and 1

17. If the forward voltage in a semiconductor diode is changed form 0.5 V to 2 V, then the forward current changes by 1.5 mA. The forward resistance of diode will be:

A. 1 KΩ

B. 2 KΩ

C. 4 KΩ

D. 8 KΩ

18. When the reverse potential in a semiconductor diode are 10 V and 20 V, then the corresponding reverse currents are 25 μA and 50 μA respectively. The reverse resistance of junction diode will be:

A. 40 Ω

B. 4×10^5 Ω

C. 40 KΩ

D. 4×10^{-5} Ω

19. The depletion region is:

A. region of opposite charges

B. neutral region

C. region of free current carriers

D. region of infinite energy

20. On increasing the reverse voltage in a PN junction diode the value of reverse current will:

A. gradually increase

B. suddently increase

C. remain constant

D. gradually decrease

21. On increasing current in a semiconductor diode, the contact potential will:

A. increase

B. decrease

C. remain constant

D. become zero

22. On reverse biasing the PN junction, its potential barrier becomes:

A. narrow

B. broad

C. zero

D. constant

23. For a junction diode the ratio of I_F and I_r is:

A. $\left(e^{qV/KT} - 1\right)$

B. $\left(e^{-qV/KT} + 1\right)$

C. $e^{V/KT}$

D. $e^{-V/KT}$

24. In depletion layer there are:
A. only holes
B. only electrons
C. both holes and electrons
D. neither electrons nor holes

25. The depletion layer in a silicon diode is 1 μm wide and its knee potential is 0.6 V, then the electric field in the depletion layer will be:
A. 0.6 V/m
B. 6×10^4 V/m
C. 6×10^5 V/m
D. zero

26. An optical fibre is a transparent rod usually made of:
A. clear plastic
B. glass
C. copper
D. Either (A) or (B)

27. An optical fibre excels at rejecting:
A. electromagnetic interference
B. radio-frequency interference
C. crosstalk
D. all of the above

28. In an optic fibre, the propagation angle of light must be equal to or less than the .... angle.
A. acceptance
B. incident
C. critical
D. refraction

29. The numerical aperture of an optical fibre depends on:
A. core refractive index
B. critical angle
C. Both of the above
D. None of the above

30. The structure of an optical fibre consists of
A. multimode step-index
B. graded index
C. single-mode step-index
D. All of the above

31. High-quality glass fibres have least losses in the light wavelength range of .... nanometre (nm).
A. 820-880
B. 1200-1320
C. 1550-1610
D. 1620-1710

32. The core diameter of a single-mode step-index fibre lies in the range of .... micrometre (μm).
A. 5-10
B. 10-20
C. 20-30
D. 30-50

33. Fibre-to-fibre connections are made by:
A. splicing
B. connectors
C. twisting
D. Either (A) or (B)

34. Fibre dispersion is usually defined in terms of time measured in:
A. nanoseconds
B. picoseconds
C. Both of the above
D. None of the above

35. Intermodal dispersion results from the fact that the light waves propagate through a fibre:
A. in different modes
B. along different routes
C. as a group of frequencies
D. with same velocity

36. Intermodal dispersion in a fibre depends on the refractive index of
A. core
B. cladding
C. surrounding medium
D. Both (A) and (B)

37. If some insulating material is filled between the plates of a parallel plate condenser, its capacity will:
A. decrease
B. increase
C. be infinite
D. remain constant

38. A variable parallel plate capacitor and an electroscope are connected parallel to a battery. The reading of the electroscope would be decreased by:
(1) increasing the area of overlap of the plates
(2) placing a block of paraffin wax between the plates
(3) decreasing the distance between the plates
(4) decreasing the battery potential
Then state:
A. If only (4) is correct
B. If only (1) and (4) are correct

C. If only (1) and (2) are correct
D. If only (1), (2) and (3) are correct

39. When a piece of aluminium wire of finite length is drawn through a series of dies to reduce its diameter to half its original value, its resistance will become:
A. sixteen times
B. nine times
C. three times
D. same

40. Resistance of a conductor increases with the rise of temperature, because:
A. electron density decreases
B. relaxation time decreases
C. electron mass increases
D. None of the above

41. In the figure given here find the resistance between points A and B. Both the circles and the diameter are made of uniform wire or resistance 10^{-4} ohm per m. The length of AB is 2 metre:

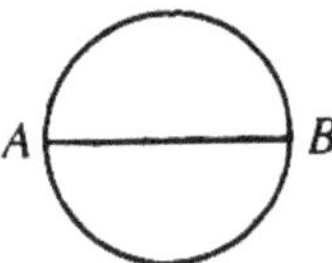

A. $0.88 \times 10^{-4}\,\Omega$
B. $0.33 \times 10^{-6}\,\Omega$
C. $6.14 \times 10^{-4}\,\Omega$
D. $9.48 \times 10^{-4}\,\Omega$

42. A stream of electrons is flowing in a solenoid as indicated below:

A. the entire solenoid behaves like the north pole of a magnet
B. face 1 behaves like south pole and face 2 like north pole
C. face 1 behaves like north pole and face 2 like south pole
D. the entire solenoid behaves like the south pole of a magnet

43. The force acting on a charge q moving with a velocity $\vec{v}$ in a magnetic field of induction $\vec{B}$ is given by:
A. $q(\vec{v} \times \vec{B})$
B. $q(\vec{v} \times \vec{B})/q$
C. $(\vec{v} \times \vec{B})/q$
D. $q/(\vec{v} \times \vec{B})$

44. Two straight horizontal parallel wires are carrying the same current in the same direction and d is the distance between the wires. You are provided with a small freely suspended magnetic needle. At which of the following positions will the orientation of the needle be independent of the magnitude of the current in the wire?
A. At a distance $d/3$ from any of the wires in the horizontal plane
B. Anywhere on the circumference of a vertical circle of radius d and centre halfway between the wires
C. At a distance $d/4$ from any of the wires
D. At points halfway between the wires in the horizontal plane

45. A galvanometer has a resistance of 5Ω and a full-scale deflection is produced by 15 milliamperes. The resistance required to enable it to read 1.5 volt will be:
A. 95 Ω in parallel
B. 15 Ω in series
C. 95 Ω in series
D. 15 Ω in parallel

46. A steel wire of length l has a magnetic moment M. It is then bent into a semicircular arc. The new magnetic moment is:
A. $3M \times l$
B. M/l
C. $2M \times \pi$
D. M/π

47. The B-H curves (A) and (B) in the adjoining figure are associated with:

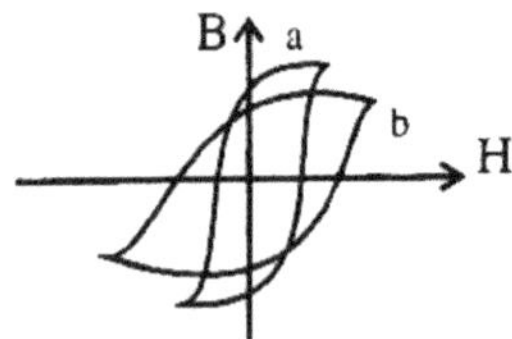

A. a paramagnetic and a ferromagnetic substance respectively
B. steel and soft iron respectively

C. soft iron and steel respectively
D. a diamagnetic and a paramagnetic substance respectively

48. The magnetic permeability is maximum for:
A. ferromagnetic
B. paramagnetic
C. diamagnetic
D. None of the above

49. The permeabilities of para and ferromagnetic materials are:
A. greater than unity
B. less than unity
C. unity
D. negative

50. Which of the following is most suitable for the core of the electromagnets?
A. Soft iron
B. Steel
C. Ag-Ni alloy
D. Silver

51. The hysteresis cycle for the material of permanent magnet is:
A. short and narrow
B. tall and wide
C. tall and narrow
D. short and wide

52. Domain formation is the necessary feature of:
A. ferromagnetic
B. paramagnetic
C. diamagnetic
D. All of the above

53. Susceptibility is positive for:
A. diamagnetic substances
B. non-magnetic substances
C. ferromagnetic substance large
D. paramagnetic substance small

54. A galvanometer of resistance 100 Ω gives a full-scale deflection for a current of 10^{-5} amp. To convert it into an ammeter capable of measuring up to one ampere connect a resistance of:
A. 9 Ω in parallel
B. 10^5 Ω in series
C. 90 Ω in series
D. 10^{-3} Ω in parallel

55. The diagram below shows two coils *A* and *B* placed parallel to each other at a very small distance. Coil *A* is connected to an AC supply. G is a very sensitive galvanometer. When the key is closed:

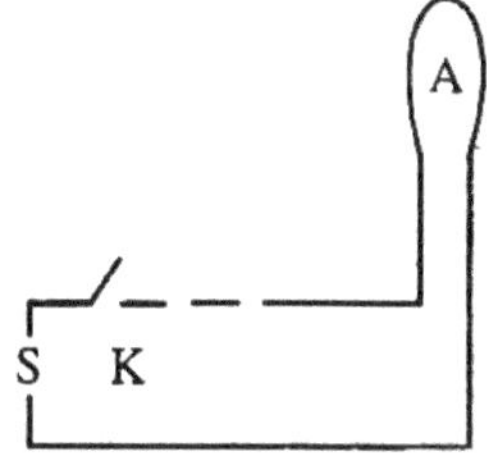

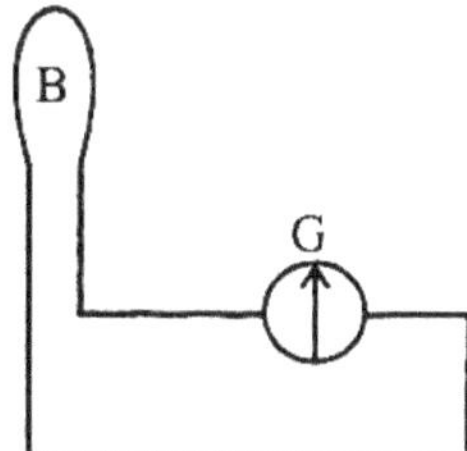

A. no variation will be observed in the galvanometer even when the input a.c. voltage is 1 or 2 Hz
B. oscillations in the galvanometer may be observed when the input a.c. voltage has a frequency of 1 to 2 Hz
C. visible small variations will be observed in the galvanometer for 50 Hz input
D. constant deflection will be observed is the galvanometer for a 50 Hz supply

56. The coefficient of self-inductance of a coil is given by:
A. $e = -Ldi/.dt$
B. $L \propto e\, dI/dt$
C. $di/dt = -L/e$
D. $-el = di/dt$

57. Two coils of self-inductance L_1 and L_2 are placed so close together that effective flux in one coil is completely linked with the other. If *M* is the mutual inductance between them then:
A. $M = 3L_1 L_2$
B. $M = \sqrt{(L_1 L_2)}$
C. $M = (L_1 L_2)^3$
D. $M = L_1 / L_2$

58. The energy of a self inductance and carrying a current *i* resides in:
A. the thermal energy due to the current
B. the electric field due to the current carriers
C. the battery that supplies the current
D. the magnetic field produced by *L*

59. The average power dissipation in a pure inductance is:
A. Li^2
B. $1/2\ Li^2$
C. $Li^2/4$
D. zero

60. Around uniformly charged infinitely long conductor, the electric field varies:
A. directly as the cube of the distance
B. inversely as the square root of the distance
C. inversely as the cube of the distance
D. inversely as the distance

61. Which of the following in motion cannot be deflected by magnetic field?
A. Alpha particles
B. Protons
C. Electrons
D. Neutrons

62. In the electromagnetic induction, the induced charge in a coil is independent of:
A. change of flux
B. resistance of the circuit
C. time
D. conductance

63. In an AC circuit containing inductance and resistance only the power factor is given by:

A. $\sqrt{R+W^2L/wL}$.

B. $\sqrt{R+wL/R}$

C. $\frac{2wL}{\sqrt{R^2+w^2+L^2}}$

D. $\frac{R}{\sqrt{w^2L^2+R^2}}$

64. When an AC flows through a pure inductance, the average power consumed is:
A. proportional to the effective current
B. proportional to the product of effective voltage and current
C. proportional to the effective resistance
D. zero

65. The resultant capacitance between A and B in the following figure is:

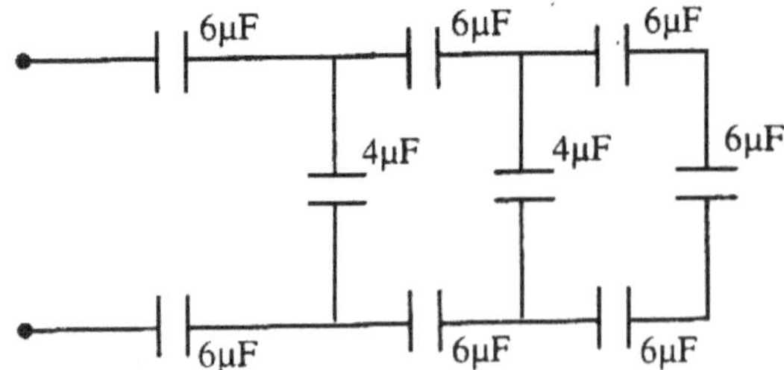

A. 3 μF
B. 2 μF
C. 9 μF
D. 8 μF

66. In the following diagram, the effective capacitance between A and B is

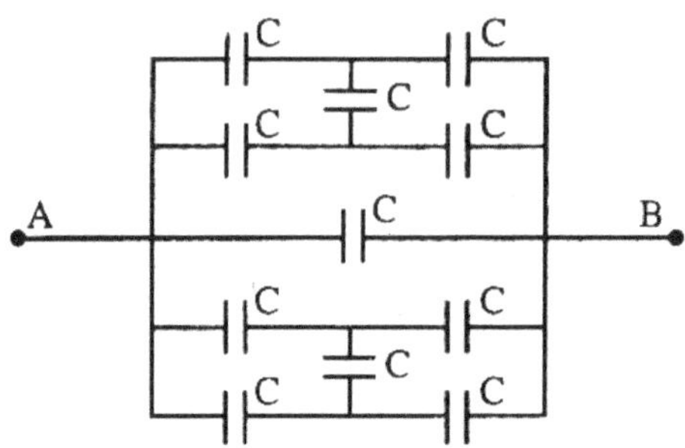

A. 12 C
B. 3 C
C. 5 C
D. 9 C

67. What is the potential difference between A and B?

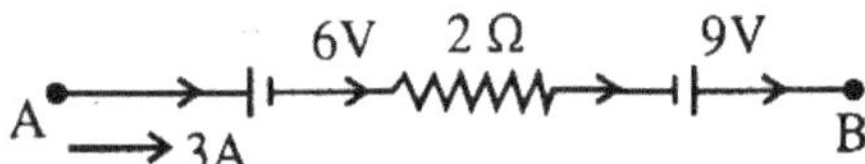

A. −9 V
B. −3 V
C. +9 V
D. +3 V

68. Three identical crossed coils A, B, and C are placed with their planes parallel to one another. Coils A and C carry equal currents as shown in the figure. Coils B and C fixed in position and coil A is moved towards B with uniform motion. Then:

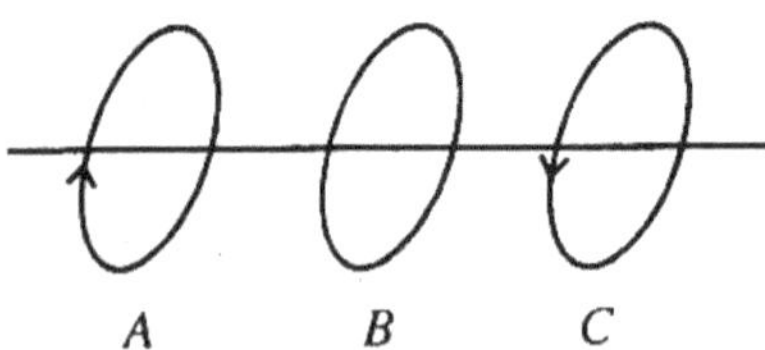

A. there is induced current in B with same direction as C
B. there is induced current in B with same direction as A
C. there is induced e.m.f. in B but no induced current
D. there is no induced current in B

69. A billion electrons are added to a pith ball. Its charge is:

A. zero

B. 1.6×10^{-10} C

C. 1.6×10^{-28} C

D. -1.6×10^{-10} C

70. In the given figure below the capacity between the points *A* and *B* is:

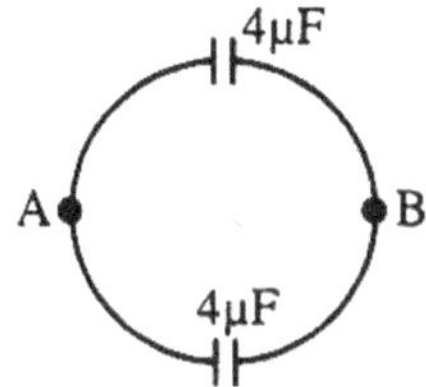

A. 16 μF

B. 8 μF

C. 7/5 μF

D. 3/8 μF

71. Capacity can be increased by connecting a number of condensers in:

A. series

B. parallel

C. in mixed grouping

D. All of the above

72. In the figure below the capacity between the points *B* and *D* is:

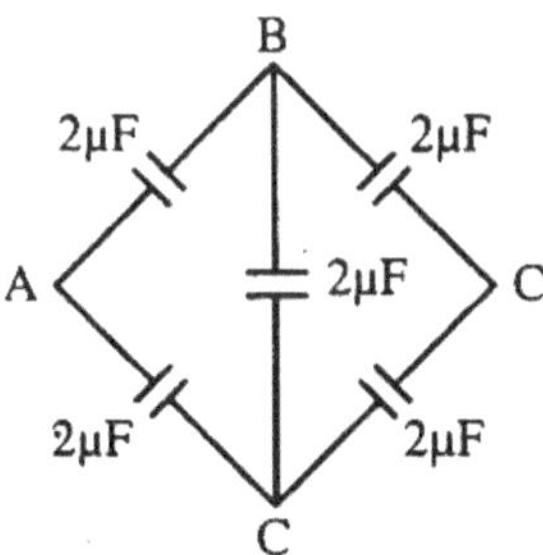

A. $\frac{3}{10}$ μF

B. $\frac{10}{3}$ μF

C. 6 μF

D. 4 μF

73. In the circuit shown in figure below, $C_1 = 3$ μF, $C_2 = 7$ μF, $E_1 = 12$ kV and $E_2 = 13$ kV. Then the p.d. across the plates of C_1 and C_2 are respectively:

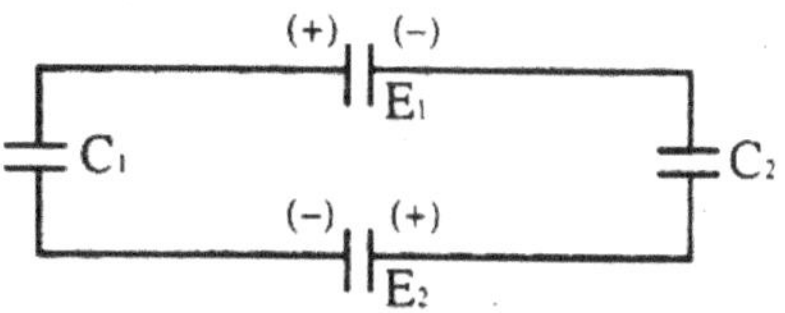

A. 22.5 kV and 22.5 kV

B. 11 kV and 17 kV

C. 14 kV and 15 kV

D. 17.5 kV and 7.5 kV

74. The resistance between the points *P* and *Q* in the figure is:

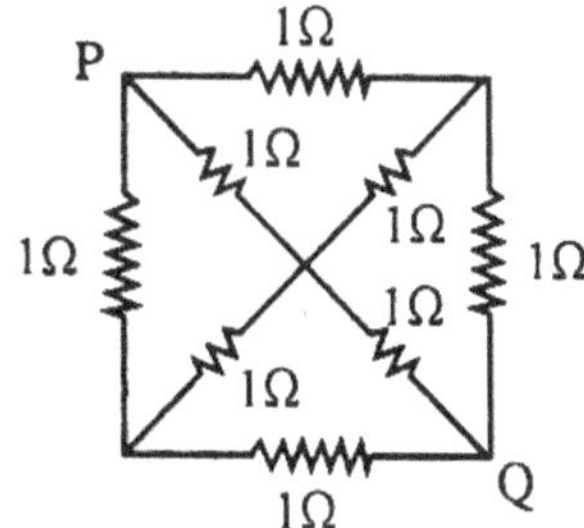

A. $\frac{2}{3}\Omega$

B. $\frac{3}{2}\Omega$

C. 5 Ω

D. 6 Ω

75. The resistance between *A* and *B* in the figure is

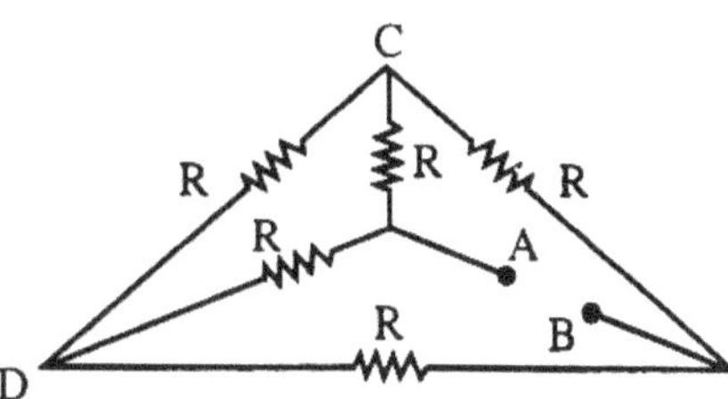

A. $\frac{R}{7}$

B. $\frac{R}{5}$

C. $7R$

D. R

76. The resistance between the points *A* and *B* in figure is:

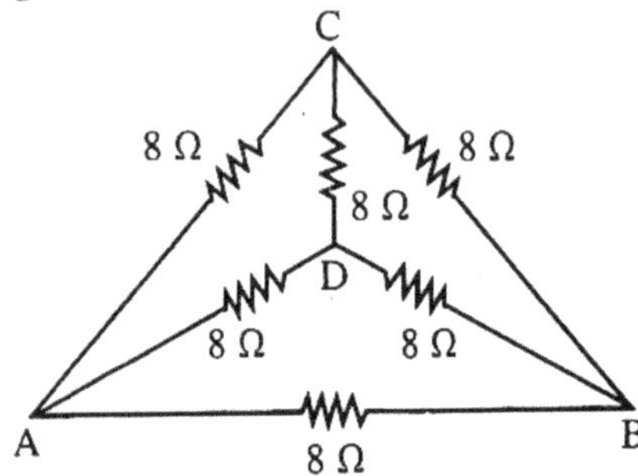

A. 36 Ω
B. 18 Ω
C. 9 Ω
D. 4 Ω

77. In figure the reading in the voltmeter is:

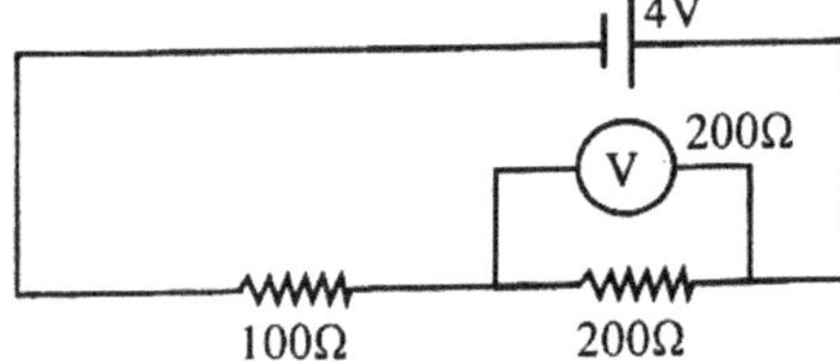

A. zero V.
B. 1.5 V
C. 3 V
D. 2 V

78. A resistance coil of 60 Ω is immersed in 42 kg of water. A current of 7 A is passed through it. The rise in temperature of water per minute is:
A. 45°C
B. 15°C
C. 0.5°C
D. 1°C

79. The strength of the magnetic field around a straight conductor
A. is directly proportional to the square of the distance from the conductor
B. is the same everywhere around the straight conductor
C. obeys inverse square law
D. None of the above

80. A straight conductor has:
A. no mutual inductance
B. no specific resistance
C. no self-inductance, no resistance
D. no resistance

81. Ferromagnetic substances have:
A. low permeability but high susceptibility
B. low permeability and susceptibility
C. very high permeability and susceptibility
D. None of the above

82. A condenser having capacity 50 microfarad is charged to 10 volts. Its energy is equal to
A. 15×10^{-2} J
B. 25×10^{-4} J
C. 12.5×10^{-2} J
D. 7.5×10^{-8} J

83. When a conductor is being charged, its potential energy:
A. goes on decreasing
B. goes on increasing
C. remains constant
D. All of the above

84. Two spherical conductors of capacities 5 cm and 2 cm are given charges to 30 e.s.u. and 20 e.s.u. respectively. They are made to share charges. Their common potential will be:
A. 3.14 e.s.u.
B. 8.14 e.s.u.
C. 2.57 e.s.u.
D. 7.14 e.s.u.

85. Some substances lose their electrical resistances at very low temperature. These substances are called:
A. semiconductors
B. non-conductors
C. good conductors
D. insulators

86. Insulators require high potentials for flow of electricity because:
A. their surfaces keep the electrons bounded together
B. there are no free electrons in them
C. there are no electron in them
D. All of the above

87. We are having three electric bulbs of 25 watts, 65 watts and 100 watts rated for 220 volts. The 25 watts lamp has the:
A. lowest resistance
B. highest resistance
C. same resistance as for 100 watts lamp
D. All of the above

88. Heat produced in a conductor depends:
A. it is yet to be decided
B. inversely on time
C. directly on time
D. Both A and B are true

89. If an insulator is to be used to support high-tension electric wire, it should preferably be made up of:
A. aluminium
B. ceramic material like porcelain
C. copper
D. wood

90. The ceramic capacitor would have .... capacitance as compared to those of other conventional capacitors.
A. lower
B. equal
C. higher
D. depends upon its size

91. Above the critical temperature of a conductor, the conductor is a:
A. normal conductor
B. super conductor
C. resistor
D. insulator

92. Superconductor can be used for many applications. One of these could be:
A. generation of power
B. transmission of power at nil losses
C. manufacture of refrigerator equipment
D. All of the above

93. The resistivity of ferrites is generally:
A. too low
B. very high
C. of moderate values
D. None of the above

94. The formula of a ferrite (having inverse spinel and cubic structure) is $MOFe_2O_3$, where M is a transition metal. Which one of the following metals would not suit this formula?
A. Mn
B. Fe
C. Co
D. Ga

95. Piezoelectric effect is not visible in the following:
A. Quartz
B. Rochelle Salt
C. Barium Titanate
D. Sodium Chloride

96. When a piezoelectric crystal is subjected to mechanical stress, it would show electric polarization. Is the reverse also true?
A. Yes
B. No
C. Depends upon the shape of the crystal
D. Depends upon the material of the crystal

97. In inverse piezo effect:
A. sound waves incident upon the crystal generate the electrical potentials
B. sound waves incident upon the crystal generate vibrations
C. sound waves do not affect the crystal
D. none of the above

98. The example of a direct piezo effect is:
A. piezoelectric microphone
B. piezoelectric loudspeaker
C. electronic watch using a piezoelectric crystal
D. All of the above

99. Which one of the following is a symbol of a piezoelectric crystal?

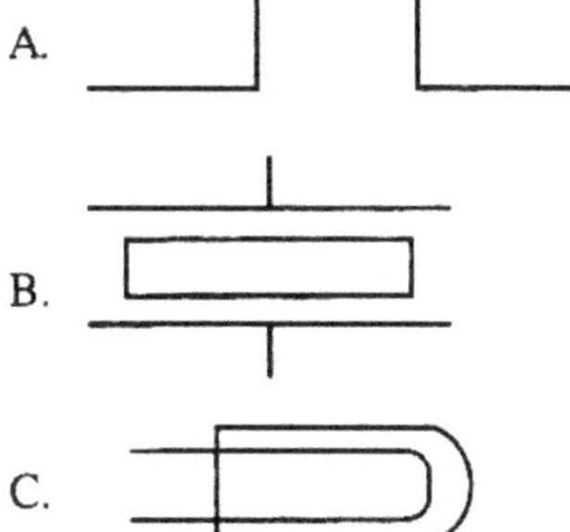

D. None of the above

100. Semiconductor materials have
A. ionic bonds
B. metallic bonds
C. covalent bonds
D. no restriction of bonding types

ANSWERS

1	2	3	4	5	6	7	8	9	10
C	C	B	D	A	B	B	C	C	A
11	**12**	**13**	**14**	**15**	**16**	**17**	**18**	**19**	**20**
D	C	D	C	B	C	A	B	A	B
21	**22**	**23**	**24**	**25**	**26**	**27**	**28**	**29**	**30**
B	B	A	D	C	D	D	C	C	D
31	**32**	**33**	**34**	**35**	**36**	**37**	**38**	**39**	**40**
C	A	D	C	A	D	B	A	A	B
41	**42**	**43**	**44**	**45**	**46**	**47**	**48**	**49**	**50**
A	B	A	D	A	C	C	A	A	A
51	**52**	**53**	**54**	**55**	**56**	**57**	**58**	**59**	**60**
C	A	C	D	B	A	B	C	D	D
61	**62**	**63**	**64**	**65**	**66**	**67**	**68**	**69**	**70**
D	C	D	D	B	B	D	B	D	B
71	**72**	**73**	**74**	**75**	**76**	**77**	**78**	**79**	**80**
B	D	D	A	D	D	D	D	A	A
81	**82**	**83**	**84**	**85**	**86**	**87**	**88**	**89**	**90**
B	C	B	D	C	B	A	C	B	C
91	**92**	**93**	**94**	**95**	**96**	**97**	**98**	**99**	**100**
B	B	B	D	D	A	A	B	B	C

Physical Electronics, Electron Devices and ICs

MULTIPLE CHOICE QUESTIONS

1. Resistivity measurements are often used to determine:
   A. carrier mobility
   B. carrier concentration in intrinsic semiconductor
   C. carrier concentration in extrinsic semiconductor
   D. lifetime of polycrystalline materials
2. When an electric field is applied to an intrinsic semiconductor at room temperature say from left to right:
   A. both electrons and holes drift to the right
   B. both electrons and holes drift to the left
   C. electrons drift to the right while the holes drift to the left
   D. electrons drift to the left while hole drift to the right
3. Energy required to break a covalent bond in a semiconductor is:
   A. always equal to 1.6 eV
   B. greater in G_e than S_i
   C. equal to the width of the forbidden energy gap
   D. is the sae in G_e as in S_i
4. In a semiconductor, rate of diffusion of charge carriers:
   A. depends on the concentration gradient and the mobility
   B. depends on the concentration gradient alone
   C. depends on the mobility alone
   D. is independent of concentration gradient and the mobility
5. Pure semiconductors are poor conductors because:
   A. they have no valence electrons
   B. all valence electrons are in electron-pairs
   C. they have a number of holes
   D. there are fewer electrons than protons
6. Forbidden energy gap in semiconductors is of the order of:
   A. 7 eV
   B. 1 eV
   C. 0.1 eV
   D. 0.05 eV
7. At 300°K, the forbidden energy gap in germanium is:
   A. 0.785 eV
   B. 1.21 eV
   C. 0.72 eV
   D. 1.1 eV
8. A 0°K, the forbidden energy gap in silicon is:
   A. 0.785 eV
   B. 1.21 eV
   C. 0.72 eV
   D. 1.1 eV
9. Mobility of a charge carrier equals:
   A. $\frac{v}{\xi}$
   B. $\frac{v^2}{\xi}$
   C. $v\,\xi$
   D. $\frac{v}{\xi^2}$

10. Mobility of holes in germanium of 300°K expressed in cm^2/V-s is:
A. 500
B. 1300
C. 1800
D. 3800

11. Mobility of holes in silicon at 300°K expressed in cm^2/V-s is:
A. 500
B. 1300
C. 1800
D. 3800

12. Diffusion constant for free electrons in silicon expressed in cm^2/s is:
A. 13
B. 40
C. 34
D. 99

13. Mobility of electrons in a semiconductor:
A. decreases as the temperature increases
B. increases as the temperature increases
C. has no effect on conductivity
D. is independent of temperature

14. Movement of charge carriers from an area of high carrier concentration to an area of low carrier concentration is called:
A. drift
B. diffusion
C. recombination
D. gradient

15. When a free electron is recaptured by a hole, the process is called:
A. recombination
B. diffusion
C. drift
D. restoration

16. Imperfections in the crystal structure result in
A. increased conductivity
B. decreased conductivity
C. increased mobility
D. decreased mobility

17. Seed crystal used in crystal growth is:
A. a small crystal formed by epitaxial growth
B. a single crystal with a specific orientation
C. a single crystal containing donor impurity
D. a single crystal containing acceptor impurity

18. The conduction band:
A. is always located at the top of the crystal
B. is the same as forbidden energy gap
C. is a range of energies corresponding to the energies of free electrons
D. is not an allowed energy band

19. A hole shifts its position basically due to:
A. a valence electron from a neighbouring atom joining the hole to fill the vacancy
B. breaking of covalent band
C. movement of atomic core
D. increase in temperature

20. Drift current in germanium is caused by
A. thermal agitation of crystal lattice
B. concentration gradient of charge carriers
C. applied electric field
D. incidence of light

21. In *n*-type germanium with boron impurity, the ionisation energy is about
A. 0.001 eV
B. 0.01 eV
C. 0.1 eV
D. 1 eV

22. If N_D and N_A are donor and acceptor concentration respectively and $N_D > N_A$, then net impurity concentration is:
A. $N_D - N_A$
B. $N_D + N_A$
C. $N_A - N_D$
D. N_D and N_A

23. In *p*-type semiconductor, with $N_A >> N_D$ minority carrier electron concentration *n* equals:
A. $N_D - N_A$
B. $N_A - N_D$
C. $\frac{n_i^2}{N_A}$
D. $\frac{n_i^2}{N_D}$

where N_D and N_A are donor and acceptor concentrations respectively.

24. If donor concentration N_D equals acceptor concentration N_A, the resulting semiconductor is:
A. *n*-type
B. *p*-type
C. intrinsic
D. may be *p* or *n* type depending on temperature

25. When a semiconductor is doped, its electrical conductivity:
A. increases
B. decreases in the direct ratio of doped material
C. decreases in the inverse ratio of doped material
D. remain unaltered

26. In any conductor, Hall voltage V_H is proportional to:
A. B
B. B^2
C. $\frac{1}{B}$
D. $\frac{1}{B^2}$
where B is the magnetic field.

27. If R_H is the Hall coefficient of a material carrying current I and subject to a transverse magnetic field B across width *w* of the specimen, the Hall voltage V_H equals:
A. $\frac{BI\,R_H}{w}$
B. $\frac{BI}{w\,R_H}$
C. $\frac{BI\,w}{R_H}$
D. $BI\,w\,R_H^2$

28. When a material carrying current I is subject to a transverse magnetic field B across width *w* of the specimen, the Hall coefficient is given by:
A. $\frac{V_H\,w\,B}{I}$
B. $\frac{V_H\,w}{BI}$
C. $\frac{V_H\,B}{w\,I}$
D. $\frac{V_H I}{w_B}$
where V_H is the Hall voltage.

29. In a given semiconductor, fermi level E_F is proportional to:
A. n
B. n^2
C. $n^{2/3}$
D. $n^{3/2}$
where *n* indicates the total number of free electrons per unit volume.

30. In an intrinsic semiconductor, the concentration of free electrons in the conduction band is given by:
A. $N_C\ \varepsilon^{-(E_C-E_F)/kT}$
B. $N_C\ \varepsilon^{-(E_F-E_C)/kT}$
C. $N_C\ \varepsilon^{-(E_C-E_F)/kT}$
D. $N_C\ \varepsilon^{-(E_F-E_C)/kT}$

31. In an intrinsic semiconductor, the concentration of charge carriers equals:
A. $A_0\,T^2\ \varepsilon^{-EG_0/kT}$
B. $A_0\,T^3\ \varepsilon^{-EG_0/kT}$
C. $A_0\,T\ \varepsilon^{-EG_0/kT}$
D. $A_0\,T^{3/2}\ \varepsilon^{-EG_0/kT}$

32. In an *n*-type semiconductor, as temperature T increases, the fermi level E_F:
A. moves towards conduction band
B. moves towards middle of forbidden energy gap
C. does not shift
D. may or may not shift depending on the concentration of donor atoms

33. In a semiconductor, the ratio of diffusion constant to mobility for a hole is proportional to:
A. T
B. T^2
C. 1/T
D. $1/T^2$
where T is the temperature in deg K.

34. In a graded semiconductor, potential of point P_2 relative to point P_1 having hole concentration p_2 and p_1 respectively equals:
A. $V_T\,l_n\,^{p_2/p_1}$
B. $V_T\,l_n\,^{p_1/p_2}$

C. $V_T\ \varepsilon^{-p_1/p_2}$

D. $V_T\ \varepsilon^{-p_2/p_1}$

where V_T is the volt equivalent of temperature.

35. In an open-circuited step graded *p-n* junction having donor concentration N_D and acceptor concentration N_A, the contact difference of potential developed is:

A. $V_T\ ln \dfrac{N_A N_D}{n_i^2}$

B. $V_T\ ln \dfrac{N_i^2}{N_A N_D}$

C. $\dfrac{1}{V_T} ln \dfrac{N_A N_D}{n_i^2}$

D. $\dfrac{1}{V_T} ln \dfrac{N_i^2}{N_A N_D}$

36. In silicon diode, reverse saturation current I_0 varies as:

A. T

B. $T^{1.5}$

C. T^2

D. $\dfrac{1}{T}$

where T is the tempeature in deg K.

37. In G_e diode, the cutin voltage is about:

A. 0.2 volt

B. 0.6 volt

C. 1.1 volt

D. 2 volt

38. In Si diode, the cutin voltage is about:

A. 0.2 volt

B. 0.6 volt

C. 1.1 volt

D. 1.75 volt

39. In a step graded reverse biased junction, the width of the depletion layer varies as:

A. $\sqrt{V_j}$

B. V_j

C. V_{j2}

D. $\dfrac{1}{V_j}$

where V_j is the magnitude of the junction potential.

40. In a forward biased *pn* diode, the injected hole current in the *n*-region is proportional to:

A. $\dfrac{1}{Q}$

B. Q

C. Q^2

D. Q^3

where Q is the total stored charge.

41. In a *pnp* transistor operating in the active region, the concentration of minority carrier holes in the *n*-region at collector junction J_e is:

A. zero

B. thermal equilibrium value p_{no} of emitter

C. thermal equilibrium concentration of hole in collector region

D. same as at J_E

42. In a G_e BJT, for usual value of the collector current, the transistor α is of the order of:

A. 0.99

B. 0.9

C. 0.5

D. 50

43. In a BJT, as the conductivity of the base region increases, the punch through voltage:

A. remains unaltered

B. increases

C. decreases

D. may increase or decrease depending on bias at J_E

44. The dynamic emitter resistance of a BJT operating in the active region is of the order of:

A. 0.01 Ω

B. 1 Ω

C. 100 Ω

D. 10 kΩ

45. In a transistor, β may be expressed in terms of α as below.

A. $\dfrac{\alpha}{1+\alpha}$

B. $\frac{\alpha}{1-\alpha}$

C. $\frac{1+\alpha}{\alpha}$

D. $\frac{1-\alpha}{\alpha}$

46. In active region operation of a transistor:
A. both junctions are reverse biased
B. both junctions are forward biased
C. Emitter junction is forward biased while collector junction is reverse biased
D. Emitter junction is reverse biased while collector junction is forward biased

47. In a BJT with $I_{CO} = 1\ \mu A$, $\alpha = 0.99$, the value of I_{CEO} is:
A. 0.01 μA
B. 0.1 μA
C. 1 μA
D. 100 μA

48. In a BJT, $I_C = 30$ mA. If $\beta = 100$, the base current approximately equals:
A. 0.03 mA
B. 3000 mA
C. 0.3 mA
D. 30 mA

49. In a BJT, the base spreading resistance is of the order of:
A. 10 Ω
B. 100 Ω
C. 1 Ω
D. 1 Ω

50. MOS transistor:
A. has only one pn-junction
B. conducts when sufficient voltage is applied to the gate electrode
C. has only two electrodes
D. has a gate elctrode in direct contact with the silicon

51. Power transistors are invariably provided with:
A. heat sink
B. metallic casing
C. soldered connections
D. fan for heat removal

52. With both junctions reverse biased the transistor operates in:
A. active region
B. cutoff region
C. saturation region
D. inverted region

53. Substrate in a monlithic IC has thickness of the order of:
A. 1 mil
B. 5 mils
C. 50 mils
D. 100 mils

54. In IC's, *npn* construction is preferred to *pnp* construction because:
A. *npn* construction is cheaper
B. *npn* construction permits higher packing of elements
C. *p*-type base is preferred
D. *n*-type collector is preferred to reduce diffusion constant

55. Solid state diffusion:
A. is an important process in fabrication of IC's
B. can be carried out at low temperature
C. can not be used to form *pn*-junction
D. result in impurities moving rapidly into silicon wafer

56. In IC technology, main merit of MOS transistor structure is:
A. lower resistance
B. lower parasitic capacitance
C. faster switching
D. higher component density and lower cost

57. Monolithic IC's:
A. are made of G*e*
B. are made using the diffused planar process
C. are made of ceramic substrate
D. use interconnection pattern formed under the final oxide

58. Extremely low power dissipation and low cost per gate can be achieved in the following IC:
A. ECL
B. CMOS
C. TTL
D. MOS

59. Thin film technology:
A. is used for fabricating active components
B. uses silk screening
C. uses vapour deposition of material on a subsrate
D. produces components cheaper than by thick film technology

60. As the complexity of a packaged IC is increased:
A. cost per gate goes on decreasing
B. package cost per gate remains the same
C. overall cost per gate reduces, reaches a minimum at some level of complexity and then increases
D. overall cost always increases with the increase of complexity

61. Very Large Scale Integration (VLSI):
A. refers to a small computer on a chip
B. can not be applied to MOS types
C. refers to IC's with complexity below 100 gates
D. does not require costly process facilities

62. One advantage of the dual-in-line package is that it:
A. is the smallest package
B. has very close lead spacing
C. was readily available when IC's were first made
D. is compatible with PCB assembly methods

63. Propagation delay t_{pd} of an IC is defined as:
A. the delay in charging the output capacitance
B. the delay in input in changing from 0 to 1 and vice-versa
C. the delay that the circuit shows between its input and output
D. the delay the transistor creates in switching from one state to the other

64. Linear MOS IC's:
A. are not possible
B. cannot use MOS load resistor
C. are difficult to fabricate
D. can be designed to use direct connection between stages

65. MOS IC's:
A. consume more power
B. occupy much less space than bipolar junction transistor
C. use load resistances
D. are preferred because of high speed

66. A typical medium gain IC Op-Amp:
A. is suitable for use at low frequencies only
B. has only one input terminal
C. has an open loop gain of about 100
D. has an open loop gain of about 2500

67. The most popularly used logic IC's are:
A. ECL and CTL
B. RTL and RCTL
C. TTL and DTL
D. simple gate types

68. CMOS input levels are given by:
A. $V_{IL} = 0; V_{IH} = 0.7\ V_{DD}$
B. $V_{IL} = 0.3\ V_{DD}; V_{IH} = 0.7\ V_{DD}$
C. $V_{IL} = 0; V_{IH} = V_{DD}$
D. $V_{IL} = 0.3\ V_{DD}; V_{IH} = V_{DD}$

69. CMOS has the following advantage over PMOS/NMOS:
A. simpler fabrication process
B. lower P_D
C. lower input capacitance
D. greater suitability for LSI

70. TTL has the following advantage over CMOS:
A. lower P_D
B. use of transistors alone as circuit elements
C. greater suitability for LSI
D. simpler fabrication process

71. The 550 IC is:
A. an Op-Amp
B. timer
C. voltage regulator
D. counter

72. The transconductance g_m of an FET in the saturation region equals:

A. $-\dfrac{2I_{DSS}}{VP}\left(1-\dfrac{V_{GS}}{V_P}\right)$

B. $-\dfrac{2I_{DSS}}{VP}\left(1-\dfrac{V_{GS}}{V_P}\right)^2$

C. $-\dfrac{2I_{DSS}}{VP}\left(1-\dfrac{V_{GS}}{V_P}\right)^{1/2}$

D. $\dfrac{I}{V_P}\ (I_{DSS} \cdot I_{DS})^{1/2}$

73. The charge carriers in a *p*-channel FET are:
A. electrons alone
B. holes alone
C. both electrons and holes
D. may be either electrons or holes

74. The dynamic drain resistance of MOSFET is of the order of:

A. 10 KΩ
B. 500 KΩ
C. 5 MΩ
D. 50 MΩ

75. The magnitude of the threshold voltage V_T for enhancement MOSFET is of the order of:
A. 4 volts
B. 10 volts
C. 40 volts
D. 100 volts

76. The input gate current of FET is:
A. a few amperes
B. a few milli-amperes
C. a few micro-amperes
D. negligibly small

77. Which of the following device has the highest input impedance:
A. CE BJT
B. CC BJT
C. JFET
D. MOSFET

78. A field-effect transistor (FET):
A. has three *pn*-junctions
B. uses a forward biased junction
C. depends on the variation of a magnetic field for its operation
D. depends on the variation of a reverse voltage its operation

79. A *pnpn* device having no gate is called:
A. UJT
B. Triac
C. Schockley diode
D. SCR

80. In a *pnpn* diode, breakover takes place when
A. $(\alpha_1 + \alpha_2) = 0.5$
B. $(\alpha_1 + \alpha_2) = 0.9$
C. $(\alpha_1 + \alpha_2) = 1.0$
D. $\alpha_1 = \alpha_2$
where α_1 and α_2 refer to the constituent transistors

81. SCR is a:
A. 2 layer device
B. 3 layer device
C. 4 layer device with one gate
D. 4 layer device with two gates

82. The advantages of SCR over SCS are:
A. larger switching time and smaller V_H
B. larger switching time and large V_H
C. smaller switching time and smaller V_H
D. smaller switching time and larger V_H

83. In an SCR, the breakover voltage V_{BO}:
A. in independent of gate current
B. increases with the increase of positive gate current
C. decreases with the increase of positive gate current
D. may increase or decrease with increase of gate current depending on temperature

84. In SCR, the turn-ON time:
A. is independent of V_g
B. decreases with increase of V_g
C. varies as V_g
D. varies as V_{g2}
where V_g is the trigger voltage amplitude.

85. In SCR, the turn-ON time:
A. is independent of ambient temperature T
B. increases with increase of T
C. varies as 1/T
D. varies as $1/T^2$

86. In SCR, the turn-OFF time:
A. increases with increase of T
B. is independent of T
C. varies as 1/T
D. varies as $1/T^2$
where T is the temperature in °K.

87. After firing an SCR, if the gate pulse is removed, the SCR current:
A. remains the same
B. reduces to zero
C. rises up
D. rises a little and then falls to zero

88. Triac is a:
A. 2 terminal bidirectional switch
B. 3 terminal bidirectional switch
C. 2 terminal unilateral switch
D. 3 terminal unilateral switch

89. Thermistor is used for measurement of power at:
A. audio frequencies
B. high frequencies
C. very high frequencies
D. microwave frequencies

90. In a photodiode, light is focussed to fall on
A. *p* region only
B. *n* region only
C. full *p* and *n* regions
D. junction region only

91. An *npn* phototransistor has typical sensitivity of the order of:
A. 25 μA/m W/cm²

B. 250 μA/m W/cm^2
C. 2.5 mA/m W/cm^2
D. 250 mA/m W/cm^2

92. LEDs have response time of the order of:
A. 0.1 ns
B. 1 ns
C. 100 ns
D. 1 μs

93. In a tunnel diode, depletion layer width is of the order of:
A. 100 Å
B. 0.1 micron
C. 1 micron
D. 5 micron

94. Tunnel diode is a *pn* diode with:
A. very high doping in *p* region
B. very high doping in *n* region
C. very high doping in both *p* and *n* regions
D. low doping in both *p* and *n* region

95. Avalanche breakdown results at applied:
A. forward bias exceeding about 6 volts
B. forward bias below 6 volts
C. reverse bias exceeding about 6 volts
D. reverse bias below 6 volts

96. In a breakdown diode, the temperature coefficient of breakdown voltage V_Z:
A. is always positive
B. is always negative
C. is always zero
D. may be positive or negative

97. At 25°C, a zener diode is rated at 2 watts. Its power rating at 50°C will be:
A. 2 watt
B. 1 watt
C. greater than 2 watts
D. much greater than 2 watts

98. LCDs have response time of the order of:
A. a few ns
B. tens of ns
C. a few milliseconds
D. hundred of milliseconds

99. Performance of the following diode is not based on its negative resistance characteristic:
A. Gunn diode
B. IMPATT diode
C. Tunnel diode
D. LSA diode

100. The main advantage of TRAPATT diode over IMPATT diode is its:
A. higher output
B. higher efficiency
C. lower noise
D. capability to operate at higher frequencies

ANSWERS

1	**2**	**3**	**4**	**5**	**6**	**7**	**8**	**9**	**10**
C	D	C	A	B	B	C	B	A	C
11	**12**	**13**	**14**	**15**	**16**	**17**	**18**	**19**	**20**
A	D	A	B	A	D	B	C	A	C
21	**22**	**23**	**24**	**25**	**26**	**27**	**28**	**29**	**30**
B	A	C	C	A	A	A	B	C	A
31	**32**	**33**	**34**	**35**	**36**	**37**	**38**	**39**	**40**
B	B	A	B	A	B	A	B	A	B
41	**42**	**43**	**44**	**45**	**46**	**47**	**48**	**49**	**50**
A	A	B	B	B	C	D	C	B	B
51	**52**	**53**	**54**	**55**	**56**	**57**	**58**	**59**	**60**
A	B	B	D	A	D	B	B	C	C
61	**62**	**63**	**64**	**65**	**66**	**67**	**68**	**69**	**70**
A	D	C	D	B	D	C	B	B	D
71	**72**	**73**	**74**	**75**	**76**	**77**	**78**	**79**	**80**
C	A	B	A	A	D	D	D	C	C
81	**82**	**83**	**84**	**85**	**86**	**87**	**88**	**89**	**90**
C	D	C	B	B	A	A	B	D	D
91	**92**	**93**	**94**	**95**	**96**	**97**	**98**	**99**	**100**
C	B	A	C	C	D	C	D	D	B

NETWORK THEORY

MULTIPLE CHOICE QUESTIONS

1. Zero of a network is the critical frequency at which network function becomes:
   A. zero
   B. unity
   C. infinite
   D. sinusoidal
2. For any impedance function *z(s)*:
   A. $z(s) = E_v\, z(s) + O_d\, z(s)$
   B. $z(s) = \dfrac{E_v z(s)}{O_d z(s)}$
   C. $z(s) = E_v\, z(s)\, O_d\, I(s)$
   D. $z(s) = \dfrac{O_d z(s)}{E_v z(s)}$
3. Poles of $E_v\, z(s)$ are:
   A. always simple
   B. always in multiple form
   C. always imaginary
   D. either simple or of higher multiplicity
4. If the impedance function $z(s)$ is positive real, then $E_v\, z(s)$ has:
   A. no zeros at origin
   B. no poles at origin
   C. no poles on imaginary axis
   D. no zeros on imaginary axis
5. An impedance function whose real part vanishes at some real frequency is called:
   A. minimum impedance function
   B. minimum reactance function
   C. minimum susceptance function
   D. minimum resistance function
6. A minimum reactance function is one which has
   A. no zeros at origin
   B. no poles at origin
   C. no zeros on imaginary axis
   D. no poles on imaginary axis
7. A polynomial $q(s)$ is Hurwitz if:
   A. $q(s)$ is real when s is real
   B. $q(s)$ is real and have real roots which are zero or negative
   C. $q(s)$ has conjugate pair of complex roots
   D. None of these
8. In an impedance function, a pole at infinity is realized by using:
   A. a capacitance in series
   B. an inductance in series
   C. an inductance in parallel with the driving point terminals
   D. None of these
9. The response of a network is decided by the location of:
   A. its zeros
   B. its poles
   C. both zeros and poles
   D. neither zeros nor poles
10. Network function $\dfrac{4s}{(s+1)(s+3)}$ has:
    A. one zero and two poles
    B. no zero and one pole
    C. two zeros and one pole
    D. one zero and no pole
11. As the poles of network shift away from the axis, the response:
    A. remains constant
    B. becomes less oscillating
    C. becomes more oscillating
    D. None of these
12. Pole of a network is a critical frequency at which network function:
    A. becomes zero
    B. becomes unity
    C. becomes infinite
    D. becomes very large

13. A resistor R of 30W is connected in series with an inductor of 9 H. The time constant of the circuits:
A. 3 second
B. 9 second
C. 27 second
D. $\frac{1}{3}$ second

14. An inductor has self-resistance of 5Ω and inductive reactance of 250Ω at the operating frequency. The quality factor of the inductor is:
A. 1250
B. 50
C. 250
D. 5

15. The Q factor of a series LCR circuit increases if:
A. R increases
B. R decreases
C. applied voltage increases
D. current decreases

16. Power factor of a purely resistive circuit is:
A. zero
B. one
C. 0.5
D. infinity

17. On increasing the Q of a coil, its power factor:
A. increases
B. decreases
C. remains unchanged
D. may increase or decrease

18. The increase the Q-factor of an inductor, it is wound with:
A. thicker wire
B. thinner wire
C. longer wire
D. wire with heavy insulation

19. In the given circuit, the equivalent inductance between terminals *a* and *b* is:

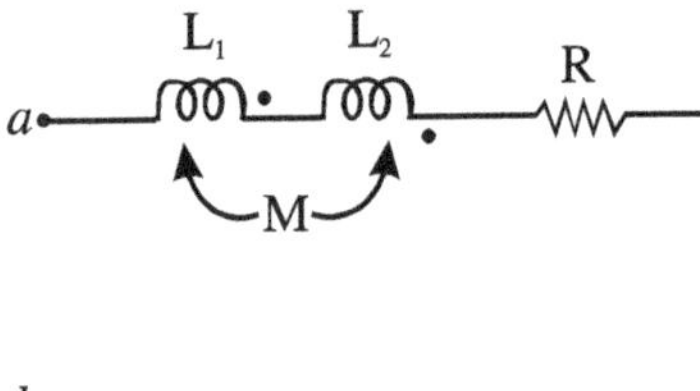

A. $L_1 + L_2$
B. $L_1 + L_2 + 2M$
C. $L_1 + L_2 - 2M$
D. $L_1 - L_2$

20. Two coupled inductors L_1 = 0.2 H and L_2 = 0.8 H have coefficient of coupling $k = 0.7$. The mutual inductance M is:
A. 0.2 H
B. 0.28 H
C. 0.112 H
D. 1 H

21. An inductor is tuned to frequency f_0 by a shunt capacitor of value 270 *p*F. The same inductor tuned to second harmonic $2f_0$ using a a capacitor of 60 *p*F. The shunt self-capacitance of the inductor is:
A. 10 pF
B. 15 pF
C. 20 pF
D. 30 pF

22. In a series LCR circuit at resonance with Q_0 = 10, and with applied voltage of 100 mV at resonance frequency, voltage across capacitor is:
A. 100 mV
B. 1 volt
C. 10 mV
D. 10 volt

23. In a resonant circuit, the power factor at resonance is:
A. zero
B. unity
C. 0.5
D. 1/2

24. A voltage of 10 mV at frequency of resonance is applied to a series LCR circuit. The Q of the circuit at resonance is 40. Then the voltage across the inductor will be:
A. 400 mV
B. 4 mV
C. 0.25 mV
D. 200 mV

25. In a series LCR circuit, the value of current at resonance is affected by the value of:
A. only L
B. only C
C. both L and C
D. only R

26. A parallel tuned LCR circuit resonates at 1000 kHz. At frequency of 990 kHz, the circuit impedance is:
A. inductive
B. capacitive
C. resistive
D. maximum

27. A parallel tuned LCR circuit resonates at 1000 kHz. At frequency of 1005 kHz, the circuit impedance is:
A. inductive
B. capacitive
C. resistive
D. maximum

28. A series LCR circuit resonates at 1000 kHz. At frequency of 995 kHz, circuit impedance is:
A. resistive
B. capacitive
C. inductive
D. minimum

29. A series LCR circuit resonates at 1000 kHz. At frequency of 1005 kHz, the circuit impedance is:
A. resistive
B. capacitive
C. inductive
D. minimum

30. In a parallel tuned circuit, f_1 and f_2 are the 3-*d*B frequencies and f_0 is the frequency of resonance. The circuit Q is given by:

A. $\frac{f_2 - f_1}{f_0}$

B. $\frac{f_2 + f_1}{\sqrt{2} f_0}$

C. $\frac{f_0}{f_2 - f_1}$

D. $\frac{2 f_0}{f_2 - f_1}$

31. A series LCR circuit resonates at 3 MHz and has 3-*d*B bandwidth of 10 kHz. The Q of the circuit at resonance is:
A. 30

B. $\frac{300}{\sqrt{2}}$

C. 300
D. $300\sqrt{2}$

32. At 3-*d*B frequencies, current in the series LCR circuit equal current at resonance multiplied by:

A. $\frac{1}{2}$

B. $\frac{1}{\sqrt{2}}$

C. $\frac{1}{4}$

D. $\frac{1}{2\sqrt{2}}$

33. A parallel tuned circuit resonating at frequency ω_0 (radians/second) has impedance at resonance equal to:

A. $\frac{\omega_0 L}{Q_0}$

B. $\omega_0 L Q_0$

C. $\frac{\omega_0 Q_0}{L}$

D. $\omega_0 L Q_0^2$

where Q_0 is the circuit at resonance.

34. In the given parallel tuned circuit, at parallel resonance, the impedance of the circuit is:

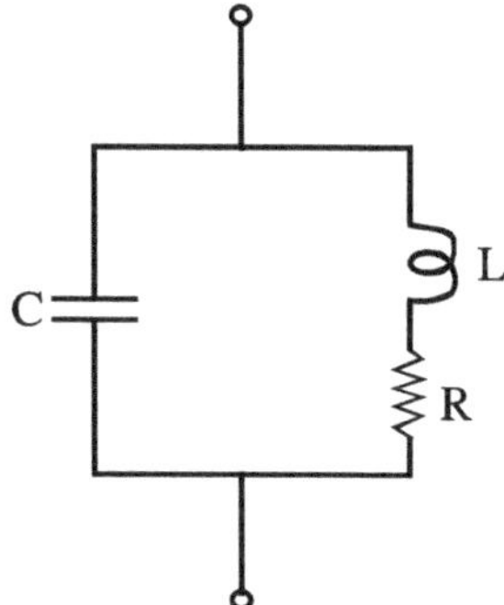

A. $\frac{L}{CR}$

B. $\frac{LC}{R}$

C. $\frac{R}{LC}$

D. LCR

35. In a series LCR circuit, 3-dB bandwidth B (Hz) equals:

A. $\frac{f_0}{Q_0}$

B. $\frac{f_0}{2\pi Q_0}$

C. $\frac{f_0}{\sqrt{2}Q_0}$

D. $\frac{f_0}{2Q_0}$

where f_0 is the frequency of resonance in Hz and Q_0 is the circuit Q at resonance.

36. In an LCR series circuit, the impedance at resonance is:

A. zero

B. R

C. infinity

D. $\frac{L}{CR}$

37. A parallel turned LC circuit resonates at 2 MHz and has Q = 100 at resonance. Its 3-dB bandwidth is:

A. 10 kHz

B. 20 kHz

C. $\frac{10}{\sqrt{2}}$ kHz

D. $\frac{20}{2\pi}$ kHz

38. A coil with reactance $X_L = 200\Omega$ at frequency of 1 MHz and internal resistance $R = 10\Omega$ has Q of:

A. 20

B. 10

C. 200

D. 2000

39. When two 2-port networks are connected in parallel, it is convenient to use:

A. z parameters

B. y parameters

C. h parameters

D. inverse-h parameters

40. When two 2-port networks are connected in parallel, it is convenient to use:

A. open circuit impedance parameters

B. short circuit admittance parameters

C. hybrid parameters

D. transmission parameters

41. The Z- parameters of the shown T-network are given below:

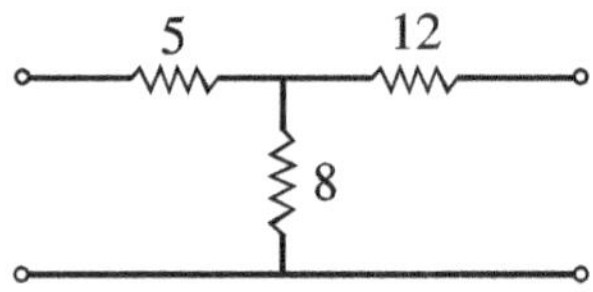

A. 13, 8, 8, 20

B. 8, 20, 13, 12

C. 5, 8, 12, 0

D. 5, 8, 8, 12

42. In a two-port network, short-circuit admittance parameters express:

A. V_1, V_2 in terms of I_1, I_2

B. I_1, I_2 in terms of V_1, V_2

C. V_1, I_1 in terms of V_2, I_2

D. V_1, I_2 in terms of V_2, I_1

43. In a two-port network, hybrid parameters express:

A. V_1, V_2 in terms of I_1, I_2

B. I_1, I_2 in terms of V_1, V_2

C. V_1, I_1 in terms of V_2, I_2

D. V_1, I_2 in terms of V_2, I_1

44. In a two-port network, open circuit impedance parameters express:

A. V_1, V_2 in terms of I_1, I_2

B. I_1, I_2 in terms of V_1, V_2

C. V_1, I_1 in terms of V_2, I_2

D. V_2, I_2 in terms of V_1, I_1

45. A two-port network is symmetrical if:

A. $Z_{12} = Z_{21}$

B. AD – BC = 1

C. $Z_{11} = Z_{22}$

D. $h_{12} = -h_{21}$

46. A source V volts having internal impedance of $(R + jx)$ is connected to the load $(R - jx)$. The power transferred to the load is:

A. $\frac{V^2}{4R}$

B. $\frac{V^2}{2R}$

C. $\frac{V^2}{R}$

D. $\frac{V^2}{4R^2}$

47. If the source impedance is capacitive, for maximum transfer of power from the source to the load, the load should be:
A. capacitive
B. resistive
C. complex conjugate of complex source impedance
D. exactly the same as the source impedance

48. In an a.c. circuit having terminal impedance of $(R + jx)$ for maximum transfer of power the load impedance should be:
A. R
B. $R - jx$
C. $R + jx$
D. x

49. A network is said to be nonlinear if it does not satisfy:
A. superposition condition
B. homogeneity condition
C. both superposition and homogeneity conditions
D. associative condition

50. In a linear system, several sources acting simultaneously produce an effect which is the sum of the separate effects caused by individual sources acting at a time. This is:
A. Reciprocity theorem
B. Millman theorem
C. Superposition theorem
D. Norton's theorem

51. The load resistance which should be conneced between terminals A and B for maximum transfer of power source to the combined load is:

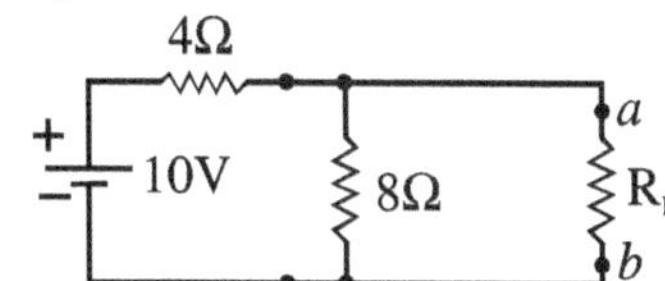

A. 4Ω
B. 8Ω
C. 12Ω
D. $\frac{8}{3}\Omega$

52. A d.c. voltage source 10 V with internal resistance of 3Ω is connected to a resistor R_L. Maximum transfer of power from source to load takes place when R_L equals:
A. 0Ω
B. 3Ω
C. 9Ω
D. Ω

53. Norton's theorem results in:
A. a current source with an impedance in parallel
B. a voltage source with an impedance in series
C. a voltage source alone
D. a current source alone

54. The superposition theorem is applicable to
A. current only
B. voltage only
C. both current and voltage
D. current, voltage and power

55. Thevenin's equivalent impedance between terminals *a* and *b* in the given circuit is:

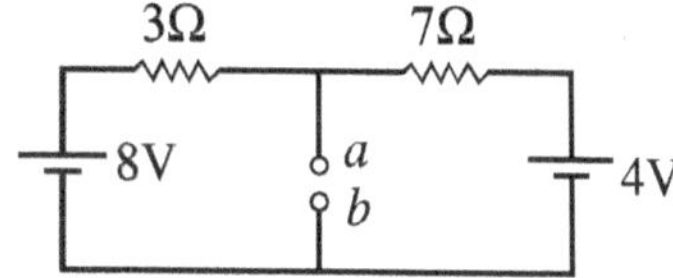

A. 10 Ω
B. 2.1 Ω
C. 5 Ω
D. 6.5 Ω

56. Thevenin's equivalent impedance of the given circuit is:

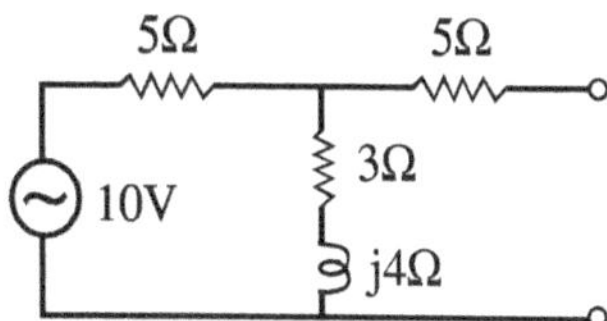

A. $7.5 + j\,1.25\ \Omega$
B. $2.5 + j\,6.25\ \Omega$
C. $10 + j\,1.25\ \Omega$
D. $13 + j\,4\ \Omega$

57. To find the Thevenin's equivalent impedance in a circuit:
A. all independent voltage sources are short-circuited and all independent current sources are open-circuited

B. all independent voltage sources are open-circuited and all independent current sources are short-circuited
C. all independent voltage and current sources are short-circuited
D. all independent voltage and current sources are open-circuited

58. A parallel LCR circuit has R = 10,000 ohms, L = 10 m H and C = 1 μF. The reasonant frequency ω_0 (rad/sec) and Q are respectively given by:
A. 10^4 and 200
B. 10^2 and 1
C. 10^4 and 100
D. 10^2 and 100

59. The elements in LCR series circuit are:
A. resistance only
B. inductance only
C. capacitance only
D. all of the above

60. The value of current at resonance in a series LCR circuit is affected by the value of:
A. R
B. C
C. L
D. all of these

61. In LCR circuits, the current at resonance is:
A. maximum in series circuit and minimum in parallel circuit
B. maximum in parallel circuit and minimum in series circuit
C. maximum in both the circuits
D. minimum in both the circuits

62. In an LCR series circuit, the impedance at resonance is:
A. maximum
B. minimum
C. infinity
D. zero

63. The power factor of a series LCR circuit at its half-power points is:
A. unity
B. lagging
C. leading
D. lagging or leading

64. Selectivities of different resonance circuits are compared in terms of their:
A. impedances
B. reactances
C. frequencies
D. bandwidths

65. A resonance curve for a series circuit is a plot of frequency versus:
A. current
B. voltage
C. impedance
D. reactance

66. Maximum power from a source having internal resistance R_i is delivered to a resistive load R_L if:
A. $R_i = R_L$
B. $R_i > R_L$
C. $R_i < R_L$
D. $R_i = R_L^2$

67. System function $H(s) = \frac{1}{(s+2)}$, for a signal cos $2t$, the steady state response is:
A. $\frac{1}{2}\cos 2t$
B. $\frac{1}{2\sqrt{2}}\cos(2t - 45)$
C. $\frac{1}{2\sqrt{2}}\cos(2t + 45)$
D. $\frac{1}{2\sqrt{2}}\cos 2t$

68. Laplace transform of shifted unit step function $u(t + a)$ is:
A. $\frac{1}{s}c^{-as}$
B. $\frac{1}{s}e^{as}$
C. $\frac{1}{s}e^{-2as}$
D. $\frac{1}{s}e^{+2as}$

69. The fourier transform of a unit step function is given by:

A. $f(j\omega) = \frac{1}{j\omega}$

B. $F(j\omega) = \frac{j}{\omega}$

C. $F(j\omega) = j\omega$

D. $F(j\omega) = -j\omega$

70. For the given circuit, the Laplace transform of current for input $v_i = 5u(t)$ is:

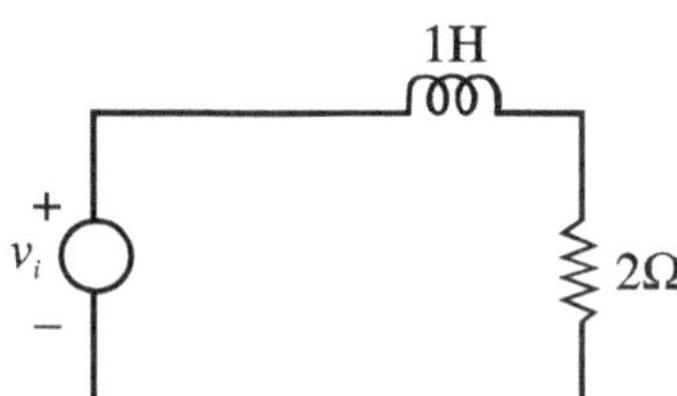

A. $\frac{5}{(s-1)(s+2)}$

B. $\frac{5}{s(s+2)}$

C. $\frac{5}{s(s+1)}$

D. $\frac{5}{(s+1)(s+2)}$

71. Laplace transform of $e^{-at} \sin \omega t$ is:

A. $\frac{\omega}{(s-a)^2 + \omega^2}$

B. $\frac{\omega}{(s+a)^2 + \omega^2}$

C. $\frac{\omega}{(s-a)^2 - \omega^2}$

D. $\frac{\omega}{(s+a)^2 - \omega^2}$

72. In an overdamped second order electrical network whose characteristic equation is $s^2 + a_1 s + a_2 = 0$, the coefficient bear the following relation:

A. $a_1^2 < 4a_2$

B. $a_1^2 = 4a_2$

C. $a_1^2 > 4a_2$

D. $a_1^2 > a_2$

73. If all the n sources in a linear network are multiplied by a factor m, the response is multiplied by:

A. n
B. m
C. mn
D. n/m

74. In a network, if the ratio of response transform to excitation transform is invariant to an interchange of their position, then the network is paid to be:

A. linear
B. non-linear
C. reciprocal
D. non-reciprocal

75. In a second order system $s^2 + 2\xi\omega_n s + \omega_n^2 = 0$. If $\xi^2 < 1$, then the roots are:

A. complex conjugate pair
B. conjugate imaginary
C. negative real and equal
D. negative real and unequal

76. The response of a critically damped second order system is of the form:

A. $e^{\sigma_1 t}(k_1 \cos \omega_1 t + k_2 \sin \omega_1 t)$

B. $k_1 e^{s_1 t} + k_2 t e^{s_2 t}$

C. $k_1 e^{s_1 t} + k_2 e^{s_2 t}$

D. $k_1 \cos \omega_1 t + k_2 \sin \omega_1 t$

77. If the roots of the characteristic equation $s^2 + a_1 s + a_2 = 0$ are negative, real and unequal, then the system is:

A. critically damped
B. underdamped
C. over damped
D. undamped

78. The frequency at which maximum voltage occurs across the capacitor in a series LCR circuit is:

A. $\frac{1}{2\pi\sqrt{LC}}$

B. $\frac{1}{2\pi}\sqrt{\frac{1}{LC}-\frac{R^2}{2C^2}}$

C. $\frac{1}{2\pi}\sqrt{\frac{1}{L^2C^2}-\frac{R^2}{2L}}$

D. $\frac{1}{2\pi\sqrt{LC}}\sqrt{\frac{R^2}{2L^2}}$

79. Frequency at which maximum voltage results across the inductor in a series LCR circuits is:

A. $\frac{1}{2\pi\sqrt{LC}}$

B. $\frac{1}{2\pi\sqrt{LC-R^2}}$

C. $\frac{1}{2\pi\sqrt{LC-\frac{C^2R^2}{2}}}$

D. $\frac{1}{2\pi\sqrt{\frac{LC-C^2R^2}{2}}}$

80. In a parallel LCR circuit, at reasonant frequency:
A. line current is maximum
B. total impedance is minimum
C. inductive branch current is minimum
D. total impedance is maximum

81. In a series LCR circuit, at the resonant frequency:
A. current is minimum
B. voltage across C is maximum
C. impedance is maximum
D. current is maximum

82. Double energy transients occur in
A. R-L circuits
B. R-C circuits
C. R-L-C circuits
D. None of these

83. Two electrical systems obeying equations of the same form are called:
A. similar systems
B. equivalent system
C. analogous system
D. identical system

84. The response of a network is of the form ke^{st} where $s = \sigma + j\omega$. Then ω is known as:
A. radian frequency
B. neper frequency
C. complex frequency
D. frequency in Hertz

85. In a series LCR circuit, complex angular frequency of an underdamped system equals:
A. $\zeta\omega_n \pm j\,\omega_d$
B. $-\omega_n \pm j\,\zeta\omega_d$
C. $\omega_d \pm j\zeta\omega_n$
D. $-\zeta\,\omega_n \pm j\,\omega_d$
where ω_n is undamped natural angular frequency and ω_d is the damped angular frequency.

86. In a series LCR circuit, the damped angular frequency of oscillation ω_d equals:
A. $\zeta\omega_n$
B. $\zeta^2\omega_n$
C. $\omega_n(1-\zeta)$
D. $\omega_n\sqrt{1-\zeta^2}$

where ω_n is undamped natural angular frequency.

87. A 5μF capacitror is charged from a 10 volt battery through a 200Ω series resistor. The capacitor will charge to its maximum value in time:
A. 1 m-s
B. 5 m-s
C. 10 m-s
D. 20 m-s

88. When an CR network is connected to a d.c. voltage source, the capacitor C gets charged to the input voltage in time period approximately equal to:
A. 2 CR
B. 5 CR

C. 10 CR

D. 20 CR

89. A capacitor charges from a d.c. source through a series resistor R. Its charging current is:

A. an exponential growth function

B. an exponential decay function

C. a linerar growth function

D. a linear decay function

90. In the given circuit, current I in the 3Ω resistor is given by:

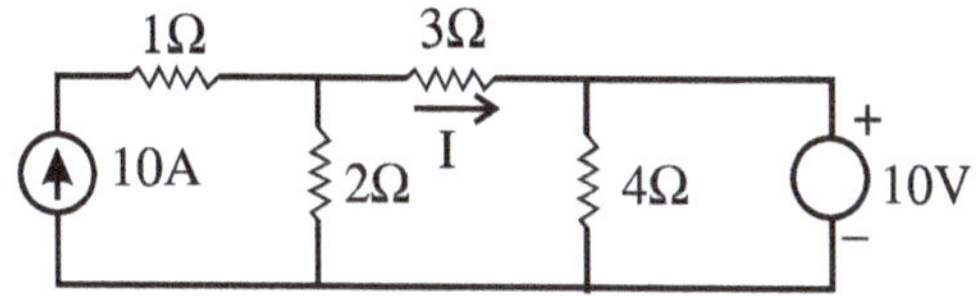

A. 2A

B. –2A

C. 6A

D. 5A

91. Transient current in a circuit results from:

A. voltage applied to the circuit

B. impedance of the circuit

C. resistance of the circuit

D. changes in the stored energy in inductors and capacitors

92. In a highly inductive circuit, a small capacitance is added in series. The angle between the applied voltage and resultant current will:

A. increase

B. decrease

C. remain absolutely unaltered

D. alter insignificantly

93. In the given circuit, switch S is opened at time $t = 0$. Then $\frac{dv}{dt}(0^+)$ is:

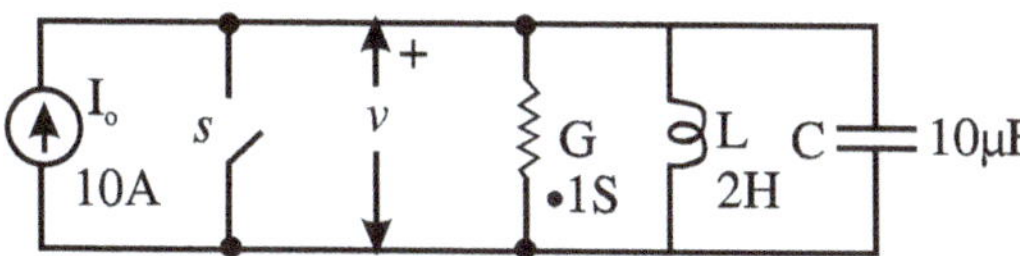

A. 10^6 volt/sec

B. 100 volt/sec

C. 10^5 volt/sec

D. 10 volt/sec

94. A capacitor C is charged to voltage V. It is connected to a resistor R at time $t = 0$. The voltage across the capacitor at time t equals:

A. $V\varepsilon^{-t/RC}$

B. $V\varepsilon^{t/RC}$

C. $V(1 - \varepsilon^{-t/RC})$

D. $V(1 - \varepsilon^{t/RC})$

95. A d.c. voltage V = 10 volts is applied to a series circuit consisting of resistor R = 2Ω and capacitor C = 10 μF. The voltage across capacitor C after one time constant is:

A. 6.3 V

B. 3.7 V

C. 3.2 V

D. 1.85 V

96. A d.c. voltage V is applied at time $t = 0$ to a series RC circuit. The voltage across capacitor C at time t is:

A. $V(1 - \varepsilon^{-t/RC})$

B. $V\,\varepsilon^{-t/RC}$

C. $V(1 - \varepsilon^{t/RC})$

D. $V\,\varepsilon^{t/RC}$

97. A d.c. voltage V is applied at time $t = 0$ to a series RC circuit. The steady state voltage across capacitor C is:

A. V

B. VC/R

C. $\frac{V}{\sqrt{R^2 + C^2}}$

D. zero

98. A d.c. voltage V is applied at time $t = 0$ to a series RC circuit. The steady state current is:

A. V/R

B. $\frac{V}{C}$

C. $\frac{V}{\sqrt{R^2 + C^2}}$

D. zero

99. A d.c. voltage V is applied to series RL circuit at time $t = 0$. The current at time t is equal to:

A. $\frac{V}{R}\varepsilon^{-\frac{Rt}{L}}$

B. $\frac{V}{R}\left(1 - \varepsilon^{-\frac{Rt}{L}}\right)$

C. $\frac{V}{R}\varepsilon^{\frac{Rt}{L}}$

D. $-\frac{V}{R}\left(1-\varepsilon^{\frac{-Rt}{L}}\right)$

100. In the circuit shown below the voltage function is $v(t) = 150 \sin \omega t$. The average power in the resistance R will be:

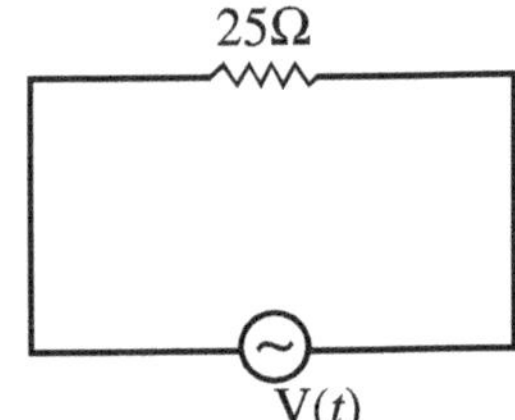

A. 300 W
B. 450 W
C. 750 W
D. 700 W

101. When the same sequence of concepts and procedures characterise both the loop and node methods, the principal quantities and concepts involved in the two methods form pairs. These are called:

A. similar
B. complementary
C. dual
D. supplementary

102. In an electrical circuit, a set of branches forming a closed path which passes through a node only once is called:

A. branch
B. node
C. loop
D. mesh

103. Graph of an electrical network has number of tree branches equal to n. The total number of branches in the network is equal to b. Then for analysis, the minimum number of link current l equals:

A. $b-n$
B. n
C. $n-b$
D. $b+n$

104. A 10 volt source is connected to a resistor R = 10Ω. How much resistance should be added in series with R in order to reduce the current to one-third its original value:

A. 10Ω
B. 20Ω
C. 30Ω
D. 40Ω

105. A battery is connected across the series combination of two resistors of value 10Ω and 30Ω. The voltage across 10Ω resistor is 20 volts. The battery voltage will be:

A. 40 volts
B. 80 volts
C. 20 volts
D. 160 volts

106. Four electric bulbs each 100 watts, 220 V are connected in series across a 220 V, 50 Hz supply. Then

A. voltage across each bulb is 220 volts
B. power dissipated in each bulb is not the same
C. current through each bulb is the same
D. total power dissipated in all the four bulbs is 1000 watts

107. A positive charge Q_1 is moved toward another positive charge Q_2 through a distance 2 m with a force of 5 newtons. The change in the potential energy of the system is:

A. 5 N-m
B. 10 N-m
C. 20 N-m
D. 50 N-m

108. A cube is constructed by soldering together twelve 1Ω resistance wires to represent the 12 edges. The resistance between any two diagonally opposite corners is:

A. 1Ω
B. 12Ω
C. 5/6Ω
D. 6Ω

109. The voltage v_1 and v_2 in the given circuit are

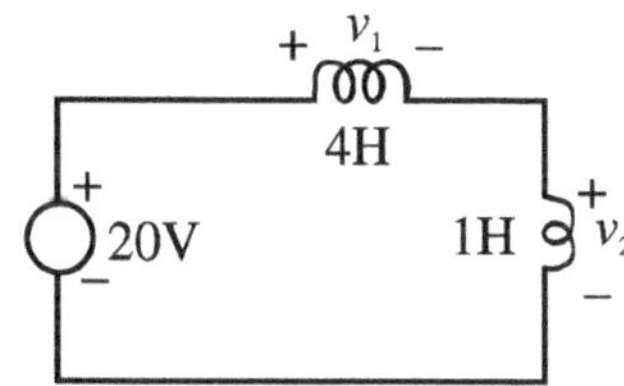

A. 16 V, 4 V
B. 4 V, 16 V
C. 10 V each
D. 20 V each

110. Two resistances are connected in parallel and each dissipates 40 watts. The total power supplied by the source equals:
A. 40 W
B. 80 W
C. 160 W
D. 20 W

111. Energy W in an inductor L carrying current i equals:

A. $\frac{1}{2}Li^2$

B. Li^2

C. $L\frac{di}{dt}$

D. Li

112. Which of the following is not a non-linear element?
A. Electric arc with unlike electrodes
B. Diode
C. Transistor
D. Heater coil

113. The following constitutes a passive circuit element:
A. FET
B. BJT
C. Diode
D. a current source

114. The following constitutes an active element:
A. a resistor
B. a capacitor
C. an inductor
D. LCR network

115. Voltages on L, C, R in a series circuit are shown below. Value of voltage source is:

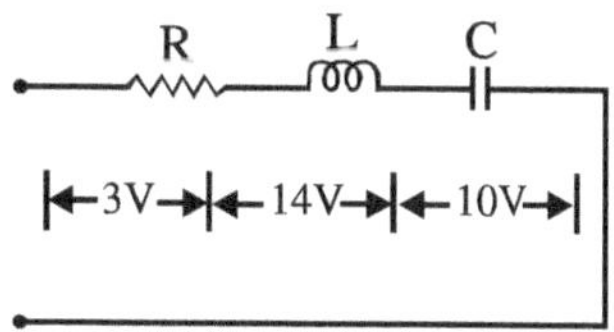

A. 10 V
B. 15 V
C. 27 V
D. 5 V

116. For the circuit shown below different time constants are given:
1. 0.5×10^{-3}
2. 2×10^{-3}
3. 0.25×10^{-3} s
4. 10^{-3} s

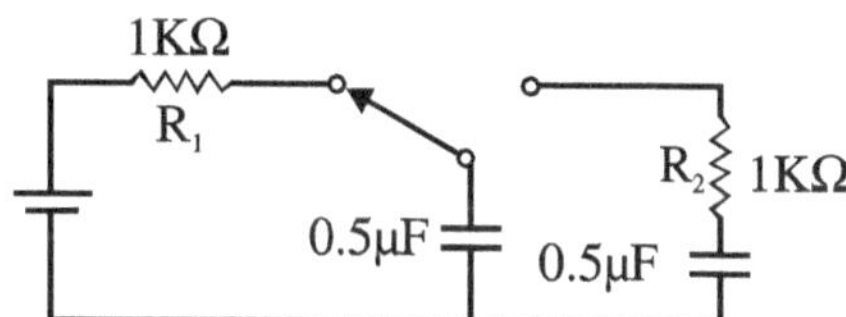

What are the changing and discharging times respectively:
A. 1, 2
B. 2, 3
C. 1, 3
D. 2, 4

117. Find the Norton's equivalent of the circuit given below:

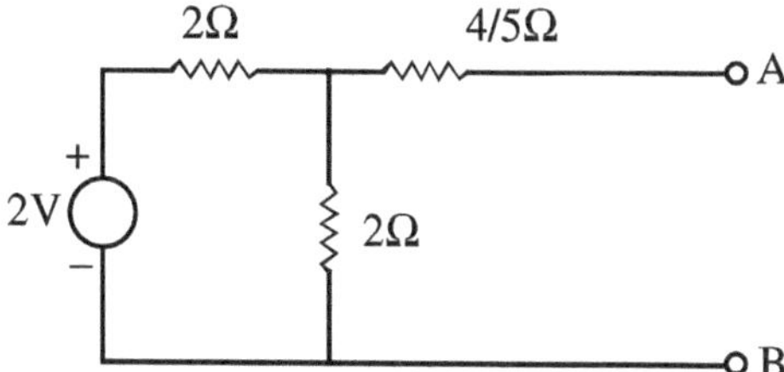

A. $\frac{5}{2}$ A, 2Ω

B. $\frac{5}{2}$ A, 1Ω

C. $\frac{2}{5}$ A, 1Ω

D. $\frac{2}{5}$ A, 2Ω

118. The given circuit represents a:

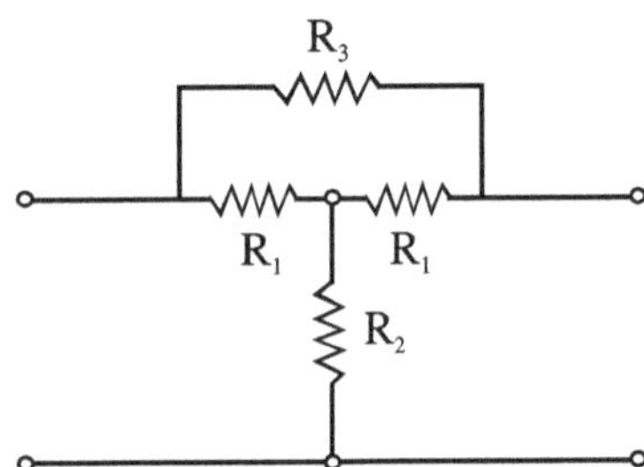

A. symmetrical T-attenuator
B. symmetrical lattice attenuator
C. symmetrical bridged-T attenuator
D. asymmetrical T-attenuator

119. A symmetrical T-attenuator uses two series arm resistors R_1 each and a shunt arm resistor R_2. If N is the attenuation ratio and R_0 is the design resistance, then R_1 equals:

A. $R_0/2$

B. $R_0 \frac{N+1}{N-1}$

C. $R_0 \frac{N-1}{N+1}$

D. $R_0 \left(\frac{N-1}{N+1}\right)^2$

120. In *m*-derived high pass T or π section filters, value *m* equals:

A. $\sqrt{1-\left(\frac{f_\infty}{f_c}\right)}$

B. $\sqrt{1-\left(\frac{f_\infty}{f_c}\right)^2}$

C. $\sqrt{1-\left(\frac{f_\infty}{f_c}\right)^2}$

D. $\sqrt{1-\left(\frac{f_\infty}{f_c}\right)^{1/2}}$

121. The following circuit represents an *m*-derived:

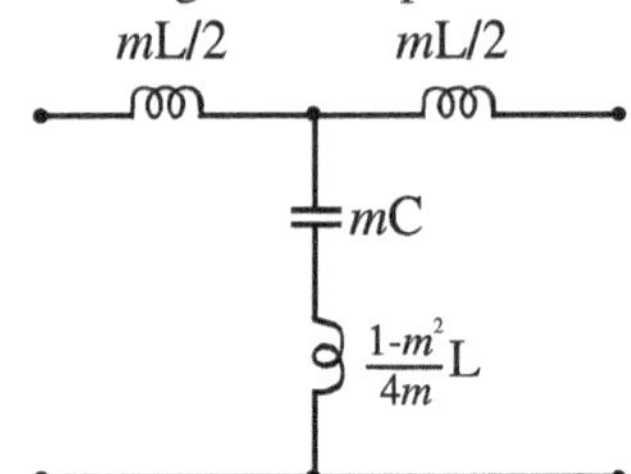

A. low pass filter
B. high pass filter
C. band pass filter
D. band stop filter

122. In *m*-derived low pass T or π section filters, value of *m* equals:

A. $\sqrt{1-\left(\frac{\omega_c}{\omega_\infty}\right)^2}$

B. $\sqrt{1-\left(\frac{\omega_c}{\omega_\infty}\right)}$

C. $\sqrt{1-\left(\frac{\omega_c}{\omega_\infty}\right)^{1/2}}$

D. $\sqrt{1-\left(\frac{\omega_c}{\omega_\infty}\right)^3}$

123. In a constant-*k* bands top filter, the shunt element is:

A. capacitor
B. inductor
C. series L-C circuit
D. shunt L-C circuit

124. In a constant-*k* T-section band stop filter, the shunt arm is a series $L_2 - C_2$ circuit where C_2 equals:

A. $\frac{(f_2 - f_1)}{R_0 f_1 f_2}$

B. $\frac{(f_2 - f_1)}{\pi R_0 f_1 f_2}$

C. $\dfrac{(f_2 - f_1)}{2\pi\, R_0 f_1 f_2}$

D. $\dfrac{(f_2 - f_1)}{4\pi\, R_0 f_1 f_2}$

125. In a constant-*k* T-section band stop filter, the shunt arm is a series $L_2 - C_2$ circuit where L_2 equals:

A. $\dfrac{R_0}{(f_2 - f_1)}$

B. $\dfrac{R_0}{\pi(f_2 - f_1)}$

C. $\dfrac{R_0}{2\pi(f_2 - f_1)}$

D. $\dfrac{R_0}{4\pi(f_2 - f_1)}$

where R_0 is the design resistance and f_1 and f_2 are the cutoff frequencies.

126. In a constant-*k* T-section band pass filter the shunt arm is a parallel $L_2 - C_2$ circuit where C_2 equals:

A. $\dfrac{1}{\pi R_0 (f_2 - f_1)}$

B. $\dfrac{1}{2\pi R_0 (f_2 - f_1)}$

C. $\dfrac{1}{4\pi R_0 (f_2 - f_1)}$

D. $\dfrac{1}{R_0 (f_2 - f_1)}$

where f_1 and f_2 are the cutoff frequencies and R_0 is the design resistance.

127. In a constant-*k* T-section bandpass filter, the shunt arm is a parallel $L_2 - C_2$ circuit where L_2 equals:

A. $\dfrac{R_0(f_1 - f_2)}{f_1 f_2}$

B. $\dfrac{R_0(f_2 - f_1)}{\pi f_1 f_2}$

C. $\dfrac{R_0(f_2 - f_1)}{2\pi f_1 f_2}$

D. $\dfrac{R_0(f_2 - f_1)}{4\pi f_1 f_2}$

where f_2 and f_1 are the cutoff frequencies and R_0 is the design resistance.

128. In a constant-*k* T-section band-pass filter, the series arm capacitance C_1 equals:

A. $\dfrac{(f_2 - f_1)}{R_0 f_1 f_2}$

B. $\dfrac{(f_2 - f_1)}{\pi\, R_0 f_1 f_2}$

C. $\dfrac{(f_2 - f_1)}{2\pi\, R_0 f_1 f_2}$

D. $\dfrac{(f_2 - f_1)}{4\pi\, R_0 f_1 f_2}$

where R_0 is the design resistance and f_2 and f_1 are the cutoff frequencies.

129. For a constant-*k* high pass filter with design resistance R_0 and cutoff frequency f_c, the shunt inductance L equals:

A. $\dfrac{R_0}{4\pi f_c}$

B. $\dfrac{R_0}{2\pi f_c}$

C. $\dfrac{R_0}{\pi f_c}$

D. $\dfrac{R_0}{f_c}$

130. For a constant-*k* high pass filter with design resistance R_0 and cutoff frequency f_c, the series capacitance C equals:

A. $\dfrac{1}{4\pi f_c R_0}$

B. $\dfrac{1}{2\pi f_c R_0}$

C. $\dfrac{1}{\pi f_c R_0}$

D. $\dfrac{1}{f_c R_0}$

131. For a constant-k low pass filter with design resistance R_0 and cutoff frequency f_c, the shunt capacitance C equals:

A. $\frac{1}{2\pi f_c R_0}$

B. $\frac{1}{\pi f_c R_0}$

C. $\frac{R_0}{\pi f_c}$

D. $\frac{R_0}{2\pi f_c}$

132. For a constant-k low pass filter with design resistance R and cutoff frequency f_c, the series inductance L equals:

A. $\frac{R_0}{2\pi f_c}$

B. $\frac{R_0}{\pi f_c}$

C. $\frac{R_0}{f_c}$

D. $\frac{2\pi R_0}{f_c}$

133. For constant-k filters:

A. $Z_{OT} \cdot Z_{0\pi} = 1$

B. $Z_{OT} \cdot Z_{0\pi} = R_0^2$

C. $Z_{OT} \cdot Z_{0\pi} = \sqrt{2}$

D. $Z_{OT} / Z_{0\pi} = \frac{1}{\sqrt{2}}$

where R_0 is the design resistance.

134. In constant-k high pass π-section filter, the characteristic impedance at any frequency f in the pass band equals:

A. $R_0 \sqrt{1-\left(\frac{f_c}{f}\right)^2}$

B. $R_0 \sqrt{1-\left(\frac{f}{f_c}\right)^2}$

C. $\frac{R_0}{\sqrt{1-\left(\frac{f_c}{f}\right)^2}}$

D. $\frac{R_0}{\sqrt{1-\left(\frac{f}{f_c}\right)^2}}$

135. In constant-k high pass T-section filter, the characteristic impedance at any frequency f in the pass band equals:

A. $R_0 \sqrt{1-\left(\frac{f_c}{f}\right)^2}$

B. $R_0 \sqrt{1-\left(\frac{f}{f_c}\right)^2}$

C. $\frac{R_0}{\sqrt{1-\left(\frac{f_c}{f}\right)^2}}$

D. $\frac{R_0}{\sqrt{1-\left(\frac{f}{f_c}\right)^2}}$

136. For a π-section constant-k low pass filter, the characteristic impedance $Z_{0\pi}$ at any frequency f in the pass band equals:

A. $R_0 \sqrt{1-\left(\frac{f}{f_c}\right)^2}$

B. $R_0 \sqrt{1-\left(\frac{f_c}{f}\right)^2}$

C. $\dfrac{R_0}{\sqrt{1-\left(\dfrac{f}{f_c}\right)^2}}$

D. $\dfrac{R_0}{1-\left(\dfrac{f_c}{f}\right)^2}$

137. For a T-section constant-*k* low pass filter, the characteristic impedance Z_{OT} at any frequency f in the pass band equals:

A. $R_0\sqrt{1-\left(\dfrac{f_c}{f}\right)^2}$

B. $R_0\sqrt{1-\left(\dfrac{f}{f_c}\right)^2}$

C. $\dfrac{R_0}{\sqrt{1-\left(\dfrac{f_c}{f}\right)^2}}$

D. $\dfrac{R_0}{\sqrt{1-\left(\dfrac{f}{f_c}\right)^2}}$

where f_c is the cutoff frequency and R_0 is the design resistance.

138. The characteristic impedance of a constant-*k* T-section filter is given by:

A. $Z_{OT}=\sqrt{Z_1 Z_2\left(1+\dfrac{Z_1}{4Z_2}\right)}$

B. $Z_{OT}=Z_1 Z_2\sqrt{1+\dfrac{Z_1}{4Z_2}}$

C. $Z_{OT}=\sqrt{1+\dfrac{Z_1}{4Z_2}}$

D. $Z_{OT}=\dfrac{Z_1 Z_2}{\sqrt{1+\dfrac{Z_1}{4Z_2}}}$

where Z_1 and Z_2 are the series and shunt impedance respective.

139. In a constant-*k* low pass filter with total series inductance L and total shunt capacitance C, the cutoff frequency f_c equals:

A. $\dfrac{1}{2\pi\sqrt{LC}}$

B. $\dfrac{1}{\sqrt{LC}}$

C. $\dfrac{1}{\pi\sqrt{LC}}$

D. $\dfrac{1}{4\pi\sqrt{LC}}$

140. The given circuit represents a:

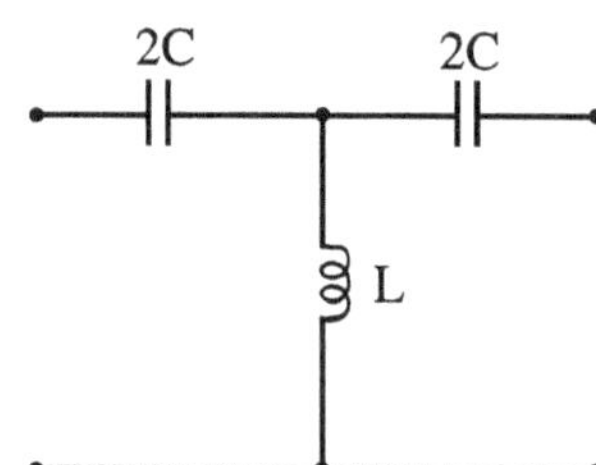

A. low pass filter
B. high pass filter
C. band-pass filter
D. band stop filter

141. The given circuit represents a:

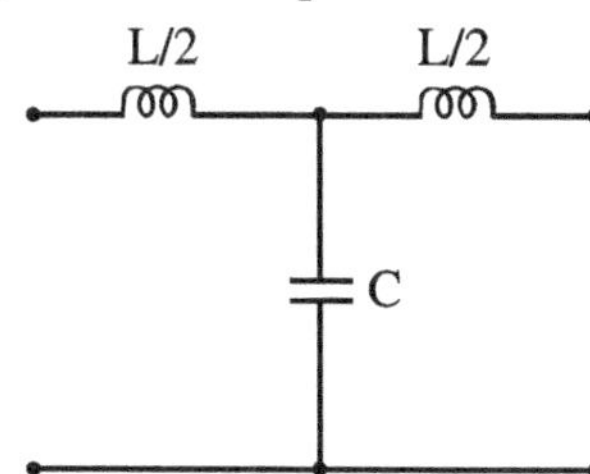

A. low pass filter
B. high pass filter
C. band-pass filter
D. band stop filter

142. A network has input and output currents of I_i and I_0 respectively. The attenuation in nepers is:

A. $10\ ln\left|\dfrac{I_i}{I_0}\right|$

B. $10 \log \left|\frac{I_i}{I_0}\right|$

C. $ln \left|\frac{I_i}{I_0}\right|$

D. $\log \left|\frac{I_i}{I_0}\right|$

143. In a prototype high pass filter, the series and shunt elements are respectively:
A. capacitive and inductive
B. inductive and capacitive
C. inductive and resistive
D. resistive and capacitive

144. In a prototype low pass filter, the series and shunt elements are respectively:
A. capacitive and inductive
B. inductive and capacitive
C. capacitive and resistive
D. resistive and inductive

145. In a constant-*k* band pass filter, the series element is:
A. capacitive
B. inductive
C. series combination of L and C
D. parallel combination of L and C

146. In a constant-*k* bandpass filter, the shunt element is:
A. capacitive
B. inductive
C. series combination of L and C
D. shunt combination of L and C

147. In a constant-*k* T-section band pass filter, the series arm inductive L_1 equals:

A. $\frac{R_0}{\pi(f_2 - f_1)}$

B. $\frac{R_0}{2\pi(f_2 - f_1)}$

C. $\frac{R_0}{4\pi(f_2 - f_1)}$

D. $\frac{R_0}{(f_2 - f_1)}$

148. The following circuit represents an *m*-derived:

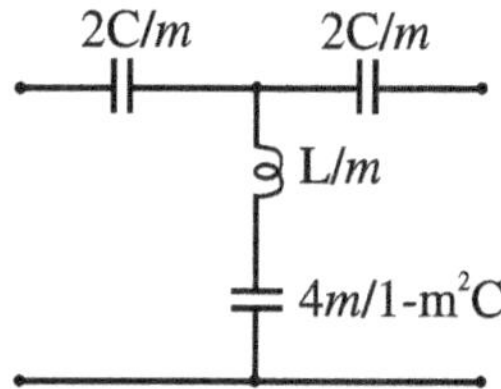

A. low pass filter
B. high pass filter
C. band pass filter
D. band stop filter

149. The given circuit represents a:

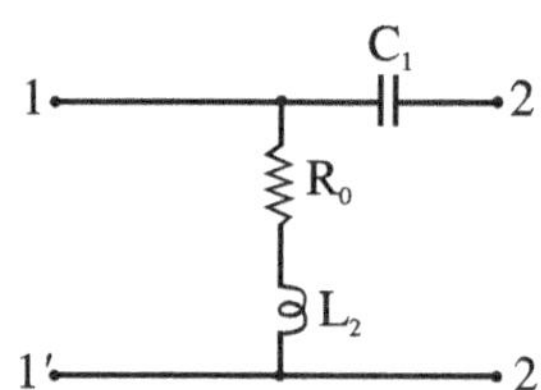

A. L-attenuator
B. full shunt equaliser
C. lattice equaliser
D. constant resistance equaliser

150. The given circuit represents a:

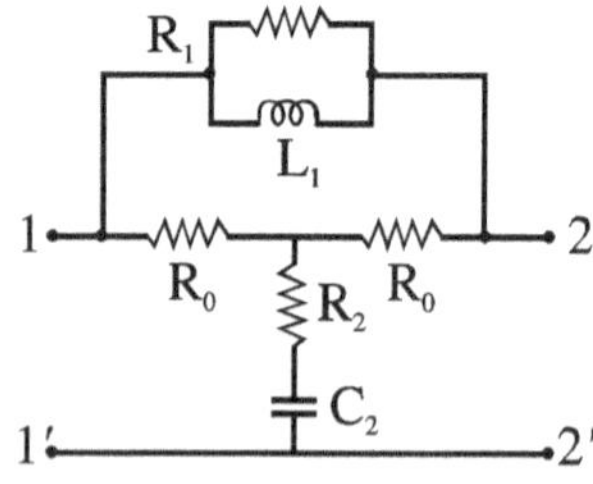

A. Bridged-T attenuation equaliser
B. Bridged-T delay equaliser
C. Lattice equaliser
D. Full shunt equaliser

ANSWERS

1	2	3	4	5	6	7	8	9	10
A	A	D	C	D	D	B	B	C	A

11	**12**	**13**	**14**	**15**	**16**	**17**	**18**	**19**	**20**
B	C	A	B	B	B	D	A	B	B
21	**22**	**23**	**24**	**25**	**26**	**27**	**28**	**29**	**30**
A	B	B	A	D	A	B	A	C	C
31	**32**	**33**	**34**	**35**	**36**	**37**	**38**	**39**	**40**
C	B	B	A	A	B	B	A	B	D
41	**42**	**43**	**44**	**45**	**46**	**47**	**48**	**49**	**50**
A	B	A	A	C	A	C	B	C	C
51	**52**	**53**	**54**	**55**	**56**	**57**	**58**	**59**	**60**
B	B	A	D	B	A	A	C	D	D
61	**62**	**63**	**64**	**65**	**66**	**67**	**68**	**69**	**70**
A	B	D	D	A	A	B	B	A	B
71	**72**	**73**	**74**	**75**	**76**	**77**	**78**	**79**	**80**
B	C	B	C	B	B	C	B	C	B
81	**82**	**83**	**84**	**85**	**86**	**87**	**88**	**89**	**90**
D	C	C	A	D	D	B	B	A	A
91	**92**	**93**	**94**	**95**	**96**	**97**	**98**	**99**	**100**
D	B	A	B	A	A	A	D	B	B
101	**102**	**103**	**104**	**105**	**106**	**107**	**108**	**109**	**110**
A	C	A	B	B	C	B	C	A	B
111	**112**	**113**	**114**	**115**	**116**	**117**	**118**	**119**	**120**
A	D	C	B	D	C	D	C	C	B
121	**122**	**123**	**124**	**125**	**126**	**127**	**128**	**129**	**130**
A	A	C	B	D	A	D	D	A	A
131	**132**	**133**	**134**	**135**	**136**	**137**	**138**	**139**	**140**
B	B	B	C	A	C	B	A	C	B
141	**142**	**143**	**144**	**145**	**146**	**147**	**148**	**149**	**150**
A	C	A	B	C	D	A	B	B	A

ELECTROMAGNETIC THEORY

MULTIPLE CHOICE QUESTIONS

1. The electric field intensity magnitude at a point (1, 2, 2) m in cartesian coordinates due to a charge (1/9) nC located at origin is equal to
   A. 3.3 V/m
   B. 0.33 V/m
   C. 0.033 V/m
   D. zero
2. The magnitude of magnetic field intensity (H) due to an infinitely long current carrying element at a distance of 1 m from the current element for a current of (2π) A is:
   A. 2π A/m
   B. 1 A/m
   C. ($1/2\pi$) A/m
   D. 0.5 A/m
3. The divergence theorem applies to:
   A. static fields only
   B. time varying fields only
   C. both static as well as time varying fields
   D. magnetic fields only
4. For a plane electromagnetic wave travelling in a linear, isotropic, homogeneous and stationary medium:
   A. E and H lie in the same plane
   B. $E \times H = 0$
   C. E and H are mutually perpendicular
   D. None of these
5. When a electromagnetic wave propagating in the z-direction, the x and y components of the electric field vector have a phase difference of 60°. The wave is:
   A. circularly polarised
   B. elliptically polarised
   C. linearly polarized
   D. None of the above
6. An electromagnetic wave propagates from a dielectric medium to the interface with free space. If the angle incidence is the critical angle of 30°, the relative permittivity of the dielectric medium is:
   A. 8
   B. 4.5
   C. 6.8
   D. 4
7. When an electromagnetic wave travelling in one medium strikes upon a second medium having a different conductivity the wave will be:
   A. neither transmitted and partially reflected
   B. totally reflected
   C. partially transmitted and partially reflected
   D. totally transmitted
8. An electromagnetic wave travelling in x direction has E_x and E_y components at 45° phase apart. The wave can be termed as:
   A. linearly polarized
   B. circularly polarized
   C. elliptically polarized
   D. horizontally polarized
9. In a perfect conductor, the incident and the reflected waves combine to produce:
   A. a stronger incident wave
   B. a stronger reflected wave
   C. a standing wave which is not progressing
   D. Both A and B
10. The polarization of dielectric materials results in:
    A. absorption of electrons
    B. release of high velocity protons
    C. creation of electric dipoles
    D. production of eddy currents
11. The absorption of radio waves by the atmosphere depends on:
    A. the polarization of atmosphere
    B. the polarization of the waves
    C. their frequency

D. their distance from the emitter

12. An infinite non-conducting sheet of charge has σ of 10^{-7} coulomb/m^2. The equipotential surfaces for a potential of 10 volts is:

A. 0.88 mm
B. 0.44 mm
C. 1.32 mm
D. 1.77 mm

13. An arbitrarily elliptically polarised wave can be broken up into:

A. two circularly polarised components rotating in the same direction
B. two oppositively rotating circularly polarised components
C. two stationary circularly polarised components
D. None of these

14. As an electromagnetic wave travels in free space, the only one of the following can happen to them:

A. reflection
B. absorption
C. attenuation
D. refraction

15. A time varying field is applied to a circuit with a capacitor:

A. The current in the circuit consists of conduction current and displacement current
B. The displacement current posses through the capacitor
C. The conduction current is equal to the displacement current
D. The conduction current density is J and displacement current density $\frac{\partial D}{\partial t}$

16. When a current carrying conductor is brought into magnetic field, the force that moves the conductor depends upon which of the following?

A. The value of current
B. The weight of the conductor
C. The direction of the conductor
D. None of the above

17. A uniform wave have components:

A. in the perpendicular direction E existing while H is zero
B. in the direction of propogation E existing while H is zero
C. E and H are zero in direction perpendicular to direction of propagation
D. E and H existing only in direction perpendicular to direction of propagation

18. For a plane good conductor, skin depth varies:

A. directly as the conductivity σ
B. inversely as σ
C. inversely as $\sqrt{\sigma}$
D. directly as $\sqrt{\sigma}$

19. Skip distance is:

A. the minimum distance from the transmitter at which a sky wave of a given frequency is returned to earth from the ionosphere
B. the maximum distance from the transmitter at which a sky wave of a given frequency is returned to earth from the ionosphere
C. $\frac{3\times10^8}{f}$ metres
D. $\frac{3\times10^8}{(f/2)}$ metres

20. Which of the following statement regarding electric flux is true?

1. Electric flux begins on positive charges and terminates on negative charges.
2. Flux is in the same direction of the electric field $\bar{E}$.
3. Flux density is proportional to the magnitude of $\bar{E}$.
4. In the SI system of units, total flux emanating from a charge of Q(C) is Q(C).
A single line will emanate from 1 C of charge

A. 1 only
B. 1 and 2 only
C. 1, 2 and 3 only
D. 1, 2, 3 and 4

21. The wavelength of an EM wave after reflection at an angle on a surface:

A. increases in the actual direction of propagation
B. decreases in the actual direction of propagation
C. remains the same as the wave length perpendicular to the surface

D. remains the same as in free space

22. Characteristic wave impedance is:

A. $\frac{\varepsilon}{\mu}$

B. $\sqrt{\varepsilon / \mu}$

C. $\sqrt{E_x / H_y}$

D. E_x/H_y

23. In a TM wave, if the electric field component E along the direction of propagation is also zero, and the EM field is entirely transverse, the principle wave is called:

A. TE Wave
B. TM Wave
C. TEM Wave
D. longitudinal Wave

24. Transverse magnetic (TM) waves have:

A. magnetic field component H in the direction of propagation and no component of electric field H in this direction
B. electric field component E in the direction of propagation and no component of magnetic field H in this direction
C. both magnetic field component H and the electric field component E in the direction of propagation
D. neither magnetic field component H nor electric field component E in the direction of propagation

25. Transverse electric (TE) waves have:

A. electric component E in the direction of propagation and no component of magnetic field H in this direction
B. magnetic field component H in the direction of propagation and no component of electric field E in this direction
C. both electric field component E and magnetic field component H in the direction of propagation
D. None of these

26. The ratio of tangential component of electric field at the surface of a conductor and the linear current density which flows as a result of this field is:

A. characteristic impedance
B. surface impedance
C. critical impedance
D. electromagnetic link impedance

27. When an EM wave is incident from a more dense medium to less dense medium at an angle equal to or exceeding the critical angle, the wave suffers:

A. total internal transmission
B. total internal reflection
C. total internal refraction
D. None of these

28. When the magnetic vector of an EM wave is parallel of the boundary surface and electric vector is parallel to the plane of incidence, the polarization is:

A. vertical
B. horizontal
C. elliptical
D. either B or C

29. As a result of reflection from a plane conducting wall, the EM waves acquire an apparent velocity greater than the velocity in free space. This velocity is called:

A. velocity of propagation
B. group velocity
C. phase velocity
D. normal velocity

30. When EM waves are incident on a perfect conductor, there results

A. a stronger incident wave
B. a stronger reflected wave
C. a weak reflected wave
D. a standing wave which is not progressing

31. The average power flow per unit area in a uniform plane wave in an electric field of maximum voltage E_0 and impedance Z_0 is:

A. $\frac{E_0^{\,2}}{2 Z_0}$

B. $\frac{E_0^{\,2}}{Z_0}$

C. $\frac{E_0^2}{Z_0^2}$

D. $I_0 E_0$

32. When EM waves are propagated in a rectangular waveguide:

A. they travel along the broader walls of the waveguide

B. they travel through the dielectric without touching the walls
C. they travel along all the four walls of the waveguide
D. they are reflected from the walls but do not travel along the walls

33. The EM field and current are concentrated close to the surface of the conductor. The phenomenon is called:
A. Faraday's effect
B. Skin effect
C. EM concentration effect
D. Ohm's effect

34. Wave propagation in plasma is called:
A. plasma dispersion
B. propagation dispersion
C. parametric dispersion
D. spherical dispersion

35. The wavelength of an electromagnetic wave in a waveguide:
A. is less than in free space
B. is greater than in free space
C. is inversely proportional to the phase velocity
D. is directly proportional to the group velocity

36. When two sinusoidally time varying vectors having different amplitudes and phases are added up, the resulting vector is:
A. linearly polarized
B. elliptically polarized
C. circularly polarized
D. None of these

37. Ponyting theorem relating the electric intensity E, magnetic intensity H and the rate of energy flow per unit area at a point gives poynting vector P as equal to:
A. $E \times H$
B. E.H
C. $\nabla.E \times H$
D. $E.\nabla \times H$

38. Potential energy W_e stored in a system of *n* point charges equals:

A. $\frac{1}{2}\sum_{i=1}^{n} Q_i V_i$

B. $\frac{1}{2}\sum_{i=1}^{n} Q_i$

C. $\frac{1}{2}\sum_{i=1}^{n} Q_i V_1^2$

D. $\frac{1}{2}\sum_{i=1}^{n} Q_i \sqrt{V_i}$

39. Maxwell's continuity equation for time varying electromagnetic field in differential form is given by:

A. $\nabla.J + \frac{d\rho}{dt} = 0$

B. $\nabla.J + \frac{\partial\rho}{\partial t} = 0$

C. $\nabla.J - \frac{\partial\rho}{\partial t} = 0$

D. $\nabla.J - \frac{d\rho}{dt} = 0$

40. Energy W_m stored in a magnetic field equals:

A. $\int_V \frac{B^2}{2\mu_0} dV$

B. $\int_V \mu_0 B^2 dV$

C. $\int_E \frac{B^2}{\mu_0} dV$

D. $\int_E \frac{B^2}{4\pi\mu_0} dV$

41. Faraday's law relating the circulation of electric field E around a closed path C and magnetic flux density B over the surface S is:

A. $\phi\, c\, E.dl = \int_S B.dS$

B. $\phi\, c\, E.dl = -\frac{d}{dt}\int_S B.dS$

C. $\phi\, c\, E.dl = \frac{d}{dt}\int_S B.dS$

D. $\phi\, c\, E.dl = -\int_S B.dS$

42. Instrinsic impedance of free space is given by:

A. $Z_0 = \mu_0 \varepsilon_0$

B. $Z_0 = \sqrt{\frac{\mu_0}{\varepsilon_0}}$

C. $Z_0 = \frac{\mu_0}{\varepsilon_0}$

D. $Z_0 = \sqrt{\mu_0 \varepsilon_0}$

43. The following is the Biot-sevart law giving the magnetic flux density B due to a wire carrying current I:

A. $B = 4\pi \varepsilon_0 I^2 r$

B. $B = \frac{1}{4\pi \varepsilon_0} \oint_c \frac{I\, dl}{r}$

C. $B = \frac{\mu_0}{4\pi} \oint_c \frac{I\, dl\, l_R}{r^2}$

D. $B = 4\pi \mu_0 \oint_c \frac{I\, dl\, l_R}{r^2}$

44. The following is the Ampere's circuital law:

A. $\oint B.dl = \mu_0 \oint_c Idl$

B. $\oint B.dl = \frac{1}{\varepsilon_o} \oint_c Idl$

C. $\oint B.dl = \varepsilon_o \oint_c Idl$

D. $\oint B.dl = \frac{1}{4\pi \varepsilon_o} \oint_c Idl$

45. Force F experienced by a test charge q moving with velocity v in a magnetic field of intensity B is given by

A. $F = qv \times B\, l_v\, l_B$

B. $F = v \times B\, l_v\, l_B$

C. $F = \frac{v \times B}{q} l_v\, l_B$

D. $F = \frac{B^2}{q} l_B$

46. In an uniform plane wave, E and H related as below:

A. $\frac{E}{H} = \sqrt{\frac{\varepsilon}{\mu}}$

B. $\frac{E}{H} = \sqrt{\frac{\mu}{\varepsilon}}$

C. $\frac{E}{H} = 1$

D. $\frac{E}{H} = \frac{\varepsilon}{\mu}$

47. A dominant wave should have:

A. no phase shift
B. no attenuation
C. highest cut off frequency
D. lowest cut off frequency

48. AC current can induce voltage because it has:

A. constant magnetic field
B. varying magnetic field
C. stronger magnetic field
D. high peak value

49. Hysteresis losses

A. generally increase with direct current in a coil
B. are caused by high frequency currents
C. can not be produced in an iron core because it is a conductor
D. None of these

50. Which of the following statements is true?

A. Alnico is commonly used in electromagnets
B. Paper cannot affect magnetic flux because it is not a magnetic material
C. Iron is generally used for permanent magnet
D. All the above

51. A magnetic material has a total flux B of 80 μ W_b with an mmf of 160 AT. The reluctance in ampereturns per weber is

A. 2×10^{-6}
B. 2×10^{6}
C. 2×10^{-8}
D. 2×10^{8}

52. If a dielectric is placed in an electric field, the field strength

A. increases
B. decreases
C. remains unaltered
D. becomes zero

53. In a parallel plate capacitor with mica and air as the media, capacitance increases on:

A. increasing the area of plates

B. increasing the thickness of mica
C. increasing the air space
D. increasing the thickness of plates

54. Displacement density D and electric field strength E are related by:
A. $D = \varepsilon E$
B. $D = E/\varepsilon$
C. $D = \varepsilon E^2$
D. None of these

55. Electric displacement density D for a spherical surface is given by:
A. $D = \frac{q}{4\pi \varepsilon r^2}$
B. $D = \frac{q}{4\pi \varepsilon}$
C. $D = \frac{q}{4\pi r^2}$
D. $D = \frac{q}{4 r^2}$

56. Force F experienced by a charge q placed in static electric field of intensity E is given by:
A. $F = \frac{E}{q}$
B. $F = Eq$
C. $F = \frac{E^2}{q}$
D. $F = Eq^2$

57. If μ and μ_r are the absolute and relative magnetic permeabilities of a medium and μ_o is the absolute permeability of free space, then:
A. $\mu = \mu_o \mu_r$
B. $\mu = \mu_o/\mu_r$
C. $\mu = \mu_r/\mu_o$
D. $\mu = \frac{1}{\mu_o \mu_r}$

58. The electric field strength at any point equals:
A. the potential gradient at the point
B. negative of the potential gradient at that point
C. the charge at that point
D. negative of the charge at that point

59. The divergence of the electric field intensity at any point equals the volume charge density at that point multiplied by:
A. 1
B. $\frac{1}{\varepsilon_0}$
C. ε_0
D. $\frac{1}{4\pi \varepsilon_0}$

60. Gauss law relates the electric field intensity E with volume charge density ρ at a point as:
A. $\nabla \times E = \varepsilon_0 \rho$
B. $\nabla . E = \frac{\rho}{\varepsilon_0}$
C. $\nabla \times E = \frac{\rho}{\varepsilon_0}$
D. $\nabla . E = \varepsilon_0 \rho$

61. Law of conservation of charge states that the sum of the current due to the flow of charges and the displacement current across any closed surface must be:
A. zero
B. infinity
C. equal to the current due to charge flow
D. equal to current due to displacement

62. Unit of electric intensity is
A. Newton
B. Newton/Amp
C. Newton/coulomb
D. volt/coulomb

63. Magnitude of electric field E at a distance from a charge q equals:
A. $\frac{q}{4 \varepsilon r^2}$
B. $\frac{q}{4\pi r^2}$
C. $\frac{q}{4\pi \varepsilon r^2}$
D. $\frac{q}{4\pi \varepsilon r}$

64. Poisson's equation relating the potential V at any point to the volume charge density ρ at that point states that:

A. $\nabla^2 V = \frac{\partial}{\rho}$

B. $\nabla^2 V = -\frac{\partial}{\rho}$

C. $\nabla^2 V = -\varepsilon\rho$

D. $\nabla^2 V = +\varepsilon\rho$

65. Laplacian of a scalar function F is:

A. gradient of F
B. divergence of F
C. gradient of divergence of F
D. divergence of gradient of F

66. The curl of a vector is:

A. a scalar
B. a vector
C. unit vector
D. may be either a scalar or a vector

67. Stoke's Theorem relating the line integration to surface integration states that:

A. $\phi_s \nabla.F\, dS = \phi_c F.dl$
B. $\phi_s (\nabla \times F).\, dS = \phi_c F.dl$
C. $\phi_s (\nabla \times F).\, dS = \phi_c F.dl$
D. $\phi_s (\nabla.F) \times dS = \phi_c F.dl$

68. Laplacian of a scalar function F is:

A. ∇F
B. $\nabla^2 F$
C. $\nabla \times F$
D. $\nabla \times \nabla F$

69. In a closed surface S enclosing a volume V and having current vector J, the current emanating from the surface is:

A. ϕ J
B. $\phi_s Jds$
C. $\phi_v Jdv$
D. d/ds J

70. Volume integral of a volume is:

A. integral of volume
B. double integral of volume
C. triple integral of volume
D. integral of weight

71. Surface integral of a surface is:

A. differential of weight
B. differential of volume
C. integral of surface
D. double integral of surface

72. Line integral around a closed path is called:

A. vector around the path
B. intensity around the path
C. circulation around the path
D. density around the path

73. Gradient of a scalar function results in:

A. scalar function
B. unit function
C. vector function
D. periodic function

74. Scalar or dot product of two vectors A and B indicated by A.B equals:

A. AB sin α
B. AB cos α
C. AB tan α
D. AB cot α

75. Unit vector is obtained by

A. dividing the vector by its magnitude
B. multiplying the vecotr by e^{j0}
C. dividing the vector by e^{j0}
D. multiplying the vector by another vector of unit length

76. Which of the following is not a wide band antenna?

A. Marconi
B. Discone
C. Helical
D. Folded dipole

77. For coupling a coaxial line to a parallel line which of the following is preferred?

A. slotted line
B. balun
C. irrotational coupler
D. quarter wave transformer

78. Antenna radiation efficiency can be increased by

A. reducing radiation resistance of the system
B. increasing radiation resistance of the system
C. providing effective earthing
D. Any of the above

79. Front to back ratio can be increased by

A. sacrificing gain

B. increasing size of conductor
C. using materials of high conductivity
D. All of the above

80. The number of lobes on each side 3λ resonant antenna will be:
A. 2
B. 3
C. 4
D. 6

81. The minimum height of outer atmosphere is:
A. 100 km
B. 150 km
C. 200 km
D. 400 km

82. The power gain of an optimum horn antenna approximately with a square aperture of 10λ on a side will be:
A. 150
B. 300
C. 450
D. 135

83. The parabolic and lens antennas are used extensively at:
A. SW
B. MW
C. microwave
D. All of the above

84. In case of lossy transmission lines, the changing SWR is represented by:
A. an arc of a circle
B. a circle
C. an inward moving sign
D. None of these

85. Higher the radius of constant SWR circle on smith chart:
A. lower is the magnitude of reflection coefficient
B. higher is the magnitude of reflection coefficient
C. higher is the permittivity of the transmission line dielectric
D. lower is the propagation velocity

86. Smith chart can be used to determine:
A. SWR for any given transmission line-load mismatch
B. magnitude and angle of reflection co-efficient
C. the length of short-circuited transmission line section of known characteristic impedance to provide the desired input impedance
D. All of the above

87. The arcs of circles on smith chart represent:
A. normalised resistances or conductors
B. different values of SWR
C. normalised reactances or susceptances
D. None of these

88. In a transmission line-load mismatch, a VSWR of 2 was observed. If the incident power is 36 watts, the reflected power would be:
A. 4 W
B. 2 W
C. 5 W
D. 3 W

89. The distance between adjacent maxima and minima of a standing wave on the transmission line is:
A. λ
B. λ/2
C. λ/4
D. None of these

90. Higher value of SWR signifies
A. higher ohmic losses
B. higher ohmic losses and possibility of insulation breakdown
C. higher radiation resistance
D. None of these

91. SWR equal to unity implies that
A. the transmission line is open-circuited
B. the transmission line is short-circuited
C. the transmission line's characteristic impedance equals load impedance
D. None of the above

92. Standing wave ratio (SWR) on a transmission line
A. is directly proportional to the load characteristic impedance and inversely proportional to the characteristic impedance
B. is directly proportional to the characteristic impedance and inversely proportional to the load impedance
C. may be defined as either (Z_L/Z_0) or (Z_0/Z_L)

D. None of these

93. The marconi antenna produces:
A. figure eight directional pattern
B. pencil beam directional pattern
C. circular directional pattern
D. None of these

94. A resonant antenna is a system with:
A. zero SWR
B. a very high SWR
C. a very low SWR
D. None of these

95. A rhombic antenna in a horizontal place produces:
A. vertically polarized waves
B. an endfire pattern
C. a broadside pattern
D. None of these

96. The transmitting antenna to be used for television broadcast is necessarily:
A. a parabolic reflector
B. a high directivity antenna
C. an omnidirectional antenna
D. None of these

97. An added capacitance:
A. increases the effective length of the antenna
B. decreases the effective length of the antenna
C. has no effect on antenna's length
D. None of these

98. An antenna behaves as a resonant circuit only when its length equals:
A. $\lambda/4$
B. $\lambda/2$
C. $\lambda/2$ or its integral multiple
D. None of these

99. Directivity of an antenna is:
A. same as its directive gain
B. directive gain expressed in dB
C. the maximum value of its directive gain
D. None of these

100. Antenna reciprocity is put to use in:
A. radar system
B. television broadcast systems
C. commercial radio broadcast systems
D. None of these

101. Range of space wave communication in meters equals:
A. $3550\ (\sqrt{h_t} + \sqrt{h_r})$
B. $3550\ (h_t + h_r)$
C. $1.23\ (\sqrt{h_t} + \sqrt{h_r})$
D. $1.23\ (h_t + h_r)$
where h_t and h_r are respectively the heights of transmitting and receiving antennas in meters.

102. EM waves travel in free space, only following can happen:
A. refraction
B. absorption
C. attenuation
D. reflection

103. Tropospheric scatter is made for communication of frequency in the range:
A. low frequencies
B. high frequencies
C. very high frequencies (VHF)
D. ultra high frequencies (UHF)

104. In Faraday rotation, the polarization phenomenon:
A. does not exist
B. is non-reciprocal
C. exists along the perpendicular direction
D. exists along the axis of the array

105. When microwave signal follows the curvature of earth, it is known as:
A. troposcatter
B. Faraday effect
C. duct propagation
D. ground wave propagation

106. Typically the rate of change of refractive index in atmosphere with height *dN/dh* approximately equals:
A. -4.3×10^{-2}/m
B. 4.3×10^{-2}/m
C. -4.3×10^{-3}/m
D. 4.3×10^{-3}/m

107. Cassegrain feed is used with a parabolic reflector to:
A. increase the gain of the system
B. increase the bandwidth of the system
C. reduce the size of the main reflector
D. allow the feed to be placed at a convenient point

108. Helical antenna is used for satellite tracking because of
A. broad bandwidth
B. good front to back ratio
C. circular polarization
D. maneuverability

109. Electromagnetic cavity resonators are energy storage devices and are used in
A. klystrons
B. band pass filters
C. wavemeters
D. All of these

110. Radiation resistance of a halfwave folded dipole antenna is about:
A. 72 Ω
B. 144 Ω
C. 288 Ω
D. 400 Ω

111. In an end fire array, maximum radiation occurs:
A. along the array
B. perpendicular to the array
C. at 30° to the array
D. at 45° to the array

112. The radiation pattern of one wavelength long centre fed antenna has number of lobes equal to:
A. 1
B. 2
C. 4
D. 8

113. The effect of ground on radiation pattern of an antenna is:
A. To produce more number of nulls
B. to cause cancellation of radiation along the ground
C. Both A and B
D. to change its effective wavelength

114. Directive gain G_d of an antenna in a given direction is the ratio of:
A. radiation intensity in that direction to the average power radiated
B. radiation intensity in that direction to the total input radiated power
C. average power to maximum power
D. peak power to rms power

115. Power gain GP of an antenna in a given direction is the ratio of:
A. radiation intensity in that direction to the average radiated power
B. radiation intensity in that direction to the total input radiated power
C. average power to maximum power
D. peak power to rms power

116. Main drawback of a two hole directional coupler is its:
A. poor directivity
B. narrow bandwidth
C. poor directional coupling
D. high standing wave ratio

117. A transmission line is required to match with the load over a wide frequency range. It is best to use:
A. a balum transformer
B. a single stub of adjustable position
C. a double stub
D. broad band directional coupler

118. A transmission line of length l, where $\lambda/4 < l < \lambda/2$, is short-circuited at the far end. The input impedance is:
A. resistive
B. inductive
C. capacitive
D. Z_o

119. A transmission line of length l, where $\lambda/4 < l < \lambda/2$, is open-circuited at the far end. The input impedance is:
A. purely resistive
B. inductive
C. capacitive
D. zero

120. Disturbance due to adjacent power lines is called:
A. cross-talk
B. crossfire
C. inductive disturbance
D. noise

121. Compared to equivalent transmission lines at low frequencies, waveguides are:
A. less loss
B. able to carry more power
C. lighter in weight
D. Both A and B

122. The length of which of the following antennas is $\lambda/2$?
A. Discone
B. Monopole
C. Hertz
D. All of above

123. The effective length of a tower radiator is:

A. $\frac{1}{2\pi}$

B. $\frac{1}{\pi}$

C. $\frac{2I}{\pi}$

D. 0.751

124. In binomial arrays, elimination of secondary lobes takes place:

A. at the cost of directivity
B. at the cost of gain
C. Both A and B
D. None of these

125. The secondary lobes in binomial arrays can be reduced by:

A. maintaining spacing between two adjacent radiating elements less than $\lambda/2$
B. maintaining the amplitudes in radiating elements from outer element to inner element, proportional to the coefficients of the successive terms of the binomial series
C. Both A and B above
D. None of the above

126. Omni-Directional Range (ODR) operates by:

A. comparing the phase between two signals
B. comparing the frequencies of two signals
C. measuring the time taken by a pulse to reach a station
D. interlocking code letter

127. As an aeroplane approaches a beacon, the aural signal:

A. disappears
B. increases in value
C. increases in volume and then disappears
D. is in a continuous tone

128. When an aeroplane is directly above the cone of silence, it means that:

A. the plane is directly over the antenna
B. the plane is very far from the range station
C. the plane is on the course
D. the plane is on the reciprocal course

129. Which of the following characteristics is shown by microwave antenna?

A. high directivity
B. high gain
C. low input impedance
D. wide frequency range

130. Compared to resonant antenna, the directive gain of a non-resonant antenna of the same length is:

A. smaller
B. greater
C. the same
D. about 12 dB more

131. If the diameter of parabolic reflector of a microwave antenna is doubled, its gain will increase by:

A. 0 dB
B. 2 dB
C. 4 dB
D. 6 dB

132. The input impedance of a folded dipole antenna is about:

A. 110 Ω
B. 220 Ω
C. 377 Ω
D. 280 Ω

133. The folded dipole antenna is quite conveniently connected to:

A. coaxial line
B. two-wire line
C. flat ribbon-type transmission line
D. shielded line

134. A Marconi antenna is required to transmit a 100 MHz signal. If the velocity factor for the antenna is 0.9, its optimum length is:

A. 0.675 meter
B. 1.35 meter
C. 0.27 meter
D. 5.4 meter

135. A half-wavelength antenna has radiation resistance equal to:

A. 700 Ω
B. 300 Ω
C. 730 Ω
D. 47 Ω

136. Both end-fire and broad-side arrays are:

A. non-linear
B. non-resonant
C. of wide bandwidth

D. not suitable for short wave transmission

137. For a wire of length λ, the number of loops in the radiation pattern are:
A. 1
B. 2
C. 3
D. 4

138. In case of surface wave propagation, the field strength at a point is directly proportional to:
A. frequency of the wave
B. height of the wave
C. distance of the point from the antenna
D. square root of effective radiated power

139. In surface wave propagation, in order to avoid short-circuiting of the electric component, the wave should be:
A. vertically polarized
B. horizontally polarized
C. circularly polarized
D. elliptically polarized

140. Which of the following effects occurs in tropospheric scatter propagation?
A. Faraday effect
B. fading
C. super - refraction
D. atmospheric storm

141. The virtual height of an ionospheric layer is:
A. more than the height a wave actually reaches
B. less than the height a wave actually reaches
C. same as the height a wave actually reaches
D. None of the above

142. The skip distance is:
A. independent of frequency
B. independent of the state of ionisation
C. independent of transmitting power
D. dependent on transmitting power

143. If the critical frequency of a layer is 9 MHz, then for angle of incidence of 60°, the maximum usable frequency is:
A. 9 MHz
B. 18 MHz
C. 10.26 MHz
D. 7.9 MHz

144. The critical frequency for a layer is:
A. the lowest frequency that can return to earth at any angle of incidence
B. the highest frequency that can return to earth at any angle of incidence
C. the lowest frequency that can return to earth for vertical incidence
D. the highest frequency than can return to earth for vertical incidence

145. Which of the following is the least important layer concerning high frequency propagation:
A. D layer
B. E layer
C. F_1 layer
D. F_2 layer

146. Reliable ship-to-ship communication may be secured with little fading by use of:
A. space diversity reception
B. frequency diversity reception
C. more directional antenna
D. broadband antenna

147. The surface waves ultimately disappear as they move away from the transmitter because of:
A. maximum single hop distance limitation
B. absorption by ground
C. interference from the sky wave
D. line-of-sight restriction

148. In ionispheric propagation, the maximum usable frequency (MUF) equals:
A. $f_c \cos\theta$
B. $\sqrt{f_c \cos\theta}$
C. $f_c \sec\theta$
D. $\sqrt{f_c \sec\theta}$

where f_c is the critical frequency and θ is the angle of incidence at the ionosphere measured with respect to the normal.

149. Refraction of radio waves in the atmosphere results due to:
A. changes in dielectric constant
B. changes in the density of air
C. changes in dielectric constant and refractive index
D. None of the above

150. Signals in UHF range use:
A. space wave propagation
B. sky wave propagation
C. surface wave propagation
D. None of the above

ANSWERS

1	**2**	**3**	**4**	**5**	**6**	**7**	**8**	**9**	**10**
B	B	C	C	B	D	C	C	C	C
11	**12**	**13**	**14**	**15**	**16**	**17**	**18**	**19**	**20**
C	D	B	C	D	A	D	D	C	D
21	**22**	**23**	**24**	**25**	**26**	**27**	**28**	**29**	**30**
A	D	C	B	B	B	C	A	D	C
31	**32**	**33**	**34**	**35**	**36**	**37**	**38**	**39**	**40**
B	B	B	C	B	B	A	A	B	A
41	**42**	**43**	**44**	**45**	**46**	**47**	**48**	**49**	**50**
B	B	C	A	A	B	D	B	B	B
51	**52**	**53**	**54**	**55**	**56**	**57**	**58**	**59**	**60**
B	B	A	D	C	B	A	B	B	B
61	**62**	**63**	**64**	**65**	**66**	**67**	**68**	**69**	**70**
A	C	C	B	D	B	B	B	B	C
71	**72**	**73**	**74**	**75**	**76**	**77**	**78**	**79**	**80**
D	C	C	B	A	A	B	B	A	D
81	**82**	**83**	**84**	**85**	**86**	**87**	**88**	**89**	**90**
D	C	C	C	B	D	C	A	C	B
91	**92**	**93**	**94**	**95**	**96**	**97**	**98**	**99**	**100**
C	C	C	B	B	C	B	C	C	A
101	**102**	**103**	**104**	**105**	**106**	**107**	**108**	**109**	**110**
A	C	D	B	C	A	D	C	D	C
111	**112**	**113**	**114**	**115**	**116**	**117**	**118**	**119**	**120**
A	B	C	A	B	B	C	C	D	C
121	**122**	**123**	**124**	**125**	**126**	**127**	**128**	**129**	**130**
D	C	C	A	C	A	C	A	C	B
131	**132**	**133**	**134**	**135**	**136**	**137**	**138**	**139**	**140**
D	D	C	A	C	C	D	D	A	B
141	**142**	**143**	**144**	**145**	**146**	**147**	**148**	**149**	**150**
A	C	B	D	A	B	B	C	C	A

ELECTRONIC MEASUREMENTS AND INSTRUMENTATION

MULTIPLE CHOICE QUESTIONS

1. Accuracy of digital voltmeter is specified as:
A. percentage of the actual reading
B. percentage of the full-scale reading
C. number of least significant digits
D. All of the above

2. An FET VM employs FETs in:
A. cascade mode
B. cascode mode
C. differential mode
D. common mode

3. A DVM measure:
A. peak value
B. rms value
C. peak-to-peak value
D. average value

4. Resolution of DVM with 4 digit display is:
A. 1/4
B. 1/1000
C. 1/10000
D. 1%

5. The best method of measuring capacitance is the:
A. CRO method
B. Transistor-Voltmeter method
C. voltmeter-ammeter method
D. a.c. bridge method

6. Transistor voltmeter is a:
A. dynamometer type instrument
B. moving iron instrument
C. moving coil instrument
D. None of these

7. Parallel connection Q-meter is used for measuring:
A. Low impedance component
B. High impedance component
C. A and B both
D. None of these

8. Transistor voltmeter is a highly sensitive instrument and can be used for measurement of:
A. energy
B. power
C. voltage and current
D. d.c. voltages only

9. Transistor voltmeter:
A. cannot measure a.c. voltage
B. cannot be designed to measure resistance as well as voltage
C. cannot measure high frequency voltage
D. can measure a.c. voltage

10. The maximum input signal frequency that can be used with a CRO is limited by:
A. input probe capacitance
B. the vertical amplifier
C. the frequency of the sync signal
D. the horizontal amplifier

11. The first stage of vertical amplifier of a CRO is a:
A. CE stage
B. CB stage
C. CS (FET) stage
D. CC stage

12. Platinum tungston thermometer can be used to measure temperature upto:
A. 400°K
B. 800°K

C. 1100°K
D. 2000°K

13. Time response of strain gauge measurement is:
A. very poor
B. poor
C. good
D. very good

14. Measurement of thickness of non-conducting coating on a magnetic material can be done by:
A. LVDT
B. capacitive pick up
C. magnetic pick up
D. inductive pick up

15. Following is used for measurement of low temperatures:
A. thermocouple
B. thermistor
C. semiconductor thermometer
D. optical pyrometer

16. Main advantage of semiconductor strain gauge over resistive strain gauge is its:
A. high gauge factor
B. high sensitivity
C. smaller size
D. better ruggedness

17. Dummy strain gauge is used to:
A. increase the efficiency
B. increase the range
C. compensate for temperature changes
D. make the bridge self-balancing

18. In a magnetostriction gauge, magnetic properties very due to
A. pressure and stress
B. pressure and temperature
C. pressure and displacement
D. force and speed

19. An active transducer is:
A. Hall effect pick-up
B. moving coil generator
C. capacitance microphone
D. potentiometer device

20. A strain gauge has a:
A. piezo-electric effect
B. piezo-resistive effect
C. piezo-capacitive effect
D. piezo-inductive effect

21. Piezo-electric quartz crystal can be used to measure:
A. temperature
B. velocity
C. acceleration
D. flow

22. Rotameter is used to measure:
A. rotation
B. flow
C. viscosity
D. specific gravity

23. In a pitot tube:
A. velocity head is converted into pressure head
B. pressure head is converted into velocity head
C. velocity head is converted into datum head
D. differential pressure is produced across the tube

24. The device best suited for measurement of temperature of hot bodies radiating energy in the visible spectrum is:
A. bolometer
B. optical pigometer
C. thermocouple
D. thermopile

25. The device can measure only level is:
A. strain gauge
B. radioactive device
C. diaphragm
D. bellow

26. A differential transformer is a:
A. variable inductance transducer
B. variable pressure transducer
C. constant displacement transducer
D. constant pressure transformer

27. Principle of Pirani gauge is based on:
A. humidity of the medium
B. thermal conductivity of the medium
C. combustibility of the medium
D. electrical resistivity of the medium

28. The following additional device is needed to measure pressure using LVDT:
A. Strain gauge
B. Pitot tube
C. Rotameter
D. Bourden tube

29. Which of the following devices can measure

pressure directly?
A. LVDT
B. Strain gauge
C. Rotameter
D. Bourden tube

30. Which of the following devices cannot be used to measure pressure?
A. LVDT
B. Strain gauge
C. Pyrometer
D. Pirani gauge

31. Thermocouple transducer is used for:
A. temperature measurement
B. velocity and vibration measurement
C. pressure measurement
D. acceleration measurement

32. Capacitor microphone transducers are used for:
A. measurement of level of noise, speech and music
B. displacement measurement
C. thickness measurement
D. liquid level measurement

33. Strain gauge is used to:
A. convert sound energy into electrical energy
B. sense temperature
C. converte electrical current into a mechanical displacement
D. to convert mechanical displacement into a change in resistance

34. LVDT:
A. converts linear motion into electrical signal
B. translates electrical signal into linear motion
C. helps measuring temperature
D. can be used to sense angular displacement

35. Transducer is a device which:
A. converts one form of power into another
B. converts one form of energy into anoher
C. helps in measuring electrical signal
D. is similar to a transformer

36. The name Gall Tinsley is associated with:
A. d.c. potentiometers
B. a.c. potentiometers
C. energy meters
D. Wheatstone bridge

37. Which of the following instruments is the best choice to measure a d.c. potential difference of about 0.5 volt, the current being 20μA?
A. Multimeter
B. Rectifier type meter
C. VTVM
D. Potentiometer

38. The resistance of a shunt for a precision grade ammeter can be best measured by:
A. Kelvin's double bridge
B. Schering's bridge
C. Maxwell's bridge
D. Wheatstone bridge

39. The Q-meter works on the principle of:
A. series resonance
B. parallel resonance
C. mutual inductance
D. self-inductance

40. Minimum number of wattmeters required to measure power in an unbalanced three-wire system is:
A. one
B. two
C. three
D. four

41. Induction wattmeter measures:
A. the apparent power
B. the reactive power
C. true power and the reactive power
D. only the true power

42. Holes are drilled in the disc of an energymeter on the opposite sides of the spindle to:
A. reduce the weight of the disc
B. avoid creeping on no load
C. increase the deflection torque
D. provide ventilation

43. Three-phase four-wire energy meter is used to measure:
A. three-phase unbalanced energy
B. three-phase balanced energy
C. two-phase energy
D. single phase energy

44. A wattmeter can measure
A. d.c. power only
B. a.c. power only
C. d.c. as well as power
D. d.c. power and a.c. power after rectification

45. Volt-box is basically a device used to:
A. measure the voltage
B. measure the power
C. extend the range of potentiometer
D. measure the current

46. Induction type single phase energymeter is:
A. an ampere-hour meter
B. true watt-hour meter
C. wattmeter
D. None of these

47. Power factor meters have:
A. only current coil
B. only voltage coil
C. both current and voltage coils
D. only inductive voltage coil

48. Which of the following is used to measure the leakage resistance of a capacitor?
A. Megger
B. Schering bridge
C. Potentiometer
D. Loss of charge method

49. The following frequency meter is suitable for measuring radio frequency:
A. Reed frequency meter
B. Electrical resonance frequency meter
C. Weston frequency meter
D. Heterodyne frequency meter

50. Reed frequency meter is essentially a:
A. recording system
B. deflection measuring system
C. vibration measuring system
D. oscillatory measuring system

51. Current effect is not used in:
A. moving coil voltmeter
B. electrodynamometer voltmeter
C. electrostatic ammeter
D. rectifier ammeter

52. Operation of Meggar is based on:
A. moving coil meter
B. moving iron meter
C. electrostatic meter
D. dynamometer

53. The main function of using springs in a measuring instrument is:
A. to conduct the current to the coils
B. to hold the pivot in position
C. to control the pointer movement
D. to minimize the vibration of the pointer

54. The function of the zero-adjust control in a multimeter is to:
A. conduct the current
B. change the sensitivity of the meter
C. to correct the zero point
D. to tighten up the moving parts to the casing

55. Meter accuracy is determined by:
A. one-tenth of full-scale deflection
B. one-fourth of full-scale deflection
C. half scale deflection
D. full-scale deflection

56. Which of the following instruments is free from hysteresis and eddy current losses?
A. M.I. instrument
B. electrostatic instrument
C. electrodynamometer type of instrument
D. All of these

57. Which of the following instruments has the poorest overload capacity?
A. moving coil insrument
B. hot wire instrument
C. induction type instrument
D. permanent magnet instrument

58. Reproducibility measurement of an instrument gives an indication of:
A. resolution
B. precision
C. reliability
D. accuracy

59. Reliability of an instrument signifies:
A. measurement of change due to temperature variation
B. degree to which repeatability continues to remain within specified limits
C. the life of the instrument
D. the extent to which the characteristics remain linear

60. Resolution of a measuring instrument is defined as:
A. the smallest measurable input change
B. consistency or the reproducibility the measurement
C. ability to reproduce the output in the same form as the input

D. ratio of the change of output signal to change in the input signal

61. Instrument efficiency is defined as:
A. the ability of the instrument to read the smallest input changes
B. the ratio of the measured quantity at full scale to the power taken by the instrument at full-scale
C. the ratio of the change in output signal to the change in input signal
D. the ability of the system to reproduce the output in the same form as the input

62. Instrument accuracy is defined as:
A. the measure of consistency or reproducibility of the measurements
B. closeness with which an instrument reading approaches the true value of the quantity being measured
C. the smallest measurable input change
D. the ratio of the change in output signal to the change in the input signal

63. The primary function of a fuse in an instrument is to:
A. open a circuit
B. close a circuit
C. protect the battery
D. prevent excessive current

64. Meggar is an instrument for:
A. measuring current
B. measuring voltage
C. testing insulation
D. measuring power

65. The phenomenon of creeping occurs in:
A. ammeters
B. voltmeters
C. watt meters
D. energy meters

66. Difference between the indicated value and the true value of a quantity is known as:
A. gross error
B. absolute error
C. dynamic error
D. relative error

67. The multiplier and the meter movement in a voltmeter are always in:
A. parallel
B. series
C. parallel-series
D. series-parallel

68. Instrument used for d.c. measurement alone is:
A. moving iron type
B. permanent magnet type
C. electrodynamic type
D. induction type

69. The measuring instruments which utilise the electrostatic effect are:
A. ammeters and voltmeters
B. voltmeters and wattmeters
C. ammeters and wattmeters
D. d.c. ampere hour meters and ammeters

70. Which of the following instruments consumes the lowest power in measurement?
A. VTVM
B. PMMC instrument
C. Electrostatic instrument
D. d' Arsonval instrument

71. The following instrument is not a current operated instrument:
A. electrodynamometer type
B. electrostatic type
C. hot wire instrument
D. moving coil instrument

72. A capacitance potential divider is generally used for extending the range of following instrument:
A. electrostatic type
B. electrodynamometer type
C. induction type
D. moving iron instrument

73. The rectifier instrument is not free from the following error:
A. temperature
B. waveshape
C. frequency
D. All of these

74. Rectifier instruments indicate:
A. rms value
B. average value
C. peak value
D. d.c. value

75. Electrostatic instrument uses the principle of variation of:
A. inductance with voltage

B. capacitance with voltage
C. capacitance with current
D. inductance with current

76. For capacitance measurement, a ballistic galvanometer should have:
A. a high critical resistance and a low period of oscillation
B. a high critical resistance and a high period of oscillation
C. a low critical resistance and a high period of oscillation
D. a low critical resistance and a low period of oscillation

77. Electrostatic instruments are not free from effect of:
A. temperature
B. frequency
C. stray magnetic field
D. stray electrostatic field

78. The following error does not result in moving iron instrument for both a.c. and d.c. measurements:
A. stray magnetic field error
B. hysteresis error
C. eddy current error
D. temperature error

79. Frequency compensation in a moving iron instrument is achieved by connecting:
A. a capacitor in series with the fixed coil
B. a capacitor across the fixed coil
C. high resistance in series with the coil
D. low resistance in series with the coil

80. When an a.c. voltage is applied to a PMMC voltmeter:
A. the meter gets damaged
B. meter reading is zero
C. pointer will oscillate to and fro
D. the pointer will not move at all

81. The main advantage of PMMC instrument is:
A. high torque weight ratio providing high accuracy
B. uniform scale
C. possibility of using a single instrument for many different current and voltage ranges by use of shunts and multipliers
D. capability of measuring both a.c. and d.c. quantities

82. The difference between the indicated value by an instrument and true value of a variable is called:
A. dead zone error
B. relative error
C. static error
D. drift error

83. The ratio of the measure and at full-scale to the power drawn by the instrument is called its:
A. efficiency
B. fidelity
C. precision
D. sensitivity

84. The following is a measure of reproducibility in a measurement system:
A. resolution
B. drift
C. precision
D. fidelity

85. Desirable static characteristics of a measuring system are:
A. accuracy and reproducibility
B. accuracy, sensitivity and reproducibility
C. low static error
D. low drift and dead zone

86. In an indicating instrument, the error when reading half-scale is:
A. much less than full-scale error
B. equal to full-scale error
C. greater than full-scale error
D. equal to three-fourth of full-scale error

87. Hysteresis in a measuring instrument implies:
A. repeatibility of the instrument
B. reliability of the instrument
C. change in the same reading when input is first increased and then decreased
D. inaccuracy due to change in temperature

88. The static error band of a recording instrument does not include:
A. hysteresis in the instrument
B. non-linearity
C. electrical drift
D. None of these

89. The static error band of a recording instrument signifies:
- A. the accuracy of the instrument
- B. the error produced when the pen is stopped at some deflection
- C. the error introduced in low varying inputs
- D. the irrepeatibility of the instrument

90. Sensitivity inaccuracy of a recording instrument signifies:
- A. the smallest signal required to produce detectable output
- B. the maximum error in sensitivity displayed by a pen
- C. the amount of input required to produce unit pen deflection
- D. degree to which the instrument is not sensitive enough to repeat readings

91. The sensitivity inaccuracy of a measuring instrument does not depend on the following factor:
- A. amplitude distortion
- B. frequency response
- C. hysteresis
- D. All of the above

92. Thereshold of sensitivity with respect to measuring instrument is:
- A. the maximum signal which can be measured
- B. the value of sensitivity on the highest scale
- C. the value of sensitivity on the lowest scale
- D. the smallest signal which results in a detectable output

93. The reliability of a measuring instrument means:
- A. the life of the instrument
- B. the extent over which the characteristics remain linear
- C. degree to which repeatibility continues to remain within specified limits
- D. All of these

94. Absolute instruments are those:
- A. which give the magnitude of the quantity under measurement in terms of the physical constants of the instruments
- B. which directly indicate the measured quantity on a display unit
- C. which are calibrated by comparison with a secondary instrument
- D. which are very popularly used

95. By resolution of an indicating instrument is meant:
- A. the smallest change in the output reading due to drifting of pointer
- B. the smallest change in applied stimulus which will indicate a detectable change in deflection
- C. the difference between various reading for the same applied stimulus
- D. None of these

96. In which part of the scale does the pointer indicate most accurately:
- A. in the first one-third of the scale
- B. in the first one-half of the scale
- C. in about the middle of the scale
- D. in the last one-third of the scale

97. As compared to indirect methods of electrical measurement, the direct methods are:
- A. less sensitive
- B. inaccurate
- C. not always feasible and practicable
- D. Both A and B

98. An ammeter is:
- A. an absolute instrument
- B. an indicating instrument
- C. a recording instrument
- D. a controlling instrument

99. Rayleigh's current balance is:
- A. an absolute instrument
- B. an indicating instrument
- C. a recording instrument
- D. a controlling instrument

100. Speedometer used for measurement of speed of an automobile is:
- A. an absolute instrument
- B. an indicating instrument
- C. a recording instrument
- D. a controlling instrument

ANSWERS

1	2	3	4	5	6	7	8	9	10
D	C	D	C	D	C	B	C	D	B
11	**12**	**13**	**14**	**15**	**16**	**17**	**18**	**19**	**20**
C	C	D	C	C	A	C	A	B	B
21	**22**	**23**	**24**	**25**	**26**	**27**	**28**	**29**	**30**
C	B	A	B	B	A	B	D	D	C
31	**32**	**33**	**34**	**35**	**36**	**37**	**38**	**39**	**40**
A	A	D	A	B	B	D	A	A	B
41	**42**	**43**	**44**	**45**	**46**	**47**	**48**	**49**	**50**
D	B	A	C	C	B	C	B	D	C
51	**52**	**53**	**54**	**55**	**56**	**57**	**58**	**59**	**60**
C	A	C	C	D	B	B	B	B	A
61	**62**	**63**	**64**	**65**	**66**	**67**	**68**	**69**	**70**
B	B	D	C	D	C	D	B	B	C
71	**72**	**73**	**74**	**75**	**76**	**77**	**78**	**79**	**80**
B	A	C	A	B	B	D	C	B	B
81	**82**	**83**	**84**	**85**	**86**	**87**	**88**	**89**	**90**
D	C	A	C	B	C	C	C	B	B
91	**92**	**93**	**94**	**95**	**96**	**97**	**98**	**99**	**100**
D	D	C	A	B	D	B	B	A	B

POWER ELECTRONICS

MULTIPLE CHOICE QUESTIONS

1. A step-up chopper is required to deliver a load voltage of 660V from a DC source of 220V. Let the non-conduction time of thyristor be 100 μ-sec. Find the required pulse width, Ton:
   A. 150 μ-sec
   B. 200 μ-sec
   C. 330 μ-sec
   D. 240 μ-sec
2. The thyristor can be turned on by applying:
   A. a positive signal to the gate
   B. a negative signal to the gate
   C. holding current to the gate
   D. None of the above
3. When two SCRs are connected in parallel configuration:
   A. the SCR with lower dynamic resistance will carry lesser value of current
   B. the SCR with lower dynamic resistance will carry higher value of current
   C. both the SCRs with carry equal values of current even if their dynamic resistances are different
   D. None of the above
4. Refer figure given below, showing a three-phase bridge invertor.

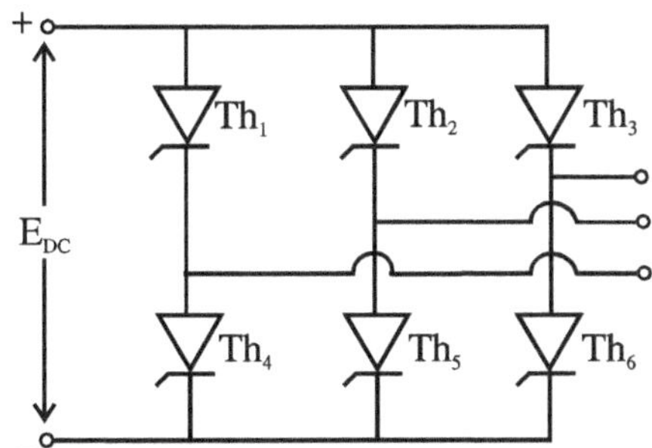

In 120°-mode operation, the maximum number of SCRs conducting at any instant is:
   A. 4
   B. 6
   C. 3
   D. 2
5. Refer figure of Q. 4, for three-phase bridge inverter. If the operation is 180°-mode, the maximum number of SCRs that conduct at any instant will be:
   A. 3
   B. 4
   C. 6
   D. 2
6. A Class-C chopper (DC to DC converter) is also known as a:
   A. one-quadrant regulator
   B. two-quadrant chopper
   C. steady state chopper
   D. All of the above
7. The load voltage curve for a Class-B chopper (DC to DC convertor) has been shown below.

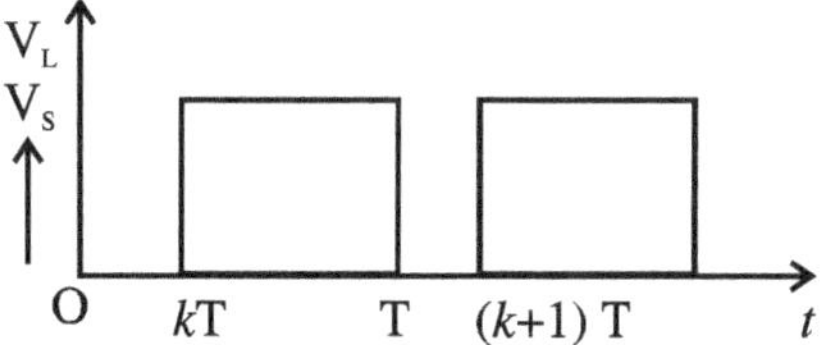

It is clear that this chopper is a first-qudrant chopper but it operates:
   A. in the first qudrant
   B. in the second qudrant
   C. by giving a sinusoidal output
   D. None of the above
8. The flyback converter topology in a DC-to-DC switched mode power supply:
   A. should be restricted to low power applications

B. can be used for high power applications
C. would be selected depending upon the DC load of the equipment in question
D. can be selected without considerations of power requirements of the equipment in question

9. In the case of an AC switched mode power supply:
A. the input-side inverter uses PWM control at a high frequency
B. the output-side inverter does not operate at output frequency
C. there is no need for a static switch
D. None of the above

10. A UPS is an example of:
A. a DC SMPS
B. an AC SMPS
C. a DC-to-DC converter
D. a voltage regulation device

11. An *n*-channel enhancement MOSFET:
A. has a physical channel
B. has no physical channel
C. is similar in construction to a *p*-channel depletion MOSFET
D. None of the above

12. The following figure is the transfer characteristic of :

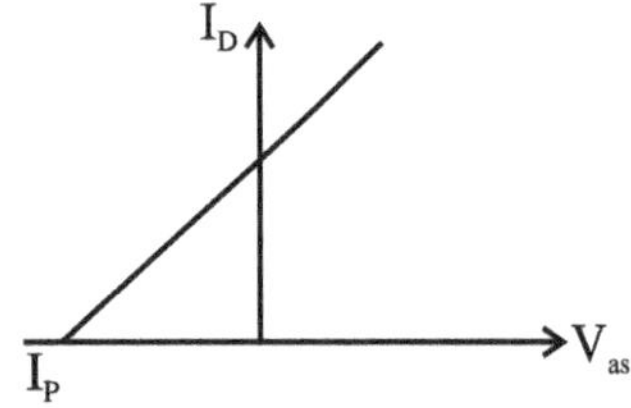

A. *n*-channel enhancement MOSFET
B. *n*-channel depletion MOSFET
C. *p*-channel depletion MOSFET
D. *p*-channel enhancement MOSFET

13. The enhancement MOSFETS respond to:
A. positive and negative gate voltages
B. positive gate voltage only
C. negative gate voltage
D. None of the above is true

14. A DC chopper (DC to DC converter) is turned on for 30 μ-sec and turned off for 10 μ-sec. What is its duty cycle?
A. 0.62
B. 0.92
C. 0.75
D. 0.85

15. In a three-phase bridge inverter:
A. firing of SCRs can be done for a period of 120° only
B. firing of SCRs can be done for a period of 180° only
C. firing of the SCRs can be done for 120° and 180° periods
D. None of the above is applicable

16. In the context of inverters, forced commutation would imply that:
A. the inverter would be cut off from the load
B. the inverter would have to be forcefully turned off through some external circuit
C. the inverter would be turned off automatically by d.c. supply
D. All of the above

17. Refer to the figure given below

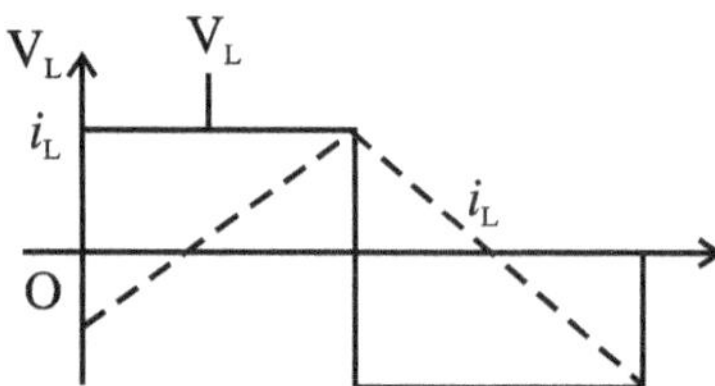

This figure shows the load voltage V_L and load current i_L for a single-phase bridge inverter. The thyristors are being fired act for continuous train of gate pulses of 180°. The output voltage is a square wave. Can you guess the type of load?
A. Resistive
B. Inductive
C. Capacitive
D. None of the above

18. A series inverter circuit has an inductor of 10 mH, a capacitor of 47 μF, connected in series with a load resistance of 5Ω. What is the resonant frequency?
A. 1292 Hz
B. 1392 Hz
C. 1492 Hz
D. 1592 Hz

19. The major part of an inverter is its:
A. step up transformer
B. oscillator circuit

C. filter
D. DC source

20. In voltage inverters:
A. load voltage wave form V_0 depends upon no load impedance Z_L but load current waveform i_L does not depend upon Z_L
B. both V_0 and i_L depend upon Z_L
C. V_0 does not depend upon Z_L but i_L depends upon Z_L
D. None of the above

21. In single-pulse modulation of PWM inverters, the pulse width is 120°. For an input voltage of 220 V DC, the RMS value of the output voltage is:
A. 280 V
B. 254.04 V
C. 179.63 V
D. None of the above

22. Which one of the following is the limitation of a current source inverter?
A. Sluggish dynamic performance
B. Stability problem
C. Unreliable
D. None of these

23. A single-phase bridge inverter delivers power to a series-connected RLC load with R = 2 Ω and $X_L = 8\Omega$. For this inverter combination, load commutation is possible case the magnitude of $\frac{1}{\omega C}$ is:
A. 20Ω
B. 12Ω
C. 10Ω
D. 14Ω

24. In a constant-source inverter, if the frequency of the output voltage is *f* Hz, then the freqeuency of voltage input to this inverter is:
A. 2*f*
B. 4*f*
C. 3*f*
D. 6*f*

25. In pulse-width modulation used in PWM inverters, the amplitudes of the reference square wave and triangular carrier wave are 1V and 2V, respectively. For generating 5 pulses per half-cycle, the pulse width should be
A. 16°
B. 24°
C. 20°
D. 18°

26. A single-phase current-source inverter has a capacitor C as the load. For a constant source current, the voltage across the capacitor is a:
A. square wave
B. step function
C. triangular wave
D. None of the above

27. In power transistors, the second breakdown can be defined as the
A. breakdown of the junction due to localising heating effects
B. breakdown due to avalanche effect
C. breakdown due to mechanical stress
D. None of the above

28. Refer to the figure given below, in which a power transistor is supplying power to an inductive load, L.

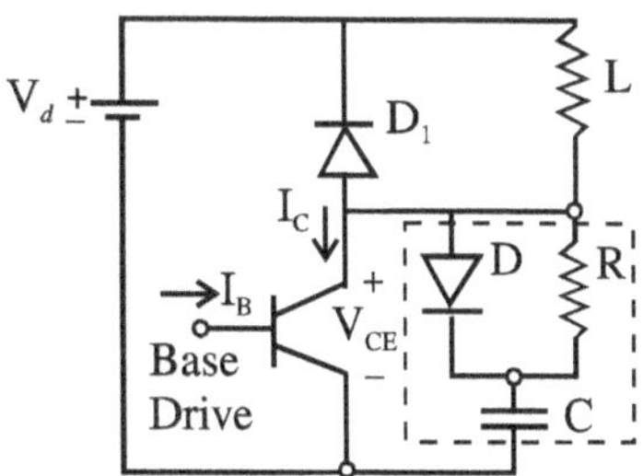

The parts in dotted box are a:
A. part of the load
B. part of the snubber circuit
C. part of the switching device
D. None of the above

29. The disadvantage of power MOSFET viz-a-viz the power transistor is:
A. nil second breakdown
B. higher conduction drop of nearly 4.5 V
C. higher peak current handling
D. more linear transfer characteristics

30. If the base current (I_B) is sufficient to drive a BJT into saturation stage, then the transistor:
A. behaves as an open switch

B. behaves as a closed switch
C. works as a switch in off and on conditions alternately
D. None of the above

31. The time required by an AC regulator to effect a change in supply voltage or load is known as:
A. time period
B. time response
C. response time
D. replanishing time

32. In a three-phase voltage regulator:
A. three SCR combinations are required per phase, each combination having 4 SCRs
B. six SCR combinations are required per phase, each combination having 5 SCRs
C. three SCR combinations are required per phase, each combination having 7 SCRs
D. None of the above

33. Refer to figure below

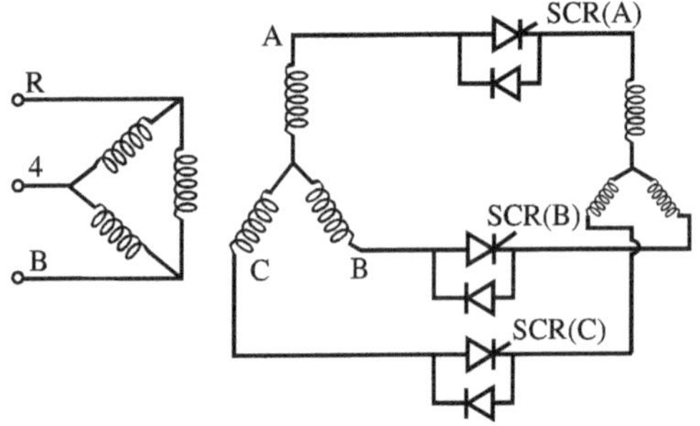

This figure represents a:
A. three-phase fully controlled AC voltage regulator
B. three-phase half-controlled AC voltage regulator
C. single-phase fully controlled AC voltage regulator
D. All of the above

34. In a single-phase voltage controller with R-L load, AC output power can be controlled if:
A. (firing angle) $\alpha > \phi$ (load phase angle) and (conduction angle) $\sqrt{\ } = \pi$
B. $\alpha < \phi$ and $\sqrt{\ } < \pi$
C. $\alpha > \phi$ and $\sqrt{\ } > \pi$
D. None of the above

35. A single-phase voltage regulator feeds power to a resistance of 10Ω. The source voltage is 200V. For a firing angle of 90°, the RMS value of thyristor current is:
A. 10 A
B. 22 A
C. 16 A
D. 4 A

36. A resistance of 10Ω is supplied power through a single-phase voltage controller from a source of 200 sin 314*t*. For a firing angle of 90°, the power delivered to the load is:
A. 2 kW
B. 0.5 kW
C. 2.5 kW
D. 1 kW

37. A single phase voltage controller is used for controlling the flow of power from 260V, 50Hz source into a load consisting of R = 5Ω and X = 12Ω. The value of maximum RHS load current and the firing angle are, respectively:
A. 20 A, 90°
B. 25A, 45°
C. 20A, 0°
D. 25A, 90°

38. When a SCR gets turned on, the gate drive:
A. should not be removed as it will turn off the SCR
B. may or may not be removed
C. should be removed
D. should be removed in order to avoid increased losses and higher junction temperature

39. The capacitive current flowing through the junction of a SCR is:

A. $I_C = C\frac{dv}{dt}$

B. $I_C = \frac{I}{C}\frac{dv}{dt}$

C. $I_C = C\frac{di}{at}$

D. $I_C = C^2\frac{di}{dt}$

40. A diac can be used to:
A. trigger a diac
B. protect a triac
C. increase the efficiency of a triac
D. None of the above

41. In a GTO, anode current begins to fall when gate current
A. is at negative peak at time $t = 0$
B. is at negative peak at t = storage time t fall time
C. is at negative peak at t = storage time
D. None of the above

42. Look at the figure given below

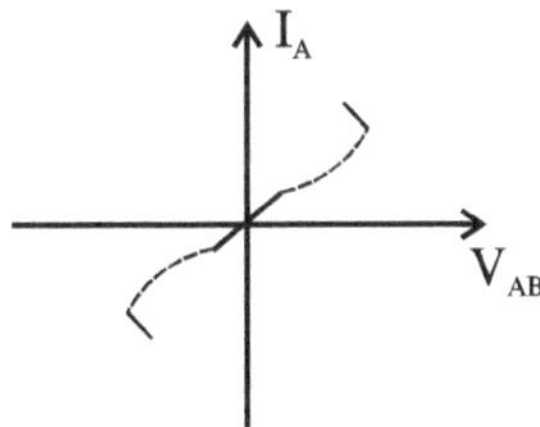

This v-i characteristic is that of a:
A. triac
B. diac
C. GTO
D. LASCR

43. The symbol of a n-channel MOSFET is:

A. 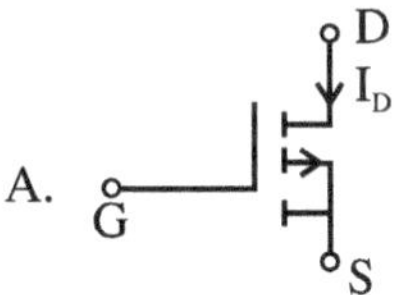

B. 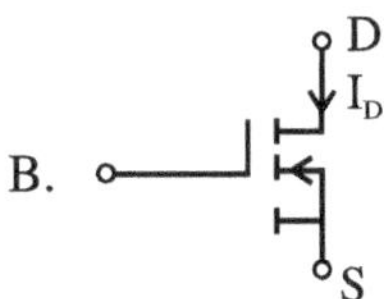

C. 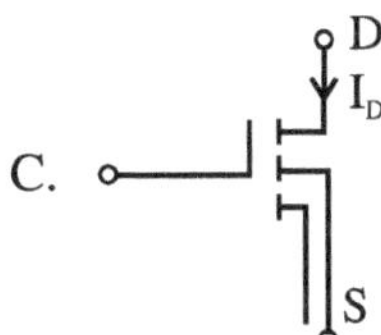

D. all are the symbols of n-channel MOSFET

44. A silicon controlled rectifier is rated for a peak inverse voltage of 650 V. Find the RMS voltage for which the device can be operated if the voltage safety factor is 2.
A. 130.4 V
B. 230.4 V
C. 200.4 V
D. 150.4 V

45. A triac can be considered to be:
A. two SCRs connected in anti-parallel with a common gate
B. two SCRs connected in parallel with common gate
C. two transistors connected in anti-parallel
D. None of these

46. Surge current rating of an SCR specifies the maximum:
A. non-repitive current with sine wave
B. non-repititive current with rectangular wave
C. repititive current with triangular wave
D. None of the above

47. In PWM technique within an inverter, the basic operation is:
A. the variation of the conduction time per cycle to alter the RMS output voltage of the inverter
B. the variation of input voltage to the inverter
C. the variation of input current to the inverter
D. All of the above are true

48. In multiple PWM control:
A. a single pulse is used in each half cycle of the output voltage
B. several equidistant pulses are used in each half cycle of the output voltage
C. Any of A or B can be executed
D. None of A or B can be executed

49. The major PWM techniques are single pulse modulation, multiple pulse modulation and:
A. skew wave modulation
B. square wave modulation
C. sinusoidal modulation
D. None of the above

50. A d.c. chopper is fed from 100 V DC. Its load voltage consists of rectangular pulses of duration1 m-sec in an overall cycle time of 3 m-sec. The average output voltage and ripple factor of this chopper are, respectively:
A. 33.3, 1.0
B. 33.8, 1.5

C. 33.33, 1.2
D. 33.33, 1.5

51. A three-phase half-wave controlled rectifier has a supply of 150 V per phase. Find the average load voltage for a firing angle of 60°. Assume a thyristor volt-drop of 1.5 V and continuous load current:
A. 46.2 V
B. 86.2 V
C. 75.4 V
D. 80.2 V

52. The forced commutation of AC to DC converters cannot be effected through:
A. input voltage control
B. sinusoidal PWM control
C. symmetrical angle control
D. extinction angle control

53. In a dual single-phase DC to AC converter:
A. two full converters are connected in parallel
B. two full converters are connected back to back
C. there is only one full converter
D. All of the above could be true

54. In a single-phase AC to DC semiconverter, the average output voltage is given by:

A. $\frac{1}{\pi}\int_{\alpha}^{\pi} V_m \cos \omega t.d\omega t$

B. $\frac{1}{\pi}\int_{\alpha}^{\pi+\alpha} V_m \cos \omega t.d\omega t$

C. $\frac{1}{\pi}\int_{\alpha-\frac{\pi}{2}}^{\alpha+\frac{\pi}{2}} V_m \cos \omega t.d\omega t$

D. $\frac{1}{\pi}\int_{\alpha-\frac{\pi}{2}}^{\pi} V_m \cos \omega t.d\omega t$

55. In a three-phase AC to DC full converter, if the load current is I and is free from ripples, then the average thyristor current is:

A. $\frac{1}{2}I$

B. $\frac{1}{3}I$

C. $\frac{1}{4}I$

D. I

56. In controlled AC to DC converters, the nature of load current:
A. does not depend upon the type of load and firing angle delay
B. depends upon the type of the load as well as the firing angle delay
C. depends only upon the load
D. depends only upon the firing angle delay

57. In a single-phase semiconverter, the extinction angle $\beta < \pi$. The conduction is discontinuous. The free wheeling diode conducts for:
A. α
B. $\beta-\pi$
C. zero degree
D. $\pi-\beta$

58. In a three-phase semiconverter, for a firing angle of 120° and extinction angle of 110°, the freewheeling diode conducts for:
A. 50°
B. 110°
C. 30°
D. None of the above

59. The three-phase AC to DC converter, which requires neutral pont connection, is one of the following:
A. Tree-phase semiconverter
B. Three-phase full converter
C. Three-phase half-wave converter
D. Three-phase full converter with diodes

60. In a three-phase full converter, the six thyristors are fired at an interval of:
A. 30°
B. 60°
C. 90°
D. 120°

61. A three-phase full-wave converter has an average output of 220 V for zero degree firing angle and for resistive load. For a firing angle of 90°, the output voltage would be:
A. zero

B. 50 V
C. 100 V
D. 26.8 V

62. For the same AC voltage and load impedance, which one of the following statements about rectifiers is true?
A. The average load current in a full-wave rectifier is π times that in a half-wave rectifier
B. Half-wave rectifier will have bigger sized transformer as compared to that of a full-wave rectifier
C. Half-wave rectifier will have a small sized transformer as compared to that of a full-wave rectifier
D. None of the above

63. A three-phase semiconverter can work as
A. a converter for $\alpha = 0°$ to $180°$
B. a converter for $\alpha = 0°$ to $90°$
C. an inverter for $\alpha = 90°$ to $180°$
D. an inverter for $\alpha = 0°$ to $90°$

64. In a single-phase full converter, if output voltage has a peak value of 325 V and average value of 133 V, then the firing angle is:
A. 40°
B. 140°
C. 50°
D. 130°

65. In a sinle-phase full converter, the number of SCRs conducting during overlap is:
A. 1
B. 2
C. 3
D. 4

66. For discontinuous load current and extinction angle of $B > \pi$, in the case of a single-phase full converter, each thyristor would conduct for an angle equal to:
A. α
B. $\beta - \alpha$
C. β
D. $\alpha + \beta$

67. In a three-phase semiconverter, the frequency of the ripple in the output voltage may be:
A. three times the supply frequency f for firing angle $\alpha < 60°$
B. $3f$ for $\alpha > 60°$
C. $6f$ for $\alpha < 60°$
D. $6f$ for $\alpha > 60°$

68. In a single phase half-wave circuit with R-L load and a free-wheeling diode across the load, extinction angle β is more than π. For a firing angle α, the thyristor and free-wheeling diode would conduct, respectively, for:
A. $(\pi - \alpha)$, β
B. $(\beta - \alpha)$, $(\pi - \alpha)$
C. $(\pi - \alpha)$, $(\beta - \pi)$
D. $(\pi - \alpha)$, $(\pi - \beta)$

69. In a circulating-current type of dual converter, the nature of voltage across the reactor is:
A. alternating
B. pulsating
C. direct
D. triangular

70. A single-wave full-wave mid-point thyristor converter uses a 230/200 V transformer with centre tap on the secondary side. The PIV for each thyristor is:
A. 100 V
B. 141.4 V
C. 200 V
D. 282.8 V

71. Look at the figure given below

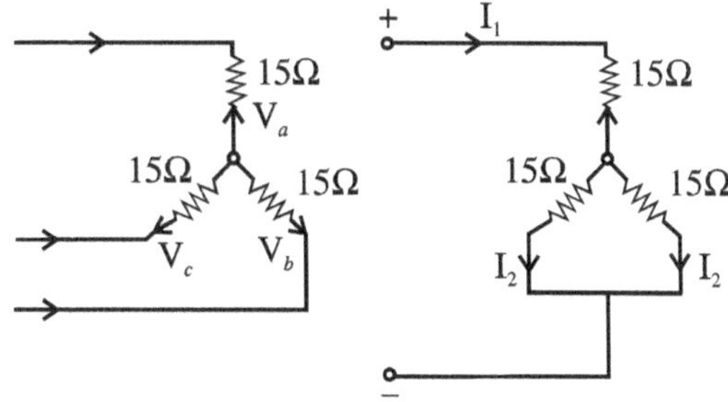

This is a circuit of three-phase voltage -sourced bridge inverter is supplied from 600 V source. For a star connected resitive load of 15Ω per phase, find the RMS value of load current.
A. 16.1 A
B. 16.33 A
C. 15.33 A
D. 14.23 A

72. In a three-phase inverter, the leading VAR

requirement can be met by a variable VAR generator using a:

A. swtiched capacitor
B. rotating synchronous condenser
C. current-fed inverter VAR generator
D. All of the above

73. So far as the harmonic current at the output voltage of the inverter is concerned, the multiple pulse modulation will provide:

A. less harmonics
B. more harmonics
C. perfect sinusoidal of fundamental frequency
D. None of the above

74. The oscillator circuit of an inverter is built up of a/an:

A. resistor
B. capacitor
C. inductor
D. All of the above

75. Which one of the following statements is correct in connection with inverters?

A. Voltage source inverter and current source inverter both need feedback diodes
B. Only current source inverter requires feedback diodes
C. GTOs can be used in current source inverters
D. Only voltage source inverter needs feedback diodes

76. PWM inverters, with multiple pulses in each half cycle of inverter output voltages, can:

A. increase power output
B. reduce harmonic content
C. improve PF of the load
D. None of the above

77. What is the disadvantage of PWM technique in inverters?

A. Harmonic content is reduced in output
B. Smaller number of commutations is required in each cycle
C. Large number of commutations is required in each cycle
D. Magnitude of fundamental component is increased

78. In PWM, all the lower order harmonics are eliminated for:

A. $V_m/V_L \leq 2$
B. $V_m/V_L \leq 1$
C. $V_m/V_L \geq 1$
D. $V_m/V_L \leq 2$

where $\frac{V_m}{V_L}$ = amplitude ratio.

79. Out of the following, which system of pulse width modulation is most superior?

A. Single pulse
B. Multiple pulse
C. Sinusoidal pulse
D. All are equally superior

80. In a single-pulse modulation used in PWM inverters, V is the input DC voltage. For eliminating the third harmonic, the magnitude of RMS value of fundamental component of output voltage and the pulse width are, respectively:

A. $\frac{2\sqrt{2}}{\pi}$ V, 120°

B. $\frac{4V}{\pi}$, 60°

C. $\frac{2\sqrt{2}}{\pi}$ V, 60°

D. $\frac{4V}{\pi}$, 120°

81. A MOSFET is easy to be put in parallel due to:

A. no second breakdown
B. positive resistance coefficient
C. larger gain
D. None of the above

82. The safe operating area of a MOSFET can be increased according to the following figure (increased in SOA shown by dotted lines).

A.

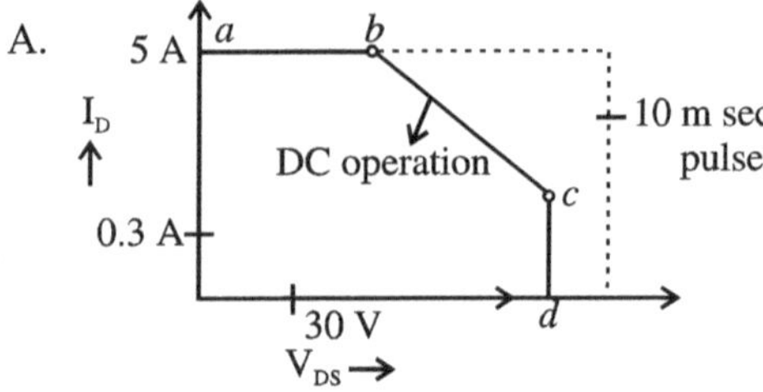

B. 5 A, a, b, 10 m sec pulse, c, I_D ↑, DC operation, 0.3 A, 30 V, d, V_{DS} →

C.

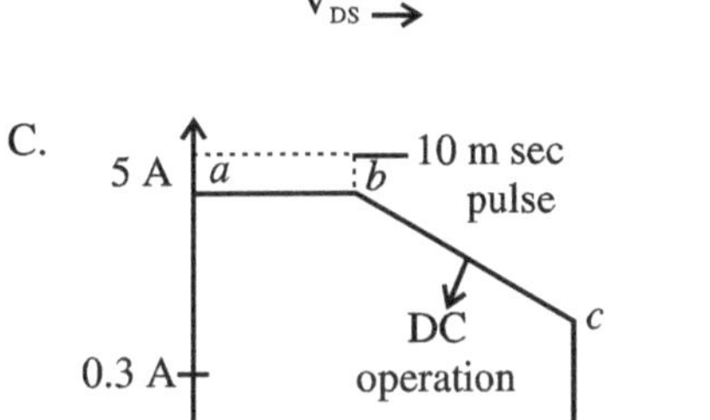

D. None of the above graphs is correct

83. The switching on/off operation of a MOSFET can be carried out in less than (nearly):

A. 100 milliseconds
B. 500 nanoseconds
C. 50 nanoseconds
D. 100 nanoseconds

84. The conduction loss of a high-voltage MOSFET is:

A. very low
B. very high
C. nil
D. depends upon the input

85. The switching loss of a MOSFET is:

A. almost negligible
B. very high
C. cannot be determined easily
D. none of the above

86. The following figure shows two transistors connected in darlington configuration.

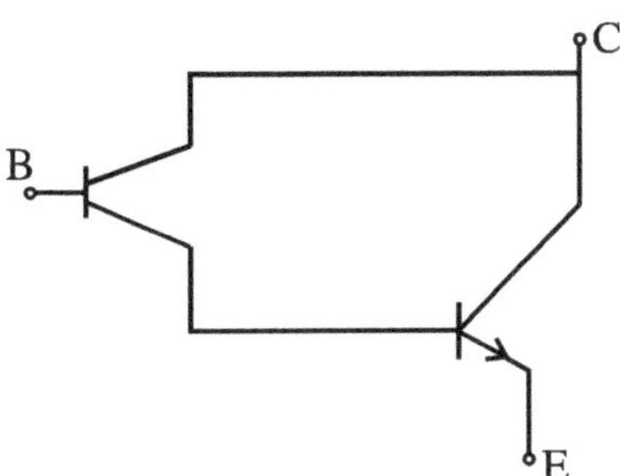

The gain of this circuit would be:

A. low
B. high
C. zero
D. Cannot predict as the input and the output are unknown

87. Diodes can be operated in parallel configuration if:

A. the load current is low
B. the input voltage is low
C. the load current is very high
D. None of the above

88. The ripple factor of a waveform is the ratio of RMS value of the ripple to the average output level. Thus the ripple factor of a half-wave rectifier is:

A. 0.246
B. 0.48
C. 1.21
D. None of the above

89. An ideal regulated power supply should have:

A. 100% regulation
B. 50% regulation
C. 0% regulation
D. None of the above

90. For a filter, no-load voltage is 300 V and full-load voltage is 280 V. The regulation of the circuit is:

A. 9.6%
B. 7.1%
C. 6.7%
D. 4.2%

91. The intrinsic standoff ratio of a UJT is given by:

A. $R_{B1} + R_{B2}$

B. $\dfrac{R_B}{R_{B1} + R_{B2}}$

C. $(R_{B1} + R_{B2})\, R_{B1}$

D. None of the above

92. For a UJT used for triggering a SCR, the stand-off ratio (n) is 0.64 and DC voltage source V_{BB} is 20 V. The UJT would trigger when the emitter voltage is:

A. 12.8 V
B. 13.1 V

C. 10 V
D. None of the above

93. Which one of the following devices cannot be triggered with the voltage of either polarity?
A. SCS
B. SUS
C. Triac
D. Diac

94. The input impedance of an IGBT is:
A. low
B. moderate
C. high
D. None of the above

95. There are four configurations of PWM inverter. Three of them are push-pull, half-bridge and full-bridge. The fourth one is:
A. start-up
B. flyback
C. start-down
D. fly-kick

96. Look at the figure given below

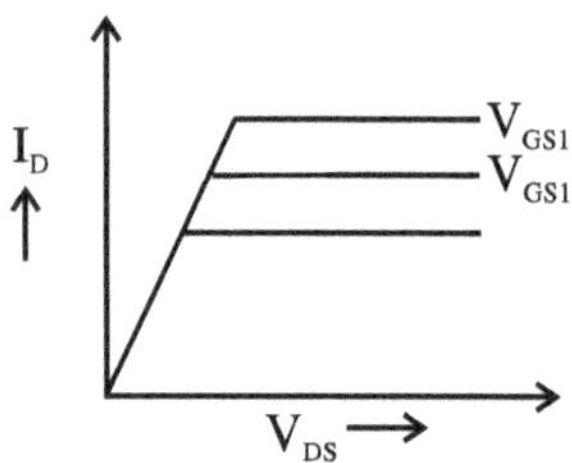

This is the *v-i* characteristic of:
A. *n*-channel MOSFET
B. *p*-channel MOSFET
C. BJT
D. UJT

97. Look at the figure given below

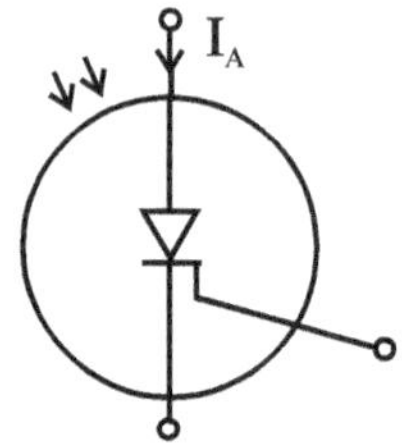

This is the symbol of a:
A. SCS
B. SUS
C. LED
D. LASCR

98. When a voltage is applied to a thyristor:
A. holes move to positive electrode
B. electrons move to negative electrode
C. the holes move to the negative electrode and electrons move to the positive electrode
D. None of the above

99. The SCR ratings $\frac{di}{dt}$ (in A per μ-sec.) and $\frac{dv}{dt}$ (in V per μ-sec) may vary, respectively between:
A. 20-500, 10-100
B. both can vary between 20 to 500
C. both can very between 10 to 100
D. 50-300, 20-500

100. In a thyristor:
A. latching current I_C is associated with the turn-off process and holding current I_H is associated with the turn-on process
B. both I_L and I_H are associated with turn-on process
C. I_H is a associated with turn-off process and I_L is associated with turn-on
D. both I_L and I_H are associated with turn-on process

ANSWERS

1	2	3	4	5	6	7	8	9	10
B	A	B	D	A	B	B	A	A	B
11	**12**	**13**	**14**	**15**	**16**	**17**	**18**	**19**	**20**
B	B	B	C	C	B	B	B	B	C
21	**22**	**23**	**24**	**25**	**26**	**27**	**28**	**29**	**30**
C	A	C	A	D	C	A	B	B	B

31	32	33	34	35	36	37	38	39	40
C	A	B	B	A	D	C	D	A	A
41	**42**	**43**	**44**	**45**	**46**	**47**	**48**	**49**	**50**
C	B	D	B	A	A	A	B	C	D
51	**52**	**53**	**54**	**55**	**56**	**57**	**58**	**59**	**60**
B	A	B	C	D	B	C	A	C	B
61	**62**	**63**	**64**	**65**	**66**	**67**	**68**	**69**	**70**
C	C	A	C	D	B	C	C	A	D
71	**72**	**73**	**74**	**75**	**76**	**77**	**78**	**79**	**80**
B	D	A	D	D	B	C	A	C	A
81	**82**	**83**	**84**	**85**	**86**	**87**	**88**	**89**	**90**
B	B	C	B	A	B	C	C	C	B
91	**92**	**93**	**94**	**95**	**96**	**97**	**98**	**99**	**100**
B	B	A	C	B	A	D	C	B	C

ELECTRONIC ANALOG CIRCUITS

MULTIPLE CHOICE QUESTIONS

1. The circuit given below forms:

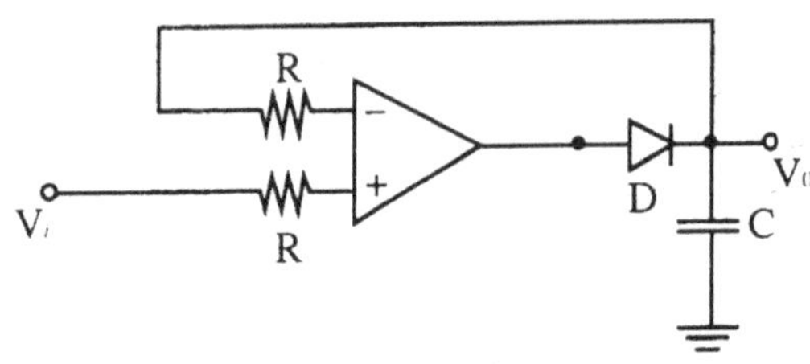

A. active positive clipper
B. active peak detector
C. active halfwave rectifier
D. sample-and-hold circuit

2. An Op-Amp clamper:

A. removes part of the output voltage above a given level
B. converts the input voltage into square wave voltage at a given level
C. clamps the output voltage at a given level
D. removes part of output voltage below a given level

3. Positive halfwave rectifier using Op-Amp is generally used to rectify signal of the order of:

A. kilovolts
B. hundreds of volts
C. tens of volts
D. a few millivolts

4. An Op-Amp positive clipper:

A. removes positive part of the input signal greater than reference voltage V_R
B. remove positive part of the input signal below the reference voltage V_R
C. produces square wave voltage at V_R
D. compresses the positive part of the input signal to level V_R

5. An Op-Amp limiter generally limits the output voltage:

A. at reference level determined by V_{CC}
B. at reference level determined by $-V_{EE}$
C. at level $(V_Z + V_D)$ and $-(V_Z + V_D)$
D. at rms values of input voltages

6. An Op-Amp Schmitt trigger is basically:

A. an Op-Amp comparator with negative feedback
B. an Op-Amp comparator with positive feedback
C. a triangle wave generator
D. a pulse generator

7. An Op-Amp comparator basically converts an input voltage into:

A. a square wave voltage
B. a triangle wave voltage
C. a ramp voltage
D. a sinusoidal votlage

8. An Op-Amp comparator:

A. is always non-inverting type
B. is always inverting type
C. is always a combination of an inverting Op-Amp and a noniverting Op-Amp
D. may be inverting type or non-inverting type

9. In the given circuit, if $V_1 = +3$ V, $V_2 = +2$ V and $V_3 = -1$ V, the output voltage V_0 will be:

A. +4 V
B. −4 V

C. +6 V

D. −6 V

10. The Op-Amp circuit given below behaves as a:

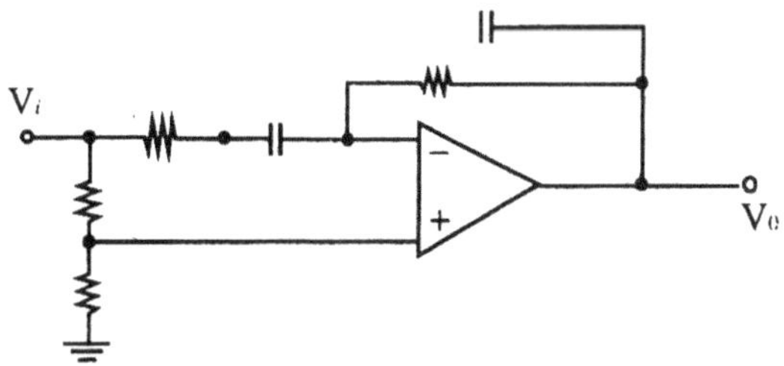

A. high pass filter

B. low pass filter

C. band-pass filter

D. notch filter

11. The Op-Amp circuit given below behaves as a:

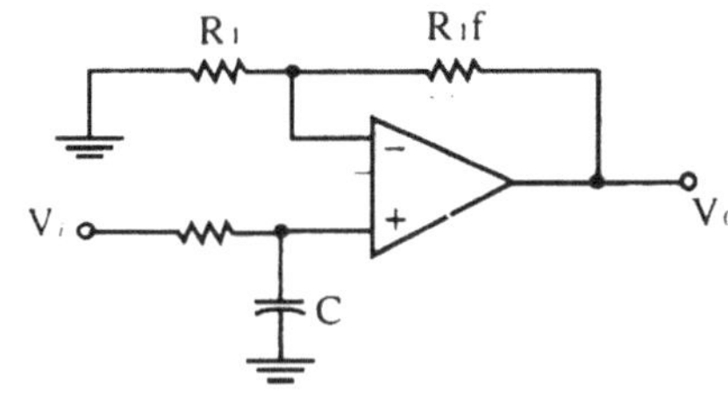

A. differentiat

B. low pass filter

C. high pass filter

D. notch filter

12. The best circuit used to eliminate power frequency (50 Hz) hum is:

A. a scratch filter

B. a speech filter

C. an octave filter

D. a rumble filter

13. An Op-Amp circuit capable of selecting sound from one out of two speakers or both is known as:

A. a crossover network

B. a tone control

C. a panning circuit

D. an octave equalizer

14. An audio Op-Amp system requires a pre-amplifier with a gain of 200 to match a 50 kΩ impedance microphone. The best amplifier configuration to use is:

A. inverting

B. non-inverting

C. signal follower

D. None of the above

15. Which of the following Op-Amp system is non-linear?

A. voltage-to-current converter

B. voltage follower

C. active filter

D. sample-and-hold circuit

16. The Op-Amp circuit given below behaves as:

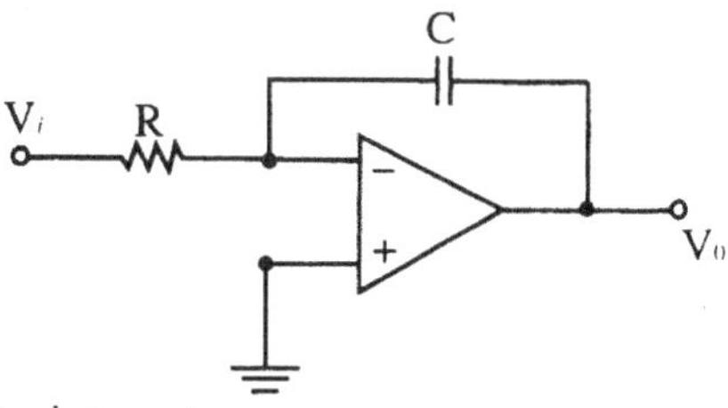

A. integrator

B. differentiator

C. adder

D. scale changer

17. The Op-Amp circuit given below behaves as:

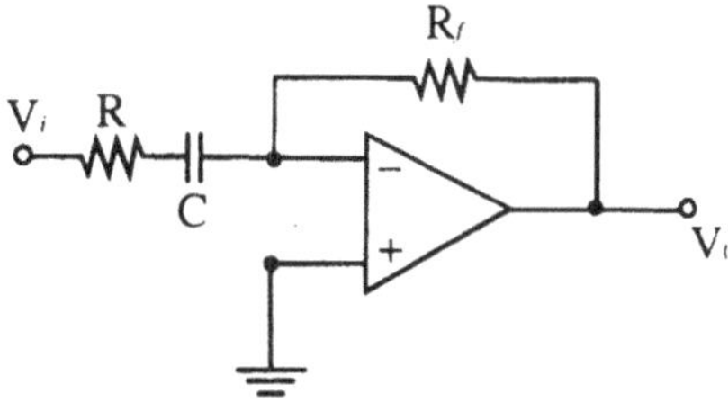

A. direct coupled amplifier

B. a.c. coupled amplifier

C. d.c. voltage follower

D. logarithmic amplifier

18. The Op-Amp circuit shown below is a:

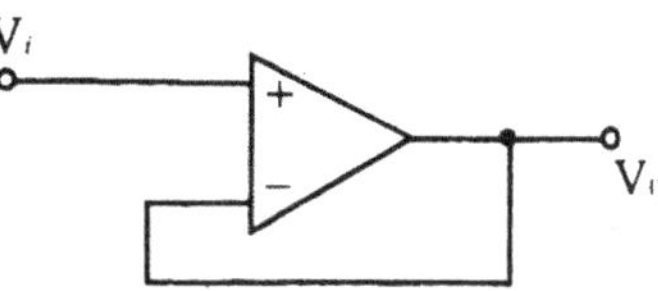

A. summer

B. voltage follower

C. non-inverting amplifier

D. voltage-to-current convertor

19. Thomas biquad is called a biquad circuit because:

A. it uses four Op-Amps

B. it involves one quadratic equation

C. it involves two quadratic equations
D. it produces two voltage in phase quadrature

20. A cascode amplifier uses:
A. two stages of CE amplifier
B. direct coupled CE-CC circuits
C. direct coupled CE-CB circuits
D. direct coupled CC-CB circuits

21. Closed loop voltage gain of an ideal inverting Op-Amp is given by:

A. $A_{vf} = \frac{R_f}{R}$

B. $A_{vf} = -\frac{R_f}{R}$

C. $A_{vf} = \frac{R + R_f}{R}$

D. $A_{vf} = -\frac{(R + R_f)}{R}$

22. Typical output offset voltage and input offset current of an Op-Amp are:
A. 1 mV, 10 mA
B. 1 mV, 10 nA
C. 10 mV, 10 nA
D. 10 mV, 10 mA

23. In an Op-Amp, the slewing rate is the maximum time rate of change of the closed loop:
A. output voltage when supplying the rated output
B. output voltage under small signal condition
C. input voltage under large signal condition
D. input voltage under small signal condition

24. Differential transconductance g_{md} of differential amplifier equals:

A. $\frac{I_0}{4V_T}$

B. $\frac{I_0}{2V_T}$

C. $\frac{\sqrt{I_0}}{4V_T}$

D. $\frac{2I}{V_T}$

25. An Op-Amp has gain of 50 and bandwidth of 100 kHz. Its unity-gain frequency is:
A. 5 MHz
B. 100 kHz
C. 10 MHz
D. 2.5 MHz

26. A non-inverting Op-Amp has Z_i = 10 kΩ, feedback resistor R_f = 120 kΩ and V_i = + 0.6 volt p–p. The output voltage will be about:
A. + 7.8 volt p–p
B. – 7.8 volt p–p
C. + 7.2 volt p–p
D. + 8.8 volt p–p

27. In an inverting Op-Amp, Z_i = 22 kΩ, Z_f = 68 kΩ, and V_i = + 0.5 volt p–p. The output voltage will be approximately:
A. + 0.5 volt p–p
B. + 15 volt p–p
C. – 1.5 volt p–p
D. zero

28. The frequency at which differential gain in Op-Amp is zero dB is known as:
A. zero dB frequency
B. lower cutoff frequency
C. upper cutoff frequency
D. unity gain crossover frequency

29. In a differential amplifier, CMMR is normally:
A. zero
B. unity
C. small
D. very large

30. A good quality Op-Amp has:
A. very large bandwidth
B. narrow bandwidth
C. high selectivity
D. low differential gain

31. The following property of Op-Amp permits voltage gain down to zero freauency:
A. capacitance coupling
B. direct coupling
C. high open loop gain
D. feedback

32. Which out of the following IC Op-Amps has the best tolerance and minimum cost:
A. 709
B. 709 A
C. 709 B
D. 709 C

33. The Op-Amp shown below has $A_v = 20{,}000$ and $Z_0 = 20\Omega$. The approximate output voltage V_0 is:

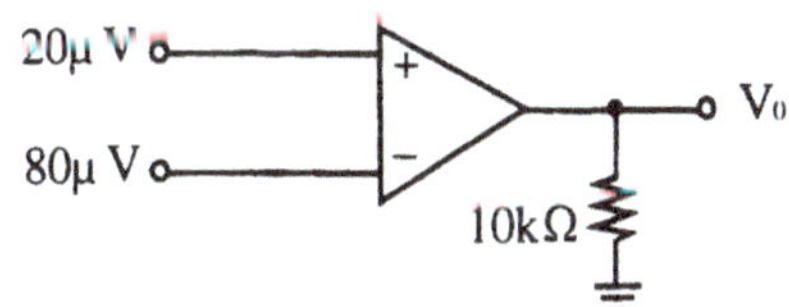

A. 0.2 V
B. 2 V
C. 2 mV
D. 20 V

34. Input bias current for 714 C Op-Amp is about:
A. 50 nA
B. 500 nA
C. 5 μA
D. 50 μA.

35. Input offset current for 714 C Op-Amp is about:
A. 20 nA
B. 200 nA
C. 2 μA
D. 20 μA

36. For precision Op-Amp 741 C, the average temperature coefficient of input offset error current $\Delta I_{io}/\Delta T$ is about:
A. 1 pA/ °C
B. 10 pA/ °C
C. 100 pA/ °C
D. 1 nA/ °C

37. The gain bandwidth product (GB) of 741 Op-Amp is about:
A. 0.1 MHz
B. 1 MHz
C. 10 MHz
D. 100 MHz

38. Which of the following is an odd man out?
A. MOSFET
B. JFET
C. Planar transistor
D. tunnel diode

39. Tuned amplifiers are generally not used in:
A. TV receivers
B. Radio receivers
C. Public address systems
D. Radar receivers

40. With $h_{fe} = 60$, the current gain of CC amplifier is approximately:
A. 1/60
B. 60
C. 61
D. 121

41. The advantage of using a junction diode as temperature stabilising element in transistor amplifier is that:
A. its temperature coefficient of resistance is negative
B. it is made of the same material as the transistor
C. its reverse current is low
D. its resistance is temperature invarient

42. The comparison unit in a d.c. voltage regulator is generally in the form of:
A. a silicon breakdown diode
B. CE amplifier
C. a potentiometer
D. a battery

43. The reference unit in a d.c. voltage regulator is generally in the form of:
A. a silicon breakdown diode
B. a transistor
C. a battery
D. a differential amplifier

44. The sampling unit in an automatic voltage regulator is generally in the form of:
A. a zener diode
B. a potentiometer
C. a transistor
D. a silicon diode

45. Stabilization ratio of a voltage regulator is defined as:

A. $\frac{dV_0}{dV_i}$

B. $\frac{dV_i}{dV_0}$

C. $\frac{dI_1}{dI_i}$

D. $\frac{dV_0}{dI_1}$

46. Wideband amplifier usually employs RC coupling instead of transformer coupling because of its:
   A. more flat frequency response in low frequency range
   B. more flat frequency response in high frequency range
   C. more flat frequency response over a wide frequency range
   D. less phase distortion

47. Which of the following circuits is used for production of delays?
   A. Astable multivibrator
   B. Bistable multivibrator
   C. Monostable multivibrator
   D. Schmitt trigger

48. Which of the following circuits may be used for converting a sine wave into a square wave?
   A. Schmitt trigger
   B. Bistable multivibrator
   C. Astable multivibrator
   D. Monostable multivibrator

49. A symmetrical astable multivibrator has R = 100Ω and C = 0.1 μF. The periodic time T is:
   A. 138 μs
   B. 69 μs
   C. 6.9 μs
   D. 13.8 μs

50. Periodic time of symmetrical free-running astable multivibrator equals:
   A. 1.38 RC
   B. 0.69 RC
   C. $\sqrt{2}$ RC
   D. RC/$\sqrt{2}$

51. In a multivibrator, commutating capacitors:
   A. reduce the transition time
   B. reduce the rise time
   C. reduced the settling time
   D. None of these

52. Schmitt trigger is also known as:
   A. squaring circuit
   B. blocking oscillator
   C. sweep circuit
   D. astable multivibrator

53. The shortest interval between pulses with which a binary can switch from one state to another is called:
   A. rise time
   B. relaxation time
   C. resolving time
   D. delay time

54. The most likely application of a crystal oscillator is:
   A. as an RF test oscillator
   B. as an electronic organ for home
   C. as a Hi-Fi test audio frequency sweep generator
   D. in commercial radio transmitter

55. In RC phase shift oscillator using FET and 3-section RC phase shift network, the condition for sustained oscillation is:
   A. $\beta > 6$
   B. $\beta > 29$
   C. $\beta > 4n + 23 + \frac{29}{n}$, $n = \frac{R_d}{R}$
   D. $\beta > 23 + \frac{29}{n}$

56. For sustained oscillations in RC phase-shift oscillator using BJT and 3-section RC network, the required transistor h_{fe} is given by:
   A. $h_{fe} > 29$
   B. $h_{fe} > \left(4n + 23 + \frac{29}{n}\right)$; $n = \frac{R_c}{R}$
   C. $h_{fe} > 4n + \frac{29}{n}$
   D. $h_{fe} > 23 + \frac{29}{n}$

57. Effective Q of the equivalent electrical circuit of quartz crystal is of the order of:
   A. 200
   B. 2,000
   C. 20,000
   D. 10^5

58. Crystal oscillator uses:
   A. silicon crystal
   B. germanium crystal
   C. crystal diode
   D. piezo-electric quartz crystal

59. FET RC phase-shift oscillator has angular frequency of oscillation ω equal to:

A. $\frac{1}{3\,RC}$

B. $\frac{1}{6\,RC}$

C. $\frac{1}{\sqrt{6}\,RC}$

D. $\frac{\sqrt{6}}{RC}$

where R and C refer to each section of three section RC phase shift network.

60. FET phase-shift oscillator uses:

A. voltage series feedback
B. voltage shunt feedback
C. current series feedback
D. current shunt feedback

61. BJT RC phase-shift oscillator has angular frequency of oscillation ω equal to:

A. $\frac{1}{RC}$

B. $\frac{3}{RC}\cdot\frac{1}{\sqrt{6+4n}}$

C. $\frac{1}{RC}\cdot\frac{1}{\sqrt{29}}$

D. $\frac{1}{RC}\cdot\frac{1}{\sqrt{6+4n}}$

where n = RC/R

62. RC phase shift oscillator using BJT employs:

A. voltage series feedback
B. voltage shunt feedback
C. current series feedback
D. current shunt feedback

63. In BJT clapp oscillator angular frequency of oscillation ω equals:

A. $\frac{1}{\sqrt{L\left(\frac{C_1C_2C_3}{C_1C_2+C_2C_3+C_3C_1}\right)}}$

B. $\frac{1}{\sqrt{L\left(\frac{C_1C_2}{C+C}\right)}}$

C. $\frac{1}{\sqrt{L\left(\frac{C_1+C_2+C_3}{C_1C_2C_3}\right)}}$

D. $\frac{1}{\sqrt{L\left(\frac{C_1C_2+C_2C_3+C_3C_1}{C_2C_2C_3}\right)}}$

64. In FET colpitts oscillator, frequency of oscillation ω:

A. $\approx \omega_0$
B. $<< \omega_0$
C. $>> \omega_0$
D. $\approx \omega_0 \frac{C_1}{C_1+C_2}$

65. In BJT colpitts oscillator, sustained oscillations are obtained when:

A. $h_{fe} \geq \frac{C_1}{C_2}$

B. $h_{fe} \geq \frac{C_1+C_2}{C_2}$

C. $h_{fe} \geq \frac{C_2}{C_1+C_2}$

D. $h_{fe} \frac{X_2}{X_1}$

where C_1 is the capacitance between base end and ground while C_2 is the capacitance between collector end and ground.

66. In BJT colpitts oscillator, angular frequency of oscillation ω equals:

A. ω_0

B. $\frac{\omega_0}{\sqrt{1-\frac{X_1X_2}{h_{ie}.h_{oe}}}}$

C. $\frac{\omega_0}{\sqrt{1+\frac{X_1X_2}{h_{ie}.h_{oe}}}}$

D. $\frac{\omega_0}{\sqrt{1+\frac{h_{ie}.h_{oe}}{X_1X_2}}}$

67. In FET Hartley oscillator, frequency oscillation ω:
A. $\approx \omega_0$
B. $<< \omega_0$
C. $>> \omega_0$
D. $\approx \omega_0 \frac{L_1}{L_2}$

where ω_0 is the frequency of series resonance of L_1, L_2 and C.

68. The condition under which a transistor amplifier is fundamentally stable is:
A. $V_{CE} < \frac{V_{CC}}{2}$
B. $V_{CE} = \frac{V_{CC}}{2}$
C. $V_{CE} > \frac{V_{CC}}{2}$
D. $V_{CE} = \frac{3\ V_{CC}}{4}$

69. The bandwidth of a double tuned transformer coupled amplifier can be adjusted by varying:
A. value of the inductance
B. the coefficient of coupling
C. value of emitter circuit biasing resistor
D. None of these

70. In a capacitance coupled single tuned amplifier, phase angle of relative voltage gain A/Ar equals:
A. arc tan ($2\,\delta\,Q_e$)
B. – arc tan ($2\,\delta\,Q_e$)
C. arc tan ($\delta\,Q_e$)
D. – arc tan ($\delta\,Q_e$)

71. In a capacitance coupled single tuned amplifier, the effective Q of the output circuit at resonance depends:
A. only on industance L
B. only on capacitance C
C. only on effective shunt resistance R_t
D. on susceptance of L (or C) and effective shunt resistance R_t

72. In a single tuned capacitance coupled amplifier. the frequency response depends on:
A. only the input circuit
B. only the output circuit
C. both input and output circuits
D. neither input circuit nor the output circuit

73. At resonance, the impedance of a parallel tuned circuit formed by capacitor C (lossless) and inductor L (with series resistance R) approximately equals:
A. $\frac{L}{CR}$
B. $\frac{LC}{R}$
C. $\frac{LR}{C}$
D. $\frac{1}{LCR}$

74. On introducing voltage series inverse feedback in an amplifier, the output resistance R_{of} and input resistance R_{if} respectively:
A. increase, increase
B. increase, decrease
C. decrease, increase
D. decrease, decrease

75. In a transconductance amplifier:
A. output voltage depends on input voltage
B. output voltage depends on input current
C. output current depends on input voltage
D. output current depends on input current

76. Negative feedback in an amplifier results in:
A. more gain, more bandwidth
B. more gain, less bandwidth
C. less gain, more bandwidth
D. less gain, less bandwidth

77. FET common source amplifier with unbypassed resistor in source-to-ground circuit forms a case of negative feedback amplifier using:
A. voltage series feedback
D. current series feedback
C. current shunt feedback
D. voltage shunt feedback

78. FET source follower is a negative feedback amplifier using:
A. voltage series feedback
B. current series feedback
C. current shunt feedback
D. voltage shunt feedback

79. In a negative feedback amplifier, series mixing:
A. tends to increase the input resistance
B. tends to decrease the input resistance
C. does not alter the input resistance
D. produces the same effect on input resistance as the shunt mixing

80. In a negative feedback amplifier, voltage sampling:
A. tends to decrease the output resistance
B. tends to increase the output resistance
C. does not alter the output resistance
D. produces the same effect on output resistance as current sampling

81. Crossover distortion takes place in:
A. tuned amplifier
B. power amplifiers
C. small signal amplifiers
D. video amplifiers

82. In a class C amplifier, full cycle conduction of the current is achieved by employing:
A. transformers
B. pushpull circuit
C. tuned circuit
D. complementary pair

83. Maximum collector dissipation $P_{D(max)}$ for class B pushpull amplifier equals

A. $\dfrac{2V^2_{CC}}{\pi^2 R_L'}$

B. $\dfrac{V^2_{CC}}{\pi^2 R_L'}$

C. $\dfrac{2V^2_{CC}}{R_L'}$

D. $\dfrac{V^2_{CC}}{R_L'}$

where $R_L' = \left(\dfrac{N_1}{N_2}\right)^2 R_L$

84. Collector circuit efficiency of class B pushpull amplifier is given by:

A. $50\dfrac{V_{max} - V_{min}}{V_{max} + V_{min}}\%$

B. $\dfrac{\pi}{2}\left(1 - \dfrac{V_{min}}{V_{CC}}\right) \times 100\%$

C. $\dfrac{\pi}{4}\left(1 - \dfrac{V_{min}}{V_{CC}}\right) \times 100\%$

D. $\pi\left(1 - \dfrac{V_{min}}{V_{CC}}\right) \times 100\%$

85. The power gain of an amplifier is 60-dB. At half power frequencies, the gain has fallen to:
A. 30 dB
B. 57 dB
C. $60/\sqrt{2}$ dB
D. 20 dB

86. Which of the following types of amplifier operations causes maximum distortion?
A. Class A
B. Class AB
C. Class B
D. Class C

87. Load impedance must match amplifier output impedance in order that:
A. minimum power is transferred to the load
B. maximum power is transferred to the load
C. collector circuit efficiency is highest
D. signal-to-noise ratio is maximum

88. Crossover distortion in class B pushpull amplifier is eliminated by
A. Class C operation
B. Class AB operation
C. elimination of output transformer
D. reducing the transistor bias

89. The reason for crossover distortion in a pushpull amplifier is that:
A. the transistors are overdriven at crossover points
B. switching of current from one transistor to the other
C. the combined transfer characteristic of the two transistors is most non-linear at zero base current
D. the input signals rise fast at their zeros

90. Higher power efficiency of a pushpull amplifier results from the fact that:
A. each transistor conducts on different half cycle of the input
B. transistors are placed in CE configuration
C. there is no collector current with zero input
D. low forward biasing voltage is needed

91. Harmonic distortion in amplifiers is due to:
A. positive feedback in the circuit
B. power supply of the amplifier
C. reactive circuit elements
D. non-linear characteristics of the active device

92. A transistor amplifier with collector circuit efficiency of 80% will certainly be:
A. Class A
B. Class AB
C. Class B
D. Class C

93. The maximum theoretical collector circuit efficiency of Class B amplifier is:
A. 15%
B. 25%
C. 50%
D. 78.5%

94. Maximum theoretical collector circuit efficiency of Class A series fed amplifier is:
A. 15%
B. 25%
C. 37.5%
D. 50%

95. In a cascade amplifier, the coupling method which is capable of providing highest gain is:
A. RC coupling
B. direct coupling
C. transformer coupling
D. impedance coupling

96. In RC coupled amplifier, the d.c. component is blocked by:
A. load resistor R_L
B. coupling capacitor C_b
C. the transistor
D. the bypass capacitor C_z in the self bias circuit

97. In an RC coupled CE amplifier, typical value of coupling capacitor is:
A. 1000 pF
B. 0.1 μF
C. 10μF
D. 0.01 μF

98. A high fidelity audio amplifier has a pass band extending from:
A. 20 Hz to 20 kHz
B. 0 Hz to 10 kHz
C. 1 kHz to 40 kHz
D. 50 Hz to 10 kHz

99. The rise time t_r of an amplifier in reproducing a square wave is related to the high 3-dB frequency f_H by the relation:

A. $t_r = \dfrac{0.35}{\sqrt{f_H}}$

B. $t_r = \dfrac{0.35}{f_H}$

C. $t_r = 0.35 f_H$

D. $t_r = 0.35\sqrt{f_H}$

100. Effect of cascading several amplifier stages is to:
A. reduce the overall gain
B. reduce the overall frequency response
C. increase the overall gain and reduce the frequency response
D. decrease the overall gain and increase the frequency response

101. In a single stage RC coupled CE amplifier, phase shift at lower 3-dB frequency is:
A. 0
B. 135°
C. 180°
D. 225°

102. In an RC coupled CE amplifier, reduction in voltage gain in the low frequency range results due to:
A. coupling capacitor C_b
B. shunt capacitance in the output circuit
C. shunt capacitance in the input circuit
D. potential divider biasing circuit from V_{CC} to ground

103. Non-linear distortion in an amplifier is caused by:
A. non-linear dynamic characteristic of the active device
B. reactive elements in the circuit
C. ripple components in the power supply
D. high temperature of operation

104. In a Class C amplifier with sinusoidal input signals, the output current flows for:
A. half the cycle
B. full cycle
C. less than half cycle
D. more than half cycle

105. In a Class AB amplifier with sinusoidal input signals, the output current flows for:
A. half the cycle
B. full cycle
C. less than half cycle
D. more than half cycle

106. The gate-to-source capacitance C_{gs} in small signal high frequency model of a JFET is of the order of:
A. 5 pF
B. 50 pF
C. 100 pF
D. 1000 pF

107. In a MOSFET, dynamic drain resistancce r_d is of the order of:
A. 1 MΩ
B. 10 kΩ
C. 10 MΩ
D. 100 MΩ

108. In a JFET, dynamic drain resistance r_d is of the order of:
A. 1 kΩ
B. 10 kΩ
C. 100 kΩ
D. 100 MΩ

109. In a JFET, transconductance g_m is of the order of:
A. 1 mS
B. 100 mS
C. 1 S
D. 100 S

110. In an FET, g_{mo} is proportional to:
A. I_{DSS}
B. $\sqrt{I_{DSS}}$
C. I^2_{DSS}
D. $1/I_{DSS}$

111. In an FET, transconductance g_m is proportional to:
A. I_{DS}
B. I^2_{DS}
C. $\sqrt{I_{DC}}$
D. $1/I_{DC}$

112. Parameter f_T of a transistor is the frequency at which the magnitude of CE short-circuit current gain is:
A. half of the midband gain
B. $1/\sqrt{2}$ of the midband gain
C. one-tenth of midband gain
D. unity

113. In the high frequency hybrid-π model of a CE amplifier, capacitance C_c accounts for:
A. emitter junction barrier capacitance
B. collector junction barrier capacitance
C. collector junction diffusion capacitance
D. emitter junction diffusion capacitance

114. In the high frequency hybrid-π model of CE transistor, typical value of resistance $r_{b'c}$ is:
A. 100 Ω
B. 1000 Ω
C. 10 kΩ
D. 4 MΩ

115. In the high frequency hybrid-π model of a CE transistor, typical value of C_c is:
A. 3 pF
B. 100 pF
C. 1000 pF
D. 0.01 μF

116. In the high frequency hybrid-π model of a CE transistor, the base spreading resistance r_{bb} equals:

A. $h_{ie} - g_m \cdot h_{re}$
B. $h_{ie} + r_{b'e}$
C. $\frac{h_{ie}}{h_{fe}}$
D. $\frac{r_{be'}}{h_{fe}}$

117. In the high frequency hybrid-π model of a CE transistor, the conductance g_{cc} is equal to:

A. $h_{ie} - g_m \cdot h_{re}$
B. $g_{b'e} - g_m \cdot h_{re}$
C. $h_{oe} - g_m h_{re}$
D. $g_m \cdot h_{re}$

118. In the high frequency hybrid-π model of a CE transistor, the conductance $g_{b'e}$ equals:

A. h_{fe}/g_m
B. g_m/h_{fe}
C. g_m/h_{ie}
D. h_{ie}/g_m

119. In CE amplifier, current gain A_I is the ratio of:

A. change in I_C to change in I_E
B. change in I_C to change in I_B
C. change in I_B to change in I_E
D. change in I_E to change in I_B

120. The input circuit of CB amplifier is:

A. emitter-base circuit
B. emitter-collector circuit
C. base-collector circuit
D. collector-base-emitter circuit

121. Compared to CB amplifier, the CE amplifier has considerably:

A. higher current gain
B. lower input resistance
C. lower voltage gain
D. lower current gain

122. CC amplifier has

A. high R_i and high R_0
B. low R_i and low R_0
C. low R_i and high R_0
D. high R_i and low R_0

where R_i is the input resistance and R_0 is the ·put resistance.

123. One characteristic of emitter follower is:

A. low input resistance
B. low current gain
C. low voltage gain
D. high output resistance

124. A transistor is said to be in quiescent state when:

A. it is unbiased
B. no signal is applied to it
C. no current is flowing in it
D. emitter junction bias is equal to collector junction bias

125. A carbon microphone having resistance of 100 ohms feeds a pre-amplifier. The best transistor configuration for pre-amplifier is:

A. CE
B. CB
C. CC
D. any of three

126. An amplifier with a frequency response of 50 Hz to 8000 Hz:

A. can reproduce intelligence speech
B. can be used as d.c. amplifier
C. is considered a high fidelity amplifier
D. constitutes inferior audio frequency amplifier

127. The output resistance of an amplifier in the CC configuration:

A. is higher than that in CE and CB configurations
F depends strongly on the source resistance
C. goes down as the source resistance increases
D. is independent of the source resistance

128. Input resistances R of ideal voltage amplifier and ideal current amplifeir are respectively:

A. $\infty, 0$
B. $0, \infty$
C. ∞, ∞
D. $0, 0$

129. Current gain of small signal transistor amplifier equals:

A. $\frac{-h_f}{1+h_0 Z_L}$

B. $\frac{+h_f}{1+h_o Z_L}$

C. $\frac{-h_i}{1+h_o Z_L}$

D. $\frac{+h_i}{1+h_o Z_L}$

130. The voltage gain of amplifier stage is lowest in:
A. CB configuration
B. CE configuration
C. CC configuration
D. same in all configurations

131. The current gain of amplifier stage is lowest in:
A. CB configuration
B. CE configuration
C. CC configuration
D. same in all configurations

132. Ignoring the biasing network, Darlington emitter follower has input resistance of the order of:
A. 20 kΩ
B. 200 kΩ
C. 2 MΩ
D. 20 MΩ

133. In a CE amplifer stage, on introducing a resistor R in the emitter circuit, the input resistance R :
A. remain unaltered
B. reduces
C. increases nominally
D. increases very much

134. With source resistance R_s of 1000 Ω, the output impedance of a typical CE amplifier stage is of the order of:
A. 500 Ω
B. 5k Ω
C. 50 Ω
D. 500 kΩ

135. With load impedance of 4 kΩ, the current gain of a typical CE amplifier stage has magnitude of the order of:
A. 0.98
B. 5
C. 50
D. 400

136. Parameter h_{fe} of a typical transistor is of the order of:
A. 5
B. 50
C. 500
D. 2000

137. In a CE amplifier, thermal runaway is unconditionally avoided if:

A. $V_{CE} = \frac{V_{CC}}{2}$

B. $V_{CE} > \frac{V_{CC}}{2}$

C. $V_{CE} < \frac{V_{CC}}{2}$

D. $V_{CE} = \frac{V_{CC}}{\sqrt{2}}$

138. To avoid permanent damage, the maximum temperature of collector junction J_C in Si transistor is generally restricted to:
A. + 60°C
B. + 100°C
C. + 175°C
D. + 225°C

139. Base-to-emitter voltage V_{BE} is a forward biased transistor decreases with increase of temperature at the following rate:
A. 25 mV/ °C
B. 0.25 mV/ °C
C. 2.5 mV/ °C
D. 0.6 mV/ °C

140. Excessive hum in a power supply is generally due to:
A. open filter choke
B. capacitor
C. defective rectifier
D. any of the above

141. DC output drops from 50 volt with no load to 48 volt with full load. Percentage load regulation is:
A. 2%
B. 4%
C. 1%
D. 8%

142. With increase of load resistance, ripple voltage of rectifier with capacitor filter:
A. decreases
B. increases
C. remains constant
D. gets multiplied

143. Bleeder resistor in a rectifier circuit is a fixed resistor connected across:
A. input of complete power supply
B. output of complete power supply
C. midway
D. None of these

144. Comparator circuits are used in:
A. address
B. integrators
C. converting sine to square wave
D. differentiator

145. A fullwave voltage doubler output has ripple frequency:
A. half the supply frequency
B. equal to the supply frequency
C. double the supply frequency
D. having no relation with the supply frequency

146. The function of centre tapping on the secondary in a fullwave rectifier is to:
A. step up the voltage
B. step down the voltage
C. isolate the load from ground
D. cause the diodes to conduct alternately

147. A capacitor filter at the output of a rectifier results in ripple which
A. increases with load resistance
B. decreases with load resistance
C. remain unaltered with increase of load resistance
D. None of these

148. Capacitor filter is ideal for currents which are:
A. small
B. medium
C. large
D. very large

149. Bleeder resistor in power supplies is used to place:
A. an infinite load across the rectifier
B. maximum load across the rectifier
C. minimum load across the rectifier
D. None of these

150. Fullwave rectifier has theorectical maximum efficiency of:
A. 40.6%
B. 81.2%
C. 78.5%
D. 50%

ANSWERS

1	**2**	**3**	**4**	**5**	**6**	**7**	**8**	**9**	**10**
B	C	D	A	C	B	A	D	B	D
11	**12**	**13**	**14**	**15**	**16**	**17**	**18**	**19**	**20**
B	D	C	B	D	A	B	B	C	C
21	**22**	**23**	**24**	**25**	**26**	**27**	**28**	**29**	**30**
B	C	A	A	A	A	C	D	D	A
31	**32**	**33**	**34**	**35**	**36**	**37**	**38**	**39**	**40**
C	A	B	B	B	B	B	D	C	C
41	**42**	**43**	**44**	**45**	**46**	**47**	**48**	**49**	**50**
B	B	A	B	B	C	C	A	D	A
51	**52**	**53**	**54**	**55**	**56**	**57**	**58**	**59**	**60**
A	A	C	D	B	B	C	D	C	A
61	**62**	**63**	**64**	**65**	**66**	**67**	**68**	**69**	**70**
D	B	A	A	D	B	A	A	B	B
71	**72**	**73**	**74**	**75**	**76**	**77**	**78**	**79**	**80**
D	B	A	B	C	C	B	A	A	A
81	**82**	**83**	**84**	**85**	**86**	**87**	**88**	**89**	**90**
B	C	A	C	B	D	B	B	C	C

91	92	93	94	95	96	97	98	99	100
D	D	D	B	C	B	C	A	B	C
101	**102**	**103**	**104**	**105**	**106**	**107**	**108**	**109**	**110**
D	A	A	C	D	A	A	C	A	A
111	**112**	**113**	**114**	**115**	**116**	**117**	**118**	**119**	**120**
C	D	B	D	A	A	C	B	B	A
121	**122**	**123**	**124**	**125**	**126**	**127**	**128**	**129**	**130**
A	D	C	B	B	A	B	A	A	C
131	**132**	**133**	**134**	**135**	**136**	**137**	**138**	**139**	**140**
A	C	D	C	C	B	C	D	A	B
141	**142**	**143**	**144**	**145**	**146**	**147**	**148**	**149**	**150**
B	A	C	C	C	D	B	A	C	B

DIGITAL ELECTRONIC CIRCUITS

MULTIPLE CHOICE QUESTIONS

1. Scratch-pad memory is a:
A. first-in first-out memory
B. last-in first-out memory
C. local permanent memory
D. local temporary memory

2. In case of bipolar or static MOS memories, the access time is about:
A. 20 to 400 ns
B. 20 to 400 μs
C. 20 to 400 ms
D. 20 to 400 s

3. Access in magnetic drum is:
A. completely random
B. sequential and cyclic
C. cyclic sequential
D. partly random and partly cyclic sequential

4. Dynamic memory cells are constructed using:
A. FETs
B. MOSFETs
C. Transistors
D. Flip-flops

5. ROM consists of:
A. a decoder followed by an encoder
B. a encoder followed by a decoder
C. a multiplexer followed by a decoder
D. None of the above

6. Semiconductor ROMs are sometimes preferred to semiconductor RAMs because:
A. ROMs are cheaper than RAMs
B. ROMs are faster
C. ROMs do not require power supply for their operation
D. programs stored in ROMs can not be altered

7. Which of the following is a permanent memory?
A. RAM
B. ROM
C. both RAM and ROM
D. None of the above

8. The following type of memory requires recharging cycle in order to retain its information:
A. static
B. dynamic
C. rechargeable
D. core

9. The storage device used to compensate the differences in the ratio of flows of data is known as:
A. auxiliary storage
B. buffer
C. main storage
D. None of the above

10. Magnetic tape:
A. is easier to handle than punched cards
B. stores data in more compact form
C. costs less than punched card for same amount of storage capacity
D. All of the above

11. Commonly used auxiliary storage devices are:
A. magnetic tape, magnetic drum and magnetic disk
B. magnetic core, magnetic tape and magnetic drum
C. magnetic disk, semiconductor memory and magnetic drum
D. magnetic tape, monolithic storage and magnetic disk

12. Transfer of information from main storage is typically n times faster than the transfer from auxiliary storage where n is about:
A. 5
B. 10

C. 100
D. 200

13. Desirable characteristics of a memory unit are:
A. compact
B. speed
C. durability
D. All of the above

14. The following logic circuit is used to change a BCD number into an equivalent decimal number:
A. decoder
B. encoder
C. multiplexer
D. code converter

15. A combinational logic circuit which generates a particular binary word or number is:
A. decoder
B. multiplexer
C. encoder
D. demultiplexer

16. Parallel adders are:
A. combinational logic circuits
B. sequential logic circuits
C. Both of the above
D. None of the above

17. How many full adders are needed to construct an m-bit parallel adder?
A. $m/2$
B. m
C. $m - 1$
D. $m + 1$

18. Which of the following counters has the highest speed?
A. asynchronous counter
B. synchronous counter
C. ripple counter
D. ring coutner

19. Output of a Moore sequential machine is a function of:
A. all present states of the machine
B. all the inputs
C. a few combinations of inputs and the present state
D. all combinations of inputs and the present state

20. Number of flip-flops required in a decade counter is:
A. 2
B. 3
C. 4
D. 5

21. A n-stage ripple counter will count upto:
A. 2^n
B. $2^n - 1$
C. n
D. 2^{n-1}

22. For an input pulse train of clock period T, the delay produced by n-stage shift register will be:
A. nT
B. $(n-1)$ T
C. $(n+1)$ T
D. $2n$ T

23. In a T-flip-flop the ratio of the frequency of the input pulse to the frequency of the output pulse is:
A. 1/2
B. 1
C. 2
D. 3

24. Switching speed of ECL logic is very high because:
A. it uses positive logic
B. it uses negative logic
C. it uses high speed transistors
D. its transistors remain unsaturated

25. The fan out TTL logic gate is about:
A. 5
B. 10
C. 20
D. 50

26. Which out of the following TTL sub-families has maximum speed?
A. standard TTL
B. Schottky-clamped TTL
C. high speed TTL
D. low power TTL

27. Which out of the following logic families has maximum fan-out capacity?
A. RTL
B. EC
C. MOS
D. CMOS

28. Which out of the following logic families is most expensive?
   A. RTL
   B. TTL
   C. DTL
   D. ECL

29. Which of the following logic gates dissipates minimum power?
   A. RTL
   B. TTL
   C. MOS
   D. ECL

30. Which of the following logics has high noise immunity?
   A. RTL
   B. TTL
   C. ECL
   D. HTL

31. The following logic has the highest speed:
   A. DTL
   B. RTL
   C. ECL
   D. TTL

32. Which of the following statements is wrong? A synchronous counter:
   A. has flip-flops that change states simultaneously
   B. is capable of operating at high frequencies
   C. requires more logic elements
   D. has large carry propagation time

33. Asynchronous sequential circuits difficult to design becuauuse:
   A. they use external clock
   B. they are more complex
   C. they have stability problem
   D. they have fast response

34. Schmitt trigger may be used to:
   A. change voltage to corresponding frequency
   B. change frequency to voltage
   C. change a slowly varying voltage into square voltage
   D. None of the above

35. Number of flip-flops needed to divide the input frequency by 32 is:
   A. 2
   B. 4
   C. 5
   D. 8

36. The functional difference between SR and JK flip-flop is that:
   A. JK flip-flop has a feedback path
   B. JK flip-flop does not require an external clock pulse
   C. JK flip-flop is faster than SR flip-flop
   D. JK flip-flop can accept both inputs at 1

37. An SR flip-flop does not accept the following input entry:
   A. both inputs zero
   B. zero at R and one at S
   C. zero at S and one at R
   D. both inputs at one

38. Simplification of function by K-map requires that the function should be in the:
   A. SOP form
   B. POS form
   C. canonical form
   D. Any of the above

39. A demultiplexer is used to:
   A. route the data from single input to one of many outputs
   B. select data from several inputs and route it to single output
   C. perform serial to parallel conversion
   D. All of the above

40. A switching function is symmetric with respect to a set of literals if and only if the function remains unchanged after:
   A. two of these literals are interchanged
   B. any permutation of the literals
   C. all the literals are changed in anticlockwise order
   D. all the literals are changed in clockwise order

41. The *dynamic hazard* problem occurs in:
   A. combinational circuit alone
   B. sequential circuit alone
   C. both sequential and combinational circuits
   D. None of the above

42. What is the form of the boolean experssion

$(A + B)\ (\overline{C} + \overline{D}) = x$?
   A. product-of-sums
   B. sum-of-products
   C. karnaugh map
   D. matrix

43. In Boolean algebra the term *product-of-sums* means:
A. AND function of several OR functions
B. OR function of several AND functions
C. OR function of several OR functions
D. AND function of several AND functions

44. Which of the following is not a sequential circuit?
A. flip-flop
B. counter
C. shift register
D. multiplexer

45. The m-bit parallel adder consists of:
A. $(m + 1)$ full adders
B. $m/2$ full adders
C. $(m - 1)$ full adders
D. m full adders

46. In Boolean algebra, the term *sum-of-products* means:
A. the AND function of several OR functions
B. the OR function of several AND functions
C. the OR function of several OR functions
D. the AND function of several AND functions

47. The circuit given below is a:

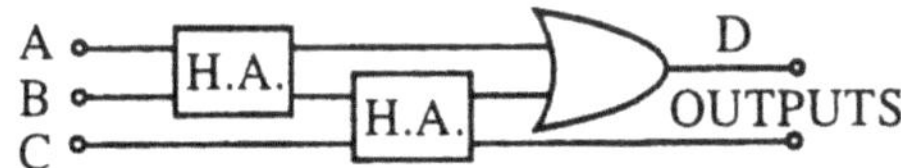

A. full adder
B. full subtractor
C. parity checker
D. None of the above

48. Which table shows the electrical state of a digital circuits output for every possible combination of electrical states in the input?
A. function table
B. truth table
C. routing table
D. ASC-II table

49. How many bits are required to encode all twenty six letters, ten symbols and ten numerals?
A. 5
B. 6
C. 7
D. 46

50. Simplified form of Boolean expression (A + B + AB) (A + C) is:
A. A + B + C
B. AB + BC
C. A + BC
D. AC + B

51. Which of the following is functionally a complete set?
A. AND, OR
B. AND, XOR
C. NOT, OR
D. AND, OR, NOT

52. Which gate is formed by adding an inverter at the output of an OR gate?
A. NOR
B. XOR
C. Equivalence
D. NAND

53. An OR gate may be imagined as:
A. switches connected in series
B. switches connected in parallel
C. MOS transistors connected in series
D. None of the above

54. Which logic function is produced by adding an inverter to each input and the output of an AND gate?
A. NAND
B. NOR
C. OR
D. XOR

55. The NAND gate can function as a NOT gate if:
A. inputs are connected together
B. inputs are left open
C. one input is set to 0
D. one input is set to 1

56. For which of the following logic gates, the output is complement of the input?
A. NOT
B. AND
C. OR
D. XOR

57. An alternative symbol for two input NAND gate is:

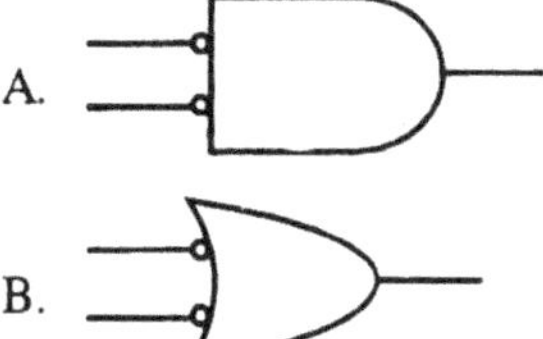

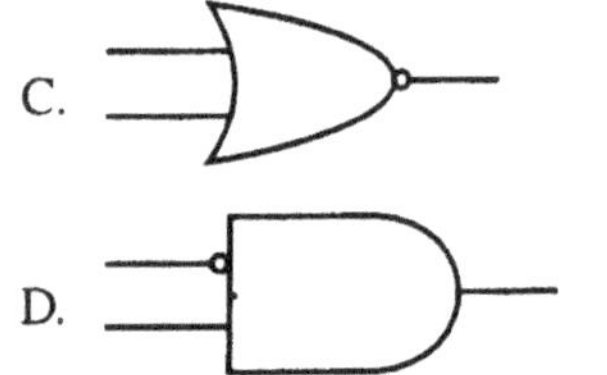

58. The circuit given below is functionally equivalent to:

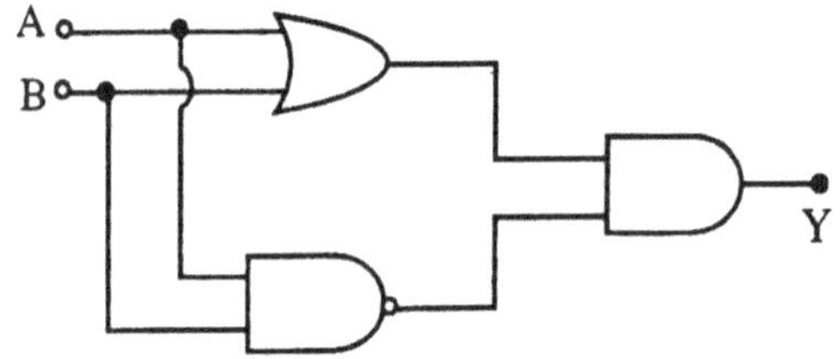

A. EX-OR gate
B. NOR gate
C. AND gate
D. NAND gate

59. Boolean expression for three input AND gate is:
A. A.B.C. = D
B. A + B + C = D
C. $\overline{A}.\overline{B}.\overline{C} = D$
D. $\overline{A} + \overline{B} + \overline{C} = D$

60. NOR gate is logically equivalent to:
A. an OR gate followed by an AND gate
B. an OR gate followed by a NAND gate
C. an AND gate followed by OR gate
D. an OR gate followed by an inverter

61. $\overline{A} + \overline{B} + \overline{C} = D$ represents a:
A. NOR gate
B. NAND gate
C. EX-OR gate
D. AND gate

62. Which of the following hexadecimal numbers represents an even decimai number?
A. EF
B. D5
C. BF
D. AC

63. Which of the following hexadecimal numbers represents an odd decimal number?
A. FF
B. EG
C. CC
D. AA

64. How many bytes are there in binary number 1011 1001 0110 1110 ?
A. 1
B. 2
C. 3
D. 4

65. BCD numbers are useful whenever:
A. binary to BCD conversion is desired
B. binary to hexadecimal conversion is desired
C. decimal information is transferred into or out of digital system
D. None of the above

66. 1 K-byte is precisely equal to:
A. 1012 bits
B. 1024 bits
C. 1000 bits
D. 1020 bits

67. Hexadecimal equivalent of binary number 10001101011001 is:
A. 2359
B. 9493
C. 8D62
D. 8DF2

68. Octal equivalent of binary number 01000100111 is:
A. 4236
B. 2117
C. 1084
D. 4136

69. Decimal equivalent of hexadecimal number 11 A is:
A. 282
B. 272
C. 200
D. 202

70. A six-bit alpha-numeric code is able to code:
A. 36 characters
B. 48 characters
C. 64 characters
D. 128 characters

71. The 2's compliment of binary number 0.01011 is:
A. 1.10101
B. 0.10101
C. 0.10100
D. 1.10100

72. Binary equivalent of gray code number 101 is:
A. 101
B. 110
C. 100
D. 111

73. Octal equivalent of decimal $(51)_{10}$ is:
A. $(63)_8$
B. $(41)_8$
C. $(67)_8$
D. $(27)_8$

74. Radix of octal number system is:
A. 2
B. 4
C. 8
D. 10

75. Decimal number 21.125 may be written in binary system as:
A. 10101.001
B. 10100.001
C. 10101.010
D. 10100.100

ANSWERS

1	2	3	4	5	6	7	8	9	10
D	A	D	B	A	D	B	B	B	D
11	**12**	**13**	**14**	**15**	**16**	**17**	**18**	**19**	**20**
A	B	D	A	B	A	B	B	C	C
21	**22**	**23**	**24**	**25**	**26**	**27**	**28**	**29**	**30**
A	B	C	D	B	B	D	A	C	D
31	**32**	**33**	**34**	**35**	**36**	**37**	**38**	**39**	**40**
C	D	C	C	C	D	D	A	B	B
41	**42**	**43**	**44**	**45**	**46**	**47**	**48**	**49**	**50**
C	A	A	D	D	B	A	A	B	C
51	**52**	**53**	**54**	**55**	**56**	**57**	**58**	**59**	**60**
A	A	B	C	A	A	B	A	A	D
61	**62**	**63**	**64**	**65**	**66**	**67**	**68**	**69**	**70**
B	D	A	B	C	B	A	B	A	C
71	**72**	**73**	**74**	**75**					
A	B	A	C	A					

CONTROL SYSTEM

MULTIPLE CHOICE QUESTIONS

1. Poles are the complex frequencies of a transfer function where the response becomes:
A. infinite
B. zero
C. oscillatory
D. decaying

2. Zeros are the complex frequencies of a transfer function where the response becomes:
A. infinite
B. zero
C. oscillatory
D. exponentially decaying

3. Which of the following is an open loop system?
A. Ward Leonard control
B. Amplidyne
C. Armature controlled d.c. motor
D. Field controlled d.c. motor

4. A driver maintains the speed of a car at 50 km per hour. Here the feedback element is:
A. steering wheel
B. needle of the speedometer
C. eyes
D. All of these

5. A mass of moment of inertia J is supported by a shaft of stiffness coefficient k and is rotated by a torque T through an angle θ. If the viscous frictional coefficient is B, the transfer function θ (s)/T (s) of this rotational mechanical system is given by (assuming zero initial condition):

A. $\frac{k}{Js+B}$

B. $\frac{B}{Js^2+k}$

C. $\frac{1}{Js^2+Bs+k}$

D. $\frac{1}{Js^2+Bs-k}$

6. Transient response of system is basically because of:
A. coupling
B. forces
C. friction
D. stored energy

7. In a critically damped system, the damping factor of the system is:
A. zero
B. unity
C. less than unity
D. greater than unity

8. The transient behaviour of an open loop system is defined to exist from initial time till a period of:
A. 2τ
B. 4τ
C. 5τ
D. 6τ

9. Transfer function of a system is characterized by:
A. zero initial condition and dependence on input excitation
B. zero initial condition and non-dependence on input excitation
C. given initial condition and dependence on input excitation
D. given initial condition and non-dependence on input excitation

10. The relation G(s) R(s)/[1 + G(s) H(s)] is basically defined for a:

A. single variable control system
B. two variable control system
C. three variable control system
D. multi-variable control system

11. With feedback, the transient response of the system as compared to that without feedback:

A. decays slowly
B. decays more quickly
C. rises at a faster rate
D. remain unaltered

12. In a second order system, if the damping factor ζ is less than unity, the system will be:

A. underdamped
B. over damped
C. critically damped
D. freely oscillating

13. The block diagram of a positional control system is shown below. The sensitivity of the closed loop transfer function with respect to the feedback element H at frequency of 1 radian/second is:

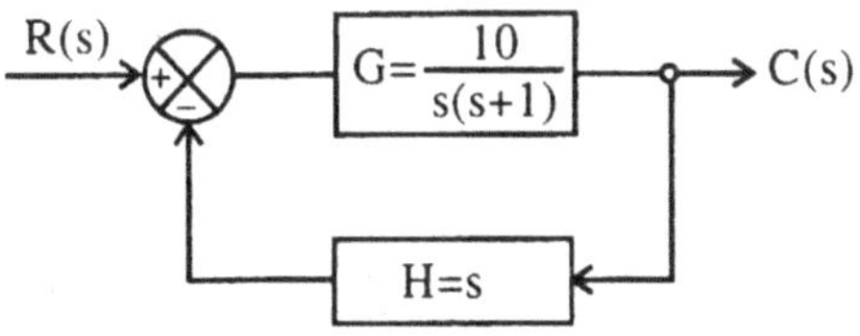

A. 0.905
B. 0.028
C. 0.084
D. 0.408

14. In the block diagram of control system given below, the effect of feedback is to shift the pole:

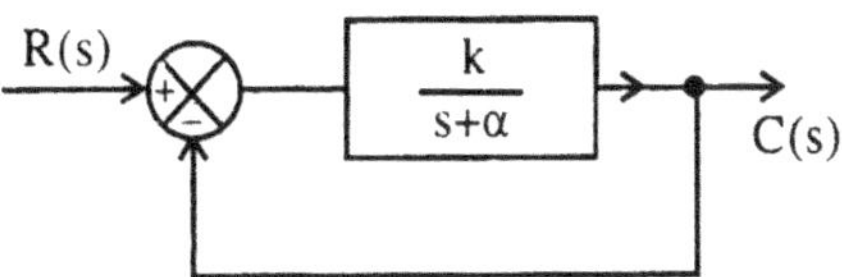

A. negatively to $s = -(\alpha + k)$ and to reduce the time constant to $(\alpha + 1/k)$
B. positively to $s = (\alpha + k)$ and reduce the time constant to $(\alpha + 1/k)$
C. negatively to $s = -(\alpha + k)$ and to increase the time constant to $(\alpha + k)$
D. positively to $s = (\alpha + k)$ and to increase the time constant to $(\alpha + k)$

15. The block diagram given below is equivalent to:

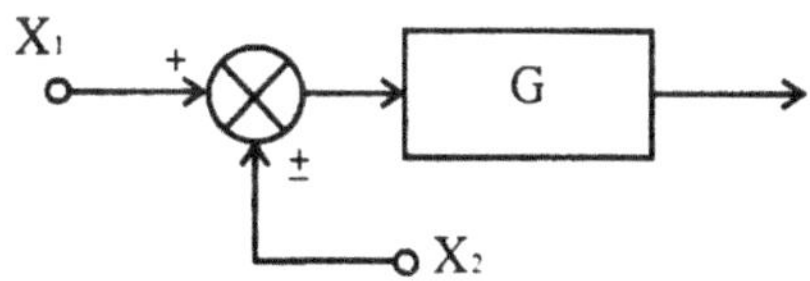

A.
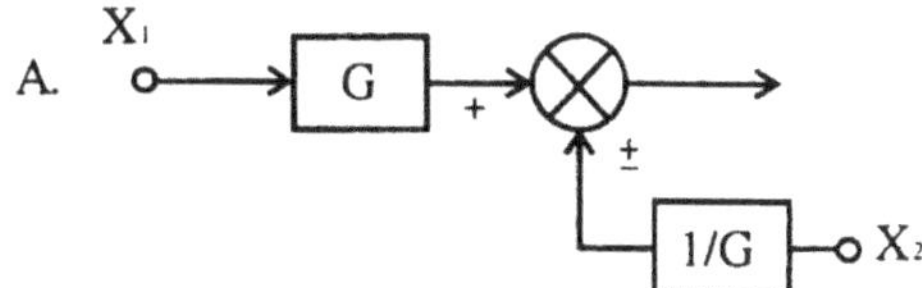

B.
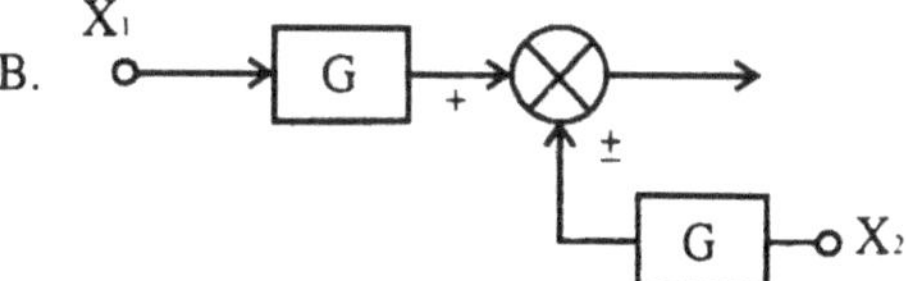

C.
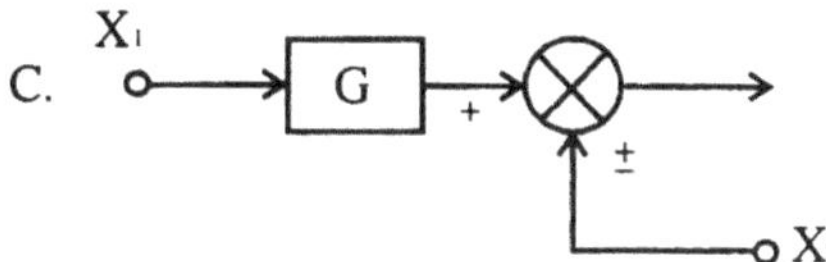

D.
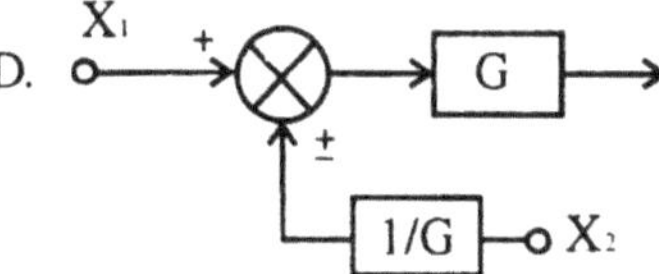

16. The block diagram given below is equivalent

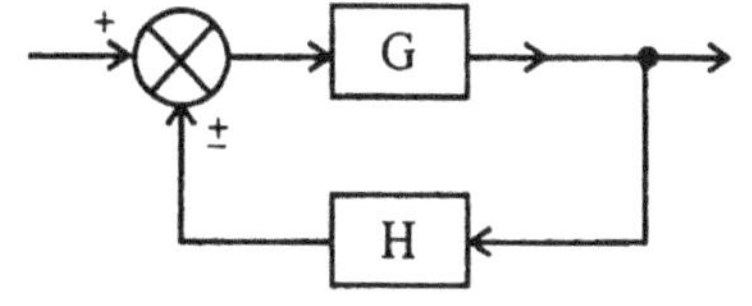

A.
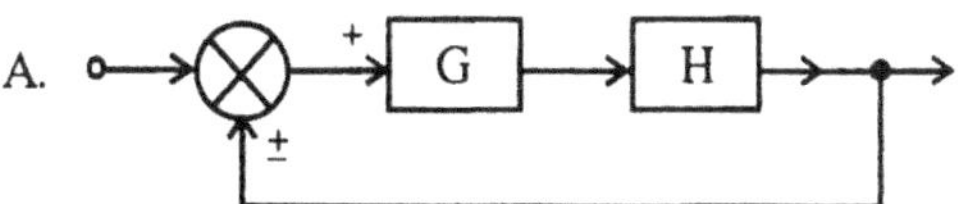

B.
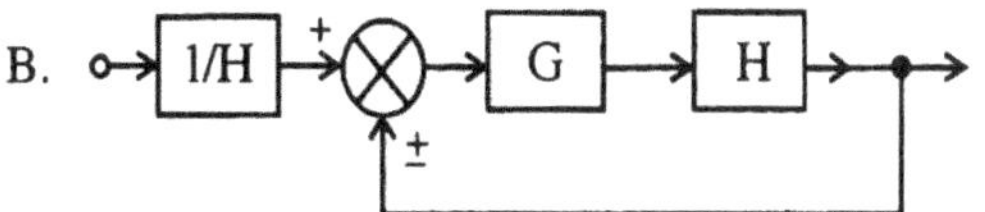

C. 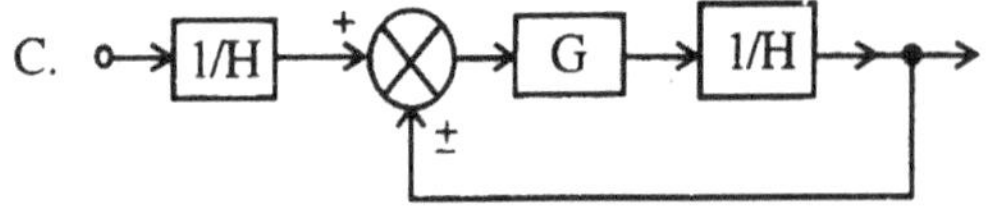

D. None of the above

17. The overall transfer function C(s)/R(s) of the system shown below is:

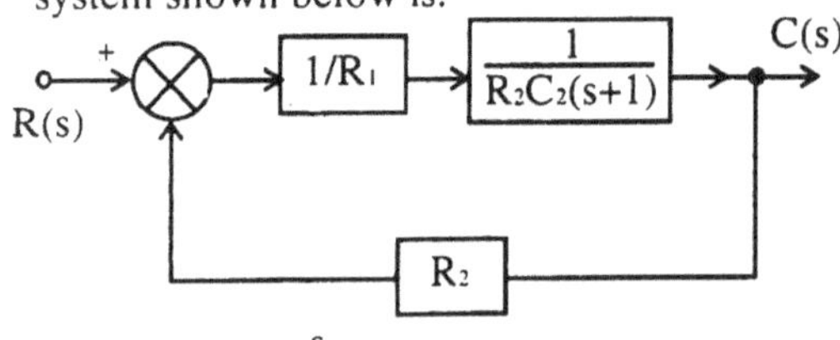

A. $\dfrac{s}{R_1 R_2 C_2 s + R_1 + R_2}$

B. $\dfrac{1}{R_1 R_2 C_2 s + R_1 + R_2}$

C. $\dfrac{s}{R_1 R_2 C_2 + R_1 + R_2}$

D. None of the above

18. The transfer function C/R of the system block shown below is given by:

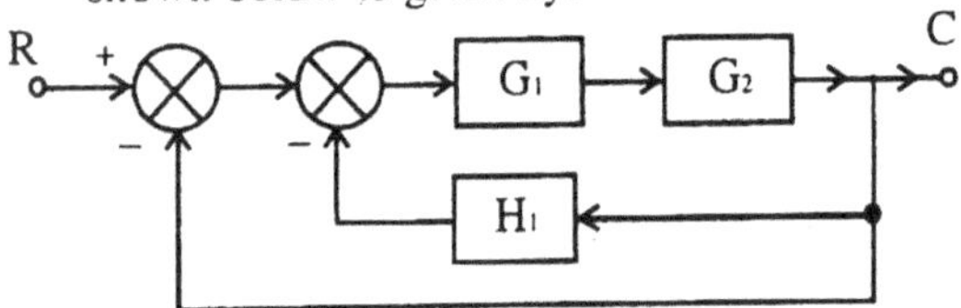

A. $\dfrac{G_1G_2}{1+H_1 G_1G_2}$

B. $\dfrac{G_1+G_2}{1+(G_1+G_2)H_1+G_1G_2}$

C. $\dfrac{G_1 G_2}{1+G_1 G_2H_1+G_1G_2}$

D. $\dfrac{G_1 G_2(1+H_1+G_1G_2)}{1+G_1 G_2 H_1+G_1 G_2}$

19. The number of forward paths in the signal flow diagram shown below are:

A. 2

B. 3

C. 4

D. 5

20. The transfer function of the signal flow diagram given below is:

R(s) —1→ [H(s) / G(s)] —1→ C(s)

A. $\dfrac{G(s)}{1+G(s)H(s)}$

B. $\dfrac{G(s)}{1-G(s)H(s)}$

C. $\dfrac{G(s)\,H(s)}{1-G(s)H(s)}$

D. $\dfrac{G(s)\,H(s)}{1+G(s)H(s)}$

21. The damping factor and undamped natural frequency for the position control system are respectively given by:

A. $2\sqrt{kJ}$, $\sqrt{kJ}$

B. $\dfrac{k}{2\,BJ}$, $\sqrt{k/J}$

C. $\dfrac{B}{2\sqrt{kJ}}$, $\sqrt{k/J}$

D. $\dfrac{J}{2\sqrt{kB}}$, $\sqrt{kJ}$

22. The rise time t_r for a type 1, second order system equals:

A. $\dfrac{\pi-\cos^{-1}\left(\dfrac{\sqrt{1-\zeta^2}}{\zeta}\right)}{\omega_n\sqrt{1-\zeta^2}}$

B. $\dfrac{\pi-1}{\omega_n\sqrt{1-\zeta^2}}$

C. $\dfrac{\pi-\cos^{-1}\zeta}{\omega_n\sqrt{1-\zeta^2}}$

D. $\dfrac{\tan^{-1}\left(\dfrac{\sqrt{1-\zeta^2}}{\zeta}\right)}{\omega_n\sqrt{1-\zeta^2}}$

23. In a type 1, second order system, the first peak overshoot occurs at a time starting from zero to:

A. $\dfrac{\pi\omega_n}{\sqrt{1-\zeta^2}}$

B. $\dfrac{\omega_n}{\pi\sqrt{1-\zeta^2}}$

C. $\dfrac{\pi}{\omega_n\sqrt{1-\zeta^2}}$

D. $\dfrac{\pi\omega_n}{(1-\zeta^2)}$

24. For a type 1 second order control system, for 2% tolerance band, the settling time is:

A: $4/(\zeta_n)$

B. $4\zeta\omega_n$

C. $\dfrac{4\zeta}{\omega_n}$

D. $\dfrac{4\omega_n}{\zeta}$

25. The number of integrators in the transfer function $G(s) = \dfrac{(s+3)}{s^2(s+5)}$ are:

A. zero

B. one

C. two

D. four

26. By increasing the gain k of the system, the steady state error of the system:

A. decreases

B. increases

C. remain unchanged

D. may increase or decrease

27. Integral error compensation is also called:

A. proportional minus integral compensation

B. proportional plus integral compensation

C. proportional plus derivative compensation

D. proportional minus derivative compensation

28. In the derivative error compensation:

A. damping increases and settling time decreases

B. dampling increases and settling time increases

C. damping decreases and settling time decreases

D. damping decreases and settling time increases

29. The expressions for Integral Square Error (ISE) and Integral of the Absolute magnitude Error (IAE) respectively are:

A. $\int_0^{\infty} e^2(t)\,dt,\ \int_0^{T} |e^2(t)|\,dt$

B. $\int_0^{\infty} e^2(t)\,dt,\ \int_0^{\infty} |e^2(t)|\,dt$

C. $\int_0^{T} e^2(t)\,dt,\ \int_0^{\infty} |e^2(t)|\,dt$

D. $\int_0^{T} e^2(t)\,dt,\ \int_0^{T} |e^2(t)|\,dt$

30. Expression of Integral Time Absolute Error (ITAE) and Integral Time Square Error (ITSE) respectively are:

A. $\int_0^{T} t|e(t)|\,dt,\ \int_0^{T} t|e^2(t)|\,dt$

B. $\int_0^{T} t|e(t)|^2\,dt,\ \int_0^{T} t|e^2(t)|\,dt$

C. $\int_0^{T} t|e(t)|^2\,dt,\ \int_0^{T} t^2|e^2(t)|\,dt$

D. $\int_0^{T} e^2(t)\,dt,\ \int_0^{T} t^2\,e^2(t)\,dt$

31. For 2% tolerance band, the settling time the second order linear system is:

A. $\tau/2$

B. τ

C. 2τ

D. 4τ

where τ is the time constant.

32. Which of the following transfer functions indicates a type 0 system?

A. $\dfrac{(s+3)(s+5)}{s\,(s+1)(s+7)}$

B. $\dfrac{(s+2)}{s^2(s^2+3)}$

C. $\dfrac{(s+1)(s+2)}{(s-3)(s-4)}$

D. $\dfrac{(s-1)(s-3)}{s^2\,(s+4)}$

33. Position error constant k_p = G (0) = ∞ or the steady state error is always zero in response to step input of displacement for:

A. type 0 system
B. type 2 system
C. type 3 system
D. type N (N > 1) system where N = 1, 2

34. A system has characteristic equation as $s^2 + 2s + 8 = 0$. The damping ratio and the natural frequency of oscillation of the system respectively are:

A. $2\sqrt{2}$, 0.5
B. 0.5, $2\sqrt{2}$
C. 0.353, $2\sqrt{2}$
D. 2, 0.353

35. The denormalized bandwidth for the particular value of ω_n and damping factor ξ is:

A. $\omega_n(1-\zeta+2\zeta^2)^{1/2}$
B. $\omega_n(1-\zeta^2)^{1/2}$
C. $\omega_n\left[1-2\zeta^2+\sqrt{2-4\zeta^2+4\zeta^4}\right]^{\frac{1}{2}}$
D. $\omega_n(1-\zeta^2)^{1/2}$

36. The Routh array of a characteristic equation is given below. The number of roots lying on the right hand side of s-plane is:

s^6	8
s^5	7
s^4	−3
s^3	2
s^2	−1
s	1
s^0	0

A. 2
B. 3
C. 4
D. 5

37. For the characteristic equation
$s^4 + 5s^3 + 5s^2 + 4s + k = 0$,
the system is stable if k lies in the range:

A. $\dfrac{84}{25} > k > 0$

B. $\dfrac{84}{25} < k < 0$

C. $3 > k > 1$
D. $4 > k > 3$

38. The Routh criterion tells us the number of roots lying:

A. in the right half of s-plane
B. in the left half of s-plane
C. on the origin of s-plane
D. None of the above

39. Which of the following characteristic equations represents a stable state:

A. $as^2 + bs + c = 0$
B. $as^4 + bs^2 + cs + d = 0$
C. $-as^2 + bs - c = 0$
D. $as^2 - bs - c = 0$

where a, b, c and d are positive integers

40. The root locus plot is symmetrical about the real axis because there are:

A. all roots occurring in pair
B. more real roots
C. more imaginary roots
D. complex roots occurring in conjugate pairs

41. The end points of root loci are:

A. open loop poles
B. closed loop poles
C. open loop zeros
D. closed loop zeros

42. For characteristic equation having more number of poles than zeros, the number of root locus branches will be equal to:

A. number of zeros
B. number of poles
C. Sum of (A) and (B)
D. Difference of (B) and (A)

43. In root locus plot, the number of asymptote is 360° divided by the number of:
A. poles
B. zeros
C. Sum of (A) and (B)
D. Difference of (A) and (B)

44. If the system gain k is increased, the roots of the system move to:
A. lower frequencies
B. higher frequencies
C. origin
D. None of the above

45. Two RC circuits each having transfer function $\left(\frac{1}{1+s\tau}\right)$ are connected in cascade. The overall transfer function of the combination is:

A. $\frac{1}{(1+s\tau)^2}$

B. $\frac{1}{1+(1+s\tau)^2}$

C. $\frac{2}{1+s\tau}$

D. $\frac{1}{1+3s\tau+s^2\tau^2}$

46. When the gain k of a system becomes zero, the roots of the loci:
A. move away from the poles
B. move away from the zeros
C. coincide with the poles
D. coincide with the zeros

47. The number of root loci, which extends to infinity equals:
A. excess of poles over zeros in G(s) H(s)
B. excess of zeros over poles in G(s) H(s)
C. number of poles in G(s) H(s)
D. number of zeros in G(s) H(s)

48. The algebraic sum of the angles of the vectors from all poles and zeros to the point on any root locus segment is:
A. always 180°
B. 180° or odd multiple thereof
C. always 90°
D. 90° or odd multiple thereof

49. The root locus of a system has four separate loci. The system can have:
A. four poles and four zeros
B. four poles or four zeros
C. two poles and two zeros
D. two poles and six zeros

50. The root locus of a system has three asymptotes. The system can have:
A. three poles
B. four poles and one zeros
C. five poles and two zeros
D. All of the above

51. At the point where 180° locus crosses the j-axis, the system is:
A. absolutely stable
B. absolutely unstable
C. conditionally stable
D. none of these

52. The function $\frac{1}{(1+j\omega T)}$ has slope of:
A. – 20 dB/decade
B. + 20 dB/decade
C. + 6 dB/decade
D. – 6 dB/decade

53. Transfer function of a system is:

$$\frac{100\ (1+0.25\ s)}{(1+0.5\ s)}$$

The corner frequencies will be:
A. – 4 and – 2
B. 4 and 2
C. 0.25 and 0.5
D. – 0.25 and 0.5

54. The transfer function of a system is

$$\frac{100\ (1+0.25\ s)}{(1+0.5\ s)}$$

The phase shift at $\omega = 0$ and $\omega = \infty$ are:
A. 0° and – 90°
B. 90° and – 90°
C. 90° and 180°
D. 0° and 180°

55. For small value of gain k, the root must be:
A. at origin
B. at infinity
C. far away from the poles of the loop transfer functions
D. near the poles of the loop transfer function

56. The minimum phase transfer function is one having:

A. poles and zeros in RHS
B. poles and zeros in LHS
C. poles in LHS and zeros in RHS
D. poles in RHS and zeros in RHS

57. An octave frequency range is given by ω_2/ω_1 equal to:

A. 2
B. 4
C. 6
D. 10

58. A decade frequency range is given by ω_2/ω_1 equal to:

A. 2
B. 4
C. 6
D. 10

59. Fro type three system, the lowest frequency asymptote will have the slope of:

A. – 20 dB/octave
B. – 6 dB/octave
C. – 12 dB/octave
D. – 18 dB/octave

60. Nichols chart is useful in determining:

A. closed loop frequency response
B. open loop frequency response
C. open and closed loop frequency responses
D. None of the above

61. Loop transfer function zeros are the same as:

A. poles of G (s) H (s)
B. zeros of G (s) H (s)
C. poles of 1 + G (s) H (s)
D. zeros of 1 + G (s) H (s)

62. Closed loop poles are:

A. zeros of G (s) H (s)
B. poles of G (s) H (s)
C. zeros of 1 + G (s) H (s)
D. poles of 1 + G (s) H (s)

63. The following technique is not applicable to non-linear system:

A. functional analysis
B. Nyquist criterion
C. Quasi lineraization
D. phase-plane representation

64. Saturation in a stable control system may cause:

A. overdamping
B. conditional stability
C. low level oscillations
D. high level oscillations

65. Noise in a control system can be kept low by:

A. reducing the bandwidth
B. attenuating such frequencies at which external signals get coupled into the system
C. Both (A) and (B)
D. None of the above

66. For a closed loop system to be stable, the Nyquist plot of G (s) H (s) must encircle the point (– 1, j0) as many times as the number of:

A. zeros of G (s) H (s) in the RHS of s-plane
B. zeros of G (s) H (s) in the LHS of s-plane
C. poles of G (s) H (s) in the RHS of s-plane
D. poles of G (s) H (s) in the LHS of s-plane

67. The function G (s) H (s) of a system has no poles in the RHS of s-plane. For the system to be stable, the Nyquist plot of G (s) must not enclose the:

A. the origin
B. point (– 1, j0)
C. point (– 1, – j1)
D. point (1, j0)

68. On adding a pole at s = 0, the Nyquist plot of a system will:

A. remain unaltered
B. rotate clockwise by 90°
C. rotate anticlockwise by 90°
D. rotate by 180°

69. The Op-Amp control components in a control system ideally draw:

A. medium current
B. small current
C. very small current
D. zero current

70. The following system provides good performance if offset is to be avoided:

A. proportional action
B. proportional plus integral action
C. proportional plus differential action
D. proportional plus integral action plus differential action

71. The following system provides excellent transient as well as steady state response:

A. proportional action
B. proportional plus integral action

C. proportional plus differential action
D. proportional plus integral action plus differential action

72. Dynamic equations are a set of equations formed on putting together:
A. input equations and output equations
B. state equations and input equations
C. output equations and state equations
D. input, output and state equations

73. Presence of non-linearities in a control system tends to introduce:
A. transient error
B. instability
C. steady state error
D. All of these

74. A system function V (s)/I (s) has a pole at s = –2 and a zero at s = – 1. For sinusoidal current excitation, the voltage response:
A. is zero
B. leads the current
C. lags behinds the current
D. is in phase with the current

75. The steady state error of a control system may be minimized by:
A. increasing the inertia of the system
B. decreasing the inertia of the system
C. decreasing the torque of servomotor
D. increasing the damping of the system

76. Gain crossover frequency is defined as one at which |G (*j*ω) H (*j*ω)| equals:
A. 0
B. 1
C. $\sqrt{2}$
D. $1/\sqrt{2}$

77. Mark the correct statement:
A. capacitance can be used to fabricate a lag-lead network
B. inductance can be used to fabricate a lag-lead network
C. capacitance and inductance both can be used to fabricate lag-lead network
D. None of the above

78. The transfer function when the Bode diagram is plotted should be of the form:
A. $\frac{1}{s+1}$
B. $\frac{s+1}{T}$
C. $(sT+1)$
D. $\frac{1}{sT+1}$

79. The circuit given below represents a:

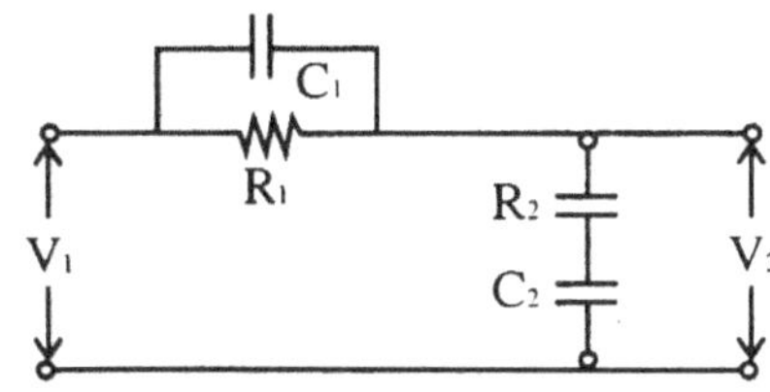

A. lag network
B. lead network
C. lag-lead network
D. None of the above

80. The bandwidth of a control system may be increased by:
A. phase lead compensator
B. phase lag compensator
C. phase lag-lead compensator
D. None of the above

81. In a feedback system, lead compensator:
A. increases the margin of stability
B. speeds up the transient response
C. increases the system error constant to some extent
D. All of the above three

82. Lead-lag compensation is needed for:
A. transient response improvement
B. steady state response improvement
C. both transient and steady state response improvement
D. None of these

83. If the Nyquist criterion for a closed-loop system is known, then the stability of the system:
A. cannot be found
B. can be found by studying the system behaviour around point (1, *j*0)
C. is the same as the stability of the loop transfer function
D. can be found by studying the behaviour of the system about a point (–1, *j*0)

84. Gain margin is the reciprocal of the gain at the frequency at which the phase angle is:

A. 0°

B. 90°

C. 180°

D. – 90°

85. The phase crossover frequency is the one at which the phase angle is:

A. 0°

B. 90°

C. – 90°

D. 180°

86. The gain margin in dB equals:

A. $-20 \log a$

B. $20 \log a$

C. $-20 \log \sqrt{a}$

D. $20 \log \sqrt{a}$

87. For stable feedback system, the phase margin:

A. is always positive

B. is always negative

C. may be positive or negative

D. None of the above

88. The phase margin Φ of a system in terms of its damping factor ζ is given by:

A. $\Phi = \tan^{-1}\left(\sqrt{1-\zeta^2}\right)$

B. $\Phi = \tan^{-1}\left[\left(1-2\zeta^2\right)^{\frac{1}{2}}\right]$

C. $\Phi = \tan^{-1}\left[2\zeta\left(\frac{1}{(4\zeta^2-1)^{1/2}-2\zeta^2}\right)^{\frac{1}{2}}\right]$

D. $\Phi = \tan^{-1}\left(\sqrt{1-\zeta}\right)$

89. The –180° phase line of Bode diagram is equal to:

A. positive imaginary axis in Nyquist plot

B. negative imaginary axis in Nyquist plot

C. positive real axis in Nyquist plot

D. negative real axis in Nyquist plot

90. The construction of a synchro-transmitter is similar to that of a:

A. three phase transformer

B. three phase altenator

C. three phase induction motor

D. None of the above

91. Which of the following elements is not used in an automatic control system?

A. sensor

B. error detector

C. oscillator

D. final control element

92. The following device works as an e: or detector:

A. synchro-transmitter

B. control transformer

C. Both (A) and (B)

D. None of the above

93. The following device is usually used for transmission of data:

A. Rotary LVDT

B. synchros

C. Both of the above

D. None of the above

94. State variable approach may be applied to:

A. time variant system

B. non-linear systems

C. linear and time variant systems

D. All of the above

95. AC systems are usually preferred to the DC systems in control applications because:

A. AC systems are cheaper

B. AC systems are more stable

C. AC systems have better performance characteristics and are smaller in size

D. All of the above

96. Time sharing of an expensive control system can be achieved by using:

A. an AC control system

B. an analog control system

C. a sampled data control system

D. None of the above

97. Zero initial condition of a control system signifies that:

A. the system is at rest but stores energy

B. the system is at rest but stores no energy

C. the reference input to working system is zero

D. the system is working but stores no energy

98. The impedance of stator and rotor windings of the control transformer are:

A. considerably lower than those of equivalent sized synchros

B. considerably higher than those of equivalent sized synchros

C. comparable with those of equivalent sized synchros

D. none of these

99. The transfer function when the Bode diagram is plotted should be of the form

A. $\frac{1}{s+1}$

B. $\frac{s+1}{T}$

C. $(sT+1)$

D. $\frac{1}{(sT+1)}$

100. Given polar plot is of the transfer function:

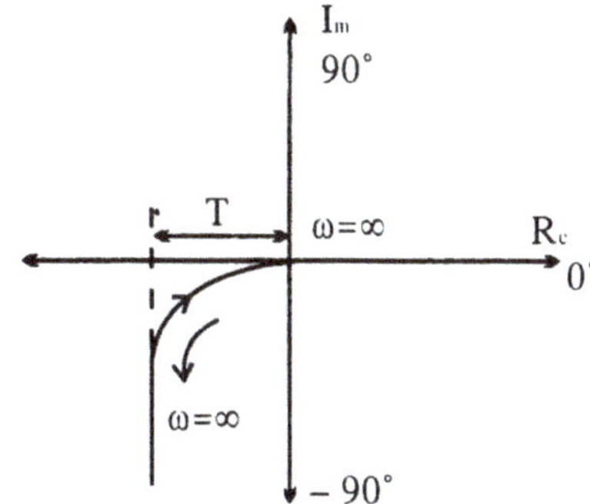

A. $\frac{1}{j\omega}$

B. $\frac{1}{(1+j\omega T)}$

C. $\frac{1}{j\omega(1+j\omega T)}$

D. None of the above

ANSWERS

1	2	3	4	5	6	7	8	9	10
A	B	D	B	C	D	B	C	B	A
11	**12**	**13**	**14**	**15**	**16**	**17**	**18**	**19**	**20**
B	A	A	A	B	C	D	C	C	B
21	**22**	**23**	**24**	**25**	**26**	**27**	**28**	**29**	**30**
C	C	C	A	C	A	B	A	A	A
31	**32**	**33**	**34**	**35**	**36**	**37**	**38**	**39**	**40**
D	C	D	C	C	B	A	A	A	D
41	**42**	**43**	**44**	**45**	**46**	**47**	**48**	**49**	**50**
C	B	D	B	C	C	A	B	B	D
51	**52**	**53**	**54**	**55**	**56**	**57**	**58**	**59**	**60**
B	A	B	A	D	B	A	D	D	A
61	**62**	**63**	**64**	**65**	**66**	**67**	**68**	**69**	**70**
B	C	B	B	C	D	B	B	D	B
71	**72**	**73**	**74**	**75**	**76**	**77**	**78**	**79**	**80**
D	C	C	B	A	B	A	C	C	B
81	**82**	**83**	**84**	**85**	**86**	**87**	**88**	**89**	**90**
D	B	D	C	D	A	A	C	D	B
91	**92**	**93**	**94**	**95**	**96**	**97**	**98**	**99**	**100**
C	B	B	D	D	C	B	B	C	C

COMMUNICATION SYSTEMS

MULTIPLE CHOICE QUESTIONS

1. In PCM if the transmission path is very long:
   A. repeater stations are used
   B. pulse width may be increased
   C. pulse mplitude is increased
   D. pulse spacing is reduced
2. Frequency shift keying is used mostly in:
   A. radio transmission
   B. telegraphy
   C. telephony
   D. None of the above
3. Leak type bias is used in a plate modulated Class C amplifier to:
   A. increase the bandwidth
   B. prevent excessive grid current
   C. prevent tuned ciruit damping
   D. prevent over modulation
4. The frequency of normal sound wave determine its:
   A. quantity
   B. echo
   C. loudness
   D. pitch
5. Which of the following is most likely to be found in an SSB transistor?
   A. Class A, RF output amplifier
   B. Class B, RF amplifier
   C. Tuned modulator
   D. Class C, audio amplifier
6. The signals contaminated with large power are demodulated by:
   A. envelope detector
   B. synchronous detector
   C. envelope detector followed by low-pass filter
   D. envelope detector followed by high-pass filter
7. A longwave AM broadcast transmitter needs:
   A. very small carrier power
   B. very large carrier power
   C. small carrier power
   D. large carrier power
8. The thermal noise is due to random:
   A. motion of atoms and molecules
   B. motion of free electrons
   C. vibrations of atom about their mean position, inside the conducting medium
   D. None of the above
9. The term noise temperature is used for:
   A. white noise source only
   B. thermal noise source only
   C. non-thermal noise source only
   D. All of the above
10. The advantages of guard time between pulses is that:
    A. it increases the efficiency of transmission
    B. it suppress the cross talk
    C. message can be reconstructed with practical filter
    D. None of the above
11. Which of the following measures cannot be used to reduce noise?
    A. By increasing the channel bandwidth
    B. By increasing the transmitted power
    C. By reducing the signaling rate
    D. By using redundancy
12. At high frequencies, which of the following types of noise becomes most important?
    A. Flicker noise
    B. Short noise
    C. Thermal noise
    D. Transit time noise

13. Which of the following is not uniformly distributed over all frequencies?
A. White noise
B. Flicker noise
C. Short noise
D. Thermal noise

14. Which of the following is not true for TDM?
A. Pulse modulated signals can be sent during wide spaces of narrow pulses
B. The transmitter and receiver are to be synchronised
C. Analog signals can also be combined
D. None of the above

15. Low speed TDM is produced by:
A. electronic switching circuits
B. delay lines
C. rotating mechanical switches
D. pulse oscillators

16. The main advantage of PCM system is lower:
A. bandwidth
B. power
C. noise
D. None of the above

17. In a FM demodulator:
A. the capacitors charged to the amplitude of FM wave
B. the frequency deviations are converted into voltage
C. simple diode is employed
D. None of the above

18. As compared to a phase discriminator a ratio detector provides:
A. double the output obtainable from a phase discriminator
B. higher stabilization against signal variations
C. all the advantages of a phase discriminator
D. lower linearity

19. The modulation index of wide-band FM signal is:
A. less than 1
B. equal to 1
C. greater than 1
D. 100%

20. The function of buffer amplifiers in transmitters is:
A. to amplify audio signals
B. to multiply frequency
C. to isolate RF stages
D. None of the above

21. The function of buffer amplifier in transmitters is to provide:
A. impeadance matching
B. frequency stability of the oscillator
C. amplification of RF signal
D. None of the above

22. The bandwidth requirement for VSB system is:
A. less than the bandwidth for SSB system
B. same as the bandwidth for SSB system
C. more than the bandwidth for SSB system
D. double the bandwidth for SSB system

23. The function of AM detector circuit is:
A. to rectify the input signal
B. to discard the carrier
C. to provide audio signal
D. All of the above

24. Which of the following cannot be used to demodulate SSB?
A. Complete phase-shift generator
B. Product detector
C. Diode balanced modulator
D. Bipolar transistor balanced modulator

25. The emphasis circuits are used for improving S/N ratio at:
A. lower frequencies
B. middle frequencies
C. higher frequencies
D. complete frequencies

26. As far the effect of thermal agitational noise is concerned a noisy resistor can be considered as a noise free resistor:
A. in series with a constant current generator
B. in parallel with a constant current generator
C. in parallel with a constant voltage generator
D. None of the above

27. Which stage contains the secondary of the last IF transformer?
A. Mixer
B. Detector
C. Oscillator
D. None of the above

28. Transistors are generally free from which of the following type of noise?
A. Resistance noise
B. Shot noise

C. Flicker noise
D. Partition noise

29. The image frequency of a superheterodyne receiver is:
A. due to insufficient adjacent channel rejection
B. created within the receiver itself
C. not rejected by the IF tube circuits
D. independent of the frequency to which the receiver is tuned

30. A trimmer is basically a:
A. insulator
B. inductor
C. capacitor
D. variable resistor

31. The function of de-emphasis circuit of an FM receiver is:
A. to control the amplitude of FM wave
B. to restore the various frequency components to their original levels
C. to reduce the high frequency component
D. none of the above

32. The function of a noise limiter in communication receivers is:
A. to eliminate noise as the signal enters the receiver
B. creates phase difference between the signal and the noise
C. to reduce the effect of noise pulses
D. limits the noise bandwidth

33. Which of the following statements is false in case of AVC in an AM receiver?
A. The gain control has to be readjusted for changing from one station to another
B. It helps to smooth out the rapid fading
C. It prevents the overloading of the final IF stage
D. None of the above

34. The major advantages of a pre-selector stage is a receiver is that is provides:
A. higher selectivity
B. higher fidelity
C. better linearity
D. larger pass-band

35. In a receiver if amplitude limiter stage does not exit, then it can be concluded that:
A. receiver is AM receiver
B. receiver is FM receiver
C. receiver is FM as well as AM type
D. None of the above

36. A padder is basically a:
A. insulator
B. variable resistor
C. capacitor
D. inductor

37. A radio receiver is tuned to 750 kHz and its oscillator frequency is 925 kHz. The image frequency would be:
A. 900 kHz
B. 1000 kHz
C. 1100 kHz
D. 1150 kHz

38. In delta modulation which of the following drawbacks are existing?
1. Slope overload
2. Serration noise
3. Granular noise
A. 1 and 2 only
B. 2 and 3 only
C. 1 and 3 only
D. All 1, 2 and 3

39. Carrier wave is generated by crystal oscillator because:
A. it provides high power output
B. no buffer stage is needed
C. tuned circuit is not required
D. it has good stability

40. In a receiver, which of the following device has IF input but RF output?
A. Demodulator
B. Loudspeaker
C. Audio amplifier
D. Frequency changer

41. In a radio detector:
A. the linearity is worse than in a phase discriminator
B. the output is twice that obtainable from a similar phase discriminator
C. stabilization against signal strength variations is provided
D. the circuit is the same as in a discriminator, except that the diode are reversed

42. The value of a resistor creating thermal noise is doubled. The noise power generated will be:
A. unchanged
B. halved
C. doubled
D. quadrupled

43. One of the following statements is wrong regarding modulation.
A. Modulation is used to ensure that intelligence can be transmitted over long distances
B. Modulation is used to ensure that different transmissions can be separated from one another
C. Modulation is used to allow the use of practicable antennas
D. Modulation is used to reduce the signal bandwidth

44. The rate at which information can be passed through a telecommunication channel depends on the:
A. carrier frequency
B. bandwidth
C. transmission loss
D. transmitter power

45. In a AM transmitter using low level modulation set up, the amplifiers following the modulator stage must be:
A. Class C amplifiers
B. Class B push pull amplifiers
C. linear devices
D. non-linear devices

46. The value of a resistor is doubled. The thermal noise power generated by this resistor is:
A. doubled
B. halved
C. unchanged
D. None of the above

47. Consider an angle modulated signal $x_c(t) = 20 \cos [(10^7)\, 2\pi t + 10 \sin 2\pi t\, (10^3)\, t]$. Maximum frequency deviation for this signal is:
A. 10 kHz
B. 10 MHz
C. 1 kHz
D. None of the above

48. The purpose of source coding is to:
A. increase the information transmission rate
B. decrease the information transmission rate
C. decrease the S/N ratio
D. decrease the probability of error

49. The bandwidth of an angle modulated signal depends both on the modulation index (β) as well as the modulating frequency (ω_m). It can be computed from:
A. $2(\beta + 1)\omega_m$
B. $2\beta\, \omega_m$
C. $(\beta + 1)\omega_m$
D. None of the above

50. The process in which the signal is first compressed before modulation and expanded after demodulation so as to allow the use of a uniform quantising rate is known as:
A. companding
B. pre-emphasis
C. de-emphasis
D. quantisation

51. Time division multiplexing (TDM) allows:
A. several subsystems to share the same transmitter facilities
B. different time slots to be used by different communicators
C. only one carrier frequency to be used to handle many communicators
D. All of the above

52. If the digital '1' level is represented by +5 V and the digital '0' is represented by 1 V, the noise margin in volts is given by:
A. 6 volts
B. 4 volts
C. 2 volts
D. None of the above

53. The envelope detector is a/an:
A. synchronous detector
B. asynchronous detector
C. product demodulator
D. coherent detector

54. A device that emits visible light is:
A. generating electromagnetic radiation
B. generating light waves whose wavelength is in millionth of meters
C. is producing energy whose sinusoidal frequency is mush higher than that of radio waves
D. All of the above

55. If the bandwidth of the modulating signal is 10 kHz, the overall bandwidth requirement of the transmitter in case of DSB amplitude modulation is:
A. 10 kHz
B. 20 kHz

C. 40 kHz
D. None of the above

56. If a modulating signal of kHz is used in a DSB of a 100 kHz carrier sinusoid, what frequency sinusoids would be present at the modulator output?
A. 100 kHz
B. 110 kHz
C. 90 kHz
D. All of the above

57. A PCM system involves following conversion:
A. voltage to frequency
B. analog to pulse widths
C. analog to binary codes
D. serial digital to parallel digital data

58. With frequency modulation, the information is placed on the carrier sinusoid by:
A. varying its frequency
B. varying its amplitude
C. turning the sinusoid on and off
D. None of the above

59. In a DM (Delta Modulation) system, the granular (idling) noise occurs when the:
A. modulation signal increases rapidly
B. pulse rate decreases
C. modulating signal remains constant
D. pulse amplitude decreases

60. In an AM system, the modulator basically:
A. causes the carrier envelope to vary in accordance with the modulating signal
B. provides the power required by the transmission system
C. multiplies the modulating signal and the carrier signal
D. (A) and (B) both

61. A hybrid communications receiver is one that:
A. employs both electronic values and solid state devices
B. is based on heterodyning principle
C. is based on 'triple-detection-receiver' principles
D. employs valves as active devices and passive components

62. In an SSB transmitter, one is most likely to find a:
A. tuned modulator
B. Class-C audio amplifier
C. Class-B RF amplifier
D. Class-A RF output amplifier

63. In a PCM system, the number of quantization levels are 16 and the maximum signal frequency is 4 kHz, the bit transmission rate is:
A. 64 K bits/sec
B. 16 K bits/sec
C. 32 K bits/sec
D. 32 bits/sec

64. In most of the grid-modulated amplitude modulators:
A. carrier is applied to the anode
B. both carrier and modulating signal are applied to the grid
C. overall efficiency is very high
D. None of the above

65. In most of plate-modulated amplitude modulators:
A. the carrier is applied to the plate of the modulator
B. the carrier is applied to the grid of the modulator
C. overall efficiency is very poor
D. None of the above

66. A crystal filter is equivalent to:
A. a high Q series reasonant circuit
B. a high Q parallel resonant circuit
C. a wideband filter
D. a low Q series resonant circuit

67. A linear-amplifier is one whose:
A. output increases with input
B. output is linearly related to input
C. output varies inversely as input in a linear fashion
D. None of the above

68. Neutralisation is generally done in:
A. pentodes
B. tetrodes
C. triodes
D. None of the above

69. If the peak transmitted power in a radar system is increased by a factor of 81 then the maximum range will be increased by a factor of:
A. 3
B. 9
C. 27
D. 81

70. A correlation receiver consists of:
A. a multiplier and an integrator
B. an integrator only
C. a multiplier only
D. an adder and integrator

71. The impulse response of filter, matched to a rectangular pulse, is:
A. an attenuator
B. a low-pass filter
C. a high-pass filter
D. an equalizer

72. The impulse response of a system is $h(t)$. When the input is $x(t)$, the output $y(t)$ will [$H(t)$ is the Fourier transform of $h(t)$]:
A. $h(t)\,x(t)$
B. $|H(f)|\,x(t)$
C. $h(t) \otimes (t)$
D. $h(t)$

73. The Fourier transform of a Gaussian time pulse is
A. uniform
B. a pair of impulses
C. gaussian
D. Rayleigh

74. A message signal band limited to 5 kHz is sampled at the minimum rate as dictated by the sampling theorem. The number of quantisation levels is 64. If the samples are encoded in binary form, the transmission rate is:
A. 60 kbps
B. 50 kbps
C. 32 kbps
D. 10 kbps

75. The probability density function of the envelope of narrow hand noise is:
A. Uniform
B. Gaussian
C. Rayleigh
D. Rician

76. A source generates four equiprobable symbols. If the source coding is adopted, the average length of code for 100 per cent efficiency is:
A. 6 bits per symbol
B. 4 bits per symbol
C. 3 bits per symbol
D. 2 bits per symbol

77. The spectral density of white noise is:
A. exponential
B. uniform
C. Poisson
D. Gaussian

78. The information capacity (bits/sec) of a channel with bandwidth W and transmission time T is given by:
A. $C \propto WT$
B. $C = W/T$
C. $C = T/W$
D. $C = W^2T$

79. The channel capacity under the Gaussian noise environment for a discrete memoryless channel with a bandwidth of 4 MHz and SNR of 31 is:
A. 20 Mbps
B. 4 Mbps
C. 8 kbps
D. 4 kbps

80. Armstrong modulator generates:
A. AM signals
B. FM signals
C. PM signals
D. None of the above

81. If each stage had a gain of 10 dB, and noise figure of 10 dB, then the overall noise figure of a two-stage cascade amplifier will be:
A. 10
B. 1.09
C. 1.0
D. 10.9

82. The saving in power in a DSBSC system modulated at 80% is:
A. Nil
B. 80%
C. 75.76%
D. 50%

83. In a PCM the amplitude levels are transmitter in a 7 unit code. The sampling is done at the rate of 10 Hz. The bandwidth should be:
A. 5 kHz
B. 35 kHz
C. 70 kHz
D. 5 MHz

84. A carrier is simultaneously modulated by two sine waves with modulation indices of 0.3 and

0.4. If the modulated power is 10 kW, what is the total modulated power:

A. 12.5 kW
B. 10 kW
C. 10.125 kW
D. 10.5 kW

85. One of the following communication system is digital. That is:

A. PCM
B. AM
C. PAM
D. FM

86. A RF carrier of 12 kV at 1 MHz is amplitude modulated by a 1 kHz signal of 6 kV peak. The modulation pattern is observed on a CRO calibrated suitably. The voltage indicated will be:

A. 18 kV
B. 1.001 kV
C. 18 V
D. 36 kV

87. Which one of the following functions qualifies to be an autocorrelation function of a real valued time signal?

A. $\sin(\omega\tau)$
B. $\exp(-\tau)$
C. $\delta(\tau)$
D. $u(\tau)$

88. A discrete zero memory informtion source has 40 symbols and each symbol is equally likely. The minimum number of bits required to code the source with uniform length code and the entropy of the source are respectively:

A. 5, 5.03
B. 6, 5.83
C. 5, 6.64
D. 6, 5.32

89. A communication channel disturbed by additive white Gaussian noise has a bandwidth of 4 kHz and SNR of 15. The highest transmission rate that such a channel can support (in K-bits/sec) is:

A. 16
B. 1.6
C. 3.2
D. 60

90. In the generation of modulated signal, a varactor diode can be used for:

A. FM generation only
B. AM generation only
C. PM generation only
D. both AM and PM generation

91. In a low level modulation AM system, the amplifiers following the modulated stage can be:

A. linear amplifiers
B. harmonic generators
C. Class C power amplifiers
D. Class B untuned amplifiers

92. A source generates four message. The entropy of the source will be maximum, when:

A. all probabilities are equal
B. one of the probabilities equals 1.0 and others are zero
C. the probabilities are 1/2, 1/4, 1/8 and 1/9
D. two of the probabilities are 1/2 each and others are zero

93. The autocorrelation function of white noise is:

A. a delta function
B. a constant
C. Gaussian
D. $\exp(-|\tau|)$ with usual notation

94. A TDM link has 20 signal channels and each channel is sampled 8000 times/sec. Each sample is represented by seven binary bits and contains an additional bit for synchronization. The total bit rate for the TDM link is:

A. 1180 k bits/sec
B. 1280 k bits/sec
C. 1480 M bits/sec
D. 1280 M bits/sec

95. PCM system require regenerative repeaters over long distances. The correct sequence of the operations which such a repeater performs is:

A. timing, equalisation and decision making
B. equalisation, timing and decision making
C. timing, thresholding (decision) and equalisation
D. thresholding, timing and equalisation

96. One Angstrom of wavelength (1 Å) equals:

A. 10^{-8} m
B. 10^{-10} m

C. 10^{-10} cm
D. 1 μm

97. The typical squelch circuit:
A. cuts off an audio-amplifier when the carrier is absent
B. eliminates the RF interference when the signal is weak
C. cuts off an IF amplifier when the AGC is maximum
D. cuts off an IF amplifier when the AGC is minimum

98. MODEM stands for:
A. modulator at transmitting side and detector at the receiving side
B. which deals with analog signals and shows digital information
C. analog to digital at transmitting side and digital to analog at a receiving side
D. a device which deals with digital signals only

99. An FM signal with a deviation δ is passed through a mixer, and has its frequency reduced five fold. The deviation in the output of the mixer is
A. 5δ
B. intermediate
C. δ/5
D. δ

100. If carrier modulated by a digital bit stream had one of the possible phases of 0, 90, 180 and 270 degrees, then the modulation is called
A. BPSK
B. QPSK
C. QAM
D. MSK

101. An amplifier operating over the frequency range of 18 to 20 MHz has a 10Ω input resistance. The RMS noise voltage at the input to the amplifier at ambient temperature of 27°C is (Assume the Boltzman constant value as 1.38×10^{-23} J/K)
A. 18.2 microvolts
B. 17.2 microvolts
C. 16.2 microvolts
D. 14.2 microvolts

102. For signal amplitude modulated to a depth of 100% by a sinusoidal signal, power is:
A. same as the power of unmodulated carrier
B. twice as the power of unmodulated carrier
C. 3/2 times the power of unmodulated carrier
D. 2/3 times the power of unmodulated carrier

103. The capacity of an analog communication channel with 4 kHz bandwidth and 15 dB SNR is approximately:
A. 20,000 bps
B. 16,000 bps
C. 10,000 bps
D. 8,000 bps

104. The entropy of a message source generating four messages with probabilities 0.5, 0.25, 0.125 and 0.125 is:
A. 1.0 bit/message
B. 1.75 bits/message
C. 3.32 bits/message
D. 5.93 bits/message

105. Frequency frogging is used in carrier system to:
A. conserve frequencies
B. reduce distortion
C. reduce cross talk
D. None of the above

106. The choice of the product RC is an envelope detector using a diode and an RC circuit is governed by:
A. both the lowest and the highest modulation frequencies
B. only the depth of modulation
C. the depth of modulation and the lowest modulation frequency
D. the depth of modulation and the highest modulation frequency

107. Typical bandwidth of an FM broadcast receiver is:
A. 20 kHz
B. 100 kHz
C. 200 kHz
D. 1 MHz

108. Armstrong modulator generates:
A. AM signals
B. phase modulated signals
C. frequency modulated signals
D. Both (B) and (C)

109. Which of the following filters is generally used in SSB generation?
A. Active filter
B. LC filter
C. Crystal filter
D. Mechanical filter

110. Term neutralisation in radio transmitter generally refers to:
   A. simultaneous modulation and demodulation
   B. modulation at low frequencies
   C. eliminating spurious oscillation due to unwanted feedback
   D. cancelling the modulation caused in modulator

111. Frequency shift keying (FSK) is used mostly in:
   A. telephony
   B. telegraphy
   C. radio broadcast
   D. radio communication

112. Over which of the following ranges is the effect of atmospheric noise most significant?
   A. Below 30 MHz
   B. From 30 MHz to 300 MHz
   C. From 300 MHz to 3,000 MHz
   D. Above 3,000 MHz

113. The best way of coupling two stages of a transmitter located physically far apart is:
   A. link coupling
   B. impedance coupling
   C. signal tuned transformer coupling
   D. double tuned transformer coupling

114. The function of buffer amplifier in transmitter is:
   A. to amplify audio signals
   B. to multiply carrier frequency
   C. to isolate master oscillator from load
   D. None of the above

115. Master oscillator in a transmitter commonly uses crystal oscillator because:
   A. it provides high power output
   B. no buffer stage is needed
   C. tuned circuit is not required
   D. it has high frequency stability

116. In the phase-shift SSB method, to get USB:
   A. the two LSBs are 180° out of phase
   B. the two LSBs are in phase
   C. the two USGs are 180° out of phase
   D. all the sidebands are in phase

117. In the phase shift method of producing SSB, the number of balanced modulators and phase shift circuits required respectively are:
   A. 1, 1
   B. 2, 1
   C. 2, 2
   D. 1, 2

118. In double sideband suppressed carrier system the receiver is costly because:
   A. synchronous detection is required
   B. it is difficult to generate local carrier at the receiver
   C. received signal is of low power
   D. None of the above

119. One important advantage of SSB-suppressed carrier system is that it provides:
   A. higher efficiency of transmission
   B. better quality of communication
   C. simpler and inexpensive circuitry
   D. longer range of communication

120. The AM broadcast band (Medium Wave band) extends from:
   A. 200 kHz to 1000 kHz
   B. 500 kHz to 1600 kHz
   C. 3 MHz to 30 MHz
   D. 30 MHz to 300 MHz

121. The Ultra High Frequency (UHF) range extends from:
   A. 3 MHz to 30 MHz
   B. 30 MHz to 300 MHz
   C. 300 MHz to 3,000 MHz
   D. 3,000 MHz to 30,000 MHz

122. The Very High Frequency (VHF) range extends from:
   A. 3 MHz to 30 MHz
   B. 30 MHz to 300 MHz
   C. 300 MHz to 3,000 MHz
   D. 3,000 MHz to 30,000 MHz

123. The High Frequency (HF) range extends from:
   A. 300 kHz to 3,000 kHz
   B. 3 MHz to 30 MHz
   C. 30 MHz to 300 MHz
   D. 300 MHz to 3,000 MHz

124. In order to separate out channels in an FDM receiver, it is necessary to use:
   A. AND gates
   B. band pass filters
   C. band stop filters
   D. integrators

125. TDM:
   A. can be used with PCM only
   B. interleaves pulses belonging to different transmissions

C. combines fine groups into a supergroup
D. stacks 24 channels in adjacent frequency slots

126. Which of the following modulation system is digital?
A. PPM
B. PCM
C. PWM
D. PFM

127. All the output pulses are at full transmitter power for a strong signal in all of the following except:
A. PWM
B. PAM
C. PFM
D. PCM

128. The principal merit of PCM system is its:
A. lower bandwidth
B. lower power requirement
C. lower noise
D. lower cost

129. For transmission of normal speech signal, the PCM channel needs a bandwidth of:
A. 64 kHz
B. 16 kHz
C. 8 kHz
D. 32 kHz

130. If PCM signal is required to be transmitted, the most common type of modulation used is:
A. AM
B. FM
C. PM
D. PAM

131. Negative peak clipping in linear diode detector results due to:
A. too high value of time constant RC
B. too low value of time constant RC
C. peaky nature of modulation voltage
D. too high frequency of modulating voltage

132. Linear diode detector utilizes:
A. linear portion of the dynamic characteristic of diode
B. square law portion of the dynamic characteristic of diode
C. rectification property of diode
D. None of the above

133. The frequency deviation in phase modulated carreir is proportional to:
A. only to amplitude of the modulating signal
B. only frequency of the modulating signal
C. amplitude as well as frequency of the modulating signal
D. None of the above

134. De-emphasis is used:
A. to attenuate high modulation frequencies
B. to attenuate low modulation frequencies
C. to attenuate midband modulation frequencies
D. to reduce overall modulation index

135. Pre-emphasis circuit is placed:
A. after modulation circuit
B. before modulation circuit
C. before detection circuit
D. after detection circuit

136. De-emphasis of high modulation frequencies is used:
A. after modulation
B. before modulation
C. before detection
D. after detection

137. Pre-emphasis is used to boost up:
A. low modulation frequencies
B. high modulation frequencies
C. both low and high modulation frequencies
D. overall modulation index

138. In an amplitude modulated carrier, the total power is 1160 watts while that of each sideband is 80 watts. The modulation index is:
A. 0.2
B. 0.4
C. 0.08
D. 0.81

139. A sinusoidal carrier voltage of amplitude 100 volts is amplitude 100 volts is amplitude modulated by a sinusoidal voltage of frequency 1 kHz resulting in maximum modulated carrier amplitude of 130 volts. The modulation index is:
A. 0.03
B. 0.3
C. 1.3
D. 0.13

140. A 2 volt, 1 kHz signal frequency modulates a carrier voltage to causes frequency deviation of

5 kHz. If the modulating voltage is changed to 20 volts, 200 Hz, the new frequency deviation is:

A. 5 kHz
B. 50 kHz
C. 0.5 kHz
D. 100 kHz

141. An 2.5 volt, 1 kHz signal frequency modulates the carrier to cause frequency deviation of 4 kHz. On increasing the modulating voltage to 10 volts, the frequency deviation becomes:

A. 8 kHz
B. 16 kHz
C. 4 kHz
D. 1 kHz

142. A 2.5 volt, 500 Hz voltage frequency modulates the carrier to cause frequency deviation of 5 kHz. The modulation index is:

A. 10
B. 25
C. 5
D. 50

143. A sinusoidal voltage of amplitude 1 kV is amplitude modulated by another sinusoidal voltage to produce 30% modulation. The amplitude of each sideband term is:

A. 300 volts
B. 150 volts
C. 500 volts
D. 250 volts

144. Which of the following is not true regarding autocorrelation function:

A. it is an average function
B. it is defined only for periodic signals
C. it is a measure of both time variation and statistical dependance
D. it is non-unique description of signals

145. For a Gaussian process, autocorrelation also implies:

A. statistical dependent
B. statistical independance
C. ergodic process
D. None of the above

146. Amount of information in a continuous signal is:

A. zero
B. 2 bits
C. 4 bits
D. infinite

147. One dit is equivalent to choice of one event from:

A. a set of 4 equiprobable events
B. a set of 8 equiprobable events
C. a set of 10 equiprobable events
D. a set of 16 equiprobable events

148. In ergodic process:

A. ensemble and time averages are identical
B. ensemble and time averages are different
C. ensemble and time averages are opposite to each other
D. None of the above

149. For a periodic function, the spectral density and the autocorrelation functions are:

A. Fourier transform pair
B. Laplace transform pair
C. one and the same thing
D. None of the above

150. If the probability pk of a message is 1/8, then the information in bits is:

A. 1 bit
B. 2 bits
C. 3 bits
D. 4 bits

ANSWERS

1	2	3	4	5	6	7	8	9	10
A	B	B	D	B	B	B	B	D	C
11	**12**	**13**	**14**	**15**	**16**	**17**	**18**	**19**	**20**
A	D	B	C	C	C	B	D	C	C
21	**22**	**23**	**24**	**25**	**26**	**27**	**28**	**29**	**30**
C	B	D	D	C	B	B	D	C	C

31	**32**	**33**	**34**	**35**	**36**	**37**	**38**	**39**	**40**
B	C	A	A	A	C	C	B	D	A
41	**42**	**43**	**44**	**45**	**46**	**47**	**48**	**49**	**50**
A	A	D	B	C	C	A	D	A	A
51	**52**	**53**	**54**	**55**	**56**	**57**	**58**	**59**	**60**
D	C	B	D	B	D	C	A	C	D
61	**62**	**63**	**64**	**65**	**66**	**67**	**68**	**69**	**70**
A	C	B	B	B	A	B	C	A	A
71	**72**	**73**	**74**	**75**	**76**	**77**	**78**	**79**	**80**
A	C	C	A	B	B	B	D	A	B
81	**82**	**83**	**84**	**85**	**86**	**87**	**88**	**89**	**90**
D	C	B	A	D	D	C	D	A	A
91	**92**	**93**	**94**	**95**	**96**	**97**	**98**	**99**	**100**
A	A	A	C	A	A	A	B	D	B
101	**102**	**103**	**104**	**105**	**106**	**107**	**108**	**109**	**110**
A	C	B	B	C	D	C	D	D	C
111	**112**	**113**	**114**	**115**	**116**	**117**	**118**	**119**	**120**
B	A	A	C	D	A	C	A	A	B
121	**122**	**123**	**124**	**125**	**126**	**127**	**128**	**129**	**130**
C	B	B	B	B	B	B	C	A	A
131	**132**	**133**	**134**	**135**	**136**	**137**	**138**	**139**	**140**
C	C	A	A	B	D	C	B	B	B
141	**142**	**143**	**144**	**145**	**146**	**147**	**148**	**149**	**150**
B	A	B	B	B	D	C	B	A	C

MICROWAVE ENGINEERING

MULTIPLE CHOICE QUESTIONS

1. A two-cavity Klystron amplifier has the following characteristics:
Input power = 5 milliwatt
Resistance of Input cavity = 30 kΩ
Calculate the input RMS voltage.
A. 12.00 V
B. 13.45 V
C. 12.25 V
D. 15.00 V

2. A two-cavity Klystron amplifier has the following specifications:
Spacing between centres of cavities = 4 cm (L)
Beam voltage = 900 V (V_0)
Find the DC transit time of electrons.
A. 2.25 sec
B. 2.25×10^{-6} sec
C. 0.225×10^{-8}
D. 0.225 sec

3. There are three types of magnetrons — the negative resistance type, the cavity type and:
A. the converting type
B. the electro-magnetic type
C. the cyclotron frequency type
D. None of the above

4. The efficiency of a depressed collector is (nearly):
A. 5 to 6 per cent
B. 5 to 20 per cent
C. 100 per cent
D. less than 5 per cent

5. In the parlance of a TWT, a depressed collector is a set of collector rings maintained at successively higher potentials to effectively slow down the electron beam that still possesses a considerable kE. If this depressed collector were used in a TWT, the efficiency of the TWT would rise to:
A. 100 per cent
B. 5 per cent
C. 30 per cent
D. 70 per cent

6. The tube life in the case of a TWT is (nearly):
A. 10,000 hours
B. 20,000 hours
C. 50,000 hours
D. 1,00,000 hours

7. The frequency range of a Backward Wave Oscillator (BWO) is:
A. 1 GHz to 200 GHz
B. 1 GHz to 1000 GHz
C. 1 GHz to 500 GHz
D. None of the above

8. A reflex Klystron operates at the peak mode of $n = 2$ with beam voltage $V_0 = 300$ V. Beam current $I_0 = 20$ mA, signal voltage $V_1 = 40$ V. Find the AC output power in watts.
A. 2.36 W
B. 1.36 W
C. 3.44 W
D. 1.16 W

9. A normal circular magnetron has the following parameters:
Inner radius = $R_a = 0.15$ m
Outer radius = $R_0 = 0.45$ m
Magnetic flux density = $B_0 = 1.2$ mWb/m^2
Find the cut-off magnetic flux density (B_C) if the beam voltage V_0 is equal to 6000 V.
A. 140.2 mWb/m^2
B. 130.595 mWb/m^2
C. 140.623 mWb/m^2
D. 130.211 mWb/m^2

10. The microwave oven used by a house-wife consists of a:
A. BWO
B. magnetron
C. TWT
D. All of the above

11. MESFETs made up of Galium Arsenide are used as the following:
A. Power oscillators
B. Driver amplifiers for high-power transmitters
C. Power amplifiers for output-stage of microwave links
D. They are used for all the three applications given above

12. Look at the figure given below, which is the equivalent circuit of a varactor diode.

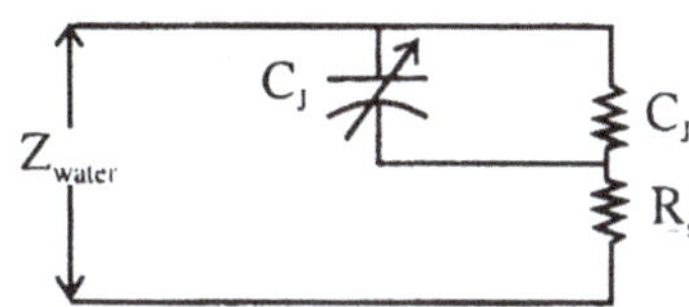

At microwave frequencies, R_J would be of the order of:
A. 1 MΩ
B. 100 MΩ
C. 10 MΩ
D. 1000 MΩ

13. The applications of varactor diodes include:
A. harmonics generation
B. active filters
C. pulse generation and pulse shaping
D. All of the above

14. In the case of a Par Amp, the output circuit, which does not require external excitation, is known as a/an:
A. excitation circuit
B. self-excited circuit
C. idler circuit
D. convertor circuit

15. The Noise Figure F, of a Parametric Up Convertor (PUC) is given by

$$F = 1 + \frac{2T_d}{T_0}\left[\frac{1}{\gamma_Q} + \frac{1}{(\gamma_Q)^2}\right]$$

In this formula, T_d stands for:
A. diode temperature in °C
B. diode temperature in °K
C. ambient temperature in °C
D. ambient temperature in °K

16. The equivalent circuit of a PIN diode is:

A.

B.
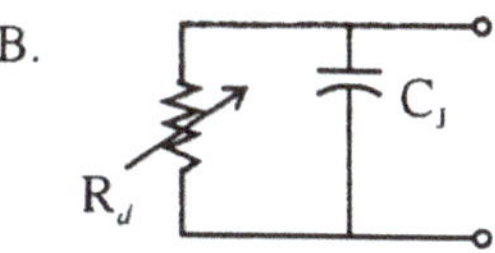

C.
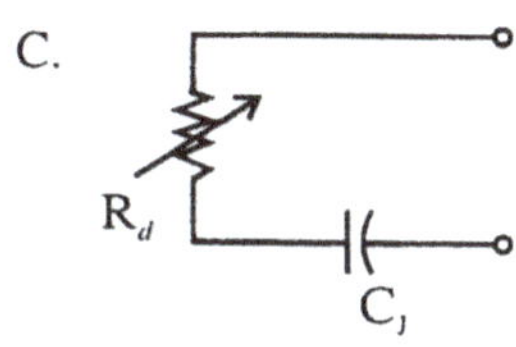

D.
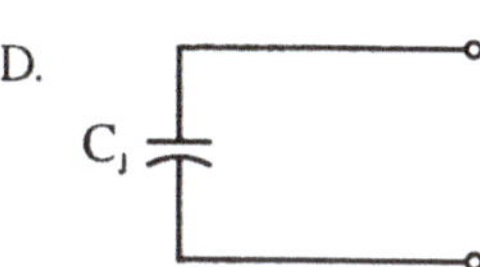

17. The PIN diode can be used as a/an:
A. switch
B. amplitude modulator
C. phase shifter
D. All of the above

18. The applications of Schottley Barrier Diodes (SBDs) include:
A. low noise mixers
B. balanced mixers in CW radars
C. microwave detectors
D. All of the above

19. The frequency range of a tunnel diode oscillator is up to:
A. 100 GHz
B. 200 GHz
C. 400 GHz
D. 50 GHz

20. In *n*-type, Ga As Gunn effect diodes:
A. electrons are transferred from lower conduction valley to the upper subsidiary valley
B. electrons are transferred from the upper subsidiary valley to lower conduction valley

C. there is no conduction of electrons to and from the upper subsidiary valley
D. None of the above

21. The efficiency range (approximate) of a Gunn diode is:
A. 10 per cent
B. 2 per.cent to 12 per cent
C. 20 per cent
D. 8 per cent to 12 per cent

22. If a Gunn diode works as a broadband linear amplifier, it can be considered to be a replacement of:
A. the TWT
B. the SBD
C. the IMPATT
D. None of the above

23. IMPATT diodes can be used as:
A. microwave generators
B. modulated output oscillators
C. receiver local oscillators
D. All of the above

24. The avalanche zone velocity, in the case of a TRAPATT diode, is given by:
A. $V_z = \frac{J}{q}$
B. $V_z = \frac{J}{qN_A}$
C. $V_z = JqN_A$
D. $V_z = Jq$
where all the terms have usual meanings.

25. The materials that can be used for MASER application are:
A. NH_3
B. Al_2O_3
C. Cesium
D. All of the above

26. An IMPATT diode has a drift length of 2 μm. Find the operating frequency of the diode if the drift velocity of Si is 10^7cm/sec.
A. 10 GHz
B. 12.5 GHz
C. 25.0 GHz
D. 33 GHz

27. A Gunn diode has operating frequency of 10 GHz, device length of 75 μm and amplified voltage pulse of 25 V. What is its threshold electric field?
A. 2.22 kV/cm
B. 3.33 kV/cm
C. 3.63 kV/cm
D. 2.63 kV/cm

28. Applications of MASER amplifiers include:
A. low noise low level amplifiers
B. radio astronomy and extra-terriestrial communications
C. radio telescopes
D. All of the above

29. If a MASER is cooled:
A. noise level is increased
B. noise level is reduced
C. noise level is not affected
D. None of the above

30. A TRAPATT has the following specifications:
Doping concentration = 2.5×10^{16} per cm^3
Current density = 33 kA/cm^2
What is the avalanche zone velocity?
A. 2.25×10^6 cm/sec
B. 8.25×10^4 cm/sec
C. 6.00×10^6 cm/sec
D. 8.25×10^6 cm/sec

31. Which one of the following is not an avalanche transit time device?
A. IMPATT
B. TRAPATT
C. BARITT
D. Gunn Diode

32. The fundamental principles, which govern the operations of LASER and MASER, can be explained on the basis of:
A. classical mechanics
B. quantum electronics and statistical mechanics
C. Newtonian mechanics
D. None of the above

33. A CD player uses a/an:
A. laser beam
B. magnetic head
C. MASER beam
D. None of the above

34. The bandwidths of MASERs are:
A. very wide
B. narrow
C. moderate
D. depend upon the crystals used as sources

35. A parametric amplifier, for its amplification process, uses:

A. a non-linear reactance or a time varying reactance
B. a simple resistance
C. a capacitor
D. None of the above

36. The applications of parametric amplifiers include:

A. satellite ground stations
B. microwave ground wave communications
C. radio telescopes and long-range radars
D. All of the above

37. As compared to waveguide circuits, microwave integrated cicruits have:

A. higher power handling abilities
B. lower power handling abilities
C. equal power handling abilities
D. None of the above

38. The chief limitations of microstrip antennas are:

A. narrow bandwidth
B. practical limitations on maximum gain (nearly 20 dB)
C. low power handling capabilities
D. All of the above

39. The guide wavelength of a microstrip line is given by:

A. $\lambda = \lambda_0 / \sqrt{E_{eff}}$
B. $\lambda = \lambda_0 \times \sqrt{E_{eff}}$
C. $\lambda = \lambda_0 / E_{eff}$
D. $\lambda = \lambda_0 \times (E_{eff})^2$

40. In parallel feed (corporate) microstrip arrays:

A. pairs of patches are connected with power splitter
B. combinations of corporate (parallel) and series feed are used
C. connected pairs are not connected with another power splitter
D. None of the above

41. Microwave communication systems are ideal in the frequency ranges of:

A. 100 MHz to 100 GHz
B. 300 MHz to 150 GHz
C. 300 MHz to 50 GHz
D. 200 MHz to 25 GHz

42. During the sky wave propagation, the refraction from ionosphere occurs only in the HF range of frequencies. If the frequency is increased beyond the critical frequency.

A. the wave is refracted back by the ionosphere
B. the wave pierces through the ionosphere
C. the wave is transmitted back to the earth through multiple hops
D. None of the above

43. The critical frequency, f_c, depends upon:

A. time of the day
B. season
C. geographical position
D. All of the above

44. What is absorption fading?.

A. It is the absorption of electro-magnetic waves when propagating over different areas like open land, hills, mountains, water, etc.
B. It is the distortion in modulator signals
C. It is a type of scatter loss in troposcatter communication links
D. None of the above

45. Selective fading is more predominant:

A. at lower frequencies for which, skywave propagation is used
B. at higher frequencies for which, skywave propagation is used
C. when Single Side Band (SSB) systems are used in skywave propagation systems
D. None of the above

46. In troposphere, the fading of the transmitted wave could be due to:

A. variation in the dielectric constant of the tropospheric region
B. local wind eddies, which weaken the signal
C. erratic variations of the refractive index and the effective earth radius factor (k)
D. All of the above

47. The uplink frequency for a satellite, in the k_n band, is:

A. 7.9-8.4 GHz
B. 14.0-14.5 GHz
C. 30-31 GHz
D. 5.9-6.4 GHz

48. The downlink frequency for a satellite, in the X band, is:

A. 3.7-4.2 GHz
B. 11.7-12.2 GHz

C. 7.25-7.75 GHz
D. 40-41 GHz

49. The frequency band of 54-58 GHz in U-band transmission signals would be ideal for:
A. military applications
B. ships navigation
C. commercial broadcasts
D. inter-satellite communications

50. What is the antenna beam angle required by a satellite antenna to provide full global coverage from a geostationary orbit?
A. 18.21°
B. 27.16°
C. 23.44°
D. 27.00°

51. Calculate the power gain of an optimum horn antenna with a square aperture of 5λ as a side dimension.
A. $G_p = 100.5$
B. $G_p = 112.5$
C. $G_p = 98.4$
D. G_p cannot be calculated

52. A geostationary communication satellite has an uplink freqeuncy of 6 GHz, with an antenna elevation of 5°. If the transmitter power is 1 kW and antenna gains of transmitter and receiver are 60 dB, respectively, then calculate the received power at the input of the satellite receiver.
A. Nearly 9×10^{-10} W
B. Nearly 9×10^{-20} W
C. Nearly 9×10^{-14} W
D. Nearly 9×10^{18} W

53. The horn antennas can be of the following type:
A. Sectoral E-plain
B. Sectoral H-plane
C. Pyramidal
D. Any of the above type is possible

54. The path that a satellite can adopt while going around the earth include:
A. geostationary
B. inclined eliptical
C. polar circular
D. All of the above are possible

55. The orbital period of a satellite is given by:

A. $T = \frac{2\pi}{60}\sqrt{\frac{r^3}{GM}}$ min

B. $T = \frac{2\pi\, r^3}{GM}$ min

C. $T = \frac{2\pi\, GM}{r^3}$

D. $T = \frac{2\pi\sqrt{GM}}{r^3}$ min

56. The field strength of a tropospheric wave is given by:

A. $F_R \simeq \frac{88\sqrt{P}\, h_t\ h_r}{d^2\lambda}$

B. $F_R \simeq \frac{88\, h_t\ h_r}{\lambda}$

C. $F_R \simeq \frac{h_t\ h_r\sqrt{P}}{d^2\ \lambda}$

D. $F_R \simeq \frac{88\sqrt{P}}{d^2\ \lambda}$

where all the terms have usual meanings.

57. Digital microwave communication is better than analog microwave communication because:
A. they do not have noise immunity for data and voice signals
B. their RF spectrum efficiencies are low
C. they are capable of using frequency bands above 10 GHz
D. All of the above are true

58. Low power can be measured through:
A. Bolometer technique
B. caloriemetric technique
C. calorimetric wattmeter technique
D. All of the above can be used

59. Two identical directional couplers are used in a waveguide to sample the incident and reflected powers. The output of the two couplers is found to be 2.5 milliwatt and 0.15 multiwatt. Find the value of VSWR in the waveguide.
A. VSWR = 2.1
B. VSWR = 1.64
C. VSWR = 1.30
D. VSWR = 1.53

60. We have two directional couplers (20 dB) in a guide to sample the incident and reflected powers. The outputs of the two couplers are 3 milliwatt and 0.1 milliwatt, respectively. What is the value of VSWR in the main waveguide?

Also what is the value of reflected power?

A. 1.44, 2 milliwatt
B. 1.44, 10 milliwatt
C. 1.32, 5 milliwatt
D. 1.32, 10 milliwatt

61. In the parlance of microwave measurements, attenuation is defined:

A. as the ratio of input power to output power in watts
B. as the ratio of input power to output power and is a mere ratio
C. as the ratio of input power to output power in dB
D. None of the above

62. The amount of attenuation may be measured by:

A. power ratio method
B. RF substitution method
C. Either of (A) or (B)
D. None of (A) and (B)

63. Impedance at microwave frequencies can be measured by using:

A. using the magic-T method
B. using the slotted line method
C. using the reflectometer
D. Either of the above can be used

64. Look at the figure given below:

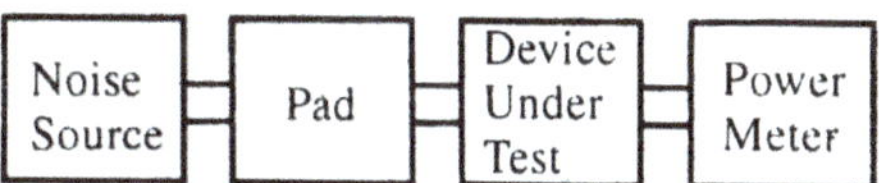

This set up can be used to measure:

A. the Q value of a cavity rasonator
B. the noise factor
C. the VSWR
D. the dielectric constant

65. VSWR, for S > 10, can be measured by using th following relationship:

A. $\text{VSWR} = \lambda_g \pi$

B. $\text{VSWR} = \dfrac{\lambda_g}{\pi(d_2 - d_1)}$

C. $\text{VSWR} = \dfrac{\lambda_g (d_2 - d_1)}{\pi}$

D. $\text{VSWR} = \lambda_g \pi (d_2 - d_1)$

ANSWERS

1	2	3	4	5	6	7	8	9	10
C	C	C	B	C	C	B	B	B	B
11	**12**	**13**	**14**	**15**	**16**	**17**	**18**	**19**	**20**
D	C	D	C	B	B	D	D	A	A
21	**22**	**23**	**24**	**25**	**26**	**27**	**28**	**29**	**30**
B	A	D	B	D	C	B	D	B	D
31	**32**	**33**	**34**	**35**	**36**	**37**	**38**	**39**	**40**
D	B	A	B	A	D	B	D	A	A
41	**42**	**43**	**44**	**45**	**46**	**47**	**48**	**49**	**50**
B	B	D	A	B	D	B	C	D	B
51	**52**	**53**	**54**	**55**	**56**	**57**	**58**	**59**	**60**
B	C	D	D	A	A	C	A	B	B
61	**62**	**63**	**64**	**65**					
C	C	D	B	B					

COMPUTER ENGINEERING

MULTIPLE CHOICE QUESTIONS

1. A computer is a device for automatically carrying out a/an:
   A. information
   B. instructions
   C. program of instructions
   D. None of the above
2. One of the more modern programming languages, is called:
   A. higher language
   B. Pascal
   C. Assembly language
   D. None of the above
3. Hardware consists of the actual:
   A. components of the computer
   B. components of the computer system
   C. components of the software
   D. None of the above
4. The component of the computer system which perform processing functions is called the:
   A. control unit
   B. main memory
   C. central processing unit (CPU)
   D. None of the above
5. The arithmetic unit (ALU) performs:
   A. mathematics computations
   B. primary memory
   C. CPU
   D. None of the above
6. Peripheral devices can be divided into:
   A. four general categories
   B. three general categories
   C. two general categories
   D. None of the above
7. Programs and data that are entered into a computer to be processed are called:
   A. output
   B. input
   C. Both (A) and (B)
   D. None of the above
8. When data is processed, or converted to some meaningful form, this is called:
   A. ALU
   B. CPU
   C. information
   D. None of the above
9. Arithmetic and logic operations are the only types of instructions that the:
   A. CPU is able to execute
   B. ALU is able to execute
   C. Hardware is able to execute
   D. None of the above
10. The primary storage unit also referred to as:
    A. main memory, primary memory
    B. internal storage
    C. Both (A) and (B)
    D. None of the above
11. In a national election, the records of all the votes cast for the office of President are data when these votes are tabulated and the final total is determined, the result is called
    A. input
    B. information
    C. ALU
    D. None of the above
12. A given computer system can have
    A. only one input device
    B. not more than two input devices
    C. many different input device
    D. None of the above

13. Output devices, allow the user to examine the computer's output. Displaying output on the screen gives the user the results in a convenient, readable form; this output is referred to as:
A. hard copy
B. soft copy
C. input devices
D. None of the above

14. As with input devices, a particular computer system can have:
A. only one output device
B. only two output devices
C. many different output devices
D. None of the above

15. Disk and tape drives are commonly used:
A. hard copy
B. soft copy
C. secondary storage devices
D. None of the above

16. In supercomputer, a machine must be capable of performing atleast:
A. 5 million arithmetic operations per second
B. 10 million arithmetic operations per second
C. 20 million arithmetic operations per second
D. None of the above

17. For most business applications, the extremely high-speed processing capabilities of a supercomputer:
A. are necessary
B. are not necessary
C. Both (A) and (B)
D. None of the above

18. Minicomputers manufactured today are more powerful than:
A. supercomputer
B. mainframes
C. microcomputer
D. None of the above

19. Which of the following computer is currently the smallest and least costly computer:
A. microcomputer
B. supercomputer
C. mainframes
D. None of the above

20. Some programs direct the computer in its own internal operations. These are called:
A. software
B. system programs
C. application programs
D. None of the above

21. Pascal, a structural programming language, was developed by:
A. Pascal, B
B. Niklaus
C. George Boole
D. None of the above

22. Pascal contains statements that allow the programmer to efficiently alter the flow of program execution by using decision:
A. structure only
B. loops only
C. structures and loops
D. None of the above

23. In block-structured language, programs are composed of subprograms or blocks, each designed to perform a:
A. decision structures
B. loop
C. specific task
D. None of the above

24. The important feature of Pascal is that it is a general-purpose:
A. programming language
B. high-level language
C. source program
D. None of the above

25. Pascal programs are usually translated into machine language by a
A. batch processing
B. computer
C. block
D. None of the above

26. There is no limit to the length of:
A. Pascal program
B. structured programming language
C. an identifier
D. None of the above

27. Before a program written in high language can be executed, it must be translated into:
A. block structural language
B. machine language
C. structural programming language
D. None of the above

28. A loop is usually the easiest way to perform a repetition task such as printing a list of:
A. a few names
B. several hundred names
C. one name only
D. None of the above

29. The syntax of a language can be described using syntax diagrams, such as those for:
A. Pascal
B. loop
C. sequence
D. None of the above

30. A program that adds three numbers together and determines the average is a:
A. loop
B. term
C. sequence
D. None of the above

31. Those languages which are strongly oriented toward the programmer rather than toward the computer, they are termed as:
A. machine language
B. high-level language
C. assembly language
D. None of the above

32. Assembly languages also referred to as:
A. high language program
B. low-level languages
C. machine language
D. None of the above

33. Peripheral devices consist of input devices, output devices and:
A. logical operation
B. secondary storage device
C. arithmetic operation
D. None of the above

34. The primary storage unit temporarily stores data, programs, and:
A. physical components
B. peripheral devices
C. processing results
D. None of the above

35. The central processing unit consists of the arithmetic/logic unit, which performs arithmetic and logical operations, and the control unit, which controls the operations of:
A. CPU
B. ALU
C. peripheral devices
D. None of the above

36. When items stored in primary storage need to be kept permanently, they can be transferred to secondary storage. Example of secondary storage media are:
A. magnetic disks and tape
B. air
C. water
D. None of the above

37. Programs are list of instructions that a computer can use to solve a problem. A program or series of programs is called:
A. hardware
B. software
C. language program
D. None of the above

38. Machine language is the only language that the computer can directly execute and is different for each type of:
A. hardware computer only
B. supercomputer only
C. computer
D. None of the above

39. The three categories of programming languages are: machine language, assembly language and:
A. structured programming language
B. high-level languages
C. subprograms
D. None of the above

40. A job is a unit of work to be processed by the:
A. CPU
B. ALU
C. software
D. None of the above

41. A computer can only execute instructions that are in:
A. assembly language
B. machine language
C. job control program
D. None of the above

42. The job control program translates the job control commands written by a programmer into:
A. high-level language
B. assembly language
C. machine language
D. None of the above

43. Application programs are the programs that meet:
A. machine language needs
B. job control program needs
C. user's needs
D. None of the above

44. When computers were first developed, machine language was the only way they could:
A. be programmed
B. not be programmed
C. Both (A) and (B)
D. None of the above

45. Often assembly language instructions will execute faster than those written in:
A. machine languge
B. high-level language
C. Pascal
D. None of the above

46. Pascal can be used to write a wide variety of programs, from those that are highly scientific to those that are strongly:
A. business oriented
B. general oriented
C. source oriented
D. None of the above

47. In Pascal, there is no limit to the length of an identifier, but some compilers only recognize the:
A. four characters in the indentifier
B. three characters in the identifier
C. eight characters in the identifier
D. None of the above

48. The formulation of user-defined identifiers should allow the rules for:
A. identifiers
B. operations
C. computer
D. None of the above

49. A storage location whose value can change during program execution is referred to as:
A. integer
B. variable
C. zero integer
D. None of the above

50. An integer is a signed or unsigned:
A. a few numbers
B. only one number
C. whole number
D. None of the above

51. The model which provides a concrete expression of these concepts by means of two fundamental ideas – logical records and links, is called a:
A. relationship model
B. network model
C. hierarhical model
D. None of the above

52. The network model has been established by:
A. CODASYL committee
B. IBM
C. EGA
D. None of the above

53. A link may be defined as the representation of an:
A. objects
B. programs
C. association
D. None of the above

54. DML is intended provide operations which operate upon data stored in the:
A. file
B. record
C. database
D. None of the above

55. In the hierarachical model a hierarchical scheme is defined by a tree whose nodes are:
A. database
B. logical record types
C. links
D. None of the above

56. The set of all values in the relations represents the contents of the:
A. database
B. record file
C. table
D. None of the above

57. In hierarchical data model, the database is represented by:
A. a directed graph
B. tree structures
C. relation
D. None of the above

58. A Database Management System (DBMS) is a software system capable of supporting and managing any number of independent:
A. results
B. relations

C. databases

D. None of the above

59. Support for data independence is arguably the most important feature of a:

A. system

B. database system

C. relation

D. None of the above

60. Distinct levels of data independence are:

A. two

B. three

C. four

D. None of the above

61. Database software may be divided into:

A. four distinct categories

B. three distinct categories

C. two distinct categories

D. None of the above

62. Any large software project may be divided into several:

A. project phases

B. sub-projects

C. hardware

D. None of the above

63. Data modelling occupies an important position in the:

A. database file

B. database life-cycle

C. relation

D. None of the above

64. The term "software engineering" implies the application of scientific knowledge and discipline to the construction of:

A. database record

B. computer hardware system

C. computer software system

D. None of the above

65. Program correctness has been a strong motivation for much of the work in software engineering and:

A. high-level language programming

B. programming language design

C. Pascal

D. None of the above

66. A data model may be defined as a pattern according to which data are:

A. organised

B. logically organised

C. well organised

D. None of the above

67. The term 'software engineering' implies the application of scientific knowledge and discipline to the construction of

A. systems

B. hardware systems

C. software systems

D. None of the above

68. In software system, the data must be shared among:

A. a very few users

B. many different users

C. two users

D. None of the above

69. One of the most important techniques for dealing with complexity in software systems is the:

A. principle of abstraction

B. principle of stepwise refinement

C. maintainability of the system

D. None of the above

70. Concurrency control in database management systems, including:

A. database procedure

B. transaction

C. alternatives to locking

D. None of the above

71. The developments in software engineering have led to the emergence of a:

A. programming language

B. new class of programming language

C. new class of system

D. None of the above

72. Program correctness has been a strong motivation for much of the work in software engineering and:

A. hardware system

B. assembly language

C. programming language design

D. None of the above

73. Modula-2 is a language which is suitable both for the implementation of:
A. database system only
B. application programs
C. database management systems and application programs
D. None of the above

74. The logical database design is a:
A. three-stage process
B. two-stage process
C. four-stage process
D. None of the above

75. The relational model whose guiding ideas were published in 1970 has dominated the area of:
A. software system
B. hardware system
C. database research and development
D. None of the above

76. A link is represented by a diagram, called a:
A. Bachman diagram
B. Pascal diagram
C. Theory diagram
D. None of the above

77. A great advantage of Modula-2 over Ada is that it is a:
A. greater language
B. small language
C. Pascal language
D. None of the above

78. Integrity constraints are predicates which must be true for data stored in the:
A. system
B. hardware system
C. database
D. None of the above

79. In a distributed DBMS one may add a third factor which is the transmission time in the:
A. relation
B. system
C. network
D. None of the above

80. A transaction is the unit of concurrency and:
A. system
B. recovery control
C. software system
D. None of the above

81. A DBMS allows data manipulation through primitive operations which can be grouped to form:
A. data transactions
B. software system
C. hardware system
D. None of the above

82. The INSERTION clause specifies the storage method as either AUTOMATIC or:
A. DBMS
B. MANUAL
C. CODASYL
D. None of the above

83. The RETENTION clause specifies the degree of freedom available to the members of a:
A. database
B. system
C. link
D. None of the above

84. The SET SELECTION clause indicates how to access to set:
A. occurrence
B. a system
C. a software
D. None of the above

85. Once complied by the DBMS, the database scheme will be used to create and to control:
A. system
B. data storage
C. a link
D. None of the above

86. The way, the occurrence of an owner is linked to the occurrence of its:
A. relation
B. system
C. members
D. None of the above

87. The SOCRATE/CLIODDL can be used to define
A. records and file:
B. relation
C. schemes
D. None of the above

88. In SOCRATE/CLIODDL one must specify the maximum numbers of occurrences for each:
A. link
B. entity type

C. scheme
D. None of the above

89. An entity defines a block structure similar to:
A. hierarchical structure
B. Pascal's record constructor
C. scheme
D. None of the above

90. An entity type may contain the declaration of another entity type, and so on, in order to define:
A. Pascal's record file
B. hierarchical structure
C. database
D. None of the above

91. Important point concerning the hierarchical model is the physical implementation of the:
A. system
B. software system
C. evaluated trees
D. None of the above

92. The data model proposed by E.F. Codd in 1970 is called the:
A. Codd Model
B. n-ary relational data model
C. network model
D. None of the above

93. The basic idea is to consider a database as a collection of time varying tables called:
A. relations
B. network model
C. Pascal's structure
D. None of the above

94. The relational algebra is a set of operators which can be used to define new relations from:
A. the old ones
B. existing ones
C. system
D. None of the above

95. Relational algebra can be used to express:
A. relations
B. system
C. queries and up dates
D. None of the above

96. SQL is a relational language which has also been implemented in several:
A. DBMSs
B. ECA
C. QUEL
D. None of the above

97. Relational algebra is one of the basic concepts for defining:
A. high-level language
B. relational language
C. machine language
D. None of the above

98. SQL is neither a pure algebraic language nor a:
A. machine language
B. assembly language
C. predicative language
D. None of the above

99. Codd shows that relational algebra is equivalent to first order:
A. differential equation
B. predicate calculus
C. differential calculus
D. None of the above

100. In SQL the statement Expand Table can be used at any time:
A. old relation
B. given relation
C. combination
D. None of the above

101. Use of microcomputer as a controller allows:
A. faster cycle time
B. simpler mechanical hardware
C. self diagnosis of the machine operation
D. all of the above

102. Integration of CAD and CAM is known as:
A. CIM
B. CIE
C. CAM alone
D. CAD alone

103. Which of the following is the positive feedback device on NC machine?
A. Linear scale
B. Shaft encoder
C. Inductosyns
D. Any of the above

104. Which of the following transmission systems provides the highest data rate to an individual device?
A. Digital PBX
B. Computer bus

C. LAN
D. Voice band modem

105. The geostationary satellite used for communication systems:
A. rotates with the earth
B. remains stationary relative to the earth
C. is positioned over equator
D. All of the above

106. A light emitting diode can couple into an optical fiber power of the order of:
A. 10 picowatts
B. 100 microwatts
C. 10 milliwatts
D. 10 watts

107. Repeated execution of simple computation may cause compounding of:
A. round-off errors
B. syntax errors
C. run-time errors
D. logic errors

108. If a graph requires *k* different colours for its proper colouring then the chromatic number of the graph is:
A. $k/2$
B. $(k-1)$
C. k
D. 1

109. A complete graph with five vertices is:
A. non-planar
B. planar
C. a non-regular graph
D. a tree

110. A circuit is a connected graph which includes every vertex of the graph is known as:
A. Eular
B. Unicursal
C. Hamiltonian
D. Clique

111. If there exists at least one path between any pair of vertices in a graph, the graph is known as:
A. complete graph
B. disconnected graph
C. connected graph
D. Eular graph

112. The maximum degree of any node in a simple graph with n vertices is:
A. $(n-1)$
B. n
C. $n/2$
D. $(n-2)$

113. A characteristic of the data which binary search uses but the linear search ignores is the:
A. order of the list
B. length of the list
C. maximum value in the list
D. mean of data values

114. The order of magnitude of the worst case performance of the linear search ever N elements is:
A. $N \log_2 N$
B. N
C. N^2
D. $\log_2 N$

115. In what order the elements of a pushdown stack are accessed?
A. First In First Out (FIFO)
B. Last In Last Out (LILO)
C. Last In First Out (LIFO)
D. None of the above

116. Which of the following sorting procedure is the slowest?
A. quick sort
B. heap sort
C. shell sort
D. bubble sort

117. A linear list in which elements can be added or removed at either end but not in the middle, is known as:
A. queue
B. deque
C. stack
D. tree

118. The following pass of an absolute loader is used to make symbol table:
A. first
B. second
C. Both first and second
D. third

119. A linker:
A. is a software program
B. combines part of a program
C. is a program used to load monitor into main memory
D. Both (A) and (B)

120. The following translates source programs into object programs:
A. assembler
B. interpreter
C. compiler
D. absolute assembler

121. The instructions of high level language are:
A. deferred instruction
B. macro-instruction
C. micro-instruction
D. mnemonic instruction

122. A bus connected between the CPU and the main memory that permits transfer of information between main memory and the CPU is know as:
A. DMA bus
B. memory bus
C. address bus
D. control bus

123. The device which is used to connect a peripheral to bus is known as:
A. control register
B. interface
C. communication protocols
D. None of the above

124. The following is used to describe the theory of control and communication in the machine and the animal:
A. inference machine
B. cybernatics
C. intelligent terminals
D. Heuristic

125. The following are used for graphical representation of the definition, analysis using symbols to represent operations, data flow and equipment:
A. algorithms
B. flow charts
C. HIPO packages
D. All of the above

126. In which addressing mode the contents of a register specified in the instruction are first decremented and then these contents are used as the effective address of the operands?
A. Index addressing
B. Indirect addressing
C. Auto-increment
D. Auto-decrement

127. System program such as compilers are designed so that they are:
A. re-enterable
B. non-reusable
C. serially usable
D. recursive

128. Which of the following operating systems uses write through catches?
A. UNIX
B. DOS
C. ULTRIX
D. XENIX

129. The operating system manages:
A. memory
B. processor
C. disk and I/O devices
D. All of the above

130. An OP code:
A. translates a mnemonic
B. instructs the CPU
C. stores data
D. All of the above

131. The ability to temporarily halt the CPU and use this time to send information on buses is called:
A. direct memory access
B. vectoring the interrupt
C. polling
D. cycle stealing

132. An interrupt which can be temporarily ignored by the counter is known as:
A. vectored interrupt
B. non-maskable interrupt
C. maskable interrupt
D. low priority interrupt

133. A bus connected between the CPU and main memory that permits transfer of information

between main memory and the CPU is known as:
A. DMA bus
B. memory bus
C. address bus
D. control bus

134. The following consists of an array of gates which can be used to patch an error in a ROM or alter the meaning of an instruction:
A. RAM
B. EAPROM
C. PLA
D. PSW

135. The three buses associated with a three-bus system are I/O bus, memory bus and the:
A. address bus
B. unibus
C. direct memory access bus
D. data bus

136. The following register is used as a working area in CPU:
A. program counter
B. instruction register
C. instruction decoder
D. accumulator

137. A technique by which more than one program can reside in primary storage at the same time and be executed concurrently is known as:
A. multiprogramming
B. multiprocessing
C. distributed processing
D. None of the above

138. The system which permits a large number of users at various remote terminals to simultaneously use a centrally located computer is known as:
A. servomechanism system
B. time sharing system
C. mainframe system
D. None of the above

139. Which of the following buses is required to interconnect all functional elements in a computer?
A. Data bus
B. Address bus
C. Control bus
D. All of the above

140. The ability of a medium sized computer to increase in data processing capability by addition of such devices as mass storage device. I/O devices, etc., is known as:
A. computer expandability
B. computer mobility
C. computer enhancement
D. computer upward compatability

141. The unit of a computer system which executes program, communicates with and often controls the operation of other subsystems of the computer is the:
A. CPU
B. control unit
C. I/O unit
D. peripheral unit

142. The register which contains the data to be written into or read out of the addressed location is known as:
A. index register
B. memory address register
C. memory data register
D. program counter

143. The register which holds the address of the location to or from which data are to be transferred is known as:
A. index register
B. instruction register
C. memory address register
D. memory data register

144. Which of the following is the fastest?
A. CPU
B. Magnetic tapes and disks
C. Video terminals
D. Sensors, mechanical controllers

145. The heart of any computer is the:
A. CPU
B. memory
C. I/O units
D. disks

146. Hard copy of a document can be obtained from:
A. card reader
B. CRT
C. Laser printer
D. paper tape

147. The hardware unit which is used to monitor computer processing is:

A. console

B. dot matrix printer

C. mouse

D. ROM

148. Which of the following storage devices can be carried around?

A. Floppy disks

B. Main memory

C. Registers

D. Core memory

149. Which of the following computers is least powerful?

A. Minicomputer

B. Microcomputer

C. Mainframe computer

D. Supercomputer

150. The control unit of a computer:

A. perform arithmetic and logical operations on the data

B. controls the operation of output devices

C. is a device for manually operating the computer

D. directs the other units of the computer

ANSWERS

1	**2**	**3**	**4**	**5**	**6**	**7**	**8**	**9**	**10**
C	B	B	C	A	B	B	C	B	C
11	**12**	**13**	**14**	**15**	**16**	**17**	**18**	**19**	**20**
B	C	B	C	C	B	B	B	A	B
21	**22**	**23**	**24**	**25**	**26**	**27**	**28**	**29**	**30**
B	C	C	A	B	C	B	B	A	C
31	**32**	**33**	**34**	**35**	**36**	**37**	**38**	**39**	**40**
B	B	B	C	A	A	B	C	B	A
41	**42**	**43**	**44**	**45**	**46**	**47**	**48**	**49**	**50**
B	C	C	A	B	A	C	A	B	C
51	**52**	**53**	**54**	**55**	**56**	**57**	**58**	**59**	**60**
B	A	B	B	B	A	B	C	B	A
61	**62**	**63**	**64**	**65**	**66**	**67**	**68**	**69**	**70**
B	A	B	C	B	B	C	B	A	C
71	**72**	**73**	**74**	**75**	**76**	**77**	**78**	**79**	**80**
B	C	C	B	C	A	B	C	C	B
81	**82**	**83**	**84**	**85**	**86**	**87**	**88**	**89**	**90**
A	B	B	A	C	C	A	B	B	B
91	**92**	**93**	**94**	**95**	**96**	**97**	**98**	**99**	**100**
C	B	A	B	C	A	B	C	B	B
101	**102**	**103**	**104**	**105**	**106**	**107**	**108**	**109**	**110**
D	A	D	B	D	B	A	C	A	C
111	**112**	**113**	**114**	**115**	**116**	**117**	**118**	**119**	**120**
C	A	A	B	C	D	B	A	D	D
121	**122**	**123**	**124**	**125**	**126**	**127**	**128**	**129**	**130**
C	B	B	B	B	D	A	B	D	B
131	**132**	**133**	**134**	**135**	**136**	**137**	**138**	**139**	**140**
D	C	B	C	C	D	A	B	D	A
141	**142**	**143**	**144**	**145**	**146**	**147**	**148**	**149**	**150**
A	C	C	A	A	C	A	A	B	D

MICROPROCESSORS

MULTIPLE CHOICE QUESTIONS

1. The purpose of the TLB located within the microprocessor is:
   A. to store the last 22 linear-to-physical address translations from the paging unit
   B. to store the first 22 linear-to-physical address translations from the paging unit
   C. to store the last 22 paging unit address translations from linear-to-physical
   D. All of the above
2. If the microprocessor sends linear address 0020000H to the paging mechanism, then the paging directory entry and page table entry accessed are:
   A. Page directory H000 and page table entry 200H
   B. Page directory H000 and page table entry H002
   C. Page directory 000H and page table entry 200H
   D. None of the above
3. 32-bit addresses stored in the page directory are:
   A. 1000
   B. 1200
   C. 1020
   D. 1024
4. Bytes found in a memory page are:
   A. 8K bytes
   B. 4K bytes
   C. 3K bytes
   D. 2K bytes
5. The program-invisible registers are:
   A. the cache portion of the segment register, the task register, and the descriptor table register
   B. the segment regiser, the task register, and the table register
   C. the cache portion of the segment register, the task register, and the table register
   D. All of the above
6. In the memory system, the local descriptor table addressed through:
   A. the descriptor table register
   B. the local descriptor table register
   C. the table register
   D. None of the above
7. A descriptor code that describes a memory segment that begins at location 03000000H and ends at location 05FFFFFFH (this memory segment is a data segment that grows upward in the memory system and can be written. The descriptor is for an 80386 microprocessor).

A.

00	30
03	01
00	00
01	10

B.

00	30
00	09
10	00
00	30

C.

03	10
09	00
00	00
30	00

D.

00	03
03	10
20	11
03	02

8. The maximum length of the global descriptor table in the Pentium microprocessor is
A. 32 K bytes
B. 64 K bytes
C. 128 K bytes
D. 256 K bytes

9. If DS = 0103H, in a protected mode system, the requested privilege lend is:
A. 3
B. 4
C. 8
D. 12

10. For a Pentium descriptor that contain a base address of 0100000H, a limit of 0FFFH, and G = 0, then the starting and ending locations addressed by this descriptor are:
A. H00000028 and 00290000H
B. 00280000H and 29000000H
C. 00280000H and 00290000H
D. 28000000H and 00290000H

11. The purpose of the segment register in protected mode memory addressing is:
A. to choose a descriptor from either the local or global descriptor table
B. to choose a register from a local descriptor
C. to choose a register from a descriptor from either the local or global descriptor
D. None of the above

12. The area of the memory in the 80286 microprocessor is allowed to access by protected mode memory addressing is any location between and including:
A. 0000H00 – FFFFFFH
B. 000H000 – FFFFFFH
C. 000000H – FFFFFFH
D. 000000H – FFFFFHF

13. The memory location address by SS = 2900H and SP = 3A00H real mode 80286 register combination is:
A. HCA002
B. 2AC00H
C. CA00H2
D. 2CA00H

14. The memory location address by DS = 2000H and SI = 1002H real mode 80286 register combination is:
A. 2100H0
B. 21003H
C. 21002H
D. 2100H2

15. The stack memory is addressed by a combination of which one of the following:
A. stack memory and stack segment
B. stack segment plus stack offset
C. stack memory plus stack offset
D. stack offset plus stack memory

16. Which register or registers are used as an offset address for string instruction destinations in the 80486 microprocessor?
A. DS : SI and ED : SI
B. ED : DS and DS : SI
C. ES : DI and DS : SI
D. ED : SI and DS : SI

17. Find the memory address of the next instruction executed by the microprocessor, when operated in the real mode CS = 1000H and IP = 2000H combinations:
A. 1000H2
B. 1200H0
C. 10002H
D. 12000H

18. Find the memory address of the next instruction executed by the microprocessor, when operated in the real mode for CS = 3456H and IP = ABCDH
A. 3F12H0
B. 3F120H
C. 4F230H
D. 8F120H

19. The purpose of a segment register in the real mode operation of the microprocessor is:
A. to locate the starts of a 64K memory segment
B. to locate the end of a 64K memory segment
C. to locate the middle of a 64K memory segment
D. None of the above

20. The flag bit controls the INTR pin on the microprocessor is:
A. the interrupt flag (IV)
B. the interrupt flag (III)
C. the interrupt flag (II)
D. the interrupt flag (I)

21. The purpose of the IP/EIP register is:
A. to locate the next instruction in a program
B. to locate the previous Instruction in a program
C. to locate the current instruction in a program
D. None of the above

22. The extended BX register is addressed as:
A. BXH
B. BXE
C. HBX
D. EBX

23. The 80286 addresses registers are:
A. 16 and 32 bit wide
B. 16 and 64 bit wide
C. 8 and 16 bit wide
D. 8 and 32 bit wide

24. The BOUND instruction interrupt a program when:
A. the boundary is outside of the boundary set by a pair of words or doublewords stored in the memory
B. the boundary is inside of the boundary set by a pair of words or double words stored in the memory
C. the boundary is inside of the boundary set by a word stored in the memory
D. None of the above

25. The personal computer interrupt services the keyboard is:
A. interrupt vector number 8
B. interrupt vector number 9
C. interrupt vector number 0
D. interrupt vector number 2

26. The instructions control the function of the INTR pin is:
A. STI
B. TSI
C. CLI
D. LCI

27. The INTO instruction only interrupts the program for:
A. if the overflow flag is added
B. if the overflow flag is removed
C. if the overflow flag is set
D. if the overflow flag is closed

28. The IRET instruction differ from the RET instructions by which one of the following:
A. IRET pops the flags. RET does not
B. IRET pops the return address, RET does not
C. IRET pops the flags, as RET does
D. None of the above

29. The purpose of interrupt vector type number 0 is:
A. not used with NMI pin
B. used with the NMI pin
C. used with markable interrupt pin
D. None of the above

30. The software instructions call an interrupt service procedure is:
A. INT
B. NMI
C. RET
D. INTR

31. The last executable instruction in a procedure must be a:
A. INT
B. RET
C. MOV
D. EDI

32. The 80486 LOOPD instruction decrements register .... and tests it for a 0 to decide if a jump occurs.
A. INT
B. RET
C. ECX
D. JMP

33. The SET instruction is used to set AL, if the flag bits indicate a zero condition is:
A. JCXZ
B. LOOP
C. DATA
D. SETZ

34. The conditional jump instructions test both the Z and C flag bits are:
A. JZT and JA
B. PWD and JA
C. JBE and JA
D. RTX and JA

35. The conditional jump instructions follow the comparison of signed number are:
A. JE, JNE, JG, JGE, JL and JLE
B. NT, NRT, GJ, JGE, LJ and NST
C. JE, JNE, NT, JGE, JL and JLE
D. JE, JNE, GJ, JGE, JL and JLE

36. The near jump modifies the program address by changing which one of the following register or registers?
  A. JMP/MP
  B. EIP/IP
  C. JNE/NE
  D. JGE/GE
37. Which one of the following type of JMP instruction assembles if the distance is 0020H bytes?
  A. Near
  B. Short
  C. Far
  D. None of the above
38. The JMP instruction, five bytes long is:
  A. A big JMP
  B. A short JMP
  C. A near JMP
  D. A far JMP
39. Which one of the following type of JMP is used when jumping to any location within the current code segment?
  A. a short JMP
  B. a near JMP
  C. a far JMP
  D. All of the above
40. What files are always generated from the source file TEST . ASM if it is processed by MASM?
  A. TEST . OBJ
  B. TEST . LST
  C. TEST . CRF
  D. None of the above
41. What does the PUBLIC directive indicate when placed in a program module?
  A. PUBLIC is used to declare that a label or even an entire segment is public to other modules.
  B. Public is used to declare that an entire segment is non-public to other modules.
  C. Public is used to declare that a labelled or even an entire segment is non-public to other modules
  D. None of the above
42. The directive appear with labels defined external are:
  A. FAR PTR, SHORT PTR, BYTE, WORD, or DWORD
  B. NEAR PTR, FAR PTR, BYTE, WORD, or DWORD
  C. SHORT PTR, NEAR PTR, BYTE, WORD, or DWORD
  D. None of the above
43. The assembler language directive delineate a macro sequence are:
  A. the MACRO directive starts a macro, and the ENDM ends it
  B. the ENDM directive starts a endm and the macro end it
  C. the macro directive starts a endm and the endm ends it
  D. All of the above
44. By which one of the following method the parameters transferred to a macro-sequence?
  A. Parameters are indicated next to the MACRO statement and then are placed in a program next to the name of the macro as operands
  B. Parameter are indicated next to the ENDUM statement and then are placed in a program next to the name of the macro as operands
  C. Parameters are indicated next to the NTRM statement and then are placed in a program next to the name of the macro as operands
  D. None of the above
45. By which of the following method the LOCAL directive used within a macro sequence?
  A. The LOCAL directive must immediately follow the MACRO statement and must contain all labels used within the macro-sequence
  B. The LOCAL directive must immediately follow the ENDM statement and must contain all labels used within the macro-sequence
  C. The LOCAL directive must immediately follow the MACRO statement and must contain all labels used within the endm sequence
  D. None of the above
46. Which one of the following INT instruction is used to access the mouse?
  A. INT 46 M
  B. INT 46 H
  C. INT 33 M
  D. INT 33 H

47. By which one of the following method, it is determined that the mouse is a serial or a bus mouse?
   A. If AH loaded with a 24H before INT 46H executes, the mouse type is returned in CH
   B. If AH loaded with a 24H before INT 33H executes, the mouse type is returned in CH
   C. If INT loaded with a 24H before AH 33H executes, the mouse type is returned in CH
   D. None of the above

48. The mouse must be disabled when data are displayed in the video display because:
   A. the mouse driver can at worst lock up the computer
   B. the mouse drive cannot at worst lock up the computer
   C. the mouse driver cannot generate a binary value
   D. None of the above

49. By which one of the following method a large number (over 100 decimal) converted from binary to BCD?
   A. By repeatedly dividing by 2
   B. By repeatedly dividing by 6
   C. By repeatedly dividing by 8
   D. By repeatedly dividing by 10

50. An ASCII coded number is converted to BCD by subtracting:
   A. 12H
   B. 16H
   C. 30H
   D. 32H

51. By which one of the following method a three digit ASCII coded number is converted to binary?
   A. A 30H is first subtracted from each digit, then most-significant digit is multiplied by 100 and the middle digit is multiplied by 10. The three are then summed to generate a binary value
   B. A 30H is first added to each digit, then most significant digit is multiplied by 100 and the middle digit is multiplied by 10. The three are then summed to generate a binary value
   C. A 30H is first subtracted from each digit, then most significant digit is divided by 100 and the middle digit is multiplied by 10. The three are then summed to generate a binary value
   D. A 30H is first subtracted from each digit, then most significant digit is multiplied by 100 and the middle digit is divided by 10. The three are then summed to generate a binary value

52. By which one of the following it can be shown to use the XLAT instruction to access a looking table called LOOK that is located in the stack segment?
   A. LOOK XLAT : SS
   B. SS XLAT : LOOK
   C. XLAT SS : LOOK
   D. LOOK SS : XLAT

53. The purpose of a Boot sector, FAT, and root directory are:
   A. Boot sector contains a program that loads DOS into the memory system. The FAT holds information that indicates which clusters are in use and which are free. The root directory is the main directory on the disk drive
   B. Boot sector contains a program that loads DOS into the operating system. The FAT holds information that indicates which cluster are in use and which are free. The root directory is the main directory on the disk drive
   C. Boot sector contain a program, the FAT holds cluster and the root directory is the disk drive
   D. None of the above

54. Which of the following is not an 8-bit microprocessor?
   A. INTEL 8085
   B. MOTOROLA 6800
   C. ZILOG Z80
   D. FAIR CHILD 9440

55. The bootstrap loader is:
   A. Program
   B. DOS
   C. Window
   D. Modem

56. The bootstrap found in:
   A. DOS
   B. Memory

C. Boot sector
D. None of the above

57. A directory entry contains an attribute byte. The following information are indicated by this byte:
1. directory name
2. file name
3. volume
A. only (1)
B. only (2)
C. (2) and (3)
D. All of the above

58. The maximum length of a file is:
A. 4G bytes
B. 5G bytes
C. 8G bytes
D. 9G bytes

59. The information appears on the address/data bus of the 8088 while ALE is active, is:
A. address bit A7 – A0
B. data bit A7 – A8
C. address bus A7
D. None of the above

60. A logic 0 on the 8086/8088 $\overline{RD}$ pin indicate:
A. memory
B. I/O
C. memory and I/O
D. the microprocessor is either reading data from memory or I/O

61. The signal that is applied to the CLK input pin of the 8086/8088 microprocessor must be:
A. a TTL.– compatible square - wave
B. a TCL – compatible square - wave
C. a TTK – compatible square - wave
D. None of the above

62. When DT/$\overline{R}$ is a logic 1, then the conditions about the operation of the 8086/8088 indicate that the microprocessor's data bus is transmitting data to:
A. the memory or I/O
B. DOS
C. Book Sector
D. All of the above

63. What three minimum mode 8086/8088 pins are decoded to discover if the processor is halted?
A. $\overline{S0}$ – $\overline{S0}$
B. $\overline{S2}$ – $\overline{S2}$
C. $\overline{S2}$ – $\overline{H2}$
D. $\overline{S2}$ – $\overline{S0}$

64. The condition that the QS_1 and QS_0 pins indicate about the 8086/8088 is
A. of the queue within the microprocessor for the arithmetic co-processor
B. of the queue with in the Boot sector for the algebric co-processor
C. of the queue within the DOS for the trigonometric co-processor
D. None of the above

65. By what factor does the 8284A clock generator divide the crystal oscillator's output frequency?
A. Two
B. Three
C. Four
D. Five

66. If the crystal oscillator is operating at 14 MHz then PCLK output of the 8184A is:
A. 5.33 MHz
B. 3.33 MHz
C. 2.33 MHz
D. 1.33 MHz

67. Which bus connections on the 8086 microprocessor are typically demultiplexed?
A. AD1 – AD0
B. AD9 – AD1
C. AD0 – AD9
D. AD19 – AD0

68. Which TTL integrated circuit is often used to demultiplex the buses on the 8086/8088?
A. An 8-bit transparent latch (74LS373)
B. An 2-bit transparent latch (74LS373)
C. An 8-bit latch (74LS373)
D. An 2-bit latch (74LS373)

69. The buffers are often required in an 8086/8088 based system because:
A. the drive current (2.0 mA) available at the output pins of the microprocessor
B. the drive current (2.0 mA) available at the input pins of the microprocessor
C. the drive current (8.0 mA) available at the input pins of the microprocessor
D. None of the above

70. A bus cycle is equal to clocking .... periods.
A. one
B. two

C. three
D. four

71. The two 8086/8088 operations occurs during a bus cycle are:
A. a read or a write
B. a read only
C. a write only
D. None of the above

72. The purpose of the BS pin on the 82C08 is to select:
A. a bank of memory
B. a Boot sector
C. a program
D. None of the above

73. If a IM × IDRAM requires 4 ms for a refresh and has 256 rows to be refreshed, no more than .... of time must pass before another row is refreshed.
A. 10.25 μs
B. 16.50 μs
C. 15.625 μs
D. 12.625 μs

74. The $\overline{RAS}$ cycle selects:
A. a new data
B. a new software
C. a old row address
D. a new row address

75. If A_0 is a logic 0, then the .... memory bank is selected.
A. high
B. low
C. middle
D. end

76. The two methods are used to select the memory in the 8086 microprocessor are:
A. separate decoders and separate control signals
B. separate data and separate boot
C. separate decoders and separate boot
D. separate boot and separate signals

77. The $\overline{BHE}$ on the 8086 microprocessor selects:
A. upper memory bank
B. lower memory bank
C. upper signal
D. lower signal

78. The A_0 pins on the 8086 microprocessor selects:
A. upper memory bank
B. lower memory bank
C. upper signal
D. lower signal

79. The 74LS636 will correct .... bits that are in error.
A. four
B. three
C. two
D. one

80. The 74LS 636 error correction and detection circuits stores a check code with each byte of data. The number of bits stored for the check code are:
A. 7
B. 6
C. 5
D. 4

81. A PROM address decoder often found in a memory system because it is more suited to:
A. memory address
B. 74LS138
C. DOS
D. None of the above

82. The time required to refresh the typical DRAM is:
A. 4-6 ms
B. 2-6 ms
C. 2-4 ms
D. None of the above

83. The address inputs to most DRAM devices are:
A. multiplexed
B. de-multiplexed
C. accessed data
D. None of the above

84. The memory access time required by the slowest 4016 is:
A. 125 ns
B. 250 ns
C. 500 ns
D. 750 ns

85. The SRAM is a device that retains data for as long as power is applied to the ....
A. memory device
B. Boot sector

C. I/O device
D. None of the above

86. The 5 MHz version of the 8088 allows 460 ns of time for the memory to access data. A 450 ns memory device will only function if the amount of time required for the address decoder and buffers in a system is ....
A. more than 20 ns
B. less than 20 ns
C. less than 10 ns
D. more than 10 ns

87. The purpose of the $\overline{WE}$ pin on a RAM is:
A. to perform a write operation, provided the $\overline{CS}$ or $\overline{CE}$ pin is also active
B. to perform a write operation, but not provided the $\overline{CS}$ or $\overline{CE}$
C. to perform a write operation
D. None of the above

88. The number of words found in each memory device for 13 address connections:
A. 256
B. 2,048
C. 4,096
D. 8,192

89. The purpose of the INTR pin on the ADC080X is to indicate that the converter has completed ....
A. a conversion
B. a operation
C. Both (A) and (B)
D. None of the above

90. If the reference voltage is –2.55V, then the step voltage at the output of the DAC0830 is:
A. 0.001V
B. 0.1V
C. 0.01V
D. 1.0V

91. The asynchronous serial data are data that are sent .... a clock pulse.
A. with
B. without
C. Both (A) and (B)
D. None of the above

92. If a 16-bit count is programmed into the 8254, then the byte of the count programmed first is:
A. least-significant byte
B. maximum-significant byte
C. Both (A) and (B)
D. None of the above

93. The 8254 internal timer functions from DC to .... Hz.
A. 16 MHz
B. 15 MHz
C. 12 MHz
D. 10 MHz

94. An overrun error occurs if the internal .... fills before the data are input to the microprocessor.
A. DRAM
B. FIFO
C. PCO
D. None of the above

95. The number of wait states are required to interface the 8279 to the 8086 microprocessor operating with an 8 MHz clock is
A. 1
B. 2
C. 3
D. 4

96. The time usually used to de-bounce a keyboard is
A. 20 ms to 15 ms
B. 20 ms to 25 ms
C. 10 ms to 20 ms
D. 10 ms to 15 ms

97. The general purpose I/O pins during mode 2 operation of the 82C55 is:
A. PC0 – PC2
B. PC2 – PC1
C. PC3 – PC2
D. PC2 – PC3

98. The two 82C55 pins accomplish internal I/O port address selection are:
A. A2 and A0
B. A0 and A2
C. A2 and A1
D. A1 and A0

99. The number of programable I/O pin connection in 82C55 have:

A. 20

B. 25

C. 24

D. 36

100. An 8-bit I/O device, located at I/O port address 0010H, is connected to which one of the following bus connections?

A. D6 – D0

B. D7 – D0

C. D8 – D0

D. D7 – D8

ANSWERS

1	2	3	4	5	6	7	8	9	10
A	C	D	B	A	B	C	B	A	C
11	**12**	**13**	**14**	**15**	**16**	**17**	**18**	**19**	**20**
A	C	D	C	B	C	D	B	A	D
21	**22**	**23**	**24**	**25**	**26**	**27**	**28**	**29**	**30**
A	D	C	A	B	A	C	A	B	A
31	**32**	**33**	**34**	**35**	**36**	**37**	**38**	**39**	**40**
B	C	D	C	A	B	B	D	B	A
41	**42**	**43**	**44**	**45**	**46**	**47**	**48**	**49**	**50**
A	B	A	A	A	D	B	A	D	C
51	**52**	**53**	**54**	**55**	**56**	**57**	**58**	**59**	**60**
A	C	A	D	A	C	D	A	A	D
61	**62**	**63**	**64**	**65**	**66**	**67**	**68**	**69**	**70**
A	A	D	A	B	C	D	A	A	D
71	**72**	**73**	**74**	**75**	**76**	**77**	**78**	**79**	**80**
A	A	C	D	B	A	A	B	D	D
81	**82**	**83**	**84**	**85**	**86**	**87**	**88**	**89**	**90**
A	C	A	B	A	C	A	D	A	C
91	**92**	**93**	**94**	**95**	**96**	**97**	**98**	**99**	**100**
B	A	D	B	B	C	A	D	C	B

YOUR SPACE

ENGLISH LANGUAGE

1. Comprehension Passages

ENGLISH LANGUAGE COMPREHENSION

The objective of language comprehension test is to ascertain the ability of the candidates to understand the passage properly. Therefore candidates are required to take notice of the following points:

1. Read the full passage very attentively and intelligently.
2. Try to comprehend the gist of it.
3. Make a mental note of all the important details and points given in the passage.
4. Read the passage for the second time in case you have not been able to understand it satisfactorily.
5. Divide the time proportionately for all the passages.
6. Answer the questions on the basis of facts, as given in the paragraph.
7. Don't waste much time in answering the questions of any one passage.
8. Check all the answers once again, very carefully, to see whether any question is left unanswered by mistake.

MODEL QUESTIONS (FOR PRACTICE)

Directions: *Each of the following passages is followed by five questions. Read the passage carefully and then answer the questions that follow each. For each question, four probable answers A, B, C and D are given. Only one out of these is correct. Choose the correct answer.*

PASSAGE-1

The use of words like 'welcome', 'thank you', 'please', etc., at the right moment reflects a polite nature. The civic sense also lies within the scope of good manners. We should not shout or talk loudly in public places like hospitals and libraries and create disturbance. We should not cheat people or make fun of them. Cleanliness is also necessary. We must not throw the waste on roads and make use of dustbins. We should not harm the public property as it belongs to all of us. While in a queue, discipline should be maintained. We must give fair chance to others.

1. Expressions like 'welcome' 'thank you' and 'please' reflect
   A. happiness B. discipline
   C. civic sense D. polite nature
2. While in a library, we should
   A. respect others B. avoid arguments
   C. talk in low tone D. be courteous
3. A public property belongs to
   A. nobody
   B. all of us
   C. government
   D. one who maintains it

4. Discipline is
   A. the rule of proper conduct or action
   B. the rule of road sense
   C. making use of dustbins
   D. forming a queue
5. The most appropriate title for this passage would be
   A. Polite Nature
   B. Courtesy
   C. Good Manners
   D. Civic Sense

PASSAGE-2

There is an old proverb 'Early to bed and early to rise makes a man healthy and wise.' I am in the habit of getting up early in the morning and have formed the habit of taking long morning walks in the past two years. It is a light exercise and best for physical fitness. The morning air which is fresh and pure is beneficial for the lungs. The early rays of the rising sun are good for healthy skin. 'Health is wealth' and doctors also recommend morning walk to their patients for gaining sound health and freshness of energy.

1. What is good for lungs?
   A. Sunrays B. Fresh air
   C. Sound sleep D. Light exercise
2. What is a light exercise?
   A. Early to bed
   B. Early to rise
   C. Morning walk
   D. Gaining sound health
3. What is good for skin?
   A. Fresh air
   B. Morning air
   C. Morning walk
   D. Rising sun's rays
4. What is best for physical fitness?
   A. Light exercise
   B. Long morning walk
   C. Early to rise
   D. Fresh and pure air
5. Long morning walk
   A. bring sound sleep
   B. ensures physical fitness
   C. ensures healthy skin
   D. keeps healthy, wealthy and wise

PASSAGE-3

Mahatma Gandhi lived a splendid long life and has set great moral standards before us. He showed to the world the true way to peace. He wished to see India prosper but he became a martyr for the noble cause of Hindu-Muslim unity at the time of partition when a religious fanatic, Nathuram Godse, shot him dead on January 30, 1948. His last words were 'Hey Ram'. He lived and died for his country and countryman.

1. Mahatma Gandhi showed the world the true way to
   A. prosperity B. love
   C. truth D. peace
2. Mahatma Gandhi became a martyr for the noble cause of
   A. truth
   B. non-violence
   C. freedom of India
   D. Hindu-Muslim unity
3. Mahatma Gandhi was shot dead
   A. before India achieved independence
   B. by a mad man
   C. by an intolerant religious person
   D. by a non-religious person
4. Mahatma Gandhi set great moral standards. It means
   A. he was a great religious teacher
   B. he was a great moralist
   C. he made India morally stronger
   D. moral was everything to him
5. Gandhiji lived and died for his country and countryman. It means
   A. he was born in India and died in India
   B. he was a patriot
   C. he was a great moralist
   D. he sacrified his life for India and her people

PASSAGE-4

On one hot day a crow felt very thirsty. He flew from one place to another in search of water. After long hours of labour he found a pitcher. Eagerly, he perched on the mouth of the pitcher. He found that the water was at the bottom of the vessel. He tried his best to dip his beak but did not succeed. He did not know what to do. Suddenly some pebbles lying nearby gave him an idea. One by one he dropped the pebbles with his beak into the pitcher. The level of water slowly came up to the mouth of the pitcher. The crow then drank the water and quenched his thirst.

1. The crow found a pitcher
   A. as it flew
   B. after many hours of labour
   C. full of water
   D. which was empty

2. What is the moral of the passage?
   A. No pains, no gains
   B. God helps those who help themselves
   C. Necessity is the mother of invention
   D. Try and try again, you will succeed at last

3. The crow flew from place to place
   A. in search of pitcher
   B. in search of pebbles
   C. in search of water
   D. in search of a vessel

4. The pitcher, the crow found
   A. was full of water
   B. was dry
   C. had little water in the bottom
   D. had water up to its mouth

5. As the crow dropped pebbles into the pitcher, what happend?
   A. The pitcher broke down
   B. The water leaked one of the pitcher
   C. The level of water into the pitcher rose up slowly
   D. Water level immediately rose to the mouth of the pitcher

PASSAGE-5

Once upon a time a crane and a fox lived in a forest. They were good friend. One day the fox invited the crane to a feast. He made a tasty food and served it before the crane on a plate. The crane could not eat anything because of the long beak. But the fox licked all his food. The crane felt insulted. He decided to teach the fox a lesson. Next day he invited the fox. He prepared the same tasty food and placed it in front of the fox inside a narrow glass. The crane ate easily while the fox looked on. Now, it was the fox's turn to remain hungry.

1. What is the moral of the passage?
   A. Beware of the wicked
   B. One good turn deserves another
   C. Be contented with what you have
   D. Tit for tat

2. The crane could not eat tasty food because the
   A. food was served in a shallow plate
   B. food was very hot
   C. food was served in a long jar
   D. crane was not hungry

3. The fox had to remain hungry because
   A. the food served was not enough in quantity
   B. the food was served inside a narrow glass
   C. the food served was not tasty
   D. the food was all liquid

4. Why did the crane feel insulted?
   A. Because he was invited to feast but he could not eat anything
   B. Because the food was served in a shallow plate and he could not eat
   C. Because the food was too hot
   D. Because the fox gulped all the food quickly

5. The crane successfully taught a lesson to the fox when he invited the fox to a feast and served the food
   A. in a narrow glass
   B. in a large plate
   C. in a broken plate
   D. in a long jar

PASSAGE-6

The family set down at the table and began to talk about the summer holidays. They had to decide a place to visit during the vacation. Should they go to their village or to a hill station? The parents preferred the village while the children wished to go the hill station. After few moments of discussion the elders decided to visit both the places. First they shall go to the village for a week and then stay at the hill station for the remaining days. For the first time the family shall be together during the holidays. The children were happy with the holiday plan.

1. The purpose for which the family set down at the table was
   A. to decide a place to visit during the vacation
   B. to educate the children how to carry articles during a visit to a hill station
   C. to decide the date when they should start their journey
   D. to tell the children that they will visit a hill station during this vacation
2. The final plan was to visit
   A. their village
   B. a hill station
   C. their village as well as a hill station
   D. their home town
3. The final decision was made by
   A. the boys B. the girls
   C. the women D. the elders
4. They decided first to go to their village and stay there for
   A. a day B. a week
   C. ten days D. a fortnight
5. Why were children happy?
   A. Because a hill station was included in their holiday plan
   B. Because a visit to their village was excluded from their holiday plan
   C. Because their choice prevailed
   D. Because they were going all alone to the hill station

PASSAGE-7

Once Govind intended to go on pilgrimage with his family. He asked Mirind to accompany. But for his trade's reason, he did not go with him. So Govind thought it safe to leave the box of his jewellery with him, as it was dangerous to leave it in a lone house or take it on the journey. So he went to him with the box. He took him to a lonely place under a tree and handed it over to him. He told Mirind, "Keep it safe with you. I shall return from the journey after six month then I shall take it back from you." Mirind said, "Don't worry, I shall keep it as safe as own."

1. Govind intended to go
   A. for a business trip
   B. to a hill station
   C. on a long journey to a sacred place
   D. to his home town for a long period
2. Why did Govind leave his box of jewellery with Mirind?
   A. Because it was not safe to take the box with him on a long journey
   B. Because Mirind was his fast friend
   C. Because the box was very heavy
   D. Because his house was unsafe
3. Why did Govind take Mirind to a lonely place?
   A. To tell him that the box contained valuable jewellery
   B. So that no third person could see box
   C. To show him what was within the box
   D. To tell him that the box will remain with him
4. Where did Govind hand over the box of jewellery to Mirind?
   A. At Mirind's house
   B. At his own house
   C. In a lonely place
   D. In a lonely place under a tree
5. It was not safe to leave the box in a lone house. Here the word 'lone house' means
   A. a house in a deserted place
   B. a house where none lives
   C. a house without door and lock
   D. a house near the forest

PASSAGE-8

Zahir-ud-din Babar was the first Mughal emperor of India. A descendent of Timur on father's side and Changez Khan on his mother's side, Babar was a brave warrior. After defeating Ibrahim Lodhi in the First Battle of Panipat in 1526 he entered Delhi and soon gained control over Agra. After many more battles with Rajputs he extended his empire over Punjab, Uttar Pradesh and north Bihar. He died at a young age of 48 years in 1530 at his capital Agra without getting much time to consolidate his victories.

1. Zahir-ud-din Babar was the first
   A. Muslim ruler of India
   B. Mughal ruler of India
   C. Afghan ruler of India
   D. Turk ruler of India
2. Babar was born in the years
   A. 1480 B. 1482
   C. 1492 D. 1962
3. Babar first occupied
   A. Punjab B. Agra
   C. Delhi D. Panipat
4. Babar was a brave warrior. Here brave warrior means
   A. courageous soldier
   B. a kind hearted soldier
   C. a clever fighter
   D. a victorious general
5. Babar extended his empire over Punjab and Uttar Pradesh after many more battles with the
   A. Afghans B. Rajputs
   C. Mughals D. Lodhies

PASSAGE-9

Our National Flag is tricolour. It has three equal horizontal strips. The strip at the top is saffron, in the middle is white and at the bottom is green. The ratio of width to length of the flag is 2 : 3. In the centre of the white strip is a wheel in navy blue. The wheel represents the *chakra.* Its design is similar to the wheel which appears on the abacus of the Sarnath Lion Capital of Ashoka. Its diameter approximates to the width of the white strip. The wheel has 24 spokes. It was adopted by Constituent Assembly on July 22, 1947. We love our national flag. We respect it. We are ready to sacrifice our life to protect its honour. It represents the nation. So it is a symbol of national honour.

1. In our national flag the wheel is located in the centre of
   A. saffron strip B. white strip
   C. green strip D. blue strip
2. In our national flag which of the strips is at the bottom in our national flag
   A. blue C. saffron
   B. white D. green
3. Why do we love our national flag?
   A. Because it is tricolour
   B. Because it has three strips
   C. Because it has a wheel at the centre
   D. Because it is a symbol of national honour
4. Our national flag was approved by
   A. President
   B. Lok Sabha
   C. Parliament
   D. Constituent Assembly
5. The diameter approximates to the width of the white strip. Here the word 'approximates' means
   A. is more or less equal
   B. is exactly equal
   C. is not equal
   D. is related

PASSAGE-10

Distance in large cities are long. All the people do not have their own means of transport. They have to depend upon the state or private buses. The number of bus users is very large. Every bus stop is, therefore, crowded. The number of buses is not adequate. Thus people suffer the torture of long wait at the bus stop. Some bus stops are quite orderly. People form queues and get into the buses turn by turn. However, often this order is forgotten and confusion spreads when the bus comes and the law of jungle prevails.

1. Why are the bus stops crowded?
   A. Because they are small is size
   B. Because the number of passengers is very large
   C. Because they are situated at some busy centre
   D. Because people do not form queues
2. Long wait at the bus stop is the result of
   A. over-crowding in the buses
   B. late running of buses
   C. shortage of buses
   D. slow speed of buses
3. Some bus stops are quite orderly where
   A. there is no crowd
   B. the number of buses is adequate

C. people do not have to wait for long
D. people form queues and enter the buses one by one

4. Most of the people who travel by buses are
A. non-working
B. do not have their own vehicles
C. have to go a long distance
D. live in large cities

5. What happens when people do not have their own transport?
A. They have to wait for a bus at a bus stop
B. They have to depend upon the state or private buses
C. They have to travel long distances
D. They form queues and get into buses one by one

PASSAGE-11

A certain king once fell ill and doctors said that only a sudden fright would restore his health but the king was not a man for anyone to play tricks on, except his fool. One day, when the fool was with him in his boat he cleverly pushed the king into water but he was rescued and put to bed. The fright, the bath and bed cured the diseased king, but he was so angry with the fool that he turned him out of the country.

1. What did the doctor say about the king?
A. Only a sudden fright would restore the king's health
B. Only fool would cure the king
C. Only a boat trick could cure the king
D. The king had suffered a sudden fright

2. He cleverly pushed the king into water but *he* was rescued and put to bed. In this sentence *he* refers to
A. the king B. the fool
C. the doctor D. the river

3. When the fool pushed the king into water they were
A. in the palace B. in the bed
C. in the garden D. in a boat

4. Who played the trick on the king?
A. The doctor B. The boatman
C. The fool D. The fright

5. The fool who cured the king was
A. rewarded
B. thrown into water
C. turned out of the country
D. put into jail

ANSWERS

	1	2	3	4	5
Passage 1.	D	C	B	A	C
Passage 2.	B	C	D	B	B
Passage 3.	D	D	C	B	D
Passage 4.	B	C	C	C	C
Passage 5.	D	A	B	B	A
Passage 6.	A	C	D	B	A
Passage 7.	C	A	B	D	B
Passage 8.	B	B	C	A	B
Passage 9.	B	D	D	D	A
Passage 10.	B	C	D	B	B
Passage 11.	A	A	D	C	C

2. English Grammar

PARTS OF SPEECH

Part of speech	Definition or Function	Examples
Noun	Name of a person, place, animal, quality or thing	Ram, boy, dog pen, sun, Delhi, truth, honesty
Pronoun	Used in place of a noun	I, you, he she, they
Articles & Determiners	Points out indefinite and definite nouns	a, an, the, few, some
Adjective	Describes a noun or pronoun	big, honest, wooden valuable, quiet, deep, soft, narrow
Adverb	Describes a verb, an adjective or another adverb	silently, widely, softly, quietly, very, carefully
Verb	Tells about action or state of something or someone	is, am, was, have, do, like, walk, work, make, throw, tell
Conjuction	Joins words, clauses or sentences	and, but, when, yet, while, else
Preposition	Links a noun or pronoun to another word	at, to, after, on for, under, over, with
Interjection	Expresses sudden feelings or emotions	Ah!, Alas!, oh!, ouch!, hi!, well!, Hurrah!

NOUNS

A word which denotes a person, a thing, an animal or a place is said to be a noun.

There are two noun numbers in English — the *Singular* and the *Plural.*

Singular Numbers : A noun that denotes one person or one thing, is said to be in the Singular number. For example — book, pencil, bird, dog, hen etc. are in singular number.

Plural Number : A noun that denotes more than one person or one thing is said to be in plural number. For example — boys, pens, lions, girls, men etc. are in plural number.

REMEMBER

Singular	*Plural*
Cat	Cats
Book	Books
Pen	Pens
Room	Rooms
Tree	Trees
Bus	Buses
Bush	Bushes
Box	Boxes
Glass	Glasses
Dish	Dishes
Judge	Judges
Tax	Taxes
Watch	Watches
Calf	Calves
Thief	Thieves
Knife	Knives

Singular	*Plural*
Scarf	Scarves
Wife	Wives
Leaf	Leaves
Wolf	Wolves
Half	Halves
Monarch	Monarchs
Roof	Roofs
Hoof	Hoofs
Gulf	Gulfs
Staff	Staffs
Radio	Radios
Bamboo	Bamboos
Folio	Folios
Hero	Heroes
Volcano	Volcanoes
Mango	Mangoes
Potato	Potatoes
Photo	Photos
Piano	Pianos
Baby	Babies
Fly	Flies
Country	Countries
Lady	Ladies
Boy	Boys
Monkey	Monkeys
Ox	Oxen
Child	Children
Man	Men
Woman	Women
Tooth	Teeth
Axis	Axes
Basis	Bases
Foot	Feet
Goose	Geese
Englishman	Englishmen
Radius	Radii
Vertex	Vertices
Stimulus	Stimuli

1. Note the plurals of the following nouns:

Singular	*Plural*	*Singular*	*Plural*
copy	copies	cry	cries
baby	babies	duty	duties
body	bodies	country	countries
family	families	diary	diaries
fly	flies	fairy	fairies
city	cities	spy	spies
army	armies	storey	storeys
bay	bays	monkey	monkeys

2. The following nouns do not undergo any change in plural form, in general.

Singular	*Plural*	*Singular*	*Plural*
deer	deer	sheep	sheep
thousand	thousand	pair	pair
hundred	hundred	score	score
dozen	dozen	gross	gross

Note: We can write—

(*a*) thousands of men; (*b*) two pairs of shoes; (*c*) dozens of mangoes; (*d*) scores of people etc.

But—

(*a*) two thousand rupees; (*b*) three hundred men; (*c*) five dozen eggs, etc.

3. The following nouns are usually used in plural forms. They take a plural verb after them—

eatables	fetters	surroundings
riches	alms	spectacles
trousers	pants	scissors
premises	thanks	annals
congratulations	goods	shorts
tongs	pains	arms
breeches	(for troubles)	

4. The following are the nouns which are plural in appearance but are usually used in singular number. They are followed by a singular verb—

news	politics	physics
mathematics	economics	ethics
politics	classics	gallows
statistics	athletics	innings
mechanics	summons	mumps

5. Collective nouns often used as plurals—

public	police	cattle
audience	clergy	folk
people	poultry	nation
elite	gentry	glitterati

6. The nouns that are usually used in singular forms—

advice	hair	rice
fuel	alphabet	machinery
offspring	issue	furniture
mischief	stationery	luggage
bedding	information	abuse

7. Material nouns are always used in singular number—

gold	copper	milk
water	silk	wool

Note: They may be used in plural with a different meaning.

copper coins (coppers), chains or fetters (irons), cans made of tin (tins).

GENDERS

The difference in sex is denoted by Gender in grammar. The various genders are as follows :

1. **Masculine Gender :** A noun that denotes a male is said to be of the masculine gender, as man, uncle, ox, boy etc.
2. **Feminine Gender :** A noun that denotes a female is said to be of feminine gender, as woman, aunt, princess, cow etc.
3. **Common Gender :** Nouns which denote both males and females are said to be of the common gender, as friend, cousin, person, parent, baby etc.
4. **Neuter Gender :** A noun that denotes the name of object without life is said to be of neuter gender, as file, table, pencil.

REMEMBER

Masculine	*Feminine*
Boy	Girl
Son	Daughter
Brother	Sister
Murderer	Murderess
Sorcerer	Sorceress
Son-in-law	Daughter-in-law
Father-in-law	Mother-in-law
Man-servant	Maid-servant
Land-lord	Land-lady
Bachelor	Maid
Gentleman	Lady
Monk	Nun
Earl	Countess
Lad	Lass
Sir	Madam
Duke	Dutchess
Emperor	Empress
Milk-man	Milk-maid
Pea-cock	Pea-hen
Step-father	Step-mother
Hero	Heroine
Viceroy	Vicerine
Mr.	Mrs.
Governor	Governess
Master	Mistress
Wizard	Witch
Heir	Heiress
Host	Hostess
Lion	Lioness
Mayor	Mayoress
Actor	Actress
Buck	Doe
Colt	Filly
Dog	Bitch
Horse	Mare
Count	Countess
Hunter	Huntress
Prince	Princess
Abbot	Abbess
God	Goddess
Author	Authoress
Ox	Cow
Widower	Widow
Grand-father	Grand-mother
He-goat	She-goat
Milk-man	Milk-woman
Bridegroom	Bride
Tiger	Tigress
Priest	Priestess
Poet	Poetess
Shepherd	Shepherdess
Nephew	Niece
Stag	Hind

PRONOUNS

The repetition of a noun in a sentence or a set of sentences is really boring. So, instead of repeating the noun, we can use a word (for that noun) called the pronoun.

"A pronoun is a word that we use instead of a noun".

Example:

This is *Sachin. He* plays cricket.

Note: *He* is the pronoun used in place of *Sachin.*

Kinds of Pronouns

1. **Personal pronouns :** A pronoun which is used instead of the name of a person is known as a 'Personal Pronoun'. A list of the 'Personal pronouns' is listed below :
   I, my, mine, me, we (First Person)
   You, your, yours (Second Person)
   He, his, him, she, her, hers, it,
   its, they, their, theirs, them (Third Person)

2. **Demonstrative, Indefinite and Distributive Pronouns :**
   (a) Demonstrative Pronouns : Pronouns used to point out the objects to which they refer are called Demonstrative Pronouns.

   Examples :
   (i) *This* is a present from my uncle.
   (ii) *These* are merely excuses.
   (iii) Bembay mangoes are better than *those* of Bangaluru.

   (b) Indefinite Pronouns : All pronouns which refer to persons or things in a general way and do not refer to any particular person or thing are called Indefinite Pronouns.

   Examples :
   (i) *Somebody* has stolen my watch.
   (ii) *Few* escaped unhurt.
   (iii) Did you ask *anybody* to come?

   (c) Distributive Pronouns : Each, either, neither are called distributive pronouns because they refer to persons or things one at a time. For this reason they are always singular and followed by the verb in singular.

   Examples :
   (i) *Each* of the men received a reward.
   (ii) *These* men received *each* a reward.
   (iii) *Either* of you can go.

3. **Relative Pronouns :** A relative pronoun refers or relates to some noun going before, which is called its Antecedent.

   Examples :
   (i) I met Hari *who* used to live here.
   (ii) I have found the pen *which* I had lost.
   (iii) Here is the book *that* you lent me.

4. **Interrogative Pronouns :** These pronouns, are used for asking questions.

   Examples :
   (i) *Whose* book is this?
   (ii) *What* will all the neighbours say?
   (iii) *Which* do you prefer, tea or coffee?

Note : Interrogative pronouns can also be used in asking indirect questions. Consider the following examples :
(i) I asked *who* was speaking.
(ii) Tell me *what* you have done.
(iii) Say *which* you would like best.

Behaviour of the Pronouns

1. If three pronouns are used together in the same sentence they are arranged in the following order :

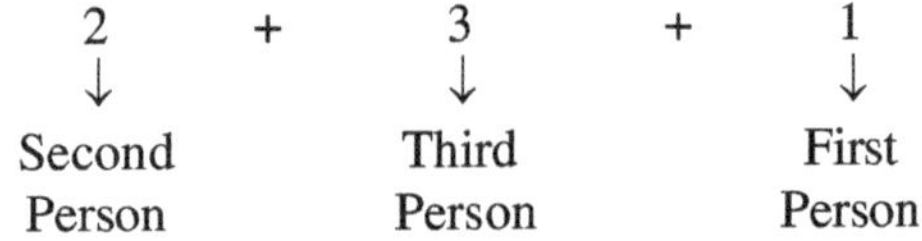

   Examples :
   I, you and he must help *that* poor man. (Incorrect)
   You, he and I must help *that* poor man. (Correct)

2. When two or more singular nouns are joined by and, the pronoun used for them should be plural.
   Examples :
   Mohan and Sohan are friends. *They* play football. *They* live at Lajpat Nagar.

3. But if these nouns joined by and refer to the same person or thing, the pronoun used should be singular.
   Examples :
   (i) Delhi, the beautiful city and the capital of India, is famous for *its* historical monuments.
   (ii) The manager and owner of the firm expressed *his* views on the demands of the workers.

4. When two nouns are used with as well as, the pronoun agrees with the first subject.

Examples :

(a) Mohan as well as his friends is doing *his* work.

(b) The students as well as their teachers are doing *their* work.

5. When two singular nouns joined by 'and' are preceded by *each* or *every,* the pronoun used must be singular and should agree in gender with the second noun.

Examples :

(a) Every man and every woman will do *her* best for the nation.

(b) Each boy and each girl went to *her* house.

6. When two nouns are joined by using 'with', the pronoun agrees with the noun coming before 'with'.

Examples :

(a) The boy with *his* parents has gone to see a movie.

(b) The children with *their* parents have gone to picnic.

7. When two different nouns are joined by either.......... or; neither .... nor, the pronoun is used according to the number and gender of the second noun.

Examples :

(a) Either your sister or you have done *your* work.

(b) Neither the students nor the teacher was in *his* class.

8. The pronoun coming after '*than*' must be in the same case as that coming before '*than*'.

Examples :

(a) She plays better than *me.* (Incorrect)
She plays better than *I.* (Correct)

(b) His elder brother is more intelligent than *him.* (Incorrect)
His elder brother is more intelligent than *he.* (Correct)

9. 'Many a' always takes a singular pronoun and singular verb.

Example :

Many a soldier has met *his* death in the battle field.

10. 'Who', 'Whose', 'Whom' are used only for persons.

Examples :

(a) *Who* is knocking at the door?

(b) *Whose* pen is this?

(c) *What* do you want?

11. 'Which' is used for things.

Example :

Which game do you like?

MULTIPLE CHOICE QUESTIONS

Directions: *In the following questions choose the correct options to fill the blanks.*

1. The place was so dirty that ..... wished to run away from there.
A. everybody B. anybody
C. few D. some

2. ..... was there to help me.
A. Somebody B. Anything
C. Anybody D. Nobody

3. Is there .... to eat?
A. some B. something
C. any D. few

4. ..... of the students were making a great noise.
A. Anyone B. Somebody
C. Many D. Nobody

5. ..... of the students can solve this sum.
A. Someone B. Anybody
C. Somebody D. None

6. ...... of us should try our best to make India a heaven.
A. Any B. Somebody
C. Anybody D. All

7. ...... of us do not know the real meaning of our lives.
A. Any B. Something
C. Several D. Many

8. My .... black.
A. hairs are B. hair is
C. hairs shall D. hair will

9. She saw two .... on the last Sunday.
A. thiefs B. theifs
C. thieves D. theives

10. My sister is a ....
A. bacheloress B. bachelor
C. unmaried D. spinster

11. One is supposed to do ....
A. our duty B. their duty
C. one's duty D. his duty

12. Take anything .... you want.
A. that B. which
C. than D. then

13. I cannot tolerate ....
A. separated you
B. your separation
C. separation from you
D. you separated

14. He is .... faithful partner.
A. Yours B. You
C. Your D. Your's

15. Ajay is more smart than ....
A. her B. hers
C. herself D. she

16. Vivek works harder than ....
A. me B. I
C. her D. his

17. They should help ....
A. the poor people B. the poor
C. the poor persons D. the poor peoples

18. .... are mad.
A. All his sons B. His all sons
C. Sons all his D. All sons his

19. The poor fellow .... to fate.
A. resigned
B. resigned himself
C. resigned itself
D. resigned themselves

20. Nobody will help you but ....
A. I B. me
C. ours D. his

21. It is a good chance, You must avail .... this opportunity.
A. of B. yourself of
C. for D. from

22. The person who is elected .... my relative.
A. is B. he is
C. his D. him

23. He made ....
A. yours mention B. mention of you
C. mention for you D. mention about you

24. .... I know, he is quite faithful.
A. As far as B. So far as
C. So far this D. So far so

25. It is a duty of a person to take .... for his family.
A. pain B. pains
C. pain-killers D. pained

26. She does not love .... husband.
A. his B. her
C. its D. their

27. Let .... work together.
A. him and me B. he and I
C. he and him D. I and me

28. Copper, Silver and Gold ....
A. each will do B. either will do
C. any one will do D. any will do

29. Jessica and Roma are very irregular .... habits.
A. in her B. in their
C. in its D. in every

30. One likes to enjoy .... who was a great poet.
A. The sonnets of Shakespeare
B. Shakespeare's sonnets
C. Sonnets
D. Shakespeare

31. That is the boy .... everybody loves.
A. whom B. who
C. that D. whose

32. That is the girl .... won the first prize.
A. whom B. who
C. whose D. which

33. That is the man .... purse was lost.
A. who B. whom
C. whose D. their

ANSWERS

1	2	3	4	5	6	7	8	9	10
A	D	B	C	D	D	D	B	C	D
11	**12**	**13**	**14**	**15**	**16**	**17**	**18**	**19**	**20**
C	A	C	C	D	B	B	A	B	B
21	**22**	**23**	**24**	**25**	**26**	**27**	**28**	**29**	**30**
B	A	B	A	B	B	A	C	B	A
31	**32**	**33**							
A	B	C							

ARTICLES

The family of the articles has only three members. They are : A, An and The. However, they fall under two groups :

(a) Definite Article *(b)* Indefinite Article

'The' is known as definite article whereas 'a' and 'an' are known as indefinite articles.

Use of the Definite Article 'The'

'The' is used before

1. The superlative degree :
   He is the ablest man of the town.
   (ablest is a superlative degree)
2. The name of states, countries etc. having a descriptive name :
   (i) The J & K is a small state. (J & K is a descriptive name)
   (ii) He lives in the U.S.A. (U.S.A. is a descriptive name)
   (But the Delhi and the America are wrong because neither Delhi nor America is a descriptive name)
3. The names of the scriptures :
   The Gita is a holy book. (Gita is a scripture)
4. Name of newspapers :
   The Tribune is published from Chandigarh.
5. Name of rivers, canals, seas, oceans, bays, gulfs, groups of islands etc. :
   (i) The Ganga is a holy river.
   (ii) The Indian Ocean is the deepest ocean.
   (iii) The Persian Gulf is a narrow gulf.
6. The name of famous buildings :
   The Taj is one of the best buildings in India.
7. The names of nationals, sects and communities:
   (i) The English defeated the Germans in the World War.
   (ii) The rich should help the poor.
   (iii) The Hindus believe in the caste system.
8. Proper nouns used as common nouns :
   (i) Kalidas is the Shakespeare of India.
   (ii) Delhi is the London of India.
9. Famous historical events :
   The Industrial Revolution changed the face of England.
10. The directions and the celestial bodies:
    The sun rises in the east.
11. Titles :
    Akbar, the Great was loved by his subjects.

Do not use 'the'

1. Before languages :
   The English is an international language. (Incorrect)
   English is an international language. (Correct)
2. Before the names of games :
   The hockey is a popular game. (Incorrect)
   Hockey is a popular game. (Correct)

Use of the Indefinite Articles 'A' and 'An'

'A' is used before :

1. All singular common nouns beginning with a consonant :
   (i) A boy sings a song.

(ii) A black and a white cow were grazing in the field.

2. If a word begins with a vowel but gives the sound of a consonant, 'a' should be used before it :
   (i) He was helped in his work by a European.
   (ii) He is a one-eyed man.
   (iii) It is a useful work.

'An' is used as follows :

1. All singular common nouns beginning with a vowel (*i.e.*, a, e, i, o, u) :
   (i) He is an artist.
   (ii) He is an old man.
   (iii) I intend to buy an umbrella.
2. If a word starts with a consonant but gives the sound of a vowel, "an" should be used before it :
   (i) Brutus is an honourable man.
   (ii) He is an honour to his profession.
   (iii) He is an L.L.B.
   (iv) He is an M.A.
   (v) You will reach there in an hour.

Demonstratives, that, these and those

1. The demonstrative adjectives and pronouns are for objects nearby the speaker:
   this (singular) those (plural)
   and for objects far away from the speaker.
   That (singular) those (plural)
2. Demonstratives are the only adjectives that agree in number with their nouns.
   That hat is nice.
   Those hats are nice.
3. When there is the idea of selection, the pronoun "one" (or "ones") often follows the demonstrative.
   I want a book. I'll get this (one).
   If the demonstrative is followed by an adjective, "one"(or "ones") must be used.
   I want a book. I'll get this big one.

MULTIPLE CHOICE QUESTIONS

Directions: *In the following questions choose the correct options to fill the blanks.*

1. .... will have to be paid for this material.
A. Half rupee B. Half a rupee
C. A half rupee D. An half rupee

2. .... is taking keen interest in India.
A. The USA B. USA
C. An USA D. A USA

3. Only .... can save our country.
A. the Hitler B. a Hitler
C. Hitler D. an Hitler

4. I can run for ....
A. hundred miles B. the hundred miles
C. a hundred miles D. an hundred miles.

5. .... man-eater has been killed.
A. The B. A
C. An D. Either A or B

6. What .... fine idea!
A. the B. an
C. a D. No article

7. .... earth is moving around the sun.
A. An B. A
C. The D. No article

8. This is .... first example while I got.
A. the B. a
C. an D. No article

9. This is .... house which was built during earthquake.
A. a B. an
C. the D. No article

10. .... America is a rich country.
A. The B. An
C. A D. No article

11. .... U.S.A. is a developed country.
A. A B. An
C. The D. No article

12. .... Bible is a holy book.
A. A B. The
C. An D. No article

13. .... rich should help the poor.
A. The B. A
C. An D. No article

14. .... Gold is a costly metal.
A. The B. A
C. An D. No article

15. Kalidas is .... Shakespeare of India.
A. a B. an
C. the D. No article

16. I cannot do .... difficult work.
A. a such B. the such
C. such the D. such a

17. How foolish .... plan it is!
A. a B. an
C. the D. No article

18. An ink is .... useful article.
A. an B. a
C. the D. No article

19. There are .... husband and wife.
A. a B. an
C. the D. No article

20. He is learning .... French
A. the B. a
C. an D. No article

ANSWERS

1	2	3	4	5	6	7	8	9	10
B	A	B	C	D	C	C	A	C	D
11	**12**	**13**	**14**	**15**	**16**	**17**	**18**	**19**	**20**
C	B	A	D	C	D	A	B	D	D

ADJECTIVES & ADVERBS

An Adjective is a word which adds something to the meaning of a noun or a pronoun.

Mridula is an *intelligent* girl.
He has a *black* goat.
He is a *brilliant* student.
She is a *clever* girl.
It is a *beautiful* picture.

In the sentences given above, the words in italics are adjectives.

An Adverb is a word which qualifies the meaning of a Verb, an Adjective or another Adverb.

(*i*) He talks *slowly.*
(*ii*) He is a *very* good student.
(*iii*) He talks *very* slowly.

In sentence (*i*), *slowly* qualifies the verb *talks.*

In sentence (*ii*), *very* qualifies the adjective *good.*

In sentence (*iii*), *very* qualifies the adverb *slowly.*

Adjectives have three degrees of comparison :

1. **Positive Degree :** It expresses the common form of an adjective.
   Example :
   Ram is a *tall* boy.
   In the above sentence *tall* is an adjective and expresses the common form.
2. **Comparative Degree :** It expresses the more of the same form.
   Example :
   Ram is *taller* than Mahesh.
   In the above sentence *taller* is an adjective that expresses the more of the common form of the adjective *tall.*

"When and How to Use" Comparative Degree?

(a) Comparative Degree is used when two persons or two groups of persons or things are compared.
Examples :
(a) He is *wiser* than his younger brother.
(b) This glass is *cleaner* than the other.

(b) When two different qualities in the same person are compared, more is used instead of 'er' to form the comparative. The formula used in this case should be :
More + Positive Degree
She is *fairer* than polite. (Incorrect)
She is *more fair* than polite. (Correct)

(c) When selection of one out of two persons or things is meant, the degree of comparison is followed by of and *the* is used before it.
Example :
Zia is abler of *the* two sisters.

(d) If two comparatives are used in the same sentence to impress upon an idea, both should be preceded by the definite article.

Examples :

(i) The higher you go, the cooler it is.

(ii) The more we get, the more we desire.

(e) When one person or thing is compared with another of the same kind, other is used after the comparative degree. In such sentences other is normally preceded by any or all.

Examples :

(i) Kalidas is greater than any dramatist. (Incorrect)

Kalidas is greater than any other dramatist. (Correct)

(ii) Lead is heavier than all metals. (Incorrect)

Lead is heavier than all other metals. (Correct)

(f) Senior, junior, superior, inferior, prior, anterior (earlier than) and posterior (later than) are always followed by 'to'.

Examples :

(i) Ram is senior *to* Mohan by three years.

(ii) That pen is inferior *to* that.

(iii) He is junior *to* me in rank.

(iv) This event was posterior *to* that.

Note: Never use *than* after the above mentioned adjectives.

Important Information

(a) 'Preferable' is also used as an adjective of the comparative degree. As such, it is always followed by *to* and not *a*.

Death is preferable than dishonour. (Incorrect)

Death is preferable *to* dishonour. (Correct)

(b) To intensify the Degree of comparison, we use *far* or *much* before the comparative.

Examples :

(i) This book is *far* better than that.

(ii) His performance was *much* better than Mohan's.

Warning : Always avoid the use of double comparatives.

Don't say : Ram is more cleverer than his younger brother.

Say: Ram is cleverer than his younger brother.

3. **Superlative Degree :** It expresses the most of the common form of an adjective.

Example :

He is the ablest man of the town.

How and when to use the Superlative Degree?

(a) The Superlative Degree is used when more than two persons or things are compared.

(b) The Superlative Degree is generally preceded by 'the' and followed by 'of' in most of the cases or otherwise.

(c) When an adjective of the superlative degree is preceded by a Possessive Adjective or a Noun in the Possessive case, 'the' should not be used before it.

Example :

Which is Kalidas' best play?

It will be a blunder to use 'the' before the Superlative Degree in such cases.

Don't say : Which is Kalidas' the best play.

(d) To intensify the degree of comparison, *by far* is used before the superlative degree.

Example :

India is *by far* the most beautiful country of the world.

Note: Always avoid the use of double superlatives.

Don't say : He is the most strongest boy in the class.

Say : He is the strongest boy in the class.

Use of some Important Adjectives

1. (a) '**Some**' is used as follows :

(i) With countable nouns where it means—a little, a small quantity.

(ii) In a question which shows some request.

Examples :

(i) There is some water in the bottle.

(ii) Some of the students were absent yesterday.

(iii) Will you have some milk?

(iv) Will you buy some fruit for me?

(b) **'Any'** is used as follows :

(i) In negative sentences.

(ii) In interrogative sentences.

(iii) After 'Hardly', 'Scarcely' and 'Barely'.
(iv) After 'If'.
Examples :
(i) There is not any sugar in the pot.
(ii) We haven't any rice in the house.
(iii) I have hardly any money.
(iv) There are scarcely any plants in this field.
(v) If there is any danger, blow the whistle.

2. (a) **Older :** Older (and oldest) are used for persons animals and things. But 'Older' and 'Oldest' refer to the persons who do not belong to the same family.
Examples :
(i) Radha is older than Shyama.
(ii) John is the oldest member of the staff. 'Older' and 'Oldest' refer to the persons who do not belong to the same family.

(b) **Elder** (and **eldest**) are used in respect of the members of the same family like sons, daughters, brothers, sisters.
Examples :
(i) My elder sister is a lecturer.
(ii) Meenakshi is the eldest of the three sisters.
Note :
(i) 'Elder' is not followed by 'than'.
(ii) 'Elder' and 'Eldest' cannot be used for things.

3. (a) **'Few'** is negative and is the opposite of 'Many'. It means 'not many'.
(b) **'A few'** is positive and means 'some at least'. It is the opposite of 'None'.
(c) **'The few'** means 'minority' and suggests 'whether there is'.
Examples :
(i) We have few holidays in school.
(ii) Only a few boys will fail in the examination.
(iii) The few poems that he wrote are very popular.

4. (a) **Further** means 'something additional'.
(b) **Farther** means 'a greater distance'.
Examples :
(i) Further discussion will be held in the office of the principal.
(ii) Amritsar is farther from Delhi than Ambala.

5. (a) **Little** is negative. It means, 'not much', or 'hardly any'.
(b) **A little** is positive. It means 'some quantity'.
(c) **The little** denotes quantity. It means, 'not much but all that is, or whatever quantity there is'.
Examples :
(i) There is little hope of his success.
(ii) He knows a little of everything.
(iii) I have spent the little money I had.
(iv) The little knowledge of shoe-making proved very useful to me.

6. (a) '**Much**' expresses 'quantity'.
(b) '**Many**' expresses 'number'.
(c) **'Many a'**—'Singular noun' and 'Singular verb' are used with 'many a'.
Examples :
(i) There is not *much* water in the jug.
(ii) *Many* boys are absent today.
(iii) *Many* a battle has been fought on the soil of India.

7. (a) '**Less**' denotes 'in a small degree'.
(b) '**Fewer**' denotes 'number'.
Examples :
(i) He devotes less time to his studies.
(ii) There are no fewer than ten chairs in this room.

8. (a) '**Each**' is used for a single number of 'two persons' or 'things'.
(b) '**Every**' is used for a single number of 'many persons' or 'things'.
Examples :
(i) Each boy must take part in games.
(ii) There are only two poets. Each poet recited his poem.
(iii) Every man dies in this world.
(iv) Every man is expected to do his duty.

9. (a) '**Either**' means one of the two or both.
(b) '**Neither**' is negative of the either.
Examples :
(i) You may buy either of these two chairs.
(ii) Neither of them could speak on the stage.

10.(a) **'Later'** expresses 'late in time'.

(b) **'Latter'** means 'second in position or order'.

Examples :

(i) My father reached later than I expected.

(ii) The latter position was better than the former.

Use of some Important Adverbs

1. (a) Also, too, enough:

(i) He taught English. Also, he edited the school magazine

(ii) He is a writer and also he is a painter.

(iii) He is too obstinate to listen to any reason.

(iv) This is too difficult a piece for the junior students.

(v) Sarla was kind enough to help the poor.

(vi) He is brave enough to help the truth.

Note: 'Too' is used in a negative sense, but enough is used in a positive sense.

(b) Fairly and rather: Both suggest the meaning 'moderately'. But, mainly 'fairly' is used with the words that denote a positive meaning and rather is used with the words that denote a negative meaning:

(i) Rita did fairly well in that competition, but her performance was rather poor in sports.

(ii) Mona is fairly rich, but she is rather stingy.

Note: 'Rather' can also be used in a positive sense.

(i) This is a rather interesting job.

(ii) That boy is rather smart.

(c) Hardly, barely, scarcely: These words mostly convey the negative suggestions and are almost similar.

(i) I have hardly any strength now.

(ii) There was barely any supply to the township,

(iii) There were scarcely a hundred guests present.

Note: With slight variance in the meaning, the words given above convey the idea of 'very little', 'not enough', 'lack of quantity and number'.

(d) Yet, Still: These adverbs can often be used to connect the sentence units:

(i) He has been defeated many times in the contest; still he wants to be a competitor.

(ii) Mona was sick; yet she went on doing her work.

(e) Alone:

(i) He alone (none else) is capable of handling that fire,

(ii) He hunted all alone in the forest. (not in any company)

Special Note:

(a) Apart from their conventional positions the adverbs might be used in different positions with different meanings and angles.

(i) He had only four books.

(ii) John only contacted his friend in need.

(iii) He greeted me only.

(iv) Only he greeted me there.

(b) Inversion: Some adverbs can be inverted *i.e.* placed in the beginning of the sentence and then be followed by an interrogative form. The most common of these adverb are: so, seldom, never, nowhere, under no circumstances, hardly, scarcely etc.

(i) So big was the bus that it could not enter the narrow lane.

(ii) Hardly had he reached the station when he received the message.

MULTIPLE CHOICE QUESTIONS

Directions: *In the following questions choose the correct options to fill the blanks.*

1. The girl whom you met is the .... sister of Ravi.

A. eldest B. elder

C. older D. oldest

2. The historical place is ....

A. seeing worth

B. worthy of seeing
C. worth seeing
D. worthy seeing

3. These flowers smell ....
A. sweet B. sweetly
C. more sweetly D. sweetest

4. .... aspirant cannot pass the entrance examination.
A. Each B. Every
C. All D. No

5. Harivansh Rai .... second Shakespeare.
A. is a B. is
C. is the D. is an

6. .... student in the class got prizes.
A. Each and every B. Every and each
C. Every D. Never

7. It is .... picture than the one we saw last Monday.
A. interesting B. much interesting
C. more interesting D. most interesting

8. She is clever ....
A. that her mother is
B. as her mother is
C. to her mother is
D. than her mother is

9. They will get ....
A. Red, green and black paper
B. Red, green black paper
C. Red and green and black paper
D. Red green black paper

10. Health is .... wealth.
A. preferable to
B. more preferable than
C. more preferable to
D. most preferable then

11. .... water that was in the jug evaporated.
A. Little B. The little
C. Small D. A small

12. He has not sung .... songs.
A. much B. most
C. more D. many

13. Srishti has searched .... office.
A. whole the B. the whole
C. a whole D. some whole

14. Premchand was .... best and .... famous writer.
A. a, the most B. the, a most
C. the, more D. the, the most

15. William Shakespeare is famous as ....
A. a poet and a dramatist
B. a poet and dramatist
C. the poet and the dramatist
D. a poet and the dramatist

16. What does .... leader suggest?
A. other B. another
C. others D. anothers

17. He .... money.
A. has few B. have few
C. has little D. have little

18. The .... boys are rewarded.
A. first two B. two first
C. firsts two D. two's first

19. He is .... brave.
A. stronger than
B. stronger then
C. more strong then
D. more strong than

20. No sooner said ....
A. so done B. and done
C. then done D. but done

21. She returned .... than I had thought.
A. quickly B. more quicker
C. more quickly D. quicker

22. He is .... foolish person.
A. rather the B. a rather
C. rather a D. rather

23. This pen .... rupees.
A. costs twenty
B. twenty costs only
C. costs only twenty
D. only costs twenty

24. It is .... pride.
A. nothing else but
B. nothing else than
C. else nothing than
D. but

25. This tea is .... to drink.
A. too hot B. very hot
C. enough hot D. much hot

ANSWERS

1	2	3	4	5	6	7	8	9	10
A	C	A	B	A	C	C	C	A	A
11	**12**	**13**	**14**	**15**	**16**	**17**	**18**	**19**	**20**
B	D	B	D	B	B	C	A	D	C
21	**22**	**23**	**24**	**25**					
C	C	C	A	A					

DETERMINERS

Determiners are actually Adjectives. They are always followed by nouns.

Determiners are of the following kinds:

1. Demonstrative Determiners

this, that, these, those

2. Possessive Determiners

my, our, your, his, her, its, their

3. Quantitative Determiners

some, any, much, enough, sufficient, whole, a little, the little, little, all, both

4. Numerical Determiners

a few, some, few, the few, any, several, many, no, etc.

One, two, three ... (Cardinals)

First, second, third ... (Ordinals)

5. Distributive Determiners

either, neither

6. Articles

Indefinite: a, an

Definite: the

MULTIPLE CHOICE QUESTIONS

Directions: *In the following questions choose the correct options to fill the blanks.*

1. Give me .... rice.
A. some B. few
C. a few D. any

2. .... sheep grazing on the slope of the hill had gone away.
A. Any B. The few
C. This D. Much

3. Have you got .... magazines to read?
A. all B. much
C. some D. little

4. I have .... money that I want to spend on shares.
A. any B. much
C. less D. some

5. There is .... owl on the branch of the tree.
A. a B. the
C. an D. some

6. My brother is .... MBA.
A. a B. an
C. the D. any

7. Have you got .... cheese?
A. some B. many
C. a few D. few

8. No, I have not got .... cheese.
A. many B. few
C. any D. some

9. There is only .... milk left in the bottle.
A. enough B. few
C. much D. a little

10. There is .... hope of his recovery.
A. any B. little
C. many D. few

11. .... dogs were barking at the strangers.
A. Some B. Any
C. Much D. Less

12. The girl bought her father .... juice.

A. few B. some

C. any D. many

13. You should take .... honey everyday.

A. any B. many

C. a little D. a few

14. .... boy was punished by the teacher.

A. Either B. All

C. Any D. Many

15. .... girl was asked to join the army.

A. None B. Neither

C. All D. Any

16. .... water in the jug has been drunk by Mohan.

A. The little B. The few

C. A few D. Few

17. I shall play .... piano at the party.

A. some B. any

C. the D. few

18. .... labourers were found dead in the mine.

A. Any B. Fewer

C. Many D. Less

19. Could I borrow .... umbrella?

A. our B. your

C. yours D. my

20. My brother is standing in the .... row.

A. any B. many

C. some D. first

ANSWERS

1	2	3	4	5	6	7	8	9	10
A	B	C	D	C	B	A	C	D	B
11	**12**	**13**	**14**	**15**	**16**	**17**	**18**	**19**	**20**
A	B	C	A	B	A	C	C	B	D

THE VERB

A Verb is a word that tells something about the action or state of or happenning to a person or thing.

A Verb tells the following:

1. What a person or thing does.

Sachin goes to school daily.
The bell *rang* loudly.
Many birds fly in the sky.
She *sang* a song.

2. What a person or thing is.

India *is* the biggest democracy in the world.
Ram Mehar *is* very rich.
They *are* happy.

3. What is done to a person or thing.

You *are liked* by all.
Two thieves *were arrested.*
Four students *were punished* by the teacher.

4. What happens to a person or thing.

His maternal uncle *died* last week.
Two ships *sank* yesterday.
Leaves *turn* yellow in autumn.

5. What a person or thing has, had, and so on.

I *have* a new car.
He *had* a scooter last year.
He *has* several cows and goats.

It goes without saying that a verb is the most important part of a sentence. No sentence is complete without a Verb.

Important Information

1. If two or more singular nouns are joined by 'and' the verb used will be plural.

   Example:

   (i) He and I were going to the market.
   (ii) Ram and Mohan are friends.

2. If two singular nouns joined by 'and' points out to the same thing or person, the verb used must be singular.

   Example:

   (i) Rice and curry is the favourite food of the Punjabis.
   (ii) The Collector and District Magistrate is away.

3. In case two subjects are joined by 'as well as' the verb agrees with the first subject.

   Example :

   (i) Kanta as well as her children is playing.
   (ii) Children as well as their mother are playing.

In the case of first sentence the verb (is) agrees with Kanta and in the case of second sentence the verb (are) agrees with the children.

4. 'Neither', 'Either', 'Every', 'Each', 'Everyone', and 'Many a' are followed by a singular verb.
   Example :
   (i) Either of the plans is to be adopted.
   (ii) Neither of the two brothers is sure to pass.
   (iii) Every student is expected to be obedient.
   (iv) Everyone of them desires this.
   (v) Many a person is drowned in the sea.

5. If two subjects are joined by 'Either .... or' / 'Neither .... nor', the verb agrees with the subject near to it.
   Example :
   (i) Either my brother or I am to do this work.
   (ii) Neither he nor they are prepared to do this work.

6. 'A great many' is always followed by a 'plural noun' and a 'plural verb'. For example :
   A great many students have been declared successful.

7. Similarly if two subjects are joined by 'with', 'together with', 'no less than', in addition to 'and not', etc. the verb agrees with the first subject.
   Example :
   (i) The boy with his parents has arrived.
   (ii) He, no less than I, is to blame.

8. Nouns, plural in form, but singular in meaning, take a singular verb.
   Example :
   This news was broadcast from television yesterday.

MULTIPLE CHOICE QUESTIONS

Directions: *In the following questions choose the correct options to fill the blanks.*

1. The bus with all its passengers .... lost.
A. were B. was
C. are D. would

2. You as well as I .... responsible for this work.
A. am B. are
C. was D. is

3. Raghava like all his companions .... a spoiled child.
A. are B. were
C. is D. will be

4. Pen and ink .... required for me.
A. are B. were
C. is D. has required

5. Every girl and every boy .... attended the seminar.
A. have B. has
C. is D. are

6. Not only she but all her sisters ...... been married.
A. has B. have
C. is D. are

7. There .... nothing but miseries in life.
A. is B. are
C. were D. will be

8. Neither prose nor poem .... given.
A. were B. was
C. has D. have

9. Either he or I .... wrong.
A. is B. are
C. am D. were

10. Either Sulekha or Rekha .... coming here.
A. are B. is
C. were D. have

11. .... the child or his parents to blame?
A. Is B. Are
C. Were D. Has

12. You and I .... neighbours.
A. am B. are
C. was D. has

13. The house with all its belongings ..... sold away.
A. were B. are
C. was D. must

14. Either water or juice .... required.
A. is B. are
C. were D. has

15. There were not as many tables as .... required.
A. was B. were
C. is D. are

16. They each .... a book.
A. have B. are
C. has D. is

17. He and I .... class friends.
A. is B. am
C. was D. are

18. She as well as I .... guilty.
A. is B. are
C. am D. must be

19. Purushottam .... not read more on this chapter.
A. needs B. has been need
C. need D. had been need

20. He came .... to his aunt.
A. run B. running
C. to run D. in run

21. She dislikes .... meat.
A. eat to B. to eat
C. eating D. to eating

22. He likes ....
A. sing to B. singing
C. to sing D. to singing

23. We are ready .... the match.
A. play to B. to playing
C. playing D. to play

24. .... is injurious to health.
A. Smoking B. To smoke
C. To smoking D. Smoke to

25. He loves .... raw vegetables.
A. eaten B. eating
C. to eating D. eat to

26. He seemed .... finished his homework.
A. have to B. to have
C. having D. to having

ANSWERS

1	2	3	4	5	6	7	8	9	10
B	B	C	C	B	B	A	B	C	B
11	**12**	**13**	**14**	**15**	**16**	**17**	**18**	**19**	**20**
A	B	C	A	B	A	D	A	C	B
21	**22**	**23**	**24**	**25**	**26**				
C	B	D	A	B	B				

CONJUNCTIONS

A conjunction is a word which connects words, clauses or sentences.

Look at the following sentences.

(i) He bought apples *and* mangoes.

(ii) God made the country *and* man made the town.

(iii) The door was open *but* there was no one in the house.

(iv) He knows that I am here *and* that I want to see him.

In the sentence (i), *and* connects two words—*apples* and *mangoes.*

In the sentence (ii), *and* connects two sentences—*God made the country* and *man made the town.*

In the sentence (iii), *but* connects two sentences—*The door was open* and *there was no one in the house.*

In the sentence (iv), *and* connects two clauses—*that I am here* and *that I want to see him.*

The main coordinating conjunctions are:

and, but, for, or, nor, also, either ..... or, neither ..... nor.

There are some conjunctions which are used in pairs. They are:

either .... or, ... neither .... nor, both .... and, though .... yet, whether .... or, not only .... but also.

Example: *Either* take it *or* leave it.

It is *neither* useful *nor* ornamental.

They *both* like *and* respect me.

Though he is suffering from high fever, *yet* he does not cry.

He does not care *whether* you go *or* stay.

He is *not only* doltish, *but also* obstinate.

The conjunctions which are used in pairs in this way, are called correlative conjunctions, or merely correlatives.

Use of Important Conjunctions

1. **As soon as :** As soon as denotes simultaneous time.

   Example : As soon as he saw his enemy, he took to his heels.

2. **No sooner .... than :**

   (a) 'No sooner' is always followed by 'than'.

   (b) Please remember that 'No sooner' is always followed by do/does/did. As such only first form of the verb should be used after the subject.

   Example :

   No sooner did he see his enemy than he took to his heels.

3. **Hardly :** Hardly is followed by when.

   Examples :

   (i) Hardly had I left the house when it started raining.

   (ii) We had hardly come into the room when his father began chastising him.

   Note :

   A. Hardly is never followed by than.

   B. 'Scarcely' can also be used in the sense and manner of 'Hardly'.

4. **Lest :** Lest is used in the sense of so that .... not. It is always followed by should. Lest is negative in sense. Hence 'not' should never be used with it.

   Example :

   Work hard lest you should fail.

   Note : 'Lest' is always followed by 'should' and not 'may'.

5. **Unless :** Unless expresses condition. It is also used in the negative sense. Use of 'not' is not allowed with unless because unless is already in the negative sense.

   Example :

   Unless you labour hard you will not pass.

6. **Until :** 'Until' expresses time. It means 'till not'.

   Example :

   Wait here until I return.

   Note : Until is in the negative sense. So 'not' should not be used with it. Example :

   Wait here until I do not return. (Incorrect)

   Wait here until I return. (Correct)

7. **As well as :** When two subjects are joined by 'as well as', the verb always agrees with the first subject.

   Examples :

   (i) The teacher as well as students is playing.

   (ii) Students as well as the teacher are playing.

   Note : 'Both' and 'as well as' cannot be used together in the same sentence.

   Examples :

   Both Sita as well as Kanta are beautiful. (Incorrect)

   Sita as well as Kanta is beautiful. (Correct)

   Both Sita and Kanta are beautiful. (Correct)

8. **As if :** 'As if' is used in the sense of pretension. While using 'as if' in a sentence, we should see that even the third person singular subject gets 'were'.

   Example :

   He talks as if he were mad.

9. **Till :** Till expresses time. Till is always used in the affirmative.

   Example :

   We did not come back till sunset.

10. **Rather than :** 'Rather than' is used in the sense of 'preference'. 'Rather' is always followed by 'than'.

    Example :

    I would rather die than submit.

11. **As long as/so long as :** Both express time during which an action or event takes place.

    Example :

    As long as there is life, there is hope.

12. **However :** It is both a subordinate and co-ordinate clause.

    Examples :

    (a) Mala worked hard, she however, failed.

    (b) However hard he may work, he cannot pass.

13. **Such as :** 'Such as' gives us the sense of 'like'. Such is always followed by 'as'.

    Example :

    Life is such a puzzle as cannot be solved.

MULTIPLE CHOICE QUESTIONS

Directions: *In the following questions choose the correct options to fill the blanks.*

1. Neither he .... his friend is good.
   A. or B. and
   C. but D. nor

2. The officer asked the peon .... why he was late.
   A. that B. if
   C. but D. No word needed

3. Both Ajay .... Vijay are intelligent.
   A. or B. nor
   C. and D. No word needed

4. No Sooner did the thief see the public .... he ran away.
   A. then B. and
   C. but D. than

5. Abhinav .... his brothers was going to Mumbai.
   A. but B. yet
   C. No word needed D. together with

6. He behaves .... he were the captain of the team.
   A. as if B. as
   C. No word needed D. that

7. Either Rupali .... Sonali is going to attend the meeting.
   A. and B. but
   C. nor D. or

8. Neither Nirmal .... Ashwinee is going to listen the speech.
   A. and B. but
   C. nor D. or

9. Ravi .... Prakash are going to Kolkata.
   A. or B. nor
   C. but D. and

10. Rice .... curry is my usual breakfast.
    A. and B. but
    C. then D. than

11. Hardly had he left .... his brother came.
    A. then B. than
    C. when D. that

12. I would rather have a copy .... a book.
    A. then B. than
    C. when D. that

13. He is no other .... my friend.
    A. then B. than
    C. when D. but

14. He saw a snake ....he awoke.
    A. then B. when
    C. than D. No word needed

15. Ten years have passed .... my grandmother died.
    A. since B. when
    C. then D. than

16. She is .... good .... bad.
    A. either, not B. neither, or
    C. neither, nor D. neither, than

17. The cellphone is both cheap .... best.
    A. than B. and
    C. then D. or

18. No sooner did the rogue see the police .... he disappeared.
    A. then B. than
    C. so D. because

19. Srishti will go .... Sanju goes.
    A. if B. than
    C. then D. although

20. She is wise .... timid.
    A. and B. yet
    C. but D. however

21. Make hay .... the sun shines.
    A. though B. while
    C. after D. before

22. He is so weak .... he cannot walk.
    A. but B. that
    C. then D. so

23. Although he is rich, ... he is unhappy.
    A. but B. yet
    C. so D. still

24. Wait here .... I come back.
    A. till B. until
    C. before D. after

25. He is my friend .... I shall help him.
A. so
B. hence
C. that is why
D. therefore

26. He must go away .... he will be beaten.
A. otherwise B. and
C. or D. else

27. God loves good men .... good men love God.
A. and B. or
C. that D. those

28. He was late .... he was not punished.
A. but B. yet
C. still D. therefore

29. Walk slowly ...., you may fall.
A. and B. or
C. so D. otherwise

30. Work hard, ... you will fail.
A. and B. or
C. otherwise D. else

ANSWERS

1	2	3	4	5	6	7	8	9	10
D	D	C	D	D	A	D	C	D	A
11	**12**	**13**	**14**	**15**	**16**	**17**	**18**	**19**	**20**
C	B	B	B	A	C	B	B	A	C
21	**22**	**23**	**24**	**25**	**26**	**27**	**28**	**29**	**30**
B	B	B	A	B	C	A	C	D	D

PREPOSITIONS

A *Preposition* is a word which is placed before a noun or a pronoun to show its relation to some other word in the sentence.

1. I saw a goat *in* the field.
2. I am fond *of* hot coffee.

In sentence 1, the word *in* shows the relation between two things—*goat* and *field.*

In sentence 2, the word *of* shows the relation between the attribute expressed by the adjective *found* and *tea.*

The words *in* and *of* are here used as prepositions.

The noun or pronoun which is used with a preposition is called its object. The noun or pronoun is in the objective case. It is governed by the preposition. Now it is absolutely clear that in sentence 1, the noun *field* is in the objective case. The word *field* is governed by the preposition *in.*

A preposition may have two or more objects.

The road runs over *hill* and *plain.*

Here, the words *hill* and *plain* are used as objects.

Use of Important Prepositions

1. Among, Between

'**Among**' is used for more than two persons or things; '**Between**' is used only for two.

Examples :

(i) Distribute these sweets *among* the poor students of the class.

(ii) Distribute these books *between* Ram and Shyam.

2. Among, In

'**Among**' is used before collective plural nouns. '**In**' is used before collective singular nouns.

Examples :

(i) I found him standing *among* the crowd.

(ii) I saw him in the crowd.

3. Beside, Besides

'**Beside**' means 'by the side of'. '**Besides**' means 'in addition to'.

Examples :

(i) The daughter was sitting *beside* her mother.

(ii) *Besides* his relatives, he invited his friends also.

4. In, Within

'**In**' means at the expiry of a period of time in future, '**Within**' means before the expiry of a period of time in any tense.

Examples :

(i) She will return *in* a week.

(ii) I shall finish my work *within* a weak.

5. On, Upon

'**On**' is used for things at rest; '**Upon**' is used for things in motion.

Examples :

(i) He is sitting *on* the floor.

(ii) The dog sprang *upon* the table.

6. By, With

'**By**' denotes the agent or doer, '**With**' denotes the instrument with which anything is done.

Examples :

(i) The bird was killed *by* the hunter with an arrow.

(ii) He beat the dog *with* a stick.

(iii) I shall reach here *by* five o'clock.

7. After, In

'**After**' means at the end of a period of time in the past. '**In**' means at the end of a period of time in future.

Examples :

(i) I shall return your book *in* a week.

(ii) He returned the book *after* a week.

8. For, From, Since

'**For**' is used before a noun denoting a period of time with all the tenses. '**From**' is used before a noun or phrase denoting a point of time, it is used in all the tenses. '**Since**' is used before a noun or phrase denoting some point of time and is always produced by a verb in the perfect continuous tense or third form of a verb.

Examples :

(i) We have been playing cards *for* two hours.

(ii) She stayed with her uncle *from* the 15th of March to the 15th of May.

(iii) I have been reading this book *since* morning.

9. Above, Over

'**Above**' means 'higher from', **Over** is used in the following four senses :

(i) In the sense of 'above' :
At noon, the sun is *over* our heads.

(ii) In the sense of 'beyond' :
I cannot get *over* my disappointment.

(iii) In the sense of 'Superiority' :
God *over* all blesses for ever more.

(iv) In the sense of 'Conclusion' :
It is all *over* with me.

10. At, Towards

'**At**' denotes the idea of aim, '**Towards**' denotes the idea of destination.

Examples :

(i) He threw the stone *at* the cat.

(ii) He went *towards* the house.

11. At, In, On

'At' is used as follows :

(i) '**At**' is used with small towns and villages.

Examples :

(a) He was born *at* Sonepat.

(b) He lives *at* village Bangra. (Bangra is a village)

(ii) '**At**' is used before a noun denoting a definite point of time.

Example :

He called on me *at* 9 p.m. yesterday.

'In' is used as follows :

(iii) '**In**' is used with the names of big cities, provinces and countries.

Examples :

(a) His father lives *in* England.

(b) His younger brother lives *in* Calcutta.

(iv) '**In**' is used before the names of months and years.

Example :

His elder sister was born *in* 1972 *in* the month of May.

'**On**' is used with dates and names of days.

Examples :

(a) I joined college *on* the 26th April.

(b) He will leave for Kolkata *on* Wednesday next.

Important Information

1. '**In**' is also used in the following phrases :
In the morning; In the evening, In winter, In summer.

2. **'In'** also denotes a place inside anything.
   He travelled *in* a crowded bus.
3. **'At'** is used in the following phrases :
   At home, *At* the station, *At* work, *At* play.

12. Below, Beneath

Below means 'of lower level in position, dignity and expectation' etc. *Beneath* means 'under'.

Examples :

(i) It is *below* my dignity to talk to her.
(ii) They rested *beneath* the shade of a tree.

13. In, Into, To

'In' expresses Rest or Motion inside anything. **'Into'** expresses Motion towards the inside of anything or change from one medium to another. **'To'** denotes motion from one place to another.

Examples :

(i) The boys are *in* the room.
(ii) Translate this passage from English *into* Hindi.
(iii) Every morning he goes *to* the temple.

14. Till, By, Of, Off

- 'Till' means upto or not earlier than.
- 'By' means not later than.
- 'Of' shows cause, source, separation, quality, contents, possession, apposition, point of reference, space in time etc.
- 'Off' shows separation at a near distance, and detached condition.

Consider the following examples:

(i) I shall work *till* 5 a.m.
(ii) Madhu died *of* cancer.
(iii) The nib *of* the pen is made *of* gold.
(iv) He presented me a bottle *of* perfume.
(v) Our principal is a man *of* principle.
(vi) He lived in the house *of* his friend.
(vii) *By* this time tomorrow, I'll have finished my job.
(viii) My house is *off* the road.
(ix) The book fell *off* the table.

MULTIPLE CHOICE QUESTIONS

Directions: *Tick the correct preposition for the blank in each of the following sentences.*

1. He applied ...... the manager.
A. for B. to
C. with D. by

2. Trust ...... God and do the right.
A. in B. for
C. to D. with

3. She is worthy ...... a prize.
A. with B. for
C. to D. of

4. Mr. Gomes has no taste ...... music.
A. of B. for
C. with D. to

5. You are hard ...... hearing.
A. at B. of
C. with D. for

6. He is sure ...... his success
A. for B. with
C. on D. of

7. Preeti was warned ...... the danger ahead.
A. for B. at
C. of D. about

8. I am thankful ...... you for a good advice.
A. for B. with
C. to D. of

9. Deepak would not surrender ...... the police.
A. with B. to
C. for D. on

10. The small plant in your lawn is very sensitive ...... touch.
A. on B. with
C. to D. about

11. Divya was sure to succeed ..... the examination.
A. for B. in
C. to D. with

12. Geeta was jealous ...... Ravina's beauty.
A. to B. with
C. for D. of

13. He was ignorant ..... what was happening there.
A. for B. of
C. to D. with

14. Your pen is inferior ...... mine.
A. than B. with
C. from D. to

15. Reenu is no match ...... Meenu.
A. to B. for
C. with D. upon

16. It is necessary ...... you to apply for this job.
A. on B. with
C. for D. to

17. Be loyal ...... your country.
A. for B. to
C. on D. with

18. Mukesh is junior ...... me.
A. than B. to
C. from D. of

19. Deepika was innocent ...... the crime.
A. of B. with
C. from D. to

20. I am desirous.... joining the Indian cricket team.
A. for B. of
C. to D. on

ANSWERS

1	2	3	4	5	6	7	8	9	10
B	A	D	B	B	D	D	D	B	D
11	**12**	**13**	**14**	**15**	**16**	**17**	**18**	**19**	**20**
B	D	B	D	B	D	B	B	A	B

SYNONYMS

A synonym is a word which conveys a meaning similar to the given word.

REMEMBER

Words	*Synonyms*
Add	Increase
Adequate	Enough
Adjust	Adapt
All	Aggregate
Allow	Permit
Abode	Dwelling
Apt	Proper
Assess	Appraise
Accuse	Calumniate
Abashed	Timid
Annoy	Displease
Ample	Enough, Sufficient
Amplify	Increase
Apathetic	Unenthusiastic
Accost	Address
Authentic	True

Words	*Synonyms*
Adjust	Fit
Approve	Assent, Allow, Accept
Adapt	Conform
Adversary	Opponent, Rival, Competitor
Beat	Whack
Benign	Kind
Breeze	Zephyr
Baffle	Puzzle
Booty	Spoil
Beauty	Charm
Beast	Animal
Bandit	Robber
Blaze	Shine
Bond	Tie
Bend	Twist
Bate	Diminish
Beg	Plead
Barbaric	Wild, Savage
Bashful	Shy, Reserved

Words	Synonyms
Begin	Start
Blend	Mix, Mingle
Bizarre	Funny
Below	Under
Bedevil	Confuse
Bemoan	Lament
Babble	Nonsense
Blame	Fault
Behaviour	Demeanour
Call	Accost
Copy	Imitate
Close	Shut
Caress	Love
Camp	Stay
Connect	Attach
Cut	Injure, Curtail
Cling	Stick
Conical	Funny
Convey	Carry
Conspicuous	Prominent
Cheerful	Happy, Pleasant
Curtail	Decrease
Cheerless	Sad, Dejected
Curious	Strange
Circumstance	Factor, Situation, Condition
Competent	Capable
Congruent	Overlapping
Cope	Deal, Endure
Confident	Sure
Complex	Intricate
Cajole	Coax, Flatter
Cunning	Crafty
Delectable	Joyful, Delightful
Devilish	Diabolical
Delicate	Soft
Devil	Fiend
Delay	Postpone
Dislike	Repugnance
Destroy	Ruin

Words	Synonyms
Dwell	Live, Dilate
Declare	Pronounce
Drunk	Flushed
Deficient	Lacking
Damn	Condemn, Curse
Decrease	Diminish
Destruction	Devastation
Efficient	Competent
Ethnic	Racial
Enthral	Enslave
Earnest	Serious
Envious	Jealous
Ending	Final
Egg	Incite
Extempore	At once
Extensive	Far-ranging
Extra	Surplus
Existence	Life
Exceed	Overstep
Enormous	Vast
Excessive	Superfluous
Free	Unhindered
Frigid	Cold
Feed	Cater
Fame	Reputation
Frame	Make
First	Initial
Frighten	Terrorise, Intimidate
Fervent	Fervid
Fall	Decline
Feeble	Frail
Fickle	Changeable
Finish	Conclude
Fraud	Deception
Forgiving	Placable
Grow	Develop
Greed	Avidity
Greet	Welcome
Grave	Serious
Group	Constellation

Words	Synonyms
Given	Bestowed
Gratitude	Thankfulness
Have	Possess
Hire	Rent
Hit	Strike
Handsome	Beautiful
Hinder	Prevent
Heap	Pile
Hope	Expect
Hard	Harsh
Help	Aid
Hymn	Song
Henpecked	Enslaved
Hoodwink	Mystify, Cheat
Humble	Polite, Urbane, Modest
Harass	Vex, Trouble
Impart	Instil
Intact	Untouched
Instal	Establish
Indict	Impeach
Imitate	Ape
Instigate	Incite
Initiate	Start, Introduce
Inimical	Unfriendly
Insufferable	Intolerable
Impartiality	Justice
Jolly	Merry
Joyful	Delectable
Join	Conjoin
Kind	Benign
Kill	Murder
Kindred	Similar
Kinship	Relationship
Keen	Sharp
Knowledge	Scholarship
Lazy	Slothful
Large	Substantial, Gargantuan
Listless	Careless, Lackadaisical
Lax	Loose
Little	Small
Lifelike	Realistic

Words	Synonyms
Lofty	High
Lenient	Soft, Gentle
Lacking	Deficient, Wanting
Lessen	Decrease
Middleclass	Bourgeois
Mitigate	Lessen, Abate
Modesty	Humility, Lowliness
Mix	Mingle, Blend
Mixture	Mingling
Mixed	Assorted
Modify	Decrease
Mean	Imply
Multifarious	Varied
Miscarry	Abort
Note	Notice
Noble	Stately
Native	Indigenous
Needful	Necessary
Notify	Declare
Nervous	Shaky, Tremulous, Timid
Natural	Spontaneous
Near	Close
Normal	Natural
Offend	Displease
Oppress	Persecute, Tyrannize
Opponent	Adversary
Obstruct	Hinder, Check
Offence	Fault
Offender	Villain
Overstep	Exceed
Overlapping	Congruent
Occult	Mystic
Profane	Unholy
Patience	Forbearance
Pornographic	Obscene
Plenitude	Abundance
Prominent	Important
Prodigal	Spender
Procrastinate	Postpone
Promote	Develop, Honour
Persecute	Tyrannise

Words	*Synonyms*
Profess	Claim
Pliant	Flexible
Plebian	Common
Polished	Sophisticated
Quake	Shake
Quit	Leave
Queer	Eccentric
Quell	Suppress
Quantify	Allot
Reply	Answer
Relinquish	Retire
Read	Peruse
Relation	Reference
Render	Do
Remainder	Residuals
Repeat	Reiterate
Repentant	Contrite
Retaliative	Retaliatory
Rumour	Hearsay
Reveal	Divulge
Ritualistic	Ceremonious
Soft	Delicate
Sort	Kind, Choose, Select
Selfish	Egoistic
Sensual	Earthly
Suppress	Quell, Check
Stimulate	Provoke
Tasteless	Insipid
Travel	Journey
True	Authentic, Faithful, Truthful
Turbulence	Turmoil
Tragedy	Calamity
Tasteful	Tasty, Delicious
Touching	Painful
Thankful	Grateful
Tremendous	Great, Huge
Tough	Strong
Terminate	Conclude, End
Theory	Doctrine
Tell	Relate
Tremble	Shake, Shiver
Urge	Spur

Words	*Synonyms*
Unbeaten	Unsubdued
Use	Utilize, Practise
Underhand	Unfair, Undue
Unfair	Unjust
Unravel	Reveal, Divulge
Unimportant	Common
Unconcerned	Apathetic
Unimitated	Inimitable
Unfortunate	Unlucky
Understand	Perceive, Comprehend
Vain	Proud, Haughty, Conceited, Shameless
Vale	Valley, Dale, Dell
Vice	Fault
Virtue	Quality
Veracity	Reality
Value	Price, Prize
Vex	Tease
Vibrate	Quiver, Shake
Violent	Excessive
Vivid	Clear, Lucid
Victory	Triumph
Vulgar	Indecent
Virtuous	Honest
Variegated	Varied, Multifarious
Well	Good
Yell	Cry, Shout
Yonder	There
Yearn	Wish, Desire
Yoke	Slavery
Zest	Earnestness, Enthusiasm
Zealous	Earnest

ANTONYMS

A antonym is a word which conveys a meaning opposite to the given word.

REMEMBER

Words	*Antonyms*
Abhor	Love
Abnormal	Normal
Able	Unable
Acceptable	Unacceptable
Adequate	Inadequate
Amusing	Boring

Words	*Antonyms*
Angry	Calm
Apex	Bottom
Attract	Repel
Bad	Good
Barren	Fertile
Beautiful	Ugly
Bitter	Sweet
Brave	Cowardly
Brief	Lengthy
Bright	Dull
Calm	Violent
Careful	Careless
Clear	Vague, Cloudy
Cold	Hot
Cruel	Kind
Dear	Cheap
Deep	Shallow
Difficult	Easy
Direct	Indirect
Dishonest	Honest
Disobey	Obey
Encourage	Discourage
Enormous	Tiny
Excellent	Bad
Expensive	Cheap
Eat	Fast
Fair	Unfair
Fake	Authentic
False	True
Famous	Notorious
Fool	Genius
Generous	Miserly
Genius	Fool
Genuine	Unauthentic
Gigantic	Tiny
Glad	Depressed
Good	Bad
Great	Little
Happy	Sad

Words	*Antonyms*
Hard	Soft
Hate	Love
Honest	Dishonest
Idle	Busy
Immoral	Moral
Include	Exclude
Incorrect	Correct
Intelligent	Unintelligent
Kind	Cruel
Like	Dislike
Long	Short
Lucid	Vague
Major	Minor
Naive	Experienced
Nadir	Apex
Neat	Clumsy
Obedient	Disobedient
Obscure	Clear
Oppose	Support
Optimistic	Pessimistic
Out	In
Patience	Impatience
Peaceful	Belligerent
Pious	Impious
Polite	Impolite
Potent	Impotent
Prominent	Unimportant
Proper	Improper
Pure	Impure
Quick	Slow
Quiet	Disturbance
Real	False, Unreal
Reject	Select, Choose
Reliable	Unreliable
Respect	Disrespect
Right	Wrong
Robust	Feeble, Weak
Sad	Happy

Words	*Antonyms*
Secret	Open
Sensible	Insensible
Severe	Mild
Sharp	Blunt
Simple	Complex
Sociable	Unsociable
Tall	Short
Tidy	Untidy
Uncanny	Canny
Violent	Calm
Vivid	Vague

Words	*Antonyms*
Strong	Weak
Big	Small
Easy	Difficult
Fast	Slow
High	Low
Catchy	Unattractive
Ugly	Handsome, Beautiful, Tidy
Tasty	Insipid
Sonorous	Harsh

MULTIPLE CHOICE QUESTIONS

Directions (Qs. 1 to 20): *In the following questions choose the word which best expresses the meaning of the given word.*

1. ABSURD
A. Foolish B. Simple
C. Courageous D. Silly

2. ABANDON
A. Lose B. Profit
C. Vacate D. Foil

3. CAJOLE
A. Pause B. Lenient
C. Blast D. Lure

4. COMBAT
A. Fight B. Conflict
C. Shoot D. Quarrel

5. LAMENT
A. Condone B. Console
C. Complain D. Contribution

6. DEBACLE
A. Disgrace B. Defeat
C. Collapse D. Decline

7. SHIVER
A. Fear B. Tremble
C. Shake D. Ache

8. TORTURE
A. Terror B. Harassment
C. Torment D. Tranquility

9. LAUDABLE
A. Lovable B. Commendable
C. Profitable D. Oblivious

10. FIXED
A. Sterile B. Static
C. Stubborn D. Parennial

11. QUEER
A. Unfamiliar B. Cute
C. Curious D. Strange

12. SUFFICIENT
A. Fit B. Proper
C. Adequate D. Vast

13. GLOSS
A. Brightness B. Soothing
C. Rubbing D. Miracle

14. LONGING
A. Prune B. Apathy
C. Curtail D. Craving

15. JEER
A. Applaud B. Magnanimity
C. Avoid D. Scoff

16. ZENITH
A. Minimum B. Nadir
C. Plant D. Peak

17. GARB
A. Distort B. Dress
C. Trivial D. Rage

18. ABHOR
A. Rude B. Reconcile
C. Crave D. Detest

19. YIELD
A. Shum B. Incisive
C. Retain D. Surrender

20. YOKE
A. Twist B. Release
C. Link D. Extra

Directions (Qs. 21 to 38): *In the following questions choose the word which best expresses the opposite of the given word.*

21. TRAGIC
A. Dramatic B. Strong
C. Gentle D. Comic

22. ORAL
A. Verbal B. Sane
C. Minor D. Written

23. ADMIRE
A. Hate B. Unlike
C. Dislike D. Enough

24. VIOLENT
A. Gentle B. Savage
C. Haughty D. Decline

25. ADVERSITY
A. Windfall B. Inprosperity
C. Prosperity D. Slave

26. GENUINE
A. Spurious B. Obscure
C. Countless D. Apathetic

27. GRUDGE
A. Essence B. Guile
C. Goodwill D. Ill-will

28. STIFF
A. Soft B. Courteous
C. Lively D. Flexible

29. VANITY
A. Conceit B. Pride
C. Ostentious D. Humility

30. FRONT
A. Upper B. Unusual
C. Back D. Rear

31. ATTRACT
A. Lured B. Longing
C. Repel D. Disguise

32. COMFORT
A. Discomfort B. Discontent
C. Uncomfort D. Miscomfort

33. WELCOME
A. Repel B. Accept
C. Resist D. Fight

34. TACTFUL
A. Naive B. Loose
C. Strict D. Uncivilized

35. DUTIFUL
A. Harmful B. Watchful
C. Forgetful D. Remiss

36. RIGID
A. Flux B. Adoptable
C. Yielding D. Adaptable

37. RARE
A. Petty B. Poor
C. Small D. Common

38. ZEAL
A. Despair B. Calmness
C. Passiveness D. Indifference

ANSWERS

1	**2**	**3**	**4**	**5**	**6**	**7**	**8**	**9**	**10**
D	C	D	A	C	C	B	C	B	B
11	**12**	**13**	**14**	**15**	**16**	**17**	**18**	**19**	**20**
D	C	A	D	D	D	B	D	D	C
21	**22**	**23**	**24**	**25**	**26**	**27**	**28**	**29**	**30**
D	D	C	A	C	A	C	D	D	D
31	**32**	**33**	**34**	**35**	**36**	**37**	**38**		
C	A	C	A	D	D	D	D		

3. Sentence Completion

It is such an exercise which starts with the primary schools and continues in the highest level of competitive examinations. One must practise it regularly to score well.

Directions (Qs. 1 to 15): *Pick out the most effective word(s) from the given words to fill in the blanks to make the sentence meaningfully complete.*

1. The student ...... that book from the library to study at home.
A. issued B. borrowed
C. hired D. lent

2. I wish I ...... a king.
A. was B. am
C. should be D. were

3. He ...... to listen to my arguments and walked away.
A. denied B. disliked
C. objected D. refused

4. The flow of blood was so ...... that the patient died.
A. intense B. adequate
C. profuse D. extensive

5. When I met her yesterday, it was the first time I ...... her since Christmas.
A. saw B. have seen
C. had seen D. have been seing

6. Can you pay ...... all these articles?
A. for B. of
C. off D. out

7. I ...... you to be at the party this evening.
A. expect B. hope
C. look forward to D. desire

8. ...... being a handicapped person, he is very cooperative and self-reliant.
A. Because B. Although
C. Since D. Despite

9. The child broke ...... from his mother and ran towards the painting.
A. away B. after
C. down D. with

10. With his ...... income, he finds it difficult to live a comfortable life.
A. brief B. sufficient
C. meagre D. huge

11. He could ...... a lot of money in such a short time by using his intelligence and working hard.
A. spend B. spoil
C. exchange D. accumulate

12. Though the brothers are twins, they look ......
A. alike B. handsome
C. indifferent D. different

13. Unfavourable weather conditions can ..... illness.
A. cure B. detect
C. treat D. enhance

14. No sooner did the bell ring, ... the actor started singing.
A. when B. than
C. after D. before

15. If I ...... realised it, I would not have acted on his advice.
A. was B. had
C. were D. have

Directions (Qs. 16 to 25): *In each question, an incomplete statement (Stem) followed by four fillers* *is given. Pick out the best one which can complete the incomplete stem correctly and meaningfully.*

16. Unless you work harder you will fail, means .....
- A. if you fail you will work harder.
- B. you must at least plan well than you will not fail.
- C. hardly you will fail if you do not desire so.
- D. if you do not put more efforts, then you will fail.

17. Even if it rains I shall come, means .....
- A. if I come it will not rain.
- B. if it rains I shall not come.
- C. I will certainly come whether it rains or not.
- D. whenever there is rain I shall come.

18. Dinesh is as stupid as he is lazy means .....
- A. Dinesh is stupid because he is lazy.
- B. Dinesh is lazy because he is stupid.
- C. Dinesh is either stupid or lazy.
- D. Dinesh is equally stupid and lazy.

19. He is so lazy that he .....
- A. cannot depend on others for getting his work done.
- B. cannot delay the schedule of completing the work.
- C. can seldom complete his work on time.
- D. dislike to postpone the work that he undertakes to do.

20. He always stammers in public meetings, but his today's speech .....
- A. was fairly audible to everyone present in the hall.
- B. was not received satisfactorily.
- C. could not be understood properly.
- D. was free from that defect.

21. In order to raise the company's profit, the employees .....
- A. demanded two additional increments.
- B. decided to go on paid holidays.
- C. requested the management to implement new welfare schemes.
- D. offered to work overtime without any compensation.

22. Although, he is reputed for making very candid statements, ...
- A. his today's speech was not fairly audible.
- B. his promises had always been realistic.
- C. his speech was very interesting.
- D. his today's statements were very ambiguous.

23. I felt somewhat more relaxed .....
- A. but tense as compared to earlier.
- B. and tense as compared to earlier.
- C. as there was already no tension at all.
- D. and tension-free as compared to earlier.

24. With great efforts his son succeeded in convincing him not to donate his entire wealth to an orphanage .....
- A. and lead the life of a wealthy merchant.
- B. but to a home for the forsaken children.
- C. and make an orphan of himself.
- D. as the orphanage needed a lot of donations.

25. Even though it is a very large house, ...
- A. there is a lot of space available in it for children.
- B. there is hardly any space available for children.
- C. there is no dearth of space for children.
- D. the servants take a long time to clean it.

ANSWERS

1	2	3	4	5	6	7	8	9	10
B	D	D	C	C	A	A	D	A	C
11	**12**	**13**	**14**	**15**	**16**	**17**	**18**	**19**	**20**
D	D	D	B	B	D	C	D	C	D
21	**22**	**23**	**24**	**25**					
D	D	D	C	B					

4. Spotting Errors

The most common errors in English are of spellings, grammar and usage of words. By regular practice, the errors can be easily spotted and minimised.

MULTIPLE CHOICE QUESTIONS

Directions: *In the following questions some of the sentences have errors and some are correct. Find out which part of a sentence has an error, the number of that part is your answer. If a sentence is free from errors, then your answer is D i.e., No error.*

1. (A) Either Ram or/(B) you is responsible/(C) for this action./(D) No error.
2. (A) The student flatly denied/(B) that he had copied/(C) in the examination hall./(D) No error.
3. (A) By the time you arrive tomorrow/(B) I have finished/(C) my work./(D) No error.
4. (A) The captain with the members of his team/ (B) are returning/(C) after a fortnight./(D) No error.
5. (A) After returning from/(B) an all-India tour/ (C) I had to describe about it./(D) No error.
6. (A) The teacher asked his students/(B) if they had gone through/(C) either of the three chapters included in the prescribed text./(D) No error.
7. (A) Do you know/(B) how old were you/(C) when you came here?/(D) No error.
8. (A) Beware of/(B) a fair-weather friend/(C) who is neither a friend in need nor a friend indeed./ (D) No error.
9. (A) Copernicus proved/(B) that Earth/(C) moves round the Sun./(D) No error.
10. (A) The property/(B) was divided/(C) among the two brothers./(D) No error.
11. (A) I am quite certain/(B) that the lady is not only greedy/(C) but miserly./(D) No error.
12. (A) The brilliant success in the examination/ (B) as well as his record in sports/(C) deserves high praise./(D) No error.
13. (A) I cannot find/(B) where has he gone/(C) though I have tried may best./(D) No error.
14. (A) If I was/(B) the Prime Minister of India/(C) I would work wonders/(D) No error.
15. (A) If it weren't/(B) for you,/(C) I wouldn't be alive today./(D) No error.
16. (A) He looked like a lion/(B) baulked from/ (C) its prey./(D) No error.
17. (A) Widespread flooding/(B) is affecting/(C) large areas of the villages./(D) No error.
18. (A) If we really set to/(B) we can get the whole house/(C) cleaned in an afternoon./(D) No error.
19. (A) It's arrogant for you/(B) to assume you'll/ (C)win every time./(D) No error.
20. (A) The two books are the same/(B) except for the fact that this/(C) has an answer in the back./(D) No error.

21. (A) Your husband doesn't/(B) believe that you are older/(C) than I./(D) No error.

22. (A) I could not/(B) answer to/(C) the question./(D) No error.

23. (A) Two years passed/(B) since/(C) my cousin died./(D) No error.

24. (A) I am learning English/(B) for ten years/(C) without much effect./(D) No error.

25. (A) Ramesh has agreed/(B) to marry with the girl/(C) of his parent's choice./ (D) No error.

26. (A) When he was arriving./(B) the party was/(C) in full swing./(D) No error.

27. (A) The most studious boy/(B) in the class/(C) was made as the captain./(D) No error.

28. (A) I am participating/(B) in the two-miles race/(C) tomorrow morning./(D) No error.

29. (A) When the boy committed a mistake/(B) the teacher made him to do/(C) the sum again./(D) No error.

30. (A) Whenever a person lost anything/(B) the poor folk around/(C) are suspected./(D) No error.

ANSWERS

1	2	3	4	5	6	7	8	9	10
B	D	B	B	C	C	D	D	B	C
11	**12**	**13**	**14**	**15**	**16**	**17**	**18**	**19**	**20**
C	D	B	A	C	C	C	A	A	C
21	**22**	**23**	**24**	**25**	**26**	**27**	**28**	**29**	**30**
C	B	A	A	B	A	C	B	B	A

EXPLANATORY ANSWERS

1. Replace 'is' by 'are'.
2. No error.
3. Replace 'have' by 'would have'.
4. Replace 'are' by 'is'.
5. Replace 'had to describe' by 'described'.
6. Replace 'either' by 'any'.
7. No error.
8. No error.
9. Omit 'that'.
10. Replace 'among' by 'between'.
11. Add 'also'.
12. No error.
13. Replace 'has he' by 'he has'.
14. Replace 'was' by 'were'.
15. Replace 'wouldn't be' by 'would not have been'.
16. Replace 'its' by 'his'.
17. Replace 'areas' by 'area'.
18. Replace 'set to' by 'set on'.
19. Replace 'for' by 'of'.
20. Replace 'in' by 'on'.
21. Replace 'I' by 'me'.
22. Omit 'to'.
23. Replace 'passed' by 'have passed'.
24. Replace 'am' by 'have been'.
25. Omit 'with'.
26. Replace 'was arriving' by 'arrived'.
27. Omit 'as'.
28. Replace 'in' by 'at'.
29. Omit 'to'.
30. Replace 'lost' by 'loses'.

5. One Word Substitution

There are many single words in English language which can be perfectly used for a number of words. These words help in expressing ideas in a short and correct manner for the right occasion. Such words not only increase the vocabulary but also enable you to economise in the use of words to a great extent.

Multiple Word Expression	*Substitution*
One who always looks towards the bright side of things	Optimist
One who always looks towards the dark side of things	Pessimist
The time when one develops from a child into an adult	Adolescence
The process of growing more plants in order to form a forest.	Afforestation
The science which deals with farming	Agriculture
From some other country or place etc.	Alien
A term, etc. giving more than one meaning	Ambiguous
A vehicle which is used to carry sick persons	Ambulance
An animal which can live both in water and on land	Amphibian
A lawless situation when there is no government	Anarchy
Belonging to the history of thousands of years old	Ancient
Once a year	Annual
A very old object but still valuable	Antique
Words of opposite meanings	Antonyms
Words of similar meanings	Synonyms
Signatures of a famous person	Autograph
A government led by one person with absolute authority	Autocracy
A written work of one's own life history	Autobiography
A person who has never been married	Bachelor
A person usually having no hair on his head	Bald
A place where one can deposit money and get interest	Bank
A person who cuts our hair	Barber
A building/group of buildings where soldiers live	Barracks
A person who makes buns and biscuits	Baker
A person who lives by asking people for food and money without doing any useful job	Beggar
The crime of having married to two persons at the same time	Bigamy
The branch of science which deals with the study of plants	Botany
Able to speak two languages	Bilingual

Multiple Word Expression	*Substitution*
Able to speak more than two languages	Polyglot
The branch of science which deals with the living organisms	Biology
A powerful snow storm	Blizzard
A great successful book or movie	Blockbuster
A short news on the radio or TV	Bulletin
A system in which the most important works are organised by the government officials	Bureaucracy
A person who has no vision in his eyes	Blind
A page or a series of pages on which the information of days, weeks, months, etc. is given	Calendar
A person who eats human flesh	Cannibal
A complete list of items often arranged alphabetically	Catalogue
A sudden disaster	Catastrophe
A period of 100 years	Century
A branch of science which deals with chemicals	Chemistry
A printed leaf usually issued by banks that we sign to carry certain financial deal	Cheque
A person who makes or mends shoes	Cobbler
A group of people who has been chosen by others to make decisions on their own	Committee
A building in which nuns live	Convent
An animal which feeds on other animals	Carnivorous
A person who does criticism	Critic
A person who cannot hear	Deaf
A condition in which one loses a lot of water from one's body because of vomiting, etc.	Dehydration
A system of government in which the people cast their votes to elect their leaders	Democracy
The study of skin problems	Dermatology
A long piece of land covered with sand	Desert
The art of managing relationships between countries	Diplomacy
A piece of information about the words in a book form	Dictionary
A piece of information about the telephone numbers of the people in a book from	Directory
A person in charge of a newspapers, magazine etc.	Editor
A person who thinks he is better than the others	Egoist
To leave your country and settle in some other country	Emigrate
A book or series of books giving almost all knowledge about an area or some persons etc.	Encyclopaedia
Study of insects	Entomology
Time when day and night are of the same duration	Equinox
To sell things out of the country	Export
To purchase things from some other country	Import
A plant or animal no longer in existence	Extinct
A situation when there is a shortage of food for a long period of time	Famine
An amount of money that we pay for some action or services	Fee
Related to women	Feminine
An animal strong and aggressive	Ferocious

Multiple Word Expression	*Substitution*
A piece of land where plants grow easily from the soil that is favourable to them	Fertile
A work of literature having some imaginary events	Fiction
A large amount of water covering certain area	Flood
A person who sells flowers	Florist
A religious ceremony for burying or cremating a dead person	Funeral
A substance which kills fungus	Fungicide
A person studying or having studied the diseases and the related things of female reproductory system	Gynaecologist
The murder of the person of the same group race or country	Genocide
A substance which kills germs	Germicide
A situation in which many people die because of fire during war	Holocaust
The act of killing a person deliberately	Homicide
A word having the pronunciation as the other one does but it differs in meaning	Homophone
A word having the same spelling as the other one does but it is pronounced in some other way	Homonym
A person who is attracted towards the person of the same sex	Homosexual
Go across and parallel to the ground	Horizontal
A substance which kills the insects	Insecticide
That cannot be corrected	Incorrigible
That cannot be defeated	Invincible
That cannot be eaten	Inedible
That cannot be seen	Invisible
A place in a school or college where books are kept for the benefit of students, teachers etc.	Library
A place in a school or college where scientific experiments are performed	Laboratory
An official who is a judge in the lowest court	Magistrate
A piece of music or a book before it is printed	Manuscript
Related to men	Masculine
One who believes in the existence of God	A theist
One who does not believe in the existence of good	An atheist
That can be believed	Credible
That cannot be believed	Incredible
That which dissolves in a solvent	Soluble
That which does not dissolves in a solvent	Insoluble
Hard writing that can be read	Legible
Hard writing that cannot be read	Illegible
A person who does jobs beneficial to mankind	Philanthropist
A person who goes on foot	Pedestrian
A person who fights for his own country	Patriot
An act of killing oneself	Suicide
A woman whose husband is dead	Widow
A man whose wife is dead	Widower
A person who eats vegetarian and non-vegetarian diets	Omnivorous

Multiple Word Expression	*Substitution*
Something which is everywhere at the same time	Omnipresent
One who knows everything	Omniscient
A child who does not have parents	Orphan
An award etc. given after the death of the person	Posthumous
The place where animals are kept for amusement and to increase the knowledge of the public	Zoo
The science which deals with the study of animals	Zoology

MULTIPLE CHOICE QUESTIONS

Directions: *In questions given below, out of the four alternatives, choose the one which can be substituted for the given words/sentences.*

1. Something that relates to everyone in the world
A. General B. Common
C. Usual D. Universal

2. An expression of mild disapproval
A. Warning B. Denigration
C. Impertinence D. Reproof

3. One who is not easily pleased by anything
A. Maiden B. Medieval
C. Precarious D. Fastidious

4. Murder of a king
A. Infanticide B. Matricide
C. Genocide D. Regicide

5. A remedy for all diseases
A. Stoic B. Marvel
C. Panacea D. Recompense

6. A dramatic performance
A. Mask B. Mosque
C. Masque D. Mascot

7. Study of birds
A. Orology B. Optology
C. Ophthalmology D. Ornithology

8. Ready to believe
A. Credulous B. Credible
C. Creditable D. Incredible

9. Incapable of being seen through
A. Ductile B. Opaque
C. Obsolete D. Potable

10. One who eats everything
A. Omnivorous B. Omniscient
C. Irresistible D. Insolvent

11. A place where bees are kept is called
A. An apiary B. A mole
C. A hive D. A sanctuary

12. One who cannot be corrected
A. Incurable B. Incorrigible
C. Hardened D. Invulnerable

13. One who is in charge of a museum
A. Curator B. Supervisor
C. Caretaker D. Warden

14. Continuing fight between parties, families, clans, etc.
A. Enmity B. Feud
C. Quarrel D. Skirmish

15. A voice loud enough to be heard
A. Audible B. Applaudable
C. Laudable D. Oral

16. A paper written by hand
A. Handicraft B. Manuscript
C. Handiwork D. Thesis

17. Habitually silent or talking little
A. Serville B. Unequivocal
C. Taciturn D. Synoptic

18. To slap with a flat object
A. Chop B. Hew
C. Gnaw D. Swat

19. A person who speaks many languages
A. Linguist B. Monolingual
C. Polyglot D. Bilingual

20. A light sailing-boat built specially for racing
A. Canoe B. Yacht
C. Frigate D. Dinghy

21. A fixed orbit in space in relation to earth
A. Geological B. Geo-synchronous
C. Geo-centric D. Geo-stationary

22. A style in which a writer makes a display of his knowledge
A. Pedantic B. Verbose
C. Pompous D. Ornate

23. A religious discourse
A. Preach B. Stanza
C. Sanctorum D. Sermon

24. A place that provides refuge
A. Asylum B. Sanatorium
C. Shelter D. Orphanage

25. Detailed plan of a journey
A. Travelogue B. Travelkit
C. Schedule D. Itinerary

26. A person who insists on something
A. Disciplinarian B. Stickler
C. Instantaneous D. Boaster

27. A drawing on transparent paper
A. Red print B. Blue print
C. Negative D. Transparency

28. One who believes that all things and events in life are predetermined is a
A. Fatalist B. Puritan
C. Egoist D. Tyrant

29. A school boy who cuts classes frequently is a
A. Defeatist B. Sycophant
C. Truant D. Martinet

30. The act of violating the sanctity of the church is
A. Blasphemy B. Heresy
C. Sacrilege D. Desecration

31. A place where monks live as a secluded community
A. Cathedral B. Diocese
C. Convent D. Monastery

32. One who is fond of fighting
A. Bellicose B. Aggressive
C. Belligerent D. Militant

33. Tending to move away from the centre or axis
A. Centrifugal B. Centripetal
C. Axiomatic D. Awry

34. Words inscribed on tomb
A. Epitome B. Epistle
C. Epilogue D. Epitaph

35. Leave or remove from a place considered dangerous
A. Evade B. Evacuate
C. Avoid D. Exterminate

36. Original inhabitants of a country
A. Abroge B. Aborger
C. Aborgory D. Aborigins

37. Government by the officials
A. Theocracy B. Plutocracy
C. Bureaucracy D. Democracy

38. Incapable of being exhausted
A. Inexhaustible B. Inaexhaustible
C. Exhaustable D. Non-tired

39. A person of good understanding, knowledge and reasoning power
A. Expert B. Intellectual
C. Snob D. Literate

40. One absorbed in his own thoughts and feelings rather than in things outside
A. Scholar B. Recluse
C. Introvert D. Intellectual

ANSWERS

1	2	3	4	5	6	7	8	9	10
D	D	D	D	C	C	D	A	B	A
11	**12**	**13**	**14**	**15**	**16**	**17**	**18**	**19**	**20**
A	B	A	B	A	B	C	D	A	B
21	**22**	**23**	**24**	**25**	**26**	**27**	**28**	**29**	**30**
D	A	D	A	D	B	D	A	C	C
31	**32**	**33**	**34**	**35**	**36**	**37**	**38**	**39**	**40**
D	A	A	D	B	B	C	A	B	C

6. Spelling Errors

There are thousands of words in English language. It is difficult to remember the spellings and meanings of all at once. Try to learn as many as you can. Use a dictionary regularly.

Directions: *Find the correctly spelt words.*

1. A. Damage B. Dammage
C. Damaige D. Dammege

2. A. Efficiant B. Effecient
C. Efficient D. Eficient

3. A. Schedule B. Schdule
C. Schedale D. Schedeule

4. A. Occurad B. Occurred
C. Ocurred D. Occured

5. A. Grieff B. Grief
C. Grieef D. Grrief

6. A. Guarantee B. Garuntee
C. Guaruntee D. Gaurantee

7. A. Meddicine B. Medicine
C. Medicene D. Medicinne

8. A. Benefeted B. Benefitted
C. Benifited D. Benefited

9. A. Acommodation B. Acomodation
C. Accomodation D. Accommodation

10. A. Querrelsome B. Quarrelsame
C. Quarrelsome D. Querralsome

11. A. Sympathetic B. Smypathetic
C. Sympothetic D. Sympethetic

12. A. Prograssive B. Progressive
C. Progresive D. Prograsive

13. A. Uncivilized B. Uncevilized
C. Uncivillized D. Uncevelized

14. A. Extravagant B. Extreragent
C. Extreregant D. Extravegent

15. A. Missunderstood B. Miesunderstood
C. Misunderstood D. Misunderstod

16. A. Belligerent B. Beligirent
C. Belligarant D. Belligerrent

17. A. Astonished B. Astronished
C. Astoneshed D. Asstonished

18. A. Sincerely B. Sencerely
C. Sincerelly D. Sincerrely

19. A. Rigourous B. Rigerous
C. Rigorous D. Regerous

20. A. Satellite B. Sattellite
C. Satelite D. Sattelite

21. A. Pesanger B. Passenger
C. Pessenger D. Pasanger

22. A. Humurous B. Humorous
C. Humoreus D. Humorrous

23. A. Exeggerate B. Exaggerate
C. Exadgerate D. Exagerate

24. A. Fariegn B. Forein
C. Foriegn D. Foreign

25. A. Excesive B. Excessive
C. Exccessive D. Exccesive

26. A. Forcaust B. Forcast
C. Forecast D. Forecaste

27. A. Paralleted B. Paralelled
C. Parralleled D. Parallelled

28. A. Ocasion B. Occassion
C. Occasion D. Ocassion

29. A. Boquet B. Bouquet
C. Bouquete D. Bouquette

30. A. Chettering B. Chaterring
C. Chattering D. Chatering

31. A. Discourage B. Disscourage
C. Discourege D. Discaurage

32. A. Curageous B. Courageous
C. Courrageous D. Couregeous

33. A. Abandon B. Abanddon
C. Abendon D. Abbandon

34. A. Embarassment
B. Emberrassement
C. Embarrassment
D. Embbaresment

35. A. Eccintric B. Eccentrie
C. Eccentric D. Eccintrie

36. A. Occasional B. Occassional
C. Occesional D. Occessional

37. A. Querrel B. Querral
C. Quarrel D. Quarel

38. A. Contrebution B. Contribution
C. Contributtion D. Conterbution

39. A. Desgrace B. Disgrece
C. Disgrice D. Disgrace

40. A. Harassment B. Herassment
C. Harasment D. Harassmient

41. A. Imaginative B. Imeginative
C. Imagenative D. Imaginetive

42. A. Suficient B. Suficiant
C. Sufficient D. Sufficiant

43. A. Adequate B. Edequate
C. Adaquete D. Edaquete

44. A. Exparienced B. Experianced
C. Experienced D. Experrienced

45. A. Flatering B. Fletering
C. Flattering D. Fletaring

46. A. Cuttiveted B. Culltrivated
C. Cultivated D. Caltivated

47. A. Praiceworthy B. Peiseworthy
C. Praiseworthy D. Praisaworthy

48. A. Profesional B. Professionel
C. Professional D. Profissional

49. A. Ameteur B. Amateur
C. Amataur D. Amateor

50. A. Unfevourable B. Unfevaurable
C. Unfavourable D. Unfivourable

ANSWERS

1	2	3	4	5	6	7	8	9	10
A	C	A	B	B	A	B	B	D	C
11	**12**	**13**	**14**	**15**	**16**	**17**	**18**	**19**	**20**
A	B	A	A	C	A	A	A	C	A
21	**22**	**23**	**24**	**25**	**26**	**27**	**28**	**29**	**30**
B	B	B	D	B	C	A	C	B	C
31	**32**	**33**	**34**	**35**	**36**	**37**	**38**	**39**	**40**
A	B	A	C	C	A	C	B	D	A
41	**42**	**43**	**44**	**45**	**46**	**47**	**48**	**49**	**50**
A	C	A	C	A	C	C	C	B	C

GENERAL AWARENESS

History

ANCIENT

HARAPPAN/INDUS CIVILIZATION (2500 BC–1750 BC)

- The most prevalent name – Indus Civilization (the largest concentration of settlement – along the Indus river valley).
- *John Marshall* was the first scholar to use the term 'Indus Civilization'.
- The Indus Civilization belongs to Proto-Historic Period (Chalcolithic Age/Bronze Age).
- **Systematic town planning** was based on grid system; burnt bricks were used to construct houses; well-managed drainage system; fortified citadel; absence of iron implements.
- The Indus Civilization was spread over Sindh, Baluchistan, Punjab, Haryana, Rajasthan, Gujarat, Western U.P. & Northern Maharashtra.

Indus Valley Civilization

Site	River	District	Province/ State	Country	Excavators
Harappa	Ravi	Montgomery	Punjab	Pakistan	Daya Ram Sahni (1921), Madho Swaroop Vatsa (1926), Wheeler (1946)
Mohenjodaro (Nakhilistan *i.e.* Oasis of Sindh	Indus	Larkana	Sindh	Pakistan	Rakhal Das Bannerji (1922), Mackay (1927) Wheeler (1930)
Chanhudaro	Indus	Nawabshah	Sindh	Pakistan	Mackay (1925), N.G. Mazumdar (1931)
Lothal	Ghaggar	Kathiyawar	Gujarat	India	S.R. Rao (1954)
Kalibangan (*i.e.*, the bangles of black colour	Ghaggar	Ganganagar	Rajasthan	India	Amalanand Ghosh (1951), B.B. Lal & B.K. Thapar (1961)
Banawali	Ghaggar	Hissar	Haryana	India	R.S. Bist (1973)
Dholavira	Luni	Kutchh	Gujarat	India	J.P. Joshi (1967-68)

Major Harappan Sites and Their Excavators

Site	Archaeological Finds
Mohenjodaro	Great Bath, Great Granery (the largest building of civilization), Assembly hall, Shell strips, Pashupati Mahadeva/Proto-Shiva (seal), Bronze Image of a nude woman dancer, Steatite image of bearded man, Human skeletons huddled together, Painted seal (Demi-God), Clay figures of Mother Goddess, A fragment of woven cotton, Brick Kilns, 2 Mesopotamian seals, 1398 seals (57% to total seals of civilization), Dice.
Chanhudaro	City without a citadel, Inkpot, Lipstick; Metal-workers', shell-ornament makers' and bead-makers' shops; Imprint of dog's paw on a brick, Terracotta model of a bullock cart, Bronze toy cart.
Kalibangan	Ploughed field surface (Pre-Harappan), 7 Fire altars, Decorated bricks, Wheels of a toy cart, Mesopotamian cylindrical seal.
Lothal	Dockyard, Rice husk; Metal-workers', shell-ornament makers' & bead-makers' shops; Fire altars, Terracotta figurine of a horse, Double burial (burying a male and a female in a single grave), Terracotta model of a ship, Dying vat, Persian/ Iranian seal, Baharainean seal, Painted jar (bird & fox).
Harappa	6 Granaries in row, Working floors, Workmen's quarters, Virgin-Goddess (seal), Cemetery (R-37, H), Stone symbols of Lingam (male sex organ) & Yoni (female sex organ), Painted pottery, Clay figures of Mother Goddess, Wheat & Barley in wooden mortar, Copper scale, Crucible for bronze, Copper-made mirror, Vanity box, Dice.
Surkotada	Bones of horse, Oval grave, Pot burials.
Banawali	Lack of chess-board or gridiron pattern town planning, Lack of systematic drainage system, Toy plough, Clay figures of Mother Goddess.
Daimabad	Bronze images (Charioteer with chariot, ox, elephant & rhinoceros)
Dholavira	A unique water harnessing system and its storm water drainage system, a large well and a bath (giant water reservoirs), Only site to be divided into 3 parts, Largest Harappan inscription used for civic purposes, A stadium.

The Vedic Period

- From the names of rivers, mountains (*Himvant i.e.,* Himalaya, *Munjavant* i.e., Hindukush) and ocean in Rig Veda we have a clear idea of the geographical area in which Rigvedic people lived.
- Rigvedic people, who called themselves Aryans, were confined the area which came to be known as *Sapta Sindhu i.e.,* land of the seven rivers. Sapta Sindhu comprises *Sindhu* and their five tributaries—*Vitasta, Asikani, Vipas, Parushni & Sutudri* and *Saraswati.*
- According to the Rig Veda, the most mentioned river—Sindhu, the most pious river—Saraswati, mention of the Ganges–1 time, mention of Yamuna–3 times.
- *The Kula* (the family) was the basis of both social & political organisations. Above the Kula were *the Grama, the Vis, the Jana* and *the Rashtra.* A group of Kula (families) formed a Grama (the village) & so on.
- Very little is known about ministers of the king. *The Purohita* or domestic priest was the first ranking official. He was the king's preceptor, friend, philosopher and guide. Other important royal officials were *Senani* (army chief) & *Gramani* (head of village).
- Rig Veda contains tribal assemblies such as the *Sabha, Samiti, Vidath, Gana.* Sabha was committee of few privileged and important individuals. Two popular assemblies, Sabha and Samiti, acted as checks on the arbitrary

rule of kings. Later Vedas record that the Sabha functioned as a court of justice.

- The Rigvedic society comprised four varnas, namely *Brahmana, Kshatriya, Vaisya* and *Shudra.* This classification of society was based on the occupational of individuals.
- The cow was already deemed *Aghanya i.e.,* not to be killed. *Indra, Agni* and *Varuna* were the most popular deities of Rigvedic Aryans.
- Money and markets were known but they were not extensively used. Cows and gold ornaments of fixed value were the media of exchange. No other coins were known.

Later Vedic Period (1000 BC – 600 BC)

- During the Later Vedic Period, the Aryan settlements covered virtually the whole of Northern India *(Aryavarta).* The centre of culture now shifted from Saraswati to Ganges *(Madhya desa).*
- The governmental machinery became more elaborate than before, as a sequel to the growth of the power of the king. New civil functionaries, besides the only civil functionary of the Rigvedic period the purohita came into existence. These were : the *Bhagadudha* (collector of taxes), the *Suta/Sarathi* (the royal herald or charioteer), the *Khasttri* (Chamberlain), the *Akshavapa* (courier).
- The military officials of the Rigvedic times, the Senani (the general) and the Gramani (the head of the village) continued to function.
- The king performed various rituals and sacrifices to strengthen his position. They include Rajasuya (consecration ceremony), Asvamedha (horse sacrifice) and Vajpeya (chariot race).
- Kingship became hereditary.
- Iron was used extensively in this period and this enabled the people to clear forests and to bring more land under cultivation. Agriculture became the chief occupation.
- A class of hereditary merchants *(vaniya)* came into existence. Vaisyas also carried on trade and commerce. The institution of *Gotra, i.e.,* the clan appeared in Later Vedic Period.

JAINISM

- According to Jain tradition, there were 24 *Thirthankaras* (literally Ford maker, across the stream of existence) the first being Rishabha-deva/Adinatha and last being Mahavira.
- *Parshwanath* was a prince of Benaras who abandoned the throne and led the life of a hermit and died at *Sammet-Shikar/ Parshwanath (Parasanath)* Hill, *Hazaribagh, Jharkhand.* His four main teachings *(Chaturthi)* were 1. *Ahimsa* (non-injury), 2. *Satya* (non-lying), 3. *Asteya* (non-stealing), 4. *Aparigraha* (non-possession). *Mahavira* adopted all these four teachings and added one more, that is *Brahmacharya* (Chastity) to it. Mahavira was born in 540 BC in a village *Kundgrama* near *Vaishali* in *Bihar.* His father *Siddhartha* was the head of the *Jnathrika Kshtriya* clan under Vajji of Vaishali and his mother *Trishala* was the sister of Chetaka, the king of Vaishali. Mahavira was also related to Bimbisara, the ruler of Magadha, who had married Chellana, the daughter of Chetaka.
- At the age of 30, after the death of his parents, he renounced his family, became an ascetic and proceeded in search of truth. He was accompained by *Makkhali Gosala,* but later due to some differences Gosala left him and founded *Ajivika* sect. At the age of 42 under a sal tree at *Jimbhikagrama* on the bank of river *Rijupalika* Mahavira attained *Kaivalya* (supreme knowledge). He delivered his first sermon at *Pava* to his 11 disciples (known as *11 Gandharas*). Later he founded a *Jain Sangha* (Jain commune) at Pava.

Jain Council	Year	Venue	Chairman	Patron	Result
1st Jain Council	300 BC	Patliputra	Sthulabhadra	Chandragupta Maurya	Compilation of 12 Angas
2nd Jain Council	512 AD	Vallabhi	Devardhi Kshmasramana	-	Final compilation of 12 Angas & 12 Upangas

BUDDHISM

- *Gautama Buddha,* founder of Buddhism, was born in 563 BC (widely accepted), on the vaisakha purnima day at *Lumbinivana* in *Kapilvastu* (now situated in the foothills of Nepal) in the *Sakya Kshatriya* clan.
- His father *Suddhodhana* was the republican king of Kapilvastu and mother *Mahamaya* was a princess of Kosala dynasty. After his mother's early death, he was brought up by his step mother and aunt *Mahaprajapati Gautami.*
- Four sights–an old man, a diseased person, a dead body and an ascetic–proved to be a turning point in his career.
- At the age of 29, he renounced home, this was his *Mahabhinishkramana* (great going forth) and became a wandering ascetic.
- His first teacher was *Alara Kalama* (Sankhya philosopher) from whom he learnt the technique of meditation.
- At the age of 35 under a pipal tree at *Uruvella (Bodh Gaya)* on the bank of river *Niranjana* (modern name Falgu) attained *Nirvana* (enlightenment) after 49 days of continuous meditation, now he was a fully enlightened (*Buddha* or *Tathagat*).
- He died at the age of 80 in 483 BC at *Kushinagar* (identical with the village Kasia in Siddharthanagar Janapada of Deoria district of U.P.).
- This is known as *Mahaparinirvana* (Final Blowing out).
- **Triratna *i.e.,* Three Jewels of Buddhism–** 1. *Buddha* (the enlightened), 2. *Dharma* (doctrine), 3. *Sangha* (commune).

Great Events of Buddha's Life Events

Events	Symbols
Janma (Birth)	Lotus & Bull
Mahabhinishkramana (Renunciation)	Horse
Nirvana/Sambodhi (Enlightenment)	Bodhi tree
Dharmachakra Pravartana (First Sermon)	Wheel
Mahaparinirvana (Death)	Stupa

Buddhist Council

Council	Year	Venue	Chairman	Patron	Result
1st Buddhist Council	483 BC	Saptaparni Cave, Rajagriha	Mahakassapa	Ajatashatru (Harayanka Dynasty)	Compilation of Sutta Pitaka and Vinaya Pitaka by Ananda and Upali respectively
2nd Buddhist Council	383 BC	Chullavanga, Vaishali	Sabbakami	Kalashoka (Shisunaga Dynasty)	(i) The monks of Vaishali wanted some change in rites. (ii) Schism into Sthaviravadins & Mahasanghikas.
3rd Buddhist Council	250 BC	Ashokarama Vihar, Patliputra	Mogaliputta Tissa	Ashoka (Maurya Dynasty)	(i) Compilation of Abhidhamma Pitaka (ii) Decision to send missionaries to various parts of the world
4th Buddhist Council	98 AD	Kundala Vana, Kashmir	Chairman- Vasumitra Vice chairman- Ashvaghosa	Kanishka (Kushana Dynasty)	(i) Compilation of Mahavibhasha shastra (Sanskrit comment on Tripitaka) (ii) The division of Buddhists into Hinayanists & Mahayanists.

MAHAJANAPADAS PERIOD (600 BC–325 BC)

- Buddhist literature *(Anguttara Nikaya, Mahavastu)* and Jain literature *(Bhagavati Sutta)* present a list of 16 Mahajanapadas with minor variation of names.
- There were two types of states–monarchical & non-monarchical/republican. Monarchical states–Anga, Magadha, Kashi, Kosala, Vatsa, Chedi, Shursena, Matsya, Avanti, Gandhara. Republican States–Vajji, Malla, Kuru, Panchal, Kamboja, Shakya, (Kapilvastu), Koliyas (Ramgrama), Moriya (Pipplivana).

Mahajanapada	Capital
Kashi	Varanasi
Anga	Champanagri
Vajji	Vaishali
Chedi	Shuktimati
Kuru	Hastinapur
Matsya	Viratnagar
Assaka	Budanya
Gandhara	Taxila
Kosala	Shravasti
Magadh	Rajagriha
Malla	Kushinagar
Vatsa	Kaushambi
Panchala	Ahichhatra
Surasena	Mathura
Avanti	Ujjain
Kamboja	Rajpur

Haryanaka Dynasty (544 BC–412 BC)

- Bimbisara was the founder of Haryana dynasty.
- Magadha came into prominence under the leadership of Bimbisara.
- He married the princesses of Kosala (Kosaldevi/Mahakosala–Sister of Kosal King Prasenjit), Lichchhavi, (Chellana-sister of Lichchhavi Head Chetaka) & Madra (Khema-daughter of Madra king), which helped him in his expansionist policy.
- Bimbisara was succeeded by his son Ajatashatru. Ajatashatru killed his father and seized the throne.
- He built the fort of Rajagriha and a watch-fort (Jaladurga) at a village called Patali on the banks of the Ganga.

Shisunaga Dynasty

- Nag-Dasak was unworthy to rule. So the people got disgusted and elected Shisunaga as the king, the minister of the last king.
- The most important achievement of Shisunaga was the destruction of Pradyota dynasty of Avanti. This brought to an end the hundred years old rivalry between Magadha and Avanti. From then on Avanti become part of the Magadha rule. Shisunaga was succeeded by Kalashoka (Kakavarna). His reign is important because he convened the second Buddhist council in Vaishali.

Nanda Dynasty

- The Shisunaga dynasty was overthrown by Mahapadma who established a new line of kings known as the Nands.
- The last king Dhanananda is possibly identical with the Agrammes or Xandrames of the Greek texts. The Nanda dynasty came to an end about 322-21 BC and was supplanted by another dynasty known as Mauryas, with Chandragupta Maurya as the founder.

Chandragupta Maurya

- Chandragupta dethroned the last Nanda ruler Dhananand and occupied Patliputra in 322 BC with the help of Kautilya.
- Chandragupta became a Jain and went to Chandragiri Hill, Sravanbelgola (Karnataka) with Bhadrabahu, where he died by slow starvation (Kaya Klesha/Salekhan)
- Under Chandragupta Maurya for the first time, the whole of Northern India was united.
- Trade flourished, agriculture was regulated, weights and measures were standardised and money came into use.

Bindusara

- Chandragupta Maurya was succeeded by his son Bindusara.
- Bindusara patronised Ajivikas.

Ashoka

- This war of succession accounts for interregnum of four years (273-269 BC), and only after securing his position on the throne, Ashoka had himself formally crowned in 269 BC.
- Ashoka fought the Kalinga war in 261 BC in 8th years of his coronation.
- The king was moved by massacre in this war and therefore, abandoned the policy of physical occupation in favour of policy of cultural conquest. In other words, Bherighosa was replaced by Dhammaghosa.

Sunga Dynasty (185-73 BC)

- Sunga Dynasty was established by Pushymitra Sunga, a Brahmin commander-in-chief of last Mauryan ruler named Brihadratha in 185 BC.
- Pushyamitra was succeeded by his son Agnimitra, the hero of Kalidasa's drama 'Malvikagnimitra'.
- In arts, the Bharhut Stupa is the most famous monument of the Sunga period.

Kanva Dynasty (73-28 BC)

- Devabhuti, the last ruler of the Sunga dynasty, was murdered by his minister Vasudeva, who usurped the throne and founded the Kanva dynasty.

Satavahana Dynasty

- Simuka was the founder of the Satvahana dynasty.
- It was Gautamiputra Satakarni (106-130 AD) who revived the Satavahana power and defeated the Saka Ksatrap Nahapana. He was the greatest Satavahan ruler.
- Yajna Sri Satakarni, its 27th ruler was the dynasty's last great ruler.
- The official language of the Satavahanas was Prakrit.
- The Satavahanas issued their coins in lead (mainly), copper, bronze and potin.

GUPTA PERIOD

Chandragupta–I (319-334 AD)

- He was the first Gupta ruler to assume the title of Maharajadhiraja.
- Chandragupta-I was able to establish his authority over Magadha, Prayaga and Saketa.
- Original type of Gold coins (Dinaras)– Chandragupta-I Kumaradevi type.

Samudragupta (335-380 AD)

- The most detailed and authentic record of his reign is preserved in the Prayaga Prasasti/ Allahabad pillar inscription, composed by his court poet Harisen.
- Samudragupta was the greatest king of Gupta dynasty.
- Samudragupta's military compaigns justify description of him as the 'Napoleon of India' by V.A. Smith.
- According to Chinese writer Wang-Hiuen-Tse, Meghavarna, king of Sri Lanka, sent on embassy to Samudragupta for his permission to build a monastery for Buddhist pilgrims at Bodh Gaya.

Chandragupta-II 'Vikramaditya' (380-414 AD)

- According to 'Devi Chandragupta' (Vishakhadatta) Samudragupta was succeeded by Ramgupta. It seems Ramgupta ruled for a very short period.
- Chandragupta-II also succeeded in killing Ramgupta and not only seized his kingdom but also married his widow Dhruvadevi.
- It was in Chandragupta's time that the Chinese pilgrim Fahien visited India.
- Original type of Gold coins (Dinaras)– Ashvarohi type, Chhatradhari type, Chakra-Vikram type etc.

Kumaragupta-I (415-455 AD)

- Chandragupta-II was succeeded by his son Kumaragupta-I.
- Kumaragupta was the worshipper of God Kartikeya.

- He founded the Nalanda Mahavidhalaya which developed into a great centre of learning.

Skandagupta (455-467 AD)

- Skandagupta, the last great ruler of the Gupta dynasty.

Pushyabhuti Dynasty (606-647 AD)

- The Pushyabhuti or Vardhana dynasty was founded at Thaneswar (Karnal District, Haryana) by Pushyabhuti probably towards the beginning of the 6th centuary.
- The first important ruler of the dynasty was Prabhakaravardhana (580-605 AD).

Harshavardhan

- After the killing of Rajyavardhana, his younger brother, Harshavardhan also known as Siladitya, ascended the Pushyabhuti throne in 606 AD and from this year started the Harsha Era.
- Harsha drove out Shashanka from Kannauj who had occupied it after killing of Rajyavardhan. He not only unfied Kannauj with Thaneswar but also made it his new capital, which made him the most powerful king of North India.
- Harshavardhan defeated Dhruvasena II, the ruler of Vallabhi. Dhruvasena-II accepted the position of a feudatary vassal. Harsha married his daughter to him. It was an important diplomatic achievement of Harsha.
- Harsha maintained diplomatic relation with China. In 641 A.D., he sent a envoy to Tai-Tsung, the Tang Emperor of China. Three Chinese missions subsequently visited his court. Hiuen-Tsang, the celebrated eight years (635-643 A.D.) in the dominions of Harsha.
- According to Hiuen-Tsang, Nalanda university, meant for Buddhist monks, was maintained by the revenue from 200 villages which granted by Harshavardhana.
- Harshavardhana was not only a patron of learning, but was himself on accomplished author. He wrote three Sanskrit plays–Nagananda, Ratnavali and Priyadarsika.

Chalukyas of Vatapi/Vadami

- Pulakesin-I was the founder of the Chalukya dynasty. He established a small kingdom with Vatapi or Badami as its capital.
- Chalukyas began the Vesara style or Deccan style in the building of structural temples, which reached culmination, however, only under the Rashtrakutas and the Hoyasalas.

Pallavas of Kanchi

- There is controversy regarding the origin of Pallavas. Possibly the Pallavas were a local tribe who established their authority in the Tondaimandalam or the land of creepers.
- Both Chalukyas and Pallavas tried to establish their supremacy over land between Krishna and Tungabhadra.
- Pallava king Narsimhavarman (630-668 AD) occupied Chalukya capital at Vatapi in about 642 AD and assumed the title Vatapikonda *i.e.,* conqueror of Vatapi.
- The development of temple architecture, particularly Dravida style, under the Pallavas can be seen in four stages:

1. Mahendravarmana Group	Mahendravarmana-I (600-630 AD)	Temple at Bhairavkona (North Arcot Distt.), Anantesvara temple at Undavalli (Guntur Distt.)
2. Mammala Group	Narsimhavarmana-I (Mammala (630-668 AD)	Mandapa temples and Ratha temples (Sapt Pagodas) at Mammalapuram (Mahabali-puram)
3. Rajasimha Group	Narsimhavarmana-II 'Rajsimha' (680-720 AD)	Kaulashnath & Vaikunth Perumal Temple at Kanchi, Shore temple at Mammalapuram.
4. Aparajit Group	Nandivarmana 'Aparajit' (879-897 AD)	Mukteshwar & Matangesvar temple at Kanchi, Parshuramesvar temple at Gudimallam.

MEDIEVAL

- In 712 A.D., Muhammad bin Quasim invaded Sind. Quasim defeated Dahir, the ruler of Sind and killed him in a well-contested battle. His capital Arar was captured.

Mahmud of Ghazni

- Mahmud of Ghazni, also known as the 'Sword of Islam', mounted, 17 plundering expeditions between 1000 and 1027 into North India, annexing Punjab as his eastern province.
- In 1025, he attacked and raided the most celebrated Hindu temple of Somnath, which is near the coast in the extreme south of Kathiawar. The temple was destroyed in 1026.

Muhammad Ghori

- Muiz-ud-din Muhammad Ghori invaded India and laid the foundation of the Muslim dominion in India.
- He may be considered as the 'founder of Muslim rule' in India.
- Prithviraj Chauhan defeated Ghori in the first battle of Tarain near Delhi in 1191 A.D.
- In the second Battle of Tarain in 1192, Muhammad Ghori thoroughly routed the army of Prithiviraj, who was captured and killed. The Battles of Tarain and Chandawar contributed to the establishment of Turkish rule in India.

DELHI SULTANATE

- Mohammed Ghori's conquests became the nucleus of a new political entity in India—the Delhi Sultanate.
- The five dynasties which ruled Delhi were:
  1. The Slave Dynasty (1206-1290)
  2. The Khilji Dynasty (1290-1320)
  3. The Tughlaq Dynasty (1320-1414)
  4. The Sayyid Dynasty (1414-1451)
  5. The Lodhi Dynasty (1451-1526)

The Slave Dynasty (1206-1290)

Qutub-ud-din Aibak

- A Turkish slave by origin, Aibak was purchased by Muhammad Ghori was later made him his governor. After the Ghori's death Aibek became the master of Hindustan and founded the Slave Dynasty in 1206. He was considered as a generous ruler and was popularly called Lakh Bakhsh (One who gives lakh).
- He constructed two mosques—Quwat-ul-Islam at Delhi and Adhai din ka Jhonpra at Ajmer. He also began the construction of Qutub Minar in the honour of famous Sufi Saint Khwaja Qutubuddin Bakhtiyar Kaki.
- Aibak was great patron of learning and patronised writers like Hasan-un-Nizami, author of Taj-ul-Massir and Fakhruddin, author of Tarikh-i-Mubarak Shahi.

Shamsuddin Iltutmish (1211-1236)

- He was a slave and son-in-law of Qutub-ud-din Aibak and occupied the throne of Delhi in 1211 after deposing Aram Baksh.
- He introduced the silver coin (Tanka) and the copper coin (Jital). He organised the Iqta system and introduced reforms in civil administration and army, which was now centrally paid and recruited. He patronised Minhaj-us-Siraj, author of Tabaqal-i-Nasiri.

Razia Sultana (1236-1239)

- She was the daughter of Iltutmish. She was the first and only Muslim lady who ever ruled India.
- In 1240 AD, Razia became the victim of a conspiracy and was assassinated near Kaithal (Haryana).

Baharam Shah (1240-42)

- After Razia, Iltutmish's third son Baharam Shah was put on throne by powerful Turkish council Chalisa.

Ghiyasuddin Balban (1266-87)

- Ghiyasuddin Balban, who was also known as Ulugh Khan served as Naib or regent to Sultan Nasiruddin Mahmud.
- He broke the power of Chalisa and restored the prestige of the crown. That was his greatest contribution towards the stability of the sultanate.
- He established the military department Diwan-i-Arz.
- The Persian court model influenced Balban's conception of Kingship. He took up the title of Zil-i-llahi.
- He introduced Sijda (prostration before the monarch) and Paibos (kissing the feet of monarch) as the normal forms of salutation.

Kaiqubad (1287-90)

- A grandson of Balban was established on the throne by Fakruddin, the Kotwal of Delhi who assumed high political authority during last days of Balban. But Kaiqubad was killed by the Khilji family, which saw end of slave dynasty and beginning of Khilji dynasty at Delhi throne.

The Khilji Dynasty (1290-1320)

Jalaluddin Khilji

- He was the founder of the Khilji Dynasty.

Alauddin Khilji

- Alauddin Khilji was a nephew of Sultan Jalaluddin, whom he killed and succeeded to the throne in 1296.
- He was the first Turkish Sultan of Delhi who separated religion from politics. He proclaimed 'Kingship know no kinship'.
- He introduced the system of Dagh (the branding of horse) and Chehra (descriptive roll of soldiers).
- The post of special officer called Mustakharaj was created for the purpose of collection of revenue.
- All goods for sale were brought to an open market called Sara-i-Adal.
- He was a patron of art & learning. Amir Khusrau, the poet-musician was his favourite court poet.
- Alauddin Khilji built a famous gateway known as Alai Darwaja.

The Tughlaq Dynasty (1320-1413 AD)

Ghiyasuddin Tughlaq

- Khusrau Khan, the last king of the Khilji dynasty was killed by Ghazi Malik, and Ghazi Malik ascended the throne assuming the title Ghiyasuddin Tughlaq.
- He died in an accident and his son Ulugh Khan, who take the title of Muhammad bin Tughlaq succeeded him.

Muhammad-bin-Tughlaq

- The writings of Barani, a historian and Ibn Batuta, a traveller from Morocco, provide information about the rule of Muhammad-bin-Tughlaq.
- Muhammad-bin-Tughlaq introduced token coins of brass and copper, which had the same value as silver coins.
- He died in Thatta while compaigning in Sindh against Taghi, a Turkish slave.

Firoz Shah Tughlaq

- He was a cousin of Mohammad-bin-Tughlaq. After his death, the nobles and theologians of the court selected Firoz Shah as the next Sultan. It was during the time of Firoz that Jizya became a separate tax. Firoz refused to exempt the Brahmanas from payment of Jizya since this was not provided for in Shariat.
- The new system of taxation was according to Quran. Four kinds of taxes sanctioned by the Quran were imposed and those were Kharaj, Zakat, Jizya and Khams. Kharaj was the land tax, which was equal to 1/10 of the produce of the land, Zakat was 2½ per cent tax on property, Jizya was levied on non-Muslims and Khams was 1/5 of the booty captured during war.
- The two pillars of Ashoka, one from Torpa (Haryana) and other from Meerut (U.P.) were brought to Delhi.

- The Sultan established at Delhi, a hospital described as Dar-ul-Shifa.
- A new department of Diwan-i-Khairat was set up to make provisions for marriage of poor girls. He had built new towns of Firozabad, Jaunpur, Hissar and Firozpur.
- Ibn Batuta was an African traveller, who visited India in 1333 during the reign of the Tughlaqs. He was appointed as chief Qazi of Delhi by the Sultan and subsequently as his ambassador to China in 1342.

The Sayyid Dynasty

- Before his departure from India, Timur appointed Khizr Khan as governor of Multan. He captured Delhi and founded the Sayyid dynasty in 1414.
- Mubarak Shah, Mohammad Shah and Alam Shah were some of the other important noteworthy rulers of Sayyid Dynasty.

The Lodhi Dynasty

- Bahlol Lodhi was an Afghan who established himself in Punjab after the invasion of Timur. He founded the Lodhi Dynasty.
- In 1504, Sikandar Lodhi founded the city of Agra and transferred his capital from Delhi to Agra.
- He introduced the Gaz-in-Sikandare of 32 digits for measuring cultivated fields.
- Babar marched against Delhi and defeated and killed Ibrahim Lodhi in the first battle of Panipat (1526).

A Literature of Delhi Sultanate

Book	Author	Historical Importance
Tahqiq-i-Hind	Alberuni	Alberuni was an Arabian scholar who wrote about the Slave dynasty
Tabaqat-i-Nasiri	Minhaj-us-Siraj	Gives an account of Iltutmish's reign
Laila-Majnu	Amir Khusrau	Court poet of Ala-ud-din Khilji
Khazain-ul-Futuh	Amir Khusrau	Describes conquests of Ala-ud-din Khilji
Tughlaq-Nama	Amir Khusrau	Gives account of Ghiyas-ud-din's reign
Nuh-Siphir	Amir Khusrau	Poetic description of Ala-ud-din Khilji
Fatawah-i-Jahandari	Ziauddin Barani	Gives an account of the Tughlaq dynasty
Tarikha-i-Firoz Shahi	Ziauddin Barani	Gives an account of Firoz Shah's reign
Fatwah-i-Firoz Shahi	Firoz Shah	Gives an account of his reign
Kitab-fi-Tahqiq	Alberuni	About Indian sciences
Qanun-e-Masudi	Alberuni	About astronomy
Jawahar-fil-Jawahir	Alberuni	About mineralogy
Qamas	Firozabadi	Arabic words dictionary
Taj-ul-Maathir	Hasan Nizami	History of Ilbaris, the slave dynasty
Chach Namah	Abu Baker	History of Sindh region
Lubab-ul-Alab	Bhukhari	Persian anthology
Khamsah	Amir Khusrau	Literature and Poems
Shah Namah	Firdausi	About Mohmud Ghazni's reign
Kitab-ul-Rehla	Ibn Batutah	A travelogue with stories
Miftah-ul-Futuh	Amir Khusrau	Jalaluddin's conquest and life
Multa-ul-Anwar	Amir Khusrau	Literary masterpieces
Ayina-i-Sikandari	Amir Khusrau	Literary masterpieces
Hasht Bihisht	Amir Khusrau	Literary masterpieces
Shirin Khusrau	Amir Khusrau	Literary masterpieces
Tarik-i-Firoz Shahi	Shams-i-Shiraj Afif	History of Tughlaqs

Central Administration

Department	Head (Founded by)
Diwan-i-Wizarat (Department of Finance)	Wazir
Diwan-i-Ariz (Military Department)	Ariz-i-Mumalik
Diwan-i-Insha (Department of Correspondence)	Dabir-i-Mumalik
Diwan-i-Risalat (Department of Appeals)	Dabir-i-Mulq
Diwan-i-Mustakharaj (Department of Arreras)	(Founded by Alauddin Khilji)
Diwan-i-Riyasat (Department of Commerce)	Rais-i-Mumalik (Founded by Alauddin Khilji)
Diwan-i-Kohi (Department of Agriculture)	(Founded by Md-bin-Tughlaq)
Diwan-i-Bandgan (Department of Slaves)	(Founded by Firoz Tughlaq)
Diwan-i-Khairat (Department of Charity)	(Founded by Firoz Tughlaq)
Diwan-i-Isthiaq (Department of Pensions)	(Founded by Firoz Tughlaq)

BHAKTI MOVEMENT

- The Bhakti movement was based on the doctrine that the relationship between God and man is through love and worship rather than through performing any ritual or religious ceremonies.
- It was in South India for the first time that Bhakti movement grew from a mere religious doctrine to a broad based popular movement based on social and religious equality.
- It was led by popular saint poets called 'Alvars', who represented emotional side of Vaishnavism through collective songs called Prabandhas.

MUGHAL PERIOD

- Babur was the founder of the Mughal Empire in India. His original name was Zahir-ud-din Muhammad.
- He was related to Timur from his father's side and to Chengez Khan through his mother.
- He adopted Tughluma and flanking party system and first to use gunpowder and artillery in India.
- Shershah introduced new silver coins called "Dam" and they were in circulation till 1835.
- Shershah also improved communications by building several highways. He built the Grand Trunk Road (G.T. Road), that runs from Calcutta to Peshawar.
- Akbar abolished the pilgrim tax and in 1562, he abolished Jaziya.
- Akbar evolved a new faith called Din-i-Illahi or Divine Faith.
- The land revenue system of Akbar was called Zabti or Bandobast system.
- It was further improved by Raja Todar Mal. It was known as Dahsala System.
- Akbar built Fatehpur Sikri, Agra fort, Lahore fort & Allahabad fort and Humayun's Tomb at Delhi.
- Tulsidas (Ramcharitamanas) also lived during Akbar's period.
- During Jahangir's reign Captain William Hawkins and Sir Thomas Roe visited his court.
- Shahjahan's reign is described by French traveller Bernier and Tavernier and the Italian traveller Nicali Manucci.
- Peter Mundi described the famine that accured during Shahjahan's time.
- Aurangazeb was also against the Sikhs and he executed the ninth Sikh Guru Tegh Bahadur. This had resulted in the transformation of Sikhs into a warring community. The tenth Sikh Guru, Guru Govind Singh built up a powerful army called the Khalsa to drive the Mughals out of India.

Important Years (Akbar)

Year	Work
1562	Visited Ajmer first time
1562	Ban on forcible conversion of war-prisoners into slaves
1563	Abolition of Pilgrimage Tax
1564	Abolition of Jaziya
1571	Foundation of Fatehpur Sikri
1574	Mansabadari System introduced
1575	Ibadatkhana was built
1578	Parliament of Religions in Ibadatkhana
1579	Proclamation of 'Mazhar' (written by Faizi)
1580	Dahsala Bandobast introduced
1582	Din-i-Ilahi/Tauhid-Illahi
1584	Ilahi Samvat *i.e.,* Calender
1587	Ilahi Gaz *i.e.,* Yard.

Important Years of Aurangzeb's religious policy

Year	Policy
1659	Forbade inscription of Kalama on the coins, celebration of Nauroj Festival; Appointment of Muhatasib (Regulator of moral character).
1663	Ban on Sati custom.
1668	Ban on Hindu festival.
1669	Ban on Jharokha darshan, Forbade music in the court.
1670	Ban on Tuladan (weighing of the emperor)
1979	Re-introduction of Jaziya.

Hindi Literature

Writer	Book
Ramanuja	Brahmsutra
Parhasarthi	Karma Mimansa
Jayadeva	Gita Govinda
Vijnansvara	Mitakshara
Jimuta Vahna	Dayabhaga
Chand Bardoi	Prithviraj Raso
Jaisi	Padmavat
Salibhadra	Bharata Bahuholi

Translation of some famous Sanskrit works into Persion during Akbar

Book	Writer
Ramayana	Badauni
Lilavati	Faizi
Atharva Veda	Hazi Ibrahim Sarhindi

Major Books of the Mughal Era

Book	Writer
Tuzuki-i-Baburi	Babur
Humayun Nama	Khwand Amair
Danun-i-Himayun	Gulbadan Begum
Akbar Nama	Sheikh Abul Fazal
Tabaqat-i-Akbari	Khwajah Nizamuddin Ahmad Baksh
Muntakhabut-ul-Tawarikh	Abdul Qadir Badauni
Tuzuk-i-Jahangiri	Jahangir
Tarikh-i-Alfi	Mulla Daud
Masnavi Nala-i-Daman	Faizi
Iqbalnama-i-Jahangiri	Muhammad Khan
Chahar Chaman	Chandrabhan Brahman
Padshahnama	Abdul Hamid Lahori
Alami-i-Salih	Mohmanad Salih
Majmua-ul-Bahrin	Dara Shikoh
Alamgirnama	Munshi Mirza Muhammad Kazin
Massir-i-Alamgiri	Saqi Mustaid Khan
Ain-e-Akbari	Abul Fazl
Muntakhab-ui-Tawarikh	Badauni
Nuriyya-e-Sultaniyya	Abdul Haqq
Waqt-e-Hyderabad	Nimat Khan Ali
Futuhat-e-Alamgiri	Ishwar Das
Nuskha-e-Dilkusha	Bhimsen
Khulasat-ul-Tawarikh	Sujan Rai Khatri
Shahjahan Namah	Muhammad Salih
Shahjahan Namah	Inayat Khan
Hamlai-Haidri	Muhammad Rafi Khan
Namah-e-Alamgiri	Aquil Khan Zafar
Salinat-ul-Auliya	Dara Shikoh
Safinat-ul-Auliya	Dara Shikoh
Najma-ul-Bahrain	Dara Shikoh
Raqqat-e-Alamgiri	Aurangzeb
Hasmat-ul-Arifin	Dara Shikoh
Muntkhbi-ul-Lubat	Kafi Khan

MARATHA STATE

- Shivaji was born at Shivner in 1627. His father was Shahji, Mother Jija Bai and religious teacher Samarth Ramdas.
- In 1674, Shivaji crowned himself at Raigarh and assumed the title Chatrapathi.

Shivaji's Ashtapradhan

1. **Peshwa**–Finance and general administration. Later he became the prime minister.
2. **Sar-i-Naubat or Senapati**–Military commander, a honorary post.
3. **Amatya**–Accountant General.
4. **Waqenavis**–Intelligence, posts and household affairs.
5. **Sachiv**–Correspondence.
6. **Sumanta**–Master of ceremonies and Foreign Minister
7. **Nyayadish**–Justice.
8. **Panditarao**–Charities and religious administration.

- Chauth and Sardeshmukhi were the taxes collected not in the Maratha Kingdom but in the neighbouring territories of the Mughal empire or Deccan Sultanates.
- Chauth was one-fourth of the land revenue paid to the Marathas in order to avoid the Maratha raids.
- Sardeshmukhi was an additional levy of ten per cent on those lands which the Marathas claimed hereditary rights.
- Rajaram created the new post of Pratinidhi, thus taking the total number of minister to nine.
- Baji Rao-I said about Mughals : 'Let us strike at the trunk of the withering tree and the branches will fall of themselves'.

MODERN

- Vasco-da-Gama, a Portuguese explorer, sailed through the route of cape of Good Hope and reached near Calicut on 20th May, 1498 AD. He was cordially received by king Zamorin, the local ruler.
- Vasco-da-Gama founded a factory at Cannanore on his second visit to India in 1501. In due course, Calicut, Cochin and Cannanore became the Portuguese trading centres.
- The United East India company of the Netherlands founded a factory at Masulipatnam in 1605. They built their first fort on the main land of India at Pulicut in 1609, near Madras (Chennai).
- The Danish East India company was established in 1616 in Denmark.
- The French East India company was established in 1664 under the inspiring and energetic leadership of Colbert, the economic advisor of the French king Louis XIV.
- In 1667, the first French factory was established at Surat by Francis Caron who was nominated as Director-General. In December 1669, another French factory was established at Masulipatnam.
- Francis Caron was replaced by Francis Martin. He founded the settlement of Pondicherry (Puducherry) in 1674 under a grant from Sher Khan Lodi, the Sultan of Bijapur. He made Pondicherry (Puducherry) as the French headquarters.
- The East India Company sent Sir William Hawkins to the court of the Mughal Emperor Jahangir in 1609 to obtain permission to eract a factory at Surat.
- In 1615, Sir Thomas Roe, another British merchant, came to Jahangir's court. He stayed for three years and succeeded in getting permission to set up their trading centres at Agra, Surat, Ahmedabad and Broach.
- Charles II leased out Bombay (Mumbai) to the East India Company. King Charles got Bombay as dowry after his marriage with Portuguese Princess.

The Advent of the Europeans

Company	Foundation
Portugese East India Company	1498
English East India Company	1600
Dutch East India Company	1602
Danish East India Company	1616
French East India Company	1664

Land Revenue Systems

Permanent Land Settlement or Zamindari System

- Introduced in Bengal, Bihar, Orissa and districts of Benaras and Northern districts of Madras by Lord Cornwallis in 1793. It declared Zamindars as the owners of the land. Hence, they could keep 1/11th of the revenue collected to themselves while the British got a fixed share of 10/11th of the revenue collected. The Zamindars were free to fix the rents.

Ryotwari System

- Introduced in Bombay and Madras. Munro (Viceroy) and Charles Reed recommended it.
- This system established a direct link between the government and the ryots or farmers.

Mahalwari System

- This was a scientific land revenue settlement implemented by Lord Bentinck.
- Revenue settlement was to made by village or estates with landlords. In Western Uttar Pradesh, a settlement was made with the village communities, which maintained from a common ownership known as Bhaichare, or with Mahals, which were groups of Village.

Mirasdari System

- In the Madras Presidency, the Palegar system was prevailent during early years of 19th century. The palegars were the landlords who exercised judicial powers and also possessed military powers also.
- In 1855, the Governor of Madras, Sir Thomas Monroe, abolished the palegar system, by curtailing their legal and military powers, and reducing them to the status of mere revenue collectors.

Peasant Movements

Movement	Place	Year	Leaders
Indigo Revolt	Bengal	1859	Bishnu and Digambar Biswas
Pabna	Bengal	1870	Ishwar Roy, Sabu Pal, Khoodi Mollah
Deccan Riots	Maharashtra	1875	-
Ramosi Movement	Maharashtra	1879	Vasudev Balwant Phadke
Bijolia	Rajasthan	1913	Sitaram Das, Vijay Pathik Singh
Champaran	Bihar	1917	Gandhiji
Kheda	Gujarat	1918	Gandhi & Vallabh Bhai Patel
Moplah	Kerala	1921	Sayyad Ali and Sayyid Fazl
Bordoli/Borsad	Gujarat	1928	Vallabh Bhai Patel
Tebhaga	Bengal	1946	Kamparan Singh, Nyamat Ali
Punnapra-Waylar	Kerala	1946	-
Telengana	Andhra Pradesh	1946	Kumaraiya & Sundaraiya
UP Kisan Sabha	UP	1918	Indra Narayan Dwivedi and Gauri Shankar Mishra
Avadh Kisan Sabha	UP	1920	Baba Ramachandra
Eka Movement	Avadh	1921	Madari Pasi
Forest Satyagrah	South India	1931	N.V. Rama Naidu, N.G. Ranga
All India Kisan Sabha	Lucknow	1936	Sahajanada Saraswati

Tribal Rovolto

Tribe	Year	Leaders	Cause
Chuar	1766-1772	Raja Jagannath	Excess Revenue demand, Bengal famine
Bhills	1817	Sewaram	Agrarian hardship
Hos	1820	-	British occupation of Singhbhum
Ramosi	1822	Chittur Singh, Pratap Singh, Dattaraya Patkar	British Rule
Kolis	1824		Dismantle of forests
Ahom	1828-1833	Gomadhar Kunwar	British occupation
Khasi	1829-1832	Tiruth Singh	British occupation
Kol	1831-1832	Buddhu Bhagat	Land transfer to outsiders
Santhals	1855-56	Sidhu and Kanhu	British Rule
Naikda	1858	Rup Singh	For Dharma Raj against ban Joria Bhagat on grazing and timber
Bhuyan & Juang	1867-68, 1891-31	Ratna Nayak, Dharni Nayak	Installation of British protege on throne
Kacha Nagas	1882	Sambhuden	British intervention
Munda (Ulgulan)	1899	Birsa Munda	Land system, Missionary activity and forced labour
Bhills	1913	Govind Guru	A temperance and purification movement
Oraons	1914	Jatra Bhagat, Tana Bhagat	Religious Reason
Chenchus	1921-22	-	British control of forests.
Koyas/Rampas	1922-24	Alluri Sitaram Raju	British Rule
Naga	1932	Jadunang (1905-31) & Rani Gaidinliu	A reformist movement later directed against excess of British rule

Lower Caste/Caste Movements and Organisations

Movement/Organisation	Year	Place	Founder
Satya Shodhak Samaj	1873	Maharashtra	Jyotibha Phule
Aravippuram Movement	1888	Aravippuram, Kerala	Shri Narayan Guru
Shri Narayan Dharma Paripalana Yogam (S.N.D.P.) Movement	1902-03	Kerala	Shri Narayan Guru, Dr. Palpu & Kumaran Asan
The Depressed Class Mission Society	1906	Bombay	V.R. Shinde
Bahujan Samaj	1910	Satara, Maharashtra	Mukundrao Patil
Justice (Party) Movement	1915-16	Madras, Tamil Nadu	C.N. Mudaliar, T.M. Nair & P. Tyagaraja Chetti
Depressed Class Welfare Institute (Bahiskrit Hitkarini Sabha)	1924	Bombay	B.R. Ambedkar
Self-Respect Movement	1925	Madras, Tamil Nadu	E.V. Ramaswami Naiker 'Periyar'
Harijan Sevak Samaj	1932	Pune	Mahatma Gandhi

THE GREAT REVOLT OF 1857

- The events that led to the Revolt began on 29 March, 1857 at Barrackpore. Mangal Pandey (a sepoy) refused to use the greased cartridges and single-handedly attacked and killed his officer. Mangal Pandey was hanged. The regiment to which he belonged was disbanded and sepoys guilty of rebellion punished.
- At Meerut in May 1857, 85 sepoys of the 3rd Cavalry regiment were sentenced to long terms of imprisonment for refusing to use the greased catridges. Therefore, on 10th May the sepoys broke out in open rebellion, shot their officers, released their fellow sepoys and headed towards Delhi.
- The leadership at Delhi was nominally in the hands of Bahadur Shah, but the real control was exercised by General Bakht Khan.
- As far as the effects of the Revolt are concerned, it brought about fundamental changes in the character of Indian administration which was transferred from the East India Company to the Crown by the Queen's Proclamation of 1 November, 1858.
- The Governor-General received the new title of Viceroy. Lord Canning had the unique opportunity to become the Governor-General as well as the first Viceroy according to the Act of 1858.

A Brief Survey of the Major Centres of the Revolt of 1857

Centre	Leader/s	British Officials who Suppressed the Revolt	Fate of the Leader
Delhi	Bahadur Shah, General Bakht Khan of Bareilly regiment	Nicholson, Lt. Willoughby, Lt. Hudson	Bahadur Shah deported to Rangoon; Bakht Khan died in battlefield
Lucknow	Begum Hazrat Mahal of Avadh	Colin Campbell	Escaped to Nepal
Kanpur	Nana Saheb, Tantia Tope, Azimullah	Colin Campbell, Henry Havelock, Henry Lawrence	Nana Saheb escaped to Nepal; Tantia Tope was hanged; Azimullah died of illness
Jhansi	Lakshmibai	Hugh Rose	Died in battlefield
Arrah	Kunwar Singh	William Tyler and Eyre	Died of wound sustained in the fight.

Comments on the Nature of 1857 Revolt

- **British Historians** – A mutiny due to the use of greased cartridges.
- **Disraeli (British Opposition Leader)** – A national revolt rooted in deep mistrust.
- **V.D. Savarkar** – First War of Independence.
- **S.B. Choudhary** – Civil rebellion.
- **R.C. Majumdar** – Neither 'first' nor 'national' nor 'a war of independence'.
- **S.N. Sen** – An effort by the conservative elements to turn the clock back.
- **Marxists** – A soldier-peasant struggle against foreign and feudal bondage.

Socio-Religious Reform Movements

- Raja Ram Mohan Roy lived during the 19th century. He was a religious and social reformer and founder of the Brahmo Samaj (1829).
- In 1815, Raja Ram Mohan Roy established the Atmiya Sabha. Later, it was developed into the Brahmo Sabha in August 1828.
- Raja Ram Mohan Roy is most remembered for helping Lord William Bentinck to declare the practice of Sati a punishable offence in 1829.
- In 1817, Raja Ram Mohan Roy founded the Hindu College (now Presidency College, Calcutta) along with David Hare, a missionary.

- Ram Mohan Roy started the first Bengali weekly *Samvad Kaumudi* and edited a Persian weekly *Mirat-ul-akhbar.*
- Henry Vivian Derozio was the founder of the Young Bengal Movement.
- The Arya Samaj was founded by Swami Dayanand Saraswati at Bombay in 1875.
- Swami Dayanand Saraswati believed the Vedas were the source of true knowledge. His motto was "Back to the Vedas". He wrote the book *Satyartha Prakash* which contains his ideas.
- The first Dayanand Anglo-Vedic (DAV) School was founded in 1886 at Lahore.
- The Prarthana Samaj was founded in 1867 in Bombay by Dr. Atmaram Pandurang. It was an off-shoot of Brahmo Samaj. Justice M.G. Ranade and R.G. Bhandarkar joined it in 1870 and infused new strength to it.
- Swami Vivekananda participated at the Parliament of Religions held in Chicago (USA) in September 1893 and raised the prestige of India and Hinduism very high.
- Swami Vivekananda founded the Ramkrishna Mission at Belur in Howrah in 1897. The objectives of this Mission are providing humanitarian relief and social work through the establishment of schools, colleges, hospitals and orphanages.
- The Theosophical Society was founded in New York (USA) in 1875 by Madam H.P. Blavatsky, a Russian lady, and Henry Steel Olcott, an American colonel. They arrived in India and established their headquarters at Adyar in Madras in 1882. Later in 1893, Mrs. Annie Besant arrived in India and took over the leadership of the Society after the death of Olcott.
- Pandit Ishwar Chandra Vidyasagar was a great educator, humanist and social reformer.
- In 1873, Jyotiba Phule founded the Satyashodak Samaj to fight against the caste system. He pioneered the widow remarriage movement in Maharashtra and worked for the education for women.
- Jyotiba Phule and his wife established the first girls' school at Poona in 1851.
- The Aligarh Movement was started by Sir Syed Ahmad Khan for the social and educational advancement of the Muslims in India.
- In 1875, Sir Syed Ahmad Khan founded a modern school at Aligarh to promote English education among the Muslims. This had later grown into the Mohammadan Anglo Oriental College and then into the Aligarh Muslim University.
- In 1890, the Khalsa college was established in Amritsar by Dayal Das who led the Sikh Movement. It opposed child marriage and Devdasi System.

National Movement

- Allan Octavian Hume, a retired civil servant in the British Government took the initiative to form an all-India organisation. Thus, the Indian National Congress was founded and its first session was held at Bombay in 1885. W.C. Banerjee was its first president. It was attended by 72 delegates from all over India.
- The second session was held in Calcutta in 1886 and the third in Madras in 1887.
- Surendranath Banerjee was called the Indian Burke. He founded the Indian Association (1876) to agitate for political reforms. He had convened the Indian National Conference (1883) which merged with the Indian National Congress in 1886.
- In 1907, Lala Hardyal came under the influence of revolutionaries working in Oxford and started a weekly paper called Gadar. His subsequent associations with more leaders culminated into the formation of the Gadar Party in 1913, in North America.
- Kashi Ram and Hardayal were the active members of the Gadar Party. During Lord Curzon's time in 1905, Bengal was divided.
- In 1911, in Lord Hardinge's time the partition of Bengal was cancelled.
- In 1911 the capital of India was shifted to Delhi from Calcutta.
- In 1916, a pact was signed between Muslim League and Congress which is known in History as the Lucknow Pact.

- In 1916, Bal Gangadhar Tilak established the Home Rule League of India.
- After Lucknow Pact, Congress and League presented the plan of political reforms based on separate electoral regions. This Pact led to an increase in communalism.
- On 30 March, 1919, Satyagrah a Day was observed in whole of India. The Satyagraha was peaceful at all places except Punjab and Delhi. Dr. Satyapal and Dr. Saifuddin, the leaders of the Punjab Satyagraha were imprisoned. In Protest, a meeting was organised at Jallianwala Bagh in Amritsar. The People who assembled here were gunned down. This is known as "Jallianwalla Bagh Massacre' of April 1919.
- The Congress joined the Muslims in Khilafat movement. On 31 August, 1919, the Khilafat Day was observed.
- Mahatma Gandhi launched the Non-cooperation Mass Movement in 1920-21. But violence broke out at Chauri Chaura then in Gorakhpur district which saddened Gandhiji. In February 1922 he announced the closure of the movement.
- In March 1922 Motilal Nehru and Deshbandhu Chitranjan Das established the Swaraj Party.
- In the elections of 1923 the Swaraj Party scored 40 seats out of 148.
- In 1927 the Bardoli Satyagraha was conducted by Sardar Vallabh Bhai Patel.
- In the 1929 Lahore Congress session held under the chairmanship of Jawaharlal Nehru, the meaning of Swaraj was declared as total independence.
- In 1930 Gandhiji broke the Salt laws by his Dandi March and he started the Civil Disobedience movement. In 1930, the Congress boycotted the first Round Table Conference.
- In 1931, after Gandhi-Irwin pact Gandhiji went to attend the second Round Table Conference along with the members of Muslim League.
- In the third Round table conference in 1932, Congress did not send any representative. Only 46 members went to participate under different categories. The British Prime Minister Ramsay Macdonald declared the communal award on 16 August, 1932.
- On 25 September, 1932, the Poona Pact was signed. Common agreement was made on two conditions for preparing the electoral regions. The representative of the Depressed classes was B.R. Ambedkar.
- In 1932 Gandhiji founded the Harijan Sewak Sangh for the uplift of the Harijans.
- On 8 May, 1933 Gandhiji declared the programme of 21 days fast for his self-purification. Gandhiji began Individual Satyagraha and Civil Disobedience on 1 August, 1933.
- The Government of India Act of 1935, the subjects were divided into three departments-Federal, Provincial and Concurrent.
- In the Haripura session of the Congress (1938), S. C. Bose was unanimously elected the President. In 1939 Bose was relected Congress President defeating Gandhi's candidate P. Sitaramayya.
- In April 1939, Subhash Chandra Bose resigned from the post of the President and started a militant party known as Forward Block. The Individual Satyagraha was started from 17 October, 1940. Acharya Vinoba Bhave was the first Satyagrahi. Gandhiji postponed it on 17 December, 1940.
- The Quit India movement resolution was passed on 14 July, 1942 in the Executive of the Congress Session held at Wardha. It was reaffirmed on 8 August, 1942.
- The interim government of free India was organized on 21 October, 1943 by Subhash Chandra Bose in Singapore.
- Muslim League observed the Direct Action Day on 16 August 1946.
- The Interim Government of India was organized under the leadership of Jawaharlal Nehru. The Cabinet took oath on 2nd September, 1946.
- The Constituent Assembly first met under the chairmanship of Dr. Rajendra Prasad on 6th December, 1946. In July 1947, the Indian Independence Act was passed by the British Parliament.
- India became independent on 15 August, 1947.

Indian Revolutionary Organisations (In India)

Name of Organisation	Place	Year	Founder
Vyayam Mandala	Poona	1896-97	Chapekar Brothers
Mitra Mela	Nasik (Later-Poona)	1901	Savarkar Brothers
Anushilan Samiti	Midnapur	1902	Jnanendra Nath Bose
Abhinava Bharata	Poona	1904	V.D. Savarkar
Swadesh Bandhav Samiti	Warisal	1905	Ashwini Kumar Datta
Anushilan Samiti	Dacca	1907	Varindra Kumar Ghosh
Bharat Mata Society	Punjab	1907	Ajit Singh & Sufi Amba Prasad
Hindustan Republican Association/Army (H.R.A.)	Kanpur	1924	Sachindra Nath Sanyal
Bharat Naujawan Sabha	Lahore	1926	Bhagat Singh
Hindustan Socialist Republican Association/Army (H.S.R.A.)	Delhi	1928	Chandrashekhar Azad
India Home Rule Society (India House)	London	1904	Shyamji Krishna Verma
Abhinav Bharat	London	1906	V.D. Savarkar
Indian Independence League	California (U.S.A.)	1907	Tarak Nath Das
Gadar Party	San Fransico	1913	Lala Hardayal
Indian Independence League	Berlin (Germany)	1914	Lala Hardayal & Virendra Nath Chattopadhyaya
Indian Independence League and Government	Kabul	1915	Raja Mahendra Pratap

Reforms/Acts

Nomenclature of the Reforms/Acts	Year	During the term of	Significance
Prohibition of Sati and Female infanticide	1829	Lord William Bentick	Supported by Raja Rammohun Roy
Doctrine of Lapse	1848	Lord Dalhousie	Adoption of sons by rulers in the absence of their natural heirs was banned
Morely-Minto Reforms	1909	Lord Minto-II	Separate electroates to widen the gulf between Hindus and Muslims.
Dyarchy	1919	Lord Chelmsford	Meaning dual system of Government
Jallianwala Bagh Massacre	1919	Lord Chelmsford	Massacre at Jallianwala Bagh in Amritsar by General Dyer
Simon commission	1928	Lord Irwin	To report working of the reforms; recommended dyarchy in provinces; India to be constituted as a federation and Indianisation of armed forces.
Cripps Mission	1942	Lord Linlithgow	Proposed Dominion status for India after the Second World War.
Cabinet Mission Plan	1946	Lord Wavell	Envisaged establishment of Constituent Assembly to frame the Constitution.
Mountbatten Plan	1947	Lord Mountbatten	Partition Plan
Indian Independence Act	1947	Lord Mountbatten	India partitioned and attained independence.

Educational Committee/Commissions

Viceroy	Committee/Commission	Year	Chairman	Objectives
Lord Ripon (1880-1884)	Hunter Commission	1882	William Hunter	To study the development in education
Lord Curzon (1899-1905)	University Commission	1902	Thomas Raleigh	To study the Universities and introduce reforms
Lord Reading (1921-1926)	Indian Disbandment Committee	1923	Lord Itchcap	To discuss the Central Committee of Education
Lord Wavell (1943-1947)	Sergeant Plan	1944	John Sergeant	To raise the standard of Education like Britain

Important Congress Session

Year	Place	Importance
1885	Bombay	At Gokuldas Tejpal Sanskrit College 72 delegates
1886	Calcutta	436 delegates
1887	Madras	Tayabji became first Muslim President
1888	Allahabad	George Yule became first English President
1889	Bombay	Congress represented all areas of British India
1890	Calcutta	Decision taken to organise a session of Congress in London
1895	Poona	Demand for a representative body only for educated class
1898	Madras	Social reform was set as the main goal.
1907	Surat	Congress split
1908	Madras	Constitution for the Congress.
1916	Lucknow	Congress merger. Pact with Muslim League, Gandhi attended
1917	Calcutta	Annie Besant became 1st women President
1920	Nagpur	Gandhian programme was adopted. Change in congress constitution
1921	Ahmedabad	Hasrat Mohani demanded for complete independence
1922	Gaya	Formation of Swaraj Party
1924	Belgaum	Gandhi became President
1925	Kanpur	Sarojini Naidu became 1st Indian women President
1927	Madras	Nehru and S.C. Bose moved resolution for independence and it was passed for the 1st time
1928	Calcutta	First All India Youth Congress
1929	Lahore	'Poona Swaraj' (complete Independence) resolution and pledge for Independence day on 26 January 1930
1931	Karachi	Resolution for Fundamental Rights and National Economic Policy
1934	Bombay	Formation of Congress Socialist Party
1936	Lucknow	Support for socialism through democracy
1937	Faizpur	Demand for Constituent Assembly
1938	Haripura	Purna Swaraj was to cover also princely states
1939	Tripuri	S.C. Bose resigned due to difference with Gandhi
1940	Ramgarh	Non-cooperation in the war effort; Demand for a national government; Resolution of August 1942 Quit India passed
1946	Meerut	Cabinet Mission was supposed to decide the fate of India; Congress wins majority in the Constitutional Assembly

❑ ❑ ❑

General Geography

UNIVERSE

- Existing matter and energy are together known as universe.
- Universe includes all the universal bodies and the complete solar family which comprises the Sun, Moon, Planets, Comets, Planetoids etc.
- A galaxy is a huge system of billions of stars and clouds of dust and gases.
- The Sun, eight planets, satellites and some other celestial bodies known as asteroids and meteoroids form the solar system.
- Mercury, Venus, Mars, Jupiter and Saturn can be seen from Earth with the naked eyes.
- The Sun is a mixture of gases. It consists of 92% hydrogen, 7.8% helium and 0.2% other gases.
- There are eight planets in our solar system. In order of their distance from the Sun, they are: Mercury, Venus, Earth, Mars, Jupiter, Saturn, Uranus, and Neptune.
- Mercury is nearest to the Sun. It takes only about 88 days to complete one round along its orbit.
- Venus is considered as 'Earth's-twin' because its size and shape are very much similar to that of the earth.
- Popularly known as Evening star and Morning star. It is seen in the east in the morning and in the west in the evening. It is not possible to see it all over the night.
- Mars has 2 satellites—Phobos and Deimos.
- Jupiter has 63 satellites (Prominent are Europa, Gannymede and Callisto). Gannymede is the largest satellite of Solar System.
- Saturn is the second largest planet in the solary system.
- Titan, the satellite of Saturn, is the largest satellite in the solar system.
- Uranus is about four times the size of the Earth. This planet appears greenish in colour because of methane gas present in its atmosphere.
- Uranus was discovered in 1781 by Sir William Hersiel.
- Uranus rotates from east to west on its axis, which is opposite to other planets.
- Asteroids are mostly found between the orbits of Mars and Jupiter.
- The star Alpha centaury (also called Proxima Centauri) is the closest star to the Earth after the Sun. It is about 4.3 light years away.

The Moon

- Earth has only one satellite, that is, the moon.
- The moon moves around the earth in about 27 days. It takes exactly the same time to complete one spin. As a result, only one side of the moon (only 59% of its surface) is visible to us on the earth.
- Moonlight takes 1.3 sec. to reach earth.
- The Moon rotates on its axis at a constant rate but its path around the earth is not perfectly circular. At some times, it is nearer the earth than at others.
- At Perigee (at its nearest) the distance is 3,64,000 kms and at Apogee, it is about 4,06,000 km.

THE EARTH

- The Earth is the fifth largest planet in the solar system.
  Diameter
  Equatorial Diameter—12,756 km
  Polar Diameter—12,714 km
  Circumference
  Equatorial Circumference—40,077 km
  Polar Circumference—40,009 km
  Total Surface Area—510 million sq. km
  29% covered by Continents;
  71% covered by Oceans
  Volume—1.083 × 10^{12} cubic km.
  Once the diameter is known, volume is calculated by using the formula $V = 4/3\pi r^3$
  Mass—5.978 × 10^{24} kg
  Using Newton's law of gravitation, mass of Earth is calculated from the orbit of the Moon.
  Density—5.52 g/cm^3 (5.2 times that of water
  Age—4.6 billion years
  Highest Land Point (Mt. Everest)—8848 m
  Lowest Land Point (Dead Sea)—397 m
  Greatest Ocean Depth (Mariana Trench)—11022 m
  Period of Rotation on Polar Axis (Sideral day)—23 hr 56 min 4.09 sec.
  Orbital Speed About Sun—29.8 km/sec
  Maximum distance from Sun (at aphelion July 4)—about 152 million km
  Minimum distance from Sun (at Peri-helion January 3)—about 147 million km
- **Escape Velocity—**11200 m/s

Rocks

- Rocks can be classified into three types based on their formation. They are: 1. Igneous 2. Sedimentary and 3. Metamorphic
- Two types of igneous rocks are basalt and granite. Today igneous rocks represent 95 per cent of the Earth's crust.
- Basalt is an extrusive rock, and granite is an intrusive rock.
- Extrusive igneous rocks are formed by magma that reaches the earth's surface along deep cracks and at volcanic vents.

Sedimentary Rocks

- Sedimentary rocks are formed by the accumulation and cementation of mud, silt, or sand derived from weathered igneous rock fragments. One of the commonest sedimentary rock is sandstone.
- Coal is a sedimentary rock of plant origin.

Metamorphic Rocks

- Metamorphic rocks are igneous or sedimentary rocks that have been altered by heat and/or pressure, either because they have been buried and folded deep in the crust, or because they have come into contact with molten igneous rock.
- The most common metamorphic rock is gneiss. It was originally granite. Due to shrinking and pressure the minerals are in distinct layers. Pressure changed bituminous (soft) coal into anthracite (hard coal). Some marble was formed by the folding and pressure on limestone. Other marble was made when water seeped through deeply buried limestone, the water dissolves the limy mineral and rearrange the particles in little crystalline blocks.
- Water also changed many sandstones by depositing quartz around the grains of sand. In time each grain become a crystal, locked to other crystals. Thus, the sandstone became quartzite.

Earthquakes and Volcanoes

- The sudden tremors or shaking of the earth's crust for a few seconds is called an earthquake. When a part of the earth's surface moves backward and forward or up and down, the earth's surface 'quakes' and these the 'earthquake'. 'Seismology' the special branch of Geology, deals with the study of earthquake.
- 'Richter scale' is the instrument to measrue/record the intensity of earthquake.
- Magma chamber is the reservoir from where volcanic materials erupt. This chamber is as close as 3 kilometers to the surface.
- Most active composite cones are located in a narrow zone, named the Ring the Fire that surrounds the Pacific ocean.

- Fujiyama in Japan, Mt. Mayean in the Philippenes, and the picturesque volcanoes of cascade range in the Northwestern United States, including Mt. St. Helens are good examples of composite cone volcanoes.
- The world's largest, active volcano in Mauna Loa in Hawaii, where famous coffee is grown in the rich volcanic soils.

Different Heat Zones of the Earth

- The region between the Tropic of cancer and Tropic of Capricorn receives direct rays of the Sun practically throughout the year.
- This area receives maximum heat and is called the Torrid (hot) zone. Longitude is the angular distance of a place east or west of the Prime meridian. In India, the Longitude of 82°30′ E, passing through Allahabad is considered the standard Meridian. This is known as the Indian Standard Time (IST).
- IST is 5 hours 30 minutes ahead of GMT.
- It is an imaginary zigzag line on the globe, approximately along, the 180° meridian of longitude. If you are travelling eastwards from Greenwich to a place A on longitude 180°, the time will be 12 hours ahead of Greenwich time, *i.e.,* 06:00 a.m. on Monday.
- If you are travelling westward from Greenwich to the same place A on longitude 180°, the time will be 12 hours behind Greenwich time, *i.e.,* 06:00 p.m. but on the previous day, Tuesday.

Layers of the Atmosphere

- There are five distinct layers of atmosphere:

Troposphere

- This is the first layer of the atmosphere.
- It extends to a hight of 18 km at the equator and 8 km at the poles. It contains more than 90% of gases in the atmosphere.

Stratosphere

- This is second layer of the atmosphere.
- It extends from the tropopause to about 50 km above the surface of the Earth. Temperature increases due to the absorption of the ultraviolet radiation of the Sun by ozone present in this layer.
- The temperature slowly increases to 4°C.
- This layer is free from clouds and associated weather Phenomena. Hence, it provides ideal flying conditions for large Jet planes.

Mesophere

- Above the stratosphere lies the mesosphere. The mesosphere extends to a hight of 80 km. Here the temperature decreases again, falling as low as –90°C.

Thermosphere

- The thermosphere lies above the mesosphere.
- This layer extends to a hight of about 90 km.
- This increase in temperature is due to the fact that the gas molecules in this layer absorb the X-rays and the short wave ultraviolet radiation of the Sun.

Major Mountain Ranges of the World

Range	Location	Highest Peak (m)	Length (km)
Andes	South America	6,960	7,200
Himalayas-Karakoram-Hindukush	South Central Asia	8,848	4,800
Rockies	North America	4,401	4,800
Great Dividing Range	East Australia	2,228	3,600
Western Ghat	Western India	2,637	1,610
Caucasus	Europe	5,642	1,200
Alaska	USA	6,194	1,130
Alps	Europe	4,808	1,050
Apennines	Europe	2,912	—
Ural	Asia	1,895	—
Atlas	North West Africa	—	1,930

Oceans of the World

Names	Area (Sq. Km.)	Greatest Depth
Pacific	16,62,40,000	Mariana Trench
Atlantic	8,65,60,000	Puerto Rico Trench
Indian	7,34,30,000	Java Trench
Arctic	1,32,30,000	—

Major Ocean Current

Current	Ocean	Type
Agulhas Current	Indian	Warm
Alaska Current	North Pacific	Warm
Benguela Current	South Atlantic	Cool
Brazil Current	South Atlantic	Warm
California Current	North Pacific	Cool
Canaries Current	North Atlantic	Cool
East Australian Current	South Pacific	Warm
Equatorial Current	Pacific	Warm
Gulf Stream	North Altantic	Warm
Humboldt (Peru) Current	South Pacific	Cool
Kuroshio (Japan) Current	North Pacific	Warm
Labrador Current	North Atlantic	Cool
North Atlantic Drift	North Atlantic	Warm
North Pacific Drift	North Pacific	Warm
Oyashio (Kamchatka) Current	North Pacific	Cool
West Australian Current	Indian	Cool
West Wind Drift	South Pacific	Cool

Important Lakes of the World

Lake	Location	Area (Sq. Km.)
Caspian	Russia and CIS	371000
Superior	Canada and USA	82414
Victoria	Tanzania (Africa)	69485
Huron	Canada and USA	59596
Michigan	USA	58016
Tanganyika	Africa	32892
Baikal	Russia (CIS)	31502
Great Bear	Canada	31080
Malawi	Malawi (Tanzania)	30044
Great Slave	Canada	28438

Famous Tribes of the World

Tribes	Place
Abhors	People of Mongolian blood living between Assam and Eastern tribes.
Afridis	Tribes residing in the North-west Frontier (Pakistan).
Bantus	Negroes living in the Central and South Africa.
Berbers	Tribals of Morocco, Algeria and Tunisia.
Boers	The Dutch settlers of South Africa.
Bushmen	Tribals of Kalahari deserts in Africa.
Cossacks	People living in the southern and eastern frontiers of Russia.
Dinka	Tribals of Sudan.
Eskimos	Inhabitants of Greenland and of Arctic regions.
Flemings	A term used for the people of Belgium.
Gaucho	Nomads of Pampas in Uruguay and Argentina.
Hamites	Inhabitants of North-West Africa.
Khirgiz	People Living in Central Asia.
Kurds	Tribes, living in Kurdistan (Iraq).
Lendu	Democratic Republic of Congo.
Magyars	Inhabitants of Hungary.
Masuds	Tribal group of Waziristan (Pakistan).
Mooris	Inhabitants of New Zealand.
Negroes	Mostly found in Africa.
Pygmies	Short sized people found in Congo basin in Africa.
Red Indians	Original inhabitants of North America.
Semites	Caucasian people of ancient times.
Swahili	Ethnic group of Kenya and Tanzania.
Yakuts	People in the Tundra region in Russia.
Zambos	People of mixed black and Indian blood inhabiting South America.
Zulus	People of South Africa living in certain part of Natal.

Highest Waterfalls of the World

Name(s) (Foreign)	Location	Source/River	High (Feet)
Angel (Salto Angel)	Canaima Nat'l Park, Venezuela	Upper trib. of Rio Caroni	3,212
Tugela	Natal Nat'l Park, South Africa	Tugela River	3,110
Utigord (Utigordsfoss)	Norway	Glacier fed	2,625
Monge (Mongefoss)	Marstein, Norway	Mongebeck	2,540
Gocta Cataracts	Chachapoyas, Peru		2,532
Mutarazi (Mtarazi)	Nyanga Nat'l Park, Zimbabwe	Mutarazi River	2,499
Yosemite	Yosemite Nat'l Park, California, U.S.	Yosemite Creek	2,425
Espelands (Espelandsfoss)	Hardanger Fjord, Norway	Opo River	2,307
Lower Mar Valley (Ostra Mardolafoss)	Eikesdal, Norway	Mardals Stream	2,151
Tyssestrengene	Odda, Norway	Tyssa River	2,123

Important Straits of the World

Straits	Water Bodies joined	Area
Bab-al-Mandeb	Red Sea & Arabian Sea	Arabia & Africa
Bering	Arctic Ocean & Bering Sea	Alaska & Asia
Bosphorus	Black Sea & Marmara Sea	Turkey
Dover	North Sea & Atlantic Ocean	England & Europe
Florida	Gulf of Mexico & Atlantic Ocean	Florida & Bahamas Islands
Gibralter	Mediterranean Sea & Atlantic Ocean	Spain & Africa
Malacca	Java Sea & Bay of Bengal	India & Indonesia
Palk	Bay of Bengal & Indian Ocean	India & Sri Lanka
Magellan	South Pacific & South Atlantic Ocean	Chile
Sunda	Java Sea & Indian Ocean	Indonesia

Major Rivers of the World

River	Origin	Falls in	Length (Km.)
Nile	Victoria lake	Mediterranean Sea	6,650
Amazon	Andes (Peru)	Atlantic Ocean	6,428
Yangtze	Tibetan Kiang Plateau	China Sea	6,300
Mississippi Missouri	Itaska lake (USA)	Gulf of Mexico (USA)	6,275
Yenisei	Tannu-Ola Mts.	Arctic Ocean	5,539
Hoang Ho	Kunlun Mts.	Gulf of Chibli	5,464
Ob	Altai Mts., Russia	Gulf of Ob	5,410
Congo	Lualaba & Luapula rivers	Atlantic Ocean	4,700
Amur	Northeast China	Sea of Okhotsk	4,444
Lena	Baikal Mts	Laptev Sea	4,400
Mekong	Tibetan Highlands	South China Sea	4,350
Mackenzie	Great Slave Lake	Beaufort Sea	4,241
Niger	Guinea	Gulf of Guinea	4,200

Major Deserts of the World

S.No.	Name	Type	Area (km^2)	Location
1.	Sahara	Subtropical	9,400,000	Algeria, Chad, Egypt, Eritrea, Libya, Mali, Mauritania, Morocco, Niger, Sudan, Tunisia and Western Sahara
2.	Arabian Desert	Subtropical	2,330,000	Iraq, Jordan, Kuwait, Oman, Qatar, Saudi Arabia, United Arab Emirates and Yemen
3.	Gobi Desert	Cold Winter	1,300,000	China and Mongolia
4.	Kalahari Desert	Subtropical	900,000	Angola, Botswana, Namibia and South Africa
5.	Patagonian Desert	Cold Winter	670,000	Argentina and Chile
6.	Great Victoria Desert	Subtropical	647,000	Australia
7.	Syrian Desert	Subtropical	520,000	Iraq, Jordan and Syria
8.	Great Sandy Desert	Subtropical	400,000	Australia
9.	Karakum Desert	Cold Winter	350,000	Turkmenistan
10.	Taklamakan Desert	Cold Winter	270,000	China
11.	Thar Desert	Subtropical	200,000	India and Pakistan
12.	Gibson Desert	Subtropical	156,000	Australia
13.	Atacama Desert	Cool Coastal	140,000	Chile and Peru
14.	Namib Desert	Cool Coastal	81,000	Angola and Namibia

Important Cities on River Banks (World)

City	Country	River	City	Country	River
Adelaide	Australia	Torrens	Amsterdam	Netherlands	Amsel
Alexandria	Egypt	Nile	Ankara	Turkey	Kazil
Bangkok	Thailand	Chao Praya	Baghdad	Iraq	Tigris
Berlin	Germany	Spree	Bonn	Germany	Rhine
Budapest	Hungary	Danube	Bristol	UK	Avon
Buenos Aires	Argentina	Laplata	Chittagong	Bangladesh	Majyani
Canton	China	Si-Kiang	Cairo	Egypt	Nile
Chung King	China	Yang-tse-kiang	Cologne	Germany	Rhine
Dandzing	Germany	Vistula	Dresden	Germany	Elbe
Dublin	Ireland	Liffy	Hamburg	Germany	Elbe
Kabul	Afghanistan	Kabul	Karachi	Pakistan	Indus
Lahore	Pakistan	Ravi	Leningrad	Russia	Neva
Lisbon	Portugal	Tagus	Liverpool	England	Messey
London	England	Thames	Moscow	Russia	Moskva
Montreal	Canada	St. Lawrence	Nanking	China	Yang-tse-kiang
New Orleans	U.S.A	Mississipi	New York	U.S.A	Hudson
Ottawa	Canada	Ottawa	Paris	France	Seine
Philadelphia	U.S.A	Delaware	Perth	Australia	Swan
Prague	Czech Republic	Vitava	Quebec	Canada	Sl. Lawrence
Rotterdam	The Netherlands	New Moss	Rome	Italy	Tiber

City	Country	River	City	Country	River
Stalingrad	Russia	Volga	Shanghai	China	Yang-tse-kiang
Sidney	Australia	Darling	Saint Louis	U.S.A	Mississippi
Tokyo	Japan	Arakava	Vienna	Austria	Danube
Warsaw	Poland	Vistula	Yangoon	Myanmar	Irawaddy

Major Gulfs of the World

Names	Area (Sq. Km.)
Gulf of Mexico	15,44,000
Gulf of St. Lawrence	2,37,000
Gulf of Hudson	12,33,000
Gulf of California	1,62,000
Arabian Gulf	2,38,000
English Channel	89,900

Important Local Winds

Wind	Type
Chinook	Hot, dry wind in Rockies, also called 'snow eater'.
Foehn	Hot, dry wind in the Alps.
Khamsin	Hot, dry wind in Egypt.
Sirocoo	Hot, moist wind from Sahara to Mediterranean Sea.
Solano	Hot, moist wind for Sahara towards Iberian Peninsula.
Harmattan	Hot, dry wind blowing outwards from the interior of West Africa, also called 'Guinea Doctor'.
Bora	Cold, dry wind blowing outwards from Hungary to the north of Italy (near Adriatic Sea).
Mistral	Very cold wind, which blows down from the Alps over France.
Punas	Cold, dry wind blowing down towards the western side of Andes.
Blizzard	Very cold winds in Tundra region.
Brickfielder	Hot wind in Australia.
Purga	Cold wind in Russian tundra
Levanter	Cold wind in Spain.
Norwester	Hot wind in New Zealand.
Santa Ana	Hot wind in Southern California in USA.

Foreign Towns Associated with Industry

Town (Country)	Associated Industry
Baku (Azerbaijan)	Petroleum
Bangkok (Thailand)	Ship building
Belfast (Ireland)	Ship building
Buenos Aires (Argentina)	Meat
Cadiz (Portugal)	Cork
Chicago (U.S.A.)	Agricultural implements, Meat
Coloigne (Germany)	Cotton and woollen industries
Dhaka (Bangladesh)	Jute
Detroit (U.S.A.)	Motor cars
Dresden (Germany)	Optical and photographic apparatus
Glasgow (Great Britain)	Machinery
Havana (Cuba)	Tobacco, Cigars
Hollywood (U.S.A.)	Film industry
Johannesburg (S. Africa)	Gold mines
Kimberlay (S. Africa)	Diamond mining
Leeds (England)	Woollen goods
Lyons (France)	Silk industries
Manchester (England)	Cotton industry
Mauritius (Indian Ocean)	Fishing, Shipping, Sugar
Milan (Italy)	Silk
Morocco (North Africa)	Leather
Munich (Germany)	Lenses
New Orleans (U.S.A)	Cotton industry
Osaka (Japan)	Cotton fabrics
Pittsburg (U.S.A)	Iron and Steel
Plymouth (England)	Ship-building
Sheffield (England)	Cutlery
Venice (Italy)	Glass manufacturing
Vienna (Austria)	Glass manufacturing
Wellington (New Zealand)	Dairy Product

INDIAN GEOGRAPHY

- India extends between 8°4' N and 37°6' N latitudes and between 68°7' E and 97°2' E longitudes.
- India, has a total geographic area of 32,87,263 sq. km. This is only 2.42 per cent of the total geographic area of the world but holds 16 per cent of the world's population.
- The 23½° N, which is the Tropic of Cancer, runs across the country. India has a length of 3214 km from north-to-south and 2933 km from east-to-west. It has a land frontier of 15200 km.
- The total length of the coastline of the mainland, Lakshadweep Islands and Andaman and Nicobar Islands is 7,516.6 km.
- India shares its common border with Afghanistan and Pakistan in the north-west, China and Bhutan in the north, and Bangladesh in the east. In the south, Sri Lanka is separated from India by a strait, known as the Palk Strait.
- 82°30' E longitude is considered as the Indian Standard Meridian. The local time of this longitude is taken as the Indian Standard Time (IST). This is 5½ hours ahead of the Greenwich Mean Time.
- India is the second most populous country in the world with a population of 1.21 billion, which is 17.44% of the world.
- India shares longest boundary with Bangladesh (4053 km), followed by China (3380 km), Pakistan (2912 km), Nepal (1690 km), Myanmar (1463 km), Bhutan (605 km) and Afghanistan (88 km).
- In India, the Tropic of Cancer (23.5° N latitude) passes through 8 states (Gujarat, Rajasthan, Madhya Pradesh, Chhattisgarh, Jharkhand, West Bengal, Tripura and Mizoram).
- Indian subcontinent is located in the Northern and Eastern hemisphere.
- Islands Andaman and Nicobar Islands in the Bay of Bengal; Lakshadweep, Amindivi and Minicoy in the Arabian Sea.
- Ocean India lies midway between the Far East and Middle East. The trans-Indian Ocean routes connecting the industrially developed countries of Europe in the West and the under developed countries of East Asia pass close by Indian subcontinent is surrounded by Arabian Sea in the South-West and Bay of Bengal in the South-East.
- India is bordered on three sides by water and on one by land, it is also a peninsula.
- There are 29 States, and 6 Union Territories and 1 National Capital Territory (Delhi) in India.

Indian States Situated on the Border

Country	Border
Pakistan (4)	Gujarat, Rajasthan, Punjab, Jammu and Kashmir
Afghanistan (1)	Jammu and Kashmir
China (5)	Jammu and Kashmir, Uttara-khand, Himachal Pradesh, Sikkim, Arunachal Pradesh
Nepal (5)	Uttar Pradesh, Uttarakhand, Bihar, Paschim Bangal, Sikkim
Bhutan (4)	Sikkim, Paschim Banga, Asom, Arunachal Pradesh
Bangladesh (5)	Paschim Banga, Asom, Meghalaya, Tripura, Mizoram

Highest Peaks of India

Highest Peak	Height (in m)	State
Mt. K2	8611	PoK (India)
Kanchenjunga	8598	Sikkim
Nanda Devi	7817	Uttarakhand
Saltoro Kangri	7742	Jammu and Kashmir
Kangto	7090	Arunachal Pradesh
Reo Purgyil	6816	Himachal Pradesh
Saramati	3841	Nagaland
Sandakphu	3636	Paschim Banga
Khayang	3114	Manipur
Anaimudi	2695	Kerala
Dodda Betta	2636	Tamil Nadu

Important Sanctuaries and National Park

Name	Location	Reserve for
Achanakmar Sanctuary	Bilaspur, Chhattisgarh	Tiger, boar, chital, sambar, bison
Bandhavgarh National Park	Shahdol, Madhya Pradesh	Tiger, panther, chital, nilgai, wild boar
Bandipur Sanctuary	Border of Karnataka and Tamil Nadu	Elephant, tigers, panther, sambar, deer, birds
Banerghatta National Park	Bangalore, Karnataka	Elephant, chital, deer, grey partridges, green pigeon
Bhadra Sanctuary	Chikmagalur, Karnataka	Elephant, chital, panther, sambar, wild boar
Bhimabandh	Monghyr, Bihar	Tiger, leopard, sambar, wild boar, chital, water birds
Bori Sanctuary	Hoshangabad, Madhya Pradesh	Tiger, panther, sambar, chital, wild boar, barking deer
Borivili National Park	Mumbai	Panther, sambar, langur, wild boar, chinkara
Chandraprabha Sanctuary	Near Varanasi, Uttar Pradesh	Famous for Gir lions, chital, sambar
Corbett National Park	Nainital, Uttarakhand	Tiger, leopard, elephant, sambar (named in memory of Jim Corbett)
Dachigam Sanctuary	Dachigam, Kashmir	Kashmiri stag
Dalma Sanctuary	Singhum, Jharkhand	Elephant, leopard, wild boar, barking deer
Dandeli Sanctuary	Dharwar, Karnataka	Tiger, panther, elephant, chital, sambar, wild boar
Dudhwa National Park	Lakhimpurkheri, Uttar Pradesh	Tiger, panther, sambar, chital, nilgai, barking deer
Gandhi Sagar Sanctuary	Mandsaur, Madhya Pradesh	Chital, sambar, chinkara, barking deer, wild birds
Garampani Sanctuary	Diphu, Assam	Elephant, leopard, wild buffalo, langur
Ghana Bird Sanctuary	Bharatpur, Rajasthan	Water birds, black-buck, chital, sambar
Gir Forest	Junagarh, Gujarat	India's biggest wildlife sanctuary famous for Gir lions
Gautam Buddha Sanctuary	Gaya, Bihar	Tiger, leopard, sambar, chital, barking deer
Hazaribagh Sanctuary	Hazaribagh, Jharkhand	Tiger, leopard, chital, nilgai, sambar, wild cat
Intangki Sanctuary	Kohima, Nagaland	Elephant, gaur, tiger, panther, barking deer, wild boar
Jaldapara Sanctuary	West Bengal	Rhinoceros
Kawal Sanctuary	Adilabad, Andhra Pradesh	Tiger, panther, gaur, chital, wild boar
Kaziranga National Park	Jorhat, Assam	Horned rhinoceros, gaur, elephant, leopard, wild buffalo
Khangchandzenda National Park	Gangtok, Sikkim	Snow leopard, musk deer, Himalayan boar
Kinnersani Sanctuary	Khamrsan, Andhra Pradesh	Tiger, panther, gaur, chital, sambar, nilgai

Name	Location	Reserve for
Kolleru Pelicanary	Elluru, Andhra Pradesh	Pelicans, painted stroke
Nogerhole National Park	Coorg, Karnataka	Elephant, tiger, panther, sambar, chital
Namdafa Sanctuary	Tirap, Arunachal Pradesh	Elephant, panther, sambar, tiger, chital, king cobra
Nawegaon National Park	Bhandara, Maharashtra	Tiger, panther, sambar, chital, nilgari
Pachmarhi Sanctuary	Hoshangabad, Madhya Pradesh	Tiger, panther, boar, sambar, nilgai, barking deer
Pakhal Sanctuary	Warangal, Andhra Pradesh	Tiger, panther, sambar chital, nilgai
Parambikulam Sanctuary	Palghat, Kerala	Tiger, leopard, gaur, elephant, nilgai, chital
Pench National Park	Nagpur, Maharashtra	Tiger, panther, gaur, sambar, chital, nilgai
Periyar Sanctuary	Idukki, Kerala	Elephant, tiger, panther, gaur, nilgai, sambar, wild beer
Ranganthittoo Bird Sanctuary	Islands in river Cauvery in Karnataka	Important bird sanctuary
Rhola National Park	Kulu, Himachal Pradesh	Snow-leopard, brown boar, musk deer, snowcock, new pigeon
Sariska Sanctuary	Alwar, Rajasthan	Tiger, panther, sambar, nilgai, chital, chinkara
Sharaswathy Valley Sanctuary	Shimoga, Karnataka	Elephant, tiger, panther, sambar, gaur, chital, wild boar
Shikari Devi Sanctuary	Mandi, Himachal Pradesh	Black boar, musk deer, panther, leopard, partridge
Shivpuri National Park	Shivpuri, Madhya Pradesh	Tiger, panther, sambar, hyena, sloth, boar, nilgai
Similipal Sanctuary	Mayurbhanj, Orrisa	Elephant, tiger, leopard, gaur, chital
Someshwara Sanctuary	Canara, Karnataka	Tiger, Panther, Wild Boar, Leopard
Sunderban Tiger Reserve	South 24 parganas, West Bengal	Tiger, deer, wild boar, crocodile, Gangetic dolphin
Sonai Rupa Sanctuary	Tezpur, Assam	Elephant, sambar, wild boar, one-horned rhinoceros
Tadoda National Park	Chandrapur, Maharashtra	Tiger, panther, sambar, nilgai, chinkara, chital
Todwai Sanctuary	Warangal, Andhra Pradesh	Tiger, panther, sambar, gaur, jungle cat
Tansa Sanctuary	Thane, Maharashtra	Panther, sambar, chital, four-horned antelope
Tungabhadra Sanctuary	Bellary, Karnataka	Panther, chital, sloth boar, four-horned antelope
Valvadore National Park	Bharnagar, Gujarat	Wolf, black buck
Vedanthangal Bird Sanctuary	Tamil Nadu	Important bird sanctuary
Wynad Sanctuary	Cannanore and Kozhikode, Kerala	Elephant, gaur, sambar, chital, wild boar, deer
Wild Ass Sanctuary	Little Rann of Kutch, Gujarat	Wild ass, wolf, nilgai, chinkara

Biosphere Reserves of India

Name of the site & Date of Notification	km²	Area in Location (State)
Nilgiri (01.08.86)	5,520	Part of Wynad, Nagarhole, Bandipur and Mudumalai, Nilambur, Silent Valley and Siruvani Hills (Tamil Nadu, Kerala and Karnataka)
Nanda Devi (18.01.88)	5,860.69	Part of Chamoli, Pithoragarh, Almora Districts (Uttaranchal)
Nokrek (01.09.88)	820	Part of Gora Hillas (Meghalaya)
Manas (14.03.89)	2,837	Part of Kokrajhar, Bongaigaon, Barpeta, Nalbari, Kamrup, and Darang district (Assam)
Sunderbans (29.03.89)	9,630	Part of delta of Ganga and Brahmaputra river system (West Bengal)
Gulf of Mannar (18.02.89)	10,500	Indian part of Gulf of Mannar, between India and Sri Lanka (Tamil Nadu)
Great Nicobar (06.01.89)	885	Southernmost islands of Andaman and Nicobar (Andaman and Nicobar islands)
Simlipal (21.06.94)	4,374	Part of Mayurbhanj district (Orissa)
Dibru-Saikhowa (28.07.97)	765	Part of Dibrugarh and Tinsukia district (Assam)
Dehand Debang (02.09.98)	5,112	Part of Siang and Debang Valley (Arunachal Pradesh)
Pachmarhi (03.03.99)	4,926.28	Parts of Betul, Hoshangabad, and Chindwara districts (Madhya Pradesh)
Kanchanjanga (07.02.00)	2,619.92	Part of Kanchanjanga Hills (Sikkim)

PROJECT TIGER

- It was launched on April 1, 1973 to ensure maintenance of viable population of the tigers in India.
- There are 39 tiger reserves in the country:

Name of Tiger Reserve	State	Name of Tiger Reserve	State
Bandipur	Karnataka	Bori, Satpura, Panchmarhi	Madhya Pradesh
Kanha	Madhya Pradesh	Katarniaghat	Uttar Pradesh
Melghat	Maharashtra	Kaziranga	Assam
Palamau	Jharkhand	Corbett	Uttarakhand
Ranthambhore	Rajasthan	Manas	Assam
Sunderbans	West Bengal	Similipal	Odisha
Periyar	Kerala	Nagarjunasagar	Andhra Pradesh
Sariska	Rajasthan	Dudhwa	Uttar Pradesh
Buxa	West Bengal	Valmiki	Bihar
Indravati	Chhattisgarh	Tadobo-Andhari	Maharashtra
Namdapha	Arunachal Pradesh	Panna	Madhya Pradesh
Kalakad-Mundanthurai	Tamil Nadu	Bhadra	Karnataka
Pench	Madhya Pradesh	Pakhui-Nameri	Arunachal-Assam
Bandhavgarh	Madhya Pradesh	Nagarhole	Karnataka
Dampha	Mizoram	Nameri	Assam
Pench	Maharashtra		

Important Projects and their Beneficiary States

Project	River	Purpose	Beneficiary States
Farakka Project	Ganga, Bhagirathi	Power, irrigation, avoid accumulation of slit to improve navigation	Paschim Banga
Beas Project	Beas	Irrigation and power	Rajasthan, Haryana, Punjab and Himachal Pradesh
Rajasthan Canal Project	Sutlej, Beas and Ravi	Irrigation	Rajasthan, Punjab and Haryana
Chambal Project	Chambal	Power and irrigation	Madhya Pradesh and Rajasthan
Kakrapara Project	Tapti	Irrigation	Gujarat
Ukai Project	Tapti	Power and irrigation	Gujarat
Ramganga Multipurpose Project	Chisot stream near Kala	Power and irrigation	Uttar Pradesh
Matatilla Project	Betwa	Multipurpose power and irrigation	Uttar Pradesh and Madhya Pradesh
Tehri Dam Project	Bhilangana, Bhagirathi	Hydroelectricity	Uttarakhand
Rihand Scheme	Rihand	Hydroelectricity	Uttar Pradesh
Kundah Project	Kundah	Hydroelectricity and irrigation	Tamil Nadu
Bhakra Nangal Project	Sutlej	Power and irrigation	Punjab, Himachal Pradesh, Haryana and Rajasthan
Damodar Valley	Damodar	Power, Irrigation and flood control	Bihar and Paschim Banga, shared by Madhya Pradesh
Hirakud	Mahanadi	Power and irrgation	Odisha
Tunghabhadra Project	Tungabhadra	Power and irrigation	Andhra Pradesh and Karnataka
Nagarjuna Sagar Project	Krishna	Power and irrigation	Andhra Pradesh
Gandak River Project	Gandak	Power and irrigation	Bihar, Uttar Pradesh, Nepal (joint venture of India and Nepal)
Kosi Project	Kosi	Flood control, Power and irrigation	Bihar
Tawa Project	Tawa (Narmada)	Irrigation	Madhya Pradesh
Poochampad Project	Godavari	Irrigation	Andhra Pradesh
Malaprabha Project	Malaprabha	Irrigation	Karnataka
Durgapur Barrage	Damodar	Irrigation and navigation	Karnataka, Paschim Banga and Jharkhand
Mahanadi Delta Project	Mahanadi	Irrigation	Odisha
Iddukki Project	Periyar	Hydroelectricity	Kerala
Koyna Project	Koyna	Hydroelectricity	Maharashtra

Mineral Resources of India

Resources	Place
Coal	West Bengal (Raniganj, Burdwan, Bankura, Purulia, Birbhum, Jalpaigudi, Darjeeling), Jharkhand (Jharia, Giridih, Kharhawadi, Bokaro, Hazaribagh, Karnapura, Rampur, Palamau), Odisha (Rampur, Hindgir, Talcher, Sambal), Madhya Pradesh and Chhattisgarh (Rewa, Pench valley, Umaria, Korba, Sohagpur, Mand river area, Kanha valley, Betul), etc. Power sector is the largest consumer of coal in India followed by steel industry, cement industry, etc.
Manganese	Maharashtra (Nagpur, Bhandara, Ratnagiri), Madhya Pradesh (Balaghat, Chhindawara), Odisha (Keonjhar, Bonai, Kalahandi), Andhra Pradesh (Kadur, Garibadi).
Copper	Madhya Pradesh (Balaghat), Rajasthan (Khetri), Jharkhand (Singhbhum, Masobani, Surda), Karnataka (Chitradurg, Hassan).
Mica	Jharkhand (Hazaribagh, Giridih, Kodarma), Bihar (Gaya, Bhagalpur), Andhra Pradesh (Guntur, Vizag, Kurnool), Rajasthan (Bhilwara, Udaipur, Jaipur).
Petroleum	Assam (Digboi, Naharkatiya, Badarpur, Masinpur and Pallharia), Gujarat (Ankleshwar, Khambat, Kalol), Mumbai High, Bassein (South of Mumbai High), etc. Recently oil has been discovered in Cauvery basin, Krishna and Godavari basin, Khambat basin, etc.
Iron	India possesses Haematite, a very high-grade iron ore. In Madhya Pradesh (Bailadila, Jabalpur), Goa (North Goa), Karnataka (Bababudan hills, Chikmangalur, Hospet), Jharkhand (Singhbhum, Noamundi), Andhra Pradesh, Odisha. India is the fifth largest exporter of iron ore in the world. Japan is the biggest buyer accounting for about 3/4th of India's total exports. Major ports handling iron ore export are Vishakhapatnam, Paradip, Marmagao and Mangalore.
Bauxite	Chief ore for producing aluminium. In Odisha (Kalahandi, Koraput, Sundargarh, Bolangir, Sambalpur), Jharkhand (Lohardaga, Gumla), Madhya Pradesh (Jabalpur, Mandla, Shahdol, Katni, Balaghat), Maharashtra, Andhra Pradesh, Gujarat, Tamil Nadu.
Gold	Karnataka (Kolar, Hutti, Raichur), Andhra Pradesh (Ramgiri and Yeppamanna goldfields in Chittor and Anantapur districts).
Silver, Zinc & Lead	Rajasthan (Zawar mines near Udaipur), Andhra Pradesh (Mysore, Chitradurg), Karnataka (Kolar mines).
Uranium	Jharkhand (Jaduguda), Rajasthan (Ajmer), Andhra Pradesh (Nellore, Nalgonda), Karnataka (Gulbarga).
Thorium	Kerala coast (From Monazite sand), rocks of Aravallis in Rajasthan

Water Transport

The Government has recognised the following National Waterways of India

NW 1	Allahabad to Haldia	1629 km
NW 2	Sadia to Dhubri on Brahmaputra river	819 km
NW 3	Kollam to Kottapuram	186 km
NW 4	Kakinada to Marakkanam along Godavari and Krishna river	1110 km
NW 5	Mangalgarhi to Paradeep and Talcher to Dhamara Mahanadi and Brahmini	1623 km

Important National Highways

NH	Connects
NH 1	New Delhi-Ambala-Jalandhar-Amritsar
NH 2	Delhi-Mathura-Agra-Kanpur-Allahabad-Varanasi-Kolkata
NH 3	Agra-Gwalior-Nasik-Mumbai
NH 4	Thane and Chennai via Pune and Belgaum
NH 5	Kolkata-Chennai
NH 6	Kolkata-Dhule
NH 7	Varanasi-Kanyakumari (2369 km)
NH 8	Delhi-Mumbai (via Jaipur, Baroda and Ahmedabad)
NH 9	Mumbai-Vijayawada
NH 10	Delhi-Fazika
NH 24	Delhi-Lucknow
NH 26	Lucknow-Varanasi

Railway Zones and Headquarters

Railway Zones	Headquarters
Central	Mumbai VT
Eastern	Kolkata
Northern	New Delhi
North Eastern	Gorakhpur
North-East Frontier	Maligaon-Guwahati
Southern	Chennai
South Central	Secunderabad
South Eastern	Kolkata
Metro Railway	Kolkata
Western	Mumbai Churchgate
East Coast	Bhubaneshwar
East Central	Hajipur
North Central	Allahabad
North Western	Jaipur
South Western	Bangalore (Hubli)
West Central	Jabalpur
South East Central	Bilaspur

Major Ports of India

Name	State	River/Strait/Ocean
Kolkata	West Bengal	Hoogly River
Mumbai (Busiest and biggest)	Maharashtra	Arabian Sea
Chennai (Oldest and artificial)	Tamil Nadu	Bay of Bengal
Cochin (Natural harbour)	Kerala	Arabian Sea
Vishakhapatnam (Deepest Port)	Andhra Pradesh	Bay of Bengal
Paradip	Orissa	Bay of Bengal
New Titucorin (Southernmost)	Tamil Nadu	Bay of Bengal
Marmagao	Goa	Arabian Sea
Kandla (Tidal Port)	Gujarat	Arabian Sea
Mangalore (exports Kudremukh iron-ore)	Karnataka	Arabian Sea
J. L. Nehru Port (Fastest Growing)	Maharashtra	Arabian Sea
Ennore (Most modern in private hands)	Tamil Nadu	Bay of Bengal

Forest Report in India

Class	Area (sq. km)	Per cent of Geographical Area
Forest Cover— Very Dense Forest	85,904	2.61
Moderate Dense Forest	3,15,374	9.59
Open Forest	3,00,395	9.14
Total Forest Cover*	**7,01,673**	**21.34**
Tree Cover	92,572	2.82
Total Forest and Tree Cover	**7,94,245**	**24.16**
Scrub	41,362	1.26
Non Forest	25,44,228	77.40
Total Geographical Area	**32,87,263**	**100.00**

Major Tribes in India

Tribes	State
Abors	Assam
Apatamis	Arunachal Pradesh
Badagas	Nilgiri (TN)
Baiga	Madhya Pradesh
Bhils	Mostly in Madhya Pradesh and Rajasthan, also in Gujarat
Bhotias	Garhwal and Kumaon regions in U.P.
Chenchus	Andhra Pradesh, Orissa
Gaddis	Himachal Pradesh
Garos	Meghalaya
Gonds	Madhya Pradesh, also in Bihar, Orissa and Andhra Pradesh
Jarawas	Little Andamans
Khas	Jaunsar-Babar area in U.P.
Khasis	Assam, Meghalaya
Khonds	Orissa
Kol	Madhya Pradesh
Kotas	Nilgiri (Tamil Nadu)
Kuki	Manipur

Tribes	State
Lepchas	Sikkim
Lushais	Tripura
Murias	Bastar region in Madhya Pradesh
Mikirs	Assam
Mundas	Jharkhand
Nagas	(Angami, Sema, Ao, Tangkul, Lahora): Nagaland, some in Assam and NEFA region
Oarons (also called Kurukh)	Bihar, Orissa
Onges	Andaman and Nicobar Islands
Santhals	Birbhum region in Bengal, Hazaribagh, Ranchi and Palamau in Jharkhand
Sentinelese	Sentinel Island, Andaman and Nicobar
Shompens	Andaman and Nicobar
Todas	Nilgiri (Tamil Nadu)
Uralis	Kerala
Warlis	Maharashtra

Mangrove Cover Report 2015

Sl. No.	States/UTs	Very Dense Mangrove	Moderately Dense Mangrove	Open Mangrove	Total Mangrove	Change with respect to ISFR 2013
1.	Andhra Pradesh	0	129	238	367	15
2.	Goa	0	20	6	26	4
3.	Gujarat	0	174	933	1107	4
4.	Karnataka	0	3	0	3	0
5.	Kerala	0	5	4	9	3
6.	Maharashtra	0	79	143	222	36
7.	Odisha	82	95	54	231	18
8.	Tamil Nadu	1	18	28	47	8
9.	West Bengal	991	700	416	2106	9
10.	A & N Islands	399	168	50	617	13
11.	Daman & Diu	0	0	3	3	1
12.	Puducherry	0	0	2	2	1
	Total	**1472**	**1391**	**1877**	**4740**	**112**

Forest and Tree Cover of India in 2015

States/UTs	Geogra-phical Area	VDF	MDF	OF	Total Forest area	Percent of geographical
Andhra Pradesh	1,60,204	375	13,093	10,956	24,424	15.25
Arunachal Pradesh	83,743	20,804	31,301	15,143	67,248	80.30
Assam	78,438	1,441	11,268	14,914	27,623	35.22
Bihar	94,163	248	3,376	3,664	7,288	7.74
Chhattisgarh	1,35,191	4,152	34,846	16,588	55,586	41.12
Delhi	1,483	6.94	57.15	124.68	188.77	12.73
Goa	3,702	542	580	1,102	2,224	60.08
Gujarat	1,96,022	376	5,220	9,064	14,660	7.48
Haryana	44,212	27	452	1,105	1,584	3.58
Himachal Pradesh	55,673	3,224	6,381	5,091	14,696	26.40
Jammu & Kashmir	2,22,236	4,061	8,815	10,112	22,988	10.34
Jharkhand	79,714	2,588	9,663	11,227	23,478	29.45
Karnataka	1,91,791	1,781	20,063	14,577	36,421	18.99
Kerala	38,863	1,523	9,301	8,415	19,239	49.50
Madhya Pradesh	3,08,245	6,629	34,902	35,931	77,462	25.13
Maharashtra	307,713	8,712	20,747	21,169	50,628	16.45
Manipur	22,327	727	5,925	10,342	16,994	76.11
Meghalaya	22,429	449	9,584	7,184	17,217	76.76
Mizoram	21,081	138	5,858	12,752	18,748	88.93
Nagaland	16,579	1,296	4,695	6,975	12,966	78.21
Odisha	1,55,707	7,023	21,470	21,861	50,354	32.34
Punjab	50,362	0	735	1,036	1,771	3.52
Rajasthan	3,42,239	76	4,426	11,669	16,171	4.73
Sikkim	7,096	500	2,160	697	3,357	47.31
Tamil Nadu	1,30,058	2,993	10,469	12,883	26,345	20.26
Telangana	1,14,865	513	12,712	8,366	21,591	18.80
Tripura	10,486	113	4,609	3,089	7,811	74.49
Uttar Pradesh	2,40,928	2,195	4,060	8,206	14,461	6.00
Uttarakhand	53,483	4,754	13,602	5,884	24,240	45.32
West Bengal	88,752	2,948	4,172	9,708	16,828	18.96
A & N Islands	8,249	5,686	685	380	6,751	81.84
Chandigarh	114	1.36	14.09	6.58	22.03	19.32
Dadra & Nagar Havelli	491	0	80	126	,206	41.96
Daman & Diu	112	1.4	5.82	12.39	19.61	17.51
Lakshdweep	32	0	17.22	9.84	27.06	84.56
Puducherry	480	0	29.68	25.70	55.38	11.54
Total	**32,87,263**	**85,904**	**3,15,374**	**3,00,395**	**7,01,673**	**21.34**

❑ ❑ ❑

Indian Polity

- The idea to have a Constitution was given by M.N. Roy (Political Philosopher).
- The Constitution was framed by the Constituent Assembly of India, set-up in December 1946, in accordance with the Cabinet Mission Plan, under the Chairmanship of Sachidanand Sinha, initially. On the demise of Sachidanand Sinha, Dr. Rajendra Prasad was elected as the President. B.N. Rau was appointed as the Constitutional Advisor.
- The Chairman of the Drafting Committee was Dr. B.R. Ambedkar, also called the Father of the Constitution.

Interim Government (1946)

Members	Portfolios Head
Jawaharlal Nehru	External Affairs & Commonwealth Relations
Sardar Vallabhbhai Patel	Home, Information & Broadcasting
Dr. Rajendra Prasad	Food & Agriculture
Dr. John Mathai	Industries & Supplies
Jagjivan Ram	Labour
Sardar Baldev Singh	Defence
C.H. Bhabha	Works, Mines & Power
Laiquat Ali Khan	Finance
Abdur Rab Nishtar	Posts & Air
Asaf Ali	Railways & Transport
C. Rajagopalachari	Education & Arts
II Chundrigar	Commerce
Ghaznafar Ali Khan	Health
Joginder Nath Mandal	Law

- The Constituent Assembly took 2 years, 11 months and 18 days to complete the Constitution.
- The Constitution, adopted on 26th November, 1949, contained 395 Articles and 8 Schedules.
- The Constitution was delayed till 26th January because, in 1929, on this day Indian National Congress demanded **Poorna Swaraj** in Lahore Session, Chaired by J.L. Nehru.

THE PREAMBLE

- The Preamble does not from an integral part of the Constitution but embodies its spirit and meaning. The 42nd Amendment (1976) added the words 'Secular', 'Socialist' and 'Integrity' now the Preamble reads as follows.
- The Idea of the Preamble was borrowed from the constitution of USA.

Sources of the Indian Constitution

UK	❑ Nominal Head-President ❑ Cabinet System of Ministers ❑ Post of Prime Minister ❑ Parliamentary Type of Government ❑ Bicameral Parliament ❑ Lower House more Powerful ❑ Single Citizenship ❑ Law making procedures
USA	❑ Written Constitution ❑ Vice-President as the Ex-officio Chairman of Upper House

	❑ Fundamental Rights ❑ Supreme Court ❑ Provision of States ❑ Independence of Judiciary and Judicial Review ❑ Preamble
Erstwhile USSR	❑ Fundamental Duties
Australia	❑ Concurrent List ❑ Language of the Preamble
Japan	❑ Law on which the Supreme Court functions
Germany	❑ Suspension of Fundamental Rights during the Emergency
Canada	❑ Scheme of federation with a strong Centre ❑ Distribution of powers between the Centre and the States and placing Residuary Powers with the Centre
Ireland	❑ Concept of Directive Principles of State Policy Method of election of the President.

IMPORTANT ARTICLES

PART-I

Union and its Territories (Article 1-4)

- The Constitution says, "India, that is Bharat, shall be a Union of States".
- The Constitution empowers the Parliament to form new States and to alter the areas, boundaries or names of existing States.
- Jammu and Kashmir has been given special status under Article 370.

PART-II

Citizenship (Article 5-11)

- The Citizenship Act of 1955 prescribes five ways to acquire citizenship of India
  1. By birth, 2. By descent, 3. By registration, 4. By naturalisation, 5. By incorporation.

PART-III

Fundamental Rights (Article 12-35)

Rights to Equality (Article 14-18)

- Equality before law (Article 14).
- Prohibition of discrimination on grounds of religion, race, caste, sex or place of birth. (Article 15).
- Equality of opportunity in matters of public employment. (Article 16)
- Abolition of untouchability (Article 17).
- Abolition of titles. (Article 18)

Rights to Freedom (Article 19-22)

- Protection of certain rights regarding freedom of speech and expression, freedom to assemble peacefully, cooperatives, etc (Article 19)
- Protection in respect of conviction for offences. (Article 20)
- Protection of life and personal liberty (Article 21).
- Protection against arrest and detention in certain cases (Article 22).

Rights to Education

- Article 21A states that the state shall provide free and compulsory education to all children of the age of 6-14 years.

Rights against Exploitation

- Prohibition of trafficking in human beings and forced labour (Article 23).
- Prohibition of employment of children below the age of 14 years in any factories, etc (Article 24).

Rights to Freedom of Religion (Article 25-28)

- Subject to public order, morality and health and to other provisions of this part. All persons are equally entitled to freedom of conscience and the right freely to profess, practise and propagate religion (Article 25).

- The right to establish and maintain its own institutions for religious and charitable purposes (Article 26).
- Provides protection to religious activity by exempting funds appropriates towards the promotion and maintenance of any particular religion from the payment of taxes (Article 27).
- Prohibits religious instruction in any educational institution wholly maintained out of state funds (Article 28).

Cultural and Educational Rights

- Protection of interest of minorities (Article 29).
- Right of minorities to establish and administer educational institutions (Article 30).

Rights to Constitutional Remedies

- Right to move to the Supreme Court (Article 32) and the High Courts (Article 226) in case of violation of the Fundamental Rights termed by B.R. Ambedkar as the *Heart and Soul of the Constitution.*
- The Supreme Court and High Courts have the power to issue writs, if fundamental rights are violated.

PART-IV

Directive Principles of State Policy (Article 36-51)

- Articles 36-37 — Definition and application of the principles contained in this part.
- Article 38 — To secure and protect a social order, which stands for the welfare of the people.
- Article 39 — Certain principles of policy to be followed by the State.
- Article 40 — To organise village panchayats as units of self-government.
- Article 41 — Rights to work, to education and to public assistance in certain cases.
- Article 42 — To secure just and humane conditions of work and maternity relief.
- Article 43 — Living wage etc for workers.
- Article 44 — Uniform Civil Code for the citizens.
- Article 45 — Provision of early childhood care and education to children below the age of 6 years.
- Article 51 — Promotion of international peace and security.

PART-IV (A)

Fundamental Duties (Article 51A)

- It was inserted by the 42nd Amendment Act in 1976 on the recommendations of Swaran Singh Committee and it was taken from erstwhile Constitution of USSR.
- The **86th Amendment Act,** 2002 inserted Article 51A (K), "each parent or guardian to provide opportunities for education to his child or ward between the age of 6 and 14 years".

PART-V

Union (Article 52-151)

THE PRESIDENT

- Executive head of the State and the first citizen of India.
- The 42nd Amendment of the Constitution has made it obligatory on the part of the President to accept the advice of the Council of Ministers. However, 44th Amendment Act amended the word 'obligatory' and added that 'President can send the advice for reconsideration'.
- **Qualification :** Must be a citizen of India; of 35 years in age; eligible to be a member of the Lok Sabha and must not hold any office of profit.
- **Election :** Indirectly elected through Electoral College consisting of elected members of both the Houses of the Parliament and elected members of the Legislative Assemblies of the States. Supreme Court decides all disputes regarding President's election.
- **Tenure :** The term is 5 years though there is no upper limit on the number of times a person can become the President (Article 57). He can give resignation to the Vice-President before the full-term.

- In case, the office of the President falls vacant due to death, resignation or removal, the Vice-President acts as the President.
- If he is not available then Chief Justice of India, if not then the senior most Judge of the Supreme Court shall act as the President of India.

Powers of President

- He is the formal head of the administration.
- The President shall have the power to appoint and remove high authorities like, the Prime Minister, other Ministers of the Union, Judges, Governors of States and appoints Chiefs of Army, Navy and Air Force.
- He appoints 12 members of special repute in the Rajya Sabha and 2 members in the Lok Sabha of the Anglo-Indian Community.
- **Under Article 72,** the President has the power to grant pardons, reprieves, respites or remission of punishment or to suspend, remit or commute the sentence of any person convicted with death sentence.
- **Under Article 123,** President can promulgate ordinances only, when the Parliament is not in session.

Types of Vetoes

Absolute Veto	Withholding the assent to the bill.
Qualified Veto	Can be overridden by the Legislature with a higher majority.
Suspensive Veto	Can be overridden by the Legislature with an ordinary majority.
Pocket Veto	Delay in giving assent to the bill. The Veto Power has been exercised only twice (a) by Dr. Rajendra Prasad of the above four the President of India is vested with three vetos absolute veto, suspensive veto and pocket veto. There is no qualified veto in the case of Indian President.

Emergency Powers

- To declare National Emergency (Article 352).
- To impose President Rule in a State (Article 356).
- To declare Financial Emergency (Article 360).
- Justice M. Hidyatullah was the first Chief Justice of India to be appointed as the President (July 1969-August 1969).

VICE-PRESIDENT

- **Article 63** of the Constitution stipulates a Vice-President for India.
- The Vice-President is the Ex-officio **Chairman** of the council of States (Rajya Sabha) as mentioned in the **Article 64.**

COUNCIL OF MINISTERS

- Article 74 of the Constitution states that there shall be a Council of Ministers, with the Prime Minister at it's head, to aid and advise the President.
- The Ministers can be **appointed by the President** on the advice of the Prime Minister under Article 75(1).
- A Minister must be a member of either House of Parliament or be elected within 6 months of assuming office under Article 75(5).
- Ministers are also responsible for their departments and can be removed from the office by the President on the advice of the Prime Minister. This is essentially an individual responsibility under Article 75(2).

PRIME MINISTER

- The Prime Minister is the head of the Government and the head of the Council of Ministers.
- The Prime Minister is appointed by the President on the basis of his being the leader of the majority party in the Lok Sabha.
- When the Lok Sabha is dissolved he can continue in office upon the request of the President.

- The Prime Minister selects other ministers who are appointed by the President on his advice; presides over Cabinet meetings; exercises a casting vote as Chairman of the Council of Ministers.
- He/she is the link between the President and the Cabinet; keeps the President informed of the decisions of the Council of Ministers (Article 78).
- The Prime Minister holds office for five years, but can be reappointed. He/she holds office with the consent of the President till a new Lok Sabha is formed; even when the Lok Sabha is dissolved by the President, he can ask the Prime Minister to hold office till new elections are held.

UNION LEGISLATURE

- Legislature of the Union is called the Parliament and consists of the Rajya Sabha (Council of States), the Lok Sabha (House of the People) and the President (Article 79).

Rajya Sabha (Council of States)

- The Rajya Sabha is the Upper House of the Parliament and the first sitting of the Rajya Sabha was held on 3rd April, 1952.
- The maximum permissible strength of the Rajya Sabha is 250. Of these, 238 members are elected indirectly from the States and Union Territories, and 12 are nominated by the President for their expertise in art, literature, science and social services (Article 80).
- Currently, the strength of the Rajya Sabha is 245. Of these, 233 members are elected from States and Union Territories and 12 are nominated members.
- The Rajya Sabha is a **Permanent House** and is not subject to dissolution and members enjoy a tenure of six years. One-third of the members retire every two years (Article 83).
- It shares legislative powers with the Lok Sabha, except in the case of **Money Bill** where the Lok Sabha has overriding powers.

Lok Sabha (People's House)

- The Lok Sabha is the **Lower House** of the Parliament and its first sitting took place on 13th May, 1952. The current Lok Sabha is the 16th Constituted Lok Sabha.
- Members 530 from States, 20 from Union Territories and 2 nominated by the President from the Anglo Indian Community.
- **Qualifications** Article 84 provides for the eligibility for membership of the Parliament. The conditions are: (a) citizen of India; (b) at least 25 years of age for the Lok Sabha and 30 years of age for the Rajya Sabha; and (c) possess such other qualifications as may be prescribed by the Parliament.

Speaker of the Lok Sabha

- He/She is the head of Lok Sabha. He/She is elected from amongst the members of Lok Sabha. The Speaker of the Lok Sabha conducts the business in the House.

SUPREME COURT

- The Supreme Court stands at the apex of the judicial system of India.
- The Supreme Court came into being on January 28, 1950.

Tenure and Qualification

- Judges of the Supreme Court are appointed by the President and retire at the age of 65.
- **Independence of Judges (Article 125) :** The salaries and allowances of Judges are charged from the Consolidated Fund of India.
- **Removal of Judges :** Judges can be removed only on the grounds of proven **misbehaviour** or **incapacity.**
- Judges can be removed only by a resolution of both Houses of Parliament supported by a majority of total membership of both the Houses and 2/3 of members present and voting. The first Judge against whom the proceedings were initiated was V. Ramaswami (1993) and the second one is Soumitra Sen (2011).

Types of Writs

Writ	Meaning	Intended Purpose
Habeas Corpus	You may have the body	To release a person who has been detained unlawfully whether in prison or in private custody.
Mandamus	We command	To secure the peformance of public duties by lower court, tribunal or public authority
Certiorari	To be certified	To quash the order already passed by an inferior court, tribunal or quasi judicial authority.
Prohibition	The act of stopping something	To prohibit an inferior court from continuing the proceedings in a particular case where it has no jurisdiction to try.
Quo Warranto	What is your authority	To restrain a person from holding a public office to which he is not entitled.

Comptroller and Auditor General (CAG) (Article 148-151)

- The Comptroller and Auditor General is appointed by the President under Article 148 of the Constitution.
- The CAG audits all receipts and expenditures of the Union and State Governments.
- The CAG submits its reports to the President (in case of accounts relating to the Union Government) or to the State Governors (for State Government Accounts).
- The CAG is also the head of the Indian Audits and Accounts Service (IA & AS). The office of the CAG was established in 1860.
- The CAG can only be removed from office in manner similar to a Judge of the Supreme Court. The CAG is not eligible for further office under the Union or State Governments. The expenses of the office of the CAG is charged to the Consolidated Fund of India.

Attorney General of India

- The Attorney General is the **highest law officer** in the country appointed by the President under Article 76 of the Constitution.
- The Attorney General can participate in proceedings of the Parliament without the Right to Vote (Article 88).
- In the performance of his duties, the Attorney-General shall have right of audience in all courts in the territory of India.

PART-VI

The States (Article 152-237)

THE GOVERNOR

- The Governor is the **Constitutional Head** of the State and the same Governor can act as Governor of more than one State (Article 153 and 154).
- The Governor is appointed by the **President** (Article 155) and Article 156 states that the Governor holds office during the pleasure of the President.
- Article 163 talks of **discretionary powers** of the Governor, which is not even provided to the President. Moreover, the courts cannot question his discretion.
- Article 171 states that the States where Legislative Councils exists, the Governor can **nominate some members** from amongst those distinguished in literature, science, arts, cooperative movement and social service.
- Article 213 empowers the Governor to issue the **ordinances** during the recess of the State Legislature.

States Legislature (Article 168-212)

- **Article 163** Council of Ministers to aid and advise the Governor.
- **Article 165** An Advocate General for each of the States.

- **Article 160** Abolition or creation of Legislative Councils in States (Currently, only seven states in India have Legislative Councils: Bihar, Jammu and Kashmir, Uttar Pradesh, Maharashtra, Karnataka, Andhra Pradesh and Telengana).
- **Article 171 Composition of Legislative Council** Unlike the members of the Legislative Assembly, the members of Legislative Council are indirectly elected. The maximum strength of the Council is fixed at one-third of the total strength of assembly and the minimum strength is fixed at 40.

HIGH COURTS (ARTICLE 214-232)

- There are 24 High Courts in India. The Calcutta High Court, established in 1862, is the oldest High Court in India.

Jurisdiction and Seats of High Courts

Court Name	Estd. Year	Territorial Jurisdiction	Seat
Bombay	1862	Maharashtra, Dadra and Nagar Haveli, Goa, Daman and Diu	Bombay (Bench at Nagpur, Panaji and Aurangabad)
Calcutta	1862	West Bengal, Andaman and Nicobar Islands	Calcutta (Circuit Bench at Port Blair)
Madras	1862	Tamil Nadu and Puducherry	Madras (Bench at Madurai)
Allahabad	1866	Uttar Pradesh	Allahabad (Bench at Lucknow)
Karnataka	1884	Karnataka	Bangalore (Circuit Benches at Hubli Dharwad and Gulbarga)
Patna	1916	Bihar	Patna
Madhya Pradesh	1936	Madhya Pradesh	Jabalpur (Benchest at Gwalior and Indore)
Jammu & Kashmir	1943	Jammu and Kashmir	Srinagar and Jammu
Punjab and Haryana	1947	Punjab, Haryana and Chandigarh	Chandigarh
Orissa	1948	Orissa	Cuttack
Guwahati	1948	Assam, Nagaland, Mizoram and Arunachal Pradesh	Guwahati (Bench at Kohima, Aizwal and Itanagar)
Rajasthan	1949	Rajasthan	Jodhpur (Bench at Jaipur)
Andhra Pradesh	1954	Andhra Pradesh, Telangana	Hyderabad
Kerala	1956	Kerala and Lakshadweep	Kochi
Gujarat	1960	Gujarat	Ahmedabad
Delhi	1966	National Capital Territory of Delhi	New Delhi
Himachal Pradesh	1971	Himachal Pradesh	Shimla
Sikkim	1975	Sikkim	Gangtok
Chhattisgarh	2000	Chhattisgarh	Bilaspur
Uttarakhand	2000	Uttarakhand	Nainital
Jharkhand	2000	Jharkhand	Ranchi
Tripura	2013	Tripura	Agartala
Manipur	2013	Manipur	Imphal
Meghalaya	2013	Meghalaya	Shillong

The Panchayats (Article 243-243 O)

- Introduced by the 73rd Amendment Act, 1992 which envisaged a three tier system of local governance. The Panchayat System exists in all states except Nagaland, Meghalaya and Mizoram. It also exists in all Union Territories except Delhi.

The Municipalities (Article 243 P-243 ZG)

- Introduced by the 74th Amendment Act, 1993 which envisages three types of urban local bodies, namely, municipality (nagar palika), city council (nagar panchayat).
- Municipal governance in India was first introduced in Madras in 1688. The Madras Municipal Corporation is the first municipal body in the whole commonwealth outside the UK. The Bombay and Calcutta Corporations were estabished in 1726. City Council administers urban areas having population greater than 30,000 and less than 1,00,000.
- A municipality administers an urban area of population 2,00,000 or less.

Relations between the Union and the States (Article 245-263)

- Article 262 Adjudication of disputes relating to waters of inter-state rivers or river valleys.
- Article 263 Inter-state council.

Finance, Property, Contracts and Suits (Article 264-300 A)

- Article 266 Consolidated Fund of India.
- Article 267 Contingency Fund of India.

PART-XIV

Services Under the Union and the States (Article 308-323)

- The Government of India Act, 1935 provided for the establishment of a Federal Public Service Commission and Provincial Public Service Commissions.
- The Chairman and other members of the UPSC are appointed by the President of India.
- The tenure of each member is six years or upto the age of 65, whichever is earlier. Members of the UPSC can be removed by the President on the charges of misbehaviour, if these charges are upheld by the Supreme Court.

Elections (Article 324-329)

- **Article 324** stipulates that the superintendence, direction and control of elections shall be vested in the Election Commission.
- **Article 325** provides for a single electoral roll for every constituency. Also stipulates that no person shall be eligible or ineligible for inclusion in electoral rolls on the basis of race, religion, caste or sex.
- **Article 326** stipulates that elections shall be held on the basis of adult suffrage. Every person, who is a citizen of India and is not less than 18 years of age shall be eligible for inclusion.

Planning Commission

- The Planning Commission was established in March 1950 by an executive resolution of the Government of India (*i.e.,* Union Cabinet) on the recommendation of the Advisory Planning Board constituted in 1946 under the Chairmanship of KC Neogi. Now, the Planning Commission has been replaced by NITI Aayog.

NITI Aayog

- NITI Aayog or National institution for transforming India Aayog is a policy think-tank of Union Government of India that replaces Planning Commission of India and aims to involve the states in economic policy-making in India. It will be providing strategic and technical advice to the central and the State Governments. Prime Minister of India heads the Aayog as its Chairperson.
- *Ex-officio members :* Rajnath Singh, Arun Jaitley, Suresh Prabhu and Radhe Mohan Singh.
- Special Invites : Nitin Gadkari, Smriti Zubin Irani and Thawar Chand Gehlot.
- Full-time members : Bibek Debray, VK Saraswat.
- Governing Council : All CMs and Lieutenant Governors of UTs.

National Development Council (NDC)

- Functions of the NDC is to review the working of national plan. The NDC was formed in 1952, to associate the states in the formulation of the plans.
- The PM is the ex-officio chairman of NDC.

Finance Commission

- Article 280 of the Constitution of India provides for a Finance Commission as a quasi-judicial body. It is constituted by the President of India every fifth year.
- It consists of Chairman and 4 other members.

Amendments of the Constitution (Article 368)

- Three methods for amendment of different provisions of the Constitution are
    1. Simple Majority in both the Houses of Parliament.
    2. Articles requiring ratification of the one-half member of the concerned Legislative Assembly and passed by the total member ship of the each House of the Parliament and majority not less than 2/3 of the members present and voting in each House.
    3. Provisions requiring passing by total membership of the each House of the Parliament and majority not less than two-third of the members present and voting in each House.
- It has been held by the Supreme Court in the **Keshavananda Bharati Case** (1973) that every provision of the Constitution is amendable under the meaning of Article 368 except the basic features of the Constitution.

Important Constitutional Amendments

- **First Amendment,** 1951 Added Ninth Schedule.
- **Seventh Amendment,** 1965 Necessitated on account of reorganisation of States on a linguistic basis.
- **Thirteenth Amendment, 1962** Created Nagaland as a State.
- **Fifteenth Amendment, 1963** Age of retirement of the Judges of High Court has been extended from 60 to 65 years.
- **Twenty Sixth Amendment, 1971** Abolished the titles and special privileges of former rulers of princely states.
- **Thirty Sixth Amendment, 1975** Made Sikkim a State.
- **Forty Third Constitutional Amendment Act, 1977** The 43rd Amendment omitted many articles inserted by 42nd Amendment. It restored the jurisdiction of the Supreme Court and the High Courts, which had been curtailed under the 42nd Amendment.
- **Forty Fourth Amendment, 1978** The Right to Property was deleted from Part III. Article 352 was amended to provide. 'Armed Rebellion' as one of the circumstances for declaration of emergency.
- **Seventy Third Constitutional Amendment Act, 1992** The institution of Panchayati Raj receive constitutional guarantee, status and legitimacy. XIth Schedule was added to deal with it. It also inserted Part IX, containing Articles, 243, 243 A to 243 O.
- **Eighty Ninth Amendment, 2003 The Act adds Article 338** A and provides for the creation of National Commission for Scheduled Tribes.
- **Ninety First Amendment, 2003** Amended the Anti-Defection Law and also made a provision that the number of ministers in the Centre and State Government, cannot the more than 15% of the strength of Lok Sabha and the respective Vidhan Sabha.
- **Ninety Third Amendment, 2005** To reserve seats for socially and educationally backward classes, besides the Scheduled Castes and the Scheduled Tribes, in private unaided institutions other than those run by minorities.
- **Ninety Fifth Amendment, 2009** Extends the reservation of seats for SC/STs in the Lok Sabha for another 10 years. In Article 334 of the Constitution, for the words 'sixty years', the words 'seventy years' was substituted.
- **Ninety Sixth Amendment, 2011** Amendment of 8th Schedule, it replaces 'Orissa' with 'Odisha'.
- **Ninety Seventh Amendment, 2011** Amendment of Article 19(1)(i), insertion of Article 43B, Insertion of Part IXB. This amendment gives constitutional status to cooperatives.
- **Ninety Eighth Amendment Act, 2012** (Insert Article 371 J). To empower the Governor of Karnataka to take steps to develop Hyderabad-Karnataka Region.

❑ ❑ ❑

Indian Economy

Nature of the Indian Economy

- **Mixed Economy :** Existence of both public and private sectors. This term was coined by J.M. Keynes.
- **Agrarian Economy :** Even after six-decades of independence 51% of the work force of India is still agriculturist and its contribution to GDP is 17.4% in 2011-12.

Broad Sectors of Indian Economy

- **Primary Sector :** Agriculture, forestry and fishing.
- **Secondary Sector :** Mining, manufacturing, electricity gas and water supply, construction.
- **Tertiary Sector :** (also called service sector) business, transport, telecommunication, banking, insurance, real estate, community and personnel services.

Planning in India

- **Planning Commission** (1950) was set up under the Chairmanship of Pandit Jawaharlal Nehru (Gulzarilal Nanda was the first Deputy Chairman).
- Basic aim of Economic Planning is to bring rapid economic growth through agriculture, industry, power and all other sectors of the economy.

Five Year Plans At a Glance

Plan	Growth Rate		Important Sector
	Target	Achieved	
First Plan (1951-56) *(Based on Harrod Domar Model)*	2.1%	3.6%	Agriculture irrigation, electricity
Second Plan (1956-61) *(Based on PC Mahalanobis two sector model)*	4.5%	4.2%	Heavy industries
Third Plan (1961-66)	5.6%	2.8%	Foodgrains, heavy industries
Plan Holiday (1966-69)			
Fourth Plan (1969-74)	5.7%	3.2%	Agriculture
Fifth Plan (1974-78)	4.4%	5.0%	Removal of poverty
Rolling Plan			
Sixth Plan (1980-85)	5.2%	5.5%	Agriculture industries
Seventh Plan (1985-90)	5.0%	6.0%	Energy, foodgrains
Annual Plans			
Eighth Plan (1992-97)	5.6%	6.6%	Human resource education
Ninth Plan (1997-2002)	6.5%	5.4%	Social Justice
Tenth Plan (2002-07)	8.1%	7.6%	Income, energy
Eleventh Plan (2007-2012)	8.0%	7.9%	Inclusive growth
Twelfth Plan (2012-2017)	8%	-	Faster, sustainable and inclusive growth

NITI Aayog

- NITI Aayog or National Institution for Transforming India Aayog came into existence on 1st January, 2015; policy-making think-tank of government that replaces planning commission and aims to involve states in economic policy making. It will be providing strategic and technical advice to the Central and the State Governments.
- The Prime Minister heads the Aayog as its Chairman. Arvind Panagariya is the Vice-Chairman of NITI Aayog of India.

NATIONAL INCOME OF INDIA

Domestic Product

- Gross Domestic Product at Market Price = Market value of final output of goods and services produced within the country's domestic economy in a period of one year.
- Net Domestic Product at Market Price = GDP_{MP} – Depreciation.
- Net Domestic Product at Factor Cost = NDP_{FC} = NDP_{MP} – Indirect Taxes + Subsidies.

National Product

- Gross National Product at Market Price = GDP_{MP} + Net Factor Income from Abroad.
- The first scientific estimate was made by Prof. VK Saraswat using output and income methods for the year 1931-32.
- National Income Committee was headed by Prof. P.C. Mahalanobis, set-up in 1949 and provided for the first time comprehensive data of national income in 1954.

Indian Tax Structure

- **Direct Tax :** The term direct tax generally means a tax paid directly to the government by the persons on whom it is imposed. *Example :* Income Tax, Corporation Tax, Capital Gain Tax, Stamp Duty, Land Tax, Estate Duty, Wealth Tax, Petroleum Revenue Tax.
- **Indirect Tax :** An indirect tax is a tax collected by an intermediary from the person who bears the ultimate economic burden of the tax. *Example :* Sales Tax or VAT, Customs Duty, Insurance Premium Tax, Excise Duties, Land-fill Tax, Electricity Duty, Climate Change Levy.
- **Corporate Income Tax :** The Government of India earns maximum from this tax.

Goods and Service Tax (GST)

- The GST as it is more commonly referred to is a system of taxation where there is a single tax in the economy for goods as well as service Indian GST was first proposed in India in the Union Budget speech in 2006-07.
- The main feature of the GST is that there is a tax credit available at each stage of the value chain.

Human Development Index (HDI)

- HDI measure was given by Pakistani Nobel Prize Winner, Mehbub-ul-Haq.
- Measured by Human Development Index (HDI), published by UNDP since, 1990.
- Three dimensions—1. Life expectancy at birth rate; 2. Education Index comprising means year of schooling and expected year of schooling. 3. GNI per capita (PPPUS $) Index.

POVERTY

- **Planning commission estimates poverty** rate based on data collected by National Sample Survey Organisation (NSSO).
- **Main Reasons for Rural Poverty** Rapid population growth, lack of capital, lack of alternate employment other than agriculture, illiteracy and lack of proper implementation of PDS.
- **Main Reasons for Urban Poverty** Migration from rural areas, lack of skilled labour, lack of housing facilities, limited job opportunities in cities. Based on 2400 calories (rural) and 2100 calories (urban) and monthly per capita consumption expenditure of ₹ 454 (rural) and 540 (urban), Planning Commission estimated poverty ratio in India in 2004-05 was 27.5% and according to the Suresh Tendulkar Committee was 37.2%. Countries whose GNI per capita is equal to or below USD 1035 is low income countries.

Various Government Schemes of India

Programme	Year of beginning	Objective
Community Development Programme (CDP)	1952	Overall development of rural areas with people's participation
Green Revolution	1966-67	To increase the foodgrains, specially wheat production
Antyodaya Yojana	1977-78	To make the poorest families of the village economically independent (only in Rajasthan State)
Training Rural Youth for Self-Employment (TRYSEM)	August 15, 1979	Programme of training rural youth for self-employment
Integrated Rural Development Programme (IRDP)	October 2, 1980	All-round development of the rural poor through a programme of asset endowment for self-employment
National Rural Employment Programme (NREP)	1980	To provide profitable employment opportunities to the rural poor
Development of Women and Children in Rural Areas (DWCRA)	September, 1982	To provide suitable opportunities of self-employment to the women belonging to the rural families who are living below the poverty line
National Fund for Rural Development (NFRD)	February, 1984	To grant 100% tax rebate to donors and also to provide financial assistance for rural development projects.
Jawahar Rozgar Yojana	April 1989	For providing employment to rural unemployed
Members of Parliament Local Area Development Scheme (MPLADS)	December 23, 1993	To sanction ₹ 1 crore per year to every Member of Parliament for various development works in their respective areas through DM of the district. The amount later increased to ₹ 5 crore.
Mahila Samridhi Yojana	2 October, 1993	To encourage the rural women to deposit in Post-office Saving Account
Ganga Kalyan Yojana	1997-98	To provide financial assistance to farmers for exploring and developing ground and surface water resources
Kasturba Gandhi Education Scheme	August 15, 1997	To establish girls schools in districts having low female literacy rate
Swarna Jayanti Shahari Rozgar Yojana (SJSRY)	December 1997	To provide gainful employment to urban unemployed and under-employed poor through self-employment or wage-employment
Bhagya Shree Bal Kalyan Policy	October 19, 1998	To uplift the girls conditions
Rajrajeshwari Mahila Kalyan Yojana	October 19, 1998	To provide insurance protection to women
Annapurna Yojana	March 1999	To provide 10 kg foodgrains to senior citizens (who did not get pension)

Programme	Year of beginning	Objective
Swarna Jayanti Gram Swarozgar Yojana	April 1999	For eliminating Rural poverty and unemployment and promoting self-employment
Antyodaya Anna Yojana	December 25, 2000	To provide food security to poor
Ashraya Bima Yojana	June 2001	To provide compensation to labourers who have lost their employment
National Food for Work Programme	2004	Providing Foodgrains to labour for the works of development
Mahatma Gandhi National Rural Employment Guarantee Act (MNREGA)	February 2, 2006	To provide at least 100 days wage-employment in rural areas
Bharat Nirman	2005	For development of rural infra-structure (Irrigation, Rural Roads, Rural Housing, Rural Drinking Water Supply, Rural Electrification and Telephone Connections).
Pradhan Mantri Jan Dhan Yojana	2014	To provide a bank account to every poor in the country (to achieve comprehensive financial inclusion).
Atal Pension Yojana	May 9, 2015	Social Sector Scheme pertaining to Pension Sector
Pradhan Mantri Suraksha Bima Yojana	May 9, 2015	Accidental Insurance with a premium of ₹ 12 per year.
Pradhan Mantri Jeevan Jyoti Bima Yojana	May 9, 2015	Life Insurance of ₹ 2 lakh with a premium of ₹ 330 per year.
HRIDAY–Heritage City Development and Augmentation Yojana	January, 2015	The scheme seeks to preserve and rejuvenate the rich cultural heritage of the country.
Sukanya Samridhi Yojana (Girl Child Prosperity Scheme)	January 2015	The scheme primarily ensures equit-able share to a girl child in resources and savings of a family in which she is generally discriminated as against a male child.
Smart Cities Mission	June 25, 2015	To enable better living and drive economic growth stressing on the need for people centric urban planning and development.
Atal Mission for Rejuvenation and Urban Transformation (AMRUT)	June 25, 2015	To enable better living and drive economic growth stressing on the need for people centric urban planning and development.
Pradhan Mantri Awas Yojana (PMAY)	June 25, 2015	To enable better living and drive economic growth stressing on the need for people centric urban planning and development.
Pradhan Mantri Ujjwala Yojana	May 1, 2016	Launched to provided free LPG connections to women from below poverty line families.

EMPLOYMENT

Estimates/Causes

- Since 1973 on the recommendation of B. Bhagwati Committee, three estimates of unemployment have been brought about by Planning Commission, viz.
  1. **Usual Principal Status :** (also called Open Unemployment) According to the NSSO, 66th Round (2009-10), rural and urban unemployment were 2% and 4%, respectively.
  2. **Current Weekly Status :** On this basis, the unemployment in 2009-10 in rural and urban areas were 3% and 4% respectively.
  3. **Current Daily Status :** The most comprehensive, measure on this basis, the unemployment for rural and urban areas were 7% and 6%, respectively (2009-10).

Women Empowerment & Child Development

Programme/Measure	Year
ICDS	1975
Mid-Day Meal Scheme	1995
Swadhar	1995
Swayam Sidha	2001
SSA	2001
Support to Training and Employment Programme for Women (STEP)	2003-04
Ujjwala	2007
Dhanlaxmi	2008
Gender Budgeting	2005-06
Integrated Child Protection Scheme	2009-10
Sabla Scheme	2010
National Mission for Empowerment of Women	2010
Bal Bandu Scheme	2011

INDUSTRIES

Navratnas, Maharatnas and Miniratnas

- **Navratnas** Bharat Electronics Ltd., BPCL, HAL, HPCL, MTNL, NALCO, National Mineral Development Corporation, Nevyeli Lignite Company Ltd., Oil India Ltd., Power Finance Company Ltd., Power Grid Corporation of India Ltd., Rashtriya Ispat Nigam Ltd., Rural Electri-fication Corporation Ltd., Shipping Shipping Corporation of India Ltd., CCIL, EIL and NBCCL.
- Miniratnas Public Sector Enterprises (PSEs) that have made profit continuously for the last three years or earned a net profit of ₹ 30 crores or more in one of three years.
- At present there are 54 in category I and 18 in Category II. NEEPCO is newly added miniratna in category I.

Major Industries in India

- **Iron and Steel :** First Steel Industry at Kulti, Paschim Banga–Bengal Iron Works Company was established in 1874. First large scale steel plant–TISCO at Jamshedpur (1907) was followed by IISCO at Burnpur (1919), Paschim Banga.

Manufactu-ring Unit	Limit	Service Unit	Limit
Micro	upto ₹ 25 lakhs	Micro	upto ₹ 10 lakhs
Small	₹ 25 lakhs ₹ 5 crore	Small	₹ 10 lakhs ₹ 2 crore
Medium	₹ 5 crore ₹ 10 crore	Medium	₹ 2 crore ₹ 5 crore

- The first public owned steel plant was Rourkela integrated Steel plant. Presently, India is the 4th largest producer of steel and comes 1st in the production of sponge iron.

Location	Assistance
Rourkela (Odisha)	Germany
Bhilai (Chhattisgarh)	Russia
Durgapur (Paschim Banga)	Britain
Bokaro (Jharkhand)	Russia
Vishakhapatnam (Andhra Pradesh)	Russia

- **Cotton and Textile Industry :** Largest organised and broad-based industry accounting for 4% of GDP, 20% of manufacture value added and one third of total exports earning.
- **Cement Industry :** First Cement producing unit was set-up at Paschim Banga, Porbandar (Gujarat), in 1914. India is the second largest producer of cement in the world.

Committees on Various Sectors of Indian Economy

Committee	Sector
A.C. Shah Committee	Non-Banking Financial Company
Bimal Jalan Committee	Market Infrastructure Instruments
Malegam Committee	Functioning of Micro Finance
Birla Committee	Corporate Governance
Kirit Parikh Committee	Rationalisation of Petroleum Product Prices
Chaturvedi Committee	Improving National Highways in India
S.R. Hashim Committee	Urban Poverty
Abhijit Sen Committee	Wholesale Price Index
C. Rangarajan Committee	Services Price Index
Damodaran Committee	Customer Service in Banks
Khandelwal Committee	Human Resource in Commercial Banks
Patil Committee	Corporate Debt
Sarangi Committee	Non-Performing Assets
Khanna Committee	Regional Rural Banks
Dantawala Committee	Lead Bank Scheme
Gadgil Committee	Financial Inclusion

BANKING AND FINANCE

- Bank of Hindustan was the first bank, established in India in 1770. First bank with limited liability managed by an Indian Board was the Oudh Commercial Bank in 1881. First purely Indian Bank is the Punjab National Bank (1894).

Nationalisation of Bank

- A step towards social banking was taken with the nationalisation of 14 Commercial Banks on 19 July, 1969. Six more banks were nationalised on 1980, total number of Public Sector banks are 27. Later on, in the year 1993, the Government merged New Bank of India with Punjab National Bank.
- Bhartiya Mahila Bank, India's first bank exclusively for women, headquarters in New Delhi. Inaugurated the bank on 19th November, 2013 and Chairman cum MD Usha Ananthasubrmaniam is the key person of the bank.
- IDBI bank is an Indian financial service company, formerly known as Industries Development Bank of India, head quartered in Mumbai, India.
- Udaya Kumar Dharmalingam is the designer of the Indian rupee sign ₹.

Establishment years of Important financial institutions

Institutions	Establishment year	Headquarter
Imperial Bank of India	1921	Mumbai
Reserve Bank of India	1 April, 1935	Mumbai
Industrial Finance Corporation of India (IFCI)	1948	Bengaluru
Industrial Credit and Investment Corporation of India (ICICI)	January 1955	Mumbai
State Bank of India (SBI)	1 July, 1955	Mumbai
Life Insurance Corporation of India (LIC)	September 1956	Mumbai
General Insurance Corporation of India (GIC)	November 1972	Mumbai
Unit Trust of India (UTI)	1 February, 1964	Mumbai
Industrial Development Bank of India (IDBI)	July, 1964	Mumbai
Regional Rural Bank (RRB)	1975	
Housing Development Finance Corporation (HDFC)	1977	Mumbai
National Bank for Agriculture and Rural Development (NABARD)	12 July, 1982	Mumbai
National Housing Bank (NHB)	1988	
Small Industries Development Bank of India (SIDBI)	2 April, 1990	Lucknow

Reserve Bank of India (RBI)

- RBI was established in 1935, according to RBI Act, 1934. RBI is the Central Bank of India. The main purpose of creating RBI was to separate currency and credit from GoI. RBI was nationalised in 1949 and its first Indian Governor was C.D. Deshmukh. Its headquarter is in Mumbai.

The RBI and Credit Control Quantitative Credit Control

- It is used to control the volume of credit and indirectly to control the inflationary and deflationary pressures. These are
  1. Cash Reserve Ratio
  2. Statutory Liquidity Ratio (SLR)
  3. Repo Rate
  4. Reverse Repo Rate
  5. Bank Rate

Functions of the RBI

- Monetary Policy, regulation and supervision of the banking and non-banking financial institutions.
- Debt and cash management for Centre and State Governments.
- Foreign exchange management current and capital account management.
- Management of foreign exchange reserves.
- Currency management; oversight of the payment and settlement systems.
- Development role.
- Research and statistics.

Operating Abroad

- 22 Indian banks (6 Private and 16 Public) operate overseas in 52 countries.
- SBI has the largest network of foreign offices followed by Bank of Baroda.
- Among private sector banks, ICICI Bank has the largest foreign presence.

14th Finance Commission

- The Fourteenth Finance Commission was constituted in accordance with the Articles 270, 275 and 280 of the Indian Constitution. The FC-XIII's overall approach was to foster inclusive and green growth promoting fiscal federalism.
- The first Finance commission was headed by K.C. Neogi and the 14th Finance Commission is headed by Yoga Venugopal Reddy.

Finance Commission of India

S. No.	Year of Constitution	Name of Commissioner	Years of Execution
1st	1951	K.C. Neogi	1952 – 1957
2nd	1956	K. Santhanam	1957 – 1962
3rd	1960	A.K. Chanda	1962 – 1966
4th	1964	Dr. P.V. Rajamannar	1966 – 1969
5th	1968	Mahavir Tyagi	1969 – 1974
6th	1972	Brahmanand Reddy	1974 – 1979
7th	1977	J.M. Shallot	1979 – 1984
8th	1983	Y.V. Chavan	1984 – 1989
9th	1987	Dr. K.P. Salve	1989 – 1995
10th	1992	K.C. Pant	1995 – 2000
11th	1998	A.M. Khusro	2000 – 2005
12th	2003	Dr. C. Rangrajan	2005 – 2010
13th	2007	Dr. V.L. Kelkar	2010 – 2015
14th	2012	Y.V. Reddy	2015 – 2020

India's Population at A Glance : 2011 (Final Data)

S.I. No.	State/UTs	Area (sq. km)	Population	Sex Ratio	Density	Literacy Rate	(%) Decadel Growth Rate (2001-2011)
1.	Jammu & Kashmir	222,236.00	1,25,41,302	889	124	67.2	23.6
2.	Himachal Pradesh	55,673.00	68,64,602	972	123	82.8	12.9
3.	Punjab	50,362.00	2,77,43,338	895	551	75.8	13.9
4.	Chandigarh	114.00	10,55,450	818	9258	86.0	17.2
5.	Uttarakhand	53483.00	1,00,86,292	963	189	78.8	18.8
6.	Haryana	44212.00	2,53,51,462	879	573	75.6	19.9
7.	Delhi	1483.00	1,67,87,941	868	11320	86.2	21.2
8.	Rajasthan	342,239.00	6,85,48,437	928	200	66.1	21.3
9.	Uttar Pradesh	240,928.00	19,98,12,341	912	829	66.7	20.2
10.	Bihar	94,163.00	10,40,99,452	918	1106	61.8	25.4
11.	Sikkim	7,096.00	6,10,577	890	86	81.4	12.9
12.	Arunachal Pradesh	83,743.00	13,83,727	938	17	65.4	26.0
13.	Nagaland	16,579.00	19,78,502	931	119	79.6	–0.6
14.	Manipur	22,327.00	28,55,794	985	128	79.2	31.79
15.	Mizoram	21,081.00	10,97,206	976	52	91.3	23.5
16.	Tripura	10,486.00	36,73,917	960	350	87.2	14.8
17.	Meghalaya	22,429.00	29,66,889	989	132	74.4	27.9
18.	Assam	78,438.00	3,12,05,576	958	398	72.2	17.1
19.	Pachim Banga	88752.00	9,12,76,115	950	1028	76.3	13.8
20.	Jharkhand	79714.00	3,29,88,134	949	414	66.4	22.4
21.	Odisha	155,707.00	4,19,74,218	979	270	72.9	14.0
22.	Chhattisgarh	135,191.00	2,55,45,198	991	189	70.3	22.6
23.	Madhya Pradesh	308,245.00	7,26,26,809	931	236	69.3	20.3
24.	Gujarat	196,024.00	6,04,39,692	919	308	78.0	19.3
25.	Daman & Diu	112.00	2,43,247	618	2191	87.1	53.8
26.	Dadra & Nagar Haveli	491.00	3,43,709	774	700	76.2	55.9
27.	Maharashtra	307,713.00	11,23,74,333	929	365	82.3	16.0
28.	Andhra Pradesh	275,045.00	8,45,80,777	993	308	67.0	11.0
29.	Karnataka	191,791.00	6,10,95,297	973	319	75.4	15.6
30.	Goa	3,702.00	14,58,545	973	394	88.7	8.2
31.	Lakshadweep	32.00	64,473	947	2149	91.8	6.3
32.	Kerala	38863.00	3,34,06,061	1084	860	94.0	4.9
33.	Tamil Nadu	130,058.00	7,21,47,030	996	555	80.1	15.6
34.	Puducherry	479.00	12,47,953	1037	2546	85.8	28.1
35.	Andaman & Nicobar Island	8249.00	3,80,581	876	46	86.6	6.9
	India	**3,287,240.00**	**1,21,08,54,977**	**943**	**382**	**73.0**	**17.7**

General Science

PHYSICS

Some Important SI Units

Quantity	Unit (SI)	Quantity	Unit (SI)
Length	Metre	Time	Second
Mass	Kilogram	Area	Square metre
Volume	Cubic metre	Velocity	Metre/second
Acceleration	Metre/second^2	Density	Kilogram/metre^3
Momentum	Kilogram metre/second	Work	Joule
Energy	Joule	Force	Newton
Pressure	Pascal or Newton/metre^2	Frequency	Hertz
Power	Watt	Weight	Newton or Kilogram
Impulse	Newton-second	Angular velocity	Radian/second
Viscosity	Poise	Surface tension	Newton/Metre
Heat	Joule	Temperature	Kelvin
Electric current	Ampere	Electromotive force	Volt
Electrical conductivity	mho/metre	Electric energy	Kilo watt hour
Electric power	Kilo watt or watt	Magnetic intensity	Orsted
Luminous flux	Candela	Intensity of sound	Decibel
Power of lens	Dioptre	Depth of sea	Fathom

Some Equipment used to Transform Energy

S.No.	Equipment	Energy Transformed
1.	Dynemo	Mechanical energy into electrical energy
2.	Candle	Chemical energy into light and heat energy
3.	Microphone	Sound energy into electrical energy
4.	Loud Speaker	Electrical energy into sound energy
5.	Solar Cells	Solar energy into electrical energy
6.	Tube light	Electrical energy into light energy
7.	Electric Bulb	Electrical energy into light and heat energy.
8.	Battery	Chemical energy into electrical energy.
9.	Electric motor	Electrical energy into mechanical energy.
10.	Sitar	Mechanical energy into sound energy.

Important Law & Principle

- **Newton's First Law :** Every body maintains its initial state of rest or motion with uniform speed on a straight line unless an external force acts on it. It is also called **Galileo's law** or **law of inertia.**
- **Newton's Second Law :** The force acting on an object is directly proportional to the product of the mass of the object and the acceleration produced on it.
- **Newton's Third Law :** To every action, there is an equal and opposite reaction. *Example :* Bogies of the trains are provided with buffers to avoid severe jerks during shunting of trains. Rocket moves up due to reaction of downward ejection of gas.

Pascal's Law of Pressure

- If gravitational attraction is negligible in equilibrium condition, pressure is same at all points in a liquid.
- The pressure exerted anywhere at a point of confined liquid is transmitted equally and undiminished in all directions throughout the liquid.
- Hydraulic lift, hydraulic pressure and hydraulic brakes are based on the Pascal's law of pressure.

Archimedes' Principle

- When a body is immersed partly or wholly in a liquid, there is an apparent loss in the weight of the body, which is equal to the weight of liquid displaced by the body.
- The weight of water displaced by an iron ball is less than its own weight. Whereas water displaced by the immersed portion of a ship is equal to its weight. So, small ball of iron ball sink in water, but large ship float.
- A fat person will quickly learn the swimming as compared to a slim person because he will displace more water. So, will be more balanced.
- Hydrogen filled balloon float in air because hydrogen is lighter than air.
- A person can lift more weight in water.

Bernoulli's Theorem

- When an incompressible and non-viscous liquid (or gas) flows in streamlined motion from one place to another, then at every point of its path the total energy per unit volume (pressure energy) + kinetic energy + potential energy) is constant.
- Venturimeter, Atomizer, filter pump motion of aeroplane are based upon the Bernoulli's theorem.

Ohm's Law

- If physical conditions like temperature, intensity of light etc. remains unchanged then electric current flowing through a conductor is directly proportional to the potential difference across its ends. If V is the potential difference across the ends of a conductor and I is the current through it, then according to ohm's law $V \propto I$ or, $V = RI$. where R is a constant called resistance of conductor.

Important Discoveries Relating to Physics

Discovery	Scientist	Year
Electron	J.J. Thomson	1897
Proton	Rutherfort	1919
Neutron	James Chadwick	1932
Atom	John Dalton	1808
Atomic Structure	Neil Bohr & Rutherford	1913
Laws of motion	Newton	1687
Radioactivity	Henry Becquerel	1896
Radium	Madam Curie	1898
Principle of Relativity	Albert Einstein	1905

Discovery	Scientist	Year
Electromagnetic Induction	Michael Faraday	1831
Raman Effect	C.V. Raman	1928
X-Rays	Roentgen	1895
Quantum theory	Max Plank	1900
Photo electric effect	Albert Einstein	1905
Photography (On Metal)	J. Neepse	1826
Photography (On paper)	W. Fox Talbot	1835
Periodic Table	Mandeleev	1888
Law of Electric resistance	G.S. Ohm	1827
Thermionic emission	Edison	–
Nuclear Reactor	Anrico Fermi	1942
Wireless Telegram	Marconi	1901
Dynamite	Alfred Nobel	1867

Scientific Instruments

Name of Instrument	Use
Altimeter	Measures altitudes (used in aircraft)
Ammeter	Measures strength of electric current
Anemometer	Measures force and velocity of wind and determiens its directions
Barometer	Measures atmospheric pressure
Callipers	Measure inner and outer diameters of bodies
Cardiogram (ECG)	Traces movements of the heart; recorded on a Cardiograph
Cathetometer	Determines heights, measurement of levels, etc., in scientific experiments
Dilatometer	Measures changes in volume of substances
Dynamometer	Measures electrical power
Electrometer	Measures very small but potential difference in electric currents
Endoscope	To examine internal parts of the body
Fathometer	Measures depth of the ocean
Lactometer	Measures the relative density of milk to determine purity
Magnetometer	Compares magnetic movements and fields
Photometer	Compares the luminous intensity of the source of light
Polygraph	Instrument that simultaneously records changes in physiological processes such as heartbeat, blood-pressure and respiration; used as a lie detector
Pyrometer	Measures very high temperature
Radar	To detect the direction and range of an approaching aeroplane by means of radiowaves, (Radio, Angle, Detection and Range)
Radio micrometer	Measures heat radiation
Stethoscope	Used by doctors to hear and analyze heart and lung sounds
Telemeter	Records physical happenings at a distant place.
Telescope	To view distant objects in space
Transponder	To receive a signal and transmit a reply immediately
Ultrasonoscope	To measure and use ultrasonic sound (beyond hearing); use to make a Ecogram to detect brain tumours, heart defects and abnormal growth
Voltmeter	To measure electric potential difference between two points

Inventions

Invention	Inventor	Country	Year
Adding machine	Pascal	France	1642
Aeroplane	Wright brothers	USA	1903
Balloon	Jacques and Joseph Montogolfier	France	1783
Ball-point pen	C. Biro	Hungary	1938
Bicycle	K. Macmillan	Sctoland	1839
Bicycle Tyre	J.B. Dunlop	Scotland	1888
Computer	Charles Babbage	Britain	1834
Cinema	A.L. and J.L. Lumiere	France	1895
Clock (machanical)	Hsing and Ling-Tsan	China	1725
Clock (pendulum)	C. Hugyens	Netherlands	1657
Diesel engine	Rudolf Diesel	Germany	1892
Dynamite	Alfred Nobel	Sweden	1867
Electric iron	H.W. Seeley	USA	1882
Electromagnet	W. Sturgeon	England	1824
Fountain Pen	LE. Waterman	USA	1884
Gas lighting	William Murdoch	Scotland	1794
Gramophone	T.A. Edison	USA	1878
Jet Engine	Sir Frank Whittle	England	1937
Machine gun	Richard Gatling	USA	1861
Match (safety)	J.E. Lurdstrom	Sweden	1855
Microphone	David Hughes	USA	1878
Motor car (petrol)	Karl Benz	Germany	1885
Motorcycle	Edward Butler	England	1884
Nylon	Dr. W.H. Carothers	USA	1937
Photography (paper)	W.H. Fox Tablot	England	1835
Printing press	J. Gutenberg	Germany	1455
Radar	Dr. A.H. Taylor and L.C. Young	USA	1922
Radium	Marie and Pierre Curie	France	1898
Razor (safety)	K.G. Gillette	USa	1895
Refrigerator	J.Harrison and A. Catling	Britain	1834
Revolver	Samuel Colt	USA	1835
Safety lamp	Sir Humphrey Davy	England	1816
Safety pin	William Hurst	USA	1849
Scooter	G. Bradshaw	England	1919
Shorthand (modem)	Sir Richard Arkwight	England	1769
Tank	Sir Ernest Swington	England	1914
Telephone	Alexander Graham Bell	USA	1876
Telescope	Hans Lippershey	Netherlands	1608
Thermometer	Galileo Galilei	Italy	1593
Transistor	Bardeen, Schockley	USA & Britain	1949
Watch	A.L. Breguet	France	1791
X-ray	Wilhelm Roentgen	Germany	1895

CHEMISTRY

- Chemistry is the branch of Science which deals with the study of matter, its physical and chemical properties, chemical composition the physical and chemical changes which it undergoes and the energy changes during these processes.

Important Metals and their Ores

Metal	Ores	Formulas
Sodium (Na)	Sodium Chloride	$NaCl$
	Sodium Carbonate	$Na_2CO_3.10H_2O$
	Sodium Nitrate	$NaNO_3$
	Sodium Sulphate	$Na_2SO_4.10H_2O$
	Borex (Sodium tetraborate)	$Na_2B_4O_7.10H_2O$
Aluminium (Al)	Bauxite	$Al_2O_3.2H_2O$
	Corundum	Al_2O_3
	Felspar	$K_2OAl_2O_3.6SiO_2$
	Cryollite	$Na_3Al_4F_6$
	Alunite	$K_2SO_4\ Al_2(SO_4)_3$ $4Al(OH)_3$
	Kaolin	$3Al_2O_3$ $6SiO_2.2H_2O$
Potassium (K)	Potassium Chloride	KCl
	Potassium Carbonate	K_2CO_3
	Potassium nitrate	KNO_3
Magnesium (Mg)	Magnesite	$MgCO_3$
	Dolomite	$MgCO_3.CaCO_3$
	Epsom salt	$MgSO_4.7H_2O$
	Carnallite	$KCl\ MgCl_2.6H_2O$
Calcium (Ca)	Dolomite	$CaCO_3.MgCO_3$
	Calcite	$CaCO_3$
	Gypsum	$CaSO_4.2H_2O$
	Fluorspar	CaF_2
	Asbestus	$CaSiO_3.MgSiO_2$
Strontium (Sr)	Strontinite	$SrCO_3$
	Silestine	$SrSO_4$
Copper (Cu)	Cuprite	Cu_2O
	Calcosite	Cu_2S
	Chalcopyrite	$CuFeS_2$
Silver (Ag)	Ruby Silver	$3Ag_2\ S\ Sb_2\ S_3$
	Horn Silver	$AgCl$
Gold (Au)	Calaverite	$AuTe_2$
	Silvenites	$[(Ag, Au)_2\ Te_2]$
Barium (Ba)	Barytes	$BaSO_4$
Zinc (Zn)	Zinc blende	ZnS
	Zincite	ZnO
	Calamine	$ZnCO_3$
Mercury (Hg)	Cinebar	HgS
Tin (Sn)	Cisiteryte	SnO_2
Lead (Pb)	Galena	PbS
Antimony (Sb)	Stibenite	SbS_2
Cadmium (Cd)	Greenocite	CdS
Bismuth (Bi)	Bismuthite	Bi_2S_3
Iron (Fe)	Haemetite	Fe_2O_3
	Magnetite	Fe_3O_4
	Ciderite	$FeCO_3$
	Iron Pyrite	FeS_2
	Chalco Pyrites	$CuFeS_2$
Cobalt (Co)	Smelite	$CoAsS_2$
Nickel (Ni)	Milarite	NiS
Magnese (Mn)	Pyrolusite	MnO_2
	Magnite	$Mn_2O_3\ 2H_2O$
Uranium (U)	Carnotite	$K(UO)_2.VO_4.3H_2O$
	Pitch blende	U_3O_8

Important Alloys and Their Uses

Alloys	Compositions	Uses
Brass	Cu + Zn (70% 30%)	Pot, Decorative purposes, screws & handles
Bronze	Cu + Sn (80% 20%)	Statues, coins & medals, utensils
German Silver	Cu + Zn + Ni (50% 30% 20%)	Ornaments, Decorative articles
Rolled gold	Cu + Al (90% 10%)	Cheap ornaments
Gun metal	Cu + Sn + Zn + Pb (88% 10% 1% 1%)	Gun, barrels, gears & bearings
Delta metal	Cu + Zn + Fe (60% 38% 2%)	Blades of aeroplane
Munz metal	Cu + Zn (60% 40%)	Coins
Dutch metal	Cu + Zn (80% 20%)	Artificial ornaments
Monel metal	Cu + Ni (70% 30%)	Pot for alkali
Rose metal	Bi + Pb + Sn (50% 28% 22%)	Automatic fuse
Solder	Pb + Sn (50% 50%)	Soldering
Magnalium	Al + Mg (95% 5%)	Light forts, Beam of a balance, Frame of aeroplane
Duralumin	Al + Cu + Mg + Mn (94% 3% 2% 1%)	Pressure Cooker, Aeroplanes Light tools
Type metal	Sn + Pb + Sb (5% 80% 15%)	Printing industry
Bell metal	Cu + Sn (80% 20%)	Casting bells, statues
Stainless steel	Fe + Cr + Ni + C (75% 15% 10% .05%)	Utensils, surgical, cutlery
Nickel steel	Fe + Ni (95% 5%)	Electrical wire, Automobile parts

Compounds of Metal and Non-metal & Their Uses

Metal & Non-metal	Chemical Name	Uses
Ferrous oxide	Fe_2O_3	In green glass, Ferrous salt
Ferric oxide	Fe_3O_4	Electroplating of ornaments
Ferric hydroxide	$Fe(OH)_3$	(i) As reagent, (ii) In Medicine
Iodine	I_2	Antiseptic, Medicine, Tincture of iodine
Bromine	Br_2	(i) Dyes industry (ii) As reagent

Metal & Non-metal	Chemical Name	Uses
Chlorine	Cl_2	(i) Mustard gas, (ii) Bleaching powder, (iii) Bleaches cloth & paper
Hydrochloric acid	HCl	(i) An aquaregia (3 HCl : 1 HNO_3), (ii) Dyes
Sulphur dioxide	SO_2	Oxidants & reductants, As bleaching agent
Ammonia	NH_3	Reagent ice factory
Producer gas	$CO + N_2$	(i) In heating furnace, (ii) Cheap fuel, (iii) In Extraction of metal
Water gas	$CO + H_2$	(i) As Fuel, (ii) Welding
Nitrous oxide	N_2O	Laughing gas, Surgery
Carbondioxide	CO_2	Sodawater, Fire extinguisher
Carbon monoxide	CO	In phosgene gas ($COCl_2$)
Alum	$K_2SO_4\ Al_2\ (SO_4)_3 . 24\ H_2O$	(i) Purification of water, (ii) Leather industry
Mercuric Chloride	$HgCl_2$	Calomel, Insecticides (Corrosive sublimate)
Zinc Sulphate or White vitriol	$ZnSO_4 . 7\ H_2O$	Lithopone, Eye ointment
Zinc Chloride	$ZnCl_2$	Textile industry
Zinc oxide	ZnO	Ointment
Zinc	Zn	In battery
Calcium carbide	CaC_2	Calcium cyanide & acetylene gas
Bleaching powder	Ca(OCl) Cl	Insecticides, Bleaching actions
Plaster of paris	$(CaSO_4)_2 . 2\ H_2O$, $CaS\ O_4\ ½\ H_2O$	Statue, Surgery
Calcium sulphate	$CaSO_4\ 2\ H_2O$	Cement industry
Magnesium Chloride	$MgCl_2 . 6\ H_2O$	Decoration of cotton
Magnesium Carbonate	$MgCO_3$	Toothpaste
Magnesium oxide	MgO	Rubber filler, Boiler
Copper sulphate	$CuSO_4 . 5\ H_2O$	Insecticides Electric cells
Cuprous Oxide	Cu_2O	Red glass, pesticides
Sodium nitrate	$NaNO_3$	Fertilizer
Heavy water	D_2O	Nuclear reactor

Different Medicines and their Examples

Medicine	Used to	Example
Analgesics	Reduce pain	Aspirin, paracetamol, morphine, phenacetin
Tranquilizers	To treat stress, mild and severe mental diseases	Equanil, valium, chlorodiazoepoxide, serotonin and meprobamate
Antiseptic	Prevent the growth of micro-organisms or kill them (applied to living tissues)	Dettol (a mixture of chloroxylenol–the antiseptic and α-terpineol), savlon, iodine tincture, boric acid, hydrogen peroxide, iodoform
Antibiotic	Destroy micro-organisms (These are obtained from micro-organisms)	Penicillin (discovered by A Fleming in 1929, ampicillin, amoxicillin, ofloxacin, chloramphenicol)
Antimalarial	Cure malaria	Chloroquine
Sulphadrugs	Alternative for antibiotics	Sulphanilamide, sulphadiazine
Antacids	Reduce acidity	Baking soda, magnesium hydroxide

Types of Batteries

Battery	Anode	Cathode	Electrolyte	Used in
Leclanche cell	Zinc	Graphite	Paste of ammonium chloride and zinc chloride	Transistors, clocks
Mercury cell	Zinc-mercury amalgam	Paste of HgO (Mercuric oxide and carbon)	Paste of KOH and ZnO	Hearing aids and camera
Lead storage battery	Lead	Lead packed in lead dioxide	38% solution of sulphuric acid	Automobiles, invertors

Some Important Fuels and Their Compositions

Fuel	Composition	Sources
Natural Gas	Methane (83%) + Ethane	From petroleum
Liquified Petroleum Gas (LPG)	Butane (C_4H_{10}) + Propane (C_3H_8)	From oil wells
Compressed Natural Gas (CNG)	Methane (CH_4) 95%	From petroleum
Biogas or Gobar Gas	Methane (CH_4) + Carbon dioxide (CO_2) + Hydrogen (H_2) + Nitrogen (N_2)	From organic wastes
Water Gas	Carbon monoxide (CO) + Hydrogen (H_2)	By passing steam over red hot coke
Producer Gas	Carbon monoxide (CO) + Nitrogen (N_2)	By passing insufficient air over red hot coke
Coal Gas	Hydrogen + Methane + Ethylene (C_2H_4) + Acetylene (C_2H_2) + CO + Nitrogen	By fractional distillation of wood

Some Fibres and Their Monomers

Fibre	Monomers	Uses
Nylon-66	Adipic acid + hexamethylene diamine	In making bristles for brushes, synthetic fibres, parachutes, as a substitute for metal in bearings
Nylon-6 or perlon	Caprolactam	In making fibres, plastics tyre cords and ropes.
Terylene	Ethylene glycol and terephthalic acid	For making wash and wear fabrics, tyre cords, safety belts, tents etc.
Kevlar	Terephthalic acid + 1, 4-diamino benzene	For making bulletproof vests.
Lexan or polycarbonate	Diethyl carbonate + bis-phenol-A	In making bulletproof windows and safety helmets
Polyurethanes	Toluene diisocyanate + ethylene glocol	For making washable and long lasting mattresses, cushions.

BIOLOGY

- Biology is the study of life and living organisms from one-celled creatures to the most complex living organism of all the human being. Biology includes the study of genes and cells that give living their special characterisitcs.
- **Biology :** Branch of science in which living beings are studied. Bios = Life & Logos = Study.
- Therefore, study of life is called Biology. The term Biology was first coined by Lamarck and Treviranus in the year 1801. Biology has two main branches:
- **Botany :** Study of different aspects of plants. Theophrastus is known as father of Botany.
- **Zoology :** Study of various aspects of animals. Aristotle is called father of Zoology as well as Biology.

Living World

- In 18th Century, Carolus Linnaeus developed binomial nomenclature for living organisms *i.e.,* scientific name consiting of genus and species.
- Whittaker (1969) classified living organisms into five kingdoms–Monera, Protista, Fungi, Plantae and Animalia.

The Cell

- According to cell theory proposed by Schleiden and Schwann (1838) cell is the structural and functional unit of living organisms.
- An organism composed of single cell (unicellular) or many cells (multicellular).
- Cells are of two types *i.e.,* prokaryotic (which lacks nucleus and membrane bound organelles) and eukaryotic (which have nucleus and membrane bound organelles).
- Prokaryotic cell is found in bacteria, mycoplasma and blue-green algae while eukaryotic cell in plants, animals and fungi.

Nucleic Acids

- These contain the genetic instructions used in the development and functioning of all know living organisms. These are of two types namely DNA and RNA. Deoxyribo Nucleic Acid (DNA). It is a long polymer made from repeating units called nucleotides. It has four bases, adenine, guanine, cytosine and thymine.
- Ribo Nucleic Acid (RNA) : It is also made up of a long chain of nucleotides. It contains uracil in place of thymine.

Major Enzymes of Digestion

Enzyme	Source	Where Active	Substrate	Main Breakdown product
Carbohydrate Digestion				
Salivary amylase	Salivary glands	Mouth	Polysaccharides	Disaccharides
Pancreatic amylase	Pancreas	Small intestine	Polysaccharides	Disaccharides
Disaccharides	Small intestine	Small intestine	Disaccharides	Monosaccharides (*e.g.,* glucose)
Protein Digestion				
Pepsin	Stomach mucosa	Stomach	Proteins	Peptide fragments
Trypsin and chymotrypsin	Pancreas	Small intestine	Proteins and polypeptide	Peptide fragments
Carboxypeptidase	Pancreas	Small intestine	Peptide fragments	Amino acids

Enzyme	Source	Where Active	Substrate	Main Breakdown product
Amino peptidase	Intestinal mucosa	Small intestine	Peptide fragments	Amino acids
Fat Digestion—Lipase	Pancreas	Small intestine	Triglycerides	Free fatty acids, Monoglycerides
Nucleic Acid Digestion				
Pancreatic nucleases	Pancreas	Small intestine	DNA, RNA	Nucleotides
Intestinal nucleases	Intestinal mucosa	Small intestine	Nucleotides	Nucleotides bases, monosaccharides

Vitamins

Vitamin (Name)	Rich Food Source	Deficiency Disease
Fat Soluble Vitamins		
A (Retinol)	Fish liver oils, dairy products, liver, most leafy vegetables and carrots contain carotene that can be converted into retinol	Dry skin, night blindness (Nyctalopia)
D (Calciferol)	Fish oils, egg yolk and butter. It can be made by the action of sunlight on skin	Rickets in children ('soft' bones that bend easily) Ostemolacia (painful bones) in adults
E (Tocopherol)	Vegetable Oils, cereal products and many other foods	Mild anaemia and sterility. Deficiency is rare in humans
K (Phylloquinone Phytomenadione or Phytonadione)	Fresh, dark green vegetables. Also made by gut bacteria	Delayed clotting time, May occur in new-born babies before their gut bacteria become established.
Water Soluble Vitamin		
B_1 (Thiamine)	Yeast, cereals, nuts, seeds, pork	Beri-beri (muscular dystrophy, stunted growth, nerve degenera-tion)
B_2 (Riboflavin)	Liver, milk, eggs, green vegetables	Cracked skin, bluured vision
B_3 (Pantothenic acid)	Liver, yeast, whole cereals, beans	Pellagra (severe skin problems, diarrhoea, dementia)
B_5 (Nicotinamide or Niacin)	Animal tissue, whole grain cereals, legumes	Pellagra, Dermatitis, Diarrhoea
B_6 (Pyridoxin)	Meat, fish, eggs, cereals bran, some vegetables	Skin problems, nerve disorder
B_{10} (Folic acid)	Liver, raw green vegetables yeast, gut bacteria	Anaemia (especially during pregnancy)
B_{12} (Cyanocobalamine)	Liver, milk, fish, yeast, None in plant foods	Pernicious anaemia, nerve disorders
C (Ascorbic acid)	Blackcurrants, peppers, sprous, citrus fruits	Scurvy, poor wound healing

Blood (Lymphatic System)

- An adult person has 5-6 litre blood.
- It is slightly alkaline having pH 7.3-7.4.
- Plasma is pale yellow transparent and constitute about 60% volume of blood.
- Plasma is composed of 90-92% water, 7% organic substances (albumin, globulin and fibrinogen protein) and 1% inorganic substances. Red blood corpuscles (Most abundant) are non-nucleated and contains haemoglobin (the respiratory pigment).
- **Eosinophils** are also called acidophils (2-8%) are phagocytic granulocytes and play important role in hypersensitivity.
- **Neutrophils** (65%) are phagocytic granulocytes and increase during bacterial infection.
- **Lymphocytes** (26%) are agranulocytes producing antibodies and increase during viral infection.
- **Rh factor** discovered by Landsteiner and Veiner in Rhesus monkey, which is responsible for erythroblastosis fetalis disease.
- Important component of blood clotting are fibrinogen, prothrombin, thromboplastin, calcium ions and Vitamin K.

Heart

- Human heart is myogenic *i.e.,* contraction is initiated by a pulse produced by SinuAtrial node (SA node) located in right atrium. It is also called pacemaker. First heart sound is lub and second heart sound is dub.
- Contraction of heart is called systole. 120 mm Hg, while relaxation is called diastole (80 mm Hg).

Central Nervous System

- **Magnetic Resonance Imaging** (MRI) : The MRI was invented by Paul C Laterbur in 1971.
- **Positron Emission Tomographic Scanning (PET) :** It was developed by Louis Sokoloff of USA in 1986.

Some Human Diseases Caused by Viruses and Bacteria

Disease	Pathogen	Symptom	Prevention/Vaccine
Chickenpox (Varicella)	Herpes zoster virus	Dark red coloured rash or pox changing into vesicles, crusts and falling	
Smallpox	Variola virus	Appearance of rash changing into pustules, scaps and falling pockmarks are left	Smallpox vaccine
Measles (Rubella disease)	Rubella virus	Rubella (skin eruptions), coughing, sneezing etc.	Measles-mumps-rubella-Varicella Combo (MMRV vaccine)
Mumps	Mumps virus	Painful enlargement of parotid and salivary glands	Mumps-vaccine, isolation
Rabies (Hydrophobia)	Rabies virus	Spasm in throat and chest muscles, fears from water, paralysis and death	Immunisation of dogs
Tuberculosis	M tuberculosis	Coughing, chest pain and bloody sputum with tuberculin	BCG vaccine
Diphtheria	C. diphtheriae	Inflammation of mucosa of nasal chamber, throat, etc. respiratory track blocked	DPT vaccine

Disease	Pathogen	Symptom	Prevention/Vaccine
Cholera	Vibrio cholerae	Acute diarrhoea and dehydration	Sanitation, boiling of water and oral cholera vaccine
Tetanus (Lock jaw)	Clostridium tetani	Degeneration of motor neurons, rigid jaw muscles, spasm and paralysis	ATS and DPT vaccines
Typhoid	Salmonella typhi	Classic typhoid fever	TAB vaccine and screening of food and water
Plague	Pasteurella pestis	Bubonic plague affects, lymph nodes, pneumonic plague affects lungs and septicemic plague causes anaemia	Killing of rats and rat fleas, plague vaccine
Pneumonia	Diplococcus pneumoniae	Decrease in respiratory efficiency	PCV 13
Swine Flu	H1 N1 flu virus (Orthomy)	Fever with or without chill, sore throat, dyspneat, myalgia, diarrhoea, vomiting, dizziness	Oseltamivir (Tamiflu), Zonamivir (Relenza) are antiviral drugs vaccines are available against this disease.
Ebola virus	Ebola virus	Haemorrhagic fever, muscle pain,	No licensed vaccine
Disease (EVD)	(Filovirus)	headache, sore throat, diarrhoea, kidney and liver dysfunction, internal and external bleeding.	available, immune, therapies are done currently.

Important Functions of Brain

Forebrain	
Olfactory region	Smell
Cerebrum	Thinking, intelligence, memory, ability to learn from experience, will power, skilled work, reasoning knowledge, consciousness and speech.
Control	Laughing, weeping, micturition (passing of urine), defecation voluntary forced breathing and voluntary muscular coordination
Diencephalon (sensation of)	Heat, cold and pain control centre of autonomic nervous system, control hunger, thirst, sweating, sleeping and sex.
Hypothalamus	Regulated body temperature so 'thermostat' of body. Appetite and safety control emotions like love, anger, pleasure and satisfication. Control metabolism of carbohydrate, fat and water.
Midbrain	Reflex centre of visual and auditory sensation.
Hindbrain	
Cerebellum	Involuntary muscular coordination, maintain posture, orientation and equilibrium of the body.
Medulla oblongata	Regulate heart rate, involuntary breathing, respiratory centre, blood pressure (vasoconstriction and vasodilation) gut peristalsis, food swallowing and vomiting gland secretion.

Pollution

- Motor vehicle contribute 60% of air pollution in major cities. Photochemical smog comprising of O_3, H_2O_2 PAN etc.
- Acid rain is composed of H_2SO_4 and HNO_3.
- Chlorofluorocarbons released into stratosphere release free chlorine atom that causes depletion of ozone.
- Bioremediation is the process of using micro-organisms to remove environmental pollutant. Example, using oil-zapper developed by TERI to prevent oil spills.
- Nitrate fertilizers caused blue baby syndrome or methemoglobinemia.

Test Tube Baby

- First attempt to produce a test tube baby was made by an Italian scientist Dr. Petrucci in 1959. The World's first test tube baby (a baby girl) named as Louise Joy Brown was born on 25th July, 1978 in Great Britain.
- India's first test tube baby was born in Mumbai on 6th August 1986. Her name is Harsha.
- **Cloning :** Dolly a sheep, the first mammal clone developed by Dr. Ian Wilmut, UK.

Largest in the world

Mammal (on land)	African elephant (Loxodonta africana)
Mammal	Blue whale (in the biosphere)
Flower	Rafflesia
Flower in India	Sapria
Bone	Femur
Bone (in frog)	Tibia-fibula
Muscles	Gluteus maximus (buttock muscle of hip)
Tooth	Tusk of elephant (upper incisor modification)
Tallest angiosperm	Eucalyptus
Coral reef	In Australia, great barrier reef
Egg or cell	Ostrich
Vein	Inferior vena cava
Cell of the body	Neuron of nerve cell
Virus	Parrot fever virus

Smallest in the World

Bacteria	Dialister Pneumosintes
Bird	Humming bird (Cuba)
Bone	Stapes
Endocrine gland	Pituitary
Flower	Wolffia microscopica (Angiosperm)
Mammal	Shrew (Suncus etruscus)
Muscles	Stapedius or arrector pili
Virus	Foot and mouth disease virus

Different Branches of Science

Branch	Concerned Field
Aeronautics	Science of flight of airplanes
Astronomy	Study of heavenly bodies
Anthology	Study of flowers
Anthropology	Study of apes and man
Apiculture	Honey industry (Bee keeping)
Biochemistry	Deals with the study of chemical reactions in relation to life activities
Cardiology	Study of heart
Cytology	Study of cells
Dermatology	Study of skin
Floriculture	Study of flower yielding plants
Horticulture	Study of garden cultivation
Myology	Study of muscles
Nephrology	Study of kidneys
Obstetrics	Branch of medicine dealing with pregnancy
Phycology	Study of algae
Pedology	Study of soils
Pathology	Study of disease causing organisms
Physiology	Science dealing with the study of functions of various parts of organisms
Pisciculture	Study of fish
Seismology	Study of earthquakes
Sericulture	Silk industry (culture of silk moth and pupa)
Serpentology	Study of snakes
Taxonomy	Study of classification of organisms
Virology	Study of virus

Important Medical Discoveries

Discovery	Made by	Country
Electrocardiogram (ECG)	William Einthoven (1903)	Dutch
CT Scan	Godfrey Hounsfield (1973)	England
Sphygmomanometer	Scipione Riva-Rocci (1898)	Italy
Stethoscope	Rene Laennee (1819)	France
Thermometer	Sir Thomas Aelburt (1867)	England
Ultrasound	Ian Donald (1950)	Ireland
X-ray	WC Roentgen (1895)	Germany
Electroencephalogram (EEG)	Hans Berger (1929)	Germany
Antibiotic	Alexender Fleming (1928)	Scotland
Antiseptic	Joseph Lister (1867)	Scotland
Blood circulation	William Harvey (1628)	Britain
Blood transfusion	Jean-Baptiste Denys (1625)	France
Cholera and TB germs	Robert Kock (1883)	Germany

Types of Antibiotic with their Source and Action

Antibiotics	Source	Action
Streptomycin	Streptomyces griseus	Meningitis, Pneumonia, Tuberculosis and Local Infection. Toxic in some through eighth cranial nerve.
Chloramphenicol	Streptomyces venezuelae,	Typhoid, Typhus, Whooping cough, Atypical
Chloromycetin	S. lavendulae, Now synthetic	Pneumonia, Bacterial Urinary Infections
Tetracyclines/ Aureomycin	Streptomyces aureofaciens	Viral pneumonia, Osteomyelitis, Whooping Cough, Eye infections.
Oxytetracycline/ Terramycin	Chlorotetracycline → Hydrogenation Streptomyces rimosus	Intestinal and Urinary Infections (Spirochaetes, Rickettsia, Viruses)
Erythromycin	Streptomyces erythreus (= S. erythraeus)	Typhoid, Common Pneumonia, Diphtheria, Whooping Cough, etc.
Gentamycin	Micromonospora purpurea	Effective against Gram (+) bacteria
Polymixin	Bacillus polymyxa	Antifungal
Pencillin	Penicillium chrysogenum, P. notatum + Phenyl Acetic Acid	Tonsilitis, Sore Throat, Gonorrhea, Rheumatic Fever, Some Pneumonia types
Griseofulvin	Penicillium griseofulvum	Antifungal, especially for Ringworm
Nystatin	Streptomyces noursei	Antifungal for Candidates and overgrowth of Intestinal Fungi during excessive antibiotic treatment.
Hamycin	Streptomyces pimprei	Antifungal for Thrush
Fumagillin	Aspergillus fumigatus	Broad spectrum antibacterial especially against Salmonella and Shigella.
Bacitracin	Bacillus Licheniformis	Syphilis, Lymphonema or Reticulosis.

Awards and Honours

INTERNATIONAL

Nobel Prize

- These prizes were instituted in 1901 to give effect to the wishes expressed by Alfred Bernhard Nobel (1833-96), the inventor of dynamite, in his will. They are awarded internationally each year for outstanding achievements in (i) chemistry, (ii) physics, (iii) medicine, (iv) literature, (v) promotion of peace, and (vi) for economics (since 1969). They are awarded each year in Oslo on Dec 10, the anniversary of benefactor Alfred Nobel's death.

Winners of Nobel Prize from India

Winner	Field	Year
Rabindranath Tagore	Literature	1913
C.V. Raman	Physics	1930
Hargobind Khorana	Medicine	1968
Mother Teresa	Peace	1979
Subrahmanyan Chandrasekhar	Physics	1983
Amartya Sen	Economics	1998
Venkatraman Ramakrishnan	Chemistry	2009
Kailash Satyarthi	Peace	2014

Kalinga Prize

- The Kalinga Prize is awarded by UNESCO for making outstanding contribution to the popularisation of science. It was established in 1951.

Magsaysay Awards

- Instituted in 1957. Named after Ramon Magsaysay, the former President of Philippines. The award is given annually on 31st August, the birth anniversary of Magsaysay, for outstanding contributions in Public service, Community Leadership, Journalism, Literature and Creative Arts and International Understanding.

Nehru Award for International understanding

- Jawaharlal Nehru Award for International Understanding was instituted by the Indian Council of Cultural Relations in 1964 to be awarded to a person who contributed the most to understanding and friendship between different countries. Angela Merkel, Chancellor of Germany is the recipient of Nehru Award 2009. (The award has not been given since 2010-2015).

Grammy Awards

- It is awarded for the outstanding achievements in the music industry by National Academy for Recording Arts and Sciencer, Amencer.
- It was started in 1959.

Indira Gandhi International Peace Prize

- The award was instituted in the memory of Mrs. Indira Gandhi to foster creative cooperation among nations of the world. The award for 2015 has been given to the office of the United Nations High Commissioner for Refugees (UNHCR). This prize carries ₹ 25 lakh and a citation.

Man Booker Prize

- Instituted in 1968, is the highest literary award of the world, set-up by the Booker Company and the British Publishers Association along the lines of the Pulitzer Prize of USA.

Pulitzer Prize

- Instituted in 1917 and named after US publisher Joseph Pulitzer. It is conferred annually in the United States for the accomplishments in Journalism, Literature and Music.

Oscar Awards

- Instituted in 1929, these awards are conferred annually by the Academy of Motion Pictures, Arts and Sciences, USA in recognition of outstanding contribution in the various fields of film making.

UN Human Rights Award

- Instituted in 1966, this award is given every 5 years for individual contributions to the establishment of human rights.

NATIONAL

Bharat Ratna

- This is India's highest civilian award. It is given for exceptional work in art, literature, science and recognition of public services of the highest order. Government servants are not eligible for it. The actual award is designed in the shape of a **peepal** leaf with Bharat Ratna inscribed in Devanagri script in the Sun Figure. It was first awarded in 1954.

Padma Vibhushan

- This award is made for exceptional and distinguished services in any field, including services rendered by Government servants.

Padma Bhushan

- This award is made for distinguished services of a high order in any field, including services rendered by Government servants.

Padma Shri

- This award is made for distinguished services in any field, including services rendered by Government servants.

Param Vir Chakra

- The highest award for bravery of some daring or pre-eminent act of valour or self-sacrifice in the presence of the enemy, whether on land, at sea or in the air.

Mahavir Chakra

- It is the second highest decoration and is awarded for acts of conspicuous gallantry in the presence of the enemy, whether on land, at sea or in the air.

Vir Chakra

- It is the third in order of awards given for acts of gallantry in the presence of the enemy, whether on land, at sea or in the air.

Ashok Chakra

- This is awarded for valour, courageous action or sacrifice, away from the battlefield. It is highest military award during peacetime.

Kirti Chakra

- The decoration is awarded for conspicuous gallantry. It is made of standard silver and is circular in shape.
- The obverse and the reverse are exactly the same as in Ashoka Chakra.

Shaurya Chakra

- The decoration is awarded for an act of gallantry during peacetime.

Rajiv Gandhi Khel Ratna

- It is instituted in 1991-92 with the objective of honouring most outstanding sports-person to enhance their general status. It is the highest award bestowed to a sports person in India. The amount of prize money is ₹ 7.5 Lakh.

Arjuna Award

- It is instituted in 1961 by the Government of India to recognise outstanding achievement in National Sports.
- The award carries a cash prize of ₹ 5 lakh, a bronze statue of Arjuna and a scroll of honour.

Dronacharya Award

- It is instituted in 1985 by the Government of India to recognise excellence in sports coaching.
- The award carries a cash prize of ₹ 5 lakh, a bronze statue of Dronacharya and a scroll of honour.

Dhyanchand Award

- It is instituted in 2002, carries a cash prize of ₹ 1.5 lakh, a plaque and a scroll of honour. This honour is given to those sportspersons who have contributed to sportspersons, and sports by their performance and continue to contribute their promotion for sports even after their retirement from active sporting career.

Gandhi Peace Prize

- Established in 2nd October, 1994, on the occasion of the 125th birthday anniversary of Mahatma Gandhi carries a cash prize of ₹ 1 crore. Indian Government instituted this annual prize to encourage and promote the significance of Gandhian values over the world.

Bharatiya Jnanpith Award

- Instituted in 22nd May, 1961, carries a cash prize of ₹ 5 lakh, a citation and a bronze replica of Vagdevi (Saraswati).
- This award is given for the best literary writting by an Indian citizen in a language listed in Eighth Schedule of the Indian Constitution.

Dadasaheb Phalke Award

- Dadasaheb Phalke is known as the Father of Indian Cinema. The highest National Film Award is named after him in 1969.
- Mrs. Devika Rani Roerich was the first person to receive Dadasaheb Phalke Award in 1969.

Borlaug Award

- Instituted in 1973, carries a cash prize of ₹ 1 lakh. Instituted to honour outstanding agricultural scientists.

Sports

Olympics

- Olympics games were started in 776 BC on Mount Olympia in the honour of Greek God, 'Zeus'. The modern Olympic Games were started in **Athens,** the capital of Greece on **6th April, 1896** with great efforts made by French nobleman, **Baron Pierre de Coubertin.**
- The Olympic Games are organised after every 4 years. The Olympic Flag is made up of white silk and contains five interwined rings as the Olympic Emblem.
- The official Olympic Motto is Citius, Altius, Fortius, a Latin phrase meaning Swifter, Higher, Stronger. The Head Office of International Olympic Committee (IOC) is at Lausanne (Switzerland).

Commonwealth Games

- The first Commonwealth Games were held in 1930 in Hamilton, Canada.
- There are currently 54 members of the Commonwealth of Nations, and 71 teams participated in the games.

The Asian Games

- The Asian Games, also called the Asiad, are a multi-sport event held every 4 years among athletes from all over Asia.
- The games are regulated by the Olympic Council of Asia (OCA), under the supervision of the International Olympic Committee (IOC). The first Asian Games were held in 1951 in New Delhi (India). 18th Asian Games of 2018 will be held at Jakarta (Indonesia).

Cricket World Cup

- The first Cricket World Cup was organised in England in 1975. A separate women's Cricket World Cup has been held every 4 years since 1973.

Hockey World Cup

- The first Hockey World Cup was organised in Barcelona (Spain) in 1971. Women's Hockey World Cup has been held since 1974. The 13th Hockey World Cup held in the Netherlands (Hague) in 2014.

Football World Cup

- The Football World Cup is organised by FIFA (Federation of International Football Association). The World Cup is called 'Jules Rimet Cup' named after the name of FIFA President Jules Rimet. The first Football World Cup was organised in Uruguay in 1930.

National Games of Some Countries

Country	Game
Australia	Cricket
China	Table Tennis
Canada	Ice Hockey
Japan	Jujutsu
Scotland	Rugby
Spain	Bull Fighting
USA	Baseball
UK	Cricket
India	Hockey
Pakistan	Hockey
Brazil	Football

Cups and trophies

Cup/Trophy	Game
Agha Khan Cup	Hockey (Mumbai)
Ashes	Cricket (England and Australia)
American Cup	International Yacht Race
Beighton Cup	Hockey (Kolkata)
Canada Cup/ World Cup	Golf (World Championship)
Colombo Cup	Football (India, Pakistan, Myanmar and Sri Lanka)
Davis Cup	Lawn Tennis (International)
Derby	Horse Racing (U.K.)
Dhyan Chand Trophy	Hockey (India)
Ezra Cup	Polo (India)
E.A. Cup	Football (India)
I.F.A. Shield	Football (India)
Jayalaxmi Cup	National Table Tennis Championship for Women (India)
Mohan Singh Trophy	Volleyball (New Delhi)
Maulana Azad Trophy	Inter-varsity Sports and Athletics (India)
Nehru Trophy	Hockey (India)
Prince of Wales Cup	Golf (U.K.)
Ramanujam Cup	Table Tennis (India)
Rangaswami Cup	National Hockey Championship (India)
Ranji Trophy	Cricket (India)
Rohinton Baria Trophy	Inter-varsity Cricket Championship (India)
Rovers Cup	Football
Ryder Cup	Golf (U.K.)
Santosh Trophy	National Football Championship (India)
Stevens Cup	Tennis (U.S.A.)
Subroto Cup	Inter School Football (India)
Thomas Cup	Badminton (World Championship)
Uber Cup	Badminton (Women's International)
Walker Cup	Golf
Westchester Cup	Polo
Wellington Trophy	Rowing (India)

Sport Terms

Badminton Deuce, Double fault, Drop, Let, Love all, Smash.

Basketball Basket, Blocking, Dribbling, Free throw, Held ball, Holding, Jump ball, Multiple throws, Pivot.

Baseball Base, Battery, Bunting, Catcher, Diamond, Hitter, Home, Infield, Outfield, Pinch, Pitcher, Put out, Short stop, strike.

Billiards Bauik line, Bolting, Cannon, Cue Hazard, In-off, Jigger, Long jenny, Pot, Scratch, Screwback, Short jenny, Spot Stroke.

Boxing Auxiliary point system, Babit punch, Break, Cut, Defence, Hook, Jab, Lying on, Knock, Slam, Upper cut, Weight in, Win by knockout.

Bridge Chicane, Dealer, Dummy, Finesse, Grand slam, Little slam, No trump, Revoke, Rubber, Ruff, Suit, Tricks, Vulnerable.

Chess Bishop, Castle, Checkmate, Gambit, King, Knight, Pawn, Queen, Rook, Stalemate.

Cricket Ashes, Boundary, Bowling, Caught, Chinaman, Cover drive, Crease, Duck, Follow on, Googly, Gully, Hat-trick, Hit Wicket, l.b.w., Legbreak, Leg bye, Maiden over, No ball, Off break, On drive, Out, Over, Pitch, Popping crease, Rubber, Run out, Sixer, Silly point, Square leg, Stone walling, Straight drive, Stumped, Wicket.

Croquet Hoops, Mallet, Peg Out

Draughts Huff

Football	Corner kick, Direct free kick, Dribble, Goal kick, Hat-trick, Off side, Penalty kick, Thrown In, Tripping.
Golf	Best-ball, Bogey, Bunker, Caddie, Dormy, Fairway, Fourball, Foursome, Greed holes, Links, Niblic, Par, Put, Rough, Stymied, Tee, Threesome.
Hockey	Bully, Corner, Dribble, Hat-trick, Offside, Roll-in, Scoop, Short Corner, Stick, Striking circle, Tiebreaker.
Horse Racing	Jockey, Punter, Steeple chase.
Polo	Bunker, Chukker, Mallet.
Rowing	Bow, Bucket, Cow, Ergometer, Feather, Paddle, Regatta.
Rugby Football	Tackle, Lines, Scrum, Touch, Try.
Shooting	Bag, Bull's eye, marksmanship, Nuzzle, Plug.
Skiing	Tobogganing.
Swimming	Breast stroke, Crawl, Dive.
Tennis	Backhand Stroke, Deuce, Double fault, Fault, Ground Stroke, Half volley, Let, Smash, Volley.
Volleyball	Blocking, Doubling, Heave, Love point, Service, Volley.
Wrestling	Half nelson, Head lock, Heave, Hold, Rebouts, Scissor.

Famous Sports and Stadium

Stadium	Place
Ambedkar Stadium	Delhi
National Stadium	Delhi
Ferozeshah Kotla Ground	Delhi
Shivaji Stadium	Delhi
Indira Gandhi Stadium	Delhi
Jawaharlal Nehru Stadium	Delhi
Wankhede Stadium	Mumbai
National Stadium	Mumbai
Brabourne Stadium	Mumbai
Eden Gardens	Kolkata
Yuva Bharti Stadium	Kolkata
Varabhati Stadium	Cuttack
Nehru (Chepak) Stadium	Chennai
Kennan Stadium	Jamshedpur
Green Park Stadium	Kanpur
Lords, Oval, Leeds	England (U.K.)
Headingley Manchester	England
Brookland	England
Twickenham	England
Trent Bridge	England
White City	England
Perth, Brisbane, Melbourne, Sydney	Australia
Sendy Lodge	Scotland
Yankee Stadium	New York (USA)
Forest Hill	USA
Flemington	Melbourn
Black Heath	London (U.K.)
Wimbledon	London

❑ ❑ ❑

Miscellaneous

Countries with Capitals and Currencies

Country	Capital	Currency
Afghanistan	Kabul	Afghani
Albania	Tirana	Lek
Algeria	Algiers	Algerian Dinar
Argentina	Beunos Aires	Peso
Australia	Canberra	Australian Dollar
Austria	Vienna	Euro
Bangladesh	Dhaka	Taka
Bhutan	Thimphu	Ngultrum
Brazil	Brasilia	Cruzeiro Real
Cambodia	Phnom-Penh	Riel
Canada	Ottawa	Canadian Dollar
Chile	Santiago	Peso
China	Beijing	Yuan, Renminbi
Denmark	Copenhagen	Krone
Egypt	Cairo	Egyptian Pound
France	Paris	Franc
Germany	Berlin	Euro
Greece	Athens	Euro
Hungary	Budapest	Forint
India	New Delhi	Rupee
Indonesia	Jakarta	Rupiah
Iran	Tehran	Riyal
Iraq	Baghdad	Dinar
Ireland	Dublin	Euro
Israel	Jerusalem	Shekel
Italy	Rome	Euro
Japan	Tokyo	Yen
Kazakhstan	Astana	Tenge
Kenya	Nairobi	Shilling
North Korea	Pyongyang	Won
Kuwait	Kuwait City	Kuwait Dinar

Country	Capital	Currency
South Korea	Seoul	Won
Malaysia	KualaLumpur	Ringgit
Maldives	Male	Rufiyaa
Mauritius	Port Louis	Rupee
Mongolia	Ulan Bator	Tugrik
Myanmar	Naypyidaw	Kyat
Nepal	Kathmandu	Rupaia
Netherlands	Amsterdam	Euro
New Zealand	Willington	NZ Dollar
Nigeria	Abuja	Naira
Pakistan	Islamabad	Rupee
Phillippines	Manila	Peso
Poland	Warsaw	Polish Zloty
Portugal	Lisbon	Euro
Qatar	Doha	Riyal
Russia	Moscow	Ruble
Saudi Arabia	Riyadh	Riyal
Singapore	Singapore	Dollar
South Africa	Pretoria	Rand
Spain	Madrid	Euro
Sri Lanka	Colombo	Rupee
Sudan	Khartoum	Sudanese Pound
Sweden	Stockholm	Krona
Switzerland	Bern	Swiss Franc
Taiwan	Taipei	New Taiwan Dollar
Thailand	Bangkok	Baht
Turkey	Ankara	Lira
Ukraine	Kiev	Hryvnia
UK	London	Pound Sterling
US	Washington DC	Dollar
Zimbabwe	Harare	Dollar

Parliaments of the World

Country Name	Parliament Name
India	Sansad
Japan	Diet
Nepal	Rashtriya Panchayat
Pakistan	National Assembly
Russia	Duma
Spain	Cortes
Afghanistan	Shora
Australia	Parliament
Bangladesh	Jatia Sansad
Bhutan	Tasongadu
Canada	Parliament
China	National People Congress
Egypt	People's Assembly
France	Parliament
Germany	Bundestag
Great Britain	Parliament
Sweden	Riksdag
South Africa	Parliament
Switzerland	Federal Assembly
USA	Congress

Major Newspapers of the World

Country	Newspaper
India	The Times of India, The Hindu, The Tribune, The Statesman, The Hindustan Times
Australia	The Sydney Morning Herald, The Age
Canada	Globe and Mail, The Gazette
France	International Herald Tribune
Germany	Die Welt
Japan	Mainichi Daily News
New Zealand	The New Zealand Herald, The Press
United Kingdom	The Times, The Scotsman, The Guardian, The Herald, The Courier,
USA	Washington Post

Intelligence/Detective Agencies of The World

Country	Detective Agency
Mossad	Israel
Egypt	Mukhabarat
Japan	Naicho
Iran	SAVAK (Sazamane Etelaatva Amniate Kechvar)
Iraq	General Security Directorate
France	DGSE (Direction General de la Securite Exterieur)
China	Central External Liaison Department
Russia	KGB/GRU
South Africa	National Intelligence Agency
UK	MI (Military Intelligence)-5 and 6, Special Branch, Joint Intelligence Organisation
Pakistan	Inter Services Intelligence (ISI)
India	Research and Analysis Wing (RAW), Intelligence Bureau (IB)
USA	Central Intelligence Agency (CIA), Federal Bureau of Investigation (FBI)

Official Books of Major Countries

Book/Paper	Country
Orange Book	An official publications of the Government of Netherlands
White Book	An official Publication of China, Germany and Portugal
White Paper	An ificial paper of the Government of Britain and India on a particular issue
Yellow Book	French official book
Blue Book	An official report of the British Government
Green Book	An official publications of Italy and Iran
Grey Book	An official reports of the Governments of Japan and Belgium

Important Symbols or Signs

Symbol/sign	Description
Lotus	Culture and civilisation
Red Cross	Medical aid and hospital
Red Flag	Revolution; also sign of danger
Black Flag	Symbol of protest
Yellow Flag	Flown on ships or vehicles carrying patients suffering from infectious diseases
Flag flown upside down	Symbol of distress
While Flag	Symbol of Truce
Pigeon or Dove	Symbol of peace
A blindfolded woman holding a balanced scale	Symbol of Justice
Black strip on face arm	Sign of mourning or protest
Wheel (Chakra)	Symbol of progress
Flag flown at half mast	Symbol of national mourning
Olive Branch	Symbol of peace

National Emblems of Major Countries

Country	Emblem
France	Lily
Germany	Corn Flower
India	Lioned Capital
Iran	Rose
Australia	Kangaroo
Bangladesh	Water Lily
Belgium	Lion
Canada	White Lily
Chile	Candor and Huemul
Pakistan	Crescent
Spain	Eagle
UK	Rose
USA	Golden Rod
Italy	White Lily
Japan	Chrysanthemum
Netherlands	Lion
New Zealand	Southern Cross, Kiwi, Fern
Norway	Lion

SUPERLATIVE (WORLD)

(The Largest, Biggest, Smallest, Longest, Highest)

- *Largest Airport :* King Abdul Khalid International Airport (Riyadh, Saudi Arabia)
- *Highest Airport :* Bangda Airport, Tibet (Now in China)
- *Tallest Building :* Burj Khalifa, Dubai United Arab Emirates (828 m)
- *Largest Bay :* Hudson Bay, Canada
- *Longest Big-ship Canal :* Suez Canal (linking Red Sea and Mediterranean Sea)
- *Longest Epic :* The Mahabharata
- *Largest Diamond :* The Cullinan (South Africa)
- *Larges Mosque :* Jama Masjid, Delhi
- *Largest Delta :* Sundarbans, India
- *Largest Lake :* Caspian Sea
- *Deepest Lake :* Baikal (Siberia)
- *Highest Lake :* Titicaca (between Boluia and Peru)
- *Largest Lake (Fresh water) :* Lake Superior, USA
- *Largest Coral formation :* The Great Barrier Reef (Australia)
- *Largest Continent :* Asia
- *Smallest Continent :* Australia
- *Highest Country :* Tibet (The Pamirs)
- *Largest Country (in Population) :* China
- *Largest Country (in area) :* Russia
- *Biggest Dome :* Gol Gumbaz (Bijapur), India
- *Tallest Minar (Free standing) :* Great Hassan II Mosque, Casablanca, Morocco
- *Largest City (in population) :* Tokyo
- *Highest City :* Wen Chuan (Tibet, China)
- *Biggest City (in area) :* Hulunbuir, China
- *Longest Bridge (Railway) :* Huex P Long Bridge (7009 m)
- *Largest Dam (Concrete) :* Grand Coulee Dam (USA)
- *Highest Straight Dam :* Bhakra Dam
- *Highest Capital City :* La Paz (Bolivia)
- *Highest Asian Desert :* Gobi, Mongolia
- *Largest Democracy :* India

- *Biggest Bell :* Great Bell at Moscow
- *Reptile which chanbges its colour :* Chameleon
- *Most intelligent Animal :* Chimpanzee
- *Highest Volcano :* Ojos del Salado, Andes, Argentine-Chile (6885 m)
- *Largest Volcano :* Mauna Loa (Hawaii Islands)
- *Longest Wall :* Great Wall of China
- *Highest Mountain Peak (world) :* Mount Everest (Nepal)
- *Highest Mountain Range :* Himalayas
- *Longest Mountain Range :* Andes Central (South America)
- *Biggest Museum :* British Museum (London)
- *Highest Waterfall :* Salto Angel Falls (Venezuela)
- *Longest Gulf :* Gulf of Mexico
- *Deepest and Biggest Ocean :* The Pacific
- *Largest Peninsula :* Arbia
- *Largest Palace :* Imperial Palace (Gugong), Beijing, China
- *Largest Park :* National Park, Greenland
- *Largest Archipelago :* Indonesia
- *Coldest Place :* Verkhoyansk (Siberia) Temperature (–89.2 °C)
- *Driest Place :* Death Valley (California)
- *Hottest Place :* Al-Aziziyah (Libya, Africa) 136°F
- *Largest Bridge (Railway) :* Danyang-Kunshan Grand Bridge (China)
- *Largest Plateau :* Tibetan Plateau
- *Largest River Basin :* Amazon Basin
- *World's Rainiest Spot :* Cherrapunji (Meghalaya)
- *Largest Gorge :* Grand Canyon on the Colorade river, USA
- *Largest & Bussiest Port :* Shanghai (China)
- *Longest Railway :* Trans-Siberian Railway
- *Longest river :* Nile (6690 km)
- *Largest Sea-Bird :* Albatross
- *Largest Sea (Inland) :* Mediterranean
- *Tallest Tower :* Sky Tree Tower, Tokyo (Japan)
- *Longest Swimming Course :* English Channel (between London and Edinburgh)
- *Longest Train Nonstop :* Flying Scotsman
- *Longest Tunnel (Railway) :* Seikan Rail Tunnel (Japan)
- *Longest and Largest Canal/Tunnel :* Le Rove Tunnel (South of France)
- *Lightest Gas :* Hydrogen
- *Lightest Metal :* Lithium
- *Highest Melting Point :* Tungstan, (34100°C)
- *Fastest Animal :* The Peregrine Falcon
- *Longest Poisonous Snake :* King Cobra
- *Largest Temple :* Angkor Vat (Combodia)

First in the World

- *First Radio Telescope Satellite Launched into Space :* Japan
- *First Space Ship Landed on Mars :* Viking-I (July 1976)
- *First Space Shuttle Launched :* Columbia
- *First country to make Constitution :* America
- *First country to start underground metro rail :* Britain
- *First unmanned mission on Moon :* Apollo-11
- *First country to do Artificial Satellite Experiment :* Russia
- *Country to give voting right to women :* New Zealand
- *First country to appoint Lokpal :* Sweden
- *First country to imposed carbon tax :* New Zealand

First in the World (Male)

- *First Asian to head the International Cricket Council :* Jagmohan Dalmiya
- *First man to climb Mount Everest :* Sherpa Tenzing Norgay and Sir Edmund Hillary (29th May, 1953)
- *First man to go into space :* Major Yuri Gagarin (USSR)
- *First person to give information about planets and their motion around the Sun :* Nicolous Copernicus
- *First man to compile Encyclopaedia :* Aspheosis (Athens)
- *First person to go on both the poles (North and South) :* Albert P. Carey

- *First man to reach North Pole :* Robert Peary
- *First man to reach South Pole :* Roald Amundsen
- *First man to climb on Mt. Everest without Oxygen :* Phu Dorji Sherpa

First in the World (Female)

- *First woman President of a country :* Maria Estela Peron (Argentina)
- *First woman in the world to cross the Strait of Gibralter :* Arti Pradhan (India)
- *First woman Cosmonaut in space :* Valentina Tereshkova (USSR)
- *First female to find asteroid :* Caroline and Derschell
- *First woman to cross the Seven Seas :* Bula Choudary
- *First non-white female to win Nobel Prize in Literature :* Toni Morrison
- *First woman to reach Antarctica :* Caroline Mikkelsen
- *First woman to have a space-walk :* Svetlana Yevgenyevna Savitskaya

Important International Organisations Groups

Name	Estd. in	Headquarter	Purpose
International Telecommunication Union (ITU)	1865	Geneva	Sets international regulations for radio telegraph, telephone and space radio communications.
International Labour Organisation (ILO)	1919	Geneva	To improve conditions and living standards of workers.
International Monetary Fund (IMF)	1945	Washington DC	Promotes international monetary cooperation.
United Nations International Children's Emergency Fund (UNICEF)	1945	New York	To promote children's welfare all over the world.
Food and Agricultural Scientific and Cultural Organisation (UNESCO)	1946	Paris	To promote collaboration among nations through education, science and culture.
World Health Organisation (WHO)	1948	Geneva	Attainment of highest possible level of health by all people.
International Atomic Energy Agency (IAEA)	1957	Vienna	To promote peaceful uses of atomic energy.
International Development Association (IDA)	1960	Washington DC	An affiliate of the World Bank, aims to help under-developed countries raise living standards.
United Nations Development Programme (UNDP)	1965	New York	Helps developing countries increase the wealth producing capabilities of their natural and human resources.
United Nations Environmental Programme (UNEP)	1972	Nairobi	Promotes international cooperation in human environment.
World Trade Organisation (WTO)	1995	Geneva	Setting rules for world trade to reduce tariffs.
United Nations Office on Drugs and Crime (UNODC)	1997	Vienna	To preventillict trafficking and abuse of drug, crime prevention.
UN Women	2010	New York City (USA)	To enable member states to achieve gen den quality and women empowerment.

Other International Organisations and Groups

Name	Estd.	Headquarter	Member	Purpose
The Commonwealth	1926	London	53	It was originally known as 'The British Commonwealth of Nations'. It is an association of sovereign and independent states which formally made up the British empire.
Asia Pacific Economic Cooperation (APEC)	1989	Singapore	21	To promote trade and investment in the Pacific basin.
Asian Development Bank (ADB)	1966	Manila	67	To promote regional economic cooperation.
Association of South-East Asian Nations (ASEAN)	1967	Jakarta	10	Regional, economic, social and cultural cooperation among the non-communist countries of South-East Asia.
Commonwealth of Independent States (CIS)	1991	Minsk (Belarus)	11	To coordinate inter-common wealth relations and to provide a mechanism for the orderly dissolution of the USSR.
Internation Olympic Committee (IOC)	1894	Lausanne (Switzerland)	105	To promote the olympic ideals and administer olympic games.
Non-Aligned Movement (NAM)	1961	-	115	Political cooperation and separate itself from both USA and USSR (in the cold-war era).
European Union	1965	Brussels (Belgium)	28	To create a united Europe in which member countries would have such strong economic and political bonds that war would cease to be a recurring fact.
North Atlantic Treaty Organisation (NATO)	1949	Brussels (Belgium)	28	Mutual defence and cooperation.
Organisation of Petroleum Exporting Countries (OPEC)	1960	Vienna (Austria)	12	Attempts to set world prices by controlling oil production and also pursues member interest in trade and development.
South Asian Association for Regional Cooperation (SAARC)	1985	Kathmandu (Nepal)	8	To promote economic, social and cultural cooperation.
Amnesty Internation (AI)	1961	London (UK)		To keep a watch over human rights violation worldwide. Got Nobel Prize in 1977 for Peace.
World Wildlife Fund (WWF) for Nature	1961	Gland (Switzerland)	All countries	To save the wildlife from extinction.

SUPERLATIVES (INDIA)

(Biggest, Highest, Largest, Longest, Smallest etc.)

- *The longest River :* The Ganga (2525 km)
- *The longest Canal :* Indira Gandhi Canal or Rajasthan Canal (Rajasthan) (649 km)
- *The longest Dam :* Hirakud Dam (Odisha) (26 km)
- *The longest Sea Beach :* Marina Beach (Chennai) (13 km)
- *The highest Lake :* Cholam Lake (Sikkim)
- *The largest Saline Water Lake :* Chilka Lake (Odisha)
- *The biggest River Islands :* Majuli, Brahmaputra river (Asom)
- *The largest Fresh Water Lake :* Wular Lake (Jammu and Kashmir)
- *The highest Dam :* Bhakhra Dam (740 ft) on Sutlej river (Punjab)
- *The highest Waterfall :* Kunchikal Falls (Karnataka) (445 m, 1493 ft)
- *The deepest River Valley :* Bhagirathi and Alaknanda
- *The longest River Bridge :* Mahatma Gandhi Setu, Patna (5,575 km)
- *The biggest Cantilever Bridge :* Rabindra Setu or Howrah Bridge (Kolkata)
- *The state with longest Coastline :* Gujarat (1600 km)
- *The longest Sea Bridge :* Bandra-Worli, Sea Link (5.6 km)
- *The largest Artificial Lake :* Govind Sagar Bhakhra Nangal (HP)
- *The longest River of Southern India :* Godavari (1465 km)
- *The longest Railway Platform :* Gorakhpur, Uttar Pradesh (1366.33 m)
- *The longest Corridor :* Corridor of Ramnath-swami Temple at Rameshwaram (Tamil Nadu)
- *The highest Road :* Road at Khardungla (in Leh-Manali Sector)
- *The highest Airport :* Leh Airport (Ladakh)
- *The largest Desert :* Thar (Rajasthan)
- *The largest Delta :* Sunderbans (Paschim Banga)
- *The state with maximum Forest Area :* Madhya Pradesh (11.24%)
- *The largest Zoo :* Zoolgoical Garden (Kolkata)
- *The biggest Stadium :* Yuva Bharti (Salt Lake) Stadium, Kolkata
- *The longest National Highway :* NH-7 (Varansi to Kanyakumari)
- *The highest Award :* Bharat Ratna
- *The largest Gurudwara :* Golden Temple, Amritsar
- *The largest Cave Temple :* Kailash Temple (Ellora, Maharashtra)
- *The highest Peak :* Godwin Austin 1, K2 (8611 m)
- *The longest Tunnel :* Jawahar Tunnel, Banihal Pass (Jammu and Kashmir)
- *The largest Auditorium :* Sri Shanmukhanand Hall (Mumbai)
- *The largest Animal Fair :* Sonepur (Bihar)
- *The largest Cave :* Amarnath (Jammu and Kashmir)
- *The highest Gate Way :* Buland Sarwaza, Fatehpur Sikri (Uttar Pradesh)
- *The tallest Statue :* Veera Abhaya Anjaneya Hanuman Swami (Andhra Pradesh, 135 ft)
- *The largest Public Sector Bank :* State Bank of India
- *The biggest Church :* Saint Cathedral at Old Goa (Goa)
- *The highest battlefield :* Siachen Glacier (5753 m)

First in India (Male)

- *First Governor-General of India :* William Bentinck (1828)
- *Last Governor-General of India :* Lord Mountbatten
- *First and last Indian Governor-General of Free India :* C. Rajgopalachari
- *First Commander-in-Chief of Free India :* General K.M. Kariappa

- *First Field Marshal of India :* General Sam Manekshaw (1971)
- *First Indian to go in Space :* Rakesh Sharma
- *First Indian to climb the Mount Everest without Oxygen :* Sherpa Ang Dorje
- *First Indian to become the Managing Director of World Bank :* Gautam Kaji
- *First Chairman of National Human Rights Commission :* Rangnath Mishra
- *First Indian to get Nobel Prize in Physics :* C.V. Raman (1930)
- *First Indian to get Nobel Prize in Literature :* Rabindranath Tagore (1913)
- *First Indian to get Nobel Prize in Economics :* Dr. Amartya Sen
- *First Indian to get Nobel Prize in Medicines (Physiology) :* Dr. Har Govind Khorana
- *First Indian to get Bharat Ratna :* Dr. S. Radhakrishnan, C. Rajgopalachari and Dr. C.V. Raman.
- *First Person to be Honoured with the Jnanpith Award :* G. Sankara Kurup (Malayalam)
- *First Cricketer to get Padma Bhushan :* C.K. Naidu
- *First Indian to get through ICS :* Satyendra Nath Tagore (1869)
- *First Indian to swim across the English Channel :* Mihir Sen (1958)

First in India (Female)

- *First Indian Female Chairperson of Indian National Congress :* Sarojini Naidu (1925)
- *First Woman to climb the Everest :* Bachendri Pal (1984)
- *First Woman Cabinet Minister :* Rajkumari Amrit Kaur (1947)
- *First Woman Chairman of the UN General Assembly :* Vijaya Laxmi Pandit
- *First Woman President of India :* Pratibha Devi Singh Patil
- *First Woman Speaker of Lok Sabha :* Meira Kumar (2009)
- *First Woman Deputy Chairman of Rajya Sabha :* Violet Alva (1962)
- *First Woman Prime Minister of India :* Indira Gandhi
- *First Woman to reach Antarctica :* Meher Moos (1976)
- *First Woman IAS Officer :* Anna Rajam George (1950)
- *First Female Chief Justice :* Leela Seth (1991)
- *First Woman to win the Jnanpith Award :* Asha Poorna Devi (1976)
- *First Woman to get the Bharat Ratna :* Indira Gandhi
- *First Female Nobel Prize Winner :* Mother Teresa (1979)
- *First Women to complete century in World Cup Cricket :* Thirush Kamini
- *First actress to be nominated to the Rajya Sabha :* Nargis Dutt
- *First Women Chairperson of UPSC :* Rose Millian Bathew (Kharbuli)

First in India

- *Newspaper :* Bengal Gazette (James Hickey)
- *Hindi Newspaper :* Udant Martand
- *Telegraph Line :* Diamond Harbour to Kolkata
- *International Telephone Service :* Mumbai to London (1851)
- *Silent Movie :* Raja Harish Chandra (Dadasaheb Phalke (1913)
- *Talkie Movie :* Alam Ara (Ardeshir Irani-1931)
- *Satellite :* Aryabhatta (19th April, 1975)
- *Satellite dedicated exclusively for Education purposes :* EDUSAT
- *Successful indigenous launch vehicle :* SLV-3
- *Nuclear Reactor :* Apsara
- *Lunar Mission :* Chandrayaan-I (October, 2008)
- *Mars Mission :* Mars Orbiter Mission (5th Novermber, 2013)
- *Hydroelectric Project :* Sivasamudram (1902)
- *Asian Games :* Delhi (1951)
- *Biosphere Reserve :* Nilgire
- *National :* Haley National Park (Jim Corbett), 1936
- *Chairman of UPSC :* Ross Barker
- *First E-court :* Ahmedabad
- *Court exclusively dedicated to women :* Malda (WB)

IMPORTANT NATIONAL & INTERNATIONAL DAYS

January

1 Global Family Day
9 NRI Day
12 National Youth Day (of Swami Vivekanand)
15 Army Day
25 National Tourism Day, Voter's Day
26 Indian Republic Day, International Customs Day
28 Data Protection Day
30 Martyr's Day (Mahatma Gandhi's Martyrdom), World Leprosy Eradication Day

February

4 World Cancer Day
20 World Day of Social Justice
24 Central Excise Day
28 National Science Day

March

8 International Women's Day
15 World Consumer Rights Day, World Disabled Day
21 World Forestry Day, International Day for the Elimination of Racial Discrimination
22 World Water Day
23 World Meteorological Day
24 World TB Day

April

5 National maritime Day, International Day for Mine Awareness
7 World Health Day
18 World Heritage Day
21 Civil Services Day
22 World Earth Day

May

1 International Labour Day (May Day)
3 International Energy Day, World Press Freedom Day
8 World Red Cross Day
21 Anti-Terrorism Day
17 World-Telecommunications Day

June

5 World Environment Day
12 World Day against Child Labour
20 World Refugee Day
21 International Yoga Day
27 International Diabetes Day

July

4 American Independence Day
7 International Day of Cooperatives
11 World Population Day

August

6 Hiroshima Day
8 World Senior Citizen's Day
12 International Youth Day
18 Day of the World's Indigenous Persons
29 National Sports Day (Dhyanchand's Birthday)

September

5 Teacher's Day (Dr. Radhakrishnan's Birthday)
14 Hindi Day, World First Aid Day
16 World Ozone Day
18 International Day of Peace
21 Biosphere Day
27 World Tourism Day

October

2 International Non-Violence Day, Lal Bahadur Shastri and Mahatma Gandhi's Birthday
3 World Habitat Day
5 World Teacher's Day
8 Indian Air Force Day
16 World Food Day
24 United Nations Day

November

9 Legal Services Day
14 Children's Day, World Diabetics Day
20 Universal Children's Day (UN)

December

1 World AIDS day
2 World Computer Literacy Day
4 Navy Day
7 Armed Forces Flag Day
10 Human Rights Day
16 Vijay Diwas
25 National Good Governance Day

BOOKS AND AUTHORS

- *The Three Musketeers* : Alexander Dumas
- *Identity and Violence : The Illusion of Destiny; The Argumentative Indian; Development as Freedom* : Amartya Sen
- *River of Smoke; Circle of Reason* : Amitav Ghose
- *Death of a City* : Amrita Pritam
- *Clear Light of the Day* : Anita Desai
- *Politics* : Aristotle
- *A Secular Agenda* : Arun Shourie
- *The Algebra of Infinite Justice; The God of Small Things; Greater Common Good* : Arundhati Roy
- *Budda Charitham* : Ashwaghosha
- *Freedom from Fear* : Aung San Suu Kyi
- *Anand Math; Durgeshnandini* : Bankim Chandra Chatterji
- *Dreams of My Father* : Barrack Obama
- *The Girl with No Name* : Barrett Lee, Marina Chapman
- *Pakistan the Gathering Storm* : Benazir Bhutto
- *The Red Saree* : Javier Moro
- *Jail Diary* : Chandrashekar
- *David Copperfield* : Charles Dickens
- *Revolution 2020 What Young India Wants; Half Girl Friend; One Night at Call Centre* : Chetan Bhagat
- *India for Sale* : Chitra Subramaniam
- *Freedom in Exile; Ethics for New Millennium* : Dalai Lama
- *A Divine Comedy* : Dante
- *Indian Economy : Essay on Money and Finance* : Dr. C. Rangarajan
- *Inspite of the Gods* : Edward Luce
- *The Luminaries* : Eleanor Catton
- *A Passage to India* : E.M. Forster
- *Kargil : From Surprise to Victory* : Gen. V.P. Malik
- *The Tin Drum* : Gunter Grass
- *Law, Lawyers and Judges* : H.R. Bhardwaj
- *Time Machine* : Herbert George Wells
- *My Truth* : Indira Gandhi
- *Ajatha Shatru* : Jai Shankar Prasad
- *Discovery of India; Glimpses of World History* : Jawaharlal Nehru
- *Prison Diary* : Jayaprakash Narayan
- *Walking with Lions; Curtain Raisers* : K. Natwar Singh
- *Straight from the Heart* : Kapil Dev
- *Das Kapital* : Karl Marx
- *Communist Manifesto* : Karl Marx and Fredrik Engles
- *Train to Pakistan* : Khushwant Singh
- *I Dare; As I See* : Kiran Bedi
- *The Inheritance of Loss* : Kiran Desai
- *Too Old to be Bold* : Kuldeep Mathur
- *A Prisoner's Scrap* : L.K. Advani
- *My Experiments with Truth* : Mahatma Gandhi
- *The Heart of India* : Mark Tully
- *Untouchable* : Mulk Raj Anand
- *Godan* : Premchand
- *The Guide* : R.K. Narayan
- *A Garland of Memories* : Ruskin Bond
- *Golden Threshold; The Broken Wings* : Sarojini Naidu
- *Arabian Nights* : Sir Richard Burton
- *Essays on Gita* : Sri Aurobindo Ghosh
- *A Brief History of Time* : Stephen Hawkings
- *Koraner Nari* : Taslima Nasreen
- Against the Day : Thomas Pynchon
- *India : A Wounded Civlisation; Letters Between a Father and Son; Half a Life; An Area of Darkness; Magic Seeds* : V.S. Naipaul
- *Bhagwad Gita* : Ved Vyas
- *Love and Longing in Bombay* : Vikram Chandra
- *An Equal Music* : Vikram Seth
- *Sachin Cricketer of the Century* : Vimal Kumar
- *As You Like it* : William Shakespeare
- *The Test of My Life* : Yuvraj Singh
- *Congress After Indira* : Zoya Hasan

PARAMILITARY AND RESERVE FORCES

Assam Rifles

- It was established in 1835 and is the oldest paramilitary force in the country.
- Its main objective is to keep vigilance of international borders in North East and counter insurgency operations in Arunachal Pradesh, Manipur, Mizoram and Nagaland.

Intelligence Bureau (IB)

- It was set-up in 1920. Its objective is to collect secret information relating to country's security.
- It was originally set-up as Central Special Branch (CSB) in 1887 and renamed IB in 1920.

Central Reserve Police Force (CRPF)

- It was set-up in 1939.
- Its main objective is to assist the State/Union Territory Police in maintenance of law and order.
- The 88th Battalion of CRPF, known as 'Mahila Battalion' (commissioned on 30th March, 1986) is the world's first paramilitary force comprising entirely of women.

Territorial Army (TA)

- It was established in 1948.
- It is a voluntary, part-time force (between 18 and 42 years), not of professional soldiers, but civilians who wish to assist in defence of the country.

National Caded Corps (NCC)

- It was established in 1948. Its main objective is to stimulate interest among the youth in the defence of the country in order to build up a reserve manpower to expand armed forces.

Home Guard

- It was established in 1962, to assist the police in maintaining security to help defence forces and to help local authorities in case of any eventuality.

Indo-Tibetan Border Police (ITBP)

- It was established in 1962, after the Chinese attack. It is basically employed in the Northern borders for monitoring the borders and also to stop smuggling and illegal immigration.

Central Bureau of Investigation (CBI)

- It was established in 1963. Its objective is to investigate cases of misconduct by public servants, cases of cheating, embezzlement and fraud. CBI is also entrusted with the investigation of international crime cases in collaboration with INTERPOL.

Border Security Force (BSF)

- It was established in 1965. It keeps a vigil over the international borders against the intrusion in the country.

Central Industrial Security Force (CISF)

- It was set-up in 1969 after the recommendations of Justice B. Mukherji.
- Its objective is to monitor the industrial complexes of Central Government.

Indian Coast Guard

- It was set-up in 1978. Its objective is to protect the maritime and other national interests in the maritime zones of India.

National Security Guards (NSG)

- It was established in 1984 to counter the surge of militancy in the country.
- It is a highly trained force deals with the militants effectively.

National Crime Records Bureau (NCRB)

- It was established in 1986. Its objective is to collect crime statistics at the national level, information of inter-state and international criminals to help investigation agencies.

Rapid Action Force (RAF)

- It was established in 1991. Under the coperational command of CRPF, 10 battalions of the CRPFs have been re-oriented for tackling communal riots in the country.

Missiles and other Weapons

Name	Class	Range	Origin (Test Fired)
Agni I	SRBM	850 km	Odisha
Agni II	MRBM	2500 km	Odisha
Agni III	IRBM	3500-5500 km	Odisha
Agni IV or Agni II Prime	IRBM	4000 km	Odisha
Agni V	ICBM	5000-6000 km	Odisha
Agni VI	ICBM	8000-10000 km	Under Development
Agni 3SL	ICBM	5200-11600 km	Under Development
Dhanush	SRBM	350 km	Odisha
Nirbhay	Subsonic Cruise Missile	1000 km	Under Development
Brahmos	Supersonic Cruise Missile	290 km	Bhubaneshwar
Brahmos 2	Hypersonic Cruise Missile	290 km	Under Development
Prithvi I	SRBM	150 km	Chandipur (Odisha)
Prithvi III	SRBM	350 km	Under Development
Sagarika	SLBM	700 km	Vishakhapatnam
Shaurya	TBM	700 km	Chandipur (Odisha)
Astra	Air to Air Missile	80-100 km	Chandipur (Odisha)

Nuclear and Space Research Centres in India

Research Centre	Place
Tata Institute of Fundamental Research	Mumbai
Saha Institute of Nuclear Physics	Kolkata
Centre of Earth Sciences Studies	Thiruvananthapuram (Kerala)
Physical Research Laboratory	Ahmedabad
Space Commission	Bengaluru
Vikram Sarabhai Space Centre	Thiruvananthapuram
Indian Space Research Organistion (ISRO)	Bengaluru
Space Application Centre	Ahmedabad
Thumba Equatorial Rocket Launching Station	Thumba (Kerala)
Indian Rare Earths Limited	Mumbai
Uranium Corporation of India Limited	Jadugoda (Jharkhand)
Atomic Energy Commission (AEC)	Mumbai
Electronics Corporation of India Limited	Hyderabad
Bhabha Atomic Research Centre (BARC)	Trombay (Mumbai)
Radio Astronomy Centre	Ootacamund (Tamil Nadu)

Indian Defence Training Institutions

Training Institution	Estd. in	Place
Rashtriya Indian Military College (RIMC)	1922	Dehradun
Army Cadet College (ACC)	1929	Dehradun
Indian Military Academy (IMA)	1932	Dehradun
National Defence Academy (NDA)	1941	Khadakwasla, Pune

High Altitude Warfare School (HAWS)	1948	Gulmarg
National Defence College (NDC)	1960	New Delhi
Officers Training Academy (OTA)	1963	Chennai
Counter Insurgency and Jungle Warfare School	1970	Vairengte (Mizoram)
College of Defence Management	1970	Secunderabad (Andhra Pradesh)
College of Combat/Army War College	1971	Mhow (Madhya Pradesh)
Army School of Physical Training (ASPT)	1978	Pune
Army Air Defence College (AADC)	1989	Gopalpur (Odisha)
Indian National Defence University		Gurgaon (Haryana)

State and Folk Dances

- **Andhra Pradesh** : Kuchipudi, Ghantamardala, Ottam Thedal
- **Assam** : Bihu, Bichhua, Natpuja, Maharas, Kaligopal, Bagurumba, Naga dance, Khel Gopal, Tabal Chongli, Canoe, Jhumura Hobjanai
- **Bihar** : Jata-Jatin, Bakho-Bakhain, Panwariya, Sama-Chakwa, Bidesia
- **Gujarat** : Garba, Dandiya Ras, Tippani Juriun, Bhavai
- **Haryana** : Jhumar, Phag, Daph, Dhamal, Loor, Gugga, Khor, Gagor
- **Himachal Pradesh** : Jhora, Jhali, Chharhi, Dhaman, Chhapeli, Mahasu, Nati, Dangi
- **Jammu and Kashmir** : Rauf, Hikat, Mandjas, Kud Dandi Nach, Damali
- **Karnataka** : Yakshagan, Huttari, Suggi, Kunitha, Karga, Lambi
- **Kerala** : Kathakali (Classical), Ottam Thulal, Mohiniattam, Kaikottikali
- **Maharashtra** : Lavani, Nakata, Koli, Lezim, Gafa, Dahikala Dasavtar or Bohada
- **Odisha** : Odissi (Classical), Savari, Ghumara, Painka, Munari, Chhau
- **Paschim Banga** : Kathi, Gambhira, Dhali, Jatra, Baul, Marasia, Mahal, Keertan.
- **Punjab** : Bhangra, Giddha, Daff, Dhaman.
- **Rajasthan** : Ghumar, Chakri, Ganagor, Jhulan Leela, Jhuma, Suisini, Ghapal, Kalbeliya
- **Tamil Nadu** : Bharatanatyam, Kumi, Kolattam, Kavadi
- **Uttar Pradesh** : Nautanki, Raslila, Kajri, Jhora, Chappeli, Jatta
- **Uttarakhand** : Garhwali, Kumayuni, Kajari, Jhora, Raslila, Chappeli

Musical Instruments and Instrumentalists

- **Been** : Asad Ali Khan, Zia Moin-ud-din Khan
- **Santoor** : Shiv Kumar Sharma
- **Sarod** : Buddhadev Dasgupta, Ali Akbar Khan, Amjad Ali Khan, Bahadur Khan, Sharan Rani, Zarin S. Sharma
- **Sarangi** : Ustand Binda Khan
- **Sitar** : Ravi Shankar, Hara Shankar Bhattacharya, Nikhil, Banerjee, Vilayat Khan, Mustaq Ali Khan
- **Surb Ahar** : Sajjad Hussain, Annapurna
- **Veena** : Soraiswamy Iyengar, Chittibabu, Emani Sankara Shastri, Dhanammal, S. Bala Chandran, K.R. Kumaraswami
- **Violin** : Gajanan Rao Joshi, M.S. Gopal Krishnan, T.N. Krishnan, Baluswamy, Dikshitar; Dwaran Venkataswamy Naidu Lalyuli G. Jayaraman, Mysore T. Chowdiah, V.G. Jog
- **Flute** : T.R. Mahalingam, N. Ramani, Hari Prasad Chaurasia, Pannalal Ghosh
- **Nadaswaran** : Sheikh Chinna Moula, Neeruswamy Pillai, Rajaratanam Pillai
- **Shehnai** : Bismillah Khan
- **Mridangam** : Palghat Mani Iyer, Karaikudi R. Mani, Palghat Raghu
- **Pakhawag** : Pt. Ayodhya Prasad, Gopal Das, Babu Ram Shanker Pagaldas
- **Tabla** : Zaikr Hussain, Nikhil Ghosh, Kishan Maharaj, Alla Rakha Khan, Pandit Samta Prasad, Kumar Bose, Laif Khan
- **Kanjira** : Pudukkotai Dakshinamurthi Pillai

❑ ❑ ❑

Multiple Choice Questions

1. Which crop was the earliest to be cultivated by the people of Indus Valley?
   A. Wheat   B. Rice
   C. Cotton   D. Rye
2. The Harappan Civilisation was chiefly concentrated in:
   A. Punjab, Rajasthan and Gujarat
   B. Sindh, Punjab and Rajasthan
   C. Haryana, Rajasthan and Tamil Nadu
   D. Punjab, Rajasthan and Assam
3. Pre-Harappan settlement was discovered in 1981 at the foot of the Bolan pass on the Bolan river at one of the following places. Identify it;
   A. Mansa   B. Manda
   C. Mehergarh   D. Alamgirpur
4. One thing is common among the following between the Indus and Mesopotamian civilisation. Identify:
   A. Larger urban complex
   B. Seals rolled on clay tablets
   C. Ornamental architecture
   D. Use of potter's wheel
5. The Indus valley belonged to:
   A. Mediterranean race
   B. Proto-Astraloid race
   C. Tibetan Mongoloid race
   D. Alpine race
6. The important public place discovered at Mohenjo-daro was:
   A. the great bath
   B. the great granary
   C. the great cemetery
   D. the acropolis
7. Indus script is:
   A. bonstrophedon   B. indeographic
   C. cuneiform   D. logographic
8. Indus seals:
   A. were purely secular in character
   B. were used as means of exchange by Harappans
   C. reveal their religions character
   D. were exported in big numbers
9. Who among the following was the first to suggest the similarity between Sanskrit language of Aryans, Greeks and Latin?
   A. Sir Williams Jones
   B. Monier Williams
   C. Max Muller
   D. None of the above
10. Which was not the cause of Aryans Victory over the Dasas or Dasyus?
    A. Use of chariots in war
    B. Use of horses in war
    C. Use of metal armours and helmets
    D. Use of elephant in war
11. The important characteristics of the vedic religion were:
    1. Goddes like 'prithvi', 'Aditi' and 'Usha'
    2. Goddes are co-equal to their male-partners
    3. Predominance of male dieties.
    4. Worship of nature.

    Codes:
    A. only 3 and 4   B. only 2 and 4
    C. only 1 and 4   D. only 1 and 3
12. Which of the following statement is not correct regarding Indra?
    A. He was the god exclusively associated with sacrifices.
    B. He was know as 'Purandara' or breaken of forts
    C. He killed the evil dragon 'Virata'
    D. He was the god of Thunder.

13. Which of the following region was called "Arya varta"?
- A. From Yamuna upto West Bengal
- B. Punjab and Delhi Region
- C. The whole of northern India
- D. The region beyond Vindhya mountain

14. In India Varna law enjoyed:
- A. Legal Sanction only
- B. Religions Sanction only
- C. Economic Sanction only
- D. The Sanction of both the state and religion.

15. Which statement an asceticism is correct?
- A. The Atharva Veda after refers to ascetics as Vratyas
- B. By the time of the Upanishads, asceticism was widespread
- C. Vratya was a priest of non-vedic fertility cult which involved ritual dancing and flagellation
- D. All of the above.

16. The speculation of the upanishads centre around one word of the Rig veda which one is that?
- A. Parjapati
- B. Varuna
- C. Brahman
- D. Indra

17. Which among the following is not be called Vedanga?
- A. Siksha
- B. Vyakarna
- C. Jyotisha
- D. Manu Smriti

18. Identify the correct sequence among the following units of political organization.
- A. kula, vis, grama, jana
- B. jana, vis, grama, kula
- C. vis, kula, jana, grama
- D. grama, jana, kula, vis

19. Which of the following are the Brahmans appended to the Rig veda?
- A. Aitareya Brahmana and Kausitiki Brahmana
- B. Gopatha Brahmana
- C. Satapatha Brahmana
- D. None of these

20. The term "Niyoga" means
- A. Marriage of a childless widow with the brother or blood relation of her deceased husband for the sake of progeny.
- B. To produce off spring by illegal means.
- C. To remain unmarried through out the life.
- D. None of the above.

21. Mahayana Buddhism originated in:
- A. Andhra Pradesh, 1st Century B.C.
- B. Bihar, 1st Century A.D.
- C. Kashmir, 3rd Century B.C.
- D. Bengal, 2nd Century A.D.

22. Where has the Rummindei Pillar been erected?
- A. The birth place of the Budha
- B. The place where he preached his sermon first
- C. The place where the Budha died
- D. The place where he attained enlightenment

23. Early Schism in Buddhism divided the Buddhist order into:
- A. Sthavirvadins and Mahasanghikas
- B. Mahasanghikas Sunyavadins
- C. Sunyavadins and Sarvastivadins
- D. Sthavirvadins and Sarvastivadins

24. The Bhabru Edict records Ashoka's faith in:
- A. Tisaranam
- B. Toleration
- C. Dharmavijya
- D. Solicitude for the Welfare of the people

25. Which one of the following statement on the social life in the Sangam era is not correct?
- A. The widows had to cut off their hair, discard all ornaments and eat only the plainest food.
- B. Civil and military offices were held by the rich peasants called velalas.
- C. Caste distinction existed.
- D. Tolkappiar calls the commercial community as vaisigas.

26. Which one of the following statement on Sangam literature is not correct?
- A. It enables us to trace a connected political history of Tamil Kingdom.
- B. It sheds light on socio-religious aspects of Tamil-Kingdom.

C. It sheds light on trade relations with "Yavanas".
D. It sheds light on the pronounced fusion that had occurred between the Sanskrit and the Tamil culture.

27. Which was the most important city under Chandragupta Maurya?
A. Pataliputra B. Kausambi
C. Ujjain D. Taxila

28. Megasthenes refers to the rule of Heracles (Vasudeva Krishna's) daughter who was assigned villages to rule in Southern region, and who was probably the founder of Pandya kingdom. Identify her:
A. Pandia B. Panlavi
C. Maitri D. Pandara

29. The term 'ahara' as an administrative united appears in:
A. Arthasastra
B. Ashokan inscriptions
C. Rig Veda
D. Epics

30. What was the name of the Chola king who performed Aswamedha sacrifice?
A. Rajendra I B. Karikala
C. Parantaka D. Rajadhiraja I

31. Name the countries which lie on either side of MacMohan Line?
A. South Vietnam and North Vietnam
B. India and China
C. India and Ceylon
D. France and Germany

32. Which is the largest Gulf in the world?
A. Gulf of Aqaba B. Gulf of Bengal
C. Gulf of Mexico D. Gulf of Aden

33. Which country is called the "Land of Morning Calm"?
A. Sweden B. Netherlands
C. Norway D. Korea

34. Which is called the Dark Continent?
A. Africa B. Australia
C. Greenland D. South America

35. What does the 38th Parallel Line separate?
A. India from Bangladesh
B. India from China
C. North Korea from South Korea
D. India from Pakistan

36. The Panama Canal Connects:
A. The Pacific and Atlantic Oceans
B. The Pacific and Arctic Oceans
C. The Atlantic and Arctic Oceans
D. The Atlantic and Indian Oceans

37. What is the habitable portion of the world land area?
A. Less than 35%
B. Less than half of it
C. 50%
D. About 75%

38. The largest of all the peninsulas in the world is the:
A. Arabian peninsula
B. Iberian peninsula
C. Indian peninsula
D. Kamehaika peninsula

39. Forbidden City is:
A. Rome B. Lhasa (Tibet)
C. Jerusalem D. Beijing

40. Rome is the:
A. Eternal city B. Forbidden city
C. City of palaces D. Cleanest city

41. The capital of Israel is:
A. Palestine B. Jerusalem
C. Vatican D. Tel Aviv

42. The City of Seven Hills is:
A. Rome B. Athens
C. Cairo D. Tokyo

43. Laplands is a group of countries located in:
A. Europe B. Australia
C. Asia D. America

44. Zanzibar is called:
A. Island of cloves B. Eternal city
C. Holy land D. Land of spices

45. The Himalayas constitute the natural boundary between:
A. China and India
B. China and Afghanistan
C. India and Nepal
D. India and Myanmar

46. Hermit Kingdom is:
A. Korea B. Palestine
C. Rome D. Haridwar

47. The highest of all the plateaus in the world is the:
A. Deccan Plateau
B. Bolivian Plateau
C. Tibetan Plateau
D. Colombian Plateau

48. Eternal City is:
A. Kolkata B. Rome
C. Athens D. Bangkok

49. The city of Khartoum is situated on the banks of:
A. Hudson B. Nile
C. Tiber D. Seine

50. In which continent some of its parts are over 1500 miles from the sea and in which the hottest as well as the coldest climates of the world are to be found:
A. America B. Asia
C. Europe D. Australia

51. Paris is situated on the river:
A. Tibea B. Seine
C. Danube D. None of the above

52. Name the place where Blue and White Nile Rivers effect a confluence:
A. Cairo
B. Constantinpole (except Greenland)
C. Khartoum
D. Teheran

53. Which place in the world is known for the least amount of rainfall?
A. Cairo (Egypt) B. Yarkand (China)
C. Atacama (Chile) D. Thar (Rajasthan)

54. Australia is the:
A. only continent in the Southern Hemisphere
B. island continent
C. smallest continent
D. all the above points are true

55. Name the place where Rhine and Spree rivers meet in Europe:
A. Lyons B. Paris
C. Hamburg D. Berlin

56. Moscow is situated on the bank of:
A. Moskva B. Spree
C. Volga D. Tagus

57. The country Congo is in the ..... continent:
A. European B. Australian
C. Asian D. African

58. China's Sorrow is:
A. Lin Piao B. Sinkiang
C. Hwang-ho D. I'sang Ho

59. Sugar Bowl of the world is:
A. Manila B. Japan
C. Cuba D. Havana

60. The City of Palaces is:
A. Peru B. Rome
C. Kolkata D. Bangkok

61. Which of the following does not come under the public sector?
A. TISCO B. IISCO
C. SAIL D. Ordnance factories

62. Term loans to industry is provided by:
A. RBI B. EXIM Bank
C. IDBI D. NABARD

63. Among the steel plants, following has been built with Russian assistance:
A Rourkela B. Bhillai
C. Durgapur D. Kothagudam

64. The "Cafeteria approach" for family planning was put forward during:
A. Third plan B. Emergency
C. Annual plans D. Janata party rule

65. The Monopolies and Restrictive Trade Practice Act was passed in:
A. 1968 B. 1969
C. 1970 D. 1971

66. Inflation is caused by:
A. decrease in general spending
B. decrease in money supply
C. increase in industrial and agriculture output
D. increase in general spending or money supply

67. Among the following, quasi-governmental institution is:
A. EXIM Bank B. IDBI
C. ICICI D. UTI

68. The main purpose of UTI is:
A. mobilise and channelise small investors' savings

B. regulate the mutual fund industry
C. act as a stabilising agent in the share market
D. coordinate activities of other financial institution

69. White revolution refers to:
A. Cement industry
B. Dairying
C. Cotton textile industry
D. Marble industry

70. In India, the major means of irrigation is:
A. Wells & Tubewalls
B. Canals
C. Tanks
D. Animal drought

71. The major function of the Food Corporation of India is:
A. to distribute high yielding variety seeds
B. to extend credit to farmers
C. to control the demand side of foodgrains
D. to check the fluctuation in foodgrain prices by controlling supply

72. The largest population among the states is in:
A. Bihar B. Bengal
C. Uttar Pradesh D. Madhya Pradesh

73. Which of the following is referred as Sunrise industry?
A. Coffee B. Hi-technology
C. Small scale D. Heavy

74. Which of the following is not a measure of economic growth?
A. Per capita income
B. Gross National Product
C. Net Domestic Product
D. Money in circulation

75. Among the following, least revenue comes from:
A. Income tax B. Excise duty
C. Custom duty D. Wealth tax

76. A blue chip company is called so because:
A. it produces computer hardware
B. it produces computer software
C. it gives high rains and has low risk
D. it exports most of its produce

77. The Command Area Development Programme was a major step in the field of:
A. irrigation
B. plant protection measures
C. farm mechanisation
D. high yield variety seeds

78. Poverty line is defined on the basis of:
A. living conditions of people
B. the number of children per family
C. per capita family income
D. the requirement of calories per person

79. The oldest large scale industry of India is in the field of:
A. iron and steel sector
B. jute sector
C. cotton textile sector
D. sugar sector

80. Seasonal unemployment is maximum in:
A. Industrial sector B. Agricultural sector
C. Service sector D. Handloom sector

81. The oldest trade union organisation in India is:
A. Hind Mazdoor Sabha
B. Indian National Trade Union Congress
C. All India Trade Union Congress
D. Central Industrial Trade Union

82. Sindri, Haldia and Paradip have:
A. Sea-ports
B. Steel plants
C. Big Multipurpose dams
D. Fertilizer plants

83. The Export-Import Bank was set up in:
A. 1969 B. 1980
C. 1982 D. 1985

84. Which among the following is India's biggest trade partner?
A. USA B. UAE
C. China D. Germany

85. Which of the following does not contribute to the development of the Indian economy?
A. Rising industrial output
B. Exports' share decreased; imports' share increased; Modern technology
C. Exports' and imports' share increased; Green Revolution
D. Population growth

86. Which of the following has largest share in India's export?
A. Manufactured goods
B. Ores and minerals
C. Agriculture and allied products
D. Others

87. Which of the following has largest share in India's imports?
A. Food and allied products
B. Capital goods
C. Fertilizers
D. Fuel

88. Special Economic Zone Act was passed on:
A. February 2004 B. May 2006
C. May 2005 D. February 2007

89. The first steel project city in India was established at
A. Bhilai B. Jamshedpur
C. Kolkata D. Bokaro

90. Rupee was made convertible on current account in:
A. 1991 B. 1994
C. 1995 D. 1997

91. The Constitution of India was adopted on:
A. 15 August, 1947
B. 26 January, 1950
C. 26 November, 1949
D. 15 August, 1950

92. Who administers the oaths of office to the Vice-President?
A. Chief Justice of India
B. President of India
C. Speaker of the Lok Sabha
D. Attorney-General of India

93. Who among the following is not appointed by the President?
A. Chairman of UPSC
B. Governor of State
C. Judge of a High Court
D. Vice-President

94. The President can dissolve the Lok Sabha before expiry of its term, on the advice of:
A. Prime Minister
B. Speaker
C. Vice-President
D. Election Commissioner

95. A presidential ordinance remains in force for:
A. three months B. one month
C. six months D. till it is revoked

96. The idea of Fundamental Duties is derived from:
A. Russian Constitution
B. British Constitution
C. American Constitution
D. German Constitution

97. The Fundamental Right, which cannot be suspended even during an emergency is:
A. Right to equality
B. Right to freedom of expression
C. Right to movement
D. Right to life

98. Whether the bill in the Parliament, is a money bill or not, is decided by:
A. Prime Minister
B. Speaker
C. President
D. Comptroller and Auditor General

99. Which of the following is presided over by a non-member:
A. Lok Sabha
B. Rajya Sabha
C. Vidhan Sabha
D. Joint sitting of Lok Sabha and Rajya Sabha

100. For the Union Territories, not having legislatures of their own, laws are enacted by:
A. Parliament
B. President
C. Administrators
D. Union Law Ministry

101. Arunachal Pradesh was granted statehood in:
A. 1972 B. 1984
C. 1985 D. 1987

102. The official language of Jammu and Kashmir is:
A. Dogri B. Kashmiri
C. Urdu D. Hindustani

103. The minimum age to qualify for the membership of Rajya Sabha is:
A. 18 years B. 21 years
C. 25 years D. 30 years

104. A minister, who is not a member of either house of the Parliament must get himself elected within:
A. a year B. six months
C. three months D. two months

105. By convention, the Governor of a state generally belongs to:
A. ruling party in the state
B. opposition party of the state
C. IAS cadre of the state
D. some other state

106. Due to the special status, Jammu and Kashmir has:
A. separate constitution
B. separate armed forces
C. separate judiciary
D. all of the above

107. Constitution describes India as a:
A. Federation of states
B. Quasi-federation
C. Union of states
D. Dominion of free states

108. Which of the following article cannot be amended by simple majority?
A. Creation of new states
B. Procedure of election of the President
C. Official language of the Union
D. Quorum in the Parliament

109. Constitution provides residual powers to the:
A. Union Government
B. State Government
C. President
D. Supreme Court

110. President of India has immunity from:
A. Criminal proceedings during the term of office
B. Civil proceedings during the term of office
C. Civil and criminal proceedings during the term of office
D. Civil and criminal proceedings after the retirement

111. Charge on a particle cannot be:
A. Positive
B. Negative
C. Neutral
D. Partly positive, partly negative

112. Current is nothing but a flow of:
A. Atoms B. Molecules
C. Electrons D. Neutrons

113. Like charges:
A. Attract each other
B. Repel each other
C. Don't affect each other
D. Depends on the charge

114. Which among the following is an insulator?
A. Iron B. Pure water
C. Nickel D. Silver

115. Which material is used to form semi-conductors?
A. Silicon B. Gold
C. Aluminium D. Gallium

116. Current passing through a conductor cannot cause:
A. Heating effect B. Gravitational effect
C. Magnetic effect D. Chemical effect

117. Lead is added to the petrol to:
A. Increase calorific value
B. Reduce pollution
C. Decrease Cracking
D. Provide lubrication to mobile parts

118. The purest form of iron is:
A. Wrought iron B. Cast iron
C. Stainless steel D. Pig iron

119. Rusting is a special case of corrosion, applied to:
A. Nickel B. Tin
C. Iron D. Silver

120. Age of a tree is calculated by:
A. Measuring its thickness
B. Measuring its length
C. Measuring the number of annual rings
D. Measuring the number of branches

121. J.G. Mendel is known for his work in the field of:
A. Biotechnology
B. Genetics
C. Parasitology
D. Medical engineering

122. Sound travels fastest in:
A. Air B. Water
C. Vacuum D. Iron

123. The maximum displacement of the particle in a wave motion is called:
A. Phase B. Frequency
C. Amplitude D. Time period

124. Each magnet has _____ poles:
A. 2 B. 3
C. 4 D. 5

125. pH factor is used to measure:
A. Colour of a liquid
B. Acidity of a liquid
C. Density of a liquid
D. Turbidity of a liquid

126. The common salt contains:
A. Sodium and sulphate
B. Sodium and nitrate
C. Sodium and chloride
D. Calcium and sulphate

127. Brass contains the metals of:
A. Copper and zinc
B. Copper and iron
C. Copper and tin
D. Sodium and iron

128. Goitre is caused by the malfunctioning of:
A. Liver B. Brain
C. Spleen D. Thyroid

129. The blood pressure of a normal man is:
A. 1,00,150 mm Hg B. 50,100 mm Hg
C. 80,120 mm Hg D. 200,300 mm Hg

130. Deficiency of Vitamin B causes:
A. Rickets
B. Scurvy
C. Beri-Beri
D. Night blindness

ANSWERS

1	**2**	**3**	**4**	**5**	**6**	**7**	**8**	**9**	**10**
C	A	C	D	A	A	D	B	A	D
11	**12**	**13**	**14**	**15**	**16**	**17**	**18**	**19**	**20**
A	A	B	D	D	C	D	B	A	A
21	**22**	**23**	**24**	**25**	**26**	**27**	**28**	**29**	**30**
A	A	A	A	D	A	A	A	B	D
31	**32**	**33**	**34**	**35**	**36**	**37**	**38**	**39**	**40**
B	C	D	A	C	A	B	C	B	A
41	**42**	**43**	**44**	**45**	**46**	**47**	**48**	**49**	**50**
D	A	A	A	A	A	C	B	B	B
51	**52**	**53**	**54**	**55**	**56**	**57**	**58**	**59**	**60**
B	B	C	D	A	A	D	C	C	C
61	**62**	**63**	**64**	**65**	**66**	**67**	**68**	**69**	**70**
A	C	B	D	B	D	C	A	B	A
71	**72**	**73**	**74**	**75**	**76**	**77**	**78**	**79**	**80**
D	C	B	D	D	C	A	D	C	B
81	**82**	**83**	**84**	**85**	**86**	**87**	**88**	**89**	**90**
C	D	C	C	D	A	D	C	D	B
91	**92**	**93**	**94**	**95**	**96**	**97**	**98**	**99**	**100**
C	B	D	A	C	A	D	B	B	A
101	**102**	**103**	**104**	**105**	**106**	**107**	**108**	**109**	**110**
D	C	D	B	D	A	C	B	A	B
111	**112**	**113**	**114**	**115**	**116**	**117**	**118**	**119**	**120**
D	C	B	D	A	B	C	A	C	C
121	**122**	**123**	**124**	**125**	**126**	**127**	**128**	**129**	**130**
B	D	C	A	B	C	A	D	C	C

❑ ❑ ❑

www.ingramcontent.com/pod-product-compliance
Lightning Source LLC
Chambersburg PA
CBHW080206170726
48282CB00030B/1023

* 9 7 8 9 3 5 0 1 2 8 3 9 8 *